EXACT SEQUENCES IN THE ALGEBRAIC THEORY OF SURGERY

by

ANDREW RANICKI

Princeton University Press

and

University of Tokyo Press

Princeton, New Jersey

1981

Published in Japan Exclusively
by University of Tokyo Press
in other parts of the world by
Princeton University Press

Printed in the United States of America
by Princeton University Press, Princeton, New Jersey

Library of Congress Cataloging in Publication Data will
be found on the last printed page of this book

for Carla

CONTENTS

Introduction

An n-dimensional algebraic Poincaré complex over a ring with involution A is an n-dimensional A-module chain complex C together with a self-dual chain equivalence

$$C^* = \mathrm{Hom}_A(C,A) \xrightarrow{\sim} C_{n-*} ,$$

so that there are induced abstract Poincaré duality A-module isomorphisms

$$H^*(C) \xrightarrow{\sim} H_{n-*}(C) .$$

A 0-dimensional algebraic Poincaré complex over A is the same as a non-singular quadratic form over A. If M is a compact n-dimensional topological manifold and $\widetilde{M}$ is a covering of M with group of covering translations π the $\mathbb{Z}[\pi]$-module chain complex $C(\widetilde{M})$ has the structure of an n-dimensional algebraic Poincaré complex over $\mathbb{Z}[\pi]$, on account of the classic Poincaré duality $H^*(\widetilde{M}) \cong H_{n-*}(\widetilde{M})$. The Poincaré-Lefschetz duality $H^*(\widetilde{M}) \cong H_{n-*}(\widetilde{M},\partial\widetilde{M})$ of a compact n-dimensional manifold with boundary $(M,\partial M)$ motivates the notion of an n-dimensional algebraic Poincaré pair over A, as a pair of chain complexes $(C,\partial C)$ together with a self-dual chain equivalence $C^* \xrightarrow{\sim} (C/\partial C)_{n-*}$. There is thus an abstract cobordism theory, with n-dimensional algebraic Poincaré complexes C,C' cobordant if $C\oplus -C' = \partial D$ is the boundary of an (n+1)-dimensional algebraic Poincaré pair $(D,\partial D)$.

In Parts I and II of a paper entitled "The algebraic theory of surgery" (Ranicki [9],[10], henceforth to be denoted I.,II.) the cobordism of algebraic Poincaré complexes with a $\begin{cases}\text{symmetric}\\ \text{quadratic}\end{cases}$ structure was used to define a sequence of covariant functors

$$\begin{cases} L^n \\ L_n \end{cases} : \{\text{rings with involution}\} \longrightarrow \{\text{abelian groups}\} \quad (n \in \mathbb{Z})$$

and to study their applications to the geometric theory of surgery on compact manifolds. In effect, this is Part III of the sequence, in which there are established various exact sequences in the algebraic L-groups, and some further applications to geometric surgery are developed.

The 0-dimensional $\begin{cases}\text{symmetric}\\ \text{quadratic}\end{cases}$ L-group $\begin{cases}L^0(A)\\ L_0(A)\end{cases}$ is the Witt group of non-singular $\begin{cases}\text{symmetric}\\ \text{quadratic}\end{cases}$ forms over A. The quadratic L-groups are 4-periodic

$$L_n(A) = L_{n+4}(A) \quad (n \in \mathbb{Z}) ,$$

and are in fact the surgery obstruction groups of Wall [4]. The higher symmetric L-groups $L^n(A)$ $(n \geqslant 0)$ are the algebraic Poincaré cobordism groups of Mishchenko [1]; they are not in general 4-periodic, $L^n(A) \neq L^{n+4}(A)$. The lower symmetric L-groups $L^n(A)$ $(n \leqslant -1)$ are defined to be such that

$$L^n(A) = L_n(A) \quad (n \leqslant -3) ,$$

with an ad hoc definition for $L^{-1}(A)$ and $L^{-2}(A)$. The symmetric L-groups are related to the quadratic L-groups by symmetrization maps

$$1+T : L_n(A) \longrightarrow L^n(A) \quad (n \in \mathbb{Z})$$

which are isomorphisms modulo 8-torsion for any A, and actually isomorphisms if 2 is a unit in A.

The principal algebraic aim here is to establish exact sequences in L-theory substantiating the assertion made in the introduction to I. that the $\left\{\begin{matrix}\text{symmetric}\\ \text{quadratic}\end{matrix}\right.$ L-groups $\left\{\begin{matrix}L^n(A)\\ L_n(A)\end{matrix}\right.$ $(n \in \mathbb{Z})$ are to the $\left\{\begin{matrix}\text{symmetric}\\ \text{quadratic}\end{matrix}\right.$ Witt group $\left\{\begin{matrix}L^0(A)\\ L_0(A)\end{matrix}\right.$ what the algebraic K-groups $K_n(A)$ $(n \in \mathbb{Z})$ are to the projective class group $K_0(A)$. It will be recalled that algebraic K-theory has to determine whether a finitely generated projective A-module is free, and if so in how many ways; similarly, algebraic L-theory has to determine whether a $\left\{\begin{matrix}\text{symmetric}\\ \text{quadratic}\end{matrix}\right.$ form is hyperbolic (= admits a maximally isotropic "lagrangian" direct summand), and if so in how many ways. The actual L-theory exact sequences obtained are listed further below, following a brief discussion of their K-theory antecedents.

The principal geometric aim is to extend the applications of algebraic surgery to topology made in II. beyond the general surgery obstruction theory for manifolds of Part 1 of Wall [4] to the theory of Part 2 arising in the classification of topological (sub)manifold structures on geometric Poincaré (sub)complexes, that is codimension q surgery obstruction theory. Exact sequences play an important role in this classification, notably the fundamental "surgery exact sequence" of the Browder-Novikov-Sullivan-Wall theory

$$L_{n+1}(\mathbb{Z}[\pi_1(X)]) \longrightarrow \mathcal{S}^{TOP}(X) \longrightarrow [X,G/TOP] \longrightarrow L_n(\mathbb{Z}[\pi_1(X)])$$

for the set $\mathcal{S}^{TOP}(X)$ of topological manifold structures on an n-dimensional geometric Poincaré complex X $(n \geqslant 5)$ with a topological reduction $\widetilde{\nu}_X : X \longrightarrow BTOP$ of the Spivak normal fibration $\nu_X : X \longrightarrow BG$. It will be recalled that an n-dimensional geometric Poincaré complex X is a finite CW complex with the Poincaré duality $H^*(\widetilde{X}) = H_{n-*}(\widetilde{X})$ of a compact n-dimensional topological manifold, but which is not required to be locally homeomorphic to Euclidean n-space $\mathbb{R}^n$. Surgery theory has to determine whether a geometric Poincaré complex is homotopy equivalent to a manifold, and if so in how many ways. The theory was first developed for smooth (= differentiable) manifolds, but it has since turned out to work just as well for topological manifolds. Moreover, the topological category has better algebraic properties, such as the homotopy-theoretic 4-periodicity of the classifying space

$$\Omega^4(G/TOP) = L_0(\mathbb{Z}) \times G/TOP \ .$$

The total surgery obstruction theory of Ranicki [7] was a tentative first step towards a purely algebraic account of the homotopy theory of compact n-dimensional topological manifolds, at least for $n \geqslant 5$, including an algebraic expression for the surgery exact sequence.

In an effort at making this book self-contained §1 recapitulates the main definitions and results of I. and II., particularly the definition of the $\left\{\begin{matrix}\text{symmetric}\\ \text{quadratic}\end{matrix}\right.$ L-groups $\left\{\begin{matrix}L^*(A)\\ L_*(A)\end{matrix}\right.$ and of the $\left\{\begin{matrix}\text{symmetric}\\ \text{quadratic}\end{matrix}\right.$ signature $\left\{\begin{matrix}\sigma^*(X) \in L^n(\mathbb{Z}[\pi_1(X)])\\ \sigma_*(f,b) \in L_n(\mathbb{Z}[\pi_1(X)])\end{matrix}\right.$ of an n-dimensional $\left\{\begin{matrix}\text{geometric Poincaré complex } X\\ \text{normal map } (f,b) : M \longrightarrow X\end{matrix}\right.$, along with the identification of the quadratic signature with the Wall surgery obstruction. The algebraic L-theory exact sequences are developed in §§2-6, and the algebraic theory of codimension q surgery is developed in §7. It should be noted that while the material of §§2-6 is in its definitive form, §7 is only a preliminary account of the applications to topology, on the level of exposition of the total surgery obstruction theory of Ranicki [7] which it extends. The full account will be spread out over the next two instalments of the series, Ranicki [11],[12].

In dealing with the algebraic K-theory motivating the algebraic L-theory it will be assumed that the reader is familiar with the definitions and basic properties of the classical algebraic K-groups $K_0(A)$ and $K_1(A)$, and their appearance in topology via the Wall finiteness obstruction and the Whitehead torsion. The algebraic K-groups $K_n(A)$ defined for $n \leqslant -1$ by Bass, for $n = 2$ by Milnor, and for $n \geqslant 3$ by Quillen are invoked only for the way in which they extend (or fail to extend) the exact sequences of classical algebraic K-theory. In particular, the algebraic K-groups

$$K_n(A) = K_n(\text{exact category of f.g. projective } A\text{-modules}) \quad (n \in \mathbb{Z})$$

are such that for a ring morphism $f:A \longrightarrow B$ there are defined relative K-groups $K_n(f)$ $(n \in \mathbb{Z})$ with a change of rings exact sequence

$$\ldots \longrightarrow K_n(A) \xrightarrow{f} K_n(B) \longrightarrow K_n(f) \longrightarrow K_{n-1}(A) \longrightarrow \ldots \quad (n \in \mathbb{Z}) .$$

Given a multiplicative subset $S \subset A$ of non-zero-divisors there is defined a ring $S^{-1}A$ inverting S, and there are defined algebraic K-groups

$$K_n(A,S) = K_{n-1}(\text{exact category of } S\text{-torsion } A\text{-modules of homological dimension } 1) \quad (n \in \mathbb{Z})$$

such that the relative K-groups $K_*(f)$ of the localization map $f: A \longrightarrow S^{-1}A$ can be identified with $K_*(A,S)$

$$K_n(f: A \longrightarrow S^{-1}A) = K_n(A,S) \quad (n \in \mathbb{Z}) .$$

The consequent expression for the change of rings exact sequence

$$\dots \longrightarrow K_n(A) \xrightarrow{f} K_n(S^{-1}A) \longrightarrow K_n(A,S) \longrightarrow K_{n-1}(A) \longrightarrow \dots \quad (n \in \mathbb{Z})$$

is the "localization exact sequence of algebraic K-theory". For a Dedekind ring R with quotient field $F = (R-\{0\})^{-1}R$ a devissage argument identifies

$$K_n(R, R-\{0\}) = \bigoplus_{\mathcal{P}} K_{n-1}(R/\mathcal{P}) \quad (n \in \mathbb{Z})$$

with $\mathcal{P}$ ranging over all the maximal ideals of R, so that the localization exact sequence for $R \longrightarrow F$ can be written as

$$\dots \longrightarrow K_n(R) \longrightarrow K_n(F) \longrightarrow \bigoplus_{\mathcal{P}} K_{n-1}(R/\mathcal{P}) \longrightarrow K_{n-1}(R) \longrightarrow \dots \quad (n \in \mathbb{Z}) .$$

An application of the localization exact sequence to the multiplicative subset $X = \{x^k \mid k \geqslant 0\} \subset A[x]$ proves the "fundamental theorem of algebraic K-theory", relating the K-groups of the polynomial extension rings $A[x]$, $A[x,x^{-1}]$ in a central indeterminate x over A ($ax = xa$, $a \in A$) by naturally split exact sequences

$$0 \longrightarrow K_n(A) \longrightarrow K_n(A[x]) \oplus K_n(A[x^{-1}]) \longrightarrow K_n(A[x,x^{-1}]) \longrightarrow K_{n-1}(A) \longrightarrow 0 \quad (n \in \mathbb{Z}) .$$

This can be generalized to the algebraic K-groups of twisted polynomial extensions $A_\alpha[x]$, $A_\alpha[x,x^{-1}]$ ($ax = x\alpha(a)$ for some automorphism $\alpha: A \longrightarrow A$), since the exact sequence for the localization $A_\alpha[x] \longrightarrow X^{-1}A_\alpha[x] = A_\alpha[x,x^{-1}]$ can be expressed as

$$\cdots \longrightarrow K_n(A_\alpha[x]) \longrightarrow K_n(A_\alpha[x,x^{-1}]) \longrightarrow K_{n-1}(A)\oplus\widetilde{Nil}_n(A,\alpha)$$
$$\longrightarrow K_{n-1}(A_\alpha[x]) \longrightarrow \cdots \qquad (n\in\mathbb{Z})$$

with the $\widetilde{Nil}$-groups such that

$$\begin{aligned} K_n(A_\alpha[x]) &= K_n(\text{exact category of f.g. projective } A\text{-modules } P \\ &\qquad \text{with an } \alpha\text{-twisted nilpotent map } \nu:P\longrightarrow P) \\ &= K_n(A)\oplus\widetilde{Nil}_n(A,\alpha^{-1}) \quad (n\in\mathbb{Z})\ . \end{aligned}$$

Given a cartesian square of rings

$$\begin{array}{ccc} A & \longrightarrow & B' \\ \downarrow & & \downarrow \\ B & \longrightarrow & A' \end{array}$$

such that $B \longrightarrow A'$ (or $B' \longrightarrow A'$) is onto there is defined a Mayer-Vietoris exact sequence of the classical algebraic K-groups

$$K_1(A) \longrightarrow K_1(B)\oplus K_1(B') \longrightarrow K_1(A') \longrightarrow K_0(A) \longrightarrow K_0(B)\oplus K_0(B')$$
$$\longrightarrow K_0(A') \longrightarrow K_{-1}(A) \longrightarrow \cdots$$

which extends on the right to the lower K-groups, but which does not in general extend to the higher K-groups on the left. However, if $S\subset A$ is a multiplicative subset of non-zero-divisors and

$$\hat{A} = \varprojlim_{s\in S} A/sA$$

is the S-adic completion of A then there is defined a cartesian square of rings

$$\begin{array}{ccc} A & \longrightarrow & S^{-1}A \\ \downarrow & & \downarrow \\ \hat{A} & \longrightarrow & \hat{S}^{-1}\hat{A} \end{array}$$

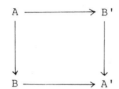

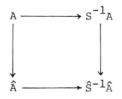

for which there is defined a Mayer-Vietoris exact sequence in all the K-groups

$$\ldots \longrightarrow K_n(A) \longrightarrow K_n(\hat{A}) \oplus K_n(S^{-1}A) \longrightarrow K_n(\hat{S}^{-1}\hat{A}) \longrightarrow K_{n-1}(A) \longrightarrow \ldots$$
$$(n \in \mathbb{Z}) .$$

It is these exact sequences of algebraic K-theory which serve as models for L-theory. The individual introductions to §§2 - 6 and §7.6 contain some further background material concerning algebraic K-theory, such as references.

In summarizing below the algebraic L-theory exact sequences obtained in §§2 - 6 the terminology will be simplified by writing $\begin{cases} L^*(A) \\ L_*(A) \end{cases}$ for all the $\begin{cases} \text{symmetric} \\ \text{quadratic} \end{cases}$ L-groups, even though the groups that actually occur are the "intermediate $\begin{cases} \varepsilon\text{-symmetric} \\ \varepsilon\text{-quadratic} \end{cases}$ L-groups" $\begin{cases} L^*_X(A,\varepsilon) \\ L^X_*(A,\varepsilon) \end{cases}$ with $X \subseteq \tilde{K}_m(A)$ $(m = 0 \text{ or } 1)$ some subgroup which is invariant under the involution of $\tilde{K}_m(A)$ determined by the involution $\bar{\ }: A \longrightarrow A; a \longmapsto \bar{a}$ of the ring A, and $\varepsilon \in A$ a central unit such that $\bar{\varepsilon} = \varepsilon^{-1} \in A$.

Following the discussion in §1 of the absolute $\begin{cases} \text{symmetric} \\ \text{quadratic} \end{cases}$ L-groups $\begin{cases} L^n(A) \\ L_n(A) \end{cases}$ $(n \in \mathbb{Z})$ of a ring with involution A there will be defined in §2 the relative $\begin{cases} \text{symmetric} \\ \text{quadratic} \end{cases}$ L-groups $\begin{cases} L^n(f) \\ L_n(f) \end{cases}$ $(n \in \mathbb{Z})$ of a morphism of rings with involution $f: A \longrightarrow B$, with a change of rings exact sequence

$$\begin{cases} \cdots \longrightarrow L^n(A) \xrightarrow{f} L^n(B) \longrightarrow L^n(f) \longrightarrow L^{n-1}(A) \longrightarrow \cdots \\ \cdots \longrightarrow L_n(A) \xrightarrow{f} L_n(B) \longrightarrow L_n(f) \longrightarrow L_{n-1}(A) \longrightarrow \cdots \end{cases} \qquad (n \in \mathbb{Z})$$

In §3 the relative $\begin{Bmatrix} \text{symmetric} \\ \text{quadratic} \end{Bmatrix}$ L-groups $\begin{Bmatrix} L^*(f) \\ L_*(f) \end{Bmatrix}$ of the localization map $f: A \longrightarrow S^{-1}A$ inverting a multiplicative subset $S \subset A$ of non-zero-divisors invariant under the involution will be identified with the cobordism groups $\begin{Bmatrix} L^*(A,S) \\ L_*(A,S) \end{Bmatrix}$ of $\begin{Bmatrix} \text{symmetric} \\ \text{quadratic} \end{Bmatrix}$ Poincaré complexes over A which become acyclic over the localization $S^{-1}A$

$$\begin{cases} L^n(f: A \longrightarrow S^{-1}A) = L^n(A,S) \\ L_n(f: A \longrightarrow S^{-1}A) = L_n(A,S) \end{cases} \qquad (n \in \mathbb{Z}) ,$$

thus obtaining the "localization exact sequence of $\begin{Bmatrix} \text{symmetric} \\ \text{quadratic} \end{Bmatrix}$ L-theory"

$$\begin{cases} \cdots \longrightarrow L^n(A) \longrightarrow L^n(S^{-1}A) \longrightarrow L^n(A,S) \longrightarrow L^{n-1}(A) \longrightarrow \cdots \\ \cdots \longrightarrow L_n(A) \longrightarrow L_n(S^{-1}A) \longrightarrow L_n(A,S) \longrightarrow L_{n-1}(A) \longrightarrow \cdots \end{cases}$$
$$(n \in \mathbb{Z}) .$$

In particular, $\begin{Bmatrix} L^0(A,S) \\ L_0(A,S) \end{Bmatrix}$ is the Witt group of non-singular $S^{-1}A/A$-valued $\begin{Bmatrix} \text{symmetric} \\ \text{quadratic} \end{Bmatrix}$ linking forms on h.d. 1 S-torsion A-modules. (See Ranicki [6] for a preliminary account of localization in quadratic L-theory). In §4 it is shown that for a ring with involution A which is an algebra over a Dedekind ring R (e.g. a group ring $A = \mathbb{Z}[\pi]$ with $R = \mathbb{Z}$)

the relative terms in the localization exact sequence for $S = R-\{0\} \subset A$ have natural direct sum decompositions

$$\begin{cases} L^n(A,S) = \bigoplus_{\mathcal{P}} L^n(A,\mathcal{P}^\infty) \\ L_n(A,S) = \bigoplus_{\mathcal{P}} L_n(A,\mathcal{P}^\infty) \end{cases} \quad (n \in \mathbb{Z})$$

with $\mathcal{P}$ ranging over all the maximal ideals of R invariant under the involution, and $\begin{cases} L^*(A,\mathcal{P}^\infty) \\ L_*(A,\mathcal{P}^\infty) \end{cases}$ defined in the same way as $\begin{cases} L^*(A,S) \\ L_*(A,S) \end{cases}$ but using only A-module chain complexes with $\mathcal{P}$-primary S-torsion homology A-modules. Furthermore, in the case A = R a symmetric L-theory devissage argument identifies

$$L^n(R,S) = \bigoplus_{\mathcal{P}} L^n(R/\mathcal{P}) \quad (n \geqslant 0) \ .$$

In general, there is no devissage in quadratic L-theory, and an example is constructed for which

$$L_n(R,S) \neq \bigoplus_{\mathcal{P}} L_n(R/\mathcal{P}) \ .$$

In §5 the localization exact sequence of §3 is applied to obtain splitting theorems for the L-groups of the α-twisted polynomial extensions $A_\alpha[x]$, $A_\alpha[x,x^{-1}]$ of a ring with involution A, with $\alpha: A \longrightarrow A$ a ring automorphism such that $\overline{\alpha(a)} = \alpha^{-1}(\bar{a}) \in A$ $(a \in A)$ and x an indeterminate over A such that

$$\bar{x} = x \ , \ ax = x\alpha(a) \ .$$

It will be shown that the $\begin{cases} \text{symmetric} \\ \text{quadratic} \end{cases}$ L-theory exact sequence for the localization inverting $X = \{x^k \mid k \geqslant 0\} \subset A_\alpha[x]$

$$A_\alpha[x] \longrightarrow X^{-1}A_\alpha[x] = A_\alpha[x,x^{-1}]$$

consists of naturally split short exact sequences of the type

$$\begin{cases} 0 \longrightarrow L^n(A_\alpha[x]) \longrightarrow L^n(A_\alpha[x,x^{-1}]) \longrightarrow L^n(A_\alpha[x],X) \longrightarrow 0 \\ 0 \longrightarrow L_n(A_\alpha[x]) \longrightarrow L_n(A_\alpha[x,x^{-1}]) \longrightarrow L_n(A_\alpha[x],X) \longrightarrow 0 \end{cases}$$

$$(n \in \mathbb{Z}) \ ,$$

and hence that there are defined naturally split exact sequences of the type

$$\begin{cases} 0 \longrightarrow L^n(A) \longrightarrow L^n(A_\alpha[x]) \oplus L^n(A_\alpha[x^{-1}]) \longrightarrow L^n(A_\alpha[x,x^{-1}]) \\ \qquad\qquad \longrightarrow L^n(A^\alpha) \longrightarrow 0 \\ 0 \longrightarrow L_n(A) \longrightarrow L_n(A_\alpha[x]) \oplus L_n(A_\alpha[x^{-1}]) \longrightarrow L_n(A_\alpha[x,x^{-1}]) \\ \qquad\qquad \longrightarrow L_n(A^\alpha) \longrightarrow 0 \end{cases} \qquad (n \in \mathbb{Z})$$

where A^α is the ring A with involution $a \longmapsto \alpha(\bar{a})$.

This "fundamental theorem of $\begin{cases}\text{symmetric}\\\text{quadratic}\end{cases}$ L-theory" is surprising in the twisted case $\alpha \neq \text{id.}$, since the corresponding localization exact sequence in algebraic K-theory

$$\ldots \longrightarrow K_n(A_\alpha[x]) \longrightarrow K_n(A_\alpha[x,x^{-1}]) \longrightarrow K_n(A_\alpha[x],X) \xrightarrow{\partial} K_{n-1}(A_\alpha[x]) \longrightarrow \ldots$$

need not break up into short exact sequences if $\alpha \neq \text{id.}$, that is $\partial \neq 0$ in general. (In §7.6 the fundamental theorem of quadratic L-theory for a group ring $A = \mathbb{Z}[\pi]$ will be given a geometric interpretation in terms of the Browder-Livesay-Wall obstruction theory for surgery on one-sided codimension 1 submanifolds.)

In §6 it will be shown that for a cartesian square of rings with involution

$$\begin{array}{ccc} A & \longrightarrow & B' \\ \downarrow & & \downarrow \\ B & \longrightarrow & A' \end{array}$$

with $B \longrightarrow A'$ (or $B' \longrightarrow A'$) onto there is defined a Mayer-Vietoris exact sequence of quadratic L-groups

$$\ldots \longrightarrow L_n(A) \longrightarrow L_n(B) \oplus L_n(B') \longrightarrow L_n(A') \longrightarrow L_{n-1}(A) \longrightarrow \ldots \quad (n \in \mathbb{Z}) .$$

In general, there is no such exact sequence in the symmetric L-groups, and an example is constructed to illustrate this failure of excision. For a localization-completion cartesian square

$$\begin{array}{ccc} A & \longrightarrow & S^{-1}A \\ \downarrow & & \downarrow \\ \hat{A} & \longrightarrow & \hat{S}^{-1}\hat{A} \end{array}$$

there will be obtained a Mayer-Vietoris exact sequence in quadratic L-theory

$$\ldots \longrightarrow L_n(A) \longrightarrow L_n(\hat{A}) \oplus L_n(S^{-1}A) \longrightarrow L_n(\hat{S}^{-1}\hat{A}) \longrightarrow L_{n-1}(A) \longrightarrow \ldots \quad (n \in \mathbb{Z}) ,$$

and if 2 is a unit in $\hat{S}^{-1}\hat{A}$ there will also be obtained such a sequence in symmetric L-theory

$$\ldots \longrightarrow L^n(A) \longrightarrow L^n(\hat{A}) \oplus L^n(S^{-1}A) \longrightarrow L^n(\hat{S}^{-1}\hat{A}) \longrightarrow L^{n-1}(A) \longrightarrow \ldots \quad (n \in \mathbb{Z}) .$$

In particular, for any group π there is a Mayer-Vietoris exact sequence in both the symmetric and the quadratic L-groups of the classical localization-completion "arithmetic square" of group rings

$$\begin{array}{ccc} \mathbb{Z}[\pi] & \longrightarrow & \mathbb{Q}[\pi] \\ \downarrow & & \downarrow \\ \hat{\mathbb{Z}}[\pi] & \longrightarrow & \hat{\mathbb{Q}}[\pi] \end{array},$$

the study of which plays such an important role in the computation of the surgery obstruction groups $L_*(\pi) \equiv L_*(\mathbb{Z}[\pi])$ of finite groups π and allied trades.

Codimension q surgery theory deals with the problem of doing surgery on a codimension q submanifold $N^{n-q} \subset M^n$ inside M, that is "ambient surgery" as opposed to "abstract surgery" on N without regard to M. For $q \geqslant 3$ the ambient and abstract surgery obstructions coincide. Besides the abstract surgery obstruction groups $L_*(\pi)$ Wall [4,§11] also introduced the codimension q surgery obstruction groups $LS_*(\Phi)$ (q = 1 or 2), by formalizing the idea due to Browder of first doing abstract surgery on the submanifold N and then fitting the result back into the supermanifold M. Given an n-dimensional geometric Poincaré complex X, a codimension q Poincaré subcomplex $Y \subset X$ with normal fibration $\xi = \nu_{Y \subset X} : Y \longrightarrow BG(q)$ (q = 1 or 2), and a homotopy equivalence $f: M \xrightarrow{\simeq} X$ from an n-dimensional manifold M there is defined an obstruction

$$s(f,Y) \in LS_{n-q}\left(\begin{array}{ccc} \pi_1(S(\xi)) & \longrightarrow & \pi_1(X-Y) \\ \downarrow & \Phi & \downarrow \\ \pi_1(Y) & \longrightarrow & \pi_1(X) \end{array}\right)$$

to deforming f by a homotopy to a map transverse at $Y \subset X$ with both the restrictions $f| : N = f^{-1}(Y) \longrightarrow Y$, $f| : M - N = f^{-1}(X-Y) \longrightarrow X - Y$ homotopy equivalences, i.e. to "splitting f along $Y \subset X$".

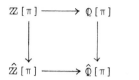

The LS-groups are defined geometrically to fit into the exact sequence

$$\ldots \longrightarrow L_{n+1}(\pi_1(X-Y) \longrightarrow \pi_1(X)) \longrightarrow LS_{n-q}(\Phi)$$

$$\longrightarrow L_{n-q}(\pi_1(Y)) \xrightarrow{\xi!} L_n(\pi_1(X-Y) \longrightarrow \pi_1(X)) \longrightarrow \ldots ,$$

the map $LS_{n-q}(\Phi) \longrightarrow L_{n-q}(\pi_1(Y))$ sending the ambient surgery obstruction $s(f,Y)$ to the abstract surgery obstruction $\sigma_*(f|:N \longrightarrow Y)$.

The expression in I. of the surgery obstruction groups $L_*(\pi)$ in terms of quadratic Poincaré complexes and the chain homotopy invariant expression in II. of the surgery obstruction $\sigma_*(f,b$ are extended in §7 to the LS-groups $LS_*(\Phi)$ and the codimension q splitting obstruction. Many authors have used geometric techniques to prove splitting theorems for manifolds, which are equivalent to vanishing theorems for the LS-groups and hence to the existence of Mayer-Vietoris exact sequences in the surgery obstruction groups. For example, the codimension 1 splitting theorem of Cappell implies that there exists such a sequence for many free products with amalgamation $\pi_1 *_\rho \pi_2$

$$\ldots \longrightarrow L_n(\rho) \longrightarrow L_n(\pi_1) \oplus L_n(\pi_2) \longrightarrow L_n(\pi_1 *_\rho \pi_2) \longrightarrow L_{n-1}(\rho) \longrightarrow \ldots$$

and for many HNN extensions $\pi *_\rho \{t\}$

$$\ldots \longrightarrow L_n(\rho) \longrightarrow L_n(\pi) \longrightarrow L_n(\pi *_\rho \{t\}) \longrightarrow L_{n-1}(\rho) \longrightarrow \ldots .$$

The next instalment of the series (Ranicki [11]) will be devoted to carrying out the programme put forward in §7.5 for an algebraic derivation of codimension q splitting theorems, using an algebraic theory of codimension q transversality. This should also apply to the symmetric L-groups $L^*(\pi)$, even though the example of Proposition 7.6.8 shows that they are not in general geometrically realizable.

This is principally a research monograph, presenting work done over a period of ten years. I should like to thank for their support the various institutions with which I have been associated in that time:

Trinity College, Cambridge (1972 - 1977)

Institute des Hautes Études Scientifiques, Bures sur Yvette (1973 - 1974)

Princeton University (*) (1977 - 1981)

Princeton,

April, 1981

(*) Including partial support from NSF Grants

§1. Absolute L-theory

In §1 we reiterate all the concepts of I. and II. which we shall be using here, particularly the definition of the n-dimensional $\left\{\begin{array}{l}\varepsilon\text{-symmetric}\\ \varepsilon\text{-quadratic}\end{array}\right.$ L-groups $\left\{\begin{array}{l}L^n(A,\varepsilon)\\ L_n(A,\varepsilon)\end{array}\right.$ $(n\in\mathbb{Z})$ of a ring with involution A as the cobordism groups of n-dimensional $\left\{\begin{array}{l}\varepsilon\text{-symmetric}\\ \varepsilon\text{-quadratic}\end{array}\right.$ Poincaré complexes over A. Also, the geometric background of the L-groups is recalled: this is important even in a purely algebraic context, since algebraic Poincaré cobordism has all the formal properties of the cobordism of manifolds (as indeed does geometric Poincaré cobordism). For example, there are cobordism exact sequences both in the geometry and in the algebra, and the relative L-groups will be defined in §2 as relative algebraic Poincaré cobordism groups.

§1 also contains some new material, specifically the triad Q-groups of §1.3, the glueing of forms and formations of §1.7 and the $\mathcal{L}$-categories of §1.8.

1.1 Q-groups

Let A be a ring with involution, that is an associative ring with 1, together with a function

$$\bar{\ } : A \longrightarrow A \; ; \; a \longmapsto \bar{a}$$

such that

$$(\overline{a+b}) = \bar{a} + \bar{b} \; , \; (\overline{ab}) = \bar{b}.\bar{a} \; , \; \bar{\bar{a}} = a \; , \; \bar{1} = 1 \in A \; (a,b \in A) .$$

Given a left A-module M let M^t be the right A-module defined by additive group of M with A acting by

$$M^t \times A \longrightarrow M^t \; ; \; (x,a) \longmapsto \bar{a}x \; .$$

Except where a right A-module structure is specified "A-module" refers to a left A-module structure.

Given A-modules M,N let $\mathrm{Hom}_A(M,N)$ be the abelian group of A-module morphisms

$$f : M \longrightarrow N \; .$$

The dual of an A-module M is the A-module

$$M^* = \mathrm{Hom}_A(M,A)$$

with A acting by

$$A \times M^* \longrightarrow M^* \; ; \; (a,f) \longmapsto (x \longmapsto f(x).\bar{a}) \quad (x \in M) .$$

The dual of an A-module morphism $f \in \mathrm{Hom}_A(M,N)$ is the A-module morphism $f^* \in \mathrm{Hom}_A(N^*,M^*)$ defined by

$$f^* : N^* \longrightarrow M^* \; ; \; g \longmapsto (x \longmapsto g(f(x))) \; .$$

If M is a f.g.(= finitely generated) projective A-module then so is M*, and there is defined a natural A-module isomorphism

$$M \longrightarrow M^{**} \; ; \; x \longmapsto (h \longmapsto \overline{h(x)}) \quad (h \in M^*)$$

which we shall use to identify M** = M.

An <u>A-module chain complex</u> C is a sequence of A-modules and A-module morphisms

$$C : \ldots \longrightarrow C_{r+1} \xrightarrow{d} C_r \xrightarrow{d} C_{r-1} \longrightarrow \ldots \quad (r \in \mathbb{Z})$$

such that

$$d^2 = 0 \in \mathrm{Hom}_A(C_r, C_{r-2}) \quad (r \in \mathbb{Z})$$

The $\left\{\begin{array}{l}\underline{\text{homology}}\\ \underline{\text{cohomology}}\end{array}\right.$ A-modules of C are defined (as usual) by

$$\left\{\begin{array}{l} H_r(C) = \ker(d:C_r \longrightarrow C_{r-1})/\mathrm{im}(d:C_{r+1} \longrightarrow C_r) \\ H^r(C) = \ker(d^*:C^r \longrightarrow C^{r+1})/\mathrm{im}(d^*:C^{r-1} \longrightarrow C^r) \end{array}\right. \quad (r \in \mathbb{Z},\ C^r = C_r^*) .$$

A <u>chain map</u> of A-module chain complexes

$$f : C \longrightarrow D$$

is a collection of A-module morphisms $\{f \in \mathrm{Hom}_A(C_r, D_r) \mid r \in \mathbb{Z}\}$ such that

$$d_D f = f d_C \in \mathrm{Hom}_A(C_r, D_{r-1}) \quad (r \in \mathbb{Z}) .$$

A <u>chain homotopy</u> of chain maps

$$g : f \simeq f' : C \longrightarrow D$$

is a collection of A-module morphisms $\{g \in \mathrm{Hom}_A(C_r, D_{r+1}) \mid r \in \mathbb{Z}\}$ such that

$$f' - f = d_D g + g d_C \in \mathrm{Hom}_A(C_r, D_r) \quad (r \in \mathbb{Z}) .$$

A <u>chain equivalence</u> is a chain map $f:C \longrightarrow D$ which admits a chain homotopy inverse, i.e. a chain map $f':D \longrightarrow C$ for which there exist chain homotopies $g:f'f \simeq 1:C \longrightarrow C$, $g':ff' \simeq 1:D \longrightarrow D$. A chain complex C is <u>chain contractible</u> if it is chain equivalent to 0; a chain homotopy $\Gamma:1 \simeq 0:C \longrightarrow C$ is a <u>chain contraction</u> of C.

An A-module chain map $f:C\longrightarrow D$ induces A-module morphisms in $\begin{cases}\text{homology}\\ \text{cohomology}\end{cases}$

$$\begin{cases} f_* : H_r(C)\longrightarrow H_r(D) \; ; \; x\longmapsto f(x) & (x\in C_r) \\ f^* : H^r(D)\longrightarrow H^r(C) \; ; \; y\longmapsto f^*y & (y\in C^r) \end{cases} \quad (r\in\mathbb{Z})$$

which depend only on the chain homotopy class of f, and are isomorphisms if f is a chain equivalence. The <u>algebraic mapping cone</u> $C(f)$ of f is the A-module chain complex defined by

$$d_{C(f)} = \begin{pmatrix} d_D & (-)^{r-1}f \\ 0 & d_C \end{pmatrix} : C(f)_r = D_r\oplus C_{r-1}\longrightarrow C(f)_{r-1} = D_{r-1}\oplus C_{r-2} \; .$$

The <u>relative</u> $\begin{cases}\underline{\text{homology}}\\ \underline{\text{cohomology}}\end{cases}$ A-modules of f are defined by

$$\begin{cases} H_r(f) = H_r(C(f)) \\ H^r(f) = H^r(C(f)) \end{cases} \quad (r\in\mathbb{Z})$$

and are such that there are defined exact sequences of A-modules

$$\begin{cases} \dots\longrightarrow H_r(C)\xrightarrow{f_*}H_r(D)\longrightarrow H_r(f)\longrightarrow H_{r-1}(C)\xrightarrow{f_*}H_{r-1}(D)\longrightarrow\dots \\ \dots\longrightarrow H^r(D)\xrightarrow{f^*}H^r(C)\longrightarrow H^r(f)\longrightarrow H^{r+1}(D)\xrightarrow{f^*}H^{r+1}(C)\longrightarrow\dots \; . \end{cases}$$

An A-module chain complex is <u>n-dimensional</u> if it is a finite complex of f.g. projective A-modules which is chain equivalent to a f.g. projective complex of the type

$$C : \dots\longrightarrow 0\longrightarrow C_n\xrightarrow{d}C_{n-1}\longrightarrow\dots\longrightarrow C_1\xrightarrow{d}C_0\longrightarrow 0\longrightarrow\dots \; .$$

For $n\leqslant -1$ n-dimensional = chain contractible, by convention. A finite-dimensional chain complex C is n-dimensional if and only if $H_r(C) = 0$ for $r<0$ and $H^r(C) = 0$ for $r>n$. A chain map $f:C\longrightarrow D$ of finite-dimensional chain complexes is a chain equivalence if and only if $H_*(f) = 0$, or equivalently if $C(f)$ is chain contractible.

Given A-module chain complexes C,D let $C^t\otimes_A D$, $\mathrm{Hom}_A(C,D)$ be the $\mathbb{Z}$-module chain complexes defined by

$$d_{C^t\otimes_A D} : (C^t\otimes_A D)_n = \sum_{p+q=n} C^t_p \,{}_A D_q \longrightarrow (C^t\otimes_A D)_{n-1} ;$$

$$x\otimes y \longmapsto x\otimes d_D(y) + (-)^q d_C(x)\otimes y ,$$

$$d_{\mathrm{Hom}_A(C,D)} : \mathrm{Hom}_A(C,D)_n = \sum_{q-p=n} \mathrm{Hom}_A(C_p,D_q) \longrightarrow \mathrm{Hom}_A(C,D)_{n-1}$$

$$f \longmapsto d_D f + (-)^q f d_C .$$

Let C^* be the A-module chain complex defined by

$$d_{C^*} = (d_C)^* : (C^*)_r = C^{-r} \longrightarrow (C^*)_{r-1} = C^{-r+1} ,$$

and let C^{n-*} ($n\in\mathbb{Z}$) be the A-module chain complex defined by

$$d_{C^{n-*}} = (-)^r (d_C)^* : (C^{n-*})_r = C^{n-r} \longrightarrow (C^{n-*})_{r-1} = C^{n-r+1} .$$

The sign conventions are such that an element $f\in H_n(\mathrm{Hom}_A(C^*,D))$ is the same as a chain homotopy class of chain maps

$$f : C^{n-*} \longrightarrow D .$$

Let $\varepsilon\in A$ be a central unit such that

$$\bar{\varepsilon} = \varepsilon^{-1} \in A$$

(e.g. $\varepsilon = \pm 1\in A$). Given a finite-dimensional A-module chain complex C let the generator $T\in\mathbb{Z}_2$ act on $\mathrm{Hom}_A(C^*,C)$ by the <u>ε-duality</u> involution

$$T_\varepsilon : \mathrm{Hom}_A(C^p,C_q) \longrightarrow \mathrm{Hom}_A(C^q,C_p) ;\ f \longmapsto \varepsilon f^*$$

$$(\varepsilon f^*(y)(x) = \varepsilon.\overline{f(y)(x)} \in A,\ x\in C^p ,\ y\in C^q) .$$

Define the $\begin{cases}\text{\underline{ε-quadratic}}\\ \text{\underline{ε-symmetric}}\\ \text{\underline{ε-hyperquadratic}}\end{cases}$ <u>Q-groups</u> $\begin{cases} Q^*(C,\varepsilon)\\ Q_*(C,\varepsilon)\\ \hat{Q}^*(C,\varepsilon)\end{cases}$ to be the

$$\left\{\begin{array}{l}\mathbb{Z}_2\text{-hypercohomology}\\ \mathbb{Z}_2\text{-hyperhomology}\\ \text{Tate } \mathbb{Z}_2\text{-hypercohomology}\end{array}\right. \quad \text{groups}$$

$$\left\{\begin{array}{l}Q^n(C,\varepsilon) = H_n(\mathrm{Hom}_{\mathbb{Z}[\mathbb{Z}_2]}(W,\mathrm{Hom}_A(C^*,C)))\\ Q_n(C,\varepsilon) = H_n(W\otimes_{\mathbb{Z}[\mathbb{Z}_2]}\mathrm{Hom}_A(C^*,C))\\ \hat{Q}^n(C,\varepsilon) = H_n(\mathrm{Hom}_{\mathbb{Z}[\mathbb{Z}_2]}(\hat{W},\mathrm{Hom}_A(C^*,C)))\end{array}\right. \qquad (n\in\mathbb{Z}) \quad ,$$

with W the standard free $\mathbb{Z}[\mathbb{Z}_2]$-resolution of $\mathbb{Z}$

$$W : \ldots \longrightarrow \mathbb{Z}[\mathbb{Z}_2] \xrightarrow{1-T} \mathbb{Z}[\mathbb{Z}_2] \xrightarrow{1+T} \mathbb{Z}[\mathbb{Z}_2] \xrightarrow{1-T} \mathbb{Z}[\mathbb{Z}_2] \longrightarrow 0$$

and $\hat{W}$ the standard complete free $\mathbb{Z}[\mathbb{Z}_2]$-resolution of $\mathbb{Z}$

$$\hat{W} : \ldots \longrightarrow \mathbb{Z}[\mathbb{Z}_2] \xrightarrow{1-T} \mathbb{Z}[\mathbb{Z}_2] \xrightarrow{1+T} \mathbb{Z}[\mathbb{Z}_2] \xrightarrow{1-T} \mathbb{Z}[\mathbb{Z}_2] \xrightarrow{1+T} \mathbb{Z}[\mathbb{Z}_2] \longrightarrow \ldots .$$

An element $\left\{\begin{array}{l}\phi \in Q^n(C,\varepsilon)\\ \psi \in Q_n(C,\varepsilon)\\ \theta \in \hat{Q}^n(C,\varepsilon)\end{array}\right.$ is an equivalence class of collections of

chains of $\mathrm{Hom}_A(C^*,C)$

$$\left\{\begin{array}{l}\{\phi_s \in \mathrm{Hom}_A(C^{n-r+s},C_r)\,|\,r\in\mathbb{Z}, s\geqslant 0\}\\ \{\psi_s \in \mathrm{Hom}_A(C^{n-r-s},C_r)\,|\,r\in\mathbb{Z}, s\geqslant 0\}\\ \{\theta_s \in \mathrm{Hom}_A(C^{n-r+s},C_r)\,|\,r\in\mathbb{Z}, s\in\mathbb{Z}\}\end{array}\right.$$

such that

$$\begin{cases} d\phi_s + (-)^r\phi_s d^* + (-)^{n+s-1}(\phi_{s-1} + (-)^s T_\varepsilon \phi_{s-1}) = 0 \\ \qquad : C^{n-r+s-1} \longrightarrow C_r \quad (s \geqslant 0,\ \phi_{-1} = 0) \\ d\psi_s + (-)^r\psi_s d^* + (-)^{n-s-1}(\psi_{s+1} + (-)^{s+1} T_\varepsilon \psi_{s+1}) = 0 \\ \qquad : C^{n-r-s-1} \longrightarrow C_r \quad (s \geqslant 0) \\ d\theta_s + (-)^r\theta_s d^* + (-)^{n+s-1}(\theta_{s-1} + (-)^s T_\varepsilon \theta_{s-1}) = 0 \\ \qquad : C^{n-r+s-1} \longrightarrow C_r \quad (s \in \mathbb{Z}) \ . \end{cases}$$

A chain map of finite-dimensional A-module chain complexes

$$f : C \longrightarrow D$$

induces a $\mathbb{Z}[\mathbb{Z}_2]$-module chain map

$$\mathrm{Hom}_A(f^*,f) : \mathrm{Hom}_A(C^*,C) \longrightarrow \mathrm{Hom}_A(D^*,D) \ ; \ \phi \longmapsto f^*\phi f \ ,$$

and hence also morphisms in the Q-groups

$$\begin{cases} f^{\%} : Q^n(C,\varepsilon) \longrightarrow Q^n(D,\varepsilon) \ ; \ \phi = \{\phi_s\} \longmapsto f^{\%}\phi = \{f^*\phi_s f\} \\ f_{\%} : Q_n(C,\varepsilon) \longrightarrow Q_n(D,\varepsilon) \ ; \ \psi = \{\psi_s\} \longmapsto f_{\%}\psi = \{f^*\psi_s f\} \\ \hat{f}^{\%} : \hat{Q}^n(C,\varepsilon) \longrightarrow \hat{Q}^n(D,\varepsilon) \ ; \ \theta = \{\theta_s\} \longmapsto \hat{f}^{\%}\theta = \{f^*\theta_s f\} \ . \end{cases}$$

An A-module chain homotopy $g:f \simeq f':C \longrightarrow D$ does not in general determine a $\mathbb{Z}[\mathbb{Z}_2]$-module chain homotopy

$$\mathrm{Hom}_A(f^*,f) \simeq \mathrm{Hom}_A(f'^*,f') : \mathrm{Hom}_A(C^*,C) \longrightarrow \mathrm{Hom}_A(D^*,D) \ .$$

Nevertheless the Q-group morphisms induced by an A-module chain map f depend only on the chain homotopy class of f (cf. Proposition 1.1.1 below). In order to account for the chain homotopy invariance of the Q-groups we define (as in §I.1) the "$\mathbb{Z}_2$-isovariant category" with objects $\mathbb{Z}[\mathbb{Z}_2]$-module chain complexes, as follows.

A $\underline{\mathbb{Z}_2\text{-isovariant chain map}}$ of $\mathbb{Z}[\mathbb{Z}_2]$-module chain complexes

$$f : C \longrightarrow D$$

is a collection of $\mathbb{Z}$- module morphisms

$$f = \{f_s \in \mathrm{Hom}_{\mathbb{Z}}(C_r, D_{r+s}) \mid r \in \mathbb{Z}, s \geqslant 0\}$$

such that

$$d_D f_s + (-)^{s-1} f_s d_C + (-)^{s-1}(f_{s-1} + (-)^s T_D f_{s-1} T_C) = 0$$

$$: C_r \longrightarrow D_{r+s-1} \quad (s \geqslant 0, f_{-1} = 0) .$$

Thus $f_0 : C \longrightarrow D$ is a $\mathbb{Z}$-module chain map, $f_1 : f_0 \simeq T_D f_0 T_C : C \longrightarrow D$ is a $\mathbb{Z}$-module chain homotopy, and $f_2, f_3, \ldots$ are higher $\mathbb{Z}$-module chain homotopies. A $\underline{\mathbb{Z}_2\text{-isovariant chain homotopy}}$ of $\mathbb{Z}_2$-isovariant chain maps

$$g : f \simeq f' : C \longrightarrow D$$

is a collection of $\mathbb{Z}$ -module morphisms

$$g = \{g_s \in \mathrm{Hom}_{\mathbb{Z}}(C_r, D_{r+s+1}) \mid r \in \mathbb{Z}, s \geqslant 0\}$$

such that

$$f'_s - f_s = d_D g_s + (-)^s g_s d_C + (-)^s (g_{s-1} + (-)^{s-1} T_D g_{s-1} T_C) = 0$$

$$: C_r \longrightarrow D_{r+s} \quad (s \geqslant 0, g_{-1} = 0) .$$

In particular, a $\mathbb{Z}_2$-isovariant chain map $f : C \longrightarrow D$ with $f_s = 0$ $(s \geqslant 1)$ is the same as a $\mathbb{Z}[\mathbb{Z}_2]$-module chain map $f_0 : C \longrightarrow D$, and a $\mathbb{Z}_2$-isovariant chain homotopy $g : f \simeq f' : C \longrightarrow D$ of such chain maps with $g_s = 0$ $(s \geqslant 1)$ is the same as a $\mathbb{Z}[\mathbb{Z}_2]$-module chain homotopy $g_0 : f_0 \simeq f'_0 : C \longrightarrow D$. The $\underline{\mathbb{Z}_2\text{-isovariant category}}$ is the category with objects $\mathbb{Z}[\mathbb{Z}_2]$-module chain complexes and morphisms the $\mathbb{Z}_2$-isovariant chain homotopy classes of $\mathbb{Z}_2$-isovariant chain maps.

A morphism is thus an element $f \in H_0(\mathrm{Hom}_{\mathbb{Z}[\mathbb{Z}_2]}(W,\mathrm{Hom}_{\mathbb{Z}}(C,D)))$, with $T \in \mathbb{Z}_2$ acting on $\mathrm{Hom}_{\mathbb{Z}}(C,D)$ by the involution

$$T : \mathrm{Hom}_{\mathbb{Z}}(C,D) \longrightarrow \mathrm{Hom}_{\mathbb{Z}}(C,D) \; ; \; h \longmapsto T_D h T_C \; .$$

The composite of the $\mathbb{Z}_2$-isovariant morphisms $f:C \longrightarrow D$, $g:D \longrightarrow E$ is the $\mathbb{Z}_2$-isovariant morphism $gf:C \longrightarrow E$ defined by

$$(gf)_s = \sum_{r=0}^{s} (-)^{r(s-r)} (T^r g_{s-r}) f_r \; : \; C_t \longrightarrow E_{s+t} \quad (s \geqslant 0,\ t \in \mathbb{Z}) \; ,$$

which is the image of $f \otimes g$ under the pairing

$$H_0(\mathrm{Hom}_{\mathbb{Z}[\mathbb{Z}_2]}(W,\mathrm{Hom}_{\mathbb{Z}}(C,D))) \otimes_{\mathbb{Z}} H_0(\mathrm{Hom}_{\mathbb{Z}[\mathbb{Z}_2]}(W,\mathrm{Hom}_{\mathbb{Z}}(D,E)))$$

$$\xrightarrow{\Delta^*} H_0(\mathrm{Hom}_{\mathbb{Z}[\mathbb{Z}_2]}(W,\mathrm{Hom}_{\mathbb{Z}}(C,D) \otimes_{\mathbb{Z}} \mathrm{Hom}_{\mathbb{Z}}(D,E)))$$

$$\xrightarrow{c^*} H_0(\mathrm{Hom}_{\mathbb{Z}[\mathbb{Z}_2]}(W,\mathrm{Hom}_{\mathbb{Z}}(C,E)))$$

defined by the composite of the product induced by the diagonal $\mathbb{Z}[\mathbb{Z}_2]$-module chain map $\Delta:W \longrightarrow W \otimes_{\mathbb{Z}} W$ given by

$$\Delta : W_s \longrightarrow (W \otimes_{\mathbb{Z}} W)_s = \sum_{r=0}^{s} W_r \otimes_{\mathbb{Z}} W_{s-r} \; ; \; 1_s \longmapsto \sum_{r=0}^{s} 1_r \otimes T^r_{s-r} \quad (s \geqslant 0)$$

and the composition pairing induced by

$$c : \mathrm{Hom}_{\mathbb{Z}}(C,D) \otimes_{\mathbb{Z}} \mathrm{Hom}_{\mathbb{Z}}(D,E) \longrightarrow \mathrm{Hom}_{\mathbb{Z}}(C,E) \; ; \; h \otimes k \longmapsto kh \; .$$

A $\mathbb{Z}_2$-isovariant morphism $f:C \longrightarrow D$ induces morphisms in the

$\left\{\begin{array}{l}\mathbb{Z}_2\text{-hypercohomology} \\ \mathbb{Z}_2\text{-hyperhomology} \\ \text{Tate } \mathbb{Z}_2\text{-hypercohomology}\end{array}\right.$ groups

$$\left\{\begin{array}{l}
f^{\%} : H_n(\mathrm{Hom}_{\mathbb{Z}[\mathbb{Z}_2]}(W,C)) \longrightarrow H_n(\mathrm{Hom}_{\mathbb{Z}[\mathbb{Z}_2]}(W,D)) \ ; \\
\quad \phi = \{\phi_s \in C_{n+s} \mid s \geqslant 0\} \longmapsto f^{\%}\phi = \{\sum_{r=0}^{s} (-)^{r(s-r)} (T^r f_{s-r})\phi_r \in D_{n+s} \mid s \geqslant 0\} \\
f_{\%} : H_n(W \otimes_{\mathbb{Z}[\mathbb{Z}_2]} C) \longrightarrow H_n(W \otimes_{\mathbb{Z}[\mathbb{Z}_2]} D) \ ; \\
\quad \psi = \{\psi_s \in C_{n-s} \mid s \geqslant 0\} \longmapsto f_{\%}\psi = \{\sum_{r=0}^{\infty} (-)^{r(s-r)} (T^r f_{r-s})\psi_r \in D_{n-s} \mid s \geqslant 0\} \\
\hat{f}^{\%} : H_n(\mathrm{Hom}_{\mathbb{Z}[\mathbb{Z}_2]}(\hat{W},C)) \longrightarrow H_n(\mathrm{Hom}_{\mathbb{Z}[\mathbb{Z}_2]}(\hat{W},D)) \ ; \\
\quad \theta = \{\theta_s \in C_{n+s} \mid s \geqslant 0\} \longmapsto \hat{f}^{\%}\theta = \{\sum_{r=-\infty}^{\infty} (-)^{r(s-r)} (T^r f_{s-r})\theta_r \in D_{n+s} \mid s \in \mathbb{Z}\} ,
\end{array}\right.$$

which are the evaluations on $f \in H_0(\mathrm{Hom}_{\mathbb{Z}[\mathbb{Z}_2]}(W,\mathrm{Hom}_{\mathbb{Z}}(C,D)))$ of the natural pairings

$$\left\{\begin{array}{l}
H_0(\mathrm{Hom}_{\mathbb{Z}[\mathbb{Z}_2]}(W,\mathrm{Hom}_{\mathbb{Z}}(C,D))) \otimes_{\mathbb{Z}} H_n(\mathrm{Hom}_{\mathbb{Z}[\mathbb{Z}_2]}(W,C)) \\
\qquad\qquad \longrightarrow H_n(\mathrm{Hom}_{\mathbb{Z}[\mathbb{Z}_2]}(W,D)) \\
H_0(\mathrm{Hom}_{\mathbb{Z}[\mathbb{Z}_2]}(W,\mathrm{Hom}_{\mathbb{Z}}(C,D)) \otimes_{\mathbb{Z}} H_n(W \otimes_{\mathbb{Z}[\mathbb{Z}_2]} C) \\
\qquad\qquad \longrightarrow H_n(W \otimes_{\mathbb{Z}[\mathbb{Z}_2]} D) \\
H_0(\mathrm{Hom}_{\mathbb{Z}[\mathbb{Z}_2]}(W,\mathrm{Hom}_{\mathbb{Z}}(C,D)) \otimes_{\mathbb{Z}} H_n(\mathrm{Hom}_{\mathbb{Z}[\mathbb{Z}_2]}(\hat{W},C)) \\
\qquad\qquad \longrightarrow H_n(\mathrm{Hom}_{\mathbb{Z}[\mathbb{Z}_2]}(\hat{W},D)) \ .
\end{array}\right.$$

Proposition 1.1.1 The Q-group morphisms induced by a chain map $f:C\longrightarrow D$ of finite-dimensional A-module chain complexes

$$\begin{cases} f^{\%}:Q^*(C,\varepsilon)\longrightarrow Q^*(D,\varepsilon) \\ f_{\%}:Q_*(C,\varepsilon)\longrightarrow Q_*(D,\varepsilon) \\ \hat{f}^{\%}:\hat{Q}^*(C,\varepsilon)\longrightarrow \hat{Q}^*(D,\varepsilon) \end{cases}$$

depend only on the chain homotopy class of f. In particular, they are isomorphisms if f is a chain equivalence.

Proof: An A-module chain homotopy

$$g\ :\ f\simeq f'\ :\ C\longrightarrow D$$

determines the $\mathbb{Z}_2$-isovariant chain homotopy

$$\mathrm{Hom}_A(g^*,g)\ :\ \mathrm{Hom}_A(f^*,f)\simeq \mathrm{Hom}_A(f'^*,f')\ :$$

$$\mathrm{Hom}_A(C^*,C)\longrightarrow \mathrm{Hom}_A(D^*,D)$$

defined by

$$\mathrm{Hom}_A(g^*,g)_s\ :\ \mathrm{Hom}_A(C^*,C)_r = \sum_{p+q=r}\mathrm{Hom}_A(C^p,C_q)$$

$$\longrightarrow \mathrm{Hom}_A(D^*,D)_{r+s+1}\ ;\ \phi\longmapsto \begin{cases} g\phi f^* + (-)^q f'\phi g^* & \text{if } s=0 \\ (-)^{q+1} g\phi g^* & \text{if } s=1 \\ 0 & \text{if } s\geqslant 2\ . \end{cases}$$

[]

The various Q-groups of a finite-dimensional A-module chain complex C are related to each other by the abelian group morphisms

$$1+T_\varepsilon : Q_n(C,\varepsilon) \longrightarrow Q^n(C,\varepsilon) \ ; \ \psi \longmapsto \{((1+T_\varepsilon)\psi)_s = \begin{cases} (1+T_\varepsilon)\psi_0 & \text{if } s=0 \\ 0 & \text{if } s \geqslant 1 \end{cases}\}$$

$$J : Q^n(C,\varepsilon) \longrightarrow \hat{Q}^n(C,\varepsilon) \ ; \ \phi \longmapsto \{(J\phi)_s = \begin{cases} \phi_s & \text{if } s \geqslant 0 \\ 0 & \text{if } s \leqslant -1 \end{cases}\}$$

$$H : \hat{Q}^n(C,\varepsilon) \longrightarrow Q_{n-1}(C,\varepsilon) \ ; \ \theta \longmapsto \{(H\theta)_s = \theta_{-s-1} \mid s \geqslant 0\} \ .$$

Proposition 1.1.2 The Q-group sequence

$$\ldots \longrightarrow \hat{Q}^{n+1}(C,\varepsilon) \xrightarrow{H} Q_n(C,\varepsilon) \xrightarrow{1+T_\varepsilon} Q^n(C,\varepsilon) \xrightarrow{J} \hat{Q}^n(C,\varepsilon) \xrightarrow{H} Q_{n-1}(C,\varepsilon) \longrightarrow \ldots$$

is exact.

Proof: See Proposition I.1.2.

[]

The $\begin{cases} \text{suspension } SC \\ \text{desuspension } \Omega C \end{cases}$ of an A-module chain complex C is the A-module chain complex defined by

$$\begin{cases} d_{SC} = d_C : (SC)_r = C_{r-1} \longrightarrow (SC)_{r-1} = C_{r-2} \\ d_{\Omega C} = d_C : (\Omega C)_r = C_{r+1} \longrightarrow (\Omega C)_{r-1} = C_r \end{cases} ,$$

so that $S\Omega C = \Omega SC = C$ and

$$\begin{cases} H_r(SC) = H_{r-1}(C) \\ H_r(\Omega C) = H_{r+1}(C) \end{cases} , \quad \begin{cases} H^r(SC) = H^{r-1}(C) \\ H^r(\Omega C) = H^{r+1}(C) \end{cases} .$$

Given a finite-dimensional A-module chain complex C define the suspension maps in the $\left\{\begin{array}{l}\varepsilon\text{-symmetric}\\ \varepsilon\text{-quadratic}\end{array}\right.$ Q-groups

$$\left\{\begin{array}{l} S : Q^n(C,\varepsilon) \longrightarrow Q^{n+1}(SC,\varepsilon) \ ; \ \phi \longmapsto S\phi \\ S : Q_n(C,\varepsilon) \longrightarrow Q_{n+1}(SC,\varepsilon) \ ; \ \psi \longmapsto S\psi \end{array}\right. \qquad (n \in \mathbb{Z})$$

by

$$\left\{\begin{array}{l} (S\phi)_s = \phi_{s-1} : (SC)^{n-r+s+1} = C^{n-r+s} \longrightarrow (SC)_r = C_{r-1} \\ (S\psi)_s = \psi_{s+1} : (SC)^{n-r-s+1} = C^{n-r-s} \longrightarrow (SC)_r = C_{r-1} \end{array}\right.$$

$$(r \in \mathbb{Z},\ s \geqslant 0,\ \phi_{-1} = 0) \ .$$

For each $p \geqslant 0$ define the $\mathbb{Z}[\mathbb{Z}_2]$-module chain complex $W[0,p]$ to be the subcomplex of W with

$$W[0,p]_s = \left\{\begin{array}{ll} W_s \ (= \mathbb{Z}[\mathbb{Z}_2]) & \text{if } 0 \leqslant s \leqslant p \\ 0 & \text{otherwise} \end{array}\right. .$$

Define the unstable <u>ε-quadratic Q-groups</u> $Q_*^{[0,p]}(C,\varepsilon)$ of a finite-dimensional A-module chain complex C by

$$Q_n^{[0,p]}(C,\varepsilon) = H_n(W[0,p] \otimes_{\mathbb{Z}[\mathbb{Z}_2]} \mathrm{Hom}_A(C^*,C)) \quad (n \in \mathbb{Z}) \ .$$

In particular, passing to the limit as $p \quad \infty$ we have

$$W[0,\infty] = W \ , \ Q_*^{[0,\infty]}(C,\varepsilon) = Q_*(C,\varepsilon) \ .$$

In the applications of the Q-groups to the algebraic theory of surgery it is useful to have available the following unstable analogue of the exact sequence of Proposition 1.1.2. For example, the unstable ε-quadratic Q-groups $Q_*^{[0,0]}(C,\varepsilon) = H_*(\mathrm{Hom}_A(C^*,C))$ appear in the algebraic theory of codimension 2 surgery outlined in §7.8 below.

Proposition 1.1.3 Given a finite-dimensional A-module chain complex C there is defined an exact sequence of Q-groups

$$\ldots \longrightarrow Q^{n+p+1}(S^pC,\varepsilon) \longrightarrow Q_n^{[0,p-1]}(C,\varepsilon)$$

$$\longrightarrow Q^n(C,\varepsilon) \xrightarrow{S^p} Q^{n+p}(S^pC,\varepsilon) \longrightarrow \ldots \quad (n\in\mathbb{Z})$$

for any $p \geqslant 1$. If C is n-dimensional and $p \geqslant n+1$ then

$$Q_*^{[0,p-1]}(C,\varepsilon) = Q_*^{[0,\infty]}(C,\varepsilon) = Q_*(C,\varepsilon) \ , \ Q^{*+p}(S^pC,\varepsilon) = \hat{Q}^*(C,\varepsilon)$$

and the sequence coincides with the exact sequence of Proposition 1.1.2

$$\ldots \longrightarrow \hat{Q}^{n+1}(C,\varepsilon) \xrightarrow{H} Q_n(C,\varepsilon) \xrightarrow{1+T_\varepsilon} Q^n(C,\varepsilon) \xrightarrow{J} \hat{Q}^n(C,\varepsilon) \longrightarrow \ldots .$$

Proof: See Proposition I.1.3.

[]

Define the relative $\left\{\begin{array}{l}\underline{\varepsilon\text{-quadratic}}\\ \underline{\varepsilon\text{-symmetric}}\\ \underline{\varepsilon\text{-hyperquadratic}}\end{array}\right.$ Q-groups $\left\{\begin{array}{l}Q^*(f,\varepsilon)\\ Q_*(f,\varepsilon)\\ \hat{Q}^*(f,\varepsilon)\end{array}\right.$

of a chain map $f:C \longrightarrow D$ of finite-dimensional A-module chain complexes by

$$\left\{\begin{array}{ll}Q^{n+1}(f,\varepsilon) = H_{n+1}(\mathrm{Hom}_{\mathbb{Z}[\mathbb{Z}_2]}(W,C(\mathrm{Hom}_A(f^*,f)))) & \\ Q_{n+1}(f,\varepsilon) = H_{n+1}(W\otimes_{\mathbb{Z}[\mathbb{Z}_2]}C(\mathrm{Hom}_A(f^*,f))) & (n\in\mathbb{Z})\\ \hat{Q}^{n+1}(f,\varepsilon) = H_{n+1}(\mathrm{Hom}_{\mathbb{Z}[\mathbb{Z}_2]}(\hat{W},C(\mathrm{Hom}_A(f^*,f)))) & ,\end{array}\right.$$

where $C(\mathrm{Hom}_A(f^*,f))$ is the algebraic mapping cone of the induced $\mathbb{Z}[\mathbb{Z}_2]$-module chain map

$$\mathrm{Hom}_A(f^*,f) : \mathrm{Hom}_A(C^*,C) \longrightarrow \mathrm{Hom}_A(D^*,D) \ ; \ \phi \longmapsto f^*\phi f \ .$$

An element $\begin{cases} (\delta\phi,\phi) \in Q^{n+1}(f,\varepsilon) \\ (\delta\psi,\psi) \in Q_{n+1}(f,\varepsilon) \\ (\delta\theta,\theta) \in \hat{Q}^{n+1}(f,\varepsilon) \end{cases}$ is an equivalence class of

collections of chains

$$\begin{cases} \{(\delta\phi,\phi)_s = (\delta\phi_s,\phi_s) \in \mathrm{Hom}_A(D^*,D)_{n+s-1} \oplus \mathrm{Hom}_A(C^*,C)_{n+s} \mid s \geqslant 0\} \\ \{(\delta\psi,\psi)_s = (\delta\psi_s,\psi_s) \in \mathrm{Hom}_A(D^*,D)_{n-s-1} \oplus \mathrm{Hom}_A(C^*,C)_{n-s} \mid s \geqslant 0\} \\ \{(\delta\theta,\theta)_s = (\delta\theta_s,\theta_s) \in \mathrm{Hom}_A(D^*,D)_{n+s-1} \oplus \mathrm{Hom}_A(C^*,C)_{n+s} \mid s \in \mathbb{Z}\} \end{cases}$$

such that

$$\begin{cases} d(\delta\phi,\phi)_s \\ \quad = (d(\delta\phi_s) + (-)^r(\delta\phi_s)d^* + (-)^{n+s}(\delta\phi_{s-1} + (-)^s T_\varepsilon \delta\phi_{s-1}) + (-)^n f\phi_s f^*, \\ \qquad d\phi_s + (-)^r \phi_s d^* + (-)^{n+s-1}(\phi_{s-1} + (-)^s T_\varepsilon \phi_{s-1})) \\ \quad = 0 \in \sum_{r=-\infty}^{\infty} \mathrm{Hom}_A(D^{n-r+s},D_r) \oplus \mathrm{Hom}_A(C^{n-r+s-1},C_r) \\ \qquad\qquad (s \geqslant 0,\ \delta\phi_{-1} = 0,\ \phi_{-1} = 0) \\ d(\delta\psi,\psi)_s \\ \quad = (d(\delta\psi_s) + (-)^r(\delta\psi_s)d^* + (-)^{n-s}(\delta\psi_{s+1} + (-)^{s+1} T_\varepsilon \delta\psi_{s+1}) + (-)^n f\psi_s f^*, \\ \qquad d\psi_s + (-)^r \psi_s d^* + (-)^{n-s-1}(\psi_{s+1} + (-)^{s+1} T_\varepsilon \psi_{s+1})) \\ \quad = 0 \in \sum_{r=-\infty}^{\infty} \mathrm{Hom}_A(D^{n-r-s},D_r) \oplus \mathrm{Hom}_A(C^{n-r-s-1},C_r) \quad (s \geqslant 0) \\ d(\delta\theta,\theta)_s \\ \quad = (d(\delta\theta_s) + (-)^r(\delta\theta_s)d^* + (-)^{n+s}(\delta\theta_{s-1} + (-)^s T_\varepsilon \delta\theta_{s-1}) + (-)^n f\theta_s f^*, \\ \qquad d\theta_s + (-)^r \theta_s d^* + (-)^{n+s-1}(\theta_{s-1} + (-)^s T_\varepsilon \theta_{s-1})) \\ \quad = 0 \in \sum_{r=-\infty}^{\infty} \mathrm{Hom}_A(D^{n-r+s},D_r) \oplus \mathrm{Hom}_A(C^{n-r+s-1},C_r) \quad (s \in \mathbb{Z}). \end{cases}$$

Proposition 1.1.4 The relative Q-groups of a chain map $f:C \longrightarrow D$ depend only on the chain homotopy class of f, and fit into the exact sequences

$$\begin{cases} \ldots \longrightarrow Q^{n+1}(C,\varepsilon) \xrightarrow{f^{\%}} Q^{n+1}(D,\varepsilon) \longrightarrow Q^{n+1}(f,\varepsilon) \longrightarrow Q^{n}(C,\varepsilon) \longrightarrow \ldots \\ \ldots \longrightarrow Q_{n+1}(C,\varepsilon) \xrightarrow{f_{\%}} Q_{n+1}(D,\varepsilon) \longrightarrow Q_{n+1}(f,\varepsilon) \longrightarrow Q_{n}(C,\varepsilon) \longrightarrow \ldots \\ \ldots \longrightarrow \hat{Q}^{n+1}(C,\varepsilon) \xrightarrow{\hat{f}^{\%}} \hat{Q}^{n+1}(D,\varepsilon) \longrightarrow \hat{Q}^{n+1}(f,\varepsilon) \longrightarrow \hat{Q}^{n}(C,\varepsilon) \longrightarrow \ldots \end{cases}$$

[]

1.2 L-groups

Let A,ε be as in §1.1 above.

An n-dimensional $\left\{\begin{array}{l}\varepsilon\text{-symmetric}\\ \varepsilon\text{-quadratic}\end{array}\right.$ complex over A $\left\{\begin{array}{l}(C,\phi)\\ (C,\psi)\end{array}\right.$ is an n-dimensional A-module chain complex C together with an element $\left\{\begin{array}{l}\phi \in Q^n(C,\varepsilon)\\ \psi \in Q_n(C,\varepsilon)\end{array}\right.$. Such a complex is Poincaré if $\left\{\begin{array}{l}\phi_0 \in H_n(\mathrm{Hom}_A(C^*,C))\\ (1+T_\varepsilon)\psi_0 \in H_n(\mathrm{Hom}_A(C^*,C))\end{array}\right.$ is a chain homotopy class of chain equivalences

$$\left\{\begin{array}{l}\phi_0 : C^{n-*} \longrightarrow C\\ (1+T_\varepsilon)\psi_0 : C^{n-*} \longrightarrow C\ .\end{array}\right.$$

The ε-symmetrization of an n-dimensional ε-quadratic complex over A (C,ψ) is the n-dimensional ε-symmetric complex over A

$$(1+T_\varepsilon)(C,\psi) = (C,(1+T_\varepsilon)\psi)\ .$$

A map (resp. homotopy equivalence) of n-dimensional $\left\{\begin{array}{l}\varepsilon\text{-symmetric}\\ \varepsilon\text{-quadratic}\end{array}\right.$ complexes over A

$$\left\{\begin{array}{l}f : (C,\phi) \longrightarrow (C',\phi')\\ f : (C,\psi) \longrightarrow (C',\psi')\end{array}\right.$$

is an A-module chain map (resp. chain equivalence) $f:C \longrightarrow C'$ such that

$$\left\{\begin{array}{l}f^{\%}(\phi) = \phi' \in Q^n(C',\varepsilon)\\ f_{\%}(\psi) = \psi' \in Q_n(C',\varepsilon)\ .\end{array}\right.$$

An <u>(n+1)-dimensional</u> $\begin{cases}\underline{\varepsilon\text{-symmetric}}\\ \underline{\varepsilon\text{-quadratic}}\end{cases}$ <u>pair over A</u>

$\begin{cases}(f:C\longrightarrow D,(\delta\phi,\phi))\\ (f:C\longrightarrow D,(\delta\psi,\psi))\end{cases}$ is a chain map $f:C\longrightarrow D$ from an n-dimensional A-module chain complex C to an (n+1)-dimensional A-module chain complex D together with an element $\begin{cases}(\delta\phi,\phi)\in Q^{n+1}(f,\varepsilon)\\ (\delta\psi,\psi)\in Q_{n+1}(f,\varepsilon)\end{cases}$.

Such a pair is <u>Poincaré</u> if the A-module chain map $D^{n+1-*}\longrightarrow C(f)$ defined (up to chain homotopy) by

$$\begin{cases}\begin{pmatrix}\delta\phi_0\\ \phi_0 f^*\end{pmatrix}: D^{n+1-r}\longrightarrow C(f)_r = D_r\oplus C_{r-1}\\ \begin{pmatrix}(1+T_\varepsilon)\delta\psi_0\\ (1+T_\varepsilon)\psi_0 f^*\end{pmatrix}: D^{n+1-r}\longrightarrow C(f)_r = D_r\oplus C_{r-1}\end{cases}$$

is a chain equivalence, in which case the <u>boundary</u> n-dimensional $\begin{cases}\varepsilon\text{-symmetric}\\ \varepsilon\text{-quadratic}\end{cases}$ complex $\begin{cases}(C,\phi\in Q^n(C,\varepsilon))\\ (C,\psi\in Q_n(C,\varepsilon))\end{cases}$ is Poincaré.

A <u>cobordism</u> of n-dimensional $\begin{cases}\varepsilon\text{-symmetric}\\ \varepsilon\text{-quadratic}\end{cases}$ Poincaré complexes over A $\begin{cases}(C,\phi),(C',\phi')\\ (C,\psi),(C',\psi')\end{cases}$ is an (n+1)-dimensional $\begin{cases}\varepsilon\text{-symmetric}\\ \varepsilon\text{-quadratic}\end{cases}$ Poincaré pair over A

$$\begin{cases}((f\ f'):C\oplus C'\longrightarrow D,(\delta\phi,\phi\oplus-\phi')\in Q^{n+1}((f\ f')),\varepsilon))\\ ((f\ f'):C\oplus C'\longrightarrow D,(\delta\psi,\psi\oplus-\psi')\in Q_{n+1}((f\ f')),\varepsilon))\end{cases}$$

with boundary $\begin{cases}(C\oplus C',\phi\oplus-\phi')\\ (C\oplus C',\psi\oplus-\psi')\end{cases}$.

Proposition 1.2.1 Cobordism is an equivalence relation on the set of n-dimensional $\begin{cases}\varepsilon\text{-symmetric}\\ \varepsilon\text{-quadratic}\end{cases}$ Poincaré complexes over A, such that homotopy equivalent Poincaré complexes are cobordant. The cobordism classes define an abelian group, the n-dimensional $\begin{cases}\varepsilon\text{-symmetric}\\ \varepsilon\text{-quadratic}\end{cases}$ L-group of A $\begin{cases}L^n(A,\varepsilon)\\ L_n(A,\varepsilon)\end{cases}$ $(n \geqslant 0)$, with addition and inverses by

$$\begin{cases}(C,\phi)+(C',\phi') = (C\oplus C',\phi\oplus\phi')\ ,\ -(C,\phi) = (C,-\phi)\in L^n(A,\varepsilon)\\ (C,\psi)+(C',\psi') = (C\oplus C',\psi\oplus\psi')\ ,\ -(C,\psi) = (C,-\psi)\in L_n(A,\varepsilon)\ .\end{cases}$$

Proof: See Proposition I.3.2.

[]

For $\varepsilon = 1 \in A$ the terminology is abbreviated:

$$\begin{cases}1\text{-symmetric} = \text{symmetric}\\ 1\text{-quadratic} = \text{quadratic}\end{cases},\ \begin{cases}Q^n(C,1) = Q^n(C)\\ Q_n(C,1) = Q_n(C)\end{cases},\ \begin{cases}L^n(A,1) = L^n(A)\\ L_n(A,1) = L_n(A)\end{cases}.$$

The ε-symmetrization maps

$$1+T_\varepsilon : L_n(A,\varepsilon) \longrightarrow L^n(A,\varepsilon)\ ;\ (C,\psi) \longmapsto (C,(1+T_\varepsilon)\psi)\quad (n \geqslant 0)$$

are isomorphisms modulo 8-torsion for any ring with involution A (Proposition I.8.2), and are actually isomorphisms if $1/2 \in A$ (I.3.3).

Define the intersection pairing of an n-dimensional ε-symmetric complex over A $(C,\phi \in Q^n(C,\varepsilon))$

$$\phi_0 : H^r(C) \times H^{n-r}(C) \longrightarrow A\ ;\ (x,y) \longmapsto \phi_0(x)(y)$$

satisfying

$$\phi_0(x,y+y') = \phi_0(x,y) + \phi_0(x,y')$$
$$\phi_0(x,ay) = a.\phi_0(x,y) \qquad (x\in H^r(C),\ y,y'\in H^{n-r}(C),a\in A)$$
$$\phi_0(y,x) = (-)^{r(n-r)}\varepsilon.\overline{\phi_0(x,y)}\quad .$$

We shall now recall from II. some of the ways in which algebraic Poincaré complexes arise in topology. (See II. for further details).

Let π be a group.

A $\underline{\pi\text{-space}}$ X is a pointed topological space with a basepoint-preserving action $\pi \times X \longrightarrow X$. The reduced singular chain complex $\dot{C}(X) = C(X,pt.)$ is then a $\mathbb{Z}[\pi]$-module chain complex. Let $T \in \mathbb{Z}_2$ act on $\dot{C}(X) \otimes_{\mathbb{Z}} \dot{C}(X)$ by the transposition involution

$$T : \dot{C}(X)_p \otimes_{\mathbb{Z}} \dot{C}(X)_q \longrightarrow \dot{C}(X)_q \otimes_{\mathbb{Z}} \dot{C}(X)_p \; ; \; x \otimes y \longmapsto (-)^{pq} y \otimes x \; .$$

Acyclic model theory equips $\dot{C}(X)$ with a canonical chain homotopy class of functorial $\mathbb{Z}[\pi]$-module chain maps ("diagonal approximations")

$$\dot{\Delta}_X : \dot{C}(X) \longrightarrow \mathrm{Hom}_{\mathbb{Z}[\mathbb{Z}_2]}(W, \dot{C}(X) \otimes_{\mathbb{Z}} \dot{C}(X))$$

with $\mathbb{Z}[\pi]$ acting on $\dot{C}(X) \otimes_{\mathbb{Z}} \dot{C}(X)$ by

$$g(x \otimes y) = gx \otimes gy \quad (x, y \in \dot{C}(X),\ g \in \pi) \quad .$$

We shall only be concerned with π-spaces X which are CW complexes with the basepoint a 0-cell, such that π acts freely by permutation on the cells of X-{pt.} with the quotient {1}-space X/π a finitely dominated CW complex - X is a "finitely dominated CWπ-complex" in the terminology of §II.3. For such X the chain complex $\dot{C}(X)$ can be replaced by a chain equivalent finite-dimensional $\mathbb{Z}[\pi]$-module chain complex, also to be denoted $\dot{C}(X)$. (If X/π is a finite CW complex $\dot{C}(X)$ can be taken to be the reduced cellular chain complex of X, which is a finite f.g. free $\mathbb{Z}[\pi]$-module chain complex.)

Given a map $w:\pi \longrightarrow \mathbb{Z}_2 = \{\pm 1\}$ (i.e. a group morphism) let $\mathbb{Z}[\pi]$ have the <u>w-twisted involution</u>

$$\bar{\ } : \mathbb{Z}[\pi] \longrightarrow \mathbb{Z}[\pi] \ ; \ \sum_{g\in\pi} n_g g \longmapsto \sum_{g\in\pi} w(g) n_g g^{-1} \quad (n_g \in \mathbb{Z}) \ .$$

Let $\mathbb{Z}^w$ denote the additive group $\mathbb{Z}$ with the right $\mathbb{Z}[\pi]$-module structure defined by

$$\mathbb{Z}^w \times \mathbb{Z}[\pi] \longrightarrow \mathbb{Z}^w \ ; \ (m, \sum_{g\in\pi} n_g g) \longmapsto m(\sum_{g\in\pi} w(g) n_g) \ .$$

Given a π-space X define the <u>reduced homology groups of X/π with w-twisted coefficients</u>

$$\dot{H}_*(X/\pi, w) = H_*(\mathbb{Z}^w \otimes_{\mathbb{Z}[\pi]} \dot{C}(X)) \ .$$

We shall write these groups as $\dot{H}_*(X/\pi)$, the contribution of w being understood. (For w = 1 these are just the usual reduced homology groups of X/π, anyway). As $\dot{C}(X)$ is a finite-dimensional $\mathbb{Z}[\pi]$-module chain complex the slant product chain map

$$\mathbb{Z}^w \otimes_{\mathbb{Z}[\pi]} (\dot{C}(X) \otimes_{\mathbb{Z}} \dot{C}(X)) = \dot{C}(X)^t \otimes_{\mathbb{Z}[\pi]} \dot{C}(X)$$
$$\longrightarrow \mathrm{Hom}_{\mathbb{Z}[\pi]}(\dot{C}(X)^*, \dot{C}(X)) \ ; \ x \otimes y \longmapsto (f \longmapsto \overline{f(x)} y)$$

is an isomorphism of $\mathbb{Z}[\mathbb{Z}_2]$-module chain complexes, with $T \in \mathbb{Z}_2$ acting on $\dot{C}(X)^t \otimes_{\mathbb{Z}[\pi]} \dot{C}(X)$ by the transposition involution $x \otimes y \longmapsto \pm y \otimes x$ and on $\mathrm{Hom}_{\mathbb{Z}[\pi]}(\dot{C}(X)^*, \dot{C}(X))$ by the duality involution $\phi \longmapsto \pm \phi^*$ $(\varepsilon = 1)$. Using this isomorphism as an identification and applying $\mathbb{Z}^w \otimes_{\mathbb{Z}[\pi]} -$ to the functorial $\mathbb{Z}[\pi]$-module chain map $\dot{\Delta}_X$ there is obtained a functorial $\mathbb{Z}$-module chain map

$$\dot{\phi}_X = 1 \otimes \dot{\Delta}_X : \mathbb{Z}^w \otimes_{\mathbb{Z}[\pi]} \dot{C}(X) \longrightarrow \mathrm{Hom}_{\mathbb{Z}[\mathbb{Z}_2]}(W, \mathrm{Hom}_{\mathbb{Z}[\pi]}(\dot{C}(X)^*, \dot{C}(X))) \ .$$

The induced morphisms in the homology groups

$$\dot{\phi}_X : \dot{H}_n(X/\pi) \longrightarrow Q^n(\dot{C}(X)) \quad (n \geqslant 0)$$

are the <u>symmetric construction on X</u> of §II.1.

If X is a finitely dominated CW complex and $\tilde{X}$ is a (regular) covering of X with group of covering translations π then the disjoint union $\tilde{X}_+ = \tilde{X} \sqcup \{pt.\}$ is a finitely dominated CWπ-complex. The symmetric construction on $\tilde{X}_+$ is written as

$$\phi_{\tilde{X}} = \dot{\phi}_{\tilde{X}_+} : H_n(X) = \dot{H}_n(X_+) \longrightarrow Q^n(C(\tilde{X})) = Q^n(\dot{C}(\tilde{X}_+))$$

with $C(\tilde{X}) = \dot{C}(\tilde{X}_+)$ the cellular chain complex of $\tilde{X}$ (up to chain equivalence), a finite-dimensional $\mathbb{Z}[\pi]$-module chain complex. The homology $\mathbb{Z}[\pi]$-modules $H_*(C(\tilde{X}))$ are the usual homology groups $H_*(\tilde{X})$ with the induced $\mathbb{Z}[\pi]$-action; the cohomology $\mathbb{Z}[\pi]$-modules $H^*(C(\tilde{X}))$ will be written as $H^*(\tilde{X})$, even though for infinite π the underlying abelian groups need not be the singular cohomology of $\tilde{X}$ (e.g. for $\tilde{S}^1 = \mathbb{R}$). Again, we have suppressed the choice of orientation map $w:\pi \longrightarrow \mathbb{Z}_2$.

An <u>n-dimensional geometric Poincaré complex</u> X (in the sense of Wall [3]) is a finitely dominated CW complex X with an <u>orientation map</u> $w(X) : \pi_1(X) \longrightarrow \mathbb{Z}_2$ and a <u>fundamental class</u> $[X] \in H_n(X)$ (defined using w(X)-twisted coefficients) such that the $\mathbb{Z}[\pi_1(X)]$-module chain map defined by the evaluation of the cap product on any cycle representative of [X]

$$[X] \cap - : C(\tilde{X})^{n-*} \longrightarrow C(\tilde{X})$$

is a chain equivalence, with $\tilde{X}$ the universal cover of X. The <u>symmetric complex</u> of X

$$\sigma^*(X) = (C(\tilde{X}), \phi_{\tilde{X}}([X]) \in Q^n(C(\tilde{X})))$$

is an n-dimensional symmetric Poincaré complex over $\mathbb{Z}[\pi_1(X)]$) such that $\phi_X([X])_0 = [X] \cap - : C(\tilde{X})^{n-*} \xrightarrow{\sim} C(\tilde{X})$. More generally, the construction applies to any <u>oriented</u> covering $\tilde{X}$ of X, that is one for which the group of covering translations π is equipped with a map $w:\pi \longrightarrow \mathbb{Z}_2$ such that $w(X)$ factors as

$$w(X) \ : \ \pi_1(X) \longrightarrow \pi \xrightarrow{w} \mathbb{Z}_2 \ ,$$

so that there is induced a morphism of rings with involution $\mathbb{Z}[\pi_1(X)] \longrightarrow \mathbb{Z}[\pi]$. The corresponding n-dimensional symmetric Poincaré complex over $\mathbb{Z}[\pi]$ is given by $\mathbb{Z}[\pi]\otimes_{\mathbb{Z}[\pi_1(X)]}\sigma^*(X) = (C(\tilde{X}),\phi)$ and is also denoted by $\sigma^*(X)$. The corresponding chain equivalence $[X] \cap - : C(\tilde{X})^{n-*} \longrightarrow C(\tilde{X})$ induces Poincaré duality $\mathbb{Z}[\pi]$-module isomorphisms

$$[X] \cap - \ : \ H^{n-*}(\tilde{X}) \xrightarrow{\sim} H_*(\tilde{X}) \ .$$

An n-dimensional symmetric complex over A (C,ϕ) is <u>finite</u> if C is a finite chain complex of f.g. free A-modules. A symmetric complex (C,ϕ) is homotopy equivalent to a finite one if and only if has vanishing reduced projective class, that is

$$[C] \equiv \sum_{r=-\infty}^{\infty} (-)^r [C_r] = 0 \in \tilde{K}_0(A) \ .$$

Similarly for quadratic complexes.

An n-dimensional symmetric Poincaré complex over A (C,ϕ) is <u>simple</u> if C is a finite chain complex of based f.g. free A-modules and

either $\tau(\phi_0:C^{n-*} \longrightarrow C) = 0 \in \tilde{K}_1(A)$

or $\quad A = \mathbb{Z}[\pi]$ and $\tau(\phi_0) = 0 \in Wh(\pi) = \tilde{K}_1(\mathbb{Z}[\pi])/\{\pi\}$.

Similarly for quadratic Poincaré complexes. In dealing with the algebraic Poincaré complexes arising in topology the Whitehead group variant is understood.

A geometric Poincaré complex X is <u>finite</u> if X is a finite CW complex; X is <u>simple</u> if it is finite and

$$\tau([X]\cap - : C(\tilde{X})^{n-*} \longrightarrow C(\tilde{X})) = 0 \in Wh(\pi_1(X)) \ .$$

A geometric Poincaré complex X is homotopy equivalent to a finite one if and only if $\sigma^*(X)$ is homotopy equivalent to a finite symmetric Poincaré complex (since $[C(\tilde{X})] \in \tilde{K}_0(\mathbb{Z}[\pi_1(X)])$ is the Wall finiteness obstruction); a finite geometric Poincaré complex X is simple if and only if $\sigma^*(X)$ is a simple symmetric Poincaré complex over $\mathbb{Z}[\pi_1(X)]$, by definition.

A compact n-dimensional topological manifold M has the structure of a simple n-dimensional geometric Poincaré complex. The intersection pairing of $\sigma^*(M)$

$$\phi_{\tilde{M}}([M])_0 \ : \ H^{n-*}(\tilde{M}) \times H^*(\tilde{M}) \longrightarrow \mathbb{Z}[\pi_1(M)]$$

agrees via the Poincaré duality isomorphisms $[M]\cap - : H^{n-*}(\tilde{M}) \xrightarrow{\sim} H_*(\tilde{M})$ with the pairing

$$H_*(\tilde{M}) \times H_{n-*}(\tilde{M}) \longrightarrow \mathbb{Z}[\pi_1(M)]$$

defined by the geometric intersection numbers of homology classes.

In dealing with the L-theoretic invariants of finitely-dominated (resp. finite, simple) geometric Poincaré complexes it is natural to consider the version of L-theory defined using f.g. projective (resp. finite, simple) algebraic Poincaré complexes. The projective theory is of greatest interest in algebra, being the most general. On the other hand, the simple L-theory is of greatest interest in topology, being the one closest to the classification theory of compact manifolds.

We continue working with the projective L-theory. We shall not spell out the analogous properties of the free and simple L-theories, except that in §1.10 the three types of L-groups are compared to each other.

An <u>(n+1)-dimensional geometric Poincaré pair</u> (Y,X) is finitely dominated CW pair (Y,X) with an <u>orientation map</u> $w(Y):\pi_1(Y)\longrightarrow\mathbb{Z}_2$ and a <u>fundamental class</u> $[Y]\in H_{n+1}(Y,X)$ such that

i) the $\mathbb{Z}[\pi_1(Y)]$-module chain map

$$[Y]\cap - \;:\; C(\tilde{Y},\tilde{X})^{n+1-*}\longrightarrow C(\tilde{Y})$$

is a chain equivalence, inducing Poincaré-Lefschetz duality isomorphisms $[Y]\cap - : H^{n+1-*}(\tilde{Y},\tilde{X})\xrightarrow{\sim}H_*(\tilde{Y})$, with $\tilde{Y}$ the universal cover of Y and $\tilde{X}$ the induced cover of X

ii) X is an n-dimensional geometric Poincaré complex (the <u>boundary</u> of (Y,X)) with orientation map

$$w(X)\;:\;\pi_1(X)\longrightarrow\pi_1(Y)\xrightarrow{w(Y)}\mathbb{Z}_2$$

and fundamental class

$$[X] = \partial[Y]\in H_n(X)\;.$$

For example, a compact (n+1)-dimensional manifold with boundary $(M,\partial M)$ has the structure of such a pair.

The <u>symmetric pair</u> of an (n+1)-dimensional geometric Poincaré pair (Y,X) with respect to an oriented covering $(\tilde{Y},\tilde{X})$ with group of covering translations π and orientation map $w:\pi\longrightarrow\mathbb{Z}_2$ is the (n+1)-dimensional symmetric Poincaré pair over $\mathbb{Z}[\pi]$ with the w-twisted involution

$$\sigma^*(Y,X) = (f:C(\tilde{X})\longrightarrow C(\tilde{Y}),\phi_{\tilde{Y},\tilde{X}}([Y])\in Q^{n+1}(C(f))\;,$$

defined using the relative symmetric construction of §II.6, with $f:C(\widetilde{X})\longrightarrow C(\widetilde{Y})$ the inclusion chain map. The boundary of $\sigma^*(Y,X)$ is the symmetric complex $\sigma^*(X)$ of X.

The <u>symmetric signature</u> of an n-dimensional geometric Poincaré complex X with respect to an oriented covering $\widetilde{X}$ with group of covering translations π is the cobordism class of the symmetric complex of X with respect to $\widetilde{X}$

$$\sigma^*(X) = (C(\widetilde{X}),\phi_{\widetilde{X}}([X])) \in L^n(\mathbb{Z}[\pi]) \quad .$$

This invariant was introduced by Mishchenko [1]. If X is the boundary of an (n+1)-dimensional geometric Poincaré pair (Y,X) and $\widetilde{X}$ extends to an oriented covering $\widetilde{Y}$ of Y then

$$\sigma^*(X) = 0 \in L^n(\mathbb{Z}[\pi]) \quad .$$

A <u>degree 1 map</u> of n-dimensional geometric Poincaré complexes

$$f : M \longrightarrow X$$

is a continuous map such that

i) $w(M) : \pi_1(M) \xrightarrow{f_*} \pi_1(X) \xrightarrow{w(X)} \mathbb{Z}_2$

ii) $f_*([M]) = [X] \in H_n(X)$.

The <u>symmetric kernel</u> of a degree 1 map $f:M\longrightarrow X$ of n-dimensional geometric Poincaré complexes with respect to an oriented cover $\widetilde{X}$ of X with group of covering translations π is the n-dimensional symmetric Poincaré complex over $\mathbb{Z}[\pi]$

$$\sigma^*(f) = (C(f^!),e^{\%}\phi_{\widetilde{M}}([M]) \in Q^n(C(f^!)))$$

with $C(f^!)$ the algebraic mapping cone of the $\mathbb{Z}[\pi]$-module <u>Umkehr chain map</u>

$$f^! : C(\widetilde{X}) \xrightarrow[\sim]{([X]\cap -)^{-1}} C(\widetilde{X})^{n-*} \xrightarrow{\widetilde{f}^*} C(\widetilde{M})^{n-*} \xrightarrow[\sim]{([M]\cap -)} C(\widetilde{M}) ,$$

with $\widetilde{M}$ the induced oriented cover of M, $\widetilde{f}:\widetilde{M}\longrightarrow\widetilde{X}$ a π-equivariant lift of f and $e:C(\widetilde{M}) \longrightarrow C(f^!)$ the inclusion. The chain equivalnce

$$\begin{pmatrix} e \\ \tilde{f} \end{pmatrix} : C(\widetilde{M}) \xrightarrow{\sim} C(f^!) \oplus C(\widetilde{X})$$

defines a homotopy equivalence of symmetric Poincaré complexes over $\mathbb{Z}[\pi]$

$$\begin{pmatrix} e \\ \tilde{f} \end{pmatrix} : \sigma^*(M) \xrightarrow{\sim} \sigma^*(f) \oplus \sigma^*(X) \quad .$$

The $\left\{\begin{matrix}\text{homology}\\ \text{cohomology}\end{matrix}\right.$ modules of $C(f^!)$ are the $\left\{\begin{matrix}\underline{\text{homology}}\\ \underline{\text{cohomology}}\end{matrix}\right.$ <u>kernels</u> of f

$$\begin{cases} K_*(M) = H_*(C(f^!)) = \ker(\tilde{f}_*:H_*(\widetilde{M}) \longrightarrow H_*(\widetilde{X})) \\ K^*(M) = H^*(C(f^!)) = \ker(f^!:H^*(\widetilde{M}) \longrightarrow H^*(\widetilde{X})) \quad , \end{cases}$$

with

$$\begin{cases} H_*(\widetilde{M}) = K_*(M) \oplus H_*(\widetilde{X}) \\ H^*(\widetilde{M}) = K^*(M) \oplus H^*(\widetilde{X}) \end{cases} \quad , \quad K^*(M) = K_{n-*}(M)$$

(up to isomorphism). If M and X are finite (resp. simple) geometric Poincaré complexes then $\sigma^*(f)$ is a finite (resp. simple) symmetric Poincaré complex. On the L-group level

$$\sigma^*(f) = \sigma^*(M) - \sigma^*(X) \in L^n(\mathbb{Z}[\pi]) \quad .$$

The <u>suspension</u> of a π-space X is the π-space

$$\Sigma X = X \wedge S^1 \quad ,$$

with π acting by

$$\pi \times \Sigma X \longrightarrow \Sigma X \ ; \ (g, x \wedge s) \longmapsto gx \wedge s \quad .$$

A <u>π-map</u> of π-spaces X,Y

$$f : X \longrightarrow Y$$

is a π-equivariant basepoint-preserving map. A <u>π-homotopy</u> of π-maps $f,f':X \longrightarrow Y$

$$H : f \simeq f' : X \longrightarrow Y$$

is a π-equivariant map

$$H : X \times I \longrightarrow Y \quad (I = [0,1])$$

such that

$$H(x,0) = f(x) \ , \ H(x,1) = f'(x) \ , \ H(pt._X,t) = pt._Y \in Y \ (x \in X, t \in I) \ .$$

Let $[X,Y]_\pi$ denote the set of π-homotopy classes of π-maps $f:X \longrightarrow Y$, and let $\{X,Y\}_\pi$ denote the abelian group of <u>stable π-homotopy</u> classes of <u>stable π-maps</u>

$$\{X,Y\}_\pi = \varinjlim_{p\to\infty} [\Sigma^p X, \Sigma^p Y]_\pi \ .$$

For $\pi = \{1\}$ the terminology is contracted in the usual manner to $[X,Y]$, $\{X,Y\}$.

The <u>quadratic construction</u> $\dot{\psi}_F$ of §II.1 associates to a stable π-map $F:\Sigma^p X \longrightarrow \Sigma^p Y$ $(p \geqslant 0)$ natural maps

$$\dot{\psi}_F : \dot{H}_*(X/\pi) \longrightarrow Q_*(\dot{C}(Y))$$

such that

$$(1+T)\dot{\psi}_F = \dot{\phi}_Y f_* - f^{\%}\dot{\phi}_X : \dot{H}_*(X/\pi) \longrightarrow Q^*(\dot{C}(Y)) \ ,$$

with $f : \dot{C}(X) = \Omega^p \dot{C}(\Sigma^p X) \xrightarrow{F} \Omega^p \dot{C}(\Sigma^p Y) = \dot{C}(Y)$ the $\mathbb{Z}[\pi]$-module chain map induced by F (up to chain homotopy). The homology groups $\dot{H}_*(X/\pi)$ are defined using w-twisted coefficients for some map $w:\pi \longrightarrow \mathbb{Z}_2$, and $\mathbb{Z}[\pi]$ is given the w-twisted involution, exactly as for the symmetric construction $\dot{\phi}_X$. There are two (equivalent) ways of obtaining $\dot{\psi}_F$, as follows.

Firstly, note that the symmetric construction $\dot{\phi}_{\Sigma X}$ on the suspension ΣX of a π-space X is the algebraic suspension $S\dot{\phi}_X$ (in the sense of §1.1) of the symmetric construction $\dot{\phi}_X$ on X

$$\dot{\phi}_{\Sigma X} : \dot{H}_*(\Sigma X/\pi) = \dot{H}_{*-1}(X/\pi) \xrightarrow{\dot{\phi}_X} Q^{*-1}(\dot{C}(X))$$

$$\xrightarrow{S} Q^*(S\dot{C}(X)) = Q^*(\dot{C}(\Sigma X)) ,$$

identifying $\dot{C}(\Sigma X) = S\dot{C}(X)$. In fact, acyclic model theory gives a functorial $\mathbb{Z}$-module chain homotopy

$$h_X : \dot{\phi}_{\Sigma X} \simeq S\dot{\phi}_X : \dot{C}(\Sigma X/\pi) = S\dot{C}(X/\pi)$$

$$\longrightarrow \mathrm{Hom}_{\mathbb{Z}[\mathbb{Z}_2]}(W, \mathrm{Hom}_{\mathbb{Z}[\pi]}(S\dot{C}(X)^*, S\dot{C}(X))) .$$

As $f : \dot{C}(X) \longrightarrow \dot{C}(Y)$ is induced by a π-map $F : \Sigma^p X \longrightarrow \Sigma^p Y$

$$\dot{\phi}_{\Sigma^p Y} f - f^{\%}\dot{\phi}_{\Sigma^p X} = 0 : \dot{C}(\Sigma^p X/\pi) = S^p\dot{C}(X/\pi)$$

$$\longrightarrow \mathrm{Hom}_{\mathbb{Z}[\mathbb{Z}_2]}(W, \mathrm{Hom}_{\mathbb{Z}[\pi]}(S^p\dot{C}(X)^*, S^p\dot{C}(X))) ,$$

by the naturality of the symmetric construction. The composite functorial $\mathbb{Z}$-module chain map

$$S^p(\dot{\phi}_Y f - f^{\%}\dot{\phi}_X) :$$

$$\dot{C}(X/\pi) \xrightarrow{\dot{\phi}_Y f - f^{\%}\dot{\phi}_X} \mathrm{Hom}_{\mathbb{Z}[\mathbb{Z}_2]}(W, \mathrm{Hom}_{\mathbb{Z}[\pi]}(\dot{C}(Y)^*, \dot{C}(Y)))$$

$$\xrightarrow{S^p} \mathrm{Hom}_{\mathbb{Z}[\mathbb{Z}_2]}(W, \mathrm{Hom}_{\mathbb{Z}[\pi]}(S^p\dot{C}(Y)^*, S^p\dot{C}(Y)))$$

is thus equipped with a functorial $\mathbb{Z}$-module chain homotopy $\dot{\psi}_F : S^p(\dot{\phi}_Y - f^{\%}\dot{\phi}_X) \simeq 0$. The chain level argument underlying the exact sequence of Proposition 1.1.3 interprets this as a functorial $\mathbb{Z}$-module chain map

$$\dot{\psi}_F : \dot{C}(X/\pi) \longrightarrow W[0,p-1] \otimes_{\mathbb{Z}[\mathbb{Z}_2]} \mathrm{Hom}_{\mathbb{Z}[\pi]}(\dot{C}(Y)^*, \dot{C}(Y))$$

inducing the <u>unstable quadratic construction</u> in homology

$$\dot{\psi}_F : \dot{H}_*(X/\pi) \longrightarrow Q_*^{[0,p-1]}(\dot{C}(Y)) .$$

Composition with the natural maps $Q_*^{[0,p-1]}(\dot{C}(Y)) \longrightarrow Q_*(\dot{C}(Y))$ (which are isomorphisms for $p >$ dimension of $\dot{C}(Y)$) gives the <u>stable quadratic construction</u>

$$\dot{\psi}_F : \dot{H}_*(X/\pi) \longrightarrow Q_*(\dot{C}(Y)) \ .$$

(Incidentally, the definition of $\dot{\psi}_F$ in Proposition II.1.5 contains a technical error in that it made use of a mythical functorial chain homotopy inverse $\Sigma_Y^{-1} : \dot{C}(\Sigma Y) \longrightarrow S\dot{C}(Y)$ of the suspension chain map $\Sigma_Y : S\dot{C}(Y) \longrightarrow \dot{C}(\Sigma Y)$ in the reduced singular chain complexes of a π-space Y and its suspension ΣY. We shall give now a new definition of $\dot{\psi}_F$ which avoids this embarrassment. The suspension Σ_Y is defined to be the composite $\mathbb{Z}[\pi]$-module chain map

$$\Sigma_Y : S\dot{C}(Y) = S\mathbb{Z}\otimes_{\mathbb{Z}}\dot{C}(Y) \xrightarrow{i\otimes 1} \dot{C}(S^1)\otimes_{\mathbb{Z}}\dot{C}(Y)$$

$$\xrightarrow{E} \dot{C}(S^1\times Y)/\dot{C}(S^1\times pt.\cup pt.\times Y) \xrightarrow{j} \dot{C}(S^1\times Y/S^1\times pt.\cup pt.\times Y)$$

$$= \dot{C}(\Sigma Y) \ ,$$

with $i : S\mathbb{Z} \longrightarrow \dot{C}(S^1)$ any $\mathbb{Z}$-module chain equivalence, E a functorial $\mathbb{Z}[\pi]$-module chain equivalence given by the relative Eilenberg-Zilber theorem, and j the $\mathbb{Z}[\pi]$-module chain map induced by the projection $S^1\times Y \longrightarrow \Sigma Y$. In general j is not a $\mathbb{Z}[\pi]$-module chain equivalence, and even if it were there might not exist a functorial $\mathbb{Z}[\pi]$-module chain homotopy inverse; a fortiori for Σ_Y. Let now $\ddot{C}(X)$ be the algebraic mapping cone of the $\mathbb{Z}[\pi]$-module chain map

$$(F\Sigma_X^p \ \Sigma_Y^p) : \Omega\dot{C}(X)\oplus\Omega\dot{C}(Y) \longrightarrow \Omega^{p+1}\dot{C}(\Sigma^p Y) \ .$$

The base point of Y is non-degenerate (by hypothesis), so that the p-fold suspension chain map $\Sigma_Y^p : S^p\dot{C}(Y) \longrightarrow \dot{C}(\Sigma^p Y)$ induces isomorphisms in homology, and hence so does $\Sigma_Y^p \otimes \Sigma_Y^p : S^p\dot{C}(Y) \otimes_{\mathbb{Z}} S^p\dot{C}(Y) \longrightarrow \dot{C}(\Sigma^p Y) \otimes_{\mathbb{Z}} \dot{C}(\Sigma^p Y)$. It follows that the projection $\ddot{C}(X) \longrightarrow \dot{C}(X)$ is also a homology equivalence, and that $\Sigma_Y^p : \dot{C}(Y) \longrightarrow \Omega^p\dot{C}(\Sigma^p Y)$ induces isomorphisms in the Q-groups. Using the terminology of p.204 of II. define a $\mathbb{Z}$-module chain map

$$\dot{\psi}_F : \mathbb{Z}^w \otimes_{\mathbb{Z}[\pi]} \ddot{C}(X) \longrightarrow C(S^p) = W[0,p-1] \otimes_{\mathbb{Z}[\mathbb{Z}_2]} (\Omega^p\dot{C}(\Sigma^p Y)^t \otimes_{\mathbb{Z}[\pi]} \Omega^p\dot{C}(\Sigma^p Y))$$

by

$$\dot{\psi}_F(x,y,z) = (\Sigma_Y^{p\%}\dot{\Delta}_Y(y) - F^{\%}\Sigma_X^p\,\dot{\Delta}_X(x), F^{\%}\Sigma_X^p(x) - \Sigma_Y^p(y) - \dot{\Delta}_{\Sigma^p Y}(z))$$

$$\in C(S^p)_n = \mathrm{Hom}_{\mathbb{Z}[\mathbb{Z}_2]}(W, \Omega^p\dot{C}(\Sigma^p Y)^t \otimes_{\mathbb{Z}[\pi]} \Omega^p\dot{C}(\Sigma^p Y))_n$$

$$\oplus \Omega^p \mathrm{Hom}_{\mathbb{Z}[\mathbb{Z}_2]}(W, \dot{C}(\Sigma^p Y)^t \otimes_{\mathbb{Z}[\pi]} \dot{C}(\Sigma^p Y))_{n+1}$$

$$((x,y,z) \in \mathbb{Z}^w \otimes_{\mathbb{Z}[\pi]} \ddot{C}(X)_n = \mathbb{Z}^w \otimes_{\mathbb{Z}[\pi]} (\dot{C}(X)_n \oplus \dot{C}(Y)_n \oplus \dot{C}(\Omega^p Y)_{n+p+1}))\ .$$

The chain map $\dot{\psi}_F$ induces the quadratic construction

$$\dot{\psi}_F : \dot{H}_*(X/\pi) = H_*(\mathbb{Z}^w \otimes_{\mathbb{Z}[\pi]} \dot{C}(X)) \longrightarrow Q_*(\Omega^p\dot{C}(\Sigma^p Y)) = Q_*(\dot{C}(Y))$$

on passing to the homology groups.)

Alternatively, for connected Y the quadratic construction ψ_F on a stable π-map $F:\Sigma^\infty X \longrightarrow \Sigma^\infty Y$ (i.e. $F:\Sigma^p X \longrightarrow \Sigma^p Y$ for some $p \geqslant 0$) may be obtained from the adjoint π-map $\mathrm{adj}(F) : X \longrightarrow \Omega^\infty\Sigma^\infty Y$ by appealing to the approximation theorem underlying infinite loop space theory

$$\Omega^\infty\Sigma^\infty Y = \bigcup_{k \geqslant 1} E\Sigma_k \times_{\Sigma_k} (\prod_k Y)/\sim$$

with Σ_k the permutation group on k letters, and setting

$$\dot{\psi}_F : \dot{H}_*(X/\pi) \xrightarrow{(\mathrm{adj}(F)/\pi)_*} \dot{H}_*(\Omega^\infty\Sigma^\infty Y/\pi) = \sum_{k=1}^{\infty} \dot{H}_*(E\Sigma_k \ltimes_{\Sigma_k} (\bigwedge_k Y)/\pi)$$

$$\xrightarrow{\text{projection}} \dot{H}_*(E\Sigma_2 \ltimes_{\Sigma_2} (Y \wedge_\pi Y)) = Q_*(\dot{C}(Y)) \quad (C(E\Sigma_2) = W) \ .$$

For disconnected π-spaces Y of type $Y = Z_+$ (for some space Z with π-action, with the added point as base) $\Omega^\infty\Sigma^\infty Y$ is approximated by the group completion of a topological monoid

$$\Omega^\infty\Sigma^\infty Y = \Omega B(\coprod_{k \geq 0} E\Sigma_k \times_{\Sigma_k} (\prod_k Z)) \ ,$$

and the quadratic construction $\dot{\psi}_F$ is given by

$$\dot{\psi}_F : \dot{H}_*(X/\pi) \xrightarrow{(\mathrm{adj}(F)/\pi)_*} \dot{H}_*(\Omega^\infty\Sigma^\infty Y/\pi)$$

$$= \mathbb{Z}[\mathbb{Z}] \otimes_{\mathbb{Z}[\mathbb{N}]} (\sum_{k=1}^{\infty} H_*(E\Sigma_k \times_{\Sigma_k} (\prod_k Z)/\pi))$$

$$\xrightarrow{\text{projection}} H_*(E\Sigma_2 \times_{\Sigma_2} (Z \times_\pi Z)) = Q_*(C(Z)) = Q_*(\dot{C}(Y)) \quad (\dot{C}(Y) = C(Z)) \ .$$

Similarly for the unstable quadratic construction, using the unstable approximation theorems.

The stable (resp. unstable) quadratic construction $\dot{\psi}_F$ on a π-map $F : \Sigma^p X \longrightarrow \Sigma^p Y$ depends only on the stable (resp. unstable) π-homotopy class of F, and $\dot{\psi}_F = 0$ if this class contains $\Sigma^p F_0$ for some π-map $F_0 : X \longrightarrow Y$.

The quadratic construction on a stable π-map $F : \Sigma^\infty X_+ \longrightarrow \Sigma^\infty Y_+$ (for some spaces with π-action X,Y) is written as

$$\psi_F = \dot{\psi}_F : H_*(X/\pi) = \dot{H}_*(X_+/\pi) \longrightarrow Q_*(C(Y)) = Q_*(\dot{C}(Y_+)) \ .$$

Given a spherical fibration $\nu: X \longrightarrow BG(k)$ over a space X

$$(D^k, S^{k-1}) \longrightarrow (E(\nu), S(\nu)) \longrightarrow X$$

and a covering $\tilde{X}$ of X with group of covering translations π define the Thom π-space of ν $T\pi(\nu)$ to be the Thom space of the pullback $\tilde{\nu}: \tilde{X} \longrightarrow X \xrightarrow{\nu} BG(k)$ with the induced π-action

$$T\pi(\nu) = T(\tilde{\nu}) \quad (= E(\tilde{\nu})/S(\tilde{\nu})) \ .$$

The quotient $\{1\}$-space $T\pi(\nu)/\pi = T(\nu)$ is just the usual Thom space of ν. If X is a finitely dominated (resp. finite) CW complex then $T\pi(\nu)$ is a finitely dominated (resp. finite) CWπ-complex. A map of spherical fibrations $b: \nu \longrightarrow \nu'$ induces a π-map of Thom π-spaces

$$T\pi(b) : T\pi(\nu) \longrightarrow T\pi(\nu') \ .$$

An n-dimensional geometric Poincaré complex X has a Spivak normal structure

$$(\nu_X : X \longrightarrow BG(k), \rho_X : S^{n+k} \longrightarrow T(\nu_X))$$

which is unique up to stable equivalence, with ν_X the Spivak normal fibration, such that $w(X) = w_1(\nu_X) : \pi_1(X) \longrightarrow \mathbb{Z}_2$ and $[X] = h(\rho_X) \cap U_{\nu_X} \in H_n(X)$ with $h: \pi_{n+k}(T(\nu_X)) \longrightarrow \dot{H}_{n+k}(T(\nu_X))$ the Hurewicz map and $U_{\nu_X} \in \dot{H}^k(T(\nu_X))$ the $w(X)$-twisted Thom class of ν_X. For finite X a Spivak normal structure (ν_X, ρ_X) may be obtained from a closed regular neighbourhood $E(\nu_X)$ of an embedding $X \subset S^{n+k}$ (k large), with $S(\nu_X) = \partial E(\nu_X)$ and

$$\rho_X : S^{n+k} \xrightarrow{\text{collapse}} S^{n+k}/\overline{S^{n+k} - E(\nu_X)} = E(\nu_X)/S(\nu_X) = T(\nu_X) \ .$$

(Similarly for finitely dominated X, using the fact that $X \times S^1$ has the homotopy type of a finite complex). We shall consider geometric Poincaré complexes X to be equipped with a particular choice of Spivak normal structure (ν_X, ρ_X). Given a covering

$\widetilde{X}$ of X (which need not be oriented) with group of covering translations π use the diagonal {1}-map

$$\Delta : T(\nu_X) = (E(\widetilde{\nu}_X)/S(\widetilde{\nu}_X))/\pi$$

$$\longrightarrow \widetilde{X}_+ \wedge_\pi T\pi(\nu_X) = ((E(\widetilde{\nu}_X) \times E(\widetilde{\nu}_X))/(E(\widetilde{\nu}_X) \times S(\widetilde{\nu}_X)))/\pi \ ;$$

$$[x] \longmapsto [x,x] \quad (x \in E(\widetilde{\nu}_X))$$

to define the <u>fundamental $S\pi$-duality map of X</u>

$$\alpha_X : S^{n+k} \xrightarrow{\rho_X} T(\nu_X) \xrightarrow{\Delta} \widetilde{X}_+ \wedge_\pi T\pi(\widetilde{\nu}_X) \ ,$$

which determines an $S\pi$-duality between the π-spaces $\widetilde{X}_+, T\pi(\nu_X)$ in the sense of the equivariant S-duality theory of §II.3. (For $\pi = \{1\}$ this is the classical Spanier-Whitehead S-duality theory for {1}-spaces). The $S\pi$-duality is characterized by the property that for any <u>π-spectrum</u> $\underline{M} = \{M_j, \Sigma M_j \longrightarrow M_{j+1} | j \geqslant 0\}$ of π-spaces and π-maps the slant products

$$\alpha_X : \dot{H}^q(T\pi(\nu_X);\underline{M}) = \varinjlim_p [\Sigma^p T\pi(\nu_X), M_{p+q}]_\pi$$

$$\longrightarrow H_{n+k-q}(X;\underline{M}) = \varinjlim_p [S^{n+k+p}, \widetilde{X}_+ \wedge_\pi M_{p+q}] \ ;$$

$$(F : \Sigma^p T\pi(\nu_X) \longrightarrow M_{p+q})$$

$$\longmapsto (S^{n+k+p} \xrightarrow{\Sigma^p \alpha_X} \widetilde{X}_+ \wedge_\pi \Sigma^p T\pi(\nu_X) \xrightarrow{1 \wedge F} \widetilde{X}_+ \wedge_\pi M_{p+q}) \quad (q \in \mathbb{Z})$$

are isomorphisms of abelian groups.

A <u>normal map</u> of n-dimensional geometric Poincaré complexes

$$(f,b) : M \longrightarrow X$$

is a degree 1 map $f: M \longrightarrow X$ together with a map of the Spivak normal fibrations $b: \nu_M \longrightarrow \nu_X$ covering f such that

$$T(b)_*(\rho_M) = \rho_X \in \pi_{n+k}(T(\nu_X)) \ .$$

Given an oriented covering $\widetilde{X}$ of X with group of covering translations π let $\widetilde{M}$ be the induced oriented covering of M and let $\tilde{f}:\tilde{M}\longrightarrow\tilde{X}$ be a π-equivariant lift of f. A <u>geometric Umkehr map</u> F of (f,b) is a stable π-map $F:\Sigma^{\infty}\tilde{X}_+\longrightarrow\Sigma^{\infty}\tilde{M}_+$ in the stable π-homotopy class $F\in\{\tilde{X}_+,\tilde{M}_+\}_{\pi}$ such that $(\Sigma^{\infty}\tilde{f}_+)F = 1\in\{\tilde{X}_+,\tilde{X}_+\}_{\pi}$ to which the composite isomorphism

$$\{T\pi(\nu_M),T\pi(\nu_X)\}_{\pi}\xrightarrow[\simeq]{\alpha_M}\{S^{n+k},\tilde{M}_+\wedge_{\pi}T\pi(\nu_X)\}\xrightarrow[\simeq]{\alpha_X^{-1}}\{\tilde{X}_+,\tilde{M}_+\}_{\pi}$$

sends $T\pi(b)\in\{T\pi(\nu_M),T\pi(\nu_X)\}_{\pi}$. A geometric Umkehr map F induces the Umkehr chain map $f^!:C(\tilde{X})\simeq C(\tilde{X})^{n-*}\xrightarrow{\tilde{f}^*}C(\tilde{M})^{n-*}\simeq C(\tilde{M})$.
The <u>quadratic kernel</u> of (f,b) is the n-dimensional quadratic Poincaré complex over $\mathbb{Z}[\pi]$

$$\sigma_*(f,b) = (C(f^!),e_{\%}\psi_F([X])\in Q_n(C(f^!)))$$

defined using the quadratic construction $\psi_F:H_n(X)\longrightarrow Q_n(C(\tilde{M}))$ with $e = \text{inclusion}: C(\tilde{M})\longrightarrow C(f^!)$. The symmetrization of the quadratic kernel is the symmetric kernel

$$(1+T)\sigma_*(f,b) = \sigma^*(f) = (C(f^!),e^{\%}\phi_{\tilde{M}}([M])\in Q^n(C(f^!)))\ .$$

The <u>quadratic signature</u> of (f,b) is the cobordism class

$$\sigma_*(f,b)\in L_n(\mathbb{Z}[\pi])$$

with symmetrization

$$(1+T)\sigma_*(f,b) = \sigma^*(f) = \sigma^*(M) - \sigma^*(X)\in L^n(\mathbb{Z}[\pi])\ .$$

A <u>topological normal structure</u> on an n-dimensional geometric Poincaré complex X is a pair

$$(\tilde{\nu}_X:X\longrightarrow \widetilde{BTOP}(k),\rho_X:S^{n+k}\longrightarrow T(\tilde{\nu}_X))$$

such that $(J\tilde{\nu}_X:X\longrightarrow BG(k),\rho_X)$ is the prescribed Spivak normal structure, i.e. it is a reduction of the Spivak normal fibration ν_X to a topological block bundle $\tilde{\nu}_X$. A compact n-dimensional topological manifold M has a canonical topological normal structure (ν_M,ρ_M) (unique up to stable equivalence) with $\nu_M:M\longrightarrow\widetilde{BTOP}(k)$ the normal bundle of an embedding $M \subset S^{n+k}$ with

$$\rho_M : S^{n+k} \longrightarrow S^{n+k}/\overline{S^{n+k} - E(\nu_M)} = E(\nu_M)/S(\nu_M) = T(\nu_M)$$

the collapsing map.

An <u>n-dimensional topological normal map</u>

$$(f,b) : M \longrightarrow X$$

is a degree 1 map $f:M \longrightarrow X$ from a compact n-dimensional topological manifold M to an n-dimensional geometric Poincaré complex X with a topological normal structure $(\tilde{\nu}_X, \rho_X)$, together with a map of bundles $b:\nu_M \longrightarrow \tilde{\nu}_X$ covering f such that

$$T(b)_*(\rho_M) = \rho_X \in \pi_{n+k}(T(\nu_X)) \quad .$$

This is a normal map in the sense of Browder [6] and Wall [4]. In fact, $(\tilde{\nu}_X, \rho_X)$ determines (f,b) up to normal bordism by the Browder-Novikov construction: make $\rho_X : S^{n+k} \longrightarrow T(\nu_X)$ topologically transverse at the zero section X $T(\nu_X)$ with respect to $\tilde{\nu}_X$ and set

$$(f = \rho_X| , b) : M = \rho_X^{-1}(X) \longrightarrow X \quad .$$

(See Ranicki [7] and §7.1 below for an algebraic treatment of topological normal maps). The <u>surgery obstruction</u> of a topological normal map (f,b) is defined to be the quadratic signature of the underlying normal map $(f,Jb):M \longrightarrow X$ of geometric Poincaré complexes

$$\sigma_*(f,b) = \sigma_*(f,Jb) \in L_n(\mathbb{Z}[\pi]) \quad .$$

<u>Proposition 1.2.2</u> The quadratic L-groups $L_*(\mathbb{Z}[\pi])$ agree with the surgery obstruction groups $L_*(\pi)$ of Wall [4], and the surgery obstruction $\sigma_*(f,b) \in L_n(\mathbb{Z}[\pi])$ of a topological normal map $(f,b):M \longrightarrow X$ agrees with the surgery obstruction $\theta(f,b) \in L_n(\pi)$.

<u>Proof</u>: See II. (Some of the details are recalled in §1.10 below).

[]

(For a topological normal map $(f,b):M \longrightarrow X$ with X finite it is possible to obtain the geometric Umkehr map $F:\Sigma^{\infty}\tilde{X}_+ \longrightarrow \Sigma^{\infty}\tilde{M}_+$ used to define the quadratic kernel

$$\sigma_*(f,b) = \sigma_*(f,Jb) = (C(f^!), e_{\%}\psi_F([X]) \in Q_n(C(f^!)))$$

directly, without appealing to the equivariant S-duality theory of §II.3, as follows. For $p \geqslant 0$ sufficiently large there exists a compact $(n+p)$-dimensional manifold with boundary $(W,\partial W)$ homotopy equivalent to $(X \times D^p, X \times S^{p-1})$ such that (f,b) is approximated by a codimension 0 embedding

$$M^n \times D^p \hookrightarrow \text{interior of } W^{n+p} \ .$$

Pass to the covers and define F using the Pontrjagin-Thom construction by

$$\begin{aligned} F \ : \ \Sigma^p\tilde{X}_+ &= (\tilde{X} \times D^p)/(\tilde{X} \times S^{p-1}) = \tilde{W}/\partial\tilde{W} \\ &\xrightarrow{\text{collapse}} \tilde{W}/(\overline{\tilde{W} - \tilde{M} \times D^p}) = (\tilde{M} \times D^p)/(\tilde{M} \times S^{p-1}) = \Sigma^p\tilde{M}_+ \ ). \end{aligned}$$

The <u>skew-suspension</u> of an n-dimensional $\begin{cases} \varepsilon\text{-symmetric} \\ \varepsilon\text{-quadratic} \end{cases}$ (Poincaré) complex over A $\begin{cases} (C,\phi \in Q^n(C,\varepsilon)) \\ (C,\psi \in Q_n(C,\varepsilon)) \end{cases}$ is the $(n+2)$-dimensional $\begin{cases} (-\varepsilon)\text{-symmetric} \\ (-\varepsilon)\text{-quadratic} \end{cases}$ (Poincaré) complex over A

$$\begin{cases} \bar{S}(C,\phi) = (SC, \bar{S}\phi \in Q^{n+2}(SC,-\varepsilon)) \\ \bar{S}(C,\psi) = (SC, \bar{S}\psi \in Q_{n+2}(SC,-\varepsilon)) \end{cases}$$

defined using the Q-group isomorphism

$$\begin{cases} \bar{S} : Q^n(C,\varepsilon) \xrightarrow{\sim} Q^{n+2}(SC,-\varepsilon) \\ \bar{S} : Q_n(C,\varepsilon) \xrightarrow{\sim} Q_{n+2}(SC,-\varepsilon) \end{cases}$$

induced by the isomorphism of $\mathbb{Z}[\mathbb{Z}_2]$-module chain complexes

$$\bar{S} : \mathrm{Hom}_A(C^*,C) \xrightarrow{\sim} \Omega^2 \mathrm{Hom}_A(SC^*,SC) \ ; \ f \longmapsto (-)^p f \quad (f \in \mathrm{Hom}_A(C^p,C_q)) \ ,$$

with $T \in \mathbb{Z}_2$ acting on $\mathrm{Hom}_A(C^*,C)$ by the ε-duality involution T_ε and on $\mathrm{Hom}_A(SC^*,SC)$ by the $(-\varepsilon)$-duality involution $T_{-\varepsilon}$.

For example, the $\begin{cases}\text{symmetric}\\ \text{quadratic}\end{cases}$ kernel of an (i-1)-connected n-dimensional $\begin{cases}\text{degree 1}\\ \text{normal}\end{cases}$ map $\begin{cases} f : M \longrightarrow X \\ (f,b) : M \longrightarrow X\end{cases}$ (i.e. one such that $K_r(M) = 0$ for $r \leqslant i-1$, with $2i \leqslant n$) is the i-fold skew-suspension of an (n-2i)-dimensional $\begin{cases}(-)^i\text{-symmetric}\\ (-)^i\text{-quadratic}\end{cases}$ Poincaré complex $\begin{cases}\sigma^i(f)\\ \sigma_i(f)\end{cases}$

$$\begin{cases} \sigma^*(f) = \bar{S}^i \sigma^i(f) \\ \sigma_*(f) = \bar{S}^i \sigma_i(f) \end{cases} .$$

The ring A is <u>m-dimensional</u> if every f.g. A-module M has a f.g. projective A-module resolution of length m

$$0 \longrightarrow P_m \longrightarrow P_{m-1} \longrightarrow \dots \longrightarrow P_1 \longrightarrow P_0 \longrightarrow M \longrightarrow 0 \quad .$$

Equivalently, A is noetherian of global dimension m.

<u>Proposition 1.2.3</u> i) $\begin{cases}\text{If } \hat{H}^1(\mathbb{Z}_2;A,\varepsilon) \equiv \{a \in A \,|\, a + \varepsilon\bar{a} = 0\}/\{b - \varepsilon\bar{b} \,|\, b \in A\} = 0 \\ \text{For all } A,\varepsilon\end{cases}$

the skew-suspension maps in the $\begin{cases}\varepsilon\text{-symmetric}\\ \varepsilon\text{-quadratic}\end{cases}$ L-groups

$$\begin{cases} \bar{S} : L^n(A,\varepsilon) \longrightarrow L^{n+2}(A,-\varepsilon) \ ; \ (C,\phi) \longmapsto \bar{S}(C,\phi) \\ \bar{S} : L_n(A,\varepsilon) \longrightarrow L_{n+2}(A,-\varepsilon) \ ; \ (C,\psi) \longmapsto \bar{S}(C,\psi) \end{cases} \quad (n \geqslant 0)$$

are isomorphisms.

ii) If A is m-dimensional the skew-suspension map

$$\bar{S} : L^n(A,\varepsilon) \longrightarrow L^{n+2}(A,-\varepsilon)$$

is an isomorphism for $n \geqslant \max(2m-2,0)$, and a monomorphism for $n = 2m-3$ (if $m \geqslant 2$).

iii) If A is 0-dimensional (i.e. if A is semisimple)

$$\begin{cases} L^{2k+1}(A,\varepsilon) = 0 \\ L_{2k+1}(A,\varepsilon) = 0 \end{cases} \quad (k \geqslant 0) \quad .$$

Proof: See Propositions I.4.3, I.4.5. (The proofs use the algebraic surgery technique summarized in §1.5 below).

[]

In particular, Proposition 1.2.3 i) shows that there are natural identifications

$$L_n(A,\varepsilon) = L_{n+2}(A,-\varepsilon) = L_{n+4}(A,\varepsilon) \quad (n \geqslant 0) \quad .$$

The periodicity of the ε-quadratic L-groups

$$L_*(A,\varepsilon) = L_{*+4}(A,\varepsilon)$$

is a generalization of the periodicity in the surgery obstruction groups of Wall [4,§9]

$$L_*(\pi) = L_{*+4}(\pi) \quad .$$

The ε-symmetric L-groups are not in general periodic,

$$L^*(A,\varepsilon) \neq L^{*+4}(A,\varepsilon) \quad ,$$

and in §I.10 some non-periodic examples were constructed. Carlsson [1],[2] has obtained an essentially cohomological description of $\mathrm{coker}(\bar{S}:L^n(A,-\varepsilon) \longrightarrow L^{n+2}(A,\varepsilon))$ $(n \geqslant 0)$ and $\mathrm{coker}(1+T_\varepsilon:L_0(A,\varepsilon) \longrightarrow L^0(A,\varepsilon))$, at least if the A-module $\hat{H}^0(\mathbb{Z}_2;A,\varepsilon) = \{a \in A \mid \varepsilon\bar{a} = a\}/\{b + \varepsilon\bar{b} \mid b \in A\}$ admits a f.g. projective A-module resolution (e.g. if A is noetherian), with A acting by

$$A \times \hat{H}^0(\mathbb{Z}_2;A,\varepsilon) \longrightarrow \hat{H}^0(\mathbb{Z}_2;A,\varepsilon) \ ; \ (a,x) \longmapsto ax\bar{a} \quad .$$

1.3 Triad Q-groups

Triads are needed for the relative L-theory of §2.2.

A *triad* Γ of A-module chain complexes

$$\Gamma : \quad \begin{array}{ccc} C & \xrightarrow{\ f\ } & D \\ {\scriptstyle g}\big\downarrow & \searrow^{k} & \big\downarrow{\scriptstyle h} \\ C' & \xrightarrow{\ f'\ } & D' \end{array}$$

consists of A-module chain maps

$$f : C \longrightarrow D \ , \ f' : C' \longrightarrow D' \ , \ g : C \longrightarrow C' \ , \ h : D \longrightarrow D'$$

and an A-module chain homotopy

$$k : hf \simeq f'g : C \longrightarrow D' \quad .$$

The *homology* A-modules $H_*(\Gamma)$ of Γ are defined by

$$H_n(\Gamma) = H_n(C(\Gamma)) \quad (n \in \mathbb{Z})$$

where $C(\Gamma)$ is the A-module chain complex given by

$$d_{C(\Gamma)} = \begin{pmatrix} d_{D'} & (-)^{r-1}h & (-)^r f' & k \\ 0 & d_D & 0 & (-)^r f \\ 0 & 0 & d_{C'} & (-)^r g \\ 0 & 0 & 0 & d_C \end{pmatrix}$$

$$: C(\Gamma)_r = D'_r \oplus D_{r-1} \oplus C'_{r-1} \oplus C_{r-2}$$

$$\longrightarrow C(\Gamma)_{r-1} = D'_{r-1} \oplus D_{r-2} \oplus C'_{r-2} \oplus C_{r-3} \ .$$

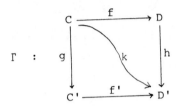

<u>Proposition 1.3.1</u> The triad homology modules $H_*(\Gamma)$ fit into a commutative diagram with exact rows and columns

$$\begin{array}{ccccccccc}
 & & \vdots & & \vdots & & \vdots & & \vdots & \\
 & & \downarrow & & \downarrow & & \downarrow & & \downarrow & \\
\cdots \longrightarrow & & H_{n+2}(\Gamma) & \longrightarrow & H_{n+1}(g) & \longrightarrow & H_{n+1}(h) & \longrightarrow & H_{n+1}(\Gamma) & \longrightarrow \cdots \\
 & & \downarrow & & \downarrow & & \downarrow & & \downarrow & \\
\cdots \longrightarrow & & H_{n+1}(f) & \longrightarrow & H_n(C) & \xrightarrow{f_*} & H_n(D) & \longrightarrow & H_n(f) & \longrightarrow \cdots \\
 & & \downarrow & & \downarrow{\scriptstyle g_*} & & \downarrow{\scriptstyle h_*} & & \downarrow & \\
\cdots \longrightarrow & & H_{n+1}(g') & \longrightarrow & H_n(C') & \xrightarrow{f'_*} & H_n(D') & \longrightarrow & H_n(f') & \longrightarrow \cdots \\
 & & \downarrow & & \downarrow & & \downarrow & & \downarrow & \\
\cdots \longrightarrow & & H_{n+1}(\Gamma) & \longrightarrow & H_n(g) & \longrightarrow & H_n(h) & \longrightarrow & H_n(\Gamma) & \longrightarrow \cdots \\
 & & \downarrow & & \downarrow & & \downarrow & & \downarrow & \\
 & & \vdots & & \vdots & & \vdots & & \vdots &
\end{array}$$

[]

Let Γ be a triad of finite-dimensional A-module chain complexes

$$\Gamma : \quad \begin{array}{ccc} C & \xrightarrow{f} & D \\ {\scriptstyle g}\downarrow & \searrow{\scriptstyle k} & \downarrow{\scriptstyle h} \\ C' & \xrightarrow{f'} & D' \end{array} .$$

The $\mathbb{Z}_2$-isovariant chain map of $\mathbb{Z}[\mathbb{Z}_2]$-module chain complexes

$$(g,h;k) : C(\mathrm{Hom}_A(f^*,f) : \mathrm{Hom}_A(C^*,C) \longrightarrow \mathrm{Hom}_A(D^*,D))$$
$$\longrightarrow C(\mathrm{Hom}_A(f'^*,f') : \mathrm{Hom}_A(C'^*,C') \longrightarrow \mathrm{Hom}_A(D'^*,D'))$$

defined by

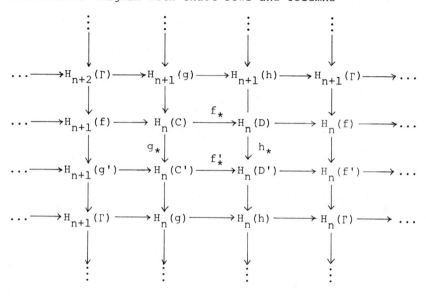

$(g,h;k)_s$:

$$C(\mathrm{Hom}_A(f^*,f))_r = \mathrm{Hom}_A(D^*,D)_r \oplus \mathrm{Hom}_A(C^*,C)_{r-1}$$

$$\longrightarrow C(\mathrm{Hom}_A(f'^*,f'))_{r+s} = \mathrm{Hom}_A(D'^*,D')_{r+s} \oplus \mathrm{Hom}_A(C'^*,C')_{r+s-1} ;$$

$$(\theta,\phi) \longmapsto \begin{cases} (h\theta h^* + (-)^r k\phi f h^* + (-)^{p+1} f' g\phi k^*, g\phi g^*) & s = 0 \\ ((-)^p k\phi k^*, 0) & s = 1 \\ 0 & s \geqslant 2 \end{cases}$$

$$((\theta,\phi) \in \sum_{p+q=r} \mathrm{Hom}_A(D^p,D_q) \oplus \mathrm{Hom}_A(C^p,C_{q-1}))$$

gives rise to a $\mathbb{Z}$-module chain map

$$\begin{cases} (g,h;k)^{\%} : \mathrm{Hom}_{\mathbb{Z}[\mathbb{Z}_2]}(W,C(\mathrm{Hom}_A(f^*,f))) \\ \qquad\qquad \longrightarrow \mathrm{Hom}_{\mathbb{Z}[\mathbb{Z}_2]}(W,C(\mathrm{Hom}_A(f'^*,f'))) ; \\ \{(\delta\phi_s,\phi_s)\} \longmapsto \{(h\delta\phi_s h^* + (-)^{n-1} k\phi_s f^* h^* + (-)^p f' g\phi_s k^* \\ \qquad\qquad + (-)^{n+p-1} kT_\varepsilon \phi_{s-1} k^*, g\phi_s g^*) \,|\, s \geqslant 0\} \\ \\ (g,h;k)_{\%} : W \otimes_{\mathbb{Z}[\mathbb{Z}_2]} C(\mathrm{Hom}_A(f^*,f))) \\ \qquad\qquad \longrightarrow W \otimes_{\mathbb{Z}[\mathbb{Z}_2]} C(\mathrm{Hom}_A(f'^*,f')) ; \\ \{(\delta\psi_s,\psi_s)\} \longmapsto \{(h\delta\psi_s h^* + (-)^{n-1} k\psi_s f^* h^* + (-)^p f' g\psi_s k^* \\ \qquad\qquad + (-)^{n+p} kT_\varepsilon \psi_{s+1} k^*, g\psi_s g^*) \,|\, s \geqslant 0\} \end{cases}$$

inducing morphisms in the relative $\begin{cases} \varepsilon\text{-symmetric} \\ \varepsilon\text{-quadratic} \end{cases}$ Q-groups

$$\begin{cases} (g,h;k)^{\%} : Q^*(f,\varepsilon) \longrightarrow Q^*(f',\varepsilon) \\ (g,h;k)_{\%} : Q_*(f,\varepsilon) \longrightarrow Q_*(f',\varepsilon) \end{cases} .$$

Define the <u>triad</u> $\begin{cases}\underline{\varepsilon\text{-symmetric}}\\ \underline{\varepsilon\text{-quadratic}}\end{cases}$ <u>Q-groups</u> $\begin{cases}Q^*(\Gamma,\varepsilon)\\ Q_*(\Gamma,\varepsilon)\end{cases}$ of a triad Γ of finite-dimensional A-module chain complexes to be the relative homology groups

$$\begin{cases}Q^*(\Gamma,\varepsilon) = H_*((g,h;k)^{\%})\\ Q_*(\Gamma,\varepsilon) = H_*((g,h;k)_{\%})\end{cases}$$

of the $\mathbb{Z}$-module chain map $\begin{cases}(g,h;k)^{\%}\\ (g,h;k)_{\%}\end{cases}$ defined above. An element $\begin{cases}(\delta\phi',\phi',\delta\phi,\phi) \in Q^{n+2}(\Gamma,\varepsilon)\\ (\delta\psi',\psi',\delta\psi,\psi) \in Q_{n+2}(\Gamma,\varepsilon)\end{cases}$ is an equivalence class of collections of chains

$$\begin{cases}\{(\delta\phi'_s,\phi'_s,\delta\phi_s,\phi_s) \in \mathrm{Hom}_A(D'^*,D')_{n+s+2}\oplus\mathrm{Hom}_A(C'^*,C')_{n+s+1}\\ \qquad\oplus\mathrm{Hom}_A(D^*,D)_{n+s+1}\oplus\mathrm{Hom}_A(C^*,C)_{n+s}\,|\,s\geqslant 0\}\\ \{(\delta\psi'_s,\psi'_s,\delta\psi_s,\psi_s) \in \mathrm{Hom}_A(D'^*,D')_{n-s+2}\oplus\mathrm{Hom}_A(C'^*,C')_{n-s+1}\\ \qquad\oplus\mathrm{Hom}_A(D^*,D)_{n-s+1}\oplus\mathrm{Hom}_A(C^*,C)_{n-s}\,|\,s\geqslant 0\}\end{cases}$$

such that

$$\begin{aligned}
&d(\delta\phi',\phi',\delta\phi,\phi)_s\\
&\equiv (d\delta\phi'_s + (-)^r\delta\phi'_s d^* + (-)^{n+s+1}(\delta\phi'_{s-1} + (-)^s T_\varepsilon\delta\phi'_{s-1})\\
&\quad + (-)^{n+1}(f'\phi'_s f'^* - h\delta\phi_s h^*) + (-)^s k\phi_s f^*h^* + (-)^{s-r} f'g\phi_s k^*\\
&\quad + (-)^{s-r+1}k\phi_{s-1}k^*,\ d\phi'_s + (-)^r\phi'_s d^* + (-)^{n+s}(\phi'_{s-1} + (-)^s T_\varepsilon\phi'_{s-1})\\
&\quad + (-)^n g\phi_s g^*,\ d\delta\phi_s + (-)^r\delta\phi_s d^* + (-)^{n+s}(\delta\phi_{s-1} + (-)^s T_\varepsilon\delta\phi_{s-1})\\
&\quad + (-)^n f\phi_s f^*,\ d\phi_s + (-)^r\phi_s d^* + (-)^{n+s-1}(\phi_{s-1} + (-)^s T_\varepsilon\phi_{s-1}))\\
&= 0 \in \mathrm{Hom}_A(D'^*,D')_{n+s+1}\oplus\mathrm{Hom}_A(C'^*,C')_{n+s}\oplus\mathrm{Hom}_A(D^*,D)_{n+s}\oplus\mathrm{Hom}_A(C^*,C)_{n+s-1}\\
&= \sum_{r=-\infty}^{\infty}\mathrm{Hom}_A(D'^{n-r+s+1},D'_r)\oplus\mathrm{Hom}_A(C'^{n-r+s},C'_r)\oplus\mathrm{Hom}_A(D^{n-r+s},D_r)\\
&\qquad\qquad\qquad\oplus\mathrm{Hom}_A(C^{n-r+s-1},C_r)\\
&(s\geqslant 0,\ (\delta\phi',\phi,\delta\phi,\phi)_{-1} = 0)
\end{aligned}$$

$$\left\{\begin{aligned}
& d(\delta\psi',\psi',\delta\psi,\psi)_s \\
&\quad \equiv (d\delta\psi'_s + (-)^r\delta\psi'_s d^* + (-)^{n-s+1}(\delta\psi'_{s+1} + (-)^{s+1}T_\varepsilon\delta\psi'_{s+1}) \\
&\qquad + (-)^{n+1}(f'\psi'f'^* - h\delta\psi_s h^*) + (-)^s k\psi_s f^*h^* + (-)^{s+r}f'g\psi_s k^* \\
&\qquad + (-)^{s+r+1}k\psi_{s+1}k^*,\ d\psi'_s + (-)^r\psi'_s d^* \\
&\qquad + (-)^{n-s}(\psi'_{s+1} + (-)^{s+1}T_\varepsilon\psi'_{s+1}) + (-)^n g\psi_s g^*, \\
&\qquad d\delta\psi_s + (-)^r\delta\psi_s d^* + (-)^{n-s}(\delta\psi_{s+1} + (-)^{s+1}T_\varepsilon\delta\psi_{s+1}) \\
&\qquad + (-)^n f\psi_s f^*,\ d\psi_s + (-)^r\psi_s d^* + (-)^{n-s-1}(\psi_{s+1} + (-)^s T_\varepsilon\psi_{s+1})) \\
&\quad = 0 \in \mathrm{Hom}_A(D'^*,D')_{n-s+1} \oplus \mathrm{Hom}_A(C'^*,C')_{n-s} \oplus \mathrm{Hom}_A(D^*,D)_{n-s} \\
&\qquad\qquad \oplus \mathrm{Hom}_A(C^*,C)_{n-s-1} \\
&\quad = \sum_{r=-\infty}^{\infty} \mathrm{Hom}_A(D'^{n-r-s+1},D'_r) \oplus \mathrm{Hom}_A(C'^{n-r-s},C'_r) \\
&\qquad\qquad \oplus \mathrm{Hom}_A(D^{n-r-s},D_r) \oplus \mathrm{Hom}_A(C^{n-r-s-1},C_r) \quad (s \geqslant 0)\ .
\end{aligned}\right.$$

<u>Proposition 1.3.2</u> The triad ε-symmetric Q-groups $Q^*(\Gamma,\varepsilon)$ fit into a commutative diagram with exact rows and columns

$$\begin{array}{ccccccccc}
 & & \vdots & & \vdots & & \vdots & & \vdots \\
 & & \downarrow & & \downarrow & & \downarrow & & \downarrow \\
\dots \to & & Q^{n+2}(\Gamma,\varepsilon) & \to & Q^{n+1}(g,\varepsilon) & \xrightarrow{(f,f';k)^\%} & Q^{n+1}(h,\varepsilon) & \to & Q^{n+1}(\Gamma,\varepsilon) \to \dots \\
 & & \downarrow & & \downarrow & & \downarrow & & \downarrow \\
\dots \to & & Q^{n+1}(f,\varepsilon) & \to & Q^n(C,\varepsilon) & \xrightarrow{f^\%} & Q^n(D,\varepsilon) & \to & Q^n(f,\varepsilon) \to \dots \\
 & & \downarrow{\scriptstyle (g,h;k)^\%} & & \downarrow{\scriptstyle g^\%} & & \downarrow{\scriptstyle h^\%} & & \downarrow{\scriptstyle (g,h;k)^\%} \\
\dots \to & & Q^{n+1}(f',\varepsilon) & \to & Q^n(C',\varepsilon) & \xrightarrow{f'^\%} & Q^n(D',\varepsilon) & \to & Q^n(f',\varepsilon) \to \dots \\
 & & \downarrow & & \downarrow & & \downarrow & & \downarrow \\
\dots \to & & Q^{n+1}(\Gamma,\varepsilon) & \to & Q^n(g,\varepsilon) & \xrightarrow{(f,f';k)^\%} & Q^n(h,\varepsilon) & \to & Q^n(\Gamma,\varepsilon) \to \dots \\
 & & \downarrow & & \downarrow & & \downarrow & & \downarrow \\
 & & \vdots & & \vdots & & \vdots & & \vdots
\end{array}\ .$$

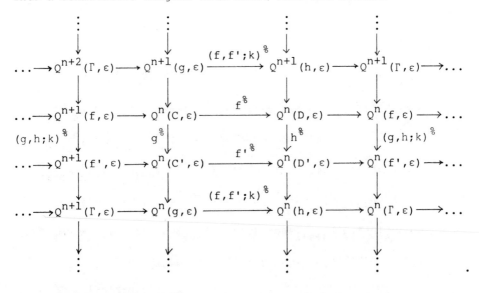

Similarly for the ε-quadratic groups $Q_*(\Gamma,\varepsilon)$.

Proof: This is a special case of Proposition 1.3.1.

[]

A homotopy equivalence of (n+1)-dimensional $\begin{cases}\varepsilon\text{-symmetric}\\ \varepsilon\text{-quadratic}\end{cases}$ pairs over A

$$\begin{cases} (g,h;k) : (f:C \longrightarrow D,(\delta\phi,\phi) \in Q^{n+1}(f,\varepsilon)) \\ \qquad\qquad \longrightarrow (f':C' \longrightarrow D',(\delta\psi',\psi') \in Q^{n+1}(f',\varepsilon)) \\ (g,h;k) : (f:C \longrightarrow D,(\delta\psi,\psi) \in Q_{n+1}(f,\varepsilon)) \\ \qquad\qquad \longrightarrow (f':C' \longrightarrow D',(\delta\psi',\psi') \in Q_{n+1}(f',\varepsilon)) \end{cases}$$

is a chain complex triad of the type

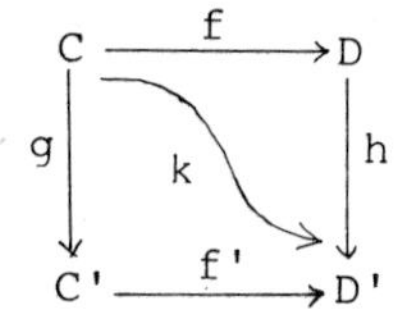

such that the chain maps $g:C \longrightarrow C'$ and $h:D \longrightarrow D'$ are chain equivalences and

$$\begin{cases} (g,h;k)^{\%}(\delta\phi,\phi) = (\delta\phi',\phi') \in Q^{n+1}(f',\varepsilon) \\ (g,h;k)_{\%}(\delta\psi,\psi) = (\delta\psi',\psi') \in Q_{n+1}(f',\varepsilon) \ . \end{cases}$$

An n-dimensional $\begin{cases}\varepsilon\text{-symmetric}\\ \varepsilon\text{-quadratic}\end{cases}$ complex over A $\begin{cases}(C,\phi \in Q^n(C,\varepsilon))\\ (C,\psi \in Q_n(C,\varepsilon))\end{cases}$ is connected (resp. contractible) if

$$\begin{cases} H_0(\psi_0:C^{n-*} \longrightarrow C) = 0 \\ H_0((1+T_\varepsilon)\psi_0:C^{n-*} \longrightarrow C) = 0 \end{cases} \quad (\text{resp. } H_*(C) = 0) \ .$$

A complex is contractible if and only if it is homotopy equivalent to 0.

The <u>algebraic Thom complex</u> of an n-dimensional $\begin{cases}\varepsilon\text{-symmetric}\\ \varepsilon\text{-quadratic}\end{cases}$ Poincaré pair over A $\begin{cases}(f:C\longrightarrow D,(\delta\phi,\phi)\in Q^n(f,\varepsilon))\\ (f:C\longrightarrow D,(\delta\psi,\psi)\in Q_n(f,\varepsilon))\end{cases}$ is the connected n-dimensional $\begin{cases}\varepsilon\text{-symmetric}\\ \varepsilon\text{-quadratic}\end{cases}$ complex over A

$$\begin{cases}(C(f),\delta\phi/\phi\in Q^n(C(f),\varepsilon)))\\ (C(f),\delta\psi/\psi\in Q_n(C(f),\varepsilon)))\end{cases}$$

defined by

$$\begin{cases}(\delta\phi/\phi)_s=\begin{pmatrix}\phi_s & 0\\ (-)^{n-r-1}\phi_s f^* & (-)^{n-r+s}T_\varepsilon\phi_{s-1}\end{pmatrix}\\ \quad : C(f)^{n-r+s}=D^{n-r+s}\oplus C^{n-r+s-1}\\ \qquad\qquad\longrightarrow C(f)_r=D_r\oplus C_{r-1}\qquad (s\geqslant 0,\phi_{-1}=0)\\ (\delta\psi/\psi)_s=\begin{pmatrix}\psi_s & 0\\ (-)^{n-r-1}\psi_s f^* & (-)^{n-r-s-1}T_\varepsilon\psi_{s+1}\end{pmatrix}\\ \quad : C(f)^{n-r-s}=D^{n-r-s}\oplus C^{n-r-s-1}\\ \qquad\qquad\longrightarrow C(f)_r=D_r\oplus C_{r-1}\qquad (s\geqslant 0)\end{cases}$$

.

For example, if $\nu:X\longrightarrow BG(k)$ is a (k-1)-spherical fibration over an (n-k)-dimensional geometric Poincaré complex X, and $\widetilde{X}$ is a covering of X with group of covering translations π, the algebraic Thom complex of the n-dimensional symmetric Poincaré pair over $\mathbb{Z}[\pi]$ associated to $(E(\nu),S(\nu))$

$$\sigma^*(E(\nu),S(\nu)) = (i:C(S(\widetilde{\nu}))\longrightarrow C(E(\widetilde{\nu})),(\delta\phi,\phi)\in Q^n(i))$$

$$(i = \text{inclusion})$$

is the connected n-dimensional symmetric complex over $\mathbb{Z}[\pi]$ associated to the Thom π-space $T\pi(\nu) = E(\widetilde{\nu})/S(\widetilde{\nu})$

$$(\dot{C}(T\pi(\nu)), \dot{\phi}_{T\pi(\nu)}(U_\nu \cap [X]) \in Q^n(\dot{C}(T\pi(\nu))))$$

up to homotopy equivalence.

The <u>algebraic Poincaré thickening</u> of a connected n-dimensional $\begin{cases} \varepsilon\text{-symmetric} \\ \varepsilon\text{-quadratic} \end{cases}$ complex over A $\begin{cases} (C, \phi \in Q^n(C,\varepsilon)) \\ (C, \psi \in Q_n(C,\varepsilon)) \end{cases}$ is the n-dimensional $\begin{cases} \varepsilon\text{-symmetric} \\ \varepsilon\text{-quadratic} \end{cases}$ Poincaré pair over A

$$\begin{cases} (i_C : \partial C \longrightarrow C^{n-*}, (0, \partial\phi) \in Q^n(i_C, \varepsilon)) \\ (i_C : \partial C \longrightarrow C^{n-*}, (0, \partial\psi) \in Q_n(i_C, \varepsilon)) \end{cases}$$

defined by

$$d_{\partial C} = \begin{cases} \begin{pmatrix} d & (-)^r \phi_0 \\ 0 & (-)^r d^* \end{pmatrix} \\ \begin{pmatrix} d & (-)^r (1+T_\varepsilon)\psi_0 \\ 0 & (-)^r d^* \end{pmatrix} \end{cases}$$

$$: \partial C_r = C_{r+1} \oplus C^{n-r} \longrightarrow \partial C_{r-1} = C_r \oplus C^{n-r+1}$$

$$i_C = (0 \quad 1) : \partial C_r = C_{r+1} \oplus C^{n-r} \longrightarrow C^{n-r}$$

$$\begin{cases} \partial\phi_0 = \begin{pmatrix} (-)^{n-r-1} T_\varepsilon \phi_1 & (-)^{r(n-r-1)} \varepsilon \\ 1 & 0 \end{pmatrix} \\ \partial\psi_0 = \begin{pmatrix} 0 & 0 \\ 1 & 0 \end{pmatrix} \end{cases}$$

$$: \partial C^{n-r-1} = C^{n-r} \oplus C_{r+1} \longrightarrow \partial C_r = C_{r+1} \oplus C^{n-r+1}$$

$$\left\{\begin{array}{l} \partial\phi_s = \begin{pmatrix} (-)^{n-r+s-1}T_\varepsilon\phi_{s+1} & 0 \\ 0 & 0 \end{pmatrix} \\ \qquad : \partial C^{n-r+s-1} = C^{n-r+s}\oplus C_{r-s+1} \longrightarrow \partial C_r = C_{r+1}\oplus C^{n-r} \\ \partial\psi_s = \begin{pmatrix} (-)^{n-r-s}T_\varepsilon\psi_{s-1} & 0 \\ 0 & 0 \end{pmatrix} \\ \qquad : \partial C^{n-r-s-1} = C^{n-r-s}\oplus C_{r+s+1} \longrightarrow \partial C_r = C_{r+1}\oplus C^{n-r} \ . \end{array}\right. \qquad (s \geqslant 1)$$

The (n-1)-dimensional $\left\{\begin{array}{l}\varepsilon\text{-symmetric}\\ \varepsilon\text{-quadratic}\end{array}\right.$ quadratic Poincaré complex over A

$$\left\{\begin{array}{l} \partial(C,\phi) = (\partial C, \partial\phi \in Q^{n-1}(\partial C,\varepsilon)) \\ \partial(C,\psi) = (\partial C, \partial\psi \in Q_{n-1}(\partial C,\varepsilon)) \end{array}\right.$$

is the __boundary__ of $\left\{\begin{array}{l}(C,\phi)\\ (C,\psi)\end{array}\right.$. For example, if $(X,\partial X)$ is an n-dimensional geometric Poincaré pair, and $(\widetilde{X},\widetilde{\partial X})$ is a covering of $(X,\partial X)$ with group of covering translations π, the n-dimensional symmetric Poincaré pair over $\mathbb{Z}[\pi]$ associated to $(X,\partial X)$

$$\sigma^*(X,\partial X) = (i = \text{inclusion} : C(\widetilde{\partial X}) \longrightarrow C(\widetilde{X}), (\phi,\partial\phi) \in Q^n(i))$$

is the algebraic Poincaré thickening of the connected n-dimensional symmetric complex over $\mathbb{Z}[\pi]$ associated to $X/\partial X$

$$(\dot{C}(\widetilde{X}/\widetilde{\partial X}), \dot{\phi}_{\widetilde{X}/\widetilde{\partial X}}(j_*[X]) \in Q^n(\dot{C}(\widetilde{X}/\widetilde{\partial X})))$$

up to homotopy equivalence, with

$$j_* = \text{projection}_* : H_n(X,\partial X) \longrightarrow \dot{H}_n(X/\partial X) \ .$$

Proposition 1.3.3 i) The algebraic Thom complex and algebraic Poincaré thickening operations are inverse to each other up to homotopy equivalence, defining a natural one-one correspondence between the homotopy equivalence classes of n-dimensional $\left\{\begin{array}{l}\varepsilon\text{-symmetric}\\ \varepsilon\text{-quadratic}\end{array}\right.$ Poincaré pairs over A and the homotopy equivalence classes of connected n-dimensional $\left\{\begin{array}{l}\varepsilon\text{-symmetric}\\ \varepsilon\text{-quadratic}\end{array}\right.$ complexes over A. The correspondence preserves boundaries; algebraic Poincaré pairs with contractible boundary correspond to algebraic Poincaré complexes.

ii) A connected n-dimensional $\left\{\begin{array}{l}\varepsilon\text{-symmetric}\\ \varepsilon\text{-quadratic}\end{array}\right.$ complex is Poincaré if and only if its boundary is a contractible (n-1)-dimensional $\left\{\begin{array}{l}\varepsilon\text{-symmetric}\\ \varepsilon\text{-quadratic}\end{array}\right.$ Poincaré complex.

iii) An n-dimensional $\left\{\begin{array}{l}\varepsilon\text{-symmetric}\\ \varepsilon\text{-quadratic}\end{array}\right.$ Poincaré complex is null-cobordant if and only if it is homotopy equivalent to the boundary of a connected (n+1)-dimensional $\left\{\begin{array}{l}\varepsilon\text{-symmetric}\\ \varepsilon\text{-quadratic}\end{array}\right.$ complex.

Proof: See Proposition I.3.4.

[]

1.4 Algebraic Wu classes

The Wu classes of an algebraic Poincaré complex over A (C,ϕ) are functions

$$v(\phi) : H^*(C) \longrightarrow \text{(subquotient groups of A)}$$

which are homotopy invariants of (C,ϕ), whose definition we shall now recall. We refer to §§II.1,5,9 for the relations between the algebraic Wu classes and the Wu classes arising in topology.

Given A,ε as in §1.1 let $T \in \mathbb{Z}_2$ act on A by the involution

$$T_\varepsilon : A \longrightarrow A \; ; \; a \longmapsto \varepsilon\bar{a} \quad ,$$

and define the $\begin{cases} \underline{\mathbb{Z}_2\text{-cohomology}} \\ \underline{\mathbb{Z}_2\text{-homology}} \\ \underline{\text{Tate } \mathbb{Z}_2\text{-cohomology}} \end{cases}$ $\underline{\text{groups of } (A,\varepsilon)}$ $\begin{cases} H^*(\mathbb{Z}_2;A,\varepsilon) \\ H_*(\mathbb{Z}_2;A,\varepsilon) \\ \hat{H}^*(\mathbb{Z}_2;A,\varepsilon) \end{cases}$

by

$$H^r(\mathbb{Z}_2;A,\varepsilon) = \begin{cases} \ker(1-T_\varepsilon : A \longrightarrow A) & r = 0 \\ \hat{H}^r(\mathbb{Z}_2;A,\varepsilon) & r \geqslant 1 \\ 0 & r \leqslant -1 \end{cases}$$

$$H_r(\mathbb{Z}_2;A,\varepsilon) = \begin{cases} \operatorname{coker}(1-T_\varepsilon : A \longrightarrow A) & r = 0 \\ \hat{H}^{r+1}(\mathbb{Z}_2;A,\varepsilon) & r \geqslant 1 \\ 0 & r \leqslant -1 \end{cases}$$

$$\hat{H}^r(\mathbb{Z}_2;A,\varepsilon) = \ker(1-(-)^r T_\varepsilon : A \longrightarrow A)/\operatorname{im}(1+(-)^r T_\varepsilon : A \longrightarrow A) \quad r \in \mathbb{Z} \; .$$

For $m \in \mathbb{Z}$ let $S^m A$ be the A-module chain complex defined by

$$(S^m A)_r = \begin{cases} A \text{ if } r = m \\ 0 \text{ if } r \neq m \; . \end{cases}$$

The cohomology classes $f \in H^m(C) = H_0(\operatorname{Hom}_A(C,S^m A))$ of an A-module chain complex C are the chain homotopy classes of chain maps $f : C \longrightarrow S^m A$.

Let C be a finite-dimensional A-module chain complex.

The <u>rth</u> $\begin{cases} \underline{\varepsilon\text{-symmetric}} \\ \underline{\varepsilon\text{-quadratic}} \\ \underline{\varepsilon\text{-hyperquadratic}} \end{cases}$ <u>Wu class</u> of an element $\begin{cases} \phi \in Q^n(C,\varepsilon) \\ \psi \in Q_n(C,\varepsilon) \\ \theta \in \hat{Q}^n(C,\varepsilon) \end{cases}$

is the function

$$\begin{cases} v_r(\phi) : H^{n-r}(C) \longrightarrow Q^n(S^{n-r}A,\varepsilon) = H^{n-2r}(\mathbb{Z}_2;A,(-)^{n-r}\varepsilon)\,;\ f \longmapsto f\phi_{n-2r}f^* \\ v^r(\psi) : H^{n-r}(C) \longrightarrow Q_n(S^{n-r}A,\varepsilon) = H_{2r-n}(\mathbb{Z}_2;A,(-)^{n-r}\varepsilon)\,;\ f \longmapsto f\psi_{2r-n}f^* \\ \hat{v}_r(\theta) : H^{n-r}(C) \longrightarrow \hat{Q}^n(S^{n-r}A,\varepsilon) = \hat{H}^r(\mathbb{Z}_2;A,\varepsilon)\ ;\ f \longmapsto f\theta_{n-2r}f^* \end{cases}$$

$(f:C_{n-r} \longrightarrow A,\ \phi_{n-2r},\psi_{2r-n},\theta_{n-2r} \in \mathrm{Hom}_A(C^{n-r},C_{n-r}),\ \mathrm{Hom}_A(A^*,A) = A)$.

The Wu classes v are quadratic functions, in the sense that

$$v(af) = a.v(f).\bar{a} \quad (a \in A, f \in H^*(C))\ .$$

<u>Proposition 1.4.1</u> i) The various Wu classes are related to each other by

$$\hat{v}_r(J\phi) : H^{n-r}(C) \xrightarrow{v_r(\phi)} H^{n-2r}(\mathbb{Z}_2;A,(-)^{n-r}\varepsilon) \xrightarrow{J} \hat{H}^r(\mathbb{Z}_2;A,\varepsilon)$$

$$v_r((1+T_\varepsilon)\psi) : H^{n-r}(C) \xrightarrow{v^r(\psi)} H_{2r-n}(\mathbb{Z}_2;A,(-)^{n-r}\varepsilon) \xrightarrow{1+T_\varepsilon} H^{n-2r}(\mathbb{Z}_2;A,(-)^{n-r}\varepsilon)$$

$$v^r(H\theta) : H^{n-r-1}(C) \xrightarrow{\hat{v}_{r+1}(\theta)} \hat{H}^{r+1}(\mathbb{Z}_2;A,\varepsilon) \xrightarrow{H} H_{2r-n+1}(\mathbb{Z}_2;A,(-)^{n-r+1}\varepsilon)$$

$(\phi \in Q^n(C,\varepsilon),\psi \in Q_n(C,\varepsilon),\theta \in \hat{Q}^n(C,\varepsilon),J\phi \in \hat{Q}^n(C,\varepsilon),(1+T_\varepsilon)\psi \in Q^n(C,\varepsilon),$
$H\theta \in Q_{n-1}(C,\varepsilon))$.

ii) The Wu classes satisfy the sum formulae

$$v_r(\phi)(f+g) - v_r(\phi)(f) - v_r(\phi)(g)$$

$$= \begin{cases} (1+T_{(-)^r\varepsilon})(g\phi_0 f^*) \in H^0(\mathbb{Z}_2;A,(-)^r\varepsilon) \\ 0 \in \hat{H}^0(\mathbb{Z}_2;A,(-)^r\varepsilon) \end{cases} \text{ if } \begin{cases} n = 2r \\ n \neq 2r \end{cases}$$

$$v^r(\psi)(f+g) - v^r(\psi)(f) - v^r(\psi)(g)$$

$$= \begin{cases} g((1+T_\varepsilon)\psi_0)f^* \in H_0(\mathbb{Z}_2;A,(-)^r\varepsilon) \\ 0 \in \hat{H}^0(\mathbb{Z}_2;A,(-)^{r+1}\varepsilon) \end{cases} \text{ if } \begin{cases} n = 2r \\ n \neq 2r \end{cases}$$

$$\hat{v}_r(\theta)(f+g) - \hat{v}_r(\theta)(f) - \hat{v}_r(\theta)(g) = 0 \in \hat{H}^0(\mathbb{Z}_2;A,(-)^r\varepsilon)$$

$$(f,g \in H^{n-r}(C)) .$$

[]

The middle-dimensional intersection pairing of a 2r-dimensional ε-symmetric complex over A $(C,\phi \in Q^{2r}(C,\varepsilon))$

$$\lambda = \phi_0 : H^r(C) \times H^r(C) \longrightarrow A \ ; \ (f,g) \longmapsto g\phi_0 f^*$$

is such that

$$\lambda(f,g_1+g_2) = \lambda(f,g_1) + \lambda(f,g_2)$$

$$\lambda(f,ag) = a\lambda(f,g)$$

$$\lambda(g,f) = (-)^r\varepsilon.\overline{\lambda(f,g)} \in A$$

$$(f,g,g_1,g_2 \in H^r(C), a \in A) ,$$

with the rth ε-symmetric Wu class given by

$$v_r(\psi) : H^r(C) \longrightarrow H^0(\mathbb{Z}_2;A,(-)^r\varepsilon) \ ; \ f \longmapsto \lambda(f,f) .$$

The rth Wu class of a 2r-dimensional ε-quadratic complex over A $(C,\psi \in Q_{2r}(C,\varepsilon))$ is a function

$$\mu = v^r(\psi) : H^r(C) \longrightarrow H_0(\mathbb{Z}_2;A,(-)^r\varepsilon)$$

such that

$$\mu(af) = a\mu(f)\overline{a}$$

$$\mu(f+g) - \mu(f) - \mu(g) = \lambda(f,g) \in H_0(\mathbb{Z}_2;A,(-)^r\varepsilon)$$

$$\lambda(f,f) = \mu(f) + \varepsilon\overline{\mu(f)} \in H^0(\mathbb{Z}_2;A,(-)^r\varepsilon)$$

$$(f,g \in H^r(C), a \in A)$$

where $\lambda = (1+T_\varepsilon)\psi_0 : H^r(C)\times H^r(C) \longrightarrow A$ is the intersection pairing of the ε-symmetrization $(C,(1+T_\varepsilon)\psi \in Q^{2r}(C,\varepsilon))$.
In particular, if $(f,b):M\longrightarrow X$ is an $(r-1)$-connected 2r-dimensional normal map the quadratic kernel

$$\sigma_*(f,b) = (C(f^!),\psi \in Q_{2r}(C(f^!)))$$

is a 2r-dimensional quadratic Poincaré complex over $\mathbb{Z}[\pi_1(X)]$ with

$$H^r(C(f^!)) = H_r(C(f^!)) = K_r(M) = \ker(\tilde{f}_*:H_r(\tilde{M})\longrightarrow H_r(\tilde{X}))$$

(up to isomorphism), and in this case the triple

$$(K_r(M),\lambda = (1+T)\psi_0 : K_r(M)\times K_r(M) \longrightarrow \mathbb{Z}[\pi_1(X)],$$

$$\mu = v^r(\psi) : K_r(M) \longrightarrow H_0(\mathbb{Z}_2;\mathbb{Z}[\pi_1(X)],(-)^r))$$

is the $(-)^r$-quadratic form $(K_r(M),\lambda,\mu)$ used by Wall [4,§5] to define the surgery obstruction $\sigma_*(f,b)\in L_{2r}(\mathbb{Z}[\pi_1(X)])$, with λ (resp. μ) the geometrically defined intersection (resp. self-intersection) form, cf. Proposition II.5.4.
We shall recall the precise relationship between $\left\{\begin{array}{l}\varepsilon\text{-symmetric}\\ \varepsilon\text{-quadratic}\end{array}\right.$ complexes and $\left\{\begin{array}{l}\varepsilon\text{-symmetric}\\ \varepsilon\text{-quadratic}\end{array}\right.$ forms in §1.6 below.

Define the <u>even ε-symmetric Q-groups</u> $Q\langle v_0\rangle^*(C,\varepsilon)$ of a finite-dimensional A-module chain complex C by

$$Q\langle v_0\rangle^n(C,\varepsilon) = \ker(\hat{v}_0:Q^n(C,\varepsilon)\longrightarrow \mathrm{Hom}_A(H^n(C),\hat{H}^0(\mathbb{Z}_2;A,\varepsilon)))\quad (n\geqslant 0)\ .$$

An n-dimensional ε-symmetric complex over A $(C,\psi\in Q^n(C,\varepsilon))$ is <u>even</u> if

$$\phi \in Q\langle v_0\rangle^n(C,\varepsilon) \subseteq Q^n(C,\varepsilon)\ .$$

For example, the ε-symmetrization $(C,(1+T_\varepsilon)\psi\in Q^n(C,\varepsilon))$ of an ε-quadratic complex $(C,\psi\in Q_n(C,\varepsilon))$ is even. The <u>relative even</u>

<u>ε-symmetric Q-groups</u> $Q\langle v_0\rangle^*(f,\varepsilon)$ of a chain map $f:C\longrightarrow D$ of finite-dimensional A-module chain complexes

$$Q\langle v_0\rangle^{n+1}(f,\varepsilon) = \ker(\hat{v}_0:Q^{n+1}(f,\varepsilon)\longrightarrow \mathrm{Hom}_A(H^{n+1}(f),\hat{H}^0(\mathbb{Z}_2;A,\varepsilon))) \quad (n\geqslant 0)$$

where the relative Wu class $\hat{v}_0$ of $(\delta\phi,\phi)\in Q^{n+1}(f,\varepsilon)$ is given by

$$\hat{v}_0(\delta\phi,\phi) \;:\; H^{n+1}(f) \longrightarrow \hat{H}^0(\mathbb{Z}_2;A,\varepsilon) \;;$$

$$(g,h)\longmapsto g(\delta\phi_{n+1})g^* + (-)^n h(\phi_n)h^*$$

$$((\delta\phi_{n+1},\phi_n)\in \mathrm{Hom}_A(D^{n+1},D_{n+1})\oplus\mathrm{Hom}_A(C^n,C_n),\ (g,h)\in D^{n+1}\oplus C^n) \ .$$

An $(n+1)$-dimensional ε-symmetric pair $(f:C\longrightarrow D,(\delta\phi,\phi))$ is <u>even</u> if

$$(\delta\phi,\phi)\in Q\langle v_0\rangle^{n+1}(f,\varepsilon)\subseteq Q^{n+1}(f,\varepsilon) \ .$$

The <u>n-dimensional even ε-symmetric L-group of A</u> $L\langle v_0\rangle^n(A,\varepsilon)$ $(n\geqslant 0)$ is the cobordism group of n-dimensional even ε-symmetric Poincaré complexes over A, where the cobordism are the $(n+1)$-dimensional even ε-symmetric Poincaré pairs over A.

By analogy with Proposition 1.2.2 i),ii) we have:

<u>Proposition 1.4.2</u> The skew-suspension maps

$$\bar{S} \;:\; L^n(A,\varepsilon)\longrightarrow L\langle v_0\rangle^{n+2}(A,-\varepsilon) \;;\; (C,\phi)\longmapsto(SC,\bar{S}\phi) \quad (n\geqslant 0)$$

are isomorphisms.

<u>Proof</u>: See Proposition I.4.4.

[]

The ε-symmetrization map in the L-groups factors through the even ε-symmetric L-groups

$$1+T_\varepsilon \;:\; L_n(A,\varepsilon)\longrightarrow L\langle v_0\rangle^n(A,\varepsilon)\longrightarrow L^n(A,\varepsilon) \quad (n\geqslant 0) \ .$$

In §1.8 below we shall recall from §I.6 the way in which the even ε-symmetric L-groups $L\langle v_0\rangle^n(A,\varepsilon)$ for $n = 0,1$ bridge the gap between the ε-quadratic and the ε-symmetric L-groups, defining a unified L-theory containing all three types of L-group.

Proposition 1.2.3 iii) also extends to the even ε-symmetric L-groups, with $L\langle v_0\rangle^1(A,\varepsilon) = 0$ for a 0-dimensional ring with involution A (cf. the proof of Proposition I.4.5).

Proposition 1.4.3 If A,ε are such that $\hat{H}^*(\mathbb{Z}_2;A,\varepsilon) = 0$ the natural maps

$$1+T_\varepsilon : L_n(A,\varepsilon) \longrightarrow L\langle v_0\rangle^n(A,\varepsilon) \ , \ L\langle v_0\rangle^n(A,\varepsilon) \longrightarrow L^n(A,\varepsilon) \quad (n \geqslant 0)$$

are isomorphisms. In particular, this is the case if there exists a central element $a \in A$ such that $a + \bar{a} = 1 \in A$ (e.g. $a = \frac{1}{2} \in A$) .

Proof: See Proposition I.3.3.

[]

Indeed, if $\hat{H}^*(\mathbb{Z}_2;A,\varepsilon) = 0$ there are natural identifications of categories

{ε-quadratic complexes over A}

= {even ε-symmetric complexes over A}

= {ε-symmetric complexes over A}

with

$$Q_*(C,\varepsilon) = Q\langle v_0\rangle^*(C,\varepsilon) = Q^*(C,\varepsilon)$$

for any finite-dimensional A-module chain complex C.

1.5 Algebraic surgery

An (n+1)-dimensional $\begin{cases} \varepsilon\text{-symmetric} \\ \varepsilon\text{-quadratic} \end{cases}$ pair over A

$\begin{cases} (f:C \longrightarrow D, (\delta\phi, \phi) \in Q^{n+1}(f,\varepsilon)) \\ (f:C \longrightarrow D, (\delta\psi, \psi) \in Q_{n+1}(f,\varepsilon)) \end{cases}$ is <u>connected</u> if

$$\begin{cases} H_0\left(\begin{pmatrix} \delta\phi_0 \\ \phi_0 f^* \end{pmatrix} : D^{n+1-*} \longrightarrow C(f)\right) = 0 \\ H_0\left(\begin{pmatrix} (1+T_\varepsilon)\delta\psi_0 \\ (1+T_\varepsilon)\psi_0 f^* \end{pmatrix} : D^{n+1-*} \longrightarrow C(f)\right) = 0 \end{cases} .$$

Define as follows the connected n-dimensional $\begin{cases} \varepsilon\text{-symmetric} \\ \varepsilon\text{-quadratic} \end{cases}$ complex over A $\begin{cases} (C', \phi' \in Q^n(C',\varepsilon)) \\ (C', \psi' \in Q_n(C',\varepsilon)) \end{cases}$ obtained from a connected n-dimensional $\begin{cases} \varepsilon\text{-symmetric} \\ \varepsilon\text{-quadratic} \end{cases}$ complex over A $\begin{cases} (C, \phi \in Q^n(C,\varepsilon)) \\ (C, \psi \in Q_n(C,\varepsilon)) \end{cases}$ by <u>surgery</u> on a connected (n+1)-dimensional $\begin{cases} \varepsilon\text{-symmetric} \\ \varepsilon\text{-quadratic} \end{cases}$ pair $\begin{cases} (f:C \longrightarrow D, (\delta\phi, \phi) \in Q^{n+1}(f,\varepsilon)) \\ (f:C \longrightarrow D, (\delta\psi, \psi) \in Q_{n+1}(f,\varepsilon)) \end{cases}$. (This is an algebraic surgery "killing $\mathrm{im}(f^*:H^*(D) \longrightarrow H^*(C))$").

In the ε-symmetric case let

$$d_{C'} = \begin{pmatrix} d_C & 0 & (-)^{n+1}\phi_0 f^* \\ (-)^r f & d_D & (-)^r \phi_0 \\ 0 & 0 & (-)^r d_D^* \end{pmatrix}$$

$$: C'_r = C_r \oplus D_{r+1} \oplus D^{n-r+1} \longrightarrow C'_{r-1} = C_{r-1} \oplus D_r \oplus D^{n-r+2}$$

$$\phi'_0 = \begin{pmatrix} \phi_0 & 0 & 0 \\ (-)^{n-r} f T_\varepsilon \phi_1 & (-)^{n-r} T_\varepsilon \delta\phi_1 & 0 \\ 0 & 1 & 0 \end{pmatrix}$$

$$: C'^{n-r} = C^{n-r} \oplus D^{n-r+1} \oplus D_{r+1} \longrightarrow C'_r = C_r \oplus D_{r+1} \oplus D^{n-r+1}$$

$$\phi'_s = \begin{pmatrix} \phi_s & 0 & 0 \\ (-)^{n-r} f T_\varepsilon \phi_s & (-)^{n-r+s} T_\varepsilon \delta\phi_{s+1} & 0 \\ 0 & 0 & 0 \end{pmatrix} \qquad (s \geqslant 1)$$

$$: C'^{n-r+s} = C^{n-r+s} \oplus D^{n-r+s+1} \oplus D_{r-s+1} \longrightarrow C'_r = C_r \oplus D_{r+1} \oplus D^{n-r+1} \quad .$$

In the ε-quadratic case let

$$d_{C'} = \begin{pmatrix} d_C & 0 & (-)^{n+1}(1+T_\varepsilon)\psi_0 f^* \\ (-)^r f & d_D & (-)^r (1+T_\varepsilon)\delta\psi_0 \\ 0 & 0 & (-)^r d_D^* \end{pmatrix}$$

$$: C'_r = C_r \oplus D_{r+1} \oplus D^{n-r+1} \longrightarrow C'_{r-1} = C_{r-1} \oplus D_r \oplus D^{n-r+2}$$

$$\psi'_0 = \begin{pmatrix} \psi_0 & 0 & 0 \\ 0 & 0 & 0 \\ 0 & 1 & 0 \end{pmatrix}$$

$$: C'^{n-r} = C^{n-r} \oplus D^{n-r+1} \oplus D_{r+1} \longrightarrow C'_r = C_r \oplus D_{r+1} \oplus D^{n-r+1}$$

$$\psi'_s = \begin{pmatrix} \psi_s & (-)^{r+s}T_\varepsilon\psi_{s-1}f^* & 0 \\ 0 & (-)^{n-r-s+1}T_\varepsilon\delta\psi_{s-1} & 0 \\ 0 & 0 & 0 \end{pmatrix}$$

$$: C'^{n-r-s} = C^{n-r-s}\oplus D^{n-r-s+1}\oplus D_{r+s+1}$$

$$\longrightarrow C'_r = C_r\oplus D_{r+1}\oplus D^{n-r+1} \qquad (s \geqslant 1)\ .$$

In §II.7 it was shown that for a $\begin{cases}\text{degree 1}\\ \text{normal}\end{cases}$ map $\begin{cases} f:M \longrightarrow X \\ (f,b):M \longrightarrow X\end{cases}$ from an n-dimensional manifold M to an n-dimensional geometric Poincaré complex X the effect on the $\begin{cases}\text{symmetric}\\ \text{quadratic}\end{cases}$ kernel $\begin{cases}\sigma^*(f) = (C,\phi)\\ \sigma_*(f,b) = (C,\psi)\end{cases}$ of an $\begin{cases}\text{oriented}\\ \text{framed}\end{cases}$ surgery on a framed embedding $S^r \subset M^n$ with a null-homotopy of $S^r \hookrightarrow M \xrightarrow{f} X$ (replacing $\begin{cases} f:M \longrightarrow X \\ (f,b):M \longrightarrow X\end{cases}$ by $\begin{cases} f':M' \longrightarrow X \\ (f',b'):M' \longrightarrow X\end{cases}$ with $M' = \overline{M\backslash S^r \times D^{n-r}} \cup D^{r+1} \times S^{n-r-1}$) is that of algebraic surgery on the $\begin{cases}\text{symmetric}\\ \text{quadratic}\end{cases}$ pair $\begin{cases}(g:C \longrightarrow S^{n-r}\mathbb{Z}[\pi_1(X)],\ (\delta\phi,\phi) \in Q^{n+1}(g))\\ (g:C \longrightarrow S^{n-r}\mathbb{Z}[\pi_1(X)],\ (\delta\psi,\psi) \in Q_{n+1}(g))\end{cases}$ determined by the commutative diagram of maps

$$\begin{array}{ccc} S^r & \hookrightarrow & M \\ \downarrow & F & \downarrow f \\ D^{r+1} & \longrightarrow & X \end{array}$$

with $g^*(1) \in H^{n-r}(C) = H_r(C) = K_r(M) = H_{r+1}(\tilde{f})$ the Hurewicz image of $F \in \pi_{r+1}(f) = \pi_{r+1}(\tilde{f})$.

Proposition 1.5.1 i) Algebraic surgery preserves the homotopy type of the boundary. In particular, surgery on an algebraic Poincaré complex results in an algebraic Poincaré complex.
ii) Algebraic Poincaré complexes x,y are cobordant if and only if x is homotopy equivalent to a complex obtained from y by surgery.

Proof: See Proposition I.4.1.

[]

Proposition 1.5.2 The skew-suspension map

$$\begin{cases} \overline{S} : L^n(A,\varepsilon) \longrightarrow L^{n+2}(A,-\varepsilon) \\ \overline{S} : L_n(A,\varepsilon) \longrightarrow L_{n+2}(A,-\varepsilon) \end{cases}$$

(for some $n \geqslant 0$) is onto (resp. one-one) if for every connected (n+2)- (resp. (n+3)-) dimensional $\begin{cases} (-\varepsilon)\text{-symmetric} \\ (-\varepsilon)\text{-quadratic} \end{cases}$ complex over A x with a boundary ∂x which is contractible (resp. a skew-suspension) it is possible to do $\begin{cases} (-\varepsilon)\text{-symmetric} \\ (-\varepsilon)\text{-quadratic} \end{cases}$ surgery on x to obtain a skew-suspension.

Proof: See Proposition I.4.2.

[]

The criterion of Proposition 1.5.2 for the skew-suspension map to be an isomorphism is always satisfied in the ε-quadratic case, cf. Proposition 1.2.3 i). It is not in general satisfied in the ε-symmetric case, cf. Proposition 1.2.3 ii) and the examples of non-periodic ε-symmetric L-groups of §I.10.

1.6 Forms and formations

Next, we recall the correspondence between n-dimensional algebraic Poincaré complexes for n = 0 (resp. 1) and quadratic forms (resp. formations), and the correspondence between the L-groups and the Witt groups.

Given a f.g. projective A-module M define the <u>ε-duality</u> involution

$$T_\varepsilon : \mathrm{Hom}_A(M,M^*) \longrightarrow \mathrm{Hom}_A(M,M^*) ;$$
$$(\varepsilon\phi^*: x \longmapsto (y \longmapsto \varepsilon\overline{\phi(y)(x)})) \quad .$$

This is just the ε-duality involution T_ε on $\mathrm{Hom}_A(C^*,C)$ with C the 0-dimensional A-module chain complex defined by

$$C_r = \begin{cases} M^* & \text{if } r = 0 \\ 0 & \text{if } r \neq 0 \end{cases} .$$

Define the $\begin{cases} \underline{\varepsilon\text{-symmetric}} \\ \underline{\text{even } \varepsilon\text{-symmetric}} \\ \underline{\varepsilon\text{-quadratic}} \\ \underline{\text{split } \varepsilon\text{-quadratic}} \end{cases}$ <u>Q-group of M</u> $\begin{cases} Q^\varepsilon(M) \\ Q\langle v_0\rangle^\varepsilon(M) \\ Q_\varepsilon(M) \\ \tilde{Q}_\varepsilon(M) \end{cases}$ by

$$\begin{cases} Q^\varepsilon(M) = Q^0(C,\varepsilon) = \ker(1-T_\varepsilon:\mathrm{Hom}_A(M,M^*) \longrightarrow \mathrm{Hom}_A(M,M^*)) \\ Q\langle v_0\rangle^\varepsilon(M) = Q\langle v_0\rangle(C,\varepsilon) = \mathrm{im}(1+T_\varepsilon:\mathrm{Hom}_A(M,M^*) \longrightarrow \mathrm{Hom}_A(M,M^*)) \\ Q_\varepsilon(M) = Q_0(C,\varepsilon) = \mathrm{coker}(1-T_\varepsilon:\mathrm{Hom}_A(M,M^*) \longrightarrow \mathrm{Hom}_A(M,M^*)) \\ \tilde{Q}_\varepsilon(M) = H_0(\mathrm{Hom}_A(C^*,C)) = \mathrm{Hom}_A(M,M^*) \quad . \end{cases}$$

The various Q-groups are related by a sequence of forgetful maps

$$\tilde{Q}_\varepsilon(M) \longrightarrow Q_\varepsilon(M) \xrightarrow{1+T_\varepsilon} Q\langle v_0\rangle^\varepsilon(M) \longrightarrow Q^\varepsilon(M)$$

with $\tilde{Q}_\varepsilon(M) \longrightarrow Q_\varepsilon(M)$ and $1+T_\varepsilon: Q_\varepsilon(M) \longrightarrow Q\langle v_0\rangle^\varepsilon(M)$ onto, and $Q\langle v_0\rangle^\varepsilon(M) \longrightarrow Q^\varepsilon(M)$ one-one.

An $\begin{cases} \underline{\varepsilon\text{-symmetric}} \\ \underline{\varepsilon\text{-quadratic}} \end{cases}$ $\underline{\text{form over } A}$ $\begin{cases} (M,\phi) \\ (M,\psi) \end{cases}$ is a f.g. projective A-module M together with an element $\begin{cases} \phi \in Q^{\varepsilon}(M) \\ \psi \in Q_{\varepsilon}(M) \end{cases}$. Such a form is *non-singular* if $\begin{cases} \phi \in \mathrm{Hom}_A(M,M^*) \\ (\psi+\varepsilon\psi^*) \in \mathrm{Hom}_A(M,M^*) \end{cases}$ is an isomorphism. A *morphism* (resp. *isomorphism*) of $\begin{cases} \varepsilon\text{-symmetric} \\ \varepsilon\text{-quadratic} \end{cases}$ forms over A

$$\begin{cases} f : (M,\phi) \longrightarrow (M',\phi') \\ f : (M,\psi) \longrightarrow (M',\psi') \end{cases}$$

is an A-module morphism (resp. isomorphism) $f \in \mathrm{Hom}_A(M,M')$ such that

$$\begin{cases} f^*\phi'f = \phi \in Q_{\varepsilon}(M) \\ f^*\psi'f = \psi \in Q_{\varepsilon}(M) \end{cases} .$$

An *even ε-symmetric form over A* (M,ϕ) is an ε-symmetric form such that

$$\phi \in Q\langle v_0\rangle^{\varepsilon}(M) \subseteq Q^{\varepsilon}(M) .$$

A *split ε-quadratic form over A* (M,ψ) is a f.g. projective A-module M together with an element $\psi \in \tilde{Q}_{\varepsilon}(M)$. A *morphism* (resp. *isomorphism*) of split ε-quadratic forms over A

$$(f,\chi) : (M,\psi) \longrightarrow (M',\psi')$$

is an A-module morphism (resp. isomorphism) $f \in \mathrm{Hom}_A(M,M')$ together with a $(-\varepsilon)$-quadratic form over A $(M,\chi \in Q_{-\varepsilon}(M))$, the *hessian* of (f,χ), such that

$$f^*\psi'f - \psi = \chi - \varepsilon\chi^* \in \tilde{Q}_{\varepsilon}(M).$$

An ε-symmetric form over A $(M,\phi \in Q^{\varepsilon}(M))$ is the same as a f.g. projective A-module M together with a pairing

$$\lambda : M \times M \longrightarrow A \ ; \ (x,y) \longmapsto \lambda(x,y) \equiv \phi(x)(y)$$

such that

$$\lambda(x,ay) = a\lambda(x,y)$$
$$\lambda(x,y+y') = \lambda(x,y) + \lambda(x,y')$$
$$\lambda(y,x) = \varepsilon\overline{\lambda(x,y)} \in A \qquad (x,y,y' \in M, a \in A) \ .$$

The form (M,ϕ) is even if for every $x \in M$ there exists $a \in A$ such that

$$\lambda(x,x) = a + \varepsilon\bar{a} \in A \ .$$

An ε-quadratic form over A $(M,\psi \in Q_{\varepsilon}(M))$ is the same (up to isomorphism) as a triple (M,λ,μ) consisting of a f.g. projective A-module M, an ε-symmetric pairing $\lambda : M \times M \longrightarrow A$ as above and a function

$$\mu : M \longrightarrow Q_{\varepsilon}(A) \equiv A/\{a-\varepsilon\bar{a} \,|\, a \in A\}$$

such that

$$\mu(ax) = a\mu(x)\bar{a}$$
$$\mu(x+y) - \mu(x) - \mu(y) = \lambda(x,y) \in Q_{\varepsilon}(A)$$
$$\lambda(x,x) = \mu(x) + \varepsilon\overline{\mu(x)} \in A \qquad (x,y,y' \in M, a \in A) \ ,$$

i.e. an ε-quadratic form in the sense of Wall [4,§5], the correspondence $(M,\psi) \longmapsto (M,\lambda,\mu)$ being given by

$$\lambda(x,y) = (\psi+\varepsilon\psi^{*})(x)(y) \in A$$
$$\mu(x) = \psi(x)(x) \in Q_{\varepsilon}(A) \qquad (x,y \in M) \ .$$

The ε-symmetrization functor

$$1+T_{\varepsilon} : \{\varepsilon\text{-quadratic forms over } A\} \longrightarrow \{\varepsilon\text{-symmetric forms over } A\} \ ;$$
$$(M,\psi) \longmapsto (M,(1+T_{\varepsilon})\psi)$$

is an isomorphism of categories if $1/2 \in A$, in which case the

ε-symmetric pairing $\lambda : M \times M \longrightarrow A ; (x,y) \longmapsto (1+T_{\varepsilon})\psi(x)(y)$ determines the ε-quadratic function $\mu : M \longrightarrow Q_{\varepsilon}(A) ; x \longmapsto \psi(x)(x)$ by

$$\mu(x) = \frac{1}{2}\lambda(x,x) \in Q_{\varepsilon}(A) \quad (x \in M) \ .$$

If A is a commutative ring with the identity involution $\bar{a} = a \in A$ $(a \in A)$ a quadratic form over A $(M, \psi \in Q_{+1}(M))$ $(\varepsilon = +1 \in A)$ is thus essentially the same as a f.g. projective A-module M together with a function

$$\mu : M \longrightarrow Q_{+1}(A) = A$$

such that

i) μ is quadratic

$$\mu(ax) = a^2\mu(x) \in A \quad (a \in A, x \in M)$$

ii) the function

$$\lambda : M \times M \longrightarrow A \ ; \ (x,y) \longmapsto (\mu(x+y) - \mu(x) - \mu(y))$$

is bilinear,

which is the classical definition of a quadratic form over a commutative ring. For any ring with involution A the forgetful functor

$$\{\text{split } \varepsilon\text{-quadratic forms over } A\}$$
$$\longrightarrow \{\varepsilon\text{-quadratic forms over } A\} \ ;$$
$$(M, \psi \in \tilde{Q}_{\varepsilon}(M)) \longmapsto (M, [\psi] \in Q_{\varepsilon}(M))$$

defines a one-one correspondence of isomorphism classes; the hessian forms appearing in the morphisms of split ε-quadratic forms are necessary for the definition of th even-dimensional relative ε-quadratic L-groups (in §2 below).

If $\begin{cases}(C,\phi) \\ (C,\psi)\end{cases}$ is a 0-dimensional $\begin{cases}\text{(even) } \varepsilon\text{-symmetric} \\ \varepsilon\text{-quadratic}\end{cases}$ complex

over A there are natural identifications

$$\begin{cases} Q^0(C,\varepsilon) = Q^{\varepsilon}(H^0(C)) \quad (Q\langle v_0\rangle^0(C,\varepsilon) = Q\langle v_0\rangle^{\varepsilon}(H^0(C))) \\ Q_0(C,\varepsilon) = Q_{\varepsilon}(H^0(C)) \quad , \end{cases}$$

so that $\begin{cases}(H^0(C),\phi) \\ (H^0(C),\psi)\end{cases}$ is an $\begin{cases}\text{(even) } \varepsilon\text{-symmetric} \\ \varepsilon\text{-quadratic}\end{cases}$ form over A such

that the 0th $\begin{cases}\varepsilon\text{-symmetric} \\ \varepsilon\text{-quadratic}\end{cases}$ Wu class of $\begin{cases}(C,\phi) \\ (C,\psi)\end{cases}$ is given by

$$\begin{cases} v_0(\phi) : H^0(C) \longrightarrow H^0(\mathbb{Z}_2;A,\varepsilon) \ ; \ x \longmapsto \phi_0(x)(x) = \lambda(x,x) \\ \qquad (\hat{v}_0(\phi) = 0) \\ v^0(\psi) : H^0(C) \longrightarrow H_0(\mathbb{Z}_2;A,\varepsilon) \ ; \ x \longmapsto \psi_0(x)(x) = \mu(x) \ . \end{cases}$$

Proposition 1.6.1 There is a natural one-one correspondence between the homotopy equivalence classes of 0-dimensional $\begin{cases}\text{(even) } \varepsilon\text{-symmetric} \\ \varepsilon\text{-quadratic}\end{cases}$ complexes over A and the isomorphism classes of $\begin{cases}\text{(even) } \varepsilon\text{-symmetric} \\ \varepsilon\text{-quadratic}\end{cases}$ forms over A. Poincaré complexes correspond to non-singular forms.

Proof: See Proposition I.5.1.

[]

A sublagrangian of an $\begin{cases}\varepsilon\text{-symmetric} \\ \varepsilon\text{-quadratic}\end{cases}$ form over A $\begin{cases}(M,\phi) \\ (M,\psi)\end{cases}$ is a direct summand L of M such that the inclusion $j \in \mathrm{Hom}_A(L,M)$ defines a morphism of $\begin{cases}\varepsilon\text{-symmetric} \\ \varepsilon\text{-quadratic}\end{cases}$ forms

$$\begin{cases} j : (L,0) \longrightarrow (M,\phi) \\ j : (L,0) \longrightarrow (M,\psi) \end{cases}$$

such that $\begin{cases} j^*\phi \in \mathrm{Hom}_A(M,L^*) \\ j^*(\psi+\varepsilon\psi^*) \in \mathrm{Hom}_A(M,L^*) \end{cases}$ is onto. The annihilator of L

$$\begin{cases} L^{\perp} = \ker(j^*\phi : M \longrightarrow L^*) \\ L^{\perp} = \ker(j^*(\psi+\varepsilon\psi^*) : M \longrightarrow L^*) \end{cases}$$

is a direct summand of M containing L as a direct summand

$$L \subseteq L^{\perp} .$$

A lagrangian is a sublagrangian which is its own annihilator

$$L^{\perp} = L ,$$

i.e. such that there is defined an exact sequence

$$\begin{cases} 0 \longrightarrow L \xrightarrow{\ j\ } M \xrightarrow{\ j^*\phi\ } L^* \longrightarrow 0 \\ 0 \longrightarrow L \xrightarrow{\ j\ } M \xrightarrow{\ j^*(\psi+\varepsilon\psi^*)\ } L^* \longrightarrow 0 . \end{cases}$$

A (sub)lagrangian (L,λ) of a split ε-quadratic form over A $(M,\tilde{\psi} \in \tilde{Q}_{\varepsilon}(M))$ is a (sub)lagrangian L of the associated ε-quadratic form $(M,[\tilde{\psi}] \in Q_{\varepsilon}(M))$, together with a hessian $(-\varepsilon)$-quadratic form $(L,\lambda \in Q_{-\varepsilon}(L))$ such that

$$j^*\tilde{\psi}j = \lambda - \varepsilon\lambda^* \in \tilde{Q}_{\varepsilon}(L) ,$$

i.e. such that there is defined a morphism of split ε-quadratic forms over A

$$(j,\lambda) : (L,0) \longrightarrow (M,\tilde{\psi}) .$$

A non-singular $\begin{cases} \text{(even) } \varepsilon\text{-symmetric} \\ \text{(split) } \varepsilon\text{-quadratic} \end{cases}$ form is hyperbolic if it admits a lagrangian.

Given $\begin{cases} \text{an } \varepsilon\text{-symmetric form over A } (L^*, \phi \in Q^{\varepsilon}(L^*)) \\ \text{a f.g. projective A-module L} \\ \quad - \text{"} - \\ \quad - \text{"} - \end{cases}$ define the

standard hyperbolic $\begin{cases} \underline{\varepsilon\text{-symmetric}} \\ \underline{\text{even } \varepsilon\text{-symmetric}} \\ \underline{\varepsilon\text{-quadratic}} \\ \underline{\text{split } \varepsilon\text{-quadratic}} \end{cases}$ form over A

$$\begin{cases} H^{\varepsilon}(L^*,\phi) = (L\oplus L^*, \begin{pmatrix} 0 & 1 \\ \varepsilon & \phi \end{pmatrix} \in Q^{\varepsilon}(L\oplus L^*)) \\ H^{\varepsilon}(L) = (L\oplus L^*, \begin{pmatrix} 0 & 1 \\ \varepsilon & 0 \end{pmatrix} \in Q\langle v_0\rangle^{\varepsilon}(L\oplus L^*)) \\ H_{\varepsilon}(L) = (L\oplus L^*, \begin{pmatrix} 0 & 1 \\ 0 & 0 \end{pmatrix} \in Q_{\varepsilon}(L\oplus L^*)) \\ \widetilde{H}_{\varepsilon}(L) = (L\oplus L^*, \begin{pmatrix} 0 & 1 \\ 0 & 0 \end{pmatrix} \in \widetilde{Q}_{\varepsilon}(L\oplus L^*)) \end{cases} .$$

The various hyperbolic forms are related to each other by

$$(1+T_{\varepsilon})H_{\varepsilon}(L) = H^{\varepsilon}(L) = H^{\varepsilon}(L^*,0) \ ,$$

with $\widetilde{H}_{\varepsilon}(L)$ a split ε-quadratic refinement of $H_{\varepsilon}(L)$.

If $(L^*, \phi \in Q\langle v_0\rangle^{\varepsilon}(L^*))$ is an even ε-symmetric form then $\phi = \psi+\varepsilon\psi^* \in Q\langle v_0\rangle^{\varepsilon}(L^*)$ for some split ε-quadratic form $(M, \psi \in \widetilde{Q}_{\varepsilon}(L^*))$ and there is defined an isomorphism of (even) ε-symmetric forms

$$\begin{pmatrix} 1 & \psi^* \\ 0 & 1 \end{pmatrix} : H^{\varepsilon}(L^*,\phi) = (L\oplus L^*, \begin{pmatrix} 0 & 1 \\ \varepsilon & \phi \end{pmatrix})$$

$$\longrightarrow H^{\varepsilon}(L) = (L\oplus L^*, \begin{pmatrix} 0 & 1 \\ \varepsilon & 0 \end{pmatrix}) \quad .$$

Proposition 1.6.2 The morphism of forms

$$\begin{cases} j : (L,0) \longrightarrow (M,\phi) \\ j : (L,0) \longrightarrow (M,\phi) \\ j : (L,0) \longrightarrow (M,\psi) \\ (j,\lambda) : (L,0) \longrightarrow (M,\tilde{\psi}) \end{cases}$$

defined by the inclusion $j \in \mathrm{Hom}_A(L,M)$ of a sublagrangian L in an

$$\begin{cases} \varepsilon\text{-symmetric} \\ \text{even } \varepsilon\text{-symmetric} \\ \varepsilon\text{-quadratic} \\ \text{split } \varepsilon\text{-quadratic} \end{cases} \text{form over A} \begin{cases} (M,\phi \in Q^{\varepsilon}(M)) \\ (M,\phi \in Q\langle v_0\rangle^{\varepsilon}(M)) \\ (M,\psi \in Q_{\varepsilon}(M)) \\ (M,\tilde{\psi} \in \tilde{Q}_{\varepsilon}(M)) \end{cases} \text{extends}$$

to an isomorphism of forms

$$\begin{cases} f : H^{\varepsilon}(L^*,\eta)\oplus(L^{\perp}/L,\phi^{\perp}/\phi) \xrightarrow{\sim} (M,\phi) \\ f : H^{\varepsilon}(L)\oplus(L^{\perp}/L,\phi^{\perp}/\phi) \xrightarrow{\sim} (M,\phi) \\ f : H_{\varepsilon}(L)\oplus(L^{\perp}/L,\psi^{\perp}/\psi) \xrightarrow{\sim} (M,\psi) \\ (f,\chi) : \tilde{H}_{\varepsilon}(L)\oplus(L^{\perp}/L,\tilde{\psi}^{\perp}/\tilde{\psi}) \xrightarrow{\sim} (M,\tilde{\psi}) \quad . \end{cases}$$

Proof: See Proposition I.2.2.

[]

In particular, Proposition 1.6.2 shows that the form $(L^{\perp}/L,\phi^{\perp}/\phi)$ is non-singular if and only if (M,ϕ) is non-singular, and similarly in the other cases.

$$\text{An}\begin{cases}\underline{\varepsilon\text{-symmetric}}\\ \underline{\varepsilon\text{-quadratic}}\end{cases}\ \underline{\text{formation over A}}\ \begin{cases}(M,\phi;F,G)\\ (M,\psi;F,G)\end{cases}\ \text{is a}$$

$$\text{non-singular}\begin{cases}\varepsilon\text{-symmetric}\\ \varepsilon\text{-quadratic}\end{cases}\ \text{form over A}\ \begin{cases}(M,\phi\in Q^{\varepsilon}(M))\\ (M,\psi\in Q_{\varepsilon}(M))\end{cases}\ \text{together}$$

with a lagrangian F and a sublagrangian G. Such a formation is <u>non-singular</u> if G is a lagrangian. An <u>isomorphism</u> of formations

$$\begin{cases}f : (M,\phi;F,G)\xrightarrow{\sim}(M',\phi';F',G')\\ f : (M,\psi;F,G)\xrightarrow{\sim}(M',\psi';F',G')\end{cases}$$

is an isomorphism of forms

$$\begin{cases}f : (M,\phi)\xrightarrow{\sim}(M',\phi')\\ f : (M,\psi)\xrightarrow{\sim}(M',\psi')\end{cases}$$

such that

$$f(F) = F'\ ,\ f(G) = G'\ .$$

A <u>stable isomorphism</u> of formations

$$\begin{cases}[f] : (M,\phi;F,G)\xrightarrow{\sim}(M',\phi';F',G')\\ [f] : (M,\psi;F,G)\xrightarrow{\sim}(M',\psi';F',G')\end{cases}$$

is an isomorphism of the type

$$\begin{cases}f : (M,\phi;F,G)\oplus(H^{\varepsilon}(P);P,P^*)\xrightarrow{\sim}(M',\phi';F',G')\oplus(H^{\varepsilon}(P');P',P'^*)\\ f : (M,\psi;F,G)\oplus(H_{\varepsilon}(P);P,P^*)\xrightarrow{\sim}(M',\psi';F',G')\oplus(H_{\varepsilon}(P');P',P'^*)\end{cases}$$

for some f.g. projective A-modules P,P'.

An <u>even ε-symmetric formation</u> $(M,\phi;F,G)$ is an ε-symmetric formation such that (M,ϕ) is an even ε-symmetric form.

A <u>split ε-quadratic formation over A</u>

$$(F,G) = (F,(\begin{pmatrix}\gamma\\ \mu\end{pmatrix},\theta)G)$$

is an ε-quadratic formation over A $(H_\varepsilon(F);F,G)$ (with $\begin{pmatrix}\gamma\\ \mu\end{pmatrix}:G \longrightarrow F\oplus F^*$ the inclusion), together with a <u>hessian</u> (-ε)-quadratic form over A $(G,\theta \in Q_{-\varepsilon}(G))$ such that

$$\gamma^*\mu = \theta - \varepsilon\theta^* \in \mathrm{Hom}_A(G,G^*) \quad ,$$

so that there is defined a morphism of split ε-quadratic forms over A

$$(\begin{pmatrix}\gamma\\ \mu\end{pmatrix},\theta) : (G,0) \longrightarrow \widetilde{H}_\varepsilon(F) = (F\oplus F^*, \begin{pmatrix}0 & 1\\ 0 & 0\end{pmatrix})$$

and (G,θ) is a sublagrangian of $\widetilde{H}_\varepsilon(F)$. The formation (F,G) is <u>non-singular</u> if the sequence

$$0 \longrightarrow G \xrightarrow{\begin{pmatrix}\gamma\\ \mu\end{pmatrix}} F\oplus F^* \xrightarrow{(\varepsilon\mu^* \ \ \gamma^*)} G^* \longrightarrow 0$$

is exact, i.e. if the underlying ε-quadratic formation $(H_\varepsilon(F);F,G)$ is non-singular. An <u>isomorphism</u> of split ε-quadratic formations

$$(\alpha,\beta,\psi) : (F,G) \xrightarrow{\sim} (F',G')$$

is a triple consisting of A-module isomorphisms $\alpha \in \mathrm{Hom}_A(F,F')$, $\beta \in \mathrm{Hom}_A(G,G')$ and a (-ε)-quadratic form $(F^*,\psi \in Q_{-\varepsilon}(F^*))$ such that

i) $\alpha\gamma + \alpha(\psi - \varepsilon\psi^*)^*\mu = \gamma'\beta \in \mathrm{Hom}_A(G,F')$

ii) $\alpha^{*-1}\mu = \mu'\beta \in \mathrm{Hom}_A(G,F'^*)$

iii) $\theta + \mu^*\psi\mu = \beta^*\theta'\beta \in Q_{-\varepsilon}(G)$.

Such an isomorphism determines a commutative diagram

$$\begin{array}{ccc} G & \xrightarrow{\beta} & G' \\ {\scriptstyle\begin{pmatrix}\gamma\\ \mu\end{pmatrix}}\downarrow & & \downarrow{\scriptstyle\begin{pmatrix}\gamma'\\ \mu'\end{pmatrix}} \\ F\oplus F^* & \xrightarrow{f} & F'\oplus F'^* \end{array}$$

with

$$f = \begin{pmatrix} \alpha & \alpha(\psi-\varepsilon\psi^*)^* \\ 0 & \alpha^{*-1} \end{pmatrix} : F\oplus F^* \xrightarrow{\sim} F'\oplus F'^* \quad ,$$

so that there is defined an isomorphism of the underlying ε-quadratic formations

$$f : (H_\varepsilon(F);F,G) \xrightarrow{\sim} (H_\varepsilon(F');F',G') \quad .$$

A <u>stable isomorphism</u> of split ε-quadratic formations

$$[\alpha,\beta,\psi] : (F,G) \longrightarrow (F',G')$$

is an isomorphism of the type

$$(\alpha,\beta,\psi) : (F,G)\oplus(P,P^*) \xrightarrow{\sim} (F',G')\oplus(P',P'^*)$$

for some f.g. projective A-modules P,P', with

$$(P,P^*) = (P,(\begin{pmatrix}0\\1\end{pmatrix},0)P^*) \quad .$$

<u>Proposition 1.6.3</u> i) Every ε-quadratic formation is isomorphic to one of the type $(H_\varepsilon(F);F,G)$.

ii) Every ε-quadratic formation of the type $(H_\varepsilon(F);F,G)$ admits a (non-unique) split ε-quadratic refinement (F,G).

iii) Every isomorphism of ε-quadratic formations of the type

$$f : (H_\varepsilon(F);F,G) \longrightarrow (H_\varepsilon(F');F',G')$$

can be refined to a (non-unique) isomorphism of split ε-quadratic formations

$$(\alpha,\beta,\psi) : (F,G) \longrightarrow (F',G') \quad .$$

Similarly for stable isomorphisms.

<u>Proof</u>: See Proposition I.2.4.

[]

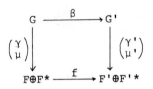

$\left\{\begin{array}{l}\text{An } \underline{\varepsilon\text{-quadratic}} \\ \text{A } \underline{\text{split } \varepsilon\text{-quadratic}}\end{array}\right.$ <u>homotopy equivalence</u> of 1-dimensional ε-quadratic complexes over A

$$f : (C,\psi) \longrightarrow (C',\psi')$$

is a chain equivalence $f:C \longrightarrow C'$ such that

$$\begin{cases} f_{\%}(\psi) - \psi' = H(\theta) \in Q_1(C',\varepsilon) \\ f_{\%}(\psi) - \psi' = 0 \in Q_1(C',\varepsilon) \end{cases}$$

for some Tate $\mathbb{Z}_2$-hypercohomology class $\theta \in \hat{Q}^2(C',\varepsilon)$ with vanishing 1st Wu class

$$\hat{v}_1(\theta) = 0 : H^1(C') \longrightarrow \hat{H}^1(\mathbb{Z}_2;A,\varepsilon) \; ; \; x \longmapsto \theta_0(x)(x)$$
$$(\theta_0 \in \mathrm{Hom}_A(C'^1,C'_1)) \; .$$

(A split ε-quadratic homotopy equivalence is the same as a homotopy equivalence).

Let C be a 1-dimensional A-module chain complex of the type

$$C : \ldots \longrightarrow 0 \longrightarrow C_1 \xrightarrow{\ d\ } C_0 \longrightarrow 0 \longrightarrow \ldots \; .$$

A $\left\{\begin{array}{l}\mathbb{Z}_2\text{-hypercohomology} \\ \mathbb{Z}_2\text{-hyperhomology}\end{array}\right.$ class $\left\{\begin{array}{l}\phi \in Q^1(C,\varepsilon) \\ \psi \in Q_1(C,\varepsilon)\end{array}\right.$ is represented by A-module morphisms

$$\begin{cases} \phi_0 : C^0 \longrightarrow C_1 \; , \; \tilde{\phi}_0 : C^1 \longrightarrow C_0 \; , \; \phi_1 : C^1 \longrightarrow C_1 \\ \psi_0 : C^0 \longrightarrow C_1 \; , \; \tilde{\psi}_0 : C^1 \longrightarrow C_0 \; , \; \psi_1 : C^0 \longrightarrow C_0 \end{cases}$$

such that

$$\begin{cases} d\phi_0 + \tilde{\phi}_0 d^* = 0 : C^0 \longrightarrow C_0 \; , \; d\phi_1 - \tilde{\phi}_0 + \varepsilon\phi_0^* = 0 : C^1 \longrightarrow C_0 \; , \\ \qquad \phi_1 - \varepsilon\phi_1^* = 0 : C^1 \longrightarrow C_1 \\ \\ d\psi_0 + \tilde{\psi}_0 d^* + \psi_1 - \varepsilon\psi_1^* = 0 : C^0 \longrightarrow C_0 \end{cases} \; .$$

A connected 1-dimensional $\begin{cases}\varepsilon\text{-symmetric} \\ \varepsilon\text{-quadratic}\end{cases}$ complex $\begin{cases}(C,\phi\in Q^1(C,\varepsilon)) \\ (C,\psi\in Q_1(C,\varepsilon))\end{cases}$

determines the $\begin{cases}\varepsilon\text{-symmetric} \\ \text{split } \varepsilon\text{-quadratic}\end{cases}$ formation

$$\begin{cases}(H^{\varepsilon}(C^1,\phi_1);C_1,C^0) = (C_1\oplus C^1, \begin{pmatrix} 0 & 1 \\ \varepsilon & \phi_1\end{pmatrix};C_1,\mathrm{im}(\begin{pmatrix}\bar{\varepsilon}\phi_0 \\ d^*\end{pmatrix}:C^0\longrightarrow C_1\oplus C^1)) \\ (C_1,C^0) = (C_1,(\begin{pmatrix}\bar{\varepsilon}\psi_0+\tilde{\psi}_0^* \\ d^*\end{pmatrix},-(\psi_1+d\psi_0))C^0)\end{cases}$$

.

Proposition 1.6.4 There is a natural one-one correspondence

between the $\begin{cases} - \\ \text{(split) } \varepsilon\text{-quadratic}\end{cases}$ homotopy equivalence classes

of connected 1-dimensional $\begin{cases}\text{(even) } \varepsilon\text{-symmetric} \\ \text{(split) } \varepsilon\text{-quadratic}\end{cases}$ complexes over A

and the stable isomorphism classes of $\begin{cases}\text{(even) } \varepsilon\text{-symmetric} \\ \text{(split) } \varepsilon\text{-quadratic}\end{cases}$

formations over A. Poincaré complexes correspond to non-singular formations.

Proof: See Propositions I.2.3, I.2.5.

[]

The boundary $\begin{cases}\partial(M,\phi;F,G) \\ \partial(M,\psi;F,G)\end{cases}$ of an $\begin{cases}\text{(even) } \varepsilon\text{-symmetric} \\ \varepsilon\text{-quadratic}\end{cases}$ formation

over A $\begin{cases}(M,\phi;F,G) \\ (M,\psi;F,G)\end{cases}$ is the non-singular $\begin{cases}\text{(even) } \varepsilon\text{-symmetric} \\ \varepsilon\text{-quadratic}\end{cases}$ form

over A

$$\begin{cases}\partial(M,\phi;F,G) = (G^{\perp}/G,\phi^{\perp}/\phi) \\ \partial(M,\psi;F,G) = (G^{\perp}/G,\psi^{\perp}/\psi)\end{cases} .$$

Proposition 1.6.2 shows that the boundary form is stably hyperbolic, with an isomorphism

$$\begin{cases} \partial(M,\phi;F,G)\oplus H^{\varepsilon}(G^*,\zeta) \longrightarrow H^{\varepsilon}(F^*,\upsilon) \\ \partial(M,\psi;F,G)\oplus H_{\varepsilon}(G) \longrightarrow H_{\varepsilon}(F) \end{cases}$$

for some ε-symmetric forms $(F^*,\zeta),(G^*,\upsilon)$ (with $\zeta=0,\upsilon=0$ if $(M,\phi;F,G)$ is even).

The <u>boundary</u> $\partial(F,G)$ of a split ε-quadratic formation (F,G) is the boundary $\partial(H_{\varepsilon}(F);F,G)$ of the underlying ε-quadratic formation $(H_{\varepsilon}(F);F,G)$.

The <u>boundary</u> $\begin{cases} \partial(M,\phi) \\ \partial(M,\phi) \\ \partial(M,\psi) \end{cases}$ of an $\begin{cases} \varepsilon\text{-symmetric} \\ \text{even } \varepsilon\text{-symmetric} \\ \varepsilon\text{-quadratic} \end{cases}$ form over A $\begin{cases} (M,\phi\in Q^{\varepsilon}(M)) \\ (M,\phi\in Q\langle v_0\rangle^{\varepsilon}(M)) \\ (M,\psi\in Q_{\varepsilon}(M)) \end{cases}$ is the non-singular $\begin{cases} \text{even } (-\varepsilon)\text{-symmetric} \\ (-\varepsilon)\text{-quadratic} \\ \text{split } (-\varepsilon)\text{-quadratic} \end{cases}$ formation over A

$$\begin{cases} \partial(M,\phi) = (H^{-\varepsilon}(M);M,\Gamma_{(M,\phi)}) \\ \partial(M,\phi) = (H_{-\varepsilon}(M);M,\Gamma_{(M,\phi)}) \\ \partial(M,\psi) = (M,(\begin{pmatrix} 1 \\ \psi+\varepsilon\psi^* \end{pmatrix},\psi)M) \end{cases} ,$$

where

$$\Gamma_{(M,\phi)} = \{(x,\phi(x))\in M\oplus M^* \mid x\in M\} \subseteq M\oplus M^*$$

is the <u>graph</u> lagrangian of (M,ϕ) in $H^{-\varepsilon}(M)$ (in $H_{-\varepsilon}(M)$ if (M,ϕ) is even).

$$\text{The } \underline{\text{Witt group of}} \left\{\begin{array}{l} \underline{\varepsilon\text{-symmetric}} \\ \underline{\text{even } \varepsilon\text{-symmetric}} \\ \underline{\varepsilon\text{-quadratic}} \end{array}\right. \underline{\text{forms over } A} \left\{\begin{array}{l} L^{\varepsilon}(A) \\ L\langle v_0\rangle^{\varepsilon}(A) \\ L_{\varepsilon}(A) \end{array}\right.$$

is the abelian group of equivalence classes of non-singular

$$\left\{\begin{array}{l} \varepsilon\text{-symmetric} \\ \text{even } \varepsilon\text{-symmetric} \\ \varepsilon\text{-quadratic} \end{array}\right. \text{forms over } A \text{ subject to the relation}$$

$(M,\phi) \sim (M',\phi')$ if there exists an isomorphism

$$(M,\phi)\oplus(H,\theta) \longrightarrow (M',\phi')\oplus(H',\theta')$$

for some hyperbolic forms $(H,\theta),(H',\theta')$.

Addition and inverses are by

$$(M,\phi) + (M',\phi') = (M\oplus M',\phi\oplus\phi') \ , \ -(M,\phi) = (M,-\phi) \ .$$

$$\text{The } \underline{\text{Witt group of}} \left\{\begin{array}{l} \underline{\varepsilon\text{-symmetric}} \\ \underline{\text{even } \varepsilon\text{-symmetric}} \\ \underline{\varepsilon\text{-quadratic}} \end{array}\right. \underline{\text{formations over } A}$$

$$\left\{\begin{array}{l} M^{\varepsilon}(A) \\ M\langle v_0\rangle^{\varepsilon}(A) \\ M_{\varepsilon}(A) \end{array}\right. \text{is the abelian group of equivalence classes of}$$

$$\text{non-singular} \left\{\begin{array}{l} \varepsilon\text{-symmetric} \\ \text{even } \varepsilon\text{-symmetric} \\ \varepsilon\text{-quadratic} \end{array}\right. \text{formations over } A \text{ subject to}$$

the relation

$(M,\phi;F,G) \sim (M',\phi';F',G')$ if there exists a stable isomorphism of the type

$$[f] : (M,\phi;F,G)\oplus(N,\nu;H,K)\oplus(N,\nu;K,L)\oplus(N',\nu';H',L')$$
$$\longrightarrow (M',\phi';F',G')\oplus(N',\nu';H',K')\oplus(N',\nu';K',L')\oplus(N,\nu;H,L) \ .$$

Addition and inverses are by

$$(M,\phi;F,G) + (M',\phi';F',G') = (M\oplus M',\phi\oplus\phi';F\oplus F',G\oplus G')$$
$$-(M,\phi;F,G) = (M,\phi;G,F) \quad (= (M,-\phi;F,G)) \qquad .$$

Proposition 1.6.5 i) For $n = 0,1$ the n-dimensional L-groups have natural expressions as Witt groups of forms and formations

$$\begin{cases} L^0(A,\varepsilon) = L^{\varepsilon}(A) \\ L\langle v_0\rangle^0(A,\varepsilon) = L\langle v_0\rangle^{\varepsilon}(A) \\ L_0(A,\varepsilon) = L_{\varepsilon}(A) \end{cases}, \quad \begin{cases} L^1(A,\varepsilon) = M^{\varepsilon}(A) \\ L\langle v_0\rangle^1(A,\varepsilon) = M\langle v_0\rangle^{\varepsilon}(A) \\ L_1(A,\varepsilon) = M_{\varepsilon}(A) \end{cases}.$$

ii) An $\begin{cases} \varepsilon\text{-symmetric} \\ \text{even } \varepsilon\text{-symmetric} \\ \varepsilon\text{-quadratic} \end{cases}$ form is non-singular if and only if its boundary formation is stably isomorphic to 0, in which case the form represents 0 in the Witt group $\begin{cases} L^{\varepsilon}(A) \\ L\langle v_0\rangle^{\varepsilon}(A) \\ L_{\varepsilon}(A) \end{cases}$ if and only if it is isomorphic to the boundary of a formation.

iii) An $\begin{cases} \varepsilon\text{-symmetric} \\ \text{even } \varepsilon\text{-symmetric} \\ \varepsilon\text{-quadratic} \end{cases}$ formation is non-singular if and only its boundary form is 0. A non-singular $\begin{cases} \text{even } \varepsilon\text{-symmetric} \\ \varepsilon\text{-quadratic} \\ \text{split } \varepsilon\text{-quadratic} \end{cases}$ formation represents 0 in the Witt group $\begin{cases} M^{\varepsilon}(A) \\ M\langle v_0\rangle^{\varepsilon}(A) \\ M_{\varepsilon}(A) \end{cases}$ if and only if it is stably isomorphic to the boundary of a form.

Proof: See Propositions I.5.1,I.5.2,I.5.4.

[]

The periodicity $L_n(A) = L_{n+2}(A,-1) = L_{n+4}(A)$ $(n \geqslant 0)$ of Proposition 1.2.3 i) combined with the expressions of Proposition 1.6.5 i) identifies the quadratic L-groups $L_n(A)$ $(n \geqslant 0)$ defined using quadratic Poincaré complexes with the quadratic L-groups $L_n(A)$ $(n \pmod 4)$ defined by Wall [4] using forms and formations.

An n-dimensional $\begin{cases} \varepsilon\text{-symmetric} \\ \varepsilon\text{-quadratic} \end{cases}$ complex $\begin{cases} (C,\phi \in Q^n(C,\varepsilon)) \\ (C,\psi \in Q_n(C,\varepsilon)) \end{cases}$ is

<u>highly-connected</u> if

for $n = 2i$: $H_r(C) = H^r(C) = 0$ $(r \neq i)$

for $n = 2i+1$: $H_r(C) = H^r(C) = 0$ $(r \neq i,i+1)$ and

$$\begin{cases} H_i(\phi_0 : C^{2i+1-*} \longrightarrow C) = 0 \\ H_i((1+T_\varepsilon)\psi_0 : C^{2i+1-*} \longrightarrow C) = 0 \ . \end{cases}$$

A highly-connected complex is connected; the boundary of a highly-connected complex is a highly-connected Poincaré complex.

<u>Proposition 1.6.6</u> For $n = 2i$ (resp. $n = 2i+1$) the homotopy equivalence classes of highly-connected n-dimensional $\begin{cases} \varepsilon\text{-symmetric} \\ \varepsilon\text{-quadratic} \end{cases}$ complexes over A are in a natural one-one correspondence with the isomorphism (resp. stable isomorphism) classes of $\begin{cases} (-)^i\varepsilon\text{-symmetric} \\ (-)^i\varepsilon\text{-quadratic} \end{cases}$ forms (resp. $\begin{cases} (-)^i\varepsilon\text{-symmetric} \\ \text{split } (-)^i\varepsilon\text{-quadratic} \end{cases}$ formations) over A. Poincaré complexes correspond to non-singular forms (resp. formations). The boundary operation on highly-connected complexes corresponds to the boundary operation on forms (resp. formations).

<u>Proof</u>: See Proposition I.5.3.

[]

1.7 Algebraic glueing

Geometric Poincaré cobordisms $(Y;X,X')$, $(Y';X',X")$ can be glued together to define a geometric Poincaré cobordism $(Y";X,X")$ with

$$Y" = Y\cup_{X'} Y' \ .$$

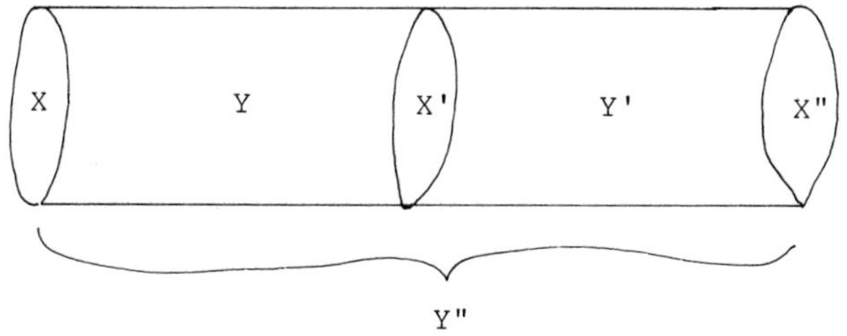

We shall now recall from §I.3 the analogue of this glueing operation for algebraic Poincaré cobordisms.

The union of adjoining (n+1)-dimensional $\begin{cases}\varepsilon\text{-symmetric}\\ \varepsilon\text{-quadratic}\end{cases}$ Poincaré cobordisms

$$\begin{cases} c = ((f_C\ f_{C'}) : C\oplus C' \longrightarrow D, (\delta\phi, \phi\oplus -\phi') \in Q^{n+1}((f_C\ f_{C'}),\varepsilon)) \\ c = ((f_C\ f_{C'}) : C\oplus C' \longrightarrow D, (\delta\psi, \psi\oplus -\psi') \in Q_{n+1}((f_C\ f_{C'}),\varepsilon)) \end{cases},$$

$$\begin{cases} c' = ((f'_{C'}\ f'_{C"}) : C'\oplus C" \longrightarrow D', (\delta\phi', \phi'\oplus -\phi") \in Q^{n+1}((f'_{C'}\ f'_{C"}),\varepsilon)) \\ c' = ((f'_{C'}\ f'_{C"}) : C'\oplus C" \longrightarrow D', (\delta\psi', \psi'\oplus -\psi") \in Q_{n+1}((f'_{C'}\ f'_{C"}),\varepsilon)) \end{cases}$$

is the (n+1)-dimensional $\begin{cases}\varepsilon\text{-symmetric}\\ \varepsilon\text{-quadratic}\end{cases}$ Poincaré cobordism

$$\begin{cases} c\cup c' = ((f"_C\ f"_{C"}) : C\oplus C" \longrightarrow D", (\delta\phi", \phi\oplus -\phi") \in Q^{n+1}((f"_C\ f"_{C"}),\varepsilon))) \\ c\cup c' = ((f"_C\ f"_{C"}) : C\oplus C" \longrightarrow D", (\delta\psi", \psi\oplus -\psi") \in Q_{n+1}((f"_C\ f"_{C"}),\varepsilon))) \end{cases}$$

defined by

$$d_{D''} = \begin{pmatrix} d_D & (-)^{r-1} f_{C'} & 0 \\ 0 & d_{C'} & 0 \\ 0 & (-)^{r-1} f'_{C'} & d_{D'} \end{pmatrix}$$

$$: D''_r = D_r \oplus C'_{r-1} \oplus D'_r \longrightarrow D''_{r-1} = D_{r-1} \oplus C'_{r-2} \oplus D'_{r-1}$$

$$f''_C = \begin{pmatrix} f_C \\ 0 \\ 0 \end{pmatrix} : C_r \longrightarrow D''_r = D_r \oplus C'_{r-1} \oplus D'_r$$

$$f''_{C''} = \begin{pmatrix} 0 \\ 0 \\ f'_{C''} \end{pmatrix} : C''_r \longrightarrow D''_r = D_r \oplus C'_{r-1} \oplus D'_r$$

$$\left\{ \begin{array}{l} \delta\phi''_s = \begin{pmatrix} \delta\phi_s & 0 & 0 \\ (-)^{n-r} \phi'_s f^*_{C'} & (-)^{n-r+s+1} T\varepsilon\phi'_{s-1} & 0 \\ 0 & (-)^s f_{C'} \phi'_s & \delta\phi'_s \end{pmatrix} \\ \quad : D''^{n-r+s+1} = D^{n-r+s+1} \oplus C'^{n-r+s} \oplus D'^{n-r+s+1} \\ \qquad \longrightarrow D''_r = D_r \oplus C'_{r-1} \oplus D'_r \qquad (s \geqslant 0, \phi'_{-1} = 0) \\ \delta\psi''_s = \begin{pmatrix} \delta\psi_s & 0 & 0 \\ (-)^{n-r} \psi'_s f^*_{C'} & (-)^{n-r-s} T\varepsilon\psi'_{s+1} & 0 \\ 0 & (-)^s f_{C'} \psi'_s & \delta\psi'_s \end{pmatrix} \\ \quad : D''^{n-r-s+1} = D^{n-r-s+1} \oplus C'^{n-r-s} \oplus D'^{n-r+s} \\ \qquad \longrightarrow D''_r = D_r \oplus C'_{r-1} \oplus D'_r \qquad (s \geqslant 0) . \end{array} \right.$$

We shall write

$$D'' = D \cup_{C'} D' \ , \quad \begin{cases} \delta\phi'' = \delta\phi \cup_{\phi'} \delta\phi' \\ \delta\psi'' = \delta\psi \cup_{\psi'} \delta\psi' \ . \end{cases}$$

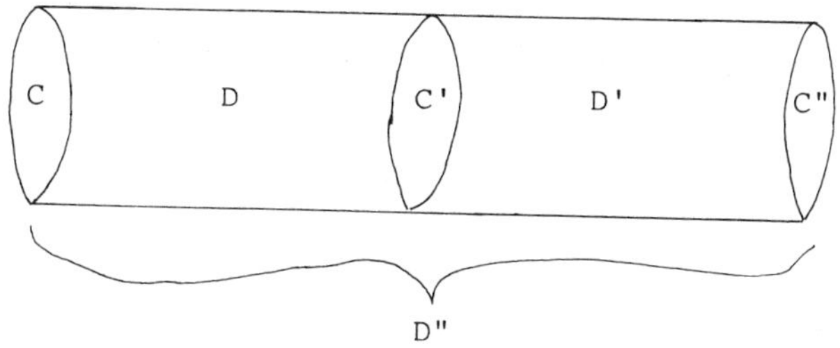

The union operation for algebraic Poincaré cobordisms has a particularly simple expression (up to homotopy equivalence) in the special case when all the chain maps involved are defined by inclusions of direct summands, as follows.

An $\begin{cases} \varepsilon\text{-symmetric} \\ \varepsilon\text{-quadratic} \end{cases}$ pair $\begin{cases} (f:C \longrightarrow D, (\delta\phi, \phi)) \\ (f:C \longrightarrow D, (\delta\psi, \psi)) \end{cases}$ is <u>direct</u> if each $f \in \mathrm{Hom}_A(C_r, D_r)$ $(r \in \mathbb{Z})$ is a split monomorphism, i.e. the inclusion of a direct summand.

The <u>direct union</u> of adjoining direct (n+1)-dimensional $\begin{cases} \varepsilon\text{-symmetric} \\ \varepsilon\text{-quadratic} \end{cases}$ Poincaré cobordisms

$$\begin{cases} c = ((f_C \ f_{C'}): C \oplus C' \longrightarrow D, (\delta\phi, \phi \oplus -\phi') \in Q^{n+1}((f_C \ f_{C'}), \varepsilon)) \\ c = ((f_C \ f_{C'}): C \oplus C' \longrightarrow D, (\delta\psi, \psi \oplus -\psi') \in Q_{n+1}((f_C \ f_{C'}), \varepsilon)) \end{cases}$$

$$\begin{cases} c' = ((f'_{C'} \ f'_{C''}): C' \oplus C'' \longrightarrow D', (\delta\phi', \phi' \oplus -\phi'') \in Q^{n+1}((f'_{C'} \ f'_{C''}), \varepsilon)) \\ c' = ((f'_{C'} \ f'_{C''}): C' \oplus C'' \longrightarrow D', (\delta\psi', \psi' \oplus -\psi'') \in Q_{n+1}((f'_{C'} \ f'_{C''}), \varepsilon)) \end{cases}$$

is the direct cobordism

$$\begin{cases} c \overline{\cup} c' = ((\bar{f}''_C \ \bar{f}''_{C''}): C \oplus C'' \longrightarrow \bar{D}'', (\overline{\delta\phi}'', \phi \oplus -\phi'') \in Q^{n+1}((\bar{f}''_C \ \bar{f}''_{C''}), \varepsilon)) \\ c \overline{\cup} c' = ((\bar{f}''_C \ \bar{f}''_{C''}): C \oplus C'' \longrightarrow \bar{D}'', (\overline{\delta\psi}'', \psi \oplus -\psi'') \in Q_{n+1}((\bar{f}''_C \ \bar{f}''_{C''}), \varepsilon)) \end{cases}$$

defined by

$$d_{\overline{D}''} = [d_D \oplus d_{D'}] : \overline{D}''_r = \operatorname{coker}(\begin{pmatrix} f_{C'} \\ f'_{C'} \end{pmatrix} : C'_r \longrightarrow D_r \oplus D'_r) \longrightarrow \overline{D}''_{r-1}$$

$$\overline{f}''_C = \begin{pmatrix} 1 \\ 0 \end{pmatrix} : C_r \longrightarrow \overline{D}''_r \ , \quad \overline{f}''_{C''} = \begin{pmatrix} 0 \\ 1 \end{pmatrix} : C''_r \longrightarrow \overline{D}''_r$$

$$\begin{cases} [\overline{\delta\phi}''] = [\delta\phi \oplus \delta\phi'] \\ [\overline{\delta\psi}''] = [\delta\psi \oplus \delta\psi'] \ . \end{cases}$$

Every $\begin{Bmatrix} \varepsilon\text{-symmetric} \\ \varepsilon\text{-quadratic} \end{Bmatrix}$ Poincaré cobordism

$\begin{cases} c = ((f\ f'):C\oplus C' \longrightarrow D, (\delta\phi, \phi\oplus -\phi')) \\ c = ((f\ f'):C\oplus C' \longrightarrow D, (\delta\psi, \psi\oplus -\psi')) \end{cases}$ is homotopy equivalent

to a direct cobordism $\begin{cases} \overline{c} = ((\overline{f}\ \overline{f}'):C\oplus C' \longrightarrow \overline{D}, (\overline{\delta\phi}, \phi\oplus -\phi')) \\ \overline{c} = ((\overline{f}\ \overline{f}'):C\oplus C' \longrightarrow \overline{D}, (\overline{\delta\psi}, \psi\oplus -\psi')) \end{cases}$,

with $\overline{D} = M(f\ f')$ the algebraic mapping cylinder of the chain map

$$(f\ f') : C\oplus C' \longrightarrow D \ .$$

(The <u>algebraic mapping cylinder</u> $M(f)$ of an A-module chain map

$$f : C \longrightarrow D$$

is the A-module chain complex defined by

$$d_{M(f)} = \begin{pmatrix} d_D & (-)^{r-1}f & 0 \\ 0 & d_C & 0 \\ 0 & (-)^r & d_C \end{pmatrix}$$

$$: M(f)_r = D_r \oplus C_{r-1} \oplus C_r \longrightarrow M(f)_{r-1} = D_{r-1} \oplus C_{r-2} \oplus C_{r-1} \ .$$

The A-module chain maps

$$\overline{f} : C \longrightarrow M(f) \ , \quad g : D \longrightarrow M(f)$$

defined by

$$\bar{f} = \begin{pmatrix} 0 \\ 0 \\ 1 \end{pmatrix} : C_r \longrightarrow M(f)_r = D_r \oplus C_{r-1} \oplus C_r$$

$$g = \begin{pmatrix} 1 \\ 0 \\ 0 \end{pmatrix} : D_r \longrightarrow M(f)_r = D_r \oplus C_{r-1} \oplus C_r$$

are such that each $\bar{f} \in \mathrm{Hom}_A(C_r, M(f)_r)$ $(r \in \mathbb{Z})$ is the inclusion of a direct summand, and $g: D \longrightarrow M(f)$ is a chain equivalence, with a chain homotopy commutative diagram

$$\begin{array}{ccc} C & \xrightarrow{\ f\ } & D \\ & \searrow^{\bar{f}} & \downarrow g \\ & & M(f) \end{array} \qquad).$$

Furthermore, if c, c' are adjoining algebraic Poincaré cobordisms there is defined a homotopy equivalence

$$c \cup c' \xrightarrow{\ \sim\ } c \,\overline{\cup}\, c'$$

from the union defined previously to the direct union.

The direct union is more obviously related to the glueing operation on geometric Poincaré cobordisms. For example, if $(Y;X,X'),(Y';X',X'')$ are adjoining geometric Poincaré cobordisms then

$$\sigma^*(Y \cup_{X'} Y';X,X'') = \sigma^*(Y;X,X') \,\overline{\cup}_{\sigma^*(X')}\, \sigma^*(Y';X',X'')$$

$$(= \sigma^*(Y;X,X') \cup_{\sigma^*(X')} \sigma^*(Y';X',X'')$$

up to homotopy equivalence) .

Similar considerations apply to the quadratic kernels of adjoining bordisms of normal maps.

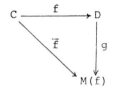

The correspondence of Proposition 1.3.3 i) shows that up to homotopy equivalence the cobordisms of n-dimensional $\begin{cases}\varepsilon\text{-symmetric}\\ \varepsilon\text{-quadratic}\end{cases}$ Poincaré complexes over A may be considered as quadruples

$$\begin{cases} c = ((D,\zeta),(C,\phi),(C',\phi'),(f\ f')) \\ c = ((D,\xi),(C,\psi),(C',\psi'),(f\ f')) \end{cases}$$

consisting of a connected (n+1)-dimensional $\begin{cases}\varepsilon\text{-symmetric}\\ \varepsilon\text{-quadratic}\end{cases}$ complex $\begin{cases}(D,\zeta)\\ (D,\xi)\end{cases}$, n-dimensional $\begin{cases}\varepsilon\text{-symmetric}\\ \varepsilon\text{-quadratic}\end{cases}$ Poincaré complexes $\begin{cases}(C,\phi)\\ (C,\psi)\end{cases}$, $\begin{cases}(C',\phi')\\ (C',\psi')\end{cases}$, and a homotopy equivalence

$$\begin{cases} (f\ f') : (C,\phi)\oplus(C',-\phi') \longrightarrow \partial(D,\zeta) \\ (f\ f') : (C,\psi)\oplus(C',-\psi') \longrightarrow \partial(D,\xi) \end{cases},$$

which we shall also called cobordisms. The union operation $(c,c') \longmapsto c \cup c'$ defined above can be written in the ε-symmetric case as

$$\begin{aligned}((D,\zeta),(C,\phi),(C',\phi'),(f\ f')) &\cup ((D',\zeta'),(C',\phi'),(C'',\phi''),(\tilde{f}'\ f''))\\ &= ((D\cup_{C'}D',\zeta\cup_{\phi'}\zeta'),(C,\phi),(C'',\phi''),(\tilde{f}\ \tilde{f}''))\end{aligned}$$

with

$$D\cup_{C'}D' = C(\begin{pmatrix} f' \\ \tilde{f}' \end{pmatrix} : C' \longrightarrow D\oplus D') \quad ,$$

and similarly in the ε-quadratic case. In particular, given connected (n+1)-dimensional ε-symmetric complexes $(D,\zeta),(D',\zeta')$ and a homotopy equivalence of the boundary n-dimensional ε-symmetric Poincaré complexes

$$g : \partial(D,\zeta) \xrightarrow{\sim} \partial(D',-\zeta')$$

we can glue (D,ζ) to (D',ζ') by g, obtaining the $(n+1)$-dimensional ε-symmetric Poincaré complex

$$(D,\zeta) \cup_g (D',\zeta')$$

appearing in the union cobordism

$$((D,\zeta),0,\partial(D,-\zeta),(0\ 1)) \cup ((D',\zeta'),\partial(D,-\zeta),0,(g\ 0))$$
$$= ((D,\zeta)\cup_g(D',\zeta'),0,0,(0\ 0)) \ ,$$

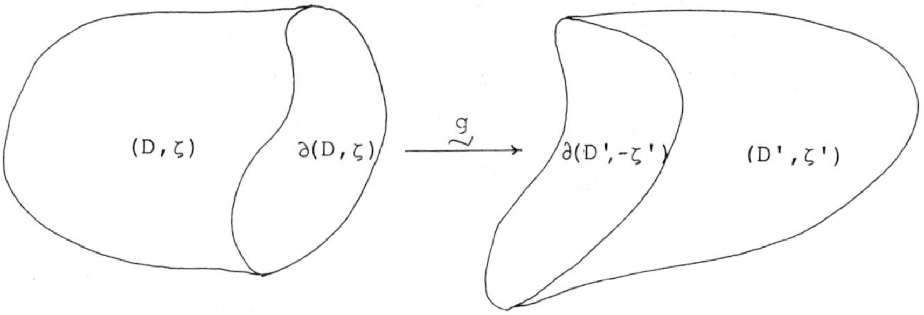

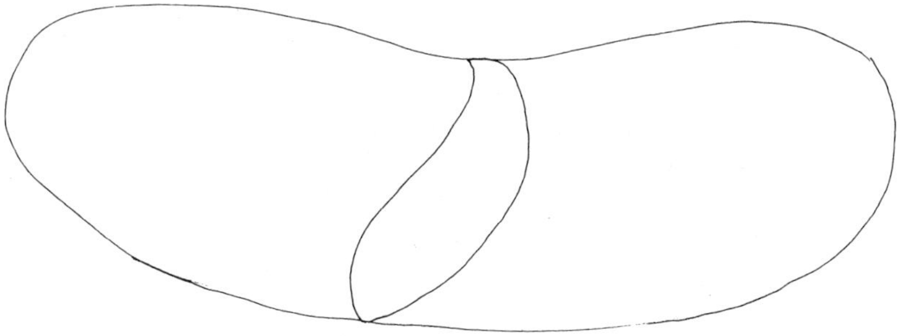

$(D,\zeta) \cup_g (D',\zeta')$,

and similarly in the ε-quadratic case.

The formulation of the union operation entirely in terms of $\begin{cases}\varepsilon\text{-symmetric}\\ \varepsilon\text{-quadratic}\end{cases}$ complexes (i.e. dispensing with pairs) has the advantage that in the low-dimensional cases n = 0,1 it translates directly into the language of forms and formations, using the correspondences of Propositions 1.6.1,1.6.4,1.6.6.

We shall now give an explicit description of the union operation for $\begin{cases}\varepsilon\text{-symmetric}\\ \varepsilon\text{-quadratic}\end{cases}$ forms and formations. In the applications (in §2 below) it is only necessary to glue along all the boundary, so that only this case will be considered.

See Ranicki [5] for further details concerning the glueing of forms and formations, at least in the ε-quadratic case.

Given $\begin{cases}\varepsilon\text{-symmetric}\\ \varepsilon\text{-quadratic}\end{cases}$ formations $\begin{cases}(M,\phi;F,G)\\ (M,\psi;F,G)\end{cases}$, $\begin{cases}(M',\phi';F',G')\\ (M',\psi';F',G')\end{cases}$ and an isomorphism of boundary non-singular $\begin{cases}\varepsilon\text{-symmetric}\\ \varepsilon\text{-quadratic}\end{cases}$ forms

$$\begin{cases} f\ :\ \partial(M,\phi;F,G) = (G^{\perp}/G,\phi^{\perp}/\phi) \xrightarrow{\ \sim\ } \partial(M',-\phi';F',G') = (G'^{\perp}/G',-\phi'^{\perp}/\phi') \\ f\ :\ \partial(M,\psi;F,G) = (G^{\perp}/G,\psi^{\perp}/\psi) \xrightarrow{\ \sim\ } \partial(M',-\psi';F',G') = (G'^{\perp}/G',-\psi'^{\perp}/\psi') \end{cases}$$

define the <u>union</u> non-singular $\begin{cases}\varepsilon\text{-symmetric}\\ \varepsilon\text{-quadratic}\end{cases}$ formation

$$\begin{cases} (M,\phi;F,G)\cup_f(M',\phi';F',G') \\ \qquad = (M\oplus M',\phi\oplus\phi';F\oplus F',G\oplus\mathrm{im}(\begin{pmatrix} j \\ j'f\end{pmatrix}:G^{\perp}/G\longrightarrow M\oplus M')\oplus G') \\ (M,\psi;F,G)\cup_f(M',\psi';F',G') \\ \qquad = (M\oplus M',\psi\oplus\psi';F\oplus F',G\oplus\mathrm{im}(\begin{pmatrix} j \\ j'f\end{pmatrix}:G^{\perp}/G\longrightarrow M\oplus M')\oplus G') \end{cases}$$

with $j\in\mathrm{Hom}_A(G^{\perp}/G,M)$, $j'\in\mathrm{Hom}_A(G'^{\perp}/G',M')$ the A-module morphisms appearing in any of the isomorphisms of $\begin{cases}\varepsilon\text{-symmetric}\\ \varepsilon\text{-quadratic}\end{cases}$ forms extending the inclusions of the sublagrangians given by Proposition 1.6.2

$$\begin{cases} (i\ j)\ :\ H^{\varepsilon}(G^*,\zeta)\oplus(G^{\perp}/G,\phi^{\perp}/\phi) \longrightarrow (M,\phi)\ , \\ \qquad\qquad (i'\ j')\ :\ H^{\varepsilon}(G'^*,\zeta')\oplus(G'^{\perp}/G',\phi'^{\perp}/\phi') \longrightarrow (M',\phi') \\ (i\ j)\ :\ H_{\varepsilon}(G)\oplus(G^{\perp}/G,\psi^{\perp}/\psi) \longrightarrow (M,\psi)\ , \\ \qquad\qquad (i'\ j')\ :\ H_{\varepsilon}(G')\oplus(G'^{\perp}/G',\psi'^{\perp}/\psi') \longrightarrow (M',\psi')\ . \end{cases}$$

The <u>union</u> of $\begin{cases}\varepsilon\text{-symmetric}\\ \varepsilon\text{-quadratic}\end{cases}$ forms $\begin{cases}(M,\phi)\\(M,\psi)\end{cases}$, $\begin{cases}(M',\phi')\\(M',\psi')\end{cases}$ along a stable isomorphism of boundary non-singular $\begin{cases}\text{even } (-\varepsilon)\text{-symmetric}\\ \text{split } (-\varepsilon)\text{-quadratic}\end{cases}$ formations

$$\begin{cases} [f]\ :\ \partial(M,\phi) = (H^{-\varepsilon}(M);M,\Gamma_{(M,\phi)}) \\ \qquad\qquad \longrightarrow \partial(M',-\phi') = (H^{-\varepsilon}(M');M',\Gamma_{(M',-\phi')}) \\ [\alpha,\beta,\sigma]\ :\ \partial(M,\psi) = (M,(\begin{pmatrix}1\\ \psi+\varepsilon\psi^*\end{pmatrix},\psi)M) \\ \qquad\qquad \longrightarrow \partial(M',-\psi') = (M',(\begin{pmatrix}1\\ -(\psi'+\varepsilon\psi'^*)\end{pmatrix},-\psi')M') \end{cases}$$

is the non-singular $\begin{cases}\varepsilon\text{-symmetric}\\ \varepsilon\text{-quadratic}\end{cases}$ form

$$\begin{cases} (M'',\phi'') = (M,\phi)\cup_{[f]}(M',\phi') \\ (M'',\psi'') = (M,\psi)\cup_{[\alpha,\beta,\sigma]}(M',\psi') \end{cases}$$

defined further below. The union operation is characterized

(up to isomorphism) by the property that $\begin{cases}(M,\phi)\\(M,\psi)\end{cases}$ and $\begin{cases}(M',\phi')\\(M',\psi')\end{cases}$ are included in the union $\begin{cases}(M'',\phi'')\\(M'',\psi'')\end{cases}$ as maximally orthogonal subforms, that is there are defined morphisms of forms

$$\begin{cases}j : (M,\phi) \longrightarrow (M'',\phi'')\\ j : (M,\psi) \longrightarrow (M'',\psi'')\end{cases}, \quad \begin{cases}j' : (M',\phi') \longrightarrow (M'',\phi'')\\ j' : (M',\psi') \longrightarrow (M'',\psi'')\end{cases}$$

with $j \in \mathrm{Hom}_A(M,M'')$, $j' \in \mathrm{Hom}_A(M',M'')$ split monomorphisms, such that the A-module sequence

$$\begin{cases}0 \longrightarrow M \xrightarrow{\ j\ } M'' \xrightarrow{\ j'^*\phi''\ } M'^* \longrightarrow 0\\ 0 \longrightarrow M \xrightarrow{\ j\ } M'' \xrightarrow{\ j'^*(\psi''+\varepsilon\psi''^*)\ } M'^* \longrightarrow 0\end{cases}$$

is exact, and such that the stable isomorphism of formations

$$\begin{cases}\partial(M,\phi) \xrightarrow{\ \sim\ } \partial(M',-\phi')\\ \partial(M,\psi) \xrightarrow{\ \sim\ } \partial(M',-\psi')\end{cases}$$

naturally associated to such inclusions is equivalent to $\begin{cases}[f]\\ [\alpha,\beta,\sigma]\end{cases}$ under the relation on stable isomorphisms corresponding (via Proposition 1.6.4) to the chain homotopy of homotopy equivalences of 1-dimensional $\begin{cases}\text{even } (-\varepsilon)\text{-symmetric}\\ (-\varepsilon)\text{-quadratic}\end{cases}$ Poincaré complexes. In particular, if the forms $\begin{cases}(M,\phi)\\(M,\psi)\end{cases}$, $\begin{cases}(M',\phi')\\(M',\psi')\end{cases}$ are non-singular the union is just the direct sum

$$\begin{cases}(M'',\phi'') = (M,\phi)\oplus(M',\phi')\\ (M'',\psi'') = (M,\psi)\oplus(M',\psi')\end{cases},$$

with j and j' the canonical inclusions.

The union operation for forms is defined as follows.

In the ε-symmetric case let

$$[f] : \partial(M,\phi) \longrightarrow \partial(M',-\phi')$$

be the stable isomorphism of even (−ε)-symmetric formations given by the isomorphism

$$f : (H^{-\varepsilon}(M);M,\Gamma_{(M,\phi)}) \oplus (H^{-\varepsilon}(P);P,P^*) \xrightarrow{\sim} (H^{-\varepsilon}(M');M',\Gamma_{(M',-\phi')}) \oplus (H^{-\varepsilon}(P');P',P'^*)$$

for some f.g. projective A-modules P,P'. Write the restrictions of f to the lagrangians as

$$\alpha = \begin{pmatrix} a & a_1 \\ a_2 & a_3 \end{pmatrix} : M\oplus P \xrightarrow{\sim} M'\oplus P'$$

$$\beta = \begin{pmatrix} b & b_1 \\ b_2 & b_3 \end{pmatrix} : M\oplus P^* \xrightarrow{\sim} M'\oplus P'^* ,$$

and let

$$\beta^{-1} = \begin{pmatrix} b' & b'_1 \\ b'_2 & b'_3 \end{pmatrix} : M'\oplus P'^* \xrightarrow{\sim} M\oplus P^* .$$

Let $(M^*\oplus P^*, \tau \in Q^{\varepsilon}(M^*\oplus P^*))$ be the unique ε-symmetric form such that there is defined a commutative square

$$\begin{array}{ccc}
M\oplus P^* & \xrightarrow[\sim]{\beta} & M'\oplus P'^* \\
\left\downarrow \begin{pmatrix} 1 & 0 \\ 0 & 0 \\ \phi & 0 \\ 0 & 1 \end{pmatrix} \right. & & \left\downarrow \begin{pmatrix} 1 & 0 \\ 0 & 0 \\ -\phi' & 0 \\ 0 & 1 \end{pmatrix} \right. \\
M\oplus P\oplus M^*\oplus P^* & \xrightarrow[\sim]{f = \begin{pmatrix} \alpha & \alpha\tau^* \\ 0 & \alpha^{*-1} \end{pmatrix}} & M'\oplus P'\oplus M'^*\oplus P'^*
\end{array}$$

and let

$$\alpha\tau\alpha^* = \begin{pmatrix} t' & t_1' \\ t_2' & t_3' \end{pmatrix} : M'^*\oplus P'^* \longrightarrow M'\oplus P' \quad .$$

The union ε-symmetric form is given by

$$(M'',\phi''\in Q^{\varepsilon}(M'')) = (M\oplus M'^*, \begin{pmatrix} \phi & a^* \\ \varepsilon a & t' \end{pmatrix}) \quad ,$$

with the canonical inclusions defined by

$$j = \begin{pmatrix} 1 \\ 0 \end{pmatrix} : (M,\phi) \longrightarrow (M'',\phi'')$$

$$j' = \begin{pmatrix} b' \\ \phi' \end{pmatrix} : (M',\phi') \longrightarrow (M'',\phi'') \quad .$$

In the ε-quadratic case let $[\alpha,\beta,\sigma]$ by the isomorphism of split ε-quadratic formations

$$(\alpha,\beta,\sigma) = (\begin{pmatrix} a & a_1 \\ a_2 & a_3 \end{pmatrix}, \begin{pmatrix} b & b_1 \\ b_2 & b_3 \end{pmatrix}, \begin{pmatrix} s & s_1 \\ s_2 & s_3 \end{pmatrix})$$

$$: (M\oplus P, (\begin{pmatrix} \begin{pmatrix} 1 & 0 \\ 0 & 0 \end{pmatrix} \\ \begin{pmatrix} \psi+\varepsilon\psi^* & 0 \\ 0 & 1 \end{pmatrix} \end{pmatrix}, \begin{pmatrix} \psi & 0 \\ 0 & 0 \end{pmatrix}) M\oplus P^*)$$

$$\longrightarrow (M'\oplus P', (\begin{pmatrix} \begin{pmatrix} 1 & 0 \\ 0 & 0 \end{pmatrix} \\ \begin{pmatrix} -(\psi'+\varepsilon\psi'^*) & 0 \\ 0 & 1 \end{pmatrix} \end{pmatrix}, \begin{pmatrix} -\psi' & 0 \\ 0 & 0 \end{pmatrix}) M'\oplus P'^*)$$

for some f.g. projective A-modules P,P'. Let

$$\alpha\sigma\alpha^* = \begin{pmatrix} s' & s' \\ s_2' & s_3' \end{pmatrix} : M'^*\oplus P'^* \longrightarrow M'\oplus P'^*$$

$$\beta^{-1} = \begin{pmatrix} b' & b_1' \\ b_2' & b_3' \end{pmatrix} : M'\oplus P'^* \longrightarrow M\oplus P^* \quad .$$

The union ε-quadratic form is given by

$$(M'',\psi'' \in Q_\varepsilon(M'')) = (M\oplus M'^*, \begin{pmatrix} \psi & 0 \\ \varepsilon a & s' \end{pmatrix}) \quad ,$$

with the canonical inclusions defined by

$$j = \begin{pmatrix} 1 \\ 0 \end{pmatrix} : (M,\psi) \longrightarrow (M'',\psi'')$$

$$j' = \begin{pmatrix} b' \\ \psi'+\varepsilon\psi'^* \end{pmatrix} : (M',\psi') \longrightarrow (M'',\psi'') \quad .$$

See Ranicki [1,4.3],[5] and Wall [8],[12] for some applications of the union of forms. Here is another:

The $\left\{\begin{array}{l}\text{(even) } \varepsilon\text{-symmetric}\\ \varepsilon\text{-quadratic}\end{array}\right.$ forms over A $\left\{\begin{array}{l}(M,\phi)\\ (M,\psi)\end{array}\right.$, $\left\{\begin{array}{l}(M',\phi')\\ (M',\psi')\end{array}\right.$

are <u>∂-equivalent</u> if there exists an isomorphism

$$\left\{\begin{array}{l} f : (M,\phi)\oplus(N,\theta) \xrightarrow{\sim} (M',\phi')\oplus(N',\theta') \\ f : (M,\psi)\oplus(N,\chi) \xrightarrow{\sim} (M',\psi')\oplus(N',\chi') \end{array}\right.$$

for some non-singular $\left\{\begin{array}{l}\text{(even) } \varepsilon\text{-symmetric}\\ \varepsilon\text{-quadratic}\end{array}\right.$ forms $\left\{\begin{array}{l}(N,\theta)\\ (N,\chi)\end{array}\right.$, $\left\{\begin{array}{l}(N',\theta')\\ (N',\chi')\end{array}\right.$,

in which case there is induced a stable isomorphism of the

boundary $\left\{\begin{array}{l}\text{even } (-\varepsilon)\text{-symmetric } ((-\varepsilon)\text{-quadratic})\\ \text{split } (-\varepsilon)\text{-quadratic}\end{array}\right.$ formations

$$\left\{\begin{array}{l} [\partial f] : \partial(M,\phi) \xrightarrow{\sim} \partial(M',\phi') \\ [\partial f] : \partial(M,\psi) \xrightarrow{\sim} \partial(M',\psi') \end{array}\right. \quad .$$

In particular, if $L\subset M$ is a sublagrangian of an $\left\{\begin{array}{l}\text{(even) } \varepsilon\text{-symmetric}\\ \varepsilon\text{-quadratic}\end{array}\right.$

form $\left\{\begin{array}{l}(M,\phi)\\ (M,\psi)\end{array}\right.$ then $\left\{\begin{array}{l}(M',\phi') = (L^\perp/L,\phi^\perp/\phi)\\ (M',\psi') = (L^\perp/L,\psi^\perp/\psi)\end{array}\right.$ is ∂-equivalent to $\left\{\begin{array}{l}(M,\phi)\\ (M,\psi)\end{array}\right.$.

Proposition 1.7.1 The boundary operations $\partial:\{\text{forms}\}\longrightarrow\{\text{formations}\}$ define natural one-one correspondences

$$\partial : \{\partial\text{-equivalence classes of } \begin{Bmatrix}\varepsilon\text{-symmetric}\\ \text{even } \varepsilon\text{-symmetric}\\ \varepsilon\text{-quadratic}\end{Bmatrix} \text{ forms over } A\}$$

$$\xrightarrow{\sim} \{\text{stable isomorphism classes of null-cobordant } \begin{Bmatrix}\text{even } (-\varepsilon)\text{-symmetric}\\ (-\varepsilon)\text{-quadratic}\\ \text{split } (-\varepsilon)\text{-quadratic}\end{Bmatrix} \text{ formations over } A\} .$$

Proof: It is sufficient to consider the ε-symmetric case, the others being entirely similar.

By Proposition 1.6.5 iii) every null-cobordant even $(-\varepsilon)$-symmetric formation is stably isomorphic to the boundary $\partial(M,\phi)$ of an ε-symmetric form (M,ϕ). Thus it remains to show that if $(M,\phi),(M',\phi')$ are ε-symmetric forms which are related by a stable isomorphism of the boundaries

$$[f] : \partial(M,\phi) \xrightarrow{\sim} \partial(M',\phi')$$

then they are ∂-equivalent. Write the union non-singular ε-symmetric form as

$$(N',\theta') = (M,\phi) \cup_{[f]} (M',-\phi') ,$$

and let

$$j' : (M',-\phi') \longrightarrow (N',\theta')$$

be the canonical inclusion. Then the submodule

$$L = \{(x,j'(x)) \in M'\oplus N' \mid x \in M'\} \subseteq M'\oplus N'$$

defines a sublagrangian of $(M',\phi')\oplus(N',\theta')$ such that

$$(L^{\perp}/L,(\phi'\oplus\theta')^{\perp}/(\phi'\oplus\theta')) = (M,\phi) .$$

Applying Proposition 1.6.2 there is obtained an isomorphism

$$f : (M,\phi)\oplus(N,\theta) \xrightarrow{\sim} (M',\phi')\oplus(N',\theta')$$

with $(N,\theta) = H^{\varepsilon}(M',\phi')$ non-singular. Thus (M,ϕ) and (M',ϕ') are ∂-equivalent.

[]

Proposition 1.7.1 is a generalization of the familiar result (cf. Kneser and Puppe [1], Wall [10] and Durfee [2]) that if $(M,\psi),(M',\psi')$ are quadratic forms over $\mathbb{Z}$ which become non-singular over $\mathbb{Q}$ then they are ∂-equivalent if and only if the boundaries $\partial(M,\psi)$, $\partial(M',\psi')$ are isomorphic as "non-singular quadratic linking forms over $(\mathbb{Z},\mathbb{Z}-\{0\})$" - see §3.4 below for the expression of $\partial(M,\psi)$ for such (M,ψ) as a non-singular quadratic linking form

$$(M^{\#}/M, \lambda : M^{\#}/M \times M^{\#}/M \longrightarrow \mathbb{Q}/\mathbb{Z}, \mu : M^{\#}/M \longrightarrow \mathbb{Q}/2\mathbb{Z}) \quad ,$$

with

$$M^{\#} = \{x \in \mathbb{Q}\otimes_{\mathbb{Z}} M \mid (\psi+\psi^*)(x)(M) \in \mathbb{Z} \subseteq \mathbb{Q}\}$$

the "dual lattice", λ a non-singular symmetric linking pairing on the finite abelian group $M^{\#}/M$, and μ a quadratic refinement of λ. The proof of Proposition 1.7.1 is a generalization of the standard proof of the Novikov additivity property for the signature: if $(M,\phi),(M',\phi')$ are symmetric forms over $\mathbb{Z}$ and $[f] : \partial(M,\phi) \xrightarrow{\sim} \partial(M',-\phi')$ is a stable isomorphism of boundary skew-symmetric formations over $\mathbb{Z}$ then the signature of the union non-singular symmetric form over $\mathbb{Z}$ $(M,\phi)\cup_{[f]}(M',\phi')$ is given by the sum of the signatures of (M,ϕ) and (M',ϕ')

$$\sigma^*((M,\phi)\cup_{[f]}(M',\phi')) = \sigma^*(M,\phi) + \sigma^*(M',\phi') \in \mathbb{Z} \,,$$

which we shall generalize in Proposition 7.3.6. Proposition 1.7.1 is generalized to complexes in Proposition 1.8.3 below.

1.8 Unified L-theory

In §I.6 there were defined lower $\begin{cases}\varepsilon\text{-symmetric}\\ \varepsilon\text{-quadratic}\end{cases}$ L-groups $\begin{cases}L^n(A,\varepsilon)\\ L_n(A,\varepsilon)\end{cases}$ $(n \leqslant -1)$, as we now recall. We shall also give a unified construction of the unified $\begin{cases}\varepsilon\text{-symmetric}\\ \varepsilon\text{-quadratic}\end{cases}$ L-groups $\begin{cases}L^n(A,\varepsilon)\\ L_n(A,\varepsilon)\end{cases}$ $(n \in \mathbb{Z})$.

Define the lower $\begin{cases}\underline{\varepsilon\text{-symmetric}}\\ \underline{\varepsilon\text{-quadratic}}\end{cases}$ $\underline{\text{L-groups of A}}$ $\begin{cases}L^n(A,\varepsilon)\\ L_n(A,\varepsilon)\end{cases}$ $(n \leqslant -1)$

by

$$\begin{cases} L^n(A,\varepsilon) = \begin{cases} L\langle v_0\rangle^{n+2}(A,-\varepsilon) & \text{if } n = -1,-2 \\ L_n(A,\varepsilon) & \text{if } n \leqslant -3 \end{cases} \\ L_n(A,\varepsilon) = L_{n+2i}(A,(-)^i\varepsilon) \quad \text{if } n \leqslant -1,\ n+2i \geqslant 0\ , \end{cases}$$

extending the semi-periodicity $\begin{cases} L^n(A,\varepsilon) = L\langle v_0\rangle^{n+2}(A,-\varepsilon) \\ L_n(A,\varepsilon) = L_{n+2}(A,-\varepsilon)\end{cases}$ $(n \geqslant 0)$

of Proposition $\begin{cases}1.4.2\\ 1.2.3\text{ i)}\end{cases}$.

Define the $\underline{\text{skew-suspension}}$ maps

$$\bar{S} : L^n(A,-\varepsilon) \longrightarrow L^{n+2}(A,\varepsilon) \quad (n \in \mathbb{Z})$$

to be the skew-suspension previously defined for $n \geqslant 0$ and $n \leqslant -4$, and to be the appropriate $\pm\varepsilon$-symmetrization maps for $-3 \leqslant n \leqslant -1$.

__Proposition 1.8.1__ If $\hat{H}^0(\mathbb{Z}_2;A,\varepsilon) = 0$ the skew-suspension maps

$$\bar{S} : L^n(A,-\varepsilon) \longrightarrow L^{n+2}(A,\varepsilon) \quad (n \in \mathbb{Z})$$

are isomorphisms.

__Proof__: See Proposition I.6.1.

[]

In particular, if there exists a central element $a \in A$ such that $a+\bar{a} = 1 \in A$ (e.g. $a = 1/2 \in A$) then $\hat{H}^*(\mathbb{Z}_2;A,\varepsilon) = 0$ and up to isomorphism

$$L_n(A,\varepsilon) = L^n(A,\varepsilon) = L^{n+2}(A,-\varepsilon) \quad (n \in \mathbb{Z})$$

cf. Proposition 1.4.3.

Define the $\left\{\begin{array}{l}\underline{\varepsilon\text{-symmetric}}\\ \underline{\varepsilon\text{-quadratic}}\end{array}\right.$ $\underline{\mathcal{L}\text{-categories}}$ $\left\{\begin{array}{l}\mathcal{L}^n(A,\varepsilon)\\ \mathcal{L}_n(A,\varepsilon)\end{array}\right.$ $(n \in \mathbb{Z})$ to be the additive categories given by

$$\left\{\begin{array}{l}
\mathcal{L}^n(A,\varepsilon) = \\
\quad\left\{\begin{array}{ll}
\{\text{connected } n\text{-dimensional } \varepsilon\text{-symmetric complexes over } A, & \\
\qquad \text{homotopy equivalences}\} & (n \geqslant 1)\\
\{\varepsilon\text{-symmetric forms over } A, \text{ isomorphisms}\} & (n = 0)\\
\{\text{even } (-\varepsilon)\text{-symmetric formations over } A, & \\
\qquad \text{stable isomorphisms}\} & (n = -1)\\
\{\text{even } (-\varepsilon)\text{-symmetric forms over } A, \text{ isomorphisms}\} & (n = -2)\\
\{(-\varepsilon)\text{-quadratic formations, stable isomorphisms}\} & (n = -3)\\
\mathcal{L}_n(A,\varepsilon) \text{ (as defined below)} & (n \leqslant -4)
\end{array}\right.\\
\mathcal{L}_n(A,\varepsilon) = \\
\quad\left\{\begin{array}{ll}
\{\text{connected } n\text{-dimensional } \varepsilon\text{-quadratic complexes over } A, & \\
\qquad \text{homotopy equivalences}\} & (n \geqslant 1)\\
\{(-)^i\varepsilon\text{-quadratic forms over } A, \text{ isomorphisms}\} & (n = 2i \leqslant 0)\\
\{\text{split } (-)^i\varepsilon\text{-quadratic formations over } A, & \\
\qquad \text{stable isomorphisms}\} & (n = 2i+1 \leqslant -1).
\end{array}\right.
\end{array}\right.$$

Note that by Proposition 1.6.4

$$\left\{\begin{array}{l}
\mathcal{L}^1(A,\varepsilon) = \{\varepsilon\text{-symmetric formations over } A, \text{ stable isomorphisms}\}\ ,\\
\mathcal{L}_1(A,\varepsilon) = \mathcal{L}_{-3}(A,\varepsilon)\ .
\end{array}\right.$$

Define the <u>orientation-reversing</u> involutions

$$\begin{cases} - : \mathcal{L}^n(A,\varepsilon) \longrightarrow \mathcal{L}^n(A,\varepsilon) \ ; \ (C,\phi) \longmapsto (C,-\phi) \\ - : \mathcal{L}_n(A,\varepsilon) \longrightarrow \mathcal{L}_n(A,\varepsilon) \ ; \ (C,\psi) \longmapsto (C,-\psi) \end{cases} \quad (n \in \mathbb{Z}) ,$$

and use the boundary operations of §§1.3,1.6 to define the <u>boundary</u> functors

$$\begin{cases} \partial : \mathcal{L}^n(A,\varepsilon) \longrightarrow \mathcal{L}^{n-1}(A,\varepsilon) \ ; \ (C,\phi) \longmapsto \partial(C,\phi) \\ \partial : \mathcal{L}_n(A,\varepsilon) \longrightarrow \mathcal{L}_{n-1}(A,\varepsilon) \ ; \ (C,\psi) \longmapsto \partial(C,\psi) \end{cases} \quad (n \in \mathbb{Z}) .$$

For any object x

$$-(-x) = x \quad , \quad \partial(-x) = -(\partial x) \quad , \quad \partial(\partial x) = 0$$

up to natural equivalence. The morphisms of the $\mathcal{L}$-categories will all be called <u>homotopy equivalences</u>; objects x,y of the same $\mathcal{L}$-category are <u>homotopy equivalent</u> $x \simeq y$ if there exists a homotopy equivalence

$$f : x \xrightarrow{\sim} y ,$$

in which case there are also defined homotopy equivalences

$$f^{-1} : y \xrightarrow{\sim} x \quad , \quad f : -x \xrightarrow{\sim} -y \quad , \quad \partial f : \partial x \xrightarrow{\sim} \partial y .$$

An object x is <u>closed</u> if $\partial x \simeq 0$, and it is a <u>boundary</u> if $x \simeq \partial y$ for some object y. In particular, boundary objects are closed, and if x is closed (resp. a boundary) then so is -x. For $n \geqslant 0$ the closed objects are precisely the algebraic Poincaré complexes, and for $n \leqslant 1$ they are precisely the non-singular forms and formations.

Given objects x,y in the same $\mathcal{L}$-category and a homotopy equivalence of the boundaries of x and -y

$$f : \partial x \xrightarrow{\sim} -\partial y$$

define the <u>union</u> $x \cup_f y$ to be the closed object of the same $\mathcal{L}$-category constructed as in §1.7. For closed objects x,y

$$x \cup_f y = x \oplus y .$$

A <u>cobordism</u> $(z;f,g)$ of objects x,y in the same n-dimensional $\mathcal{L}$-category is a triple consisting of an object z of the corresponding $(n+1)$-dimensional $\mathcal{L}$-category, and homotopy equivalences

$$f : \partial x \xrightarrow{\simeq} \partial y \quad , \quad g : x \cup_f -y \xrightarrow{\simeq} \partial z \ .$$

For closed objects x,y this is just the cobordism of §§1.1,1.7.

A <u>surgery</u> on an object x of an n-dimensional $\mathcal{L}$-category is an operation

$$x \longmapsto x'$$

sending x to an object x' of the same $\mathcal{L}$-category; for $n \geqslant 0$ this is to be surgery on complexes as defined in §1.5, and for $n \leqslant 1$ it is the translation of this surgery from the language of complexes to that of forms and formations. For example, if (M,ϕ) is an ε-symmetric form over A and $L \subset M$ is a sublagrangian the operation

$$(M,\phi) \longmapsto (M',\phi') = (L^{\perp}/L, \phi^{\perp}/\phi)$$

is a surgery on (M,ϕ).

<u>Proposition 1.8.2</u> i) Cobordism is the equivalence relation on the set of objects of $\begin{cases} \mathcal{L}^n(A,\varepsilon) \\ \mathcal{L}_n(A,\varepsilon) \end{cases}$ $(n \in \mathbb{Z})$ generated by surgery and homotopy equivalence. The cobordism classes of closed objects form an abelian group with respect to the direct sum $\oplus$, namely $\begin{cases} L^n(A,\varepsilon) \\ L_n(A,\varepsilon) \end{cases}$ $(n \in \mathbb{Z})$.

ii) If x,y,z are objects of $\begin{cases} \mathcal{L}^n(A,\varepsilon) \\ \mathcal{L}_n(A,\varepsilon) \end{cases}$ and $f:\partial x \xrightarrow{\simeq} \partial y, g:\partial y \xrightarrow{\simeq} \partial z$ are homotopy equivalences then

$$(x \cup_f -y) \oplus (y \cup_g -z) = (x \cup_{gf} -z) \in \begin{cases} L^n(A,\varepsilon) \\ L_n(A,\varepsilon) \end{cases} .$$

Proof: i) Immediate from Propositions 1.3.3, 1.5.1 and 1.6.5.

ii) It is possible to obtain $x \cup_{gf} -z$ from $(x \cup_f -y) \oplus (y \cup_g -z)$ by surgery (as in the proof of Proposition 1.8.3 below).

[]

The homotopy equivalence classes of the null-cobordant objects of $\begin{cases} \mathcal{L}^n(A,\varepsilon) \\ \mathcal{L}_n(A,\varepsilon) \end{cases}$ (i.e those representing 0 in $\begin{cases} L^n(A,\varepsilon) \\ L_n(A,\varepsilon) \end{cases}$) are in one-one correspondence with the following equivalence classes of objects of $\begin{cases} \mathcal{L}^{n+1}(A,\varepsilon) \\ \mathcal{L}_{n+1}(A,\varepsilon) \end{cases}$.

Let ∂-equivalence be the equivalence relation on the objects of $\begin{cases} \mathcal{L}^{n+1}(A,\varepsilon) \\ \mathcal{L}_{n+1}(A,\varepsilon) \end{cases}$ $(n \in \mathbb{Z})$ generated by the elementary operations:

i) $x \longmapsto x'$ if x' is homotopy equivalent to x

ii) $x \longmapsto x'$ if x' is obtained from x by surgery

iii) $x \longmapsto x'$ if $x' = x \oplus y$ for some closed object y.

Note that $\partial x \simeq \partial x'$ in each case, so that the homotopy type of ∂x is an invariant of the ∂-equivalence class of an object x. For $n+1 = 0$ ∂-equivalence is just the ∂-equivalence relation on forms defined in §1.7 above. Proposition 1.7.1 is the special case $n+1 = 0$ of:

Proposition 1.8.3 The boundary operation defines a natural one-one correspondence

$$\partial : \{\partial\text{-equivalence classes of objects } x \text{ of } \begin{cases} \mathcal{L}^{n+1}(A,\varepsilon) \\ \mathcal{L}_{n+1}(A,\varepsilon) \end{cases}\}$$

$$\xrightarrow{\sim} \{\text{homotopy equivalence classes of null-cobordant objects } \partial x \text{ of } \begin{cases} \mathcal{L}^{n}(A,\varepsilon) \\ \mathcal{L}_{n}(A,\varepsilon) \end{cases}\} \quad (n \in \mathbb{Z}) .$$

Proof: Given connected (n+1)-dimensional ε-symmetric complexes over A $(C,\phi),(C',\phi')$ (for some $n \geqslant 0$) and a homotopy equivalence of the boundary n-dimensional ε-symmetric Poincaré complexes over A

$$f : \partial(C,\phi) \xrightarrow{\simeq} \partial(C',\phi')$$

there is defined a union (n+1)-dimensional ε-symmetric ε-symmetric Poincaré complex over A

$$(C'',\phi'') = (C \cup_f C', \phi \cup -\phi') \ .$$

Surgery on $(C',\phi')\oplus(C'',\phi'')$ by the connected (n+2)-dimensional ε-symmetric pair $(g:C'\oplus C'' \longrightarrow C',(0,\phi'\oplus\phi''))$ with

$$g = (1 \quad 0 \quad 0 \quad 1)$$
$$: (C'\oplus C'')_r = C'_r \oplus C_r \oplus \partial C_{r-1} \oplus C'_r \longrightarrow C'_r$$

results in an (n+1)-dimensional ε-symmetric complex homotopy equivalent to (C,ϕ), so that (C,ϕ) and (C',ϕ') are ∂-equivalent. Similarly for the other cases.

[]

The matrix identity of Wall [4,p.63] was used to prove that the odd-dimensional surgery obstruction group $L_{2i+1}(\pi)$ defined as the quotient of the stable $(-)^i$-unitary group of $\mathbb{Z}[\pi]$ by the subgroup generated by the elementary $(-)^i$-unitary matrices is in fact abelian. Proposition 1.8.3 is a generalization of this identity, and also of the related normal forms of Sharpe [1] and Wall [11] for the elementary $(-)^i$-unitary group. The normal forms may in fact be deduced from the ε-quadratic case for $n+1 = 0$ (as has already been done in Proposition I.9.2 iii)). The sum formula of Proposition 1.8.2 ii) is an L-theoretic analogue of the Whitehead lemma of algebraic K-theory.

1.9 Products

The tensor product $A\otimes_{\mathbb{Z}}B$ of rings with involution A,B is a ring with involution

$$\bar{\ } : A\otimes_{\mathbb{Z}}B \longrightarrow A\otimes_{\mathbb{Z}}B \ ; \ a\otimes b \longmapsto \overline{a\otimes b} = \bar{a}\otimes\bar{b} \quad .$$

If $\varepsilon \in A$, $\eta \in B$ are central units such that

$$\bar{\varepsilon} = \varepsilon^{-1} \in A \quad , \quad \bar{\eta} = \eta^{-1} \in B$$

then $\varepsilon\otimes\eta \in A\otimes_{\mathbb{Z}}B$ is a central unit such that

$$(\overline{\varepsilon\otimes\eta}) = (\varepsilon\otimes\eta)^{-1} \in A\otimes_{\mathbb{Z}}B \quad .$$

If C is a p-dimensional A-module chain complex and D is a q-dimensional B-module chain complex then $C\otimes_{\mathbb{Z}}D$ is a (p+q)-dimensional $A\otimes_{\mathbb{Z}}B$-module chain complex, with $A\otimes_{\mathbb{Z}}B$ acting by

$$A\otimes_{\mathbb{Z}}B \times C\otimes_{\mathbb{Z}}D \longrightarrow C\otimes_{\mathbb{Z}}D \ ; \ (a\otimes b, x\otimes y) \longmapsto ax\otimes by \quad .$$

As in §I.8 there are defined products in the Q-groups

$$\otimes : Q^m(C,\varepsilon)\otimes_{\mathbb{Z}}Q^n(D,\eta) \longrightarrow Q^{m+n}(C\otimes_{\mathbb{Z}}D,\varepsilon\otimes\eta) \ ;$$

$$\{\phi_s \in \mathrm{Hom}_A(C^*,C)_{m+s} \,|\, s \geqslant 0\}\otimes\{\theta_s \in \mathrm{Hom}_B(D^*,D)_{n+s} \,|\, s \geqslant 0\}$$

$$\longmapsto \{(\phi\otimes\theta)_s = \sum_{r=0}^{s} (-)^{(m+r)s}\phi_r\otimes T_\eta^r\theta_{s-r}$$

$$\in \mathrm{Hom}_{A\otimes_{\mathbb{Z}}B}((C\otimes_{\mathbb{Z}}D)^*, C\otimes_{\mathbb{Z}}D))_{m+n+s}$$

$$= \sum_{r=-\infty}^{\infty} \mathrm{Hom}_A(C^*,C)_{m+r}\otimes_{\mathbb{Z}}\mathrm{Hom}_B(D^*,D)_{n-r+s} \,|\, s \geqslant 0\}$$

$$\otimes : Q^m(C,\varepsilon)\otimes_{\mathbb{Z}} Q_n(D,\eta) \longrightarrow Q_{m+n}(C\otimes_{\mathbb{Z}} D,\varepsilon\otimes\eta) \ ;$$

$$\{\phi_s \in \mathrm{Hom}_A(C^*,C)_{m+s} \mid s \geqslant 0\}\otimes\{\psi_s \in \mathrm{Hom}_B(D^*,D)_{n-s} \mid s \geqslant 0\}$$

$$\longmapsto \{(\phi\otimes\psi)_s = \sum_{r=0}^{\infty} (-)^{(m+r)s} \phi_r \otimes T_\eta^r \psi_{s+r}$$

$$\in \mathrm{Hom}_{A\otimes_{\mathbb{Z}} B}((C\otimes_{\mathbb{Z}} D)^*, C\otimes_{\mathbb{Z}} D)_{m+n-s}$$

$$= \sum_{r=-\infty}^{\infty} \mathrm{Hom}_A(C^*,C)_{m+r} \otimes_{\mathbb{Z}} \mathrm{Hom}_B(D^*,D)_{m+n-s} \mid s \geqslant 0\}$$

which extend to the L-groups:

Proposition 1.9.1 Given A,B,ε,η as above there are defined external products in the L-groups

$$\otimes : L^m(A,\varepsilon)\otimes_{\mathbb{Z}} L^n(B,\eta) \longrightarrow L^{m+n}(A\otimes_{\mathbb{Z}} B,\varepsilon\otimes\eta) \ ;$$

$$(C,\phi)\otimes(D,\theta) \longmapsto (C\otimes_{\mathbb{Z}} D,\phi\otimes\theta) \quad ,$$

$$\otimes : L^m(A,\varepsilon)\otimes_{\mathbb{Z}} L_n(B,\eta) \longrightarrow L_{m+n}(A\otimes_{\mathbb{Z}} B,\varepsilon\otimes\eta) \ ;$$

$$(C,\phi)\otimes(D,\psi) \longmapsto (C\otimes_{\mathbb{Z}} D,\phi\otimes\psi)$$

for all $m,n \in \mathbb{Z}$.

Proof: See Proposition I.8.1.

[]

Given rings with involution A,R we shall say that A is an R-module if there is given a morphism of rings with involution

$$R\otimes_{\mathbb{Z}} A \longrightarrow A \ ; \ r\otimes a \longmapsto ra \quad .$$

(We are anticipating here the definition in §2.2 below of a morphism of rings with involution).

Proposition 1.9.2 If A is an R-module there are defined internal products in the L-groups

$$\begin{cases} \otimes : L^m(R,\rho)\otimes_{\mathbb{Z}} L^n(A,\varepsilon) \longrightarrow L^{m+n}(A,\rho\varepsilon) \ ; \ (C,\phi)\otimes(D,\theta) \longmapsto (C\otimes_R D, \phi\otimes\theta) \\ \otimes : L^m(R,\rho)\otimes_{\mathbb{Z}} L_n(A,\varepsilon) \longrightarrow L_{m+n}(A,\rho\varepsilon) \ ; \ (C,\phi)\otimes(D,\psi) \longmapsto (C\otimes_R D, \phi\otimes\psi) \end{cases}$$

for any $m,n \in \mathbb{Z}$. In particular, the symmetric Witt group $L^0(R)$ acts on the $\begin{cases}\varepsilon\text{-symmetric}\\ \varepsilon\text{-quadratic}\end{cases}$ L-groups of A $\begin{cases}L^*(A,\varepsilon)\\ L_*(A,\varepsilon)\end{cases}$

$$\begin{cases} \otimes : L^0(R)\otimes_{\mathbb{Z}} L^*(A,\varepsilon) \longrightarrow L^*(A,\varepsilon) \\ \otimes : L^0(R)\otimes_{\mathbb{Z}} L_*(A,\varepsilon) \longrightarrow L_*(A,\varepsilon) \end{cases}$$

with the element

$$(R, 1: R \longrightarrow R^*; r \longmapsto (s \longmapsto s\bar{r})) \in L^0(R)$$

acting by the identity.

Proof: Compose the external products given by Proposition 1.9.1 with the L-group morphisms induced by $R\otimes_{\mathbb{Z}} A \longrightarrow A$, defining

$$\begin{cases} \otimes : L^m(R,\rho)\otimes_{\mathbb{Z}} L^n(A,\varepsilon) \xrightarrow{\otimes} L^{m+n}(R\otimes_{\mathbb{Z}} A, \rho\otimes\varepsilon) \longrightarrow L^{m+n}(A,\rho\varepsilon) \\ \otimes : L^m(R,\rho)\otimes_{\mathbb{Z}} L_n(A,\varepsilon) \xrightarrow{\otimes} L_{m+n}(R\otimes_{\mathbb{Z}} A, \rho\otimes\varepsilon) \longrightarrow L_{m+n}(A,\rho\varepsilon) \ . \end{cases}$$

[]

The symmetric Witt group $L^0(R)$ of a commutative ring R (with any involution) is a commutative ring with respect to the internal product $L^0(R)\otimes_{\mathbb{Z}} L^0(R) \longrightarrow L^0(R)$, with unit $(R,1) \in L^0(R)$, and the $\begin{cases}\varepsilon\text{-symmetric}\\ \varepsilon\text{-quadratic}\end{cases}$ L-groups $\begin{cases}L^*(A,\varepsilon)\\ L_*(A,\varepsilon)\end{cases}$ of an R-module A are all $L^0(R)$-modules.

The external L-group products appear in the product formula of Proposition II.8.1 for the $\begin{cases}\text{symmetric}\\ \text{quadratic}\end{cases}$ signature of the cartesian product of an m-dimensional $\begin{cases}\text{geometric Poincaré complex } X\\ \text{normal map } (f,b):M \longrightarrow X\end{cases}$ and an n-dimensional $\begin{cases}\text{geometric Poincaré complex } Y\\ \text{normal map } (g,c):N \longrightarrow Y\end{cases}$

$$\begin{cases}\sigma^*(X\times Y) = \sigma^*(X)\otimes\sigma^*(Y) \in L^{m+n}(\mathbb{Z}[\pi_1(X\times Y)])\\ \sigma_*((f\times g, b\times c):M\times N \longrightarrow X\times Y)\\ \qquad = \sigma_*(f,b)\otimes\sigma_*(g,c) + \sigma^*(X)\otimes\sigma_*(g,c) + \sigma_*(f,b)\otimes\sigma^*(Y)\\ \qquad\qquad \in L_{m+n}(\mathbb{Z}[\pi_1(X\times Y)])\end{cases} ,$$

identifying $\pi_1(X\times Y) = \pi_1(X)\times\pi_1(Y)$ and

$$\mathbb{Z}[\pi_1(X\times Y)] = \mathbb{Z}[\pi_1(X)]\otimes_{\mathbb{Z}}\mathbb{Z}[\pi_1(Y)] .$$

1.10 Change of K-theory

Given a ring with involution A define the duality involution in the reduced $\begin{cases}\text{projective class}\\ \text{torsion}\end{cases}$ group $\begin{cases}\widetilde{K}_0(A) = K_0(A)/K_0(\mathbb{Z})\\ \widetilde{K}_1(A) = K_1(A)/K_1(\mathbb{Z})\end{cases}$ of the underlying ring A

$$\begin{cases} * : \widetilde{K}_0(A) \longrightarrow \widetilde{K}_0(A) \ ; \ x = [P] \longmapsto x^* = [P^*] \\ * : \widetilde{K}_1(A) \longrightarrow \widetilde{K}_1(A) \ ; \\ \qquad x = \tau(f:M \xrightarrow{\sim} N) \longmapsto x^* = \tau(f^*:N^* \xrightarrow{\sim} M^*) \end{cases}$$

with $\begin{cases}P \text{ a f.g. projective A-module}\\ f \in \mathrm{Hom}_A(M,N) \text{ an isomorphism of based f.g. free A-modules}\end{cases}$.

A *-invariant subgroup $X \subseteq \widetilde{K}_m(A)$ $(m = 0,1)$ is a subgroup X of $\widetilde{K}_m(A)$ such that $x^* \in X$ for all $x \in X$.

The projective class of an n-dimensional $\begin{cases}\varepsilon\text{-symmetric}\\ \varepsilon\text{-quadratic}\end{cases}$ complex over A $\begin{cases}(C,\phi)\\ (C,\psi)\end{cases}$ is the projective Euler class of C

$$[C] = \sum_{r=-\infty}^{\infty} (-)^r [C_r] \in \widetilde{K}_0(A) \quad ,$$

which for a Poincaré complex is such that

$$[C]^* = (-)^n [C] \in \widetilde{K}_0(A) \ .$$

The projective class is a homotopy invariant such that $[C] = 0 \in \widetilde{K}_0(A)$ if and only if $\begin{cases}(C,\phi)\\ (C,\psi)\end{cases}$ is homotopy equivalent to a complex such that each C_r $(r \in \mathbb{Z})$ is a f.g. free A-module (of which all but a finite number are 0, by hypothesis).

An $\begin{cases}\varepsilon\text{-symmetric}\\ \varepsilon\text{-quadratic}\end{cases}$ complex over A $\begin{cases}(C,\phi)\\ (C,\psi)\end{cases}$ is <u>based</u> if each C_r $(r \in \mathbb{Z})$ is a based f.g. free A-module.

The <u>torsion</u> of a based n-dimensional $\begin{cases}\varepsilon\text{-symmetric}\\ \varepsilon\text{-quadratic}\end{cases}$ Poincaré complex over A $\begin{cases}(C,\phi \in Q^n(C,\varepsilon))\\ (C,\psi \in Q_n(C,\varepsilon))\end{cases}$ is the torsion of the Poincaré duality chain equivalence

$$\begin{cases}\tau = \tau(\phi_0 : C^{n-*} \longrightarrow C) \in \tilde{K}_1(A)\\ \tau = \tau((1+T_\varepsilon)\psi_0 : C^{n-*} \longrightarrow C) \in \tilde{K}_1(A)\end{cases} ,$$

which is such that

$$\tau^* = (-)^n \tau \in \tilde{K}_1(A) .$$

In dealing with the torsion of based complexes we shall assume that

$$\tau(\varepsilon : A \longrightarrow A) \in X \subseteq \tilde{K}_1(A) ,$$

which is automatically the case if $\varepsilon = \pm 1 \in A$.

As in §I.9, given a $*$-invariant subgroup $X \subseteq \tilde{K}_m(A)$ $(m = 0,1)$ define the <u>intermediate</u> $\begin{cases}\underline{\varepsilon\text{-symmetric}}\\ \underline{\varepsilon\text{-quadratic}}\end{cases}$ <u>L-groups of A</u> $\begin{cases}L^n_X(A,\varepsilon)\\ L^X_n(A,\varepsilon)\end{cases}$ $(n \in \mathbb{Z})$ in the same way as $\begin{cases}L^n(A,\varepsilon)\\ L_n(A,\varepsilon)\end{cases}$ but using algebraic Poincaré complexes <u>with K-theory in X</u>, meaning the projective class if $m = 0$, and the torsion if $m = 1$ (in which case all the complexes are to be based). In particular, for $X = \tilde{K}_0(A)$ we have

$$\begin{cases}L^*_{\tilde{K}_0(A)}(A,\varepsilon) = L^*(A,\varepsilon)\\ L^{\tilde{K}_0(A)}_*(A,\varepsilon) = L_*(A,\varepsilon)\end{cases} .$$

The <u>Tate $\mathbb{Z}_2$-cohomology groups</u> $\hat{H}^*(\mathbb{Z}_2;G)$ of a $\mathbb{Z}[\mathbb{Z}_2]$-module G are defined by

$$\hat{H}^n(\mathbb{Z}_2;G) = \{g \in G \mid Tg = (-)^n g\}/\{h + (-)^n Th \mid h \in G\} \quad (n(\mathrm{mod}\ 2)) .$$

<u>Proposition 1.10.1</u> Given $*$-invariant subgroups $X \subseteq Y \subseteq \tilde{K}_m(A)$ $(m = 0,1)$ there is defined an exact sequence of the intermediate

$\begin{cases} \varepsilon\text{-symmetric} \\ \varepsilon\text{-quadratic} \end{cases}$ L-groups

$$\begin{cases} \dots \longrightarrow L^n_X(A,\varepsilon) \longrightarrow L^n_Y(A,\varepsilon) \xrightarrow{\kappa} \hat{H}^n(\mathbb{Z}_2;Y/X) \longrightarrow L^{n-1}_X(A,\varepsilon) \longrightarrow \dots \\ \dots \longrightarrow L^X_n(A,\varepsilon) \longrightarrow L^Y_n(A,\varepsilon) \xrightarrow{\kappa} \hat{H}^n(\mathbb{Z}_2;Y/X) \longrightarrow L^X_{n-1}(A,\varepsilon) \longrightarrow \dots \end{cases}$$

$$(n \in \mathbb{Z})$$

with $T \in \mathbb{Z}_2$ acting on Y/X by the duality involution, with κ the map associating to an algebraic Poincaré complex the Tate $\mathbb{Z}_2$-cohomology class of its K-theory.

<u>Proof</u>: See Proposition I.9.1.

[]

As in §I.9 we introduce the following terminology for the intermediate L-groups

$$\begin{cases} L^n_X(A,\varepsilon) = U^n_X(A,\varepsilon) \\ L^X_n(A,\varepsilon) = U^X_n(A,\varepsilon) \end{cases} \text{for } X \subseteq \tilde{K}_0(A) , \quad \begin{cases} L^n_X(A,\varepsilon) = V^n_X(A,\varepsilon) \\ L^X_n(A,\varepsilon) = V^X_n(A,\varepsilon) \end{cases} \text{for } X \subseteq \tilde{K}_1(A)$$

$$\begin{cases} U^n_{\tilde{K}_0(A)}(A,\varepsilon) = U^n(A,\varepsilon) \quad (= L^n(A,\varepsilon)) \\ U^{\tilde{K}_0(A)}_n(A,\varepsilon) = U_n(A,\varepsilon) \quad (= L_n(A,\varepsilon)) \end{cases}$$

$$\begin{cases} V^n_{\tilde{K}_1(A)}(A,\varepsilon) = U^n_{\{0\}}(A,\varepsilon) = V^n(A,\varepsilon) \\ V^{\tilde{K}_1(A)}_n(A,\varepsilon) = U^{\{0\}}_n(A,\varepsilon) = V_n(A,\varepsilon) \end{cases} .$$

For $\varepsilon = 1 \in A$ the notation is contracted in the usual fashion, for example

$$L_X^*(A,1) = L_X^*(A) \quad .$$

The original surgery obstruction groups of Wall [4] are the simple quadratic L-groups of a group ring $\mathbb{Z}[\pi]$ with a w-twisted involution

$$L_*^s(\pi,w) = V_*^{\{\pi\}}(\mathbb{Z}[\pi]) \quad .$$

The L^h-groups of Shaneson [1] are the free quadratic L-groups

$$L_*^h(\pi,w) = V_*(\mathbb{Z}[\pi]) \ ,$$

and the Rothenberg exact sequence

$$\cdots \longrightarrow L_n^s(\pi,w) \longrightarrow L_n^h(\pi,w) \longrightarrow \hat{H}^n(\mathbb{Z}_2;Wh(\pi)) \longrightarrow L_{n-1}^s(\pi,w) \longrightarrow \cdots$$

is the special case of the exact sequence of Proposition 1.10.1 for the intermediate quadratic L-groups associated to

$$X = \{\pi\} \subseteq Y = \tilde{K}_1(\mathbb{Z}[\pi]) \quad ,$$

since $Y/X = Wh(\pi)$ is the Whitehead group of π. The $\begin{cases} \text{simple} \\ \text{finite} \end{cases}$ L-groups $\begin{cases} L_*^s(\pi,w) \\ L_*^h(\pi,w) \end{cases}$ are the obstruction groups for surgery to $\begin{cases} \text{simple} \\ - \end{cases}$ homotopy equivalence on topological normal maps $(f,b):M \longrightarrow X$ from compact manifolds M to $\begin{cases} \text{simple} \\ \text{finite} \end{cases}$ geometric Poincaré complexes X. The projective quadratic L-groups originally introduced by Novikov [1]

$$L_*^p(\pi,w) = U_*(\mathbb{Z}[\pi])$$

have two distinct geometric interpretations: either as

the obstruction groups for surgery to proper homotopy equivalence on normal maps from paracompact manifolds to infinite locally finite CW complexes with the Poincaré duality of such manifolds, as in Maumary [1] and Taylor [1], or else as the obstruction groups for surgery to homotopy equivalence on topological normal maps from compact manifolds to finitely dominated geometric Poincaré complexes (i.e. Poincaré complexes in the sense of Wall [3]) as in Pedersen and Ranicki [1]. See Hambleton [1] and Taylor and Williams [3] for applications of projective L-theory to the description of the surgery obstructions of topological normal maps of closed manifolds with finite fundamental groups.

<u>Proposition 1.10.2</u> The surgery obstruction of an n-dimensional topological normal map $(f,b):M \longrightarrow X$ with X

$$\begin{cases} \text{simple} \\ \text{finite} \\ \text{finitely dominated} \end{cases} \text{ is such that } \begin{cases} \sigma_*(f,b) = 0 \in L_n^s(\pi_1(X),w(X)) \\ \sigma_*(f,b) = 0 \in L_n^h(\pi_1(X),w(X)) \\ \sigma_*(f,b) = 0 \in L_n^p(\pi_1(X),w(X)) \end{cases}$$

if (and for $n \geqslant 5$ only if) $\begin{cases} (f,b):M \longrightarrow X \\ (f,b):M \longrightarrow X \\ (f,b)\times 1:M\times S^1 \longrightarrow X\times S^1 \end{cases}$ is normal

bordant to a $\begin{cases} \text{simple} \\ - \\ - \end{cases}$ homotopy equivalence.

[]

Given a topological normal map of n-dimensional pairs

$$((f,b),(\partial f,\partial b)) : (M,\partial M) \longrightarrow (X,\partial X)$$

such that $\partial f:\partial M \longrightarrow \partial X$ is a homotopy equivalence there is defined a <u>rel∂ surgery obstruction</u> $\sigma_*(f,b) \in L_n(\mathbb{Z}[\pi_1(X)])$ such that the analogue of Proposition 1.10.2 holds for topological normal bordism rel $(\partial f,\partial b)$. By the realization theorems of Wall [4,§§5,6] every element of $L_n(\mathbb{Z}[\pi])$ $(n \geqslant 6)$ for a finitely presented group π is the rel∂ surgery obstruction $\sigma_*(f,b)$ of such a topological normal map $(f,b):(M,\partial M) \longrightarrow (X,\partial X)$.

§2. Relative L-theory

Bass [1] related the projective class group $K_0(A)$ of a ring A to the torsion group $K_1(A) = GL(A)/E(A)$, associating to a morphism of rings

$$f : A \longrightarrow B$$

a change of rings exact sequence

$$K_1(A) \xrightarrow{f} K_1(B) \longrightarrow K_1(f) \longrightarrow K_0(A) \xrightarrow{f} K_0(B)$$

with the relative K-group $K_1(f)$ defined to be the Grothendieck group of triples (P,Q,h) consisting of f.g. projective A-modules P,Q and an isomorphism $h \in \mathrm{Hom}_B(B\otimes_A P, B\otimes_A Q)$.

The sequence extends on the right to the lower K-groups $K_n(A)$ $(n \leq -1)$ of Bass [2,XII] and on the left to the higher K-groups $K_n(A)$ $(n \geq 2)$ of Quillen [1],[2] (with $K_2(A)$ the K_2-group of Milnor [4])

$$\ldots \longrightarrow K_n(A) \xrightarrow{f} K_n(B) \longrightarrow K_n(f) \longrightarrow K_{n-1}(A) \longrightarrow \ldots \quad (n \in \mathbb{Z}) \quad .$$

Gersten [2] constructed a spectrum $\underline{\mathbb{K}}(A)$ such that

$$K_n(A) = \pi_n(\underline{\mathbb{K}}(A)) \quad (n \in \mathbb{Z}) \quad ,$$

so that the relative K-groups $K_*(f)$ can be defined to be the relative homotopy groups of the induced map of spectra

$$K_n(f) = \pi_n(f:\underline{\mathbb{K}}(A) \longrightarrow \underline{\mathbb{K}}(B)) \quad (n \in \mathbb{Z}) \quad .$$

Wall [4] used the geometric interpretation of the surgery obstruction groups $L_*(\pi)$ for a finitely presented group π as bordism groups of normal maps to geometrically define the relative L-groups $L_*(f)$ of a morphism $f:\pi \longrightarrow \pi'$ of such groups as relative bordism groups, fitting into an exact sequence

$$\ldots \longrightarrow L_n(\pi) \xrightarrow{f} L_n(\pi') \longrightarrow L_n(f) \longrightarrow L_{n-1}(\pi) \longrightarrow \ldots \quad (n \pmod 4)) .$$

Wall [4,§7] also gave an algebraic definition of the odd-dimensional relative L-groups $L_{2i+1}(f)$, as the Witt groups of pairs

(non-singular $(-)^i$-quadratic form over $\mathbb{Z}[\pi]$ (M,ψ),

lagrangian L of the induced form over $\mathbb{Z}[\pi']$ $\mathbb{Z}[\pi']\otimes_{\mathbb{Z}[\pi]}(M,\psi)$) .

Sharpe [2] gave an algebraic definition of the even-dimensional relative L-groups $L_{2i}(f)$ (which however only applies to the simple L-groups, since it is based on the unitary Steinberg group relations of Sharpe [1]).

Following the definition in §2.1 of algebraic Poincaré triads we shall define in §2.2 the relative $\begin{cases}\varepsilon\text{-symmetric}\\ \varepsilon\text{-quadratic}\end{cases}$ L-groups $\begin{cases}L^n(f,\varepsilon)\\ L_n(f,\varepsilon)\end{cases}$ $(n\in\mathbb{Z})$ of a morphism of rings with involution $f:A\longrightarrow B$, to fit into an exact sequence

$$\begin{cases}\dots\longrightarrow L^n(A,\varepsilon)\xrightarrow{f} L^n(B,\varepsilon)\longrightarrow L^n(f,\varepsilon)\longrightarrow L^{n-1}(A,\varepsilon)\longrightarrow\dots\\ \dots\longrightarrow L_n(A,\varepsilon)\xrightarrow{f} L_n(B,\varepsilon)\longrightarrow L_n(f,\varepsilon)\longrightarrow L_{n-1}(A,\varepsilon)\longrightarrow\dots\end{cases}\quad (n\in\mathbb{Z}) .$$

For $n\geqslant 1$ $\begin{cases}L^n(f,\varepsilon)\\ L_n(f,\varepsilon)\end{cases}$ is defined to be the relative cobordism group of pairs

((n-1)-dimensional $\begin{cases}\varepsilon\text{-symmetric}\\ \varepsilon\text{-quadratic}\end{cases}$ Poincaré complex over A $\begin{cases}(C,\phi)\\ (C,\psi)\end{cases}$,

n-dimensional $\begin{cases}\varepsilon\text{-symmetric}\\ \varepsilon\text{-quadratic}\end{cases}$ Poincaré pair over B

$\begin{cases}(g:B\otimes_A C\longrightarrow D,(\delta\phi,1\otimes\phi))\\ (g:B\otimes_A C\longrightarrow D,(\delta\psi,1\otimes\psi))\end{cases}$ with boundary $\begin{cases}B\otimes_A(C,\phi)\\ B\otimes_A(C,\psi)\end{cases}$)

in evident analogy with the definition of relative geometric cobordism groups. For $n \leqslant 0$ $\begin{cases} L^n(f,\varepsilon) \\ L_n(f,\varepsilon) \end{cases}$ is defined in terms of forms and formations. (In Ranicki [12] there will be defined spectra $\begin{cases} \underline{\mathbb{L}}^0(A,\varepsilon) \\ \underline{\mathbb{L}}_0(A,\varepsilon) \end{cases}$ such that

$$\begin{cases} \pi_*(\underline{\mathbb{L}}^0(A,\varepsilon)) = L^*(A,\varepsilon) \\ \pi_*(\underline{\mathbb{L}}_0(A,\varepsilon)) = L_*(A,\varepsilon) \end{cases}$$

using algebraic Poincaré n-ads, allowing the relative L-groups to be defined as relative homotopy groups

$$\begin{cases} L^*(f,\varepsilon) = \pi_*(f:\underline{\mathbb{L}}^0(A,\varepsilon) \longrightarrow \underline{\mathbb{L}}^0(B,\varepsilon)) \\ L_*(f,\varepsilon) = \pi_*(f:\underline{\mathbb{L}}_0(A,\varepsilon) \longrightarrow \underline{\mathbb{L}}_0(B,\varepsilon)) \end{cases} .$$

See Ranicki [7] for a brief discussion of the algebraic $\mathbb{L}$-spectra. The ε-quadratic $\mathbb{L}$-spectrum $\underline{\mathbb{L}}_0(A,\varepsilon)$ may be defined using forms and formations, as was in fact done in Ranicki [5]). In §2.3 the construction is extended to some of the other types of relative L-groups arising in topology, such as the ε-hyperquadratic L-groups $\hat{L}^n(A,\varepsilon)$ $(n \in \mathbb{Z})$ which fit into the exact sequence

$$\dots \longrightarrow L_n(A,\varepsilon) \xrightarrow{1+T_\varepsilon} L^n(A,\varepsilon) \xrightarrow{J} \hat{L}^n(A,\varepsilon) \xrightarrow{H} L_{n-1}(A,\varepsilon) \longrightarrow \dots$$
$$(n \in \mathbb{Z}) .$$

In §2.4 we shall define the Γ-groups $\begin{cases} \Gamma^n(f:A \longrightarrow B,\varepsilon) \\ \Gamma_n(f:A \longrightarrow B,\varepsilon) \end{cases}$ $(n \geqslant 0)$ of cobordism classes of n-dimensional $\begin{cases} \varepsilon\text{-symmetric} \\ \varepsilon\text{-quadratic} \end{cases}$ complexes over A which become Poincaré over B, for some morphism of rings with involution $f:A \longrightarrow B$. The quadratic Γ-groups $\Gamma_*(f) \equiv \Gamma_*(f,1)$

will be identified in §2.5 with the homology surgery obstruction groups originally defined by Cappell and Shaneson [1]. (The related homology surgery theory will be discussed in §7.7 below).

We shall also define lower Γ-groups $\left\{\begin{array}{l}\Gamma^n(f,\varepsilon)\\ \Gamma_n(f,\varepsilon)\end{array}\right.$ $(n \leqslant -1)$, using forms and formations. Given a commutative square of rings with involution

$$\begin{array}{ccc} A & \longrightarrow & A' \\ f\downarrow & F & \downarrow f' \\ B & \longrightarrow & B' \end{array}$$

there will also be defined relative Γ-groups $\left\{\begin{array}{l}\Gamma^n(F,\varepsilon)\\ \Gamma_n(F,\varepsilon)\end{array}\right.$ $(n \in \mathbb{Z})$ to fit into a long exact sequence

$$\left\{\begin{array}{l} \dots \longrightarrow \Gamma^n(f,\varepsilon) \longrightarrow \Gamma^n(f',\varepsilon) \longrightarrow \Gamma^n(F,\varepsilon) \longrightarrow \Gamma^{n-1}(f,\varepsilon) \longrightarrow \dots \\ \dots \longrightarrow \Gamma_n(f,\varepsilon) \longrightarrow \Gamma_n(f',\varepsilon) \longrightarrow \Gamma_n(F,\varepsilon) \longrightarrow \Gamma_{n-1}(f,\varepsilon) \longrightarrow \dots \ . \end{array}\right.$$

The relative Γ-groups $\left\{\begin{array}{l}\Gamma^*(F,\varepsilon)\\ \Gamma_*(F,\varepsilon)\end{array}\right.$ in the special case $1 : A \longrightarrow A' = A$ will be expressed as the cobordism groups of the $\left\{\begin{array}{l}\varepsilon\text{-symmetric}\\ \varepsilon\text{-quadratic}\end{array}\right.$ complexes over A which become Poincaré over B and contractible over B'. This expression will then be used in §3 for the commutative square

$$\begin{array}{ccc} A & \xrightarrow{\ 1\ } & A \\ 1\downarrow & & \downarrow \\ A & \longrightarrow & S^{-1}A \end{array}$$

associated to a localization map $A \longrightarrow S^{-1}A$ inverting a multiplicative subset $S \subset A$, allowing the relative L-groups

$\begin{cases} L^*(A \longrightarrow S^{-1}A, \varepsilon) \\ L_*(A \longrightarrow S^{-1}A, \varepsilon) \end{cases}$ (of the appropriate intermediate type) to be identified with the L-groups $\begin{cases} L^*(A,S,\varepsilon) \\ L_*(A,S,\varepsilon) \end{cases}$ of $\begin{cases} \varepsilon\text{-symmetric} \\ \varepsilon\text{-quadratic} \end{cases}$ Poincaré complexes over A which become contractible over $S^{-1}A$.

2.1 Algebraic Poincaré triads

An (n+2)-dimensional $\left\{\begin{array}{l}\varepsilon\text{-symmetric}\\ \varepsilon\text{-quadratic}\end{array}\right.$ triad over A $\left\{\begin{array}{l}(\Gamma,\Phi)\\ (\Gamma,\Psi)\end{array}\right.$ $(n \geqslant 0)$

is a triad of finite-dimensional A-module chain complexes

$$\Gamma : \begin{array}{ccc} C & \xrightarrow{\ f\ } & D \\ {\scriptstyle f'}\big\downarrow & \searrow^{h} & \big\downarrow{\scriptstyle g} \\ D' & \xrightarrow[\ g'\ ]{} & C' \end{array}$$

such that C is n-dimensional, D and D' are (n+1)-dimensional, C' is (n+2)-dimensional, together with an element

$$\left\{\begin{array}{l}\Phi = (\phi',\delta\phi',\delta\phi,\phi) \in Q^{n+2}(\Gamma,\varepsilon)\\ \Psi = (\psi',\delta\psi',\delta\psi,\psi) \in Q_{n+2}(\Gamma,\varepsilon)\end{array}\right.$$

of the $\left\{\begin{array}{l}\varepsilon\text{-symmetric}\\ \varepsilon\text{-quadratic}\end{array}\right.$ triad Q-group defined in §1.3. Such a triad is Poincaré if

i) the (n+1)-dimensional $\left\{\begin{array}{l}\varepsilon\text{-symmetric}\\ \varepsilon\text{-quadratic}\end{array}\right.$ pairs over A

$$\left\{\begin{array}{l}(f:C \longrightarrow D,\ (\delta\phi,\phi) \in Q^{n+1}(f,\varepsilon))\\ (f:C \longrightarrow D,\ (\delta\psi,\psi) \in Q_{n+1}(f,\varepsilon))\end{array}\right. ,$$

$$\left\{\begin{array}{l}(f':C \longrightarrow D',\ (\delta\phi',\phi) \in Q^{n+1}(f',\varepsilon))\\ (f':C \longrightarrow D',\ (\delta\psi',\psi) \in Q_{n+1}(f',\varepsilon))\end{array}\right.$$

are Poincaré

ii) the A-module chain map

$$\left\{\begin{array}{l}\Phi_0 : C'^{\,n+2-*} \longrightarrow C(\Gamma)\\ (1+T_\varepsilon)\Psi_0 : C'^{\,n+2-*} \longrightarrow C(\Gamma)\end{array}\right.$$

defined by

$$\left\{\begin{array}{l} \Phi_0 = \begin{pmatrix} \phi_0 \\ (-)^{n-r}\delta\phi_0 g^* \\ f'\phi_0 h^* + (-)^{n-r}\delta\phi_0' g_0'^* \\ \phi_0 f^* g^* \end{pmatrix} \\ \qquad : C'^{n+2-r} \longrightarrow C(\Gamma)_r = C_r' \oplus D_{r-1} \oplus D_{r-1}' \oplus C_{r-2} \\ (1+T_\varepsilon)\Psi_0 = \begin{pmatrix} (1+T_\varepsilon)\psi_0 \\ (-)^{n-r}(1+T_\varepsilon)\delta\psi_0 g^* \\ f'(1+T_\varepsilon)\psi_0 h^* + (-)^{n-r}(1+T_\varepsilon)\delta\psi_0' g'^* \\ (1+T_\varepsilon)\psi_0 f^* g^* \end{pmatrix} \\ \qquad : C'^{n+2-r} \longrightarrow C(\Gamma)_r = C_r' \oplus D_{r-1} \oplus D_{r-1}' \oplus C_{r-2} \end{array}\right.$$

is a chain equivalence.

Proposition 2.1.1 There is a natural one-one correspondence between quadruples

(n-dimensional $\left\{\begin{array}{l}\varepsilon\text{-symmetric} \\ \varepsilon\text{-quadratic}\end{array}\right.$ Poincaré complex over A $\left\{\begin{array}{l}(C,\phi) \\ (C,\psi)\end{array}\right.$,

(n+1)-dimensional $\left\{\begin{array}{l}\varepsilon\text{-symmetric} \\ \varepsilon\text{-quadratic}\end{array}\right.$ Poincaré pairs over A

$\left\{\begin{array}{l}(f:C \longrightarrow D,(\delta\phi,\phi)) \\ (f:C \longrightarrow D,(\delta\psi,\psi))\end{array}\right.$, $\left\{\begin{array}{l}(f':C \longrightarrow D',(\delta\phi',\phi)) \\ (f':C \longrightarrow D',(\delta\psi',\psi))\end{array}\right.$ bounding $\left\{\begin{array}{l}(C,\phi) \\ (C,\psi)\end{array}\right.$,

(n+2)-dimensional $\left\{\begin{array}{l}\varepsilon\text{-symmetric} \\ \varepsilon\text{-quadratic}\end{array}\right.$ Poincaré pair over A

$\left\{\begin{array}{l}(e:D \cup_C D' \longrightarrow C',(\phi',\delta\phi \cup_\phi \delta\phi')) \\ (e:D \cup_C D' \longrightarrow C',(\psi',\delta\psi \cup_\psi \delta\psi'))\end{array}\right.$ with boundary the union

$\left\{\begin{array}{l}(D \cup_C D',\delta\phi \cup_\phi \delta\phi') \\ (D \cup_C D',\delta\psi \cup_\psi \delta\psi')\end{array}\right.$)

and (n+2)-dimensional $\begin{cases} \varepsilon\text{-symmetric} \\ \varepsilon\text{-quadratic} \end{cases}$ Poincaré triads over A $\begin{cases} (\Gamma,\Phi) \\ (\Gamma,\Psi) \end{cases}$, under which

$$\Gamma : \quad \begin{array}{ccc} C & \xrightarrow{\ f\ } & D \\ {\scriptstyle f'}\big\downarrow & \searrow^{h} & \big\downarrow{\scriptstyle g'} \\ D' & \xrightarrow[\ g\ ]{} & C' \end{array}$$

$$\begin{cases} \Phi = (\phi',\delta\phi',\delta\phi,\phi) \in Q^{n+2}(\Gamma,\varepsilon) \\ \Psi = (\psi',\delta\psi',\delta\psi,\psi) \in Q_{n+2}(\Gamma,\varepsilon) \end{cases}$$

$$e = (g \ \ (-)^{r-1}h \ \ -g')$$

$$: (D \cup_C D')_r = D_r \oplus C_{r-1} \oplus D'_r \longrightarrow C'_r \quad .$$

[]

A <u>cobordism</u> of (n+1)-dimensional $\begin{cases} \varepsilon\text{-symmetric} \\ \varepsilon\text{-quadratic} \end{cases}$ Poincaré pairs over A $\begin{cases} (f:C \longrightarrow D,(\delta\phi,\phi)) \\ (f:C \longrightarrow D,(\delta\psi,\psi)) \end{cases}$, $\begin{cases} (f':C' \longrightarrow D',(\delta\phi',\phi')) \\ (f':C' \longrightarrow D',(\delta\psi',\psi')) \end{cases}$ is an (n+2)-dimensional $\begin{cases} \varepsilon\text{-symmetric} \\ \varepsilon\text{-quadratic} \end{cases}$ Poincaré triad over A $\begin{cases} (\Gamma,\Phi) \\ (\Gamma,\Psi) \end{cases}$ such that Γ is defined by

$$\begin{array}{ccc} C\oplus C' & \xrightarrow{\begin{pmatrix} f & 0 \\ 0 & f' \end{pmatrix}} & D\oplus D' \\ {\scriptstyle (g\ g')}\big\downarrow & \searrow^{(k\ k')} & \big\downarrow{\scriptstyle (h\ h')} \\ \delta C & \xrightarrow[\ \delta f\ ]{} & \delta D \end{array}$$

and

$$\begin{cases} \Phi = (\delta\nu, \nu, \delta\phi\oplus -\delta\phi', \phi\oplus -\phi') \in Q^{n+2}(\Gamma,\varepsilon) \\ \Psi = (\delta\chi, \chi, \delta\psi\oplus -\delta\psi', \psi\oplus -\psi') \in Q_{n+2}(\Gamma,\varepsilon) \ . \end{cases}$$

As it stands this cobordism relation is trivial (i.e. with a single equivalence class), since every $\begin{cases}\varepsilon\text{-symmetric}\\ \varepsilon\text{-quadratic}\end{cases}$ Poincaré pair $\begin{cases}(f:C \longrightarrow D, (\delta\phi,\phi))\\ (f:C \longrightarrow D, (\delta\psi,\psi))\end{cases}$ is cobordant to 0 by the $\begin{cases}\varepsilon\text{-symmetric}\\ \varepsilon\text{-quadratic}\end{cases}$ Poincaré triad $\begin{cases}(\Gamma, (0, \delta\phi, \delta\phi, \phi))\\ (\Gamma, (0, \delta\psi, \delta\psi, \psi))\end{cases}$ with Γ defined by

$$\Gamma \ : \ \begin{array}{ccc} C & \xrightarrow{\ f\ } & D \\ {\scriptstyle f}\big\downarrow & \searrow^{0} & \big\downarrow{\scriptstyle 1} \\ D & \xrightarrow[\ ]{\ 1\ } & D \end{array} \quad .$$

However, in the applications we shall be considering the cobordism of algebraic Poincaré pairs in which the boundary is restricted in some way. The above null-cobordism will not in general be restricted in that sense, so that the restricted cobordism need not be trivial. In verifying that such restricted cobordisms are in fact equivalence relations we shall make use of the following algebraic glueing operation, which is an evident generalization of the union of algebraic Poincaré cobordisms of §1.5. (The glueing is required for the verification of transitivity; reflexitivity and symmetry are clear).

Let

$$\begin{cases}(f:C \longrightarrow D,(\delta\phi,\phi)),(f':C' \longrightarrow D',(\delta\phi',\phi')),(f'':C'' \longrightarrow D'',(\delta\phi'',\phi'')) \\ (f:C \longrightarrow D,(\delta\psi,\psi)),(f':C' \longrightarrow D',(\delta\psi',\psi')),(f'':C'' \longrightarrow D'',(\delta\psi'',\psi''))\end{cases}$$

be (n+1)-dimensional $\begin{cases}\varepsilon\text{-symmetric}\\ \varepsilon\text{-quadratic}\end{cases}$ Poincaré pairs over A.

The <u>union</u> of adjoining $\begin{cases}\varepsilon\text{-symmetric}\\ \varepsilon\text{-quadratic}\end{cases}$ Poincaré cobordisms of pairs

$$\begin{cases}(\Gamma,\Phi = (\delta\nu,\nu,\delta\phi\oplus-\delta\phi',\phi\oplus-\phi')),\ (\Gamma',\Phi' = (\delta\nu',\nu',\delta\phi'\oplus-\delta\phi'',\phi'\oplus-\phi'')) \\ (\Gamma,\Psi = (\delta\chi,\chi,\delta\psi\oplus-\delta\psi',\psi\oplus-\psi')),\ (\Gamma',\Psi' = (\delta\chi',\chi',\delta\psi'\oplus-\delta\psi'',\psi'\oplus-\psi''))\end{cases}$$

with

$$\Gamma : \quad \begin{array}{ccc} C\oplus C' & \xrightarrow{f\oplus f'} & D\oplus D' \\ {\scriptstyle (g\ g')}\big\downarrow & \overset{(k\ k')}{\searrow} & \big\downarrow{\scriptstyle (h\ h')} \\ \delta C & \xrightarrow{\delta f} & \delta D \end{array}$$

$$\Gamma' : \quad \begin{array}{ccc} C'\oplus C'' & \xrightarrow{f'\oplus f''} & D'\oplus D'' \\ {\scriptstyle (\tilde{g}'\ g'')}\big\downarrow & \overset{(k'\ k'')}{\searrow} & \big\downarrow{\scriptstyle (\tilde{h}'\ h'')} \\ \delta C' & \xrightarrow{\delta f'} & \delta D' \end{array}$$

is the cobordism

$$\begin{cases}(\Gamma'',\Phi'' = (\delta\nu'',\nu'',\delta\phi\oplus-\delta\phi'',\phi\oplus-\phi'')) = (\Gamma\cup_f\Gamma',\Phi\cup_{(\delta\phi',\phi')}\Phi') \\ (\Gamma'',\Psi'' = (\delta\chi'',\chi'',\delta\psi\oplus-\delta\psi'',\psi\oplus-\psi'')) = (\Gamma\cup_f\Gamma',\Psi\cup_{(\delta\psi',\psi')}\Psi')\end{cases}$$

with

$$\Gamma'' : \quad \begin{array}{ccc} C\oplus C'' & \xrightarrow{f\oplus f''} & D\oplus D'' \\ {\scriptstyle (\tilde{g}\ \tilde{g}'')}\big\downarrow & \overset{(\tilde{k}\ \tilde{k}'')}{\searrow} & \big\downarrow{\scriptstyle (\tilde{h}\ \tilde{h}'')} \\ \delta C'' & \xrightarrow{\delta f''} & \delta D'' \end{array}$$

defined by

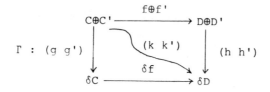

$$d_{\delta C''} = \begin{pmatrix} d_{\delta C} & (-)^{r-1}g' & 0 \\ 0 & d_{C'} & 0 \\ 0 & (-)^{r-1}\tilde{g}' & d_{\delta C'} \end{pmatrix}$$

$$: \delta C''_r = \delta C_r \oplus C'_{r-1} \oplus \delta C'_r \longrightarrow \delta C''_{r-1} = \delta C_{r-1} \oplus C'_{r-2} \oplus \delta C'_{r-1}$$

$$d_{\delta D''} = \begin{pmatrix} d_{\delta D} & (-)^{r-1}h' & 0 \\ 0 & d_{D'} & 0 \\ 0 & (-)^{r-1}\tilde{h}' & d_{\delta D'} \end{pmatrix}$$

$$: \delta D''_r = \delta D_r \oplus D'_{r-1} \oplus \delta D'_r \longrightarrow \delta D''_{r-1} = \delta D_{r-1} \oplus D'_{r-2} \oplus \delta D'_{r-1}$$

$$(\tilde{g}\ \tilde{g}'') = \begin{pmatrix} g & 0 \\ 0 & 0 \\ 0 & g'' \end{pmatrix} : C_r \oplus C''_r \longrightarrow \delta C''_r = \delta C_r \oplus C'_{r-1} \oplus \delta C'_r$$

$$(\tilde{h}\ \tilde{h}'') = \begin{pmatrix} h & 0 \\ 0 & 0 \\ 0 & h'' \end{pmatrix} : D_r \oplus D''_r \longrightarrow \delta D''_r = \delta D_r \oplus D'_{r-1} \oplus \delta D'_r$$

$$(\tilde{k}\ \tilde{k}'') = \begin{pmatrix} k & 0 \\ 0 & 0 \\ 0 & k'' \end{pmatrix} : C_r \oplus C''_r \longrightarrow \delta D''_{r+1} = \delta D_{r+1} \oplus D'_r \oplus \delta D'_r$$

$$\left\{ \begin{array}{l} \nu''_s = \begin{pmatrix} \nu_s & 0 & 0 \\ (-)^{n-r}\phi'_s g'^* & (-)^{n-r+s+1}T_\varepsilon \phi'_{s-1} & 0 \\ 0 & (-)^s g'\phi'_s & \nu'_s \end{pmatrix} \\ \quad : \delta C''^{n-r+s+1} = \delta C^{n-r+s+1} \oplus C'^{n-r+s} \oplus \delta C'^{n-r+s+1} \\ \qquad\qquad \longrightarrow \delta C''_r = \delta C_r \oplus C'_{r-1} \oplus \delta C'_r \end{array} \right.$$

$$(s \geqslant 0,\ \phi'_{-1} = 0)$$

$$\left.\begin{array}{l} \chi''_s = \begin{pmatrix} \chi_s & 0 & 0 \\ (-)^{n-r+1}\psi'_s g'^* & (-)^{n-r-s}T_\varepsilon \psi'_{s+1} & 0 \\ 0 & (-)^s g'\psi'_s & \chi'_s \end{pmatrix} \\ \quad : \delta C''^{n-r-s+1} = \delta C^{n-r-s+1} \oplus C'^{n-r-s} \oplus \delta C'^{n-r-s+1} \end{array}\right\}$$

$$\longrightarrow \delta C''_r = \delta C_r \oplus C'_{r-1} \oplus \delta C'_r$$

$$\left\{\begin{array}{l} \delta\nu''_s = \begin{pmatrix} \delta\nu_s & 0 & 0 \\ (-)^{n-r+1}\nu'_s h'^* & (-)^{n-r+s+2}T_\varepsilon \nu'_{s-1} & 0 \\ 0 & (-)^s \tilde{h}'\nu'_s & \delta\nu'_s \end{pmatrix} \\ \quad : \delta D''^{n-r+s+2} = \delta D^{n-r+s+2} \oplus D'^{n-r+s+1} \oplus \delta D'^{n-r+s+1} \\ \qquad \longrightarrow \delta D''_r = \delta D_r \oplus D'_{r-1} \oplus \delta D'_r \\ \delta\chi''_s = \begin{pmatrix} \delta\chi_s & 0 & 0 \\ (-)^{n-r+1}\chi'_s h'^* & (-)^{n-r+s+1}T_\varepsilon \chi'_{s+1} & 0 \\ 0 & (-)^s \tilde{h}'\chi'_s & \delta\chi'_s \end{pmatrix} \\ \quad : \delta D''^{n-r-s+2} = \delta D^{n-r-s+2} \oplus D'^{n-r-s+1} \oplus \delta D'^{n-r-s+2} \end{array}\right. \qquad (s \geqslant 0)$$

$$\longrightarrow \delta D''_r = \delta D_r \oplus D'_{r-1} \oplus \delta D'_r$$

$$\delta f'' = \begin{pmatrix} \delta f & (-)^{r-1}k' & 0 \\ 0 & f' & 0 \\ 0 & (-)^{r-1}k' & \delta f' \end{pmatrix}$$

$$: \delta C''_r = \delta C_r \oplus C'_{r-1} \oplus \delta C'_r \longrightarrow \delta D''_r = \delta D_r \oplus D'_{r-1} \oplus \delta D'_r \quad .$$

2.2 Change of rings

Let A,B be rings with involution.

A morphism

$$f : A \longrightarrow B$$

is a function such that

$$f(a+a') = f(a)+f(a') \ , \ f(aa') = f(a)f(a') \ ,$$

$$f(a) = f(a) \ , \ f(1) = 1 \in B \qquad\qquad (a,a' \in A) \ .$$

Regard B as a (B,A)-bimodule by

$$B \times B \times A \longrightarrow B \ ; \ (b,x,a) \longmapsto b.x.f(a) \ .$$

An A-module M induces a B-module $B \otimes_A M$, with $B \otimes_A B = B$. If N is another A-module there is defined a morphism of abelian groups

$$\mathrm{Hom}_A(M,N) \longrightarrow \mathrm{Hom}_B(B\otimes_A M, B\otimes_A N) \ ; \ g \longmapsto (1\otimes g : b\otimes x \longmapsto b\otimes g(x)) \ .$$

If M is a f.g. projective A-module then $B\otimes_A M$ is a f.g. projective B-module, and there is defined a natural B-module isomorphism

$$B\otimes_A(M^*) \longrightarrow (B\otimes_A M)^* \ ;$$

$$b\otimes g \longmapsto (c\otimes y \longmapsto c.f(g(y)).\bar{b}) \quad ,$$

allowing us to write

$$B\otimes_A M^* = B\otimes_A(M^*) = (B\otimes_A M)^* \ .$$

If C is an (n-dimensional) A-module chain complex then $B\otimes_A C$ is an (n-dimensional) B-module chain complex, and the $\mathbb{Z}$-module chain map

$$\begin{cases} C \longrightarrow B\otimes_A C \ ; \ x \longmapsto 1\otimes x \\ C^* \longrightarrow B\otimes_A C^* \ ; \ g \longmapsto (b\otimes x \longmapsto b.f(g(x))) \end{cases}$$

induces a change of rings $\mathbb{Z}$-module morphisms in

$\begin{cases} \text{homology} \\ \text{cohomology} \end{cases}$

$$\begin{cases} f : H_*(C) \longrightarrow H_*(B\otimes_A C) \\ f : H^*(C) \longrightarrow H^*(B\otimes_A C) \ . \end{cases}$$

Let $\varepsilon_A \in A$, $\varepsilon_B \in B$ be central units such that

$$\bar{\varepsilon}_A = \varepsilon_A^{-1} \in A\ ,\ \bar{\varepsilon}_B = \varepsilon_B^{-1} \in B\ ,\ f(\varepsilon_A) = \varepsilon_B \in B\ .$$

Given a finite-dimensional A-module chain complex C let $T \in \mathbb{Z}_2$ act on $\mathrm{Hom}_A(C^*,C)$ by the ε_A-duality involution T_{ε_A} and on $\mathrm{Hom}_B(B\otimes_A C^*, B\otimes_A C)$ by the ε_B-duality involution T_{ε_B}.

The $\mathbb{Z}[\mathbb{Z}_2]$-module chain map

$$f : \mathrm{Hom}_A(C^*,C) \longrightarrow \mathrm{Hom}_B(B\otimes_A C^*, B\otimes_A C)\ ;$$

$$\phi \longmapsto (b\otimes x \longmapsto (c\otimes y \longmapsto c.f(\overline{x(\phi(y))}).\bar{b}))$$

$$(b,c \in B,\ x,y \in C^*,\ B\otimes_A C = (B\otimes_A C^*)^*)$$

induces a natural transformation of the long exact sequences of Q-groups given by Proposition 1.1.2

$$\begin{array}{ccccccccc} \dots \longrightarrow & \hat{Q}^{n+1}(C,\varepsilon) & \xrightarrow{H} & Q_n(C,\varepsilon) & \xrightarrow{1+T_\varepsilon} & Q^n(C,\varepsilon) & \xrightarrow{J} & \hat{Q}^n(C,\varepsilon) & \longrightarrow \dots \\ & \downarrow f & & \downarrow f & & \downarrow f & & \downarrow f & \\ \dots \longrightarrow & \hat{Q}^{n+1}(B\otimes_A C,\varepsilon) & \xrightarrow{H} & Q_n(B\otimes_A C,\varepsilon) & \xrightarrow{1+T_\varepsilon} & Q^n(B\otimes_A C,\varepsilon) & \xrightarrow{J} & \hat{Q}^n(B\otimes_A C,\varepsilon) & \longrightarrow \dots \end{array}$$

denoting both ε_A and ε_B by ε. It follows that the various algebraic Wu classes of §1.4 are invariant under the change of rings. For example, the ε-symmetric Wu classes $v^*(\phi)$ of an element $\phi \in Q^n(C,\varepsilon)$ are such that there is defined a commutative diagram

$$\begin{array}{ccc}
H^{n-r}(C) & \xrightarrow{f} & H^{n-r}(B\otimes_A C) \\
\downarrow{\scriptstyle v_r(\phi)} & & \downarrow{\scriptstyle v_r(1\otimes\phi)} \\
H^{n-2r}(\mathbb{Z}_2;A,(-)^{n-r}\varepsilon) & \xrightarrow{f} & H^{n-2r}(\mathbb{Z}_2;B,(-)^{n-r}\varepsilon)
\end{array} .$$

An n-dimensional $\begin{cases}\varepsilon\text{-symmetric} \\ \varepsilon\text{-quadratic}\end{cases}$ (Poincaré) complex over A $\begin{cases}(C,\phi) \\ (C,\psi)\end{cases}$ induces an n-dimensional $\begin{cases}\varepsilon\text{-symmetric} \\ \varepsilon\text{-quadratic}\end{cases}$ (Poincaré) complex over B

$$\begin{cases} B\otimes_A(C,\phi) = (B\otimes_A C, 1\otimes\phi) \\ B\otimes_A(C,\psi) = (B\otimes_A C, 1\otimes\psi) \end{cases} ,$$

and similarly for pairs.

<u>Proposition 2.2.1</u> A morphism of rings with involution

$$f : A \longrightarrow B$$

induces morphisms in the $\begin{cases}\varepsilon\text{-symmetric} \\ \varepsilon\text{-quadratic}\end{cases}$ L-groups

$$\begin{cases} f : L^n(A,\varepsilon) \longrightarrow L^n(B,\varepsilon) \ ; \ (C,\phi) \longmapsto B\otimes_A(C,\phi) \\ f : L_n(A,\varepsilon) \longrightarrow L_n(B,\varepsilon) \ ; \ (C,\psi) \longmapsto B\otimes_A(C,\psi) \end{cases} \quad (n \in \mathbb{Z}) .$$

[]

Define the <u>(n+1)-dimensional relative</u> $\begin{cases}\underline{\varepsilon\text{-symmetric}} \\ \underline{\varepsilon\text{-quadratic}}\end{cases}$ <u>L-group</u> $\begin{cases}L^{n+1}(f,\varepsilon) \\ L_{n+1}(f,\varepsilon)\end{cases}$ $(n \geqslant 0)$ of a morphism of rings with involution $f:A \longrightarrow B$ to be the abelian group of equivalence classes of pairs

$$\text{(n-dimensional } \begin{cases} \varepsilon\text{-symmetric} \\ \varepsilon\text{-quadratic} \end{cases} \text{ Poincaré complex over } A \begin{cases} (C,\phi \in Q^n(C,\varepsilon)) \\ (C,\psi \in Q_n(C,\varepsilon)) \end{cases},$$

$$(n+1)\text{-dimensional } \begin{cases} \varepsilon\text{-symmetric} \\ \varepsilon\text{-quadratic} \end{cases} \text{ Poincaré pair over } B$$

$$\begin{cases} (g:B\otimes_A C \longrightarrow D,(\delta\phi,1\otimes\phi) \in Q^{n+1}(g,\varepsilon)) \\ (g:B\otimes_A C \longrightarrow D,(\delta\psi,1\otimes\psi) \in Q_{n+1}(g,\varepsilon)) \end{cases} \text{ with boundary } \begin{cases} B\otimes_A(C,\phi) \\ B\otimes_A(C,\psi) \end{cases})$$

under the <u>relative cobordism</u> equivalence relation

$$\begin{cases} ((C,\phi),(g:B\otimes_A C \longrightarrow D,(\delta\phi,1\otimes\phi)) \sim ((C',\phi'),(g':B\otimes_A C' \longrightarrow D',(\delta\phi',1\otimes\phi')) \\ ((C,\psi),(g:B\otimes_A C \longrightarrow D,(\delta\psi,1\otimes\psi)) \sim ((C',\psi'),(g':B\otimes_A C' \longrightarrow D',(\delta\psi',1\otimes\psi')) \end{cases}$$

if there exists a pair

$$((n+1)\text{-dimensional } \begin{cases} \varepsilon\text{-symmetric} \\ \varepsilon\text{-quadratic} \end{cases} \text{ Poincaré pair over } A$$

$$\begin{cases} ((h\ h'):C\oplus C' \longrightarrow E,(\nu,\phi\oplus-\phi') \in Q^{n+1}((h\ h'),\varepsilon)) \\ ((h\ h'):C\oplus C' \longrightarrow E,(\chi,\psi\oplus-\psi') \in Q_{n+1}((h\ h'),\varepsilon)) \end{cases},$$

$$(n+2)\text{-dimensional } \begin{cases} \varepsilon\text{-symmetric} \\ \varepsilon\text{-quadratic} \end{cases} \text{ Poincaré triad over } B$$

$$\begin{cases} (\Gamma,(\delta\nu,1\otimes\nu,\delta\phi\oplus-\delta\phi',1\otimes(\phi\oplus-\phi')) \in Q^{n+2}(\Gamma,\varepsilon)) \\ (\Gamma,(\delta\chi,1\otimes\chi,\delta\psi\oplus-\delta\psi',1\otimes(\psi\oplus-\psi')) \in Q_{n+2}(\Gamma,\varepsilon)) \end{cases}$$

with

$$\Gamma \ : \ \begin{array}{ccc} B\otimes_A(C\oplus C') & \xrightarrow{g\oplus g'} & D\oplus D' \\ {\scriptstyle 1\otimes(h\ h')}\downarrow & \searrow & \downarrow \\ B\otimes_A E & \longrightarrow & F \end{array} \qquad) \quad .$$

The verification that relative cobordism is an equivalence relation proceeds as in the absolute case in §I.3, with transitivity requiring the union operation defined in §2.1 above. Addition in the relative L-groups is by the direct sum $\oplus$, and inverses are given by changing signs, as in the absolute case.

Proposition 2.2.2 The relative $\begin{cases}\varepsilon\text{-symmetric}\\ \varepsilon\text{-quadratic}\end{cases}$ L-groups $\begin{cases}L^*(f,\varepsilon)\\ L_*(f,\varepsilon)\end{cases}$

fit into a change of rings exact sequence

$$\begin{cases}\ldots\longrightarrow L^{n+1}(f,\varepsilon)\longrightarrow L^n(A,\varepsilon)\xrightarrow{f} L^n(B,\varepsilon)\longrightarrow L^n(f,\varepsilon)\longrightarrow\ldots\longrightarrow L^0(B,\varepsilon)\\ \ldots\longrightarrow L_{n+1}(f,\varepsilon)\longrightarrow L_n(A,\varepsilon)\xrightarrow{f} L_n(B,\varepsilon)\longrightarrow L_n(f,\varepsilon)\longrightarrow\ldots\longrightarrow L_0(B,\varepsilon)\end{cases}$$

involving the forgetful maps

$$\begin{cases}L^{n+1}(B,\varepsilon)\longrightarrow L^{n+1}(f,\varepsilon)\ ;\ (D,\delta\phi)\longmapsto(0,(0:0\longrightarrow D,(\delta\phi,0)))\\ L_{n+1}(B,\varepsilon)\longrightarrow L_{n+1}(f,\varepsilon)\ ;\ (D,\delta\psi)\longmapsto(0,(0:0\longrightarrow D,(\delta\psi,0)))\end{cases}$$

$$\begin{cases}L^{n+1}(f,\varepsilon)\longrightarrow L^n(A,\varepsilon)\ ; \qquad (n\geqslant 0)\\ \qquad((C,\phi),(g:B\otimes_A C\longrightarrow D,(\delta\phi,1\otimes\phi)))\longmapsto(C,\phi)\\ L_{n+1}(f,\varepsilon)\longrightarrow L_n(A,\varepsilon)\ ;\\ \qquad((C,\psi),(g:B\otimes_A C\longrightarrow D,(\delta\psi,1\otimes\psi)))\longmapsto(C,\psi)\end{cases}.$$

Proof: Exactness is obvious at $\begin{cases}L^n(A,\varepsilon)\\ L_n(A,\varepsilon)\end{cases}$ $(n\geqslant 0)$ and $\begin{cases}L^n(B,\varepsilon)\\ L_n(B,\varepsilon)\end{cases}$ $(n\geqslant 1)$.

As for $\begin{cases}L^{n+1}(f,\varepsilon)\\ L_{n+1}(f,\varepsilon)\end{cases}$ $(n\geqslant 0)$ consider

$$\begin{cases}((C,\phi),(g:B\otimes_A C\longrightarrow D,(\delta\phi,1\otimes\phi)))\in\ker(L^{n+1}(f,\varepsilon)\longrightarrow L^n(A,\varepsilon))\\ ((C,\psi),(g:B\otimes_A C\longrightarrow D,(\delta\psi,1\otimes\psi)))\in\ker(L_{n+1}(f,\varepsilon)\longrightarrow L_n(A,\varepsilon))\end{cases},$$

so that there exists a null-cobordism $\begin{cases}(h:C\longrightarrow E,(\nu,\phi))\\ (h:C\longrightarrow E,(\chi,\psi))\end{cases}$ over A of $\begin{cases}(C,\phi)\\(C,\psi)\end{cases}$. Write the union $(n+1)$-dimensional $\begin{cases}\varepsilon\text{-symmetric}\\\varepsilon\text{-quadratic}\end{cases}$ Poincaré complex over B as

$$\begin{cases}(D',\delta\phi') = (D\cup_{B\otimes_A C}B\otimes_A E,\ \delta\phi\cup_{1\otimes\phi}1\otimes\nu)\\(D',\delta\psi') = (D\cup_{B\otimes_A C}B\otimes_A E,\ \delta\psi\cup_{1\otimes\psi}1\otimes\chi)\end{cases},$$

and define an $(n+2)$-dimensional $\begin{cases}\varepsilon\text{-symmetric}\\\varepsilon\text{-quadratic}\end{cases}$ Poincaré triad over B

$$\begin{cases}(\Gamma,(0,1\otimes\nu,\delta\phi\oplus-\delta\phi',1\otimes\phi))\\(\Gamma,(0,1\otimes\chi,\delta\psi\oplus-\delta\psi',1\otimes\psi))\end{cases}$$

by

$$\Gamma\ :\ \begin{array}{ccc} B\otimes_A C & \xrightarrow{\begin{pmatrix}g\\0\end{pmatrix}} & D\oplus D' \\ {\scriptstyle 1\otimes h}\downarrow & \searrow{\scriptstyle k} & \downarrow{\scriptstyle (i\ \ 1)} \\ B\otimes_A E & \xrightarrow{\ \ j\ \ } & D' \end{array},$$

where

$$i = \begin{pmatrix}1\\0\\0\end{pmatrix} : D_r \longrightarrow D'_r = D_r\oplus B\otimes_A C_{r-1}\oplus B\otimes_A E_r$$

$$j = \begin{pmatrix}0\\0\\1\end{pmatrix} : B\otimes_A E_r \longrightarrow D'_r = D_r\oplus B\otimes_A C_{r-1}\oplus B\otimes_A E_r$$

$$k = \begin{pmatrix}0\\(-)^r\\0\end{pmatrix} : B\otimes_A C_r \longrightarrow D'_{r+1} = D_{r+1}\oplus B\otimes_A C_r\oplus B\otimes_A E_{r+1}\ .$$

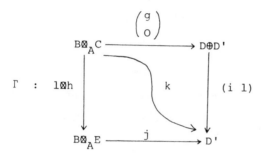

It follows that $\begin{cases} ((C,\phi),(g:B\otimes_A C \longrightarrow D,(\delta\phi,1\otimes\phi))) \in L^{n+1}(f,\varepsilon) \\ ((C,\psi),(g:B\otimes_A C \longrightarrow D,(\delta\psi,1\otimes\psi))) \in L_{n+1}(f,\varepsilon) \end{cases}$

is the image of $\begin{cases} (D',\delta\phi') \in L^{n+1}(B,\varepsilon) \\ (D',\delta\psi') \in L_{n+1}(B,\varepsilon) \end{cases}$ under the natural map.

[]

Define the <u>(n+1)-dimensional relative even ε-symmetric L-groups</u> $L\langle v_0\rangle^{n+1}(f,\varepsilon)$ $(n \geqslant 0)$ of a morphism of rings with involution $f:A \longrightarrow B$ to be the relative cobordism groups of pairs

(n-dimensional even ε-symmetric Poincaré complex over A (C,ϕ),

(n+1)-dimensional even ε-symmetric Poincaré pair over B

$(g:B\otimes_A C \longrightarrow D,(\delta\phi,1\otimes\phi)))$,

where the (n+2)-dimensional ε-symmetric Poincaré triads appearing in the relative cobordisms are even in the sense that all the ε-symmetric Poincaré complexes and pairs associated to them by Proposition 2.1.1 are even.

<u>Proposition 2.2.3</u> i) The relative even ε-symmetric L-groups $L\langle v_0\rangle^*(f,\varepsilon)$ fit into a change of rings exact sequence

$$\ldots \longrightarrow L\langle v_0\rangle^{n+1}(f,\varepsilon) \longrightarrow L\langle v_0\rangle^{n}(A,\varepsilon) \xrightarrow{f} L\langle v_0\rangle^{n}(B,\varepsilon)$$
$$\longrightarrow L\langle v_0\rangle^{n}(f,\varepsilon) \longrightarrow \ldots \longrightarrow L\langle v_0\rangle^{0}(B,\varepsilon) .$$

ii) The skew-suspension maps

$$\begin{cases} \bar{S} : L^n(f,\varepsilon) \longrightarrow L\langle v_0\rangle^{n+2}(f,-\varepsilon) ; \\ ((C,\phi),(g:B\otimes_A C \longrightarrow D,(\delta\phi,1\otimes\phi))) \\ \qquad \longmapsto ((SC,\bar{S}\phi),(g:B\otimes_A SC \longrightarrow SD,\bar{S}(\delta\phi,1\otimes\phi))) \end{cases}$$

$(n \geqslant 1)$

$$\left\{\begin{array}{l}\bar{S} : L_n(f,\varepsilon) \longrightarrow L_{n+2}(f,-\varepsilon) \ ; \\ \quad ((C,\psi),(g:B\otimes_A C \longrightarrow D,(\delta\psi,1\otimes\psi))) \\ \qquad \longmapsto ((SC,\bar{S}\psi),(g:B\otimes_A SC \longrightarrow SD,\bar{S}(\delta\psi,1\otimes\psi)))\end{array}\right.$$

are isomorphisms.

Proof: i) By analogy with Proposition 2.2.2.

ii) This follows from Proposition $\left\{\begin{array}{l}1.4.2\\1.2.3\ \text{i)}\end{array}\right.$ by applying the 5-lemma to the skew-suspension morphism of the change of rings exact sequences.

[]

Define the lower relative $\left\{\begin{array}{l}\underline{\varepsilon\text{-symmetric}}\\ \underline{\varepsilon\text{-quadratic}}\end{array}\right.$ L-groups $\left\{\begin{array}{l}L^n(f,\varepsilon)\\L_n(f,\varepsilon)\end{array}\right.$ $(n\leqslant 0)$ of a morphism of rings with involution $f:A\longrightarrow B$ by

$$\left\{\begin{array}{l} L^n(f,\varepsilon) = \left\{\begin{array}{ll} L\langle v_0\rangle^{n+2}(f,-\varepsilon) & n=0,-1 \\ \mathrm{coker}((\ker(1+T_{-\varepsilon}:L_0(B,-\varepsilon)\longrightarrow L\langle v_0\rangle^0(B,-\varepsilon)) & \\ \qquad \longrightarrow L_0(f,-\varepsilon)) & n=-2 \\ L_n(f,\varepsilon)\ \text{(as defined below)} & n\leqslant -3 \end{array}\right. \\ L_n(f,\varepsilon) = L_{n+2i}(f,(-)^i\varepsilon) \quad (n\leqslant 0,\ n+2i\geqslant 1)\ . \end{array}\right.$$

Proposition 2.2.4 The relative $\left\{\begin{array}{l}\varepsilon\text{-symmetric}\\ \varepsilon\text{-quadratic}\end{array}\right.$ L-groups $\left\{\begin{array}{l}L^n(f,\varepsilon)\\L_n(f,\varepsilon)\end{array}\right.$ $(n\in\mathbb{Z})$ fit into a long exact sequence

$$\begin{cases} \ldots \longrightarrow L^{n+1}(f,\varepsilon) \longrightarrow L^n(A,\varepsilon) \xrightarrow{f} L^n(B,\varepsilon) \longrightarrow L^n(f,\varepsilon) \longrightarrow \ldots \\ \ldots \longrightarrow L_{n+1}(f,\varepsilon) \longrightarrow L_n(A,\varepsilon) \xrightarrow{f} L_n(B,\varepsilon) \longrightarrow L_n(f,\varepsilon) \longrightarrow \ldots \end{cases}$$

[]

In the range $-\infty \leqslant n \leqslant 1$ the change of rings exact sequence of Proposition 2.2.4 can be expressed entirely in terms of the Witt groups of forms and formations defined in the absolute case in §1.6

$$\begin{aligned} \ldots &\longrightarrow M^{\varepsilon}(A) \xrightarrow{f} M^{\varepsilon}(B) \longrightarrow M^{\varepsilon}(f) \longrightarrow L^{\varepsilon}(A) \xrightarrow{f} L^{\varepsilon}(B) \longrightarrow L^{\varepsilon}(f) \\ &\longrightarrow M\langle v_0\rangle^{-\varepsilon}(A) \xrightarrow{f} M\langle v_0\rangle^{-\varepsilon}(B) \longrightarrow M\langle v_0\rangle^{-\varepsilon}(f) \\ &\longrightarrow L\langle v_0\rangle^{-\varepsilon}(A) \xrightarrow{f} L\langle v_0\rangle^{-\varepsilon}(B) \longrightarrow L\langle v_0\rangle^{-\varepsilon}(f) \\ &\longrightarrow M_{\varepsilon}(A) \xrightarrow{f} M_{\varepsilon}(B) \longrightarrow M_{\varepsilon}(f) \longrightarrow L_{\varepsilon}(A) \xrightarrow{f} L_{\varepsilon}(B) \longrightarrow L_{\varepsilon}(f) \\ &\longrightarrow M_{-\varepsilon}(A) \xrightarrow{f} M_{-\varepsilon}(B) \longrightarrow M_{-\varepsilon}(f) \longrightarrow L_{-\varepsilon}(A) \longrightarrow \ldots . \end{aligned}$$

The relative Witt groups of forms and formations are defined as follows.

The full force of the equivalence relation of Proposition 1.3.3 i) (between the homotopy equivalence classes of $\begin{cases}\varepsilon\text{-symmetric}\\ \varepsilon\text{-quadratic}\end{cases}$ Poincaré pairs and those of connected $\begin{cases}\varepsilon\text{-symmetric}\\ \varepsilon\text{-quadratic}\end{cases}$ complexes) allows the higher relative $\begin{cases}\varepsilon\text{-symmetric}\\ \varepsilon\text{-quadratic}\end{cases}$ L-groups $\begin{cases}L^{n+1}(f,\varepsilon)\\ L_{n+1}(f,\varepsilon)\end{cases}$ $(n \geqslant 0)$ to be expressed as the cobordism groups of triples

$$(n\text{-dimensional} \begin{cases} \varepsilon\text{-symmetric} \\ \varepsilon\text{-quadratic} \end{cases} \text{Poincaré complex over } A \begin{cases} (C,\phi) \\ (C,\psi) \end{cases},$$

$$\text{connected } (n+1)\text{-dimensional} \begin{cases} \varepsilon\text{-symmetric} \\ \varepsilon\text{-quadratic} \end{cases} \text{complex over } B \begin{cases} (D,\nu) \\ (D,\chi) \end{cases},$$

$$\text{homotopy equivalence} \begin{cases} g: B\otimes_A (C,\phi) \longrightarrow \partial(D,\nu) \\ g: B\otimes_A (C,\psi) \longrightarrow \partial(D,\chi) \end{cases})\quad .$$

A <u>cobordism</u> between two such triples $\begin{cases} ((C,\phi),(D,\nu),g) \\ ((C,\psi),(D,\chi),g) \end{cases}$,

$\begin{cases} ((C',\phi'),(D',\nu'),g') \\ ((C',\psi'),(D',\chi'),g') \end{cases}$ is a quadruple

$$(\text{connected } (n+1)\text{-dimensional} \begin{cases} \varepsilon\text{-symmetric} \\ \varepsilon\text{-quadratic} \end{cases} \text{complex over } A \begin{cases} (E,\delta\phi) \\ (E,\delta\psi) \end{cases},$$

$$\text{homotopy equivalence} \begin{cases} h : \partial(E,\delta\phi) \longrightarrow (C,-\phi)\oplus(C',\phi') \\ h : \partial(E,\delta\psi) \longrightarrow (C,-\psi)\oplus(C',\psi') \end{cases},$$

$$\text{connected } (n+2)\text{-dimensional} \begin{cases} \varepsilon\text{-symmetric} \\ \varepsilon\text{-quadratic} \end{cases} \text{complex over } B \begin{cases} (F,\delta\nu) \\ (F,\delta\chi) \end{cases},$$

homotopy equivalence

$$\begin{cases} B\otimes_A (E,\delta\phi) \cup_{(g\oplus g')(1\otimes h)} ((D,\nu)\oplus(D',-\nu')) \longrightarrow \partial(F,\delta\nu) \\ B\otimes_A (E,\delta\psi) \cup_{(g\oplus g')(1\otimes h)} ((D,\chi)\oplus(D',-\chi')) \longrightarrow \partial(F,\delta\chi) \end{cases})\quad .$$

In the low-dimensional cases this formulation translates directly into the language of forms and formations:

Proposition 2.2.5 The 0- (resp. 1-) dimensional relative

$$\left\{\begin{array}{l}\varepsilon\text{-symmetric}\\ \text{even } \varepsilon\text{-symmetric L-group}\\ \varepsilon\text{-quadratic}\end{array}\right. \left\{\begin{array}{l}L^0(f,\varepsilon)\\ L\langle v_0\rangle^0(f,\varepsilon)\\ L_0(f,\varepsilon)\end{array}\right. \text{(resp. } \left\{\begin{array}{l}L^1(f,\varepsilon)\\ L\langle v_0\rangle^1(f,\varepsilon))\\ L_1(f,\varepsilon)\end{array}\right.$$

is naturally isomorphic to the relative Witt group

$$\left\{\begin{array}{l}L^\varepsilon(f)\\ L\langle v_0\rangle^\varepsilon(f)\\ L_\varepsilon(f)\end{array}\right. \text{(resp. } \left\{\begin{array}{l}M^\varepsilon(f)\\ M\langle v_0\rangle^\varepsilon(f))\\ M_\varepsilon(f)\end{array}\right. \text{of cobordism classes of triples}$$

$$\text{(non-singular } \left\{\begin{array}{l}\text{even } (-\varepsilon)\text{-symmetric}\\ (-\varepsilon)\text{-quadratic}\\ \text{split } (-\varepsilon)\text{-quadratic}\end{array}\right. \text{formation}$$

$$\text{(resp. } \left\{\begin{array}{l}\varepsilon\text{-symmetric}\\ \text{even } \varepsilon\text{-symmetric form) over A } x,\\ \varepsilon\text{-quadratic}\end{array}\right.$$

$$\left\{\begin{array}{l}\varepsilon\text{-symmetric}\\ \text{even } \varepsilon\text{-symmetric form (resp. formation) over B } y,\\ \varepsilon\text{-quadratic}\end{array}\right.$$

$$\text{stable isomorphism of } \left\{\begin{array}{l}\text{even } (-\varepsilon)\text{-symmetric}\\ (-\varepsilon)\text{-quadratic}\\ \text{split } (-\varepsilon)\text{-quadratic}\end{array}\right. \text{formations}$$

$$\text{(resp. isomorphism of } \left\{\begin{array}{l}\varepsilon\text{-symmetric}\\ \text{even } \varepsilon\text{-symmetric forms) over B}\\ \varepsilon\text{-quadratic}\end{array}\right.$$

$$g : B\otimes_A x \longrightarrow \partial y \qquad) ,$$

where two such triples $(x,y,g),(x',y',g')$ are cobordant if there exists a quadruple

($\begin{cases} \varepsilon\text{-symmetric} \\ \text{even } \varepsilon\text{-symmetric} \\ \varepsilon\text{-quadratic} \end{cases}$ form (resp. formation) over A z,

stable isomorphism of formations (resp. isomorphism of forms) over A

$$h : \partial z \longrightarrow x\oplus -x' \quad ,$$

$\begin{cases} \varepsilon\text{-symmetric} \\ \text{even } \varepsilon\text{-symmetric} \\ \text{split } \varepsilon\text{-quadratic} \end{cases}$ formation (resp.

$\begin{cases} \text{connected 2-dimensional } \varepsilon\text{-symmetric} \\ (-\varepsilon)\text{-symmetric form} \\ \text{even } (-\varepsilon)\text{-symmetric form} \end{cases}$) over B w,

isomorphism of forms (resp. stable isomorphism of formations) over B

$$k : \partial w \longrightarrow (B\otimes_A z) \cup_{(g\oplus g')(1\otimes h)} (y\oplus -y') \quad).$$

[]

(In the cobordism relation for $L^1(f,\varepsilon) = L^{\varepsilon}(f)$ we are using Proposition 1.6.4 to identify the boundary of the connected 2-dimensional ε-symmetric complex z, a 1-dimensional ε-symmetric Poincaré complex ∂z, with the corresponding non-singular ε-symmetric formation).

In §§3,4 we shall need the following extension to the relative L-groups of the products of §1.9.

<u>Proposition 2.2.6</u> Let R be a ring with involution, and let

$$f : A \longrightarrow B$$

be a morphism of rings with involution which are R-modules, with f also an R-module morphism ($f(ra) = rf(a) \in B$ for all $r \in R, a \in A$). There are then defined products

$$\begin{cases} \otimes : L^m(R,\rho)\otimes_{\mathbb{Z}} L^n(f:A \longrightarrow B,\varepsilon) \longrightarrow L^{m+n}(f:A \longrightarrow B,\rho\varepsilon) \\ \otimes : L^m(R,\rho)\otimes_{\mathbb{Z}} L_n(f:A \longrightarrow B,\varepsilon) \longrightarrow L_{m+n}(f:A \longrightarrow B,\rho\varepsilon) \end{cases} \quad (m,n \in \mathbb{Z}) .$$

<u>Proof</u>: Immediate from Proposition 1.9.2 and the definition of the relative L-groups.

[]

In particular, the symmetric Witt group $L^0(R)$ of a commutative ring R (with any involution) is a ring with $1 = (R,1) \in L^0(R)$, so that the relative $\begin{cases} \varepsilon\text{-symmetric} \\ \varepsilon\text{-quadratic} \end{cases}$ L-groups $\begin{cases} L^*(f,\varepsilon) \\ L_*(f,\varepsilon) \end{cases}$ of an R-module morphism of rings with involution $f:A \longrightarrow B$ are all $L^0(R)$-modules, and the change of rings exact sequence of Proposition 2.2.4

$$\begin{cases} \dots \longrightarrow L^n(A,\varepsilon) \xrightarrow{f} L^n(B,\varepsilon) \longrightarrow L^n(f,\varepsilon) \longrightarrow L^{n-1}(A,\varepsilon) \longrightarrow \dots \\ \dots \longrightarrow L_n(A,\varepsilon) \xrightarrow{f} L_n(B,\varepsilon) \longrightarrow L_n(f,\varepsilon) \longrightarrow L_{n-1}(A,\varepsilon) \longrightarrow \dots \end{cases}$$

$$(n \in \mathbb{Z})$$

is an exact sequence of $L^0(R)$-modules.

2.3 Change of categories

The unified L-groups of §1.8 were constructed using the $\mathcal{L}$-categories and the ∂-functors. We shall now define relative L-groups for a ∂-preserving functor of the $\mathcal{L}$-categories, which include the change of rings relative L-groups of §2.2 as a special case.

Let A,B be rings with involution, and let $\varepsilon_A \in A$, $\varepsilon_B \in B$ be central units such that

$$\bar{\varepsilon}_A = \varepsilon_A^{-1} \in A \ , \ \bar{\varepsilon}_B = \varepsilon_B^{-1} \in B \quad .$$

As in §2.2 both ε_A and ε_B will be denoted by ε.

An $\begin{cases} \underline{\varepsilon\text{-symmetric}} \\ \underline{\varepsilon\text{-quadratic}} \\ \underline{\varepsilon\text{-hyperquadratic}} \end{cases}$ chain functor

$$\begin{cases} F : \mathcal{L}^*(A,\varepsilon) \longrightarrow \mathcal{L}^*(B,\varepsilon) \\ F : \mathcal{L}_*(A,\varepsilon) \longrightarrow \mathcal{L}_*(B,\varepsilon) \\ F : \mathcal{L}_*(A,\varepsilon) \longrightarrow \mathcal{L}^*(B,\varepsilon) \end{cases}$$

is a collection of additive functors $\begin{cases} \{F:\mathcal{L}^n(A,\varepsilon) \longrightarrow \mathcal{L}^n(B,\varepsilon) \mid n \in \mathbb{Z}\} \\ \{F:\mathcal{L}_n(A,\varepsilon) \longrightarrow \mathcal{L}_n(B,\varepsilon) \mid n \in \mathbb{Z}\} \\ \{F:\mathcal{L}_n(A,\varepsilon) \longrightarrow \mathcal{L}^n(B,\varepsilon) \mid n \in \mathbb{Z}\} \end{cases}$

such that $\partial F = F\partial$, $-F = F-$ (up to natural equivalence). There are induced abelian group morphisms in the cobordism groups

$$\begin{cases} F : L^n(A,\varepsilon) \longrightarrow L^n(B,\varepsilon) \\ F : L_n(A,\varepsilon) \longrightarrow L_n(B,\varepsilon) \\ F : L_n(A,\varepsilon) \longrightarrow L^n(B,\varepsilon) \end{cases} \quad (n \in \mathbb{Z}) \quad .$$

Define the relative L-groups of F $\begin{cases} L^n(F,\varepsilon) \\ L_n(F,\varepsilon) \\ \hat{L}^n(F,\varepsilon) \end{cases}$ $(n \in \mathbb{Z})$ to be the abelian groups of cobordism classes of triples (x,y,f) consisting of a closed object x of $\begin{cases} \mathcal{L}^{n-1}(A,\varepsilon) \\ \mathcal{L}_{n-1}(A,\varepsilon) \\ \mathcal{L}^{n-1}(A,\varepsilon) \end{cases}$, an object y of $\begin{cases} \mathcal{L}^{n}(B,\varepsilon) \\ \mathcal{L}_{n}(B,\varepsilon) \\ \mathcal{L}^{n}(B,\varepsilon) \end{cases}$, and a homotopy equivalence $f:F(x) \longrightarrow \partial y$. Two such pairs (x,y,f), (x',y',f') are cobordant if there exists a quadruple (z,g,w,h) consisting of an object z of $\begin{cases} \mathcal{L}^{n}(A,\varepsilon) \\ \mathcal{L}_{n}(A,\varepsilon) \\ \mathcal{L}^{n}(A,\varepsilon) \end{cases}$, a homotopy equivalence $g:\partial z \longrightarrow x\oplus -x'$, an object w of $\begin{cases} \mathcal{L}^{n+1}(B,\varepsilon) \\ \mathcal{L}_{n+1}(B,\varepsilon) \\ \mathcal{L}^{n+1}(B,\varepsilon) \end{cases}$, and a homotopy equivalence $h : F(z) \cup_{(f\oplus -f')F(g)} (-y\oplus y') \longrightarrow \partial w$, where

$$\begin{aligned}(f\oplus -f')F(g) \ : \ \partial F(z) = F\partial(z) &\xrightarrow{F(g)} F(x\oplus -x') = F(x)\oplus F(x') \\ &\xrightarrow{f\oplus -f'} \partial y\oplus -\partial y' = \partial(y\oplus -y') \ .\end{aligned}$$

Addition and inverses are given by

$$(x,y,f) + (x',y',f') = (x\oplus x',y\oplus y',f\oplus f') \ , \ -(x,y,f) = (-x,-y,-f) \ .$$

<u>Proposition 2.3.1</u> The relative L-groups $\begin{cases} L^*(F,\varepsilon) \\ L_*(F,\varepsilon) \\ \hat{L}^*(F,\varepsilon) \end{cases}$ of an $\begin{cases} \varepsilon\text{-symmetric} \\ \varepsilon\text{-quadratic} \\ \varepsilon\text{-hyperquadratic} \end{cases}$ chain functor $\begin{cases} F:\mathcal{L}^*(A,\varepsilon) \longrightarrow \mathcal{L}^*(B,\varepsilon) \\ F:\mathcal{L}_*(A,\varepsilon) \longrightarrow \mathcal{L}_*(B,\varepsilon) \\ F:\mathcal{L}_*(A,\varepsilon) \longrightarrow \mathcal{L}^*(B,\varepsilon) \end{cases}$ fit into the <u>change of categories</u> exact sequence

$$\begin{cases} \dots \longrightarrow L^n(A,\varepsilon) \xrightarrow{F} L^n(B,\varepsilon) \longrightarrow L^n(F,\varepsilon) \longrightarrow L^{n-1}(A,\varepsilon) \longrightarrow \dots \\ \dots \longrightarrow L_n(A,\varepsilon) \xrightarrow{F} L_n(B,\varepsilon) \longrightarrow L_n(F,\varepsilon) \longrightarrow L_{n-1}(A,\varepsilon) \longrightarrow \dots \\ \dots \longrightarrow L_n(A,\varepsilon) \xrightarrow{F} L^n(B,\varepsilon) \longrightarrow \hat{L}^n(F,\varepsilon) \longrightarrow L_{n-1}(A,\varepsilon) \longrightarrow \dots \end{cases}$$

$(n \in \mathbb{Z})$.

Proof: By analogy with Proposition 2.2.4 (which is a special case).

[]

The following examples of relative L-groups arise in topology:

i) A morphism of rings with involution

$$f : A \longrightarrow B$$

induces an $\begin{cases} \varepsilon\text{-symmetric} \\ \varepsilon\text{-quadratic} \end{cases}$ chain functor

$$\begin{cases} f : \mathcal{L}^*(A,\varepsilon) \longrightarrow \mathcal{L}^*(B,\varepsilon) \ ; \ x \longmapsto B\otimes_A x \\ f : \mathcal{L}_*(A,\varepsilon) \longrightarrow \mathcal{L}_*(B,\varepsilon) \ ; \ x \longmapsto B\otimes_A x \end{cases} .$$

The relative L-groups $\begin{cases} L^*(f,\varepsilon) \\ L_*(f,\varepsilon) \end{cases}$ for this change of categories are just the relative L-groups for the change of rings defined in §2.2 above. The methods of II. associate to an n-dimensional $\begin{cases} \text{geometric Poincaré pair } (X,\partial X) \\ \text{normal map of pairs } (g,c):(M,\partial M) \longrightarrow (X,\partial X) \end{cases}$ the relative $\begin{cases} \text{symmetric} \\ \text{quadratic} \end{cases}$ signature

$$\begin{cases} \sigma^*(X,\partial X) \in L^n(\mathbb{Z}[\pi_1(\partial X)] \longrightarrow \mathbb{Z}[\pi_1(X)]) \\ \sigma_*(g,c) \in L_n(\mathbb{Z}[\pi_1(\partial X)] \longrightarrow \mathbb{Z}[\pi_1(X)]) \end{cases} .$$

(The terminology is contracted in the usual fashion for $\varepsilon = 1$).

The relative quadratic signature is the obstruction for framed surgery to a homotopy equivalence of pairs. The relative quadratic L-groups $L_*(f)$ were first defined by Wall [4] using geometric methods; the $\begin{cases}\text{odd-}\\\text{even-}\end{cases}$ dimensional relative quadratic L-groups $\begin{cases}L_{2*+1}(f)\\L_{2*}(f)\end{cases}$ were first obtained algebraically by $\begin{cases}\text{Wall [4,§7]}\\\text{Sharpe [2]}\end{cases}$.

ii) Given an integer $m \geqslant 1$ define an $\begin{cases}\varepsilon\text{-symmetric}\\\varepsilon\text{-quadratic}\end{cases}$ functor

$$\begin{cases} m : \mathcal{L}^*(A,\varepsilon) \longrightarrow \mathcal{L}^*(A,\varepsilon) \ ; \ x \longmapsto mx = x\oplus x\oplus\ldots\oplus x \text{ (m times)} \\ m : \mathcal{L}_*(A,\varepsilon) \longrightarrow \mathcal{L}_*(A,\varepsilon) \ ; \ x \longmapsto mx = x\oplus x\oplus\ldots\oplus x \text{ (m times)} \end{cases} .$$

The relative L-groups are the $\begin{cases}\underline{\varepsilon\text{-symmetric}}\\\underline{\varepsilon\text{-quadratic}}\end{cases}$ <u>mod m L-groups of A</u>

$$\begin{cases} L^*(A,\varepsilon;\mathbb{Z}_m) = L^*(m:\mathcal{L}^*(A,\varepsilon) \longrightarrow \mathcal{L}^*(A,\varepsilon)) \\ L_*(A,\varepsilon;\mathbb{Z}_m) = L_*(m:\mathcal{L}_*(A,\varepsilon) \longrightarrow \mathcal{L}_*(A,\varepsilon)) \end{cases} ,$$

and fit into the exact sequence

$$\begin{cases} \ldots \longrightarrow L^n(A,\varepsilon) \xrightarrow{m} L^n(A,\varepsilon) \longrightarrow L^n(A,\varepsilon;\mathbb{Z}_m) \longrightarrow L^{n-1}(A,\varepsilon) \longrightarrow \ldots \\ \ldots \longrightarrow L_n(A,\varepsilon) \xrightarrow{m} L_n(A,\varepsilon) \longrightarrow L_n(A,\varepsilon;\mathbb{Z}_m) \longrightarrow L_{n-1}(A,\varepsilon) \longrightarrow \ldots \end{cases} \quad (n \in \mathbb{Z})$$

A <u>geometric $\mathbb{Z}_m$-Poincaré complex</u> (resp. <u>$\mathbb{Z}_m$-manifold</u>) $(X,\partial X)$ is a geometric Poincaré pair (resp. manifold with boundary) such that the boundary is the disjoint union of m copies of the <u>Bockstein</u> geometric Poincaré complex (resp. manifold) δX

$$\partial X = \coprod_m \delta X \quad .$$

The <u>mod m</u> $\begin{cases}\underline{\text{symmetric}}\\\underline{\text{quadratic}}\end{cases}$ <u>signature</u> of an n-dimensional

$$\begin{cases} \text{geometric } \mathbb{Z}_m\text{-Poincaré complex } (X,\partial X) \\ \text{normal map } (f,b):(M,\partial M) \longrightarrow (X,\partial X) \text{ from a } \mathbb{Z}_m\text{-manifold } (M,\partial M) \\ \quad \text{to a geometric } \mathbb{Z}_m\text{-Poincaré complex } (X,\partial X) \ ((\partial f,\partial b) = \bigsqcup_m (\delta f,\delta b)) \end{cases}$$

is an element

$$\begin{cases} \sigma^*(X;\mathbb{Z}_m) \in L^n(\mathbb{Z}[\pi_1(X)];\mathbb{Z}_m) \\ \sigma_*(f,b;\mathbb{Z}_m) \in L_n(\mathbb{Z}[\pi_1(X)];\mathbb{Z}_m) \end{cases} \qquad (\varepsilon = 1)$$

defined using the methods of II. exactly as in the case $m = 1$. The mod m quadratic signature is the obstruction to surgery to a homotopy equivalence of $\mathbb{Z}_m$-objects. Surgery on $\mathbb{Z}_m$- manifolds plays an important role in the characteristic variety theorem of Sullivan [2], and in the subsequent work of Morgan and Sullivan [1], Wall [13], Jones [2] and Taylor and Williams [1] on characteristic classes for the surgery obstructions of normal maps of closed manifolds.

iii) The ε-symmetrization is an ε-hyperquadratic chain functor

$$1+T_\varepsilon : \mathcal{L}_*(A,\varepsilon) \longrightarrow \mathcal{L}^*(A,\varepsilon) .$$

The relative L-groups of $1+T_\varepsilon$ are the <u>ε-hyperquadratic L-groups of A</u> $\hat{L}^*(A,\varepsilon)$, which fit into the exact sequence

$$\ldots \longrightarrow L_n(A,\varepsilon) \xrightarrow{1+T_\varepsilon} L^n(A,\varepsilon) \xrightarrow{J} \hat{L}^n(A,\varepsilon) \xrightarrow{H} L_{n-1}(A,\varepsilon) \longrightarrow \ldots \qquad (n \in \mathbb{Z})$$

and are 8-torsion groups (by Proposition I.8.2).
In §7.4 the hyperquadratic L-groups $\hat{L}^*(A)$ $(\varepsilon = 1)$ will be used to define a "hyperquadratic signature" invariant $\hat{\sigma}^*(X) \in \hat{L}^n(\mathbb{Z}[\pi_1(X)])$ for an n-dimensional normal space X in the sense of Quinn [3]. In particular, given an (n+1)-dimensional degree 1 map of geometric Poincaré pairs

$$g : (N,M) \longrightarrow (Y,X)$$

such that the restriction $f = g| : M \longrightarrow X$ underlies a normal map

$$(f,b) : M \longrightarrow X$$

there is defined a <u>hyperquadratic signature</u>

$$\hat{\sigma}^*(g,f,b) \in \hat{L}^{n+1}(\mathbb{Z}[\pi_1(Y)])$$

such that

$$H\hat{\sigma}^*(g,f,b) = \sigma_*(f,b) \in L_n(\mathbb{Z}[\pi_1(Y)]) \ ,$$

as follows. (This is the hyperquadratic signature $\hat{\sigma}^*(N\cup_f -Y)$ of the $(n+1)$-dimensional normal space obtained from (N,M) and (Y,X) by glueing along $f:M \longrightarrow X$).

In the first instance, recall from §II.9 (and see also §7.3 below) that a stable spherical fibration $p:X \longrightarrow BG$ over a finitely dominated CW complex X has associated to it a Tate $\mathbb{Z}_2$- hypercohomology class

$$\theta(p) \in \hat{Q}^0(C(\tilde{X})^{-*}) \quad ,$$

with $\tilde{X}$ the universal cover of X (say). The hyperquadratic Wu classes of $\theta(p)$ are the <u>equivariant Wu classes of p</u>

$$v_*(p) = \hat{v}_*(\theta(p)) : H_*(\tilde{X}) \longrightarrow \hat{H}^*(\mathbb{Z}_2;\mathbb{Z}[\pi_1(X)]) \ .$$

The equivariant Wu classes are stable fibre homotopy invariants which are generalizations of the familiar mod 2 Wu classes $v_*(p) \in H^*(X;\mathbb{Z}_2)$. Let (p,q,r) be a triple consisting of two stable spherical fibrations $p,q:X \longrightarrow BG$ over X and a stable fibre homotopy equivalence $r:p|_Y \xrightarrow{\sim} q|_Y$ of their restrictions to a subcomplex Y of X, which is classified by a homotopy

$$r : p|_Y \simeq q|_Y : Y \longrightarrow BG \ .$$

The relative version of the above construction associates to (p,q,r) a Tate $\mathbb{Z}_2$-hypercohomology class

$$\theta(p,q,r) \in \hat{Q}^0(C(\tilde{X},\tilde{Y})^{-*})$$

with image $\theta(p) - \theta(q) \in \hat{Q}^0(C(\tilde{X})^{-*})$ under the map induced by the projection $C(\tilde{X}) \longrightarrow C(\tilde{X},\tilde{Y})$. If r extends to a stable fibre homotopy equivalence $p \xrightarrow{\simeq} q$ then $\theta(p,q,r) = 0$. The <u>relative equivariant Wu classes of (p,q,r)</u> $v_*(p,q,r)$ are the hyperquadratic Wu classes of $\theta(p,q,r)$

$$v_*(p,q,r) = \hat{v}_*(\theta(p,q,r)) : H_*(\tilde{X},\tilde{Y}) \longrightarrow \hat{H}^*(\mathbb{Z}_2;\mathbb{Z}[\pi_1(X)]) \ ,$$

and are such that

$$v_*(p) - v_*(q) : H_*(\tilde{X}) \longrightarrow H_*(\tilde{X},\tilde{Y}) \xrightarrow{v_*(p,q,r)} \hat{H}^*(\mathbb{Z}_2;\mathbb{Z}[\pi_1(X)]) \ .$$

A stable fibre homotopy self equivalence

$$c : \nu \longrightarrow \nu$$

of a stable spherical fibration $\nu : X \longrightarrow BG$ over X is classified by a map $c : X \longrightarrow G = \Omega BG$.

The <u>equivariant suspended Wu classes of c</u> $\sigma v_*(c)$ are defined by

$$\sigma v_*(c) = v_*(p,q,r) \equiv \hat{v}_*(\theta_{\nu,c}) \in H_*(\tilde{X}\times I,\tilde{X}\times\{0,1\}) = H_{*-1}(\tilde{X})$$
$$\longrightarrow \hat{H}^*(\mathbb{Z}_2;\mathbb{Z}[\pi_1(X)]) \ ,$$

with $\theta_{\nu,c} \in \hat{Q}^0(C(\tilde{X}\times I,\tilde{X}\times\{0,1\})^{-*})$ defined by

$$\theta_{\nu,c} = \theta(p = \text{adjoint of } c : X\times I \longrightarrow BG \ , \ q : X\times I \longrightarrow * \longrightarrow BG \ ,$$
$$r = \text{id.} : p|_{X\times\{0,1\}} = \varepsilon^\infty \longrightarrow q|_{X\times\{0,1\}} = \varepsilon^\infty)$$

The equivariant suspended Wu classes were defined in §II.9 in connection with a formula for the change in the quadratic kernel $\sigma_*(f,b)$ of a normal map $(f,b):M \longrightarrow X$ caused by a change in the bundle map $b:\nu_M \longrightarrow \nu_X$, which we shall generalize in Proposition 2.3.2 below to the quadratic signature $\sigma_*(f,b) \in L_n(\mathbb{Z}[\pi_1(X)])$.

Given a chain map $f:C \longrightarrow D$ of finite-dimensional A-module chain complexes define the $\underline{\tilde{Q}\text{-groups}}$ $\tilde{Q}^{n+1}(f,\varepsilon)$ $(n \in \mathbb{Z})$ to be the relative groups appearing in the exact sequence

$$\ldots \longrightarrow Q^{n+1}(D,\varepsilon) \longrightarrow \tilde{Q}^{n+1}(f,\varepsilon) \longrightarrow Q_n(C,\varepsilon) \xrightarrow{f^{\%}(1+T_\varepsilon)} Q^n(D,\varepsilon) \longrightarrow \ldots .$$

For example, if $f = 1 : C \longrightarrow D = C$ then $\tilde{Q}^*(f,\varepsilon) = \hat{Q}^*(C,\varepsilon)$. An element $(\phi,\psi) \in \tilde{Q}^{n+1}(f,\varepsilon)$ is an equivalence class of collections of chains

$$\{(\phi_s,\psi_s) \in \mathrm{Hom}_A(D^{n+1-r+s},D_r) \oplus \mathrm{Hom}_A(C^{n-r-s},C_r) \mid r \in \mathbb{Z},\ s \geqslant 0\}$$

such that

$$d\phi_s + (-)^r\phi_s d^* + (-)^{n+s}(\phi_{s-1} + (-)^s T_\varepsilon \phi_{s-1})$$
$$= \begin{cases} (1+T_\varepsilon)f\psi_0 f^* : D^{n-r} \longrightarrow D_r \\ 0 : D^{n-r+s} \longrightarrow D_r \end{cases} \text{if } \begin{cases} s = 0 \\ s \geqslant 1 \end{cases} \quad (\phi_{-1} = 0)$$

$$d\psi_s + (-)^r\psi_s d^* + (-)^{n-s-1}(\psi_{s+1} + (-)^{s+1}T_\varepsilon\psi_{s+1}) = 0$$
$$: C^{n-r-s-1} \longrightarrow C_r \quad (s \geqslant 0) .$$

The ε-hyperquadratic L-groups $\hat{L}^{n+1}(A,\varepsilon)$ $(n \geqslant 0)$ can be viewed as the cobordism groups of objects $(f:C \longrightarrow D, (\phi,\psi) \in \tilde{Q}^{n+1}(f,\varepsilon))$ such that $(f:C \longrightarrow D, (\phi,(1+T_\varepsilon)\psi) \in Q^{n+1}(f,\varepsilon))$ is an $(n+1)$-dimensional ε-symmetric Poincaré pair over A.

Let now $g:(N,M) \longrightarrow (Y,X)$ be a degree 1 map of $(n+1)$-dimensional geometric Poincaré pairs such that $g| = f : M \longrightarrow X$ is part of an n-dimensional normal map $(f,b) : M \longrightarrow X$. Let $\tilde{Y}$ be the universal cover of Y, and let $\tilde{M},\tilde{N},\tilde{X}$ be the induced covers of M,N,X. There is then defined a commutative diagram of $\mathbb{Z}[\pi_1(Y)]$-module chain complexes and chain maps

$$\begin{array}{ccccc}
C(\tilde{X}) & \xrightarrow{f^!} & C(\tilde{M}) & \xrightarrow{e} & C(f^!) \\
\downarrow & & \downarrow & & \downarrow i \\
C(\tilde{Y}) & \xrightarrow{g^!} & C(\tilde{N}) & \xrightarrow{h} & C(g^!) \\
\downarrow & & \downarrow & & \downarrow \\
C(\tilde{Y},\tilde{X}) & \xrightarrow{(g,f)^!} & C(\tilde{N},\tilde{M}) & \xrightarrow{(h,e)} & C((g,f)^!)
\end{array} \quad .$$

Now $b:\nu_M \longrightarrow \nu_X$ defines a stable fibre homotopy equivalence of the restrictions to $M \subset N$ of $\nu_N:N \longrightarrow BG$ and $g^*\nu_Y:N \longrightarrow BG$

$$b \;:\; \nu_N|_M = \nu_M \xrightarrow{\;\sim\;} g^*\nu_Y|_M = f^*\nu_X \quad ,$$

so that by the above construction there is defined an element

$$\theta(\nu_N, g^*\nu_Y, b) \in \hat{Q}^0(C(\tilde{N},\tilde{M})^{-*}) = \hat{Q}^{n+1}(C(\tilde{N},\tilde{M})^{n+1-*}) = \hat{Q}^{n+1}(C(\tilde{N}))$$

with image

$$\begin{aligned}\theta(\nu_N) - (g,f)^{!\%}\theta(\nu_Y) \in \hat{Q}^0(C(\tilde{N})^{-*}) &= \hat{Q}^{n+1}(C(\tilde{N})^{n+1-*}) \\ &= \hat{Q}^{n+1}(C(\tilde{N},\tilde{M})) \ .\end{aligned}$$

Let $F:\Sigma^\infty\tilde{M}_+ \longrightarrow \Sigma^\infty\tilde{X}_+$ be the geometric Umkehr map associated to (f,b), so that $\psi_F([X]) \in Q_n(C(\tilde{M}))$ (ψ_F = quadratic construction) is such that

$$(1+T)\psi_F([X]) = \phi_{\tilde{M}}([M]) - f^{!\%}\phi_{\tilde{X}}([X]) \in Q^n(C(\tilde{M}))$$

and

$$\sigma_*(f,b) = (C(f^!), e_\%\psi_F([X]) \in Q_n(C(f^!))) \in L_n(\mathbb{Z}[\pi_1(Y)]) \ .$$

Now $\tilde{Q}^{n+1}(i)$ fits into a commutative braid of exact sequences

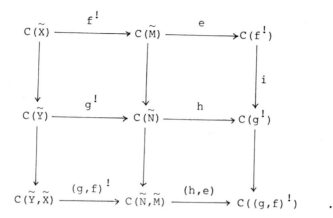

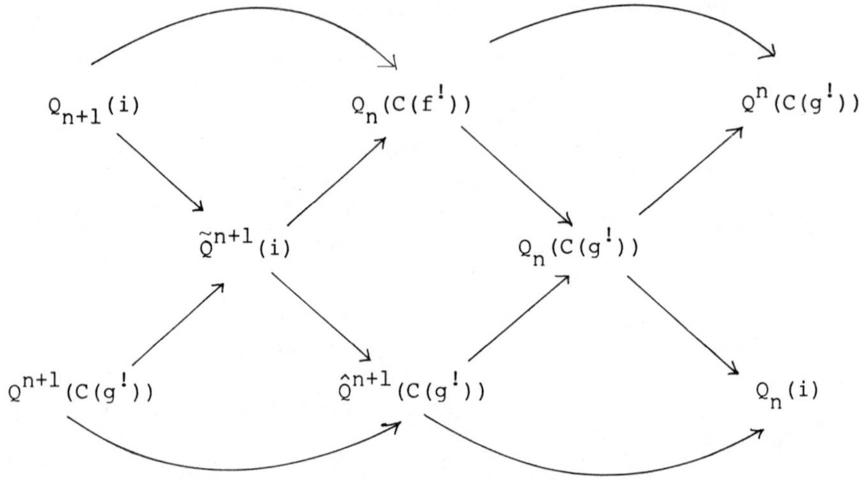

The elements $e_{\%}\psi_F([X]) \in Q_n(C(f^!))$, $\hat{h}^{\%}\theta(\nu_N, g^*\nu_Y, b) \in \hat{Q}^{n+1}(C(g^!))$ have the same image in $Q_n(C(g^!))$, and in fact there is defined an element

$$\theta(g,f,b) \in \tilde{Q}^{n+1}(i)$$

with images $e_{\%}\psi_F([X])$, $\hat{h}^{\%}\theta(\nu_N, g^*\nu_Y, b)$. The <u>hyperquadratic signature</u> of $(g:(N,M) \longrightarrow (Y,X), (f,b):M \longrightarrow X)$ is defined to be

$$\hat{\sigma}^*(g,f,b) = (i:C(f^!) \longrightarrow C(g^!), \theta(g,f,b)) \in \hat{L}^{n+1}(\mathbb{Z}[\pi_1(Y)]) ,$$

and has image $\sigma_*(f,b) \in L_n(\mathbb{Z}[\pi_1(Y)])$. If $b:\nu_M \longrightarrow \nu_X$ extends to $c:\nu_N \longrightarrow \nu_Y$ then $\theta(\nu_N, g^*\nu_Y, b) = 0$ and $\hat{\sigma}^*(g,f,b) = 0$.

(I should like to thank Jean Lannes for his suggestion that I apply the algebraic theory of surgery to normal maps which bound as degree 1 maps).

Let $(f,b):M \longrightarrow X, (f,b'):M \longrightarrow X$ be n-dimensional normal maps with the same underlying degree 1 map $f:M \longrightarrow X$, so that $b' = bc : \nu_M \longrightarrow \nu_X$ for some stable fibre self homotopy equivalence $c : \nu_M \longrightarrow \nu_M$. In Proposition II.9.10 the difference of the $\mathbb{Z}_2$-hyperhomology classes ψ, ψ' appearing in the quadratic kernels

$$\sigma_*(f,b) = (C(f^!), \psi \in Q_n(C(f^!)))$$
$$\sigma_*(f,b') = (C(f^!), \psi' \in Q_n(C(f^!)))$$

was expressed in terms of $\theta_{\nu_M, c} \in \hat{Q}^{n+1}(C(\tilde{M}))$ as

$$\psi - \psi' = e_{\%}H(\theta_{\nu_M, c}) \in Q_n(C(f^!)) \ .$$

<u>Proposition 2.3.2</u> The difference of quadratic signatures is given by

$$\sigma_*(f,b) - \sigma_*(f,b') = H\hat{\sigma}^*(g, f \cup f, b \cup -b') \in L_n(\mathbb{Z}[\pi_1(X)]) \ ,$$

with

$$g = f \times 1 : (M \times I, M \times \{0,1\}) \longrightarrow (X \times I, X \times \{0,1\})$$
$$f \cup f = g| \ : M \times \{0,1\} \longrightarrow X \times \{0,1\} \ .$$

The hyperquadratic signature $\hat{\sigma}^*(g, f \cup f, b \cup -b') \in \hat{L}^{n+1}(\mathbb{Z}[\pi_1(X)])$ is represented by

$$\begin{aligned} &\hat{\sigma}^*(g, f \cup f', b \cup -b') \\ &\quad = (i = (1\ 1) : C(f^!) \oplus C(f^!) \longrightarrow C(g^!) = C(f^!), \\ &\qquad \theta(g, f \cup f, b \cup -b') = (\psi, \hat{e}^{\%}(\theta_{\nu_M, c}), 0) \in \tilde{Q}^{n+1}(i) \\ &\qquad = Q_n(C(f^!)) \oplus \hat{Q}^{n+1}(C(f)^!) \oplus H_n(C(f^!)^t \otimes_{\mathbb{Z}[\pi_1(X)]} C(f^!)) \ ) \ . \end{aligned}$$

[]

2.4 <u>Γ-groups</u>

Let $f:A \longrightarrow B$ be a morphism of rings with involution, as before.

An n-dimensional $\begin{cases} \varepsilon\text{-symmetric} \\ \varepsilon\text{-quadratic} \end{cases}$ complex over A $\begin{cases} (C,\phi) \\ (C,\psi) \end{cases}$ is <u>B-Poincaré</u> if $\begin{cases} B\otimes_A(C,\phi) \\ B\otimes_A(C,\psi) \end{cases}$ is an n-dimensional $\begin{cases} \varepsilon\text{-symmetric} \\ \varepsilon\text{-quadratic} \end{cases}$ Poincaré complex over B. Similarly for pairs and triads.

The <u>n-dimensional</u> $\begin{cases} \underline{\varepsilon\text{-symmetric}} \\ \underline{\text{even } \varepsilon\text{-symmetric}} \\ \underline{\varepsilon\text{-quadratic}} \end{cases}$ <u>Γ-group of f</u>

$\begin{cases} \Gamma^n(f,\varepsilon) \\ \Gamma\langle v_0\rangle^n(f,\varepsilon) \\ \Gamma_n(f,\varepsilon) \end{cases}$ $(n \geqslant 0)$ is the B-Poincaré cobordism group of

n-dimensional $\begin{cases} \varepsilon\text{-symmetric} \\ \text{even } \varepsilon\text{-symmetric} \\ \varepsilon\text{-quadratic} \end{cases}$ B-Poincaré complexes over A.

In particular

$$\begin{cases} \Gamma^*(1:A \longrightarrow A,\varepsilon) = L^*(A,\varepsilon) \\ \Gamma\langle v_0\rangle^*(1:A \longrightarrow A,\varepsilon) = L\langle v_0\rangle^*(A,\varepsilon) \\ \Gamma_*(1:A \longrightarrow A,\varepsilon) = L_*(A,\varepsilon) \end{cases} .$$

The quadratic Γ-groups $\Gamma_*(f) \equiv \Gamma_*(f,1)$ are projective analogues of the original Γ-groups of Cappell and Shaneson [1].

The morphism $f:A \longrightarrow B$ is <u>locally epic</u> if for every finite subset $B_0 \subseteq B$ there exists a unit $u \in B$ such that

$$uB_0 \in \operatorname{im}(f:A \longrightarrow B) \subseteq B .$$

(This definition is due to Cappell and Shaneson [1]).

For example, if $f:A \longrightarrow B$ is onto it is locally epic; also, a localization map $f:A \longrightarrow S^{-1}A$ is locally epic - see §3 below for the application of the Γ-groups to the L-theory of localization. In dealing with Γ-groups we shall always assume that $f:A \longrightarrow B$ is locally epic. (It is in fact possible to develop Γ-theory for more general morphisms - see Vogel [3] and the discussion in §3.2 below).

An A-module morphism $g \in \mathrm{Hom}_A(M,N)$ is a B-isomorphism if $1\otimes g \in \mathrm{Hom}_B(B\otimes_A M, B\otimes_A N)$ is a B-module isomorphism.

Proposition 2.4.1 Let $f:A \longrightarrow B$ be a locally epic morphism, and let M,N be f.g. free A-modules. A morphism $g \in \mathrm{Hom}_A(M,N)$ is such that $1\otimes g \in \mathrm{Hom}_B(B\otimes_A M, B\otimes_A N)$ is onto if and only if there exists an A-module morphism $h \in \mathrm{Hom}_A(N,M)$ such that $gh \in \mathrm{Hom}_A(N,N)$ is a B-isomorphism.

Proof: Assume that $1\otimes g \in \mathrm{Hom}_B(B\otimes_A M, B\otimes_A N)$ is onto, so that there exists $b \in \mathrm{Hom}_B(B\otimes_A N, B\otimes_A M)$ right inverse to $1\otimes g$, with $(1\otimes g)b = 1 \in \mathrm{Hom}_B(B\otimes_A N, B\otimes_A N)$. Choose bases for M and N, and let (b_{ij}) $(b_{ij} \in B)$ be the corresponding matrix of b. As f is locally epic there exist a matrix (a_{ij}) with entries $a_{ij} \in A$ and a unit $u \in B$ such that

$$f(a_{ij}) = b_{ij}u \in B \ .$$

Let $h \in \mathrm{Hom}_A(N,M)$ be the A-module morphism with matrix (a_{ij}). Then $1\otimes gh = u \in \mathrm{Hom}_B(B\otimes_A N, B\otimes_A N)$ is a B-module isomorphism, so that $gh \in \mathrm{Hom}_A(N,N)$ is a B-isomorphism.

The converse is obvious.

[]

An A-module chain map $g:C \longrightarrow D$ is a <u>B-equivalence</u> if

$$1\otimes g : B\otimes_A C \longrightarrow B\otimes_A D$$

is a B-module chain equivalence.

If (C,ϕ) is an n-dimensional ε-symmetric B-Poincaré complex over A and $g:C \longrightarrow D$ is a B-equivalence with D an n-dimensional A-module chain complex then $(D,g^{\%}(\phi))$ is also an n-dimensional ε-symmetric B-Poincaré complex. Furthermore, $((g\ 1):C\oplus D \longrightarrow D,(0,\phi\oplus -g^{\%}(\phi)))$ is an $(n+1)$-dimensional ε-symmetric B-Poincaré pair over A, so that

$$(C,\phi) = (D,g^{\%}(\phi)) \in \Gamma^n(f,\varepsilon) \ .$$

Similarly for the ε-quadratic and even ε-symmetric cases.

The semi-periodicities of the L-groups given by Propositions 1.2.3 i), 1.4.2 extend to the Γ-groups:

<u>Proposition 2.4.2</u> If $f:A \longrightarrow B$ is locally epic the skew-suspension maps

$$\begin{cases} \bar{S} : \Gamma^n(f,\varepsilon) \longrightarrow \Gamma\langle v_0\rangle^{n+2}(f,-\varepsilon) \ ; \ (C,\phi) \longmapsto (SC,\bar{S}\phi) \\ \bar{S} : \Gamma_n(f,\varepsilon) \longrightarrow \Gamma_{n+2}(f,-\varepsilon) \ ; \ (C,\psi) \longmapsto (SC,\bar{S}\psi) \end{cases} \quad (n \geqslant 0)$$

are isomorphisms.

<u>Proof</u>: Given an n-dimensional $\begin{cases} \varepsilon\text{-symmetric} \\ \varepsilon\text{-quadratic} \end{cases}$ B-Poincaré complex over A $\begin{cases} (C,\phi \in Q^n(C,\varepsilon)) \\ (C,\psi \in Q_n(C,\varepsilon)) \end{cases}$ and an $(n+3)$-dimensional $\begin{cases} \text{even } (-\varepsilon)\text{-symmetric} \\ (-\varepsilon)\text{-quadratic} \end{cases}$ B-Poincaré pair over A $\begin{cases} (g:SC \longrightarrow D,(\delta\phi,\bar{S}\phi) \in Q\langle v_0\rangle^{n+3}(g,-\varepsilon)) \\ (g:SC \longrightarrow D,(\delta\psi,\bar{S}\psi) \in Q_{n+3}(g,-\varepsilon)) \end{cases}$ we shall define an

(n+1)-dimensional $\begin{cases}\varepsilon\text{-symmetric}\\ \varepsilon\text{-quadratic}\end{cases}$ B-Poincaré pair over A

$$\begin{cases}(g':C \longrightarrow D',(\delta\phi',\phi) \in Q^{n+1}(g',\varepsilon))\\ (g':C \longrightarrow D',(\delta\psi',\psi) \in Q_{n+1}(g',\varepsilon))\end{cases}$$

as follows.

Without loss of generality it may be assumed that

$$C_r = 0 \ (r<0, r>n) \ , \ D_r = 0 \ (r<0, r>n+3)$$

and that in the ε-symmetric case

$$\delta\phi_{n+3} = 0 : D^{n+3} \longrightarrow D_{n+3} \ .$$

Define an (n+3)-dimensional $\begin{cases}\text{even } (-\varepsilon)\text{-symmetric}\\ (-\varepsilon)\text{-quadratic}\end{cases}$ B-Poincaré pair over A $\begin{cases}(g'':SC \longrightarrow D'',(\delta\phi'',\bar{S}\phi) \in Q\langle v_0\rangle^{n+3}(g'',-\varepsilon))\\ (g'':SC \longrightarrow D'',(\delta\psi'',\bar{S}\psi) \in Q_{n+3}(g',-\varepsilon))\end{cases}$ by

$$g'' = \begin{cases} g : C_{r-1} \longrightarrow D''_r = D_r \quad (2 \leqslant r \leqslant n+1)\\ \begin{pmatrix} g \\ 0\end{pmatrix} : C_0 \longrightarrow D''_1 = D_1 \oplus D^{n+3}\end{cases} ,$$

$$D'' : \ldots \longrightarrow 0 \longrightarrow D_{n+3} \xrightarrow{\begin{pmatrix} d_D \\ (-)^{n+3}\end{pmatrix}} D_{n+2} \oplus D_{n+3} \xrightarrow{(d_D \ \ 0)} D_{n+1} \xrightarrow{d_D} \ldots$$

$$\longrightarrow D_2 \xrightarrow{\begin{pmatrix} d_D \\ 0\end{pmatrix}} D_1 \oplus D^{n+3} \xrightarrow{\left(d_D \ \begin{cases}-\delta\phi_0\\ -(1+T_{-\varepsilon})\delta\psi_0\end{cases}\right)} D_0 \longrightarrow 0 \longrightarrow \ldots ,$$

with $\begin{cases}\delta\phi''_s = \delta\phi_s \oplus 0\\ \delta\psi''_s = \delta\psi_s \oplus 0\end{cases}$ $(s \geqslant 0)$ except for

$$\delta\phi''_0 = \begin{cases} \begin{pmatrix} \delta\phi_0 & 0 \\ 0 & (-)^{n+3}\varepsilon \end{pmatrix} : D''^1 = D^1 \oplus D_{n+3} \longrightarrow D''_{n+2} = D_{n+2} \oplus D^{n+3} \\ \begin{pmatrix} \delta\phi_0 & 0 \\ 0 & 1 \end{pmatrix} : D''^{n+2} = D^{n+2} \oplus D^{n+3} \longrightarrow D''_1 = D_1 \oplus D^{n+3} \end{cases}$$

$$\delta\psi''_0 = \begin{pmatrix} \delta\psi_0 & 0 \\ 0 & 1 \end{pmatrix} : D''^{n+2} = D^{n+2} \oplus D^{n+3} \longrightarrow D''_1 = D_1 \oplus D^{n+3} \quad .$$

Now $1\otimes d_{D''} \in \mathrm{Hom}_B(B\otimes_A D''_1, B\otimes_A D''_0)$ is onto. Stabilizing if necessary it may be assumed that D''_1 and D''_0 are f.g. free A-modules, so that by Proposition 2.4.1 there exists an A-module morphism $e \in \mathrm{Hom}_A(D''_0, D''_1)$ such that $d_{D''}e \in \mathrm{Hom}_A(D''_0, D''_0)$ is a B-isomorphism. Define an $(n+1)$-dimensional A-module chain complex D' and a B-equivalence

$$h : D'' \longrightarrow SD'$$

by

$$d_{D'} = d_{D''} : D'_r = D''_{r+1} \longrightarrow D'_{r-1} = D''_r \quad (r \neq 0,1,2,3)$$

$$h = 1 : D''_r \longrightarrow D'_{r-1} = D''_r \quad (r \neq 0,1,2)$$

$$\begin{array}{ccccccccccc} D'' : \dots \longrightarrow & D''_3 & \xrightarrow{d_{D''}} & D''_2 & \xrightarrow{d_{D''}} & D''_1 & \xrightarrow{d_{D''}} & D''_0 & \longrightarrow & 0 & \longrightarrow \dots \\ h \big\downarrow & \big\downarrow 1 & & \big\downarrow \begin{pmatrix} 1 \\ 0 \end{pmatrix} & & \big\downarrow 1 & & \big\downarrow & & \big\downarrow & \\ SD' : \dots \longrightarrow & D''_3 & \xrightarrow{\begin{pmatrix} d_{D''} \\ 0 \end{pmatrix}} & D''_2 \oplus D''_0 & \xrightarrow{(d_{D''} \;\; e)} & D''_1 & \longrightarrow & 0 & \longrightarrow & 0 & \longrightarrow \dots \end{array}$$

Then

$$g' = hg'' : C \longrightarrow D' \ , \quad \begin{cases} (\delta\phi', \phi) = (h^{\%}\delta\phi'', \phi) \in Q^{n+1}(g', \varepsilon) \\ (\delta\psi', \psi) = (h_{\%}\delta\psi'', \psi) \in Q_{n+1}(g', \varepsilon) \end{cases}$$

define an (n+1)-dimensional $\begin{cases}\varepsilon\text{-symmetric}\\ \varepsilon\text{-quadratic}\end{cases}$ B-Poincaré pair over A

$\begin{cases}(g':C \longrightarrow D',(\delta\phi',\phi))\\ (g':C \longrightarrow D',(\delta\psi',\psi))\end{cases}$, as required.

The above construction shows that the skew-suspension map $\begin{cases}\bar{S}:\Gamma^n(f,\varepsilon) \longrightarrow \Gamma\langle v_0\rangle^{n+2}(f,-\varepsilon)\\ \bar{S}:\Gamma_n(f,\varepsilon) \longrightarrow \Gamma_{n+2}(f,-\varepsilon)\end{cases}$ is one-one; to see that it is also onto set $\begin{cases}(C,\phi) = 0\\ (C,\psi) = 0\end{cases}$ in the construction, which now associates to an (n+3)-dimensional $\begin{cases}\text{even } (-\varepsilon)\text{-symmetric}\\ (-\varepsilon)\text{-quadratic}\end{cases}$ B-Poincaré complex over A $\begin{cases}(D,\delta\phi)\\ (D,\delta\psi)\end{cases}$ $(n \geqslant -1)$ a B-Poincaré cobordant skew-suspension $\begin{cases}\bar{S}(D',\delta\phi')\\ \bar{S}(D',\delta\psi')\end{cases}$.

[]

Define the <u>lower</u> $\begin{cases}\underline{\varepsilon\text{-symmetric}}\\ \underline{\varepsilon\text{-quadratic}}\end{cases}$ <u>Γ-groups</u> $\begin{cases}\Gamma^n(f,\varepsilon)\\ \Gamma_n(f,\varepsilon)\end{cases}$ $(n \leqslant -1)$ by

$$\begin{cases}\Gamma^n(f,\varepsilon) = \begin{cases}\Gamma\langle v_0\rangle^{n+2}(f,-\varepsilon)\\ \Gamma_n(f,\varepsilon) \text{ (as defined below)}\end{cases} \text{ if } \begin{cases}n = -1,-2\\ n \leqslant -3\end{cases}\\ \Gamma_n(f,\varepsilon) = \Gamma_{n+2i}(f,(-)^i\varepsilon) \quad (n+2i \geqslant 0)\end{cases},$$

thus extending the semi-periodicity $\begin{cases}\Gamma^n(f,\varepsilon) = \Gamma\langle v_0\rangle^{n+2}(f,-\varepsilon)\\ \Gamma_n(f,\varepsilon) = \Gamma_{n+2}(f,-\varepsilon)\end{cases}$ $(n \geqslant 0)$ of Proposition 2.4.2.

We shall justify the above definitions of the unified Γ-groups $\begin{cases}\Gamma^n(f,\varepsilon) \\ \Gamma_n(f,\varepsilon)\end{cases}$ $(n \in \mathbb{Z})$ by extending the definition of the relative L-groups in §2.2 to relative Γ-groups. First, however, we shall express the Γ-groups for $n \leqslant 1$ in terms of forms and formations, extending the expressions of the L-groups for $n \leqslant 1$ as Witt groups in §1.6.

An $\begin{cases}\varepsilon\text{-symmetric} \\ \varepsilon\text{-quadratic}\end{cases}$ form over A $\begin{cases}(M,\phi \in Q^{\varepsilon}(M)) \\ (M,\psi \in Q_{\varepsilon}(M))\end{cases}$ is <u>B-non-singular</u> if $\begin{cases}\phi \in \mathrm{Hom}_A(M,M^*) \\ (1+T_{\varepsilon})\psi \in \mathrm{Hom}_A(M,M^*)\end{cases}$ is a B-isomorphism, i.e. if $\begin{cases}B\otimes_A(M,\phi) \\ B\otimes_A(M,\psi)\end{cases}$ is a non-singular $\begin{cases}\varepsilon\text{-symmetric} \\ \varepsilon\text{-quadratic}\end{cases}$ form over B.

A <u>B-lagrangian</u> of a B-non-singular $\begin{cases}\varepsilon\text{-symmetric} \\ \varepsilon\text{-quadratic}\end{cases}$ form over A $\begin{cases}(M,\phi) \\ (M,\psi)\end{cases}$ is a morphism of $\begin{cases}\varepsilon\text{-symmetric} \\ \varepsilon\text{-quadratic}\end{cases}$ forms over A

$$\begin{cases} j : (L,0) \longrightarrow (M,\phi) \\ j : (L,0) \longrightarrow (M,\psi) \end{cases}$$

which becomes the inclusion of a lagrangian over B, i.e. such that the sequence of A-modules

$$\begin{cases} 0 \longrightarrow L \xrightarrow{\;\;j\;\;} M \xrightarrow{\;\;j^*\phi\;\;} L^* \longrightarrow 0 \\ 0 \longrightarrow L \xrightarrow{\;\;j\;\;} M \xrightarrow{\;\;j^*(1+T_{\varepsilon})\psi\;\;} L^* \longrightarrow 0 \end{cases}$$

induces an exact sequence of B-modules. A B-non-singular $\begin{cases}\varepsilon\text{-symmetric} \\ \varepsilon\text{-quadratic}\end{cases}$ form over A is <u>B-hyperbolic</u> if it admits a B-lagrangian.

The $\begin{cases}\underline{\varepsilon\text{-symmetric}} \\ \underline{\text{even } \varepsilon\text{-symmetric}} \\ \underline{\varepsilon\text{-quadratic}}\end{cases}$ $\underline{\text{Witt } \Gamma\text{-group}}$ $\begin{cases}\Gamma^{\varepsilon}(f) \\ \Gamma\langle v_0\rangle^{\varepsilon}(f) \\ \Gamma_{\varepsilon}(f)\end{cases}$ of a locally epic morphism $f:A \longrightarrow B$ is the abelian group of equivalence classes of B-non-singular $\begin{cases}\varepsilon\text{-symmetric} \\ \text{even } \varepsilon\text{-symmetric} \\ \varepsilon\text{-quadratic}\end{cases}$ forms over A subject to the relation

$(M,\phi) \sim (M',\phi')$ if there exists an isomorphism of forms

$$g : (M,\phi)\oplus(H,\theta) \xrightarrow{\sim} (M',\phi')\oplus(H',\theta')$$

for some B-hyperbolic forms $(H,\theta),(H',\theta')$.

<u>Proposition 2.4.3</u> i) There is a natural one-one correspondence between the homotopy equivalence classes of 0-dimensional $\begin{cases}\text{(even) } \varepsilon\text{-symmetric} \\ \varepsilon\text{-quadratic}\end{cases}$ B-Poincaré complexes over A and the isomorphism classes of B-non-singular $\begin{cases}\text{(even) } \varepsilon\text{-symmetric} \\ \varepsilon\text{-quadratic}\end{cases}$ forms over A.

ii) There is a natural identification of the 0-dimensional Γ-groups of $f:A \longrightarrow B$ with the Witt groups of B-non-singular forms over A

$$\begin{cases} \Gamma^0(f,\varepsilon) = \Gamma^{\varepsilon}(f) \\ \Gamma\langle v_0\rangle^0(f,\varepsilon) = \Gamma\langle v_0\rangle^{\varepsilon}(f) \\ \Gamma_0(f,\varepsilon) = \Gamma_{\varepsilon}(f) \end{cases} .$$

<u>Proof</u>: i) Immediate from Proposition 1.6.1.

ii) Given a 1-dimensional $\begin{cases}\text{(even) } \varepsilon\text{-symmetric} \\ \varepsilon\text{-quadratic}\end{cases}$ B-Poincaré pair

over A $\begin{cases}(g:C \longrightarrow D, (\delta\phi,\phi) \in Q^1(g,\varepsilon)) \\ (g:C \longrightarrow D, (\delta\psi,\psi) \in Q_1(g,\varepsilon))\end{cases}$ such that

$$C_r = 0 \ (r \neq 0) \ , \ D_r = 0 \ (r \neq),1)$$

there is defined a B-lagrangian

$$\begin{cases} \begin{pmatrix} g^* \\ d_D^* \\ T_\varepsilon \delta\phi_0 \end{pmatrix} : (D^0,0) \longrightarrow (C^0,\phi_0) \oplus (D^1 \oplus D_1, \begin{pmatrix} T_\varepsilon \delta\phi_1 & \varepsilon \\ 1 & 0 \end{pmatrix}) \\ \begin{pmatrix} g^* \\ d_D^* \\ (1+T_\varepsilon)\delta\psi_0 \end{pmatrix} : (D^0,0) \longrightarrow (C_0,\psi_0) \oplus (D^1 \oplus D_1, \begin{pmatrix} 0 & 0 \\ 1 & 0 \end{pmatrix}) \end{cases} ,$$

so that $\begin{cases}(C^0,\phi_0) \\ (C^0,\psi_0)\end{cases}$ is a stably B-hyperbolic $\begin{cases}\text{(even) } \varepsilon\text{-symmetric} \\ \varepsilon\text{-quadratic}\end{cases}$

form over A, representing 0 in $\begin{cases}\Gamma^\varepsilon(f) \ \ (\ \Gamma\langle v_0\rangle^\varepsilon(f)) \\ \Gamma_\varepsilon(f)\end{cases}$. Conversely, stably B-hyperbolic forms correspond to the boundaries of 1-dimensional B-Poincaré pairs under the correspondence of i).

[]

A <u>B-non-singular</u> $\begin{cases}\underline{\text{(even) } \varepsilon\text{-symmetric}} \\ \underline{\varepsilon\text{-quadratic}}\end{cases}$ <u>formation over A</u>

$\begin{cases}(M,\phi;F,G) \\ (M,\psi;F,G)\end{cases}$ is a non-singular $\begin{cases}\text{(even) } \varepsilon\text{-symmetric} \\ \varepsilon\text{-quadratic}\end{cases}$ form over A

$\begin{cases}(M,\phi) \\ (M,\psi)\end{cases}$ together with a lagrangian F and a B-lagrangian

$\begin{cases}(G,0) \longrightarrow (M,\phi) \\ (G,0) \longrightarrow (M,\psi)\end{cases}$. Thus $\begin{cases}B\boxtimes_A(M,\phi;F,G) \\ B\boxtimes_A(M,\psi;F,G)\end{cases}$ is a non-singular

$\begin{cases} \text{(even) } \varepsilon\text{-symmetric} \\ \varepsilon\text{-quadratic} \end{cases}$ formation over B. There are evident notions of <u>isomorphism</u> and <u>stable isomorphism</u> for B-non-singular formations, generalizing the case $f = 1 : A \longrightarrow B = A$ (already treated in §1.6).

<u>Proposition 2.4.4</u> i) There is a natural one-one correspondence between the $\begin{cases} - \\ \varepsilon\text{-quadratic} \end{cases}$ homotopy equivalence classes of 1-dimensional $\begin{cases} \text{(even) } \varepsilon\text{-symmetric} \\ \varepsilon\text{-quadratic} \end{cases}$ B-Poincaré complexes over A and the stable isomorphism classes of B-non-singular $\begin{cases} \text{(even) } \varepsilon\text{-symmetric} \\ \varepsilon\text{-quadratic} \end{cases}$ formations over A.

ii) The 1-dimensional Γ-groups $\begin{cases} \Gamma^1(f,\varepsilon) \quad (\Gamma\langle v_0\rangle^1(f,\varepsilon)) \\ \Gamma_1(f,\varepsilon) \end{cases}$ have natural expressions as Witt groups of B-non-singular $\begin{cases} \text{(even) } \varepsilon\text{-symmetric} \\ \varepsilon\text{-quadratic} \end{cases}$ formations over A.

iii) The forgetful map

$$\begin{cases} \Gamma\langle v_0\rangle^1(f,\varepsilon) \longrightarrow L\langle v_0\rangle^1_X(B,\varepsilon) \ ; \ (M,\phi;F,G) \longmapsto B\otimes_A(M,\phi;F,G) \\ \Gamma_1(f,\varepsilon) \longrightarrow L^X_1(B,\varepsilon) \ ; \ (M,\psi;F,G) \longmapsto B\otimes_A(M,\psi;F,G) \end{cases}$$

is one-one, where

$$X = \operatorname{im}(\tilde{K}_0(A) \longrightarrow \tilde{K}_0(B)) \subseteq \tilde{K}_0(B) \ .$$

<u>Proof</u>: i) A straightforward generalization of Proposition 1.6.4.

ii) Immediate from i).

iii) Let $\begin{cases}(C,\phi\in Q\langle v_0\rangle^1(C,\varepsilon))\\(C,\psi\in Q_1(C,\varepsilon))\end{cases}$ be a 1-dimensional $\begin{cases}\text{even } \varepsilon\text{-symmetric}\\ \varepsilon\text{-quadratic}\end{cases}$

B-Poincaré complex over A such that $C_r = 0$ $(r \neq 0,1)$,

C_1 is f.g. free and

$$\begin{cases}B\otimes_A(C,\phi) = 0 \in L\langle v_0\rangle^1_X(B,\varepsilon)\\ B\otimes_A(C,\psi) = 0 \in L_1^X(B,\varepsilon)\end{cases} .$$

By Proposition 1.6.5 iii) there exists a 2-dimensional

$\begin{cases}\text{even } \varepsilon\text{-symmetric}\\ \varepsilon\text{-quadratic}\end{cases}$ Poincaré pair over B

$\begin{cases}(g:B\otimes_A C \longrightarrow D,(0,1\otimes\phi)\in Q\langle v_0\rangle^2(g,\varepsilon))\\(g:B\otimes_A C \longrightarrow D,(0,1\otimes\psi)\in Q_2(g,\varepsilon))\end{cases}$ with $D_r = 0$ $(r \neq 1)$

and $[D_1]\in X\subseteq \widetilde{K}_0(B)$. Stabilizing if necessary it may be assumed that $D_1 = B\otimes_A D_1'$ for some f.g. projective A-module D_1'. Let D_1'' be a f.g. projective A-module such that $D_1'\oplus D_1''$ is a f.g. free A-module, and let $\begin{pmatrix} g\\ 0\end{pmatrix} : B\otimes_A C_1 \longrightarrow B\otimes_A(D_1'\oplus D_1'')$ have matrix representation (b_{ij}) $(b_{ij}\in B)$ with respect to the B-module bases induced from A-module bases of C_1 and $D_1'\oplus D_1''$.
As $f:A\longrightarrow B$ is locally epic there exists a unit $u\in B$ such that $ub_{ij} = f(a_{ij})\in B$ for some matrix (a_{ij}) with entries $a_{ij}\in A$. Define an A-module morphism $g'\in \mathrm{Hom}_A(C_1,D_1')$ by

$$g' : C_1 \xrightarrow{(a_{ij})} D_1'\oplus D_1'' \xrightarrow{(1\ 0)} D_1' .$$

Then $\begin{cases}(g':C\longrightarrow D',(0,\phi)\in Q\langle v_0\rangle^2(g',\varepsilon))\\(g':C\longrightarrow D',(0,\psi)\in Q_2(g',\varepsilon))\end{cases}$ is a 2-dimensional

$\begin{cases}\text{even } \varepsilon\text{-symmetric}\\ \varepsilon\text{-quadratic}\end{cases}$ B-Poincaré pair over A (with $D_r' = 0$ for $r\neq 1$),

so that

$$\begin{cases} (C,\phi) = 0 \in \Gamma\langle v_0\rangle^1(f,\varepsilon) \\ (C,\psi) = 0 \in \Gamma_1(f,\varepsilon) \end{cases} .$$

[]

Let F be a commutative square of rings with involution

$$\begin{array}{ccc} A & \longrightarrow & A' \\ f\downarrow & F & \downarrow f' \\ B & \longrightarrow & B' \end{array} ,$$

with $f:A \longrightarrow B$ and $f':A' \longrightarrow B'$ locally epic morphisms.

Define the <u>(n+1)-dimensional relative</u> $\begin{cases} \text{\underline{(even) } \varepsilon\text{-symmetric}} \\ \text{\underline{\varepsilon-quadratic}} \end{cases}$

<u>Γ-group</u> $\begin{cases} \Gamma^{n+1}(F,\varepsilon) \quad (\Gamma\langle v_0\rangle^{n+1}(F,\varepsilon)) \\ \Gamma_{n+1}(F,\varepsilon) \end{cases}$ $(n \geqslant 0)$ to be the relative

cobordiam group of pairs

(n-dimensional $\begin{cases} \text{(even) } \varepsilon\text{-symmetric} \\ \varepsilon\text{-quadratic} \end{cases}$ B-Poincaré complex

over A x, (n+1)-dimensional $\begin{cases} \text{(even) } \varepsilon\text{-symmetric} \\ \varepsilon\text{-quadratic} \end{cases}$

B'-Poincaré pair over A' with boundary $A'\otimes_A x$) .

As usual, the skew-suspension maps

$$\begin{cases} \bar{S} : \Gamma^n(F,\varepsilon) \longrightarrow \Gamma\langle v_0\rangle^{n+2}(F,-\varepsilon) \\ \bar{S} : \Gamma_n(F,\varepsilon) \longrightarrow \Gamma_{n+2}(F,-\varepsilon) \end{cases} \quad (n \geqslant 1)$$

are isomorphisms. The <u>lower relative Γ-groups</u> are defined by

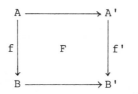

$$\Gamma^n(F,\varepsilon) = \begin{cases} \Gamma\langle v_0\rangle^{n+2}(F,-\varepsilon) & (n = 0,-1) \\ \text{coker}((\ker(1+T_{-\varepsilon}:\Gamma_0(f',-\varepsilon)\longrightarrow \Gamma\langle v_0\rangle^0(f',-\varepsilon)) \\ \qquad\longrightarrow \Gamma_0(F,-\varepsilon)) & (n = -2) \\ \Gamma_n(F,\varepsilon) \text{ (as defined below)} & (n \leqslant -3) \end{cases}$$

$$\Gamma_n(F,\varepsilon) = \Gamma_{n+2i}(F,(-)^i\varepsilon) \quad (n \leqslant 0,\ n+2i \geqslant 1) \quad ,$$

generalizing the definition of the lower relative L-groups in §2.2 (the case $f = 1 : A \longrightarrow B = A$, $f' = 1 : A' \longrightarrow B' = A'$).

Proposition 2.4.5 The relative Γ-groups $\begin{cases} \Gamma^*(F,\varepsilon) \\ \Gamma_*(F,\varepsilon) \end{cases}$ fit into a change of rings exact sequence

$$\begin{cases} \ldots \longrightarrow \Gamma^{n+1}(f',\varepsilon) \longrightarrow \Gamma^{n+1}(F,\varepsilon) \longrightarrow \Gamma^n(f,\varepsilon) \longrightarrow \Gamma^n(f',\varepsilon) \longrightarrow \ldots \\ \ldots \longrightarrow \Gamma_{n+1}(f',\varepsilon) \longrightarrow \Gamma_{n+1}(F,\varepsilon) \longrightarrow \Gamma_n(f,\varepsilon) \longrightarrow \Gamma_n(f',\varepsilon) \longrightarrow \ldots \end{cases}$$

$(n \in \mathbb{Z})$.

[]

Given a morphism of rings with involution $f:A \longrightarrow B$ we shall say that an A-module chain complex is B-acyclic if

$$H_*(B\otimes_A C) = 0 \ .$$

A finite-dimensional A-module chain complex C is B-acyclic if and only if $B\otimes_A C$ is a chain contractible B-module chain complex.

An $\begin{cases} \varepsilon\text{-symmetric} \\ \varepsilon\text{-quadratic} \end{cases}$ complex (resp. pair) over A $\begin{cases} (C,\phi) \\ (C,\psi) \end{cases}$ (resp. $\begin{cases} (C \longrightarrow D,(\delta\phi,\phi)) \\ (C \longrightarrow D,(\delta\psi,\psi)) \end{cases}$) is B-acyclic if C (resp. C,D) are B-acyclic A-module chain complexes.

In Propositions 2.4.6,2.4.7 below we shall express the relative Γ-groups $\left\{\begin{array}{l}\Gamma^*(F,\varepsilon)\\ \Gamma_*(F,\varepsilon)\end{array}\right.$ for a commutative square of the type

$$\begin{array}{ccc} A & \xrightarrow{1} & A \\ \downarrow & F & \downarrow \\ B & \longrightarrow & B' \end{array}$$

as the cobordism groups of algebraic B-Poincaré B'-acyclic complexes over A. In §3 this expression will be used in the special case

$$\begin{array}{ccc} A & \xrightarrow{1} & A \\ 1\downarrow & F & \downarrow \\ A & \longrightarrow & S^{-1}A \end{array}$$

to obtain the localization exact sequence in algebraic L-theory. We shall give a geometric interpretation of this expression in Proposition 7.7.2.

<u>Proposition 2.4.6</u> Let F be a commutative square of rings with involution of the type

$$\begin{array}{ccc} A & \xrightarrow{1} & A \\ f\downarrow & F & \downarrow f' \\ B & \longrightarrow & B' \end{array},$$

with $f:A\longrightarrow B$ and $f':A\longrightarrow B'$ locally epic.

i) The relative $\left\{\begin{array}{l}\varepsilon\text{-symmetric}\\ \varepsilon\text{-quadratic}\end{array}\right.$ Γ-group $\left\{\begin{array}{l}\Gamma^n(F,\varepsilon)\\ \Gamma_n(F,\varepsilon)\end{array}\right.$ $(n\geqslant 1)$ is naturally isomorphic to the cobordism group of connected (n+1)-dimensional $\left\{\begin{array}{l}\text{even } (-\varepsilon)\text{-symmetric}\\ (-\varepsilon)\text{-quadratic}\end{array}\right.$ B-Poincaré B'-acyclic complexes over A.

The maps appearing in the Γ-group change of rings exact sequence are given by

$$\begin{cases} \Gamma^n(f',\varepsilon) \longrightarrow \Gamma^n(F,\varepsilon) \ ; \ (C,\phi) \longmapsto \bar{S}\partial(C,\phi) \\ \Gamma_n(f',\varepsilon) \longrightarrow \Gamma_n(F,\varepsilon) \ ; \ (C,\psi) \longmapsto \bar{S}\partial(C,\psi) \end{cases} \quad (n \geqslant 1)$$

$$\begin{cases} \Gamma^n(F,\varepsilon) \longrightarrow \Gamma^{n-1}(f,\varepsilon) = \Gamma\langle v_0\rangle^{n+1}(f,-\varepsilon) \, ; \, (C,\phi) \longmapsto (C,\phi) \\ \Gamma_n(F,\varepsilon) \longrightarrow \Gamma_{n-1}(f,\varepsilon) = \Gamma_{n+1}(f,-\varepsilon) \ ; \ (C,\psi) \longmapsto (C,\psi) \end{cases} \quad (n \geqslant 1) \ .$$

ii) $\begin{cases} \Gamma^0(F,\varepsilon) \\ \Gamma_0(F,\varepsilon) \end{cases}$ is naturally isomorphic to the cobordism group of 1-dimensional $\begin{cases} \text{even } (-\varepsilon)\text{-symmetric} \\ (-\varepsilon)\text{-quadratic} \end{cases}$ B-Poincaré B'-acyclic complexes over A.

iii) $\Gamma_n(F,\varepsilon)$ $(n \geqslant 2)$ is naturally isomorphic to the cobordism group of $(n-1)$-dimensional ε-quadratic B-Poincaré B'-acyclic complexes over A.

<u>Proof</u>: i) An element of $\begin{cases} \Gamma^n(F,\varepsilon) \\ \Gamma_n(F,\varepsilon) \end{cases}$ $(n \geqslant 1)$ is the cobordism class of a pair

$((n-1)$-dimensional $\begin{cases} \varepsilon\text{-symmetric} \\ \varepsilon\text{-quadratic} \end{cases}$ B-Poincaré complex

over A $\begin{cases} (C,\phi) \\ (C,\psi) \end{cases}$, n-dimensional $\begin{cases} \varepsilon\text{-symmetric} \\ \varepsilon\text{-quadratic} \end{cases}$ B'-Poincaré

pair over A $\begin{cases} (g:C \longrightarrow D,(\delta\phi,\phi)) \\ (g:C \longrightarrow D,(\delta\psi,\psi)) \end{cases}$) ,

by definition. Let $\begin{cases} (C',\phi') \\ (C',\psi') \end{cases}$ be the connected $(n+1)$-dimensional

$\begin{cases}\text{even } (-\varepsilon)\text{-symmetric} \\ (-\varepsilon)\text{-quadratic}\end{cases}$ B-Poincaré B'-acyclic complex over A

obtained from the skew-suspension $\begin{cases}\bar{S}(C,\phi) \\ \bar{S}(C,\psi)\end{cases}$ by surgery

on the connected (n+2)-dimensional $\begin{cases}\text{even } (-\varepsilon)\text{-symmetric} \\ (-\varepsilon)\text{-quadratic}\end{cases}$

pair $\begin{cases}\bar{S}(g:C \longrightarrow D,(\delta\phi,\phi)) \\ \bar{S}(g:C \longrightarrow D,(\delta\psi,\psi))\end{cases}$. Thus $\begin{cases}((C,\phi),(g:C\longrightarrow D,(\delta\phi,\phi))) \\ ((C,\psi),(g:C\longrightarrow D,(\delta\psi,\psi)))\end{cases}$

determines an element $\begin{cases}(C',\phi') \\ (C',\psi')\end{cases}$ of the cobordism group of

connected (n+1)-dimensional $\begin{cases}\text{even } (-\varepsilon)\text{-symmetric} \\ (-\varepsilon)\text{-quadratic}\end{cases}$ B-Poincaré

B'-acyclic complexes over A.

Conversely, let $\begin{cases}(C',\phi') \\ (C',\psi')\end{cases}$ be a connected (n+1)-dimensional

$\begin{cases}\text{even } (-\varepsilon)\text{-symmetric} \\ (-\varepsilon)\text{-quadratic}\end{cases}$ B-Poincaré B'-acyclic complex over A

such that $C'_r = 0$ $(r < 0, r > n+1)$, as is the case up to homotopy equivalence. Surgery on the connected (n+2)-dimensional

$\begin{cases}\text{even } (-\varepsilon)\text{-symmetric} \\ (-\varepsilon)\text{-quadratic}\end{cases}$ pair over A $\begin{cases}(g':C \longrightarrow D',(0,\phi')) \\ (g':C \longrightarrow D',(0,\psi'))\end{cases}$

defined by

$$g' = 1 : C'_{n+1} \longrightarrow D'_{n+1} = C'_{n+1} \ , \ D'_r = 0 \ (r \neq n+1)$$

results in the skew-suspension $\begin{cases}\bar{S}(C,\phi) \\ \bar{S}(C,\psi)\end{cases}$ of an (n-1)-dimensional

$\begin{cases}\varepsilon\text{-symmetric} \\ \varepsilon\text{-quadratic}\end{cases}$ B-Poincaré complex over A $\begin{cases}(C,\phi) \\ (C,\psi)\end{cases}$.

The n-dimensional $\begin{cases}\varepsilon\text{-symmetric}\\ \varepsilon\text{-quadratic}\end{cases}$ pair over A $\begin{cases}(g:C \longrightarrow D,(0,\phi))\\ (g:C \longrightarrow D,(0,\psi))\end{cases}$

defined by

$$g = (0\ 1) : C_0 = C'_1 \oplus C'^{n+1} \longrightarrow D_0 = C'^{n+1} \quad , \quad D_r = 0 \ (r \neq 0)$$

is B'-Poincaré. Thus $\begin{cases}(C',\phi')\\ (C',\psi')\end{cases}$ determines an element

$$\begin{cases}((C,\phi),(g:C \longrightarrow D,(\delta\phi,\phi)) \in \Gamma^n(F,\varepsilon)\\ ((C,\psi),(g:C \longrightarrow D,(\delta\psi,\psi)) \in \Gamma_n(F,\varepsilon)\end{cases} .$$

ii) An element of $\begin{cases}\Gamma^0(F,\varepsilon)\\ \Gamma_0(F,\varepsilon)\end{cases}$ is the cobordism class of a pair

(1-dimensional $\begin{cases}\text{even } (-\varepsilon)\text{-symmetric}\\ (-\varepsilon)\text{-quadratic}\end{cases}$ B-Poincaré complex

over A $\begin{cases}(C,\phi)\\ (C,\psi)\end{cases}$, 2-dimensional $\begin{cases}\text{even } (-\varepsilon)\text{-symmetric}\\ (-\varepsilon)\text{-quadratic}\end{cases}$

B'-Poincaré pair over A $\begin{cases}(g:C \longrightarrow D,(\delta\phi,\phi))\\ (g:C \longrightarrow D,(\delta\psi,\psi))\end{cases}$) ,

by definition. As in the proof of Proposition 2.4.4 iii) it may be assumed that

$$C_r = 0 \ (r \neq 0,1) \ , \ D_r = 0 \ (r \neq 1) \ .$$

It follows that the result of surgery on $\begin{cases}(g:C \longrightarrow D,(\delta\phi,\phi))\\ (g:C \longrightarrow D,(\delta\psi,\psi))\end{cases}$

is a 1-dimensional $\begin{cases}\text{even } (-\varepsilon)\text{-symmetric}\\ (-\varepsilon)\text{-quadratic}\end{cases}$ B-Poincaré B'-acyclic

complex over A $\begin{cases}(C',\phi')\\ (C',\psi')\end{cases}$.

Conversely, given a 1-dimensional $\begin{cases}\text{even } (-\varepsilon)\text{-symmetric} \\ (-\varepsilon)\text{-quadratic}\end{cases}$

B-Poincaré B'-acyclic complex over A $\begin{cases}(C',\phi') \\ (C',\psi')\end{cases}$ there is defined

an element

$$\begin{cases}((C',\phi'),(0:C'\longrightarrow 0,(0,\phi'))) \in \Gamma^0(F,\varepsilon) \\ ((C',\psi'),(0:C'\longrightarrow 0,(0,\psi'))) \in \Gamma_0(F,\varepsilon)\end{cases}$$

iii) Given a connected $(n+1)$-dimensional B-Poincaré B'-acyclic $(-\varepsilon)$-quadratic complex over A (C,ψ) we shall define an $(n-1)$-dimensional B-Poincaré B'-acyclic ε-quadratic complex over A (C',ψ') such that

$$(C,\psi) = \bar{S}(C',\psi') \in \Gamma_{n+1}(F,-\varepsilon)$$

and $(C',\psi') = (C'',\psi'')$ if $(C,\psi) = \bar{S}(C'',\psi'')$, as follows.

Without loss of generality it may be assumed that $C_r = 0$ $(r < 0, r > n+1)$ and that C_n, C_{n+1} are f.g. free A-modules. By Proposition 2.4.2 there exists an A-module morphism $h \in \mathrm{Hom}_A(C_n, C_{n+1})$ such that $hd \in \mathrm{Hom}_A(C_n, C_n)$ is a B'-isomorphism. Define an A-module chain complex D and an A-module chain map $g:C \longrightarrow D$ by

$$\begin{array}{ccccccccccc}
C: & \ldots \longrightarrow & 0 & \longrightarrow & C_{n+1} & \xrightarrow{d} & C_n & \xrightarrow{d} & C_{n-1} & \longrightarrow & \ldots \\
\big\downarrow g & & \big\downarrow & & \big\downarrow 1 & & \big\downarrow h & & \big\downarrow & & \\
D: & \ldots \longrightarrow & 0 & \longrightarrow & C_{n+1} & \xrightarrow{hd} & C_{n+1} & \longrightarrow & 0 & \longrightarrow & \ldots
\end{array}$$

The complex obtained from (C,ψ) by surgery on the connected $(n+2)$-dimensional B'-acyclic $(-\varepsilon)$-quadratic pair over A $(g:C\longrightarrow D,(0,\psi))$ is the skew-suspension $\bar{S}(C',\psi')$ of an $(n-1)$-dimensional B-Poincaré B'-acyclic ε-quadratic complex over A (C',ψ').

[]

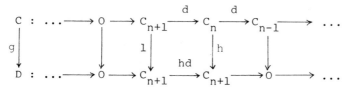

The low-dimensional relative Γ-groups $\begin{cases}\Gamma^n(F,\varepsilon)\\ \Gamma_n(F,\varepsilon)\end{cases}$ $(n \leqslant 1)$ of a commutative square of rings with involution of the type

$$\begin{array}{ccc} A & \xrightarrow{\ 1\ } & A \\ \big\downarrow & F & \big\downarrow \\ B & \longrightarrow & B' \end{array}$$

can be expressed in terms of forms and formations, as follows.

A <u>B-non-singular</u> $\begin{cases}\underline{\text{(even) }\varepsilon\text{-symmetric}}\\ \underline{\varepsilon\text{-quadratic}}\end{cases}$ <u>B'-form over A</u> $\begin{cases}(M,\phi;L)\\ (M,\psi;L)\end{cases}$ is a B-non-singular $\begin{cases}\text{(even) }\varepsilon\text{-symmetric}\\ \varepsilon\text{-quadratic}\end{cases}$ form over A $\begin{cases}(M,\phi)\\ (M,\psi)\end{cases}$ together with a B'-lagrangian $\begin{cases}(L,0)\longrightarrow(M,\phi)\\ (L,0)\longrightarrow(M,\psi)\end{cases}$.

A <u>B-non-singular</u> $\begin{cases}\underline{\text{(even) }\varepsilon\text{-symmetric}}\\ \underline{\varepsilon\text{-quadratic}}\end{cases}$ <u>B'-formation over A</u> $\begin{cases}(M,\phi;F,G)\\ (M,\psi;F,G)\end{cases}$ is a non-singular $\begin{cases}\text{(even) }\varepsilon\text{-symmetric}\\ \varepsilon\text{-quadratic}\end{cases}$ form over A $\begin{cases}(M,\phi)\\ (M,\psi)\end{cases}$ together with a lagrangian F and a B-lagrangian $\begin{cases}(G,0)\longrightarrow(M,\phi)\\ (G,0)\longrightarrow(M,\psi)\end{cases}$ such that the projection $G\longrightarrow M/F$ is a B'-isomorphism.

A <u>B-non-singular split ε-quadratic B'-formation over A</u> (F,G) is a morphism of split ε-quadratic forms over A

$$\left(\begin{pmatrix}\gamma\\ \mu\end{pmatrix},\theta\right)\ :\ (G,0)\longrightarrow \widetilde{H}_{\varepsilon}(F)$$

defining a B-lagrangian of $\widetilde{H}_{\varepsilon}(F)$, and such that $\mu\in \mathrm{Hom}_A(G,F^*)$ is a B'-isomorphism.

The boundary of a B'-non-singular $\begin{cases}\text{(even) }\varepsilon\text{-symmetric}\\ \varepsilon\text{-quadratic}\end{cases}$ formation over A $\begin{cases}(M,\phi;F,G)\\ (M,\psi;F,G)\end{cases}$ is the B-non-singular $\begin{cases}\text{(even) }\varepsilon\text{-symmetric}\\ \varepsilon\text{-quadratic}\end{cases}$ B'-form over A

$$\begin{cases}\partial(M,\phi;F,G) = (M,\phi;G)\\ \partial(M,\psi;F,G) = (M,\psi;G)\ .\end{cases}$$

The boundary of a B'-non-singular $\begin{cases}\varepsilon\text{-symmetric}\\ \text{even }\varepsilon\text{-symmetric}\\ \varepsilon\text{-quadratic}\end{cases}$ form over A $\begin{cases}(M,\phi\in Q^{\varepsilon}(M))\\ (M,\phi\in Q\langle v_0\rangle^{\varepsilon}(M))\\ (M,\psi\in Q_{\varepsilon}(M))\end{cases}$ is the B-non-singular $\begin{cases}\text{even }(-\varepsilon)\text{-symmetric}\\ (-\varepsilon)\text{-quadratic}\\ \text{split }(-\varepsilon)\text{-quadratic}\end{cases}$ B'-formation over A

$$\begin{cases}\partial(M,\phi) = (H^{-\varepsilon}(M);M,\begin{pmatrix}1\\ \phi\end{pmatrix}:(M,0)\longrightarrow H^{-\varepsilon}(M))\\ \partial(M,\phi) = (H_{-\varepsilon}(M);M,\begin{pmatrix}1\\ \phi\end{pmatrix}:(M,0)\longrightarrow H_{-\varepsilon}(M))\\ \partial(M,\psi) = (M,(\begin{pmatrix}1\\ \psi+\varepsilon\psi^*\end{pmatrix},\psi)M)\ .\end{cases}$$

Proposition 2.4.7 Let F be a commutative square of rings with involution of the type

$$\begin{array}{ccc} A & \xrightarrow{1} & A \\ {\scriptstyle f}\downarrow & F & \downarrow{\scriptstyle f'} \\ B & \longrightarrow & B' \end{array}$$

with f and f' locally epic.

i) $\left\{\begin{array}{l}\Gamma^0(F,\varepsilon)\\ \Gamma^{-2}(F,-\varepsilon)\\ \Gamma_0(F,\varepsilon)\end{array}\right.$ is naturally isomorphic to the Witt group

of B-non-singular $\left\{\begin{array}{l}\text{even } (-\varepsilon)\text{-symmetric}\\ (-\varepsilon)\text{-quadratic}\\ \text{split } (-\varepsilon)\text{-quadratic}\end{array}\right.$ B'-formations over A,

with

$$\left\{\begin{array}{l}\Gamma^0(f',\varepsilon) \longrightarrow \Gamma^0(F,\varepsilon) \; ; \; (M,\phi) \longmapsto \partial(M,\phi)\\ \Gamma^{-2}(f',-\varepsilon) \longrightarrow \Gamma^{-2}(F,-\varepsilon) \; ; \; (M,\phi) \longmapsto \partial(M,\phi)\\ \Gamma_0(f',\varepsilon) \longrightarrow \Gamma_0(F,\varepsilon) \; ; \; (M,\psi) \longmapsto \partial(M,\psi)\end{array}\right. .$$

ii) $\left\{\begin{array}{l}\Gamma^1(F,\varepsilon)\\ \Gamma^{-1}(F,-\varepsilon)\\ \Gamma_1(F,\varepsilon)\end{array}\right.$ is naturally isomorphic to the Witt group

of B-non-singular $\left\{\begin{array}{l}\varepsilon\text{-symmetric}\\ \text{even } \varepsilon\text{-symmetric}\\ \varepsilon\text{-quadratic}\end{array}\right.$ B'-forms over A, with

$$\left\{\begin{array}{l}\Gamma^1(f',\varepsilon) \longrightarrow \Gamma^1(F,\varepsilon) \; ; \; (M,\phi;F,G) \longmapsto \partial(M,\phi;F,G)\\ \Gamma^{-1}(f',-\varepsilon) \longrightarrow \Gamma^{-1}(F,-\varepsilon) \; ; \; (M,\phi;F,G) \longmapsto \partial(M,\phi;F,G)\\ \Gamma_1(f',\varepsilon) \longrightarrow \Gamma_1(F,\varepsilon) \; ; \; (M,\psi;F,G) \longmapsto \partial(M,\psi;F,G)\end{array}\right. .$$

<u>Proof</u>: The expression of the low-dimensional relative Γ-groups in terms of forms and formations follows from Proposition 2.4.6 and the following generalizations of the correspondences of Propositions 1.6.1,1.6.4:

i) the homotopy equivalence classes of 1-dimensional $\left\{\begin{array}{l}\text{(even) } \varepsilon\text{-symmetric}\\ \varepsilon\text{-quadratic}\end{array}\right.$ B-Poincaré B'-acyclic complexes over A are in a natural one-one correspondence with equivalence classes of B-non-singular $\left\{\begin{array}{l}\text{(even) } \varepsilon\text{-symmetric}\\ \text{split } \varepsilon\text{-quadratic}\end{array}\right.$ B'-formations over A,

ii) the homotopy equivalence classes of connected 2-dimensional $\left\{\begin{array}{l}\text{(even) } \varepsilon\text{-symmetric}\\ \varepsilon\text{-quadratic}\end{array}\right.$ B-Poincaré B'-acyclic complexes over A are in a natural one-one correspondence with equivalence classes of B-non-singular $\left\{\begin{array}{l}\text{(even) } (-\varepsilon)\text{-symmetric}\\ (-\varepsilon)\text{-quadratic}\end{array}\right.$ B'-forms over A.

(We shall give a more detailed account of these correspondences in §3 below, in the special case

$$\begin{array}{ccc} A & \xrightarrow{\ 1\ } & A \\ {\scriptstyle f=1}\big\downarrow & F & \big\downarrow{\scriptstyle f'=\text{inclusion}} \\ A & \longrightarrow & S^{-1}A \end{array}$$

).

[]

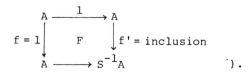

2.5 Change of K-theory

There are evident extensions of all the results of §§2.1 - 2.4 to the intermediate L-groups of §1.10 and their intermediate Γ-group analogues. Here, we shall only state the extensions for which we shall need a reference.

Given a morphism of rings with involution

$$f : A \longrightarrow B$$

and *-invariant subgroups $X \subseteq \tilde{K}_m(A)$, $Y \subseteq \tilde{K}_m(B)$ $(m = 0,1)$ such that

$$B \otimes_A X \subseteq Y \subseteq \tilde{K}_m(B)$$

define the relative intermediate $\begin{cases} \underline{\varepsilon\text{-symmetric}} \\ \underline{\varepsilon\text{-quadratic}} \end{cases}$ L-groups

$\begin{cases} L^n_{Y,X}(f,\varepsilon) \\ L_n^{Y,X}(f,\varepsilon) \end{cases}$ $(n \in \mathbb{Z})$ in the same way as the relative L-groups

$\begin{cases} L^*(f,\varepsilon) \\ L_*(f,\varepsilon) \end{cases}$ (which are the special case $X = \tilde{K}_0(A)$, $Y = \tilde{K}_0(B)$)

but using only algebraic Poincaré complexes over A with K-theory in X and algebraic Poincaré cobordisms over B with K-theory in Y.

Given a morphism of $\mathbb{Z}[\mathbb{Z}_2]$-modules

$$f : G \longrightarrow H$$

define the relative Tate $\mathbb{Z}_2$-cohomology groups $\hat{H}^*(\mathbb{Z}_2;f)$ by

$$\hat{H}^n(\mathbb{Z}_2;f) = \frac{\{(x,y) \in G \oplus H \mid Tx = (-)^{n-1}x,\ fx = y + (-)^{n-1}Ty\}}{\{(u + (-)^{n-1}Tu, fu + v + (-)^n Tv) \mid (u,v) \in G \oplus H\}} \quad (n \pmod 2),$$

to fit into the long exact sequence

$$\ldots \longrightarrow \hat{H}^{n+1}(\mathbb{Z}_2;H) \longrightarrow \hat{H}^{n+1}(\mathbb{Z}_2;f) \longrightarrow \hat{H}^n(\mathbb{Z}_2;G) \xrightarrow{f} \hat{H}^n(\mathbb{Z}_2;H) \longrightarrow \ldots .$$

<u>Proposition 2.5.1</u> Given a morphism of rings with involution

$$f : A \longrightarrow B$$

and $*$-invariant subgroups $X \subseteq X' \subseteq \tilde{K}_m(A)$, $Y \subseteq Y' \subseteq \tilde{K}_m(B)$ $(m = 0,1)$ such that

$$B\otimes_A X \subseteq Y \subseteq \tilde{K}_m(B)\ ,\ B\otimes_A X' \subseteq Y' \subseteq \tilde{K}_m(B)$$

there is defined a commutative diagram of abelian groups with exact rows and columns

$$\begin{array}{ccccccccc}
 & \vdots & & \vdots & & \vdots & & \vdots & \\
 & \downarrow & & \downarrow & & \downarrow & & \downarrow & \\
\cdots\longrightarrow & L^n_X(A,\varepsilon) & \xrightarrow{f} & L^n_Y(B,\varepsilon) & \longrightarrow & L^n_{Y,X}(f,\varepsilon) & \longrightarrow & L^{n-1}_X(A,\varepsilon) & \longrightarrow\cdots \\
 & \downarrow & & \downarrow & & \downarrow & & \downarrow & \\
\cdots\longrightarrow & L^n_{X'}(A,\varepsilon) & \xrightarrow{f} & L^n_{Y'}(B,\varepsilon) & \longrightarrow & L^n_{Y',X'}(f,\varepsilon) & \longrightarrow & L^{n-1}_{X'}(A,\varepsilon) & \longrightarrow\cdots \\
 & \downarrow & & \downarrow & & \downarrow & & \downarrow & \\
\cdots\longrightarrow & \hat{H}^n(\mathbb{Z}_2;X'/X) & \xrightarrow{f} & \hat{H}^n(\mathbb{Z}_2;Y'/Y) & \longrightarrow & \hat{H}^n(\mathbb{Z}_2;f) & \longrightarrow & \hat{H}^{n-1}(\mathbb{Z}_2;X'/X) & \longrightarrow\cdots \\
 & \downarrow & & \downarrow & & \downarrow & & \downarrow & \\
\cdots\longrightarrow & L^{n-1}_X(A,\varepsilon) & \xrightarrow{f} & L^{n-1}_Y(B,\varepsilon) & \longrightarrow & L^{n-1}_{Y,X}(f,\varepsilon) & \longrightarrow & L^{n-2}_X(A,\varepsilon) & \longrightarrow\cdots \\
 & \downarrow & & \downarrow & & \downarrow & & \downarrow & \\
 & \vdots & & \vdots & & \vdots & & \vdots &
\end{array}$$

Similarly for the ε-quadratic L-groups L_*.

<u>Proof</u>: Immediate from Propositions 1.10.1, 2.2.4.

[]

Given a locally epic morphism of rings with involution

$$f : A \longrightarrow B$$

and a $*$-invariant subgroup $X \subseteq \tilde{K}_m(B)$ $(m = 0,1)$ define the

<u>intermediate</u> $\left\{\begin{array}{l}\underline{\varepsilon\text{-symmetric}}\\ \underline{\varepsilon\text{-quadratic}}\end{array}\right.$ <u>Γ-groups</u> $\left\{\begin{array}{l}\Gamma^n_X(f,\varepsilon)\\ \Gamma_n^X(f,\varepsilon)\end{array}\right.$ $(n \in \mathbb{Z})$ in the

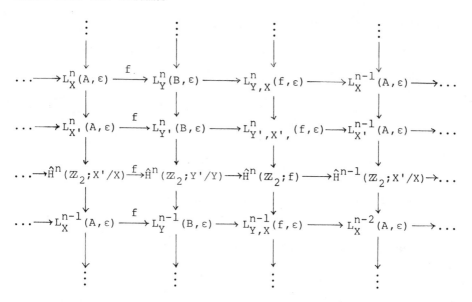

same way as $\begin{cases}\Gamma^n(f,\varepsilon)\\ \Gamma_n(f,\varepsilon)\end{cases}$ (the special case $X = \widetilde{K}_0(B)$) but using algebraic B-Poincaré complexes over A (based if m = 1) such that the induced algebraic Poincaré complexes over B have K-theory in X.

<u>Proposition 2.5.2</u> The intermediate Γ-groups associated to $\star$-invariant subgroups $X \subseteq Y \subseteq \mathrm{im}(f:\widetilde{K}_m(A)\longrightarrow \widetilde{K}_m(B))$ $(m = 0,1)$ are such that there is defined an exact sequence

$$\begin{cases} \dots \longrightarrow \Gamma^n_X(f,\varepsilon) \longrightarrow \Gamma^n_Y(f,\varepsilon) \longrightarrow \hat{H}^n(\mathbb{Z}_2;Y/X) \longrightarrow \Gamma^{n-1}_X(f,\varepsilon) \longrightarrow \dots \\ \dots \longrightarrow \Gamma_n^X(f,\varepsilon) \longrightarrow \Gamma_n^Y(f,\varepsilon) \longrightarrow \hat{H}^n(\mathbb{Z}_2;Y/X) \longrightarrow \Gamma_{n-1}^X(f,\varepsilon) \longrightarrow \dots \end{cases}$$
$$(n \in \mathbb{Z}) \ .$$

<u>Proof</u>: As for Proposition 1.10.1 (the special case $f = 1 : A \longrightarrow B = A$).

[]

It follows from the intermediate analogues of Propositions 2.4.3,2.4.4 that the original Γ-groups of Cappell and Shaneson [1] are the intermediate quadratic Γ-groups

$$\Gamma_*(f) = \Gamma_*^{\{f(\pi)\}}(f:\mathbb{Z}[\pi]\longrightarrow B) \qquad (\varepsilon = 1)$$

of a locally epic morphism $f:\mathbb{Z}[\pi]\longrightarrow B$, with $\{f(\pi)\} \subseteq \widetilde{K}_1(B)$. Similarly, the P-groups of Matsumoto [1] are the intermediate t-quadratic Γ-groups

$$P_*(\mathcal{E}) = \Gamma_*^{\{\pi'\}}(f:\mathbb{Z}[\pi]\longrightarrow \mathbb{Z}[\pi'],t)$$

associated to a group extension

$$\mathcal{E} : \{1\} \longrightarrow C \longrightarrow \pi \xrightarrow{\ f\ } \pi' \longrightarrow \{1\}$$

with C a cyclic group and $t \in \pi$ the image of a generator of C. See §7.8 for a discussion of the geometric significance as codimension 2 surgery obstruction groups of the Γ- and P-groups.

§3. Localization

Let A be a ring with involution, and let $S \subset A$ be a multiplicative subset of non-zero-divisors such that the ring with involution $S^{-1}A$ inverting S is defined - this is the "localization of A away from S". We shall now apply the theory of §§1,2 to express the relative $\left\{\begin{matrix}\varepsilon\text{-symmetric}\\ \varepsilon\text{-quadratic}\end{matrix}\right.$ L-groups $\left\{\begin{matrix}L^*(A \longrightarrow S^{-1}A,\varepsilon)\\ L_*(A \longrightarrow S^{-1}A,\varepsilon)\end{matrix}\right.$ of the inclusion $A \longrightarrow S^{-1}A$ as the cobordism groups of algebraic Poincaré complexes over A which become contractible over $S^{-1}A$.

Our role model here is the localization exact sequence of algebraic K-theory, which identifies the relative K-groups $K_*(A \longrightarrow S^{-1}A)$ appearing in the change of rings exact sequence

$$\ldots \longrightarrow K_n(A) \longrightarrow K_n(S^{-1}A) \longrightarrow K_n(A \longrightarrow S^{-1}A) \longrightarrow K_{n-1}(A) \longrightarrow \ldots \quad (n \in \mathbb{Z})$$

(where $K_n(A) = K_n$(exact category of f.g. projective A-modules)) with the K-groups

$K_n(A,S) = K_{n-1}$(exact category of S-torsion A-modules of homological dimension 1) $(n \in \mathbb{Z})$,

that is

$$K_n(A \longrightarrow S^{-1}A) = K_n(A,S) \quad (n \in \mathbb{Z}) \ .$$

This identification was first obtained for central S ($as = sa$ for all $a \in A, s \in S$) by Bass [2,IX] for $n = 1$, and then extended to $n \geqslant 2$ by Quillen (Grayson [1]), and to $n \leqslant 0$ by Carter [1]. The extension to eccentric localizations $A \longrightarrow S^{-1}A$ (i.e. those in which S is not necessarily central in A) is due to Grayson [2].

The "S-adic completion of A" is the inverse limit

$$\hat{A} = \varprojlim_{s\in S} A/sA \quad ,$$

which fits into the cartesian square of rings

$$\begin{array}{ccc} A & \longrightarrow & S^{-1}A \\ \downarrow & & \downarrow \\ \hat{A} & \longrightarrow & \hat{S}^{-1}\hat{A} \end{array} \quad .$$

The functor

$$\{\text{h.d. } 1 \text{ } S\text{-torsion } A\text{-modules}\} \longrightarrow \{\text{h.d. } 1 \text{ } \hat{S}\text{-torsion } \hat{A}\text{-modules}\};$$

$$M \longmapsto \hat{A}\otimes_A M$$

is an isomorphism of exact categories (an observation due to Karoubi [2]), so that it induces excision isomorphisms in the relative K-groups

$$K_*(A,S) \xrightarrow{\sim} K_*(\hat{A},\hat{S})$$

and there is defined a Mayer-Vietoris exact sequence

$$\ldots \longrightarrow K_n(A) \longrightarrow K_n(S^{-1}A) \oplus K_n(\hat{A}) \longrightarrow K_n(\hat{S}^{-1}\hat{A}) \longrightarrow K_{n-1}(A) \longrightarrow \ldots \quad (n \in \mathbb{Z}) \quad .$$

In particular, this applies to the "arithmetic square"

$$\begin{array}{ccc} \mathbb{Z}[\pi] & \longrightarrow & \mathbb{Q}[\pi] \\ \downarrow & & \downarrow \\ \hat{\mathbb{Z}}[\pi] & \longrightarrow & \hat{\mathbb{Q}}[\pi] \end{array}$$

associated to a group ring $A = \mathbb{Z}[\pi]$ with $S = \mathbb{Z}-\{0\} \subset A$,

$$\hat{\mathbb{Z}} = \varprojlim_{s} \mathbb{Z}/s\mathbb{Z} = \prod_{p} \hat{\mathbb{Z}}_p \quad (p \text{ prime})$$

the profinite completion of $\mathbb{Z}$, and

$$\hat{\mathbb{Q}} = \hat{S}^{-1}\hat{\mathbb{Z}} = \coprod_{p} (\hat{\hat{\mathbb{Q}}}_p, \hat{\mathbb{Z}}_p)$$

the ring of finite adèles of $\mathbb{Z}$.

Following some generalities in §3.1 on the localization of rings with involution we shall define in §3.2 the $\begin{cases}\varepsilon\text{-symmetric}\\ \varepsilon\text{-quadratic}\end{cases}$ L-groups $\begin{cases}L^n(A,S,\varepsilon)\\ L_n(A,S,\varepsilon)\end{cases}$ $(n \in \mathbb{Z})$ of $S^{-1}A$-acyclic algebraic Poincaré complexes over A. In §3.3 the algebraic Wu classes of §1.4 will be generalized to linking Wu classes, the analogues of the Wu classes appropriate to $S^{-1}A$-acyclic complexes over A. In §§3.4,3.5,3.6 we shall show that there are natural identifications

$$\begin{cases} L^*_S(A \longrightarrow S^{-1}A,\varepsilon) = L^*(A,S,\varepsilon) \\ L^S_*(A \longrightarrow S^{-1}A,\varepsilon) = L_*(A,S,\varepsilon) \end{cases},$$

the groups on the left being the relative intermediate $\begin{cases}\varepsilon\text{-symmetric}\\ \varepsilon\text{-quadratic}\end{cases}$ L-groups of the localization map

$$A \longrightarrow S^{-1}A$$

associated to the *-invariant subgroup

$$S = \operatorname{im}(\tilde{K}_0(A) \longrightarrow \tilde{K}_0(S^{-1}A)) \subseteq \tilde{K}_0(S^{-1}A) \quad ,$$

so that there is obtained a localization exact sequence in algebraic L-theory

$$\begin{cases} \ldots \longrightarrow L^n(A,\varepsilon) \longrightarrow L^n_S(S^{-1}A,\varepsilon) \longrightarrow L^n(A,S,\varepsilon) \longrightarrow L^{n-1}(A,\varepsilon) \longrightarrow \ldots \\ \ldots \longrightarrow L_n(A,\varepsilon) \longrightarrow L^S_n(S^{-1}A,\varepsilon) \longrightarrow L_n(A,S,\varepsilon) \longrightarrow L_{n-1}(A,\varepsilon) \longrightarrow \ldots \end{cases}$$
$$(n \in \mathbb{Z}) \ .$$

(Special cases of these sequences have been obtained by many previous authors, listed below). In §§3.4,3.5 the low-dimensional $\begin{cases}\varepsilon\text{-symmetric}\\ \varepsilon\text{-quadratic}\end{cases}$ L-groups $\begin{cases}L^n(A,S,\varepsilon)\\ L_n(A,S,\varepsilon)\end{cases}$ $(n \leqslant 1)$ will be interpreted as Witt groups of non-singular $S^{-1}A/A$-valued linking forms and linking formations involving S-torsion A-modules of homological dimension 1. It will thus be possible to express the lower-dimensional ε-symmetric L-theory localization exact sequence as a localization exact sequence of Witt groups

$$\begin{aligned}\ldots &\longrightarrow L^2(A,S,\varepsilon) \longrightarrow M^{\varepsilon}(A) \longrightarrow M^{\varepsilon}_S(S^{-1}A) \longrightarrow M\langle v_0\rangle^{\varepsilon}(A,S) \longrightarrow L^{\varepsilon}(A)\\
&\longrightarrow L^{\varepsilon}_S(S^{-1}A) \longrightarrow L\langle v_0\rangle^{\varepsilon}(A,S) \longrightarrow M\langle v_0\rangle^{-\varepsilon}(A) \longrightarrow M\langle v_0\rangle^{-\varepsilon}_S(S^{-1}A)\\
&\longrightarrow M_{-\varepsilon}(A,S) \longrightarrow L\langle v_0\rangle^{-\varepsilon}(A) \longrightarrow L\langle v_0\rangle^{-\varepsilon}_S(S^{-1}A) \longrightarrow L_{-\varepsilon}(A,S)\\
&\longrightarrow M_{\varepsilon}(A) \longrightarrow M^S_{\varepsilon}(S^{-1}A) \longrightarrow \widetilde{M}_{\varepsilon}(A,S) \longrightarrow L_{\varepsilon}(A) \longrightarrow L^S_{\varepsilon}(S^{-1}A)\\
&\longrightarrow \widetilde{L}_{\varepsilon}(A,S) \longrightarrow M_{-\varepsilon}(A) \longrightarrow M^S_{-\varepsilon}(S^{-1}A) \longrightarrow \widetilde{M}_{-\varepsilon}(A,S) \longrightarrow \ldots\end{aligned}$$

which extends to the left as the localization exact sequence in the higher-dimensional ε-symmetric L-groups (non-periodic in general) and to the right as the 12-periodic localization exact sequence in the ±ε-quadratic Witt groups. Here,

$$\left\{\begin{array}{l} L^{\varepsilon}(A) = L^0(A,\varepsilon) \\ L\langle v_0\rangle^{\varepsilon}(A) = L^{-2}(A,-\varepsilon) \\ L_{\varepsilon}(A) = L_0(A,\varepsilon)\end{array}\right. \text{(resp.} \left\{\begin{array}{l} M^{\varepsilon}(A) = L^1(A,\varepsilon) \\ M\langle v_0\rangle^{\varepsilon}(A) = L^{-1}(A,-\varepsilon)\text{)} \\ M_{\varepsilon}(A) = L_1(A,\varepsilon)\end{array}\right. \text{is}$$

the Witt group of non-singular $\left\{\begin{array}{l}\varepsilon\text{-symmetric} \\ \text{even } \varepsilon\text{-symmetric} \\ \varepsilon\text{-quadratic}\end{array}\right.$ forms

(resp. formations) over A, and $\left\{\begin{array}{l} L^{\varepsilon}_S(S^{-1}A) = L^0_S(S^{-1}A,\varepsilon) \\ L\langle v_0\rangle^{\varepsilon}_S(S^{-1}A) = L^{-2}_S(S^{-1}A,-\varepsilon) \\ L^S_{\varepsilon}(S^{-1}A) = L^S_0(S^{-1}A,\varepsilon)\end{array}\right.$

(resp. $\left\{\begin{array}{l} M^{\varepsilon}_S(S^{-1}A) = L^1_S(S^{-1}A,\varepsilon) \\ M\langle v_0\rangle^{\varepsilon}_S(S^{-1}A) = L^{-1}_S(S^{-1}A,-\varepsilon)\text{)} \\ M^S_{\varepsilon}(S^{-1}A) = L^S_1(S^{-1}A,\varepsilon)\end{array}\right.$ is the Witt group of

$$\text{non-singular}\left\{\begin{array}{l}\varepsilon\text{-symmetric}\\ \text{even } \varepsilon\text{-symmetric forms (resp. formations) over}\\ \varepsilon\text{-quadratic}\end{array}\right.$$

$S^{-1}A$ involving only the f.g. projective $S^{-1}A$-modules induced from f.g. projective A-modules. The relative L-group

$$\left\{\begin{array}{l}L\langle v_0\rangle^{\varepsilon}(A,S) = L^0(A,S,\varepsilon)\\ L_{\varepsilon}(A,S) = L^{-2}(A,S,-\varepsilon)\\ \widetilde{L}_{\varepsilon}(A,S) = L_0(A,S,\varepsilon)\end{array}\right. \quad \left(\text{resp.} \left\{\begin{array}{l}M\langle v_0\rangle^{\varepsilon}(A,S) = L^1(A,S,\varepsilon)\\ M_{\varepsilon}(A,S) = L^{-1}(A,S,-\varepsilon)\\ \widetilde{M}_{\varepsilon}(A,S) = L_1(A,S,\varepsilon)\end{array}\right.\right)$$

$$\text{is the Witt group of non-singular}\left\{\begin{array}{l}\text{even } \varepsilon\text{-symmetric}\\ \varepsilon\text{-quadratic}\\ \text{split } \varepsilon\text{-quadratic}\end{array}\right.$$

linking forms (resp. formations) over (A,S).

A localization exact sequence for Witt groups of the type

$$\begin{aligned}\ldots \longrightarrow M^{\varepsilon}(A) \longrightarrow M^{\varepsilon}_S(S^{-1}A) \longrightarrow M^{\varepsilon}(A,S) \longrightarrow L^{\varepsilon}(A) \longrightarrow L^{\varepsilon}_S(S^{-1}A)\\ \longrightarrow L^{\varepsilon}(A,S) \longrightarrow M^{-\varepsilon}(A) \longrightarrow M^{-\varepsilon}_S(S^{-1}A) \longrightarrow \ldots\end{aligned}$$

for arbitrary rings with involution A was first obtained by Karoubi [2],[3] in the case $1/2 \in A$ (when the various categories of linking forms over (A,S) coincide), following on from the work of earlier authors for Dedekind rings A - see §4 below for a discussion of the L-theory of Dedekind rings.

A localization exact sequence for the surgery obstruction groups of the type

$$\ldots \longrightarrow L_n(\mathbb{Z}[\pi]) \longrightarrow L^S_n(\mathbb{Q}[\pi]) \longrightarrow L_n(\mathbb{Z}[\pi],S) \longrightarrow L_{n-1}(\mathbb{Z}[\pi]) \longrightarrow \ldots$$

$$(n \pmod 4),\ S = \mathbb{Z}-\{0\} \subset \mathbb{Z}[\pi])$$

was first obtained by Pardon [1],[2],[3] for finite groups π, following on from the earlier work on linking forms in

odd-dimensional surgery obstruction theory of Wall [2], Passman and Petrie [1], and Connolly [1]. The algebraic methods of Pardon [2] apply to the quadratic L-groups of more general localizations, provided that $1/2 \in S^{-1}A$ (e.g. if $2 \in S$).
The localization exact sequence of Witt groups

$$L^{\varepsilon}(A) \longrightarrow L^{\varepsilon}_S(S^{-1}A) \longrightarrow L\langle v_0\rangle^{\varepsilon}(A,S) \longrightarrow M\langle v_0\rangle^{-\varepsilon}(A) \longrightarrow M\langle v_0\rangle^{-\varepsilon}_S(S^{-1}A)$$

has also been obtained by Carlsson and Milgram [3].

In §3.6 we shall apply the localization exact sequence in the $\begin{cases}\varepsilon\text{-symmetric}\\ \varepsilon\text{-quadratic}\end{cases}$ L-groups to prove that

$\begin{cases}\text{if } \mathrm{im}(\hat{H}^0(\mathbb{Z}_2;S^{-1}A/A,\varepsilon) \longrightarrow \hat{H}^1(\mathbb{Z}_2;A,\varepsilon)) = 0\\ \text{for all } A,S,\varepsilon\end{cases}$ there are defined

excision isomorphisms in the relative L-groups

$$\begin{cases} L^n(A,S,\varepsilon) \xrightarrow{\ \sim\ } L^n(\hat{A},\hat{S},\varepsilon) \\ L_n(A,S,\varepsilon) \xrightarrow{\ \sim\ } L_n(\hat{A},\hat{S},\varepsilon) \end{cases} \quad (n \in \mathbb{Z}) \quad ,$$

giving rise to a Mayer-Vietoris exact sequence in the absolute L-groups

$$\begin{cases} \dots \longrightarrow L^n(A,\varepsilon) \longrightarrow L^n_S(S^{-1}A,\varepsilon) \oplus L^n(\hat{A},\varepsilon) \longrightarrow L^n_S(\hat{S}^{-1}\hat{A},\varepsilon) \longrightarrow L^{n-1}(A,\varepsilon) \longrightarrow \dots \\ \dots \longrightarrow L_n(A,\varepsilon) \longrightarrow L^S_n(S^{-1}A,\varepsilon) \oplus L_n(\hat{A},\varepsilon) \longrightarrow L^S_n(\hat{S}^{-1}\hat{A},\varepsilon) \longrightarrow L_{n-1}(A,\varepsilon) \longrightarrow \dots \end{cases}$$
$$(n \in \mathbb{Z}) \quad .$$

Such a Mayer-Vietoris exact sequence was first obtained by Wall [8] for the quadratic L-groups of a finitely generated ring A with $S = \mathbb{Z}-\{0\} \subset A$, using arithmetic methods such as the strong approximation theorem for algebraic groups over $\mathbb{Q}$. Karoubi [2] obtained such a sequence for more genral localizations $A \longrightarrow S^{-1}A$, but with the restriction $1/2 \in A$. Bak [2] has obtained a similar sequence in the context of the unitary algebraic K-theory of Bass [3].

In §3.6 we shall also use the localization exact sequence for $S = \mathbb{Z}-\{0\} \subset A$ and the natural action of the symmetric Witt ring $L^0(\hat{\mathbb{Z}})$ (which is of exponent 8) on the relative L-groups

$$\begin{cases} L^*(\hat{A},\hat{S},\varepsilon) = L^*(A,S,\varepsilon) \\ L_*(\hat{A},\hat{S},\varepsilon) = L_*(A,S,\varepsilon) \end{cases}$$

to prove that the natural maps

$$\begin{cases} L^*(A,\varepsilon) \longrightarrow L^*_S(S^{-1}A,\varepsilon) \\ L_*(A,\varepsilon) \longrightarrow L^S_*(S^{-1}A,\varepsilon) \end{cases}$$

are isomorphisms modulo 8-torsion for any torsion-free ring with involution A (e.g. a group ring $A = \mathbb{Z}[\pi]$, in which case $S^{-1}A = \mathbb{Q}[\pi]$). Results of this type were first obtained for the surgery obstruction groups $L_*(\mathbb{Z}[\pi])$ of finite groups π. Taking for granted the result that the natural maps

$$L^S_{2i}(\mathbb{Q}[\pi]) \longrightarrow L_{2i}(\mathbb{R}[\pi]) \quad (i \pmod 2,\ \pi \text{ finite})$$

are isomorphisms modulo 2-primary torsion it is possible to interpret Theorems 13A.3, 13A.4 i) of Wall [4] as stating that the natural maps

$$L_{2i}(\mathbb{Z}[\pi]) \longrightarrow L^S_{2i}(\mathbb{Q}[\pi]) \quad (i \pmod 2,\ \pi \text{ finite})$$

are isomorphisms modulo 2-primary torsion. Passman and Petrie [1] and Connolly [1] showed that the natural maps

$$L_{2i+1}(\mathbb{Z}[\pi]) \longrightarrow L^S_{2i+1}(\mathbb{Q}[\pi]) \quad (i \pmod 2,\ \pi \text{ finite})$$

are isomorphisms modulo 2^j-torsion, $j \leqslant 3$. (Actually, they were working with the simple quadratic L-groups). Karoubi [2] obtained similar results for the L-groups of arbitrary torsion-free rings with involution A such that $1/2 \in A$.

The localization exact sequence and the Mayer-Vietoris exact sequence associated to a localization-completion square are key tools in the computations of the surgery obstruction groups $L_*(\mathbb{Z}[\pi])$ of finite groups π due to Wall [9], Bak [2], Pardon [5], Carlsson and Milgram [1],[2], Kolster [1],[2], Bak and Kolster [1], Hambleton and Milgram [2].

The localization exact sequence for the quadratic L-groups $L_*(R[\pi])$ of group rings $R[\pi]$ $(R = S^{-1}\mathbb{Z} \subseteq \mathbb{Q},\ S \subset \mathbb{Z}-\{0\})$ has a geometric interpretation involving homotopy-theoretic localization, which is discussed in §7.7. below.

3.1 Localization and completion

We refer to Chapter II of Stenström [1] for the general theory of localization in noncommutative rings.

Let A be a ring with involution.

A subset $S \subset A$ is multiplicative if

i) $st \in S$ for all $s,t \in S$

ii) if $sa = 0 \in A$ for some $s \in S, a \in A$ then $a = 0 \in A$

iii) $\bar{s} \in S$ for all $s \in S$

iv) for all $a \in A, s \in S$ there exist $b,b' \in A$, $t,t' \in S$ such that $at = sb$, $t'a = b's \in A$ ("the two-sided Ore condition")

v) $1 \in S$.

The localization of A away from S $S^{-1}A$ is the ring with involution defined by the equivalence classes of pairs

$$(a,s) \in A \times S$$

under the relation

$(a,s) \sim (b,t)$ if there exist $c,d \in A$ such that

$$ca = db \in A \ , \ cs = dt \in S \subset A \ ,$$

with

$$(a,s) + (a',s') = (b'a + ba',t) \text{ if } b,b' \in A \text{ are such that } t = b's = bs' \in S \subset A,$$

$$(a,s).(a',s') = (ba',ts) \text{ if } b \in A,\ t \in S \text{ are such that } ta = bs' \in A,$$

$$\overline{(a,s)} = (b,t) \text{ if } b \in A,\ t \in S \text{ are such that } t\bar{a} = b\bar{s} \in A \ .$$

The equivalence class of $(a,s) = (1,s).(a,1)$ will be denoted by

$$\frac{a}{s} \in S^{-1}A \qquad ,$$

as usual. The injection

$$A \longrightarrow S^{-1}A \ ; \ a \longmapsto \frac{a}{1}$$

is a locally epic morphism of rings with involution.

An A-module chain complex C is <u>S-acyclic</u> if $H_*(S^{-1}A \otimes_A C) = 0$, that is if it is $S^{-1}A$-acyclic in the sense of §2.4.

Here are some important examples of localization:

i) if A is an algebra over an integral domain R, then

$$S = R-\{0\} \subset R \subset A$$

is a multiplicative subset of both R and A. The localization $S^{-1}R = F$ is the quotient field of R and $S^{-1}A = F \otimes_R A$ is the induced algebra over F.

ii) if A is an algebra over a commutative ring R, and $\mathcal{P}$ is a prime ideal of R, then

$$S = R - \mathcal{P} \subset R \subseteq A$$

is a multiplicative subset of both R and A. The localization $S^{-1}R = R_{\mathcal{P}}$ is the "localization of A at $\mathcal{P}$", and $S^{-1}A = A_{\mathcal{P}}$ is the "localization of A at $\mathcal{P}$".

(The L-theory of localizations of type i) and ii) will be studied in §4 in the case when R is a Dedekind ring).

iii) if A is a ring with involution and $\alpha : A \longrightarrow A$ is a ring automorphism such that $\overline{\alpha(a)} = \alpha^{-1}(\bar{a}) \in A$ for all $a \in A$ (e.g. $\alpha = 1$) let x be an indeterminate over A such that

$$ax = x\alpha(a) \quad (a \in A) \ .$$

The "α-twisted polynomial extension of A" $A_\alpha[x]$ is then defined, a ring with involution

$$- : A_\alpha[x] \longrightarrow A_\alpha[x] \ ; \ \sum_{j=0}^{\infty} a_j x^j \longmapsto \sum_{j=0}^{\infty} x^j \bar{a}_j \ .$$

The multiplicative subset

$$X = \{x^k \mid k \geqslant 0\} \subset A_\alpha[x]$$

is such that the localization

$$X^{-1}A_\alpha[x] = A_\alpha[x,x^{-1}]$$

is the "α-twisted Laurent polynomial extension of A".
The L-theory of such polynomial extensions will be dealt with in §5 below.

iv) if $A = \mathbb{Z}[\pi]$ is the group ring of a group π which is an extension of a finitely generated torsion-free nilpotent group by a finite extension ρ of a polycyclic group then

$$S = \{1+i \mid i \in \ker(\mathbb{Z}[\pi] \longrightarrow \mathbb{Z}[\rho])\} \subset A$$

is a multiplicative subset, such that a finite-dimensional A-module chain complex is $\mathbb{Z}[\rho]$-acyclic if and only if it is S-acyclic. This example is due to Smith [1],[2].
We shall consider a particular case of this type of localization in §7.9 below, for $\pi = \mathbb{Z}$, $\rho = \{1\}$, in connection with the algebraic theory of knot cobordism.

A multiplicative subset $S \subset A$ is <u>central</u> if

$$as = sa \in A \text{ for all } a \in A, s \in S \ .$$

For central $S \subset A$ it is possible to express the localization $S^{-1}A$ in the familiar way as the set of equivalence classes of pairs $(a,s) \in A \times S$ under the relation

$$(a,s) \sim (b,t) \text{ if } at = bs \in A \ ,$$

with

$$(a,s) + (a',s') = (as' + a's, ss')$$

$$(a,s).(a',s') = (aa', ss')$$

$$\overline{(a,s)} = (\bar{a},\bar{s}) \ .$$

We shall now develop some general properties of modules and chain complexes over a ring with involution A and the localization $S^{-1}A$ of A away from a multiplicative subset $S \subset A$ (which in general will not be assumed to be central).

An A-module M induces an $S^{-1}A$-module

$$S^{-1}M = S^{-1}A \otimes_A M \ .$$

The elements of $S^{-1}M$ can be regarded as the equivalence classes $\frac{x}{s}$ of pairs $(x,s) \in M \times S$ under the relation

$$(x,s) \sim (y,t) \text{ if there exist } c,d \in A \text{ such that } cx = dy \in M,\ cs = dt \in S \subset A$$

with

$$(x,s) + (x',s') = (b'x + bx',t) \text{ if } b,b' \in A \text{ are such that } t = b's = bs' \in S \subset A$$

$$(a,s)(y,t) = (by,us) \text{ if } b \in A, u \in S \text{ are such that } ua = bt \in A \ .$$

(Again, if $S \subset A$ is central this can be simplified to

$$(x,s) \sim (y,t) \text{ if } tx = sy \in M$$

$$(x,s) + (x',s') = (s'x + sx',ss')$$

$$(a,s)(y,t) = (ay,st) \qquad).$$

If M is a f.g. projective A-module then $S^{-1}M$ is a f.g. projective $S^{-1}A$-module, and there is defined a natural $S^{-1}A$-module isomorphism

$$S^{-1}(M^*) = S^{-1}\mathrm{Hom}_A(M,A) \longrightarrow (S^{-1}M)^* = \mathrm{Hom}_{S^{-1}A}(S^{-1}M,S^{-1}A) \ ;$$

$$\frac{f}{s} \longmapsto (\frac{x}{t} \longmapsto \frac{f(x)}{t}\cdot\frac{1}{s}) \quad ,$$

allowing us to write

$$S^{-1}M^* = S^{-1}(M^*) = (S^{-1}M)^* \ .$$

An A-module morphism $f \in \mathrm{Hom}_A(M,N)$ induces an $S^{-1}A$-module morphism

$$S^{-1}f : S^{-1}M \longrightarrow S^{-1}N \; ; \; \frac{x}{s} \longmapsto \frac{f(x)}{s} \quad .$$

An <u>S-isomorphism</u> is an A-module morphism $f \in \mathrm{Hom}_A(M,N)$ such that $S^{-1}f \in \mathrm{Hom}_{S^{-1}A}(S^{-1}M, S^{-1}N)$ is an $S^{-1}A$-module isomorphism, i.e. f is an $S^{-1}A$-isomorphism in the sense of §2.4.

An A-module M is <u>S-torsion</u> if

$$S^{-1}M = 0 \; ,$$

that is if for every $x \in M$ there exists $s \in S$ such that

$$sx = 0 \in M \; .$$

An <u>(A,S)-module</u> M is an S-torsion A-module of homological dimension 1, that is an A-module which admits a f.g. projective A-module resolution of length 1

$$0 \longrightarrow P_1 \xrightarrow{\;d\;} P_0 \xrightarrow{\;h\;} M \longrightarrow 0$$

with $d \in \mathrm{Hom}_A(P_1, P_0)$ an S-isomorphism.

The <u>S-dual</u> $M^\wedge$ of an (A,S)-module M is the (A,S)-module

$$M^\wedge = \mathrm{Hom}_A(M, S^{-1}A/A) \quad ,$$

with A acting by

$$A \times M^\wedge \longrightarrow M^\wedge \; ; \; (a,f) \longmapsto (x \longmapsto f(x).\bar{a}) \quad .$$

The S-dual has f.g. projective A-module resolution

$$0 \longrightarrow P_0^* \xrightarrow{\;d^*\;} P_1^* \xrightarrow{\;Th\;} M^\wedge \longrightarrow 0$$

with

$$Th : P_1^* \longrightarrow M^\wedge \; ; \; f \longmapsto ([x] \longmapsto \frac{f(y)}{s})$$

$$(x \in P_0, \; [x] \in M, \; s \in S, \; y \in P_1, \; sx = dy \in P_0) \; .$$

The natural A-module isomorphism

$$M \longrightarrow M^{\wedge\wedge} \ ; \ x \longmapsto (f \longmapsto \overline{f(x)})$$

will be used to identify

$$M^{\wedge\wedge} = M \ .$$

If M,N are (A,S)-modules there is defined an <u>S-duality</u> isomorphism of abelian groups

$$\mathrm{Hom}_A(M,N) \longrightarrow \mathrm{Hom}_A(N^\wedge,M^\wedge) \ ; \ f \longmapsto (f^\wedge : g \longmapsto (x \longmapsto gf(x))) \ .$$

For example, a $(\mathbb{Z},\mathbb{Z}-\{0\})$-module M is the same as a finite abelian group and the $(\mathbb{Z}-\{0\})$-dual

$$M^\wedge = \mathrm{Hom}_{\mathbb{Z}}(M,\mathbb{Q}/\mathbb{Z})$$

is the character group.

An <u>n-dimensional (A,S)-module chain complex</u> is an A-module chain complex

$$C : 0 \longrightarrow C_n \xrightarrow{d} C_{n-1} \xrightarrow{d} \cdots \xrightarrow{d} C_1 \xrightarrow{d} C_0 \longrightarrow 0$$

such that each C_r $(0 \leqslant r \leqslant n)$ is an (A,S)-module. The S-dual A-module chain complex

$$C^{n-\wedge} : 0 \longrightarrow C_0^\wedge \xrightarrow{d^\wedge} C_1^\wedge \xrightarrow{d^\wedge} \cdots \xrightarrow{d^\wedge} C_{n-1}^\wedge \xrightarrow{d^\wedge} C_n^\wedge \longrightarrow 0$$

is also an n-dimensional (A,S)-module chain complex. The homology A-modules $H_*(C)$ are S-torsion (but not in general (A,S)-modules), since localization is exact

$$S^{-1}H_*(C) = H_*(S^{-1}C) = 0 \ .$$

The <u>S-dual cohomology</u> $H_S^*(C)$ are the S-torsion A-modules defined by

$$H_S^r(C) = H_{n-r}(C^{n-\wedge}) = \ker(d^\wedge : C_r^\wedge \longrightarrow C_{r+1}^\wedge)/\mathrm{im}(d^\wedge : C_{r-1}^\wedge \longrightarrow C_r^\wedge) \quad (0 \leqslant r \leqslant n) \ .$$

If $\varepsilon \in A$ is a central unit such that

$$\bar{\varepsilon} = \varepsilon^{-1} \in A$$

then $\frac{\varepsilon}{1} \in S^{-1}A$ is a central unit (also to be denoted by ε) such that

$$\overline{(\tfrac{\varepsilon}{1})} = (\tfrac{\varepsilon}{1})^{-1} \in S^{-1}A \ .$$

Further below we shall define the $\begin{cases} \varepsilon\text{-symmetric} \\ \text{even } \varepsilon\text{-symmetric} \\ \varepsilon\text{-quadratic} \end{cases}$

Q_S-groups $\begin{cases} Q_S^*(C,\varepsilon) \\ Q\langle v_0\rangle_S^*(C,\varepsilon) \\ Q_*^S(C,\varepsilon) \end{cases}$ of a finite-dimensional (A,S)-module

chain complex C, generalizing the Q-groups of a finite-dimensional A-module chain complex defined in §1.1. (Indeed, the Q_S-groups of C will be defined to be the Q-groups of a finite-dimensional A-module chain complex D such that $H_*(D) = H_*(C)$, $H^*(D) = H_S^{*-1}(C)$). The localization exact sequence of §3.2 will identify the relative $\begin{cases} \varepsilon\text{-symmetric} \\ \varepsilon\text{-quadratic} \end{cases}$ L-group

$\begin{cases} L_S^n(A \longrightarrow S^{-1}A,\varepsilon) \\ L_n^S(A \longrightarrow S^{-1}A,\varepsilon) \end{cases}$ $(n \geqslant 0)$ with the cobordism group of

"n-dimensional $\begin{cases} \text{even } \varepsilon\text{-symmetric} \\ \varepsilon\text{-quadratic} \end{cases}$ Poincaré complexes over (A,S)"

$\begin{cases} (C,\phi \in Q\langle v_0\rangle_S^n(C,\varepsilon)) \\ (C,\psi \in Q_n^S(C,\varepsilon)) \end{cases}$ with C an n-dimensional (A,S)-module chain

complex and $\begin{cases} \phi \\ \psi \end{cases}$ such that there are defined Poincaré duality

isomorphisms of S-torsion A-modules

$$\begin{cases} \phi_0^S : H_S^*(C) \xrightarrow{\sim} H_{n-*}(C) \\ (1+T_\varepsilon)\psi_0^S : H_S^*(C) \xrightarrow{\sim} H_{n-*}(C) \end{cases} .$$

In §3.4 (resp. §3.5) we shall identify the n-dimensional

$\begin{cases} \text{(even) } \varepsilon\text{-symmetric} \\ \varepsilon\text{-quadratic} \end{cases}$ Poincaré complexes over (A,S) for

$n = 0$ (resp. $n = 1$) with the "non-singular $\begin{cases} \text{(even) } \varepsilon\text{-symmetric} \\ \text{split } \varepsilon\text{-quadratic} \end{cases}$

linking forms (resp. formations) over (A,S)", going on in §3.6

to identify the relative L-groups $\begin{cases} L^n_S(A \longrightarrow S^{-1}A,\varepsilon) & (-\infty < n \leqslant 1) \\ L^S_n(A \longrightarrow S^{-1}A,\varepsilon) & (n \in \mathbb{Z}) \end{cases}$

with the Witt groups of such objects by analogy with the identifications of §1.6 of the absolute L-groups

$\begin{cases} L^n(A,\varepsilon) & (-\infty < n \leqslant 1) \\ L_n(A,\varepsilon) & (n \in \mathbb{Z}) \end{cases}$ with the Witt groups of forms and formations

over A. A "linking form over (A,S)" is an (A,S)-module M together with a pairing

$$M \times M \longrightarrow S^{-1}A/A \quad ,$$

and a "linking formation over (A,S)" is a linking form over (A,S) together with a lagrangian and a sublagrangian. The familiar equivalence of categories

$$\{\text{S-acyclic 1-dimensional A-module chain complexes}\}$$
$$\xrightarrow{\sim} \{(A,S)\text{-modules}\} \; ; \; C \longmapsto H_0(C) \quad .$$

will be generalized to equivalences

$$\{\text{S-acyclic algebraic Poincaré complexes over A}\}$$
$$\xrightarrow{\sim} \{\text{algebraic Poincaré complexes over } (A,S)\} \quad .$$

The <u>maximal S-torsion submodule</u> T_SM of an A-module M is the submodule

$$\begin{aligned} T_SM &= \{x \in M \mid sx = 0 \in M \text{ for some } s \in S\} \\ &= \ker(M \longrightarrow S^{-1}M; x \longmapsto \tfrac{x}{s}) \subseteq M \quad . \end{aligned}$$

The A-module M is S-torsion if and only if

$$T_SM = M \quad .$$

The <u>linking pairing</u> ϕ_0^S of an n-dimensional ε-symmetric complex over A $(C,\phi \in Q^n(C,\varepsilon))$ is defined by

$$\phi_0^S : T_SH^r(C) \times T_SH^{n-r+1}(C) \longrightarrow S^{-1}A/A \ ;$$
$$(x,y) \longmapsto \frac{1}{s}\phi_0(x)(z)$$

$$(x \in C^r,\ y \in C^{n-r+1},\ z \in C^{n-r},\ s \in S,\ d^*z = sy \in C^{n-r+1})\ ,$$

and satisfies

$$\text{i)}\ \phi_0^S(x,y+y') = \phi_0^S(x,y) + \phi_0^S(x,y')$$
$$\text{ii)}\ \phi_0^S(x,ay) = a\phi_0^S(x,y)$$
$$\text{iii)}\ \phi_0^S(y,x) = (-)^{r(n-r+1)}\overline{\varepsilon\phi_0^S(x,y)}$$

$$(x \in T_SH^r(C),\ y,y' \in T_SH^{n-r+1}(C),\ a \in A)\ .$$

The name arises as follows.

Let M be a compact n-dimensional manifold, and let $\widetilde{M}$ be a covering of M with group of covering translations π such that the orientation map of M factors as

$$w(M) : \pi_1(M) \longrightarrow \pi \xrightarrow{\ w\ } \mathbb{Z}_2$$

for some group morphism w, so that there is defined a symmetric Poincaré complex over $\mathbb{Z}[\pi]$ with the w-twisted involution

$$\sigma^*(M) = (C(\widetilde{M}),\phi \in Q^n(C(\widetilde{M})))$$

(as recalled from II. in §1.2 above). Define a multiplicative subset

$$S = \mathbb{Z}-\{0\} \subset \mathbb{Z}[\pi]\ .$$

The linking pairing of $\sigma^*(M)$

$$\phi_0^S : T_SH^r(\widetilde{M}) \times T_SH^{n-r+1}(\widetilde{M}) \longrightarrow \mathbb{Q}[\pi]/\mathbb{Z}[\pi]$$

agrees via the Poincaré duality $H^*(\widetilde{M}) \cong H_{n-*}(\widetilde{M})$ with the pairing

$$T_S H_{n-r}(\widetilde{M}) \times T_S H_{r-1}(\widetilde{M}) \longrightarrow \mathbb{Q}[\pi]/\mathbb{Z}[\pi]$$

defined by the geometric linking numbers of torsion homology classes, as originally studied by deRham [1] and Seifert [1] (for $\pi = \{1\}$) and more recently by Kervaire and Milnor [1], Wall [2] and Pardon [3] (for π finite) in connection with odd-dimensional surgery obstruction theory.

In §4.2 below we shall identify the cobordism class $\begin{cases}(C,\phi) \in L^{2i}(A,\varepsilon) \\ (C,\phi) \in L^{2i-1}(A,\varepsilon)\end{cases}$ of a $\begin{cases}2i \\ 2i-1\end{cases}$-dimensional ε-symmetric Poincaré complex $\begin{cases}(C,\phi \in Q^{2i}(C,\varepsilon)) \\ (C,\phi \in Q^{2i-1}(C,\varepsilon))\end{cases}$ over a Dedekind ring A with a cobordism class of the non-singular $(-)^i\varepsilon$-symmetric $\begin{cases}\text{intersection} \\ \text{linking}\end{cases}$ pairing

$$\begin{cases}\phi_0 : H^i(C)/T_S H^i(C) \times H^i(C)/T_S H^i(C) \longrightarrow A \\ \phi_0^S : T_S H^i(C) \times T_S H^i(C) \longrightarrow S^{-1}A/A\end{cases}$$

(and similarly for the ε-quadratic case) with

$$S = A - \{0\} \subset A \ .$$

The expression in §3.2 below of the relative L-groups of a localization map $A \longrightarrow S^{-1}A$ as the cobordism groups of S-acyclic algebraic Poincaré complexes over A will be based on the following results:

Proposition 3.1.1 i) An n-dimensional $S^{-1}A$-module chain complex D with projective class

$$[D] \in \mathrm{im}(\widetilde{K}_0(A) \longrightarrow \widetilde{K}_0(S^{-1}A)) \subseteq \widetilde{K}_0(S^{-1}A)$$

has the chain homotopy type of $S^{-1}C = S^{-1}A \otimes_A C$ for some n-dimensional A-module chain complex C.

ii) An S-acyclic finite-dimensional A-module chain complex C is chain equivalent to a complex C' for which there exist A-module morphisms $e \in \mathrm{Hom}_A(C'_r, C'_{r+1})$ ($r \in \mathbb{Z}$) such that the A-module morphisms

$$s = de + ed : C'_r \longrightarrow C'_r \quad (r \in \mathbb{Z})$$

are S-isomorphisms. (If $S \subset A$ is central can take $C' = C, s \in S$).

Proof: Clear denominators.

[]

Localization is exact, so that for any A-module chain complex C there are natural identifications of $S^{-1}A$-modules

$$H_*(S^{-1}C) = S^{-1}H_*(C)$$
$$H^*(S^{-1}C) = S^{-1}H^*(C) \ .$$

Thus C is S-acyclic if and only if the homology A-modules $H_*(C)$ are S-torsion; similarly for $C^*, H^*(C)$.

A chain map of A-module chain complexes

$$f : C \longrightarrow C'$$

is a homology equivalence if it induces isomorphisms in the homology A-modules

$$f_* : H_*(C) \xrightarrow{\sim} H_*(C') \ .$$

In particular, a chain equivalence is a homology equivalence. A homology equivalence of finite-dimensional chain complexes is a chain equivalence, but in general homology equivalences are not chain equivalences.

A resolution (D,g) of an n-dimensional (A,S)-module chain complex C consists of an (n+1)-dimensional A-module chain complex D together with a homology equivalence

$$g : D \longrightarrow C \ .$$

The S-dual chain complex $C^{n-\wedge}$ admits a dual resolution (D^{n+1-*}, Tg) inducing the A-module isomorphisms

$$Tg_* : H_r(D^{n+1-*}) = H^{n+1-r}(D) \longrightarrow H_r(C^{n-\wedge}) = H_S^{n-r}(C) \; ;$$

$$f \longmapsto (g(x) \longmapsto \frac{f(y)}{s})$$

$$(f \in D^{n+1-r},\ x \in D_{n-r},\ y \in D_{n+1-r},\ s \in S,\ sx = dy \in D_{n-r}) \ .$$

For example, a resolution (D,g) of a 0-dimensional (A,S)-module chain complex C is a f.g. projective A-module resolution of the (A,S)-module C_0

$$0 \longrightarrow D_1 \xrightarrow{\quad d \quad} D_0 \xrightarrow{\quad g \quad} C_0 \longrightarrow 0$$

with $d \in \mathrm{Hom}_A(D_1, D_0)$ an S-isomorphism, and (D^{1-*}, Th) is the dual resolution of $C_0^{\wedge}$

$$0 \longrightarrow D^0 \xrightarrow{\quad d^* \quad} D^1 \xrightarrow{\quad Tg \quad} C_0 \longrightarrow 0$$

defined above.

A <u>resolution</u> (h,k) of a chain map of n-dimensional (A,S)-module chain complexes

$$f : C \longrightarrow C'$$

is a triad of A-module chain complexes

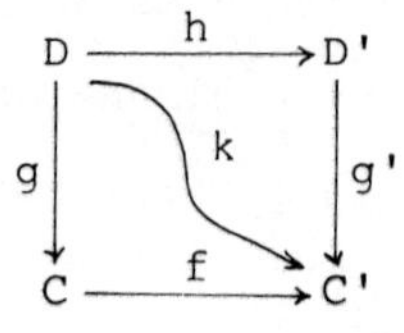

(i.e. a chain map $h: D \longrightarrow D'$ together with a chain homotopy $k: fg \simeq g'h: D \longrightarrow C'$) such that (D,g) is a resolution of C and (D',g') is a resolution of C'. Note that f is a homology equivalence if and only if h is a chain equivalence.

Proposition 3.1.2 i) Every n-dimensional (A,S)-module chain complex C admits a resolution (D,g), and every chain map $f:C \longrightarrow C'$ of n-dimensional (A,S)-module chain complexes admits a resolution (h,k).

ii) There are natural identifications of sets of homology equivalence classes

{n-dimensional (A,S)-module chain complexes}

= {S-acyclic (n+1)-dimensional A-module chain complexes}

$(n \geq 0)$.

Proof: i) Given an n-dimensional (A,S)-module chain complex C write a f.g. projective A-module resolution of C_r $(0 \leq r \leq n)$ as

$$0 \longrightarrow P_r \xrightarrow{f} Q_r \xrightarrow{h} C_r \longrightarrow 0 \quad ,$$

and resolve $d \in \mathrm{Hom}_A(C_r, C_{r-1})$ $(1 \leq r \leq n)$ by

$$\begin{array}{ccccccccc} 0 & \longrightarrow & P_r & \xrightarrow{f} & Q_r & \xrightarrow{h} & C_r & \longrightarrow & 0 \\ & & \downarrow i & & \downarrow j & & \downarrow d & & \\ 0 & \longrightarrow & P_{r-1} & \xrightarrow{f} & Q_{r-1} & \xrightarrow{h} & C_{r-1} & \longrightarrow & 0 \end{array} \qquad (fi = jf, hj = dh) \quad .$$

As $d^2 = 0$ there exist chain homotopies $k \in \mathrm{Hom}_A(Q_r, P_{r-2})$ $(2 \leq r \leq n)$ such that

$$i^2 = kf \in \mathrm{Hom}_A(P_r, P_{r-2}) \ , \ j^2 = fk \in \mathrm{Hom}_A(Q_r, Q_{r-2}) \ .$$

Define a resolution (D,g) of C by

$$d_D = \begin{pmatrix} i & (-)^r k \\ f & (-)^r j \end{pmatrix} : D_r = P_{r-1} \oplus Q_r \longrightarrow D_{r-1} = P_{r-2} \oplus Q_{r-1}$$

$$(1 \leq r \leq n+1, \ P_{-1} = Q_{n+1} = 0)$$

$$g = (0 \ \ (-)^r h) : D_r = P_{r-1} \oplus Q_r \longrightarrow C_r \quad (0 \leq r \leq n) \ .$$

Similarly for chain maps.

ii) Given an S-acyclic (n+1)-dimensional A-module chain complex D define an n-dimensional (A,S)-module chain complex C with resolution (D,g), as follows. Since $S^{-1}D$ is a chain contractible $S^{-1}A$-module chain complex there exist A-module morphisms $e \in \mathrm{Hom}_A(D_r, D_{r+1})$ $(0 \leqslant r \leqslant n)$ such that the A-module morphisms

$$s = de + ed : D_r \longrightarrow D_r \quad (0 \leqslant r \leqslant n)$$

are S-isomorphisms, by Proposition 3.1.1 ii). Define a collection of f.g. projective A-modules and A-module morphisms

$$(P,Q,f,h,i,j,k)$$

as in i) by

$$f = \begin{cases} \begin{pmatrix} d & 0 & 0 & . \\ e & d & 0 & . \\ 0 & e & d & . \\ . & . & . & . \end{pmatrix} : P_0 = D_1 \oplus D_3 \oplus D_5 \oplus \ldots \longrightarrow Q_0 = D_0 \oplus D_2 \oplus D_4 \oplus \ldots \\ \begin{pmatrix} s & 0 & 0 & . \\ e^2 & s & 0 & . \\ 0 & e^2 & s & . \\ . & . & . & . \end{pmatrix} : P_r = D_{r+1} \oplus D_{r+3} \oplus D_{r+5} \oplus \ldots \\ \qquad \longrightarrow Q_r = D_{r+1} \oplus D_{r+3} \oplus D_{r+5} \oplus \ldots \quad (r \geqslant 1) \end{cases}$$

$$i = \begin{pmatrix} d & 0 & 0 & . \\ e & d & 0 & . \\ 0 & e & d & . \\ . & . & . & . \end{pmatrix} : P_r = D_{r+1} \oplus D_{r+3} \oplus D_{r+5} \oplus \ldots$$

$$\longrightarrow P_{r-1} = D_r \oplus D_{r+2} \oplus D_{r+4} \oplus \ldots \quad (r \geqslant 0)$$

$$j = \begin{cases} \begin{pmatrix} 0 & 0 & 0 & . \\ 1 & 0 & 0 & . \\ 0 & 1 & 0 & . \\ . & . & . & . \end{pmatrix} : Q_1 = D_2 \oplus D_4 \oplus D_6 \oplus \dots \longrightarrow Q_0 = D_0 \oplus D_2 \oplus D_4 \oplus \dots \\ \begin{pmatrix} d & 0 & 0 & . \\ e & d & 0 & . \\ 0 & e & d & . \\ . & . & . & . \end{pmatrix} : Q_r = D_{r+1} \oplus D_{r+3} \oplus D_{r+5} \oplus \dots \\ \qquad \longrightarrow Q_{r-1} = D_r \oplus D_{r+2} \oplus D_{r+4} \oplus \dots \quad (r \geqslant 2) \end{cases}$$

$$k = \begin{pmatrix} 0 & 0 & 0 & . \\ 1 & 0 & 0 & . \\ 0 & 1 & 0 & . \\ . & . & . & . \end{pmatrix} : Q_r = D_{r+1} \oplus D_{r+3} \oplus D_{r+5} \oplus \dots$$

$$\longrightarrow P_{r-2} = D_{r-1} \oplus D_{r+1} \oplus D_{r+3} \oplus \dots \quad (r \geqslant 2)$$

$$g = \text{projection} : Q_r \longrightarrow C_r = \text{coker}(f: P_r \longrightarrow Q_r) \quad (r \geqslant 0) .$$

The n-dimensional (A,S)-module chain complex C has a resolution (D',g') with

$$D'_r = P_{r-1} \oplus Q_r \quad (r \geqslant 0)$$

(as in i)) such that D' is chain equivalent to D. Thus (D,g) is also a resolution of C.

[]

Given a finite-dimensional (A,S)-module chain complex C define the $\left\{\begin{array}{l}\underline{\varepsilon\text{-symmetric}}\\ \underline{\text{even } \varepsilon\text{-symmetric}}\\ \underline{\varepsilon\text{-quadratic}}\end{array}\right.$ $\underline{Q^S\text{-groups}}$ $\left\{\begin{array}{l}Q_S^*(C,\varepsilon)\\ Q\langle v_0\rangle_S^*(C,\varepsilon)\\ Q_*^S(C,\varepsilon)\end{array}\right.$ by

$$\left\{\begin{array}{l}Q_S^n(C,\varepsilon) = Q^{n+1}(D,-\varepsilon)\\ Q\langle v_0\rangle_S^n(C,\varepsilon) = Q\langle v_0\rangle^{n+1}(D,-\varepsilon) \quad (n \in \mathbb{Z})\\ Q_n^S(C,\varepsilon) = Q_{n+1}(D,-\varepsilon)\end{array}\right.$$

for any resolution (D,g) of C. The Q^S-groups are independent of the choice of resolution, on account of the chain homotopy invariance of the Q-groups. As already noted above the relative L-groups of a localization map $A \longrightarrow S^{-1}A$ will be expressed in §§3.2 - 3.6 as the cobordism groups of algebraic Poincaré complexes over (A,S) defined using the Q^S-groups.

(It does not in general seem to possible to express the Q^S-groups of a finite-dimensional (A,S)-module chain complex C directly in terms of C, although there are natural candidates for such expressions: let $\mathrm{Hom}_A(C^\wedge,C)$ be the $\mathbb{Z}[\mathbb{Z}_2]$-module chain complex defined by

$$d : \mathrm{Hom}_A(C^\wedge,C)_r = \sum_{p+q=r} \mathrm{Hom}_A(C_p^\wedge,C_q) \longrightarrow \mathrm{Hom}_A(C^\wedge,C)_{r-1} ;$$

$$f \longmapsto d_C f + (-)^q f d_C^\wedge$$

$$T_\varepsilon : \mathrm{Hom}_A(C_p^\wedge,C_q) \longrightarrow \mathrm{Hom}_A(C_q^\wedge,C_p) ;$$

$$(\varepsilon\phi^\wedge : x \longmapsto (y \longmapsto \varepsilon\overline{\phi(y)(x)})) \quad (C_p^{\wedge\wedge} = C_p) ,$$

and define the $\left\{\begin{array}{l}\underline{\varepsilon\text{-symmetric}}\\ \underline{\text{even } \varepsilon\text{-symmetric}}\\ \underline{\varepsilon\text{-quadratic}}\end{array}\right.$ $\underline{\bar{Q}^S\text{-groups}}$ of C

$$\left\{\begin{array}{l}\bar{Q}^n_S(C,\varepsilon) = H_n(\mathrm{Hom}_{\mathbb{Z}[\mathbb{Z}_2]}(W,\mathrm{Hom}_A(C^\wedge,C)))\\ \bar{Q}\langle v_0\rangle^n_S(C,\varepsilon) = \ker(\hat{v}_0:\bar{Q}^n_S(C,\varepsilon)\longrightarrow \mathrm{Hom}_A(H^n_S(C),\hat{H}^1(\mathbb{Z}_2;A,\varepsilon)))\quad (n\in\mathbb{Z})\\ \bar{Q}^S_n(C,\varepsilon) = H_n(W\otimes_{\mathbb{Z}[\mathbb{Z}_2]}\mathrm{Hom}_A(C^\wedge,C))\end{array}\right.$$

by analogy with the Q-groups of §1.1. In particular, if C is a 0-dimensional (A,S)-module chain complex then

$$\left\{\begin{array}{l}\bar{Q}^0_S(C,\varepsilon) = \ker(1-T_\varepsilon:\mathrm{Hom}_A(H^0_S(C),H^0_S(C)^\wedge)\longrightarrow \mathrm{Hom}_A(H^0_S(C),H^0_S(C)^\wedge))\\ \bar{Q}\langle v_0\rangle^0_S(C,\varepsilon) = \ker(\hat{v}_0:\bar{Q}^0_S(C,\varepsilon)\longrightarrow \mathrm{Hom}_A(H^0_S(C),\hat{H}^1(\mathbb{Z}_2;A,\varepsilon)))\\ \bar{Q}^S_0(C,\varepsilon) = \mathrm{coker}(1-T_\varepsilon:\mathrm{Hom}_A(H^0_S(C),H^0_S(C)^\wedge)\longrightarrow \mathrm{Hom}_A(H^0_S(C),H^0_S(C)^\wedge))\end{array}\right.$$

and in §3.4 we shall show that there are natural identifications

$$\begin{array}{l}\bar{Q}^0_S(C,\varepsilon) = Q^0_S(C,\varepsilon)\\ \bar{Q}\langle v_0\rangle^0_S(C,\varepsilon) = Q\langle v_0\rangle^0_S(C,\varepsilon)\quad .\end{array}$$

Also, we shall identify $Q^S_0(C,\varepsilon)$ with the additive group of pairs (λ,ν) of functions

$$\lambda : H^0_S(C)\times H^0_S(C)\longrightarrow S^{-1}A/A$$

$$\nu : H^0_S(C)\longrightarrow H_0(\mathbb{Z}_2;S^{-1}A/A,\varepsilon) = \mathrm{coker}(1-T_\varepsilon:S^{-1}A/A\longrightarrow S^{-1}A/A)$$

satisfying

$$\lambda(x,y+y') = \lambda(x,y) + \lambda(x,y')\ ,\ \lambda(x,ay) = a\lambda(x,y)\ ,$$

$$\lambda(y,x) = \varepsilon\overline{\lambda(x,y)} \in S^{-1}A/A\ ,\ \lambda(x,x) = \nu(x) + \varepsilon\overline{\nu(x)} \in S^{-1}A/A$$

$$\nu(ax) = a\nu(x)\bar{a}\ ,\ \nu(x+y) - \nu(x) - \nu(y) = \lambda(x,y) \in H_0(\mathbb{Z}_2;S^{-1}A/A,\varepsilon)$$

$$(x,y,y'\in H^0_S(C),\ a\in A)\ .$$

The triple $(H^0_S(C),\lambda,\nu)$ is a "split ε-quadratic linking form over (A,S)" in the terminology of §3.4. There is defined a natural map

$$\overline{Q}^S_0(C,\varepsilon) \longrightarrow Q^S_0(C,\varepsilon) \ ; \ \psi \longmapsto (\lambda,\nu)$$

sending a representative $\psi \in \mathrm{Hom}_A(H^0_S(C),H^0_S(C)^{\wedge})$ to the pair (λ,ν) defined by

$$\lambda(x,y) = \psi(x)(y) + \varepsilon\overline{\psi(y)(x)} \in S^{-1}A/A$$
$$\nu(x) = \psi(x)(x) \in H_0(\mathbb{Z}_2;S^{-1}A/A,\varepsilon) \quad (x,y \in H^0_S(C)) .$$

In general this map is not an isomorphism, so that

$$\overline{Q}^S_0(C,\varepsilon) \neq Q^S_0(C,\varepsilon) .$$

For example, in the case

$$S = \{2^k \mid k \geqslant 0\} \subset A = \mathbb{Z} , \quad C_r = \begin{cases} \mathbb{Z}_2 & \text{if } r=0 \\ 0 & \text{if } r \neq 0 \end{cases}$$

we have that $H^0_S(C) = \mathbb{Z}_2$, and that the generator $1 = (\lambda,\nu) \in Q^S_0(C) = \mathbb{Z}_4$ $(\varepsilon = 1)$ is represented by the pair (λ,ν) defined by

$$\lambda(1,1) = \tfrac{1}{2} \in S^{-1}A/A = \mathbb{Z}[\tfrac{1}{2}]/\mathbb{Z}$$
$$\nu(1) = \tfrac{1}{4} \in H_0(\mathbb{Z}_2;S^{-1}A/A) = \mathbb{Z}[\tfrac{1}{2}]/\mathbb{Z} ;$$

the natural map $\overline{Q}^S_0(C) \longrightarrow Q^S_0(C)$ sends the generator $1 \in \overline{Q}^S_0(C) = \mathrm{Hom}_{\mathbb{Z}}(\mathbb{Z}_2,\hat{\mathbb{Z}_2}) = \mathbb{Z}_2$ to $2 = (2\lambda,2\nu) \in Q^S_0(C) = \mathbb{Z}_4$ $(2\lambda(1,1) = 0,\ 2\nu(1) = \tfrac{1}{2})$ and

$$\overline{Q}^S_0(C) = \mathbb{Z}_2 \neq Q^S_0(C) = \mathbb{Z}_4 .$$

Vogel [2,2.4] has shown that for any finite (A,S)-module chain complex C there is defined a long exact sequence of Q-groups

$$\ldots \longrightarrow \bar{Q}^S_n(C,\varepsilon) \longrightarrow Q^S_n(C,\varepsilon) \longrightarrow Q_{n+1}(C,-\varepsilon) \longrightarrow \bar{Q}^S_{n-1}(C,\varepsilon) \longrightarrow \ldots$$

in which the groups $Q_*(C,\varepsilon)$ are defined by

$$Q_n(C,\varepsilon) = H_n(W \otimes_{\mathbb{Z}[\mathbb{Z}_2]} (C^t \otimes_A C)) \quad (n \in \mathbb{Z})$$

with $T \in \mathbb{Z}_2$ acting on $C^t \otimes_A C$ by the <u>ε-transposition involution</u>

$$T_\varepsilon : C^t_p \otimes_A C_q \longrightarrow C^t_q \otimes_A C_p \ ; \ x \otimes y \longmapsto (-)^{pq} y \otimes \varepsilon x$$

exactly as in the original definition of the Q-groups in §I.1.1. The maps $Q^S_n(C,\varepsilon) = Q_{n+1}(D,-\varepsilon) \longrightarrow Q_{n+1}(C,-\varepsilon)$ are the ones naturally induced by $g^t \otimes g : C^t \otimes_A C \longrightarrow D^t \otimes_A D$ for any resolution $g : D \longrightarrow C$ of C by a finite f.g. projective A-module chain complex D, using the <u>slant</u> isomorphism of $\mathbb{Z}[\mathbb{Z}_2]$-module chain complexes

$$\backslash : D^t \otimes_A D \longrightarrow \mathrm{Hom}_A(D^*,D) \ ; \ x \otimes y \longmapsto (f \longmapsto \overline{f(x)}.y)$$

to identify

$$Q_*(D,\varepsilon) = H_*(W \otimes_{\mathbb{Z}[\mathbb{Z}_2]} (D^t \otimes_A D)) \ .$$

A chain map of finite (A,S)-module chain complexes

$$f : C \longrightarrow C'$$

induces a natural transformation of exact sequences

$$\begin{array}{ccccccccc}
\ldots \longrightarrow & \bar{Q}^S_n(C,\varepsilon) & \longrightarrow & Q^S_n(C,\varepsilon) & \longrightarrow & Q_{n+1}(C,-\varepsilon) & \longrightarrow & \bar{Q}^S_{n-1}(C,\varepsilon) & \longrightarrow \ldots \\
& \downarrow \bar{f}^S_\% & & \downarrow f^S_\% & & \downarrow f_\% & & \downarrow \bar{f}^S_\% & \\
\ldots \longrightarrow & \bar{Q}^S_n(C',\varepsilon) & \longrightarrow & Q^S_n(C',\varepsilon) & \longrightarrow & Q_{n+1}(C',-\varepsilon) & \longrightarrow & \bar{Q}^S_{n-1}(C',\varepsilon) & \longrightarrow \ldots
\end{array}$$

However, a homology equivalence $f : C \longrightarrow C'$ need not induce isomorphisms $\overline{f}^S_{\%}, f_{\%}$ (although the maps $f^S_{\%}$ are isomorphisms), since already the $\mathbb{Z}[\mathbb{Z}_2]$-module chain map inducing $\overline{f}^S_{\%}$

$$f^t \otimes f : C^t \otimes_A C \longrightarrow C'^t \otimes_A C'$$

need not be a homology equivalence. For example, if

$$S = \{2^k | k \geqslant 0\} \subset A = \mathbb{Z}$$

and $f: C \longrightarrow C'$ is defined by

$$\begin{array}{ccccccccccc} C: & \dots \longrightarrow & 0 & \longrightarrow & \mathbb{Z}_2 & \longrightarrow & 0 & \longrightarrow & 0 & \longrightarrow & \dots \\ \downarrow f & & & & \downarrow f & & \downarrow & & & & \\ C': & \dots \longrightarrow & 0 & \longrightarrow & \mathbb{Z}_4 & \xrightarrow{d'} & \mathbb{Z}_2 & \longrightarrow & 0 & \longrightarrow & \dots \end{array}$$

with

$$f : C_0 = \mathbb{Z}_2 \longrightarrow C'_0 = \mathbb{Z}_4 ; 1 \longmapsto 2$$

$$d' : C'_0 = \mathbb{Z}_4 \longrightarrow C'_{-1} = \mathbb{Z}_2 ; 1 \longmapsto 1$$

it is the case that

$$f \otimes f = 0 : C \otimes_{\mathbb{Z}} C \longrightarrow C' \otimes_{\mathbb{Z}} C' \quad .$$

Vogel [2,§3] has also shown that for every "n-dimensional ε-quadratic complex over (A,S)" $(C, \psi \in Q^S_n(C,\varepsilon))$ there exists a finite (but not necessarily n-dimensional) (A,S)-module chain complex C' with a homology equivalence

$$f : C \longrightarrow C'$$

such that

$$f^S_{\%}(\psi) \in \operatorname{im}(\overline{Q}^S_n(C',\varepsilon) \longrightarrow Q^S_n(C',\varepsilon)) \quad ,$$

and hence that the ε-quadratic L-groups $L_n(A,S,\varepsilon)$ $(n \geqslant 0)$ defined in §3.2 below using n-dimensional ε-quadratic Poincaré complexes over (A,S) $(C, \psi \in Q^S_n(C,\varepsilon))$ with C n-dimensional are isomorphic to the ε-quadratic L-groups $\overline{L}_n(A,S,\varepsilon)$ $(n \geqslant 0)$ defined using ε-quadratic Poincaré complexes over (A,S) of type $(C,\psi\in\bar{Q}^S_n(C,\varepsilon))$ with C finite. For example, with $f:C\longrightarrow C'$ as in the special case above $(S=\{2^k\}\subset A=\mathbb{Z}$, $C_0=\mathbb{Z}_2$ etc.) the map

$$\bar{Q}^S_0(C')=\mathbb{Z}_4\longrightarrow Q^S_0(C')=\mathbb{Z}_4\quad(\varepsilon=1)$$

is an isomorphism, whereas the map

$$\bar{Q}^S_0(C)=\mathbb{Z}_2\longrightarrow Q^S_0(C)=\mathbb{Z}_4$$

is not an isomorphism. Similar considerations apply in the $\begin{cases}\varepsilon\text{-symmetric}\\ \text{even } \varepsilon\text{-symmetric}\end{cases}$ case, with an exact sequence

$$\begin{cases}\ldots\longrightarrow\bar{Q}^n_S(C,\varepsilon)\longrightarrow Q^n_S(C,\varepsilon)\longrightarrow Q^{n+1}(C,-\varepsilon)\longrightarrow\bar{Q}^{n-1}_S(C,\varepsilon)\longrightarrow\ldots\\ \ldots\longrightarrow\bar{Q}\langle v_0\rangle^n_S(C,\varepsilon)\longrightarrow Q\langle v_0\rangle^n_S(C,\varepsilon)\longrightarrow Q^{n+1}(C,-\varepsilon)\longrightarrow\bar{Q}\langle v_0\rangle^{n-1}_S(C,\varepsilon)\\ \qquad\longrightarrow\ldots\end{cases}$$

for any finite (A,S)-module chain complex C).

The <u>S-adic completion of A</u> is the inverse limit

$$\hat{A}=\varprojlim_{s\in S}A/sA$$

of the inverse system of abelian groups $\{A/sA\,|\,s\in S\}$, with S partially ordered by

$$s\leqslant s'\text{ if there exists } t\in S \text{ such that } s'=st\in S\ ,$$

the structure maps being the projections

$$A/stA\longrightarrow A/sA\quad(s,t\in S)\ .$$

Thus an element $\hat{a}\in\hat{A}$ is a sequence

$$\hat{a}=\{a_s\in A/sA\,|\,s\in S\}$$

such that

$$a_s=[a_{st}]\in A/sA\ (s,t\in S)\ .$$

In dealing with completions we shall always assume that S is central in A, so that each A/sA $(s \in S)$ inherits a ring structure from A, and $\hat{A}$ is a ring with involution

$$\bar{\ } : \hat{A} \longrightarrow \hat{A} \ ; \ \hat{a} = \{a_s \in A/sA \mid s \in S\} \longmapsto \bar{\hat{a}} = \{\overline{a_{\bar{s}}} \in A/sA \mid s \in S\} \ .$$

For example, the ring of m-adic integers

$$\hat{\mathbb{Z}}_m = \varprojlim_k \mathbb{Z}/m^k\mathbb{Z} \qquad (k \geqslant 0, m \geqslant 2)$$

is the $\{m^k\}$-adic completion of $\mathbb{Z}$. The inclusion

$$i : A \longrightarrow \hat{A} \ ; \ a \longmapsto \{[a] \in A/sA \mid s \in S\}$$

is a morphism of rings with involution such that

$$\hat{S} = i(S) \subset \hat{A}$$

is a multiplicative subset.

A commutative square of rings with involution

$$\begin{array}{ccc} A & \longrightarrow & B \\ \downarrow & & \downarrow \\ B' & \longrightarrow & A' \end{array}$$

is <u>cartesian</u> if it gives rise to an exact sequence of abelian groups with involution

$$0 \longrightarrow A \longrightarrow B \oplus B' \longrightarrow A' \longrightarrow 0 \ .$$

In particular, the <u>localization-completion</u> square

$$\begin{array}{ccc} A & \longrightarrow & S^{-1}A \\ \downarrow & & \downarrow \\ \hat{A} & \longrightarrow & \hat{S}^{-1}\hat{A} \end{array}$$

is cartesian. As described in the introduction to §3 such a square gives rise to excision isomorphisms in the relative K-groups

$$K_*(A,S) \xrightarrow{\sim} K_*(\hat{A},\hat{S})$$

(which follows from the isomorphism of exact categories

$$i : \{(A,S)\text{-modules}\} \xrightarrow{\sim} \{(\hat{A},\hat{S})\text{-modules}\} ;$$

$$M \longmapsto \hat{M} = \hat{A}\otimes_A M = \varprojlim_{s\in S} M/sM \quad)$$

and a Mayer-Vietoris exact sequence in the absolute K-groups

$$\ldots \longrightarrow K_n(A) \longrightarrow K_n(\hat{A}) \oplus K_n(S^{-1}A) \longrightarrow K_n(\hat{S}^{-1}\hat{A}) \longrightarrow K_{n-1}(A) \longrightarrow \ldots \quad (n \in \mathbb{Z}) .$$

In §§3.2,3.6 below we shall identify the relative $\begin{cases} \varepsilon\text{-symmetric} \\ \varepsilon\text{-quadratic} \end{cases}$ L-groups $\begin{cases} L^*_S(A \longrightarrow S^{-1}A,\varepsilon) \\ L^S_*(A \longrightarrow S^{-1}A,\varepsilon) \end{cases}$ with the cobordism groups $\begin{cases} L^*(A,S,\varepsilon) \\ L_*(A,S,\varepsilon) \end{cases}$ of $\begin{cases} \text{even } \varepsilon\text{-symmetric} \\ \varepsilon\text{-quadratic} \end{cases}$ Poincaré complexes over (A,S) $\begin{cases} (C,\phi \in Q\langle v_0\rangle^n_S(C,\varepsilon)) \\ (C,\psi \in Q^S_n(C,\varepsilon)) \end{cases}$ with C an n-dimensional (A,S)-module chain complexes. The functors

$$i : \{\text{n-dimensional } (A,S)\text{-module chain complexes}\}$$
$$\longrightarrow \{\text{n-dimensional } (\hat{A},\hat{S})\text{-module chain complexes}\} ;$$
$$C \longmapsto \hat{C} = \hat{A}\otimes_A C \qquad (n \geqslant 0)$$

are isomorphisms of categories. Thus if the induced maps

$$\begin{cases} i : Q\langle v_0\rangle^*_S(C,\varepsilon) \longrightarrow Q\langle v_0\rangle^*_{\hat{S}}(\hat{C},\varepsilon) \\ i : Q^S_*(C,\varepsilon) \longrightarrow Q^{\hat{S}}_*(\hat{C},\varepsilon) \end{cases}$$

are isomorphisms there are defined excision isomorphisms in the relative L-groups

$$\begin{cases} L^*(A,S,\varepsilon) \xrightarrow{\sim} L^*(\hat{A},\hat{S},\varepsilon) \\ L_*(A,S,\varepsilon) \xrightarrow{\sim} L_*(\hat{A},\hat{S},\varepsilon) \end{cases}$$

and there is defined a Mayer-Vietoris exact sequence in the absolute L-groups

$$\begin{cases} \ldots \longrightarrow L^n(A,\varepsilon) \longrightarrow L^n(\hat{A},\varepsilon) \oplus L^n_S(S^{-1}A,\varepsilon) \longrightarrow L^n_{\hat{S}}(\hat{S}^{-1}\hat{A},\varepsilon) \longrightarrow L^{n-1}(A,\varepsilon) \longrightarrow \ldots \\ \ldots \longrightarrow L_n(A,\varepsilon) \longrightarrow L_n(\hat{A},\varepsilon) \oplus L^S_n(S^{-1}A,\varepsilon) \longrightarrow L^{\hat{S}}_n(\hat{S}^{-1}\hat{A},\varepsilon) \longrightarrow L_{n-1}(A,\varepsilon) \longrightarrow \ldots \end{cases}$$

$$(n \in \mathbb{Z}) \quad .$$

Use the cartesian property of the localization-completion square to define the abelian group morphism

$$\hat{\delta} \; : \; \hat{H}^0(\mathbb{Z}_2;\hat{S}^{-1}\hat{A},\varepsilon) \longrightarrow \hat{H}^0(\mathbb{Z}_2;\hat{S}^{-1}\hat{A}/\hat{A},\varepsilon) \cong \hat{H}^0(\mathbb{Z}_2;S^{-1}A/A,\varepsilon)$$

$$\xrightarrow{\;\;\delta\;\;} \hat{H}^1(\mathbb{Z}_2;A,\varepsilon) \; .$$

In Proposition 3.1.3 ii) we shall show that $\begin{cases} \text{if } \hat{\delta} = 0 \\ \text{for all } A,S,\varepsilon \end{cases}$ the completion map $i:(A,S) \longrightarrow (\hat{A},\hat{S})$ does induce isomorphisms in the $\begin{cases} Q\langle v_0\rangle^*_S \\ Q^S_* \end{cases}$-groups. The conclusions regarding excision isomorphisms and Mayer-Vietoris exact sequences in the L-groups will be drawn in §3.6.

The property of the completion map i implying excision in the K- and L-groups can be abstracted as follows.

Let (B,T) be another pair such as (A,S), with B a ring with involution and $T \subset B$ a multiplicative subset. A <u>morphism</u> of such pairs

$$f \; : \; (A,S) \longrightarrow (B,T)$$

is a morphism of rings with involution

$$f \; : \; A \longrightarrow B$$

such that

$$f(S) \subseteq T \subseteq B \; .$$

If C is an n-dimensional (A,S)-module chain complex then $B\otimes_A C$ is an n-dimensional (B,T)-module chain complex; if (D,g) is a resolution of C then $(B\otimes_A D, 1\otimes g)$ is a resolution of $B\otimes_A C$.

The morphism $f:(A,S) \longrightarrow (B,T)$ is <u>cartesian</u> if

i) $f| : S \longrightarrow T$ is a bijection

ii) for each $s \in S$ the abelian group morphism

$$A/sA \longrightarrow B/f(s)B \ ; \ [a] \longmapsto [f(a)] \quad (a \in A)$$

is an isomorphism .

Cartesian morphisms were introduced by Karoubi [2]. In particular, the completion map

$$i : (A,S) \longrightarrow (\hat{A},\hat{S})$$

is a cartesian morphism.

Define a direct system of abelian groups

$$\{A/sA \mid s \in S\}$$

by giving S the partial ordering

$$s \leqslant s' \text{ if } s' = ts \in S \text{ for some } t \in S \ ,$$

and defining the structure maps by

$$A/sA \longrightarrow A/s'A \ ; \ [a] \longmapsto [ta] \ .$$

Use the abelian group morphisms

$$A/sA \longrightarrow S^{-1}A/A \ ; \ [a] \longmapsto \frac{a}{s} \quad (s \in S)$$

to identify

$$\varinjlim_{s \in S} A/sA = S^{-1}A/A \ .$$

It follows from this identification that a cartesian morphism $f:(A,S) \longrightarrow (B,T)$ induces isomorphisms

$$f : \varinjlim_{s \in S} A/sA = S^{-1}A/A \longrightarrow \varinjlim_{t \in T} B/tB = T^{-1}B/B$$

and hence that the commutative square of rings with involution

$$\begin{array}{ccc} A & \longrightarrow & S^{-1}A \\ {\scriptstyle f}\downarrow & & \downarrow \\ B & \longrightarrow & T^{-1}B \end{array}$$

is cartesian. There is thus defined a short exact sequence of $\mathbb{Z}[\mathbb{Z}_2]$-modules

$$0 \longrightarrow A \longrightarrow B \oplus S^{-1}A \longrightarrow T^{-1}B \longrightarrow 0$$

inducing a long exact sequence of Tate $\mathbb{Z}_2$-cohomology groups

$$\ldots \longrightarrow \hat{H}^n(\mathbb{Z}_2;A,\varepsilon) \longrightarrow \hat{H}^n(\mathbb{Z}_2;B,\varepsilon) \oplus \hat{H}^n(\mathbb{Z}_2;S^{-1}A,\varepsilon)$$
$$\longrightarrow \hat{H}^n(\mathbb{Z}_2;T^{-1}B,\varepsilon) \xrightarrow{\hat{\delta}} \hat{H}^{n+1}(\mathbb{Z}_2;A,\varepsilon) \longrightarrow \ldots \quad (n \in \mathbb{Z}) .$$

Proposition 3.1.3 i) A cartesian morphism

$$f : (A,S) \longrightarrow (B,T)$$

induces an isomorphism of exact categories

$$f : \{(A,S)\text{-modules}\} \xrightarrow{\sim} \{(B,T)\text{-modules}\} \ ; \ M \longmapsto B \otimes_A M .$$

If M,N are (A,S)-modules there are defined $\mathbb{Z}$-module isomorphisms

$$M \xrightarrow{\sim} B \otimes_A M \ ; \ x \longmapsto 1 \otimes x$$

$$\mathrm{Hom}_A(M,N) \xrightarrow{\sim} \mathrm{Hom}_B(B \otimes_A M, B \otimes_A N) \ ; \ g \longmapsto (b \otimes x \longmapsto b \otimes g(x))$$

$$M\hat{} = \mathrm{Hom}_A(M,S^{-1}A/A) \xrightarrow{\sim} (B \otimes_A M)\hat{} = \mathrm{Hom}_B(B \otimes_A M, T^{-1}B/B) \ ;$$
$$g \longmapsto (b \otimes x \longmapsto b.f(g(x))) .$$

ii) If $f:(A,S) \longrightarrow (B,T)$ is a cartesian morphism and C is a finite-dimensional (A,S)-module chain complex the induced abelian group morphisms

$$\begin{cases} f : Q_S^*(C,\varepsilon) \longrightarrow Q_T^*(B \otimes_A C,\varepsilon) \\ f : Q\langle v_0\rangle_S^*(C,\varepsilon) \longrightarrow Q\langle v_0\rangle_T^*(B \otimes_A C,\varepsilon) \\ f : Q_*^S(C,\varepsilon) \longrightarrow Q_*^T(B \otimes_A C,\varepsilon) \end{cases}$$

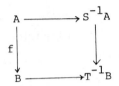

are $\left\{\begin{array}{l}\text{isomorphisms}\\ \text{monomorphisms. If}\\ \text{isomorphisms}\end{array}\right.$

$$\hat{\delta} = 0 : \hat{H}^0(\mathbb{Z}_2;T^{-1}B,\varepsilon) \longrightarrow \hat{H}^1(\mathbb{Z}_2;A,\varepsilon)$$

the maps $f : Q\langle v_0\rangle^*_S(C,\varepsilon) \longrightarrow Q\langle v_0\rangle^*_S(B\otimes_A C,\varepsilon)$ are also isomorphisms.

Proof: i) See Appendix 5 of Karoubi [2].

ii) Let (D,g) be a resolution of C, and consider the commutative diagram of abelian group chain complexes

$$\begin{array}{ccc}
\mathrm{Hom}_A(D^*,D) & \xrightarrow{\mathrm{Hom}_A(1,g)} & \mathrm{Hom}_A(D^*,C) \\
\Big\downarrow f & & \Big\downarrow f \\
\mathrm{Hom}_B(B\otimes_A D^*,B\otimes_A D) & \xrightarrow{\mathrm{Hom}_B(1,1\otimes_A g)} & \mathrm{Hom}_B(B\otimes_A D^*,B\otimes_A C)
\end{array} \quad .$$

As $g:D \longrightarrow C$ is a homology equivalence (by definition) and D^* is a f.g. projective A-module chain complex the chain maps $\mathrm{Hom}_A(1,g)$, $\mathrm{Hom}_B(1,1\otimes_A g)$ are also homology equivalences.

As $f:(A,S) \longrightarrow (B,T)$ is cartesian and D^* is a f.g. projective A-module chain complex the chain map

$$f : \mathrm{Hom}_A(D^*,C) \longrightarrow \mathrm{Hom}_B(B\otimes_A D^*,B\otimes_A C)$$

is an isomorphism of abelian group chain complexes. It now follows from the commutativity of the above diagram that the $\mathbb{Z}[\mathbb{Z}_2]$-module chain map

$$f : \mathrm{Hom}_A(D^*,D) \longrightarrow \mathrm{Hom}_B(B\otimes_A D^*,B\otimes_A D)$$

is a homology equivalence, so that it induces isomorphisms

in the $\left\{\begin{array}{l}\mathbb{Z}_2\text{-hypercohomology}\\ \mathbb{Z}_2\text{-hyperhomology}\end{array}\right.$ groups

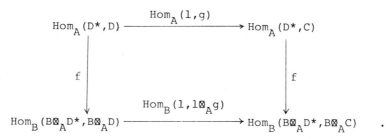

$$\begin{cases} f : Q_S^n(C,\varepsilon) = Q^{n+1}(D,-\varepsilon) \xrightarrow{\sim} Q_T^n(B\otimes_A D,\varepsilon) = Q^{n+1}(B\otimes_A D,-\varepsilon) \\ f : Q_n^S(C,\varepsilon) = Q_{n+1}(D,-\varepsilon) \xrightarrow{\sim} Q_n^T(B\otimes_A D,\varepsilon) = Q_{n+1}(B\otimes_A D,-\varepsilon) \end{cases}$$

$$(n \in \mathbb{Z}) \quad .$$

As D is S-acyclic, for every $\phi \in Q^{n+1}(D,-\varepsilon)$

$$\mathrm{im}(\hat{v}_0(\phi) : H^{n+1}(D) \longrightarrow \hat{H}^1(\mathbb{Z}_2;A,\varepsilon))$$
$$\subseteq \mathrm{im}(\hat{\delta} : \hat{H}^0(\mathbb{Z}_2;S^{-1}A/A,\varepsilon) \longrightarrow \hat{H}^1(\mathbb{Z}_2;A,\varepsilon))$$

(cf. the definition in §3.3 below of the "linking Wu class"

$$\hat{v}_0^S(\phi) \; : \; H^{n+1}(D) \longrightarrow \hat{H}^0(\mathbb{Z}_2;S^{-1}A/A,\varepsilon)$$

such that $\hat{\delta}\hat{v}_0^S(\phi) = \hat{v}_0(\phi)$). We can thus identify

$$Q\langle v_0\rangle_S^n(C,\varepsilon) = Q\langle v_0\rangle^{n+1}(D,-\varepsilon)$$
$$= \ker(\hat{v}_0 : Q^{n+1}(D,-\varepsilon) \longrightarrow \mathrm{Hom}_A(H^{n+1}(D), \mathrm{im}(\hat{H}^0(\mathbb{Z}_2;S^{-1}A/A,\varepsilon)$$
$$\longrightarrow \hat{H}^1(\mathbb{Z}_2;A,\varepsilon))) \quad .$$

It follows from the exact sequence

$$0 \longrightarrow \mathrm{im}(\hat{\delta} : \hat{H}^0(\mathbb{Z}_2;T^{-1}B,\varepsilon) \longrightarrow \hat{H}^1(\mathbb{Z}_2;A,\varepsilon))$$
$$\longrightarrow \mathrm{im}(\hat{H}^0(\mathbb{Z}_2;S^{-1}A/A,\varepsilon) \longrightarrow \hat{H}^1(\mathbb{Z}_2;A,\varepsilon))$$
$$\longrightarrow \mathrm{im}(\hat{H}^0(\mathbb{Z}_2;T^{-1}B/B,\varepsilon) \longrightarrow \hat{H}^1(\mathbb{Z}_2;B,\varepsilon)) \longrightarrow 0$$

that if $\hat{\delta} = 0$ there are also induced isomorphisms

$$f : Q\langle v_0\rangle_S^n(C,\varepsilon) = Q\langle v_0\rangle^{n+1}(D,-\varepsilon)$$
$$\xrightarrow{\sim} Q\langle v_0\rangle_T^n(B\otimes_A C,\varepsilon) = Q\langle v_0\rangle^{n+1}(B\otimes_A D,-\varepsilon)$$
$$(n \in \mathbb{Z}) \quad .$$

[]

3.2 The localization exact sequence ($n \geqslant 0$)

Let A, S, ε be as in §3.1 above.

Let

$$S = \mathrm{im}(\tilde{K}_0(A) \longrightarrow \tilde{K}_0(S^{-1}A)) \subseteq \tilde{K}_0(S^{-1}A)$$

be the $*$-invariant subgroup of the projective classes $[S^{-1}P]$ of the f.g. projective $S^{-1}A$-modules $S^{-1}P$ induced from f.g. projective A-modules P.

Let $\begin{cases} L^n_S(A \longrightarrow S^{-1}A, \varepsilon) \\ L^S_n(A \longrightarrow S^{-1}A, \varepsilon) \end{cases}$ ($n \in \mathbb{Z}$) be the relative $\begin{cases} \varepsilon\text{-symmetric} \\ \varepsilon\text{-quadratic} \end{cases}$ L-groups appearing in the exact sequence

$$\begin{cases} \dots \longrightarrow L^n(A,\varepsilon) \longrightarrow L^n_S(S^{-1}A,\varepsilon) \longrightarrow L^n_S(A \longrightarrow S^{-1}A,\varepsilon) \longrightarrow L^{n-1}(A,\varepsilon) \longrightarrow \dots \\ \dots \longrightarrow L_n(A,\varepsilon) \longrightarrow L^S_n(S^{-1}A,\varepsilon) \longrightarrow L^S_n(A \longrightarrow S^{-1}A,\varepsilon) \longrightarrow L_{n-1}(A,\varepsilon) \longrightarrow \dots \end{cases}.$$

An <u>n-dimensional</u> $\begin{cases} \underline{\text{(even) } \varepsilon\text{-symmetric}} \\ \underline{\varepsilon\text{-quadratic}} \end{cases}$ <u>complex over (A,S)</u> $\begin{cases} (C,\phi) \\ (C,\psi) \end{cases}$ is an n-dimensional (A,S)-module chain complex C together with an element $\begin{cases} \phi \in Q^n_S(C,\varepsilon) \quad (\phi \in Q\langle v_0\rangle^n_S(C,\varepsilon)) \\ \psi \in Q^S_n(C,\varepsilon) \end{cases}.$ Such a complex is <u>Poincaré</u> if the A-module morphisms

$$\begin{cases} \phi_0 \ : \ H^*_S(C) \longrightarrow H_{n-*}(C) \\ (1+T_\varepsilon)\psi_0 \ : \ H^*_S(C) \longrightarrow H_{n-*}(C) \end{cases}$$

are isomorphisms. There is a corresponding notion of <u>pair</u>.

Define the <u>n-dimensional</u> $\begin{cases} \underline{\varepsilon\text{-symmetric}} \\ \underline{\varepsilon\text{-quadratic}} \end{cases}$ <u>L-group of (A,S)</u> $\begin{cases} L^n(A,S,\varepsilon) \\ L_n(A,S,\varepsilon) \end{cases}$ ($n \geqslant 0$) to be the cobordism group of n-dimensional

$\begin{cases} \text{even } \varepsilon\text{-symmetric} \\ \varepsilon\text{-quadratic} \end{cases}$ Poincaré complexes over (A,S).

<u>Proposition 3.2.1</u> A cartesian morphism $f:(A,S)\longrightarrow(B,T)$

$\begin{cases} \text{such that } \hat{\delta} = 0 : \hat{H}^0(\mathbb{Z}_2;T^{-1}B,\varepsilon)\longrightarrow\hat{H}^1(\mathbb{Z}_2;A,\varepsilon) \\ - \end{cases}$ induces

isomorphisms in the $\begin{cases} \varepsilon\text{-symmetric} \\ \varepsilon\text{-quadratic} \end{cases}$ L-groups

$$\begin{cases} f : L^n(A,S,\varepsilon) \xrightarrow{\sim} L^n(B,T,\varepsilon) \\ f : L_n(A,S,\varepsilon) \xrightarrow{\sim} L_n(B,T,\varepsilon) \end{cases} \quad (n \geqslant 0) \quad .$$

<u>Proof</u>: Immediate from Proposition 3.1.3 ii).

[]

In Proposition 3.2.3 below we shall apply the algebraic Γ-theory of §2.4 to identify

$$\begin{cases} L^n_S(A\longrightarrow S^{-1}A,\varepsilon) = L^n(A,S,\varepsilon) \\ L^S_n(A\longrightarrow S^{-1}A,\varepsilon) = L_n(A,S,\varepsilon) \end{cases} \quad (n \geqslant 0) \quad .$$

In §3.6 this will be extended to the range $n \leqslant -1$, and these identifications will be used together with Proposition 3.2.1 to obtain Mayer-Vietoris exact sequences for the L-groups of the rings with involution appearing in the cartesian square

$$\begin{array}{ccc} A & \longrightarrow & S^{-1}A \\ \big\downarrow & & \big\downarrow \\ B & \longrightarrow & T^{-1}B \end{array}$$

associated to a cartesian morphism $(A,S)\longrightarrow(B,T)$.

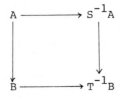

<u>Proposition 3.2.2</u> i) The homotopy equivalence classes of

n-dimensional $\begin{cases}\text{(even) } \varepsilon\text{-symmetric} \\ \varepsilon\text{-quadratic}\end{cases}$ (Poincaré) complexes over (A,S)

are in a natural one-one correspondence with the homotopy equivalence classes of S-acyclic (n+1)-dimensional

$\begin{cases}\text{(even) } (-\varepsilon)\text{-symmetric} \\ (-\varepsilon)\text{-quadratic}\end{cases}$ (Poincaré) complexes over A.

Similarly for pairs.

ii) $\begin{cases}L^n(A,S,\varepsilon) \\ L_n(A,S,\varepsilon)\end{cases}$ $(n \geqslant 0)$ is naturally isomorphic to the cobordism

group of S-acyclic (n+1)-dimensional $\begin{cases}\text{even } (-\varepsilon)\text{-symmetric} \\ (-\varepsilon)\text{-quadratic}\end{cases}$

Poincaré complexes over A.

<u>Proof</u>: i) Immediate from Proposition 3.1.2.

ii) Immediate from i).

[]

We shall be mainly working with the characterization of the L-groups $\begin{cases}L^*(A,S,\varepsilon) \\ L_*(A,S,\varepsilon)\end{cases}$ as the cobordism groups of S-acyclic algebraic Poincaré complexes (Proposition 3.2.2 ii)), because all the A-module chain complex manipulations developed in §1 in connection with the L-groups $\begin{cases}L^*(A,\varepsilon) \\ L_*(A,\varepsilon)\end{cases}$ specialize to manipulations of S-acyclic complexes. In particular, if it is insisted that all the A-module chain complexes involved be S-acyclic there is obtained from §1.5 an algebraic S-acyclic surgery theory with which to analyze S-acyclic algebraic Poincaré cobordism. (Localization in geometric surgery theory will be discussed more fully in §7.7 below. For the present we note that if $(f,b):M\longrightarrow X$ is an n-dimensional normal map which is a rational homotopy equivalence ($\pi_*(f)\otimes\mathbb{Q} = 0$) then the quadratic kernel

$$\sigma_*(f,b) = (C(f^!),\psi)$$

is an S-acyclic n-dimensional quadratic Poincaré complex over $\mathbb{Z}[\pi_1(X)]$, with

$$S = \mathbb{Z}-\{0\} \subset \mathbb{Z}[\pi_1(X)] .$$

The S-acyclic cobordism class of the skew-suspension

$$\sigma_*^S(f,b) = \bar{S}\sigma_*(f,b) \in L_{n+1}(\mathbb{Z}[\pi_1(X)],S) \quad (\varepsilon = 1)$$

is the obstruction to making (f,b) normal bordant to a homotopy equivalence by a bordism which is also a rational homotopy equivalence, i.e. it is the "local surgery obstruction" in the sense of Pardon [3]. The chain level effect of a "local surgery on a conglomerate Moore space" in the sense of Pardon [3] is that of an S-acyclic surgery on a connected S-acyclic (n+1)-dimensional quadratic pair over $\mathbb{Z}[\pi_1(X)]$ $(g:C(f^!)\longrightarrow D,(\delta\psi,\psi))$ with $D_r = 0$ $(r \neq k,k+1)$ for some k, $0 \leqslant k \leqslant n+1)$.

<u>Proposition 3.2.3</u> i) There are natural identifications

$$\begin{cases} L_S^n(A\longrightarrow S^{-1}A,\varepsilon) = L^n(A,S,\varepsilon) \\ L_n^S(A\longrightarrow S^{-1}A,\varepsilon) = L_n(A,S,\varepsilon) \end{cases} \quad (n \geqslant 0)$$

under which the maps appearing in the localization exact sequence

$$\begin{cases} \dots\longrightarrow L^n(A,\varepsilon)\longrightarrow L_S^n(S^{-1}A,\varepsilon)\longrightarrow L_S^n(A\rightarrow S^{-1}A,\varepsilon)\longrightarrow L^{n-1}(A,\varepsilon)\longrightarrow \dots \\ \dots\longrightarrow L_n(A,\varepsilon)\longrightarrow L_n^S(S^{-1}A,\varepsilon)\longrightarrow L_n^S(A\rightarrow S^{-1}A,\varepsilon)\longrightarrow L_{n-1}(A,\varepsilon)\longrightarrow \dots \end{cases}$$

are given by

$$\begin{cases} \partial : L^n_S(S^{-1}A,\varepsilon) \longrightarrow L^n(A,S,\varepsilon) \; ; \; S^{-1}(C,\phi) \longmapsto \partial\bar{S}(C,\phi) \\ \partial : L_n^S(S^{-1}A,\varepsilon) \longrightarrow L_n(A,S,\varepsilon) \; ; \; S^{-1}(C,\psi) \longmapsto \partial\bar{S}(C,\psi) \end{cases}$$

$$\begin{cases} L^n(A,S,\varepsilon) \longrightarrow L^{n-1}(A,\varepsilon) = L\langle v_0\rangle^{n+1}(A,-\varepsilon) \; ; \; (C,\phi) \longmapsto (C,\phi) \\ L_n(A,S,\varepsilon) \longrightarrow L_{n-1}(A,\varepsilon) = L_{n+1}(A,-\varepsilon) \; ; \; (C,\psi) \longmapsto (C,\psi) \end{cases}$$

$$(n \geqslant 0) \; .$$

ii) The skew-suspension maps in the $\pm\varepsilon$-quadratic L-groups

$$\bar{S} : L_n(A,S,\varepsilon) \longrightarrow L_{n+2}(A,S,-\varepsilon) \; ; \; (C,\psi) \longmapsto (SC,\bar{S}\psi) \quad (n \geqslant 0)$$

are isomorphisms.

<u>Proof</u>: i) It follows from Proposition 3.1.1 i) that the maps

$$\begin{cases} \Gamma^n(A \longrightarrow S^{-1}A,\varepsilon) \longrightarrow L^n_S(S^{-1}A,\varepsilon) \; ; \; (C,\phi) \longmapsto S^{-1}(C,\phi) \\ \Gamma_n(A \longrightarrow S^{-1}A,\varepsilon) \longrightarrow L_n^S(S^{-1}A,\varepsilon) \; ; \; (C,\psi) \longmapsto S^{-1}(C,\psi) \end{cases} \quad (n \geqslant 0)$$

are isomorphisms, so that there are natural identifications

$$\begin{cases} L^n_S(A \longrightarrow S^{-1}A,\varepsilon) = \Gamma^n(F,\varepsilon) \\ L_n^S(A \longrightarrow S^{-1}A,\varepsilon) = \Gamma_n(F,\varepsilon) \end{cases} \quad (n \geqslant 0) \; ,$$

the groups on the right being the relative $\begin{cases} \varepsilon\text{-symmetric} \\ \varepsilon\text{-quadratic} \end{cases}$ Γ-groups

of the commutative square of rings with involution

$$\begin{array}{ccc} A & \xrightarrow{\;1\;} & A \\ {\scriptstyle 1}\big\downarrow & F & \big\downarrow \\ A & \longrightarrow & S^{-1}A \end{array} \quad .$$

By Proposition 3.2.2 ii) $\begin{cases} L^n(A,S,\varepsilon) \\ L_n(A,S,\varepsilon) \end{cases}$ $(n \geqslant 0)$ is naturally isomorphic

to the cobordism group of S-acyclic (n+1)-dimensional

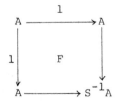

$\begin{cases}\text{even } (-\varepsilon)\text{-symmetric} \\ (-\varepsilon)\text{-quadratic}\end{cases}$ Poincaré complexes over A, which is just the expression obtained for $\begin{cases}\Gamma^n(F,\varepsilon) \\ \Gamma_n(F,\varepsilon)\end{cases}$ $(n \geqslant 0)$ in Proposition 2.4.6.

We can thus identify

$$\begin{cases} L^n_S(A \longrightarrow S^{-1}A,\varepsilon) = \Gamma^n(F,\varepsilon) = L^n(A,S,\varepsilon) \\ L_n^S(A \longrightarrow S^{-1}A,\varepsilon) = \Gamma_n(F,\varepsilon) = L_n(A,S,\varepsilon)\end{cases} \quad (n \geqslant 0) \quad .$$

Explicitly, the isomorphism

$$\begin{cases} L^n_S(A \longrightarrow S^{-1}A,\varepsilon) \xrightarrow{\sim} L^n(A,S,\varepsilon) \\ L_n^S(A \longrightarrow S^{-1}A,\varepsilon) \xrightarrow{\sim} L_n(A,S,\varepsilon)\end{cases}$$

sends the element

$$\begin{cases} ((C,\phi \in Q^{n-1}(C,\varepsilon)),S^{-1}(f:C \longrightarrow D,(\delta\phi,\phi) \in Q^n(f,\varepsilon))) \in L^n_S(A \longrightarrow S^{-1}A,\varepsilon) \\ ((C,\psi \in Q_{n-1}(C,\varepsilon)),S^{-1}(f:C \longrightarrow D,(\delta\psi,\psi) \in Q_n(f,\varepsilon))) \in L_n^S(A \longrightarrow S^{-1}A,\varepsilon)\end{cases}$$

to the cobordism class $\begin{cases}(C',\phi') \in L^n(A,S,\varepsilon) \\ (C',\psi') \in L_n(A,S,\varepsilon)\end{cases}$ of the S-acyclic (n+1)-dimensional $\begin{cases}\text{even } (-\varepsilon)\text{-symmetric} \\ (-\varepsilon)\text{-quadratic}\end{cases}$ Poincaré complex over A $\begin{cases}(C',\phi') \\ (C',\psi')\end{cases}$ obtained from the skew-suspension $\begin{cases}\overline{S}(C',\phi') \\ \overline{S}(C',\psi')\end{cases}$ by surgery on the connected (n+2)-dimensional $\begin{cases}\text{even } (-\varepsilon)\text{-symmetric} \\ (-\varepsilon)\text{-quadratic}\end{cases}$ $S^{-1}A$-Poincaré pair over A $\begin{cases}\overline{S}(f:C \longrightarrow D,(\delta\phi,\phi)) \\ \overline{S}(f:C \longrightarrow D,(\delta\psi,\psi))\end{cases}$. In particular, for $C = 0$ $\begin{cases}(C',\phi') = \partial\overline{S}(D,\delta\phi) \\ (C',\psi') = \partial\overline{S}(D,\delta\psi)\end{cases}$, and $\begin{cases}S^{-1}(D,\delta\phi) \in L^n_S(S^{-1}A,\varepsilon) \\ S^{-1}(D,\delta\psi) \in L_n^S(S^{-1}A,\varepsilon)\end{cases}$.

ii) Immediate from i) and Proposition 2.2.3 ii).

[]

The pair (A,S) is <u>m-dimensional</u> if every f.g. S-torsion A-module M has a f.g. projective A-module resolution of length m+1

$$0 \longrightarrow P_{m+1} \longrightarrow P_m \longrightarrow \ldots \longrightarrow P_1 \longrightarrow P_0 \longrightarrow M \longrightarrow 0 \ .$$

For example, if A is m-dimensional (in the sense of §1.2) then (A,S) is m-dimensional; if π is a finite group and p is a prime such that $p \nmid |\pi|$ then $(\mathbb{Z}[\pi],\{p^k | k \geqslant 0\})$ is 0-dimensional.
By analogy with Proposition 1.2.3 ii) we have:

<u>Proposition 3.2.4</u> If (A,S) is m-dimensional the skew-suspension maps in the $\pm\varepsilon$-symmetric L-groups

$$\bar{S} : L^n(A,S,\varepsilon) \longrightarrow L^{n+2}(A,S,-\varepsilon) \quad (n \geqslant 2m+1)$$

are isomorphisms, and there are natural identifications

$$\begin{cases} L^{2i}(A,S,\varepsilon) \\ L^{2i-1}(A,S,\varepsilon) \end{cases} = \text{the cobordism group of S-acyclic } \begin{cases} 2m+1 \\ 2m \end{cases}\text{-dimensional } (-)^{i-m-1}\varepsilon\text{-symmetric Poincaré complexes over A} \quad (i \geqslant m+1)$$

under which $L^n(A,S,\varepsilon) \longrightarrow L^{n-1}(A,\varepsilon)$ $(n \geqslant 2m+1)$ becomes the forgetful map

$$\begin{cases} L^{2i}(A,S,\varepsilon) \longrightarrow L^{2i-1}(A,\varepsilon) \ ; \ (C,\phi) \longmapsto \bar{S}^{i-m-1}(C,\phi) \\ L^{2i-1}(A,S,\varepsilon) \longrightarrow L^{2i-2}(A,\varepsilon) \ ; \ (C,\phi) \longmapsto \bar{S}^{i-m-1}(C,\phi) \end{cases} \quad (i \geqslant m+1) \ .$$

In particular, for m = 0

$$L^{2i-1}(A,S,\varepsilon) = 0 \quad (i \geqslant 1) \ .$$

<u>Proof</u>: In order to identify $\begin{cases} L^{2i}(A,S,\varepsilon) \\ L^{2i-1}(A,S,\varepsilon) \end{cases}$ $(i \geqslant m+1)$ with the cobordism group of S-acyclic $\begin{cases} (2m+1)- \\ 2m- \end{cases}$ dimensional

$(-)^{i-m-1}\varepsilon$-symmetric Poincaré complexes over A it suffices (by the S-acyclic counterpart of Proposition 1.4.2) to show that it is possible to perform S-acyclic surgery on a connected S-acyclic (n+1)-dimensional even $(-\varepsilon)$-symmetric Poincaré complex over A $(C,\phi \in Q\langle v_0\rangle^{n+1}(C,-\varepsilon))$ $(n \geqslant 2m+1)$ so as to obtain a skew-suspension, killing $H^{n+1}(C)$. Working exactly as in the proof (in I.) of Proposition 1.2.3 use a f.g. projective resolution of the f.g. S-torsion A-module $H_0(C)$

$$0 \longrightarrow D_{m+1} \longrightarrow D_m \longrightarrow \dots \longrightarrow D_1 \longrightarrow D_0 \longrightarrow H_0(C) \longrightarrow 0$$

to define a connected S-acyclic (n+2)-dimensional even $(-\varepsilon)$-symmetric pair over A $(f:C \longrightarrow D,(0,\phi) \in Q\langle v_0\rangle^{n+2}(f,-\varepsilon))$ with which to perform such a surgery.

In particular, if (A,S) is 0-dimensional we have that $L^{2i-1}(A,S,\varepsilon)$ $(i \geqslant 1)$ is the cobordism group of S-acyclic 0-dimensional $(-)^{i-1}\varepsilon$-symmetric Poincaré complexes over A $(C,\phi \in Q^0(C,(-)^{i-1}\varepsilon))$. Now $H_0(C)$ is an S-torsion f.g. projective A-module, and S consists of non-zero-divisors, so that $H_0(C) = 0$ and $L^{2i-1}(A,S,\varepsilon) = 0$ $(i \geqslant 1)$.

[]

Let

$$f : A \longrightarrow B$$

be a morphism of rings with involution for which there exists a multiplicative subset $S \subset A$ such that f factors through the localization $S^{-1}A$

$$f : A \longrightarrow S^{-1}A \longrightarrow B$$

with the property

¶ a finite-dimensional A-module chain complex C is B-acyclic if and only if C is S-acyclic.

It then follows that

$$\begin{aligned}\Gamma^n(f:A \longrightarrow B,\varepsilon) &= \text{the cobordism group of } n\text{-dimensional} \\ &\quad \varepsilon\text{-symmetric } B\text{-Poincaré complexes over } A \\ &= \text{the cobordism group of } n\text{-dimensional} \\ &\quad \varepsilon\text{-symmetric } S^{-1}A\text{-Poincaré complexes over } A \\ &= \Gamma^n(A \longrightarrow S^{-1}A,\varepsilon) = L^n_S(S^{-1}A,\varepsilon) \quad (n \geqslant 0)\ ,\end{aligned}$$

and similarly for the ε-quadratic case. The connection between the Γ-groups and the L-groups of localizations has been investigated in the quadratic case by Smith [1] (following some preliminary work of Cappell and Shaneson in the commutative case). In particular, Smith showed that if

$$f : A = \mathbb{Z}[\pi] \longrightarrow B = \mathbb{Z}[\rho]$$

is the morphism of rings with involution induced by a surjective group morphism $f:\pi \longrightarrow \rho$ such that ρ is a finite extension of a polycyclic group and $\ker(f:\pi \longrightarrow \rho)$ is a finitely generated nilpotent group then the multiplicative subset

$$S = \{1+i \mid i \in \ker(f:A \longrightarrow B)\} \subset A$$

is such that the evident factorization

$$f : A \longrightarrow S^{-1}A \longrightarrow B$$

does indeed have the property ¶, and hence that

$$\Gamma_*(f:A \longrightarrow B) = L^S_*(S^{-1}A)\ .$$

The case of the projection induced by $f:\pi = \mathbb{Z} \longrightarrow \rho = \{1\}$

$$f : A = \mathbb{Z}[\mathbb{Z}] \longrightarrow B = \mathbb{Z}$$

is of particular interest, since the groups

$$\Gamma_*(f:\mathbb{Z}[\mathbb{Z}] \longrightarrow \mathbb{Z}) = \Gamma_*(\mathbb{Z}[\mathbb{Z}] \longrightarrow S^{-1}\mathbb{Z}[\mathbb{Z}]) = L^S_*(S^{-1}\mathbb{Z}[\mathbb{Z}])$$

are closely related to the high-dimensional knot cobordism groups C_*, as described in §7.9 below. More recently, Vogel [3] has obtained natural identifications of the type

$$\Gamma_*(f:A \longrightarrow B) = L_*^X(\Lambda)$$

for any locally epic morphism $f:A \longrightarrow B$ with a factorization

$$f : A \longrightarrow \Lambda \longrightarrow B$$

satisfying the property

¶ a finite-dimensional A-module chain complex C is B-acyclic if and only if C is Λ-acyclic

universally, with $\Lambda \longrightarrow B$ onto.

3.3 Linking Wu classes

The linking Wu classes are the S-acyclic counterparts of the algebraic Wu classes of §1.4.

Let $T \in \mathbb{Z}_2$ act on the additive groups A, $S^{-1}A$, $S^{-1}A/A$ by $T: x \longmapsto \varepsilon\bar{x}$ in each case. Define the $\begin{cases} \mathbb{Z}_2\text{-cohomology} \\ \mathbb{Z}_2\text{-homology} \\ \text{Tate } \mathbb{Z}_2\text{-cohomology} \end{cases}$ groups $\begin{cases} H^r(\mathbb{Z}_2;G,\varepsilon) \\ H_r(\mathbb{Z}_2;G,\varepsilon) \\ \hat{H}^r(\mathbb{Z}_2;G,\varepsilon) \end{cases}$ $(r \in \mathbb{Z})$ for $G = S^{-1}A$, $S^{-1}A/A$ by analogy with the case $G = A$ considered in §1.4. The short exact sequence of $\mathbb{Z}[\mathbb{Z}_2]$-modules

$$0 \longrightarrow A \longrightarrow S^{-1}A \longrightarrow S^{-1}A/A \longrightarrow 0$$

induces a long exact sequence of abelian groups

$$\begin{cases} \dots \longrightarrow H^r(\mathbb{Z}_2;A,\varepsilon) \longrightarrow H^r(\mathbb{Z}_2;S^{-1}A,\varepsilon) \longrightarrow H^r(\mathbb{Z}_2;S^{-1}A/A,\varepsilon) \\ \qquad\qquad \longrightarrow H^{r+1}(\mathbb{Z}_2;A,\varepsilon) \longrightarrow \dots \\ \dots \longrightarrow H_r(\mathbb{Z}_2;A,\varepsilon) \longrightarrow H_r(\mathbb{Z}_2;S^{-1}A,\varepsilon) \longrightarrow H_r(\mathbb{Z}_2;S^{-1}A/A,\varepsilon) \\ \qquad\qquad \longrightarrow H_{r-1}(\mathbb{Z}_2;A,\varepsilon) \longrightarrow \dots \\ \dots \longrightarrow \hat{H}^r(\mathbb{Z}_2;A,\varepsilon) \longrightarrow \hat{H}^r(\mathbb{Z}_2;S^{-1}A,\varepsilon) \longrightarrow \hat{H}^r(\mathbb{Z}_2;S^{-1}A/A,\varepsilon) \\ \qquad\qquad \longrightarrow \hat{H}^{r+1}(\mathbb{Z}_2;A,\varepsilon) \longrightarrow \dots \end{cases} \quad (r \in \mathbb{Z}) \; .$$

Let C be an S-acyclic finite-dimensional A-module chain complex. The <u>rth</u> $\begin{cases} \underline{\varepsilon\text{-symmetric}} \\ \underline{\varepsilon\text{-quadratic}} \\ \underline{\varepsilon\text{-hyperquadratic}} \end{cases}$ <u>linking Wu class</u> $\begin{cases} v_r^S(\phi) \\ v_S^r(\psi) \\ \hat{v}_r^S(\theta) \end{cases}$ of an element $\begin{cases} \phi \in Q^{n+1}(C,\varepsilon) \\ \psi \in Q_{n+1}(C,\varepsilon) \\ \theta \in \hat{Q}^{n+1}(C,\varepsilon) \end{cases}$ is the function

$$v^S_r(\phi) : H^{n-r+1}(C) \longrightarrow H^{n-2r}(\mathbb{Z}_2;S^{-1}A/A,(-)^{n-r+1}\varepsilon) ;$$

$$x \longmapsto (\tfrac{1}{s}).(\phi_{n-2r-1} + (-)^{n-r}\phi_{n-2r}d^*)(y)(y).\overline{(\tfrac{1}{s})}$$

$$v^r_S(\psi) : H^{n-r+1}(C) \longrightarrow H_{2r-n}(\mathbb{Z}_2;S^{-1}A/A,(-)^{n-r+1}\varepsilon) :$$

$$x \longmapsto (\tfrac{1}{s}).(\psi_{2r-n+1} + (-)^{n-r}\psi_{2r-n}d^*)(y)(y).\overline{(\tfrac{1}{s})}$$

$$\hat{v}^S_r(\theta) : H^{n-r+1}(C) \longrightarrow \hat{H}^{r-1}(\mathbb{Z}_2;S^{-1}A/A,\varepsilon) ;$$

$$x \longmapsto (\tfrac{1}{s}).(\theta_{n-2r-1} + (-)^{n-r}\theta_{n-2r}d^*)(y)(y).\overline{(\tfrac{1}{s})}$$

$$(x \in C^{n-r+1}, y \in C^{n-r}, s \in S, sx = d^*y \in C^{n-r+1}) .$$

Motivation: the cohomology classes $x \in H^m(C)$ of an S-acyclic A-module chain complex C are in a natural one-one correspondence with the chain homotopy classes of A-module chain maps

$$x : C \longrightarrow C_m(A,S) ,$$

where $C_m(A,S)$ is the S-acyclic A-module chain complex defined by

$$d : C_m(A,S)_m = A \longrightarrow C_m(A,S)_{m-1} = S^{-1}A ; a \longmapsto \frac{a}{1}$$

$$C_m(A,S)_i = 0 \quad (i \neq m-1,m) .$$

The linking Wu classes are such that

$$\begin{cases} v^S_r(\phi)(x) = x^\%(\phi) \in Q^{n+1}(C_{n-r+1}(A,S),\varepsilon) = H^{n-2r}(\mathbb{Z}_2;S^{-1}A/A,(-)^{n-r+1}\varepsilon) \\ v^r_S(\psi)(x) = x_\%(\psi) \in Q_{n+1}(C_{n-r+1}(A,S),\varepsilon) = H_{2r-n}(\mathbb{Z}_2;S^{-1}A/A,(-)^{n-r+1}\varepsilon) \\ \hat{v}^S_r(\theta)(x) = \hat{x}^\%(\theta) \in \hat{Q}^{n+1}(C_{n-r+1}(A,S),\varepsilon) = \hat{H}^{r+1}(\mathbb{Z}_2;S^{-1}A/A,\varepsilon) \end{cases} .$$

Now $C_m(A,S)$ is the direct limit

$$C_m(A,S) = \varinjlim_{s \in S} C_m(A,s)$$

of the directed system $\{C_m(A,s) \mid s \in S\}$ of finite-dimensional S-acyclic A-module chain complexes defined by

$$d : C_m(A,s)_m = A \longrightarrow C_m(A,s)_{m-1} = A \ ; \ a \longmapsto as$$

$$C_m(A,S)_i = 0 \ (i \neq m-1,m)$$

with $s \leqslant s'$ if there exists $t \in S$ such that $s' = st \in S$ and

$$C_m(A,s)_i \longrightarrow C_m(A,s')_i \ ; \ a \longmapsto \begin{cases} at & i = m-1 \\ a & i = m \end{cases} .$$

The $\begin{cases} \varepsilon\text{-symmetric} \\ \varepsilon\text{-quadratic} \end{cases}$ linking Wu class $\begin{cases} v_r^S(\phi)(x) \\ v_S^r(\psi)(x) \end{cases}$ is the obstruction to killing $x \in H^{n-r+1}(C)$ (= $H_r(C)$ if $\begin{cases} (C,\phi) \\ (C,\psi) \end{cases}$ is Poincaré) by S-acyclic surgery on an $\begin{cases} \varepsilon\text{-symmetric} \\ \varepsilon\text{-quadratic} \end{cases}$ pair of the type

$$\begin{cases} (x:C \longrightarrow C_{n-r+1}(A,s),(\delta\phi,\phi) \in Q^{n+2}(x,\varepsilon)) \\ (x:C \longrightarrow C_{n-r+1}(A,s),(\delta\psi,\psi) \in Q_{n+2}(x,\varepsilon)) \end{cases} \quad (s \in S) \quad .$$

By analogy with Proposition 1.4.1 we have:

<u>Proposition 3.3.1</u> i) The linking Wu classes are related to each other by

$$\hat{v}_r^S(J\phi) \ : \ H^{n-r+1}(C) \xrightarrow{v_r^S(\phi)} H^{n-2r}(\mathbb{Z}_2;S^{-1}A/A,(-)^{n-r+1}\varepsilon)$$
$$\xrightarrow{J} \hat{H}^{r+1}(\mathbb{Z}_2;S^{-1}A/A,\varepsilon)$$

$$v_r^S((1+T_\varepsilon)\psi) \ : \ H^{n-r+1}(C) \xrightarrow{v_S^r(\psi)} H_{2r-n}(\mathbb{Z}_2;S^{-1}A/A,(-)^{n-r+1}\varepsilon)$$
$$\xrightarrow{1+T_\varepsilon} H^{n-2r}(\mathbb{Z}_2;S^{-1}A/A,(-)^{n-r+1}\varepsilon)$$

$$v_S^{r-1}(H\theta) \ : \ H^{n-r+1}(C) \xrightarrow{\hat{v}_r^S(\theta)} \hat{H}^{r-1}(\mathbb{Z}_2;S^{-1}A/A,\varepsilon)$$
$$\xrightarrow{H} H_{2r-n-1}(\mathbb{Z}_2;S^{-1}A/A,(-)^{n-r+1}\varepsilon)$$

$$(\phi \in Q^{n+1}(C,\varepsilon), \ \psi \in Q_{n+1}(C,\varepsilon), \ \theta \in \hat{Q}^{n+1}(C,\varepsilon)) \ .$$

ii) The linking Wu classes are related to the algebraic Wu classes of §1.4 by

$$v_r(\phi)\ :\ H^{n-r+1}(C) \xrightarrow{v_r^S(\phi)} H^{n-2r}(\mathbb{Z}_2;S^{-1}A/A,(-)^{n-r+1}\varepsilon) \xrightarrow{\delta} H^{n-2r+1}(\mathbb{Z}_2;A,(-)^{n-r+1}\varepsilon)$$

$$v^r(\psi)\ :\ H^{n-r+1}(C) \xrightarrow{v_S^r(\psi)} H_{2r-n}(\mathbb{Z}_2;S^{-1}A/A,(-)^{n-r+1}\varepsilon) \xrightarrow{\partial} H_{2r-n-1}(\mathbb{Z}_2;A,(-)^{n-r+1}\varepsilon)$$

$$\hat{v}_r(\theta)\ :\ H^{n-r+1}(C) \xrightarrow{\hat{v}_r^S(\theta)} \hat{H}^{r-1}(\mathbb{Z}_2;S^{-1}A/A,\varepsilon) \xrightarrow{\hat{\delta}} \hat{H}^r(\mathbb{Z}_2;A,\varepsilon)\ .$$

iii) The linking Wu classes satisfy the sum formulae

$$v_r^S(\phi)(x+y) - v_r^S(\phi)(x) - v_r^S(\phi)(y)$$

$$= \begin{cases} \phi_0^S(x,y) + \phi_0^S(y,x) \in H^0(\mathbb{Z}_2;S^{-1}A/A,(-)^{r+1}\varepsilon) & (n = 2r) \\ 0 \in H^{n-2r}(\mathbb{Z}_2;S^{-1}A/A,(-)^{n-r+1}\varepsilon) & (n \neq 2r) \end{cases}$$

$$v_S^r(\psi)(x+y) - v_S^r(\psi)(x) - v_S^r(\psi)(y)$$

$$= \begin{cases} (1+T_\varepsilon)\psi_0^S(x,y) \in H_0(\mathbb{Z}_2;S^{-1}A/A,(-)^{r+1}\varepsilon) & (n = 2r) \\ 0 \in H_{2r-n}(\mathbb{Z}_2;S^{-1}A/A,(-)^{n-r+1}\varepsilon) & (n \neq 2r) \end{cases}$$

$$\hat{v}_r^S(\theta)(x+y) - \hat{v}_r^S(\theta)(x) - \hat{v}_r^S(\theta)(y) = 0 \in \hat{H}^{r-1}(\mathbb{Z}_2;S^{-1}A/A,\varepsilon)$$

$$(x,y \in H^{n-r+1}(C))$$

with $\begin{cases} \phi_0^S \\ (1+T_\varepsilon)\psi_0^S \end{cases} : H^{r+1}(C) \times H^{r+1}(C) \longrightarrow S^{-1}A/A$ the linking pairing of $\begin{cases} (C,\phi \in Q^{2r+1}(C,\varepsilon)) \\ (C,\psi \in Q_{2r+1}(C,\varepsilon)) \end{cases}$ $(n = 2r)$. Furthermore,

$$v_r^S(\phi)(x) = \phi_0^S(x,x) \in H^0(\mathbb{Z}_2;S^{-1}A/A,(-)^{r+1}\varepsilon) \quad (n = 2r)\ .$$

[]

As a first application of the linking Wu classes we have the following S-acyclic analogues of Proposition 1.2.3 i):

Proposition 3.3.2 i) If A,S,ε are such that

$$\ker(\hat{\delta}:\hat{H}^0(\mathbb{Z}_2;S^{-1}A/A,\varepsilon)\longrightarrow\hat{H}^1(\mathbb{Z}_2;A,\varepsilon)) = 0$$

there is a natural identification

$$L^n(A,S,\varepsilon) = \text{the cobordism group of S-acyclic } (n-1)\text{-dimensional } \varepsilon\text{-symmetric Poincaré complexes over } A \quad (n\geqslant 2)\ ,$$

under which $L^n(A,S,\varepsilon)\longrightarrow L^{n-1}(A,\varepsilon)$ becomes the forgetful map. In particular, this is the case if $\hat{H}^0(\mathbb{Z}_2;S^{-1}A,\varepsilon) = 0$ (e.g. if $1/2\in S^{-1}A$).

ii) If A,S,ε are such that

$$\hat{H}^0(\mathbb{Z}_2;A,\varepsilon)\longrightarrow\hat{H}^0(\mathbb{Z}_2;S^{-1}A,\varepsilon)$$

is an isomorphism then the skew-suspension maps

$$\overline{S}\ :\ L^n(A,S,\varepsilon)\longrightarrow L^{n+2}(A,S,-\varepsilon)\quad(n\geqslant 0)$$

are isomorphisms. In particular, this is the case if $1/2\in A$.

Proof: i) By the S-acyclic counterpart of Proposition 1.5.2 it suffices to show that it is possible to perform S-acyclic surgery on a connected S-acyclic $(n+1)$-dimensional even $(-\varepsilon)$-symmetric complex over A $(C,\phi\in Q\langle v_0\rangle^{n+1}(C,-\varepsilon))$ $(n\geqslant 2)$ so as to obtain a skew-suspension, killing $H^{n+1}(C)$. For any element $x\in H^{n+1}(C)$ we have

$$\hat{\delta}v_0^S(\phi)(x) = v_0(\phi)(x) = 0\in\hat{H}^1(\mathbb{Z}_2;A,\varepsilon)\ ,$$

so that

$$v_0^S(\phi)(x)\in\ker(\hat{\delta}:\hat{H}^0(\mathbb{Z}_2;S^{-1}A/A,\varepsilon)\longrightarrow\hat{H}^1(\mathbb{Z}_2;A,\varepsilon)) = 0\ .$$

It follows that $x\in H^{n+1}(C)$ may be represented by an A-module chain map

$$x : C \longrightarrow C_{n+1}(A,s)$$

for some $s \in S$ (with $C_{n+1}(A,s)$ as defined above) such that there is defined a connected S-acyclic (n+2)-dimensional even $(-\varepsilon)$-symmetric pair over A

$$(x:C \longrightarrow C_{n+1}(A,s),(\delta\phi,\phi) \in Q\langle v_0\rangle^{n+2}(x,-\varepsilon)) \quad .$$

Surgery on this pair results in a connected S-acyclic (n+1)-dimensional even $(-\varepsilon)$-symmetric complex over A $(C',\phi' \in Q\langle v_0\rangle^{n+1}(C',-\varepsilon))$ such that

$$H^{n+1}(C') = H^{n+1}(C)/(x) \ .$$

Now $H^{n+1}(C)$ is a f.g. S-torsion A-module, so that it is possible to kill $H^{n+1}(C)$ in (C,ϕ) by successively killing off a finite set of generators.

ii) Consider the exact sequence of abelian groups

$$\hat{H}^1(\mathbb{Z}_2;S^{-1}A/A,\varepsilon) \xrightarrow{\hat{\delta}} \hat{H}^0(\mathbb{Z}_2;A,\varepsilon) \longrightarrow \hat{H}^0(\mathbb{Z}_2;S^{-1}A,\varepsilon)$$
$$\longrightarrow \hat{H}^0(\mathbb{Z}_2;S^{-1}A/A,\varepsilon) \xrightarrow{\hat{\delta}} \hat{H}^1(\mathbb{Z}_2;A,\varepsilon) \ .$$

If $\hat{H}^0(\mathbb{Z}_2;A,\varepsilon) \longrightarrow \hat{H}^0(\mathbb{Z}_2;S^{-1}A,\varepsilon)$ is onto then

$$\ker(\hat{\delta}:\hat{H}^0(\mathbb{Z}_2;S^{-1}A/A,\varepsilon) \longrightarrow \hat{H}^1(\mathbb{Z}_2;A,\varepsilon)) = 0$$

and by i) we can identify

$L^{n+2}(A,S,\varepsilon)$ = the cobordism group of S-acyclic (n+1)-dimensional ε-symmetric Poincaré complexes over A $\quad (n \geq 0)$.

If $\hat{H}^0(\mathbb{Z}_2;A,\varepsilon) \longrightarrow \hat{H}^0(\mathbb{Z}_2;S^{-1}A,\varepsilon)$ is one-one then

$$\mathrm{im}(\hat{\delta}:\hat{H}^1(\mathbb{Z}_2;S^{-1}A/A,\varepsilon) \longrightarrow \hat{H}^0(\mathbb{Z}_2;A,\varepsilon)) = 0$$

and every S-acyclic ε-symmetric complex (or pair) over A is even. Thus if $\hat{H}^0(\mathbb{Z}_2;A,\varepsilon) \longrightarrow \hat{H}^0(\mathbb{Z}_2;S^{-1}A,\varepsilon)$ is an isomorphism we can identify

$$L^{n+2}(A,S,\varepsilon) = \text{the cobordism group of } S\text{-acyclic } (n+1)\text{-dimensional even } \varepsilon\text{-symmetric Poincaré complexes over } A$$

$$= L^{n}(A,S,-\varepsilon) \quad (n \geqslant 0) \ .$$

[]

3.4 Linking forms

In the first instance we define some subquotient groups of $S^{-1}A$, which are needed to define the various types of linking form that arise in the localization exact sequences of Witt groups.

Write $Q^{\varepsilon}(S^{-1}A/A)$ for the $\mathbb{Z}_2$-cohomology group

$$Q^{\varepsilon}(S^{-1}A/A) = H^{0}(\mathbb{Z}_2;S^{-1}A/A,\varepsilon) = \{b\in S^{-1}A \,|\, b-\varepsilon\bar{b}\in A\}/A$$

and let $Q^{\varepsilon}(A,S)$ be the subgroup of $Q^{\varepsilon}(S^{-1}A/A)$ defined by

$$\begin{aligned} Q^{\varepsilon}(A,S) &= \operatorname{im}(H^{0}(\mathbb{Z}_2;S^{-1}A,\varepsilon)\longrightarrow H^{0}(\mathbb{Z}_2;S^{-1}A/A,\varepsilon)) \\ &= \{b\in S^{-1}A \,|\, b-\varepsilon\bar{b} = a-\varepsilon\bar{a}, a\in A\}/A \subseteq Q^{\varepsilon}(S^{-1}A/A)\ . \end{aligned}$$

Write $Q_{\varepsilon}(S^{-1}A/A)$ for the $\mathbb{Z}_2$-homology group

$$Q_{\varepsilon}(S^{-1}A/A) = H_{0}(\mathbb{Z}_2;S^{-1}A/A,\varepsilon) = S^{-1}A/\{a+b-\varepsilon\bar{b} \,|\, a\in A, b\in S^{-1}A\}\ ,$$

and define also the abelian group

$$\begin{aligned} Q_{\varepsilon}(A,S) &= \operatorname{coker}(1+T_{\varepsilon}:H_{0}(\mathbb{Z}_2;A,\varepsilon)\longrightarrow H^{0}(\mathbb{Z}_2;S^{-1}A,\varepsilon)) \\ &= \{b\in S^{-1}A \,|\, b=\varepsilon\bar{b}\}/\{a+\varepsilon\bar{a} \,|\, a\in A\}\ . \end{aligned}$$

The ε-symmetrization map

$$1+T_{\varepsilon}\ :\ Q_{\varepsilon}(S^{-1}A/A)\longrightarrow Q^{\varepsilon}(S^{-1}A/A)\ ;\ x\longmapsto x+\varepsilon\bar{x}$$

factorizes as

$$1+T_{\varepsilon}\ :\ Q_{\varepsilon}(S^{-1}A/A)\xrightarrow{\ p\ }Q_{\varepsilon}(A,S)\xrightarrow{\ q\ }Q^{\varepsilon}(A,S)\xrightarrow{\ r\ }Q^{\varepsilon}(S^{-1}A/A)$$

with

$$p\ :\ Q_{\varepsilon}(S^{-1}A/A)\longrightarrow Q_{\varepsilon}(A,S)\ ;\ x\longmapsto x+\varepsilon\bar{x}$$

$$q\ :\ Q_{\varepsilon}(A,S)\longrightarrow Q^{\varepsilon}(A,S)\ ;\ x\longmapsto x$$

$$r\ :\ Q^{\varepsilon}(A,S)\longrightarrow Q^{\varepsilon}(S^{-1}A/A)\ ;\ x\longmapsto x\ .$$

An <u>ε-symmetric linking form over (A,S)</u> (M,λ) is an (A,S)-module M together with an A-module morphism $\lambda \in \mathrm{Hom}_A(M,M^{\wedge})$ such that

$$\varepsilon\lambda^{\wedge} = \lambda \in \mathrm{Hom}_A(M,M^{\wedge}) \ .$$

Equivalently, $\lambda \in \mathrm{Hom}_A(M,M^{\wedge})$ can be regarded as a pairing

$$\lambda : M \times M \longrightarrow S^{-1}A/A \ ; \ (x,y) \longmapsto \lambda(x,y) \equiv \lambda(x)(y)$$

such that

$$\text{i)}\ \lambda(x,y+y') = \lambda(x,y) + \lambda(x,y')$$

$$\text{ii)}\ \lambda(x,ay) = a\lambda(x,y)$$

$$\text{iii)}\ \lambda(y,x) = \varepsilon\overline{\lambda(x,y)}$$

$$(x,y,y' \in M, a \in A) \ .$$

For example, an ε-symmetric linking form over $(\mathbb{Z},\mathbb{Z}-\{0\})$ (M,λ) is the same as a finite abelian group M together with a bilinear ε-symmetric pairing

$$\lambda : M \times M \longrightarrow \mathbb{Q}/\mathbb{Z} \ .$$

If (M,λ) is an ε-symmetric linking form over (A,S) then

$$\lambda(x,x) \in Q^{\varepsilon}(S^{-1}A/A) \quad (x \in M) \ .$$

The linking form (M,λ) is <u>even</u> if

$$\lambda(x,x) \in Q^{\varepsilon}(A,S) \subseteq Q^{\varepsilon}(S^{-1}A/A) \quad (x \in M) \ .$$

An <u>ε-quadratic linking form over (A,S)</u> (M,λ,μ) is an even ε-symmetric linking form over (A,S) (M,λ) together with a function

$$\mu : M \longrightarrow Q_{\varepsilon}(A,S)$$

such that

$$\text{i)}\ \mu(ax) = a\mu(x)\bar{a} \in Q_{\varepsilon}(A,S)$$

$$\text{ii)}\ \mu(x+y) - \mu(x) - \mu(y) = \lambda(x,y) + \varepsilon\overline{\lambda(x,y)} \in Q_{\varepsilon}(A,S)$$

$$\text{iii)}\ q\mu(x) = \lambda(x,x) \in Q^{\varepsilon}(A,S)$$

$$(x,y \in M, a \in A) \ .$$

This definition is due to Wall [2] (in the special case $(A,S) = (\mathbb{Z}[\pi], \mathbb{Z}-\{0\})$ arising in odd-dimensional surgery obstruction theory). If A is a commutative ring with the identity involution $\bar{a} = a \in A$ $(a \in A)$ and $1/2 \in S^{-1}A$ a quadratic linking form over (A,S) (M,λ,μ) $(\varepsilon = 1 \in A)$ consists of an (A,S)-module M together with a function

$$\mu : M \longrightarrow Q_{+1}(A,S) = S^{-1}A/2A$$

such that

i) $\mu(ax) = a^2\mu(x) \in S^{-1}A/2A \quad (x \in M, a \in A)$

ii) the function

$$\lambda : M \times M \longrightarrow S^{-1}A/A \ ; \ (x,y) \longmapsto \tfrac{1}{2}(\mu(x+y) - \mu(x) - \mu(y))$$

is bilinear.

A <u>split ε-quadratic linking form over (A,S)</u> (M,λ,ν) is an even ε-symmetric linking form over (A,S) (M,λ) together with a function

$$\nu : M \longrightarrow Q_{\varepsilon}(S^{-1}A/A)$$

such that

i) $\nu(ax) = a\nu(x)\bar{a} \in Q_{\varepsilon}(S^{-1}A/A)$

ii) $\nu(x+y) - \nu(x) - \nu(y) = \lambda(x,y) \in Q_{\varepsilon}(S^{-1}A/A)$

iii) $qp\nu(x) = \lambda(x,x) \in Q^{\varepsilon}(A,S)$

$(x,y \in M, a \in A)$,

in which case the function

$$\mu : M \longrightarrow Q_{\varepsilon}(A,S) \ ; \ x \longmapsto p\nu(x)$$

defines an ε-quadratic linking form (M,λ,μ). This definition is due to Karoubi [2]. In Proposition 3.4.2 below we shall show that every ε-quadratic linking form (M,λ,μ) has a split

ε-quadratic refinement (M,λ,μ) (with $\mu = p\nu$), and that if $1/2 \in S^{-1}A$ there is no difference between ε-quadratic and split ε-quadratic linking forms over (A,S). If A is a commutative ring with the identity involution a split quadratic linking form over (A,S) (M,λ,ν) $(\varepsilon = 1 \in A)$ consists of an (A,S)-module M together with a function

$$\nu : M \longrightarrow Q_{+1}(S^{-1}A/A) = S^{-1}A/A$$

such that

i) $\nu(ax) = a^2\nu(x) \in S^{-1}A/A \quad (x \in M, a \in A)$

ii) the function

$$\lambda : M \times M \longrightarrow S^{-1}A/A \ ; \ (x,y) \longmapsto (\nu(x+y) - \nu(x) - \nu(y))$$

is bilinear.

The associated quadratic linking form $(M,\lambda,\mu = p\nu)$ is obtained by composing ν with

$$p = 2 : Q_{+1}(S^{-1}A/A) = S^{-1}A/A \longrightarrow Q_{+1}(A,S) = S^{-1}A/2A \ ; \quad b \longmapsto 2b$$

(which is an isomorphism if $1/2 \in S^{-1}A$).

An $\begin{cases} \text{(even) } \varepsilon\text{-symmetric} \\ \text{(split) } \varepsilon\text{-quadratic} \end{cases}$ linking form over (A,S) $\begin{cases} (M,\lambda) \\ (M,\lambda,\mu) \ ((M,\lambda,\nu)) \end{cases}$ is <u>non-singular</u> if $\lambda \in \mathrm{Hom}_A(M,M^\wedge)$ is an isomorphism.

A <u>morphism</u> (resp. <u>isomorphism</u>) of $\begin{cases} \text{(even) } \varepsilon\text{-symmetric} \\ \varepsilon\text{-quadratic} \\ \text{split } \varepsilon\text{-quadratic} \end{cases}$ linking forms over (A,S)

$$\begin{cases} f : (M,\lambda) \longrightarrow (M',\lambda') \\ f : (M,\lambda,\mu) \longrightarrow (M',\lambda',\mu') \\ f : (M,\lambda,\nu) \longrightarrow (M',\lambda',\nu') \end{cases}$$

is an A-module morphism (resp. isomorphism) $f \in \mathrm{Hom}_A(M,M')$ such that

$$\lambda : M \times M \xrightarrow{f \times f} M' \times M' \xrightarrow{\lambda'} S^{-1}A/A$$

and

$$\begin{cases} \mu : M \xrightarrow{f} M' \xrightarrow{\mu'} Q_\varepsilon(A,S) \\ \nu : M \xrightarrow{f} M' \xrightarrow{\nu'} Q_\varepsilon(S^{-1}A/A) \end{cases} .$$

In Proposition 3.4.1 below the isomorphism classes of (non-singular) linking forms over (A,S) will be identified with the homotopy equivalence classes of S-acyclic 1-dimensional (Poincaré) complexes over A. In Proposition 3.4.7 this will be extended to an identification of $\begin{cases} L^2(A,S,-\varepsilon) & ((A,S) \text{ 0-dimensional}) \\ L^0(A,S,\varepsilon) \\ L^{-2}(A,S,-\varepsilon) \\ L_0(A,S,\varepsilon) \end{cases}$

with the Witt group $\begin{cases} L^\varepsilon(A,S) \\ L\langle v_0\rangle^\varepsilon(A,S) \\ L_\varepsilon(A,S) \\ \widetilde{L}_\varepsilon(A,S) \end{cases}$ of non-singular $\begin{cases} \varepsilon\text{-symmetric} \\ \text{even } \varepsilon\text{-symmetric} \\ \varepsilon\text{-quadratic} \\ \text{split } \varepsilon\text{-quadratic} \end{cases}$

linking forms over (A,S). Thus the even-dimensioanal ε-quadratic L-groups of (A,S)

$$L_{2i}(A,S,\varepsilon) = L_0(A,S,(-)^i\varepsilon)$$

are the Witt groups of the split $(-)^i\varepsilon$-quadratic linking forms over (A,S), rather then the Witt groups of $(-)^i\varepsilon$-quadratic linking forms. However, if $1/2 \in S^{-1}A$ (e.g. if $(A,S) = (\mathbb{Z}[\pi],\mathbb{Z}-\{0\})$) it will be shown in Proposition 3.4.2 below that the forgetful functor

$$\{\text{split } \varepsilon\text{-quadratic linking forms over } (A,S)\} \longrightarrow \{\varepsilon\text{-quadratic linking forms over } (A,S)\}$$

is an isomorphism of categories, so that the Witt groups are also isomorphic. See Ranicki [6,§6] and §5.1 below for an example of a pair (A,S) (with $1/2 \notin S^{-1}A$) for which the Witt groups are not isomorphic.

An $\begin{cases} \underline{\varepsilon\text{-symmetric}} \\ \underline{\varepsilon\text{-quadratic}} \\ \underline{\text{split } \varepsilon\text{-quadratic}} \end{cases}$ $\underline{\text{map}}$ (resp. $\underline{\text{homotopy equivalence}}$)

of S-acyclic 1-dimensional $\begin{cases} \varepsilon\text{-symmetric} \\ \varepsilon\text{-quadratic} \\ \varepsilon\text{-quadratic} \end{cases}$ complexes over A

$$\begin{cases} f : (C,\phi) \longrightarrow (C',\phi') \\ f : (C,\psi) \longrightarrow (C',\psi') \\ f : (C,\psi) \longrightarrow (C',\psi') \end{cases}$$

is an A-module chain map (resp. chain equivalence)

$$f : C \longrightarrow C'$$

such that

$$\begin{cases} f^{\%}(\phi) - \phi' = 0 \in Q^1(C',\varepsilon) \\ f_{\%}(\psi) - \psi' = H(\theta) \in Q_1(C',\varepsilon) \\ f_{\%}(\psi) - \psi' = H(\theta) \in Q_1(C',\varepsilon) \end{cases}$$

for some Tate $\mathbb{Z}_2$-cohomology class $\theta \in \hat{Q}^2(C',\varepsilon)$ such that

$$\begin{cases} \hat{v}_1(\theta) = 0 : H^1(C') \longrightarrow \hat{H}^1(\mathbb{Z}_2;A,\varepsilon) \\ \hat{v}_1^S(\theta) = 0 : H^1(C') \longrightarrow \hat{H}^0(\mathbb{Z}_2;S^{-1}A/A,\varepsilon) \end{cases} .$$

An ε-quadratic homotopy equivalence in this sense is the same as an ε-quadratic homotopy equivalence in the sense of §1.6.

An $\left\{\begin{array}{l}\varepsilon\text{-quadratic}\\ \text{split } \varepsilon\text{-quadratic}\end{array}\right.$ map $f:(C,\psi)\longrightarrow(C',\psi')$ determines an $\left\{\begin{array}{l}\varepsilon\text{-symmetric}\\ \varepsilon\text{-quadratic}\end{array}\right.$ map

$$\left\{\begin{array}{l} f\ :\ (C,(1+T_\varepsilon)\psi)\longrightarrow(C',(1+T_\varepsilon)\psi') \\ f\ :\ (C,\psi)\longrightarrow(C',\psi') \end{array}\right.$$

since

$$\left\{\begin{array}{l} f^{\%}((1+T_\varepsilon)\psi)\ -\ (1+T_\varepsilon)\psi'\ =\ (1+T_\varepsilon)H(\theta)\ =\ 0\in Q^1(C',\varepsilon) \\ \hat{v}_1^S(\theta)\ =\ 0 \\ \hat{v}_1(\theta)\ :\ H^1(C')\longrightarrow\hat{H}^0(\mathbb{Z}_2;S^{-1}A/A,\varepsilon)\xrightarrow{\ \hat{\delta}\ }\hat{H}^1(\mathbb{Z}_2;A,\varepsilon)\ . \end{array}\right.$$

<u>Proposition 3.4.1</u> The category of $\left\{\begin{array}{l}\varepsilon\text{-symmetric}\\ \text{even } \varepsilon\text{-symmetric}\\ \varepsilon\text{-quadratic}\\ \text{split } \varepsilon\text{-quadratic}\end{array}\right.$ linking forms over (A,S) is naturally equivalent to the opposite of the category of S-acyclic 1-dimensional $\left\{\begin{array}{l}(-\varepsilon)\text{-symmetric}\\ \text{even } (-\varepsilon)\text{-symmetric}\\ (-\varepsilon)\text{-quadratic}\\ (-\varepsilon)\text{-quadratic}\end{array}\right.$ complexes over A and $\left\{\begin{array}{l}(-\varepsilon)\text{-symmetric}\\ (-\varepsilon)\text{-symmetric}\\ (-\varepsilon)\text{-quadratic}\\ \text{split } (-\varepsilon)\text{-quadratic}\end{array}\right.$ maps.

Isomorphisms of linking forms correspond to homotopy equivalences of complexes. Non-singular linking forms correspond to Poincaré complexes.

<u>Proof</u>: The linking pairing of an S-acyclic 1-dimensional $(-\varepsilon)$-symmetric complex over A $(C,\phi\in Q^1(C,-\varepsilon))$

$$\phi_0^S : H^1(C)\times H^1(C) \longrightarrow S^{-1}A/A \ ; \ (x,y)\longmapsto \frac{1}{s}\phi_0(x)(z)$$

$$(x,y\in C^1, z\in C^0, s\in S, d^*z = sy\in C^1)$$

defines an ε-symmetric linking form over (A,S)

$$(M,\lambda) = (H^1(C),\phi_0^S) \ .$$

The 0th $(-\varepsilon)$-symmetric Wu class of (C,ϕ) factors as

$$v_0(\phi) : H^1(C) \xrightarrow{v_0^S(\phi)} H^0(\mathbb{Z}_2;S^{-1}A/A,\varepsilon) \xrightarrow{\delta} H^1(\mathbb{Z}_2;A,\varepsilon)$$

and

$$\ker(\delta) = Q^{\varepsilon}(A,S) \subseteq H^0(\mathbb{Z}_2;S^{-1}A/A,\varepsilon) = Q^{\varepsilon}(S^{-1}A/A) \ ,$$

so that the complex (C,ϕ) is even $(v_0(\phi) = 0)$ if and only if the linking form (M,λ) is even $(\lambda(x,x) \equiv v_0^S(\phi)(x) \in Q^{\varepsilon}(A,S) \subseteq Q^{\varepsilon}(S^{-1}A/A)$ for all $x\in M = H^1(C))$.

The 0th $(-\varepsilon)$-quadratic linking Wu class of an S-acyclic 1-dimensional $(-\varepsilon)$-quadratic complex over A $(C,\psi\in Q_1(C,-\varepsilon))$

$$v_S^0(\psi) : H^1(C) \longrightarrow H_0(\mathbb{Z}_2;S^{-1}A/A,\varepsilon) = Q_{\varepsilon}(S^{-1}A/A) \ ;$$

$$y\longmapsto (\tfrac{1}{s}).(\psi_1+\psi_0 d^*)(z)(z).\overline{(\tfrac{1}{s})}$$

$$(y\in C^1, z\in C^0, s\in S, d^*z = sy\in C^1)$$

defines a split ε-quadratic linking form over (A,S)

$$(M,\lambda,\nu) = (H^1(C),(1+T_{-\varepsilon})\psi_0^S, v_S^0(\psi))$$

with associated ε-quadratic linking form over (A,S)

$$(M,\lambda,\mu) = (H^1(C),(1+T_{-\varepsilon})\psi_0^S, pv_S^0(\psi) : H^1(C)\longrightarrow Q_{\varepsilon}(A,S)) \ .$$

A map of S-acyclic 1-dimensional $(-\varepsilon)$-symmetric complexes over A

$$f : (C,\phi) \longrightarrow (C',\phi')$$

induces contravariantly a morphism of the associated ε-symmetric linking forms over (A,S)

$$f^* : (H^1(C'),\phi'^S_0) \longrightarrow (H^1(C),\phi^S_0) \quad .$$

Conversely, every morphism of the associated ε-symmetric linking forms is induced by a map of complexes.

A map of the $(-\varepsilon)$-symmetrizations

$$f : (C,(1+T_{-\varepsilon})\psi) \longrightarrow (C',(1+T_{-\varepsilon})\psi')$$

of the S-acyclic 1-dimensional $(-\varepsilon)$-quadratic complexe over A $(C,\psi),(C',\psi')$ induces contravariantly a morphism of the associated $\begin{cases}\varepsilon\text{-quadratic} \\ \text{split } \varepsilon\text{-quadratic}\end{cases}$ linking forms over (A,S)

$$\begin{cases} f^* : (H^1(C'),(1+T_{-\varepsilon})\psi'^S_0,pv^O_S(\psi')) \longrightarrow (H^1(C),(1+T_{-\varepsilon})\psi^S_0,pv^O_S(\psi)) \\ f^* : (H^1(C'),(1+T_{-\varepsilon})\psi'^S_0,v^O_S(\psi')) \longrightarrow (H^1(C),(1+T_{-\varepsilon})\psi^S_0,v^O_S(\psi)) \end{cases}$$

if and only if $f:(C,\psi) \longrightarrow (C',\psi')$ is a $\begin{cases}(-\varepsilon)\text{-quadratic} \\ \text{split } (-\varepsilon)\text{-quadratic}\end{cases}$ map, since by the exact sequence of Proposition 1.1.3 there exists an element $\theta \in \hat{Q}^2(C',-\varepsilon)$ such that

$$f_\%(\psi) - \psi' = H(\theta) \in Q_1(C',-\varepsilon)$$

and there is defined a commutative diagram

$$\hat{v}_1(\theta) : H^1(C') \longrightarrow \hat{H}^0(\mathbb{Z}_2;A,\varepsilon) , \quad \hat{v}_1^S(\theta) : H^1(C') \longrightarrow \hat{H}^1(\mathbb{Z}_2;S^{-1}A/A,\varepsilon) , \quad v_S^0(f_\%(\psi)) - v_S^0(\psi') = v_S^0(H\theta) : H^1(C') \longrightarrow Q_\varepsilon(S^{-1}A/A)$$

$$\begin{CD}
\hat{H}^1(\mathbb{Z}_2;S^{-1}A/A,\varepsilon) @>H>> Q_\varepsilon(S^{-1}A/A) @>qp>> Q^\varepsilon(A,S) \\
@V{\hat{\delta}}VV @V{p}VV @| \\
\hat{H}^0(\mathbb{Z}_2;A,\varepsilon) @>j>> Q_\varepsilon(A,S) @>q>> Q^\varepsilon(A,S)
\end{CD}$$

involving the exact sequences

$$0 \longrightarrow \hat{H}^1(\mathbb{Z}_2;S^{-1}A/A,\varepsilon) \xrightarrow{\ H\ } Q_\varepsilon(S^{-1}A/A) \xrightarrow{\ qp\ } Q^\varepsilon(A,S)$$

$$0 \longrightarrow \hat{H}^0(\mathbb{Z}_2;A,\varepsilon) \xrightarrow{\ j\ } Q_\varepsilon(A,S) \xrightarrow{\ q\ } Q^\varepsilon(A,S)$$

with

$$H : \hat{H}^1(\mathbb{Z}_2;S^{-1}A/A,\varepsilon) \longrightarrow Q_\varepsilon(S^{-1}A/A) \ ; \ x \longmapsto x$$

$$j : \hat{H}^0(\mathbb{Z}_2;A,\varepsilon) \longrightarrow Q_\varepsilon(A,S) \ ; \ a \longmapsto \frac{a}{1} \ .$$

Conversely, given an $\begin{cases} \text{(even) } \varepsilon\text{-symmetric} \\ \varepsilon\text{-quadratic} \\ \text{split } \varepsilon\text{-quadratic} \end{cases}$ linking form over (A,S) $\begin{cases} (M,\lambda) \\ (M,\lambda,\mu) \\ (M,\lambda,\nu) \end{cases}$ we shall construct an S-acyclic 1-dimensional $\begin{cases} \text{(even) } (-\varepsilon)\text{-symmetric} \\ (-\varepsilon)\text{-quadratic} \\ (-\varepsilon)\text{-quadratic} \end{cases}$ complex over A $\begin{cases} (C,\phi \in Q^1(C,-\varepsilon)) \\ (C,\psi \in Q_1(C,-\varepsilon)) \\ (C,\psi \in Q_1(C,-\varepsilon)) \end{cases}$

such that

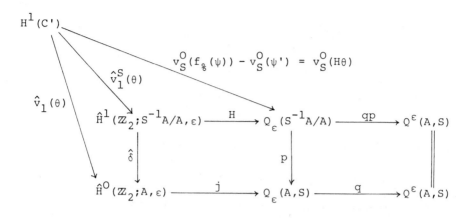

$$\begin{cases} (H^1(C),\phi_0^S) = (M,\lambda) \\ (H^1(C),(1+T_{-\varepsilon})\psi_0^S,pv_S^0(\psi)) = (M,\lambda,\mu) \\ (H^1(C),(1+T_{-\varepsilon})\psi_0^S,v_S^0(\psi)) = (M,\lambda,\nu) \end{cases} ,$$

as follows.

Given an ε-symmetric linking form over (A,S) (M,λ) let

$$0 \longrightarrow C_1 \xrightarrow{\ d\ } C_0 \longrightarrow M\hat{} \longrightarrow 0$$

be a f.g. projective A-module resolution of the S-dual (A,S)-module $M\hat{}$ of M. The A-module morphism $\lambda \in \mathrm{Hom}_A(M,M\hat{})$ can be resolved by a chain map

$$\phi_0 : C^{1-*} \longrightarrow C$$

such that there is defined a commutative diagram

$$\begin{array}{ccccccccc} 0 & \longrightarrow & C^0 & \xrightarrow{\ d^*\ } & C^1 & \longrightarrow & M & \longrightarrow & 0 \\ & & \big\downarrow{\scriptstyle \phi_0} & & \big\downarrow{\scriptstyle -\tilde{\phi}_0} & & \big\downarrow{\scriptstyle \lambda} & & \\ 0 & \longrightarrow & C_1 & \xrightarrow{\ d\ } & C_0 & \longrightarrow & M\hat{} & \longrightarrow & 0 \end{array} .$$

We thus have A-module morphisms

$$\phi_0 : C^0 \longrightarrow C_1 \ , \ \tilde{\phi}_0 : C^1 \longrightarrow C_0$$

such that

$$d\phi_0 + \tilde{\phi}_0 d^* = 0 : C^0 \longrightarrow C_0$$

and

$$\lambda : M = \mathrm{coker}(d^*:C^0 \longrightarrow C^1) \longrightarrow M\hat{} \ ;$$

$$x \longmapsto (y \longmapsto \tfrac{1}{s}\phi_0(x)(z))$$

$$(x,y \in C^1,\ z \in C^0,\ s \in S,\ d^*z = sy \in C^1) .$$

The relation $T_\varepsilon\lambda = \lambda \in \mathrm{Hom}_A(M,M\hat{\ })$ is resolved by a chain homotopy

$$\phi_1 : T_{-\varepsilon}\phi_0 \simeq \phi_0 : C^{1-*} \longrightarrow C \ ,$$

as defined by an A-module morphism

$$\phi_1 : C^1 \longrightarrow C_1$$

such that

$$\phi_0 + \varepsilon\tilde{\phi}_0^* = -\phi_1 d^* : C^0 \longrightarrow C_1 \ ,$$

$$\tilde{\phi}_0 + \varepsilon\phi_0^* = d\phi_1 : C^1 \longrightarrow C_0 \ .$$

Now

$$d(\phi_1 + \varepsilon\phi_1^*) = (\tilde{\phi}_0 + \varepsilon\phi_0^*) - \varepsilon(\phi_0 + \varepsilon\tilde{\phi}_0^*)^* = 0 : C^1 \longrightarrow C_0$$

and $d \in \mathrm{Hom}_A(C_1,C_0)$ is a monomorphism, so that

$$\phi_1 + \varepsilon\phi_1^* = 0 : C^1 \longrightarrow C_1 \ .$$

The S-acyclic 1-dimensional $(-\varepsilon)$-symmetric complex over A $(C,\phi \in Q^1(C,-\varepsilon))$ is such that

$$(H^1(C),\phi_S^0) = (M,\lambda) \ ,$$

by construction. The chain map $\phi_0 : C^{1-*} \longrightarrow C$ is a chain equivalence if and only if it induces an A-module isomorphism

$$(\phi_0)_* = \lambda : H^1(C) = M \longrightarrow H_0(C) = M\hat{\ } \ ,$$

so that the complex (C,ϕ) is Poincaré if and only if the linking form (M,λ) is non-singular.

Given an ε-quadratic linking form over (A,S) (M,λ,μ) let

$$0 \longrightarrow C_1 \xrightarrow{\ d\ } C_0 \longrightarrow M\hat{\ } \longrightarrow 0$$

be a f.g. projective A-module resolution of the S-dual $M\hat{\ }$ (as above), stabilized so as to have C_1 a f.g. free A-module. Write the dual resolution for the double S-dual $(M\hat{\ })\hat{\ } = M$ as

$$0 \longrightarrow C^0 \xrightarrow{d^*} C^1 \xrightarrow{e} M \longrightarrow 0 \quad .$$

Choose a base $\{x_i | 1 \leqslant i \leqslant n\}$ for $C^1 = C_1^*$ and let $\{y_{ij} \in S^{-1}A | 1 \leqslant i,j \leqslant n\}$ be such that

$$\text{i)}\ y_{ij} = \varepsilon\overline{y_{ji}} \in S^{-1}A \quad (1 \leqslant i,j \leqslant n)$$

$$\text{ii)}\ \lambda(ex_i)(ex_j) = y_{ij} \in S^{-1}A/A \quad (1 \leqslant i < j \leqslant n)$$

$$\text{iii)}\ \mu(ex_i) = y_{ii} \in Q_\varepsilon(A,S) \quad (1 \leqslant i \leqslant n) \ .$$

Define an A-module structure on $\mathrm{Hom}_A(C^1, S^{-1}A)$ by

$$A \times \mathrm{Hom}_A(C^1, S^{-1}A) \longrightarrow \mathrm{Hom}_A(C^1, S^{-1}A) \ ;$$
$$(a,f) \longmapsto (x \longmapsto f(x)\bar{a}) \quad .$$

The A-module morphism

$$\alpha : C^1 \longrightarrow \mathrm{Hom}_A(C^1, S^{-1}A) \ ;$$

$$\sum_{i=1}^n a_i x_i \longmapsto (\sum_{j=1}^n b_j x_j \longmapsto \sum_{1\leqslant i,j\leqslant n} b_j y_{ij} \overline{a_i}) \quad (a_i, b_j \in A)$$

is such that

$$\text{i)}\ \alpha(y)(x) = \varepsilon\overline{\alpha(x)(y)} \in S^{-1}A/A$$

$$\text{ii)}\ \lambda(ex)(ey) = \alpha(x)(y) \in S^{-1}A/A$$

$$\text{iii)}\ \mu(ex) = \alpha(x)(x) \in Q_\varepsilon(A,S)$$

$$(x,y \in C^1) .$$

Now

$$(d^*z)(y) \in A \subseteq S^{-1}A$$
$$(d^*z)(d^*z) \in \mathrm{im}(1+T_\varepsilon : A \longrightarrow A ; a \longmapsto a+\varepsilon\bar{a}) \subseteq S^{-1}A$$
$$(y \in C^1,\ z \in C^0) \ ,$$

so that there is a well-defined A-module morphism

$$\psi_0 : C^0 \longrightarrow (C^1)^* = C_1 \ ; \ z \longmapsto (y \longmapsto \alpha d^*(z)(y))$$

such that for some $\psi_1 \in \mathrm{Hom}_A(C^0, C_0)$

$$d\psi_0 + \psi_1 + \varepsilon\psi_1^* = 0 : C^0 \longrightarrow C_0 \quad .$$

The S-acyclic 1-dimensional $(-\varepsilon)$-quadratic complex over A $(C,\psi \in Q_1(C,-\varepsilon))$ is such that

$$(H^1(C),(1+T_{-\varepsilon})\psi_0^S, pv_S^0(\psi)) = (M,\lambda,\mu) ,$$

by construction.

Given a split ε-quadratic linking form over (A,S) (M,λ,ν) let (C,ψ) be the S-acyclic 1-dimensional $(-\varepsilon)$-quadratic complex over A constructed as above, but with $\psi_1 \in \mathrm{Hom}_A(C^0,C_0)$ determined by $\nu:M \longrightarrow Q_\varepsilon(S^{-1}A/A)$, as follows.
Let $\{z_i \in S^{-1}A | 1 \leqslant i \leqslant n\}$ be such that

$$\nu(ex_i) = z_i \in Q_\varepsilon(S^{-1}A/A) \quad (1 \leqslant i \leqslant n) ,$$

and define an A-module morphism

$$\beta : C^1 \longrightarrow \mathrm{Hom}_A(C^1,S^{-1}A) ;$$
$$\sum_{i=1}^n a_i x_i \longmapsto (\sum_{j=1}^n b_j x_j \longmapsto \sum_{1 \leqslant i<j \leqslant n} b_j y_{ij} \overline{a_i} + \sum_{i=1}^n b_i z_i \overline{a_i})$$
$$(a_i, b_j \in A)$$

such that

$$\alpha(x)(y) = \beta(x)(y) + \varepsilon\overline{\beta(y)(x)} \in S^{-1}A$$
$$\nu(ex) = \beta(x)(x) \in Q_\varepsilon(S^{-1}A/A) \qquad (x,y \in C^1).$$

Let $\psi_1 \in \mathrm{Hom}_A(C^0,C_0)$ be an A-module morphism such that

$$\psi_1(z)(z) = -\beta(d^*z)(d^*z) \in Q_\varepsilon(S^{-1}A) \quad (z \in C^0)$$
$$d\psi_0 + \psi_1 + \varepsilon\psi_1^* = 0 : C^0 \longrightarrow C_0 .$$

The S-acyclic 1-dimensional $(-\varepsilon)$-quadratic complex over A $(C,\psi \in Q_1(C,-\varepsilon))$ is such that

$$(H^1(C),(1+T_{-\varepsilon})\psi_0^S, v_S^0(\psi)) = (M,\lambda,\nu) ,$$

by construction.

[]

Proposition 3.4.2 i) Every ε-quadratic linking form over (A,S) (M,λ,μ) admits a split ε-quadratic linking form (M,λ,ν) with ν a refinement of μ,

$$\mu : M \xrightarrow{\nu} Q_\varepsilon(S^{-1}A/A) \xrightarrow{p} Q_\varepsilon(A,S) \quad .$$

ii) If A,S,ε are such that

$$\begin{cases} \operatorname{im}(\hat{\delta}:\hat{H}^0(\mathbb{Z}_2;S^{-1}A/A,\varepsilon) \longrightarrow \hat{H}^1(\mathbb{Z}_2;A,\varepsilon)) = 0 \\ \hat{H}^0(\mathbb{Z}_2;A,\varepsilon) \longrightarrow \hat{H}^0(\mathbb{Z}_2;S^{-1}A,\varepsilon) \quad \text{is an isomorphism} \\ \operatorname{im}(\hat{H}^1(\mathbb{Z}_2;S^{-1}A,\varepsilon) \longrightarrow \hat{H}^1(\mathbb{Z}_2;S^{-1}A/A,\varepsilon)) = 0 \end{cases}$$

then the forgetful functor

$$\left(\begin{cases} \text{even } \varepsilon\text{-symmetric} \\ \varepsilon\text{-quadratic} \\ \text{split } \varepsilon\text{-quadratic} \end{cases} \text{linking forms over } (A,S)\right)$$

$$\longrightarrow \left(\begin{cases} \varepsilon\text{-symmetric} \\ \text{even } \varepsilon\text{-symmetric} \\ \varepsilon\text{-quadratic} \end{cases} \text{linking forms over } (A,S)\right)$$

is an isomorphism of categories. In particular, this is the case if $1/2 \in A$; if $1/2 \in S^{-1}A$ (e.g. if $2 \in S$) the forgetful functor

(split ε-quadratic linking forms over (A,S))

$\longrightarrow$ (ε-quadratic linking forms over (A,S))

is an isomorphism of categories.

Proof: i) Immediate from Proposition 3.4.1.

ii) Let $\tilde{Q}_\varepsilon(A,S)$ be the subgroup of $Q_\varepsilon(A,S)$ defined by

$$\tilde{Q}_\varepsilon(A,S) = \{b + \varepsilon\bar{b} \mid b \in S^{-1}A\}/\{a + \varepsilon\bar{a} \mid a \in A\} \quad ,$$

and define abelian group morphisms

$$\tilde{p} = p| \;:\; Q_\varepsilon(S^{-1}A/A) \longrightarrow \tilde{Q}_\varepsilon(A,S) \;;\; b \longmapsto b + \varepsilon\bar{b}$$

$$\tilde{q} = q| \;:\; \tilde{Q}_\varepsilon(A,S) \longrightarrow Q^\varepsilon(A,S) \;;\; x \longmapsto x \;.$$

By i) we have that for every ε-quadratic linking form over (A,S) (M,λ,μ)

$$\mu(x) \in \tilde{Q}_\varepsilon(A,S) \subseteq Q_\varepsilon(A,S) \quad (x \in M) \;.$$

The isomorphisms of categories of linking forms may now be deduced from the correspondences of Proposition 3.4.1 and the exact sequences

$$\begin{cases} 0 \longrightarrow Q^\varepsilon(A,S) \xrightarrow{\ r\ } Q^\varepsilon(S^{-1}A/A) \longrightarrow \mathrm{im}(\hat{\delta}:\hat{H}^0(\mathbb{Z}_2;S^{-1}A/A,\varepsilon) \longrightarrow \hat{H}^1(\mathbb{Z}_2;A,\varepsilon)) \\ \qquad\qquad\qquad\qquad\qquad\qquad \longrightarrow 0 \\ 0 \longrightarrow \ker(\hat{H}^0(\mathbb{Z}_2;A,\varepsilon) \longrightarrow \hat{H}^0(\mathbb{Z}_2:S^{-1}A,\varepsilon)) \longrightarrow \tilde{Q}_\varepsilon(A,S) \xrightarrow{\ \tilde{q}\ } Q^\varepsilon(A,S) \\ \qquad\qquad \mathrm{coker}(\hat{H}^0(\mathbb{Z}_2;A,\varepsilon) \longrightarrow \hat{H}^0(\mathbb{Z}_2;S^{-1}A,\varepsilon)) \longrightarrow 0 \\ 0 \longrightarrow \mathrm{im}(\hat{H}^1(\mathbb{Z}_2;A,\varepsilon) \longrightarrow \hat{H}^1(\mathbb{Z}_2;S^{-1}A/A,\varepsilon)) \\ \qquad\qquad \longrightarrow Q_\varepsilon(S^{-1}A/A) \xrightarrow{\ \tilde{p}\ } \tilde{Q}_\varepsilon(A,S) \longrightarrow 0 \end{cases}$$

(which are valid for any A,S,ε).

[]

Proposition 3.4.1 related linking forms over (A,S) to S-acyclic 1-dimensional complexes over A. Proposition 1.6.4 relates such complexes to formations over A which become stably isomorphic to 0 over $S^{-1}A$. We shall now establish the direct connection between linking forms and such formations - such a connection was first observed by Wall [1] in the case $(A,S) = (\mathbb{Z},\mathbb{Z}-\{0\})$.

An <u>S-lagrangian</u> L of an $\begin{cases}\varepsilon\text{-symmetric}\\ \varepsilon\text{-quadratic}\end{cases}$ form over A

$\begin{cases}(K,\alpha\in Q^{\varepsilon}(K))\\ (K,\beta\in Q_{\varepsilon}(K))\end{cases}$ is a f.g. projective A-submodule L of K

(not necessarily a direct summand) such that the inclusion $j\in \mathrm{Hom}_A(L,K)$ defines a morphism of forms over A

$$\begin{cases} j\ :\ (L,0)\longrightarrow (K,\alpha)\\ j\ :\ (L,0)\longrightarrow (K,\beta)\end{cases}$$

which becomes the inclusion of a lagrangian over $S^{-1}A$. (An S-lagrangian is an $S^{-1}A$-lagrangian in the sense of §2.4).

An $\begin{cases}\underline{\text{(even) }\varepsilon\text{-symmetric}}\\ \underline{\varepsilon\text{-quadratic}}\end{cases}$ <u>S-formation over A</u> $\begin{cases}(Q,\phi;F,G)\\ (Q,\psi;F,G)\end{cases}$

is a non-singular $\begin{cases}\text{(even) }\varepsilon\text{-symmetric}\\ \varepsilon\text{-quadratic}\end{cases}$ form over A $\begin{cases}(Q,\phi)\\ (Q,\psi)\end{cases}$

together with a lagrangian F and an S-lagrangian G, such that $S^{-1}F$ and $S^{-1}G$ are complementary lagrangians in $\begin{cases}S^{-1}(Q,\phi)\\ S^{-1}(Q,\psi)\end{cases}$

$$S^{-1}Q = S^{-1}F\oplus S^{-1}G\ .$$

It follows that $F\cap G = \{0\}$, and that $Q/(F+G) = \mathrm{coker}(G\longrightarrow Q/F)$ is an (A,S)-module supporting an $\begin{cases}\text{(even) }(-\varepsilon)\text{-symmetric}\\ (-\varepsilon)\text{-quadratic}\end{cases}$ linking form over (A,S) (as made precise in Proposition 3.4.3 below). The S-formation is <u>non-singular</u> if G is a lagrangian. (An S-formation is an $S^{-1}A$-formation in the sense of §2.4).

An <u>isomorphism</u> of $\begin{cases}\text{(even) }\varepsilon\text{-symmetric}\\ \varepsilon\text{-quadratic}\end{cases}$ S-formations over A

$$\begin{cases} f : (Q,\phi;F,G) \longrightarrow (Q',\phi';F',G') \\ f : (Q,\psi;F,G) \longrightarrow (Q',\psi';F',G') \end{cases}$$

is an isomorphism of the $\begin{cases}\text{(even) } \varepsilon\text{-symmetric} \\ \varepsilon\text{-quadratic}\end{cases}$ forms

$$\begin{cases} f : (Q,\phi) \longrightarrow (Q',\phi') \\ f : (Q,\psi) \longrightarrow (Q',\psi') \end{cases}$$

such that

$$f(F) = F' \ , \ f(G) = G' \ .$$

A <u>stable isomorphism</u> of $\begin{cases}\text{(even) } \varepsilon\text{-symmetric} \\ \varepsilon\text{-quadratic}\end{cases}$ S-formations over A

$$\begin{cases} [f] : (Q,\phi;F,G) \longrightarrow (Q',\phi';F',G') \\ [f] : (Q,\psi;F,G) \longrightarrow (Q',\psi';F',G') \end{cases}$$

is an isomorphism of the type

$$\begin{cases} f : (Q,\phi;F,G)\oplus(H^{\varepsilon}(P);P,P^*) \longrightarrow (Q',\phi';F',G')\oplus(H^{\varepsilon}(P');P',P'^*) \\ f : (Q,\psi;F,G)\oplus(H_{\varepsilon}(P);P,P^*) \longrightarrow (Q',\psi';F',G')\oplus(H_{\varepsilon}(P');P',P'^*) \end{cases}$$

for some f.g. projective A-modules P,P'.

A <u>split ε-quadratic S-formation over A</u>

$$(F,G) = (F,(\binom{\gamma}{\mu},\theta)G)$$

is an ε-quadratic S-formation $(H_{\varepsilon}(F);F,G)$, with $\binom{\gamma}{\mu}:G \longrightarrow F\oplus F^*$ the inclusion, together with a <u>hessian</u> $(-\varepsilon)$-quadratic form over A $(G,\theta \in Q_{-\varepsilon}(G))$ such that

$$\gamma^*\mu = \theta - \varepsilon\theta^* \in \mathrm{Hom}_A(G,G^*) \ .$$

Then $\mu \in \mathrm{Hom}_A(G,F^*)$ is an S-isomorphism, and

$$(F\oplus F^*)/(F+G) = \mathrm{coker}(\mu:G \longrightarrow F^*)$$

is an (A,S)-module supporting a split $(-\varepsilon)$-quadratic linking form over (A,S) (as made precise in Proposition 3.4.3 below). The S-formation (F,G) is <u>non-singular</u> if G is a lagrangian, that is such that the sequence

$$0 \longrightarrow G \xrightarrow{\begin{pmatrix}\gamma\\ \mu\end{pmatrix}} F\oplus F^* \xrightarrow{(\varepsilon\mu^*\ \gamma^*)} G^* \longrightarrow 0$$

is exact.

An <u>isomorphism</u> of split ε-quadratic S-formations over A

$$(\alpha,\beta,\psi) : (F,(\begin{pmatrix}\gamma\\ \mu\end{pmatrix},\theta)G) \longrightarrow (F',(\begin{pmatrix}\gamma'\\ \mu'\end{pmatrix},\theta')G')$$

is a triple $(\alpha \in \mathrm{Hom}_A(F,F'), \beta \in \mathrm{Hom}_A(G,G'), \theta \in Q_{-\varepsilon}(F^*))$ such that α and β are isomorphisms, and such that

i) $\mu'\beta = \alpha^{*-1}\mu \in \mathrm{Hom}_A(G,F'^*)$

ii) $\gamma'\beta = \alpha\gamma + \alpha(\psi - \varepsilon\psi^*)^*\mu \in \mathrm{Hom}_A(G,F')$

iii) $\beta^*\theta'\beta - \theta - \mu^*\psi\mu \in \ker(S^{-1}: Q_{-\varepsilon}(G) \longrightarrow Q_{-\varepsilon}(S^{-1}G))$.

A <u>stable isomorphism</u> of split ε-quadratic S-formations over A

$$[\alpha,\beta,\psi] : (F,G) \longrightarrow (F',G')$$

is an isomorphism of the type

$$(\alpha,\beta,\psi) : (F,G)\oplus(P,P^*) \longrightarrow (F',G')\oplus(P',P'^*)$$

for some f.g. projective A-modules P,P', with $(P,P^*) = (P,(\begin{pmatrix}0\\ 1\end{pmatrix},0)P^*)$.

<u>Proposition 3.4.3</u> The isomorphism classes of $\begin{cases}\text{(even) } \varepsilon\text{-symmetric}\\ \varepsilon\text{-quadratic}\\ \text{split } \varepsilon\text{-quadratic}\end{cases}$ linking forms over (A,S) are in a natural one-one correspondence with the stable isomorphism classes of $\begin{cases}\text{(even) } (-\varepsilon)\text{-symmetric}\\ (-\varepsilon)\text{-quadratic}\\ \text{split } (-\varepsilon)\text{-quadratic}\end{cases}$ S-formations over A. Non-singular linking forms correspond to non-singular S-formations.

<u>Proof</u>: Proposition 3.4.1 gives a natural one-one correspondence between the isomorphism classes of (non-singular)

$$\left\{\begin{array}{l}\text{(even) } \varepsilon\text{-symmetric} \\ \varepsilon\text{-quadratic} \\ \text{split } \varepsilon\text{-quadratic}\end{array}\right. \quad \text{linking forms over } (A,S) \text{ and the}$$

$$\left\{\begin{array}{l}(-\varepsilon)\text{-symmetric} \\ (-\varepsilon)\text{-quadratic} \\ \text{split } (-\varepsilon)\text{-quadratic}\end{array}\right. \quad \text{homotopy equivalence classes of S-acyclic}$$

$$\text{1-dimensional} \left\{\begin{array}{l}\text{(even) } (-\varepsilon)\text{-symmetric} \\ (-\varepsilon)\text{-quadratic} \\ (-\varepsilon)\text{-quadratic}\end{array}\right. \quad \text{(Poincaré) complexes over A.}$$

A straightforward modification of the proof of Proposition 1.6.4 shows that the latter are in a natural one-one correspondence with the stable isomorphism classes of (non-singular)

$$\left\{\begin{array}{l}\text{(even) } (-\varepsilon)\text{-symmetric} \\ (-\varepsilon)\text{-quadratic} \\ \text{split } (-\varepsilon)\text{-quadratic}\end{array}\right. \quad \text{S-formations over A. Explicitly, a}$$

$$\text{(non-singular)} \left\{\begin{array}{l}\text{(even) } (-\varepsilon)\text{-symmetric} \\ (-\varepsilon)\text{-quadratic} \\ \text{split } (-\varepsilon)\text{-quadratic}\end{array}\right. \quad \text{S-formation over A}$$

$$\left\{\begin{array}{l}(Q,\phi;F,G) \\ (Q,\psi;F,G) \\ (F,(\binom{\gamma}{\mu},\theta)G)\end{array}\right. \quad \text{corresponds to the (non-singular)}$$

$$\left\{\begin{array}{l}\text{(even) } \varepsilon\text{-symmetric} \\ \varepsilon\text{-quadratic} \\ \text{split } \varepsilon\text{-quadratic}\end{array}\right. \quad \text{linking form over } (A,S) \left\{\begin{array}{l}(M,\lambda) \\ (M,\lambda,\mu) \\ (M,\lambda,\nu)\end{array}\right.$$

defined by

$$\begin{cases} \lambda : M = Q/(F+G) \longrightarrow M\hat{} \ ; \ x \longmapsto (y \longmapsto \frac{1}{s}\phi(x)(g)) \\ \qquad (x,y \in Q, g \in G, s \in S, sy - g \in F) \\ \begin{cases} \lambda : M = Q/(F+G) \longrightarrow M\hat{} \ ; \ x \longmapsto (y \longmapsto \frac{1}{s}(\psi - \varepsilon\psi^*)(x)(g)) \\ \mu : M = Q/(F+G) \longrightarrow Q_{\varepsilon}(A,S) \ ; \ y \longmapsto \frac{1}{s}(\psi - \varepsilon\psi^*)(y)(g) - \psi(y)(y) \\ \qquad (x,y \in Q, g \in G, s \in S, sy - g \in F) \end{cases} \\ \begin{cases} \lambda : M = \mathrm{coker}(\mu : G \longrightarrow F^*) \longrightarrow M\hat{} \ ; \ x \longmapsto (y \longmapsto \frac{1}{s}\gamma^*(x)(g)) \\ \nu : M = \mathrm{coker}(\mu : G \longrightarrow F^*) \longrightarrow Q_{\varepsilon}(S^{-1}A/A) \ ; \\ \qquad y \longmapsto (\frac{1}{s}).\theta(g)(g).\overline{(\frac{1}{s})} \\ \qquad (x,y \in F^*, g \in G, s \in S, sy = \mu g \in F^*) \ . \end{cases} \end{cases}$$

[]

An $\begin{cases} \varepsilon\text{-symmetric} \\ \varepsilon\text{-quadratic} \end{cases}$ form over A $\begin{cases} (K, \alpha \in Q^{\varepsilon}(K)) \\ (K, \beta \in Q_{\varepsilon}(K)) \end{cases}$ is <u>S-non-singular</u> if $\begin{cases} S^{-1}(K,\alpha) \\ S^{-1}(K,\beta) \end{cases}$ is a non-singular $\begin{cases} \varepsilon\text{-symmetric} \\ \varepsilon\text{-quadratic} \end{cases}$ form over $S^{-1}A$, that is if $\begin{cases} \alpha \in \mathrm{Hom}_A(K,K^*) \\ \beta + \varepsilon\beta^* \in \mathrm{Hom}_A(K,K^*) \end{cases}$ is an S-isomorphism. (Thus an S-non-singular form is an $S^{-1}A$-non-singular form in the sense of §2.4. S-non-singular forms were called "non-degenerate" in Ranicki [6], but an explicit reference to the multiplicative subset $S \subset A$ now seems preferable).

We shall now use the correspondence of Proposition 3.4.1 to characterize the non-singular $\begin{cases} \text{even } \varepsilon\text{-symmetric} \\ \text{split } \varepsilon\text{-quadratic} \end{cases}$ linking forms over (A,S) representing 0 in $\begin{cases} L^0(A,S,\varepsilon) \\ L_0(A,S,\varepsilon) \end{cases}$ in terms of S-non-singular $\begin{cases} \varepsilon\text{-symmetric} \\ \varepsilon\text{-quadratic} \end{cases}$ forms over A.

The <u>boundary</u> of an S-non-singular $\begin{cases}\varepsilon\text{-symmetric}\\ \text{even } \varepsilon\text{-symmetric}\\ \varepsilon\text{-quadratic}\end{cases}$

form over A $\begin{cases}(K,\alpha \in Q^{\varepsilon}(K))\\ (K,\alpha \in Q\langle v_0\rangle^{\varepsilon}(K))\\ (K,\beta \in Q_{\varepsilon}(K))\end{cases}$ is the non-singular

$\begin{cases}\text{even } \varepsilon\text{-symmetric}\\ \varepsilon\text{-quadratic}\\ \text{split } \varepsilon\text{-quadratic}\end{cases}$ linking form over (A,S)

$$\begin{cases}\partial(K,\alpha) = (\partial K,\lambda)\\ \partial(K,\alpha) = (\partial K,\lambda,\mu)\\ \partial(K,\beta) = (\partial K,\lambda,\nu)\end{cases}$$

defined by

$$\lambda : \partial K = \mathrm{coker}(\alpha : K \longrightarrow K^*) \longrightarrow \partial K^{\wedge} \; ; \; x \longmapsto (y \longrightarrow \frac{x(z)}{s})$$

$$\mu : \partial K = \mathrm{coker}(\alpha : K \text{——} K^*) \longrightarrow Q_{\varepsilon}(A,S) \; ; \; y \longmapsto \frac{y(z)}{s}$$

$$\nu : \partial K = \mathrm{coker}(\alpha : K \longrightarrow K^*) \longrightarrow Q_{\varepsilon}(S^{-1}A/A) \; ;$$

$$y \longmapsto (\frac{1}{s}).\beta(z)(z).\overline{(\frac{1}{s})}$$

$(x,y \in K^*, z \in K, s \in S, sy = \alpha(z) \in K^*, \alpha = \beta+\varepsilon\beta^*$ in the ε-quadratic case).

The boundary linking form corresponds (via Proposition 3.4.3)

to the boundary $\begin{cases}\text{even } (-\varepsilon)\text{-symmetric}\\ (-\varepsilon)\text{-quadratic}\\ \text{split } (-\varepsilon)\text{-quadratic}\end{cases}$ S-formation over A

$$\begin{cases}\partial(K,\alpha) = (H^{-\varepsilon}(K);K,\Gamma_{(K,\alpha)})\\ \partial(K,\alpha) = (H_{-\varepsilon}(K);K,\Gamma_{(K,\alpha)})\\ \partial(K,\beta) = (K,(\begin{pmatrix}1\\ \beta+\varepsilon\beta^*\end{pmatrix},\beta)K) \; .\end{cases} \qquad (\Gamma_{(K,\alpha)} = \{(x,\alpha(x)) \in K\oplus K^* \mid x \in K\})$$

The boundary operations

$$\partial : (\text{S-non-singular forms}) \longrightarrow (\text{linking forms})$$

are thus seen to be special cases of the boundary operations

$$\partial : (\text{forms}) \longrightarrow (\text{formations})$$

defined in §1.6. (The boundary operations on S-non-singular forms can also be expressed in terms of the "dual lattice" construction familiar in the classical theory of quadratic forms over Dedekind rings (particularly in the case $(A,S) = (\mathbb{Z},\mathbb{Z}-\{0\})$, when $S^{-1}A = \mathbb{Q}$), as follows.

A lattice in a non-singular $\begin{cases}\text{(even) } \varepsilon\text{-symmetric} \\ \varepsilon\text{-quadratic}\end{cases}$ form over $S^{-1}A$ $\begin{cases}(Q,\phi) \\ (Q,\psi)\end{cases}$ is an S-non-singular $\begin{cases}\text{(even) } \varepsilon\text{-symmetric} \\ \varepsilon\text{-quadratic}\end{cases}$ form over A $\begin{cases}(K,\alpha) \\ (K,\beta)\end{cases}$ with K a f.g. projective A-submodule of Q, such that the inclusion $j \in \mathrm{Hom}_A(K,Q)$ extends to an isomorphism of forms over $S^{-1}A$

$$\begin{cases} j : S^{-1}(K,\alpha) \longrightarrow (Q,\phi) \\ j : S^{-1}(K,\beta) \longrightarrow (Q,\psi) \end{cases} .$$

A non-singular $\begin{cases}\text{(even) } \varepsilon\text{-symmetric} \\ \varepsilon\text{-quadratic}\end{cases}$ form over $S^{-1}A$ $\begin{cases}(Q,\phi) \\ (Q,\psi)\end{cases}$ admits such lattices if and only if Q is isomorphic to $S^{-1}K$ for some f.g. projective A-module K. The dual lattice $K^{\#}$ of a lattice $\begin{cases}(K,\alpha) \subseteq (Q,\phi) \\ (K,\beta) \subseteq (Q,\psi)\end{cases}$ is the A-submodule

$$\begin{cases} K^{\#} = \{x \in Q \mid \phi(x)(K) \subseteq A \subseteq S^{-1}A\} \subseteq Q \\ K^{\#} = \{x \in Q \mid (\psi+\varepsilon\psi^*)(x)(K) \subseteq A \subseteq S^{-1}A\} \subseteq Q \end{cases} .$$

The A-module isomorphism

$$\begin{cases} K^{\#} \longrightarrow K^* \ ; \ x \longmapsto (y \longmapsto \phi(x)(y)) \\ K^{\#} \longrightarrow K^* \ ; \ x \longmapsto (y \longmapsto (\psi+\varepsilon\psi^*)(x)(y)) \end{cases}$$

sends $K \subseteq K^{\#}$ to $\begin{cases} \mathrm{im}(\alpha : K \longrightarrow K^*) \subseteq K^* \\ \mathrm{im}(\beta+\varepsilon\beta^* : K \longrightarrow K^*) \subseteq K^* \end{cases}$, so there is induced an isomorphism of (A,S)-modules

$$\begin{cases} K^{\#}/K \longrightarrow \partial K = \mathrm{coker}(\alpha : K \longrightarrow K^*) \\ K^{\#}/K \longrightarrow \partial K = \mathrm{coker}(\beta+\varepsilon\beta^* : K \longrightarrow K^*) \ . \end{cases}$$

Given a non-singular $\begin{cases} \varepsilon\text{-symmetric} \\ \text{even } \varepsilon\text{-symmetric} \\ \varepsilon\text{-quadratic} \end{cases}$ form over $S^{-1}A$ $\begin{cases} (Q,\phi) \\ (Q,\phi) \\ (Q,\psi) \end{cases}$

and a lattice $\begin{cases} (K,\alpha) \subseteq (Q,\phi) \\ (K,\alpha) \subseteq (Q,\phi) \\ (K,\beta) \subseteq (Q,\psi) \end{cases}$ define a non-singular

$\begin{cases} \text{even } \varepsilon\text{-symmetric} \\ \varepsilon\text{-quadratic} \\ \text{split } \varepsilon\text{-quadratic} \end{cases}$ linking form over (A,S) $\begin{cases} (K^{\#}/K,\lambda) \\ (K^{\#}/K,\lambda,\mu) \\ (K^{\#}/K,\lambda,\nu) \end{cases}$ by

$$\lambda : K^{\#}/K \longrightarrow (K^{\#}/K)\hat{}\ \ ; \ x \longmapsto (y \longmapsto \phi(x)(y))$$

$$\mu : K^{\#}/K \longrightarrow Q_{\varepsilon}(A,S) \ ; \ x \longmapsto \phi(x)(x)$$

$$\nu : K^{\#}/K \longrightarrow Q_{\varepsilon}(S^{-1}A/A) \ ; \ x \longmapsto \psi(x)(x)$$

$(x,y \in K^{\#}, \phi = \psi+\varepsilon\psi^*$ in the ε-quadratic case) .

The isomorphism of (A,S)-modules $K^{\#}/K \xrightarrow{\sim} \partial K$ defined above

actually defines an isomorphism of $\begin{cases} \text{even } \varepsilon\text{-symmetric} \\ \varepsilon\text{-quadratic} \\ \text{split } \varepsilon\text{-quadratic} \end{cases}$

linking forms over (A,S)

$$\left\{\begin{array}{l}(K^{\#}/K,\lambda) \xrightarrow{\ \sim\ } \partial(K,\alpha) \\ (K^{\#}/K,\lambda,\mu) \xrightarrow{\ \sim\ } \partial(K,\alpha) \\ (K^{\#}/K,\lambda,\nu) \xrightarrow{\ \sim\ } \partial(K,\beta)\end{array}\right. \quad).$$

An S-non-singular $\left\{\begin{array}{l}\varepsilon\text{-symmetric} \\ \varepsilon\text{-quadratic}\end{array}\right.$ form over A $\left\{\begin{array}{l}(K,\alpha) \\ (K,\beta)\end{array}\right.$ is <u>S-hyperbolic</u> if it admits an S-lagrangian, or equivalently if $\left\{\begin{array}{l}S^{-1}(K,\alpha) \\ S^{-1}(K,\beta)\end{array}\right.$ is a hyperbolic $\left\{\begin{array}{l}\varepsilon\text{-symmetric} \\ \varepsilon\text{-quadratic}\end{array}\right.$ form over $S^{-1}A$ with a lagrangian isomorphic to $S^{-1}L$ for some f.g. projective A-module L. (Thus an S-hyperbolic form is the same as an $S^{-1}A$-hyperbolic form in the sense of §2.4).

<u>Proposition 3.4.4</u> Let $\left\{\begin{array}{l}(C,\phi \in Q\langle v_0\rangle^1(C,-\varepsilon)) \\ (C,\psi \in Q_1(C,-\varepsilon))\end{array}\right.$ be an S-acyclic 1-dimensional $\left\{\begin{array}{l}\text{even } (-\varepsilon)\text{-symmetric} \\ (-\varepsilon)\text{-quadratic}\end{array}\right.$ Poincaré complex over A, with associated non-singular $\left\{\begin{array}{l}\text{even } \varepsilon\text{-symmetric} \\ \text{split } \varepsilon\text{-quadratic}\end{array}\right.$ linking form over (A,S) $\left\{\begin{array}{l}(M,\lambda) = (H^1(C),\phi_0^S) \\ (M,\lambda,\nu) = (H^1(C),(1+T_{-\varepsilon})\psi_0^S,v_S^0(\psi))\end{array}\right.$.

i) The S-acyclic cobordism class $\left\{\begin{array}{l}(C,\phi) \in L^0(A,S,\varepsilon) \\ (C,\psi) \in L_0(A,S,\varepsilon)\end{array}\right.$ depends only on the isomorphism class of $\left\{\begin{array}{l}(M,\lambda) \\ (M,\lambda,\nu)\end{array}\right.$.

ii) $\left\{\begin{array}{l}(C,\phi) = 0 \in L^0(A,S,\varepsilon) \\ (C,\psi) = 0 \in L_0(A,S,\varepsilon)\end{array}\right.$ if and only if $\left\{\begin{array}{l}(M,\lambda) \\ (M,\lambda,\nu)\end{array}\right.$ is isomorphic to the boundary $\left\{\begin{array}{l}\partial(K,\alpha) \\ \partial(K,\beta)\end{array}\right.$ of an S-hyperbolic

S-non-singular $\begin{cases}\varepsilon\text{-symmetric}\\ \varepsilon\text{-quadratic}\end{cases}$ form over A $\begin{cases}(K,\alpha)\\ (K,\beta)\end{cases}$.

<u>Proof</u>: By the S-acyclic counterpart of Proposition 1.3.3 iii) an S-acyclic 1-dimensional $\begin{cases}\text{even } (-\varepsilon)\text{-symmetric}\\ (-\varepsilon)\text{-quadratic}\end{cases}$ Poincaré complex over A $\begin{cases}(C,\phi)\\ (C,\psi)\end{cases}$ represents 0 in $\begin{cases}L^0(A,S,\varepsilon)\\ L_0(A,S,\varepsilon)\end{cases}$ if and only if it is homotopy equivalent to the boundary $\begin{cases}\partial(D,\eta)\\ \partial(D,\zeta)\end{cases}$ of a connected S-acyclic 2-dimensional $\begin{cases}\text{even } (-\varepsilon)\text{-symmetric}\\ (-\varepsilon)\text{-quadratic}\end{cases}$ complex over A $\begin{cases}(D,\eta\in Q\langle v_0\rangle^2(D,-\varepsilon))\\ (D,\zeta\in Q_2(D,-\varepsilon))\end{cases}$ with D a f.g. projective A-module chain complex of the type

$$D\ :\ \ldots\longrightarrow 0\longrightarrow D_2\xrightarrow{d}D_1\xrightarrow{d}D_0\longrightarrow 0\longrightarrow\ldots\ .$$

Let $\begin{cases}(C,\phi)=\partial(D,\eta)\\ (C,\psi)=\partial(D,\zeta)\end{cases}$ be an S-acyclic boundary, as above.

The associated $\begin{cases}\text{even } \varepsilon\text{-symmetric}\\ \text{split } \varepsilon\text{-quadratic}\end{cases}$ linking form over (A,S) is the boundary

$$\begin{cases}(H^1(C),\phi_0^S)=\partial(K,\alpha)\\ (H^1(C),(1+T_{-\varepsilon})\psi_0^S,v_S^0(\psi))=\partial(K,\beta)\end{cases}$$

of the S-non-singular $\begin{cases}\varepsilon\text{-symmetric}\\ \varepsilon\text{-quadratic}\end{cases}$ form over A

$$(K,\alpha) = (\operatorname{coker}\begin{pmatrix} d^* \\ \eta_0 \end{pmatrix} : D^0 \longrightarrow D^1 \oplus D_2),\ \begin{bmatrix} \eta_0 + d\eta_1 & d \\ d^* & 0 \end{bmatrix} \in Q^{\varepsilon}(K))$$

$$(K,\beta) = (\operatorname{coker}\begin{pmatrix} d^* \\ (1+T_{-\varepsilon})\zeta_0 \end{pmatrix} : D^0 \longrightarrow D^1 \oplus D_2),\ \begin{bmatrix} \zeta_0 & d \\ 0 & 0 \end{bmatrix} \in Q_{\varepsilon}(K))$$

(which is obtained from $\begin{cases} (D,\eta) \\ (D,\zeta) \end{cases}$ by a surgery killing $H^2(D)$).

Moreover, the morphism of $\begin{cases} \varepsilon\text{-symmetric} \\ \varepsilon\text{-quadratic} \end{cases}$ forms over A

$$\begin{cases} \begin{bmatrix} 0 \\ 1 \end{bmatrix} : (D_2, 0) \longrightarrow (K,\alpha) \\ \begin{bmatrix} 0 \\ 1 \end{bmatrix} : (D_2, 0) \longrightarrow (K,\beta) \end{cases}$$

is the inclusion of an S-lagrangian, so that $\begin{cases} (K,\alpha) \\ (K,\beta) \end{cases}$ is an S-hyperbolic form.

Conversely, let $\begin{cases} (K,\alpha) \\ (K,\beta) \end{cases}$ be an S-hyperbolic S-non-singular $\begin{cases} \varepsilon\text{-symmetric} \\ \varepsilon\text{-quadratic} \end{cases}$ form over A, and let

$$\begin{cases} j : (L,0) \longrightarrow (K,\alpha) \\ j : (L,0) \longrightarrow (K,\beta) \end{cases}$$

be the inclusion of an S-lagrangian. Define a connected S-acyclic 2-dimensional $\begin{cases} \text{even } (-\varepsilon)\text{-symmetric} \\ (-\varepsilon)\text{-quadratic} \end{cases}$ complex over A $\begin{cases} (D, \eta \in Q\langle v_0\rangle^2(D,-\varepsilon)) \\ (D, \zeta \in Q_2(D,-\varepsilon)) \end{cases}$ by

$$d = j^* : D_1 = K^* \longrightarrow D_0 = L^*$$

$$d = \begin{cases} \alpha j \\ (\beta+\varepsilon\beta^*)j \end{cases} : D_2 = L \longrightarrow D_1 = K^*$$

$$\eta_0 = \begin{cases} 1 : D^0 = L \longrightarrow D_2 = L \\ \alpha : D^1 = K \longrightarrow D_1 = K^* \\ -\varepsilon : D^2 = L^* \longrightarrow D_0 = L^* \end{cases} , \ \eta_s = 0 \ (s \geqslant 1)$$

$$\zeta_0 = \begin{cases} 1 : D^0 = L \longrightarrow D_2 = L \\ \beta : D^1 = K \longrightarrow D_1 = K^* \\ 0 : D^2 = L^* \longrightarrow D_0 = L^* \end{cases}$$

$$\zeta_1 = \begin{cases} \varepsilon\beta^*j : D^0 = L \longrightarrow D_1 = K^* \\ 0 : D^1 = K \longrightarrow D_0 = L^* \end{cases}$$

$$\zeta_2 = \chi : D^0 = L \longrightarrow D_0 = L^*$$

for any $\beta \in \mathrm{Hom}_A(K,K^*)$ representing $\beta \in Q_\varepsilon(K)$, and any $\chi \in \mathrm{Hom}_A(L,L^*)$ such that

$$j^*\beta j = \chi - \varepsilon\chi^* \in \mathrm{Hom}_A(L,L^*) \ .$$

The boundary $\begin{cases} \partial(K,\alpha) \\ \partial(K,\beta) \end{cases}$ is the non-singular $\begin{cases} \text{even } \varepsilon\text{-symmetric} \\ \text{split } \varepsilon\text{-quadratic} \end{cases}$ linking form over (A,S) associated by Proposition 3.4.1 to the boundary S-acyclic 1-dimensional $\begin{cases} \text{even } (-\varepsilon)\text{-symmetric} \\ (-\varepsilon)\text{-quadratic} \end{cases}$ Poincaré complex over A $\begin{cases} (C,\phi) = \partial(D,\eta) \\ (C,\psi) = \partial(D,\zeta) \end{cases}$,

$$\begin{cases} \partial(K,\alpha) = (H^1(C),\phi_0^S) \\ \partial(K,\beta) = (H^1(C),(1+T_{-\varepsilon})\psi_0^S,v_S^0(\psi)) \end{cases} .$$

It remains to show that if $\begin{cases}(C,\phi)\\(C,\psi)\end{cases}$, $\begin{cases}(C',\phi')\\(C',\psi')\end{cases}$ are S-acyclic 1-dimensional $\begin{cases}\text{even } (-\varepsilon)\text{-symmetric}\\ (-\varepsilon)\text{-quadratic}\end{cases}$ Poincaré complexes over A which are related by an isomorphism of the associated non-singular $\begin{cases}\text{even } \varepsilon\text{-symmetric}\\ \text{split } \varepsilon\text{-quadratic}\end{cases}$ linking forms over (A,S)

$$\begin{cases}(H^1(C),\phi_0^S) \longrightarrow (H^1(C'),\phi_0'^S)\\ (H^1(C),(1+T_{-\varepsilon})\psi_0^S,v_S^0(\psi)) \longrightarrow (H^1(C'),(1+T_{-\varepsilon})\psi_0'^S,v_S^0(\psi'))\end{cases}$$

then

$$\begin{cases}(C,\phi) = (C',\phi') \in L^0(A,S,\varepsilon)\\ (C,\psi) = (C',\psi') \in L_0(A,S,\varepsilon)\ .\end{cases}$$

Proposition 3.4.1 associates to such an isomorphism a $\begin{cases}(-\varepsilon)\text{-symmetric}\\ \text{split } (-\varepsilon)\text{-quadratic}\end{cases}$ homotopy equivalence

$$\begin{cases}f : (C,\phi) \xrightarrow{\ \sim\ } (C',\phi')\\ f : (C,\psi) \xrightarrow{\ \sim\ } (C',\psi')\ ,\end{cases}$$

so that

$$\begin{cases}f^{\%}(\phi) = \phi' \in Q\langle v_0\rangle^1(C',-\varepsilon)\\ f_{\%}(\psi) = \psi' + H(\theta) \in Q_1(C',-\varepsilon)\end{cases}$$

for some $\theta \in \hat{Q}^2(C',-\varepsilon)$ such that

$$\hat{v}_1^S(\theta) = 0 : H^1(C') \longrightarrow \hat{H}^0(\mathbb{Z}_2;S^{-1}A/A,\varepsilon)\ .$$

Now $\begin{cases}(C,\phi)\oplus(C',-\phi')\\ (C,\psi)\oplus(C',-(\psi'+H(\theta)))\end{cases}$ is the boundary of an S-acyclic 2-dimensional $\begin{cases}\text{even } (-\varepsilon)\text{-symmetric}\\ (-\varepsilon)\text{-quadratic}\end{cases}$ Poincaré pair over A

$$\begin{cases} ((f\ 1):C\oplus C' \longrightarrow C', (\delta\phi, \phi\oplus -\phi') \in Q\langle v_0\rangle^2((f\ 1),-\varepsilon)) \\ ((f\ 1):C\oplus C' \longrightarrow C', (\delta\psi, \psi\oplus -(\psi'+H(\theta))) \in Q_2((f\ 1),-\varepsilon)) \end{cases}$$

so that

$$\begin{cases} (C,\phi) = (C',\phi') \in L^0(A,S,\varepsilon) \\ (C,\psi) = (C',\psi'+H(\theta)) \in L_0(A,S,\varepsilon) \ . \end{cases}$$

We shall prove that

$$(C',\psi'+H(\theta)) = (C',\psi') \in L_0(A,S,\varepsilon)$$

using the language of S-formations (Proposition 3.4.3), as follows.

Given non-singular split $(-\varepsilon)$-quadratic S-formations over A $(F,(\binom{\gamma}{\mu},\theta)G),(F,(\binom{\gamma}{\mu},\theta')G)$ such that

$$\theta' - \theta \in \ker(S^{-1}:Q_\varepsilon(G) \longrightarrow Q_\varepsilon(S^{-1}G))$$

we have to show that the non-singular split $(-\varepsilon)$-quadratic formation over A $(F,(\binom{\gamma}{\mu},\theta')G)\oplus(F,(\binom{-\gamma}{\mu},-\theta)G)$ is stably isomorphic to the boundary $\partial(K,\beta) = (K,(\begin{pmatrix} 1 \\ \beta+\varepsilon\beta^* \end{pmatrix},\beta)K)$ of an S-hyperbolic S-non-singular ε-quadratic form over A $(K,\beta \in Q_\varepsilon(K))$. By Proposition 1.6.2 the inclusion of the lagrangian

$$\begin{pmatrix} \gamma \\ \mu \end{pmatrix} : (G,0) \longrightarrow H_{-\varepsilon}(F)$$

extends to an isomorphism of hyperbolic $(-\varepsilon)$-quadratic forms over A

$$\begin{pmatrix} \gamma & \tilde{\gamma} \\ \mu & \tilde{\mu} \end{pmatrix} : H_{-\varepsilon}(G) = (G\oplus G^*, \begin{pmatrix} 0 & 1 \\ 0 & 0 \end{pmatrix}) \xrightarrow{\sim} H_{-\varepsilon}(F) = (F\oplus F^*, \begin{pmatrix} 0 & 1 \\ 0 & 0 \end{pmatrix}) \ .$$

Define an S-non-singular ε-quadratic form over A

$$(K,\beta) = (G\oplus F, \begin{pmatrix} -\theta & 0 \\ \mu & 0 \end{pmatrix} + \begin{pmatrix} -\gamma^*\tilde{\mu} & -\varepsilon\mu^*\tilde{\gamma} \\ \tilde{\mu} & \tilde{\mu} \end{pmatrix}\begin{pmatrix} 0 & 0 \\ 0 & \theta'-\theta \end{pmatrix}\begin{pmatrix} -\tilde{\mu}^*\gamma & \tilde{\mu}^* \\ -\varepsilon\tilde{\gamma}^*\mu & \tilde{\mu}^* \end{pmatrix}$$

$$\in Q_\varepsilon(G\oplus F)) \ .$$

For some S-isomorphism $s \in \mathrm{Hom}_A(F,F)$ there is defined a morphism of ε-quadratic forms over A

$$\begin{pmatrix} 0 \\ s \end{pmatrix} : (F,0) \longrightarrow (K,\beta) = (G\oplus F,\beta)$$

which is the inclusion of an S-lagrangian, so that (K,β) is an S-hyperbolic form. The isomorphism of non-singular split $(-\varepsilon)$-quadratic formations over A

$$(a,b,c) = \left(\begin{pmatrix} \gamma & -1 & -\varepsilon\tilde{\gamma} & 0 \\ 0 & 1 & 0 & 0 \\ \mu & 0 & \tilde{\mu} & 1 \\ \mu & 0 & \tilde{\mu} & 0 \end{pmatrix}, \begin{pmatrix} -\tilde{\mu}^*\gamma & \tilde{\mu}^* & 1 & 0 \\ -\bar{\varepsilon}\tilde{\gamma}^*\mu & \tilde{\mu}^* & 1 & 0 \\ 0 & 0 & 0 & 1 \\ -\gamma\tilde{\gamma}^*\mu & \varepsilon\tilde{\gamma}\mu^* & \gamma & -1 \end{pmatrix}, \right.$$

$$\left. \begin{pmatrix} 0 & \tilde{\gamma}^* & 0 & 0 \\ 0 & 0 & 0 & 0 \\ 0 & \gamma^* & \theta'-\theta & 0 \\ 0 & 0 & 0 & 0 \end{pmatrix} \right)$$

$$: \partial(G\oplus F,\beta)\oplus(G^*\oplus F^*,G\oplus F)$$

$$\longrightarrow (F,(\begin{pmatrix} -\gamma \\ \mu \end{pmatrix},-\theta)G)\oplus(F,(\begin{pmatrix} \gamma \\ \mu \end{pmatrix},\theta')G)\oplus(F^*\oplus F^*,F\oplus F)$$

defines a stable isomorphism

$$[a,b,c] : \partial(K,\beta) \longrightarrow (F,(\begin{pmatrix} -\gamma \\ \mu \end{pmatrix},- \quad)G)\oplus(F,(\begin{pmatrix} \gamma \\ \mu \end{pmatrix},\theta')G) \quad .$$

It follows that the S-acyclic 1-dimensional $(-\varepsilon)$-quadratic Poincaré complexes over A associated to $(F,(\begin{pmatrix} \gamma \\ \mu \end{pmatrix},\theta)G)$ and $(F,(\begin{pmatrix} \gamma \\ \mu \end{pmatrix},\theta')G)$ are S-acyclic cobordant

$$(C,\psi) = (C,\psi') \in L_0(A,S,\varepsilon) \quad .$$

[]

A <u>sublagrangian</u> of an $\begin{cases} \varepsilon\text{-symmetric} \\ \varepsilon\text{-quadratic} \\ \text{split } \varepsilon\text{-quadratic} \end{cases}$ linking form

over (A,S) $\begin{cases} (M,\lambda) \\ (M,\lambda,\mu) \\ (M,\lambda,\nu) \end{cases}$ is a submodule L of M such that

i) L and M/L are (A,S)-modules

ii) the inclusion $j \in \mathrm{Hom}_A(L,M)$ defines a morphism of

$\begin{cases} \varepsilon\text{-symmetric} \\ \varepsilon\text{-quadratic} \\ \text{split } \varepsilon\text{-quadratic} \end{cases}$ linking forms over (A,S)

$$\begin{cases} j : (L,0) \longrightarrow (M,\lambda) \\ j : (L,0,0) \longrightarrow (M,\lambda,\mu) \\ j : (L,0,0) \longrightarrow (M,\lambda,\nu) \end{cases}$$

iii) the A-module morphism

$$[\lambda] : M/L \longrightarrow L^{\wedge} \; ; \; x \longmapsto (y \longmapsto \lambda(x)(y))$$

is onto.

The <u>annihilator</u> of a sublagrangian L is the submodule

$$L^{\perp} = \ker(j^{\wedge}\lambda : M \longrightarrow L^{\wedge} ; x \longmapsto (y \longmapsto \lambda(x)(y))) \subseteq M$$

which contains L

$$L \subseteq L^{\perp} \ .$$

Both $L^{\perp}$ and $L^{\perp}/L$ are (A,S)-modules, where

$$L^{\perp}/L = \ker([\lambda] : M/L \longrightarrow L^{\wedge}) \ .$$

A <u>lagrangian</u> is a sublagrangian L such that $[\lambda] \in \mathrm{Hom}_A(M/L, L^{\perp})$ is an isomorphism, that is

$$L^{\perp} = L \ .$$

A non-singular $\begin{cases}\text{(even) } \varepsilon\text{-symmetric} \\ \text{(split) } \varepsilon\text{-quadratic}\end{cases}$ linking form is

<u>hyperbolic</u> if it admits a lagrangian.

<u>Proposition 3.4.5</u> i) Given a sublagrangian L of a non-singular $\begin{cases}\text{(even) } \varepsilon\text{-symmetric} \\ \varepsilon\text{-quadratic} \\ \text{split } \varepsilon\text{-quadratic}\end{cases}$ linking form over (A,S) $\begin{cases}(M,\lambda) \\ (M,\lambda,\mu) \\ (M,\lambda,\nu)\end{cases}$ there is

defined a non-singular $\begin{cases}\text{(even) } \varepsilon\text{-symmetric} \\ \varepsilon\text{-quadratic} \\ \text{split } \varepsilon\text{-quadratic}\end{cases}$ linking form over (A,S)

$\begin{cases}(L^{\perp}/L,\lambda^{\perp}/\lambda) \\ (L^{\perp}/L,\lambda^{\perp}/\lambda,\mu^{\perp}/\mu) \\ (L^{\perp}/L,\lambda^{\perp}/\lambda,\nu^{\perp}/\nu)\end{cases}$ such that $\begin{cases}(M,\lambda)\oplus(L^{\perp}/L,-\lambda^{\perp}/\lambda) \\ (M,\lambda,\mu)\oplus(L^{\perp}/L,-\lambda^{\perp}/\lambda,-\mu^{\perp}/\mu) \\ (M,\lambda,\nu)\oplus(L^{\perp}/L,-\lambda^{\perp}/\lambda,-\nu^{\perp}/\nu)\end{cases}$ is

hyperbolic, with lagrangian

$$\Delta = \{(x,[x]) \in M\oplus L^{\perp}/L \mid x \in L^{\perp}\} \subseteq M\oplus L^{\perp}/L\ .$$

ii) A non-singular $\begin{cases}\text{(even) } \varepsilon\text{-symmetric} \\ \text{(split) } \varepsilon\text{-quadratic}\end{cases}$ linking form over (A,S) is

hyperbolic if and only if the associated $\begin{cases}(-\varepsilon)\text{-symmetric} \\ \text{(split) } (-\varepsilon)\text{-quadratic}\end{cases}$

homotopy equivalence class of S-acyclic 1-dimensional

$\begin{cases}\text{(even) } (-\varepsilon)\text{-symmetric} \\ (-\varepsilon)\text{-quadratic}\end{cases}$ Poincaré complexes over A contains the

boundary $\begin{cases}(C,\phi \in Q^1(C,-\varepsilon)) \\ (C,\psi \in Q_1(C,-\varepsilon))\end{cases}$ of an S-acyclic 2-dimensional

$\begin{cases}\text{(even) } (-\varepsilon)\text{-symmetric} \\ (-\varepsilon)\text{-quadratic}\end{cases}$ Poincaré pair over A

$$\begin{cases} (f:C \longrightarrow D, (\delta\phi,\phi) \in Q^2(f,-\varepsilon)) \\ (f:C \longrightarrow D, (\delta\psi,\psi) \in Q_2(f,-\varepsilon)) \end{cases} \text{ with } H^2(D) = 0.$$

Proof: i) Trivial.

ii) Let $\begin{cases} (f:C \longrightarrow D, (\delta\phi,\phi)) \\ (f:C \longrightarrow D, (\delta\psi,\psi)) \end{cases}$ be an S-acyclic 2-dimensional

$\begin{cases} \text{(even) } (-\varepsilon)\text{-symmetric} \\ (-\varepsilon)\text{-quadratic} \end{cases}$ Poincaré pair over A such that $H^2(D) = 0$.

The non-singular $\begin{cases} \text{(even) } \varepsilon\text{-symmetric} \\ \text{split } \varepsilon\text{-quadratic} \end{cases}$ linking form over (A,S)

$\begin{cases} (H^1(C), \phi_0^S) \\ (H^1(C), (1+T_{-\varepsilon})\psi_0^S, v_S^0(\psi)) \end{cases}$ associated to the boundary $\begin{cases} (C,\phi) \\ (C,\psi) \end{cases}$

is hyperbolic, with lagrangian

$$L = \operatorname{im}(f^*:H^1(D) \longrightarrow H^1(C)) \subseteq H^1(C) \ .$$

The correspondence of Proposition 3.4.1 associates to a hyperbolic (even) ε-symmetric linking form over (A,S) (M,λ) woth a lagrangian L a map of S-acyclic 1-dimensional (even) $(-\varepsilon)$-symmetric complexes over A

$$f : (C,\phi) \longrightarrow (D,0) \ ,$$

with $f:C \longrightarrow D$ a chain map of f.g. projective A-module chain complexes

$$\begin{array}{ccccccccc} C : \ldots \longrightarrow 0 \longrightarrow & C_1 & \xrightarrow{d} & C_0 & \longrightarrow 0 \longrightarrow \ldots \\ f\big\downarrow & f\big\downarrow & & \big\downarrow\tilde{f} & \\ D : \ldots \longrightarrow 0 \longrightarrow & D_1 & \xrightarrow{\tilde{d}} & D_0 & \longrightarrow 0 \longrightarrow \ldots \end{array}$$

resolving

$$f^* = \text{inclusion} : H^1(D) = L \longrightarrow H^1(C) = M \ .$$

From the exact sequence of Proposition 1.1.4 we have

$$\phi \in \ker(f^{\%}:Q^1(C,-\varepsilon) \longrightarrow Q^1(D,-\varepsilon)) = \mathrm{im}(\partial:Q^2(f,-\varepsilon) \longrightarrow Q^1(C,-\varepsilon)) ,$$

so that there exists an S-acyclic 2-dimensional (even) $(-\varepsilon)$-symmetric Poincaré pair over A $(f:C \longrightarrow D,(\delta\phi,\phi) \in Q^2(f,-\varepsilon))$ such that $H^2(D) = 0$, with boundary (C,ϕ). Thus a non-singular (even) ε-symmetric linking form over (A,S) (M,λ) is hyperbolic if and only if an associated S-acyclic 1-dimensional (even) $(-\varepsilon)$-symmetric Poincaré complex over A (C,ϕ) is such a boundary.

The correspondence of Proposition 3.4.1 also associates to a hyperbolic $\begin{Bmatrix} \varepsilon\text{-quadratic} \\ \text{split } \varepsilon\text{-quadratic} \end{Bmatrix}$ linking form over (A,S) $\begin{Bmatrix} (M,\lambda,\mu) \\ (M,\lambda,\nu) \end{Bmatrix}$ with lagrangian L a $\begin{Bmatrix} (-\varepsilon)\text{-quadratic} \\ \text{split } (-\varepsilon)\text{-quadratic} \end{Bmatrix}$ map of S-acyclic 1-dimensional $(-\varepsilon)$-quadratic complexes over A

$$f : (C,\psi) \longrightarrow (D,0)$$

with $f:C \longrightarrow D$ exactly as in the ε-symmetric case dealt with above. It is possible to choose resolutions such that $\tilde{f} \in \mathrm{Hom}_A(C_0,D_0)$ is an isomorphism. (Explicitly, given a f.g. projective A-module resolution of $M\hat{}$

$$0 \longrightarrow C_1 \xrightarrow{d} C_0 \longrightarrow M\hat{} \longrightarrow 0$$

write the dual resolution of $(M\hat{})\hat{} = M$ as

$$0 \longrightarrow C^0 \xrightarrow{d^*} C^1 \xrightarrow{e} M \longrightarrow 0 .$$

Define a f.g. projective A-module

$$P = e^{-1}(L) \subseteq C^1 ,$$

let $g \in \mathrm{Hom}_A(P,C^1)$ be the inclusion, and let $h \in \mathrm{Hom}_A(C^0,P)$ be the restriction of $d^* \in \mathrm{Hom}_A(C^0,C^1)$, so that

$$d^* = gh \in \mathrm{Hom}_A(C^0,C^1) .$$

The S-acyclic 1-dimensional f.g. projective A-module chain complex D defined by

$$\tilde{d} = h^* : D_1 = P^* \longrightarrow D_0 = C_0 \ , \ D_r = 0 \ (r \neq 0,1)$$

is a resolution of $L^\wedge$

$$0 \longrightarrow D_1 \xrightarrow{\tilde{d}} D_0 \longrightarrow L^\wedge \longrightarrow 0 \ .$$

The A-module chain map $f:C \longrightarrow D$ defined by

$$f = g^* : C_1 \longrightarrow D_1 = P^*$$

$$\tilde{f} = 1 : C_0 \longrightarrow D_0 = C_0$$

is a resolution of

$$f^* = \text{inclusion} : H^1(D) = L \longrightarrow H^1(C) = M$$

with $\tilde{f} \in \mathrm{Hom}_A(C_0,D_0)$ an isomorphism). By the definition of a $\begin{cases} (-\varepsilon)\text{-quadratic} \\ \text{split } (-\varepsilon)\text{-quadratic} \end{cases}$ map we have that

$$f_{\%}(\psi) = H(\theta) \in Q_1(D,-\varepsilon)$$

for some element $\theta \in \hat{Q}^2(D,-\varepsilon)$ such that

$$\begin{cases} \hat{v}_1(\theta) = 0 : H^1(D) \longrightarrow \hat{H}^0(\mathbb{Z}_2;A,\varepsilon) \\ \hat{v}_1^S(\theta) = 0 : H^1(D) \longrightarrow \hat{H}^1(\mathbb{Z}_2;S^{-1}A/A,\varepsilon) \end{cases} .$$

On the chain level the elements $\psi \in Q_1(C,-\varepsilon)$, $\theta \in \hat{Q}^2(D,-\varepsilon)$ are represented by A-module morphisms

$$\psi_0 : C^0 \longrightarrow C_1 \ , \ \tilde{\psi}_0 : C^1 \longrightarrow C_0 \ , \ \psi_1 : C^0 \longrightarrow C_0$$

$$\theta_0 : D^1 \longrightarrow D_1 \ , \ \theta_{-1} : D^0 \longrightarrow D_{-1} \ , \ \tilde{\theta}_{-1} : D^1 \longrightarrow D_0 \ ,$$

$$\theta_{-2} : D^0 \longrightarrow D_0$$

such that

$$d\psi_0 + \tilde{\psi}_0 d^* + \psi_1 + \varepsilon\psi_1^* = 0 \ , \ \tilde{d}\theta_{-1} + \tilde{\theta}_{-1}\tilde{d}^* + \theta_{-2} + \varepsilon\theta_{-2}^* = 0 \ ,$$

$$\theta_0 - \varepsilon\theta_0^* = 0 \ , \ \tilde{d}\theta_0 - \theta_{-1} + \varepsilon\tilde{\theta}_{-1}^* = 0 \ , \ \theta_0\tilde{d}^* + \tilde{\theta}_{-1} - \varepsilon\theta_{-1}^* = 0 \ ,$$

$$f\psi_0\tilde{f}^* = \theta_{-1} \ , \ \tilde{f}\tilde{\psi}_0 f^* = \tilde{\theta}_{-1} \ , \ \tilde{f}\psi_1\tilde{f}^* = \theta_{-2} \ .$$

The vanishing of the $(-\varepsilon)$-hyperquadratic $\begin{cases} - \\ \text{linking} \end{cases}$ Wu class

$$\begin{cases} \hat{v}_1(\theta) = 0 : H^1(D) \longrightarrow \hat{H}^0(\mathbb{Z}_2;A,\varepsilon) \ ; \ x \longmapsto \theta_0(x)(x) \\ \hat{v}_1^S(\theta) = 0 : H^1(D) \longrightarrow \hat{H}^1(\mathbb{Z}_2;S^{-1}A/A,\varepsilon) \ ; \\ \qquad x \longmapsto (\tfrac{1}{s}).(\theta_{-2}+\tilde{\theta}_{-1}\tilde{d}^*)(y)(y).\overline{(\tfrac{1}{s})} \end{cases}$$

$$(x \in D^1, y \in D^0, s \in S, sx = d^*y \in D^1)$$

implies that there exists an A-module morphism $\chi \in \mathrm{Hom}_A(D^1, D_1)$ such that

$$\begin{cases} \theta_0 = \chi + \varepsilon\chi^* \in Q\langle v_0\rangle^{\varepsilon}(D^1) \subseteq Q^{\varepsilon}(D^1) \\ \tilde{f}(\psi_1 + \tilde{\psi}_0 d^*)\tilde{f}^* - \tilde{d}\chi\tilde{d}^* \in \ker(S^{-1}:Q_{\varepsilon}(D^0) \longrightarrow Q_{\varepsilon}(S^{-1}D^0)) \ . \end{cases}$$

Define $\theta' \in Q^2(C,-\varepsilon)$ by

$$\theta'_{-2} = \tilde{f}^{-1}\tilde{d}\chi\tilde{d}^*\tilde{f}^{*-1} - (\psi_1 + \tilde{\psi}_0 d^*) \ : \ C^0 \longrightarrow C_0$$

$$\theta'_s = 0 \ \ (s \neq -2) \ ,$$

and let

$$\psi' = \psi + H(\theta') \in Q_1(C,-\varepsilon) \ .$$

Then

$$f_{\%}(\psi') = 0 \in Q_1(D,-\varepsilon)$$

and

$$\begin{cases} v_1(\theta') = 0 : H^1(C) \longrightarrow \hat{H}^0(\mathbb{Z}_2;A,\varepsilon) \\ v_1^S(\theta') = 0 : H^1(C) \longrightarrow \hat{H}^1(\mathbb{Z}_2;S^{-1}A/A,\varepsilon) \ . \end{cases}$$

Now (C,ψ') is an S-acyclic 1-dimensional $(-\varepsilon)$-quadratic Poincaré complex over A which is the boundary of an S-acyclic 2-dimensional $(-\varepsilon)$-quadratic Poincaré pair over A $(f:C \longrightarrow D,(\delta\psi',\psi') \in Q_2(f,-\varepsilon))$ and such that there is defined a $\begin{cases} (-\varepsilon)\text{-quadratic} \\ \text{split } (-\varepsilon)\text{-quadratic} \end{cases}$ homotopy equivalence

$$1 : (C,\psi) \longrightarrow (C,\psi')$$

with (C,ψ) a complex associated to the hyperbolic $\begin{cases} \varepsilon\text{-quadratic} \\ \text{split } \varepsilon\text{-quadratic} \end{cases}$ linking form over (A,S) $\begin{cases} (M,\lambda,\mu) \\ (M,\lambda,\nu) \end{cases}$.

[]

Next, we shall relate the (sub)lagrangians of the boundary linking form over (A,S) of an S-non-singular form over A to morphisms of S-non-singular forms over A which become isomorphisms over $S^{-1}A$. This relationship will then be used in Proposition 3.4.7 to identify the relative L-group $\begin{cases} L^0(A,S,\varepsilon) \\ L_0(A,S,\varepsilon) \end{cases}$ with the Witt group of non-singular $\begin{cases} \text{even } \varepsilon\text{-symmetric} \\ \text{split } \varepsilon\text{-quadratic} \end{cases}$ linking forms over (A,S).

An <u>S-isomorphism</u> of S-non-singular $\begin{cases} \varepsilon\text{-symmetric} \\ \varepsilon\text{-quadratic} \end{cases}$ forms over A

$$\begin{cases} f : (K,\alpha) \longrightarrow (K',\alpha') \\ f : (K,\beta) \longrightarrow (K',\beta') \end{cases}$$

is an S-isomorphism of A-modules $f \in \mathrm{Hom}_A(K,K')$ such that

$$\begin{cases} f^*\alpha' f - \alpha \in \ker(S^{-1}:Q^{\varepsilon}(K)\longrightarrow Q^{\varepsilon}(S^{-1}K)) = \{0\}\subseteq Q^{\varepsilon}(K) \\ f^*\beta' f - \beta \in \ker(S^{-1}:Q_{\varepsilon}(K)\longrightarrow Q_{\varepsilon}(S^{-1}(K))\subseteq Q_{\varepsilon}(K) \end{cases}$$

$$(\neq \{0\}, \text{ in general}) .$$

Then

$$\begin{cases} S^{-1}f : S^{-1}(K,\alpha) \longrightarrow S^{-1}(K',\alpha') \\ S^{-1}f : S^{-1}(K,\beta) \longrightarrow S^{-1}(K',\beta') \end{cases}$$

is an isomorphism of non-singular $\begin{cases}\varepsilon\text{-symmetric} \\ \varepsilon\text{-quadratic}\end{cases}$ forms over $S^{-1}A$.

Note that if $1/2 \in S^{-1}A$ there is a natural identification of sets of S-isomorphism classes

(S-non-singular ε-quadratic forms over A)

= (S-non-singular even ε-symmetric forms over A) .

(Specifically, if $(K,\beta),(K',\beta')$ are S-non-singular ε-quadratic forms over A which are related by an S-isomorphism of the ε-symmetrizations

$$f : (K,\beta+\varepsilon\beta^*) \longrightarrow (K',\beta'+\varepsilon\beta'^*)$$

then

$$S^{-1}(f^*\beta' f - \beta) = \tfrac{1}{2}(f^*\beta' f - \beta) - \tfrac{1}{2}\varepsilon(f^*\beta' f - \beta)^* = 0 \in Q_{\varepsilon}(S^{-1}K) ,$$

so that there is also defined an S-isomorphism of ε-quadratic forms

$$f : (K,\beta) \longrightarrow (K',\beta') \quad).$$

An <u>equivalence</u> of S-isomorphisms of S-non-singular

$\begin{cases}\varepsilon\text{-symmetric} \\ \varepsilon\text{-quadratic}\end{cases}$ forms over A

$$\begin{cases} (g,g') : (f:(K,\alpha)\longrightarrow (K',\alpha')) \longrightarrow (\tilde{f}:(\tilde{K},\tilde{\alpha}) \longrightarrow (\tilde{K}',\tilde{\alpha}')) \\ (g,g') : (f:(K,\beta)\longrightarrow (K',\beta')) \longrightarrow (\tilde{f}:(\tilde{K},\tilde{\beta}) \longrightarrow (\tilde{K}',\tilde{\beta}')) \end{cases}$$

is defined by S-isomorphisms of forms

$$\begin{cases} g : (K,\alpha) \longrightarrow (\widetilde{K},\widetilde{\alpha}) \ , \ g' : (K',\alpha') \longrightarrow (\widetilde{K}',\widetilde{\alpha}') \\ g : (K,\beta) \longrightarrow (\widetilde{K},\widetilde{\beta}) \ , \ g' : (K',\beta') \longrightarrow (\widetilde{K}',\widetilde{\beta}') \end{cases}$$

with $g \in \mathrm{Hom}_A(K,K)$, $g' \in \mathrm{Hom}_A(K',K')$ isomorphisms, and such that the there is defined a commutative diagram

$$\begin{array}{ccc} K & \xrightarrow{f} & K' \\ {\scriptstyle g}\downarrow & & \downarrow{\scriptstyle g'} \\ \widetilde{K} & \xrightarrow{\widetilde{f}} & \widetilde{K}' \end{array} \quad .$$

A non-singular $\begin{cases} \text{(even) } \varepsilon\text{-symmetric} \\ \text{(split) } \varepsilon\text{-quadratic} \end{cases}$ linking form X is <u>stably hyperbolic</u> if there exists an isomorphism of such linking forms

$$X \oplus Y \xrightarrow{\sim} Y'$$

with Y,Y' hyperbolic.

<u>Proposition 3.4.6</u> i) Let $\begin{cases} (K,\alpha \in Q^{\varepsilon}(K)) \\ (K,\alpha \in Q\langle v_0\rangle^{\varepsilon}(K)) \\ (K,\beta \in Q_{\varepsilon}(K)) \end{cases}$ be an S-non-singular $\begin{cases} \varepsilon\text{-symmetric} \\ \text{even } \varepsilon\text{-symmetric} \\ \varepsilon\text{-quadratic} \end{cases}$ form over A. The sublagrangians L of the boundary $\begin{cases} \text{even } \varepsilon\text{-symmetric} \\ \varepsilon\text{-quadratic} \\ \text{split } \varepsilon\text{-quadratic} \end{cases}$ linking form over (A,S) $\begin{cases} \partial(K,\alpha) = (M,\lambda) \\ \partial(K,\alpha) = (M,\lambda,\mu) \\ \partial(K,\beta) = (M,\lambda,\nu) \end{cases}$ are in a natural one-one correspondence with the equivalence classes of S-isomorphisms of S-non-singular

$$\left\{\begin{array}{l}\varepsilon\text{-symmetric}\\ \text{even }\varepsilon\text{-symmetric}\\ \varepsilon\text{-quadratic}\end{array}\right. \text{ forms over } A$$

$$\left\{\begin{array}{l} f : (K,\alpha) \longrightarrow (K',\alpha') \\ f : (K,\alpha) \longrightarrow (K',\alpha') \\ f : (K,\beta) \longrightarrow (K',\beta')\ , \end{array}\right.$$

under which

$$L = \operatorname{coker}(f:K \longrightarrow K') \subseteq M = \operatorname{coker}(\alpha:K \longrightarrow K^*)$$

(with $\alpha = \beta+\varepsilon\beta^*$ in the ε-quadratic case)

and

$$\left\{\begin{array}{l} (L^\perp/L, \lambda^\perp/\lambda) = \partial(K',\alpha') \\ (L^\perp/L, \lambda^\perp/\lambda, \mu^\perp/\mu) = \partial(K',\alpha') \\ (L^\perp/L, \lambda^\perp/\lambda, \nu^\perp/\nu) = \partial(K',\beta')\ . \end{array}\right.$$

Lagrangians L correspond to S-isomorphisms with $\left\{\begin{array}{l}(K',\alpha')\\(K',\alpha')\\(K',\beta')\end{array}\right.$ non-singular.

ii) A non-singular $\left\{\begin{array}{l}\text{even }\varepsilon\text{-symmetric}\\ \varepsilon\text{-quadratic}\\ \text{split }\varepsilon\text{-quadratic}\end{array}\right.$ linking form over (A,S)

$\left\{\begin{array}{l}(M,\lambda)\\(M,\lambda,\mu)\\(M,\lambda,\nu)\end{array}\right.$ is stably hyperbolic if and only if it is isomorphic

to the boundary $\left\{\begin{array}{l}\partial(K,\alpha)\\ \partial(K,\alpha)\\ \partial(K,\beta)\end{array}\right.$ of an S-hyperbolic S-non-singular

$\left\{\begin{array}{l}\varepsilon\text{-symmetric}\\ \text{even }\varepsilon\text{-symmetric}\\ \varepsilon\text{-quadratic}\end{array}\right.$ form over A $\left\{\begin{array}{l}(K,\alpha)\\(K,\alpha)\ .\\(K,\beta)\end{array}\right.$

Proof: i) Given an S-isomorphism of S-non-singular $\left\{\begin{array}{l}\text{(even) } \varepsilon\text{-symmetric}\\ \varepsilon\text{-quadratic}\end{array}\right.$ forms over A

$$\left\{\begin{array}{l} f : (K,\alpha) \longrightarrow (K',\alpha') \\ f : (K,\beta) \longrightarrow (K',\beta') \end{array}\right.$$

define a sublagrangian L of the boundary $\left\{\begin{array}{l}\text{even } \varepsilon\text{-symmetric } (\varepsilon\text{-quadratic})\\ \text{split } \varepsilon\text{-quadratic}\end{array}\right.$ linking form over (A,S) $\left\{\begin{array}{l}\partial(K,\alpha)\\ \partial(K,\beta)\end{array}\right.$ by the resolution

$$\begin{array}{ccccccccc} 0 & \longrightarrow & K & \xrightarrow{\;f\;} & K' & \longrightarrow & L & \longrightarrow & 0 \\ & & \big\downarrow 1 & & \big\downarrow f^*\alpha' & & \big\downarrow & & \\ 0 & \longrightarrow & K & \xrightarrow{\;\alpha\;} & K^* & \longrightarrow & M & \longrightarrow & 0 \end{array}$$

with $\alpha = \beta + \varepsilon\beta^* \in \mathrm{Hom}_A(K,K^*)$, $\alpha' = \beta' + \varepsilon\beta'^* \in \mathrm{Hom}_A(K',K'^*)$ in the ε-quadratic case. An equivalence of S-isomorphisms of S-non-singular $\left\{\begin{array}{l}\text{(even) } \varepsilon\text{-symmetric}\\ \varepsilon\text{-quadratic}\end{array}\right.$ forms over A

$$\left\{\begin{array}{l} (g,g') : (f:(K,\alpha) \longrightarrow (K',\alpha')) \longrightarrow (\tilde{f}:(\tilde{K},\tilde{\alpha}) \longrightarrow (\tilde{K}',\tilde{\alpha}')) \\ (g,g') : (f:(K,\beta) \longrightarrow (K',\beta')) \longrightarrow (\tilde{f}:(\tilde{K},\tilde{\beta}) \longrightarrow (\tilde{K}',\tilde{\beta}')) \end{array}\right.$$

induces an isomorphism of $\left\{\begin{array}{l}\text{even } \varepsilon\text{-symmetric } (\varepsilon\text{-quadratic})\\ \text{split } \varepsilon\text{-quadratic}\end{array}\right.$ linking forms over (A,S)

$$\left\{\begin{array}{l} h : \partial(K,\alpha) \xrightarrow{\;\sim\;} \partial(\tilde{K},\tilde{\alpha}) \\ h : \partial(K,\beta) \xrightarrow{\;\sim\;} \partial(\tilde{K},\tilde{\beta}) \end{array}\right.$$

such that

$$h(L) = \tilde{L} \subsetneq \tilde{M} ,$$

where $h \in \mathrm{Hom}_A(M,\tilde{M})$ is the isomorphism with resolution

$$\begin{array}{ccccccccc} 0 & \longrightarrow & K & \xrightarrow{\alpha} & K^* & \longrightarrow & M & \longrightarrow & 0 \\ & & \big\downarrow g & & \big\downarrow g^{*-1} & & \big\downarrow h & & \\ 0 & \longrightarrow & \widetilde{K} & \xrightarrow{\widetilde{\alpha}} & \widetilde{K}^* & \longrightarrow & \widetilde{M} & \longrightarrow & 0 \end{array} \quad .$$

Conversely, given an S-non-singular $\begin{cases} \varepsilon\text{-symmetric} \\ \text{even } \varepsilon\text{-symmetric} \\ \varepsilon\text{-quadratic} \end{cases}$

form over A $\begin{cases} (K,\alpha) \\ (K,\alpha) \\ (K,\beta) \end{cases}$ and a sublagrangian L of the boundary

$\begin{cases} \partial(K,\alpha) \\ \partial(K,\alpha) \\ \partial(K,\beta) \end{cases}$ define an S-isomorphism $\begin{cases} f:(K,\alpha) \longrightarrow (K',\alpha') \\ f:(K,\alpha) \longrightarrow (K',\alpha') \\ f:(K,\beta) \longrightarrow (K',\beta') \end{cases}$

as follows.

In the first instance, define an S-acyclic 1-dimensional $\begin{cases} \text{even } (-\varepsilon)\text{-symmetric} \\ (-\varepsilon)\text{-quadratic} \\ (-\varepsilon)\text{-quadratic} \end{cases}$ Poincaré complex over A $\begin{cases} (C,\phi \in Q\langle v_0\rangle^1(C,-\varepsilon)) \\ (C,\psi \in Q_1(C,-\varepsilon)) \\ (C,\psi \in Q_1(C,-\varepsilon)) \end{cases}$

with associated non-singular $\begin{cases} \text{even } \varepsilon\text{-symmetric} \\ \varepsilon\text{-quadratic} \\ \text{split } \varepsilon\text{-quadratic} \end{cases}$ linking form

over (A,S) $\begin{cases} \partial(K,\alpha) \\ \partial(K,\alpha) \\ \partial(K,\beta) \end{cases}$ by

$$d = \begin{cases} \alpha \\ \alpha \\ \beta + \varepsilon\beta^* \end{cases} : C_1 = K \longrightarrow C_0 = K^* \ , \ C_r = 0 \ (r \neq 0,1)$$

$$\begin{cases} \phi_0 = \begin{cases} 1 : C^0 = K \longrightarrow C_1 = K \\ -\varepsilon : C^1 = K^* \longrightarrow C_0 = K^* \end{cases}, & \phi_1 = 0 : C^1 = K^* \longrightarrow C_1 = K \\ \psi_1 = \begin{cases} 1 : C^0 = K \longrightarrow C_1 = K \\ 0 : C^1 = K^* \longrightarrow C_0 = K^* \end{cases}, & \psi_1 = -\widetilde{\beta} : C^0 = K \longrightarrow C_0 = K^* \end{cases}$$

for any $\widetilde{\beta} \in \mathrm{Hom}_A(K,K^*)$ such that $\begin{cases} \alpha = \widetilde{\beta} + \varepsilon\widetilde{\beta}^* \in \mathrm{Hom}_A(K,K^*) \\ \beta = \widetilde{\beta} \in Q_\varepsilon(K) \end{cases}$.

Let $e \in \mathrm{Hom}_A(K^*,M)$ be the natural projection

$$e : K^* \longrightarrow M = \begin{cases} \mathrm{coker}(\alpha : K \longrightarrow K^*) \\ \mathrm{coker}(\alpha : K \longrightarrow K^*) \\ \mathrm{coker}(\beta+\varepsilon\beta^* : K \longrightarrow K^*) \end{cases},$$

define a f.g. projective A-module

$$K' = e^{-1}(L) \subseteq K^*$$

and let $f \in \mathrm{Hom}_A(K,K')$, $g \in \mathrm{Hom}_A(K',K^*)$ be defined by

$$f = \begin{cases} \alpha| \\ \alpha| \\ \beta+\varepsilon\beta^*| \end{cases} : K \longrightarrow K'$$

$$g = \text{inclusion} : K' \longrightarrow K^* .$$

The A-module chain complex D and the A-module chain map

$$h : C \longrightarrow D$$

defined by

$$\begin{array}{ccccccccccc} C : & \dots \longrightarrow & 0 & \longrightarrow & K & \xrightarrow{\alpha} & K^* & \longrightarrow & 0 & \longrightarrow & \dots \\ h \big\downarrow & & & & g^* \big\downarrow & & \big\downarrow 1 & & & & \\ D : & \dots \longrightarrow & 0 & \longrightarrow & K'^* & \xrightarrow{f^*} & K^* & \longrightarrow & 0 & \longrightarrow & \dots \end{array}$$

(with $\alpha = \beta+\varepsilon\beta^* \in \mathrm{Hom}_A(K,K^*)$ in the ε-quadratic case) are such that

$$h_* = (\text{inclusion})\hat{} : H_0(C) = M\hat{} \longrightarrow H_0(D) = L\hat{} .$$

The morphism of $\begin{cases} \text{even } \varepsilon\text{-symmetric} \\ \varepsilon\text{-quadratic} \\ \text{split } \varepsilon\text{-quadratic} \end{cases}$ linking forms over (A,S)

defined by the inclusion

$$\begin{cases} (L,0) \longrightarrow \partial(K,\alpha) \\ (L,0,0) \longrightarrow \partial(K,\alpha) \\ (L,0,0) \longrightarrow \partial(K,\beta) \end{cases}$$

is associated by Proposition 3.4.1 to an $\begin{cases} \varepsilon\text{-symmetric} \\ \varepsilon\text{-quadratic} \\ \text{split } \varepsilon\text{-quadratic} \end{cases}$

map

$$\begin{cases} h : (C,\phi) \longrightarrow (D,0) \\ h : (C,\psi) \longrightarrow (D,0) \\ h : (C,\psi) \longrightarrow (D,0) \ , \end{cases}$$

so that

$$\begin{cases} h^{\%}(\phi) = 0 \in Q\langle v_0\rangle^1(D,-\varepsilon) \\ h_{\%}(\psi) = H(\theta) \in Q_1(D,-\varepsilon) \\ h_{\%}(\psi) = H(\theta) \in Q_1(D,-\varepsilon) \end{cases}$$

for some $\theta \in \hat{Q}^2(D,-\varepsilon)$ such that

$$\begin{cases} \hat{v}_1(\theta) = 0 : H^1(D) = L \longrightarrow \hat{H}^0(\mathbb{Z}_2;A,\varepsilon) \\ \hat{v}_1^S(\theta) = 0 : H^1(D) = L \longrightarrow \hat{H}^1(\mathbb{Z}_2;S^{-1}A/A,\varepsilon) \ . \end{cases}$$

Working exactly as in the proof of Proposition 3.4.5 it is possible to replace $\psi \in Q_1(C,-\varepsilon)$ by $\psi + H(\theta') \in Q_1(C,-\varepsilon)$ for some $\theta' \in \hat{Q}^2(C,-\varepsilon)$ such that $\begin{cases} \hat{v}_1(\theta') = 0 \\ \hat{v}_1^S(\theta') = 0 \end{cases}$, to ensure that

$$h_{\%}(\psi) = 0 \in Q_1(D,-\varepsilon) \ .$$

It follows from $\begin{cases} h^{\%}(\phi) = 0 \in Q\langle v_0\rangle^1(D,-\varepsilon) \\ h_{\%}(\psi) = 0 \in Q_1(D,-\varepsilon) \\ h_{\%}(\psi) = 0 \in Q_1(D,-\varepsilon) \end{cases}$ that there exists a

connected S-acyclic 2-dimensional $\begin{cases} \text{even } (-\varepsilon)\text{-symmetric} \\ (-\varepsilon)\text{-quadratic} \\ (-\varepsilon)\text{-quadratic} \end{cases}$ pair

$\begin{cases} (h:C \longrightarrow D, (\delta\phi,\phi) \in Q\langle v_0\rangle^2(h,-\varepsilon)) \\ (h:C \longrightarrow D, (\delta\psi,\psi) \in Q_2(h,-\varepsilon)) \\ (h:C \longrightarrow D, (\delta\psi,\psi) \in Q_2(h,-\varepsilon)) \end{cases}$. Define an S-non-singular

$\begin{cases} \varepsilon\text{-symmetric} \\ \text{even } \varepsilon\text{-symmetric form over } A \\ \varepsilon\text{-quadratic} \end{cases}$ $\begin{cases} (K', \alpha' \in Q^{\varepsilon}(K')) \\ (K', \alpha' \in Q\langle v_0\rangle^{\varepsilon}(K')) \\ (K', \beta' \in Q_{\varepsilon}(K')) \end{cases}$ by

$$\begin{cases} \alpha' = -\delta\phi_0 \\ \alpha' = -(\delta\psi_0 + \varepsilon\delta\psi_0^*) \\ \beta' = -\delta\psi_0 \end{cases} : D^1 = K' \longrightarrow D_1 = K'^* .$$

The S-isomorphism of S-non-singular forms

$$\begin{cases} f : (K,\alpha) \longrightarrow (K',\alpha') \\ f : (K,\alpha) \longrightarrow (K',\alpha') \\ f : (K,\beta) \longrightarrow (K',\beta') \end{cases}$$

determines the sublagrangian L of $\begin{cases} \partial(K,\alpha) \\ \partial(K,\alpha) \\ \partial(K,\beta) \end{cases}$, with

$L = \text{coker}(f:K \longrightarrow K')$ etc.

ii) We need a preliminary result.

<u>Lemma</u> An S-isomorphism of S-non-singular $\begin{cases} \varepsilon\text{-symmetric} \\ \text{even } \varepsilon\text{-symmetric} \\ \varepsilon\text{-quadratic} \end{cases}$

forms over A

$$\begin{cases} f : (K,\alpha) \longrightarrow (K',\alpha') \\ f : (K,\alpha) \longrightarrow (K',\alpha') \\ f : (K,\beta) \longrightarrow (K',\beta') \end{cases}$$

determines a lagrangian L of the boundary

$\begin{cases} \text{even } \varepsilon\text{-symmetric} \\ \varepsilon\text{-quadratic} \\ \text{split } \varepsilon\text{-quadratic} \end{cases}$ linking form over (A,S) $\begin{cases} \partial(K\oplus K',\alpha\oplus-\alpha') \\ \partial(K\oplus K',\alpha\oplus-\alpha') . \\ \partial(K\oplus K',\beta\oplus-\beta') \end{cases}$

<u>Proof</u>: The S-isomorphism of S-non-singular $\begin{cases} \text{(even) } \varepsilon\text{-symmetric} \\ \varepsilon\text{-quadratic} \end{cases}$

forms over A

$$\begin{cases} \begin{pmatrix} \alpha'f & 0 \\ \varepsilon f & 1 \end{pmatrix} : (K,\alpha)\oplus(K',-\alpha') \longrightarrow (K'^*\oplus K', \begin{pmatrix} 0 & 1 \\ \varepsilon & -\alpha' \end{pmatrix}) \\ \begin{pmatrix} (\beta'+\varepsilon\beta'^*)f & 0 \\ \varepsilon f & 1 \end{pmatrix} : (K,\beta)\oplus(K',-\beta') \longrightarrow (K'^*\oplus K', \begin{pmatrix} 0 & 1 \\ 0 & -\beta' \end{pmatrix}) \end{cases}$$

has non-singular range, so that it determines a lagrangian L of

$\begin{cases} \partial(K\oplus K',\alpha\oplus-\alpha') \\ \partial(K\oplus K',\beta\oplus-\beta') \end{cases}$ by i).

[]

Let now $\begin{cases} (K,\alpha) \\ (K,\beta) \end{cases}$ be an S-hyperbolic S-non-singular

$\begin{cases} \text{(even) } \varepsilon\text{-symmetric} \\ \varepsilon\text{-quadratic} \end{cases}$ form over A, and let $j \in \mathrm{Hom}_A(L,K)$ be the

inclusion of an S-lagrangian L. As $\begin{cases} j^*\alpha \in \mathrm{Hom}_A(K,L^*) \\ j^*(\beta+\varepsilon\beta^*) \in \mathrm{Hom}_A(K,L^*) \end{cases}$

becomes onto over $S^{-1}A$ there exists $k \in \mathrm{Hom}_A(L^*,K)$ such that

$$\begin{cases} s = j^*\alpha k \in \mathrm{Hom}_A(L^*,L^*) \\ s = j^*(\beta+\varepsilon\beta^*)k \in \mathrm{Hom}_A(L^*,L^*) \end{cases}$$

is an S-isomorphism. Applying the Lemma to the S-isomorphism of S-non-singular $\begin{cases}\text{(even) } \varepsilon\text{-symmetric} \\ \varepsilon\text{-quadratic}\end{cases}$ forms over A

$$\begin{cases} (j\ k) : (K',\alpha') = (L\oplus L^*, \begin{pmatrix} 0 & s \\ \varepsilon s^* & k^*\alpha k \end{pmatrix}) \longrightarrow (K,\alpha) \\ (j\ k) : (K',\beta') = (L\oplus L^*, \begin{pmatrix} 0 & 0 \\ 0 & k^*\beta k \end{pmatrix}) \longrightarrow (K,\beta) \end{cases}$$

we have that $\begin{cases} \partial(K,\alpha)\oplus\partial(K',-\alpha') \\ \partial(K,\beta)\oplus\partial(K',-\beta') \end{cases}$ is a hyperbolic linking form over (A,S). Furthermore, there is defined an S-isomorphism of S-non-singular forms

$$\begin{cases} \begin{pmatrix} s^* & 0 \\ 0 & 1 \end{pmatrix} : (K',\alpha') \longrightarrow (L\oplus L^*, \begin{pmatrix} 0 & 1 \\ \varepsilon & k^*\alpha k \end{pmatrix}) \\ \begin{pmatrix} s^* & 0 \\ 0 & 1 \end{pmatrix} : (K',\beta') \longrightarrow (L\oplus L^*, \begin{pmatrix} 0 & 1 \\ 0 & k^*\beta k \end{pmatrix}) \end{cases}$$

with non-singular range, so that $\begin{cases} \partial(K',\alpha') \\ \partial(K',\beta') \end{cases}$ is hyperbolic by i).

We have just shown that $\begin{cases} \partial(K,\alpha) \\ \partial(K,\beta) \end{cases}$ is a stably hyperbolic linking form.

It remains to prove the converse, that a stably hyperbolic $\begin{cases}\text{even } \varepsilon\text{-symmetric} \\ \varepsilon\text{-quadratic} \\ \text{split } \varepsilon\text{-quadratic}\end{cases}$ linking form over (A,S) is isomorphic to the boundary of an S-hyperbolic $\left\{\begin{array}{l}\varepsilon\text{-symmetric}\\ \text{even }\varepsilon\text{-symmetric}\\ \varepsilon\text{-quadratic}\end{array}\right.$ form over A.

Let $\left\{\begin{array}{l}(M,\lambda)\\ (M,\lambda,\nu)\end{array}\right.$ be a non-singular $\left\{\begin{array}{l}\text{even }\varepsilon\text{-symmetric}\\ \text{split }\varepsilon\text{-quadratic}\end{array}\right.$ linking form over (A,S). By Proposition 3.4.1 there exists an S-acyclic 1-dimensional $\left\{\begin{array}{l}\text{even }(-\varepsilon)\text{-symmetric}\\ (-\varepsilon)\text{-quadratic}\end{array}\right.$ Poincaré complex over A $\left\{\begin{array}{l}(C,\phi\in Q\langle v_0\rangle^1(C,-\varepsilon))\\ (C,\psi\in Q_1(C,-\varepsilon))\end{array}\right.$ such that

$$\left\{\begin{array}{l}(H^1(C),\phi_0^S) = (M,\lambda)\\ (H^1(C),(1+T_{-\varepsilon})\psi_0^S,v_S^0(\psi)) = (M,\lambda,\nu)\ .\end{array}\right.$$

The S-acyclic cobordism class $\left\{\begin{array}{l}(C,\phi)\in L^0(A,S,\varepsilon)\\ (C,\psi)\in L_0(A,S,\varepsilon)\end{array}\right.$ depends only on the isomorphism class of $\left\{\begin{array}{l}(M,\lambda)\\ (M,\lambda,\nu)\end{array}\right.$ (Proposition 3.4.4 i)), vanishing if $\left\{\begin{array}{l}(M,\lambda)\\ (M,\lambda,\nu)\end{array}\right.$ is hyperbolic (Proposition 3.4.5 ii)). It follows that if $\left\{\begin{array}{l}(M,\lambda)\\ (M,\lambda,\nu)\end{array}\right.$ is stably hyperbolic then

$$\left\{\begin{array}{l}(C,\phi) = 0\in L^0(A,S,\varepsilon)\\ (C,\psi) = 0\in L_0(A,S,\varepsilon)\end{array}\right.$$

and hence (by Proposition 3.4.4 ii)) that $\left\{\begin{array}{l}(M,\lambda)\\ (M,\lambda,\nu)\end{array}\right.$ is isomorphic to the boundary $\left\{\begin{array}{l}\partial(K,\alpha)\\ \partial(K,\beta)\end{array}\right.$ of an S-hyperbolic S-non-singular $\left\{\begin{array}{l}\varepsilon\text{-symmetric}\\ \varepsilon\text{-quadratic}\end{array}\right.$ form over A $\left\{\begin{array}{l}(K,\alpha)\\ (K,\beta)\end{array}\right.$. It may be verified that if (M,λ) is the ε-symmetrization of a stably hyperbolic ε-quadratic linking form over (A,S) (M,λ,μ) then the S-hyperbolic S-non-singular ε-symmetric form (K,α) arising is even, and that (M,λ,μ) is isomorphic to the boundary $\partial(K,\alpha)$.

[]

Define the Witt group of $\left\{\begin{array}{l}\underline{\varepsilon\text{-symmetric}}\\ \underline{\text{even }\varepsilon\text{-symmetric}}\\ \underline{\varepsilon\text{-quadratic}}\\ \underline{\text{split }\varepsilon\text{-quadratic}}\end{array}\right.$ linking

forms over (A,S) $\left\{\begin{array}{l}L^{\varepsilon}(A,S)\\ L\langle v_0\rangle^{\varepsilon}(A,S)\\ L_{\varepsilon}(A,S)\\ \widetilde{L}_{\varepsilon}(A,S)\end{array}\right.$ to be the abelian group of

stable isomorphism classes of non-singular $\left\{\begin{array}{l}\varepsilon\text{-symmetric}\\ \text{even }\varepsilon\text{-symmetric}\\ \varepsilon\text{-quadratic}\\ \text{split }\varepsilon\text{-quadratic}\end{array}\right.$

linking forms over (A,S), the stability being with respect to the hyperbolic linking forms (i.e. a stable isomorphism of linking forms X,X' is an isomorphism $X\oplus Y \longrightarrow X'\oplus Y'$ for some hyperbolic linking forms Y,Y'). Addition is by the direct sum $\oplus$, and inverses are given by

$$\left\{\begin{array}{l}-(M,\lambda) = (M,-\lambda) \in L^{\varepsilon}(A,S)\\ -(M,\lambda) = (M,-\lambda) \in L\langle v_0\rangle^{\varepsilon}(A,S)\\ -(M,\lambda,\mu) = (M,-\lambda,-\mu) \in L_{\varepsilon}(A,S)\\ -(M,\lambda,\nu) = (M,-\lambda,-\nu) \in \widetilde{L}_{\varepsilon}(A,S)\ ,\end{array}\right.$$

since the diagonal $\Delta = \{(x,x) \in M\oplus M \mid x \in M\} \subsetneq M\oplus M$ is a lagrangian of $X\oplus -X$ for any non-singular linking form X, by Proposition 3.4.5 i).

There are evident forgetful maps

$$L\langle v_0\rangle^{\varepsilon}(A,S) \longrightarrow L^{\varepsilon}(A,S) \ ; \ (M,\lambda) \longmapsto (M,\lambda)$$

$$L_{\varepsilon}(A,S) \longrightarrow L\langle v_0\rangle^{\varepsilon}(A,S) \ ; \ (M,\lambda,\mu) \longmapsto (M,\lambda)$$

$$\widetilde{L}_{\varepsilon}(A,S) \longrightarrow L_{\varepsilon}(A,S) \ ; \ (M,\lambda,\nu) \longmapsto (M,\lambda,p\nu) \ .$$

A non-singular $\begin{cases} \text{(even) } \varepsilon\text{-symmetric} \\ \varepsilon\text{-quadratic} \end{cases}$ form over $S^{-1}A$ $\begin{cases} (Q,\phi) \\ (Q,\psi) \end{cases}$

with projective class

$$[Q] \in S = \mathrm{im}(\widetilde{K}_0(A) \longrightarrow \widetilde{K}_0(S^{-1}A)) \subseteq \widetilde{K}_0(S^{-1}A)$$

is stably isomorphic to $\begin{cases} S^{-1}(K,\alpha) \\ S^{-1}(K,\beta) \end{cases}$ for some S-non-singular

$\begin{cases} \text{(even) } \varepsilon\text{-symmetric} \\ \varepsilon\text{-quadratic} \end{cases}$ form over A $\begin{cases} (K,\alpha) \\ (K,\beta) \end{cases}$. It follows from Proposition 3.4.6 ii) that the boundary operations

$$\partial : (\text{S-non-singular forms over A}) \longrightarrow (\text{linking forms over } (A,S))$$

give rise to well-defined abelian group morphisms

$$\partial : L_S^{\varepsilon}(S^{-1}A) \longrightarrow L\langle v_0\rangle^{\varepsilon}(A,S) \ ; \ S^{-1}(K,\alpha) \longmapsto \partial(K,\alpha)$$

$$\partial : L\langle v_0\rangle_S^{\varepsilon}(S^{-1}A) \longrightarrow L_{\varepsilon}(A,S) \ ; \ S^{-1}(K,\alpha) \longmapsto \partial(K,\alpha)$$

$$\partial : L_{\varepsilon}^S(S^{-1}A) \longrightarrow \widetilde{L}_{\varepsilon}(A,S) \ ; \ S^{-1}(K,\beta) \longmapsto \partial(K,\beta) \ .$$

There is also defined a morphism

$$\partial : L_S^{\varepsilon}(S^{-1}A) \longrightarrow L^{\varepsilon}(A,S) \ ; \ S^{-1}(K,\alpha) \longmapsto \partial(K,\alpha) \ ,$$

namely the composite

$$L_S^{\varepsilon}(S^{-1}A) \xrightarrow{\ \partial\ } L\langle v_0\rangle^{\varepsilon}(A,S) \longrightarrow L^{\varepsilon}(A,S) \ .$$

The correspondence of Proposition 3.4.3 associates to a non-singular $\begin{cases}\text{(even) }\varepsilon\text{-symmetric}\\ \varepsilon\text{-quadratic}\end{cases}$ linking form over (A,S) $\begin{cases}(M,\lambda)\\ (M,\lambda,\mu)\end{cases}$ a stable isomorphism class of non-singular $\begin{cases}\text{(even) }(-\varepsilon)\text{-symmetric}\\ (-\varepsilon)\text{-quadratic}\end{cases}$ formations over A $\begin{cases}(Q,\phi;F,G)\\ (Q,\psi;F,G)\end{cases}$ (i.e. the associated S-formations, regarded as formations), and it follows from Proposition 3.4.6 ii) that there are well-defined abelian group morphisms

$$L^{\varepsilon}(A,S) \longrightarrow M^{-\varepsilon}(A) \; ; \; (M,\lambda) \longmapsto (Q,\phi;F,G)$$

$$L\langle v_0\rangle^{\varepsilon}(A,S) \longrightarrow M\langle v_0\rangle^{-\varepsilon}(A) \; ; \; (M,\lambda) \longmapsto (Q,\phi;F,G)$$

$$L_{\varepsilon}(A,S) \longrightarrow M_{-\varepsilon}(A) \; ; \; (M,\lambda,\mu) \longmapsto (Q,\psi;F,G)$$

from the Witt groups of linking forms over (A,S) to the Witt groups of formations over A defined in §1.6 above. There is also defined an abelian group morphism

$$\tilde{L}_{\varepsilon}(A,S) \longrightarrow M_{-\varepsilon}(A) \; ; \; (M,\lambda,\nu) \longmapsto (Q,\psi;F,G) \; ,$$

namely the composite

$$\tilde{L}_{\varepsilon}(A,S) \longrightarrow L_{\varepsilon}(A,S) \longrightarrow M_{-\varepsilon}(A) \; .$$

Define the <u>lower even-dimensional</u> $\begin{cases}\underline{\varepsilon\text{-symmetric}}\\ \underline{\varepsilon\text{-quadratic}}\end{cases}$ <u>L-groups of (A,S)</u> $\begin{cases}L^{2k}(A,S,\varepsilon)\\ L_{2k}(A,S,\varepsilon)\end{cases}$ $(k \leqslant -1)$ by

$$\begin{cases} L^{2k}(A,S,\varepsilon) = \begin{cases} L_{-\varepsilon}(A,S) & (k=-1) \\ L_{2k}(A,S,\varepsilon) & (k \leqslant -2) \end{cases} \\ L_{2k}(A,S,\varepsilon) = L_{2k+2i}(A,S,(-)^{i}\varepsilon) \quad (k \leqslant -1, k+i \geqslant 0) \; . \end{cases}$$

<u>Proposition 3.4.7</u> i) The localization exact sequence of algebraic Poincaré cobordism groups

$$L^{2k}(A,(-)^k\varepsilon) \longrightarrow L^{2k}_S(S^{-1}A,(-)^k\varepsilon) \longrightarrow L^{2k}(A,S,(-)^k\varepsilon)$$
$$\longrightarrow L^{2k-1}(A,(-)^k\varepsilon) \longrightarrow L^{2k-1}_S(S^{-1}A,(-)^k\varepsilon) \qquad (*)_{2k}$$

is naturally isomorphic for $\begin{cases} k = 0 \\ k = -1 \\ k \leqslant -2 \end{cases}$ to a localization exact sequence of Witt groups

$$\begin{cases} L^{\varepsilon}(A) \longrightarrow L^{\varepsilon}_S(S^{-1}A) \xrightarrow{\partial} L\langle v_0\rangle^{\varepsilon}(A,S) \longrightarrow M\langle v_0\rangle^{-\varepsilon}(A) \longrightarrow M\langle v_0\rangle^{-\varepsilon}_S(S^{-1}A) \\ L\langle v_0\rangle^{\varepsilon}(A) \longrightarrow L\langle v_0\rangle^{\varepsilon}_S(S^{-1}A) \xrightarrow{\partial} L_{\varepsilon}(A,S) \longrightarrow M_{-\varepsilon}(A) \longrightarrow M^S_{-\varepsilon}(S^{-1}A) \\ L_{\varepsilon}(A) \longrightarrow L^S_{\varepsilon}(S^{-1}A) \xrightarrow{\partial} \widetilde{L}_{\varepsilon}(A,S) \longrightarrow M_{-\varepsilon}(A) \longrightarrow M^S_{-\varepsilon}(S^{-1}A) \end{cases} .$$

ii) There are defined natural abelian group morphisms

$$L^{\varepsilon}(A,S) \longrightarrow L^{2k}(A,S,(-)^k\varepsilon) \quad (k \geqslant 1)$$

for all A,S,ε. If $\begin{cases} (A,S) \text{ is 0-dimensional} \\ \ker(\hat{\delta}:\hat{H}^1(\mathbb{Z}_2;S^{-1}A/A,\varepsilon) \longrightarrow \hat{H}^0(\mathbb{Z}_2;A,\varepsilon)) = 0 \end{cases}$

then for $\begin{cases} k \geqslant 1 \\ k = 1 \end{cases}$ these are isomorphisms, and $(*)_{2k}$ is naturally isomorphic to a localization exact sequence of Witt groups

$$\begin{cases} L^{\varepsilon}(A) \longrightarrow L^{\varepsilon}_S(S^{-1}A) \xrightarrow{\partial} L^{\varepsilon}(A,S) \longrightarrow M^{-\varepsilon}(A) \longrightarrow M^{-\varepsilon}_S(S^{-1}A) \\ L^2(A,-\varepsilon) \longrightarrow L^2_S(S^{-1}A,-\varepsilon) \xrightarrow{\partial} L^{\varepsilon}(A,S) \longrightarrow M^{-\varepsilon}(A) \longrightarrow M^{-\varepsilon}_S(S^{-1}A) \end{cases} .$$

iii) For all A,S,ε the forgetful map of Witt groups

$$\widetilde{L}_{\varepsilon}(A,S) \longrightarrow L_{\varepsilon}(A,S) \; ; \; (M,\lambda,\nu) \longmapsto (M,\lambda,p\nu)$$

is onto, and there are natural identifications

$$\mathrm{coker}(\partial:L^S_\varepsilon(S^{-1}A)\longrightarrow \tilde{L}_\varepsilon(A,S)) = \mathrm{coker}(\partial:L\langle v_0\rangle^\varepsilon_S(S^{-1}A)\longrightarrow L_\varepsilon(A,S))$$
$$= \ker(M_{-\varepsilon}(A)\longrightarrow M^S_{-\varepsilon}(S^{-1}A)) \ .$$

If (A,S) is 0-dimensional

$$\ker(L^\varepsilon(A,S)\longrightarrow M^{-\varepsilon}(A)) = \ker(L\langle v_0\rangle^\varepsilon(A,S)\longrightarrow M\langle v_0\rangle^{-\varepsilon}(A))$$
$$= \mathrm{coker}(L^\varepsilon(A)\longrightarrow L^\varepsilon_S(S^{-1}A)) \ .$$

If $\begin{cases} \mathrm{im}(\hat{\delta}:\hat{H}^0(\mathbb{Z}_2;S^{-1}A/A,\varepsilon)\longrightarrow \hat{H}^1(\mathbb{Z}_2;A,\varepsilon)) = 0 \\ \hat{H}^0(\mathbb{Z}_2;A,\varepsilon)\longrightarrow \hat{H}^0(\mathbb{Z}_2;S^{-1}A,\varepsilon) \text{ is an isomorphism} \\ \mathrm{im}(\hat{H}^1(\mathbb{Z}_2;S^{-1}A,\varepsilon)\longrightarrow \hat{H}^1(\mathbb{Z}_2;S^{-1}A/A,\varepsilon)) = 0 \end{cases}$

the forgetful maps identify

$$\begin{cases} L\langle v_0\rangle^\varepsilon(A,S) = L^\varepsilon(A,S) \\ L_\varepsilon(A,S) = L\langle v_0\rangle^\varepsilon(A,S) \\ \tilde{L}_\varepsilon(A,S) = L_\varepsilon(A,S) \ . \end{cases}$$

In particular, if $1/2\in S^{-1}A$

$$\tilde{L}_\varepsilon(A,S) = L_\varepsilon(A,S) \ ,$$

and if $1/2\in A$ then

$$\tilde{L}_\varepsilon(A,S) = L_\varepsilon(A,S) = L\langle v_0\rangle^\varepsilon(A,S) = L^\varepsilon(A,S) \ .$$

<u>Proof</u>: i) An S-acyclic 1-dimensional $\begin{cases}\text{even } (-\varepsilon)\text{-symmetric} \\ (-\varepsilon)\text{-quadratic}\end{cases}$

Poincaré complex over A $\begin{cases}(C,\phi\in Q\langle v_0\rangle^1(C,-\varepsilon)) \\ (C,\psi\in Q_1(C,-\varepsilon))\end{cases}$ represents 0 in

$\begin{cases}L^0(A,S,\varepsilon) \\ L_0(A,S,\varepsilon)\end{cases}$ if and only if the associated non-singular

$\begin{cases}\text{even } \varepsilon\text{-symmetric} \\ \text{split } \varepsilon\text{-quadratic}\end{cases}$ linking form over (A,S)

$$\begin{cases}(H^1(C),\phi_0^S) \\ (H^1(C),(1+T_{-\varepsilon})\psi_0^S,v_S^0(\psi))\end{cases} \text{ represents } 0 \text{ in } \begin{cases}L\langle v_0\rangle^{\varepsilon}(A,S) \\ \widetilde{L}_{\varepsilon}(A,S)\end{cases},$$

by Propositions 3.4.4 ii), 3.4.6 ii). It follows that the correspondence of Proposition 3.4.1 gives rise to isomorphisms of abelian groups

$$\begin{cases}L^0(A,S,\varepsilon) \longrightarrow L\langle v_0\rangle^{\varepsilon}(A,S) \ ; \ (C,\phi) \longmapsto (H^1(C),\phi_0^S) \\ L_0(A,S,\varepsilon) \longrightarrow \widetilde{L}_{\varepsilon}(A,S) \ ; \ (C,\psi) \longmapsto (H^1(C),(1+T_{-\varepsilon})\psi_0^S,v_S^0(\psi)) \ .\end{cases}$$

The exactness of the Witt group sequences

$$\begin{cases}L^{\varepsilon}(A) \longrightarrow L_S^{\varepsilon}(S^{-1}A) \xrightarrow{\partial} L\langle v_0\rangle^{\varepsilon}(A,S) \longrightarrow M\langle v_0\rangle^{-\varepsilon}(A) \longrightarrow M\langle v_0\rangle_S^{-\varepsilon}(S^{-1}A) \\ L_{\varepsilon}(A) \longrightarrow L_{\varepsilon}^S(S^{-1}A) \xrightarrow{\partial} \widetilde{L}_{\varepsilon}(A,S) \longrightarrow M_{-\varepsilon}(A) \longrightarrow M_{-\varepsilon}^S(S^{-1}A)\end{cases}$$

can now be deduced from the exactness of $(*)_0$ and $(*)_{-4}$ (which is given by Proposition 3.2.3 i)), or else may be established directly using Proposition 3.4.6 ii). The direct method also applies to the exactness of $(*)_{-2}$

$$L\langle v_0\rangle^{\varepsilon}(A) \longrightarrow L\langle v_0\rangle_S^{\varepsilon}(S^{-1}A) \xrightarrow{\partial} L_{\varepsilon}(A,S) \longrightarrow M_{-\varepsilon}(A) \longrightarrow M_{-\varepsilon}^S(S^{-1}A) \ .$$

ii) Define abelian group morphisms

$$L^{\varepsilon}(A,S) \longrightarrow L^{2k}(A,S,(-)^k\varepsilon) \ ; \ (M,\lambda) \longmapsto \bar{S}^k(C,\phi) \quad (k \geqslant 1)$$

by sending a non-singular ε-symmetric linking form over (A,S) (M,λ) to the k-fold skew-suspension

$$\bar{S}^k(C,\phi) = (S^kC,\bar{S}^k\phi \in Q\langle v_0\rangle^{2k+1}(S^kC,(-)^{k+1}\varepsilon))$$

of an S-acyclic 1-dimensional $(-\varepsilon)$-symmetric Poincaré complex over A $(C,\phi \in Q^1(C,-\varepsilon))$ such that

$$(H^1(C),\phi_0^S) = (M,\lambda) ,$$

as given by Proposition 3.4.1. The S-acyclic cobordism class

$\bar{S}^k(C,\phi) \in L^{2k}(A,S,(-)^k\varepsilon)$ depends only on the isomorphism class of (M,λ) (which may be proved exactly as was done in Proposition 3.4.4 i) in the even ε-symmetric case), and vanishes if (M,λ) is stably hyperbolic (Proposition 3.4.5 ii)), so that the morphisms are well-defined. If

$$\begin{cases} (A,S) \text{ is 0-dimensional} \\ \ker(\hat{\delta}:\hat{H}^1(\mathbb{Z}_2;S^{-1}A/A,\varepsilon) \longrightarrow \hat{H}^0(\mathbb{Z}_2;A,\varepsilon)) = 0 \end{cases}$$

then by Proposition $\begin{cases} 3.2.4 \\ 3.3.2 \text{ i)} \end{cases}$ there are natural identifications for $\begin{cases} k \geqslant 1 \\ k = 1 \end{cases}$

$L^{2k}(A,S,(-)^k\varepsilon)$ = the cobordism group of S-acyclic 1-dimensional $(-\varepsilon)$-symmetric Poincaré complexes over A ,

so that the morphisms are onto. Moreover, if $(M,\lambda) \in \ker(L^{\varepsilon}(A,S) \longrightarrow L^2(A,S,-\varepsilon))$ then (C,ϕ) is homotopy equivalent to the boundary $\partial(D,\eta)$ of a connected S-acyclic 2-dimensional $(-\varepsilon)$-symmetric complex over A $(D,\eta \in Q^2(D,-\varepsilon))$, and the proof of Proposition 3.4.6 ii) generalizes to show that (M,λ) is stably hyperbolic, so that the morphisms are also one-one, and hence isomorphisms.

iii) Immediate from i), ii) and Proposition 3.4.2 ii).

[]

3.5 Linking formations

A "non-singular linking formation over (A,S)" is a linking form over (A,S) together with an ordered pair of lagrangians. In Proposition 3.5.2 below we shall show that the homotopy equivalence classes of S-acyclic 2-dimensional algebraic Poincaré complexes over A are in one-one correspondence with the "stable equivalence" classes of non-singular linking formations over (A,S), and in Proposition 3.5.5 the cobordism groups of such complexes will be identified with Witt groups of linking formations. There is an evident analogy between the theory of forms and formations set out in §1.6 and the theory of linking forms and linking formations.

An $\begin{cases} \underline{\text{(even) } \varepsilon\text{-symmetric}} \\ \underline{\varepsilon\text{-quadratic}} \end{cases}$ linking formation over (A,S)

$\begin{cases} (M,\lambda;F,G) \\ (M,\lambda,\mu;F,G) \end{cases}$ is a non-singular $\begin{cases} \text{(even) } \varepsilon\text{-symmetric} \\ \varepsilon\text{-quadratic} \end{cases}$ linking form over (A,S) $\begin{cases} (M,\lambda) \\ (M,\lambda,\mu) \end{cases}$ together with a lagrangian F and a sublagrangian G. The linking formation is non-singular if G is a lagrangian.

An isomorphism of $\begin{cases} \text{(even) } \varepsilon\text{-symmetric} \\ \varepsilon\text{-quadratic} \end{cases}$ linking formations over (A,S)

$$\begin{cases} f : (M,\lambda;F,G) \longrightarrow (M',\lambda';F',G') \\ f : (M,\lambda,\mu;F,G) \longrightarrow (M',\lambda',\mu';F'.G') \end{cases}$$

is an isomorphism of the $\begin{cases} \text{(even) } \varepsilon\text{-symmetric} \\ \varepsilon\text{-quadratic} \end{cases}$ linking forms over (A,S)

$$\begin{cases} f : (M,\lambda) \longrightarrow (M',\lambda') \\ f : (M,\lambda,\mu) \longrightarrow (M',\lambda',\mu') \end{cases}$$

such that

$$f(F) = F' \ , \ f(G) = G' \ .$$

A <u>sublagrangian</u> H of an $\begin{cases}\text{(even) } \varepsilon\text{-symmetric} \\ \varepsilon\text{-quadratic}\end{cases}$ linking formation over (A,S) $\begin{cases}(M,\lambda;F,G) \\ (M,\lambda,\mu;F,G)\end{cases}$ is a sublagrangian H of $\begin{cases}(M,\lambda) \\ (M,\lambda,\mu)\end{cases}$ such that

i) $H \subseteq G$, with G/H an (A,S)-module

ii) $F \cap H = \{0\}$, $M = F + H^{\perp}$.

An <u>elementary equivalence</u> of $\begin{cases}\text{(even) } \varepsilon\text{-symmetric} \\ \varepsilon\text{-quadratic}\end{cases}$ linking formations over (A,S) is the transformation

$$\begin{cases} (M,\lambda;F,G) \longmapsto (M',\lambda';F',G') \\ (M,\lambda,\mu;F,G) \longmapsto (M',\lambda',\mu';F',G') \end{cases}$$

determined by a sublagrangian H of $\begin{cases}(M,\lambda;F,G) \\ (M,\lambda,\mu;F,G)\end{cases}$, with

$$\begin{cases} (M',\lambda';F',G') = (H^{\perp}/H,\lambda^{\perp}/\lambda;F\cap H^{\perp},G/H) \\ (M',\lambda',\mu';F',G') = (H^{\perp}/H,\lambda^{\perp}/\lambda,\mu^{\perp}/\mu;F\cap H^{\perp},G/H) \end{cases}$$

(where $F\cap H^{\perp}$ stands for the image of the natural injection $F\cap H^{\perp} \longrightarrow H^{\perp}/H; x \longmapsto [x]$). Note that there are natural identifications of S-torsion A-modules

$$F'\cap G' = F\cap G \quad , \quad M'/(F'+G') = M/(F+G) \quad , \quad G'^{\perp}/G' = G^{\perp}/G$$

- in general, only $G^{\perp}/G$ is an (A,S)-module.

Elementary equivalences and isomorphisms generate an equivalence relation on the set of $\begin{cases}\text{(even) } \varepsilon\text{-symmetric} \\ \varepsilon\text{-quadratic}\end{cases}$

linking formations over (A,S), called <u>stable equivalence</u>. Note that $\begin{cases}(M,\lambda;F,G)\\(M,\lambda,\mu;F,G)\end{cases}$ is stably equivalent to 0 if and only if

$$M = F\oplus G \ .$$

In Proposition 3.5.2 ii) below the stable equivalence classes of (even) ε-symmetric linking formations over (A,S) will be shown to be in one-one correspondence with the homotopy equivalence classes of connected S-acyclic 2-dimensional (even) $(-\varepsilon)$-symmetric complexes over A, with non-singular linking formations corresponding to Poincaré complexes.

Given an (A,S)-module L define the <u>standard hyperbolic</u> $\begin{cases}\text{\underline{even } } \varepsilon\text{\underline{-symmetric}}\\ \varepsilon\text{\underline{-quadratic}}\\ \text{\underline{split }}\varepsilon\text{\underline{-quadratic}}\end{cases}$ <u>linking form over (A,S)</u>

$$\begin{cases} H^{\varepsilon}(L) = (L\oplus L^{\wedge},\lambda : L\oplus L^{\wedge}\times L\oplus L^{\wedge}\longrightarrow S^{-1}A/A; \\ \qquad\qquad ((x,y),(x',y'))\longmapsto y(x')+\varepsilon\overline{y'(x)}) \\ H_{\varepsilon}(L) = (L\oplus L^{\wedge},\lambda,\mu : L\oplus L^{\wedge}\longrightarrow Q_{\varepsilon}(A,S); \\ \qquad\qquad (x,y)\longmapsto y(x)+\varepsilon\overline{y(x)}) \\ \widetilde{H}_{\varepsilon}(L) = (L\oplus L^{\wedge},\lambda,\nu : L\oplus L^{\wedge}\longrightarrow Q_{\varepsilon}(S^{-1}A/A);(x,y)\longmapsto y(x)) \ , \end{cases}$$

for which both L and $L^{\wedge}$ are lagrangians.

A <u>split ε-quadratic linking formation over (A,S)</u>

$$(F,G) = (F,(\begin{pmatrix}\gamma\\ \mu\end{pmatrix},\theta)G)$$

is an ε-quadratic linking formation over (A,S) of the type $(H_{\varepsilon}(F);F,G)$, with $\begin{pmatrix}\gamma\\ \mu\end{pmatrix}:G\longrightarrow F\oplus F^{\wedge}$ the inclusion, together with a function

$$\theta : G\longrightarrow Q_{-\varepsilon}(A,S)$$

such that $(G,\gamma^{\wedge}\mu\in \mathrm{Hom}_A(G,G^{\wedge}),\theta)$ is a $(-\varepsilon)$-quadratic linking form over (A,S), the <u>hessian</u> of (F,G). (Such objects were first considered by Pardon [2]). Note that the existence of the hessian θ ensures that G is a sublagrangian of the hyperbolic split ε-quadratic linking form $\widetilde{H}_{\varepsilon}(F)$. The linking formation (F,G) is <u>non-singular</u> if G is a lagrangian, that is if the sequence

$$0 \longrightarrow G \xrightarrow{\begin{pmatrix}\gamma\\ \mu\end{pmatrix}} F\oplus F^{\wedge} \xrightarrow{(\varepsilon\mu^{\wedge}\ \gamma^{\wedge})} G^{\wedge} \longrightarrow 0$$

is exact.

An <u>isomorphism</u> of split ε-quadratic linking formations over (A,S)

$$(\alpha,\beta,\phi,\psi)\ :\ (F,G) \longrightarrow (F',G')$$

is a quadruple

$$(\alpha \in \mathrm{Hom}_A(F,F'), \beta \in \mathrm{Hom}_A(G,G'), \phi \in \mathrm{Hom}_A(F^{\wedge},F), \psi : F^{\wedge} \longrightarrow Q_{-\varepsilon}(A,S))$$

with α,β isomorphisms and $(F^{\wedge},\phi,\psi)$ a $(-\varepsilon)$-quadratic linking form over (A,S), such that

$$\text{i)}\ \ \alpha^{\wedge-1}\mu = \mu'\beta \in \mathrm{Hom}_A(G,F'^{\wedge})$$

$$\text{ii)}\ \ \alpha\gamma + \alpha\phi^{\wedge}\mu = \gamma'\beta \in \mathrm{Hom}_A(G.F')$$

$$\text{iii)}\ \ \theta + \psi\mu = \theta'\beta\ :\ G \longrightarrow Q_{-\varepsilon}(A,S)\quad .$$

The isomorphism of (A,S)-modules

$$f = \begin{pmatrix}\alpha & \alpha\phi^{\wedge}\\ 0 & \alpha^{\wedge-1}\end{pmatrix}\ :\ F\oplus F^{\wedge} \longrightarrow F^{\wedge}\oplus F$$

defines an isomorphism of the underlying ε-quadratic linking formations over (A,S)

$$f\ :\ (H_{\varepsilon}(F);F,G) \longrightarrow (H_{\varepsilon}(F');F',G')\ .$$

Conversely, every such isomorphism arises from a triple (α,β,ϕ) satisfying i) and ii).

A <u>sublagrangian</u> H of a split ε-quadratic linking formation over (A,S) (F,G) is a sublagrangian H of the underlying ε-quadratic linking formation $(H_{\varepsilon}(F);F,G)$ such that

i) $\theta j = 0 : H \longrightarrow Q_{-\varepsilon}(A,S)$, where $j \in \mathrm{Hom}_A(H,G)$ is the inclusion,

ii) $\gamma j = 0 \in \mathrm{Hom}_A(H,F)$, i.e. $H \subseteq F^{\wedge} \subseteq F \oplus F^{\wedge}$.

An <u>elementary equivalence</u> of split ε-quadratic linking formations over (A,S) is the transformation

$$(F,G) \longmapsto (F',G')$$

determined by a sublagrangian H of (F,G), with

$$F' = F \cap H^{\perp} = \ker(j^{\wedge}\mu^{\wedge}:F \longrightarrow H^{\wedge})$$

$$G' = G/H = \mathrm{coker}(j:H \longrightarrow G)$$

$$\gamma' : G' \longrightarrow F' \; ; \; [x] \longmapsto \gamma(x)$$

$$\mu' : G' \longrightarrow F'^{\wedge} \; ; \; [x] \longmapsto (y \longmapsto \mu(x)(y))$$

$$\theta' : G' \longrightarrow Q_{-\varepsilon}(A,S) \; ; \; [x] \longmapsto \theta(x) \qquad (x \in G, y \in F').$$

(The ε-quadratic linking formation $(H_{\varepsilon}(F');F',G')$ underlying (F',G') is then obtained from $(H_{\varepsilon}(F);F,G)$ by an elementary equivalence of ε-quadratic linking formations).

Elementary equivalences and isomorphisms generate an equivalence relation on the set of split ε-quadratic linking formations over (A,S), called <u>stable equivalence</u>. Note that (F,G) is stably equivalent to 0 if and only if $\mu \in \mathrm{Hom}_A(G,F^{\wedge})$ is an isomorphism. In Proposition 3.5.2 iii) below the stable equivalence classes of split ε-quadratic linking formations over (A,S) will be shown to be in one-one correspondence with the appropriate equivalence classes of connected S-acyclic 2-dimensional $(-\varepsilon)$-quadratic complexes over A, with non-singular linking formations corresponding to Poincaré complexes.

Prior to such an identification we need some preliminary results on the homotopy classification of 2-dimensional complexes.

A 2-dimensional A-module chain complexes C is in <u>normal form</u> if $C_r = 0$ $(r \neq 0,1,2)$ and each C_r $(r = 0,1,2)$ is a f.g. projective A-module,

$$C : \ldots \longrightarrow 0 \longrightarrow C_2 \xrightarrow{d} C_1 \xrightarrow{d} C_0 \longrightarrow 0 \longrightarrow \ldots .$$

A connected 2-dimensional $\left\{\begin{array}{l}\varepsilon\text{-symmetric}\\ \varepsilon\text{-quadratic}\end{array}\right.$ complex over A $\left\{\begin{array}{l}(C,\phi)\\ (C,\psi)\end{array}\right.$ is in <u>normal form</u> if C is in normal form and $\left\{\begin{array}{l}\phi \in Q^2(C,\varepsilon)\\ \psi \in Q_2(C,\varepsilon)\end{array}\right.$ has a chain representative

$$\left\{\begin{array}{l}\phi \in \mathrm{Hom}_{\mathbb{Z}[\mathbb{Z}_2]}(W,\mathrm{Hom}_A(C^*,C))_2\\ \psi \in W\otimes_{\mathbb{Z}[\mathbb{Z}_2]}\mathrm{Hom}_A(C^*,C)_2\end{array}\right. \text{ such that}$$

i) $\left\{\begin{array}{l}\phi_0 \in \mathrm{Hom}_A(C^0,C_2)\\ \psi_0 \in \mathrm{Hom}_A(C^0,C_2)\end{array}\right.$ is an isomorphism

ii) $\left\{\begin{array}{l}\phi_1 = 0 \in \mathrm{Hom}_A(C^2,C_1)\\ \psi_1 = 0 \in \mathrm{Hom}_A(C^1,C_0),\ \psi_0 = 0 \in \mathrm{Hom}_A(C^2,C_0)\ .\end{array}\right.$

An $\left\{\begin{array}{l}\varepsilon\text{-symmetric}\\ \varepsilon\text{-quadratic}\end{array}\right.$ complex $\left\{\begin{array}{l}(C,\phi)\\ (C,\psi)\end{array}\right.$ in normal form is Poincaré if and only if $\left\{\begin{array}{l}\phi_0 \in \mathrm{Hom}_A(C^1,C_1)\\ (1+T_\varepsilon)\psi_0 \in \mathrm{Hom}_A(C^1,C_1)\end{array}\right.$ is an isomorphism.

A <u>stable isomorphism</u> of connected 2-dimensional $\begin{cases}\varepsilon\text{-symmetric}\\ \varepsilon\text{-quadratic}\end{cases}$ complexes over A in normal form

$$\begin{cases}[f] : (C,\phi) \longrightarrow (C',\phi') \\ [f] : (C,\psi) \longrightarrow (C',\psi')\end{cases}$$

is an isomorphism of $\begin{cases}\varepsilon\text{-symmetric}\\ \varepsilon\text{-quadratic}\end{cases}$ complexes

$$\begin{cases} f : (C,\phi)\oplus C^{\varepsilon}(P) \longrightarrow (C',\phi')\oplus C^{\varepsilon}(P') \\ f : (C,\psi)\oplus C_{\varepsilon}(P) \longrightarrow (C',\psi')\oplus C_{\varepsilon}(P')\end{cases}$$

for some f.g. projective A-modules P,P', with

$$\begin{cases} C^{\varepsilon}(P) = (D,\eta \in Q^2(D,\varepsilon)) \\ C_{\varepsilon}(P) = (D,\zeta \in Q_2(D,\varepsilon))\end{cases}$$

the contractible 2-dimensional $\begin{cases}\varepsilon\text{-symmetric}\\ \varepsilon\text{-quadratic}\end{cases}$ complex over A in normal form defined by

$$D : \dots \longrightarrow 0 \longrightarrow P \xrightarrow{\begin{pmatrix} 0 \\ -\varepsilon \end{pmatrix}} P^*\oplus P \xrightarrow{(1\ 0)} P^* \longrightarrow 0 \longrightarrow \dots$$

$$\begin{cases}
\begin{cases}
\eta_0 = \begin{cases} 1 : D^0 = P \longrightarrow D_2 = P \\ \begin{pmatrix} 0 & 1 \\ -\varepsilon & 0 \end{pmatrix} : D^1 = P\oplus P^* \longrightarrow D_1 = P^*\oplus P \\ \varepsilon : D^2 = P^* \longrightarrow D_0 = P^* \end{cases} \\
\eta_s = 0 : D^{2-r+s} \longrightarrow D_r \ (s \geqslant 1)
\end{cases} \\
\begin{cases}
\zeta_0 = \begin{cases} 1 : D^0 = P \longrightarrow D_2 = P \\ \begin{pmatrix} 0 & 0 \\ -\varepsilon & 0 \end{pmatrix} : D^1 = P\oplus P^* \longrightarrow D_1 = P^*\oplus P \\ 0 : D^2 = P^* \longrightarrow D_0 = P^* \end{cases} \\
\zeta_s = 0 : D^{2-r-s} \longrightarrow D_r \ (s \geqslant 1)
\end{cases}
\end{cases}$$

and similarly for $\begin{cases} C^{\varepsilon}(P') \\ C_{\varepsilon}(P') \end{cases}$.

Proposition 3.5.1 The homotopy equivalence classes of connected 2-dimensional $\begin{cases} \varepsilon\text{-symmetric} \\ \varepsilon\text{-quadratic} \end{cases}$ complexes over A are in a natural one-one correspondence with the stable isomorphism classes of connected 2-dimensional $\begin{cases} \varepsilon\text{-symmetric} \\ \varepsilon\text{-quadratic} \end{cases}$ complexes over A in normal form.

Proof: A stable isomorphism is a homotopy equivalence. Therefore it is sufficient to prove that every connected 2-dimensional $\begin{cases} \varepsilon\text{-symmetric} \\ \varepsilon\text{-quadratic} \end{cases}$ complex is homotopy equivalent to one in normal form, and that homotopy equivalent complexes determine stably isomorphic complexes in normal form.

Every 2-dimensional $\begin{cases} \varepsilon\text{-symmetric} \\ \varepsilon\text{-quadratic} \end{cases}$ complex over A $\begin{cases} (C,\phi \in Q^2(C,\varepsilon)) \\ (C,\psi \in Q_2(C,\varepsilon)) \end{cases}$ is homotopy equivalent to one in which the chain complex C is in normal form, and for such C the class $\begin{cases} \phi \\ \psi \end{cases}$ is represented by A-module morphisms

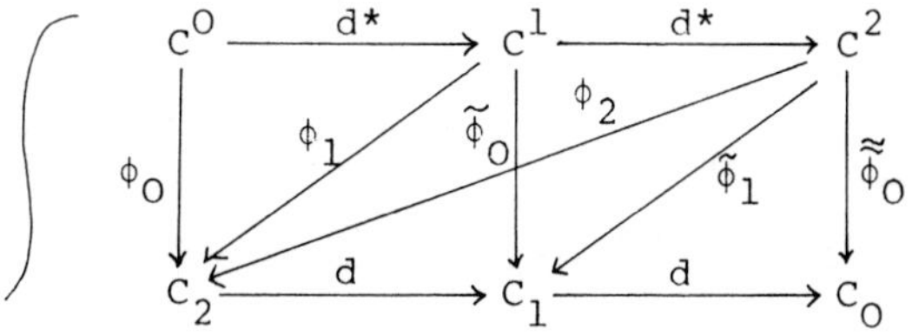

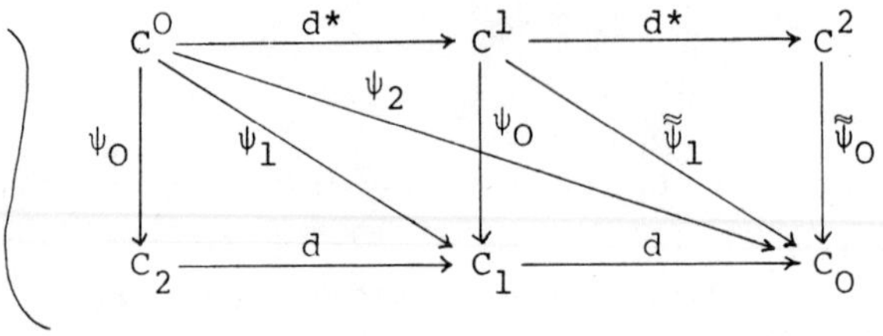

such that

$$\begin{cases} \begin{cases} d\phi_0 - \tilde{\phi}_0 d^* = 0 : C^0 \longrightarrow C_1 \ , \ d\tilde{\phi}_0 + \tilde{\tilde{\phi}}_0 d^* = 0 : C^1 \longrightarrow C_0 \ , \\ \phi_1 d^* + \phi_0 - \varepsilon\tilde{\tilde{\phi}}{}_0^* = 0 : C^0 \longrightarrow C_2 \ , \\ d\phi_1 - \tilde{\phi}_1 d^* + \tilde{\phi}_0 + \varepsilon\tilde{\phi}_0^* = 0 : C^1 \longrightarrow C_2 \ , \\ d\tilde{\phi}_1 + \tilde{\tilde{\phi}}_0 - \varepsilon\phi_0^* = 0 : C^2 \longrightarrow C_0 \ , \\ \phi_2 d^* - \phi_1 - \varepsilon\tilde{\phi}_1^* = 0 : C^1 \longrightarrow C_2 \ , \\ d\phi_2 - \tilde{\phi}_1 - \varepsilon\phi_1^* = 0 : C^2 \longrightarrow C_1 \ , \ \phi_2 - \varepsilon\phi_2^* = 0 : C^2 \longrightarrow C_2 \end{cases} \\ \begin{cases} d\psi_0 - \tilde{\psi}_0 d^* - \psi_1 + \varepsilon\tilde{\psi}_1^* = 0 : C^0 \longrightarrow C_1 \ , \\ d\tilde{\psi}_0 + \tilde{\tilde{\psi}}_0 d^* - \tilde{\psi}_1 + \varepsilon\psi_1^* = 0 : C^1 \longrightarrow C_0 \ , \\ d\psi_1 + \tilde{\psi}_1 d^* + \psi_2 + \varepsilon\psi_2^* = 0 : C^0 \longrightarrow C_0 \ . \end{cases} \end{cases}$$

Such a complex $\begin{cases} (C,\phi) \\ (C,\psi) \end{cases}$ is connected if and only if the A-module morphism

$$\begin{cases} (d \quad \tilde{\tilde{\phi}}_0) : C_1 \oplus C^2 \longrightarrow C_0 \\ (d \quad (\tilde{\tilde{\psi}}_0 + \varepsilon\psi_0^*)) : C_1 \oplus C^2 \longrightarrow C_0 \end{cases}$$

is onto, in which case we shall construct a homotopy equivalent complex $\begin{cases} (C',\phi' \in Q^2(C',\varepsilon)) \\ (C',\psi' \in Q_2(C',\varepsilon)) \end{cases}$ in normal form, as follows.

Define a connected 2-dimensional ε-symmetric complex over A $(C',\phi' \in Q^2(C',\varepsilon))$ in normal form by

$$d' = \begin{cases} \begin{pmatrix} d \\ 0 \end{pmatrix} : C'_2 = C_2 \longrightarrow C'_1 = \ker((d \ \ -\phi_0^*) : C_1 \oplus C^2 \longrightarrow C_0) \ , \\ (0 \ \ 1) \ : C'_1 \longrightarrow C'_0 = C^2 \ , \end{cases}$$

$$\phi'_0 = 1 : C'^2 = C^2 \longrightarrow C'_0 = C^2 \ ,$$

$$\tilde{\phi}'_0 = \begin{pmatrix} \tilde{\phi}_0 - \tilde{\phi}_1 d^* & d \\ -\varepsilon d^* & 0 \end{pmatrix} : C'^1 = \operatorname{coker}\left(\begin{pmatrix} d^* \\ -\phi_0 \end{pmatrix} : C^0 \longrightarrow C^1 \oplus C_2\right) \longrightarrow C'_1 \ ,$$

$$\tilde{\tilde{\phi}}'_0 = \varepsilon : C'^0 \longrightarrow C'_2 = C_2 \ , \ \phi'_1 = [\phi_2 d^* \ 0] : C'^1 \longrightarrow C'_2 = C_2 \ ,$$

$$\phi'_1 = 0 : C'^2 = C_2 \longrightarrow C'_1 \ , \ \phi'_2 = \phi_2 : C'^2 = C^2 \longrightarrow C'_2 = C_2 \ .$$

The chain equivalence

$$f : C' \longrightarrow C$$

given by

$$\begin{array}{ccccccccccc} C' : \ldots \longrightarrow & 0 & \longrightarrow & C'_2 & \xrightarrow{d'} & C'_1 & \xrightarrow{d'} & C'_0 & \longrightarrow & 0 & \longrightarrow \ldots \\ \downarrow f & & & \downarrow 1 & & \downarrow (1 \ 0) & & \downarrow \phi_0^* & & & \\ C \ : \ldots \longrightarrow & 0 & \longrightarrow & C_2 & \xrightarrow{d} & C_1 & \xrightarrow{d} & C_0 & \longrightarrow & 0 & \longrightarrow \ldots \end{array}$$

defines a homotopy equivalence of 2-dimensional ε-symmetric complexes over A

$$f : (C',\phi') \longrightarrow (C,\phi) \ .$$

Given (C,ψ) as above we define first an auxiliary 2-dimensional ε-quadratic complex over A $(C'',\psi'' \in Q_2(C'',\varepsilon))$ by

$$d'' = \begin{cases} \begin{pmatrix} d \\ 0 \end{pmatrix} : C''_2 = C_2 \longrightarrow C''_1 = C_1 \oplus C^2 \ , \\ \begin{pmatrix} d & -(\psi_0 + \varepsilon \tilde{\tilde{\psi}}_0^*) \\ 0 & 1 \end{pmatrix} : C''_1 = C_1 \oplus C^2 \longrightarrow C''_0 = C_0 \oplus C^2 \end{cases}$$

$$\psi''_0 = (0 \ \ 1) : C''^0 = C^0 \oplus C_2 \longrightarrow C''_2 = C_2$$

$$\tilde{\psi}''_0 = \begin{pmatrix} \tilde{\psi}_0 & 0 \\ -\epsilon d^* & 0 \end{pmatrix} : C''^1 = C^1 \oplus C_2 \longrightarrow C''_1 = C_1 \oplus C^2$$

$$\tilde{\tilde{\psi}}''_0 = 0 : C''^2 = C^2 \longrightarrow C''_0 = C_0 \oplus C^2$$

$$\psi''_1 = \begin{pmatrix} -\tilde{\psi}_0 d^* & d \\ 0 & 0 \end{pmatrix} : C''^0 = C^0 \oplus C_2 \longrightarrow C''_1 = C_1 \oplus C^2$$

$$\tilde{\psi}''_1 = 0 : C''^1 = C^1 \oplus C_2 \longrightarrow C''_0 = C_0 \oplus C^2$$

$$\psi''_2 = \begin{pmatrix} \psi_2 + \tilde{\psi}_1 d^* & 0 \\ 0 & 0 \end{pmatrix} : C''^0 = C^0 \oplus C_2 \longrightarrow C''_0 = C_0 \oplus C^2 \quad .$$

The chain equivalence

$$f'' : C \longrightarrow C''$$

given by

$$\begin{array}{ccccccccccccc} C & : & \ldots \longrightarrow & 0 & \longrightarrow & C_2 & \xrightarrow{d} & C_1 & \xrightarrow{d} & C_0 & \longrightarrow & 0 & \longrightarrow \ldots \\ \downarrow f'' & & & & & \downarrow 1 & & \downarrow \begin{pmatrix} 1 \\ 0 \end{pmatrix} & & \downarrow \begin{pmatrix} 1 \\ 0 \end{pmatrix} & & & \\ C'' & : & \ldots \longrightarrow & 0 & \longrightarrow & C''_2 & \xrightarrow{d''} & C''_1 & \xrightarrow{d''} & C''_0 & \longrightarrow & 0 & \longrightarrow \ldots \end{array}$$

defines a homotopy equivalence of 2-dimensional ϵ-quadratic complexes over A

$$f'' : (C,\psi) \longrightarrow (C'',\psi'') \quad .$$

Defien a 2-dimensional A-module chain complex C' in normal form by

$$d' = \begin{cases} \begin{pmatrix} d \\ 0 \end{pmatrix} : C'_2 = C_2 \longrightarrow C'_1 = \ker((d\ \tilde{\tilde{\psi}}_0 + \epsilon\psi^*_0) : C_1 \oplus C^2 \longrightarrow C_0) \\ (0 \quad 1) : C'_1 \longrightarrow C'_0 = C^2 \quad . \end{cases}$$

Choose a splitting map

$$\begin{pmatrix} j \\ k \end{pmatrix} : C_0 \longrightarrow C_1 \oplus C^2$$

for $(d\ \tilde{\tilde{\psi}}_0 + \epsilon\psi^*_0) : C_1 \oplus C^2 \longrightarrow C_0$, so that

$$(d\ \tilde{\tilde{\psi}}_0+\varepsilon\psi_0^*)\begin{pmatrix} j \\ k\end{pmatrix} = dj + (\tilde{\tilde{\psi}}_0+\varepsilon\psi_0^*)k = 1 : C_0 \longrightarrow C_0 \quad ,$$

and define a chain equivalence

$$f' : C'' \longrightarrow C'$$

by

$$\begin{array}{ccccccccccccc}
C'' & : & \ldots \longrightarrow & 0 & \longrightarrow & C''_2 & \xrightarrow{d''} & C''_1 & \xrightarrow{d''} & C''_0 & \longrightarrow & 0 & \longrightarrow \ldots \\
{\scriptstyle f'}\big\downarrow & & & & & \big\downarrow{\scriptstyle f'_2} & & \big\downarrow{\scriptstyle f'_1} & & \big\downarrow{\scriptstyle f'_0} & & & \\
C' & : & \ldots \longrightarrow & 0 & \longrightarrow & C'_2 & \xrightarrow{d'} & C'_1 & \xrightarrow{d'} & C'_0 & \longrightarrow & 0 & \longrightarrow \ldots
\end{array} ,$$

with

$$f'_0 = (-k \quad -\bar{\varepsilon}) : C''_0 = C_0 \oplus C^2 \longrightarrow C'_0 = C^2$$

$$f'_1 = \begin{pmatrix} 1 - jd & \bar{\varepsilon}j(\tilde{\tilde{\psi}}_0+\varepsilon\psi_0^*) \\ -kd & \bar{\varepsilon}k(\tilde{\tilde{\psi}}_0+\varepsilon\psi_0^*) - \bar{\varepsilon}\end{pmatrix}$$

$$: C''_1 = C_1 \oplus C^2 \longrightarrow C'_1 = \ker((d\ \ \tilde{\tilde{\psi}}_0+\varepsilon\psi_0^*) : C_1 \oplus C^2 \longrightarrow C_0)$$

$$f'_2 = 1 : C''_2 = C_2 \longrightarrow C'_2 = C_2 \ .$$

The connected 2-dimensional ε-quadratic complex over A (C',ψ') defined by

$$\psi' = f'_\%(\psi'') \in Q_2(C',\varepsilon)$$

is in normal form, and there is defined a homotopy equivalence

$$f = f'f'' : (C,\psi) \longrightarrow (C',\psi') \ .$$

Tha above procedure associates to an isomorphism class of connected 2-dimensional $\left\{\begin{array}{l}\varepsilon\text{-symmetric}\\ \varepsilon\text{-quadratic}\end{array}\right.$ complexes over A $\left\{\begin{array}{l}(C,\phi)\\ (C,\psi)\end{array}\right.$ with the chain complex C in normal form an isomorphism class of connected 2-dimensional $\left\{\begin{array}{l}\varepsilon\text{-symmetric}\\ \varepsilon\text{-quadratic}\end{array}\right.$ complexes over A $\left\{\begin{array}{l}(C',\phi')\\ (C',\psi')\end{array}\right.$ in normal form. The association preserves homotopy types, and also the direct sum $\oplus$. In particular, if C is a chain contractible 2-dimensional A-module chain complex in normal form it is isomorphic to one of the type

$$C : \ldots \longrightarrow 0 \longrightarrow P \xrightarrow{\begin{pmatrix}1\\0\end{pmatrix}} P\oplus Q \xrightarrow{(0\ 1)} Q \longrightarrow 0 \longrightarrow \ldots$$

for some f.g. projective A-modules P,Q, so that $\begin{cases}(C',\phi')\\(C',\psi')\end{cases}$ is isomorphic to $\begin{cases}C^{\varepsilon}(P)\\C_{\varepsilon}(P)\end{cases}$, and hence is stably isomorphic to 0. It follows from the Lemma below that homotopy equivalent complexes $\begin{cases}(C,\phi)\\(C,\psi)\end{cases}$, $\begin{cases}(\tilde{C},\tilde{\phi})\\(\tilde{C},\tilde{\psi})\end{cases}$ with C and $\tilde{C}$ in normal form determine stably isomorphic complexes in normal form $\begin{cases}(C',\phi')\\(C',\psi')\end{cases}$, $\begin{cases}(\tilde{C}',\tilde{\phi}')\\(\tilde{C}',\tilde{\psi}')\end{cases}$.

<u>Lemma</u> Let $\begin{cases}(C,\phi)\\(C,\psi)\end{cases}$, $\begin{cases}(\tilde{C},\tilde{\phi})\\(\tilde{C},\tilde{\psi})\end{cases}$ be 2-dimensional $\begin{cases}\varepsilon\text{-symmetric}\\\varepsilon\text{-quadratic}\end{cases}$ complexes over A with C,$\tilde{C}$ in normal form. There exists a homotopy equivalence

$$\begin{cases}f : (C,\phi) \longrightarrow (\tilde{C},\tilde{\phi})\\ f : (C,\psi) \longrightarrow (\tilde{C},\tilde{\psi})\end{cases}$$

if and only if there exists an isomorphism

$$\begin{cases}(C,\phi)\oplus(D,0) \longrightarrow (\tilde{C},\tilde{\phi})\oplus(\tilde{D},0)\\ (C,\psi)\oplus(D,0) \longrightarrow (\tilde{C},\tilde{\psi})\oplus(\tilde{D},0)\end{cases}$$

for some contractible 2-dimensional A-module chain complexes D,$\tilde{D}$ in normal form.

<u>Proof</u>: This is a special case of Proposition I.1.5.

[]

[]

An $\begin{cases} \underline{\text{(even) } \varepsilon\text{-symmetric}} \\ \underline{\varepsilon\text{-quadratic}} \end{cases}$ $\underline{\text{S-form over A}}$ $\begin{cases} (K,\alpha;L) \\ (K,\beta;L) \end{cases}$ is an S-non-singular $\begin{cases} \text{(even) } \varepsilon\text{-symmetric} \\ \varepsilon\text{-quadratic} \end{cases}$ form over A $\begin{cases} (K,\alpha \in Q^{\varepsilon}(K)) \\ (K,\beta \in Q_{\varepsilon}(K)) \end{cases}$ together with an S-lagrangian L. Such an S-form is <u>non-singular</u> if $\begin{cases} (K,\alpha) \\ (K,\beta) \end{cases}$ is a non-singular form. (An S-form is an $S^{-1}A$-form in the sense of §2.4).

An <u>isomorphism</u> of $\begin{cases} \text{(even) } \varepsilon\text{-symmetric} \\ \varepsilon\text{-quadratic} \end{cases}$ S-forms over A

$$\begin{cases} f : (K,\alpha;L) \xrightarrow{\sim} (K',\alpha';L') \\ f : (K,\beta;L) \xrightarrow{\sim} (K',\beta';L') \end{cases}$$

is an isomorphism of the $\begin{cases} \text{(even) } \varepsilon\text{-symmetric} \\ \varepsilon\text{-quadratic} \end{cases}$ forms over A

$$\begin{cases} f : (K,\alpha) \xrightarrow{\sim} (K',\alpha') \\ f : (K,\beta) \xrightarrow{\sim} (K',\beta') \end{cases}$$

such that

$$f(L) = L' .$$

A <u>stable isomorphism</u> of $\begin{cases} \text{(even) } \varepsilon\text{-symmetric} \\ \varepsilon\text{-quadratic} \end{cases}$ S-forms over A

$$\begin{cases} [f] : (K,\alpha;L) \xrightarrow{\sim} (K',\alpha';L') \\ [f] : (K,\beta;L) \xrightarrow{\sim} (K',\beta';L') \end{cases}$$

is an isomorphism of S-forms

$$\begin{cases} f : (K,\alpha;L)\oplus(M,\phi;N) \xrightarrow{\sim} (K',\alpha';L')\oplus(M',\phi';N') \\ f : (K,\beta;L)\oplus(M,\phi;N) \xrightarrow{\sim} (K',\beta';L')\oplus(M',\psi';N') \end{cases}$$

for some non-singular $\begin{cases} \text{(even) } \varepsilon\text{-symmetric} \\ \varepsilon\text{-quadratic} \end{cases}$ S-forms over A

$\begin{cases}(M,\phi;N)\\(M,\psi;N)\end{cases}$, $\begin{cases}(M',\phi';N')\\(M',\psi';N')\end{cases}$ such that N is a lagrangian of $\begin{cases}(M,\phi)\\(M,\psi)\end{cases}$

and N' is a lagrangian of $\begin{cases}(M',\phi')\\(M',\psi')\end{cases}$.

Proposition 3.5.2 i) The stable equivalence classes of

$\begin{cases}\text{even } \varepsilon\text{-symmetric}\\ \varepsilon\text{-quadratic}\\ \text{split } \varepsilon\text{-quadratic}\end{cases}$ linking formations over (A,S) are in a

natural one-one correspondence with the stable isomorphism

classes of $\begin{cases}\varepsilon\text{-symmetric}\\ \text{even } \varepsilon\text{-symmetric}\\ \varepsilon\text{-quadratic}\end{cases}$ S-forms over A. Non-singular

linking formations correspond to non-singular S-forms.

ii) The stable equivalence classes of (even) ε-symmetric linking formations over (A,S) $(M,\lambda;F,G)$ are in a natural one-one correspondence with the homotopy equivalence classes of connected S-acyclic 2-dimensional (even) $(-\varepsilon)$-symmetric complexes over A $(C,\phi\in Q^2(C,-\varepsilon))$. Under this correspondence the exact sequence of S-torsion A-modules

$$0\longrightarrow H^1(C)\xrightarrow{\phi_0}H_1(C)\longrightarrow H_1(\phi_0)\longrightarrow H^2(C)\xrightarrow{\phi_0}H_0(C)\longrightarrow 0$$

can be identified with

$$0\longrightarrow F\cap G\longrightarrow F\cap G^{\perp}\longrightarrow G^{\perp}/G\longrightarrow M/(F+G)\longrightarrow M/(F+G^{\perp})\longrightarrow 0\ ,$$

and

$$v_0^S(\phi)\ :\ H^2(C)=M/(F+G)\longrightarrow \hat{H}^0(\mathbb{Z}_2;S^{-1}A/A,\varepsilon)\ ;\ x\longmapsto \lambda(x)(x)\ .$$

Non-singular linking formations correspond to Poincaré complexes.

iii) There is a natural projection of the set of homotopy equivalence classes of connected S-acyclic 2-dimensional $(-\varepsilon)$-quadratic complexes over A $(C,\psi \in Q_2(C,-\varepsilon))$ onto the set of stable equivalence classes of split ε-quadratic linking formations over (A,S) $(F,(\binom{\gamma}{\mu},\theta)G)$. If the complexes (C,ψ), (C',ψ') project to the same stable equivalence class then (C',ψ') is homotopy equivalent to a complex obtained from (C,ψ) by an S-acyclic $(-\varepsilon)$-quadratic surgery preserving the $(-\varepsilon)$-symmetric homotopy type, and

$$pv_S^1(\psi) = pv_S^1(\psi') : H^1(C) = \ker(\mu:G\longrightarrow F^\wedge) \longrightarrow Q_{-\varepsilon}(A,S) \ ;$$
$$x \longmapsto \theta(x) \ .$$

[]

(Before embarking on the proof of Proposition 3.5.2 we remark on the similarity between these correspondences and those of

(linking forms over (A,S)) $\longleftrightarrow$

(S-acyclic 1-dimensional complexes over A) (Proposition 3.4.1)

(linking forms over (A,S)) $\longleftrightarrow$

(S-formations over A) (Proposition 3.4.3)

(formations over A) $\longleftrightarrow$

(1-dimensional complexes over A) (Proposition 1.6.4).

In particular, given a connected 1-dimensional $\begin{cases}\varepsilon\text{-symmetric}\\ \varepsilon\text{-quadratic}\end{cases}$ complex over A $\begin{cases}(C,\phi \in Q^1(C,\varepsilon))\\ (C,\psi \in Q_1(C,\varepsilon))\end{cases}$ with a corresponding $\begin{cases}\varepsilon\text{-symmetric}\\ \text{split } \varepsilon\text{-quadratic}\end{cases}$ formation over A $\begin{cases}(M,\alpha;F,G)\\ (F,(\binom{\gamma}{\mu},\theta)G)\end{cases}$ the exact sequence of A-modules

$$0 \longrightarrow H^0(C) \xrightarrow{\phi_0} H_1(C) \longrightarrow H_1(\phi_0) \longrightarrow H^1(C) \xrightarrow{\phi_0} H_0(C) \longrightarrow 0$$

(with $\phi_0 = (1+T_\varepsilon)\psi_0$ in the ε-quadratic case) may be identified with the exact sequence

$$0 \longrightarrow F \cap G \longrightarrow F \cap G^\perp \longrightarrow G^\perp/G \longrightarrow M/(F+G) \longrightarrow M/(F+G^\perp) \longrightarrow 0$$

and

$$v_0(\phi) \ : \ H^1(C) = M/(F+G) \longrightarrow \hat{H}^0(\mathbb{Z}_2;A,\varepsilon) \ ; \ [x] \longmapsto \alpha(x)(x) \quad (x \in M)$$

$$v^1(\psi) \ : \ H^0(C) = F \cap G = \ker(\mu : G \longrightarrow F^*) \longrightarrow \hat{H}^0(\mathbb{Z}_2;A,\varepsilon) \ ;$$
$$y \longmapsto \theta(y)(y) \ ) .$$

<u>Proof</u>: i) Given an $\begin{cases} \text{even } \varepsilon\text{-symmetric} \\ \varepsilon\text{-quadratic} \end{cases}$ linking formation over (A,S) $\begin{cases} (M,\lambda;F,G) \\ (M,\lambda,\mu;F,G) \end{cases}$ we have from Proposition 3.4.6 that the $\begin{cases} \text{even } \varepsilon\text{-symmetric} \\ \varepsilon\text{-quadratic} \end{cases}$ linking form $\begin{cases} (M,\lambda) \\ (M,\lambda,\mu) \end{cases}$ is isomorphic to the boundary $\partial(K,\alpha)$ of an S-hyperbolic S-non-singular $\begin{cases} \varepsilon\text{-symmetric} \\ \text{even } \varepsilon\text{-symmetric} \end{cases}$ form over A $\begin{cases} (K,\alpha \in Q^\varepsilon(K)) \\ (K,\alpha \in Q\langle v_0\rangle^\varepsilon(K)) \end{cases}$, and that

$$F = \mathrm{coker}(f:K \longrightarrow K_F) \subseteq M = \mathrm{coker}(\alpha:K \longrightarrow K^*)$$

$$G = \mathrm{coker}(g:K \longrightarrow K_G) \subseteq M = \mathrm{coker}(\alpha:K \longrightarrow K^*)$$

for some S-isomorphisms of S-non-singular $\begin{cases} \varepsilon\text{-symmetric} \\ \text{even } \varepsilon\text{-symmetric} \end{cases}$ forms over A

$$f : (K,\alpha) \longrightarrow (K_F,\alpha_F) \ , \ g : (K,\alpha) \longrightarrow (K_G,\alpha_G)$$

with (K_F,α_F) non-singular. The $\begin{cases} \varepsilon\text{-symmetric} \\ \text{even } \varepsilon\text{-symmetric} \end{cases}$ S-form over A associated to $\begin{cases} (M,\lambda;F,G) \\ (M,\lambda,\mu;F,G) \end{cases}$ is defined to be

$$(K_F \oplus K_G, \begin{pmatrix} \alpha_F & 0 \\ 0 & -\alpha_G \end{pmatrix} \in Q^{\varepsilon}(K_F \oplus K_G); \mathrm{im}(\begin{pmatrix} f \\ g \end{pmatrix} : K \longrightarrow K_F \oplus K_G)) \quad .$$

We defer to ii) the proof that the stable isomorphism class of this S-form is independent of the choice of S-non-singular form (K,α) such that

$$\begin{cases} (M,\lambda) = \partial(K,\alpha) \\ (M,\lambda,\mu) = \partial(K,\alpha) \quad . \end{cases}$$

We shall now prove that the S-forms associated to stably equivalent linking formations are stably isomorphic.

Given an $\begin{cases} \text{even } \varepsilon\text{-symmetric} \\ \varepsilon\text{-quadratic} \end{cases}$ linking formation over (A,S) $\begin{cases} (M,\lambda{:}F,G) \\ (M,\lambda,\mu;F,G) \end{cases}$ and a sublagrangian H write the linking formation obtained by elementary equivalence as

$$\begin{cases} (M',\lambda';F',G') = (H^{\perp}/H,\lambda^{\perp}/\lambda;F\cap H^{\perp},G/H) \\ (M',\lambda',\mu';F',G') = (H^{\perp}/H,\lambda^{\perp}/\lambda,\mu^{\perp}/\mu;F\cap H^{\perp},G/H) \quad . \end{cases}$$

Continuing with the previous terminology, let

$$h : (K,\alpha) \longrightarrow (K',\alpha')$$

be the S-isomorphism of S-non-singular $\begin{cases} \varepsilon\text{-symmetric} \\ \text{even } \varepsilon\text{-symmetric} \end{cases}$ forms over A associated to H by Proposition 3.4.6 i), with

$$H = \mathrm{coker}(h{:}K \longrightarrow K') \subsetneq M = \mathrm{coker}(\alpha{:}K \longrightarrow K^*)$$

$$\begin{cases} (M',\lambda') = \partial(K',\alpha') \\ (M',\lambda',\mu') = \partial(K',\alpha') \quad . \end{cases}$$

As $H \subset G$ there is also defined an S-isomorphism

$$g' : (K',\alpha') \longrightarrow (K_G,\alpha_G)$$

$$g = g'h : (K,\alpha) \xrightarrow{h} (K',\alpha') \xrightarrow{g'} (K_G,\alpha_G) .$$

The composite

$$F \xrightarrow{[\text{inclusion}]} M/H^{\perp} \xrightarrow{[\lambda]} H^{\wedge}$$

is onto, with resolution

$$\begin{array}{ccccccccc} 0 & \longrightarrow & K & \xrightarrow{f} & K_F & \longrightarrow & F & \longrightarrow & 0 \\ & & \downarrow{\scriptstyle \alpha'h} & & \downarrow{\scriptstyle f^*\alpha_F} & & \downarrow & & \\ 0 & \longrightarrow & K'^* & \xrightarrow{h^*} & K^* & \longrightarrow & H^{\wedge} & \longrightarrow & 0 \end{array} .$$

Thus the (A,S)-module $F' = \ker(F \longrightarrow H^{\wedge})$ has f.g. projective A-module resolution

$$0 \longrightarrow K \xrightarrow{e} J \longrightarrow F' \longrightarrow 0$$

with

$$e = \begin{pmatrix} f \\ \alpha'h \end{pmatrix} : K \longrightarrow J = \ker((f^*\alpha_F \ -h^*) : K_F \oplus K'^* \longrightarrow K^*) .$$

Define a non-singular $\begin{cases} \varepsilon\text{-symmetric} \\ \text{even } \varepsilon\text{-symmetric} \end{cases}$ form over A

$$(R,\rho) = (K_F \oplus K'^* \oplus K', \begin{pmatrix} \alpha_F & 0 & 0 \\ 0 & 0 & 1 \\ 0 & \varepsilon & \alpha' \end{pmatrix}) \in Q^{\varepsilon}(K_F \oplus K'^* \oplus K')) ,$$

and let L be the sublagrangian of (R,ρ) defined by

$$L = \operatorname{im}(\begin{pmatrix} f \\ -\alpha'^*h \\ h \end{pmatrix} : K \longrightarrow K_F \oplus K'^* \oplus K') \subseteq R ,$$

so that

$$(L^{\perp}/L, \lambda^{\perp}/\lambda) = (K_{F'}, \alpha_{F'})$$

is also a non-singular $\begin{cases}\varepsilon\text{-symmetric}\\ \text{even } \varepsilon\text{-symmetric}\end{cases}$ form over A.

The S-isomorphism of f.g. projective A-modules

$$f' = \begin{pmatrix} 0 \\ 0 \\ 1 \end{pmatrix} : K' \longrightarrow K_{F'} = \frac{\ker((f^*\alpha_F \quad \varepsilon h^* \quad 0):K_F \oplus K'^* \oplus K' \longrightarrow K^*)}{\operatorname{im}(\begin{pmatrix} f \\ -\alpha'^* h \\ h \end{pmatrix} : K \longrightarrow K_F \oplus K'^* \oplus K')}$$

defines an S-isomorphism of S-non-singular $\begin{cases}\varepsilon\text{-symmetric}\\ \text{even } \varepsilon\text{-symmetric}\end{cases}$ forms over A

$$f' : (K',\alpha') \longrightarrow (K_{F'},\alpha_{F'})$$

such that

$$\operatorname{coker}(f':K' \longrightarrow K_{F'}) = \operatorname{coker}(e:K \longrightarrow J) = F'$$

is the associated lagrangian of $\partial(K',\alpha') = \begin{cases}(M',\lambda')\\ (M',\lambda',\mu')\end{cases}$.

Thus the $\begin{cases}\varepsilon\text{-symmetric}\\ \text{even } \varepsilon\text{-symmetric}\end{cases}$ S-form over A associated to $\begin{cases}(M',\lambda';F',G')\\ (M',\lambda',\mu';F',G')\end{cases}$ is given by

$$(K_{F'} \oplus K_G, \begin{pmatrix} \alpha_{F'} & 0 \\ 0 & -\alpha_G \end{pmatrix} \in Q^{\varepsilon}(K_{F'} \oplus K_G); \operatorname{im}(\begin{pmatrix} f' \\ g' \end{pmatrix} : K' \longrightarrow K_{F'} \oplus K_G)) .$$

Define an $\begin{cases}\varepsilon\text{-symmetric}\\ \text{even } \varepsilon\text{-symmetric}\end{cases}$ S-form over A

$$(Q,\phi;P) = (K_F \oplus K'^* \oplus K' \oplus K_G, \begin{pmatrix} \alpha_F & 0 & 0 & 0 \\ 0 & 0 & \varepsilon & 0 \\ 0 & 1 & -\alpha' & 0 \\ 0 & 0 & 0 & -\alpha_G \end{pmatrix};$$

$$\mathrm{im}(\begin{pmatrix} f & 0 \\ -\alpha' h & -\alpha' \\ 0 & 1 \\ g & g' \end{pmatrix} : K \oplus K' \longrightarrow K_F \oplus K'^* \oplus K' \oplus K_G)) \quad .$$

By Proposition 1.6.2 the inclusion of the sublagrangian

$$\begin{pmatrix} f \\ -\alpha' h \\ h \\ 0 \end{pmatrix} : (K,0) \longrightarrow (Q,\phi) = (K_F \oplus K'^* \oplus K' \oplus K_G, \phi)$$

extends to an isomorphism of $\begin{cases} \varepsilon\text{-symmetric} \\ \text{even } \varepsilon\text{-symmetric} \end{cases}$ forms over A

$$H^{\varepsilon}(K,\theta) \oplus (K^{\perp}/K, \phi^{\perp}/\phi) = (K \oplus K^*, \begin{pmatrix} 0 & 1 \\ \varepsilon & \theta \end{pmatrix}) \oplus (K_{F'} \oplus K_G, \begin{pmatrix} \alpha_{F'} & 0 \\ 0 & -\alpha_G \end{pmatrix})$$

$$\longrightarrow (Q,\phi)$$

sending $K \oplus \mathrm{im}(\begin{pmatrix} f' \\ g' \end{pmatrix} : K' \longrightarrow K_{F'} \oplus K_G)$ to P, for some $\begin{cases} \theta \in Q^{\varepsilon}(K^*) \\ \theta \in Q\langle v_0 \rangle^{\varepsilon}(K^*) \end{cases}$.

Thus there are defined isomorphisms of $\begin{cases} \varepsilon\text{-symmetric} \\ \text{even } \varepsilon\text{-symmetric} \end{cases}$

S-forms over A

$$(H^{\varepsilon}(K,\theta);K) \oplus (K_{F'} \oplus K_G, \begin{pmatrix} \alpha_{F'} & 0 \\ 0 & -\alpha_G \end{pmatrix}; \mathrm{im}(\begin{pmatrix} f' \\ g' \end{pmatrix} : K' \longrightarrow K_{F'} \oplus K_G))$$

$$\longrightarrow (Q,\phi;P)$$

$$\begin{pmatrix} 1 & 0 & 0 & 0 \\ 0 & 0 & g' & 1 \\ 0 & 0 & 1 & 0 \\ 0 & 1 & 0 & g'^*\alpha_G \end{pmatrix}$$

$$: (Q,\phi;P) \longrightarrow (K_F\oplus K_G, \begin{pmatrix} \alpha_F & 0 \\ 0 & -\alpha_G \end{pmatrix}; \mathrm{im}(\begin{pmatrix} f \\ g \end{pmatrix}: K \longrightarrow K_F\oplus K_G)) \oplus (H^{\varepsilon}(K');K')$$

The $\begin{cases} \varepsilon\text{-symmetric} \\ \text{even } \varepsilon\text{-symmetric} \end{cases}$ S-forms over A associated to the stably

equivalent $\begin{cases} \text{even } \varepsilon\text{-symmetric} \\ \varepsilon\text{-quadratic} \end{cases}$ linking formations over (A,S)

$\begin{cases} (M,\lambda;F,G) \\ (M,\lambda,\mu;F,G) \end{cases}$, $\begin{cases} (M',\lambda';F',G') \\ (M',\lambda',\mu';F',G') \end{cases}$ are therefore related by a

stable isomorphism

$$(K_F\oplus K_G, \begin{pmatrix} \alpha_F & 0 \\ 0 & -\alpha_G \end{pmatrix}; \mathrm{im}(\begin{pmatrix} f \\ g \end{pmatrix}: K \longrightarrow K_F\oplus K_G))$$

$$\longrightarrow (K_{F'}\oplus K_{G'}, \begin{pmatrix} \alpha_{F'} & 0 \\ 0 & -\alpha_{G'} \end{pmatrix}; \mathrm{im}(\begin{pmatrix} f' \\ g' \end{pmatrix}: K' \longrightarrow K_{F'}\oplus K_{G'})) \ .$$

Given a split ε-quadratic linking formation over (A,S) $(F,(\binom{\gamma}{\mu},\theta)G)$ we shall obtain an ε-quadratic S-form over A $(K,\beta;L)$, as follows. Let $u \in \mathrm{Hom}_A(L',L^*)$ be an S-isomorphism of f.g. projective A-modules defining a resolution of F by

$$0 \longrightarrow L \xrightarrow{u^*} L'^* \longrightarrow F \longrightarrow 0 \ .$$

Let $e \in \mathrm{Hom}_A(L^*\oplus L'^*, F\oplus F^{\wedge})$ be the projection appearing in the corresponding resolution of $F\oplus F^{\wedge}$

$$0 \longrightarrow L \oplus L' \xrightarrow{\begin{pmatrix} 0 & u \\ \varepsilon u^* & 0 \end{pmatrix}} L^* \oplus L'^* \xrightarrow{\ e\ } F \oplus F\hat{} \longrightarrow 0 \ ,$$

define a f.g. projective A-module

$$K = e^{-1}(G) \subseteq L^* \oplus L'^* \quad ,$$

and write the inclusion as

$$(j \ \ k) \ : \ L \oplus L' \longrightarrow K \ .$$

There is then a natural identification

$$\mathrm{coker}((j \ \ k) : L \oplus L' \longrightarrow K) = G \ ,$$

and there exists an S-non-singular ε-quadratic form over A $(K, \beta \in Q_\varepsilon(K))$ such that the inclusion $\binom{\gamma}{\mu} : G \longrightarrow F \oplus F\hat{}$ is resolved by

$$\begin{array}{ccccccccc}
0 & \longrightarrow & L \oplus L' & \xrightarrow{(j \ \ k)} & K & \xrightarrow{e|} & G & \longrightarrow & 0 \\
 & & \Big\downarrow 1 & & \Big\downarrow \begin{pmatrix} j^*(\beta+\varepsilon\beta^*) \\ k^*(\beta+\varepsilon\beta^*) \end{pmatrix} & & \Big\downarrow \begin{pmatrix} \gamma \\ \mu \end{pmatrix} & & \\
0 & \longrightarrow & L \oplus L' & \xrightarrow{\begin{pmatrix} 0 & u \\ \varepsilon u^* & 0 \end{pmatrix}} & L^* \oplus L'^* & \longrightarrow & F \oplus F\hat{} & \longrightarrow & 0
\end{array} \quad .$$

(As in Proposition 3.4.6 i) (K, β) is only determined by G up to S-isomorphism, i.e. only the coset

$$[\beta] \in Q_\varepsilon(K)/\ker(S^{-1} : Q_\varepsilon(K) \longrightarrow Q_\varepsilon(S^{-1}K))$$

is determined). Proposition 3.4.3 associates to the $(-\varepsilon)$-quadratic linking form over (A,S)

$$(G, \gamma\hat{}\mu \in \mathrm{Hom}_A(G, G\hat{}), \theta : G \longrightarrow Q_{-\varepsilon}(A,S))$$

the ε-quadratic S-formation over A

$$(K^*\oplus K,\begin{pmatrix}0 & 1\\ 0 & \beta\end{pmatrix}\in Q_{\varepsilon}(K^*\oplus K);K^*,\mathrm{im}(\begin{pmatrix}-(\beta+\varepsilon\beta^*)^*j & 0\\ j & k\end{pmatrix}:L\oplus L'\longrightarrow K^*\oplus K))$$

with

$$\theta : G = \mathrm{coker}((j\ k):L\oplus L'\longrightarrow K)\longrightarrow Q_{-\varepsilon}(A,S)\ ;$$

$$x\longmapsto(\frac{1}{s}).\varepsilon u^*(y)(y').\overline{(\frac{1}{s})} - \beta(x)(x)$$

$$(x\in K,s\in S,y\in L,y'\in L',sx = j(y)+k(y')\in K)\ ,$$

for a unique ε-quadratic form over A $(K,\beta\in Q_{\varepsilon}(K))$ in the prescribed S-isomorphism class. The ε-quadratic S-form over A associated to (F,G) is defined to be

$$(K,\beta\in Q_{\varepsilon}(K);\mathrm{im}(j:L\longrightarrow K))\ .$$

The verification that stably equivalent split ε-quadratic linking formations over (A,S) determine stably isomorphic ε-quadratic S-forms over A proceeds as in the (even) ε-symmetric case.

Conversely, given an $\begin{cases}\varepsilon\text{-symmetric}\\ \text{even } \varepsilon\text{-symmetric}\\ \varepsilon\text{-quadratic}\end{cases}$ S-form over A $\begin{cases}(K,\alpha\in Q^{\varepsilon}(K);L)\\ (K,\alpha\in Q\langle v_0\rangle^{\varepsilon}(K);L)\\ (K,\beta\in Q_{\varepsilon}(K);L)\end{cases}$ we shall define an $\begin{cases}\text{even } \varepsilon\text{-symmetric}\\ \varepsilon\text{-quadratic}\\ \text{split } \varepsilon\text{-quadratic}\end{cases}$ linking formation over (A,S) $\begin{cases}(M,\lambda;F,G)\\ (M,\lambda,\mu;F,G)\\ (F,G)\end{cases}$, as follows.

Given an $\begin{cases}\text{(even) } \varepsilon\text{-symmetric}\\ \varepsilon\text{-quadratic}\end{cases}$ S-form over A $\begin{cases}(K,\alpha;L)\\ (K,\beta;L)\end{cases}$ let $j\in \mathrm{Hom}_A(L,K)$ be the inclusion, and apply Proposition 1.6.2

to extend the inclusion of the lagrangian

$$\begin{cases} j : (S^{-1}L,0) \longrightarrow S^{-1}(K,\alpha) \\ j : (S^{-1}L,0) \longrightarrow S^{-1}(K,\beta) \end{cases}$$

to an isomorphism of non-singular $\begin{cases}\text{(even) } \varepsilon\text{-symmetric} \\ \varepsilon\text{-quadratic}\end{cases}$ forms over $S^{-1}A$

$$\begin{cases} (j\ j') : (S^{-1}L \oplus S^{-1}L^*, \begin{pmatrix} 0 & 1 \\ \varepsilon & j'^*\alpha j' \end{pmatrix}) \longrightarrow S^{-1}(K,\alpha) \\ (j\ j') : (S^{-1}L \oplus S^{-1}L^*, \begin{pmatrix} 0 & 1 \\ 0 & 0 \end{pmatrix}) \longrightarrow S^{-1}(K,\beta) \end{cases}$$

for some $j' \in \mathrm{Hom}_{S^{-1}A}(S^{-1}L^*, S^{-1}K)$ such that

$$\begin{cases} j^*\alpha j' = 1 \in \mathrm{Hom}_{S^{-1}A}(S^{-1}L^*, S^{-1}L^*) \\ j^*(\beta+\varepsilon\beta^*)j' = 1 \in \mathrm{Hom}_{S^{-1}A}(S^{-1}L^*, S^{-1}L^*) \ . \end{cases}$$

By Proposition 3.1.1 there exists an S-isomorphism $s \in \mathrm{Hom}_A(L^*,L^*)$ such that

$$j's = k \in \mathrm{Hom}_A(L^*,K) \subseteq \mathrm{Hom}_{S^{-1}A}(S^{-1}L^*, S^{-1}K)$$

(stabilizing $\begin{cases}(K,\alpha;L) \\ (K,\beta;L)\end{cases}$ if necessary). In the ε-quadratic case

$$j'^*\beta j' = 0 \in Q_\varepsilon(S^{-1}L^*) \ ,$$

so that $k^*\beta k \in \ker(S^{-1}:Q_\varepsilon(L^*) \longrightarrow Q_\varepsilon(S^{-1}L^*))$ and there exists an $S^{-1}A$-module morphism $\chi \in \mathrm{Hom}_{S^{-1}A}(S^{-1}L^*, S^{-1}L)$ such that

$$k^*\beta k = \chi - \varepsilon\chi^* \in \mathrm{Hom}_{S^{-1}A}(S^{-1}L^*, S^{-1}L) \ .$$

Applying Proposition 3.1.1 again let $t \in \mathrm{Hom}_A(L^*,L^*)$ be an S-isomorphism such that

$$\chi t \in \mathrm{Hom}_A(L^*,L) \subset \mathrm{Hom}_{S^{-1}A}(S^{-1}L^*, S^{-1}L) \ .$$

Replacing s,k,χ by $st,kt,t^*\chi t$ ensures that

$$k^*\beta k = 0 \in Q_\varepsilon(L^*) \ .$$

Define an S-isomorphism of f.g. projective A-modules

$$\begin{cases} u = j^*\alpha k : L^* \longrightarrow L^* \\ u = j^*(\beta+\varepsilon\beta^*)k : L^* \longrightarrow L^* \ . \end{cases}$$

The S-isomorphism of S-non-singular $\begin{cases}\text{(even) } \varepsilon\text{-symmetric} \\ \varepsilon\text{-quadratic}\end{cases}$ forms

over A

$$\begin{cases} \begin{pmatrix} u^* & 0 \\ 0 & 1 \end{pmatrix} : (L\oplus L^*, \begin{pmatrix} 0 & u \\ \varepsilon u^* & k^*\alpha k \end{pmatrix}) \longrightarrow (L\oplus L^*, \begin{pmatrix} 0 & 1 \\ \varepsilon & k^*\alpha k \end{pmatrix}) \\ \begin{pmatrix} u^* & 0 \\ 0 & 1 \end{pmatrix} : (L\oplus L^*, \begin{pmatrix} 0 & u \\ 0 & 0 \end{pmatrix}) \longrightarrow (L\oplus L^*, \begin{pmatrix} 0 & 1 \\ 0 & 0 \end{pmatrix}) \end{cases}$$

has non-singular range, corresponding by Proposition 3.4.6 i) to a lagrangian

$$F = \mathrm{coker}(u^*: L \longrightarrow L) \subseteq M$$

of the boundary $\begin{cases}\text{even } \varepsilon\text{-symmetric } (\varepsilon\text{-quadratic}) \\ \text{split } \varepsilon\text{-quadratic}\end{cases}$ linking form

over (A,S)

$$\begin{cases} \partial(L\oplus L^*, \begin{pmatrix} 0 & u \\ \varepsilon u^* & k^*\alpha k \end{pmatrix}) = (M,\lambda) \ (= (M,\lambda,\mu)) \\ \partial(L\oplus L^*, \begin{pmatrix} 0 & u \\ 0 & 0 \end{pmatrix}) = \widetilde{H}_\varepsilon(F) = (M,\lambda,\nu) \ . \end{cases}$$

The inclusion of the lagrangian

$$F \hookrightarrow M$$

is resolved by

$$\begin{array}{ccccccccc}
0 & \longrightarrow & L\oplus L^* & \xrightarrow{\begin{pmatrix} u^* & 0 \\ 0 & 1 \end{pmatrix}} & L\oplus L^* & \longrightarrow & F & \longrightarrow & 0 \\
 & & \Big\downarrow 1 & \begin{pmatrix} 0 & u \\ u^* & k^*\alpha k \end{pmatrix} & \Big\downarrow \begin{pmatrix} 0 & u \\ \varepsilon & k^*\alpha k \end{pmatrix} & & \Big\downarrow & & \\
0 & \longrightarrow & L\oplus L^* & \longrightarrow & L^*\oplus L & \longrightarrow & M & \longrightarrow & 0
\end{array}$$

with $k^*\alpha k = k^*(\beta+\varepsilon\beta^*)k = 0 \in \mathrm{Hom}_A(L^*,L)$ in the ε-quadratic case. The S-isomorphism of S-non-singular $\begin{cases} \text{(even) } \varepsilon\text{-symmetric} \\ \varepsilon\text{-quadratic} \end{cases}$ forms over A

$$\begin{cases} (j\ k) : (L\oplus L^*, \begin{pmatrix} 0 & u \\ \varepsilon u^* & k^*\alpha k \end{pmatrix}) \longrightarrow (K,\alpha) \\ (j\ k) : (L\oplus L^*, \begin{pmatrix} 0 & u \\ 0 & 0 \end{pmatrix}) \longrightarrow (K,\beta) \end{cases}$$

corresponds by Proposition 3.4.6 i) to a sublagrangian

$$G = \mathrm{coker}((j\ k) : L\oplus L^* \longrightarrow K) \subseteq M$$

of the $\begin{cases} \text{even } \varepsilon\text{-symmetric } (\varepsilon\text{-quadratic}) \\ \text{split } \varepsilon\text{-quadratic} \end{cases}$ linking form over (A,S) $\begin{cases} (M,\lambda)\ \ ((M,\lambda,\mu)) \\ (M,\lambda,\nu) = \widetilde{H}_\varepsilon(F) \end{cases}$, such that the inclusion

$$G \hookrightarrow M$$

has resolution

$$\begin{array}{ccccccccc}
0 & \longrightarrow & L\oplus L^* & \xrightarrow{(j\ k)} & K & \longrightarrow & G & \longrightarrow & 0 \\
 & & \Big\downarrow 1 & \begin{pmatrix} 0 & u \\ \varepsilon u^* & k^*\alpha k \end{pmatrix} & \Big\downarrow \begin{pmatrix} j^*\alpha \\ k^*\alpha \end{pmatrix} & & \Big\downarrow & & \\
0 & \longrightarrow & L\oplus L^* & \longrightarrow & L^*\oplus L & \longrightarrow & M & \longrightarrow & 0
\end{array}$$

with $\alpha = \beta+\varepsilon\beta^*$, $k^*\beta k = 0$ in the ε-quadratic case.

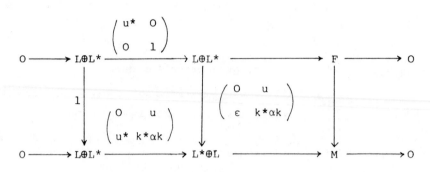

The above procedure associates an $\begin{cases} \varepsilon\text{-symmetric} \\ \text{even } \varepsilon\text{-symmetric} \\ \varepsilon\text{-quadratic} \end{cases}$

S-form over A $\begin{cases} (K,\alpha;L) \\ (K,\alpha;L) \\ (K,\beta;L) \end{cases}$ the $\begin{cases} \text{even } \varepsilon\text{-symmetric} \\ \varepsilon\text{-quadratic} \\ \text{split } \varepsilon\text{-quadratic} \end{cases}$ linking

formation over (A,S)

$$\begin{cases} (M,\lambda;F,G) \\ (M,\lambda,\mu;F,G) \\ (F,(\binom{\gamma}{\mu}),\theta)G) \end{cases} ,$$

where the hessian $(G,\gamma\hat{}\mu \in \mathrm{Hom}_A(G,G\hat{}),\theta:G \longrightarrow Q_{-\varepsilon}(A,S))$ is the $(-\varepsilon)$-quadratic linking form over (A,S) associated by Proposition 3.4.3 to the ε-quadratic S-formation over A

$$(H^{\varepsilon}(K^*);K^*,\mathrm{im}(\begin{pmatrix} -\bar{\varepsilon}\beta j & \beta^* k \\ j & k \end{pmatrix}:L\oplus L' \longrightarrow K^*\oplus K)) .$$

(For an even ε-symmetric S-form $(K,\alpha;L)$ Proposition 1.6.2 actually gives an extension of $j \in \mathrm{Hom}_{S^{-1}A}(S^{-1}L,S^{-1}K)$ to an isomorphism of non-singular even ε-symmetric forms over $S^{-1}A$

$$(j\ j') : H^{\varepsilon}(S^{-1}L) = (S^{-1}L\oplus S^{-1}L^*,\begin{pmatrix} 0 & 1 \\ \varepsilon & 0 \end{pmatrix}) \longrightarrow S^{-1}(K,\alpha) ,$$

leading to an S-isomorphism of S-non-singular even ε-symmetric forms over A

$$(j\ k) : (L\oplus L^*,\begin{pmatrix} 0 & u \\ \varepsilon u^* & 0 \end{pmatrix}) \longrightarrow (K,\alpha) .$$

In this way it can be proved that every ε-quadratic linking formation over (A,S) $(M,\lambda,\mu;F,G)$ is stably equivalent to one of the type $(H_{\varepsilon}(F);F,G)$).

It remains to show that the $\begin{cases}\text{even } \varepsilon\text{-symmetric } (\varepsilon\text{-quadratic}) \\ \text{split } \varepsilon\text{-quadratic}\end{cases}$

linking formations over (A,S) $\begin{cases}(M_r,\lambda_r;F_r,G_r) \ \ ((M_r,\lambda_r,\mu_r;F_r,G_r)) \\ (F_r,G_r)\end{cases}$

$(r = 1,2)$ associated to an $\begin{cases}\text{(even) } \varepsilon\text{-symmetric} \\ \varepsilon\text{-quadratic}\end{cases}$ S-form over A

$\begin{cases}(K,\alpha;L) \\ (K,\beta;L)\end{cases}$ using two different choices $(k_1,u_1),(k_2,u_2)$ of the pair

$$(k \in \mathrm{Hom}_A(K,L^*), u \in \mathrm{Hom}_A(L^*,L^*))$$

(that is two different extensions of the inclusion $j \in \mathrm{Hom}_A(L,K)$ to an S-isomorphism of S-non-singular $\begin{cases}\text{(even) } \varepsilon\text{-symmetric} \\ \varepsilon\text{-quadratic}\end{cases}$ forms over A

$$\begin{cases}(j\ k_r) : (L\oplus L^*, \begin{pmatrix} 0 & u_r \\ \varepsilon u_r^* & k_r^*\alpha k_r\end{pmatrix}) \longrightarrow (K,\alpha) \\ \\ (j\ k_r) : (L\oplus L^*, \begin{pmatrix} 0 & u_r \\ 0 & 0\end{pmatrix}) \longrightarrow (K,\beta)\end{cases} \quad (r = 1,2) \quad)$$

are stably equivalent. The two choices are related by an A-module morphism $h \in \mathrm{Hom}_A(L^*,L)$ and S-isomorphisms $v_1,v_2 \in \mathrm{Hom}_A(L^*,L^*)$ such that

$$u_1v_1 = u_2v_2 \in \mathrm{Hom}_A(L^*,L^*)$$

$$k_2v_2 - k_1v_1 = jh \in \mathrm{Hom}_A(L^*,K)$$

$$\begin{cases} v_1^*u_1^*h \in Q^{-\varepsilon}(L^*) \\ v_1^*u_1h \in Q\langle v_0\rangle^{-\varepsilon}(L^*)\ .\end{cases}$$

We shall consider separately the effects of the transformations

$$(k_1,u_1) \longmapsto (k_1v_1,u_1v_1) \longmapsto (k_1v_1+jh,u_1v_1) = (k_2v_2,u_2v_2) \longmapsto (k_2,u_2)\,.$$

If the choices $(k_1,u_1),(k_2,u_2)$ are related by an S-isomorphism $v \in \mathrm{Hom}_A(L^*,L^*)$ such that

$$u_1 = u_2 v \in \mathrm{Hom}_A(L^*,L^*) \ , \ k_1 = k_2 v \in \mathrm{Hom}_A(L^*,K)$$

then the sublagrangian H of $\begin{cases} (M_1,\lambda_1;F_1,G_1) \quad ((M_1,\lambda_1,\mu_1;F_1,G_1)) \\ (F_1,(\begin{pmatrix}\gamma_1\\ \mu_1\end{pmatrix},\theta_1)G_1) \end{cases}$

defined by the resolution

$$\begin{array}{ccccccccc}
0 & \longrightarrow & L\oplus L^* & \xrightarrow{\begin{pmatrix}1 & 0\\ 0 & v\end{pmatrix}} & L\oplus L^* & \longrightarrow & H & \longrightarrow & 0\\
 & & \Big\downarrow 1 & & \Big\downarrow (j\ k_2) & & \Big\downarrow & & \\
0 & \longrightarrow & L\oplus L^* & \xrightarrow{(j\ k_1)} & K & \longrightarrow & G_1 & \longrightarrow & 0
\end{array}$$

is such that

$$\begin{cases} (H^\perp/H,\lambda_1^\perp/\lambda_1;F_1\cap H^\perp,G_1/H) = (M_2,\lambda_2;F_2,G_2) \\ \qquad ((H^\perp/H,\lambda_1^\perp/\lambda,\mu_1^\perp/\mu;F_1\cap H^\perp,G_1/H) = (M_2,\lambda_2,\mu_2;F_2,G_2)) \\ (F_1\cap H^\perp,(\begin{pmatrix}[\gamma_1]\\ [\mu_1]\end{pmatrix},[\theta_1])G_1/H) = (F_2,(\begin{pmatrix}\gamma_2\\ \mu_2\end{pmatrix},\theta_2)G_2) \end{cases} .$$

Thus the linking formations associated to the choices (k_1,u_1), (k_2,u_2) are related by an elementary equivalence.

If the choices $(k_1,u_1),(k_2,u_2)$ are related by

$$u_1 = u_2 \in \mathrm{Hom}_A(L^*,L^*) \ , \ k_2 = k_1 + jh \in \mathrm{Hom}_A(L^*,K)$$

for some $h \in \mathrm{Hom}_A(L^*,L)$ such that

$$\begin{cases} u_1^* h \in Q^{-\varepsilon}(L^*) \\ u_1^* h \in Q\langle v_0\rangle^{-\varepsilon}(L^*) \end{cases}$$

there is defined an isomorphism of $\begin{cases} \text{even } \varepsilon\text{-symmetric } (\varepsilon\text{-quadratic}) \\ \text{split } \varepsilon\text{-quadratic} \end{cases}$

linking formations over (A,S)

$$\begin{cases} f : (M_1,\lambda_1;F_1,G_1) \longrightarrow (M_2,\lambda_2;F_2,G_2) \\ \qquad (f:(M_1,\lambda_1,\mu_1;F_1,G_1) \longrightarrow (M_2,\lambda_2,\mu_2;F_2,G_2)) \\ (1,g,\phi,\psi) : (F_1,G_1) \longrightarrow (F_2,G_2) \end{cases}$$

with

$$\begin{cases} \begin{array}{ccccccccc} 0 & \longrightarrow & L\oplus L^* & \xrightarrow{\begin{pmatrix} 0 & u \\ \varepsilon u_1^* & k_1^*\alpha k_1 \end{pmatrix}} & L^*\oplus L & \longrightarrow & M_1 & \longrightarrow & 0 \\ & & \Big\downarrow{\scriptstyle\begin{pmatrix} 1 & -h \\ 0 & 1 \end{pmatrix}} & & \Big\downarrow{\scriptstyle\begin{pmatrix} 1 & 0 \\ h & 0 \end{pmatrix}} & & \Big\downarrow{\scriptstyle f} & & \\ 0 & \longrightarrow & L\oplus L^* & \xrightarrow{\begin{pmatrix} 0 & u_2 \\ \varepsilon u_1^* & k_2^*\alpha k_2 \end{pmatrix}} & L^*\oplus L & \longrightarrow & M_2 & \longrightarrow & 0 \\ & & & & & & & & \\ 0 & \longrightarrow & L\oplus L^* & \xrightarrow{(j\ k_1)} & K & \longrightarrow & G_1 & \longrightarrow & 0 \\ & & \Big\downarrow{\scriptstyle\begin{pmatrix} 1 & -h \\ 0 & 1 \end{pmatrix}} & & \Big\downarrow{\scriptstyle 1} & & \Big\downarrow{\scriptstyle g} & & \\ 0 & \longrightarrow & L\oplus L^* & \xrightarrow{(j\ k_2)} & K & \longrightarrow & G_2 & \longrightarrow & 0 \end{array} \end{cases}$$

and $(F_1^{\wedge} = \mathrm{coker}(u_1:L^* \longrightarrow L^*), \phi \in \mathrm{Hom}_A(F_1^{\wedge},F_1), \psi:F_1^{\wedge} \longrightarrow Q_{-\varepsilon}(A,S))$ the $(-\varepsilon)$-quadratic linking form over (A,S) associated by Proposition 3.4.3 to the ε-quadratic S-formation over A

$$(H_\varepsilon(L);L,\mathrm{im}(\begin{pmatrix} -\bar{\varepsilon}h \\ u_1 \end{pmatrix}: L^* \longrightarrow L\oplus L^*)) \ .$$

This completes the verification that the stable equivalence class of the linking formation $\begin{cases} (M,\lambda;F,G) \quad ((M,\lambda,\mu;F,G)) \\ (F,G) \end{cases}$

associated to the S-form $\begin{cases}(K,\alpha;L)\\(K,\beta;L)\end{cases}$ is independent of the choice of (k,u).

ii) A connected S-acyclic 2-dimensional (even) $(-\varepsilon)$-symmetric complex over A $(C,\phi \in Q^2(C,-\varepsilon))$ is homotopy equivalent to one in normal form, by Proposition 3.5.1. Given such a complex in normal form we shall construct an (even) ε-symmetric linking formation over (A,S) $(M,\lambda;F,G)$, as follows.

Choose a cycle representative $\phi \in \mathrm{Hom}_{\mathbb{Z}[\mathbb{Z}_2]}(W,\mathrm{Hom}_A(C^*,C))_2$ in normal form, i.e. such that $\phi_0 \in \mathrm{Hom}_A(C^0,C_2)$ is an isomorphism (which we shall use as an identification), and $\tilde{\phi}_1 = 0 \in \mathrm{Hom}_A(C^2,C_1)$. It is thus possible to write the diagram of f.g. projective A-modules and A-module morphisms

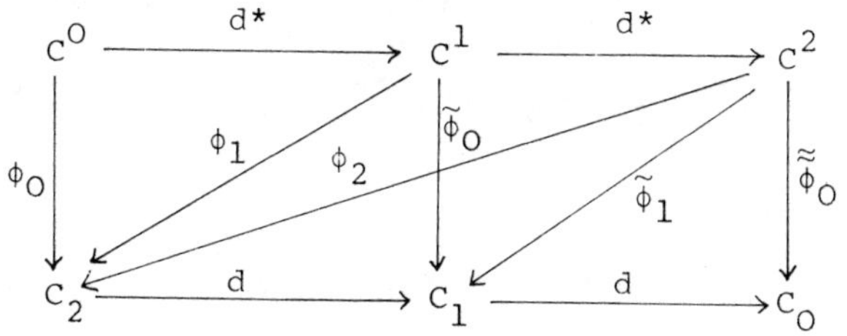

as

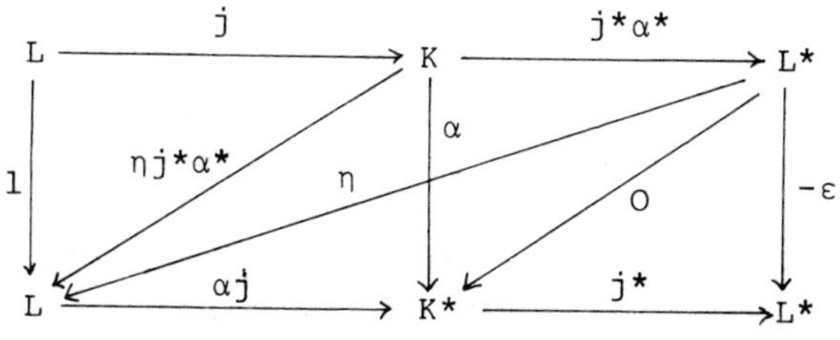

with $j \in \mathrm{Hom}_A(L,K)$, $\alpha \in \mathrm{Hom}_A(K,K^*)$, $\eta \in \mathrm{Hom}_A(L^*,L)$ such that

$$j^*\alpha j = 0 \in \mathrm{Hom}_A(L,L^*)$$

$$\alpha - \varepsilon\alpha^* + \alpha j\eta j^*\alpha^* = 0 \in \mathrm{Hom}_A(K,K^*)$$

$$\eta + \varepsilon\eta^* = 0 \in \mathrm{Hom}_A(L^*,L) \qquad .$$

The sequence of f.g. projective A-modules and A-module morphisms

$$0 \longrightarrow L \xrightarrow{\quad j \quad} K \xrightarrow{\quad j^*\alpha \quad} L^* \longrightarrow 0$$

becomes exact over $S^{-1}A$, so that there exists an A-module morphism $k \in \mathrm{Hom}_A(L^*,K)$ such that the A-module morphism

$$u = j^*\alpha k : L^* \longrightarrow L^*$$

is an S-isomorphism. Let (M,λ) be the non-singular (even) ε-symmetric linking form over (A,S) associated by Proposition 3.4.3 to the non-singular (even) $(-\varepsilon)$-symmetric S-formation over A

$$(L\oplus L^*\oplus L^*\oplus L, \begin{pmatrix} 0 & 0 & 1 & 0 \\ 0 & 0 & 0 & 1 \\ -\varepsilon & 0 & \eta & 0 \\ 0 & -\varepsilon & 0 & 0 \end{pmatrix}; \mathrm{im}(\begin{pmatrix} 1 & 0 \\ 0 & 1 \\ 0 & 0 \\ 0 & 0 \end{pmatrix} : L\oplus L^* \longrightarrow L\oplus L^*\oplus L^*\oplus L),$$

$$\mathrm{im}(\begin{pmatrix} 1 & 0 \\ 0 & 1 \\ 0 & u \\ \varepsilon u^* & k^*\alpha k \end{pmatrix} : L\oplus L^* \longrightarrow L\oplus L^*\oplus L^*\oplus L)) \qquad .$$

Define a lagrangian F and a sublagrangian G of (M,λ) by the resolutions

$$\begin{array}{ccccccccc}
0 & \longrightarrow & L\oplus L^* & \xrightarrow{\begin{pmatrix} u^* & 0 \\ 0 & 1 \end{pmatrix}} & L^*\oplus L & \longrightarrow & F & \longrightarrow & 0 \\
 & & \Big\downarrow{\scriptstyle 1} & & \Big\downarrow{\scriptstyle \begin{pmatrix} 0 & u \\ \varepsilon & k^*\alpha k \end{pmatrix}} & & \Big\downarrow & & \\
0 & \longrightarrow & L\oplus L^* & \xrightarrow{\begin{pmatrix} 0 & u \\ \varepsilon u^* & k^*\alpha k \end{pmatrix}} & L^*\oplus L & \longrightarrow & M & \longrightarrow & 0 \\
 & & \Big\uparrow{\scriptstyle 1} & & \Big\uparrow{\scriptstyle \begin{pmatrix} j^*\alpha \\ k^*\alpha \end{pmatrix}} & & \Big\uparrow & & \\
0 & \longrightarrow & L\oplus L^* & \xrightarrow{(j\ \ k)} & K & \longrightarrow & G & \longrightarrow & 0
\end{array} \quad .$$

Then $(M,\lambda;F,G)$ is the (even) ε-symmetric linking formation over (A,S) associated to the complex (C,ϕ).

Replacing ϕ by a different cycle representative $\phi' \in \mathrm{Hom}_{\mathbb{Z}[\mathbb{Z}_2]}(W,\mathrm{Hom}_A(C^*,C))_2$ of $\phi \in Q^2(C,-\varepsilon)$ replaces (α,η) by $(\alpha' \in \mathrm{Hom}_A(K,K^*),\eta' \in \mathrm{Hom}_A(L^*,L))$ such that for some $\chi \in \mathrm{Hom}_A(L^*,L)$

$$\alpha' - \alpha = \alpha j\chi j^*\alpha^* \in \mathrm{Hom}_A(K,K^*)$$

$$\eta' - \eta = -\chi + \varepsilon\chi^* \in \mathrm{Hom}_A(L^*,L) \ .$$

The A-module isomorphism $f \in \mathrm{Hom}_A(M,M')$ given by the resolution

$$\begin{array}{ccccccccc}
0 & \longrightarrow & L\oplus L^* & \xrightarrow{\begin{pmatrix} 0 & u \\ \varepsilon u^* & k^*\alpha k \end{pmatrix}} & L^*\oplus L & \longrightarrow & M & \longrightarrow & 0 \\
 & & \Big\downarrow{\scriptstyle 1} & \begin{pmatrix} 0 & u \\ \varepsilon u^* & k^*\alpha' k \end{pmatrix} & \Big\downarrow{\scriptstyle \begin{pmatrix} 1 & 0 \\ u^*\chi & 1 \end{pmatrix}} & & \Big\downarrow{\scriptstyle f} & & \\
0 & \longrightarrow & L\oplus L^* & \longrightarrow & L^*\oplus L & \longrightarrow & M' & \longrightarrow & 0
\end{array}$$

defines an isomorphism of the associated (even) ε-symmetric linking formations over (A,S)

$$f : (M,\lambda;F,G) \longrightarrow (M',\lambda';F',G') \ .$$

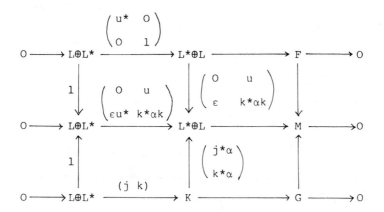

The verification that the stable equivalence class of $(M,\lambda;F,G)$ is independent of the choice of $(k \in \mathrm{Hom}_A(L^*,K), u \in \mathrm{Hom}_A(L^*,L^*))$ proceeds exactly as in the proof of i) above - indeed, if $(C,\phi \in Q^2(C,-\varepsilon))$ is even then $\phi \in Q\langle v_0\rangle^2(C,-\varepsilon)$ has a cycle representative with

$$\phi_2 \equiv \eta = 0 \in \mathrm{Hom}_A(L^*,L) \qquad (L \equiv C_2) \quad ,$$

in which case $(K,\alpha \in Q^{\varepsilon}(K);\mathrm{im}(j:L \longrightarrow K))$ is an ε-symmetric S-form over A and $(M,\lambda;F,G)$ is the associated even ε-symmetric linking formation over (A,S). Moreover, if $(C,\phi) = C^{-\varepsilon}(P)$ for some f.g. projective A-module P we can take

$$(k,u) = (\begin{pmatrix}1\\0\end{pmatrix}:P^* \longrightarrow P^*\oplus P, 1:P^* \longrightarrow P^*)$$

so that the associated even ε-symmetric linking formation is $(M,\lambda;F,G) = 0$.

We have shown that the stable equivalence class of the (even) ε-symmetric linking formation over (A,S) $(M,\lambda;F,G)$ associated to a connected S-acyclic 2-dimensional (even) $(-\varepsilon)$-symmetric complex over A (C,ϕ) in normal form depends only on the stable isomorphism class of (C,ϕ), which by Proposition 3.5.1 is just the homotopy equivalence class of (C,ϕ).

Conversely, given an (even) ε-symmetric linking formation over (A,S) $(M,\lambda;F,G)$ we shall construct a connected S-acyclic 2-dimensional (even) $(-\varepsilon)$-symmetric complex over A (C,ϕ) in normal form, such that $(M,\lambda;F,G)$ is in the stable equivalence class determined by (C,ϕ), as follows.

Let $(D,\eta \in Q^1(D,-\varepsilon))$ be an S-acyclic 1-dimensional (even) $(-\varepsilon)$-symmetric Poincaré complex over A associated to the non-singular (even) ε-symmetric linking form over (A,S) (M,λ) by Proposition 3.4.1, with D an S-acyclic 1-dimensional f.g. projective A-module chain complex

$$D : \ldots \longrightarrow 0 \longrightarrow D_1 \xrightarrow{d} D_0 \longrightarrow 0 \longrightarrow \ldots ,$$

such that

$$(H^1(D), \eta_0^S) = (M, \lambda) \quad .$$

Let $e \in \mathrm{Hom}_A(D^1, M)$ be the projection appearing in the resolution

$$0 \longrightarrow D^0 \xrightarrow{d^*} D^1 \xrightarrow{e} M \longrightarrow 0 ,$$

and define f.g. projective A-modules

$$D_1' = (e^{-1}(F))^* , \ D_1'' = (e^{-1}(G))^* .$$

Define A-module morphisms $f' \in \mathrm{Hom}_A(D_1', D_0)$, $f'' \in \mathrm{Hom}_A(D_1'', D_0)$ to be such that their duals are the inclusions

$$f'^* : D'^1 = e^{-1}(F) \longrightarrow D^1 , \ f''^* : D''^1 = e^{-1}(G) \longrightarrow D^1 ,$$

and let $d' \in \mathrm{Hom}_A(D_1', D_0)$, $d'' \in \mathrm{Hom}_A(D_1'', D_0)$ be the duals of the restrictions of $d^* \in \mathrm{Hom}_A(D^0, D^1)$

$$d'^* = d^*| : D^0 \longrightarrow D'^1 , \ d''^* = d^*| : D^0 \longrightarrow D''^1$$

(which are well-defined since

$$\mathrm{im}(d^* : D^0 \longrightarrow D^1) = e^{-1}(0) \subseteq e^{-1}(F) \cap e^{-1}(G) \quad).$$

Let D',D" be the S-acyclic 1-dimensional f.g. projective A-module chain complexes defined by

$$d_{D'} = d' : D_1' \longrightarrow D_0' = D_0 , \ D_r' = 0 \ (r \neq 0,1)$$

$$d_{D''} = d'' : D_1'' \longrightarrow D_0'' = D_0 , \ D_r'' = 0 \ (r \neq 0,1)$$

and let

$$f' : D \longrightarrow D' , \ f'' : D \longrightarrow D''$$

be the A-module chain maps defined by

$$\begin{array}{ccccccccccc}
D: & \dots \longrightarrow & 0 & \longrightarrow & D_1 & \xrightarrow{d} & D_0 & \longrightarrow & 0 & \longrightarrow & \dots \\
\downarrow f' & & & & \downarrow f' & & \downarrow 1 & & & & \\
D': & \dots \longrightarrow & 0 & \longrightarrow & D'_1 & \xrightarrow{d'} & D'_0 & \longrightarrow & 0 & \longrightarrow & \dots
\end{array}$$

$$\begin{array}{ccccccccccc}
D: & \dots \longrightarrow & 0 & \longrightarrow & D_1 & \xrightarrow{d} & D_0 & \longrightarrow & 0 & \longrightarrow & \dots \\
\downarrow f'' & & & & \downarrow f'' & & \downarrow 1 & & & & \\
D'': & \dots \longrightarrow & 0 & \longrightarrow & D''_1 & \xrightarrow{d''} & D''_0 & \longrightarrow & 0 & \longrightarrow & \dots
\end{array},$$

so that

$$f'^* = \text{inclusion} : H^1(D') = F \longrightarrow H^1(D) = M$$

$$f''^* = \text{inclusion} : H^1(D'') = G \longrightarrow H^1(D) = M \quad .$$

These inclusions define morphisms of (even) ε-symmetric linking forms over (A,S)

$$(F,0) \longrightarrow (M,\lambda) \quad , \quad (G,0) \longrightarrow (M,\lambda)$$

which by Proposition 3.4.1 correspond to maps of S-acyclic 1-dimensional (even) $(-\varepsilon)$-symmetric complexes over A

$$f' : (D,\eta) \longrightarrow (D',0) \quad , \quad f'' : (D,\eta) \longrightarrow (D'',0) \quad .$$

Thus there are defined an S-acyclic 2-dimensional (even) $(-\varepsilon)$-symmetric Poincaré pair over A

$$(f':D \longrightarrow D', (\delta\eta',\eta) \in Q^2(f',-\varepsilon))$$

and a connected S-acyclic 2-dimensional (even) $(-\varepsilon)$-symmetric pair over A

$$(f'':D \longrightarrow D'', (\delta\eta'',\eta) \in Q^2(f'',-\varepsilon)) \quad .$$

The union

$$(C,\phi) = (D' \cup_D D'', -\delta\eta' \cup_\eta \delta\eta'' \in Q^2(D' \cup_D D'',-\varepsilon))$$

(as defined in §1.7) is a connected S-acyclic 2-dimensional (even) $(-\varepsilon)$-symmetric complex over A. Next, we show how to recover the stable equivalence class of $(M,\lambda;F,G)$ from (C,ϕ).

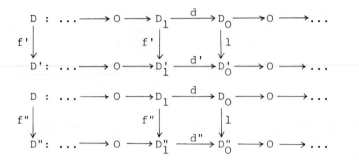

The relative $\mathbb{Z}_2$-hypercohomology classes $(\delta\eta',\eta) \in Q^2(f',-\varepsilon)$, $(\delta\eta'',\eta) \in Q^2(f'',-\varepsilon)$ are represented by A-module morphism

$$\eta_0 : D^0 \longrightarrow D_1 \ , \ \tilde{\eta}_0 : D^1 \longrightarrow D_0 \ , \ \eta_1 : D^1 \longrightarrow D_1$$

$$\delta\eta_0' : D'^1 \longrightarrow D_1' \ , \ \delta\eta_0'' : D''^1 \longrightarrow D_1''$$

such that

$$d\eta_0 + \tilde{\eta}_0 d^* = 0 : D^0 \longrightarrow D_0 \ , \ d\eta_1 - \tilde{\eta}_0 - \varepsilon\eta_0^* = 0 : D^1 \longrightarrow D_1 \ ,$$

$$\eta_1 + \varepsilon\eta_1^* = 0 : D^1 \longrightarrow D_1 \ , \ \eta_1 d^* + \eta_0 + \varepsilon\tilde{\eta}_0^* = 0 : D^0 \longrightarrow D_1 \ ,$$

$$f'\eta_0 = -\delta\eta_0' d'^* : D^0 \longrightarrow D_1' \ , \ \tilde{\eta}_0 f'^* = d'\delta\eta_0' : D'^1 \longrightarrow D_0 \ ,$$

$$f'\eta_1 f'^* = \delta\eta_0' - \varepsilon\delta\eta_0'^* : D'^1 \longrightarrow D_1' \ ,$$

$$f''\eta_0 = -\delta\eta_0'' d''^* : D^0 \longrightarrow D_1'' \ , \ \tilde{\eta}_0 f''^* = d''\delta\eta_0'' : D''^1 \longrightarrow D_0,$$

$$f''\eta_1 f''^* = \delta\eta_0'' - \varepsilon\delta\eta_0''^* : D''^1 \longrightarrow D_1'' \ .$$

Define a connected S-acyclic 2-dimensional (even) $(-\varepsilon)$-symmetric complex over A $(C',\phi' \in Q^2(C',-\varepsilon))$ by

$$d_{C'} = \begin{cases} \begin{pmatrix} f' \\ f'' \end{pmatrix} : C_2' = D_1 \longrightarrow C_1' = D_1' \oplus D_1'' \\ (d' \ \ -d'') : C_1' = D_1' \oplus D_1'' \longrightarrow C_0' = D_0 \end{cases} , \ C_r' = 0 \ (r \neq 0,1,2)$$

$$\phi_0' = \begin{cases} \eta_0 : C'^0 = D^0 \longrightarrow C_2' = D_1 \\ \begin{pmatrix} -\delta\eta_0' & 0 \\ 0 & \delta\eta_0'' \end{pmatrix} : C'^1 = D'^1 \oplus D''^1 \longrightarrow C_1' = D_1' \oplus D_1'' \\ \tilde{\eta}_0 : C'^2 = D^1 \longrightarrow C_0' = D_0 \end{cases}$$

$$\phi_1' = \begin{cases} (\eta_1 f'^* \ \ 0) : C'^1 = D'^1 \oplus D''^1 \longrightarrow C_2' = D_1 \\ \begin{pmatrix} 0 \\ f''\eta_1 \end{pmatrix} : C'^2 = D^1 \longrightarrow C_1' = D_1' \oplus D_1'' \end{cases}$$

$$\phi_2' = \eta_1 : C'^2 = D^1 \longrightarrow C_2' = D_1 \ .$$

There is defined a homotopy equivalence

$$h : (C,\phi) \longrightarrow (C',\phi') \ ,$$

with $h:C \longrightarrow C'$ the A-module chain equivalence given by

$$\begin{array}{ccccccccccc}
C : & \ldots \longrightarrow 0 \longrightarrow & D_1 & \xrightarrow{\begin{pmatrix} -f' \\ d \\ -f'' \end{pmatrix}} & D_1'\oplus D_0\oplus D_1'' & \xrightarrow{\begin{pmatrix} d' & 1 & 0 \\ 0 & 1 & d'' \end{pmatrix}} & D_0\oplus D_0 & \longrightarrow 0 \longrightarrow \ldots \\
\Big\downarrow h & & \Big\downarrow -1 & & \Big\downarrow \begin{pmatrix} 1 & 0 & 0 \\ 0 & 0 & 1 \end{pmatrix} & & \Big\downarrow (1\ \ -1) & \\
C' : & \ldots \longrightarrow 0 \longrightarrow & D_1 & \xrightarrow{\begin{pmatrix} f' \\ f'' \end{pmatrix}} & D_1'\oplus D_1'' & \xrightarrow{(d'\ \ -d'')} & D_0 & \longrightarrow 0 \longrightarrow \ldots \ .
\end{array}$$

Now apply the method of the proof of Proposition 3.5.1 to obtain from (C',ϕ') a homotopy equivalent S-acyclic 2-dimensional (even) $(-\varepsilon)$-symmetric complex over A $(C'',\phi'' \in Q^2(C'',-\varepsilon))$ with

$$d_{C''} = \begin{cases} \begin{pmatrix} f' \\ f'' \\ 0 \end{pmatrix} : C_2'' = D_1 \longrightarrow C_1'' = \ker((d'\ \ -d''\ \ -\eta_0^*) : D_1'\oplus D_1''\oplus D^1 \longrightarrow D_0) \\ (0\ \ 0\ \ 1) : C_1'' \longrightarrow C_0'' = D^1 \end{cases}$$

$$\phi_0'' = \begin{cases} 1 : C''^0 = D_1 \longrightarrow C_2'' = D_1 \\ \begin{pmatrix} -\delta\eta_0' & 0 & f' \\ -f''\eta_1 f'^* & \varepsilon\delta\eta_0''^* & f'' \\ \varepsilon f'^* & \varepsilon f''^* & 0 \end{pmatrix} : C''^1 = \mathrm{coker}\left(\begin{pmatrix} d'^* \\ -d''^* \\ -\eta_0 \end{pmatrix} : D^0 \longrightarrow D'^1\oplus D''^1\oplus D_1\right) \longrightarrow C_1'' \\ -\varepsilon : C''^2 = D^1 \longrightarrow C_0'' = D^1 \end{cases}$$

$$\phi_1'' = \begin{cases} (\eta_1 f'^* \quad \eta_1 f''^* \quad 0) : C''^1 \longrightarrow C_2'' = D^1 \\ 0 : C''^2 \longrightarrow C_1'' \end{cases}$$

$$\phi_2'' = \eta_1 : C''^2 = D^1 \longrightarrow C_2'' = D_1 \ .$$

As before, write

$$d^*_{C''} = j : C''^0 = L \longrightarrow C''^1 = K$$

$$\phi_0'' = \alpha : C''^1 = K \longrightarrow C_1'' = K^* \quad ,$$

and let $s \in \mathrm{Hom}_A(D_1, D_1)$ be an S-isomorphism such that

$$sf'^{-1} = i \in \mathrm{Hom}_A(D_1', D_1) \subseteq \mathrm{Hom}_{S^{-1}A}(S^{-1}D_1', S^{-1}D_1) \quad ,$$

so that the A-module morphism

$$k = \begin{pmatrix} i^* \\ 0 \\ 0 \end{pmatrix} : L^* = D^1 \longrightarrow K = \mathrm{coker}\left(\begin{pmatrix} d'^* \\ -d''^* \\ -\eta_0 \end{pmatrix} : D^0 \longrightarrow D'^1 \oplus D''^1 \oplus D_1\right)$$

is such that there is defined an S-isomorphism

$$u \equiv j^*\alpha k = \varepsilon s^* : L^* \longrightarrow L^* \ .$$

The (even) ε-symmetric linking formation over (A,S) $(M',\lambda';F',G')$ associated to the complex (C,ϕ) (via (C'',ϕ'')) is thus described by the resolutions

$$\begin{array}{ccccccccc}
0 & \longrightarrow & L\oplus L^* & \xrightarrow{\begin{pmatrix} \bar{\varepsilon} s & 0 \\ 0 & 1 \end{pmatrix}} & L\oplus L^* & \longrightarrow & F' & \longrightarrow & 0 \\
 & & \Big\downarrow 1 & \begin{pmatrix} 0 & \varepsilon s^* \\ s & k^*\alpha k \end{pmatrix} & \Big\downarrow \begin{pmatrix} 0 & \varepsilon s^* \\ \varepsilon & k^*\alpha k \end{pmatrix} & & \Big\downarrow & & \\
0 & \longrightarrow & L\oplus L^* & \longrightarrow & L^*\oplus L & \longrightarrow & M' & \longrightarrow & 0 \\
 & & \Big\uparrow 1 & & \Big\uparrow \begin{pmatrix} j^*\alpha \\ k^*\alpha \end{pmatrix} & & \Big\uparrow & & \\
0 & \longrightarrow & L\oplus L^* & \xrightarrow{(j \ \ k)} & K & \longrightarrow & G' & \longrightarrow & 0
\end{array} \ .$$

Let H' be the sublagrangian of $(M',\lambda';F',G')$ with resolution

$$\begin{array}{ccccccc}
0 & \longrightarrow & D_1\oplus D^1 & \xrightarrow{\begin{pmatrix}1 & 0\\ 0 & i^*\end{pmatrix}} & D_1\oplus D'^1 & \longrightarrow & H' & \longrightarrow & 0 \\
 & & \Big\downarrow 1 & & \Big\downarrow \begin{pmatrix}0 & 1\\ 0 & 0\\ 1 & 0\end{pmatrix} & & \Big\downarrow & & \\
0 & \longrightarrow & D_1\oplus D^1 & \xrightarrow{(j\ k)} & K & \longrightarrow & G' & \longrightarrow & 0
\end{array}$$

There is defined an isomorphism of (even) ε-symmetric linking formations over (A,S)

$$(M,\lambda;F,G) \longrightarrow (H'^{\perp}/H',\lambda'^{\perp}/\lambda';F'\cap H'^{\perp},G'/H') \quad ,$$

so that $(M',\lambda';F',G')$ is stably equivalent to $(M,\lambda;F,G)$.

Next, we consider the effect on the complex

$$(C,\phi) = (D'\cup_D D'',-\delta\eta'\cup_\eta\delta\eta'')$$

of the elementary equivalence

$$(M,\lambda;F,G) \longmapsto (\overline{M},\overline{\lambda};\overline{F},\overline{G}) = (H^{\perp}/H,\lambda^{\perp}/\lambda;F\cap H^{\perp},G/H)$$

determined by a sublagrangian H of $(M,\lambda;F,G)$. Let the inclusion $j\in \mathrm{Hom}_A(H,G)$ have resolution

$$\begin{array}{ccccccccc}
0 & \longrightarrow & D^0 & \xrightarrow{d'''^*} & D'''^1 & \longrightarrow & H & \longrightarrow & 0 \\
 & & \Big\downarrow 1 & & \Big\downarrow g^* & & \Big\downarrow j & & \\
0 & \longrightarrow & D^0 & \xrightarrow{d''^*} & D''^1 & \longrightarrow & G & \longrightarrow & 0
\end{array}$$

with $g^*\in \mathrm{Hom}_A(D'''^1,D''^1)$ the inclusion of $D'''^1 = e^{-1}(H)\subseteq D^1$ in $D''^1 = e^{-1}(G)\subseteq D^1$, where $e\in \mathrm{Hom}_A(D^1,M)$ is the projection (as above). The A-module chain map

$$f''' : D \longrightarrow D'''$$

defined by

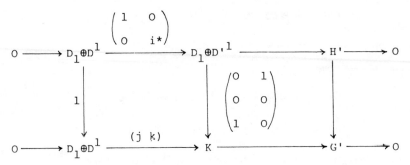

$$\begin{array}{ccccccccccc} D & : & \dots \longrightarrow & 0 & \longrightarrow & D_1 & \xrightarrow{d} & D_0 & \longrightarrow & 0 & \longrightarrow \dots \\ \downarrow f''' & & & & & \downarrow gf'' & & \downarrow 1 & & & \\ D''' & : & \dots \longrightarrow & 0 & \longrightarrow & D_1''' & \xrightarrow{d'''} & D_0 & \longrightarrow & 0 & \longrightarrow \dots \end{array}$$

is such that there exists a connected S-acyclic 2-dimensional (even) $(-\varepsilon)$-symmetric pair over A

$$(f''':D \longrightarrow D''', (\delta\eta''', \eta) \in Q^2(f''', -\varepsilon)) \ .$$

The S-acyclic 1-dimensional (even) $(-\varepsilon)$-symmetric Poincaré complex over A $(\bar{D}, \bar{\eta} \in Q^1(\bar{D}, -\varepsilon))$ obtained from (D,η) by S-acyclic surgery on $(f''':D \longrightarrow D''', (\delta\eta''', \eta))$ has associated non-singular (even) ε-symmetric linking form over (A,S)

$$(H^1(\bar{D}), \bar{\eta}_0^S) = (\bar{M}, \bar{\lambda}) \ .$$

Define S-acyclic 2-dimensional A-module chain complexes $\bar{D}', \bar{D}''$ by

$$\bar{D}' = C(f'\eta_0 f'''^*:D'''^{1-*} \longrightarrow D') \ ,$$

$$\bar{D}'' = \Omega C(g:D'' \longrightarrow D''') \ ,$$

and let

$$\bar{f}' : \bar{D} \longrightarrow \bar{D}' \ , \ \bar{f}'' : \bar{D} \longrightarrow \bar{D}''$$

be the A-module chain maps defined by

$$\bar{f}' = \begin{pmatrix} f' & 0 & 0 \\ 0 & 0 & 1 \end{pmatrix} : \bar{D}_r = D_r \oplus D_{r+1}''' \oplus D'''^{2-r} \longrightarrow \bar{D}_r' = D_r' \oplus D'''^{2-r}$$

$$\bar{f}'' = \begin{pmatrix} f'' & 0 & 0 \\ 0 & 1 & 0 \end{pmatrix} : \bar{D}_r = D_r \oplus D_{r+1} \oplus D'''^{2-r} \longrightarrow \bar{D}_r'' = D_r'' \oplus D_{r+1}'''$$

$$(r \in \mathbb{Z}) \ ,$$

so that

$$\bar{f}'^* = \text{inclusion} : H^1(\bar{D}') = \bar{F} \longrightarrow H^1(\bar{D}) = \bar{M}$$

$$\bar{f}''^* = \text{inclusion} : H^1(\bar{D}'') = \bar{G} \longrightarrow H^1(\bar{D}) = \bar{M} \ .$$

There exist connected S-acyclic 2-dimensional (even) $(-\varepsilon)$-symmetric pairs over A

$$(\bar{f}':\bar{D} \longrightarrow \bar{D}', (\overline{\delta\eta}', \bar{\eta}) \in Q^2(\bar{f}', -\varepsilon)) ,$$

$$(\bar{f}'':\bar{D} \longrightarrow \bar{D}'', (\overline{\delta\eta}'', \bar{\eta}) \in Q^2(\bar{f}'', -\varepsilon))$$

such that the union

$$(\bar{C}, \bar{\phi}) = (\bar{D}' \cup_{\bar{D}} \bar{D}'', -\overline{\delta\eta}' \cup_{\bar{\eta}} \overline{\delta\eta}'' \in Q^2(\bar{D}' \cup_{\bar{D}} \bar{D}'', -\varepsilon))$$

is a connected S-acyclic 2-dimensional (even) $(-\varepsilon)$-symmetric complex over A associated to $(\bar{M}, \bar{\lambda}; \bar{F}, \bar{G})$. It may be verified that $(\bar{C}, \bar{\phi})$ is homotopy equivalent to (C, ϕ), the complex associated to $(M, \lambda; F, G)$.

This completes the proof of ii). It remains to complete the proof of i). Given an $\begin{cases} \text{even } \varepsilon\text{-symmetric} \\ \varepsilon\text{-quadratic} \end{cases}$ linking formation over (A,S) $\begin{cases} (M,\lambda;F,G) \\ (M,\lambda,\mu;F,G) \end{cases}$ let $(K,\alpha),(K',\alpha')$ be S-non-singular $\begin{cases} \varepsilon\text{-symmetric} \\ \text{even } \varepsilon\text{-symmetric} \end{cases}$ forms over A such that

$$(M,\lambda) = \partial(K,\alpha) = \partial(K',\alpha')$$

(up to isomorphism), so that

$$F = \mathrm{coker}(f:K \longrightarrow K_F) = \mathrm{coker}(f':K' \longrightarrow K'_{F'})$$

$$G = \mathrm{coker}(g:K \longrightarrow K_G) = \mathrm{coker}(g':K' \longrightarrow K'_G)$$

for some S-isomorphisms of S-non-singular $\begin{cases} \varepsilon\text{-symmetric} \\ \text{even } \varepsilon\text{-symmetric} \end{cases}$ forms over A

$$f : (K,\alpha) \longrightarrow (K_F,\alpha_F) \ , \ f' : (K',\alpha') \longrightarrow (K'_F,\alpha'_F)$$

$$g : (K,\alpha) \longrightarrow (K_G,\alpha_G) \ , \ g' : (K',\alpha') \longrightarrow (K'_G,\alpha'_G)$$

with $(K_F,\alpha_F),(K'_F,\alpha'_F)$ non-singular. We have to show that the associated $\begin{cases}\varepsilon\text{-symmetric} \\ \text{even } \varepsilon\text{-symmetric}\end{cases}$ S-forms over A

$$(K_F \oplus K_G, \begin{pmatrix} \alpha_F & 0 \\ 0 & -\alpha_G \end{pmatrix} \in Q^{\varepsilon}(K_F \oplus K_G); \mathrm{im}(\begin{pmatrix} f \\ g \end{pmatrix}: K \longrightarrow K_F \oplus K_G))$$

$$(K'_F \oplus K'_G, \begin{pmatrix} \alpha'_F & 0 \\ 0 & -\alpha'_G \end{pmatrix} \in Q^{\varepsilon}(K'_F \oplus K'_G); \mathrm{im}(\begin{pmatrix} f' \\ g' \end{pmatrix}: K' \longrightarrow K'_F \oplus K'_G))$$

are stably isomorphic. The S-acyclic connected 2-dimensional $\begin{cases}(-\varepsilon)\text{-symmetric} \\ \text{even } (-\varepsilon)\text{-symmetric}\end{cases}$ complexes over A in normal form obtained from the S-forms (as in i)) are homotopy equivalent, since they correspond to the same linking formation, and are therefore stably isomorphic (by Proposition 3.5.1). It follows that the S-forms are stably isomorphic.

iii) A connected S-acyclic 2-dimensional $(-\varepsilon)$-quadratic complex over A $(C, \psi \in Q_2(C, -\varepsilon))$ is homotopy equivalent to one in normal form (by Proposition 3.5.1). Given such a complex in normal form we shall construct a split ε-quadratic linking formation over (A,S)

$$(F,G) = (F, ((\begin{smallmatrix}\gamma \\ \mu\end{smallmatrix}), \theta)G) \quad ,$$

as follows. Choose a cycle representative $\psi \in (W \otimes_{\mathbb{Z}[\mathbb{Z}_2]} \mathrm{Hom}_A(C^*, C))_2$ in normal form, i.e. such that $\psi_0 \in \mathrm{Hom}_A(C^0, C_2)$ is an isomorphism (which we shall use as an identification), $\tilde{\tilde{\psi}}_0 = 0 \in \mathrm{Hom}_A(C^2, C_0)$, $\psi_1 = 0 \in \mathrm{Hom}_A(C^1, C_0)$. It is thus possible to write the diagram of f.g. projective A-modules and A-module morphisms

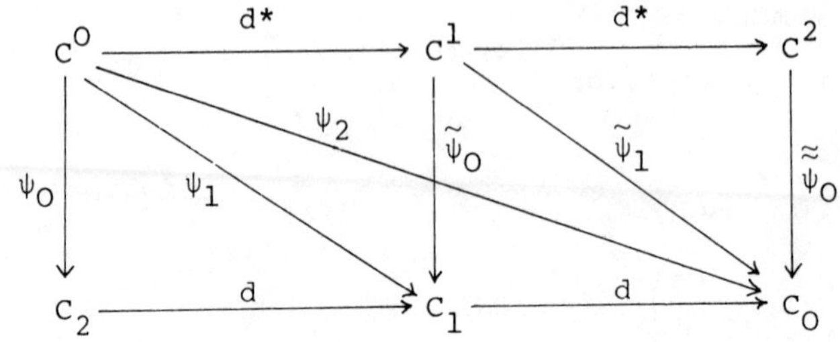

as

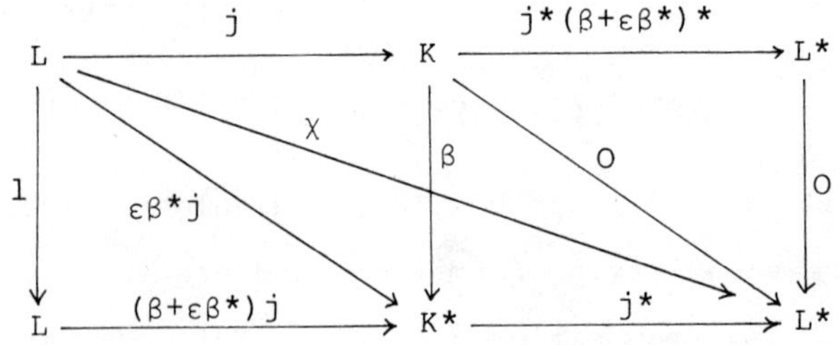

,

with $j \in \mathrm{Hom}_A(L,K)$, $\beta \in \mathrm{Hom}_A(K,K^*)$, $\chi \in \mathrm{Hom}_A(L,L^*)$ such that

$$j^*\beta j = \chi - \varepsilon\chi^* \in \mathrm{Hom}_A(L,L^*) \ .$$

Let (F,G) be the split ε-quadratic linking formation over (A,S) associated by i) to the ε-quadratic S-form over A

$$(K,\beta \in Q_\varepsilon(K);\mathrm{im}(j:L \longrightarrow K)) \ .$$

Replacing ψ by a different cycle representative $\psi' \in (W\otimes_{\mathbb{Z}[\mathbb{Z}_2]}\mathrm{Hom}_A(C^*,C))_2$ replaces β,χ by β',χ' such that

$$\beta' - \beta = \omega - \varepsilon\omega^* \in \mathrm{Hom}_A(K,K^*)$$

$$\chi' - \chi = j^*\omega j + \eta + \varepsilon\eta^* \in \mathrm{Hom}_A(L,L^*)$$

for some $\omega \in \mathrm{Hom}_A(K,K^*)$, $\eta \in \mathrm{Hom}_A(L,L^*)$. Neither the ε-quadratic S-form $(K,\beta;L)$ nor the split ε-quadratic linking formation (F,G) are affected by such a change.

In particular, if P is a f.g. projective A-module the ε-quadratic S-form over A $(K,\beta;L)$ associated to the contractible S-acyclic 2-dimensional $(-\varepsilon)$-quadratic complex over A $C_{-\varepsilon}(P)$ is given by

$$(K,\beta;L) = (P\oplus P^*, \begin{pmatrix} 0 & 1 \\ 0 & 0 \end{pmatrix}; P) \quad ,$$

corresponding by i) to a split ε-quadratic linking formation (F,G) stably equivalent to 0 (take $k = \begin{pmatrix} 1 \\ 0 \end{pmatrix} : L = P \longrightarrow K = P\oplus P^*$).

Thus the stable equivalence class of the linking formation (F,G) associated to (C,ψ) depends only on the stable isomorphism class of (C,ψ), which by Proposition 3.5.1 is the same as the homotopy equivalence class of (C,ψ).

Conversely, given a split ε-quadratic linking formation over (A,S) (F,G) we shall construct a connected S-acyclic 2-dimensional $(-\varepsilon)$-quadratic complex over A (C,ψ) in normal form, such that (F,G) is in the stable equivalence class determined by (C,ψ), as follows.

Let $(K,\beta;L)$ be an ε-quadratic S-form over A associated by i) to (F,G), and let $j \in \mathrm{Hom}_A(L,K)$ be the inclusion. For any lift $\tilde{\beta} \in \mathrm{Hom}_A(K,K^*)$ of $\beta \in Q_\varepsilon(K)$ there exists $\chi \in \mathrm{Hom}_A(L,L^*)$ such that

$$j^*\tilde{\beta}j = \chi - \varepsilon\chi^* \in \mathrm{Hom}_A(L,L^*) \quad .$$

Given such a choice $(\tilde{\beta},\chi) \in \mathrm{Hom}_A(K,K^*)\oplus\mathrm{Hom}_A(L,L^*)$ define $(C,\psi \in Q_2(C,-\varepsilon))$ by

$$d = \begin{cases} (\tilde{\beta}+\varepsilon\tilde{\beta}^*)j : C_2 = L \longrightarrow C_1 = K^* \\ j^* : C_1 = K^* \longrightarrow C_0 = L^* \end{cases} \quad , \quad C_r = 0 \ (r \neq 0,1,2)$$

$$\psi_0 = 1 : C^0 = L \longrightarrow C_2 = L \ , \ \tilde{\psi}_0 = \tilde{\beta} : C^1 = K \longrightarrow C_1 = K^* \ ,$$

$$\tilde{\tilde{\psi}}_0 = 0 : C^2 = L^* \longrightarrow C_0 = L^* \ , \ \psi_1 = \varepsilon\tilde{\beta}^* j : C^0 = L \longrightarrow C_1 = K^*$$

$$\tilde{\psi}_1 = 0 : C^1 = K \longrightarrow C_0 = L^* \ , \ \psi_2 = \chi : C^0 = L \longrightarrow C_0 = L^* \ .$$

The method of proof of i) shows that the homotopy equivalence class of (C,ψ) depends only on the stable equivalence class of (F,G) together with a choice of hessian $(\tilde{\beta},\chi) \in Q_\varepsilon(K,L)$ for the S-lagrangian L of $(K,\beta \in Q_\varepsilon(K))$, where

$$Q_\varepsilon(K,L) = \frac{\{(\tilde{\beta},\chi) \in \mathrm{Hom}_A(K,K^*) \oplus \mathrm{Hom}_A(L,L^*) \mid j^*\tilde{\beta}j = \chi - \varepsilon\chi^*\}}{\{(\omega - \varepsilon\omega^*, j^*\omega j + \eta + \varepsilon\eta^*) \mid (\omega,\eta) \in \mathrm{Hom}_A(K,K^*) \oplus \mathrm{Hom}_A(L,L^*)\}} \ .$$

(Define a <u>split ε-quadratic S-form over A</u> $(K,\tilde{\beta};L,\chi)$ to be an S-non-singular split ε-quadratic form over A $(K,\tilde{\beta} \in \tilde{Q}_\varepsilon(K))$ together with an S-lagrangian L and a choice of hessian $\chi \in Q_{-\varepsilon}(L)$. The homotopy equivalence classes of connected S-acyclic 2-dimensional $(-\varepsilon)$-quadratic complexes over A are in a natural one-one correspondence with the stable isomorphism classes of split ε-quadratic S-forms over A).

It remains to show that if $(C,\psi),(C,\bar{\psi})$ are the complexes associated to two different choices $\chi,\bar{\chi} \in \mathrm{Hom}_A(L,L^*)$ such that

$$j^*\tilde{\beta}j = \chi - \varepsilon\chi^* = \bar{\chi} - \varepsilon\bar{\chi}^* \in \mathrm{Hom}_A(L,L^*)$$

then $(C,\bar{\psi})$ is homotopy equivalent to a complex obtained from (C,ψ) by an S-acyclic $(-\varepsilon)$-quadratic surgery. As before, let $k \in \mathrm{Hom}_A(L^*,K)$ be such that

i) $u = j^*(\beta+\varepsilon\beta^*)k \in \mathrm{Hom}_A(L^*,L^*)$ is an S-isomorphism

ii) $k^*\beta k = 0 \in Q_\varepsilon(L^*)$.

Also, let $\chi' \in \mathrm{Hom}_A(L^*,L)$ be such that

$$k^*\tilde{\beta}k = \chi' - \varepsilon\chi'^* \in \mathrm{Hom}_A(L^*,L) \ ,$$

and let $(C',\psi' \in Q_2(C',-\varepsilon))$ be the connected S-acyclic 2-dimensional $(-\varepsilon)$-quadratic complex over A in normal form associated to the ε-quadratic S-form over A

$$(K,\beta \in Q_{\varepsilon}(K);\mathrm{im}(k:L^* \longrightarrow K))$$

with choice of hessian $(\widetilde{\beta},\chi') \in Q_{\varepsilon}(K,L^*)$, corresponding by i) to the split ε-quadratic linking formation over (A,S)

$$(F^{\wedge},(\begin{pmatrix} \mu \\ -\varepsilon\gamma \end{pmatrix},\theta)G) \quad .$$

Let $(C'',\psi'' \in Q_2(C'',-\varepsilon))$ be the connected S-acyclic 2-dimensional $(-\varepsilon)$-quadratic complex over A obtained from (C,ψ) by surgery on the connected S-acyclic 3-dimensional $(-\varepsilon)$-quadratic pair over A $(f:C \longrightarrow D,(\delta\psi,\psi) \in Q_3(f,-\varepsilon))$ defined by

$$d_D = \varepsilon u^* : D_2 = L \longrightarrow D_1 = L \quad , \quad D_r = 0 \ (r \neq 1,2) \ ,$$

$$f = \begin{cases} 1 : C_2 = L \longrightarrow D_2 = L \\ k^* : C_1 = K^* \longrightarrow D_1 = L \end{cases} ,$$

$$\delta\psi_1 = -\chi' : D^1 = L^* \longrightarrow D_1 = L \ ,$$

$$\delta\psi_0 = 0 : D^r \longrightarrow D_{3-r} \quad (r = 1,2) \ .$$

The A-module chain equivalence

$$h : C'' \longrightarrow C'$$

given by

$$\begin{array}{ccccccc}
C'':\ldots\longrightarrow 0 \longrightarrow & L\oplus L^* & \xrightarrow{\begin{pmatrix} (\beta+\varepsilon\beta^*)j & -(\beta+\varepsilon\beta^*)k \\ 1 & 0 \\ 0 & \overline{\varepsilon} u \end{pmatrix}} & K^*\oplus L\oplus L^* & \xrightarrow{\begin{pmatrix} j^* & 0 & \varepsilon \\ -k^* & \varepsilon u^* & 0 \end{pmatrix}} & L^*\oplus L & \longrightarrow 0 \longrightarrow \ldots \\
h \downarrow \qquad & \downarrow (0\ 1) & & \downarrow (-1\ \ (\beta+\varepsilon\beta^*)j\ \ 0) & & \downarrow (0\ 1) & \\
C':\ldots\longrightarrow 0 \longrightarrow & L^* & \xrightarrow{(\beta+\varepsilon\beta^*)k} & K^* & \xrightarrow{k^*} & L & \longrightarrow 0 \longrightarrow \ldots
\end{array}$$

defines a homotopy equivalence

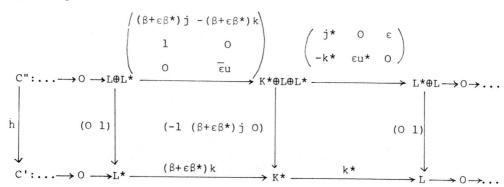

$$h : (C'',\psi'') \longrightarrow (C',\psi') .$$

Now (C,ψ) is homotopy equivalent to a complex obtained from (C'',ψ'') by S-acyclic surgery (since (C'',ψ'') is obtained from (C,ψ) by S-acyclic surgery), so that (C,ψ) is also homotopy equivalent to a complex obtained from (C',ψ') by S-acyclic surgery. The complex (C',ψ') is independent of the choices $\chi,\bar{\chi} \in \mathrm{Hom}_A(L,L^*)$, and the effect of successive S-acyclic surgeries may be composed (cf. Proposition I.4.7), so that $(C,\bar{\psi})$ is homotopy equivalent to a complex obtained from (C,ψ) by S-acyclic surgery.

[]

An <u>S-non-singular</u> $\left\{\begin{array}{l}\underline{\text{(even) } \varepsilon\text{-symmetric}} \\ \underline{\varepsilon\text{-quadratic}}\end{array}\right.$ <u>formation over A</u> $\left\{\begin{array}{l}(K,\alpha;I,J) \\ (K,\beta;I,J)\end{array}\right.$ ia non-singular $\left\{\begin{array}{l}\text{(even) } \varepsilon\text{-symmetric} \\ \varepsilon\text{-quadratic}\end{array}\right.$ form over A $\left\{\begin{array}{l}(K,\alpha) \\ (K,\beta)\end{array}\right.$ together with a lagrangian I and an S-lagrangian J. (An S-non-singular formation over A is an $S^{-1}A$-non-singular formation over A in the sense of §2.4). The induced $\left\{\begin{array}{l}\text{(even) } \varepsilon\text{-symmetric} \\ \varepsilon\text{-quadratic}\end{array}\right.$ formation over $S^{-1}A$ $\left\{\begin{array}{l}S^{-1}(K,\alpha;I,J) \\ S^{-1}(K,\beta;I,J)\end{array}\right.$ is non-singular, and it is stably isomorphic to 0 precisely when $\left\{\begin{array}{l}(K,\alpha;I,J) \\ (K,\beta;I,J)\end{array}\right.$ is an S-formation (i.e. $S^{-1}K = S^{-1}I \oplus S^{-1}J$).

The S-non-singular formation $\left\{\begin{array}{l}(K,\alpha;I,J) \\ (K,\beta;I,J)\end{array}\right.$ is <u>non-singular</u> if J is a lagrangian of $\left\{\begin{array}{l}(K,\alpha) \\ (K,\beta)\end{array}\right.$.

The boundary of an S-non-singular $\begin{cases} \varepsilon\text{-symmetric} \\ \text{even } \varepsilon\text{-symmetric} \\ \varepsilon\text{-quadratic} \end{cases}$

formation over A $\begin{cases} (K,\alpha \in Q^{\varepsilon}(K);I,J) \\ (K,\alpha \in Q\langle v_0\rangle^{\varepsilon}(K);I,J) \\ (K,\beta \in Q_{\varepsilon}(K);I,J) \end{cases}$ is the non-singular

$\begin{cases} \text{even } \varepsilon\text{-symmetric} \\ \varepsilon\text{-quadratic} \\ \text{split } \varepsilon\text{-quadratic} \end{cases}$ linking formation over A

$$\begin{cases} \partial(K,\alpha;I,J) = (M,\lambda;F,G) \\ \partial(K,\alpha;I,J) = (M,\lambda,\mu;F,G) \\ \partial(K,\beta;I,J) = (F,G) \end{cases}$$

associated (uniquely up to stable equivalence) to the

non-singular $\begin{cases} \varepsilon\text{-symmetric} \\ \text{even } \varepsilon\text{-symmetric} \\ \varepsilon\text{-quadratic} \end{cases}$ S-form over A $\begin{cases} (K,\alpha;J) \\ (K,\alpha;J) \\ (K,\beta;J) \end{cases}$.

An S-non-singular formation is non-singular if and only if its boundary is stably equivalent to 0.

(The boundary operations

$$\partial : (\text{S-non-singular formations over A}) \longrightarrow (\text{non-singular linking formations over (A,S)})$$

can also be expressed in terms of the "dual lattice" construction, by analogy with the corresponding expression in §3.4 for the boundary operations

$$\partial : (\text{S-non-singular forms over A}) \longrightarrow (\text{non-singular linking forms over (A,S)}).$$

A lattice $\begin{cases} (K,\alpha) \\ (K,\beta) \end{cases}$ in a non-singular $\begin{cases} \text{(even) } \varepsilon\text{-symmetric} \\ \varepsilon\text{-quadratic} \end{cases}$ form over $S^{-1}A \begin{cases} (Q,\phi) \\ (Q,\psi) \end{cases}$ (which is an $\begin{cases} \text{(even) } \varepsilon\text{-symmetric} \\ \varepsilon\text{-quadratic} \end{cases}$ form over A

such that $\begin{cases} S^{-1}(K,\alpha) = (Q,\phi) \\ S^{-1}(K,\beta) = (Q,\psi) \end{cases}$) is <u>non-singular</u> if it is a non-singular form over A, or equivalently if the lattice $K \subseteq Q$ is <u>self-dual</u>

$$\begin{cases} K^{\#} \equiv \{x \in Q \mid \phi(x)(K) \subseteq A \subseteq S^{-1}A\} = K \\ K^{\#} \equiv \{x \in Q \mid (\psi+\varepsilon\psi^*)(x)(K) \subseteq A \subseteq S^{-1}A\} = K \end{cases} .$$

Given an S-non-singular $\begin{cases} \text{(even) } \varepsilon\text{-symmetric} \\ \varepsilon\text{-quadratic} \end{cases}$ formation over A

$\begin{cases} (K,\alpha;I,J) \\ (K,\beta;I,J) \end{cases}$ there exist

i) a non-singular lattice $\begin{cases} (K_I,\alpha_I) \\ (K_I,\beta_I) \end{cases}$ in $\begin{cases} S^{-1}(K,\alpha) \\ S^{-1}(K,\beta) \end{cases}$ such that

$$K_I \cap S^{-1}I = I \subseteq S^{-1}K$$

is a lagrangian of $\begin{cases} (K_I,\alpha_I) \\ (K_I,\beta_I) \end{cases}$,

ii) a lattice $\begin{cases} (K_J,\alpha_J) \\ (K_J,\beta_J) \end{cases}$ in $\begin{cases} S^{-1}(K,\alpha) \\ S^{-1}(K,\beta) \end{cases}$ such that

$$K_J \cap S^{-1}J = J \subseteq S^{-1}K$$

is an S-lagrangian of $\begin{cases} (K_J,\alpha_J) \\ (K_J,\beta_J) \end{cases}$,

iii) a lattice $\begin{cases} (K',\alpha') \\ (K',\beta') \end{cases}$ in $\begin{cases} S^{-1}(K,\alpha) \\ S^{-1}(K,\beta) \end{cases}$ such that

$$K' \subseteq K_I \cap K_J \subseteq S^{-1}K \quad .$$

The boundary $\begin{cases}\text{even } \varepsilon\text{-symmetric } (\varepsilon\text{-quadratic}) \\ \text{split } \varepsilon\text{-quadratic}\end{cases}$ linking formation over (A,S) is given by

$$\begin{cases}\partial(K,\alpha;I,J) = (\partial(K',\alpha');K_I/K',K_J/K') \\ \partial(K,\beta;I,J) = (K_I/K',K_J/K')\end{cases},$$

using Proposition 3.4.6 i) to translate the S-isomorphisms

S-non-singular $\begin{cases}\text{(even) } \varepsilon\text{-symmetric} \\ \varepsilon\text{-quadratic}\end{cases}$ forms over A

$$\begin{cases}(K',\alpha') \longrightarrow (K_I,\alpha_I) \\ (K',\beta') \longrightarrow (K_I,\beta_I)\end{cases}, \quad \begin{cases}(K',\alpha') \longrightarrow (K_J,\alpha_J) \\ (K',\beta') \longrightarrow (K_J,\beta_J)\end{cases}$$

defined by the inclusions into the lagrangian K_I/K' of the boundary $\begin{cases}\text{even } \varepsilon\text{-symmetric } (\varepsilon\text{-quadratic}) \\ \text{split } \varepsilon\text{-quadratic}\end{cases}$ linking form over A(A,S)

$$\begin{cases}\partial(K',\alpha') = (K'^{\#}/K',\alpha'^{\#}/\alpha') \\ \partial(K',\beta') = (K'^{\#}/K',\beta'^{\#}/\beta') = \widetilde{H}_{\varepsilon}(K_I/K')\end{cases}$$

and the sublagrangian K_J/K).

The <u>boundary</u> $\begin{cases}\partial(M,\lambda) \\ \partial(M,\lambda) \\ \partial(M,\lambda,\mu)\end{cases}$ of an $\begin{cases}\varepsilon\text{-symmetric} \\ \text{even } \varepsilon\text{-symmetric linking} \\ \varepsilon\text{-quadratic}\end{cases}$ form over (A,S) $\begin{cases}(M,\lambda) \\ (M,\lambda) \\ (M,\lambda,\mu)\end{cases}$ is the non-singular $\begin{cases}\text{even } (-\varepsilon)\text{-symmetric} \\ (-\varepsilon)\text{-quadratic} \\ \text{split } (-\varepsilon)\text{-quadratic}\end{cases}$ linking formation over (A,S)

$$\begin{cases}\partial(M,\lambda) = (H^{-\varepsilon}(M);M,\Gamma_{(M,\lambda)}) \\ \partial(M,\lambda) = (H_{-\varepsilon}(M);M,\Gamma_{(M,\lambda)}) \\ \partial(M,\lambda,\mu) = (M,(\begin{pmatrix}1 \\ \lambda\end{pmatrix},\mu)M)\end{cases}$$

where

$$\Gamma_{(M,\lambda)} = \{(x,\lambda(x)) \in M\oplus M^{\wedge} \mid x \in M\} \subseteq M\oplus M^{\wedge}$$

is the <u>graph</u> lagrangian of (M,λ) in $H^{-\varepsilon}(M)$ (in $H_{-\varepsilon}(M)$ if (M,λ) is even).

The boundary operations on S-non-singular formations and linking forms are related by the factorization

$$\partial : \{\text{linking forms over } (A,S)\} = \{\text{S-formations over } A\} \longrightarrow \{\text{S-non-singular formations over } A\} \xrightarrow{\partial} \{\text{linking formations over } (A,S)\} .$$

Thus if the $\begin{cases} \text{(even) } (-\varepsilon)\text{-symmetric} \\ (-\varepsilon)\text{-quadratic} \end{cases}$ S-formation over A $\begin{cases} (K,\alpha;I,J) \\ (K,\beta;I,J) \end{cases}$ associated by Proposition 3.4.3 to an $\begin{cases} \text{(even) } \varepsilon\text{-symmetric} \\ \varepsilon\text{-quadratic} \end{cases}$ linking form over (A,S) $\begin{cases} (M,\lambda) \\ (M,\lambda,\mu) \end{cases}$ is regarded as an S-non-singular $\begin{cases} \text{(even) } (-\varepsilon)\text{-symmetric} \\ (-\varepsilon)\text{-quadratic} \end{cases}$ formation over A there is a natural identification (up to stable equivalence) of the boundary $\begin{cases} \text{even } (-\varepsilon)\text{-symmetric } ((-\varepsilon)\text{-quadratic}) \\ \text{split } (-\varepsilon)\text{-quadratic} \end{cases}$ linking formations over (A,S)

$$\begin{cases} \partial(K,\alpha;I,J) = \partial(M,\lambda) \\ \partial(K,\beta;I,J) = \partial(M,\lambda,\mu) . \end{cases}$$

(There is an evident analogy between the boundary operations

$$\partial : \{\text{linking forms}\} \longrightarrow \{\text{linking formations}\}$$

and the boundary operations of §1.6

$$\partial : \{\text{forms}\} \longrightarrow \{\text{formations}\} .$$

To complete the analogy we can also define boundary operations

$$\partial : \{\text{linking formations}\} \longrightarrow \{\text{linking forms}\}$$

corresponding to the boundary operations of §1.6

$$\partial : \{\text{formations}\} \longrightarrow \{\text{forms}\} .$$

The <u>boundary</u> of an $\begin{cases} \text{(even) } \varepsilon\text{-symmetric} \\ \varepsilon\text{-quadratic} \\ \text{split } \varepsilon\text{-quadratic} \end{cases}$ linking formation over (A,S)

$\begin{cases} (M,\lambda;F,G) \\ (M,\lambda,\mu;F,G) \\ (F,G) \end{cases}$ is the non-singular $\begin{cases} \text{(even) } \varepsilon\text{-symmetric} \\ \varepsilon\text{-quadratic} \\ \text{split } \varepsilon\text{-quadratic} \end{cases}$ linking form over (A,S)

$$\begin{cases} \partial(M,\lambda;F,G) = (G^{\perp}/G,\lambda^{\perp}/\lambda) \\ \partial(M,\lambda,\mu;F,G) = (G^{\perp}/G,\lambda^{\perp}/\lambda,\mu^{\perp}/\mu) \\ \partial(F,G) = (G^{\perp}/G,\lambda^{\perp}/\lambda,\nu^{\perp}/\nu) \quad (\widetilde{H}_{\varepsilon}(F) = (F\oplus F^{\wedge},\lambda,\nu)) \end{cases} .$$

An $\begin{cases} \varepsilon\text{-symmetric} \\ \text{even } \varepsilon\text{-symmetric} \\ \varepsilon\text{-quadratic} \\ \text{split } \varepsilon\text{-quadratic} \end{cases}$ linking form over (A,S) is non-singular

(resp. represents 0 in the Witt group $\begin{cases} L^{\varepsilon}(A,S) \\ L\langle v_0\rangle^{\varepsilon}(A,S) \\ L_{\varepsilon}(A,S) \\ \widetilde{L}_{\varepsilon}(A,S) \end{cases}$) if and only

if its boundary $\begin{cases} \text{even } (-\varepsilon)\text{-symmetric} \\ (-\varepsilon)\text{-quadratic} \\ \text{split } (-\varepsilon)\text{-quadratic} \\ \text{split } (-\varepsilon)\text{-quadratic} \end{cases}$ linking formation

over (A,S) is stably equivalent to 0 (resp. if it is isomorphic to the boundary of an $\begin{cases} \varepsilon\text{-symmetric} \\ \text{even } \varepsilon\text{-symmetric} \\ \varepsilon\text{-quadratic} \\ \text{split } \varepsilon\text{-quadratic} \end{cases}$ linking formation

over (A,S))).

<u>Proposition 3.5.3</u> Let $\begin{cases} (C,\phi \in Q\langle v_0\rangle^2(C,-\varepsilon)) \\ (C,\psi \in Q_2(C,-\varepsilon)) \end{cases}$ be an S-acyclic

2-dimensional $\begin{cases} \text{even } (-\varepsilon)\text{-symmetric} \\ (-\varepsilon)\text{-quadratic} \end{cases}$ Poincaré complex over A,

and let $\begin{cases} (M,\lambda;F,G) \\ (F,G) \end{cases}$ be an associated non-singular

$\begin{cases} \text{even } \varepsilon\text{-symmetric} \\ \text{split } \varepsilon\text{-quadratic} \end{cases}$ linking formation over (A,S).

i) The S-acyclic cobordism class $\begin{cases} (C,\phi) \in L^1(A,S,\varepsilon) \\ (C,\psi) \in L_1(A,S,\varepsilon) \end{cases}$

depends only on the stable equivalence class of $\begin{cases} (M,\lambda;F,G) \\ (F,G) \end{cases}$.

ii) $\begin{cases} (C,\phi) = 0 \in L^1(A,S,\varepsilon) \\ (C,\psi) = 0 \in L_1(A,S,\varepsilon) \end{cases}$ if and only if $\begin{cases} (M,\lambda;F,G) \\ (F,G) \end{cases}$ is

stably equivalent to the boundary $\begin{cases} \partial(K,\alpha;I,J) \\ \partial(K,\beta;I,J) \end{cases}$ of an

S-non-singular $\begin{cases} \varepsilon\text{-symmetric} \\ \varepsilon\text{-quadratic} \end{cases}$ formation over A $\begin{cases} (K,\alpha;I,J) \\ (K,\beta;I,J) \end{cases}$

such that

$$\begin{cases} S^{-1}(K,\alpha;I,J) = 0 \in M_S^{\varepsilon}(S^{-1}A) = L_S^1(S^{-1}A,\varepsilon) \\ S^{-1}(K,\beta;I,J) = 0 \in M_{\varepsilon}^S(S^{-1}A) = L_1^S(S^{-1}A,\varepsilon) \end{cases}.$$

$\begin{cases} \text{If } \ker(\hat{\delta}:\hat{H}^0(\mathbb{Z}_2;S^{-1}A/A,\varepsilon)\longrightarrow \hat{H}^1(\mathbb{Z}_2;A,\varepsilon)) = 0 \\ \text{For all } A,S,\varepsilon \end{cases}$ it is possible

to choose $\begin{cases}(K,\alpha;I,J)\\(K,\beta;I,J)\end{cases}$ to be an $\begin{cases}\varepsilon\text{-symmetric}\\ \varepsilon\text{-quadratic}\end{cases}$ S-formation over A

(i.e. such that $S^{-1}K = S^{-1}I\oplus S^{-1}J$), so that

$\begin{cases}(C,\phi) = 0 \in L^1(A,S,\varepsilon)\\(C,\psi) = 0 \in L_1(A,S,\varepsilon)\end{cases}$ if and only if $\begin{cases}(M,\lambda;F,G)\\(F,G)\end{cases}$ is stably

equivalent to the boundary $\begin{cases}\partial(N,\xi)\\ \partial(N,\xi,\rho)\end{cases}$ of a $\begin{cases}(-\varepsilon)\text{-symmetric}\\(-\varepsilon)\text{-quadratic}\end{cases}$

linking form over (A,S) $\begin{cases}(N,\xi)\\(N,\xi,\rho)\end{cases}$.

<u>Proof</u>: i) Immediate from Proposition 3.5.2 $\begin{cases}\text{ii)}\\\text{iii)}\end{cases}$.

ii) By the S-acyclic counterpart of Proposition 1.2.2 iii) an S-acyclic 2-dimensional $\begin{cases}\text{even } (-\varepsilon)\text{-symmetric}\\(-\varepsilon)\text{-quadratic}\end{cases}$ Poincaré complex over A $\begin{cases}(C,\phi\in Q\langle v_0\rangle^2(C,-\varepsilon))\\(C,\psi\in Q_2(C,-\varepsilon))\end{cases}$ represents 0 in $\begin{cases}L^1(A,S,\varepsilon)\\L_1(A,S,\varepsilon)\end{cases}$ if and only if it is homotopy equivalent to the boundary $\begin{cases}\partial(D,\eta)\\\partial(D,\zeta)\end{cases}$ of a connected S-acyclic 3-dimensional $\begin{cases}\text{even } (-\varepsilon)\text{-symmetric}\\(-\varepsilon)\text{-quadratic}\end{cases}$ complex over A $\begin{cases}(D,\eta\in Q\langle v_0\rangle^3(D,-\varepsilon))\\(D,\zeta\in Q_3(D,-\varepsilon))\end{cases}$ with D a f.g. projective A-module chain complex of the type

$$D : \ldots\longrightarrow 0\longrightarrow D_3\xrightarrow{d} D_2\xrightarrow{d} D_1\xrightarrow{d} D_0\longrightarrow 0\longrightarrow\ldots .$$

Let then $\begin{cases}(C,\phi) = \partial(D,\eta) \\ (C,\psi) = \partial(D,\zeta)\end{cases}$ be the boundary of such a complex $\begin{cases}(D,\eta) \\ (D,\zeta)\end{cases}$, and let $\begin{cases}(D',\eta' \in Q\langle v_0\rangle^3(D',-\varepsilon)) \\ (D',\zeta' \in Q_3(D',-\varepsilon))\end{cases}$ be the connected 3-dimensional $\begin{cases}\text{even } (-\varepsilon)\text{-symmetric} \\ (-\varepsilon)\text{-quadratic}\end{cases}$ complex obtained from $\begin{cases}(D,\eta) \\ (D,\zeta)\end{cases}$ by surgery on the connected 4-dimensional $\begin{cases}\text{even } (-\varepsilon)\text{-symmetric} \\ (-\varepsilon)\text{-quadratic}\end{cases}$ pair over A $\begin{cases}(f:D \longrightarrow \delta D,(0,\eta) \in Q\langle v_0\rangle^4(f,-\varepsilon)) \\ (f:D \longrightarrow \delta D,(0,\zeta) \in Q_4(f,-\varepsilon))\end{cases}$ defined by

$$f = 1 : D_3 \longrightarrow \delta D_3 = D_3 \ , \ \delta D_r = 0 \ (r \neq 3) \ .$$

Then $\begin{cases}(D',\eta') = \bar{S}(D'',\eta'') \\ (D',\zeta') = \bar{S}(D'',\zeta'')\end{cases}$ is the skew-suspension of a 1-dimensional $\begin{cases}\varepsilon\text{-symmetric} \\ \varepsilon\text{-quadratic}\end{cases}$ complex over A $\begin{cases}(D'',\eta'' \in Q^1(D'',\varepsilon)) \\ (D'',\zeta'' \in Q_1(D'',\varepsilon))\end{cases}$ such that $\begin{cases}S^{-1}(D'',\eta'') \\ S^{-1}(D'',\zeta'')\end{cases}$ is Poincaré and null-cobordant over $S^{-1}A$.

The homotopy equivalence classes of 1-dimensional $\begin{cases}\varepsilon\text{-symmetric} \\ \varepsilon\text{-quadratic}\end{cases}$ $S^{-1}A$-Poincaré complexes over A are in a natural one-one correspondence with the stable isomorphism classes of S-non-singular $\begin{cases}\varepsilon\text{-symmetric} \\ \text{split } \varepsilon\text{-quadratic}\end{cases}$ formations over A (by a straightforward generalization of Proposition 1.6.4).

In particular, the S-non-singular $\begin{cases}\varepsilon\text{-symmetric} \\ \varepsilon\text{-quadratic}\end{cases}$ formation over A associated to $\begin{cases}(D'',\eta'')\\(D'',\zeta'')\end{cases}$ is given up to stable isomorphism by

$$\begin{cases}(K,\alpha;I,J) = (D_2\oplus D^2, \begin{pmatrix}0 & 1\\ \varepsilon & \eta_1\end{pmatrix};D_2, \mathrm{im}(\begin{pmatrix}d & \eta_0\\ 0 & d^*\end{pmatrix}:D_3\oplus D^1 \longrightarrow D_2\oplus D^2))\\ (K,\beta;I,J) = (D_2\oplus D^2, \begin{pmatrix}0 & 1\\ 0 & 0\end{pmatrix};D_2, \mathrm{im}(\begin{pmatrix}d & (1+T_{-\varepsilon})\zeta_0\\ 0 & d^*\end{pmatrix}:D_3\oplus D^1 \longrightarrow D_2\oplus D^2))\end{cases}$$

and is such that

$$\begin{cases}S^{-1}(K,\alpha;I,J) = S^{-1}(D'',\eta'') = 0 \in M^{\varepsilon}_S(S^{-1}A) = L^1_S(S^{-1}A,\varepsilon)\\ S^{-1}(K,\beta;I,J) = S^{-1}(D'',\zeta'') = 0 \in M^S_{\varepsilon}(S^{-1}A) = L^S_1(S^{-1}A,\varepsilon)\ .\end{cases}$$

The non-singular $\begin{cases}\text{even } \varepsilon\text{-symmetric}\\ \text{split } \varepsilon\text{-quadratic}\end{cases}$ linking formation over (A,S) associated to $\begin{cases}(C,\phi)\\(C,\psi)\end{cases}$ is the boundary

$$\begin{cases}(M,\lambda;F,G) = \partial(K,\alpha;I,J)\\ (F,G) = \partial(K,\beta;I,J)\end{cases}\ .$$

Since D is S-acyclic there exists an A-module morphism $g \in \mathrm{Hom}_A(D_2,D_3)$ such that the composite

$$s = gd : D_3 \xrightarrow{d} D_2 \xrightarrow{g} D_3$$

is an S-isomorphism. Let

$$\widetilde{f} : D \longrightarrow \delta\widetilde{D}$$

be the A-module chain map defined by

$$\begin{array}{ccccccccccccc} D: & \ldots \longrightarrow & 0 & \longrightarrow & D_3 & \xrightarrow{d} & D_2 & \xrightarrow{d} & D_1 & \xrightarrow{d} & D_0 & \longrightarrow 0 \longrightarrow & \ldots \\ \downarrow{\scriptstyle \widetilde{f}} & & & & \downarrow{\scriptstyle 1} & & \downarrow{\scriptstyle g} & & \downarrow & & \downarrow & & \\ \delta\widetilde{D}: & \ldots \longrightarrow & 0 & \longrightarrow & D_3 & \xrightarrow{s} & D_3 & \longrightarrow & 0 & \longrightarrow & 0 & \longrightarrow 0 \longrightarrow & \ldots ,\end{array}$$

so that $\begin{cases} \text{if } v_0^S(\eta) = 0 : H^3(D) \longrightarrow \hat{H}^0(\mathbb{Z}_2;S^{-1}A/A,\varepsilon) \\ - \end{cases}$ then

$$\begin{cases} \tilde{f}^{\%}(\eta) = 0 \in Q\langle v_0\rangle^3(\delta\tilde{D},-\varepsilon) \\ \tilde{f}_{\%}(\zeta) = 0 \in Q_3(\delta\tilde{D},-\varepsilon) \end{cases} .$$

The connected S-acyclic 3-dimensional $\begin{cases} \text{even } (-\varepsilon)\text{-symmetric} \\ (-\varepsilon)\text{-quadratic} \end{cases}$

complex over A $\begin{cases} (\tilde{D}',\tilde{\eta}' \in Q\langle v_0\rangle^3(\tilde{D}',-\varepsilon)) \\ (\tilde{D}',\tilde{\zeta}' \in Q_3(\tilde{D}',-\varepsilon)) \end{cases}$ obtained from $\begin{cases} (D,\eta) \\ (D,\zeta) \end{cases}$

by surgery on the pair $\begin{cases} (\tilde{f}:D \longrightarrow \delta\tilde{D},(0,\eta) \in Q\langle v_0\rangle^4(\tilde{f},-\varepsilon)) \\ (\tilde{f}:D \longrightarrow \delta\tilde{D},(0,\zeta) \in Q_4(\tilde{f},-\varepsilon)) \end{cases}$ is

the skew-suspension $\begin{cases} (\tilde{D}',\tilde{\eta}') = \bar{S}(\tilde{D}'',\tilde{\eta}'') \\ (\tilde{D}',\tilde{\zeta}') = \bar{S}(\tilde{D}'',\tilde{\zeta}'') \end{cases}$ of an S-acyclic

1-dimensional $\begin{cases} \varepsilon\text{-symmetric} \\ \varepsilon\text{-quadratic} \end{cases}$ complex over A $\begin{cases} (\tilde{D}'',\tilde{\eta}'' \in Q^1(\tilde{D}'',\varepsilon)) \\ (\tilde{D}'',\tilde{\zeta}'' \in Q_1(\tilde{D}'',\varepsilon)) \end{cases}$.

The $\begin{cases} (-\varepsilon)\text{-symmetric} \\ (-\varepsilon)\text{-quadratic} \end{cases}$ linking form over (A,S) associated to $\begin{cases} (\tilde{D}'',\tilde{\eta}'') \\ (\tilde{D}'',\tilde{\zeta}'') \end{cases}$

$$\begin{cases} (N,\xi) = (H^1(\tilde{D}''),\tilde{\eta}''^S_0) \\ (N,\xi,\rho) = (H^1(\tilde{D}''),(1+T_\varepsilon)\tilde{\zeta}''^S_0,pv_S^0(\tilde{\zeta}'')) \end{cases}$$

is such that up to stable equivalence

$$\begin{cases} (M,\lambda;F,G) = \partial(N,\xi) \\ (F,G) = \partial(N,\xi,\rho) \end{cases} .$$

If $\ker(\hat{\delta}:\hat{H}^0(\mathbb{Z}_2;S^{-1}A/A,\varepsilon) \longrightarrow \hat{H}^1(\mathbb{Z}_2;A,\varepsilon)) = 0$ then for any S-acyclic 3-dimensional even $(-\varepsilon)$-symmetric complex over A $(D,\eta \in Q\langle v_0\rangle^3(D,-\varepsilon))$ we have

$$v_0(\eta) = 0 : H^3(D) \xrightarrow{v_0^S(\eta)} \hat{H}^0(\mathbb{Z}_2;S^{-1}A/A,\varepsilon) \xrightarrow{\hat{\delta}} \hat{H}^1(\mathbb{Z}_2;A,\varepsilon)$$

(by Proposition 3.3.1 ii)), and so $v_0^S(\eta) = 0$.

Conversely, given an S-non-singular $\begin{cases}\varepsilon\text{-symmetric}\\ \varepsilon\text{-quadratic}\end{cases}$ formation over A $\begin{cases}(K,\alpha;I,J)\\ (K,\beta;I,J)\end{cases}$ such that

$$\begin{cases} S^{-1}(K,\alpha;I,J) = 0 \in M^{\varepsilon}_{S}(S^{-1}A) \\ S^{-1}(K,\beta;I,J) = 0 \in M^{S}_{\varepsilon}(S^{-1}A) \end{cases}$$

we have to show that the S-acyclic 2-dimensional $\begin{cases}\text{even } (-\varepsilon)\text{-symmetric}\\ (-\varepsilon)\text{-quadratic}\end{cases}$ Poincaré complex over A $\begin{cases}(C,\phi \in Q\langle v_0\rangle^2(C,-\varepsilon))\\ (C,\psi \in Q_2(C,-\varepsilon))\end{cases}$ associated to the boundary $\begin{cases}\text{even } \varepsilon\text{-symmetric}\\ \text{split } \varepsilon\text{-quadratic}\end{cases}$ linking formation over (A,S) $\begin{cases}\partial(K,\alpha;I,J) = (M,\lambda;F,G)\\ \partial(K,\beta;I,J) = (F,G)\end{cases}$ is an S-acyclic boundary. As I is a lagrangian of $\begin{cases}(K,\alpha)\\ (K,\beta)\end{cases}$ we can identify

$$\begin{cases} (K,\alpha) = (I\oplus I^*, \begin{pmatrix} 0 & 1 \\ \varepsilon & \theta \end{pmatrix}) \\ (K,\beta) = (I\oplus I^*, \begin{pmatrix} 0 & 1 \\ 0 & 0 \end{pmatrix}) \end{cases}$$

for some ε-symmetric form $(I^*,\theta \in Q^{\varepsilon}(I^*))$ (by Proposition 1.6.2). Write the inclusion of J in $K = I\oplus I^*$ as

$$\begin{pmatrix} j \\ k \end{pmatrix} : J \longrightarrow I\oplus I^* \quad ,$$

such that in the ε-quadratic case

$$j^*k = \chi - \varepsilon\chi^* : J \longrightarrow J^*$$

for some $(-\varepsilon)$-quadratic form $(J,\chi \in Q_{-\varepsilon}(J))$. Define a 1-dimensional $\begin{cases}\varepsilon\text{-symmetric}\\ \varepsilon\text{-quadratic}\end{cases}$ $S^{-1}A$-Poincaré complex over A $\begin{cases}(D,\eta \in Q^1(D,\varepsilon))\\ (D,\zeta \in Q_1(D,\varepsilon))\end{cases}$ by

$$d = k^* : D_1 = I \longrightarrow D_0 = J^* \quad , \quad D_r = 0 \ (r \neq 0,1)$$

$$\begin{cases} \eta_0 = \begin{cases} \varepsilon j : D^0 = J \longrightarrow D_1 = I \\ j^* + k^*\theta : D^1 = I^* \longrightarrow D_0 = J^* \end{cases} , \\ \qquad \eta_1 = \theta : D^1 = I^* \longrightarrow D_1 = I \\ \zeta_0 = \begin{cases} \varepsilon j : D^0 = J \longrightarrow D_1 = I \\ 0 : D^1 = I^* \longrightarrow D_0 = J^* \end{cases} , \ \zeta_1 = -\chi : D^0 = J \longrightarrow D_0 = J^* . \end{cases}$$

Now

$$\begin{cases} S^{-1}(D,\eta) = S^{-1}(K,\alpha;I,J) = 0 \in L^1_S(S^{-1}A,\varepsilon) = M^\varepsilon_S(S^{-1}A) \\ S^{-1}(D,\zeta) = S^{-1}(K,\beta;I,J) = 0 \in L^S_1(S^{-1}A,\varepsilon) = M^S_\varepsilon(S^{-1}A) \end{cases} ,$$

so that there exists a 2-dimensional $\begin{cases} \varepsilon\text{-symmetric} \\ \varepsilon\text{-quadratic} \end{cases}$ $S^{-1}A$-Poincaré

pair ovet A $\begin{cases} (f:D \longrightarrow \delta D, (\delta\eta,\eta) \in Q^2(f,\varepsilon)) \\ (f:D \longrightarrow \delta D, (\delta\zeta,\zeta) \in Q_2(f,\varepsilon)) \end{cases}$. Let

$\begin{cases} (D',\eta' \in Q\langle v_0\rangle^3(D',-\varepsilon)) \\ (D',\zeta' \in Q_3(D',-\varepsilon)) \end{cases}$ be the connected S-acyclic 3-dimensional

$\begin{cases} \text{even } (-\varepsilon)\text{-symmetric} \\ (-\varepsilon)\text{-quadratic} \end{cases}$ complex over A obtained from the

skew-suspension $\begin{cases} \bar{S}(D,\eta) \\ \bar{S}(D,\zeta) \end{cases}$ by surgery on the skew-suspension

$\begin{cases} \bar{S}(f:D \longrightarrow \delta D, (\delta\eta,\eta)) \\ \bar{S}(f:D \longrightarrow \delta D, (\delta\zeta,\zeta)) \end{cases}$. The boundary $\begin{cases} \text{even } \varepsilon\text{-symmetric} \\ \text{split } \varepsilon\text{-quadratic} \end{cases}$

linking formation over (A,S) $\begin{cases} \partial(K,\alpha;I,J) \\ \partial(K,\beta;I,J) \end{cases}$ is the linking

formation associated to the S-acyclic boundary

$\begin{cases} (C,\phi \in Q\langle v_0\rangle^2(C,-\varepsilon)) = \partial(D',\eta') \\ (C,\psi \in Q_2(C,-\varepsilon)) = \partial(D',\zeta') \end{cases}$ (= $\begin{cases} \partial\bar{S}(D,\eta) \\ \partial\bar{S}(D,\zeta) \end{cases}$, up to homotopy

equivalence).

[]

Define the <u>Witt group of</u> $\left\{\begin{array}{l}\underline{\varepsilon\text{-symmetric}}\\ \underline{\text{even }\varepsilon\text{-symmetric}}\\ \underline{\varepsilon\text{-quadratic}}\\ \underline{\text{split }\varepsilon\text{-quadratic}}\end{array}\right.$ <u>linking</u>

<u>formations over (A,S)</u> $\left\{\begin{array}{l}M^{\varepsilon}(A,S)\\ M\langle v_0\rangle^{\varepsilon}(A,S)\\ M_{\varepsilon}(A,S)\\ \widetilde{M}_{\varepsilon}(A,S)\end{array}\right.$ to be the abelian group

with one generator for each isomorphism class of non-singular

$$\left\{\begin{array}{l}\varepsilon\text{-symmetric}\\ \text{even }\varepsilon\text{-symmetric}\\ \varepsilon\text{-quadratic}\\ \text{split }\varepsilon\text{-quadratic}\end{array}\right. \text{ linking formations over } (A,S) \left\{\begin{array}{l}(M,\lambda;F,G)\\ (M,\lambda;F,G)\\ (M,\lambda,\mu;F,G)\\ (F,G)\end{array}\right.$$

subject to the relations:

in the (even) ε-symmetric case

$$(M,\lambda;F,G) + (M',\lambda';F',G') = (M\oplus M',\lambda\oplus\lambda';F\oplus F',G\oplus G')$$

$$(M,\lambda;F,G) + (M,\lambda;G,H) = (M,\lambda;F,H)$$

$$(M,\lambda;F,G) = (L^{\perp}/L,\lambda^{\perp}/\lambda;F\cap L^{\perp},G/L)$$

if L is a sublagrangian of $(M,\lambda;F,G)$

$$(M,\lambda;F,G) = (L^{\perp}/L,\lambda^{\perp}/\lambda;F/L,G/L)$$

if L is a sublagrangian of $(M,\lambda;F,G)$ such that $L \subseteq F\cap G$,

similarly in the ε-quadratic case,

in the split ε-quadratic case,

$$(F,G) + (F',G') = (F\oplus F',G\oplus G')$$

$(F,G) = (F\cap L^{\perp},G/L)$ if L is a sublagrangian of (F,G)

$\partial(M,\lambda,\mu) = 0$ if (M,λ,μ) is a $(-\varepsilon)$-quadratic linking form over (A,S).

In particular, stably equivalent linking formations represent the same element in the Witt group. There are defined forgetful maps

$$M\langle v_0\rangle^{\varepsilon}(A,S) \longrightarrow M^{\varepsilon}(A,S) \ ; \ (M,\lambda;F,G) \longmapsto (M,\lambda;F,G)$$

$$M_{\varepsilon}(A,S) \longrightarrow M\langle v_0\rangle^{\varepsilon}(A,S) \ ; \ (M,\lambda,\mu;F,G) \longmapsto (M,\lambda;F,G)$$

$$\widetilde{M}_{\varepsilon}(A,S) \longrightarrow M_{\varepsilon}(A,S) \ ; \ (F,G) \longmapsto (M,\lambda,\mu;F,G) \quad .$$

In order to verify that $\widetilde{M}_{\varepsilon}(A,S) \longrightarrow M_{\varepsilon}(A,S)$ is well-defined we have to show that

$$\partial(M,\lambda) = 0 \in M_{\varepsilon}(A,S)$$

for any $(-\varepsilon)$-quadratic linking form over (A,S) (M,λ,μ); for any non-singular ε-quadratic linking formation $(H_{\varepsilon}(F);F,G)$ we have

$$\begin{aligned}(H_{\varepsilon}(F);F,G) &= (H_{\varepsilon}(F);F,F^{\wedge}) \oplus (H_{\varepsilon}(F);F^{\wedge},G)\\ &= (H_{\varepsilon}(F);F^{\wedge},G) \in M_{\varepsilon}(A,S) \quad ,\end{aligned}$$

so that for any even $(-\varepsilon)$-symmetric linking form (M,λ)

$$\begin{aligned}\partial(M,\lambda) = (H_{\varepsilon}(M);M,\Gamma_{(M,\lambda)}) &= (H_{\varepsilon}(M);M^{\wedge},\Gamma_{(M,\lambda)})\\ &= 0 \in M_{\varepsilon}(A,S) .\end{aligned}$$

The following result is the analogue for linking formations of Proposition 1.6.5 iii) (a formation represents 0 in the Witt group if and and only if it is stably isomorphic to the boundary of a form).

<u>Proposition 3.5.4</u> A non-singular $\left\{\begin{array}{l}\text{even } \varepsilon\text{-symmetric}\\ \varepsilon\text{-quadratic}\\ \text{split } \varepsilon\text{-quadratic}\end{array}\right.$ linking formation over (A,S) $\left\{\begin{array}{l}(M,\lambda;F,G)\\ (M,\lambda,\mu;F,G)\\ (F,G)\end{array}\right.$ represents 0 in the Witt group

$\begin{cases} M\langle v_0\rangle^{\varepsilon}(A,S) \\ M_{\varepsilon}(A,S) \\ M_{\varepsilon}(A,S) \end{cases}$ if and only if it is stably equivalent to the

boundary $\begin{cases} \partial(K,\alpha;I,J) \\ \partial(K,\alpha;I,J) \\ \partial(K,\beta;I,J) \end{cases}$ of an S-non-singular $\begin{cases} \varepsilon\text{-symmetric} \\ \text{even } \varepsilon\text{-symmetric} \\ \varepsilon\text{-quadratic} \end{cases}$

formation over A $\begin{cases} (K,\alpha;I,J) \\ (K,\alpha;I,J) \\ (K,\beta;I,J) \end{cases}$ such that

$$\begin{cases} S^{-1}(K,\alpha;I,J) = 0 \in M_S^{\varepsilon}(S^{-1}A) \\ S^{-1}(K,\alpha;I,J) = 0 \in M\langle v_0\rangle_S^{\varepsilon}(S^{-1}A) \\ S^{-1}(K,\beta;I,J) = 0 \in M_{\varepsilon}^S(S^{-1}A) \end{cases} .$$

$\begin{cases} \text{If } \ker(\hat{\delta}:\hat{H}^0(\mathbb{Z}_2;S^{-1}A/A,\varepsilon) \longrightarrow \hat{H}^1(\mathbb{Z}_2;A,\varepsilon)) = 0 \\ \text{For all } A,S,\varepsilon \\ \text{For all } A,S,\varepsilon \end{cases}$ it is possible

to choose $\begin{cases} (K,\alpha;I,J) \\ (K,\alpha;I,J) \\ (K,\beta;I,J) \end{cases}$ to be an S-formation (i.e. such that

$S^{-1}K = S^{-1}I \oplus S^{-1}J$), so that $\begin{cases} (M,\lambda;F,G) = 0 \in M\langle v_0\rangle^{\varepsilon}(A,S) \\ (M,\lambda,\mu;F,G) = 0 \in M_{\varepsilon}(A,S) \\ (F,G) = 0 \in \widetilde{M}_{\varepsilon}(A,S) \end{cases}$

if and only if $\begin{cases} (M,\lambda;F,G) \\ (M,\lambda,\mu;F,G) \\ (F,G) \end{cases}$ is stably equivalent to the

boundary $\begin{cases} \partial(N,\xi) \\ \partial(N,\xi) \\ \partial(N,\xi,\rho) \end{cases}$ of an $\begin{cases} (-\varepsilon)\text{-symmetric} \\ \text{even } (-\varepsilon)\text{-symmetric linking} \\ (-\varepsilon)\text{-quadratic} \end{cases}$

form over (A,S) $\begin{cases} (N,\xi) \\ (N,\xi) \\ (N,\xi,\rho) \end{cases}$.

Proof: It is convenient to introduce the following construction, which associates an element

$$[(Q,\phi),f,(Q',\phi')] \in M\langle v_0\rangle^{\varepsilon}(A,S)$$

to an isomorphism of the non-singular ε-symmetric forms over $S^{-1}A$

$$f : S^{-1}(Q,\phi) \longrightarrow S^{-1}(Q',\phi')$$

induced from non-singular ε-symmetric forms over A $(Q,\phi),(Q',\phi')$.

Let $u \in \mathrm{Hom}_A(P,Q)$ be an S-isomorphism of f.g. projective A-modules such that

$$fu \in \mathrm{Hom}_A(P,Q') \subseteq \mathrm{Hom}_{S^{-1}A}(S^{-1}P,S^{-1}Q') .$$

(Such u exist for $P = Q$). Let (P,θ) be the ε-symmetric form over A defined by

$$\theta = u^*\phi u : P \longrightarrow P^* .$$

The S-isomorphisms of S-non-singular ε-symmetric forms over A

$$u : (P,\theta) \longrightarrow (Q,\phi) \quad , \quad fu : (P,\theta) \longrightarrow (Q',\phi')$$

correspond by Proposition 3.4.6 i) to lagrangians

$$F = \mathrm{coker}(u:P \longrightarrow Q) \quad , \quad G = \mathrm{coker}(fu:P \longrightarrow Q')$$

of the boundary even ε-symmetric linking form over (A,S)

$$(M,\lambda) = \partial(P,\theta) .$$

Set

$$[(Q,\phi),f,(Q',\phi')] = (M,\lambda;F,G) \in M\langle v_0\rangle^{\varepsilon}(A,S) .$$

Lemma 1 The Witt class $[(Q,\phi),f,(Q',\phi')] \in M\langle v_0\rangle^{\varepsilon}(A,S)$ is independent of the choice of S-isomorphism $u \in \mathrm{Hom}_A(P,Q)$.

Proof: If $\tilde{u} \in \mathrm{Hom}_A(\tilde{P},Q)$ is another choice of S-isomorphism there exist a f.g. projective A-module $\tilde{\tilde{P}}$ and S-isomorphisms $v \in \mathrm{Hom}_A(\tilde{\tilde{P}},P)$, $\tilde{v} \in \mathrm{Hom}_A(\tilde{\tilde{P}},\tilde{P})$ such that

$$uv = \tilde{u}\tilde{v} \in \mathrm{Hom}_A(\tilde{\tilde{P}},Q) \quad .$$

Therefore it is sufficient to consider the effect of replacing $u \in \mathrm{Hom}_A(P,Q)$ by $\tilde{u} = uv \in \mathrm{Hom}_A(\tilde{P},Q)$ for some S-isomorphism $v \in \mathrm{Hom}_A(\tilde{P},P)$. The non-singular even ε-symmetric linking formation over (A,S)

$$(M,\lambda;F,G) = (\partial(P,\theta);\mathrm{coker}(u:P \longrightarrow Q),\mathrm{coker}(fu:P \longrightarrow Q))$$

is replaced by

$$(\tilde{M},\tilde{\lambda};\tilde{F},\tilde{G}) = (\partial(\tilde{P},\tilde{\theta});\mathrm{coker}(\tilde{u}:\tilde{P} \longrightarrow Q),\mathrm{coker}(f\tilde{u}:\tilde{P} \longrightarrow Q))$$

with

$$\tilde{\theta} = \tilde{u}^*\phi\tilde{u} = v^*\theta v \in \mathrm{Hom}_A(\tilde{P},\tilde{P}^*) \quad .$$

By Proposition 3.4.6 i) the S-isomorphism of S-non-singular ε-symmetric forms over A

$$v : (\tilde{P},\tilde{\theta}) \longrightarrow (P,\theta)$$

determines the sublagrangian

$$H = \mathrm{coker}(v:\tilde{P} \longrightarrow P)$$

of $(\tilde{M},\tilde{\lambda}) = \partial(\tilde{P},\tilde{\theta})$. Now $H \subseteq \tilde{F} \cap \tilde{G}$, and there is defined an isomorphism of non-singular even ε-symmetric linking formations over (A,S)

$$(M,\lambda;F,G) \xrightarrow{\sim} (H^{\perp}/H,\tilde{\lambda}^{\perp}/\tilde{\lambda};\tilde{F}/H,\tilde{G}/H) \quad ,$$

so that

$$\begin{aligned}(M,\lambda;F,G) &= (H^{\perp}/H,\tilde{\lambda}^{\perp}/\tilde{\lambda};\tilde{F}/H,\tilde{G}/H) \\ &= (\tilde{M},\tilde{\lambda};\tilde{F},\tilde{G}) \in M\langle v_0\rangle^{\varepsilon}(A,S) \quad .\end{aligned}$$

[]

<u>Lemma 2</u> Given non-singular ε-symmetric forms over A $(Q,\phi),(Q',\phi')$, (Q'',ϕ'') and isomorphisms of the induced forms over $S^{-1}A$

$$f : S^{-1}(Q,\phi) \xrightarrow{\sim} S^{-1}(Q',\phi') \quad , \quad f' : S^{-1}(Q',\phi') \xrightarrow{\sim} S^{-1}(Q'',\phi'')$$

the composite isomorphism

$$f'f : S^{-1}(Q,\phi) \longrightarrow S^{-1}(Q'',\phi'')$$

is such that

$$[(Q,\phi),f'f,(Q'',\phi'')] = [(Q,\phi),f,(Q',\phi')]\oplus[(Q',\phi'),f',(Q'',\phi'')] \in M\langle v_0\rangle^{\varepsilon}(A,S) \ .$$

Proof: Let $u \in \mathrm{Hom}_A(P,Q)$ be an S-isomorphism such that

$$fu \in \mathrm{Hom}_A(P,Q') \subseteq \mathrm{Hom}_{S^{-1}A}(S^{-1}P,S^{-1}Q')$$

$$f'fu \in \mathrm{Hom}_A(P,Q'') \subseteq \mathrm{Hom}_{S^{-1}A}(S^{-1}P,S^{-1}Q'') \ ,$$

and let $\theta = u^*\phi u \in Q^{\varepsilon}(P)$ (as before). Let F,G,H be the lagrangians of $\partial(P,\theta) = (M,\lambda)$ associated by Proposition 3.4.6 i) to the S-isomorphisms of S-non-singular ε-symmetric forms over A

$$u : (P,\theta) \longrightarrow (Q,\phi)$$

$$fu : (P,\theta) \longrightarrow (Q',\phi')$$

$$f'fu : (P,\theta) \longrightarrow (Q'',\phi'') \ .$$

Then

$$\begin{aligned}[(Q,\phi),f,(Q',\phi')]&\oplus[(Q',\phi'),f',(Q'',\phi'')]\\ &= (M,\lambda;F,G)\oplus(M,\lambda;G,H)\\ &= (M,\lambda;F,H) = [(Q,\phi),f'f,(Q'',\phi'')] \in M\langle v_0\rangle^{\varepsilon}(A,S) \ .\end{aligned}$$

[]

Lemma 3 Let $(Q,\phi),(Q',\phi')$ be hyperbolic ε-symmetric forms over A, with lagrangians L,L'. If an isomorphism of the induced hyperbolic ε-symmetric forms over $S^{-1}A$

$$f : S^{-1}(Q.\phi) \longrightarrow S^{-1}(Q',\phi')$$

is such that

$$f(S^{-1}L) = S^{-1}L' \subseteq S^{-1}Q'$$

then

$$[(Q,\phi),f,(Q',\phi')] = 0 \in M\langle v_0\rangle^{\varepsilon}(A,S) \ .$$

<u>Proof</u>: Choose direct complements to L in Q and to L' in Q', so that

$$(Q,\phi) = (L\oplus L^*, \begin{pmatrix} 0 & 1 \\ \varepsilon & \alpha \end{pmatrix} \in Q^{\varepsilon}(L\oplus L^*))$$

$$(Q',\phi') = (L'\oplus L'^*, \begin{pmatrix} 0 & 1 \\ \varepsilon & \alpha' \end{pmatrix} \in Q^{\varepsilon}(L'\oplus L'^*))$$

for some ε-symmetric forms over A $(L^*,\alpha \in Q^{\varepsilon}(L^*))$, $(L'^*,\alpha' \in Q^{\varepsilon}(L'^*))$. There exist S-isomorphisms $s \in \mathrm{Hom}_A(L,L)$, $t \in \mathrm{Hom}_A(L^*,L)$ such that

$$f\begin{pmatrix} s & 0 \\ 0 & t \end{pmatrix} = \begin{pmatrix} g & k \\ 0 & g' \end{pmatrix}$$

$$\in \mathrm{Hom}_A(L\oplus L^*, L'\oplus L'^*) \subseteq \mathrm{Hom}_{S^{-1}A}(S^{-1}L\oplus S^{-1}L^*, S^{-1}L'\oplus S^{-1}L'^*)$$

for some S-isomorphisms $g \in \mathrm{Hom}_A(L,L')$, $g' \in \mathrm{Hom}_A(L^*,L')$ and some $k \in \mathrm{Hom}_A(L^*,L')$. The S-isomorphism of S-non-singular ε-symmetric forms over A

$$u = \begin{pmatrix} s & 0 \\ 0 & t \end{pmatrix} : (P,\theta) = (L\oplus L^*, \begin{pmatrix} 0 & s^*t \\ \varepsilon t^*s & t^*\alpha t \end{pmatrix})$$

$$\longrightarrow (Q,\phi) = (L\oplus L^*, \begin{pmatrix} 0 & 1 \\ \varepsilon & \alpha \end{pmatrix})$$

determines a non-singular even ε-symmetric linking formation over (A,S)

$$(M,\lambda;F,G) = (\partial(P,\theta);\mathrm{coker}(f:P\longrightarrow Q),\mathrm{coker}(fu:P\longrightarrow Q'))$$

such that

$$[(Q,\phi),f,(Q',\phi')] = (M,\lambda;F,G) \in M\langle v_0\rangle^{\varepsilon}(A,S) \ .$$

The S-isomorphisms of S-non-singular ε-symmetric forms over A

$$h = \begin{pmatrix} t^*s & 0 \\ 0 & 1 \end{pmatrix} : (P,\theta) = (L\oplus L^*, \begin{pmatrix} 0 & s^*t \\ \varepsilon t^*s & t^*\alpha t \end{pmatrix})$$

$$\longrightarrow (L\oplus L^*, \begin{pmatrix} 0 & 1 \\ \varepsilon & t^*\alpha t \end{pmatrix})$$

$$i = \begin{pmatrix} s & 0 \\ 0 & 1 \end{pmatrix} : (P,\theta) = (L\oplus L^*, \begin{pmatrix} 0 & s^*t \\ \varepsilon t^*s & t^*\alpha t \end{pmatrix})$$

$$\longrightarrow (L\oplus L^*, \begin{pmatrix} 0 & t \\ \varepsilon t^* & t^*\alpha t \end{pmatrix})$$

$$j = \begin{pmatrix} g & 0 \\ 0 & 1 \end{pmatrix} : (P,\theta) = (L\oplus L^*, \begin{pmatrix} 0 & s^*t \\ \varepsilon t^*s & t^*\alpha t \end{pmatrix})$$

$$\longrightarrow (L'\oplus L^*, \begin{pmatrix} 0 & g' \\ \varepsilon g'^* & t^*\alpha t \end{pmatrix})$$

correspond by Proposition 3.4.6 i) to a lagrangian

$$H = \mathrm{coker}(h : L\oplus L^* \longrightarrow L\oplus L^*) \subseteq M = \mathrm{coker}(\theta : P \longrightarrow P^*)$$

and the sublagrangians

$$I = \mathrm{coker}(i : L\oplus L^* \longrightarrow L\oplus L^*) \subseteq M = \mathrm{coker}(\theta : P \longrightarrow P^*)$$

$$J = \mathrm{coker}(j : L\oplus L^* \longrightarrow L'\oplus L^*) \subseteq M = \mathrm{coker}(\theta : P \longrightarrow P^*)$$

of the boundary even ε-symmetric linking form over (A,S) $(M,\lambda) = \partial(P,\theta)$. Now $I \subseteq F \cap H$, $J \subseteq G \cap H$ and the even ε-symmetric linking formations over (A,S) $(I^\perp/I, \lambda^\perp/\lambda; F/I, H/I)$, $(J^\perp/J, \lambda^\perp/\lambda; H/J, G/J)$ are stably equivalent to 0, so that

$$\begin{aligned}[(Q,\phi),f,(Q',\phi')] &= (M,\lambda;F,G) \\ &= (M,\lambda;F,H)\oplus(M,\lambda;H,G) \\ &= (I^\perp/I,\lambda^\perp/\lambda;F/I,H/I)\oplus(J^\perp/J,\lambda^\perp/\lambda;H/J,G/J) \\ &= 0 \in M\langle v_0\rangle^\varepsilon(A,S) \ .\end{aligned}$$

[]

The Witt class $\partial(K,\alpha;I,J) \in M\langle v_0\rangle^{\varepsilon}(A,S)$ of the boundary even ε-symmetric linking formation over (A,S) of an S-non-singular ε-symmetric formation over A $(K,\alpha;I,J)$ may be described as follows. Choose a direct complement to the lagrangian I in K, so that

$$(K,\alpha) = (I\oplus I^*, \begin{pmatrix} 0 & 1 \\ \varepsilon & \theta \end{pmatrix} \in Q^{\varepsilon}(I\oplus I^*))$$

for some $\theta \in Q^{\varepsilon}(I^*)$. The inclusion of the S-lagrangian

$$\begin{pmatrix} j \\ k \end{pmatrix} : (J,0) \longrightarrow (I\oplus I^*, \begin{pmatrix} 0 & 1 \\ \varepsilon & \theta \end{pmatrix})$$

extends to an S-isomorphism of S-non-singular ε-symmetric forms over A

$$\begin{pmatrix} j & \tilde{j} \\ k & \tilde{k} \end{pmatrix} : (J\oplus J^*, \begin{pmatrix} 0 & s^* \\ \varepsilon s & \phi \end{pmatrix}) \longrightarrow (I\oplus I^*, \begin{pmatrix} 0 & 1 \\ \varepsilon & \theta \end{pmatrix})$$

for some S-isomorphism $s \in \mathrm{Hom}_A(J,J)$. Define an isomorphism of hyperbolic ε-symmetric forms over $S^{-1}A$

$$f = \begin{pmatrix} js^{-1} & \tilde{j} \\ ks^{-1} & \tilde{k} \end{pmatrix} : S^{-1}(J\oplus J^*, \begin{pmatrix} 0 & 1 \\ \varepsilon & \phi \end{pmatrix})$$

$$\xrightarrow{\sim} S^{-1}(I\oplus I^*, \begin{pmatrix} 0 & 1 \\ \varepsilon & \theta \end{pmatrix}) .$$

Then

$$(K,\alpha;I,J) = [(J\oplus J^*, \begin{pmatrix} 0 & 1 \\ \varepsilon & \phi \end{pmatrix}), f, (I\oplus I^*, \begin{pmatrix} 0 & 1 \\ \varepsilon & \theta \end{pmatrix})] \in M\langle v_0\rangle^{\varepsilon}(A,S) .$$

(To verify that this is the linking formation associated to the non-singular ε-symmetric S-form over A $(K,\alpha;J)$ used to define $\partial(K,\alpha;I,J)$ use the S-isomorphism

$$u = \begin{pmatrix} s & 0 \\ 0 & 1 \end{pmatrix} : P = J\oplus J^* \longrightarrow J\oplus J^*$$

in the construction of $[(J\oplus J^*, \begin{pmatrix} 0 & 1 \\ \varepsilon & \phi \end{pmatrix}), f, (I\oplus I^*, \begin{pmatrix} 0 & 1 \\ \varepsilon & \theta \end{pmatrix})])$.

Lemma 4 If $(K,\alpha;I,J),(K',\alpha';I',J')$ are S-non-singular ε-symmetric formations over A such that the induced non-singular ε-symmetric formations over $S^{-1}A$ $S^{-1}(K,\alpha;I,J),S^{-1}(K',\alpha';I',J')$ are isomorphic then

$$\partial(K,\alpha;I,J) = \partial(K',\alpha';I',J') \in M\langle v_O\rangle^{\varepsilon}(A,S) .$$

Proof: As above, let

$$f : S^{-1}(J\oplus J^*, \begin{pmatrix} 0 & 1 \\ \varepsilon & \phi \end{pmatrix}) \longrightarrow S^{-1}(I\oplus I^*, \begin{pmatrix} 0 & 1 \\ \varepsilon & \theta \end{pmatrix}) = S^{-1}(K,\alpha)$$

$$f' : S^{-1}(J'\oplus J'^*, \begin{pmatrix} 0 & 1 \\ \varepsilon & \phi' \end{pmatrix})$$

$$\longrightarrow S^{-1}(I'\oplus I'^*, \begin{pmatrix} 0 & 1 \\ \varepsilon & \theta' \end{pmatrix}) = S^{-1}(K',\alpha')$$

be isomorphisms of the induced hyperbolic ε-symmetric forms over $S^{-1}A$. Let

$$g : S^{-1}(K,\alpha;I,J) \longrightarrow S^{-1}(K',\alpha';I',J')$$

be an isomorphism of the induced non-singular ε-symmetric formations over $S^{-1}A$. The isomorphisms of hyperbolic ε-symmetric forms over $S^{-1}A$

$$g : S^{-1}(K,\alpha) \longrightarrow S^{-1}(K',\alpha')$$

$$h = f'^{-1}gf : S^{-1}(J\oplus J^*, \begin{pmatrix} 0 & 1 \\ \varepsilon & \phi \end{pmatrix}) \longrightarrow S^{-1}(J'\oplus J'^*, \begin{pmatrix} 0 & 1 \\ \varepsilon & \phi' \end{pmatrix})$$

are such that

$$g(S^{-1}I) = S^{-1}I' \subseteq S^{-1}K'$$

$$h(S^{-1}J) = S^{-1}J' \subseteq S^{-1}(J'\oplus J'^*) .$$

Applying Lemma 3, we have

$$[(K,\alpha),g,(K',\alpha')] = 0 \in M\langle v_O\rangle^{\varepsilon}(A,S)$$

$$[(J\oplus J^*,\begin{pmatrix} 0 & 1 \\ \varepsilon & \phi \end{pmatrix}),h,(J'\oplus J'^*,\begin{pmatrix} 0 & 1 \\ \varepsilon & \phi' \end{pmatrix})] = 0 \in M\langle v_O\rangle^{\varepsilon}(A,S) .$$

Applying Lemma 2, we have

$$\begin{aligned}\partial(K,\alpha;I,J) &= [(J\oplus J^*,\begin{pmatrix}0 & 1\\ \varepsilon & \phi\end{pmatrix}),f,(I\oplus I^*,\begin{pmatrix}0 & 1\\ \varepsilon & \theta\end{pmatrix})]\\ &= [(J\oplus J^*,\begin{pmatrix}0 & 1\\ \varepsilon & \phi\end{pmatrix}),h,(J'\oplus J'^*,\begin{pmatrix}0 & 1\\ \varepsilon & \phi'\end{pmatrix})]\\ &\quad\oplus[(J'\oplus J'^*,\begin{pmatrix}0 & 1\\ \varepsilon & \phi'\end{pmatrix}),f',(I'\oplus I'^*,\begin{pmatrix}0 & 1\\ \varepsilon & \theta'\end{pmatrix})]\\ &\quad\oplus[(K',\alpha'),g^{-1},(K,\alpha)]\\ &= [(J'\oplus J'^*,\begin{pmatrix}0 & 1\\ \varepsilon & \phi'\end{pmatrix}),f',(I'\oplus I'^*,\begin{pmatrix}0 & 1\\ \varepsilon & \theta'\end{pmatrix})]\\ &= \partial(K',\alpha';I',J') \in M\langle v_0\rangle^{\varepsilon}(A,S)\ .\end{aligned}$$

[]

<u>Lemma 5</u> If $(K,\alpha;I,J)$ is an S-non-singular ε-symmetric formation over A such that

$$S^{-1}(K,\alpha;I,J) = 0 \in M_S^{\varepsilon}(S^{-1}A)$$

then

$$\partial(K,\alpha;I,J) = 0 \in M\langle v_0\rangle^{\varepsilon}(A,S)\ .$$

<u>Proof</u>: Let $(D,\eta\in Q^1(D,\varepsilon))$ be the 1-dimensional ε-symmetric $S^{-1}A$-Poincaré complex associated to $(K,\alpha;I,J)$, with

$$d = k^* : D_1 = I \longrightarrow D_0 = J^*\quad,\quad D_r = 0\ (r\neq 0,1)$$

$$\eta_0 = \begin{cases}\varepsilon j : D^0 = J \longrightarrow D_1 = I\\ j^* + k^* \ : D^1 = I^* \longrightarrow D_0 = J^*\end{cases}$$

$$\eta_1 = \theta : D^1 = I^* \longrightarrow D_1 = I$$

for some ε-symmetric form over A $(I^*,\theta\in Q^{\varepsilon}(I^*))$ such that

$$(K,\alpha) = (I\oplus I^*,\begin{pmatrix}0 & 1\\ \varepsilon & \theta\end{pmatrix})\quad ,$$

with $\begin{pmatrix} j \\ k \end{pmatrix} : J \longrightarrow I\oplus I^*$ the inclusion. Now

$$S^{-1}(D,\eta) = S^{-1}(K,\alpha;I,J) = 0$$

$$\in L^1_S(S^{-1}A,\varepsilon) = M^{\varepsilon}_S(S^{-1}A) = \Gamma^1(A \longrightarrow S^{-1}A,\varepsilon) ,$$

so that there exists a 2-dimensional ε-symmetric $S^{-1}A$-Poincaré pair over A $(f:D \longrightarrow \delta D, (\delta\eta,\eta) \in Q^2(f,\varepsilon))$ with δD a f.g. projective A-module chain complex such that $\delta D_r = 0$ $(r \neq 0,1,2)$. Define an A-module chain complex $\delta D'$ and an A-module chain map

$$g : \delta D \longrightarrow \delta D'$$

by

$$\begin{array}{ccccccccccc} \delta D : & \dots \to 0 & \longrightarrow & \delta D_2 & \xrightarrow{d} & \delta D_1 & \xrightarrow{d} & \delta D_0 & \longrightarrow & 0 \longrightarrow \dots \\ \Big\downarrow g & & & \Big\downarrow 1 & & \Big\downarrow 1 & & \Big\downarrow & & \\ \delta D' : & \dots \to 0 & \longrightarrow & \delta D_2 & \xrightarrow{d} & \delta D_1 & \longrightarrow & 0 & \longrightarrow & 0 \longrightarrow \dots \end{array} ,$$

and let $(D',\eta' \in Q^1(D',\varepsilon))$ be the 1-dimensional ε-symmetric $S^{-1}A$-Poincaré complex over A obtained from (D,η) by surgery on the 2-dimensional ε-symmetric pair over A

$$(gf:D \longrightarrow \delta D', (g,1)_{\%}(\delta\eta,\eta) \in Q^2(gf,\varepsilon))$$

(which becomes connected over $S^{-1}A$). The S-non-singular ε-symmetric formation over A associated to (D',η') is given by

$$(K',\alpha';I',J') = (D'_1\oplus D'^1, \begin{pmatrix} 0 & 1 \\ \varepsilon & \eta'_1 \end{pmatrix} ; D'_1, \mathrm{im}(\begin{pmatrix} \bar{\varepsilon}\eta'_0 \\ d'^* \end{pmatrix} : D'^0 \longrightarrow D'_1\oplus D'^1)) .$$

The boundary even ε-symmetric linking formation over (A,S) $\partial(K',\alpha';I',J')$ is stably equivalent to $\partial(K,\alpha;I,J)$, by Propositions 1.5.1 i), 3.5.2 ii), since $\partial(K,\alpha;I,J)$ corresponds to the S-acyclic 2-dimensional even $(-\varepsilon)$-symmetric Poincaré complex over A $\partial\bar{S}(D,\eta)$, and $\partial\bar{S}(D,\eta)$ is homotopy equivalent to $\partial\bar{S}(D',\eta')$. Define an A-module chain complex $\delta\tilde{D}'$ and an A-module chain map

$$g' : D' \longrightarrow \delta\tilde{D}'$$

by

$$\begin{array}{ccccccccccc} D' & : & \dots \longrightarrow & 0 & \longrightarrow & D'_1 & \xrightarrow{d'} & D'_0 & \longrightarrow & 0 & \longrightarrow \dots \\ \big\downarrow g' & & & & & \big\downarrow & & \big\downarrow g' & & & \\ \delta\tilde{D}' & : & \dots \longrightarrow & 0 & \longrightarrow & 0 & \longrightarrow & \delta D_0 & \longrightarrow & 0 & \longrightarrow \dots \end{array} ,$$

with

$$d' = \begin{pmatrix} d & 0 & \eta_0 f^* \\ -f & d & -\delta\eta_0 \\ 0 & 0 & -d^* \end{pmatrix}$$

$$: D'_1 = D_1 \oplus \delta D_2 \oplus \delta D^1 \longrightarrow D'_0 = D_0 \oplus \delta D_1 \oplus \delta D^2 ,$$

$$g' = (f \quad d \quad -\delta\eta_0) : D'_0 = D_0 \oplus \delta D_1 \oplus \delta D^2 \longrightarrow \delta\tilde{D}'_0 = \delta D_0 .$$

The 2-dimensional ε-symmetric pair over A

$$(g' : D' \longrightarrow \delta\tilde{D}', (0,\eta') \in Q^2(g',\varepsilon))$$

is $S^{-1}A$-Poincaré, so that there exists an A-module morphism $i \in \mathrm{Hom}_A(D'_1, \delta D^0)$ such that the composite A-module morphism

$$s = i\eta'_0 g'^* : \delta D^0 \xrightarrow{g'^*} D'^0 \xrightarrow{\eta'_0} D'_1 \xrightarrow{i} \delta D^0$$

is an S-isomorphism. Define a 1-dimensional A-module chain complex D'' and an S-equivalence

$$h : D' \longrightarrow D''$$

by

$$\begin{array}{ccccccccccc} D' & : & \dots \longrightarrow & 0 & \longrightarrow & D'_1 & \xrightarrow{d'} & D'_0 & \longrightarrow & 0 & \longrightarrow \dots \\ \big\downarrow h & & & & & \big\downarrow i & & \big\downarrow g' & & & \\ D'' & : & \dots \longrightarrow & 0 & \longrightarrow & \delta D^0 & \xrightarrow{0} & \delta D_0 & \longrightarrow & 0 & \longrightarrow \dots \end{array} .$$

Let

$$\eta'' = h^{\%}(\eta') \in Q^1(D'',\varepsilon) \quad ,$$

so that (D'',η'') is a 1-dimensional ε-symmetric $S^{-1}A$-Poincaré complex over A corresponding to the S-non-singular ε-symmetric formation over A

$$(K'',\alpha'';I'',J'') = (D_1''\oplus D''^1, \begin{pmatrix} 0 & 1 \\ \varepsilon & \eta_1'' \end{pmatrix}; D_1'', \mathrm{im}(\begin{pmatrix} \bar{\varepsilon}\eta_0'' \\ d''^* \end{pmatrix}: D''^0 \longrightarrow D_1''\oplus D''^1))$$

$$= (\delta D^0 \oplus \delta D_0, \begin{pmatrix} 0 & 1 \\ \varepsilon & i\eta_1' i^* \end{pmatrix}; \delta D^0, \mathrm{im}(s: \delta D^0 \longrightarrow \delta D^0))$$

with

$$S^{-1}I'' = S^{-1}J'' \subseteq S^{-1}K'' \ .$$

Proposition 1.6.4 translates the homotopy equivalence of 1-dimensional ε-symmetric Poincaré complexes over $S^{-1}A$

$$h : S^{-1}(D',\eta') \longrightarrow S^{-1}(D'',\eta'')$$

into an isomorphism of non-singular ε-symmetric formations over $S^{-1}A$

$$h : S^{-1}(K',\alpha';I',J') \oplus S^{-1}(H^{\varepsilon}(J''^*);J''^*,J'')$$
$$\longrightarrow S^{-1}(K'',\alpha'';I'',J'') \oplus S^{-1}(H^{\varepsilon}(J'^*);J'^*,J') \ .$$

Applying Lemma 4, we have

$$\partial(K,\alpha;I,J) = \partial(K',\alpha';I',J')$$
$$= \partial(K'',\alpha'';I'',J'') = \partial(K'',\alpha'';I'',I'') = 0 \in M\langle v_0\rangle^{\varepsilon}(A,S) \ .$$

[]

It follows from Proposition 3.5.3 ii) and Lemma 5 that the correspondence of Proposition 3.5.2 ii)

(S-acyclic 2-dimensional even $(-\varepsilon)$-symmetric Poincaré complexes over A (C,ϕ))

$\longleftrightarrow$ (non-singular even ε-symmetric linking

can be used to define an abelian group morphism

$$L^1(A,S,\varepsilon) \longrightarrow M\langle v_0\rangle^{\varepsilon}(A,S) \ ; \ (C,\phi) \longmapsto (M,\lambda;F,G) \quad .$$

We shall prove that this is in fact an isomorphism, so that applying Proposition 3.5.3 ii) again it will follow that a non-singular even ε-symmetric linking formation over (A,S) $(M,\lambda;F,G)$ representing 0 in $M\langle v_0\rangle^{\varepsilon}(A,S)$ is stably equivalent to the boundary $\partial(K,\alpha;I,J)$ of an S-non-singular ε-symmetric formation over A $(K,\alpha;I,J)$ such that $S^{-1}(K,\alpha;I,J) = 0 \in M_S^{\varepsilon}(S^{-1}A)$. In order to verify that the correspondence of Proposition 3.5.2 ii) also defines an abelian group morphism

$$M\langle v_0\rangle^{\varepsilon}(A,S) \longrightarrow L^1(A,S,\varepsilon) \ ; \ (M,\lambda;F,G) \longmapsto (C,\phi)$$

we have to show that the S-acyclic 2-dimensional even $(-\varepsilon)$-symmetric Poincaré complex over A associated to the non-singular even ε-symmetric linking formation over (A,S)

$$\begin{cases} (M,\lambda;F,G)\oplus(M,\lambda;G,H) \\ (M,\lambda;F,G) \\ (M,\lambda;F,G) \end{cases}$$ is S-acyclic cobordant to the complex

associated to $\begin{cases} (M,\lambda;F,H) \\ (L^{\perp}/L,\lambda^{\perp}/\lambda;F\cap L^{\perp},G/L)\text{, for any non-singular} \\ (L^{\perp}/L,\lambda^{\perp}/\lambda;F/L,G/L) \end{cases}$

even ε-symmetric linking form over (A,S) (M,λ) and lagrangians

F,G together with a $\begin{cases} \text{lagrangian } H \text{ of } (M,\lambda) \\ \text{sublagrangian } L \text{ of } (M,\lambda;F,G) \\ \text{sublagrangian } L \text{ of } (M,\lambda) \text{ such that } L\subseteq F\cap G \end{cases}$.

We shall consider the three cases separately.

Recall from the proof of Proposition 3.5.2 ii) that the S-acyclic 2-dimensional even $(-\varepsilon)$-symmetric Poincaré complex over A $(C,\phi \in Q\langle v_0\rangle^2(C,-\varepsilon))$ associated to the even ε-symmetric linking formation over (A,S) $(M,\lambda;F,G)$ is the union

$$(C,\phi) = (\delta D \cup_D \delta D', -\delta\eta \cup_\eta \delta\eta' \in Q\langle v_0\rangle^2(\delta D\cup_D \delta D',-\varepsilon))$$

of the S-acyclic null-cobordisms $(f:D \longrightarrow \delta D, (\delta\eta,\eta) \in Q\langle v_0\rangle^2(f,-\varepsilon))$, $(f':D \longrightarrow \delta D', (\delta\eta',\eta) \in Q\langle v_0\rangle^2(f',-\varepsilon))$ associated to the lagrangians F,G by Proposition 3.4.5 ii), with $(D,\eta \in Q\langle v_0\rangle^1(D,-\varepsilon))$ the S-acyclic complex associated to the linking form (M,λ) by Proposition 3.4.1. Let $(f'':D \longrightarrow \delta D'', (\delta\eta'',\eta) \in Q\langle v_0\rangle^2(f'',-\varepsilon))$ be the S-acyclic null-cobordism of (D,η) corresponding to the lagrangian H of (M,λ), so that the S-acyclic complexes associated to the linking formations $(M,\lambda;G,H)$, $(M,\lambda;F,H)$ are the unions

$$(C',\phi') = (\delta D' \cup_D \delta D'', -\delta\eta' \cup_\eta \delta\eta'' \in Q\langle v_0\rangle^2(\delta D' \cup_D \delta D'',-\varepsilon))$$

$$(C'',\phi'') = (\delta D \cup_D \delta D'', -\delta\eta \cup_\eta \delta\eta'' \in Q\langle v_0\rangle^2(\delta D \cup_D \delta D'',-\varepsilon)) \quad .$$

Now (C'',ϕ'') is homotopy equivalent to the S-acyclic complex obtained from $(C,\phi)\oplus(C',\phi')$ by surgery on the connected S-acyclic 3-dimensional even $(-\varepsilon)$-symmetric pair over A

$$((g\ g'):C\oplus C' \longrightarrow \delta D', (0,\phi\oplus\phi') \in Q\langle v_0\rangle^3((g\ g'),-\varepsilon)) \ ,$$

where

$$\left.\begin{array}{l} g = (0\quad 0\quad 1) : C_r = \delta D_r \oplus D_{r-1} \oplus \delta D'_r \longrightarrow \delta D'_r \\ g' = (1\quad 0\quad 0) : C'_r = \delta D'_r \oplus D_{r-1} \oplus \delta D''_r \longrightarrow \delta D'_r \end{array}\right. \quad (r \in \mathbb{Z}) \quad .$$

It follows from the S-acyclic counterpart of Proposition 1.4.2 that

$$(C,\phi)\oplus(C',\phi') = (C'',\phi'') \in L^1(A,S,\varepsilon) \ .$$

The S-acyclic complexes associated to the stably equivalent even ε-symmetric linking formations $(M,\lambda;F,G)$, $(L^{\perp}/L,\lambda^{\perp}/\lambda;F\cap L^{\perp},G/L)$ are homotopy equivalent (by Proposition 3.5.2 ii)) and hence represent the same element of $L^1(A,S,\varepsilon)$.

Given a non-singular even ε-symmetric linking formation over (A,S) $(M,\lambda;F,G)$ let $(K,\alpha;J)$ be a non-singular ε-symmetric S-form over A associated to it by Proposition 3.5.2 i). Let $j\in \mathrm{Hom}_A(J,K)$ be the inclusion, and let

$$(j\ k)\ :\ (J\oplus J^*,\begin{pmatrix} 0 & s \\ \varepsilon s^* & k^*\alpha k\end{pmatrix})\longrightarrow (K,\alpha)$$

be an extension of the inclusion to an S-isomorphism of S-non-singular ε-symmetric forms over A, with

$$s = j^*\alpha k\ :\ J^*\longrightarrow J^*$$

an S-isomorphism. Given a sublagrangian L of (M,λ) such that $L\subseteq F\cap G$ there exist a f.g. projective A-module J', an A-module morphism $j'\in \mathrm{Hom}_A(J',K)$ and S-isomorphisms $u\in \mathrm{Hom}_A(J',J)$, $v\in \mathrm{Hom}_A(J',J)$ such that the inclusions $L\longrightarrow F$, $L\longrightarrow G$ have resolutions

$$\begin{array}{ccccccccc}
0 & \longrightarrow & J\oplus J^* & \xrightarrow{\begin{pmatrix} u & 0 \\ 0 & 1\end{pmatrix}} & J'\oplus J^* & \longrightarrow & L & \longrightarrow & 0 \\
 & & \Big\downarrow{\scriptstyle 1} & \begin{pmatrix} s^* & 0 \\ 0 & 1\end{pmatrix} & \Big\downarrow & \begin{pmatrix} v & 0 \\ 0 & 1\end{pmatrix} & \Big\downarrow & & \\
0 & \longrightarrow & J\oplus J^* & \longrightarrow & J\oplus J^* & \longrightarrow & F & \longrightarrow & 0
\end{array},$$

$$\begin{array}{ccccccccc}
0 & \longrightarrow & J\oplus J^* & \xrightarrow{\begin{pmatrix} u & 0 \\ 0 & 1 \end{pmatrix}} & J'\oplus J^* & \longrightarrow & L & \longrightarrow & 0 \\
 & & \Big\downarrow{\scriptstyle 1} & & \Big\downarrow{\scriptstyle (j'\ k)} & & \Big\downarrow & & \\
0 & \longrightarrow & J\oplus J^* & \xrightarrow{(j\ \ k)} & K & \longrightarrow & G & \longrightarrow & 0
\end{array} \quad .$$

Let $(C,\phi\in Q\langle v_0\rangle^2(C,-\varepsilon))$ be the S-acyclic 2-dimensional even $(-\varepsilon)$-symmetric Poincaré complex over A in normal form associated to the S-form $(K,\alpha;J)$ (as in the proof of Proposition 3.5.2 ii)). Define a 3-dimensional S-acyclic A-module chain complex D and a chain map

$$f : C \longrightarrow D$$

by

$$\begin{array}{cccccccccccc}
C: & \dots & \longrightarrow & 0 & \longrightarrow & J & \xrightarrow{\alpha^* j} & K^* & \xrightarrow{j^*} & J^* & \longrightarrow & 0 \longrightarrow \dots \\
\Big\downarrow{\scriptstyle f} & & & & & \Big\downarrow & & \Big\downarrow{\scriptstyle j'^*} & & \Big\downarrow{\scriptstyle 1} & & \\
D: & \dots & \longrightarrow & 0 & \longrightarrow & 0 & \longrightarrow & J'^* & \xrightarrow{u^*} & J^* & \longrightarrow & 0 \longrightarrow \dots
\end{array} \quad .$$

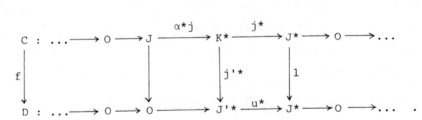

Let $(C',\phi'\in Q\langle v_0\rangle^2(C',-\varepsilon))$ be the S-acyclic 2-dimensional even $(-\varepsilon)$-symmetric Poincaré complex over A obtained from (C,ϕ) by surgery on the connected S-acyclic 3-dimensional even $(-\varepsilon)$-symmetric pair over A $(f:C\longrightarrow D,(0,\phi)\in Q\langle v_0\rangle^3(f,-\varepsilon))$. Let $(C'',\phi''\in Q\langle v_0\rangle^2(C'',-\varepsilon))$ be the S-acyclic 2-dimensional even $(-\varepsilon)$-symmetric Poincaré complex over A in normal form associated by Proposition 3.5.2 ii) to the non-singular ε-symmetric S-form over A $(K,\alpha;\mathrm{im}(j':J'\longrightarrow K))$, which corresponds by Proposition 3.5.2 i) to the non-singular even ε-symmetric linking formation over (A,S) $(L^{\perp}/L,\lambda^{\perp}/\lambda;F/L,G/L)$. The chain equivalence

$$h : C' \longrightarrow C''$$

given by

$$\begin{array}{ccccccccccccc}
C' : \ldots \to 0 \to & J & \xrightarrow{\begin{pmatrix}\varepsilon\\ u\end{pmatrix}} & J\oplus J' & \xrightarrow{(\alpha^*j\ \ -\alpha j')} & K^* & \xrightarrow{\begin{pmatrix}j^*\\ -j'^*\end{pmatrix}} & J^*\oplus J'^* & \xrightarrow{(1\ \ u^*)} & J^* & \to 0 \to \ldots \\
\Big\downarrow h & \Big\downarrow & & \Big\downarrow (u\ \ -\varepsilon) & & \Big\downarrow 1 & & \Big\downarrow (0\ \ -1) & & \Big\downarrow & \\
C'' : \ldots \to 0 \to & 0 & \longrightarrow & J' & \xrightarrow{\alpha^*j'} & K^* & \xrightarrow{j'^*} & J'^* & \longrightarrow & 0 & \to 0 \to \ldots
\end{array}$$

defines a homotopy equivalence

$$h : (C',\phi') \longrightarrow (C'',\phi'') .$$

It follows that

$$(C,\phi) = (C',\phi') = (C'',\phi'') \in L^1(A,S,\varepsilon) ,$$

verifying that the S-acyclic complexes associated to the linking formations $(M,\lambda;F,G)$, $(L^{\perp}/L,\lambda^{\perp}/\lambda;F/L,G/L)$ are S-acyclic cobordant.

This completes the identification

$$L^1(A,S,\varepsilon) = M\langle v_0\rangle^{\varepsilon}(A,S) .$$

The verification that a non-singular $\begin{cases}\varepsilon\text{-quadratic}\\ \text{split } \varepsilon\text{-quadratic}\end{cases}$ linking formation over (A,S) $\begin{cases}(M,\lambda,\mu;F,G)\\ (F,G)\end{cases}$ represents 0 in the Witt group $\begin{cases}M_{\varepsilon}(A,S)\\ \widetilde{M}_{\varepsilon}(A,S)\end{cases}$ if and only if it is stably equivalent to the boundary $\begin{cases}\partial(K,\alpha;I,J)\\ \partial(K,\beta;I,J)\end{cases}$ of an S-non-singular $\begin{cases}\text{even } \varepsilon\text{-symmetric}\\ \varepsilon\text{-quadratic}\end{cases}$ formation over A $\begin{cases}(K,\alpha;I,J)\\ (K,\beta;I,J)\end{cases}$ such that

$$\begin{cases} S^{-1}(K,\alpha;I,J) = 0 \in M\langle v_0\rangle^{\varepsilon}_{S}(S^{-1}A) \\ S^{-1}(K,\beta;I,J) = 0 \in M^{S}_{\varepsilon}(S^{-1}A) \end{cases}$$

proceeds by analogy with the case of even ε-symmetric linking formations dealt with above.

It remains to prove that

$$\left\{\begin{array}{l} \text{if } \ker(\hat{\delta}:\hat{H}^0(\mathbb{Z}_2;S^{-1}A/A,\varepsilon)\longrightarrow\hat{H}^1(\mathbb{Z}_2;A,\varepsilon)) = 0 \\ \text{for all } A,S,\varepsilon \\ \text{for all } A,S,\varepsilon \end{array}\right. \quad \text{a non-singular}$$

$$\left\{\begin{array}{l} \text{even } \varepsilon\text{-symmetric} \\ \varepsilon\text{-quadratic} \\ \text{split } \varepsilon\text{-quadratic} \end{array}\right. \text{ linking formation over } (A,S) \left\{\begin{array}{l} (M,\lambda;F,G) \\ (M,\lambda,\mu;F,G) \\ (F,G) \end{array}\right.$$

$$\text{represents 0 in the Witt group } \left\{\begin{array}{l} M\langle v_0\rangle^{\varepsilon}(A,S) \\ M_{\varepsilon}(A,S) \\ \widetilde{M}_{\varepsilon}(A,S) \end{array}\right. \text{ if and only if}$$

$$\text{it is stably equivalent to the boundary } \left\{\begin{array}{l} \partial(N,\xi) \\ \partial(N,\xi) \\ \partial(N,\xi,\rho) \end{array}\right. \text{ of an}$$

$$\left\{\begin{array}{l} (-\varepsilon)\text{-symmetric} \\ \text{even } (-\varepsilon)\text{-symmetric} \\ (-\varepsilon)\text{-quadratic} \end{array}\right. \text{ linking form over } (A,S) \left\{\begin{array}{l} (N,\xi) \\ (N,\xi) \\ (N,\xi,\rho) \end{array}\right. .$$

For $\left\{\begin{array}{l} \text{even } \varepsilon\text{-symmetric} \\ \text{split } \varepsilon\text{-quadratic} \end{array}\right.$ linking formations this follows from Proposition 3.5.3 ii). (The projection of Proposition 3.5.2 iii)

(S-acyclic 2-dimensional $(-\varepsilon)$-quadratic Poincaré complexes over A (C,ψ))

$\longrightarrow$ (non-singular split ε-quadratic linking formations over (A,S) (F,G))

can thus be used to define an isomorphism of abelian groups

$$L_1(A,S,\varepsilon) \longrightarrow \widetilde{M}_\varepsilon(A,S) \;;\; (C,\psi) \longmapsto (F,G) \;).$$

The method of proof of Proposition 3.5.2 ii) is readily modified to give the corresponding result for ε-quadratic linking formations.

[]

A non-singular $\begin{cases} \text{(even) } \varepsilon\text{-symmetric} \\ \varepsilon\text{-quadratic} \end{cases}$ formation over $S^{-1}A$ $\begin{cases} (Q,\phi;F,G) \\ (Q,\psi;F,G) \end{cases}$ with projective class

$$[G] - [F^*] \in S = \mathrm{im}(\widetilde{K}_0(A) \longrightarrow \widetilde{K}_0(S^{-1}A)) \subseteq \widetilde{K}_0(S^{-1}A)$$

is stably isomorphic to $\begin{cases} S^{-1}(K,\alpha;I,J) \\ S^{-1}(K,\beta;I,J) \end{cases}$ for some S-non-singular $\begin{cases} \text{(even) } \varepsilon\text{-symmetric} \\ \varepsilon\text{-quadratic} \end{cases}$ formation over A $\begin{cases} (K,\alpha;I,J) \\ (K,\beta;I,J) \end{cases}$. It follows from Proposition 3.5.4 that the boundary operations

$$\partial : (\text{S-non-singular formations over A}) \longrightarrow (\text{linking formations over } (A,S))$$

can be used to define abelian group morphisms

$$\partial : M_S^\varepsilon(S^{-1}A) \longrightarrow M\langle v_0\rangle^\varepsilon(A,S) \;;\; S^{-1}(K,\alpha;I,J) \longmapsto \partial(K,\alpha;I,J)$$

$$\partial : M\langle v_0\rangle_S^\varepsilon(S^{-1}A) \longrightarrow M_\varepsilon(A,S) \;;\; S^{-1}(K,\alpha;I,J) \longmapsto \partial(K,\alpha;I,J)$$

$$\partial : M_\varepsilon^S(S^{-1}A) \longrightarrow \widetilde{M}_\varepsilon(A,S) \;;\; S^{-1}(K,\beta;I,J) \longmapsto \partial(K,\beta;I,J) \;.$$

There is also defined a morphism

$$\partial : M_S^\varepsilon(S^{-1}A) \longrightarrow M^\varepsilon(A,S) \;;\; S^{-1}(K,\alpha;I,J) \longmapsto \partial(K,\alpha;I,J) \;,$$

namely the composite

$$M_S^\varepsilon(S^{-1}A) \xrightarrow{\;\partial\;} M\langle v_0\rangle^\varepsilon(A,S) \longrightarrow M^\varepsilon(A,S) \;.$$

The correspondence of Proposition 3.5.2 i) associates to a

$$\text{non-singular}\begin{cases}\text{even } \varepsilon\text{-symmetric}\\ \varepsilon\text{-quadratic}\\ \text{split } \varepsilon\text{-quadratic}\end{cases}\text{linking formation over } (A,S)$$

$$\begin{cases}(M,\lambda;F,G)\\ (M,\lambda,\mu;F,G)\\ (F,G)\end{cases}\text{a stable isomorphism class of non-singular}$$

$$\begin{cases}\varepsilon\text{-symmetric}\\ \text{even } \varepsilon\text{-symmetric}\\ \varepsilon\text{-quadratic}\end{cases}\text{S-forms over A}\begin{cases}(K,\alpha;L)\\ (K,\alpha;L)\\ (K,\beta;L)\end{cases}$$, and it follows

from Proposition 3.5.4 that there are well-defined abelian group morphisms

$$\begin{cases}M\langle v_0\rangle^{\varepsilon}(A,S)\longrightarrow L^{\varepsilon}(A) \ ; \ (M,\lambda;F,G)\longmapsto(K,\alpha)\\ M_{\varepsilon}(A,S)\longrightarrow L\langle v_0\rangle^{\varepsilon}(A) \ ; \ (M,\lambda,\mu;F,G)\longmapsto(K,\alpha)\\ \widetilde{M}_{\varepsilon}(A,S)\longrightarrow L_{\varepsilon}(A) \ ; \ (F,G)\longmapsto(K,\beta)\end{cases}$$

from the Witt groups of linking formations over (A,S) to the Witt groups of forms over A defined in §1.6 above.

Define the $\underline{\text{lower odd-dimensional}}\begin{cases}\underline{\varepsilon\text{-symmetric}}\\ \underline{\varepsilon\text{-quadratic}}\end{cases}\underline{\text{L-groups}}$

$\underline{\text{of }(A,S)}\begin{cases}L^{2k+1}(A,S,\varepsilon)\\ L_{2k+1}(A,S,\varepsilon)\end{cases}(k\leqslant -1)$ by

$$\begin{cases}L^{2k+1}(A,S,\varepsilon) = \begin{cases}M_{-\varepsilon}(A,S) & (k=-1)\\ L_{2k+1}(A,S,\varepsilon) & (k\leqslant -2)\end{cases}\\ L_{2k+1}(A,S,\varepsilon) = L_{2k+2i+1}(A,S,(-)^i\varepsilon) \quad (k\leqslant -1, k+i\geqslant 0)\ .\end{cases}$$

Proposition 3.5.5 i) The localization exact sequence of algebraic Poincaré cobordism groups

$$L^{2k+1}(A,(-)^k\varepsilon) \longrightarrow L_S^{2k+1}(S^{-1}A,(-)^k\varepsilon) \longrightarrow L^{2k+1}(A,S,(-)^k\varepsilon)$$
$$\longrightarrow L^{2k}(A,(-)^k\varepsilon) \longrightarrow L_S^{2k}(S^{-1}A,(-)^k\varepsilon) \qquad (*)_{2k+1}$$

is naturally isomorphic for $\begin{cases} k=0 \\ k=-1 \\ k \leqslant -2 \end{cases}$ to a localization exact sequence of Witt groups

$$\begin{cases} M^\varepsilon(A) \longrightarrow M_S^\varepsilon(S^{-1}A) \xrightarrow{\partial} M\langle v_0\rangle^\varepsilon(A,S) \longrightarrow L^\varepsilon(A) \longrightarrow L_S^\varepsilon(S^{-1}A) \\ M\langle v_0\rangle^\varepsilon(A) \longrightarrow M\langle v_0\rangle_S^\varepsilon(S^{-1}A) \xrightarrow{\partial} M_\varepsilon(A,S) \longrightarrow L\langle v_0\rangle^\varepsilon(A) \longrightarrow L\langle v_0\rangle_S^\varepsilon(S^{-1}A) \\ M_\varepsilon(A) \longrightarrow M_\varepsilon^S(S^{-1}A) \xrightarrow{\partial} \widetilde{M}_\varepsilon(A,S) \longrightarrow L_\varepsilon(A) \longrightarrow L_\varepsilon^S(S^{-1}A) \ . \end{cases}$$

ii) There are defined natural abelian group morphisms

$$M^\varepsilon(A,S) \longrightarrow L^{2k+1}(A,S,(-)^k\varepsilon) \quad (k \geqslant 1)$$

for all A,S,ε. If $\begin{cases} (A,S) \text{ is 1-dimensional} \\ \ker(\hat{\delta}:\hat{H}^1(\mathbb{Z}_2;S^{-1}A/A,\varepsilon) \longrightarrow \hat{H}^0(\mathbb{Z}_2;A,\varepsilon)) = 0 \end{cases}$

then for $\begin{cases} k \geqslant 1 \\ k = 1 \end{cases}$ these are isomorphisms.

iii) If (A,S) is 0-dimensional then

$$M_\varepsilon(A,S) = M\langle v_0\rangle^\varepsilon(A,S) = M^\varepsilon(A,S) = 0 \ ,$$

and there are defined localization exact sequences of Witt groups

$$0 \longrightarrow L^\varepsilon(A) \longrightarrow L_S^\varepsilon(S^{-1}A) \xrightarrow{\partial} L^\varepsilon(A,S)$$
$$\longrightarrow M^{-\varepsilon}(A) \longrightarrow M_S^{-\varepsilon}(S^{-1}A) \longrightarrow 0 \ ,$$

$$0 \longrightarrow L^{\varepsilon}(A) \longrightarrow L_S^{\varepsilon}(S^{-1}A) \xrightarrow{\partial} L\langle v_0\rangle^{\varepsilon}(A,S)$$
$$\longrightarrow M\langle v_0\rangle^{-\varepsilon}(A) \longrightarrow M\langle v_0\rangle_S^{-\varepsilon}(S^{-1}A) \longrightarrow 0 \ ,$$

$$0 \longrightarrow L\langle v_0\rangle^{\varepsilon}(A) \longrightarrow L\langle v_0\rangle_S^{\varepsilon}(S^{-1}A) \xrightarrow{\partial} L_{\varepsilon}(A,S) \longrightarrow M_{-\varepsilon}(A)$$
$$\longrightarrow M_{-\varepsilon}^S(S^{-1}A) \xrightarrow{\partial} \widetilde{M}_{-\varepsilon}(A,S) \longrightarrow L_{-\varepsilon}(A) \longrightarrow L_{-\varepsilon}^S(S^{-1}A)$$
$$\xrightarrow{\partial} \widetilde{L}_{-\varepsilon}(A,S) \longrightarrow M_{\varepsilon}(A) \longrightarrow M_{\varepsilon}^S(S^{-1}A) \xrightarrow{\partial} \widetilde{M}_{\varepsilon}(A,S) \longrightarrow \ldots \quad .$$

iv) If $\begin{cases} \mathrm{im}(\hat{\delta}:\hat{H}^0(\mathbb{Z}_2;S^{-1}A/A,\varepsilon) \longrightarrow \hat{H}^1(\mathbb{Z}_2;A,\varepsilon)) = 0 \\ \hat{H}^0(\mathbb{Z}_2;A,\varepsilon) \longrightarrow \hat{H}^0(\mathbb{Z}_2;S^{-1}A,\varepsilon) \text{ is an isomorphism} \\ \hat{H}^1(\mathbb{Z}_2;A,\varepsilon) \longrightarrow \hat{H}^1(\mathbb{Z}_2;S^{-1}A,\varepsilon) \text{ is an isomorphism} \end{cases}$

there is a natural identification of Witt groups of linking formations over (A,S)

$$\begin{cases} M\langle v_0\rangle^{\varepsilon}(A,S) = M^{\varepsilon}(A,S) \\ M_{\varepsilon}(A,S) = M\langle v_0\rangle^{\varepsilon}(A,S) \\ \widetilde{M}_{\varepsilon}(A,S) = M_{\varepsilon}(A,S) \quad . \end{cases}$$

In particular, if $1/2 \in A$

$$\widetilde{M}_{\varepsilon}(A,S) = M_{\varepsilon}(A,S) = M\langle v_0\rangle^{\varepsilon}(A,S) = M^{\varepsilon}(A,S) \quad .$$

<u>Proof</u>: i) It has already been verified (in the course of the proof of Proposition 3.5.4) that there are natural identifications

$$\begin{cases} L^1(A,S,\varepsilon) = M\langle v_0\rangle^{\varepsilon}(A,S) \\ L_1(A,S,\varepsilon) = \widetilde{M}_{\varepsilon}(A,S) \ . \end{cases}$$

The localization exact sequence of L-groups $\begin{cases} (*)_1 \\ (*)_{-3} \end{cases}$ can thus be identified with the localization sequence of Witt groups

$$\begin{cases} M^{\varepsilon}(A) \longrightarrow M_S^{\varepsilon}(S^{-1}A) \xrightarrow{\partial} M\langle v_0\rangle_S^{\varepsilon}(A,S) \longrightarrow L^{\varepsilon}(A) \longrightarrow L_S^{\varepsilon}(S^{-1}A) \\ M_{\varepsilon}(A) \longrightarrow M_{\varepsilon}^S(S^{-1}A) \xrightarrow{\partial} \widetilde{M}_{\varepsilon}(A,S) \longrightarrow L_{\varepsilon}(A) \longrightarrow L_{\varepsilon}^S(S^{-1}A) \ . \end{cases}$$

The exactness of the Witt group sequence can also be established directly, using Proposition 3.5.4. The direct method applies also to the verification of the exactness of $(*)_{-1}$

$$M\langle v_0\rangle^{\varepsilon}(A) \longrightarrow M\langle v_0\rangle^{\varepsilon}_S(S^{-1}A) \longrightarrow M_{\varepsilon}(A,S)$$
$$\longrightarrow L\langle v_0\rangle^{\varepsilon}(A) \longrightarrow L\langle v_0\rangle^{\varepsilon}_S(S^{-1}A) \ .$$

ii) Define abelian group morphisms

$$M^{\varepsilon}(A,S) \longrightarrow L^{2k+1}(A,S,(-)^k\varepsilon) \ ; \ (M,\lambda;F,G) \longmapsto \bar{S}^k(C,\phi) \quad (k \geqslant 1)$$

by sending a non-singular ε-symmetric linking formation over (A,S) $(M,\lambda;F,G)$ to the k-fold skew-suspension $\bar{S}^k(C,\phi)$ of an S-acyclic 2-dimensional $(-\varepsilon)$-symmetric Poincaré complex over A $(C,\phi \in Q^2(C,-\varepsilon))$ associated to $(M,\lambda;F,G)$ by Proposition 3.5.2 ii). The S-acyclic cobordism class $\bar{S}^k(C,\phi) \in L^{2k+1}(A,S,(-)^k\varepsilon)$ depends only on the stable equivalence class of $(M,\lambda;F,G)$ (proved exactly as in Proposition 3.5.3 i)), vanishing if $(M,\lambda;F,G) = 0 \in M^{\varepsilon}(A,S)$ (proved exactly as in Proposition 3.5.4), so that the morphisms are well-defined. If $\begin{cases} (A,S) \text{ is 1-dimensional} \\ \ker(\hat{\delta}:\hat{H}^1(\mathbb{Z}_2;S^{-1}A/A,\varepsilon) \longrightarrow \hat{H}^0(\mathbb{Z}_2;A,\varepsilon)) = 0 \end{cases}$ then for $\begin{cases} k \geqslant 1 \\ k = 1 \end{cases}$ $L^{2k+1}(A,S,(-)^k\varepsilon)$ is the cobordism group of S-acyclic 2-dimensional $(-\varepsilon)$-symmetric Poincaré complexes over A (by Proposition $\begin{cases} 3.2.4 \\ 3.3.2 \end{cases}$), so that the morphisms are onto.

If $(M,\lambda;F,G) \in \ker(M^{\varepsilon}(A,S) \longrightarrow L^{2k+1}(A,S,(-)^k\varepsilon))$ the S-acyclic complex (C,ϕ) is homotopy equivalent to the boundary $\partial(D,\eta)$ of a connected S-acyclic 3-dimensional $(-\varepsilon)$-symmetric complex over A $(D,\eta \in Q^3(D,-\varepsilon))$, and the proof of Proposition 3.5.4 generalizes to show that $(M,\lambda;F,G) = 0 \in M^{\varepsilon}(A,S)$, so that the morphisms are also one-one.

iii) If (A,S) is 0-dimensional Proposition 3.2.4 shows that

$$L^1(A,S,\varepsilon) = L^3(A,S,-\varepsilon) = 0 \quad ,$$

so that

$$M\langle v_0\rangle^{\varepsilon}(A,S) = M^{\varepsilon}(A,S) = 0 \ .$$

The proof of Proposition 3.2.4 generalizes to also show that

$$L^{-1}(A,S,-\varepsilon) = M_{\varepsilon}(A,S) = 0 \ .$$

iv) If $\begin{cases} \mathrm{im}(\hat{\delta}:\hat{H}^0(\mathbb{Z}_2;S^{-1}A/A,\varepsilon)\longrightarrow\hat{H}^1(\mathbb{Z}_2;A,\varepsilon)) = 0 \\ \hat{H}^0(\mathbb{Z}_2;A,\varepsilon)\longrightarrow\hat{H}^0(\mathbb{Z}_2;S^{-1}A,\varepsilon) \text{ is an isomorphism} \end{cases}$

Proposition 3.4.2 ii) gives an identification of categories

$$\begin{cases} \text{(even } \varepsilon\text{-symmetric linking forms over (A,S))} \\ \qquad = (\varepsilon\text{-symmetric linking forms over (A,S))} \\ (\varepsilon\text{-quadratic linking forms over (A,S))} \\ \qquad = \text{(even } \varepsilon\text{-symmetric linking forms over (A,S))} \end{cases}$$

There is thus also an identification of categories

$$\begin{cases} \text{(even } \varepsilon\text{-symmetric linking formations over (A,S))} \\ \qquad = (\varepsilon\text{-symmetric linking formations over (A,S))} \\ (\varepsilon\text{-quadratic linking formations over (A,S))} \\ \qquad = \text{(even } \varepsilon\text{-symmetric linking formations over (A,S))} \ , \end{cases}$$

giving rise to an identification of the Witt groups

$$\begin{cases} M\langle v_0\rangle^{\varepsilon}(A,S) = M^{\varepsilon}(A,S) \\ M_{\varepsilon}(A,S) = M\langle v_0\rangle^{\varepsilon}(A,S) \ . \end{cases}$$

If $\hat{H}^1(\mathbb{Z}_2;A,\varepsilon)\longrightarrow\hat{H}^1(\mathbb{Z}_2;S^{-1}A,\varepsilon)$ is an isomorphism Proposition 3.4.2 ii) gives identifications of categories

$$\text{(split } \varepsilon\text{-quadratic linking forms over (A,S))}$$
$$= (\varepsilon\text{-quadratic linking forms over (A,S))}$$

((−ε)-quadratic linking forms over (A,S))

= (even (−ε)-symmetric linking forms over (A,S)),

so that there is an identification of stable equivalence classes

(split ε-quadratic linking formations over (A,S))

= (ε-quadratic linking formations over (A,S)) ,

giving rise to an identification of the Witt groups

$$\widetilde{M}_{\varepsilon}(A,S) = M_{\varepsilon}(A,S) \ .$$

[]

3.6 The localization exact sequence ($n \in \mathbb{Z}$)

In the course of §§3.4,3.5 the definition of the

$\left\{\begin{array}{l}\varepsilon\text{-symmetric}\\ \varepsilon\text{-quadratic}\end{array}\right.$ L-groups $\left\{\begin{array}{l}L^n(A,S,\varepsilon)\\ L_n(A,S,\varepsilon)\end{array}\right.$ $(n \geq 0)$ of §3.2 was extended

to the range $n \leq -1$, by S-acyclic analogy with the lower

L-groups $\left\{\begin{array}{l}L^n(A,\varepsilon)\\ L_n(A,\varepsilon)\end{array}\right.$ $(n \leq -1)$ of §1.8. Combining the results of

Propositions 3.4.7,3.5.5 we have:

Proposition 3.6.1 i) There is defined a localization exact

sequence of $\left\{\begin{array}{l}\varepsilon\text{-symmetric}\\ \varepsilon\text{-quadratic}\end{array}\right.$ L-groups

$$\left\{\begin{array}{l}\ldots \longrightarrow L^n(A,\varepsilon) \longrightarrow L^n_S(S^{-1}A,\varepsilon) \xrightarrow{\partial} L^n(A,S,\varepsilon) \longrightarrow L^{n-1}(A,\varepsilon) \longrightarrow \ldots\\ \ldots \longrightarrow L_n(A,\varepsilon) \longrightarrow L^S_n(S^{-1}A,\varepsilon) \xrightarrow{\partial} L_n(A,S,\varepsilon) \longrightarrow L_{n-1}(A,\varepsilon) \longrightarrow \ldots\end{array}\right.$$

$(n \in \mathbb{Z})$.

ii) The localization exact sequence of ε-quadratic L-groups is 12-periodic, all the groups involved being 4-periodic in n, and it is naturally isomorphic to the localization exact sequence of $\pm\varepsilon$-quadratic Witt groups

$$\begin{aligned}\ldots \longrightarrow \widetilde{L}_{-\varepsilon}(A,S) \longrightarrow M_\varepsilon(A) \longrightarrow M^S_\varepsilon(S^{-1}A) \xrightarrow{\partial} \widetilde{M}_\varepsilon(A,S)\\ \longrightarrow L_\varepsilon(A) \longrightarrow L^S_\varepsilon(S^{-1}A) \xrightarrow{\partial} \widetilde{L}_\varepsilon(A,S) \longrightarrow M_{-\varepsilon}(A) \longrightarrow \ldots\ .\end{aligned}$$

iii) In the range $n \leq 2$ the localization exact sequence of ε-symmetric L-groups is naturally isomorphic to the localization exact sequence of Witt groups

$$(\ldots \longrightarrow L^2(A,S,\varepsilon) \longrightarrow)\ M^{\varepsilon}(A) \longrightarrow M^{\varepsilon}_S(S^{-1}A) \xrightarrow{\partial} M\langle v_0\rangle^{\varepsilon}(A,S)$$

$$\longrightarrow L^{\varepsilon}(A) \longrightarrow L^{\varepsilon}_S(S^{-1}A) \xrightarrow{\partial} L\langle v_0\rangle^{\varepsilon}(A,S) \longrightarrow M\langle v_0\rangle^{-\varepsilon}(A)$$

$$\longrightarrow M\langle v_0\rangle^{-\varepsilon}_S(S^{-1}A) \xrightarrow{\partial} M_{-\varepsilon}(A,S) \longrightarrow L\langle v_0\rangle^{-\varepsilon}(A) \longrightarrow L\langle v_0\rangle^{-\varepsilon}_S(S^{-1}A)$$

$$\xrightarrow{\partial} L_{-\varepsilon}(A,S) \longrightarrow M_{\varepsilon}(A) \longrightarrow M^S_{\varepsilon}(S^{-1}A) \xrightarrow{\partial} \widetilde{M}_{\varepsilon}(A,S) \longrightarrow L_{\varepsilon}(A)$$

$$\longrightarrow L^S_{\varepsilon}(S^{-1}A) \xrightarrow{\partial} \widetilde{L}_{\varepsilon}(A,S) \longrightarrow M_{-\varepsilon}(A) \longrightarrow M^S_{-\varepsilon}(S^{-1}A)$$

$$\xrightarrow{\partial} \widetilde{M}_{-\varepsilon}(A,S) \longrightarrow \ldots ,$$

becoming the 12-periodic ε-quadratic sequence on the right.

iv) If $\hat{H}^0(\mathbb{Z}_2;A,\varepsilon) \longrightarrow \hat{H}^0(\mathbb{Z}_2;S^{-1}A,\varepsilon)$ is an isomorphism the skew-suspension maps

$$\bar{S} : L^n(A,S,\varepsilon) \longrightarrow L^{n+2}(A,S,-\varepsilon) \quad (n \in \mathbb{Z})$$

are isomorphisms.

v) If (A,S) is 0-dimensional

$$L^{2k}(A,S,(-)^k\varepsilon) = L^{\varepsilon}(A,S) \quad (k \geqslant 1)$$

$$L^{2k+1}(A,S,(-)^k\varepsilon) = M_{\varepsilon}(A,S) = M\langle v_0\rangle^{\varepsilon}(A,S) = M^{\varepsilon}(A,S) = 0 \ (k \geqslant -1).$$

If (A,S) is 1-dimensional

$$L^{2k+1}(A,S,(-)^k\varepsilon) = M^{\varepsilon}(A,S) \quad (k \geqslant 1) \ .$$

If $\ker(\hat{\delta}:\hat{H}^1(\mathbb{Z}_2;S^{-1}A/A,\varepsilon) \longrightarrow \hat{H}^0(\mathbb{Z}_2;A,\varepsilon)) = 0$

$$L^2(A,S,-\varepsilon) = L^{\varepsilon}(A,S)$$

$$L^3(A,S,-\varepsilon) = M^{\varepsilon}(A,S) \ .$$

[]

(Note that Proposition 3.6.1 iv) is an S-acyclic analogue of the result of Proposition 1.8.1 that if $\hat{H}^0(\mathbb{Z}_2;A,\varepsilon) = 0$ then the skew-suspension maps $\bar{S}:L^n(A,\varepsilon) \longrightarrow L^{n+2}(A,-\varepsilon)$ $(n \in \mathbb{Z})$ are isomorphisms).

As promised in §3.1 we shall now apply the localization exact sequence to obtain excision isomorphisms and Mayer-Vietoris exact sequences for the L-groups of the rings with involution appearing in the cartesian square

$$\begin{array}{ccc} A & \longrightarrow & S^{-1}A \\ {\scriptstyle f}\big\downarrow & & \big\downarrow \\ B & \longrightarrow & T^{-1}B \end{array}$$

associated to a cartesian morphism

$$f : (A,S) \longrightarrow (B,T)$$

of rings with involution and multiplicative subsets.

In the first instance we consider the Witt groups of linking $\left\{\begin{array}{l}\text{forms}\\ \text{formations}\end{array}\right.$:

Proposition 3.6.2 A cartesian morphism

$$f : (A,S) \longrightarrow (B,T)$$

induces isomorphisms of Witt groups

$$\left\{\begin{array}{l} f : L^{\varepsilon}(A,S) \longrightarrow L^{\varepsilon}(B,T) \\ f : M^{\varepsilon}(A,S) \longrightarrow M^{\varepsilon}(B,T) \end{array}\right. , \quad \left\{\begin{array}{l} f : \tilde{L}_{\varepsilon}(A,S) \longrightarrow \tilde{L}_{\varepsilon}(B,T) \\ f : \tilde{M}_{\varepsilon}(A,S) \longrightarrow \tilde{M}_{\varepsilon}(B,T) \end{array}\right. .$$

If $\hat{\delta} = 0 : \hat{H}^0(\mathbb{Z}_2;T^{-1}B,\varepsilon) \longrightarrow \hat{H}^1(\mathbb{Z}_2;A,\varepsilon)$ there are also induced isomorphisms

$$\left\{\begin{array}{l} f : L\langle v_0\rangle^{\varepsilon}(A,S) \longrightarrow L\langle v_0\rangle^{\varepsilon}(B,T) \\ f : M\langle v_0\rangle^{\varepsilon}(A,S) \longrightarrow M\langle v_0\rangle^{\varepsilon}(B,T) \end{array}\right. ,$$

$$\left\{\begin{array}{l} f : L_{-\varepsilon}(A,S) \longrightarrow L_{-\varepsilon}(B,T) \\ f : M_{-\varepsilon}(A,S) \longrightarrow M_{-\varepsilon}(B,T) \end{array}\right. .$$

Proof: The cartesian morphism $f:(A,S)\longrightarrow(B,T)$ induces an isomorphism of exact categories

$$f : \{(A,S)\text{-modules}\} \longrightarrow \{(B,T)\text{-modules}\} \ ; \ M \longmapsto B\otimes_A M$$

(Proposition 3.1.3 i)), so that it also induces an isomorphism of categories

$$f : \{\varepsilon\text{-symmetric linking} \begin{cases}\text{forms}\\ \text{formations}\end{cases} \text{over } (A,S)\}$$

$$\longrightarrow \{\varepsilon\text{-symmetric linking} \begin{cases}\text{forms}\\ \text{formations}\end{cases} \text{over } (B,T)\},$$

and hence also isomorphisms of the corresponding Witt groups. Although the functor

$$f : \{\text{split } \varepsilon\text{-quadratic linking} \begin{cases}\text{forms}\\ \text{formations}\end{cases} \text{over } (A,S)\}$$

$$\longrightarrow \{\text{split } \varepsilon\text{-quadratic linking} \begin{cases}\text{forms}\\ \text{formations}\end{cases} \text{over } (B,T)\}$$

need not be an isomorphism of categories in the linking formation case it does induce isomorphisms in the corresponding Witt groups, since it induces isomorphisms

$$f : Q_*^S(C,\varepsilon) \longrightarrow Q_*^T(B\otimes_A C,\varepsilon)$$

for any finite-dimensional (A,S)-module chain complex C (Proposition 3.1.3 ii)). It follows from the exact sequences

$$0 \longrightarrow Q^{\varepsilon}(A,S) \xrightarrow{\ f\ } Q^{\varepsilon}(B,T) \longrightarrow \operatorname{im}(\hat{\delta}) \longrightarrow 0$$

$$0 \longrightarrow \operatorname{im}(\hat{\delta}) \longrightarrow \tilde{Q}_{-\varepsilon}(A,S) \xrightarrow{\ f\ } \tilde{Q}_{-\varepsilon}(B,T) \longrightarrow 0$$

(with the $\tilde{Q}$-groups as defined in the proof of Proposition 3.4.2 ii)) that if $\hat{\delta} = 0 : \hat{H}^0(\mathbb{Z}_2;T^{-1}B,\varepsilon) \longrightarrow \hat{H}^1(\mathbb{Z}_2;A,\varepsilon)$ then $f:(A,S)\longrightarrow(B,T)$ induces isomorphisms of categories

$$f : \{\text{even } \varepsilon\text{-symmetric linking } \left\{\begin{matrix}\text{forms}\\ \text{formations}\end{matrix}\right. \text{ over } (A,S)\}$$

$$\xrightarrow{\sim} \{\text{even } \varepsilon\text{-symmetric linking } \left\{\begin{matrix}\text{forms}\\ \text{formations}\end{matrix}\right. \text{ over } (B,T)\} ,$$

$$f : \{(-\varepsilon)\text{-quadratic linking } \left\{\begin{matrix}\text{forms}\\ \text{formations}\end{matrix}\right. \text{ over } (A,S)\}$$

$$\xrightarrow{\sim} \{(-\varepsilon)\text{-quadratic linking } \left\{\begin{matrix}\text{forms}\\ \text{formations}\end{matrix}\right. \text{ over } (B,T)\}$$

(as well as an isomorphism

$$f : \{\text{split } \varepsilon\text{-quadratic linking } \left\{\begin{matrix}\text{forms}\\ \text{formations}\end{matrix}\right. \text{ over } (A,S)\}$$

$$\xrightarrow{\sim} \{\text{split } \varepsilon\text{-quadratic linking } \left\{\begin{matrix}\text{forms}\\ \text{formations}\end{matrix}\right. \text{ over } (B,T)\}),$$

and hence also isomorphisms of the corresponding Witt groups.

[]

Next, we consider the excision properties of the L-groups:

Proposition 3.6.3 i) A cartesian morphism

$$f : (A,S) \longrightarrow (B,T)$$

induces excision isomorphisms in the ε-quadratic L-groups

$$f : L_n(A,S,\varepsilon) \longrightarrow L_n(B,T,\varepsilon) \quad (n \in \mathbb{Z}) ,$$

and there is defined a Mayer-Vietoris exact sequence

$$\ldots \longrightarrow L_n(A,\varepsilon) \longrightarrow L_n^S(S^{-1}A,\varepsilon) \oplus L_n(B,\varepsilon) \longrightarrow L_n(T^{-1}B,\varepsilon)$$

$$\xrightarrow{\partial} L_{n-1}(A,\varepsilon) \longrightarrow \ldots \quad (n \in \mathbb{Z}) .$$

ii) A cartesian morphism

$$f : (A,S) \longrightarrow (B,T)$$

such that $\left\{\begin{matrix}(A,S) \text{ is 0-dimensional}\\ \hat{\delta} = 0 : \hat{H}^0(\mathbb{Z}_2;T^{-1}B,\varepsilon) \longrightarrow \hat{H}^1(\mathbb{Z}_2;A,\varepsilon)\end{matrix}\right.$ induces excision isomorphisms in the ε-symmetric L-groups

$$f : L^n(A,S,\varepsilon) \longrightarrow L^n(B,T,\varepsilon) \quad \left(\begin{cases} n \geqslant 1 \\ n \in \mathbb{Z} \end{cases}\right)$$

and there is defined a Mayer-Vietoris exact sequence

$$\ldots \longrightarrow L^n(A,\varepsilon) \longrightarrow L^n_S(S^{-1}A,\varepsilon) \oplus L^n(B,\varepsilon) \longrightarrow L^n_T(T^{-1}B,\varepsilon)$$
$$\longrightarrow L^{n-1}(A,\varepsilon) \longrightarrow \ldots$$

for $\begin{cases} n \geqslant 1, \text{ with } L^{2k+1}(A,S,\varepsilon) = L^{2k+1}(B,T,\varepsilon) = 0 \ (k \geqslant 0) \text{ and} \\ n \in \mathbb{Z} \end{cases}$

$$\begin{cases} \hat{\delta} = 0 : L^{2k+1}_T(T^{-1}B,\varepsilon) \longrightarrow L^{2k}(A,\varepsilon) \quad (k \geqslant 0) \\ - \end{cases} .$$

<u>Proof</u>: Immediate from Propositions 3.2.1, 3.6.1 and 3.6.2.

[]

In particular, for a central multiplicative subset $S \subset A$ there is defined a cartesian morphism

$$(A,S) \longrightarrow (\hat{A},\hat{S})$$

with $\hat{A} = \varprojlim_{s \in S} A/sA$ the S-adic completion of A, giving rise to the cartesian square of rings with involution

$$\begin{array}{ccc} A & \longrightarrow & S^{-1}A \\ \downarrow & & \downarrow \\ \hat{A} & \longrightarrow & \hat{S}^{-1}\hat{A} \end{array} .$$

Proposition 3.6.3 $\begin{cases} \text{ii)} \\ \text{i)} \end{cases}$ shows that

$\begin{cases} \text{if } \hat{\delta} = 0 : \hat{H}^0(\mathbb{Z}_2;\hat{S}^{-1}\hat{A},\varepsilon) \longrightarrow \hat{H}^1(\mathbb{Z}_2;A,\varepsilon) \text{ (e.g. if } 1/2 \in \hat{S}^{-1}\hat{A}) \\ \text{for all } A,S,\varepsilon \end{cases}$

there are defined excision isomorphisms in the $\begin{cases} \varepsilon\text{-symmetric} \\ \varepsilon\text{-quadratic} \end{cases}$

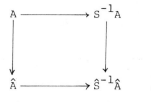

L-groups

$$\begin{cases} L^n(A,S,\varepsilon) \longrightarrow L^n(\hat{A},\hat{S},\varepsilon) \\ L_n(A,S,\varepsilon) \longrightarrow L_n(\hat{A},\hat{S},\varepsilon) \end{cases} \quad (n \in \mathbb{Z})$$

and a Mayer-Vietoris exact sequence

$$\begin{cases} \ldots \longrightarrow L^n(A,\varepsilon) \longrightarrow L^n_S(S^{-1}A,\varepsilon) \oplus L^n(\hat{A},\varepsilon) \longrightarrow L^n_{\hat{S}}(\hat{S}^{-1}\hat{A},\varepsilon) \\ \qquad \xrightarrow{\partial} L^{n-1}(A,\varepsilon) \longrightarrow \ldots \\ \ldots \longrightarrow L_n(A,\varepsilon) \longrightarrow L^S_n(S^{-1}A,\varepsilon) \oplus L_n(\hat{A},\varepsilon) \longrightarrow L^{\hat{S}}_n(\hat{S}^{-1}\hat{A},\varepsilon) \\ \qquad \xrightarrow{\partial} L_{n-1}(A,\varepsilon) \longrightarrow \ldots \end{cases}$$

$$(n \in \mathbb{Z}) \quad .$$

A ring with involution A is <u>m-torsion-free</u> for some integer $m \geqslant 2$ if $m1 \in A$ is a non-zero-divisor of A, in which case

$$S = \{m^k \mid k \geqslant 0\} \subset A$$

is a central multiplicative subset of A. The <u>localization of A away from m</u> is the localization

$$A[\tfrac{1}{m}] = S^{-1}A \ .$$

The <u>m-adic completion of A</u> is the S-adic completion of A

$$\hat{A}_m = \varprojlim_k A/m^kA \quad .$$

The completion $\hat{A}_m$ is an m-torsion-free ring with involution which is a module over the ring of m-adic integers $\hat{\mathbb{Z}}_m = \varprojlim_k \mathbb{Z}/m^k\mathbb{Z}$, and the localization of the completion

$$\hat{A}_m[\tfrac{1}{m}] = \hat{S}^{-1}\hat{A}_m = \hat{\mathbb{Q}}_m \otimes_{\hat{\mathbb{Z}}_m} \hat{A}_m$$

is a vector space over the field of m-adic numbers $\hat{\mathbb{Q}}_m = \hat{\mathbb{Z}}_m[\tfrac{1}{m}]$.

A ring with involution A is <u>torsion-free</u> if it is m-torsion-free for each integer $m \geqslant 2$, in which case

$$S = \mathbb{Z}-\{0\} \subset A$$

is a central multiplicative subset of A. The <u>localization of A at 0</u> is the localization

$$A_{(0)} = S^{-1}A = \mathbb{Q}\otimes_{\mathbb{Z}}A \quad ,$$

which is a vector space over the field of rational numbers $\mathbb{Q}$. The <u>profinite completion of A</u> is the S-adic completion

$$\hat{A} = \varprojlim_{m} A/mA \quad ,$$

which is a module over $\hat{\mathbb{Z}} = \varprojlim_{m} \mathbb{Z}/m\mathbb{Z}$. Furthermore, $\hat{A}$ is a torsion-free ring with involution, and the localization of the completion

$$\hat{A}_{(0)} = \hat{S}^{-1}\hat{A} = \hat{\mathbb{Q}}\otimes_{\hat{\mathbb{Z}}}\hat{A}$$

is a module over the ring of finite adèles $\hat{\mathbb{Q}} = \hat{S}^{-1}\hat{\mathbb{Z}}$ of $\mathbb{Z}$.

As in Ranicki [6,§4] define for each integer $m \geqslant 2$ the number

$$\hat{\psi}(m) = \text{the exponent of } L^0(\hat{\mathbb{Z}}_m)$$

$$= \begin{cases} 2 \text{ if } m \text{ is a product of odd primes } p \equiv 1 \pmod 4 \\ 4 \text{ if } m \text{ is a product of odd primes at least} \\ \qquad \text{one of which is } p \equiv 3 \pmod 4 \\ 8 \text{ if } m \text{ is even ,} \end{cases}$$

and note that $L^0(\hat{\mathbb{Z}})$ has exponent 8. (In fact, $L^0(\hat{\mathbb{Z}}_m)$ and $L^0(\hat{\mathbb{Z}})$ are given by

$$L^0(\hat{\mathbb{Z}}_m) = \sum_{i=1}^{r} L^0(\hat{\mathbb{Z}}_{p_i}) \text{ if } m = p_1^{k_1}p_2^{k_2}\dots p_r^{k_r} \text{ is the}$$

factorization of m into prime powers,

$$L^0(\hat{\mathbb{Z}}) = \prod_{p \text{ prime}} L^0(\hat{\mathbb{Z}}_p) \quad ,$$

with

$$L^0(\hat{\mathbb{Z}}_p) = \begin{cases} \mathbb{Z}_8 \oplus \mathbb{Z}_2 & \text{if } p = 2 \\ \mathbb{Z}_2 \oplus \mathbb{Z}_2 & \text{if } p \equiv 1 \pmod 4 \\ \mathbb{Z}_4 & \text{if } p \equiv 3 \pmod 4 \end{cases} \quad).$$

Proposition 3.6.4 Let A be a ring with involution which is m-torsion-free (resp. torsion-free), and let

$$S = \{m^k 1 \mid k \geqslant 0\} \subset A \quad (\text{resp. } S = \mathbb{Z}-\{0\} \subset A) \ ,$$

so that the cartesian square of rings with involution

$$\begin{array}{ccc} A & \longrightarrow & S^{-1}A \\ \downarrow & & \downarrow \\ \hat{A} & \longrightarrow & \hat{S}^{-1}\hat{A} \end{array}$$

is given by

$$\begin{array}{ccc} A & \longrightarrow & A[\frac{1}{m}] \\ \downarrow & & \downarrow \\ \hat{A}_m & \longrightarrow & \hat{A}_m[\frac{1}{m}] \end{array} \qquad \left(\text{resp. } \begin{array}{ccc} A & \longrightarrow & A_{(0)} \\ \downarrow & & \downarrow \\ \hat{A} & \longrightarrow & \hat{A}_{(0)} \end{array} \right) .$$

i) There is defined a Mayer-Vietoris exact sequence of $\begin{Bmatrix} \varepsilon\text{-symmetric} \\ \varepsilon\text{-quadratic} \end{Bmatrix}$ L-groups

$$\begin{cases} \ldots \longrightarrow L^n(A,\varepsilon) \longrightarrow L^n_S(S^{-1}A,\varepsilon) \oplus L^n(\hat{A},\varepsilon) \longrightarrow L^n_{\hat{S}}(\hat{S}^{-1}\hat{A},\varepsilon) \longrightarrow L^{n-1}(A,\varepsilon) \longrightarrow \ldots \\ \ldots \longrightarrow L_n(A,\varepsilon) \longrightarrow L^S_n(S^{-1}A,\varepsilon) \oplus L_n(\hat{A},\varepsilon) \longrightarrow L^{\hat{S}}_n(\hat{S}^{-1}\hat{A},\varepsilon) \longrightarrow L_{n-1}(A,\varepsilon) \longrightarrow \ldots \end{cases}$$

$(n \in \mathbb{Z})$.

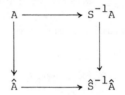

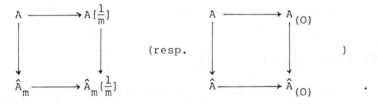

ii) The localization maps

$$\begin{cases} L^n(A,\varepsilon) \longrightarrow L^n_S(S^{-1}A,\varepsilon) \\ L^n(A,\varepsilon) \longrightarrow L_n^S(S^{-1}A,\varepsilon) \end{cases} \quad (n \in \mathbb{Z})$$

are isomorphisms modulo $\hat{\psi}(m)$- (resp. 8-) torsion, and the L-groups $\begin{cases} L^n(\hat{A},\varepsilon) \\ L_n(\hat{A},\varepsilon) \end{cases}$, $\begin{cases} L^n_{\hat{S}}(\hat{S}^{-1}\hat{A},\varepsilon) \\ L_n^{\hat{S}}(\hat{S}^{-1}\hat{A},\varepsilon) \end{cases}$ $(n \in \mathbb{Z})$ have exponents dividing $\hat{\psi}(m)$ (resp. 8).

Proof: i) The Mayer-Vietoris exact sequence in the ε-quadratic case is just that given by Proposition 3.6.3 i) for the cartesian morphism

$$(A,S) \longrightarrow (\hat{A},\hat{S})$$

defined by the inclusion. In the ε-symmetric case it is the sequence given by Proposition 3.6.3 ii) - this applies here since $1/2 \in \hat{S}^{-1}\hat{A}$, so that

$$\hat{\delta} = 0 : \hat{H}^0(\mathbb{Z}_2;\hat{S}^{-1}\hat{A},\varepsilon) = 0 \longrightarrow \hat{H}^1(\mathbb{Z}_2;A,\varepsilon) .$$

ii) The maps

$$\begin{cases} L^n(A,S,\varepsilon) \longrightarrow L^n(\hat{A},\hat{S},\varepsilon) \\ L_n(A,S,\varepsilon) \longrightarrow L_n(\hat{A},\hat{S},\varepsilon) \end{cases} \quad (n \in \mathbb{Z})$$

are isomorphisms (by Proposition 3.6.3 again). Now the localization map

$$\hat{A} \longrightarrow \hat{S}^{-1}\hat{A}$$

is a morphism of $\hat{\mathbb{Z}}_m$- (resp. $\hat{\mathbb{Z}}$-) modules, so that by Proposition 2.2.6 the localization exact sequence

$$\begin{cases} \ldots \longrightarrow L^n(\hat{A},\varepsilon) \longrightarrow L^n_{\hat{S}}(\hat{S}^{-1}\hat{A},\varepsilon) \longrightarrow L^n(\hat{A},\hat{S},\varepsilon) \longrightarrow L^{n-1}(\hat{A},\varepsilon) \longrightarrow \ldots \\ \ldots \longrightarrow L_n(\hat{A},\varepsilon) \longrightarrow L_n^{\hat{S}}(\hat{S}^{-1}\hat{A},\varepsilon) \longrightarrow L_n(\hat{A},\hat{S},\varepsilon) \longrightarrow L_{n-1}(\hat{A},\varepsilon) \longrightarrow \ldots \end{cases}$$

$$(n \in \mathbb{Z})$$

is a sequence of $L^0(\hat{\mathbb{Z}}_m)$- (resp. $L^0(\hat{\mathbb{Z}})$-) modules.

[]

In particular, for any group π the group ring $\mathbb{Z}[\pi]$ is torsion-free, so that by Proposition 3.6.4 ii) the localization maps

$$\begin{cases} L^n(\mathbb{Z}[\pi]) \longrightarrow L^n_S(\mathbb{Q}[\pi]) \\ L_n(\mathbb{Z}[\pi]) \longrightarrow L^S_n(\mathbb{Q}[\pi]) \end{cases} \quad (n \in \mathbb{Z})$$

are isomorphisms modulo 8-torsion, with

$$S = \mathbb{Z}-\{0\} \subset \mathbb{Z}[\pi] \ , \ S^{-1}\mathbb{Z}[\pi] = \mathbb{Q}[\pi] \ .$$

For each prime p define the multiplicative subset

$$S_p = \{p^k 1 \,|\, k \geqslant 0\} \subset \mathbb{Z}[\pi] \ .$$

In §4.1 below the L-groups of $(\mathbb{Z}[\pi],S)$ will be expressed as direct sums

$$\begin{cases} L^n(\mathbb{Z}[\pi],S) = \bigoplus\limits_p L^n(\mathbb{Z}[\pi],S_p) = \bigoplus\limits_p L^n(\hat{\mathbb{Z}}_p[\pi],\hat{S}_p) \\ L_n(\mathbb{Z}[\pi],S) = \bigoplus\limits_p L_n(\mathbb{Z}[\pi],S_p) = \bigoplus\limits_p L_n(\hat{\mathbb{Z}}_p[\pi],\hat{S}_p) \end{cases} \quad (n \in \mathbb{Z})$$

in which the p-components are $L^0(\hat{\mathbb{Z}}_p)$-modules, and hence of exponent dividing $\hat{\psi}(p)$.

Returning to general rings with involution, we have the following result (which is needed for §4.1):

<u>Proposition 3.6.5</u> If $S,T \subset A$ are multiplicative subsets such that

$$S^{-1}A = T^{-1}A$$

(in the sense that there exists an isomorphism of rings with involution $S^{-1}A \longrightarrow T^{-1}A$ which is the identity on A) there are defined natural identifications

$$\begin{cases} L^n(A,S,\varepsilon) = L^n(A,T,\varepsilon) \\ L_n(A,S,\varepsilon) = L_n(A,T,\varepsilon) \end{cases} \quad (n \in \mathbb{Z}) \quad .$$

Proof: Immediate from the definitions and the identification of exact categories

$$\{(A,S)\text{-modules}\} = \{(A,T)\text{-modules}\} \quad .$$

[]

Given central multiplicative subsets $S,T \subset A$ in a ring with involution A define a central multiplicative subset

$$ST = \{st \mid s \in S, t \in T\} \subset A \ ,$$

such that

$$(ST)^{-1}A = S^{-1}(T^{-1}A) = T^{-1}(S^{-1}A) \ .$$

The central multiplicative subsets $S,T \subset A$ are coprime if for all $s \in S, t \in T$ the ideals sA, tA of A are coprime, that is if there exist $a,b \in A$ such that

$$as + bt = 1 \in A \ .$$

It follows that the inclusion defines a cartesian morphism

$$(A,S) \longrightarrow (T^{-1}A,S) \ ,$$

giving rise to the cartesian square of rings with involution

$$\begin{array}{ccc} A & \longrightarrow & S^{-1}A \\ \big\downarrow & & \big\downarrow \\ T^{-1}A & \longrightarrow & (ST)^{-1}A \end{array} \quad .$$

Proposition 3.6.6 Let $S,T \subset A$ be coprime central multiplicative subsets in a ring with involution A.

i) For all A,S,T,ε there is defined a Mayer-Vietoris exact sequence of ε-quadratic L-groups

$$\ldots \longrightarrow L_n(A,\varepsilon) \longrightarrow L_n^S(S^{-1}A,\varepsilon) \oplus L_n^T(T^{-1}A,\varepsilon) \longrightarrow L_n^{ST}((ST)^{-1}A,\varepsilon)$$
$$\xrightarrow{\ \partial\ } L_{n-1}(A,\varepsilon) \longrightarrow \ldots \qquad (n \in \mathbb{Z}) \quad .$$

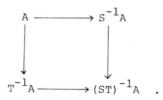

ii) If $\hat{\delta} = 0 : \hat{H}^0(\mathbb{Z}_2;(ST)^{-1}A,\varepsilon) \longrightarrow \hat{H}^1(\mathbb{Z}_2;A,\varepsilon)$ (e.g. if the involution on A restricts to the identity on S and T) there is defined a Mayer-Vietoris exact sequence of ε-symmetric L-groups

$$\ldots \longrightarrow L^n(A,\varepsilon) \longrightarrow L^n_S(S^{-1}A,\varepsilon) \oplus L^n_T(T^{-1}A,\varepsilon) \longrightarrow L^n_{ST}((ST)^{-1}A,\varepsilon)$$
$$\xrightarrow{\partial} L^{n-1}(A,\varepsilon) \longrightarrow \ldots \quad (n \in \mathbb{Z}) .$$

<u>Proof</u>: By Proposition 3.6.3 $\begin{cases} \text{ii)} \\ \text{i)} \end{cases}$ there are defined excision isomorphisms of $\begin{cases} \varepsilon\text{-symmetric} \\ \varepsilon\text{-quadratic} \end{cases}$ L-groups

$$\begin{cases} L^n(A,S,\varepsilon) \longrightarrow L^n(T^{-1}A,S,\varepsilon) \\ L_n(A,S,\varepsilon) \longrightarrow L_n(T^{-1}A,S,\varepsilon) \end{cases} \quad (n \in \mathbb{Z}) .$$

It follows from the exact sequence

$$\ldots \longrightarrow \tilde{K}_1((ST)^{-1}A) \longrightarrow \tilde{K}_0(A) \longrightarrow \tilde{K}_0(S^{-1}A) \oplus \tilde{K}_0(T^{-1}A)$$
$$\longrightarrow \tilde{K}_0((ST)^{-1}A) \longrightarrow K_{-1}(A) \longrightarrow \ldots$$

that the natural map

$$\tilde{K}_0(T^{-1}A)/T \longrightarrow S/ST$$

is an isomorphism, where S,T,ST are the $*$-invariant subgroups

$$S = \mathrm{im}(\tilde{K}_0(T^{-1}A) \longrightarrow \tilde{K}_0((ST)^{-1}A)) \subseteq \tilde{K}_0((ST)^{-1}A)$$
$$T = \mathrm{im}(\tilde{K}_0(A) \longrightarrow \tilde{K}_0(T^{-1}A)) \subseteq \tilde{K}_0(T^{-1}A)$$
$$ST = \mathrm{im}(\tilde{K}_0(A) \longrightarrow \tilde{K}_0((ST)^{-1}A)) \subseteq \tilde{K}_0((ST)^{-1}A) .$$

Proposition 2.5.1 shows that the natural maps

$$\begin{cases} L^n(T^{-1}A,S,\varepsilon) = L^n_{S,\tilde{K}_0(T^{-1}A)}(T^{-1}A \longrightarrow (ST)^{-1}A,\varepsilon) \\ \qquad\qquad \longrightarrow L^n_{ST,S}(T^{-1}A \longrightarrow (ST)^{-1}A,\varepsilon) \end{cases}$$

$$\left\{\begin{array}{l} L_n(A,S,\varepsilon) = L_n^{S,\tilde{K}_0(A)}(A \longrightarrow S^{-1}A,\varepsilon) \\ \qquad\qquad \longrightarrow L_n^{ST,S}(T^{-1}A \longrightarrow (ST)^{-1}A,\varepsilon) \end{array}\right. \quad (n \in \mathbb{Z})$$

are isomorphisms, so that the natural maps

$$\left\{\begin{array}{l} L^n(A,S,\varepsilon) = L^n_{S,\tilde{K}_0(A)}(A \longrightarrow S^{-1}A,\varepsilon) \\ \qquad\qquad \longrightarrow L^n_{ST,S}(T^{-1}A \longrightarrow (ST)^{-1}A,\varepsilon) \\ L_n(A,S,\varepsilon) = L_n^{S,\tilde{K}_0(A)}(A \longrightarrow S^{-1}A,\varepsilon) \\ \qquad\qquad \longrightarrow L_n^{ST,S}(T^{-1}A \longrightarrow (ST)^{-1}A,\varepsilon) \end{array}\right. \quad (n \in \mathbb{Z})$$

are excision isomorphisms and give rise to the Mayer-Vietoris exact sequences claimed in the statement.

[]

Given disjoint collections of primes in $\mathbb{N}$

$$P = \{p_1,p_2,\ldots\} \quad , \quad Q = \{q_1,q_2,\ldots\}$$

(one of which may be empty) such that

$$P \cup Q = \{\text{all primes in } \mathbb{N}\}$$

there are defined coprime multiplicative subsets

$$S = \{p_1^{k_1}p_2^{k_2}\ldots p_r^{k_r} \mid k_1,k_2,\ldots,k_r \geqslant 0, r \geqslant 0\} \subset A$$

$$T = \{q_1^{j_1}q_2^{j_2}\ldots q_s^{j_s} \mid j_1,j_2,\ldots,j_s \geqslant 0, s \geqslant 0\} \subset A$$

for any torsion-free ring with involution A. The <u>localization of A away from P</u>, or equivalently the <u>localization of A at Q</u>, is defined to be the ring with involution

$$S^{-1}A = A[\tfrac{1}{P}] = A_{(Q)} \ .$$

The localizations at and away from P are related by a cartesian square

$$\begin{array}{ccc} A & \longrightarrow & A_{(P)} \\ \downarrow & & \downarrow \\ A[\frac{1}{P}] & \longrightarrow & A_{(O)} \end{array} ,$$

for which Proposition 3.6.6 gives a Mayer-Vietoris exact sequence

$$\begin{cases} \dots \longrightarrow L^n(A,\varepsilon) \longrightarrow L^n_S(A[\frac{1}{P}],\varepsilon) \oplus L^n_T(A_{(P)},\varepsilon) \longrightarrow L^n_{ST}(A_{(O)},\varepsilon) \\ \qquad \xrightarrow{\partial} L^{n-1}(A,\varepsilon) \longrightarrow \dots \\ \dots \longrightarrow L_n(A,\varepsilon) \longrightarrow L^S_n(A[\frac{1}{P}],\varepsilon) \oplus L^T_n(A_{(P)},\varepsilon) \longrightarrow L^{ST}_n(A_{(O)},\varepsilon) \\ \qquad \xrightarrow{\partial} L_{n-1}(A,\varepsilon) \longrightarrow \dots \end{cases}$$

$$(n \in \mathbb{Z}) \ .$$

In particular, there is defined such a Mayer-Vietoris exact sequence of the $\begin{cases} \varepsilon\text{-symmetric} \\ \varepsilon\text{-quadratic} \end{cases}$ L-groups of the rings with involution appearing in the cartesian square

$$\begin{array}{ccc} \mathbb{Z}[\pi] & \longrightarrow & \mathbb{Z}_{(p)}[\pi] \\ \downarrow & & \downarrow \\ \mathbb{Z}[\frac{1}{p}][\pi] & \longrightarrow & \mathbb{Q}[\pi] \end{array}$$

with $A = \mathbb{Z}[\pi]$ a group ring and $P = \{p\}$ for some prime p.

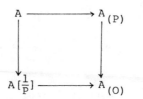

3.7 Change of K-theory

The localization exact sequence of §3.6 will now be extended to the intermediate L-groups of §1.10. In fact there are two such extensions, one indexed by the *-invariant subgroups $X \subseteq K_1(A,S)$ and one which is indexed by the *-invariant subgroups $X \subseteq \tilde{K}_m(A)$ $(m = 0,1)$. The generalizations may be proved in the same way as the original sequence, or else may be deduced from it using the comparison exact sequences of §1.10.

In the first instance it is necessary to consider the action of the duality involutions * on the localization exact sequence of algebraic K-theory

$$\tilde{K}_1(A) \xrightarrow{i_1} \tilde{K}_1(S^{-1}A) \xrightarrow{\partial} K_1(A,S) \xrightarrow{j} \tilde{K}_0(A) \xrightarrow{i_0} \tilde{K}_0(S^{-1}A)$$

for a localization $A \longrightarrow S^{-1}A$ of rings with involution. The duality involution

$$* : \tilde{K}_m(A) = \tilde{K}_m(\underline{\underline{P}}(A)) \longrightarrow \tilde{K}_m(A) = \tilde{K}_m(\underline{\underline{P}}(A)) \quad (m = 0,1)$$

is induced by the duality involution on the exact category $\underline{\underline{P}}(A)$ of f.g. projective A-modules

$$* : \underline{\underline{P}}(A) \longrightarrow \underline{\underline{P}}(A) \ ; \ P \longmapsto P^* = \mathrm{Hom}_A(P,A) \quad ,$$

and similarly for $*: \tilde{K}_m(S^{-1}A) \longrightarrow \tilde{K}_m(S^{-1}A)$. The morphisms $i_m : \tilde{K}_m(A) \longrightarrow \tilde{K}_m(S^{-1}A)$ are induced by a functor of categories with involution

$$\underline{\underline{P}}(A) \longrightarrow \underline{\underline{P}}(S^{-1}A) \ ; \ P \longmapsto S^{-1}P = S^{-1}A \otimes_A P \ ,$$

so that the diagrams

$$\begin{array}{ccc}
\tilde{K}_m(A) & \xrightarrow{i_m} & \tilde{K}_m(A) \\
\downarrow * & & \downarrow * \\
\tilde{K}_m(A) & \xrightarrow{i_m} & \tilde{K}_m(A)
\end{array}$$

commute. The duality involution

$$* : K_1(A,S) = K_0(\underline{\underline{P}}(A,S)) \longrightarrow K_1(A,S) = K_0(\underline{\underline{P}}(A,S))$$

is induced by the S-duality involution on the exact category $\underline{\underline{P}}(A,S)$ of (A,S)-modules

$$* : \underline{\underline{P}}(A,S) \longrightarrow \underline{\underline{P}}(A,S) \ ; \ M \longmapsto M\hat{} = \mathrm{Hom}_A(M, S^{-1}A/A) \ .$$

If an (A,S)-module M has f.g. projective A-module resolution

$$0 \longrightarrow P_1 \xrightarrow{d} P_0 \longrightarrow M \longrightarrow 0$$

the S-dual $M\hat{}$ has resolution

$$0 \longrightarrow P_0^* \xrightarrow{d^*} P_1^* \longrightarrow M\hat{} \longrightarrow 0 \quad .$$

It follows that the morphisms

$$\partial : \tilde{K}_1(S^{-1}A) \longrightarrow K_1(A,S) \ ;$$

$$\tau(s^{-1}f : S^{-1}P \longrightarrow S^{-1}P) \longmapsto [P/f(P)] - [P/s(P)]$$

(with $f,s \in \mathrm{Hom}_A(P,P)$ S-automorphisms of a f.g. projective A-module P) and

$$j : K_1(A,S) \longrightarrow \tilde{K}_0(A) \ ; \ [M] \longmapsto [P_0] - [P_1]$$

(with P_0, P_1 the f.g. projective A-modules appearing in a resolution $0 \longrightarrow P_1 \longrightarrow P_0 \longrightarrow M \longrightarrow 0$ of an (A,S)-module M) are well-defined and such that the diagram

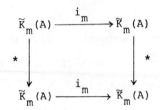

$$\begin{array}{ccccc} \tilde{K}_1(S^{-1}A) & \longrightarrow & K_1(A,S) & \xrightarrow{\ j\ } & \tilde{K}_0(A) \\ \Big\downarrow{\scriptstyle *} & & \Big\downarrow{\scriptstyle *} & & \Big\downarrow{\scriptstyle -*} \\ \tilde{K}_1(S^{-1}A) & \longrightarrow & K_1(A,S) & \xrightarrow{\ j\ } & \tilde{K}_0(A) \end{array}$$

commutes.

Define the *S-projective class* of an (A,S)-module M to be the element

$$[M] \in K_1(A,S) = K_0(\underline{\underline{P}}(A,S)) \ .$$

More generally, the S-projective class of an n-dimensional (A,S)-module chain complex C is defined to be

$$[C] = \sum_{r=0}^{n} (-)^r [C_r] \in K_1(A,S) \ ,$$

and is such that

$$[C^{n-\wedge}] = (-)^n [C]^* \in K_1(A,S) \ .$$

A short exact sequence of (A,S)-modules

$$0 \longrightarrow M \longrightarrow M' \longrightarrow M'' \longrightarrow 0$$

is an acyclic 2-dimensional (A,S)-module chain complex with S-projective class

$$[M] - [M'] + [M''] = 0 \in K_1(A,S) \ .$$

Given a $\mathbb{Z}[\mathbb{Z}_2]$-module G let G^- denote the $\mathbb{Z}[\mathbb{Z}_2]$-module with the same additive group, but with $T \in \mathbb{Z}_2$ acting by

$$T_{G^-} : G \longrightarrow G \ ; \ x \longmapsto -T_G(x) \ .$$

The Tate $\mathbb{Z}_2$-cohomology groups are such that

$$\hat{H}^*(\mathbb{Z}_2;G^-) = \hat{H}^{*-1}(\mathbb{Z}_2;G) \ .$$

Given a $*$-invariant subgroup $X \subseteq K_1(A,S)$ let

$\begin{cases} L^n_X(A,S,\varepsilon) \\ L^X_n(A,S,\varepsilon) \end{cases}$ $(n \in \mathbb{Z})$ be the L-groups defined in the same way as

$\begin{cases} L^n(A,S,\varepsilon) \\ L_n(A,S,\varepsilon) \end{cases}$ (the special case $X = K_1(A,S)$) but using only

(A,S)-module chain complexes C with S-projective class

$$[C] \in X \subseteq K_1(A,S) \ .$$

Define $*$-invariant subgroups

$$jX = \operatorname{im}(j| : X \longrightarrow \tilde{K}_0(A)) \subseteq \tilde{K}_0(A) \ ,$$

$$X^{\partial} = \partial^{-1}(X) \subseteq \tilde{K}_1(S^{-1}A) \ ,$$

so that there is defined a short exact sequence of

$\mathbb{Z}[\mathbb{Z}_2]$-modules

$$0 \longrightarrow X^{\partial}/\ker(\partial : \tilde{K}_1(S^{-1}A) \longrightarrow K_1(A,S)) \xrightarrow{\ \partial\ } X \xrightarrow{\ j\ } jX^{-} \longrightarrow 0$$

inducing a long exact sequence of Tate $\mathbb{Z}_2$-cohomology groups

$$\ldots \longrightarrow \hat{H}^n(\mathbb{Z}_2;X^{\partial}/\ker(\partial)) \xrightarrow{\ \partial\ } \hat{H}^n(\mathbb{Z}_2;X) \xrightarrow{\ j\ } \hat{H}^{n-1}(\mathbb{Z}_2;jX)$$

$$\longrightarrow \hat{H}^{n-1}(\mathbb{Z}_2;X^{\partial}/\ker(\partial)) \longrightarrow \ldots \ .$$

The exact sequences of Proposition 2.5.1, 3.6.1

generalize to the intermediate $\begin{cases} \varepsilon\text{-symmetric} \\ \varepsilon\text{-quadratic} \end{cases}$ L-groups

$\begin{cases} L^*_X(A,S,\varepsilon) \\ L^X_*(A,S,\varepsilon) \end{cases}$ $(X \subseteq K_1(A,S))$ as follows:

<u>Proposition 3.7.1</u> Given $*$-invariant subgroups $X \subseteq Y \subseteq K_1(A,S)$ there is defined a commutative diagram of abelian groups with exact rows and columns

$$\begin{array}{ccccccccc}
 & & \vdots & & \vdots & & \vdots & & \vdots & \\
 & & \downarrow & & \downarrow & & \downarrow & & \downarrow & \\
\dots \longrightarrow & & L^n_{jX}(A,\varepsilon) & \longrightarrow & L^n_{X^\partial}(S^{-1}A,\varepsilon) & \xrightarrow{\partial} & L^n_X(A,S,\varepsilon) & \longrightarrow & L^{n-1}_{jX}(A,\varepsilon) & \longrightarrow \dots \\
 & & \downarrow & & \downarrow & & \downarrow & & \downarrow & \\
\dots \longrightarrow & & L^n_{jY}(A,\varepsilon) & \longrightarrow & L^n_{Y^\partial}(S^{-1}A,\varepsilon) & \xrightarrow{\partial} & L^n_Y(A,S,\varepsilon) & \longrightarrow & L^{n-1}_{jY}(A,\varepsilon) & \longrightarrow \dots \\
 & & \downarrow & & \downarrow & & \downarrow & & \downarrow & \\
\dots \longrightarrow & & \hat{H}^n(\mathbb{Z}_2;jY/jX) & \longrightarrow & \hat{H}^n(\mathbb{Z}_2;Y^\partial/X^\partial) & \xrightarrow{\partial} & \hat{H}^n(\mathbb{Z}_2;Y/X) & \longrightarrow & \hat{H}^{n-1}(\mathbb{Z}_2;jY/jX) & \longrightarrow \dots \\
 & & \downarrow & & \downarrow & & \downarrow & & \downarrow & \\
\dots \longrightarrow & & L^{n-1}_{jX}(A,\varepsilon) & \longrightarrow & L^{n-1}_{X^\partial}(S^{-1}A,\varepsilon) & \xrightarrow{\partial} & L^{n-1}_X(A,S,\varepsilon) & \longrightarrow & L^{n-2}_{jX}(A,\varepsilon) & \longrightarrow \dots \\
 & & \downarrow & & \downarrow & & \downarrow & & \downarrow & \\
 & & \vdots & & \vdots & & \vdots & & \vdots &
\end{array}$$

$(n \in \mathbb{Z})$.

Similarly for the ε-quadratic L-groups L_n.

[]

Given a $*$-invariant subgroup $X \subseteq \tilde{K}_0(A)$ define $*$-invariant subgroups

$$S^{-1}X = \operatorname{im}(i_0| : X \longrightarrow \tilde{K}_0(S^{-1}A)) \subseteq \tilde{K}_0(S^{-1}A) \ ,$$

$$X^j = j^{-1}(X) \subseteq K_1(A,S) \ ,$$

so that there is defined a short exact sequence of $\mathbb{Z}[\mathbb{Z}_2]$-modules

$$0 \longrightarrow (X^j/\ker(j:K_1(A,S) \longrightarrow \tilde{K}_0(A)))^- \xrightarrow{\ j\ } X \xrightarrow{\ i_0\ } S^{-1}X \longrightarrow 0$$

inducing a long exact sequence of Tate $\mathbb{Z}_2$-cohomology groups

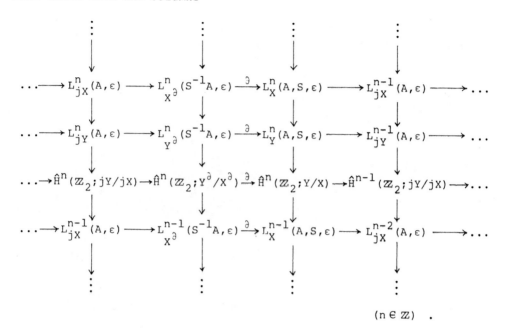

$$\cdots \longrightarrow \hat{H}^{n+1}(\mathbb{Z}_2;X^j/\ker(j)) \xrightarrow{\ j\ } \hat{H}^n(\mathbb{Z}_2;X) \xrightarrow{\ i_0\ } \hat{H}^n(\mathbb{Z}_2;S^{-1}X)$$
$$\longrightarrow \hat{H}^n(\mathbb{Z}_2;X^j/\ker(j)) \longrightarrow \cdots \ .$$

<u>Proposition 3.7.2</u> Given $*$-invariant subgroups $X \subseteq Y \subseteq \tilde{K}_0(A)$ there is defined a commutative diagram of abelian groups with exact rows and columns

$$\begin{array}{ccccccccc}
 & & \vdots & & \vdots & & \vdots & & \vdots \\
 & & \downarrow & & \downarrow & & \downarrow & & \downarrow \\
\cdots \longrightarrow & L^n_X(A,\varepsilon) & \longrightarrow & L^n_{S^{-1}X}(S^{-1}A,\varepsilon) & \xrightarrow{\partial} & L^n_{X^j}(A,S,\varepsilon) & \longrightarrow & L^{n-1}_X(A,\varepsilon) & \longrightarrow \cdots \\
 & \downarrow & & \downarrow & & \downarrow & & \downarrow & \\
\cdots \longrightarrow & L^n_Y(A,\varepsilon) & \longrightarrow & L^n_{S^{-1}Y}(S^{-1}A,\varepsilon) & \xrightarrow{\partial} & L^n_{Y^j}(A,S,\varepsilon) & \longrightarrow & L^{n-1}_Y(A,\varepsilon) & \longrightarrow \cdots \\
 & \downarrow & & \downarrow & & \downarrow & & \downarrow & \\
\cdots \longrightarrow & \hat{H}^n(\mathbb{Z}_2;Y/X) & \longrightarrow & \hat{H}^n(\mathbb{Z}_2;S^{-1}Y/S^{-1}X) & \xrightarrow{\partial} & \hat{H}^n(\mathbb{Z}_2;Y^j/X^j) & \longrightarrow & \hat{H}^{n-1}(\mathbb{Z}_2;Y/X) & \longrightarrow \cdots \\
 & \downarrow & & \downarrow & & \downarrow & & \downarrow & \\
\cdots \longrightarrow & L^{n-1}_X(A,\varepsilon) & \longrightarrow & L^{n-1}_{S^{-1}X}(S^{-1}A,\varepsilon) & \xrightarrow{\partial} & L^{n-1}_{X^j}(A,S,\varepsilon) & \longrightarrow & L^{n-2}_X(A,\varepsilon) & \longrightarrow \cdots \\
 & \downarrow & & \downarrow & & \downarrow & & \downarrow & \\
 & \vdots & & \vdots & & \vdots & & \vdots &
\end{array}$$

Similarly for the ε-quadratic L-groups L_n.

[]

The localization exact sequence of Proposition 3.7.1 for $X = \mathrm{im}(\partial:\tilde{K}_1(S^{-1}A)\longrightarrow K_1(A,S)) = \ker(j:K_1(A,S)\longrightarrow \tilde{K}_0(A)) \subseteq K_1(A,S)$

$$\cdots \longrightarrow L^n_{jX}(A,\varepsilon) \longrightarrow L^n_{X^\partial}(S^{-1}A,\varepsilon) \xrightarrow{\ \partial\ } L^n_X(A,S,\varepsilon) \longrightarrow L^{n-1}_{jX}(A,\varepsilon) \longrightarrow \cdots$$

coincides with the localization exact sequence of Proposition 3.7.2 for $Y = \{0\} \subseteq \tilde{K}_0(A)$

$$\ldots \longrightarrow L^n_Y(A,\varepsilon) \longrightarrow L^n_{S^{-1}Y}(S^{-1}A,\varepsilon) \xrightarrow{\partial} L^n_{Y^j}(A,S,\varepsilon) \longrightarrow L^{n-1}_Y(A,\varepsilon) \longrightarrow \ldots .$$

This sequence can be written as

$$\ldots \longrightarrow V^n(A,\varepsilon) \longrightarrow V^n(S^{-1}A,\varepsilon) \xrightarrow{\partial} V^n(A,S,\varepsilon) \longrightarrow V^{n-1}(A,\varepsilon) \longrightarrow \ldots$$

with $V^*(A,\varepsilon)$ (resp. $V^*(S^{-1}A,\varepsilon)$) the V-groups of §1.10, i.e. the analogues of the L-groups $L^*(A,\varepsilon)$ defined using only f.g. free A- (resp. $S^{-1}A$-) modules, and $V^*(A,S,\varepsilon)$ the analogues of the L-groups $L^*(A,S,\varepsilon)$ defined using only (A,S)-modules with a f.g. free A-module resolution of length 1.

An (A,S)-module M is <u>S-based</u> is there is given a f.g. free A-module resolution of length 1

$$0 \longrightarrow P_1 \xrightarrow{d} P_0 \longrightarrow M \longrightarrow 0$$

with P_0 and P_1 based. The <u>S-torsion</u> of M is then defined to be

$$\tau_S(M) = \tau(S^{-1}d:S^{-1}P_1 \longrightarrow S^{-1}P_0) \in \tilde{K}_1(S^{-1}A) .$$

The S-dual (A,S)-module $M^\wedge$ is also S-based (using the dually based A-modules P_0^*,P_1^*), with S-torsion

$$\tau_S(M^\wedge) = \tau(S^{-1}d^*:S^{-1}P_0^* \longrightarrow S^{-1}P_1^*) = \tau_S(M)^* \in \tilde{K}_1(S^{-1}A) .$$

More generally, the S-torsion of an n-dimensional S-based (A,S)-module chain complex C is defined to be

$$\tau_S(C) = \sum_{r=0}^{n} (-)^r \tau_S(C_r) \in \tilde{K}_1(S^{-1}A) .$$

If D is an (n+1)-dimensional S-acyclic based A-module chain complex resolving C (with the S-bases of the (A,S)-modules C_r determined by the bases of the A-modules D_r) then $S^{-1}D$ is an (n+1)-dimensional acyclic based $S^{-1}A$-module chain complex such that

$$\tau_S(C) = \tau(S^{-1}D) \in \tilde{K}_1(S^{-1}A) .$$

It follows from the definitions that

$$\tau_S(C^{n-\wedge}) = \tau(S^{-1}D^{n+1-*}) = (-)^n\tau_S(C)^* \in \tilde{K}_1(S^{-1}A) .$$

The <u>torsion</u> of an n-dimensional acyclic S-based (A,S)-module chain complex C is defined to be

$$\tau(C) = \tau(D) \in \tilde{K}_1(A)$$

with D an (n+1)-dimensional acyclic based A-module chain complex resolving C. It follows from the definitions that

$$S^{-1}\tau(C) = \tau_S(C) \in \tilde{K}_1(S^{-1}A) ,\quad \tau(C^{n-\wedge}) = \tau(D^{n+1-*}) = (-)^n\tau(C)^* \in \tilde{K}_1(A) .$$

The torsion of a homology equivalence

$$f : C \longrightarrow C'$$

of n-dimensional S-based (A,S)-module chain complexes is defined to be

$$\tau(f) = \tau(C(f)) \in \tilde{K}_1(A) ,$$

and is such that

$$\tau(f^{n-\wedge}:C'^{n-\wedge}\longrightarrow C^{n-\wedge}) = (-)^{n+1}\tau(f)^* \in \tilde{K}_1(A)$$

$$S^{-1}\tau(f) = \tau_S(C) - \tau_S(C') \in \tilde{K}_1(S^{-1}A) .$$

The <u>torsion</u> of an n-dimensional S-based ε-symmetric Poincaré complex over (A,S) $(C,\phi\in Q^n_S(C,\varepsilon))$ is defined to be

$$\tau(C,\phi) = (\tau(\phi_0:C^{n-\wedge}\longrightarrow C),\tau_S(C))$$

$$\in \ker(\begin{pmatrix} 1+(-)^nT & 0 \\ -S^{-1} & 1-(-)^nT \end{pmatrix} : \tilde{K}_1(A)\oplus\tilde{K}_1(S^{-1}A)\longrightarrow \tilde{K}_1(A)\oplus\tilde{K}_1(S^{-1}A)) ,$$

with $T:\tilde{K}_1\longrightarrow\tilde{K}_1;\tau\longmapsto\tau^*$ the duality involutions. The torsion of an n-dimensional S-based ε-quadratic Poincaré complex over (A,S) $(C,\psi\in Q_n^S(C,\varepsilon))$ is defined to be the torsion of the ε-symmetrization

$$\tau(C,\psi) = \tau(C,(1+T_\varepsilon)\psi \in Q_S^n(C,\varepsilon)) .$$

Given a $*$-invariant subgroup $X \subseteq \tilde{K}_1(A)$ define a $*$-invariant subgroup

$$S^{-1}X = \mathrm{im}(i_1| : X \longrightarrow \tilde{K}_1(S^{-1}A)) \subseteq \tilde{K}_1(S^{-1}A) .$$

Given $*$-invariant subgroups $X \subseteq \tilde{K}_1(A), Y \subseteq \tilde{K}_1(S^{-1}A)$ such that

$$S^{-1}X \subseteq Y \subseteq \tilde{K}_1(S^{-1}A)$$

let $\begin{cases} L^n_{X,Y}(A,S,\varepsilon) \\ L^{X,Y}_n(A,S,\varepsilon) \end{cases}$ $(n \in \mathbb{Z})$ be the L-groups defined in the same way as $\begin{cases} L^n(A,S,\varepsilon) \\ L_n(A,S,\varepsilon) \end{cases}$ $(n \in \mathbb{Z})$ but using only S-based algebraic Poincaré complexes over (A,S) with torsion in

$$\{(x,y) \in X \oplus Y \mid x^* = (-)^n x,\ S^{-1}x = y + (-)^{n-1}y^*\}$$

$$\subseteq \ker\left(\begin{pmatrix} 1+(-)^n T & 0 \\ -S^{-1} & 1-(-)^n T \end{pmatrix} : \tilde{K}_1(A) \oplus \tilde{K}_1(S^{-1}A) \longrightarrow \tilde{K}_1(A) \oplus \tilde{K}_1(S^{-1}A)\right) .$$

For $X = \tilde{K}_1(A)$, $Y = \tilde{K}_1(S^{-1}A)$ these are the free L-groups $\begin{cases} V^n(A,S,\varepsilon) \\ V_n(A,S,\varepsilon) \end{cases}$.

As in §2.5 define the relative Tate $\mathbb{Z}_2$-cohomology groups of a morphism of $\mathbb{Z}[\mathbb{Z}_2]$-modules

$$f : G \longrightarrow H$$

by

$$\hat{H}^n(\mathbb{Z}_2;f) = \frac{\{(x,y) \in G \oplus H \mid x^* = (-)^{n-1}x,\ fx = y + (-)^{n-1}y^*\}}{\{(u + (-)^{n-1}u^*,\ fu + v + (-)^n v^*) \mid (u,v) \in G \oplus H\}} \quad (n \pmod 2) ,$$

and note that there is defined a long exact sequence

$$\ldots \longrightarrow \hat{H}^n(\mathbb{Z}_2;G) \xrightarrow{f} \hat{H}^n(\mathbb{Z}_2;H) \longrightarrow \hat{H}^n(\mathbb{Z}_2;f) \longrightarrow \hat{H}^{n-1}(\mathbb{Z}_2;G) \longrightarrow \ldots .$$

The exact sequences of Propositions 1.10.1, 2.5.1 and 3.6.1 generalize to the intermediate torsion L-groups as follows:

<u>Proposition 3.7.3</u> Given $*$-invariant subgroups $X \subseteq X' \subseteq \tilde{K}_1(A)$, $Y \subseteq Y' \subseteq \tilde{K}_1(S^{-1}A)$ such that $S^{-1}X \subseteq Y$, $S^{-1}X' \subseteq Y'$ there is defined a commutative diagram of abelian groups with exact rows and columns

$$\begin{array}{ccccccccc}
 & \vdots & & \vdots & & \vdots & & \vdots & \\
 & \downarrow & & \downarrow & & \downarrow & & \downarrow & \\
\ldots \longrightarrow & L^n_X(A,\varepsilon) & \longrightarrow & L^n_Y(S^{-1}A,\varepsilon) & \longrightarrow & L^n_{X,Y}(A,S,\varepsilon) & \longrightarrow & L^{n-1}_X(A,\varepsilon) & \longrightarrow \ldots \\
 & \downarrow & & \downarrow & & \downarrow & & \downarrow & \\
\ldots \longrightarrow & L^n_{X'}(A,\varepsilon) & \longrightarrow & L^n_{Y'}(S^{-1}A,\varepsilon) & \longrightarrow & L^n_{X',Y'}(A,S,\varepsilon) & \longrightarrow & L^{n-1}_{X'}(A,\varepsilon) & \longrightarrow \ldots \\
 & \downarrow & & \downarrow & & \downarrow & & \downarrow & \\
\ldots \longrightarrow & \hat{H}^n(\mathbb{Z}_2;X'/X) & \longrightarrow & \hat{H}^n(\mathbb{Z}_2;Y'/Y) & \longrightarrow & \hat{H}^n(\mathbb{Z}_2;X'/X \to Y'/Y) & \longrightarrow & \hat{H}^{n-1}(\mathbb{Z}_2;X'/X) & \longrightarrow \ldots \\
 & \downarrow & & \downarrow & & \downarrow & & \downarrow & \\
\ldots \longrightarrow & L^{n-1}_X(A,\varepsilon) & \longrightarrow & L^{n-1}_Y(S^{-1}A,\varepsilon) & \longrightarrow & L^{n-1}_{X,Y}(A,S,\varepsilon) & \longrightarrow & L^{n-2}_X(A,\varepsilon) & \longrightarrow \ldots \\
 & \downarrow & & \downarrow & & \downarrow & & \downarrow & \\
 & \vdots & & \vdots & & \vdots & & \vdots &
\end{array}$$

$$(n \in \mathbb{Z}) \ .$$

Similarly for the ε-quadratic L-groups L_n.

[]

The generalizations to the intermediate L-groups of the excision isomorphisms and Mayer-Vietoris exact sequences of §3.6 will be dealt with in §6.3 below.

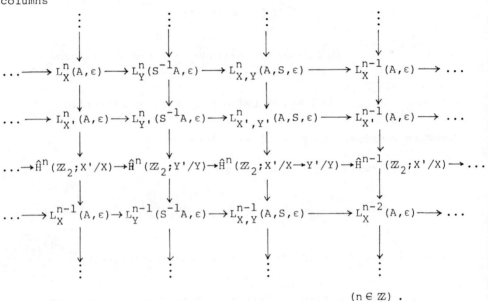

§4. Arithmetic L-theory

Localization has long been a key tool in the unstable classification of quadratic forms over rings of arithmetic type - cf. the work of Gauss, Minkowski, Hasse et. al. The classification over a global ring such as an algebraic number field (e.g. $\mathbb{Q}$) is reduced to the classifications over local rings such as the completions at the various valuations (e.g. the p-adic fields $\hat{\mathbb{Q}}_p$ and the reals $\mathbb{R}$ for $\mathbb{Q}$). This reduction can also be used for the classification over an order such as the ring of algebraic integers (e.g. $\mathbb{Z}$ in $\mathbb{Q}$). See O'Meara [1], Milnor and Husemoller [1] and Cassels [1] for modern accounts of the arithmetic theory of quadratic forms.

Many authors have used the localization techniques of algebraic number theory to obtain localization exact sequences for the Witt groups of quadratic forms over rings of arithmetic type and more general Dedekind rings, notably Kneser, Milnor, Wall [6], Fröhlich [1], Knebusch and Scharlau [1], Durfee [1] and Barge, Lannes, Latour and Vogel [1]. The arithmetic approach has been extended to more general orders in semi-simple algebras (e.g. $\mathbb{Z}[\pi]$ in $\mathbb{Q}[\pi]$ for a finite group π) by Wall [8], Bak and Scharlau [1], and Bak [2].

We shall now apply the localization exact sequence of §3 to the L-theory of rings with involution which are algebras over a Dedekind ring. As usual, we start with some K-theory.

Let R be a Dedekind ring, and let A be a ring which is an algebra over R. Then $S = R-\{0\} \subset A$ is a multiplicative subset of A such that the localization $S^{-1}A = F\otimes_R A$ is the induced algebra over the quotient field $F = S^{-1}R$.

An (A,S)-module M is "$\mathcal{P}$-primary" if the annihilator of M is

$$\{r \in R \mid rM = 0\} = \mathcal{P}^k \triangleleft R$$

for some maximal ideal $\mathcal{P} \triangleleft R$, with $k \geqslant 1$. Every (A,S)-module M has a canonical decomposition as a direct sum of $\mathcal{P}$-primary (A,S)-modules

$$M = \bigoplus_{\mathcal{P}} M_{\mathcal{P}} \quad ,$$

with $\mathcal{P}$ ranging over all the maximal ideals of R. The resulting identification of exact categories

$$\{(A,S)\text{-modules}\} = \bigoplus_{\mathcal{P}} \{\mathcal{P}\text{-primary } (A,S)\text{-modules}\}$$

gives rise to an identification of algebraic K-groups

$$K_n(A,S) = \bigoplus_{\mathcal{P}} K_n(A,\mathcal{P}^\infty) \quad (n \in \mathbb{Z}) \quad ,$$

so that the algebraic K-theory localization exact sequence can be written as

$$\ldots \longrightarrow K_n(A) \longrightarrow K_n(S^{-1}A) \longrightarrow \bigoplus_{\mathcal{P}} K_n(A,\mathcal{P}^\infty) \longrightarrow K_{n-1}(A) \longrightarrow \ldots \; .$$

In the case A = R a devissage argument (due to Bass [2] for $n \leqslant 1$, and to Quillen [1] for $n \geqslant 2$) identifies

$$K_n(R,\mathcal{P}^\infty) = K_{n-1}(R/\mathcal{P}) \quad (n \in \mathbb{Z}) \quad ,$$

so that the sequence can also be written as

$$\ldots \longrightarrow K_n(R) \longrightarrow K_n(F) \longrightarrow \bigoplus_{\mathcal{P}} K_{n-1}(R/\mathcal{P}) \longrightarrow K_{n-1}(R) \longrightarrow \ldots \; .$$

In §4.1 we shall deal with with the algebraic L-theory localization exact sequence for a ring with involution A which is an algebra over a Dedekind ring R, with $S = R-\{0\} \subset A$. The decomposition of (A,S)-modules into $\mathcal{P}$-primary components will be used to obtain natural direct sum decompositions of L-groups

$$\begin{cases} L^n(A,S,\varepsilon) = \bigoplus_{\mathcal{P}} L^n(A,\mathcal{P}^\infty,\varepsilon) \\ L_n(A,S,\varepsilon) = \bigoplus_{\mathcal{P}} L_n(A,\mathcal{P}^\infty,\varepsilon) \end{cases} \quad (n \in \mathbb{Z})$$

with $\mathcal{P}$ ranging over all the maximal ideals of R which are invariant under the involution, $\bar{\mathcal{P}} = \mathcal{P} \triangleleft R$. For $A = R$, $1/2 \in A$ such decompositions have been previously obtained by Karoubi [3].

In §4.2 the results of §4.1 are specialized to the L-theory of a Dedekind ring, with $A = R$. In particular, an L-theoretic devissage argument will be used to identify

$$L^0(R,\mathcal{P}^\infty,\varepsilon) = L^0(R/\mathcal{P},\varepsilon) \quad ,$$

thus recovering the localization exact sequence of Milnor and Husemoller [1,IV.3.3] relating the symmetric Witt groups of a Dedekind ring R and its quotient field F

$$0 \longrightarrow L^0(R) \longrightarrow L^0(F) \longrightarrow \bigoplus_{\mathcal{P}} L^0(R/\mathcal{P}) \quad (\longrightarrow L^{-1}(R) \longrightarrow 0) \quad ,$$

extending it to the right by the map onto $L^{-1}(R)$.

In §4.3 the results of §4.2 are applied to obtain the L-groups of $\mathbb{Z}$ and $\mathbb{Q}$.

4.1 Dedekind algebra

We refer to Zariski and Samuel [1,§V.6] for the basic properties of Dedekind rings.

A Dedekind algebra with involution (A,S) is a ring with involution A together with a central multiplicative subset $S \subset A$ such that $R = S \cup \{0\}$ is a Dedekind ring with respect to the ring operations inherited from A. The localization away from S

$$S^{-1}A = F \otimes_R A$$

is the induced algebra over the quotient field $F = S^{-1}R$. For example, a torsion-free ring with involution A is the same as a Dedekind algebra $(A, \mathbb{Z}-\{0\})$, and a Dedekind ring with involution R is the same as a Dedekind algebra $(R, R-\{0\})$.

Let (A,S) be a Dedekind algebra with involution, and let max(R) be the spectrum of maximal ideals of the Dedekind ring $R = S \cup \{0\}$, that is the set of maximal ideals (= non-zero prime ideals) of R.

The annihilator of an A-module M is the ideal of R defined by

$$\mathrm{ann}(M) = \{r \in R \mid rx = 0 \in M \text{ for all } x \in M\} \lhd R \ .$$

By the classical ideal theory of Dedekind rings this has a unique factorization as a product of powers of maximal ideals $\mathcal{P}_1, \mathcal{P}_2, \ldots, \mathcal{P}_q \in \max(R)$

$$\mathrm{ann}(M) = \mathcal{P}_1^{k_1}\mathcal{P}_2^{k_2}\ldots\mathcal{P}_q^{k_q} \quad (k_i \geqslant 1) \ .$$

A non-zero A-module M is S-torsion if and only if ann(M) is a proper ideal of R. If M is an (A,S)-module then

$$\mathrm{ann}(M\hat{\ }) = \overline{\mathrm{ann}(M)} \lhd R$$

(since $\overline{\mathrm{ann}(M)} \subseteq \mathrm{ann}(M\hat{\ })$ for any A-module M and $M\hat{\ }\hat{\ } = M$ for an (A,S)-module M).

An S-torsion A-module M is <u>$\mathcal{P}$-primary</u> if

$$\mathrm{ann}(M) = \mathcal{P}^k$$

for some $\mathcal{P} \in \max(R)$, $k \geqslant 1$. An <u>$(A,\mathcal{P}^\infty)$-module</u> is an (A,S)-module which is $\mathcal{P}$-primary, that is a $\mathcal{P}$-primary S-torsion A-module of homological dimension 1. An <u>n-dimensional $(A,\mathcal{P}^\infty)$-module chain complex</u> C is an n-dimensional (A,S)-module chain complex

$$C : \ldots \longrightarrow 0 \longrightarrow C_n \xrightarrow{d} C_{n-1} \longrightarrow \ldots \longrightarrow C_1 \xrightarrow{d} C_0 \longrightarrow 0 \longrightarrow \ldots$$

such that each C_r $(0 \leqslant r \leqslant n)$ is an $(A,\mathcal{P}^\infty)$-module. An A-module chain complex D is <u>$\mathcal{P}^\infty$-acyclic</u> if it is S-acyclic and the homology S-torsion A-modules $H_*(D)$ are $\mathcal{P}$-primary.

<u>Proposition 4.1.1</u> There are natural identifications of sets of homology equivalence classes

$$\{\text{n-dimensional } (A,\mathcal{P}^\infty)\text{-module chain complexes}\}$$

$$= \{\mathcal{P}^\infty\text{-acyclic (n+1)-dimensional A-module chain complexes}\}$$

$$(n \in \mathbb{Z}, \mathcal{P} \in \max(R)) \ .$$

<u>Proof</u>: Immediate from Proposition 3.1.2.

[]

Let $\mathcal{P} \in \max(R)$. The <u>localization of A at $\mathcal{P}$</u> is the ring obtained from A by inverting $R-\mathcal{P} \subset A$

$$A_{\mathcal{P}} = (R-\mathcal{P})^{-1}A \ .$$

If $\overline{\mathcal{P}} = \mathcal{P}$ then $R-\mathcal{P} \subset A$ is a multiplicative subset in the sense of §3.1, and $A_{\mathcal{P}}$ is a ring with involution

$$\overline{} : A_{\mathcal{P}} \longrightarrow A_{\mathcal{P}} \ ; \ x = \frac{a}{r} \longmapsto \overline{x} = \frac{\overline{a}}{\overline{r}} \quad (a \in A, r \in R-\mathcal{P}) \ .$$

(If $\overline{\mathcal{P}} \neq \mathcal{P}$ then $A_{\mathcal{P}} \times A_{\overline{\mathcal{P}}}$ is a ring with involution

$$\overline{} : A_{\mathcal{P}} \times A_{\overline{\mathcal{P}}} \longrightarrow A_{\mathcal{P}} \times A_{\overline{\mathcal{P}}} \ ; \ (\frac{a}{r},\frac{b}{s}) \longmapsto (\frac{\overline{b}}{\overline{s}},\frac{\overline{a}}{\overline{r}}) \qquad).$$

If M is an (A,S)-module the localization of M at $\mathcal{P}$ is the $(A,\mathcal{P}^\infty)$-module

$$M_{\mathcal{P}} = A_{\mathcal{P}}\otimes_A M \ (= R_{\mathcal{P}}\otimes_R M) ,$$

that is

$$M_{\mathcal{P}} = \{\tfrac{x}{r}\in S^{-1}M \mid x\in M, r\in R-\mathcal{P}\subseteq S\}\subseteq S^{-1}M .$$

If $\mathrm{ann}(M) = \mathcal{P}_1^{k_1}\mathcal{P}_2^{k_2}\ldots\mathcal{P}_q^{k_q}$ (as above) there are natural identifications

$$M_{\mathcal{P}} = \begin{cases} \mathcal{P}_1^{k_1}\mathcal{P}_2^{k_2}\ldots\mathcal{P}_{i-1}^{k_{i-1}}\mathcal{P}_{i+1}^{k_{i+1}}\ldots\mathcal{P}_q^{k_q}M & \text{if } \mathcal{P} = \mathcal{P}_i \text{ for some } i,\ 1\leq i\leq q \\ 0 & \text{if } \mathcal{P}\notin\{\mathcal{P}_1,\mathcal{P}_2,\ldots,\mathcal{P}_q\} , \end{cases}$$

$$M = \bigoplus_{i=1}^{q} M_{\mathcal{P}_i} ,$$

and if M' is another (A,S)-module

$$\mathrm{Hom}_A(M,M') = \bigoplus_{\mathcal{P}\in\max(R)} \mathrm{Hom}_A(M_{\mathcal{P}},M'_{\mathcal{P}}) .$$

There is thus an identification of exact categories

$$\{(A,S)\text{-modules}\} = \bigoplus_{\mathcal{P}\in\max(R)} \{(A,\mathcal{P}^\infty)\text{-modules}\} .$$

The S-duality involution

$$\{(A,S)\text{-modules}\}\longrightarrow\{(A,S)\text{-modules}\} ;\ M\longmapsto M\hat{}$$

sends the $\mathcal{P}$-primary component $M_{\mathcal{P}}$ of an (A,S)-module M to the $\overline{\mathcal{P}}$-primary component $(M\hat{})_{\overline{\mathcal{P}}}$ of the S-dual $M\hat{}$, that is

$$(M\hat{})_{\overline{\mathcal{P}}} = (M_{\mathcal{P}})\hat{} .$$

Define $\overline{\max}(R)$ to be the subset of max(R) consisting of the maximal ideals of R which are invariant under the involution

$$\overline{\max}(R) = \{\mathcal{P}\in\max(R)\mid\overline{\mathcal{P}} = \mathcal{P}\triangleleft R\} .$$

For each $\mathcal{P} \in \overline{\max}(R)$ define the <u>n-dimensional</u> $\left\{\begin{array}{l}\underline{\varepsilon\text{-symmetric}}\\ \underline{\varepsilon\text{-quadratic}}\end{array}\right.$ <u>L-group</u>

<u>of $(A,\mathcal{P}^\infty)$</u> $\left\{\begin{array}{l}L^n(A,\mathcal{P}^\infty,\varepsilon)\\ L_n(A,\mathcal{P}^\infty,\varepsilon)\end{array}\right.$ $(n \in \mathbb{Z})$ in exactly the same way as

$\left\{\begin{array}{l}L^n(A,S,\varepsilon)\\ L_n(A,S,\varepsilon)\end{array}\right.$ $(n \in \mathbb{Z})$ but using only $(A,\mathcal{P}^\infty)$-module chain complexes, or equivalently $\mathcal{P}^\infty$-acyclic A-module chain complexes.

<u>Proposition 4.1.2</u> The L-groups of a Dedekind algebra (A,S) have natural direct sum decompositions

$$\left\{\begin{array}{l}L^n(A,S,\varepsilon) = \bigoplus_{\mathcal{P}\in\overline{\max}(R)} L^n(A,\mathcal{P}^\infty,\varepsilon)\\ L_n(A,S,\varepsilon) = \bigoplus_{\mathcal{P}\in\overline{\max}(R)} L_n(A,\mathcal{P}^\infty,\varepsilon)\end{array}\right. \quad (n \in \mathbb{Z}) \quad .$$

<u>Proof</u>: In the first instance recall from Proposition I.1.4 that for any finite-dimensional A-module chain complexes C,D there are natural direct sum decompositions

$$Q^n(C\oplus D,\varepsilon) = Q^n(C,\varepsilon)\oplus Q^n(D,\varepsilon)\oplus H_n(\mathrm{Hom}_A(C^*,D))$$

$$Q\langle v_0\rangle^n(C\oplus D,\varepsilon) = Q\langle v_0\rangle^n(C,\varepsilon)\oplus Q\langle v_0\rangle^n(D,\varepsilon)\oplus H_n(\mathrm{Hom}_A(C^*,D))$$

$$Q_n(C\oplus D,\varepsilon) = Q_n(C,\varepsilon)\oplus Q_n(D,\varepsilon)\oplus H_n(\mathrm{Hom}_A(C^*,D))$$

$$(n \in \mathbb{Z}) \quad .$$

By Proposition 4.1.1 an S-acyclic (n+1)-dimensional A-module chain complex C is chain equivalent to the direct sum $\bigoplus_{\mathcal{P}\in\max(R)} C(\mathcal{P})$ of $\mathcal{P}^\infty$-acyclic (n+1)-dimensional A-module chain complexes $C(\mathcal{P})$. If $\mathcal{P}_1,\mathcal{P}_2 \in \max(R)$ are such that $\overline{\mathcal{P}}_1 \neq \mathcal{P}_2$ then $\mathrm{Hom}_A(C(\mathcal{P}_1)^*,C(\mathcal{P}_2))$ is an acyclic $\mathbb{Z}$-module chain complex, so that in particular

$$H_{n+1}(\mathrm{Hom}_A(C(\mathcal{P}_1)^*,C(\mathcal{P}_2))) = 0 \ ;$$

if $\mathcal{Q} \in \max(R)$ is such that $\overline{\mathcal{Q}} \neq \mathcal{Q}$ then $\mathrm{Hom}_A(C(\mathcal{Q})^*,C(\mathcal{Q}))$ is an acyclic $\mathbb{Z}[\mathbb{Z}_2]$-module chain complex, so that in particular

$$\begin{cases} Q\langle v_0\rangle^{n+1}(C(\mathcal{P}),-\varepsilon) = 0 \\ Q_{n+1}(C(\mathcal{P}),-\varepsilon) = 0 . \end{cases}$$

Choose a decomposition of $\max(R) - \overline{\max}(R)$ as a disjoint union

$$\max(R) - \overline{\max}(R) = \{Q\} \cup \{\overline{Q}\} .$$

Applying the above sum formula there is thus a direct sum decomposition

$$\begin{cases} Q\langle v_0\rangle^{n+1}(C,-\varepsilon) = \bigoplus_{\mathcal{P}} Q\langle v_0\rangle^{n+1}(C(\mathcal{P}),-\varepsilon) \oplus \bigoplus_{Q} H_{n+1}(\mathrm{Hom}_A(C(\overline{Q})^*,C(Q))) \\ Q_{n+1}(C,-\varepsilon) = \bigoplus_{\mathcal{P}} Q_{n+1}(C(\mathcal{P}),-\varepsilon) \oplus \bigoplus_{Q} H_{n+1}(\mathrm{Hom}_A(C(\overline{Q})^*,C(Q))) \end{cases}$$

with $\mathcal{P}$ ranging over $\overline{\max}(R)$. An S-acyclic (n+1)-dimensional $\begin{cases}\text{even } (-\varepsilon)\text{-symmetric} \\ (-\varepsilon)\text{-quadratic}\end{cases}$ Poincaré complex over A $\begin{cases}(C,\phi \in Q\langle v_0\rangle^{n+1}(C,-\varepsilon)) \\ (C,\psi \in Q_{n+1}(C,-\varepsilon))\end{cases}$ is thus homotopy equivalent to a direct sum $\begin{cases}\bigoplus_{\mathcal{P}}(C(\mathcal{P}),\phi(\mathcal{P})) \oplus \bigoplus_{Q}(C(Q)\oplus C(\overline{Q}),\phi(Q,\overline{Q})) \\ \bigoplus_{\mathcal{P}}(C(\mathcal{P}),\psi(\mathcal{P})) \oplus \bigoplus_{Q}(C(Q)\oplus C(\overline{Q}),\psi(Q,\overline{Q}))\end{cases}$, with each $\begin{cases}(C(\mathcal{P}),\phi(\mathcal{P}) \in Q\langle v_0\rangle^{n+1}(C(\mathcal{P}),-\varepsilon)) \\ (C(\mathcal{P}),\psi(\mathcal{P}) \in Q_{n+1}(C(\mathcal{P}),-\varepsilon))\end{cases}$ a $\mathcal{P}^\infty$-acyclic (n+1)-dimensional $\begin{cases}\text{even } (-\varepsilon)\text{-symmetric} \\ (-\varepsilon)\text{-quadratic}\end{cases}$ Poincaré complex over A and

$$\begin{cases}\phi(Q,\overline{Q}) \in Q\langle v_0\rangle^{n+1}(C(Q)\oplus C(\overline{Q}),-\varepsilon) = H_{n+1}(\mathrm{Hom}_A(C(\overline{Q})^*,C(Q))) \\ \psi(Q,\overline{Q}) \in Q_{n+1}(C(Q)\oplus C(\overline{Q}),-\varepsilon) = H_{n+1}(\mathrm{Hom}_A(C(\overline{Q})^*,C(Q)))\end{cases}$$

a chain homotopy class of chain equivalences

$$\begin{cases}\phi(Q,\overline{Q}) : C(\overline{Q})^{n+1-*} \longrightarrow C(Q) \\ \psi(Q,\overline{Q}) : C(\overline{Q})^{n+1-*} \longrightarrow C(Q) \end{cases} .$$

The (n+2)-dimensional S-acyclic $\begin{cases} \text{even } (-\varepsilon)\text{-symmetric} \\ (-\varepsilon)\text{-quadratic} \end{cases}$ Poincaré

pair over A

$$\begin{cases} ((1\ 0):C(Q)\oplus C(\bar{Q}) \longrightarrow C(Q),\ (0,\phi(Q,\bar{Q})) \in Q\langle v_0\rangle^{n+2}((1\ 0),-\varepsilon)) \\ ((1\ 0):C(Q)\oplus C(\bar{Q}) \longrightarrow C(Q),\ (0,\psi(Q,\bar{Q})) \in Q_{n+2}((1\ 0),-\varepsilon)) \end{cases}$$

shows that for each Q

$$\begin{cases} (C(Q)\oplus C(\bar{Q}),\phi(Q,\bar{Q})) = 0 \in L^n(A,S,\varepsilon) \\ (C(Q)\oplus C(\bar{Q}),\psi(Q,\bar{Q})) = 0 \in L_n(A,S,\varepsilon) \end{cases},$$

and so

$$\begin{cases} (C,\phi) = \bigoplus_{\mathcal{P}} (C(\mathcal{P}),\phi(\mathcal{P})) \in L^n(A,S,\varepsilon) = \bigoplus_{\mathcal{P}} L^n(A,\mathcal{P}^\infty,\varepsilon) \\ (C,\psi) = \bigoplus_{\mathcal{P}} (C(\mathcal{P}),\psi(\mathcal{P})) \in L_n(A,S,\varepsilon) = \bigoplus_{\mathcal{P}} L_n(A,\mathcal{P}^\infty,\varepsilon) \end{cases}$$

$$(\mathcal{P} \in \overline{\max}(R),\ n \geqslant 0)\ .$$

Similarly for the lower L-groups.

[]

A multiplicative subset $P \subset A$ is <u>characteristic</u> for $\mathcal{P} \in \overline{\max}(R)$ if there is an identity of categories

$$\{(A,\mathcal{P}^\infty)\text{-modules}\} = \{(A,P)\text{-modules}\}\ .$$

For example, if some power $\mathcal{P}^k$ $(k \geqslant 1)$ of $\mathcal{P}$ is a principal ideal of R, with generator $\pi \in R$

$$\mathcal{P}^k = \pi R \triangleleft R\ ,$$

then $\bar{\pi} = \pi u \in R$ for some unit $u \in R$ such that $u\bar{u} = 1$ and the multiplicative subset

$$S_\pi = \{\pi^m u^n \mid m \geqslant 0, n \in \mathbb{Z}\} \subset A$$

is characteristic for $\mathcal{P}$.

Proposition 4.1.3 If $\mathcal{P} \in \overline{\max}(R)$ has a characteristic multiplicative subset $P \subset A$ there are natural identifications of L-groups

$$\begin{cases} L^n(A,\mathcal{P}^\infty,\varepsilon) = L^n(A,P,\varepsilon) \\ L_n(A,\mathcal{P}^\infty,\varepsilon) = L_n(A,P,\varepsilon) \end{cases} \quad (n \in \mathbb{Z}) \quad .$$

Proof: Immediate from the definitions and Proposition 3.6.5.

[]

Let $\mathcal{P} \in \overline{\max}(R)$. The localization of (A,S) at $\mathcal{P}$ is the Dedekind algebra with involution $(A_\mathcal{P},S_\mathcal{P})$ defined by

$$S_\mathcal{P} = \{\tfrac{s}{r} \mid r \in R-\mathcal{P}, s \in S\} \subset A_\mathcal{P} = (R-\mathcal{P})^{-1}A \ ,$$

with

$$S_\mathcal{P} \cup \{0\} = R_\mathcal{P} \ , \ S_\mathcal{P}^{-1}A_\mathcal{P} = S^{-1}A$$

Now $R_\mathcal{P}$ is a local ring, with unique maximal ideal

$$\mathcal{P}_\mathcal{P} = \mathcal{P} R_\mathcal{P} \in \overline{\max}(R_\mathcal{P}) \quad ,$$

and $S_\mathcal{P} \subset A_\mathcal{P}$ is a characteristic multiplicative subset for $\mathcal{P}_\mathcal{P}$ so that by Proposition 4.1.3 there are natural identifications

$$\begin{cases} L^n(A_\mathcal{P},\mathcal{P}_\mathcal{P}^\infty,\varepsilon) = L^n(A_\mathcal{P},S_\mathcal{P},\varepsilon) \\ L_n(A_\mathcal{P},\mathcal{P}_\mathcal{P}^\infty,\varepsilon) = L_n(A_\mathcal{P},S_\mathcal{P},\varepsilon) \end{cases} \quad (n \in \mathbb{Z}) \quad .$$

The functor

$$\{(A,\mathcal{P}^\infty)\text{-modules}\} \longrightarrow \{(A_\mathcal{P},S_\mathcal{P})\text{-modules}\} = \{(A_\mathcal{P},\mathcal{P}_\mathcal{P}^\infty)\text{-modules}\} \ ;$$
$$M \longmapsto M_\mathcal{P} \ (= M \text{ as an } A\text{-module})$$

is an isomorphism of categories.

Proposition 4.1.4 i) For every $\mathcal{P} \in \overline{\max}(R)$ there are natural identifications

$$L_n(A,\mathcal{P}^\infty,\varepsilon) = L_n(A_\mathcal{P},S_\mathcal{P},\varepsilon) \quad (n \in \mathbb{Z}) \ .$$

ii) If $\mathcal{P} \in \overline{\max}(R)$ is such that

either the map $\hat{H}^1(\mathbb{Z}_2;A,\varepsilon) \longrightarrow \hat{H}^1(\mathbb{Z}_2;A_{\mathcal{P}},\varepsilon)$ is one-one

or there exists a characteristic multiplicative subset $P \subset A$ for $\mathcal{P}$ and the map

$$\hat{H}^1(\mathbb{Z}_2;A,\varepsilon) \longrightarrow \hat{H}^1(\mathbb{Z}_2;A_{\mathcal{P}},\varepsilon) \oplus \hat{H}^1(\mathbb{Z}_2;P^{-1}A,\varepsilon)$$

is one-one

(e.g. if some power $\mathcal{P}^k$ $(k \geqslant 1)$ is principal and $\hat{H}^0(\mathbb{Z}_2;F,\varepsilon) = 0$)

then there are natural identifications

$$L^n(A,\mathcal{P}^\infty,\varepsilon) = L^n(A_{\mathcal{P}},S_{\mathcal{P}},\varepsilon) \quad (n \in \mathbb{Z}) \quad .$$

<u>Proof</u>: Consider first the case $n \geqslant 0$.

Let C be a $\mathcal{P}^\infty$-acyclic $(n+1)$-dimensional A-module chain complex, so that $A_{\mathcal{P}} \otimes_A C$ is an $S_{\mathcal{P}}$-acyclic $(n+1)$-dimensional $A_{\mathcal{P}}$-module chain complex. Working as in the proof of Proposition 3.1.4 we can identify

$$\begin{cases} Q^{n+1}(C,-\varepsilon) = Q^{n+1}(A_{\mathcal{P}} \otimes_A C,-\varepsilon) \\ Q_{n+1}(C,-\varepsilon) = Q_{n+1}(A_{\mathcal{P}} \otimes_A C,-\varepsilon) \quad . \end{cases}$$

Also, there is defined an exact sequence

$$0 \longrightarrow Q\langle v_0\rangle^{n+1}(C,-\varepsilon) \longrightarrow Q\langle v_0\rangle^{n+1}(A_{\mathcal{P}} \otimes_A C,-\varepsilon)$$
$$\longrightarrow \mathrm{Hom}_A(H^{n+1}(C),\ker(\hat{H}^1(\mathbb{Z}_2;A,\varepsilon) \longrightarrow \hat{H}^1(\mathbb{Z}_2;A_{\mathcal{P}},\varepsilon)))$$

so that if $\ker(\hat{H}^1(\mathbb{Z}_2;A,\varepsilon) \longrightarrow \hat{H}^1(\mathbb{Z}_2;A_{\mathcal{P}},\varepsilon)) = 0$ we can also identify

$$Q\langle v_0\rangle^{n+1}(C,-\varepsilon) = Q\langle v_0\rangle^{n+1}(A_{\mathcal{P}} \otimes_A C,-\varepsilon) \quad .$$

As for the Q-groups, so for the L-groups.

For the case $n \leqslant -1$ we need only consider $n = -1,-2$. In the first instance, we show that if the map $\hat{H}^1(\mathbb{Z}_2;A,\varepsilon) \longrightarrow \hat{H}^1(\mathbb{Z}_2;A_{\mathcal{P}},\varepsilon)$ is one-one there are identifications of categories

$$\{(-\varepsilon)\text{-quadratic linking forms (resp. formations) over } (A,\mathcal{P}^\infty)\}$$
$$= \{(-\varepsilon)\text{-quadratic linking forms (resp. formations) over } (A_{\mathcal{P}},S_{\mathcal{P}})\}$$

where a linking form (resp. formation) over $(A,\mathcal{P}^\infty)$ is defined to be a linking form (resp. formation) over (A,S) involving only $(A,\mathcal{P}^\infty)$-modules. By the above identifications of Q-groups there are identifications of categories

$$\{(-\varepsilon)\text{-symmetric (resp. split } (-\varepsilon)\text{-quadratic) linking forms over } (A,\mathcal{P}^\infty)\}$$
$$= \{(-\varepsilon)\text{-symmetric (resp. split } (-\varepsilon)\text{-quadratic) linking forms over } (A_{\mathcal{P}},S_{\mathcal{P}})\} \ .$$

By Proposition 3.4.2 i) every $(-\varepsilon)$-quadratic linking form over $(A_{\mathcal{P}},S_{\mathcal{P}})$

$$(M,\lambda:M\times M \longrightarrow S^{-1}A/A,\mu:M \longrightarrow Q_{-\varepsilon}(A_{\mathcal{P}},S_{\mathcal{P}}))$$

can be lifted to a split $(-\varepsilon)$-quadratic linking form over $(A_{\mathcal{P}},S_{\mathcal{P}})$, and hence to a $(-\varepsilon)$-quadratic linking form over $(A,\mathcal{P}^\infty)$

$$(M,\lambda_1:M\times M \longrightarrow S^{-1}A/A,\mu_1:M \longrightarrow Q_{-\varepsilon}(A,S))$$

with λ_1 uniquely determined by λ. If (M,λ_1,μ_2) is another such lifting of (M,λ,μ) then for each $x \in M$

$$\begin{aligned}\mu_1(x)-\mu_2(x) &\in \{a\in A \mid a = b-\varepsilon\bar{b} \text{ for some } b\in A_{\mathcal{P}}\}/\{c-\varepsilon\bar{c} \mid c\in A\}\\ &= \ker(\hat{H}^1(\mathbb{Z}_2;A,\varepsilon)\longrightarrow \hat{H}^1(\mathbb{Z}_2;A_{\mathcal{P}},\varepsilon))\\ &\subsetneq \ker(Q_{-\varepsilon}(A,S)\longrightarrow Q_{-\varepsilon}(A_{\mathcal{P}},S_{\mathcal{P}})) \ .\end{aligned}$$

Thus if $\ker(\hat{H}^1(\mathbb{Z}_2;A,\varepsilon)\longrightarrow\hat{H}^1(\mathbb{Z}_2;A_{\mathcal{P}},\varepsilon)) = 0$ there are identifications of categories as claimed above and $L^n(A,\mathcal{P}^\infty,\varepsilon) = L^n(A_{\mathcal{P}},S_{\mathcal{P}},\varepsilon)$ for $n = -2$ (resp. $n = -1$).

If there exists a characteristic multiplicative subset $P \subset A$ for $\mathcal{P} \in \overline{\max}(R)$ there is defined a cartesian morphism

$$(A,P) \longrightarrow (A_{\mathcal{P}},P) \ .$$

There is an identity of categories

$$\{(A_{\mathcal{P}},P)\text{-modules}\} = \{(A_{\mathcal{P}},S_{\mathcal{P}})\text{-modules}\} \ (= \{(A,\mathcal{P}^{\infty})\text{-modules}\}),$$

so that by Proposition 3.6.5 there are identifications

$$L^n(A_{\mathcal{P}},P,\varepsilon) = L^n(A_{\mathcal{P}},S_{\mathcal{P}},\varepsilon) \quad (n \in \mathbb{Z}) \ .$$

If $\hat{H}^1(\mathbb{Z}_2;A,\varepsilon) \longrightarrow \hat{H}^1(\mathbb{Z}_2;A_{\mathcal{P}},\varepsilon) \oplus \hat{H}^1(\mathbb{Z}_2;P^{-1}A,\varepsilon)$ is one-one then by Proposition 3.6.3 ii) there are also identifications

$$L^n(A,P,\varepsilon) = L^n(A_{\mathcal{P}},P,\varepsilon) \quad (n \in \mathbb{Z}) \ ,$$

and by Proposition 4.1.3

$$L^n(A,\mathcal{P}^{\infty},\varepsilon) = L^n(A,P,\varepsilon) \quad (n \in \mathbb{Z}) \ .$$

[]

Given $\mathcal{P} \in \max(R)$ define the <u>$\mathcal{P}$-adic completion of A</u> to be the ring

$$\hat{A}_{\mathcal{P}} = \varprojlim_k A/\mathcal{P}^k A$$

which is also the $\mathcal{P}_{\mathcal{P}}$-adic completion $\widehat{(A_{\mathcal{P}})}_{\mathcal{P}_{\mathcal{P}}} = \varprojlim_k A_{\mathcal{P}}/\mathcal{P}^k A_{\mathcal{P}}$ of the localization $A_{\mathcal{P}}$ of A at $\mathcal{P}$. If $\mathcal{P} \in \overline{\max}(R) \subseteq \max(R)$ the completion $\hat{A}_{\mathcal{P}}$ is a ring with involution, and

$$\hat{S}_{\mathcal{P}} = \hat{R}_{\mathcal{P}} - \{0\} \subset \hat{A}_{\mathcal{P}}$$

is a multiplicative subset such that $(\hat{A}_{\mathcal{P}},\hat{S}_{\mathcal{P}})$ is a Dedekind algebra with involution. The quotient field of $\hat{R}_{\mathcal{P}}$

$$\hat{F}_{\mathcal{P}} = \hat{S}_{\mathcal{P}}^{-1}\hat{R}_{\mathcal{P}}$$

is the <u>$\mathcal{P}$-adic field of R</u>, and is such that

$$\hat{S}_{\mathcal{P}}^{-1}\hat{A}_{\mathcal{P}} = \hat{F}_{\mathcal{P}} \otimes_{\hat{R}_{\mathcal{P}}} \hat{A}_{\mathcal{P}} \ .$$

The S-adic completion $\hat{A} = \varprojlim_{s \in S} A/sA$ is the unrestricted product of the $\mathcal{P}$-adic completions

$$\hat{A} = \prod_{\mathcal{P}\in\max(R)} \hat{A}_{\mathcal{P}} ,$$

and the localization of the completion is the restricted product

$$\hat{S}^{-1}\hat{A} = \coprod_{\mathcal{P}\in\max(R)} (\hat{S}_{\mathcal{P}}^{-1}\hat{A}_{\mathcal{P}},\hat{A}_{\mathcal{P}})$$

consisting of collections $\{x_{\mathcal{P}} \in \hat{S}_{\mathcal{P}}^{-1}\hat{A}_{\mathcal{P}} | \mathcal{P}\in \max(R)\}$ such that $x_{\mathcal{P}} \in \hat{A}_{\mathcal{P}} \subseteq \hat{S}_{\mathcal{P}}^{-1}\hat{A}_{\mathcal{P}}$ for all but a finite number of $\mathcal{P} \in \max(R)$. Thus the cartesian square of rings with involution associated to (A,S)

$$\begin{array}{ccc} A & \longrightarrow & S^{-1}A \\ \downarrow & & \downarrow \\ \hat{A} & \longrightarrow & \hat{S}^{-1}\hat{A} \end{array}$$

can be written as

$$\begin{array}{ccc} A & \longrightarrow & S^{-1}A \\ \downarrow & & \downarrow \\ \prod_{\mathcal{P}} \hat{A}_{\mathcal{P}} & \longrightarrow & \coprod_{\mathcal{P}}(\hat{S}_{\mathcal{P}}^{-1}\hat{A}_{\mathcal{P}},\hat{A}_{\mathcal{P}}) \end{array} .$$

If $P\subset A$ is a characteristic multiplicative subset for $\mathcal{P} \in \overline{\max}(R)$ the $\mathcal{P}$-adic completion of A is just the P-adic completion of A

$$\hat{A} = \varprojlim_{p\in P} A/pA .$$

For example, the ring of p-adic integers $\hat{\mathbb{Z}}_p = \varprojlim_k \mathbb{Z}/p^k\mathbb{Z}$ is the (p)-adic completion of $\mathbb{Z}$, with $(p) = p\mathbb{Z} \in \overline{\max}(\mathbb{Z})$ (p prime), and $\hat{S}_{(p)}^{-1}\hat{\mathbb{Z}}_p = \hat{\mathbb{Q}}_p$ is the field of p-adic numbers.

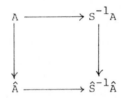

Given $\mathcal{P} \in \overline{\max}(R)$ let

$$\hat{\mathcal{P}}_{\mathcal{P}} = \mathcal{P}\hat{R}_{\mathcal{P}} \in \overline{\max}(\hat{R}_{\mathcal{P}})$$

be the unique maximal ideal of the complete local ring $\hat{R}_{\mathcal{P}}$.

The multiplicative subset

$$\hat{S}_{\mathcal{P}} = \hat{R}_{\mathcal{P}} - \{0\} \subset \hat{A}_{\mathcal{P}}$$

is characteristic for $\hat{\mathcal{P}}_{\mathcal{P}} \in \overline{\max}(\hat{R}_{\mathcal{P}})$, and there are natural identifications of exact categories

$$\begin{aligned}\{(A,\mathcal{P}^{\infty})\text{-modules}\} &= \{(A_{\mathcal{P}},\mathcal{P}^{\infty})\text{-modules}\} \\ &= \{(\hat{A}_{\mathcal{P}},\hat{\mathcal{P}}_{\mathcal{P}}^{\infty})\text{-modules}\} = \{(\hat{A}_{\mathcal{P}},\hat{S}_{\mathcal{P}})\text{-modules}\} .\end{aligned}$$

<u>Proposition 4.1.5</u> i) There are natural identifications of ε-quadratic L-groups

$$L_n(A,\mathcal{P}^{\infty},\varepsilon) = L_n(\hat{A}_{\mathcal{P}},\hat{S}_{\mathcal{P}},\varepsilon) \quad (n \in \mathbb{Z}) ,$$

giving rise to a Mayer-Vietoris exact sequence

$$\begin{aligned}\ldots \longrightarrow L_n(A,\varepsilon) \longrightarrow L_n^S(S^{-1}A,\varepsilon) \oplus \prod_{\mathcal{P}} L_n(\hat{A}_{\mathcal{P}},\varepsilon) \\ \longrightarrow \coprod_{\mathcal{P}} (L_n^{\hat{S}_{\mathcal{P}}}(\hat{S}_{\mathcal{P}}^{-1}\hat{A}_{\mathcal{P}},\varepsilon), L_n(\hat{A}_{\mathcal{P}},\varepsilon)) \longrightarrow L_{n-1}(A,\varepsilon) \longrightarrow \ldots \quad (n \in \mathbb{Z}) ,\end{aligned}$$

with $\mathcal{P}$ ranging over $\overline{\max}(R)$.

ii) If $\mathcal{P} \in \overline{\max}(R)$ is such that

either the maps $\hat{H}^1(\mathbb{Z}_2;A,\varepsilon) \longrightarrow \hat{H}^1(\mathbb{Z}_2;\hat{A}_{\mathcal{P}},\varepsilon)$ and

$\hat{H}^1(\mathbb{Z}_2;A_{\mathcal{P}},\varepsilon) \longrightarrow \hat{H}^1(\mathbb{Z}_2;\hat{A}_{\mathcal{P}},\varepsilon) \oplus \hat{H}^1(\mathbb{Z}_2;\hat{S}_{\mathcal{P}}^{-1}\hat{A}_{\mathcal{P}},\varepsilon)$ are one-one

or there exists a characteristic multiplicative subset $P \subset A$ for $\mathcal{P}$ and the map

$\hat{H}^1(\mathbb{Z}_2;A,\varepsilon) \longrightarrow \hat{H}^1(\mathbb{Z}_2;\hat{A}_{\mathcal{P}},\varepsilon) \oplus \hat{H}^1(\mathbb{Z}_2;P^{-1}A,\varepsilon)$

is one-one

(e.g. if some power $\mathcal{P}^k$ $(k \geqslant 1)$ is principal and

$\hat{H}^0(\mathbb{Z}_2;\hat{F}_{\mathcal{P}},\varepsilon) = 0$)

there are natural identifications of ε-symmetric L-groups

$$L^n(A,\mathcal{P}^\infty,\varepsilon) = L^n(\hat{A}_{\mathcal{P}},\hat{S}_{\mathcal{P}},\varepsilon) \quad (n \in \mathbb{Z}) \ .$$

If one of these conditions is satisfied for each $\mathcal{P} \in \overline{\max}(R)$ there is defined a Mayer-Vietoris exact sequence

$$\ldots \longrightarrow L^n(A,\varepsilon) \longrightarrow L^n_S(S^{-1}A,\varepsilon) \oplus \prod_{\mathcal{P}} L^n(\hat{A}_{\mathcal{P}},\varepsilon)$$

$$\longrightarrow \coprod_{\mathcal{P}} (L^n_{\hat{S}}(\hat{S}_{\mathcal{P}}^{-1}\hat{A}_{\mathcal{P}},\varepsilon), L^n(\hat{A}_{\mathcal{P}},\varepsilon)) \longrightarrow L^{n-1}(A,\varepsilon) \longrightarrow \ldots \quad (n \in \mathbb{Z}) .$$

iii) For $A,\mathcal{P},\varepsilon$ as in $\begin{cases} \text{ii)} \\ \text{i)} \end{cases}$ the groups $\begin{cases} L^*(A,\mathcal{P}^\infty,\varepsilon) = L^*(\hat{A}_{\mathcal{P}},\hat{S}_{\mathcal{P}},\varepsilon) \\ L_*(A,\mathcal{P}^\infty,\varepsilon) = L_*(\hat{A}_{\mathcal{P}},\hat{S}_{\mathcal{P}},\varepsilon) \end{cases}$

are $L^0(\hat{R}_{\mathcal{P}})$-modules.

<u>Proof</u>: i),ii) Immediate from Propositions 3.6.3,3.6.5 and 4.1.4 using the cartesian morphisms $(A_{\mathcal{P}},S_{\mathcal{P}}) \longrightarrow (\hat{A}_{\mathcal{P}},S_{\mathcal{P}}),(A,P) \longrightarrow (\hat{A}_{\mathcal{P}},P)$. (The restricted product $\coprod_{\mathcal{P}}(G_{\mathcal{P}},H_{\mathcal{P}})$ of a collection of abelian group morphisms $H_{\mathcal{P}} \longrightarrow G_{\mathcal{P}}$ indexed by a set $\{\mathcal{P}\}$ is the direct limit

$$\coprod_{\mathcal{P}} (G_{\mathcal{P}},H_{\mathcal{P}}) = \varinjlim_{I} \left(\prod_{\mathcal{P}\in I} G_{\mathcal{P}} \times \prod_{\mathcal{P}\notin I} H_{\mathcal{P}} \right)$$

taken over all the finite subsets I of $\{\mathcal{P}\}$).

iii) Immediate from i),ii) and Proposition 2.2.6.

[]

The hypotheses of Propositions 4.1.4 ii), 4.1.5 ii) are satisfied if the Dedekind ring R is of characteristic ≠ 2 and has finite reduced projective class group $\tilde{K}_0(R)$ (= the ideal class group), such as is the case for the ring of integers R in an algebraic number field F. In particular, the hypotheses are satisfied if (A,S) is the Dedekind algebra with involution defined by a torsion-free ring with involution A, with

$$S = \mathbb{Z}-\{0\} \subset A \ , \ R = S \cup \{0\} = \mathbb{Z} \ ,$$

for which Proposition 4.1.5 gives identifications

$$\begin{cases} L^n(A,S,\varepsilon) = \bigoplus_p L^n(A,(p)^\infty,\varepsilon) = \bigoplus_p L^n(\hat{A}_p,(\hat{p})^\infty,\varepsilon) \\ L_n(A,S,\varepsilon) = \bigoplus_p L_n(A,(p)^\infty,\varepsilon) = \bigoplus_p L_n(\hat{A}_p,(\hat{p})^\infty,\varepsilon) \end{cases}$$

$$(n \in \mathbb{Z},\ p \text{ prime},\ (p) = p\mathbb{Z} \triangleleft \mathbb{Z})$$

and a Mayer-Vietoris exact sequence

$$\begin{cases} \dots \longrightarrow L^n(A,\varepsilon) \longrightarrow L^n_S(S^{-1}A,\varepsilon) \oplus \prod_p L^n(\hat{A}_p,\varepsilon) \\ \qquad \longrightarrow \coprod_p (L^n_{\hat{S}_p}(\hat{S}_p^{-1}\hat{A}_p,\varepsilon), L^n(\hat{A}_p,\varepsilon)) \longrightarrow L^{n-1}(A,\varepsilon) \longrightarrow \dots \\ \dots \longrightarrow L_n(A,\varepsilon) \longrightarrow L^S_n(S^{-1}A,\varepsilon) \oplus \prod_p L_n(\hat{A}_p,\varepsilon) \\ \qquad \longrightarrow \coprod_p (L^{\hat{S}_p}_n(\hat{S}_p^{-1}\hat{A}_p,\varepsilon), L_n(\hat{A}_p,\varepsilon)) \longrightarrow L_{n-1}(A,\varepsilon) \longrightarrow \dots , \end{cases}$$

where $S^{-1}A = \mathbb{Q}\otimes_{\mathbb{Z}}A$, $\hat{A}_p = \varprojlim_k A/p^kA = \hat{\mathbb{Z}}_p\otimes_{\mathbb{Z}}A$, $\hat{S}_p^{-1}\hat{A}_p = \hat{\mathbb{Q}}_p\otimes_{\mathbb{Z}}A$.

Moreover, the L-groups $\begin{cases} L^*(A,(p)^\infty,\varepsilon) \\ L_*(A,(p)^\infty,\varepsilon) \end{cases}$ are $L^0(\hat{\mathbb{Z}}_p)$-modules, and hence of exponent $\hat{\psi}(p)$. (See §3.6 for the definition of $\hat{\psi}(p)$).

4.2 <u>Dedekind rings</u>

We shall now specialize the results of §4.1 to the case of a Dedekind algebra with involution (A,S) with $A = R = S\cup\{0\}$, i.e. A is itself the underlying Dedekind ring with involution.

Let then R be a Dedekind with involution, and let

$$S = R-\{0\}\subset R \ ,$$

so that the quotient field of R is given by

$$F = S^{-1}R \ .$$

Recall from §3.1 the definition of the maximal S-torsion submodule of an R-module M

$$T_SM = \{x\in M | sx = 0\in M \text{ for some } s\in S\}\subseteq M \ .$$

An R-module M is $\begin{cases}\underline{\text{torsion}}\\ \underline{\text{torsion-free}}\end{cases}$ if

$$\begin{cases}T_SM = M\\ T_SM = \{0\} \ .\end{cases}$$

An R-module is $\begin{cases}\text{an } (R,S)\text{-module}\\ \text{a f.g. projective R-module}\end{cases}$ if and only if it is f.g. and $\begin{cases}\text{torsion}\\ \text{torsion-free}\end{cases}$.

Given a finite-dimensional R-module chain complex C let

$$\begin{cases}T_r(C) = T_SH_r(C)\\ F_r(C) = H_r(C)/T_SH_r(C)\end{cases} \quad (\text{resp.} \begin{cases}T^r(C) = T_SH^r(C)\\ F^r(C) = H^r(C)/T_SH^r(C)\end{cases}) \quad (r\in\mathbb{Z})$$

be the $\begin{cases}\text{maximal torsion submodule}\\ \text{minimal torsion-free quotient module}\end{cases}$ of $H_r(C)$ (resp. $H^r(C)$), which is $\begin{cases}\text{an } (R,S)\text{-module}\\ \text{a f.g. projective R-module}\end{cases}$. The universal coefficient theorem gives natural R-module isomorphisms

$$T_r(C) \longrightarrow T^{r+1}(C)^\wedge = \mathrm{Hom}_R(T^{r+1}(C),F/R) \;;\; x \longmapsto (f \longmapsto \overline{(\frac{f(y)}{s})})$$

$$(x \in C_r, y \in C_{r+1}, s \in S, sx = dy \in C_r, f \in C^{r+1})$$

$$F_r(C) \longrightarrow F^r(C)^* = \mathrm{Hom}_R(F^r(C),R) \;;\; x \longmapsto (f \longmapsto \overline{f(x)})$$

$$(x \in C_r, f \in C^r) \;.$$

<u>Proposition 4.2.1</u> The L-groups of a Dedekind ring with involution R and of the quotient field $F = S^{-1}R$ are such that

i) The skew-suspension maps in the $\pm\varepsilon$-symmetric L-groups

$$\begin{cases} \overline{S} : L^n(R,\varepsilon) \longrightarrow L^{n+2}(R,-\varepsilon) & (n \geqslant 0) \\ \overline{S} : L^n(R,S,\varepsilon) \longrightarrow L^{n+2}(R,S,-\varepsilon) & (n \geqslant 1) \end{cases}$$

are isomorphisms.

ii) The Witt group of $\begin{cases} \varepsilon\text{-symmetric} \\ \text{even } \varepsilon\text{-symmetric} \\ \varepsilon\text{-quadratic} \end{cases}$ formations over F

vanishes

$$\begin{cases} M^\varepsilon(F) = L^1(F,\varepsilon) = 0 \\ M\langle v_0\rangle^\varepsilon(F) = L^{-1}(F,-\varepsilon) = 0 \\ M_\varepsilon(F) = L_1(F,\varepsilon) = 0 \;, \end{cases}$$

as does the Witt group of $\begin{cases} \varepsilon\text{-symmetric} \\ \text{even } \varepsilon\text{-symmetric} \\ \varepsilon\text{-quadratic} \end{cases}$ linking formations

over (R,S)

$$\begin{cases} M^\varepsilon(R,S) = L^3(R,S,-\varepsilon) = 0 \\ M\langle v_0\rangle^\varepsilon(R,S) = L^1(R,S,\varepsilon) = 0 \\ M_\varepsilon(R,S) = L^{-1}(R,S,-\varepsilon) = 0 \;. \end{cases}$$

iii) There are defined localization exact sequences of Witt groups

$$0 \longrightarrow L^{\varepsilon}(R) \longrightarrow L^{\varepsilon}(F) \xrightarrow{\partial} L^{\varepsilon}(R,S) \longrightarrow M^{-\varepsilon}(R) \longrightarrow 0$$

$$0 \longrightarrow L^{\varepsilon}(R) \longrightarrow L^{\varepsilon}(F) \xrightarrow{\partial} L\langle v_0\rangle^{\varepsilon}(R,S) \longrightarrow M\langle v_0\rangle^{-\varepsilon}(R) \longrightarrow 0$$

$$0 \longrightarrow L\langle v_0\rangle^{\varepsilon}(R) \longrightarrow L\langle v_0\rangle^{\varepsilon}(F) \xrightarrow{\partial} L_{\varepsilon}(R,S) \longrightarrow M_{-\varepsilon}(R) \longrightarrow 0$$

$$0 \longrightarrow \tilde{M}_{\varepsilon}(R,S) \longrightarrow L_{\varepsilon}(R) \longrightarrow L_{\varepsilon}(F) \xrightarrow{\partial} \tilde{L}_{\varepsilon}(R,S) \longrightarrow M_{-\varepsilon}(R) \longrightarrow 0 \ .$$

In particular, there are natural identifications of Witt groups of formations over R with quotients of Witt groups of linking forms over (R,S)

$$M^{\varepsilon}(R) = \frac{(\text{non-singular } (-\varepsilon)\text{-symmetric linking forms over } (R,S))}{\begin{matrix}(\text{boundaries of } S\text{-non-singular} \\ (-\varepsilon)\text{-symmetric forms over } R\,)\end{matrix} + (\text{hyperbolics})}$$

$$M\langle v_0\rangle^{\varepsilon}(R) = \frac{(\text{non-singular even } (-\varepsilon)\text{-symmetric linking forms over } (R,S))}{(\text{boundaries of } S\text{-non-singular } (-\varepsilon)\text{-symmetric forms over } R)}$$

$$M_{\varepsilon}(R) = \frac{(\text{non-singular } (-\varepsilon)\text{-quadratic linking forms over } (R,S))}{\begin{matrix}(\text{boundaries of } S\text{-non-singular even} \\ (-\varepsilon)\text{-symmetric forms over } R)\end{matrix}}$$

$$= \frac{(\text{non-singular split } (-\varepsilon)\text{-quadratic linking forms over } (R,S))}{(\text{boundaries of } S\text{-non-singular } (-\varepsilon)\text{-quadratic forms over } R)} \ .$$

(Note that a stably hyperbolic $\begin{cases}\text{even } (-\varepsilon)\text{-symmetric} \\ (-\varepsilon)\text{-quadratic} \\ \text{split } (-\varepsilon)\text{-quadratic}\end{cases}$ linking form over (R,S) is isomorphic to the boundary of an S-non-singular $\begin{cases}(-\varepsilon)\text{-symmetric} \\ \text{even } (-\varepsilon)\text{-symmetric} \\ (-\varepsilon)\text{-quadratic}\end{cases}$ form over R, by Proposition 3.4.6 ii)).

iv) For $n = 2i$ (resp. $n = 2i+1$) the isomorphism

$$\begin{cases} \bar{S}^{-i} : L^n(R,\varepsilon) \longrightarrow L^{n-2i}(R,(-)^i\varepsilon) \\ \bar{S}^{-i} : L_n(R,\varepsilon) \longrightarrow L_{n-2i}(R,(-)^i\varepsilon) \end{cases} \quad (i \geqslant 0)$$

sends the cobordism class of an n-dimensional $\begin{cases} \varepsilon\text{-symmetric} \\ \varepsilon\text{-quadratic} \end{cases}$

Poincaré complex over R $\begin{cases} (C,\phi \in Q^n(C,\varepsilon)) \\ (C,\psi \in Q_n(C,\varepsilon)) \end{cases}$ to the class in

$$\begin{cases} L^0(R,(-)^i\varepsilon) = L^{(-)^i\varepsilon}(R) \\ L_0(R,(-)^i\varepsilon) = L_{(-)^i\varepsilon}(R) \end{cases} \text{(resp.} \begin{cases} L^1(R,(-)^i\varepsilon) = M^{(-)^i\varepsilon}(R) \\ L_1(R,(-)^i\varepsilon) = M_{(-)^i\varepsilon}(R) \end{cases})$$

of the non-singular $\begin{cases} (-)^i\varepsilon\text{-symmetric} \\ (-)^i\varepsilon\text{-quadratic} \end{cases}$ form over R

$$\begin{cases} (F^i(C),\phi_0 : F^i(C) \times F^i(C) \longrightarrow R) \\ (F^i(C),(1+T_\varepsilon)\psi_0,v^i(\psi) : F^i(C) \longrightarrow Q_{(-)^i\varepsilon}(R)) \end{cases} \text{(resp. of the}$$

non-singular $\begin{cases} (-)^{i+1}\varepsilon\text{-symmetric} \\ (-)^{i+1}\varepsilon\text{-quadratic} \end{cases}$ linking form over (R,S)

$$\begin{cases} (T^{i+1}(C),\phi_0^S : T^{i+1}(C) \times T^{i+1}(C) \longrightarrow F/R) \\ (T^{i+1}(C),(1+T_\varepsilon)\psi_0^S,pv_S^i(\psi) : T^{i+1}(C) \longrightarrow Q_{(-)^{i+1}\varepsilon}(R,S)) \end{cases}).$$

v) There are natural direct sum decompositions of Witt groups

$$W(R,S) = \bigoplus_{\mathcal{P} \in \overline{\max}(R)} W(R,\mathcal{P}^\infty)$$

for $W = L^\varepsilon, L\langle v_0\rangle^\varepsilon, L_\varepsilon, \tilde{L}_\varepsilon, \tilde{M}_\varepsilon$. (The Witt groups $W(R,\mathcal{P}^\infty)$ are defined in the same way as W(R,S) but using only $(R,\mathcal{P}^\infty)$-modules).

vi) There are natural identifications

$$L^\varepsilon(R,\mathcal{P}^\infty) = L^\varepsilon(R/\mathcal{P}) \quad (\mathcal{P} \in \overline{\max}(R)) \quad ,$$

and the localization exact sequence of Witt groups can be

expressed as

$$0 \longrightarrow L^{\varepsilon}(R) \longrightarrow L^{\varepsilon}(F) \longrightarrow \bigoplus_{\mathcal{P}\in \overline{\max}(R)} L^{\varepsilon}(R/\mathcal{P}) \longrightarrow M^{-\varepsilon}(R) \longrightarrow 0 \ .$$

<u>Proof</u>: i) - v) Immediate from Propositions 1.2.3,3.6.1 and 4.1.2, since a Dedekind ring is 1-dimensional and the quotient field F is 0-dimensional.

vi) Define an <u>$(R,\mathcal{P}^k)$-module</u> M $(k \geqslant 1)$ to be a $\mathcal{P}$-primary (R,S)-module with annihilator

$$\mathrm{ann}(M) = \mathcal{P}^j \triangleleft R$$

for some $j \leqslant k$. Let $L^{\varepsilon}(R,\mathcal{P}^k)$ $(k \geqslant 1)$ be the Witt group of non-singular ε-symmetric linking forms over $(R,\mathcal{P}^k)$, that is non-singular ε-symmetric linking forms over (R,S) (M,λ) with M an $(R,\mathcal{P}^k)$-module. The natural maps

$$L^{\varepsilon}(R,\mathcal{P}^k) \longrightarrow L^{\varepsilon}(R,\mathcal{P}^{k+1}) \ ; \ (M,\lambda) \longmapsto (M,\lambda) \quad (k \geqslant 1)$$

are isomorphisms, for if (M,λ) is a non-singular ε-symmetric linking form over $(R,\mathcal{P}^{k+1})$ then

$$L = \mathcal{P}^k M \subseteq M$$

is a sublagrangian of (M,λ) such that L is an $(R,\mathcal{P}^{k+1})$-module and $(L^{\perp}/L,\lambda^{\perp}/\lambda)$ is a non-singular ε-symmetric linking form over $(R,\mathcal{P}^k)$, so that there are defined inverses

$$L^{\varepsilon}(R,\mathcal{P}^{k+1}) \longrightarrow L^{\varepsilon}(R,\mathcal{P}^k) \ ; \ (M,\lambda) \longmapsto (L^{\perp}/L,\lambda^{\perp}/\lambda) \quad (k > 1) \ .$$

We can thus identify

$$L^{\varepsilon}(R,\mathcal{P}^{\infty}) = \varinjlim_k L^{\varepsilon}(R,\mathcal{P}^k) = L^{\varepsilon}(R,\mathcal{P}) \ .$$

There is a natural identification of categories

$\{(R,\mathcal{P})\text{-modules}\}$

$= \{$finite-dimensional vector spaces over the residue class field $R/\mathcal{P}\}$.

Choose an element $\pi \in \mathcal{P} - \mathcal{P}^2$, so that $\pi \in R_{\mathcal{P}}$ is a generator of the unique maximal ideal $\mathcal{P}_{\mathcal{P}} = \pi R_{\mathcal{P}} \in \overline{\max}(R_{\mathcal{P}})$ of the local ring $R_{\mathcal{P}}$, and note that for any $(R,\mathcal{P})$-module M there is defined an R-module isomorphism

$$M^* = \mathrm{Hom}_{R/\mathcal{P}}(M,R/\mathcal{P}) \longrightarrow M^{\wedge} = \mathrm{Hom}_{R_{\mathcal{P}}}(M,F/R_{\mathcal{P}}) \ ;$$
$$f \longmapsto (x \longmapsto \frac{f(x)}{\pi}) \ .$$

Moreover, the ε-duality involution T_ε on $\mathrm{Hom}_{R/\mathcal{P}}(M,M^*)$ corresponds under this isomorphism to the ε-duality involution T_ε on $\mathrm{Hom}_{R_{\mathcal{P}}}(M,M^{\wedge})$. (We are assuming here that $\bar{\pi} = \pi \in \mathcal{P}$).
The natural R-module morphism

$$\mathrm{Hom}_R(M,F/R) \longrightarrow \mathrm{Hom}_{R_{\mathcal{P}}}(M,F/R_{\mathcal{P}}) \ ; \ f \longmapsto (x \longmapsto f(x))$$

is an isomorphism, so that we have identifications of categories

$$\begin{aligned}&\{\varepsilon\text{-symmetric linking forms over } R/\mathcal{P}\}\\ &\qquad = \{\varepsilon\text{-symmetric linking forms over } (R_{\mathcal{P}},\mathcal{P}_{\mathcal{P}})\}\\ &\qquad = \{\varepsilon\text{-symmetric linking forms over } (R,\mathcal{P})\}\end{aligned}$$

and hence also of ε-symmetric Witt groups

$$L^\varepsilon(R/\mathcal{P}) = L^\varepsilon(R_{\mathcal{P}},\mathcal{P}_{\mathcal{P}}) = L^\varepsilon(R,\mathcal{P}) = L^\varepsilon(R,\mathcal{P}^\infty)$$

(with only the identification $L^\varepsilon(R/\mathcal{P}) = L^\varepsilon(R_{\mathcal{P}},\mathcal{P}_{\mathcal{P}})$ depending on the choice of uniformizer π).

[]

The L-theoretic devissage argument used to identify $L^\varepsilon(R,\mathcal{P}^\infty) = L^\varepsilon(R,\mathcal{P})$ in the proof of Proposition 4.2.1 vi) above breaks down in the ε-quadratic case. Given a non-singular ε-quadratic linking form over (R,S) (M,λ,μ) such that

$$\mathrm{ann}(M) = \mathcal{P}^{k+1} \quad (k \geqslant 1)$$

for some $\mathcal{P} \in \overline{\max}(R)$ it need not be the case that $L = \mathcal{P}^k M \subset M$ is a sublagrangian of (M,λ,μ) as well as of (M,λ).

For example, consider the non-singular quadratic linking form over $(\mathbb{Z},\mathbb{Z}-\{0\})$ (M,λ,μ) defined by

$$M = \mathbb{Z}_4 \quad (\text{so that } \mathrm{ann}(M) = (2)^2 \lhd \mathbb{Z})$$

$$\lambda : M \times M \longrightarrow \mathbb{Q}/\mathbb{Z} ; \ (m,n) \longmapsto \frac{1}{4}mn$$

$$\mu : M \longrightarrow Q_{+1}(\mathbb{Z},\mathbb{Z}-\{0\}) = \mathbb{Q}/2\mathbb{Z} ; \ m \longmapsto \frac{1}{4}m^2 .$$

Then $L = 2M \subset M$ is a lagrangian of the symmetrization (M,λ) but not of (M,λ,μ), since

$$\mu(2) = 1 \neq 0 \in \mathbb{Q}/2\mathbb{Z} .$$

In fact, the kernel of the symmetrization map of Witt groups

$$1+T : L_{+1}(\mathbb{Z},(2)^\infty) = \mathbb{Z}_8 \oplus \mathbb{Z}_2 \longrightarrow L^{+1}(\mathbb{Z},(2)^\infty) = L^{+1}(\mathbb{Z}_2) = \mathbb{Z}_2 ;$$

$$(a,b) \longmapsto b$$

is generated by $(M,\lambda,\mu) = (1,0) \in \ker(1+T) = \mathbb{Z}_8$, and the non-singular quadratic linking form over $(\mathbb{Z},\mathbb{Z}-\{0\})$ (M',λ',μ') defined by

$$M' = \mathbb{Z}_2$$

$$\lambda' : M' \times M' \longrightarrow \mathbb{Q}/\mathbb{Z} ; \ (m,n) \longmapsto \frac{1}{2}mn$$

$$\mu' : M' \longrightarrow \mathbb{Q}/2\mathbb{Z} ; \ m \longmapsto \frac{1}{2}m^2$$

represents $(M',\lambda',\mu') = (0,1) \in L_{+1}(\mathbb{Z},(2)^\infty)$. Furthermore, $L_{+1}(\mathbb{Z}_2) = \mathbb{Z}_2$ (generated by the non-singular quadratic form over $\mathbb{Z}_2$

$(\mathbb{Z}_2 \oplus \mathbb{Z}_2, \begin{pmatrix} 1 & 1 \\ 0 & 1 \end{pmatrix} \in Q_{+1}(\mathbb{Z}_2 \oplus \mathbb{Z}_2))$ of Arf invariant 1), so that

$$L_{+1}(\mathbb{Z},(2)^\infty) \neq L_{+1}(\mathbb{Z}_2) .$$

In Proposition 4.3.3 below we shall relate this failure of devissage in quadratic L-theory to a failure of reduction modulo a complete ideal (= Hensel's lemma) in symmetric L-theory.

The ε-symmetric Witt group localization exact sequence of Proposition 4.2.1 vi)

$$0 \longrightarrow L^{\varepsilon}(R) \longrightarrow L^{\varepsilon}(F) \longrightarrow \bigoplus_{\mathcal{P}\in\overline{\max}(R)} L^{\varepsilon}(R/\mathcal{P})$$

was first obtained by Milnor (cf. Corollary IV.3.3 of Milnor and Husemoller [1]) in the case $\varepsilon = +1 \in R$, with R a Dedekind ring of characteristic $\neq 2$. The identifications of Proposition 4.2.1 iii),vi)

$$M^{-\varepsilon}(R) = \mathrm{coker}(\partial : L^{\varepsilon}(F) \longrightarrow L^{\varepsilon}(R,S)) \ ,$$

$$L^{\varepsilon}(R,S) = \bigoplus_{\mathcal{P}} L^{\varepsilon}(R/\mathcal{P})$$

were first obtained by Karoubi [3], in the case $1/2 \in R$. Example IV.3.5 of Milnor and Husemoller [1] can be interpreted as stating that for the coordinate ring of the circle

$$R = \mathbb{R}[x,y]/(x^2+y^2-1)$$

the Witt group of non-singular skew-symmetric formations over R is given by

$$M^{-1}(R) = \mathbb{Z} \ ,$$

generated by the formation

$$(R\oplus R^*, \begin{pmatrix} 0 & 1 \\ -1 & 0 \end{pmatrix}; R, \mathrm{im}(\begin{pmatrix} x \\ y \end{pmatrix} : R \longrightarrow R\oplus R^*)) \ ,$$

corresponding to the symplectic automorphism

$$\begin{pmatrix} x & -y \\ y & x \end{pmatrix} \in SL_2(R) = \mathrm{Aut}(R\oplus R^*, \begin{pmatrix} 0 & 1 \\ -1 & 0 \end{pmatrix}) \ .$$

Given a Dedekind ring with involution R and $\mathcal{P} \in \overline{\max}(R)$ together with a choice of uniformizer $\pi \in \mathcal{P} - \mathcal{P}^2$ such that $\bar{\pi} = \pi$ there is defined a non-singular skew-symmetric S-formation over R

$$(R\oplus R^*, \begin{pmatrix} 0 & 1 \\ -1 & 0 \end{pmatrix}; R, \mathrm{im}(\begin{pmatrix} i \\ \pi i \end{pmatrix}: \mathcal{P} \longrightarrow R\oplus R^*))$$

with $i \in \mathrm{Hom}_R(\mathcal{P},R)$ the inclusion, corresponding by Proposition 3.4.3 to the non-singular symmetric linking form over $(R,\mathcal{P})$

$$(R/\mathcal{P}, \lambda : R/\mathcal{P} \times R/\mathcal{P} \longrightarrow F/R ; (x,y) \longmapsto \frac{\bar{x}y}{\pi}) \quad .$$

The inclusion of the lagrangian $\begin{pmatrix} i \\ \pi i \end{pmatrix}: \mathcal{P} \longrightarrow R\oplus R^*$ extends by Proposition 1.6.2 to an R-module isomorphism

$$\mathcal{P} \oplus \mathcal{P}^* \longrightarrow R\oplus R^* \ .$$

Thus if R has the identity involution ($\bar{r} = r$ for all $r \in R$) the duality involution on the reduced projective class group $\tilde{K}_0(R)$ (= the ideal class group of R) is given by

$$* : \tilde{K}_0(R) \longrightarrow \tilde{K}_0(R) \ ; \ [\mathcal{P}] \longmapsto [\mathcal{P}]^* = -[\mathcal{P}]$$

and

$$\hat{H}^m(\mathbb{Z}_2;\tilde{K}_0(R)) = \begin{cases} \ker(2:\tilde{K}_0(R) \longrightarrow \tilde{K}_0(R)) \\ \mathrm{coker}(2:\tilde{K}_0(R) \longrightarrow \tilde{K}_0(R)) \end{cases} \text{if} \begin{cases} m \equiv 0 \pmod 2 \\ m \equiv 1 \pmod 2 \end{cases} .$$

Now $V^0(R,-1) = 0$ (by Corollary I.3.5 of Milnor and Husemoller [1]), so that a portion of the relevant exact sequence of Proposition 1.10.1 can be written as

$$\begin{aligned} \ldots \longrightarrow U^2(R,-1) \longrightarrow \hat{H}^2(\mathbb{Z}_2;\tilde{K}_0(R)) \longrightarrow V^1(R,-1) \\ \longrightarrow U^1(R,-1) \longrightarrow \hat{H}^1(\mathbb{Z}_2;\tilde{K}_0(R)) \longrightarrow 0 \ . \end{aligned}$$

The map

$$U^2(R,-1) = U^0(R) \longrightarrow \hat{H}^2(\mathbb{Z}_2;\tilde{K}_0(R)) \ ; \ (M,\phi) \longmapsto [M]$$

is onto: if I is an ideal of R such that $I^2 = rR$ is a principal ideal, with generator $r \in I^2$, then

$$(I, \phi: I \times I \longrightarrow R \ ; \ (x,y) \longmapsto \frac{xy}{r})$$

is a non-singular symmetric form over R with projective class $[I] \in \widetilde{K}_0(R)$. Thus the map $\hat{H}^1(\mathbb{Z}_2;\widetilde{K}_0(R)) \longrightarrow V^1(R,-1)$ is 0, and there is defined a short exact sequence

$$0 \longrightarrow V^1(R,-1) \longrightarrow U^1(R,-1) \longrightarrow \hat{H}^1(\mathbb{Z}_2;\widetilde{K}_0(R)) \longrightarrow 0 .$$

If R is the ring of integers in an algebraic number field F then by Milnor [4,Cor.16.3]

$$V^1(R,-1) = SK_1(R) = 0 ,$$

so that there are identifications

$$U^1(R,-1) = M^{-1}(R) = \hat{H}^1(\mathbb{Z}_2;\widetilde{K}_0(R)) .$$

The consequent identification

$$\mathrm{coker}(\partial : L^{+1}(F) \longrightarrow \bigoplus_{\mathfrak{P} \in \max(R)} L^{+1}(R/\mathfrak{P})) = \hat{H}^1(\mathbb{Z}_2;\widetilde{K}_0(R))$$

appears as Example IV.3.4 of Milnor and Husemoller [1] - in this connection see also Knebusch and Scharlau [1].

See Pardon [6],[7] for an extension of the localization exact sequence of L-groups of Dedekind rings to more general regular rings, and for an application of the algebraic theory of surgery to the conjecture that for a regular local ring R with quotient field $F = (R-\{0\})^{-1}R$ the natural map of symmetric Witt groups $L^0(R) \longrightarrow L^0(F)$ is injective.

4.3 Integral and rational L-theory

The results of §4.2 will now be applied to obtain the L-groups of the Dedekind ring $R = \mathbb{Z}$ and of its quotient field $S^{-1}R = \mathbb{Q}$ (with $S = \mathbb{Z}-\{0\}\subset\mathbb{Z}$), in the sense of reducing the computation to the well-known stable classifications of forms over $\mathbb{Z}$ and $\mathbb{Q}$. In the first instance we recall the classical invariants of forms over $\mathbb{Z}$.

A symmetric form over $\mathbb{Z}$ (M,ϕ) induces a symmetric form over $\mathbb{R}$ which can be expressed as

$$\mathbb{R}\otimes_{\mathbb{Z}}(M,\phi) = \bigoplus_{p}(\mathbb{R},1)\oplus\bigoplus_{q}(\mathbb{R},-1)\oplus\bigoplus_{r}(\mathbb{R},0) \quad (p,q,r \geqslant 0)$$

up to isomorphism. The <u>signature</u> of (M,ϕ) is defined by

$$\sigma^*(M,\phi) = p-q \in \mathbb{Z} \ .$$

If (M,ϕ) is even ($\phi(x)(x) \equiv 0 \pmod 2$ for each $x\in M$) then

$$\sigma^*(M,\phi) \equiv 0 \pmod 8 \ .$$

The <u>deRham invariant</u> of a non-singular skew-symmetric linking form over $(\mathbb{Z},S)$ (M,λ) is defined by

$$\sigma^*(M,\lambda) = |M| - 1 \in \mathbb{Z}_2 \ ,$$

or equivalently the mod 2 reduction of the number of summands in the decomposition of M as a direct sum of cyclic groups of type $\mathbb{Z}_{p^k}$ (p prime, $k \geqslant 1$). If (M,λ) is even ($\lambda(x)(x) = 0 \in Q^{-1}(\mathbb{Q}/\mathbb{Z}) = \mathbb{Z}_2$ for each $x\in M$, e.g. if M is of odd order) then

$$\sigma^*(M,\lambda) = 0 \in \mathbb{Z}_2 \ .$$

A non-singular skew-quadratic form over $\mathbb{Z}$ (M,ψ) induces a form over $\mathbb{Z}_2$ which can be expressed as

$$\mathbb{Z}_2\otimes_{\mathbb{Z}}(M,\psi) = \bigoplus_b(\mathbb{Z}_2\oplus\mathbb{Z}_2,\begin{pmatrix}0 & 1\\ 0 & 0\end{pmatrix})\oplus\bigoplus_c(\mathbb{Z}_2\oplus\mathbb{Z}_2,\begin{pmatrix}1 & 1\\ 0 & 1\end{pmatrix})$$

$$(b\geqslant 0, c = 0 \text{ or } 1)$$

up to isomorphism. The <u>Arf invariant</u> of (M,ψ) is defined by

$$\sigma_*(M,\psi) = c \in \mathbb{Z}_2 .$$

<u>Proposition 4.3.1</u> The symmetric and quadratic L-groups of $\mathbb{Z}$ are given by

$$L^n(\mathbb{Z}) = \begin{cases}\mathbb{Z}\\ \mathbb{Z}_2\\ 0\\ 0\end{cases} ,\quad L_n(\mathbb{Z}) = \begin{cases}\mathbb{Z}\\ 0\\ \mathbb{Z}_2\\ 0\end{cases} \quad \text{if } n\equiv\begin{cases}0\\ 1\\ 2\\ 3\end{cases} \pmod 4$$

$$(n\geqslant 0)$$

$$L_n(\mathbb{Z}) = L_{n+4k}(\mathbb{Z}) \quad (n\leqslant -1,\ n+4k\geqslant 0)$$

$$L^n(\mathbb{Z}) = \begin{cases}0\\ L_n(\mathbb{Z})\end{cases} \text{if } \begin{cases}n = -1,-2\\ n\leqslant -3\end{cases} .$$

The invariants are given by

$$L^{4k}(\mathbb{Z})\longrightarrow\mathbb{Z}\ ;\ (C,\phi\in Q^{4k}(C))\longmapsto \text{signature of } (F^{2k}(C),\phi_0)$$

$$L^{4k+1}(\mathbb{Z})\longrightarrow\mathbb{Z}_2\ ;$$

$$(C,\phi\in Q^{4k+1}(C))\longmapsto \text{deRham invariant of } (T^{2k+1}(C),\phi_0^S)$$

$$L_{4k}(\mathbb{Z})\longrightarrow\mathbb{Z}\ ;$$

$$(C,\psi\in Q_{4k}(C))\longmapsto\frac{1}{8}(\text{signature of } (F^{2k}(C),(1+T)\psi_0))$$

$$L_{4k+2}(\mathbb{Z})\longrightarrow\mathbb{Z}_2\ ;$$

$$(C,\psi\in Q_{4k+2}(C))\longmapsto \text{Arf invariant of } (F^{2k+1}(C),\psi_0) .$$

The hyperquadratic L-groups $\hat{L}^*(\mathbb{Z})$ (as defined in §2.3) are given by

$$\hat{L}^n(\mathbb{Z}) = \begin{cases} \begin{cases}\mathbb{Z}_8\\ \mathbb{Z}_2\\ 0\\ \mathbb{Z}_2\end{cases} & \text{if } n \equiv \begin{cases}0\\ 1\\ 2\\ 3\end{cases} \pmod 4 \quad (n \leqslant -2) \\ 0 & \text{if } n \leqslant -3 \ . \end{cases}$$

<u>Proof</u>: Proposition 4.2.1 reduces the computation of the $\begin{cases}\text{even-}\\ \text{odd-}\end{cases}$ dimensional L-groups of $\mathbb{Z}$ to the stable classification of non-singular $\begin{cases}\text{forms over } \mathbb{Z}\\ \text{linking forms over } (\mathbb{Z},\mathbb{Z}-\{0\})\end{cases}$, for which we refer to $\begin{cases}\text{Arf [1], Milnor and Husemoller [1]}\\ \text{deRham [1], Wall [1]}\end{cases}$.

The generator of the $\begin{cases}\text{symmetric}\\ \text{quadratic}\end{cases}$ L-group $\begin{cases}L^0(\mathbb{Z}) = \mathbb{Z}\\ L_0(\mathbb{Z}) = \mathbb{Z}\end{cases}$ is represented by the non-singular $\begin{cases}\text{symmetric}\\ \text{quadratic}\end{cases}$ form over $\mathbb{Z}$ $\begin{cases}(\mathbb{Z}, 1 \in Q^{+1}(\mathbb{Z}))\\ (\mathbb{Z}^8, E_8 \in Q_{+1}(\mathbb{Z}))\end{cases}$ of signature $\begin{cases}1 \in \mathbb{Z}\\ 8 \in \mathbb{Z}\end{cases}$. The generator of $\begin{cases}L^1(\mathbb{Z}) = L^2(\mathbb{Z},\mathbb{Z}-\{0\}) = L^2(\mathbb{Z},(2)^\infty) = \mathbb{Z}_2\\ L_2(\mathbb{Z}) = L_3(\mathbb{Z},\mathbb{Z}-\{0\}) = L_3(\mathbb{Z},(2)^\infty) = \mathbb{Z}_2\end{cases}$ is represented by the non-singular $\begin{cases}\text{symmetric formation}\\ \text{skew-quadratic form}\end{cases}$ over $\mathbb{Z}$ of $\begin{cases}\text{deRham}\\ \text{Arf}\end{cases}$ invariant $1 \in \mathbb{Z}_2$

$$\begin{cases} \left(\mathbb{Z}\oplus\mathbb{Z}, \begin{pmatrix} 0 & 1 \\ 1 & 1 \end{pmatrix} \in Q^{+1}(\mathbb{Z}\oplus\mathbb{Z}) ; \operatorname{im}(\begin{pmatrix} 1 \\ 0 \end{pmatrix} : \mathbb{Z} \longrightarrow \mathbb{Z}\oplus\mathbb{Z}), \right. \\ \qquad\qquad\qquad\qquad \left. \operatorname{im}(\begin{pmatrix} 1 \\ -2 \end{pmatrix} : \mathbb{Z} \longrightarrow \mathbb{Z}\oplus\mathbb{Z})\right) \\ \left(\mathbb{Z}\oplus\mathbb{Z}, \begin{pmatrix} 1 & 1 \\ 0 & 1 \end{pmatrix} \in Q_{-1}(\mathbb{Z}\oplus\mathbb{Z})\right) \end{cases} ,$$

corresponding to the non-singular $\begin{cases} \text{skew-symmetric linking form} \\ \text{split skew-quadratic linking} \end{cases}$ $\begin{cases} - \\ \text{formation} \end{cases}$ over $(\mathbb{Z},(2)^\infty)$

$$\begin{cases} (\mathbb{Z}_2, \lambda : \mathbb{Z}_2 \times \mathbb{Z}_2 \longrightarrow \mathbb{Q}/\mathbb{Z} ; (m,n) \longmapsto \frac{1}{2}mn) \\ (\mathbb{Z}_4, (\begin{pmatrix} (2\ 0) \\ (0\ 2) \end{pmatrix}, \theta) \mathbb{Z}_2\oplus\mathbb{Z}_2), \text{ with} \\ \theta : \mathbb{Z}_2\oplus\mathbb{Z}_2 \longrightarrow Q_{+1}(\mathbb{Z},\mathbb{Z}-\{0\}) = \mathbb{Q}/2\mathbb{Z} ; \ (m,n) \longmapsto m^2+n^2 . \end{cases}$$

[]

Of course, the computation of the simply-connected surgery obstruction groups $L_*(\mathbb{Z})$ is well-known, going back to Kervaire and Milnor [1].

<u>Proposition 4.3.2</u> The L-groups of $\mathbb{Q}$ are given by

$$L^n(\mathbb{Q}) = L_n(\mathbb{Q}) = \begin{cases} \mathbb{Z}\oplus\mathbb{Z}_2^\infty\oplus\mathbb{Z}_4^\infty & \text{if } n \equiv 0 \pmod 4 \\ 0 & \text{if } n \not\equiv 0 \pmod 4 \end{cases} .$$

The $\begin{cases} \text{symmetric} \\ \text{quadratic} \end{cases}$ L-theory localization exact sequence

$$\begin{cases} 0 \longrightarrow L^0(\mathbb{Z}) \longrightarrow L^0(\mathbb{Q}) \xrightarrow{\partial} \bigoplus_p L^0(\mathbb{Z},(p)^\infty) \longrightarrow 0 \\ 0 \longrightarrow L_0(\mathbb{Z}) \longrightarrow L_0(\mathbb{Q}) \xrightarrow{\partial} \bigoplus_p L_0(\mathbb{Z},(p)^\infty) \longrightarrow 0 \end{cases} \quad (p \text{ prime}) ,$$

$\left\{\begin{array}{l}\text{splits}\\ \text{does not split}\end{array}\right.$, with

$$L^0(\mathbb{Z},(p)^\infty) = L_0(\mathbb{Z},(p)^\infty) = L^0(\mathbb{Z}_p) = \begin{cases}\mathbb{Z}_2\oplus\mathbb{Z}_2 \\ \mathbb{Z}_4\end{cases} \text{if} \begin{cases}p \equiv 1 \pmod 4 \\ p \equiv 3 \pmod 4\end{cases}$$

$$\begin{cases}L^0(\mathbb{Z},(2)^\infty) = L^0(\mathbb{Z}_2) = \mathbb{Z}_2 \\ L_0(\mathbb{Z},(2)^\infty) = \mathbb{Z}_8\oplus\mathbb{Z}_2\end{cases} .$$

[]

The computation of $L^*(\mathbb{Q})$ is also well-known, cf. §IV.2 of Milnor and Husemoller [1].

The element $\left\{\begin{array}{l}1 \in L^0(\mathbb{Z},(2)^\infty) = \mathbb{Z}_2\\ (1,0),(0,1) \in L_0(\mathbb{Z},(2)^\infty) = \mathbb{Z}_8\oplus\mathbb{Z}_2\end{array}\right.$ is the image under $\left\{\begin{array}{l}\partial : L^0(\mathbb{Q}) \longrightarrow L^0(\mathbb{Z},(2)^\infty)\\ \partial : L_0(\mathbb{Q}) \longrightarrow L_0(\mathbb{Z},(2)^\infty)\end{array}\right.$ of the Witt class of the non-singular $\left\{\begin{array}{l}\text{symmetric}\\ \text{quadratic}\end{array}\right.$ form over $\mathbb{Q}$ $\left\{\begin{array}{l}(\mathbb{Q},2 \in Q^{+1}(\mathbb{Q}))\\ (\mathbb{Q},2 \in Q_{+1}(\mathbb{Q})),(\mathbb{Q},1 \in Q_{+1}(\mathbb{Q}))\end{array}\right.$, corresponding to the non-singular $\left\{\begin{array}{l}\text{symmetric}\\ \text{quadratic}\end{array}\right.$ linking form over $(\mathbb{Z},(2)^\infty)$ $\left\{\begin{array}{l}(\mathbb{Z}_2,\lambda')\\ (\mathbb{Z}_4,\lambda,\mu),\ (\mathbb{Z}_2,\lambda',\mu')\end{array}\right.$ defined in §4.2 above.

By contrast with Proposition 4.3.2 both the symmetric and quadratic localization exact sequences for the Witt group of the 2-adic field $\hat{\mathbb{Q}}_2$

$$\begin{cases}0 \longrightarrow L^0(\hat{\mathbb{Z}}_2) \longrightarrow L^0(\hat{\mathbb{Q}}_2) \xrightarrow{\partial} L^0(\hat{\mathbb{Z}}_2,(\hat{2})^\infty) \longrightarrow 0 \\ 0 \longrightarrow L_0(\hat{\mathbb{Z}}_2) \longrightarrow L_0(\hat{\mathbb{Q}}_2) \xrightarrow{\partial} L_0(\hat{\mathbb{Z}}_2,(\hat{2})^\infty) \longrightarrow 0\end{cases}$$

split, with

$$L^0(\hat{\mathbb{Q}}_2) = L_0(\hat{\mathbb{Q}}_2) = \mathbb{Z}_8\oplus\mathbb{Z}_2\oplus\mathbb{Z}_2 ,$$

$$\begin{cases} L^0(\hat{\mathbb{Z}}_2,(\hat{2})^\infty) = L^0(\mathbb{Z},(2)^\infty) = L^0(\mathbb{Z}_2) = \mathbb{Z}_2 \ , \ L^0(\hat{\mathbb{Z}}_2) = \mathbb{Z}_8 \oplus \mathbb{Z}_2 \\ L_0(\hat{\mathbb{Z}}_2,(\hat{2})^\infty) = L_0(\mathbb{Z},(2)^\infty) = \mathbb{Z}_8 \oplus \mathbb{Z}_2 \ , \ L_0(\hat{\mathbb{Z}}_2) = L_0(\mathbb{Z}_2) = \mathbb{Z}_2 \ . \end{cases}$$

The element $\begin{cases} (1,0),(0,1) \in L^0(\hat{\mathbb{Z}}_2) = \mathbb{Z}_8 \oplus \mathbb{Z}_2 \\ 1 \in L_0(\hat{\mathbb{Z}}_2) = \mathbb{Z}_2 \end{cases}$ is represented by

the non-singular $\begin{cases} \text{symmetric} \\ \text{quadratic} \end{cases}$ form over the 2-adic ring $\hat{\mathbb{Z}}_2$

$$\begin{cases} (\hat{\mathbb{Z}}_2, 1 \in Q^{+1}(\hat{\mathbb{Z}}_2)) \quad , \quad (\hat{\mathbb{Z}}_2 \oplus \hat{\mathbb{Z}}_2, \begin{pmatrix} 2 & 1 \\ 1 & 2 \end{pmatrix} \in Q^{+1}(\hat{\mathbb{Z}}_2 \oplus \hat{\mathbb{Z}}_2)) \\ (\hat{\mathbb{Z}}_2 \oplus \hat{\mathbb{Z}}_2, \begin{pmatrix} 1 & 1 \\ 0 & 1 \end{pmatrix} \in Q_{+1}(\hat{\mathbb{Z}}_2 \oplus \hat{\mathbb{Z}}_2)) \end{cases} .$$

In identifying $L_0(\hat{\mathbb{Z}}_2) = L_0(\mathbb{Z}_2)$ we are dealing with a special case of the result of Wall [7] concerning reduction modulo a complete ideal in quadratic L-theory: if R is a ring with involution which is complete in the I-adic topology, i.e. such that the canonical map

$$R \longrightarrow \hat{R} = \varprojlim_k R/I^k$$

is an isomorphism, for some 2-sided ideal I in R such that $\bar{I} = I$, then the projection $R \longrightarrow R/I$ induces isomorphisms in the quadratic L-groups

$$L_*(R) \longrightarrow L_*(R/I)$$

- an L-theoretic version of Hensel's lemma. In particular, $\hat{\mathbb{Z}}_2$ is complete in the $(\hat{2})$-adic topology, with $(\hat{2}) = 2\hat{\mathbb{Z}}_2 \triangleleft \hat{\mathbb{Z}}_2$ and $\hat{\mathbb{Z}}_2/(\hat{2}) = \mathbb{Z}_2$.

Proposition 4.3.3 i) Reduction modulo a complete ideal fails for the symmetric L-groups, since

$$L^0(\hat{\mathbb{Z}}_2) = \mathbb{Z}_8 \oplus \mathbb{Z}_2 \neq L^0(\mathbb{Z}_2) = \mathbb{Z}_2 \ .$$

ii) Devissage fails for the quadratic L-groups, since

$$L_0(\hat{\mathbb{Z}}_2, (\hat{2})^\infty) = \mathbb{Z}_8 \oplus \mathbb{Z}_2 \neq L_0(\mathbb{Z}_2) = \mathbb{Z}_2 \ .$$

[]

The result of Proposition 4.3.3 i) is a direct consequence of the well-known failure of Hensel's lemma for symmetric forms at the prime 2, which is remedied by reducing modulo $(2)^3 = (8)$ instead of (2) - cf. Weyl [1,§III.5]. In particular, the natural map $L^0(\hat{\mathbb{Z}}_2) \longrightarrow L^0(\hat{\mathbb{Z}}_2/(\hat{2})^3) = L^0(\mathbb{Z}_8) = \mathbb{Z}_8 \oplus \mathbb{Z}_2$ is an isomorphism.

See Proposition 4.2.1 vi) for devissage in the symmetric L-groups of Dedekind rings.

§5. Polynomial extensions ($\bar{x} = x$)

We shall now study the L-theory of the polynomial extensions $A_\alpha[x]$, $A_\alpha[x,x^{-1}]$ of a ring with involution A, with

$$ax = x\alpha(a) \quad (a \in A)$$

for some ring automorphism $\alpha: A \longrightarrow A$ such that $\overline{\alpha(a)} = \alpha^{-1}(\bar{a}) \in A$ for all $a \in A$ (e.g. $\alpha = \text{id.} : A \longrightarrow A$), with the involution extended by

$$\bar{x} = x \ .$$

(See Ranicki [2],[3] and §7.6 below for the L-theory of polynomial extensions $A_\alpha[z,z^{-1}]$ with $\bar{z} = z^{-1}$). As usual, we start with a discussion of the relevant algebraic K-theory.

Given a central indeterminate x over a ring A let A[x] be the ring of polynomials $\sum_{j=0}^{\infty} a_j x^j$ in x with coefficients $a_j \in A$, only a finite number of which are to be non-zero. The central multiplicative subset

$$X = \{x^k \mid k \geqslant 0\} \subset A[x]$$

is then such that the localization

$$X^{-1}A[x] = A[x,x^{-1}]$$

is the ring of polynomials $\sum_{j=-\infty}^{\infty} a_j x^j$ in an invertible central indeterminate x with coefficients $a_j \in A$, only a finite number of which are to be non-zero. Bass, Heller and Swan [1] (for $n = 1$), Bass [2,XII] (for $n \leqslant 0$) and Quillen (for $n \geqslant 2$, cf. Grayson [1]) used the linearization trick of Higman [1] and the isomorphism of exact categories

$$\underline{\underline{\mathrm{Nil}}}(A) = (\text{f.g. projective } A\text{-modules } P \text{ with a nilpotent } A\text{-module morphism } \nu \in \mathrm{Hom}_A(P,P) \text{ (i.e. } \nu^k = 0 \text{ for some } k \geq 0))$$

$$\longrightarrow ((A[x],X)\text{-modules}) \ ; \ (P,\nu) \longmapsto (P, x = \nu : P \longrightarrow P)$$

to identify

$$K_n(A[x],X) = K_{n-1}(\underline{\underline{\mathrm{Nil}}}(A)) = K_{n-1}(A) \oplus \widetilde{\mathrm{Nil}}_n(A)$$

$$K_n(A[x]) = K_n(A) \oplus \widetilde{\mathrm{Nil}}_n(A) \qquad (n \in \mathbb{Z})$$

with $\widetilde{\mathrm{Nil}}_n(A) = K_{n-1}(\widetilde{\underline{\underline{\mathrm{Nil}}}}(A))$ and $\widetilde{\underline{\underline{\mathrm{Nil}}}}(A)$ the fibre of the forgetful functor of exact categories

$$\underline{\underline{\mathrm{Nil}}}(A) \longrightarrow \underline{\underline{P}}(A) = (\text{f.g. projective } A\text{-modules}) \ ; \ (P,\nu) \longmapsto P \ .$$

The algebraic K-theory localization exact sequence

$$\ldots \longrightarrow K_n(A[x]) \longrightarrow K_n(A[x,x^{-1}]) \xrightarrow{\partial} K_n(A[x],X) \longrightarrow K_{n-1}(A[x]) \longrightarrow \ldots$$

was shown to be made up of naturally split short exact sequences

$$0 \longrightarrow K_n(A) \oplus \widetilde{\mathrm{Nil}}_n(A) \longrightarrow K_n(A) \oplus K_{n-1}(A) \oplus \widetilde{\mathrm{Nil}}_n(A) \oplus \widetilde{\mathrm{Nil}}_n(A) \xrightarrow{\partial} K_{n-1}(A) \oplus \widetilde{\mathrm{Nil}}_n(A) \longrightarrow 0 \ ,$$

and the "fundamental theorem of algebraic K-theory" was proved, the naturally split exact sequences

$$0 \longrightarrow K_n(A) \longrightarrow K_n(A[x]) \oplus K_n(A[x^{-1}]) \longrightarrow K_n(A[x,x^{-1}]) \longrightarrow K_{n-1}(A) \longrightarrow 0 \ .$$

These results were extended for $n = 1$ to the α-twisted polynomial extensions $A_\alpha[x], A_\alpha[x,x^{-1}]$ of a ring A by Farrell and Hsiang [1],[2] and Siebenmann [1], with x no longer a central indeterminate over A but such that

$$ax = x\alpha(a) \quad (a \in A)$$

for some automorphism $\alpha : A \longrightarrow A$. (These results were obtained in connection with the obstruction theory of Farrell [1] for the problem of fibring a manifold over S^1, and the codimension 1 splitting obstruction theory of Farrell and Hsiang [1] (resp. [3]) for homotopy equivalences of finite CW complexes (resp. compact manifolds) with fundamental group $\pi\times_\alpha\mathbb{Z}$ the α-twisted extension of a group π by $\mathbb{Z}$ for some automorphism $\alpha:\pi\longrightarrow\pi$ - cf. the discussion of codimension 1 splitting in §7.6 below).

The multiplicative subset

$$X = \{x^k \mid k \geqslant 0\} \subset A_\alpha[x]$$

is such that

$$X^{-1}A_\alpha[x] = A_\alpha[x,x^{-1}] ,$$

and the functor

$$\underline{\underline{\mathrm{Nil}}}(A,\alpha) = (\text{pairs } (P,\nu) \text{ consisting of a f.g. projective } A\text{-module } P \text{ and a function } \nu:P\longrightarrow P \text{ such that } \nu(y+z) = \nu(y)+\nu(z),\ \nu(ay) = \alpha^{-1}(a)\nu(y)\in P\ (y,z\in P, a\in A) \text{ and } \nu^k = 0 : P\longrightarrow P \text{ for some } k\geqslant 1)$$

$$\longrightarrow ((A_\alpha[x],X)\text{-modules}) \ ;\ (P,\nu)\longmapsto(P, x = \nu : P\longrightarrow P)$$

is an isomorphism of categories, so that the algebraic K-theory eccentric localization exact sequence of Grayson [2]

$$\ldots\longrightarrow K_n(A_\alpha[x])\longrightarrow K_n(A_\alpha[x,x^{-1}])\xrightarrow{\partial} K_n(A_\alpha[x],X)$$
$$\longrightarrow K_{n-1}(A_\alpha[x])\longrightarrow\ldots \qquad (n\in\mathbb{Z})$$

is naturally isomorphic to the exact sequence

$$\ldots \longrightarrow K_n(A) \oplus \widetilde{Nil}_n(A,\alpha^{-1}) \xrightarrow{\begin{pmatrix} \beta & 0 \\ 0 & 0 \\ 0 & 1 \end{pmatrix}} K_n(A,\alpha) \oplus \widetilde{Nil}_n(A,\alpha) \oplus \widetilde{Nil}_n(A,\alpha^{-1})$$

$$\xrightarrow{\begin{pmatrix} \gamma & 0 & 0 \\ 0 & 1 & 0 \end{pmatrix}} K_{n-1}(A) \oplus \widetilde{Nil}_n(A,\alpha) \xrightarrow{\begin{pmatrix} 1-\alpha & 0 \\ 0 & 0 \end{pmatrix}} K_{n-1}(A) \oplus Nil_{n-1}(A,\alpha^{-1}) \longrightarrow \ldots ,$$

with $\widetilde{Nil}_n(A,\alpha) = K_{n-1}(\underline{\underline{\widetilde{Nil}}}(A,\alpha))$ $(n \in \mathbb{Z})$ the algebraic K-groups of the fibre $\underline{\underline{\widetilde{Nil}}}(A,\alpha)$ of the forgetful functor of exact categories

$$\underline{\underline{Nil}}(A,\alpha) \longrightarrow \underline{\underline{P}}(A) \ ; \ (P,\nu) \longmapsto P \quad ,$$

and $K_*(A,\alpha),\beta,\gamma$ the abelian groups and morphisms appearing in the exact sequence

$$\ldots \longrightarrow K_n(A) \xrightarrow{1-\alpha} K_n(A) \xrightarrow{\beta} K_n(A,\alpha) \xrightarrow{\gamma} K_{n-1}(A) \xrightarrow{1-\alpha} K_{n-1}(A) \longrightarrow \ldots .$$

In §5.1 we shall study the algebraic L-groups of the α-twisted polynomial extensions $A_\alpha[x]$, $A_\alpha[x,x^{-1}]$ of a ring with involution A, with $\alpha:A \longrightarrow A$ a ring automorphism such that $\overline{\alpha(a)} = \alpha^{-1}(\bar{a}) \in A$ $(a \in A)$ and $\bar{x} = x \in A_\alpha[x]$, so that

$$X = \{x^k \,|\, k \geqslant 0\} \subset A_\alpha[x]$$

is a multiplicative susbset in the sense of §3.1 and

$$X^{-1}A_\alpha[x] = A_\alpha[x,x^{-1}]$$

as a ring with involution. We shall show that the $\begin{cases} \varepsilon\text{-symmetric} \\ \varepsilon\text{-quadratic} \end{cases}$ L-theory localization exact sequence given for $(A_\alpha[x],X)$ by §3.6

$$\left\{\begin{array}{l} \ldots \longrightarrow L^n_K(A_\alpha[x],\varepsilon) \longrightarrow L^n_K(A_\alpha[x,x^{-1}],\varepsilon) \xrightarrow{\partial} L^n(A_\alpha[x],X,\varepsilon) \\ \qquad\qquad\qquad\qquad\qquad\qquad \longrightarrow L^{n-1}_K(A_\alpha[x],\varepsilon) \longrightarrow \ldots \\ \ldots \longrightarrow L_n^K(A_\alpha[x],\varepsilon) \longrightarrow L_n^K(A_\alpha[x,x^{-1}],\varepsilon) \xrightarrow{\partial} L_n(A_\alpha[x],X,\varepsilon) \\ \qquad\qquad\qquad\qquad\qquad\qquad \longrightarrow L_{n-1}^K(A_\alpha[x],\varepsilon) \longrightarrow \ldots \end{array}\right.$$

$$(K = \mathrm{im}(\tilde{K}_0(A) \longrightarrow \tilde{K}_0(B)) \subseteq \tilde{K}_0(B),\ B = A_\alpha[x], A_\alpha[x,x^{-1}],\ n \in \mathbb{Z})$$

is made up of naturally split short exact sequences

$$\left\{\begin{array}{l} 0 \longrightarrow L^n_K(A_\alpha[x],\varepsilon) \longrightarrow L^n_K(A_\alpha[x,x^{-1}],\varepsilon) \xrightarrow{\partial} L^n(A_\alpha[x],X,\varepsilon) \longrightarrow 0 \\ 0 \longrightarrow L_n^K(A_\alpha[x],\varepsilon) \longrightarrow L_n^K(A_\alpha[x,x^{-1}],\varepsilon) \xrightarrow{\partial} L_n(A_\alpha[x],X,\varepsilon) \longrightarrow 0 \end{array}\right. \quad (n \in \mathbb{Z})$$

(by contrast with the corresponding localization exact sequence in algebraic K-theory, which need not split if $\alpha \neq \mathrm{id.}$). Furthermore, we shall show that each of these short exact sequences is naturally isomorphic to

$$\left\{\begin{array}{l} 0 \longrightarrow L^n(A,\varepsilon) \oplus \widetilde{LNil}^n(A,\alpha^{-1},\varepsilon) \xrightarrow{\begin{pmatrix} 1 & 0 \\ 0 & 0 \\ 0 & 0 \\ 0 & 1 \end{pmatrix}} \\ \qquad L^n(A,\varepsilon) \oplus L^n(A^\alpha,\varepsilon) \oplus \widetilde{LNil}^n(A,\alpha,\varepsilon) \oplus \widetilde{LNil}^n(A,\alpha^{-1},\varepsilon) \\ \xrightarrow{\begin{pmatrix} 0 & 1 & 0 & 0 \\ 0 & 0 & 1 & 0 \end{pmatrix}} L^n(A^\alpha,\varepsilon) \oplus \widetilde{LNil}^n(A,\alpha,\varepsilon) \longrightarrow 0 \\ \text{(the } \varepsilon\text{-quadratic analogue)}, \end{array}\right.$$

with the L-groups $\left\{\begin{array}{l} \widetilde{LNil}^*(A,\alpha,\varepsilon) \\ \widetilde{LNil}_*(A,\alpha,\varepsilon) \end{array}\right.$ cobordism groups of chain complexes in $\underline{\underline{\mathrm{Nil}}}(A,\alpha)$ with an α-twisted $\left\{\begin{array}{l} \varepsilon\text{-symmetric} \\ \varepsilon\text{-quadratic} \end{array}\right.$ Poincaré duality, and A^α the ring with involution defined by giving the ring underlying A the involution

$$\bar{\ } : A^{\alpha} \longrightarrow A^{\alpha} \ ; \ a \longmapsto \alpha(\bar{a}) \quad .$$

In §5.2 the results of §5.1 will be extended to more general intermediate L-groups of the α-twisted polynomial extensions $A_{\alpha}[x]$, $A_{\alpha}[x,x^{-1}]$ of a ring with involution A. (In §7.6 we shall outline a geometric interpretation of an appropriately intermediate version of the decomposition

$$L^K_*(A_{\alpha}[x,x^{-1}]) = L_*(A) \oplus L_*(A^{\alpha}) \oplus \widetilde{LNil}_*(A,\alpha) \oplus \widetilde{LNil}_*(A,\alpha^{-1}) \quad (\varepsilon = 1)$$

for a group ring $A = \mathbb{Z}[\pi]$). In particular, in the untwisted case $\alpha = \text{id.} : A \longrightarrow A$ there will be obtained the "fundamental theorem of $\begin{cases} \varepsilon\text{-symmetric} \\ \varepsilon\text{-quadratic} \end{cases}$ L-theory", the naturally split exact sequence

$$\begin{cases} 0 \longrightarrow V^n(A,\varepsilon) \longrightarrow V^n(A[x],\varepsilon) \oplus V_n(A[x^{-1}],\varepsilon) \\ \qquad \longrightarrow V^n(A[x,x^{-1}],\varepsilon) \longrightarrow U^n(A,\varepsilon) \longrightarrow 0 \\ 0 \longrightarrow V_n(A,\varepsilon) \longrightarrow V_n(A[x],\varepsilon) \oplus V_n(A[x^{-1}],\varepsilon) \\ \qquad \longrightarrow V_n(A[x,x^{-1}],\varepsilon) \longrightarrow U_n(A,\varepsilon) \longrightarrow 0 \end{cases} \quad (n \in \mathbb{Z})$$

relating the free $\begin{cases} \varepsilon\text{-symmetric} \\ \varepsilon\text{-quadratic} \end{cases}$ L-groups $\begin{cases} V^*(A,\varepsilon) \\ V_*(A,\varepsilon) \end{cases}$ (as defined in §1.10) to the projective $\begin{cases} \varepsilon\text{-symmetric} \\ \varepsilon\text{-quadratic} \end{cases}$ L-groups $\begin{cases} U^*(A,\varepsilon) = L^*(A,\varepsilon) \\ U_*(A,\varepsilon) = L_*(A,\varepsilon) \end{cases}$.

The ε-quadratic L-theory fundamental theorem was obtained by Karoubi [2],[3] for $1/2 \in A$ (when $\widetilde{LNil}_*(A) = 0$ by Karoubi [1]) using localization, and by Ranicki [4] using the techniques developed by Novikov [1] and Ranicki [2] in the proof of the splitting theorem $V_n(A[z,z^{-1}]) = V_n(A) \oplus U_{n-1}(A)$ $(n \in \mathbb{Z}, \bar{z} = z^{-1})$.

5.1 L-theory of polynomial extensions

Let A be a ring with involution, and let

$$\alpha : A \longrightarrow A$$

be a ring automorphism such that

$$\overline{\alpha(a)} = \alpha^{-1}(\bar{a}) \in A \quad (a \in A) .$$

The *α-twisted polynomial extension of A* $A_\alpha[x]$ is the ring of finite polynomials $\sum_{j=0}^{\infty} a_j x^j$ in an indeterminate x over A such that

$$ax = x\alpha(a) \quad (a \in A) ,$$

with the involution on A extended to $A_\alpha[x]$ by

$$\bar{x} = x .$$

Thus addition and multiplication in $A_\alpha[x]$ are given by

$$\sum_{j=0}^{\infty} a_j x^j + \sum_{j=0}^{\infty} b_j x^j = \sum_{j=0}^{\infty} (a_j+b_j) x^j$$

$$(\sum_{j=0}^{\infty} a_j x^j)(\sum_{k=0}^{\infty} b_k x^k) = \sum_{j=0}^{\infty}\sum_{k=0}^{\infty} a_j \alpha^{-k}(b_k) x^{j+k} ,$$

and the involution is given by

$$\overline{(\sum_{j=0}^{\infty} a_j x^j)} = \sum_{j=0}^{\infty} \alpha^{-j}(\bar{a}_j) x^j \qquad (a_j, b_j \in A) .$$

Define the multiplicative subset

$$X = \{x^k \mid k \geqslant 0\} \subset A_\alpha[x] .$$

The localization away from X

$$X^{-1}A_\alpha[x] = A_\alpha[x,x^{-1}]$$

is the *α-twisted Laurent polynomial extension of A*, the ring of finite polynomials $\sum_{j=-\infty}^{\infty} a_j x^j$ $(a_j \in A)$ with involution by $\bar{x} = x$, containing $A_\alpha[x]$ as a subring with involution.

Given an A-module M and $j \in \mathbb{Z}$ let x^jM be the A-module with elements x^jy $(y \in M)$, addition by

$$x^jy + x^jy' = x^j(y+y') \in x^jM \quad (y,y' \in M)$$

and A acting by

$$a(x^jy) = x^j(\alpha^j(y)) \in x^jM \quad (y \in M) .$$

(The automorphism of the projective class group $K_0(A)$ induced by the ring automorphism $\alpha^j : A \longrightarrow A$ is given by

$$\alpha^j : K_0(A) \longrightarrow K_0(A) \; ; \; [M] \longmapsto [x^{-j}M] \;).$$

An A-module morphism $f \in \mathrm{Hom}_A(x^jM, x^kN)$ (for some A-modules M,N and $j,k \in \mathbb{Z}$) is a function

$$f : M \longrightarrow N$$

such that

$$\text{i)} \; f(y+y') = f(y) + f(y') \in N$$

$$\text{ii)} \; f(ay) = \alpha^{k-j}(a)f(y) \in N$$

$$(a \in A, \; y,y' \in M)$$

with

$$f : x^jM \longrightarrow x^kN \; ; \; x^jy \longmapsto x^kf(y) .$$

In particular, there is defined an A-module isomorphism

$$(x^jM)^* = \mathrm{Hom}_A(x^jM, A) \longrightarrow x^{-j}(M^*) = x^{-j}\mathrm{Hom}_A(M,A) \; ;$$

$$(f : x^jM \longrightarrow A) \longmapsto x^{-j}(y \longmapsto \alpha^{-j}(f(x^jy))) \quad (y \in M)$$

with inverse

$$x^{-j}(M^*) \longrightarrow (x^jM)^* \; ; \; x^{-j}g \longmapsto (x^jy \longmapsto \alpha^j(g(y))) \quad (g \in M^*, y \in M) .$$

We shall write $x^{-j}M$ as Mx^j,

$$x^{-j}M = Mx^j \quad (j \in \mathbb{Z}) .$$

With this terminology there is a natural identification of A-modules

$$(x^jM)^* = M^*x^j \quad (j \in \mathbb{Z}) .$$

For any A-module M the induced $A_\alpha[x]$-module

$$M_\alpha[x] = A_\alpha[x]\otimes_A M$$

consists of finite polynomials $\sum_{j=0}^{\infty} x^j y_j$ $(y_j \in M)$. As an A-module it can be expressed as a direct sum

$$M_\alpha[x] = \sum_{j=0}^{\infty} x^j M \quad .$$

For any A-modules M,N there is a natural identification of abelian groups

$$\mathrm{Hom}_{A_\alpha[x]}(M_\alpha[x],N_\alpha[x]) = \sum_{j=0}^{\infty} \mathrm{Hom}_A(M,x^j N) \quad .$$

Similarly, the induced $A_\alpha[x,x^{-1}]$-module

$$M_\alpha[x,x^{-1}] = A_\alpha[x,x^{-1}]\otimes_A M$$

consists of finite polynomials $\sum_{j=-\infty}^{\infty} x^j y_j$ $(y_j \in M)$, it can be expressed as a direct sum of A-modules

$$M_\alpha[x,x^{-1}] = \sum_{j=-\infty}^{\infty} x^j M \ ,$$

and there is a natural identification

$$\mathrm{Hom}_{A_\alpha[x,x^{-1}]}(M_\alpha[x,x^{-1}],N_\alpha[x,x^{-1}]) = \sum_{j=-\infty}^{\infty} \mathrm{Hom}_A(M,x^j N) \quad .$$

An A-module morphism $\nu \in \mathrm{Hom}_A(M,Mx)$ is <u>nilpotent</u> if the composite A-module morphism

$$\nu^k : M \xrightarrow{\nu} Mx \xrightarrow{\nu} Mx^2 \longrightarrow \ldots \longrightarrow Mx^{k-1} \xrightarrow{\nu} Mx^k$$

is 0, $\nu^k = 0 \in \mathrm{Hom}_A(M,Mx^k)$, for some $k \geqslant 1$. (An A-module morphism $f \in \mathrm{Hom}_A(M,N)$ induces A-module morphisms $f \in \mathrm{Hom}_A(x^j M,x^j N)$ $(j \in \mathbb{Z})$ by

$$f : x^j M \longrightarrow x^j N \ ; \ x^j y \longmapsto x^j f(y) \quad (y \in M) \ ).$$

Equivalently, ν can be regarded as a function $\nu : M \longrightarrow M$ such that

$$\text{i)} \ \nu(y+y') = \nu(y) + \nu(y')$$

$$\text{ii)} \ \nu(ay) = \alpha^{-1}(a)\nu(y)$$

$$\text{iii)} \ \nu^k(y) = 0 \text{ for some } k \geqslant 1$$

$$(a \in A, \ y, y' \in M) \ .$$

An <u>α-twisted nilmodule over A</u> is a pair

(f.g. projective A-module M, nilpotent morphism $\nu \in \mathrm{Hom}_A(M, Mx)$) .

A <u>morphism</u> of α-twisted nilmodules over A

$$f : (M,\nu) \longrightarrow (M',\nu')$$

is an A-module morphism $f \in \mathrm{Hom}_A(M,M')$ such that there is defined a commutative diagram

$$\begin{array}{ccc} M & \xrightarrow{\ f\ } & M' \\ {\scriptstyle\nu}\big\downarrow & & \big\downarrow{\scriptstyle\nu'} \\ Mx & \xrightarrow{\ f\ } & M'x \end{array} \ .$$

Define the <u>duality</u> involution on the category of α-twisted nilmodules over A

$$* : \underline{\underline{\mathrm{Nil}}}(A,\alpha) = (\alpha\text{-twisted nilmodules over } A) \longrightarrow \underline{\underline{\mathrm{Nil}}}(A,\alpha) \ ;$$

$$(M,\nu) \longmapsto (M,\nu)^* = (\ (Mx)^* = x(M^*), \nu^* : (Mx)^* \longrightarrow (Mx)^*x = M) \ .$$

An <u>n-dimensional α-twisted nilcomplex over A</u> (C,ν) is an n-dimensional chain complex of α-twisted nilmodules over A

$$(C,\nu) : (C_n,\nu) \xrightarrow{\ d\ } (C_{n-1},\nu) \longrightarrow \dots \longrightarrow (C_1,\nu) \xrightarrow{\ d\ } (C_0,\nu) \ .$$

Equivalently, we have that C is an n-dimensional A-module chain complex such that $C_r = 0$ for $r < 0$ and $r > n$ together with a nilpotent A-module chain map

$$\nu : C \longrightarrow Cx \ .$$

Note that (C^{n-*},ν^*) is also an n-dimensional α-twisted nilcomplex over A.

Proposition 5.1.1 i) There is a natural isomorphism of exact categories

$$((A_\alpha[x],X)\text{-modules}) \longrightarrow \underline{\underline{\text{Nil}}}(A,\alpha) \ ;$$

$$M \longmapsto (M,\nu : M \longrightarrow Mx = x^{-1}M \ ; \ y \longmapsto x^{-1}(xy)) \quad (y,xy \in M)$$

under which the X-duality involution

$$((A_\alpha[x],X)\text{-modules}) \longrightarrow ((A_\alpha[x],X)\text{-modules}) \ ;$$

$$M \longmapsto M\hat{} = \text{Hom}_{A_\alpha[x]}(M,A_\alpha[x,x^{-1}]/A_\alpha[x])$$

corresponds to the duality involution on the category of α-twisted nilmodules over A

$$* : \underline{\underline{\text{Nil}}}(A,\alpha) \longrightarrow \underline{\underline{\text{Nil}}}(A,\alpha) \ ; \ (M,\nu) \longmapsto (M,\nu)^* = (xM^*,\nu^*) \ ,$$

with a natural $A_\alpha[x]$-module isomorphism

$$xM^* = \text{Hom}_A(Mx,A) \longrightarrow M\hat{} = \text{Hom}_{A_\alpha[x]}(M,A_\alpha[x,x^{-1}]/A_\alpha[x]) \ ;$$

$$g \longmapsto (y \longmapsto \sum_{j=-\infty}^{-1} x^j\alpha(g(\nu^{-j-1}yx))) \qquad (y \in M) \ .$$

ii) For each $n \geqslant 0$ there is a natural identification of exact categories

(n-dimensional $(A_\alpha[x],X)$-module chain complexes)

= (n-dimensional α-twisted nilcomplexes over A) .

Proof: i) This isomorphism was first established by Bass [2,XII.6.4] in the untwisted case $\alpha = \text{id.} : A \longrightarrow A$. (See also Proposition 3.10 of Karoubi [2]). The extension to the α-twisted case is due to Farrell and Hsiang [1].

In particular, an α-twisted nilmodule over A (M,ν) determines an $(A_\alpha[x],X)$-module M by

$$A_\alpha[x] \times M \longrightarrow M \; ; \; (\sum_{j=0}^{\infty} a_j x^j, y) \longmapsto \sum_{j=0}^{\infty} a_j \nu^j(y) \; ,$$

with a canonical f.g. projective $A_\alpha[x]$-module resolution

$$0 \longrightarrow (xM)_\alpha[x,x^{-1}] \xrightarrow{x-\nu} M_\alpha[x,x^{-1}] \xrightarrow{h} M \longrightarrow 0 \quad ,$$

where

$$x - \nu : (xM)_\alpha[x,x^{-1}] \longrightarrow M_\alpha[x,x^{-1}] \; ;$$
$$\sum_{j=0}^{\infty} x^j(xy_j) \longmapsto \sum_{j=0}^{\infty} (x^{j+1}y_j - x^j(\nu(y_j)))$$

$$h : M_\alpha[x,x^{-1}] \longrightarrow M \; ; \; \sum_{j=0}^{\infty} x^j y_j \longmapsto \sum_{j=0}^{\infty} \nu^j(y_j)$$

$$(y_j \in M) \; .$$

ii) Immediate from i).

[]

We shall now use the identifications of Proposition 5.1.1 to express the relative $\begin{cases} \varepsilon\text{-symmetric} \\ \varepsilon\text{-quadratic} \end{cases}$ L-groups $\begin{cases} L^*(A_\alpha[x],X,\varepsilon) \\ L_*(A_\alpha[x],X,\varepsilon) \end{cases}$ appearing in the localization exact sequence of Proposition 3.6.1

$$\begin{cases} \ldots \longrightarrow L^n(A_\alpha[x],\varepsilon) \longrightarrow L^n_X(A_\alpha[x,x^{-1}],\varepsilon) \xrightarrow{\partial} L^n(A_\alpha[x],X,\varepsilon) \\ \qquad\qquad \longrightarrow L^{n-1}(A_\alpha[x],\varepsilon) \longrightarrow \ldots \\ \ldots \longrightarrow L_n(A_\alpha[x],\varepsilon) \longrightarrow L^X_n(A_\alpha[x,x^{-1}],\varepsilon) \xrightarrow{\partial} L_n(A_\alpha[x],X,\varepsilon) \\ \qquad\qquad \longrightarrow L_{n-1}(A_\alpha[x],\varepsilon) \longrightarrow \ldots \end{cases}$$
$$(n \in \mathbb{Z})$$

as the L-groups $\begin{cases} LNil^*(A,\alpha,\varepsilon) \\ LNil_*(A,\alpha,\varepsilon) \end{cases}$ of finite-dimensional α-twisted nilcomplexes over A with an $\begin{cases}\varepsilon\text{-symmetric} \\ \varepsilon\text{-quadratic}\end{cases}$ Poincaré duality structure, and to prove that the maps ∂ are split surjections.

In dealing with $\begin{cases}\varepsilon\text{-symmetric} \\ \varepsilon\text{-quadratic}\end{cases}$ complexes over the α-twisted polynomial extensions $A_\alpha[x], A_\alpha[x,x^{-1}]$ we shall assume that

$$\alpha(\varepsilon) = \varepsilon \in A\ .$$

This is automatically the case if $\varepsilon = \pm 1 \in A$, for example.

Define the <u>α-twisted</u> $\begin{cases}\underline{\varepsilon\text{-symmetric}} \\ \underline{\varepsilon\text{-quadratic}}\end{cases}$ <u>Q-groups</u> $\begin{cases}Q^*(C,\alpha,\varepsilon) \\ Q_*(C,\alpha,\varepsilon)\end{cases}$ of a finite-dimensional A-module chain complex C to be

$$\begin{cases} Q^n(C,\alpha,\varepsilon) = H_n(\mathrm{Hom}_{\mathbb{Z}[\mathbb{Z}_2]}(W,\mathrm{Hom}_A(C^*,xC))) \\ Q_n(C,\alpha,\varepsilon) = H_n(W\otimes_{\mathbb{Z}[\mathbb{Z}_2]}\mathrm{Hom}_A(C^*,xC)) \end{cases} \quad (n \in \mathbb{Z})\ ,$$

with $T \in \mathbb{Z}_2$ acting on $\mathrm{Hom}_A(C^*,xC)$ by the <u>α-twisted ε-duality</u> involution

$$\begin{aligned} T_{\alpha,\varepsilon}\ :\ \mathrm{Hom}_A(C^p,xC_q) &\longrightarrow \mathrm{Hom}_A((xC_q)^*,C_p) = \mathrm{Hom}_A(C^q,xC_p)\ ; \\ \phi &\longmapsto (-)^{pq}\varepsilon\phi^*\ . \end{aligned}$$

An <u>n-dimensional α-twisted</u> $\begin{cases}\underline{\varepsilon\text{-symmetric}} \\ \underline{\varepsilon\text{-quadratic}}\end{cases}$ <u>complex over A</u> $\begin{cases}(C,\phi) \\ (C,\psi)\end{cases}$ is an n-dimensional A-module chain complex C together with an element $\begin{cases}\phi \in Q^n(C,\alpha,\varepsilon) \\ \psi \in Q_n(C,\alpha,\varepsilon)\end{cases}$. Such a complex is <u>Poincaré</u> if the A-module chain map

$$\begin{cases} \phi_0\ :\ C^{n-*} \longrightarrow xC \\ (1+T_{\alpha,\varepsilon})\psi_0\ :\ C^{n-*} \longrightarrow xC \end{cases}$$

is a chain equivalence, inducing A-module isomorphisms

$$H^{n-*}(C) \xrightarrow{\sim} H_*(xC) = xH_*(C) \ .$$

The <u>n-dimensional α-twisted</u> $\left\{\begin{array}{l}\underline{\varepsilon\text{-symmetric}}\\ \underline{\varepsilon\text{-quadratic}}\end{array}\right.$ <u>L-group of A</u>

$\left\{\begin{array}{l}L^n(A,\alpha,\varepsilon)\\ L_n(A,\alpha,\varepsilon)\end{array}\right.$ $(n \geqslant 0)$ is the cobordism group of n-dimensional

α-twisted $\left\{\begin{array}{l}\varepsilon\text{-symmetric}\\ \varepsilon\text{-quadratic}\end{array}\right.$ Poincaré complexes over A. Note that

there are defined isomorphisms

$$\left\{\begin{array}{l}L^n(A,\alpha,\varepsilon) \xrightarrow{\sim} L^n(A,\alpha^{-1},\varepsilon) \ ; \ (C,\phi) \longmapsto (C^{n-*},(\phi_0{}^{-1})^{\%}(\phi))\\ L_n(A,\alpha,\varepsilon) \xrightarrow{\sim} L_n(A,\alpha^{-1},\varepsilon) \ ; \ (C,\psi) \longmapsto (C^{n-*},((1+T_{\alpha,\varepsilon})\psi_0{}^{-1})_{\%}(\psi))\end{array}\right.$$

$$(n \geqslant 0) \ .$$

In the untwisted case $\alpha = 1 : A \longrightarrow A$ the canonical isomorphism of A-module chain complexes

$$xC \longrightarrow C \ ; \ xy \longmapsto y$$

can be used to identify

$$\mathrm{Hom}_A(C^*,xC) = \mathrm{Hom}_A(C^*,C) \ ,$$

$$\left\{\begin{array}{l}Q^*(C,1,\varepsilon) = Q^*(C,\varepsilon)\\ Q_*(C,1,\varepsilon) = Q_*(C,\varepsilon) \ ,\end{array}\right.$$

$$\left\{\begin{array}{l}L^*(A,1,\varepsilon) = L^*(A,\varepsilon)\\ L_*(A,1,\varepsilon) = L_*(A,\varepsilon) \ .\end{array}\right.$$

Let A^α denote the ring with involution with the ring structure of A, but with the involution

$$\bar{\ } : A^\alpha \longrightarrow A^\alpha \ ; \ a \longmapsto \alpha(\bar{a}) \ .$$

Given an A-module M let M^α denote M regarded as an A^α-module, so that $M^{\alpha t}$ is the right A^α-module with the additive group of M and A^{α} acting by

$$M^{\alpha t} \times A^{\alpha} \longrightarrow M^{\alpha t} \; ; \; (y,a) \longmapsto \alpha(\bar{a})y \; ,$$

and $M^{\alpha *}$ is the A^{α}-module with additive group $\mathrm{Hom}_A(M,A)$ and A^{α} acting by

$$A^{\alpha} \times M^{\alpha *} \longrightarrow M^{\alpha *} \; ; \; (a,f) \longmapsto (y \longmapsto f(y)\alpha(\bar{a})) \; .$$

<u>Proposition 5.1.2</u> There are natural identifications of L-groups

$$\begin{cases} L^n(A,\alpha,\varepsilon) = L^n(A^{\alpha},\varepsilon) \\ L_n(A,\alpha,\varepsilon) = L_n(A^{\alpha},\varepsilon) \end{cases} \quad (n \geqslant 0) \; .$$

<u>Proof</u>: For any finite-dimensional A-module chain complexes C,D there is defined an isomorphism of $\mathbb{Z}$-module chain complexes

$$C^t \otimes_A D = C \otimes_{\mathbb{Z}} D/\{y \otimes az - \bar{a}y \otimes z \,|\, a \in A, y \in C, z \in D\} \longrightarrow \mathrm{Hom}_A(C^*,D) \; ;$$
$$u \otimes v \longmapsto (f \longmapsto \overline{f(u)}v)$$

which can be used to identify

$$C^t \otimes_A D = \mathrm{Hom}_A(C^*,D) \; .$$

In particular, for D = Cx there are identifications of $\mathbb{Z}$-module chain complexes

$$\begin{aligned} \mathrm{Hom}_A(C^*,Cx) &= C \otimes_{\mathbb{Z}} C/\{y \otimes \alpha^{-1}(a)z - \bar{a}y \otimes z \,|\, a \in A,\, y,z \in C\} \\ &= C \otimes_{\mathbb{Z}} C/\{y \otimes az - \alpha(\bar{a})y \otimes z \,|\, a \in A,\, y,z \in C\} \\ &= \mathrm{Hom}_{A^{\alpha}}(C^{\alpha *},C^{\alpha}) \; . \end{aligned}$$

Furthermore, the α-twisted ε-duality involution $T_{\alpha,\varepsilon}$ on $\mathrm{Hom}_A(C^*,Cx)$ can be identified with the ε-duality involution T_{ε} on $\mathrm{Hom}_{A^{\alpha}}(C^{\alpha *},C^{\alpha})$, so that

$$\begin{cases} Q^n(C,\alpha,\varepsilon) = Q^n(C^{\alpha},\varepsilon) \\ Q_n(C,\alpha,\varepsilon) = Q_n(C^{\alpha},\varepsilon) \end{cases} \quad (n \geqslant 0)$$

and similarly for the L-groups.

[]

Given a finite-dimensional α-twisted nilcomplex over A $(C,\nu:C \longrightarrow Cx)$ define a $\mathbb{Z}[\mathbb{Z}_2]$-module chain map

$$\Gamma_\nu : \mathrm{Hom}_A(C^*,xC) \longrightarrow \overline{\mathrm{Hom}}_A(C^*,C) \; ; \; \phi \longmapsto \nu\phi - \phi\nu^* \; ,$$

with $T \in \mathbb{Z}_2$ acting on $\mathrm{Hom}_A(C^*,xC)$ by $T_{\alpha,\varepsilon}$ (as above) and $\overline{\mathrm{Hom}}_A(C^*,C)$ the $\mathbb{Z}[\mathbb{Z}_2]$-module chain complex defined by the $\mathbb{Z}$-module chain complex $\mathrm{Hom}_A(C^*,C)$ with $T \in \mathbb{Z}_2$ acting by the $(-\varepsilon)$-duality involution $T_{-\varepsilon}$. Define the $\begin{cases} \underline{\varepsilon\text{-symmetric}} \\ \underline{\varepsilon\text{-quadratic}} \end{cases}$ $\underline{\text{QNil-groups of } (C,\nu)}$

$\begin{cases} \mathrm{QNil}^*(C,\nu,\varepsilon) \\ \mathrm{QNil}_*(C,\nu,\varepsilon) \end{cases}$ to be the relative groups appearing in the long exact sequence of abelian groups

$$\begin{cases} \dots \longrightarrow Q^{n+1}(C,-\varepsilon) \longrightarrow \mathrm{QNil}^n(C,\nu,\varepsilon) \longrightarrow Q^n(C,\alpha,\varepsilon) \xrightarrow{\Gamma_\nu} Q^n(C,-\varepsilon) \longrightarrow \dots \\ \dots \longrightarrow Q_{n+1}(C,-\varepsilon) \longrightarrow \mathrm{QNil}_n(C,\nu,\varepsilon) \longrightarrow Q_n(C,\alpha,\varepsilon) \xrightarrow{\Gamma_\nu} Q_n(C,-\varepsilon) \longrightarrow \dots \end{cases} \quad (n \in \mathbb{Z})$$

An element $\begin{cases} (\delta\phi,\phi) \in \mathrm{QNil}^n(C,\nu,\varepsilon) \\ (\delta\psi,\psi) \in \mathrm{QNil}_n(C,\nu,\varepsilon) \end{cases}$ is an equivalence class of collections of A-module morphisms

$$\begin{cases} \{(\delta\phi_s,\phi_s) \in \mathrm{Hom}_A(C^{n-r+s+1},C_r) \oplus \mathrm{Hom}_A(C^{n-r+s},xC_r) \,|\, r \in \mathbb{Z}, s \geqslant 0\} \\ \{(\delta\psi_s,\psi_s) \in \mathrm{Hom}_A(C^{n-r-s+1},C_r) \oplus \mathrm{Hom}_A(C^{n-r-s},xC_r) \,|\, r \in \mathbb{Z}, s \geqslant 0\} \end{cases}$$

such that

$$\left\{\begin{array}{l}\left\{\begin{array}{l} d\phi_s + (-)^r \phi_s d^* + (-)^{n+s-1}(\phi_{s-1} + (-)^s T_{\alpha,\varepsilon}\phi_{s-1}) = 0 \\ \qquad : C^{n-r+s-1} \longrightarrow xC_r \\ d(\delta\phi_s) + (-)^r(\delta\phi_s)d^* + (-)^{n+s}(\delta\phi_{s-1} + (-)^s T_{-\varepsilon}\delta\phi_{s-1}) \\ \qquad + (-)^n(\nu\phi_s - \phi_s\nu^*) = 0 : C^{n-r+s} \longrightarrow C_r \\ \qquad\qquad (s \geqslant 0,\ \phi_{-1} = 0\ ,\ \delta\phi_{-1} = 0) \end{array}\right. \\ \left\{\begin{array}{l} d\psi_s + (-)^r \psi_s d^* + (-)^{n-s-1}(\psi_{s+1} + (-)^{s+1} T_{\alpha,\varepsilon}\psi_{s+1}) = 0 \\ \qquad : C^{n-r-s+1} \longrightarrow xC_r \\ d(\delta\psi_s) + (-)^r(\delta\psi_s)d^* + (-)^{n-s}(\delta\psi_{s+1} + (-)^{s+1} T_{-\varepsilon}\delta\psi_{s+1}) \\ \qquad + (-)^n(\nu\psi_s - \psi_s\nu^*) = 0 : C^{n-r-s} \longrightarrow C_r \\ \qquad\qquad (s \geqslant 0) \end{array}\right. \end{array}\right.$$

An <u>n-dimensional α-twisted</u> $\left\{\begin{array}{l}\underline{\varepsilon\text{-symmetric}} \\ \underline{\varepsilon\text{-quadratic}}\end{array}\right.$ <u>nilcomplex over A</u> $\left\{\begin{array}{l}(C,\nu,\delta\phi,\phi) \\ (C,\nu,\delta\psi,\psi)\end{array}\right.$ $(n \geqslant 0)$ is an n-dimensional α-twisted nilcomplex over A (C,ν) together with an element $\left\{\begin{array}{l}(\delta\phi,\phi) \in QNil^n(C,\nu,\varepsilon) \\ (\delta\psi,\psi) \in QNil_n(C,\nu,\varepsilon)\end{array}\right.$. Such a complex is <u>Poincaré</u> if $\left\{\begin{array}{l}(C,\phi \in Q^n(C,\alpha,\varepsilon)) \\ (C,\psi \in Q_n(C,\alpha,\varepsilon))\end{array}\right.$ is an n-dimensional α-twisted $\left\{\begin{array}{l}\varepsilon\text{-symmetric} \\ \varepsilon\text{-quadratic}\end{array}\right.$ Poincaré complex over A.

The <u>n-dimensional α-twisted</u> $\left\{\begin{array}{l}\underline{\varepsilon\text{-symmetric}} \\ \underline{\varepsilon\text{-quadratic}}\end{array}\right.$ <u>LNil-group of A</u> $\left\{\begin{array}{l}LNil^n(A,\alpha,\varepsilon) \\ LNil_n(A,\alpha,\varepsilon)\end{array}\right.$ is the cobordism group of n-dimensional $\left\{\begin{array}{l}\varepsilon\text{-symmetric} \\ \varepsilon\text{-quadratic}\end{array}\right.$ Poincaré nilcomplexes over A $(n \geqslant 0)$.

<u>Proposition 5.1.3</u> i) For each $n \geqslant 0$ there is a natural identification of categories

$$(\text{n-dimensional} \begin{cases} \text{(even) } \varepsilon\text{-symmetric} \\ \varepsilon\text{-quadratic} \end{cases} \text{(Poincaré) complexes over } (A_\alpha[x],X))$$

$$= (\text{n-dimensional } \alpha\text{-twisted} \begin{cases} \varepsilon\text{-symmetric} \\ \varepsilon\text{-quadratic} \end{cases} \text{(Poincaré) nilcomplexes over A)} \ .$$

ii) There are natural identifications of L-groups

$$\begin{cases} L^n(A_\alpha[x],X,\varepsilon) = LNil^n(A,\alpha,\varepsilon) \\ L_n(A_\alpha[x],X,\varepsilon) = LNil_n(A,\alpha,\varepsilon) \end{cases} \quad (n \geqslant 0) \ .$$

iii) The $\begin{cases} \varepsilon\text{-symmetric} \\ \varepsilon\text{-quadratic} \end{cases}$ L-theory localization exact sequence of $(A_\alpha[x],X)$ in the range $n \geqslant 0$ is made up of naturally split short exact sequences

$$\begin{cases} 0 \longrightarrow L^n(A_\alpha[x],\varepsilon) \longrightarrow L^n_X(A_\alpha[x,x^{-1}],\varepsilon) \xrightarrow{\partial} LNil^n(A,\alpha,\varepsilon) \longrightarrow 0 \\ 0 \longrightarrow L_n(A_\alpha[x],\varepsilon) \longrightarrow L_n^X(A_\alpha[x,x^{-1}],\varepsilon) \xrightarrow{\partial} LNil_n(A,\alpha,\varepsilon) \longrightarrow 0 \ . \end{cases}$$

<u>Proof</u>: i) This follows from Proposition 5.1.1 i), provided we can show that if D is an X-acyclic $(n+1)$-dimensional $A_\alpha[x]$-module chain complex resolving an n-dimensional $(A_\alpha[x],X)$-module chain complex (= α-twisted nilcomplex over A) (C,ν) that

$$\begin{cases} Q^{n+1}(D,-\varepsilon) = Q\langle v_0\rangle^{n+1}(D,-\varepsilon) = QNil^n(C,\nu,\varepsilon) \\ Q_{n+1}(D,-\varepsilon) = QNil_n(C,\nu,\varepsilon) \end{cases} \quad (n \geqslant 0) \ .$$

The exact sequence of $\mathbb{Z}[\mathbb{Z}_2]$-modules

$$0 \longrightarrow A_\alpha[x] \longrightarrow A_\alpha[x,x^{-1}] \longrightarrow A_\alpha[x,x^{-1}]/A_\alpha[x] \longrightarrow 0$$

splits, with $T \in \mathbb{Z}_2$ acting in each case by

$$T_\varepsilon : \sum_j a_j x^j \longmapsto \varepsilon(\overline{\sum_j a_j x^j}) = \sum_j \varepsilon\alpha^{-j}(\bar{a}_j)x^j$$

$$(a_j \in A) \quad .$$

It follows that every X-acyclic (n+1)-dimensional $(-\varepsilon)$-symmetric complex over $A_\alpha[x]$ $(D,\phi \in Q^{n+1}(D,-\varepsilon))$ is even, since

$$\hat{v}_0(\phi) : H^{n+1}(D) \xrightarrow{\hat{v}_0^X(\phi)} \hat{H}^0(\mathbb{Z}_2;A_\alpha[x,x^{-1}]/A_\alpha[x],\varepsilon)$$
$$\xrightarrow{\hat{\delta} = 0} \hat{H}^1(\mathbb{Z}_2;A_\alpha[x],\varepsilon)$$

and

$$Q\langle v_0\rangle^{n+1}(D,-\varepsilon) = Q^{n+1}(D,-\varepsilon) \quad .$$

An n-dimensional α-twisted nilcomplex over A (C,ν) can be regarded as an n-dimensional $(A_\alpha[x],X)$-module chain complex by Proposition 5.1.1 ii), and as such has a canonical resolution $(D,h:D \longrightarrow C)$ with D the X-acyclic (n+1)-dimensional $A_\alpha[x]$-module chain complex defined by

$$d_D = \begin{pmatrix} d_C & (-)^{r-1}(x-\nu) \\ 0 & d_C \end{pmatrix}$$

$$: D_r = (C_r)_\alpha[x] \oplus (xC_{r-1})_\alpha[x]$$

$$\longrightarrow D_{r-1} = (C_{r-1})_\alpha[x] \oplus (xC_{r-2})_\alpha[x]$$

and $h:D \longrightarrow C$ the homology equivalence of $A_\alpha[x]$-module chain complexes defined by

$$h : D_r = (C_r)_\alpha[x]\oplus(xC_{r-1})_\alpha[x] \longrightarrow C_r \ ;$$

$$(\sum_{j=0}^{\infty} x^j y_j, \sum_{j=0}^{\infty} x^j(xz_j)) \longmapsto \sum_{j=0}^{\infty} \nu^j(y_j) \qquad (y_j \in C_r, z_j \in C_{r-1}) .$$

The $\mathbb{Z}$-module chain map

$$h : \mathrm{Hom}_{A_\alpha[x]}(D^*,D) \longrightarrow \mathrm{Hom}_{A_\alpha[x]}(D^*,C) \ ; \ g \longmapsto hg$$

is also a homology equivalence (working as in the proof of Proposition 3.1.3 ii)). Now

$$\begin{aligned}\mathrm{Hom}_{A_\alpha[x]}(D^*,C)_r &= \sum_{p+q=r} \mathrm{Hom}_{A_\alpha[x]}(D^p,C_q) \\ &= \sum_{p+q=r} \mathrm{Hom}_A(C^p\oplus C^{p-1}x,C_q) \\ &= \overline{\mathrm{Hom}}_A(C^*,C)_r\oplus\mathrm{Hom}_A(C^*,xC)_{r-1} \qquad (r\in\mathbb{Z}) \ ,\end{aligned}$$

allowing $\mathrm{Hom}_{A_\alpha[x]}(D^*,C)$ to be identified with the algebraic mapping cone $C(\Gamma_\nu)$ of the $\mathbb{Z}[\mathbb{Z}_2]$-module chain map

$$\Gamma_\nu : \overline{\mathrm{Hom}}_A(C^*,C) \longrightarrow \mathrm{Hom}_A(C^*,xC)$$

used to define the QNil-groups. Thus

$$h : \mathrm{Hom}_{A_\alpha[x]}(D^*,D) \longrightarrow \mathrm{Hom}_{A_\alpha[x]}(D^*,C) = C(\Gamma_\nu)$$

is a $\mathbb{Z}[\mathbb{Z}_2]$-module chain map inducing isomorphisms in homology, and hence also in the $\begin{cases}\mathbb{Z}_2\text{-hypercohomology} \\ \mathbb{Z}_2\text{-hyperhomology}\end{cases}$ groups

$$\begin{cases} h^\% : Q^{n+1}(D,-\varepsilon) = H_{n+1}(\mathrm{Hom}_{\mathbb{Z}[\mathbb{Z}_2]}(W,\mathrm{Hom}_{A_\alpha[x]}(D^*,D))) \\ \qquad\qquad \longrightarrow H_{n+1}(\mathrm{Hom}_{\mathbb{Z}[\mathbb{Z}_2]}(W,C(\Gamma_\nu))) = \mathrm{QNil}^n(C,\nu,\varepsilon) \\ h_\% : Q_{n+1}(D,-\varepsilon) = H_{n+1}(W\otimes_{\mathbb{Z}[\mathbb{Z}_2]}\mathrm{Hom}_{A_\alpha[x]}(D^*,D)) \\ \qquad\qquad \longrightarrow H_{n+1}(W\otimes_{\mathbb{Z}[\mathbb{Z}_2]}C(\Gamma_\nu)) = \mathrm{QNil}_n(C,\nu,\varepsilon) \end{cases} .$$

ii) The identifications $\begin{cases} L^n(A_\alpha[x],X,\varepsilon) = LNil^n(A,\alpha,\varepsilon) \\ L_n(A_\alpha[x],X,\varepsilon) = LNil_n(A,\alpha,\varepsilon) \end{cases}$ $(n \geqslant 0)$

are immediate from i).

iii) The abelian group morphisms

$$\begin{cases} \Delta : LNil^n(A,\alpha,\varepsilon) \longrightarrow L^n_X(A_\alpha[x,x^{-1}],\varepsilon) \ ; \\ \qquad (C,\nu,\delta\phi,\phi) \longmapsto (C_\alpha[x,x^{-1}],[\nu,\delta\phi,\phi]) \\ \Delta : LNil_n(A,\alpha,\varepsilon) \longrightarrow L^X_n(A_\alpha[x,x^{-1}],\varepsilon) \\ \qquad (C,\nu,\delta\psi,\psi) \longmapsto (C_\alpha[x,x^{-1}],[\nu,\delta\psi,\psi]) \end{cases} \quad (n \geqslant 0)$$

defined by

$$\begin{cases} [\nu,\delta\phi,\phi]_s = (x-\nu)\phi_s - T_{-\varepsilon}\delta\phi_{s-1} \\ \qquad : (C^{n-r+s})_\alpha[x,x^{-1}] \longrightarrow (C_r)_\alpha[x,x^{-1}] \\ [\nu,\delta\psi,\psi]_s = (x-\nu)\psi_s + T_{-\varepsilon}\delta\psi_{s+1} \\ \qquad : (C^{n-r-s})_\alpha[x,x^{-1}] \longrightarrow (C_r)_\alpha[x,x^{-1}] \\ \qquad (r \in \mathbb{Z},\ s \geqslant 0,\ \delta\phi_{-1} = 0) \end{cases}$$

are right inverses ($\partial\Delta = 1$) for the morphisms

$$\begin{cases} \partial : L^n_X(A_\alpha[x,x^{-1}],\varepsilon) \longrightarrow L^n(A_\alpha[x],X,\varepsilon) = LNil^n(A,\alpha,\varepsilon) \\ \partial : L^X_n(A_\alpha[x,x^{-1}],\varepsilon) \longrightarrow L_n(A_\alpha[x],X,\varepsilon) = LNil_n(A,\alpha,\varepsilon) \end{cases} \quad (n \geqslant 0)$$

appearing in the localization exact sequence, since

$\begin{cases} \partial(C_\alpha[x,x^{-1}],[\nu,\delta\phi,\phi]) \\ \partial(C_\alpha[x,x^{-1}],[\nu,\delta\psi,\psi]) \end{cases}$ is homotopy equivalent to $\begin{cases} (C,\nu,\delta\phi,\phi) \\ (C,\nu,\delta\psi,\psi) \end{cases}$.

(The underlying $(n+1)$-dimensional X-acyclic $A_\alpha[x]$-module chain complex $\begin{cases} C((x-\nu)\phi_0 : (C^{n-*})_\alpha[x] \longrightarrow C_\alpha[x]) \\ C((x-\nu)(1+T_{\alpha,\varepsilon})\psi_0 : (C^{n-*})_\alpha[x] \longrightarrow C_\alpha[x]) \end{cases}$ is chain

equivalent to the canonical resolution $D = C(x-\nu : (xC)_\alpha[x] \longrightarrow C_\alpha[x])$

of C, since $\begin{cases} \phi_0 : C^{n-*} \longrightarrow xC \\ (1+T_{\alpha,\varepsilon})\psi_0 : C^{n-*} \longrightarrow xC \end{cases}$ is a chain equivalence of A-module chain complexes).

[]

Define the <u>n-dimensional α-twisted</u> $\begin{cases} \underline{\varepsilon\text{-symmetric}} \\ \underline{\varepsilon\text{-quadratic}} \end{cases}$ <u>$\widetilde{\text{LNil}}$-group</u> $\begin{cases} \widetilde{LNil}^n(A,\alpha,\varepsilon) \\ \widetilde{LNil}_n(A,\alpha,\varepsilon) \end{cases}$ $(n \geqslant 0)$ to be the group appearing in the natural direct sum decomposition

$$\begin{cases} LNil^n(A,\alpha,\varepsilon) = L^n(A,\alpha,\varepsilon) \oplus \widetilde{LNil}^n(A,\alpha,\varepsilon) \\ LNil_n(A,\alpha,\varepsilon) = L_n(A,\alpha,\varepsilon) \oplus \widetilde{LNil}_n(A,\alpha,\varepsilon) \end{cases}$$

obtained from the natural injection

$$\begin{cases} \bar{\eta} : L^n(A,\alpha,\varepsilon) \longrightarrow LNil^n(A,\alpha,\varepsilon) \; ; \; (C,\phi) \longmapsto (C,0,0,\phi) \\ \bar{\eta} : L_n(A,\alpha,\varepsilon) \longrightarrow LNil_n(A,\alpha,\varepsilon) \; ; \; (C,\psi) \longmapsto (C,0,0,\psi) \end{cases}$$

and the natural projection

$$\begin{cases} \eta : LNil^n(A,\alpha,\varepsilon) \longrightarrow L^n(A,\alpha,\varepsilon) \; ; \; (C,\nu,\delta\phi,\phi) \longmapsto (C,\phi) \\ \eta : LNil_n(A,\alpha,\varepsilon) \longrightarrow L_n(A,\alpha,\varepsilon) \; ; \; (C,\nu,\delta\psi,\psi) \longmapsto (C,\psi) \end{cases}$$

(which are such that $\eta\bar{\eta} = 1$).

Let $A_\alpha[x^{-1}]$ be the ring with involution defined in the same way as $A_\alpha[x]$ but with α^{-1}, x^{-1} in place of α, x so that

$$ax^{-1} = x^{-1}\alpha^{-1}(a) \in A_\alpha[x^{-1}] \quad (a \in A)$$

$$\overline{x^{-1}} = x^{-1} \in A_\alpha[x^{-1}] \quad .$$

The function

$$A_{\alpha^{-1}}[x] \longrightarrow A_\alpha[x^{-1}] \; ; \; \sum_{j=0}^{\infty} a_j x^j \longmapsto \sum_{j=0}^{\infty} a_j x^{-j}$$

is an isomorphism of rings with involution. The multiplicative subset

$$X_- = \{x^{-k} \mid k \geqslant 0\} \subset A_\alpha[x^{-1}]$$

is such that

$$(X_-)^{-1}A_\alpha[x^{-1}] = A_\alpha[x,x^{-1}] = (X_+)^{-1}A_\alpha[x] \ ,$$

where $X_+ = X = \{x^k \mid k \geqslant 0\} \subset A_\alpha[x]$ is the multiplicative subset dealt with above. The inclusions

$$\bar{e}_\pm : A \longrightarrow A_\alpha[x^{\pm 1}] \ ; \ a \longmapsto a$$

are split by the projections

$$e_\pm : A_\alpha[x^{\pm 1}] \longrightarrow A \ ; \ \sum_{j=0}^{\infty} a_j x^{\pm j} \longmapsto a_0 \ ,$$

with $e_\pm \bar{e}_\pm = 1$. The inclusions

$$\bar{E}_\pm : A_\alpha[x^{\pm 1}] \longrightarrow A_\alpha[x,x^{-1}] \ ;$$
$$\sum_{j=0}^{\infty} a_j x^{\pm j} \longmapsto \sum_{j=0}^{\infty} a_j x^{\pm j}$$

do not split.

Let $K \subseteq \tilde{K}_0(A_\alpha[x^{\pm 1}])$ (resp. $\tilde{K}_0(A_\alpha[x,x^{-1}])$) be the $*$-invariant subgroup of the projective classes $[P_\alpha[x^{\pm 1}]]$ (resp. $[P_\alpha[x,x^{-1}]]$) of the modules induced from f.g. projective A-modules P by $\bar{e}_\pm : A \longrightarrow A_\alpha[x^{\pm 1}]$ (resp. $\bar{E}_+\bar{e}_+ = \bar{E}_-\bar{e}_- : A \longrightarrow A_\alpha[x,x^{-1}]$).

<u>Proposition 5.1.4</u> For each $n \geqslant 0$ there is defined a commutative braid of naturally split exact sequences of $\begin{cases} \varepsilon\text{-symmetric} \\ \varepsilon\text{-quadratic} \end{cases}$ L-groups

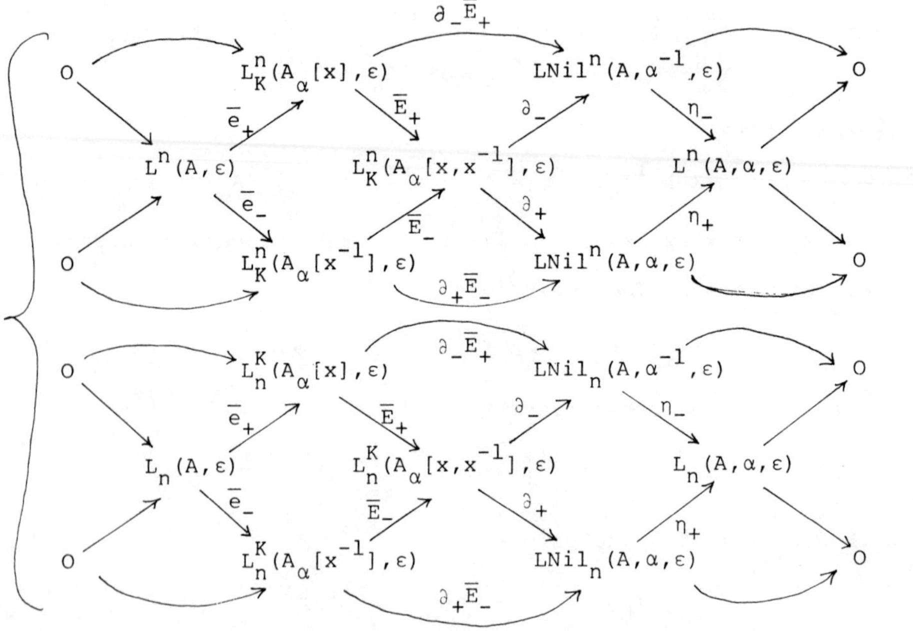

and there are defined natural direct sum decompositions

$$\left\{\begin{array}{l} \left\{\begin{array}{l} L^n_K(A_\alpha[x],\varepsilon) = L^n(A,\varepsilon)\oplus\widetilde{LNil}^n(A,\alpha^{-1},\varepsilon) \\ L^n_K(A_\alpha[x^{-1}],\varepsilon) = L^n(A,\varepsilon)\oplus\widetilde{LNil}^n(A,\alpha,\varepsilon) \\ L^n_K(A_\alpha[x,x^{-1}],\varepsilon) = L^n(A,\varepsilon)\oplus\widetilde{LNil}^n(A,\alpha,\varepsilon)\oplus\widetilde{LNil}^n(A,\alpha^{-1},\varepsilon)\oplus L^n(A,\alpha,\varepsilon) \end{array}\right. \\ \left\{\begin{array}{l} L^K_n(A_\alpha[x],\varepsilon) = L_n(A,\varepsilon)\oplus\widetilde{LNil}_n(A,\alpha^{-1},\varepsilon) \\ L^K_n(A_\alpha[x^{-1}],\varepsilon) = L_n(A,\varepsilon)\oplus\widetilde{LNil}_n(A,\alpha,\varepsilon) \\ L^K_n(A_\alpha[x,x^{-1}],\varepsilon) = L_n(A,\varepsilon)\oplus\widetilde{LNil}_n(A,\alpha,\varepsilon)\oplus\widetilde{LNil}_n(A,\alpha^{-1},\varepsilon)\oplus L_n(A,\alpha,\varepsilon) \end{array}\right. \end{array}\right.$$

$(n \geqslant 0)$.

<u>Proof</u>: In the first instance note that the sequence of ε-symmetric L-groups

$$0 \longrightarrow L^n_K(A_\alpha[x],\varepsilon) \xrightarrow{\overline{E}_+} L^n_K(A_\alpha[x,x^{-1}],\varepsilon) \xrightarrow{\partial_+} LNil^n(A,\alpha,\varepsilon) \longrightarrow 0$$

and its ε-quadratic counterpart are both naturally split exact:

this may be deduced from Proposition 5.1.3 iii) using the appropriate comparison exact sequences of Proposition 1.10.1, noting that

$$\overline{E}_+ : \tilde{K}_0(A_\alpha[x])/\mathrm{im}(\overline{e}_+ : \tilde{K}_0(A) \longrightarrow \tilde{K}_0(A_\alpha[x]))$$

$$\longrightarrow \mathrm{im}(\overline{E}_+ : \tilde{K}_0(A_\alpha[x]) \longrightarrow \tilde{K}_0(A_\alpha[x,x^{-1}]))/\mathrm{im}(\overline{E}_+\overline{e}_+ : \tilde{K}_0(A) \longrightarrow \tilde{K}_0(A_\alpha[x,x^{-1}]))$$

is an isomorphism, or else may be obtained directly from the corresponding intermediate L-theory localization exact sequence of Proposition 3.7.2 by constructing a splitting map Δ_+ for ∂_+ as in the proof of Proposition 5.1.3 iii). Let

$$E_+ : L^n_K(A_\alpha[x,x^{-1}],\varepsilon) \longrightarrow L^n_K(A_\alpha[x],\varepsilon) \quad (n \geqslant 0)$$

be the split surjections associated to the split injections

$$\Delta_+ : LNil^n(A,\alpha,\varepsilon) \longrightarrow L^n_K(A_\alpha[x,x^{-1}],\varepsilon) \quad (n \geqslant 0) \quad ,$$

and similarly in the ε-quadratic case. By Lemma 1.1 of Ranicki [4] it now suffices to prove that the diagrams of ε-symmetric L-groups

$$\begin{array}{ccc} LNil^n(A,\alpha,\varepsilon) & \xrightarrow{\eta_+} & L^n(A,\alpha,\varepsilon) \\ \Big\downarrow{\scriptstyle \Delta_+} & & \Big\downarrow{\scriptstyle \overline{\eta}_-} \\ L^n_K(A_\alpha[x,x^{-1}],\varepsilon) & \xrightarrow{\partial_-} & LNil^n(A,\alpha^{-1},\varepsilon) \end{array}$$

$$\begin{array}{ccc} L^n_K(A_\alpha[x],\varepsilon) & \xrightarrow{e_+} & L^n(A,\varepsilon) \\ \Big\downarrow{\scriptstyle \overline{E}_+} & & \Big\downarrow{\scriptstyle \overline{e}_-} \\ L^n_K(A_\alpha[x,x^{-1}],\varepsilon) & \xrightarrow{E_-} & L^n_K(A_\alpha[x^{-1}],\varepsilon) \end{array}$$

and their ε-quadratic analogues are commutative in order to establish the split exactness of the other sequences in the braids, and hence to obtain the direct sum decompositions.

The relation $\overline{\eta}_-\eta_+ = \partial_-\Delta_+$ is easy to verify directly. The relation $E_-\overline{E}_+ = \overline{e}_-e_+$ was verified explicitly in Ranicki [4,§4] in the ε-quadratic case for $\alpha = 1 : A \longrightarrow A$ (taking into account that the splitting maps E_+ defined there are slghtly different from those defined here, being geared to the splitting maps of rings $A[x^{\pm 1}] \longrightarrow A$ for $\overline{e}_\pm : A \longrightarrow A[x^{\pm 1}]$ given by $x^{\pm 1} \longmapsto 1$ rather than $e_\pm : x^{\pm 1} \longmapsto 0$). The verification of $E_-\overline{E}_+ = \overline{e}_-e_+$ in the general case requires a symmetric L-theory Higman linearization trick, and is deferred to Ranicki [11].

[]

We shall now identify linking forms (resp. formations) over $(A_\alpha[x],X)$ with α-twisted forms (resp. formations) over A together with a nilpotent structure. This will allow us to express the LNil-groups $\begin{cases} LNil^n(A,\alpha,\varepsilon) \\ LNil_n(A,\alpha,\varepsilon) \end{cases}$ for $n = 0$ (resp. $n = 1$) as the Witt groups of such objects, and also to define lower LNil-groups $\begin{cases} LNil^n(A,\alpha,\varepsilon) \\ LNil_n(A,\alpha,\varepsilon) \end{cases}$ $(n \leqslant -1)$, which appear in the extensions to the lower L-groups of the results of Propositions 5.1.2, 5.1.3 and 5.1.4.

Given a f.g. projective A-module M define the <u>α-twisted ε-duality</u> involution

$$T_{\alpha,\varepsilon} : \mathrm{Hom}_A(M,xM^*) \longrightarrow \mathrm{Hom}_A(M,xM^*) \ ; \ \phi \longmapsto \varepsilon\phi^* \ .$$

An element $\phi \in \mathrm{Hom}_A(M,xM^*)$ is the same as a pairing

$$\phi : M \times M \longrightarrow A \ ; \ (y,z) \longmapsto \phi(y,z) \equiv \phi(y)(zx)$$

(identifying $xM^* = (Mx)^*$) such that

i) $\phi(y+y',z) = \phi(y,z) + \phi(y',z)$

ii) $\phi(y,z+z') = \phi(y,z) + \phi(y,z')$

iii) $\phi(ay,bz) = \alpha(b)\phi(y,z)\bar{a} \in A$

$(y,y',z,z' \in M,\ a,b \in A)$.

The α-twisted ε-dual $T_{\alpha,\varepsilon}\phi \in \mathrm{Hom}_A(M,M^*x)$ is the pairing defined by

$$T_{\alpha,\varepsilon}\phi(y,z) = \varepsilon\alpha(\overline{\phi(z,y)}) \in A \quad (y,z \in M) .$$

(Working as in the proof of Proposition 5.1.2 it is possible to identify

$$(\mathrm{Hom}_A(M,xM^*),T_{\alpha,\varepsilon}) = (\mathrm{Hom}_{A^\alpha}((Mx)^\alpha,(Mx)^{\alpha*}),T_\varepsilon) \quad).$$

Define the <u>α-twisted Q-groups of M</u>

$$\begin{cases} Q^\varepsilon(M,\alpha) = \ker(1-T_{\alpha,\varepsilon}:\mathrm{Hom}_A(M,xM^*) \longrightarrow \mathrm{Hom}_A(M,xM^*)) \\ Q\langle v_0\rangle^\varepsilon(M,\alpha) = \mathrm{im}(1+T_{\alpha,\varepsilon}:\mathrm{Hom}_A(M,xM^*) \longrightarrow \mathrm{Hom}_A(M,xM^*)) \subseteq Q^\varepsilon(M,\alpha) \\ Q_\varepsilon(M,\alpha) = \mathrm{coker}(1-T_{\alpha,\varepsilon}:\mathrm{Hom}_A(M,xM^*) \longrightarrow \mathrm{Hom}_A(M,xM^*)) \end{cases} .$$

An <u>α-twisted</u> $\begin{cases} \text{(even) } \varepsilon\text{-symmetric} \\ \varepsilon\text{-quadratic} \end{cases}$ <u>form over A</u> $\begin{cases} (M,\phi) \\ (M,\psi) \end{cases}$ is a f.g. projective A-module M together with an element $\begin{cases} \phi \in Q^\varepsilon(M,\alpha) \quad (\phi \in Q\langle v_0\rangle^\varepsilon(M,\alpha)) \\ \psi \in Q_\varepsilon(M,\alpha) \end{cases}$. Such a form is <u>non-singular</u> if $\begin{cases} \phi \in \mathrm{Hom}_A(M,xM^*) \\ (1+T_{\alpha,\varepsilon})\psi \in \mathrm{Hom}_A(M,xM^*) \end{cases}$ is an isomorphism. There are evident notions of <u>morphism</u>, <u>(sub)lagrangian</u>, <u>hyperbolic</u>, <u>Witt group</u> for α-twisted forms. The Witt group of non-singular α-twisted $\begin{cases} \text{(even) } \varepsilon\text{-symmetric} \\ \varepsilon\text{-quadratic} \end{cases}$ forms over A is denoted by $\begin{cases} L^\varepsilon(A,\alpha) \quad (L\langle v_0\rangle^\varepsilon(A,\alpha)) \\ L_\varepsilon(A,\alpha) \end{cases}$. There are identifications of categories

$$(\text{0-dimensional } \alpha\text{-twisted } \begin{cases}\text{(even) } \varepsilon\text{-symmetric} \\ \varepsilon\text{-quadratic}\end{cases} \text{(Poincaré) complexes over } A)$$

$$= ((\text{non-singular}) \ \alpha\text{-twisted } \begin{cases}\text{(even) } \varepsilon\text{-symmetric} \\ \varepsilon\text{-quadratic}\end{cases} \text{ forms over } A)$$

and hence also identifications of groups

$$\begin{cases} L^0(A,\alpha,\varepsilon) = L^\varepsilon(A,\alpha) \\ L_0(A,\alpha,\varepsilon) = L_\varepsilon(A,\alpha) \ . \end{cases}$$

Given an α-twisted nilmodule over A $(M,\nu \in \mathrm{Hom}_A(M,Mx))$ define the <u>QNil-groups of (M,ν)</u> by

$$\begin{cases} \mathrm{QNil}^\varepsilon(M,\nu) = \\ \quad \{\phi \in \mathrm{Hom}_A(M,xM^*) \,|\, \varepsilon\phi^* = \phi \in \mathrm{Hom}_A(M,xM^*), \nu^*\phi = \phi\nu \in \mathrm{Hom}_A(M,M^*)\} \\ \mathrm{QNil}\langle v_0\rangle^\varepsilon(M,\nu) = \\ \quad \{\phi \in \mathrm{Hom}_A(M,xM^*) \,|\, \phi = \psi + \varepsilon\psi^* \text{ for some } \psi \in \mathrm{Hom}_A(M,xM^*) \\ \qquad \text{such that } \nu^*\psi - \psi\nu = \delta\psi + \varepsilon\delta\psi^* \in \mathrm{Hom}_A(M,M^*) \\ \qquad \text{some } \delta\psi \in \mathrm{Hom}_A(M,M^*)\} \subseteq \mathrm{QNil}^\varepsilon(M,\nu) \\ \mathrm{QNil}_\varepsilon(M,\nu) = \\ \quad \dfrac{\{(\delta\psi,\psi) \in \mathrm{Hom}_A(M,M^*) \oplus \mathrm{Hom}_A(M,xM^*) \,|\, \nu^*\psi - \psi\nu = \delta\psi + \varepsilon\delta\psi^* \in \mathrm{Hom}_A(M,M^*)\}}{\{(\delta\chi - \varepsilon\delta\chi^* + \nu^*\chi - \chi\nu, \chi - \varepsilon\chi^*) \,|\, (\delta\chi,\chi) \in \mathrm{Hom}_A(M,M^*) \oplus \mathrm{Hom}_A(M,xM^*)\}} \end{cases} .$$

An <u>α-twisted</u> $\begin{cases}\underline{\text{(even) } \varepsilon\text{-symmetric}} \\ \underline{\varepsilon\text{-quadratic}}\end{cases}$ <u>nilform over A</u> $\begin{cases}(M,\nu,\phi) \\ (M,\nu,\delta\psi,\psi)\end{cases}$ is an α-twisted nilmodule over A (M,ν) together with an element $\begin{cases}\phi \in \mathrm{QNil}^\varepsilon(M,\nu) \ \ (\phi \in \mathrm{QNil}\langle v_0\rangle^\varepsilon(M,\nu)) \\ (\delta\psi,\psi) \in \mathrm{QNil}_\varepsilon(M,\nu)\end{cases}$. Such a nilform is <u>non-singular</u> if $\begin{cases}(M,\phi \in Q^\varepsilon(M,\alpha)) \ \ (M,\phi \in Q\langle v_0\rangle^\varepsilon(M,\alpha)) \\ (M,\psi \in Q_\varepsilon(M,\alpha))\end{cases}$ is a non-singular α-twisted $\begin{cases}\text{(even) } \varepsilon\text{-symmetric} \\ \varepsilon\text{-quadratic}\end{cases}$ form over A.

There are evident notions of <u>morphism</u>, <u>(sub)lagrangian</u>, <u>hyperbolic</u>, <u>Witt group</u> for nilforms. The Witt group of

non-singular α-twisted $\begin{cases}\text{(even) } \varepsilon\text{-symmetric} \\ \varepsilon\text{-quadratic}\end{cases}$ nilforms over A

is denoted by $\begin{cases} LNil^{\varepsilon}(A,\alpha) \ \ (LNil\langle v_0\rangle^{\varepsilon}(A,\alpha)) \\ LNil_{\varepsilon}(A,\alpha)\end{cases}$.

<u>Proposition 5.1.5</u> i) There are natural identifications of categories

(0-dimensional α-twisted $\begin{cases}\varepsilon\text{-symmetric} \\ \varepsilon\text{-quadratic}\end{cases}$ (Poincaré) nilcomplexes over A)

= ((non-singular) α-twisted $\begin{cases}\varepsilon\text{-symmetric} \\ \varepsilon\text{-quadratic}\end{cases}$ nilforms over A)

= ((non-singular) $\begin{cases}\varepsilon\text{-symmetric} \\ \text{split } \varepsilon\text{-quadratic}\end{cases}$ linking forms over $(A_{\alpha}[x],X)$)

((non-singular) α-twisted even ε-symmetric nilforms over A)

= ((non-singular) even ε-symmetric linking forms over $(A_{\alpha}[x],X)$) .

ii) There are natural identifications of L-groups

$$LNil^0(A,\alpha,\varepsilon) = LNil^{\varepsilon}(A,\alpha) = L^{\varepsilon}(A_{\alpha}[x],X) = L\langle v_0\rangle^{\varepsilon}(A_{\alpha}[x],X) = L^0(A_{\alpha}[x],X,\varepsilon)$$

$$LNil_0(A,\alpha,\varepsilon) = LNil_{\varepsilon}(A,\alpha) = \tilde{L}_{\varepsilon}(A_{\alpha}[x],X) = L_0(A_{\alpha}[x],X,\varepsilon)$$

$$LNil\langle v_0\rangle^{\varepsilon}(A,\varepsilon) = L_{\varepsilon}(A_{\alpha}[x],X) = L^{-2}(A_{\alpha}[x],X,-\varepsilon) .$$

<u>Proof</u>: i) Given an α-twisted nilmodule over A $(M,\nu \in \mathrm{Hom}_A(M,Mx))$ define a O-dimensional α-twisted nilcomplex over A $(C,\nu^*:C \longrightarrow Cx)$ by

$$C_r = \begin{cases} M^* & \text{if } r = 0 \\ 0 & \text{if } r \neq 0 \end{cases} ,$$

and note that

$$\begin{cases} QNil^0(C,\nu^*,\varepsilon) = QNil^{\varepsilon}(M,\nu) \\ QNil_0(C,\nu^*,\varepsilon) = QNil_{\varepsilon}(M,\nu) \end{cases} .$$

This gives the identification of O-dimensional α-twisted $\begin{cases}\varepsilon\text{-symmetric}\\ \varepsilon\text{-quadratic}\end{cases}$ nilcomplexes over A with α-twisted $\begin{cases}\varepsilon\text{-symmetric}\\ \varepsilon\text{-quadratic}\end{cases}$ nilforms over A. The correspondence between nilforms and linking forms now follows from Propositions 3.4.1, 5.1.3 i).

In particular, an α-twisted $\begin{cases}\varepsilon\text{-symmetric}\\ \varepsilon\text{-quadratic}\end{cases}$ nilform over A $\begin{cases}(M,\nu,\phi)\\ (M,\nu,\delta\psi,\psi)\end{cases}$ determines the $\begin{cases}\varepsilon\text{-symmetric}\\ \text{split } \varepsilon\text{-quadratic}\end{cases}$ linking form over $(A_\alpha[x],X)$ $\begin{cases}((M,\nu),\xi)\\ ((M,\nu),\xi,\zeta)\end{cases}$ defined by

$$\begin{cases} \xi : M \times M \longrightarrow A_\alpha[x,x^{-1}]/A_\alpha[x] \; ; \\ \qquad (y,z) \longmapsto \sum\limits_{j=-\infty}^{-1} \phi(\nu^{-j-1}y)(z)x^j \\ \begin{cases} \xi : M \times M \longrightarrow A_\alpha[x,x^{-1}]/A_\alpha[x] \; ; \\ \qquad (y,z) \longmapsto \sum\limits_{j=-\infty}^{-1} (\psi+\varepsilon\psi^*)(\nu^{-j-1}y)(z)x^j \end{cases} \end{cases}$$

$$\left.\left.\begin{array}{l}\zeta : M \longrightarrow Q_{\varepsilon}(X^{-1}A_{\alpha}[x]/A_{\alpha}[x]) \\ \qquad\qquad = H_0(\mathbb{Z}_2;A_{\alpha}[x,x^{-1}]/A_{\alpha}[x],\varepsilon)\ ; \\ \qquad y \longmapsto \psi(y)(y)x^{-1} + (\psi\nu + \delta\psi)(y)(y)x^{-2} + \ldots\end{array}\right)\right)$$

and every $\begin{cases}\varepsilon\text{-symmetric} \\ \text{split } \varepsilon\text{-quadratic}\end{cases}$ linking form over $(A_{\alpha}[x],X)$ can be expressed in this way.

By definition, an ε-quadratic linking form over $(A_{\alpha}[x],X)$ $((M,\nu),\xi,\rho)$ is an ε-symmetric linking form over $(A_{\alpha}[x],X)$ $((M,\nu),\xi)$ together with a function

$$\begin{aligned}\rho : M \longrightarrow \widetilde{Q}_{\varepsilon}(A_{\alpha}[x],X) \\ &= \frac{\{b+\varepsilon\bar{b} \in A_{\alpha}[x,x^{-1}]\,|\,b \in A_{\alpha}[x,x^{-1}]\}}{\{c+\varepsilon\bar{c}\,|\,c \in A_{\alpha}[x]\}} \\ &= \{d+\varepsilon\bar{d} \in A_{\alpha}[x,x^{-1}]/A_{\alpha}[x]\,|\,d \in A_{\alpha}[x]\}\end{aligned}$$

(with $\widetilde{Q}_{\varepsilon}$ as in the proof of Proposition 3.4.2 ii)) satisfying, among others,

$$\rho(y) = \xi(y,y) \in H^0(\mathbb{Z}_2;A_{\alpha}[x,x^{-1}]/A_{\alpha}[x],\varepsilon) \quad (y \in M)\ .$$

The natural map

$$\widetilde{Q}_{\varepsilon}(A_{\alpha}[x],X) \longrightarrow H^0(\mathbb{Z}_2;A_{\alpha}[x,x^{-1}]/A_{\alpha}[x],\varepsilon)\,;\ d+\varepsilon\bar{d} \longmapsto d+\varepsilon\bar{d}$$

is injective, so that ρ (if it exists) is determined by ξ. By Proposition 2.4.1 i) every ε-quadratic linking form admits a split ε-quadratic refinement. Thus we can also identify α-twisted even ε-symmetric nilforms over A with ε-quadratic linking forms over $(A_{\alpha}[x],X)$.

ii) Immediate from i) and Propositions 3.4.7 i), 5.1.3 ii).

[]

An $\underline{\alpha\text{-twisted}} \begin{cases} \underline{\text{(even) } \varepsilon\text{-symmetric}} \\ \underline{\varepsilon\text{-quadratic}} \end{cases}$ $\underline{\text{formation over A}}$

$\begin{cases} (M,\phi;F,G) \\ (M,\psi;F,G) \end{cases}$ is an α-twisted $\begin{cases} \text{(even) } \varepsilon\text{-symmetric} \\ \varepsilon\text{-quadratic} \end{cases}$ form over A $\begin{cases} (M,\phi \in Q^{\varepsilon}(M,\alpha)) \quad ((M,\phi \in Q\langle v_0\rangle^{\varepsilon}(M,\alpha))) \\ (M,\psi \in Q_{\varepsilon}(M,\alpha)) \end{cases}$ together with a lagrangian F and a sublagrangian G. Such a formation is $\underline{\text{non-singular}}$ if G is a lagrangian. There are evident notions of $\underline{\text{(stable) isomorphism}}$ and $\underline{\text{Witt group}}$ for α-twisted formations. The Witt group of non-singular α-twisted $\begin{cases} \text{(even) } \varepsilon\text{-symmetric} \\ \varepsilon\text{-quadratic} \end{cases}$ formations over A is denoted by $\begin{cases} M^{\varepsilon}(A,\alpha) \quad (M\langle v_0\rangle^{\varepsilon}(A,\alpha)) \\ M_{\varepsilon}(A,\alpha) \end{cases}$. There are identifications of groups

$$\begin{cases} L^1(A,\alpha,\varepsilon) = M^{\varepsilon}(A,\alpha) \\ L_1(A,\alpha,\varepsilon) = M_{\varepsilon}(A,\alpha) \end{cases} .$$

An $\underline{\alpha\text{-twisted}} \begin{cases} \underline{\text{(even) } \varepsilon\text{-symmetric}} \\ \underline{\varepsilon\text{-quadratic}} \end{cases}$ $\underline{\text{nilformation over A}}$

$\begin{cases} (M,\nu,\phi;F,G) \\ (M,\nu,\delta\psi,\psi;F,G) \end{cases}$ is an α-twisted $\begin{cases} \text{(even) } \varepsilon\text{-symmetric} \\ \varepsilon\text{-quadratic} \end{cases}$ nilform over A $\begin{cases} (M,\nu \in \mathrm{Hom}_A(M,Mx),\phi \in QNil^{\varepsilon}(M,\nu)) \quad ((M,\nu,\phi \in QNil\langle v_0\rangle^{\varepsilon}(M,\nu))) \\ (M,\nu \in \mathrm{Hom}_A(M,Mx),\psi \in QNil_{\varepsilon}(M,\nu)) \end{cases}$

together with a lagrangian F and a sublagrangian G, and such that in the ε-quadratic case

$$(M,\nu,\delta\psi,\psi) = (F\oplus xF^*, \begin{pmatrix} \omega & \overline{\varepsilon}(\lambda+\varepsilon\lambda^*) \\ 0 & \omega^* \end{pmatrix}, \begin{pmatrix} 0 & 0 \\ 0 & \lambda \end{pmatrix}, \begin{pmatrix} 0 & 1 \\ 0 & 0 \end{pmatrix})$$

for some nilpotent map $\omega \in \mathrm{Hom}_A(F,Fx)$ and some ε-quadratic form over A $(xF^*, \lambda \in Q_\varepsilon(xF^*))$, with $(G, \nu|_G, \psi|_G)$ an α-twisted even $(-\varepsilon)$-symmetric nilform over A. Such a nilformation is non-singular if G is a lagrangian. There are evident notions of (stable) isomorphism and Witt group for α-twisted nilformations. The Witt group of non-singular α-twisted

$\begin{cases} \text{(even) } \varepsilon\text{-symmetric} \\ \varepsilon\text{-quadratic} \end{cases}$ nilformations over A is denoted by

$$\begin{cases} MNil^\varepsilon(A,\alpha) \quad (MNil\langle v_0\rangle^\varepsilon(A,\alpha)) \\ MNil_\varepsilon(A,\alpha) \end{cases} .$$

A 1-dimensional α-twisted $\begin{cases} \varepsilon\text{-symmetric} \\ \varepsilon\text{-quadratic} \end{cases}$ nilcomplex over A

$\begin{cases} (C, \nu: C \longrightarrow Cx, (\delta\phi, \phi) \in QNil^1(C, \nu, \varepsilon)) \\ (C, \nu: C \longrightarrow Cx, (\delta\psi, \psi) \in QNil_1(C, \nu, \varepsilon)) \end{cases}$ is connected if

$$\begin{cases} H_0(\phi_0 : C^{1-*} \longrightarrow xC) = 0 \\ H_0((1+T_{\alpha,\varepsilon})\psi_0 : C^{1-*} \longrightarrow xC) = 0 \end{cases} .$$

Proposition 5.1.6 i) There are natural identifications of sets of equivalence classes

(connected 1-dimensional α-twisted $\begin{cases} \varepsilon\text{-symmetric} \\ \varepsilon\text{-quadratic} \end{cases}$ complexes over A)

= (α-twisted $\begin{cases} \varepsilon\text{-symmetric} \\ \varepsilon\text{-quadratic} \end{cases}$ nilformations over A)

= ($\begin{cases} \text{(even) } \varepsilon\text{-symmetric} \\ \text{split } \varepsilon\text{-quadratic} \end{cases}$ linking formations over $(A_\alpha[x], X)$) ,

(α-twisted even ε-symmetric nilformations over A)

= (ε-quadratic linking formations over $(A_\alpha[x], X)$) .

Poincaré nilcomplexes correspond to non-singular nilformations, which in turn correspond to non-singular linking formations.

ii) There are natural identifications of L-groups

$$\begin{aligned} LNil^1(A,\alpha,\varepsilon) &= MNil^{\varepsilon}(A,\alpha) = M^{\varepsilon}(A_\alpha[x],X) \\ &= M\langle v_0\rangle^{\varepsilon}(A_\alpha[x],X) = L^1(A_\alpha[x],X,\varepsilon) \end{aligned}$$

$$LNil_1(A,\alpha,\varepsilon) = MNil_{\varepsilon}(A,\alpha) = \widetilde{M}_{\varepsilon}(A_\alpha[x],X) = L_1(A_\alpha[x],X,\varepsilon)$$

$$MNil\langle v_0\rangle^{\varepsilon}(A,\alpha) = M^{\varepsilon}(A_\alpha[x],X) = L^{-1}(A_\alpha[x],X,-\varepsilon) .$$

Proof: By analogy with Proposition 5.1.5.

In particular, given a connected 1-dimensional α-twisted $\begin{cases}\varepsilon\text{-symmetric} \\ \varepsilon\text{-quadratic}\end{cases}$ nilcomplex over A $\begin{cases}(C,\nu,\delta\phi,\phi) \\ (C,\nu,\delta\psi,\psi)\end{cases}$ there is defined an α-twisted $\begin{cases}\varepsilon\text{-symmetric} \\ \varepsilon\text{-quadratic}\end{cases}$ nilformation over A, as follows.

The nilcomplex $(C,\nu:C\longrightarrow Cx)$ is defined by a morphism of α-twisted nilmodules over A

$$d : (C_1,\nu) \longrightarrow (C_0,\nu) .$$

The class $\begin{cases}(\delta\phi,\phi) \in QNil^1(C,\nu,\varepsilon) \\ (\delta\psi,\psi) \in QNil_1(C,\nu,\varepsilon)\end{cases}$ is represented by a collection of A-module morphisms

$$\begin{cases} \begin{cases} \phi_0 : C^0 \longrightarrow xC_1 , \ \tilde{\phi}_0 : C^1 \longrightarrow xC_0 , \ \phi_1 : C^1 \longrightarrow xC_1 , \\ \delta\phi_0 : C^1 \longrightarrow C_1 \end{cases} \\ \begin{cases} \psi_0 : C^0 \longrightarrow xC_1 , \ \tilde{\psi}_0 : C^1 \longrightarrow xC_0 , \ \psi_1 : C^0 \longrightarrow xC_0 , \\ \delta\psi_0 : C^1 \longrightarrow C_1 , \ \delta\psi_1 : C^0 \longrightarrow C_1 , \ \delta\tilde{\psi}_1 : C^1 \longrightarrow C_0 , \\ \delta\psi_2 : C^0 \longrightarrow C_0 \end{cases} \end{cases}$$

satisfying

$$\left\{\begin{array}{l}\begin{cases} d\phi_0 + \tilde{\phi}_0 d^* = 0\ ,\ d\phi_1 - \tilde{\phi}_0 + \varepsilon\phi_0^* = 0\ ,\ \phi_1 - \varepsilon\phi_1^* = 0\ , \\ \nu\phi_0 - \phi_0\nu^* + \delta\phi_0 d^* = 0\ ,\ \nu\tilde{\phi}_0 - \tilde{\phi}_0\nu^* - d\delta\phi_0 = 0\ , \\ \nu\phi_1 - \phi_1\nu^* - \delta\phi_0 + \varepsilon\delta\phi_0^* = 0 \end{cases} \\ \begin{cases} d\psi_0 + \tilde{\psi}_0 d^* + \psi_1 - \varepsilon\psi_1^* = 0\ ,\ \nu\psi_0 - \psi_0\nu^* - \delta\psi_0 d^* - \delta\psi_1 - \varepsilon\delta\tilde{\psi}_1^* = 0\ , \\ \nu\tilde{\psi}_0 - \tilde{\psi}_0\nu^* - d\delta\psi_0 + \delta\tilde{\psi}_1 + \varepsilon\delta\psi_1^* = 0\ , \\ \nu\psi_1 - \psi_1\nu^* - d\delta\psi_1 - \delta\tilde{\psi}_1 d^* - \delta\psi_2 + \varepsilon\delta\psi_2^* = 0 \end{cases}\end{array}\right. .$$

The α-twisted $\begin{cases}\varepsilon\text{-symmetric}\\ \varepsilon\text{-quadratic}\end{cases}$ nilformation over A associated to $\begin{cases}(C,\nu,\delta\phi,\phi)\\ (C,\nu,\delta\psi,\psi)\end{cases}$ is defined by

$$\left\{\begin{array}{l}(xC_1\oplus C^1, \begin{pmatrix}\nu & -\bar{\varepsilon}\delta\phi_0\\ 0 & \nu^*\end{pmatrix}, \begin{pmatrix}0 & 1\\ \varepsilon & \phi_1\end{pmatrix}; xC_1, \mathrm{im}(\begin{pmatrix}\bar{\varepsilon}\phi_0\\ d^*\end{pmatrix}: C^0 \longrightarrow xC_1\oplus C^1)) \\ (xC_1\oplus C^1, \begin{pmatrix}\nu & -\bar{\varepsilon}(\delta\psi_0+\varepsilon\delta\psi_0^*)\\ 0 & \nu^*\end{pmatrix}, \begin{pmatrix}0 & 0\\ 0 & -\delta\psi_0\end{pmatrix}, \begin{pmatrix}0 & 1\\ 0 & 0\end{pmatrix}; \\ \qquad\qquad xC_1, \mathrm{im}(\begin{pmatrix}\bar{\varepsilon}(\psi_0+\varepsilon\tilde{\psi}_0^*)\\ d^*\end{pmatrix}: C^0 \longrightarrow xC_1\oplus C^1))\end{array}\right. .$$

[]

Define the <u>n-dimensional α-twisted</u> $\begin{cases}\underline{\varepsilon\text{-symmetric}}\\ \underline{\varepsilon\text{-quadratic}}\end{cases}$ <u>L- (resp. LNil-) groups</u> $\begin{cases}L^n(A,\alpha,\varepsilon)\\ L_n(A,\alpha,\varepsilon)\end{cases}$ (resp. $\begin{cases}LNil^n(A,\alpha,\varepsilon)\\ LNil_n(A,\alpha,\varepsilon)\end{cases}$) for $n \leqslant -1$ by

$$\left\{\begin{array}{l} L^n(A,\alpha,\varepsilon) = \begin{cases} M\langle v_0\rangle^{-\varepsilon}(A,\alpha) \\ L\langle v_0\rangle^{-\varepsilon}(A,\alpha) \\ L_n(A,\alpha,\varepsilon) \end{cases}\end{array}\right.$$

$$(\text{resp. } LNil^n(A,\alpha,\varepsilon) = \begin{cases} MNil\langle v_0\rangle^{-\varepsilon}(A,\alpha) & n=-1 \\ LNil\langle v_0\rangle^{-\varepsilon}(A,\alpha)) \text{ if} & n=-2 \\ LNil_n(A,\alpha,\varepsilon) & n\leqslant -3 \end{cases}$$

$$L_n(A,\alpha,\varepsilon) = L_{n+2i}(A,\alpha,(-)^i\varepsilon)$$

$$(\text{resp. } LNil_n(A,\alpha,\varepsilon) = LNil_{n+2i}(A,\alpha,(-)^i\varepsilon)) \quad (n\leqslant -1, n+2i\geqslant 0).$$

Let $\begin{cases}\widetilde{LNil}^n(A,\alpha,\varepsilon) \\ \widetilde{LNil}_n(A,\alpha,\varepsilon)\end{cases}$ $(n\leqslant -1)$ be the groups appearing in the natural direct sum decompositions

$$\begin{cases} LNil^n(A,\alpha,\varepsilon) = L^n(A,\alpha,\varepsilon)\oplus\widetilde{LNil}^n(A,\alpha,\varepsilon) \\ LNil_n(A,\alpha,\varepsilon) = L_n(A,\alpha,\varepsilon)\oplus\widetilde{LNil}_n(A,\alpha,\varepsilon) \end{cases}$$

as in the case $n\geqslant 0$ dealt with above.

<u>Proposition 5.1.7</u> For each $n\in\mathbb{Z}$ there is defined a commutative braid of naturally split exact sequences of ε-symmetric L-groups

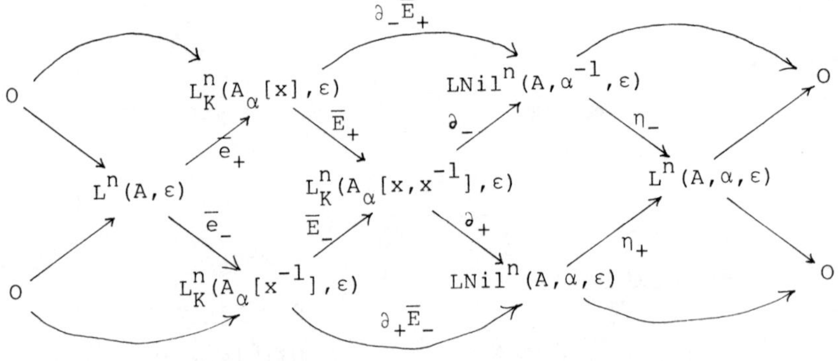

and there are defined natural direct sum decompositions

$$\begin{cases} L^n_K(A_\alpha[x],\varepsilon) = L^n(A,\varepsilon)\oplus\widetilde{LNil}^n(A,\alpha^{-1},\varepsilon) \\ L^n_K(A_\alpha[x^{-1}],\varepsilon) = L^n(A,\varepsilon)\oplus\widetilde{LNil}^n(A,\alpha,\varepsilon) \\ L^n_K(A_\alpha[x,x^{-1}],\varepsilon) = L^n(A,\varepsilon)\oplus\widetilde{LNil}^n(A,\alpha,\varepsilon)\oplus\widetilde{LNil}^n(A,\alpha^{-1},\varepsilon)\oplus L^n(A,\alpha,\varepsilon), \end{cases}$$

as well as natural identifications

$$L^n(A,\alpha,\varepsilon) = L^n(A,\alpha^{-1},\varepsilon) = L^n(A^\alpha,\varepsilon) .$$

Similarly for the ε-quadratic L-groups.

<u>Proof</u>: By analogy with the case $n \geqslant 0$ (Propositions 5.1.2, 5.1.3 and 5.1.4).

[]

In the untwisted case $\alpha = 1 : A \longrightarrow A$ the terminology involving α is contracted, for example

$$LNil^*(A,1,\varepsilon) = LNil^*(A,\varepsilon) .$$

We shall now reiterate the example given in Ranicki [6,§6] of a pair (A,S) for which the natural projection of Witt groups

$$\tilde{L}_\varepsilon(A,S) \longrightarrow L_\varepsilon(A,S) \ ; \ (M,\lambda,\nu) \longmapsto (M,\lambda,p\nu)$$

is not injective, showing that in general split ε-quadratic linking forms over (A,S) carry more information than ε-quadratic linking forms over (A,S). Namely, let

$$\varepsilon = -1 \in A = \mathbb{Z}[x]$$

and note that the non-singular skew-quadratic nilform over $\mathbb{Z}$

$$c = (\mathbb{Z}\oplus\mathbb{Z}, 0, (0, \begin{pmatrix} 1 & 1 \\ 0 & 1 \end{pmatrix}) \in QNil_\varepsilon(\mathbb{Z}\oplus\mathbb{Z}))$$

represents the element

$$\begin{aligned} c = (1,0) \in \tilde{L}_\varepsilon(\mathbb{Z}[x],X) &= LNil_\varepsilon(\mathbb{Z}) \\ &= L_\varepsilon(\mathbb{Z}) \oplus \widetilde{LNil}_\varepsilon(\mathbb{Z}) , \end{aligned}$$

where $1 = (\mathbb{Z}\oplus\mathbb{Z}, \begin{pmatrix} 1 & 1 \\ 0 & 1 \end{pmatrix} \in Q_\varepsilon(\mathbb{Z}\oplus\mathbb{Z})) \in L_\varepsilon(\mathbb{Z}) = \mathbb{Z}_2$ is the Arf invariant 1 element, and that the natural map

$$\tilde{L}_{\varepsilon}(\mathbb{Z}[x],X) \longrightarrow L_{\varepsilon}(\mathbb{Z}[x],X) = L\langle v_0\rangle^{\varepsilon}(\mathbb{Z}) \oplus \widetilde{LNil}\langle v_0\rangle^{\varepsilon}(\mathbb{Z})$$

sends this element to 0, since $L\langle v_0\rangle^{\varepsilon}(\mathbb{Z}) = 0$ (Proposition 4.3.1).

5.2 Change of K-theory

The results of §5.1 will now be extended to the intermediate L-groups of the α-twisted polynomial extensions $A_\alpha[x]$, $A_\alpha[x,x^{-1}]$ of a ring with involution A, using the intermediate L-theory localization exact sequences of §3.7.

As in §3.7 we start by considering the action of the duality involutions $*$ on the algebraic K-theory localization exact sequence, which in this case is

$$\tilde{K}_1(A_\alpha[x]) \xrightarrow{\bar{E}_+} \tilde{K}_1(A_\alpha[x,x^{-1}]) \xrightarrow{\partial} K_1(A_\alpha[x],X)$$
$$\xrightarrow{j} \tilde{K}_0(A_\alpha[x]) \xrightarrow{\bar{E}_+} \tilde{K}_0(A_\alpha[x,x^{-1}])$$

for a ring with involution A and a ring automorphism $\alpha: A \longrightarrow A$ such that $\overline{\alpha(a)} = \alpha^{-1}(\bar{a}) \in A$ $(a \in A)$. The action is such that

$$*\bar{E}_+ = \bar{E}_+* \quad , \quad *\partial = \partial* \quad , \quad *j = -j* \quad .$$

By the results of Farrell and Hsiang [1],[2] and Siebenmann [1] this sequence can be expressed as

$$\tilde{K}_1(A)\oplus\widetilde{Nil}_1(A,\alpha^{-1}) \xrightarrow{\bar{E}_+ = \begin{pmatrix} i_1 & 0 \\ 0 & 0 \\ 0 & 1 \end{pmatrix}} \tilde{K}_1(A,\alpha)\oplus\widetilde{Nil}_1(A,\alpha)\oplus\widetilde{Nil}_1(A,\alpha^{-1})$$

$$\xrightarrow{\partial = \begin{pmatrix} \partial_1 & 0 & 0 \\ 0 & 1 & 0 \end{pmatrix}} K_0(A)\oplus\widetilde{Nil}_1(A,\alpha) \xrightarrow{j = \begin{pmatrix} 1-\alpha & 0 \\ 0 & 0 \end{pmatrix}} \tilde{K}_0(A)\oplus\widetilde{Nil}_1(A,\alpha^{-1})$$

$$\xrightarrow{\bar{E}_+ = \begin{pmatrix} i_0 & 0 \\ 0 & 0 \\ 0 & 1 \end{pmatrix}} \tilde{K}_0(A,\alpha)\oplus\widetilde{Nil}_0(A,\alpha)\oplus\widetilde{Nil}_0(A,\alpha^{-1}) \quad ,$$

with

$$K_m(A_\alpha[x],X) = K_{m-1}(\underline{\underline{\mathrm{Nil}}}(A,\alpha)) = K_{m-1}(A)\oplus\widetilde{\mathrm{Nil}}_m(A,\alpha) \quad (m=0,1)$$

and $\widetilde{K}_m(A,\alpha)$ $(m=0,1)$ the relative K-groups appearing in the exact sequence

$$\ldots \longrightarrow K_1(A) \xrightarrow{i_1} \widetilde{K}_1(A,\alpha) \xrightarrow{\partial_1} K_0(A) \xrightarrow{1-\alpha} \widetilde{K}_0(A)$$
$$\xrightarrow{i_0} \widetilde{K}_0(A,\alpha) \longrightarrow \ldots .$$

The duality involution on the exact category $\underline{\underline{\mathrm{Nil}}}(A,\alpha)$ of α-twisted nilmodules over A

$$* : \underline{\underline{\mathrm{Nil}}}(A,\alpha) \longrightarrow \underline{\underline{\mathrm{Nil}}}(A,\alpha) ;$$
$$(P,\nu:P \longrightarrow Px) \longmapsto (xP^*,\nu^*:xP^* \longrightarrow P^*)$$

induces the duality involution

$$* = \begin{pmatrix} *\alpha & 0 \\ 0 & * \end{pmatrix}$$
$$: K_m(A_\alpha[x],X) = K_{m-1}(A)\oplus\widetilde{\mathrm{Nil}}_m(A,\alpha)$$
$$\longrightarrow K_m(A_\alpha[x],X) = K_{m-1}(A)\oplus\widetilde{\mathrm{Nil}}_m(A,\alpha) \quad (m=0,1)$$

with $*\alpha$ the composite of the automorphism of the K-group

$$\alpha : K_{m-1}(A) = K_{m-1}(\underline{\underline{P}}(A)) \longrightarrow K_{m-1}(A)$$

induced by the automorphism of the exact category $\underline{\underline{P}}(A)$

$$\alpha : \underline{\underline{P}}(A) = (\text{f.g. projective A-modules}) \longrightarrow \underline{\underline{P}}(A) ;$$
$$P \longmapsto Px$$

and the duality involution

$$* : K_{m-1}(A) \longrightarrow K_{m-1}(A)$$

induced by

$$* : \underline{\underline{P}}(A) \longrightarrow \underline{\underline{P}}(A) ; \; P \longmapsto P^* = \mathrm{Hom}_A(P,A) .$$

Note that $*\alpha : K_{m-1}(A) \longrightarrow K_{m-1}(A)$ is just the duality involution $*:K_{m-1}(A^\alpha) \longrightarrow K_{m-1}(A^\alpha)$ associated to the ring with involution A^α

defined in §5.1 above, with the ring structure of A and involution

$$^{-} : A^{\alpha} \longrightarrow A^{\alpha} \ ; \ a \longmapsto \alpha(\bar{a}) \quad .$$

The duality involution

$$* : \tilde{K}_m(A,\alpha) \longrightarrow \tilde{K}_m(A,\alpha)$$

which $\tilde{K}_m(A,\alpha)$ inherits from $\tilde{K}_1(A_\alpha[x,x^{-1}])$ is such that in the diagram

$$\begin{array}{ccccccccc}
\tilde{K}_1(A) & \xrightarrow{i_1} & \tilde{K}_1(A,\alpha) & \xrightarrow{\partial_1} & K_0(A^{\alpha}) & \xrightarrow{1-\alpha} & \tilde{K}_0(A) & \xrightarrow{i_0} & \tilde{K}_0(A,\alpha) \\
\downarrow * & & \downarrow * & & \downarrow * & & \downarrow * & & \downarrow * \\
\tilde{K}_1(A) & \xrightarrow{i_1} & \tilde{K}_1(A,\alpha) & \xrightarrow{\partial_1} & K_0(A^{\alpha}) & \xrightarrow{1-\alpha} & \tilde{K}_0(A) & \xrightarrow{i_0} & \tilde{K}_0(A,\alpha)
\end{array}$$

we have the relations

$$*i_m = i_m* \ (m=0,1) \ , \ *\partial_1 = \partial_1* \ , \ *(1-\alpha) = -(1-\alpha)* \quad .$$

The duality involutions on the remaining groups of the algebraic K-theory localization exact sequence are given by

$$* = \begin{pmatrix} * & 0 \\ 0 & * \end{pmatrix} :$$

$$\tilde{K}_m(A_\alpha[x]) = \tilde{K}_m(A) \oplus \widetilde{Nil}_m(A,\alpha^{-1}) \longrightarrow \tilde{K}_m(A) \oplus \widetilde{Nil}_m(A,\alpha^{-1})$$

$$* = \begin{pmatrix} * & 0 & 0 \\ 0 & * & 0 \\ 0 & 0 & * \end{pmatrix} :$$

$$\tilde{K}_m(A_\alpha[x,x^{-1}]) = \tilde{K}_m(A,\alpha) \oplus \widetilde{Nil}_m(A,\alpha) \oplus \widetilde{Nil}_m(A,\alpha^{-1})$$

$$\longrightarrow \tilde{K}_m(A,\alpha) \oplus \widetilde{Nil}_m(A,\alpha) \oplus \widetilde{Nil}_m(A,\alpha^{-1})$$

$$(m=0,1) \quad .$$

Given a $*$-invariant subgroup $Y \subseteq K_1(A_\alpha[x],X)$ define the

$$\underline{\text{intermediate}} \left\{ \begin{array}{l} \underline{\varepsilon\text{-symmetric}} \\ \underline{\varepsilon\text{-quadratic}} \end{array} \right. \underline{\text{LNil-groups}} \left\{ \begin{array}{l} \mathrm{LNil}^n_Y(A,\alpha,\varepsilon) \\ \mathrm{LNil}^Y_n(A,\alpha,\varepsilon) \end{array} \right. \quad (n \in \mathbb{Z})$$

in the same way as the groups $\left\{ \begin{array}{l} \mathrm{LNil}^*(A,\alpha,\varepsilon) \\ \mathrm{LNil}_*(A,\alpha,\varepsilon) \end{array} \right.$ of §5.1 (the special case $Y = K_1(A_\alpha[x],X)$) but using only α-twisted nilcomplexes over A with X-projective class in $Y \subseteq K_1(A_\alpha[x],X)$. The proof of Proposition 5.1.3 ii) gives natural identifications

$$\left\{ \begin{array}{l} L^*_Y(A_\alpha[x],X,\varepsilon) = \mathrm{LNil}^*_Y(A,\alpha,\varepsilon) \\ L^Y_*(A_\alpha[x],X,\varepsilon) = \mathrm{LNil}^Y_*(A,\alpha,\varepsilon) \end{array} \right. ,$$

the groups on the left hand side being defined as in §3.7. If

$$Y = Y_0 \oplus Y_1 \subseteq K_1(A_\alpha[x],X) = K_0(A^\alpha) \oplus \widetilde{\mathrm{Nil}}_1(A,\alpha)$$

for some $*$-invariant subgroups $Y_0 \subseteq K_0(A^\alpha)$, $Y_1 \subseteq \widetilde{\mathrm{Nil}}_1(A,\alpha)$ there are natural direct sum decompositions

$$\left\{ \begin{array}{l} \mathrm{LNil}^*_Y(A,\alpha,\varepsilon) = L^*_{\widetilde{Y}_0}(A,\alpha,\varepsilon) \oplus \widetilde{\mathrm{LNil}}^*_{Y_1}(A,\alpha,\varepsilon) \\ \mathrm{LNil}^Y_*(A,\alpha,\varepsilon) = L^{\widetilde{Y}_0}_*(A,\alpha,\varepsilon) \oplus \widetilde{\mathrm{LNil}}^{Y_1}_*(A,\alpha,\varepsilon) \end{array} \right.$$

with $\widetilde{Y}_0 \subseteq \widetilde{K}_0(A^\alpha)$ the image of Y_0 under the natural projection $K_0(A^\alpha) \longrightarrow \widetilde{K}_0(A^\alpha)$.

Given a $*$-invariant subgroup $Y \subseteq \widetilde{K}_0(A^\alpha)$ define $*$-invariant subgroups

$$Y^\partial = \partial_1^{-1}(Y) \subseteq \widetilde{K}_1(A,\alpha) \subseteq \widetilde{K}_1(A_\alpha[x,x^{-1}])$$

$$(1-\alpha)(Y) \subseteq \widetilde{K}_0(A) \subseteq \widetilde{K}_0(A_\alpha[x]) \quad .$$

The natural map

$$\iota : Y \longrightarrow \widetilde{K}_0(A)/(1-\alpha)(Y) \; ; \; [P] \longmapsto (1-\alpha)(Y) + [P]$$

is a $\mathbb{Z}[\mathbb{Z}_2]$-module morphism, with $T \in \mathbb{Z}_2$ acting on Y by the α-twisted duality involution it inherits from $\tilde{K}_0(A)$ (i.e. the duality involution on $\tilde{K}_0(A^\alpha)$ with respect to which Y is invariant) and by the duality involution inherited from $\tilde{K}_0(A)$ on $\tilde{K}_0(A)/(1-\alpha)(Y)$. Let $\tilde{L}_{Y^\partial}(A,\alpha,\varepsilon)$ $(n \in \mathbb{Z})$ be the relative ε-symmetric L-groups appearing in the exact sequence

$$\ldots \longrightarrow L^n_{(1-\alpha)Y}(A,\varepsilon) \xrightarrow{\beta} \tilde{L}^n_{Y^\partial}(A,\alpha,\varepsilon) \xrightarrow{\gamma} L^n_Y(A^\alpha,\varepsilon)$$

$$\xrightarrow{\delta} L^{n-1}_{(1-\alpha)Y}(A,\varepsilon) \xrightarrow{\beta} \tilde{L}^{n-1}_{Y^\partial}(A,\alpha,\varepsilon) \longrightarrow \ldots$$

with δ the composite

$$\delta : L^n_Y(A^\alpha,\varepsilon) \longrightarrow \hat{H}^n(\mathbb{Z}_2;Y) \xrightarrow{\iota} \hat{H}^n(\mathbb{Z}_2;\tilde{K}_0(A)/(1-\alpha)Y)$$

$$\longrightarrow L^{n-1}_{(1-\alpha)Y}(A,\varepsilon) \ .$$

Define similarly relative ε-quadratic L-groups $\tilde{L}^{Y^\partial}_*(A,\alpha,\varepsilon)$.

<u>Proposition 5.2.1</u> Given $*$-invariant subgroups $Y \subseteq \tilde{K}_0(A^\alpha)$, $Z_\pm \subseteq \widetilde{\mathrm{Nil}}_1(A,\alpha^{\pm 1})$ there is a natural identification of the intermediate ε-symmetric L-theory localization exact sequence

$$\ldots \longrightarrow L^n_{(1-\alpha)Y\oplus Z_-}(A_\alpha[x],\varepsilon) \longrightarrow L^n_{Y^\partial\oplus Z_+\oplus Z_-}(A_\alpha[x,x^{-1}],\varepsilon)$$

$$\xrightarrow{\partial} L^n_{Y\oplus Z_+}(A_\alpha[x],X,\varepsilon) \longrightarrow L^{n-1}_{(1-\alpha)Y\oplus Z_-}(A_\alpha[x],\varepsilon) \longrightarrow \ldots$$

$$(n \in \mathbb{Z})$$

with the exact sequence

$$\cdots \longrightarrow L^n_{(1-\alpha)Y}(A,\varepsilon) \oplus \widetilde{LNil}^n_{Z_-}(A,\alpha^{-1},\varepsilon)$$

$$\xrightarrow{\begin{pmatrix} \beta & 0 \\ 0 & 0 \\ 0 & 1 \end{pmatrix}} \widetilde{L}^n_{Y^\partial}(A,\alpha,\varepsilon) \oplus \widetilde{LNil}^n_{Z_+}(A,\alpha,\varepsilon) \oplus \widetilde{LNil}^n_{Z_-}(A,\alpha^{-1},\varepsilon)$$

$$\xrightarrow{\begin{pmatrix} \gamma & 0 & 0 \\ 0 & 1 & 0 \end{pmatrix}} L^n_Y(A^\alpha,\varepsilon) \oplus \widetilde{LNil}^n_{Z_+}(A,\alpha,\varepsilon)$$

$$\xrightarrow{\begin{pmatrix} \delta & 0 \\ 0 & 0 \end{pmatrix}} L^{n-1}_{(1-\alpha)Y}(A,\varepsilon) \oplus \widetilde{LNil}^{n-1}_{Z_-}(A,\alpha^{-1},\varepsilon) \longrightarrow \cdots \quad .$$

If $Y = \{0\} \subseteq \widetilde{K}_0(A^\alpha)$ or if $\alpha = 1 : A \longrightarrow A$ then $\delta = 0$ and the exact sequence is naturally split.

Similarly for the ε-quadratic L-groups L_*.

Proof: By analogy with Propositions 3.7.1, 5.1.4.

If $Y = \{0\}$ or $\alpha = 1$ define splitting maps

$$\Delta_+ : L^n_{Y\oplus Z_+}(A_\alpha[x],X,\varepsilon) \longrightarrow L^n_{Y^\partial \oplus Z_+ \oplus Z_-}(A_\alpha[x,x^{-1}],\varepsilon)$$

exactly as in the proof of Proposition 5.1.3 iii).

[]

Given a $*$-invariant subgroup $Y \subseteq \widetilde{K}_m(A)$ ($m = 0$ or 1) define $*$-invariant subgroups

$$(1-\alpha)^{-1}Y = \{w \in \widetilde{K}_m(A^\alpha) \mid (1-\alpha)(w) \in Y\} \subseteq \widetilde{K}_m(A^\alpha)$$

$$x^{-1}Y = \begin{cases} i_0(Y) \subseteq \widetilde{K}_0(A_\alpha[x,x^{-1}]) \text{ if } m = 0 \\ i_1(Y) + \{\tau(x^k : A_\alpha[x,x^{-1}] \longrightarrow A_\alpha[x,x^{-1}]) \mid k \geqslant 0\} \\ \qquad \subseteq \widetilde{K}_1(A_\alpha[x,x^{-1}]) \text{ if } m = 1 \end{cases} .$$

The natural map

$$\iota : (1-\alpha)^{-1}Y \longrightarrow \widetilde{K}_m(A)/Y \; ; \; w \longmapsto Y + w$$

is a $\mathbb{Z}[\mathbb{Z}_2]$-module morphism. Let $\widetilde{L}^n_{X^{-1}Y}(A,\alpha,\epsilon)$ be the relative ϵ-symmetric L-groups appearing in the exact sequence

$$\ldots \longrightarrow L^n_Y(A,\epsilon) \xrightarrow{\beta} \widetilde{L}^n_{X^{-1}Y}(A,\alpha,\epsilon) \xrightarrow{\gamma} L^n_{(1-\alpha)^{-1}Y}(A^\alpha,\epsilon)$$
$$\xrightarrow{\delta} L^{n-1}_Y(A,\epsilon) \longrightarrow \ldots ,$$

with δ the composite

$$\delta : L^n_{(1-\alpha)^{-1}Y}(A^\alpha,\epsilon) \longrightarrow \hat{H}^n(\mathbb{Z}_2;(1-\alpha)^{-1}Y) \xrightarrow{\iota} \hat{H}^n(\mathbb{Z}_2;\widetilde{K}_m(A)/Y)$$
$$\longrightarrow L^{n-1}_Y(A,\epsilon) \; .$$

<u>Proposition 5.2.2</u> Let $Y \subseteq \widetilde{K}_m(A)$, $Z_\pm \subseteq \widetilde{Nil}_m(A,\alpha^{\pm 1})$ ($m = 0$ or 1) be $*$-invariant subgroups, with $Z_\pm = 0$ if $m = 0$. There is a natural identification of the intermediate ϵ-symmetric L-theory localization exact sequence

$$\ldots \longrightarrow L^n_{Y\oplus Z_-}(A_\alpha[x],\epsilon) \longrightarrow L^n_{X^{-1}Y\oplus Z_+\oplus Z_-}(A_\alpha[x,x^{-1}],\epsilon)$$
$$\xrightarrow{\partial} L^n_{(1-\alpha)^{-1}Y\oplus Z_+}(A_\alpha[x],X,\epsilon) \longrightarrow L^{n-1}_{Y\oplus Z_-}(A_\alpha[x],\epsilon) \longrightarrow \ldots$$
$$(n \in \mathbb{Z})$$

with the exact sequence

$$\ldots \longrightarrow L^n_Y(A,\varepsilon)\oplus\widetilde{LNil}^n_{Z_-}(A,\alpha^{-1},\varepsilon)$$

$$\xrightarrow{\begin{pmatrix}\beta & 0\\ 0 & 0\\ 0 & 1\end{pmatrix}} \tilde{L}^n_{X^{-1}Y}(A,\alpha,\varepsilon)\oplus\widetilde{LNil}^n_{Z_+}(A,\alpha,\varepsilon)\oplus\widetilde{LNil}^n_{Z_-}(A,\alpha^{-1},\varepsilon)$$

$$\xrightarrow{\begin{pmatrix}\gamma & 0 & 0\\ 0 & 1 & 0\end{pmatrix}} L^n_{(1-\alpha)^{-1}Y}(A^\alpha,\varepsilon)\oplus\widetilde{LNil}^n_{Z_+}(A,\alpha,\varepsilon)$$

$$\xrightarrow{\begin{pmatrix}\delta & 0\\ 0 & 0\end{pmatrix}} L^{n-1}_Y(A,\varepsilon)\oplus\widetilde{LNil}^{n-1}_{Z_-}(A,\alpha^{-1},\varepsilon)\longrightarrow \ldots ,$$

where $\widetilde{LNil}^*_{Z_\pm}(A,\alpha^{\pm 1},\varepsilon) \equiv \widetilde{LNil}^*(A,\alpha^{\pm 1},\varepsilon)$ if $m = 0$. If $Y = \tilde{K}_m(A)$ or if $\alpha = 1 : A \longrightarrow A$ then $\delta = 0$ and the exact sequence is naturally split.

Similarly for the ε-quadratic L-groups L_*.

Proof: As for Proposition 5.2.1, but using 3.7.2 and 3.7.3 instead of 3.7.1.

[]

In the special case $\alpha = 1 : A \longrightarrow A$, $Y = \{0\} \subseteq \tilde{K}_0(A)$ Proposition 5.2.2 gives the "fundamental theorem of $\begin{cases}\varepsilon\text{-symmetric}\\ \varepsilon\text{-quadratic}\end{cases}$ L-theory", the naturally split exact sequences

$$\begin{cases} 0 \longrightarrow V^n(A,\varepsilon) \longrightarrow V^n(A[x],\varepsilon)\oplus V^n(A[x^{-1}],\varepsilon) \longrightarrow V^n(A[x,x^{-1}],\varepsilon) \\ \qquad\qquad \longrightarrow U^n(A,\varepsilon) \longrightarrow 0 \\ 0 \longrightarrow V_n(A,\varepsilon) \longrightarrow V_n(A[x],\varepsilon)\oplus V_n(A[x^{-1}],\varepsilon) \longrightarrow V_n(A[x,x^{-1}],\varepsilon) \\ \qquad\qquad \longrightarrow U_n(A,\varepsilon) \longrightarrow 0 \end{cases} \quad (n \in \mathbb{Z}) .$$

(The ε-quadratic case was previously obtained in Ranicki [4]).

§6. Mayer-Vietoris sequences

We shall now investigate the existence or otherwise of a Mayer-Vietoris exact sequence of intermediate $\begin{cases}\varepsilon\text{-symmetric}\\ \varepsilon\text{-quadratic}\end{cases}$ L-groups

$$\begin{cases} \dots \longrightarrow L^n_X(A,\varepsilon) \xrightarrow{\begin{pmatrix} f \\ f' \end{pmatrix}} L^n_Y(B,\varepsilon)\oplus L^n_{Y'}(B',\varepsilon) \xrightarrow{(g\ \ -g')} L^n_{X'}(A',\varepsilon) \\ \qquad \xrightarrow{\partial} L^{n-1}_X(A,\varepsilon) \longrightarrow L^{n-1}_Y(B,\varepsilon)\oplus L^{n-1}_{Y'}(B',\varepsilon) \longrightarrow \dots \\ \dots \longrightarrow L_n^X(A,\varepsilon) \xrightarrow{\begin{pmatrix} f \\ f' \end{pmatrix}} L_n^Y(B,\varepsilon)\oplus L_n^{Y'}(B',\varepsilon) \xrightarrow{(g\ \ -g')} L_n^{X'}(A',\varepsilon) \\ \qquad \xrightarrow{\partial} L_{n-1}^X(A,\varepsilon) \longrightarrow L_{n-1}^Y(B,\varepsilon)\oplus L_{n-1}^{Y'}(B',\varepsilon) \longrightarrow \dots \end{cases} \quad (n\in\mathbb{Z})$$

for a commutative square of rings with involution

$$\begin{array}{ccc} A & \xrightarrow{f} & B \\ {\scriptstyle f'}\downarrow & \Phi & \downarrow{\scriptstyle g} \\ B' & \xrightarrow{g'} & A' \end{array}$$

and a commutative square of $*$-invariant subgroups

$$\kappa = \begin{pmatrix} X & \longrightarrow & Y \\ \downarrow & & \downarrow \\ Y' & \longrightarrow & X' \end{pmatrix} \subseteq \begin{pmatrix} \tilde{K}_m(A) & \longrightarrow & \tilde{K}_m(B) \\ \downarrow & & \downarrow \\ \tilde{K}_m(B') & \longrightarrow & \tilde{K}_m(A') \end{pmatrix}$$

for $m = 0$ or 1. As usual, we start with a review of the relevant algebraic K-theory.

A Mayer-Vietoris exact sequence of classical algebraic K-groups

$$K_1(A) \xrightarrow{\begin{pmatrix} f \\ f' \end{pmatrix}} K_1(B) \oplus K_1(B') \xrightarrow{(g\ -g')} K_1(A')$$

$$\xrightarrow{\partial} K_0(A) \xrightarrow{\begin{pmatrix} f \\ f' \end{pmatrix}} K_0(B) \oplus K_0(B') \xrightarrow{(g\ -g')} K_0(A')$$

has been obtained for three types of commutative square of rings Φ (as above):

I) Φ is cartesian, i.e. the sequence of additive groups

$$0 \longrightarrow A \xrightarrow{\begin{pmatrix} f \\ f' \end{pmatrix}} B \oplus B' \xrightarrow{(g\ -g')} A' \longrightarrow 0$$

is exact, and $g: B \longrightarrow A'$ (or $g': B' \longrightarrow A'$) is onto,

II) Φ is the cartesian localization-completion square

$$\begin{array}{ccc} A & \longrightarrow & S^{-1}A \\ \downarrow & & \downarrow \\ \hat{A} & \longrightarrow & \hat{S}^{-1}\hat{A} \end{array}$$

associated to a multiplicative subset $S \subset A$, with $\hat{A} = \varprojlim_{s \in S} A/sA$, or some abstraction thereof (e.g the cartesian square

$$\begin{array}{ccc} A & \longrightarrow & S^{-1}A \\ \downarrow & & \downarrow \\ B & \longrightarrow & T^{-1}B \end{array}$$

associated to a cartesian morphism $(A,S) \longrightarrow (B,T)$),

III) Φ is a pushout square with

$$A' = B *_A B'$$

the free product of B and B' amalgamated along A, with the morphisms $f: A \longrightarrow B$, $f': A \longrightarrow B'$ injective, and satisfying some extra conditions.

The first such exact sequence was obtained by Milnor [4,§4], who showed that for a cartesian square of rings Φ of type I there is indeed a Mayer-Vietoris exact sequence of the type

$$K_1(A) \longrightarrow K_1(B) \oplus K_1(B') \longrightarrow K_1(A') \xrightarrow{\partial} K_0(A) \longrightarrow K_0(B) \oplus K_0(B') \longrightarrow K_0(A') \ .$$

Bass [2,XII] defined the lower algebraic K-groups $K_n(A)$ $(n \leqslant -1)$ inductively by

$$K_n(A) = \mathrm{coker}(K_{n+1}(A[x]) \oplus K_{n+1}(A[x^{-1}]) \longrightarrow K_{n+1}(A[x,x^{-1}]))$$

and extended this sequence to the right by

$$\ldots \longrightarrow K_0(B) \oplus K_0(B') \longrightarrow K_0(A') \xrightarrow{\partial} K_{-1}(A) \longrightarrow K_{-1}(B) \oplus K_{-1}(B') \longrightarrow K_{-1}(A') \longrightarrow \ldots \ .$$

Swan [1] showed that there does not exist a K_2-functor extending the sequence to the left for all squares Φ of type I. However, Milnor [4,§§5,6] defined $K_2(A)$ using Steinberg relations such that for the squares Φ of type I with both $g:B \longrightarrow A'$ and $g':B' \longrightarrow A'$ onto there is an extension of the sequence to the left by

$$K_2(A) \longrightarrow K_2(B) \oplus K_2(B') \longrightarrow K_2(A') \xrightarrow{\partial} K_1(A) \longrightarrow K_1(B) \oplus K_1(B') \longrightarrow \ldots \ .$$

Quillen [1],[2] defined the higher K-groups $K_n(A)$ $(n \geqslant 3)$ of a ring A to be the homotopy groups of a space $BGL(A)^+$, with

$$K_n(A) = \pi_n(BGL(A)^+) \quad (n \geqslant 1) \ .$$

Gersten [2] extended this definition to the lower K-groups, constructing a spectrum $\underline{\mathbb{K}}(A)$ such that

$$K_n(A) = \pi_n(\underline{\mathbb{K}}(A)) \quad (n \in \mathbb{Z}) \ .$$

The triad K-groups $K_*(\Phi)$ of a commutative square of rings

$$\begin{array}{ccc} A & \xrightarrow{f} & B \\ {\scriptstyle f'}\downarrow & \Phi & \downarrow{\scriptstyle g} \\ B' & \xrightarrow{g'} & A' \end{array}$$

can thus be defined by

$$K_n(\Phi) = \pi_n \left(\begin{array}{ccc} \underline{\mathbb{K}}(A) & \xrightarrow{f} & \underline{\mathbb{K}}(B) \\ {\scriptstyle f'}\downarrow & & \downarrow{\scriptstyle g} \\ \underline{\mathbb{K}}(B') & \xrightarrow{g'} & \underline{\mathbb{K}}(A') \end{array} \right) \qquad (n \in \mathbb{Z})$$

and are such that there is defined a commutative diagram of abelian groups with exact rows and columns

$$\begin{array}{ccccccccccc}
 & & \vdots & & \vdots & & \vdots & & \vdots & & \vdots & \\
 & & \downarrow & & \downarrow & & \downarrow & & \downarrow & & \downarrow & \\
\cdots & \to & K_n(A) & \xrightarrow{f} & K_n(B) & \to & K_n(f) & \to & K_{n-1}(A) & \xrightarrow{f} & K_{n-1}(B) & \to \cdots \\
 & & {\scriptstyle f'}\downarrow & & \downarrow{\scriptstyle g} & & \downarrow{\scriptstyle (f',g)} & & {\scriptstyle f'}\downarrow & & \downarrow{\scriptstyle g} & \\
\cdots & \to & K_n(B') & \xrightarrow{g'} & K_n(A') & \to & K_n(g') & \to & K_{n-1}(B') & \xrightarrow{g'} & K_{n-1}(A') & \to \cdots \\
 & & \downarrow & & \downarrow & & \downarrow & & \downarrow & & \downarrow & \\
\cdots & \to & K_n(f') & \xrightarrow{(f,g')} & K_n(g) & \to & K_n(\Phi) & \to & K_{n-1}(f') & \xrightarrow{(f,g')} & K_{n-1}(g) & \to \cdots \\
 & & \downarrow & & \downarrow & & \downarrow & & \downarrow & & \downarrow & \\
\cdots & \to & K_{n-1}(A) & \xrightarrow{f} & K_{n-1}(B) & \to & K_{n-1}(f) & \to & K_{n-2}(A) & \xrightarrow{f} & K_{n-2}(B) & \to \cdots \\
 & & \downarrow & & \downarrow & & \downarrow & & \downarrow & & \downarrow & \\
 & & \vdots & & \vdots & & \vdots & & \vdots & & \vdots &
\end{array}$$

$$(n \in \mathbb{Z}) .$$

The triad K-groups vanish

$$K_*(\Phi) = 0$$

if and only if the natural maps

$$(f',g) : K_*(f) \longrightarrow K_*(g')$$

(or equivalently $(f,g'):K_*(f') \longrightarrow K_*(g)$) are isomorphisms, in which case they are called "excision isomorphisms". If in fact $K_*(\Phi) = 0$ the above diagram collapses to a commutative braid of exact sequences

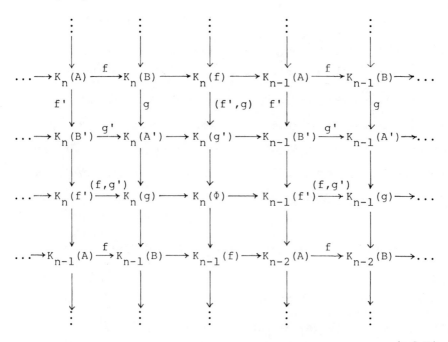

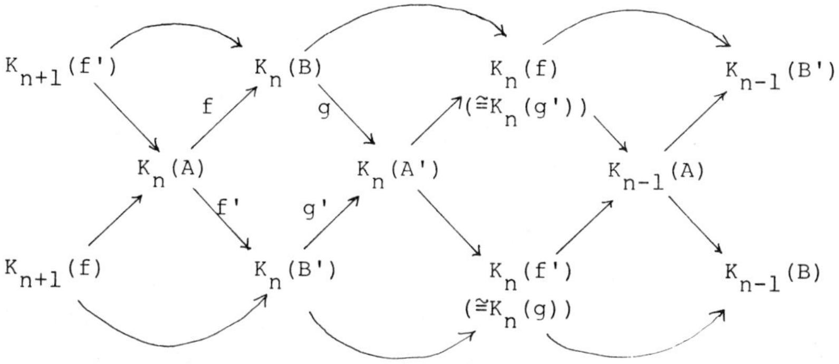

and there is defined a Mayer-Vietoris exact sequence

$$\cdots \longrightarrow K_n(A) \xrightarrow{\begin{pmatrix} f \\ f' \end{pmatrix}} K_n(B) \oplus K_n(B') \xrightarrow{(g \ -g')} K_n(A')$$

$$\xrightarrow{\ \partial\ } K_{n-1}(A) \xrightarrow{\begin{pmatrix} f \\ f' \end{pmatrix}} K_{n-1}(B) \oplus K_{n-1}(B') \longrightarrow \cdots \qquad (n \in \mathbb{Z})$$

with the connecting maps ∂ given by

$$\partial : K_n(A') \longrightarrow K_n(g') \xrightarrow[\sim]{(f',g)^{-1}} K_n(f) \longrightarrow K_{n-1}(A)$$

(or equivalently

$$\partial : K_n(A') \longrightarrow K_n(f') \xrightarrow[\sim]{(f,g')^{-1}} K_n(g) \longrightarrow K_{n-1}(A) \ ).$$

In particular, for a cartesian square Φ of type II (localization-completion) it is the case that $K_*(\Phi) = 0$, since the identification of exact categories

$$((A,S)\text{-modules}) = ((\hat{A},\hat{S})\text{-modules})$$

of Karoubi [2,App.5] (cf. Proposition 3.1.3 i) above) gives that

$$\begin{aligned} K_*(A \longrightarrow S^{-1}A) &= K_{*-1}((A,S)\text{-modules}) \\ &= K_{*-1}((\hat{A},\hat{S})\text{-modules}) = K_*(\hat{A} \longrightarrow \hat{S}^{-1}\hat{A}), \end{aligned}$$

so that there is defined a Mayer-Vietoris exact sequence

$$\ldots \longrightarrow K_n(A) \longrightarrow K_n(S^{-1}A) \oplus K_n(\hat{A}) \longrightarrow K_n(\hat{S}^{-1}\hat{A}) \xrightarrow{\partial} K_{n-1}(A) \longrightarrow \ldots \quad (n \in \mathbb{Z}) \ .$$

We shall only consider the K- and L-theory Mayer-Vietoris sequences for squares of type I and II in §6, leaving type III (pushout) to §7, on account of the close connections with topology.

In §6.1 we shall define the $\left\{\begin{array}{l}\varepsilon\text{-symmetric}\\ \varepsilon\text{-quadratic}\end{array}\right.$ triad L-groups $\left\{\begin{array}{l}L^n(\Phi,\varepsilon)\\ L_n(\Phi,\varepsilon)\end{array}\right.$ $(n \in \mathbb{Z})$ of a commutative square of rings with involution

$$\begin{array}{ccc} A & \xrightarrow{f} & B \\ {\scriptstyle f'}\big\downarrow & \Phi & \big\downarrow{\scriptstyle g} \\ B' & \xrightarrow{g'} & A' \end{array} \ .$$

The necessary and sufficient condition

$$\left\{\begin{array}{l} L^*(\Phi,\varepsilon) = 0 \\ L_*(\Phi,\varepsilon) = 0 \end{array}\right.$$

for there to be excision isomorphisms in the relative $\left\{\begin{array}{l}\varepsilon\text{-symmetric}\\ \varepsilon\text{-quadratic}\end{array}\right.$ L-groups

$$\left\{\begin{array}{l} (f',g) : L^*(f,\varepsilon) \xrightarrow{\sim} L^*(g',\varepsilon) \\ (f',g) : L_*(f,\varepsilon) \xrightarrow{\sim} L_*(g',\varepsilon) \end{array}\right.$$

and a Mayer-Vietoris exact sequence in the absolute $\left\{\begin{array}{l}\varepsilon\text{-symmetric}\\ \varepsilon\text{-quadratic}\end{array}\right.$ L-groups

$$\left\{\begin{array}{l} \ldots \longrightarrow L^n(A,\varepsilon) \xrightarrow{\begin{pmatrix} f \\ f' \end{pmatrix}} L^n(B,\varepsilon) \oplus L^n(B',\varepsilon) \xrightarrow{(g\ \ -g')} L^n(A',\varepsilon) \\ \qquad \xrightarrow{\partial} L^{n-1}(A,\varepsilon) \xrightarrow{\begin{pmatrix} f \\ f' \end{pmatrix}} L^{n-1}(B,\varepsilon) \oplus L^{n-1}(B',\varepsilon) \longrightarrow \ldots \\ \ldots \longrightarrow L_n(A,\varepsilon) \xrightarrow{\begin{pmatrix} f \\ f' \end{pmatrix}} L_n(B,\varepsilon) \oplus L_n(B',\varepsilon) \xrightarrow{(g\ \ -g')} L_n(A',\varepsilon) \\ \qquad \xrightarrow{\partial} L_{n-1}(A,\varepsilon) \xrightarrow{\begin{pmatrix} f \\ f' \end{pmatrix}} L_{n-1}(B,\varepsilon) \oplus L_{n-1}(B',\varepsilon) \longrightarrow \ldots \end{array}\right.$$

$$(n \in \mathbb{Z})$$

will be interpreted in §6.1 in terms of Mayer-Vietoris splittings of algebraic Poincaré complexes over A' with respect to Φ, using the algebraic glueing operations of §1.7. In §6.2 the theory will be extended to the intermediate L-groups of §1.10, since in practice there are only such excision isomorphisms and Mayer-Vietoris exact sequences for the intermediate L-groups associated to a commutative square of $*$-invariant subgroups

$$\kappa = \begin{pmatrix} X & \longrightarrow & Y \\ \downarrow & & \downarrow \\ Y' & \longrightarrow & X' \end{pmatrix} \subseteq \begin{pmatrix} \tilde{K}_m(A) & \longrightarrow & \tilde{K}_m(B) \\ \downarrow & & \downarrow \\ \tilde{K}_m(B') & \longrightarrow & \tilde{K}_m(A') \end{pmatrix} \quad (m = 0 \text{ or } 1)$$

such that

i) $I_m = \ker\left(\begin{pmatrix} f \\ f' \end{pmatrix} : \tilde{K}_m(A) \longrightarrow \tilde{K}_m(B) \oplus \tilde{K}_m(B')\right) \subseteq X \subseteq \tilde{K}_m(A)$

ii) the sequence

$$0 \longrightarrow X/I_m \xrightarrow{\begin{pmatrix} f \\ f' \end{pmatrix}} Y \oplus Y' \xrightarrow{(g\ \ -g')} X' \longrightarrow 0$$

is exact.

Moreover, if Φ is such that there is defined a Mayer-Vietoris exact sequence in the reduced classical algebraic K-groups

$$\tilde{K}_1(A) \xrightarrow{\begin{pmatrix} f \\ f' \end{pmatrix}} \tilde{K}_1(B)\oplus\tilde{K}_1(B') \xrightarrow{(g\ -g')} \tilde{K}_1(A')$$

$$\xrightarrow{\partial} \tilde{K}_0(A) \xrightarrow{\begin{pmatrix} f \\ f' \end{pmatrix}} \tilde{K}_0(B)\oplus\tilde{K}_0(B') \xrightarrow{(g\ -g')} \tilde{K}_0(A')$$

and there is defined a Mayer-Vietoris exact sequence of intermediate $\begin{cases}\varepsilon\text{-symmetric} \\ \varepsilon\text{-quadratic}\end{cases}$ L-groups

$$\begin{cases} \ldots \longrightarrow L^n_X(A,\varepsilon) \longrightarrow L^n_Y(B,\varepsilon)\oplus L^n_{Y'}(B',\varepsilon) \longrightarrow L^n_{X'}(A',\varepsilon) \xrightarrow{\partial} L^{n-1}_X(A,\varepsilon) \longrightarrow \ldots \\ \ldots \longrightarrow L^X_n(A,\varepsilon) \longrightarrow L^Y_n(B,\varepsilon)\oplus L^{Y'}_n(B',\varepsilon) \longrightarrow L^{X'}_n(A',\varepsilon) \xrightarrow{\partial} L^X_{n-1}(A,\varepsilon) \longrightarrow \ldots \end{cases}$$

$$(n \in \mathbb{Z})$$

for one such square κ then there is defined such a sequence for all squares κ satisfying i) and ii). At any rate, for any commutative squares Φ,κ there are defined intermediate $\begin{cases}\varepsilon\text{-symmetric} \\ \varepsilon\text{-quadratic}\end{cases}$ triad L-groups $\begin{cases} L^n_\kappa(\Phi,\varepsilon) \\ L^\kappa_n(\Phi,\varepsilon)\end{cases}$ $(n \in \mathbb{Z})$ such that if

$$\begin{cases} L^*_\kappa(\Phi,\varepsilon) = 0 \\ L^\kappa_*(\Phi,\varepsilon) = 0 \end{cases}$$

then there is defined a Mayer-Vietoris exact sequence in the corresponding intermediate $\begin{cases}\varepsilon\text{-symmetric} \\ \varepsilon\text{-quadratic}\end{cases}$ L-groups.

(The generalities of §§6.1,6.2 apply equally well to L-theory Mayer-Vietoris sequences for squares of type III as to those of types I and II). In §6.3 we shall show that for squares κ satisfying i) and ii) it is indeed the case that $L^\kappa_*(\Phi,\varepsilon) = 0$ if Φ is either a cartesian square of type I ($g:B \longrightarrow A'$ or

$g':B'\longrightarrow A'$ is onto) or a cartesian square of type II (localization-completion), thus obtaining a Mayer-Vietoris exact sequence in the corresponding intermediate ε-quadratic L-groups. Furthermore, we shall show that $L^*_\kappa(\Phi,\varepsilon) = 0$ for κ satisfying i) and ii) with Φ of type II satisfying the extra condition

$$\hat{\delta} = 0 : \hat{H}^0(\mathbb{Z}_2;\hat{S}^{-1}\hat{A},\varepsilon) \longrightarrow \hat{H}^1(\mathbb{Z}_2;A,\varepsilon) \quad ,$$

thus obtaining a Mayer-Vietoris exact sequence in the corresponding intermediate ε-symmetric L-groups. (Special cases of the localization-completion Mayer-Vietoris sequences have already been obtained in §3.6 above). In §6.4 we shall consider the excision properties of the L-groups of cartesian squares of type I associated to ideals. In particular, an example will be constructed for which

$$L^*_\kappa(\Phi,\varepsilon) \neq 0$$

with Φ of type I (with both $g:B\longrightarrow A'$ and $g':B'\longrightarrow A'$ onto) and κ satisfying i) and ii). Thus the ε-symmetric L-groups do not have as good excision as the ε-quadratic L-groups.

Quadratic L-theory Mayer-Vietoris exact sequences for cartesian squares of types I and II have also been obtained by Bass [3], Wall [8], Karoubi [2] and Bak [2], in various special cases.

6.1 Triad L-groups

We shall now define the $\left\{\begin{array}{l}\varepsilon\text{-symmetric}\\ \varepsilon\text{-quadratic}\end{array}\right.$ triad L-groups $\left\{\begin{array}{l}L^*(\Phi,\varepsilon)\\ L_*(\Phi,\varepsilon)\end{array}\right.$ of a commutative square of rings with involution

$$\begin{array}{ccc} A & \longrightarrow & B \\ \big\downarrow & \Phi & \big\downarrow \\ B' & \longrightarrow & A' \end{array}$$

using the algebraic Poincaré triads of §1.3. The condition for excision $\left\{\begin{array}{l}L^*(\Phi,\varepsilon) = 0\\ L_*(\Phi,\varepsilon) = 0\end{array}\right.$ will be interpreted in terms of algebraic Poincaré splittings with respect to Φ of algebraic Poincaré complexes over A'. (The connections with geometric Poincaré splittings will be explored in §7.5 below). In order to do this it is convenient to use the unified L-theory of §1.8 to adopt the following terminology for algebraic Poincaré complexes, pairs and triads, which is a straightforward adaptation of the familiar terminology for geometric Poincaré complexes, pairs and triads. A more detailed account of this terminology will appear in Ranicki [11].

An <u>n-dimensional</u> $\left\{\begin{array}{l}\underline{\varepsilon\text{-symmetric}}\\ \underline{\varepsilon\text{-quadratic}}\end{array}\right.$ <u>Poincaré complex over A</u> ($n \in \mathbb{Z}$) is a closed object x of $\left\{\begin{array}{l}\mathcal{L}^n(A,\varepsilon)\\ \mathcal{L}_n(A,\varepsilon)\end{array}\right.$. For $n \geqslant 0$ this is exactly the same as an algebraic Poincaré complex of this type in the sense of §1.1.

An *n-dimensional* $\left\{\begin{array}{l}\underline{\varepsilon\text{-symmetric}}\\ \underline{\varepsilon\text{-quadratic}}\end{array}\right.$ *Poincaré pair over A* (x,y)

$(n \in \mathbb{Z})$ is defined by an object x of $\left\{\begin{array}{l}\mathcal{L}^n(A,\varepsilon)\\ \mathcal{L}_n(A,\varepsilon)\end{array}\right.$ and an object y of

$\left\{\begin{array}{l}\mathcal{L}^{n-1}(A,\varepsilon)\\ \mathcal{L}_{n-1}(A,\varepsilon)\end{array}\right.$ together with a homotopy equivalence

$$f : \partial x \xrightarrow{\ \sim\ } y \ ,$$

which will be used to identify $y = \partial x$.

An *n-dimensional* $\left\{\begin{array}{l}\underline{\varepsilon\text{-symmetric}}\\ \underline{\varepsilon\text{-quadratic}}\end{array}\right.$ *Poincaré triad over A*

$(x;\partial_+x,\partial_-x;\partial\partial_+x)$ $(n \in \mathbb{Z})$ is defined by an object x of $\left\{\begin{array}{l}\mathcal{L}^n(A,\varepsilon)\\ \mathcal{L}_n(A,\varepsilon)\end{array}\right.$

and objects ∂_+x,∂_-x of $\left\{\begin{array}{l}\mathcal{L}^{n-1}(A,\varepsilon)\\ \mathcal{L}_{n-1}(A,\varepsilon)\end{array}\right.$ together with homotopy

equivalences

$$f : \partial\partial_+x \xrightarrow{\ \sim\ } \partial\partial_-x \quad , \quad g : \partial_+x \cup_f -\partial_-x \xrightarrow{\ \sim\ } \partial x$$

which will be used as identifications.

The algebraic glueing operation of §1.7 is readily generalized to define the *union* of adjoining n-dimensional

$\left\{\begin{array}{l}\varepsilon\text{-symmetric}\\ \varepsilon\text{-quadratic}\end{array}\right.$ Poincaré triads over A $(x;y,y';z),(x';y',y'';z)$

as an n-dimensional $\left\{\begin{array}{l}\varepsilon\text{-symmetric}\\ \varepsilon\text{-quadratic}\end{array}\right.$ Poincaré triad over A

$$(x;y,y';z) \cup (x';y',y'';z) = (x \cup_{y'} x';y,y'';z) \quad .$$

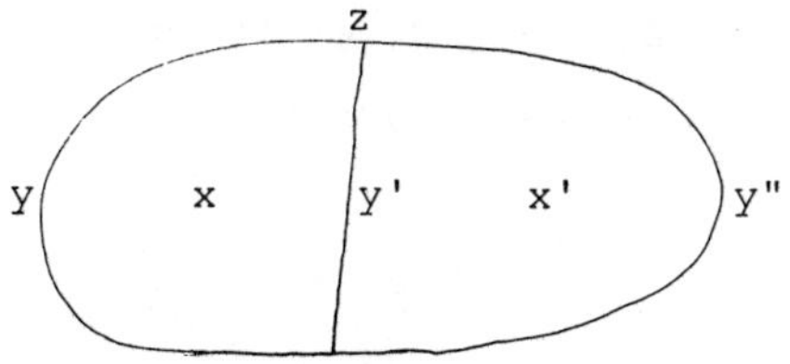

Given a morphism of rings with involution

$$f : A \longrightarrow B$$

let $\varepsilon \in A$ be a central unit such that $\bar{\varepsilon} = \varepsilon^{-1} \in A$ (as usual) and such that $f(\varepsilon) \in B$ is a central unit, also to be denoted ε.

An <u>n-dimensional</u> $\left\{\begin{array}{l} \underline{\varepsilon\text{-symmetric}} \\ \underline{\varepsilon\text{-quadratic}} \end{array}\right.$ <u>Poincaré pair over f</u> (y,x) $(n \in \mathbb{Z})$

consists of

i) an $(n-1)$-dimensional $\left\{\begin{array}{l} \varepsilon\text{-symmetric} \\ \varepsilon\text{-quadratic} \end{array}\right.$ Poincaré complex

over A x

ii) an n-dimensional $\left\{\begin{array}{l} \varepsilon\text{-symmetric} \\ \varepsilon\text{-quadratic} \end{array}\right.$ Poincaré pair over B

$(y, B\otimes_A x)$.

The relative $\left\{\begin{array}{l} \varepsilon\text{-symmetric} \\ \varepsilon\text{-quadratic} \end{array}\right.$ L-group $\left\{\begin{array}{l} L^n(f,\varepsilon) \\ L_n(f,\varepsilon) \end{array}\right.$ $(n \in \mathbb{Z})$ defined in §2.2 is the cobordism group of n-dimensional $\left\{\begin{array}{l} \varepsilon\text{-symmetric} \\ \varepsilon\text{-quadartic} \end{array}\right.$

Poincaré pairs over f.

Let Φ be a commutative square of rings with involution

$$\begin{array}{ccc} A & \xrightarrow{\quad f \quad} & B \\ {\scriptstyle f'}\big\downarrow & \Phi & \big\downarrow{\scriptstyle g} \\ B' & \xrightarrow{\quad g' \quad} & A' \end{array} \quad ,$$

and let $\varepsilon \in A$ be a central unit such that $\bar{\varepsilon} = \varepsilon^{-1} \in A$, and such that the elements $f(\varepsilon) \in B, f'(\varepsilon) \in B', gf(\varepsilon) = g'f'(\varepsilon) \in A'$ are also central units, all to be denoted by ε.

An n-dimensional $\left\{\begin{array}{l}\varepsilon\text{-symmetric}\\ \varepsilon\text{-quadratic}\end{array}\right.$ Poincaré triad over Φ

$(x';y,y';x)$ $(n \in \mathbb{Z})$ consists of

i) an (n-2)-dimensional $\left\{\begin{array}{l}\varepsilon\text{-symmetric}\\ \varepsilon\text{-quadratic}\end{array}\right.$ Poincaré complex

over A x

ii) an (n-1)-dimensional $\left\{\begin{array}{l}\varepsilon\text{-symmetric}\\ \varepsilon\text{-quadratic}\end{array}\right.$ Poincaré pair

over B $(y, B\otimes_A x)$

iii) an (n-1)-dimensional $\left\{\begin{array}{l}\varepsilon\text{-symmetric}\\ \varepsilon\text{-quadratic}\end{array}\right.$ Poincaré pair

over B' $(y', B'\otimes_A x)$

iv) an n-dimensional $\left\{\begin{array}{l}\varepsilon\text{-symmetric}\\ \varepsilon\text{-quadratic}\end{array}\right.$ Poincaré pair

over A' $(x', A'\otimes_B y \cup_{A'\otimes_A(-x)} A'\otimes_{B'}(-y'))$.

In particular, $(x';A'\otimes_B y, A'\otimes_{B'} y';A'\otimes_A x)$ is an n-dimensional

$\left\{\begin{array}{l}\varepsilon\text{-symmetric}\\ \varepsilon\text{-quadratic}\end{array}\right.$ Poincaré triad over A'.

Define the n-dimensional $\left\{\begin{array}{l}\varepsilon\text{-symmetric}\\ \varepsilon\text{-quadratic}\end{array}\right.$ triad L-group of Φ

$\left\{\begin{array}{l}L^n(\Phi,\varepsilon)\\ L_n(\Phi,\varepsilon)\end{array}\right.$ $(n \in \mathbb{Z})$ to be the cobordism group of n-dimensional

$\left\{\begin{array}{l}\varepsilon\text{-symmetric}\\ \varepsilon\text{-quadratic}\end{array}\right.$ Poincaré triads over Φ.

Proposition 6.1.1 i) The ε-symmetric triad L-groups $L^*(\Phi,\varepsilon)$ fit into a commutative diagram of abelian groups with exact rows and columns

$$
\begin{array}{ccccccccc}
 & & \vdots & & \vdots & & \vdots & & \vdots & \\
 & & \downarrow & & \downarrow & & \downarrow & & \downarrow & \\
\dots \longrightarrow & L^n(A,\varepsilon) & \xrightarrow{f} & L^n(B,\varepsilon) & \longrightarrow & L^n(f,\varepsilon) & \longrightarrow & L^{n-1}(A,\varepsilon) & \longrightarrow \dots \\
 & \downarrow{\scriptstyle f'} & & \downarrow{\scriptstyle g} & & \downarrow{\scriptstyle (f',g)} & & \downarrow{\scriptstyle f'} & \\
\dots \longrightarrow & L^n(B',\varepsilon) & \xrightarrow{g'} & L^n(A',\varepsilon) & \longrightarrow & L^n(g',\varepsilon) & \longrightarrow & L^{n-1}(B',\varepsilon) & \longrightarrow \dots \\
 & \downarrow & & \downarrow & & \downarrow & & \downarrow & \\
\dots \longrightarrow & L^n(f',\varepsilon) & \xrightarrow{(f,g')} & L^n(g,\varepsilon) & \longrightarrow & L^n(\Phi,\varepsilon) & \longrightarrow & L^{n-1}(f',\varepsilon) & \longrightarrow \dots \\
 & \downarrow & & \downarrow & & \downarrow & & \downarrow & \\
\dots \longrightarrow & L^{n-1}(A,\varepsilon) & \xrightarrow{f} & L^{n-1}(B,\varepsilon) & \longrightarrow & L^{n-1}(f,\varepsilon) & \longrightarrow & L^{n-2}(A,\varepsilon) & \longrightarrow \dots \\
 & \downarrow & & \downarrow & & \downarrow & & \downarrow & \\
 & \vdots & & \vdots & & \vdots & & \vdots &
\end{array}
$$

$(n \in \mathbb{Z})$.

Similarly for the ε-quadratic triad L-groups $L_*(\Phi,\varepsilon)$.

ii) If Φ,ε are such that $\begin{cases} L^*(\Phi,\varepsilon) = 0 \\ L_*(\Phi,\varepsilon) = 0 \end{cases}$ then there are defined excision isomorphisms of relative $\begin{cases} \varepsilon\text{-symmetric} \\ \varepsilon\text{-quadratic} \end{cases}$ L-groups

$$
\begin{cases} (f',g) : L^*(f,\varepsilon) \xrightarrow{\simeq} L^*(g',\varepsilon) \\ (f',g) : L_*(f,\varepsilon) \xrightarrow{\simeq} L_*(g',\varepsilon) \end{cases}
$$

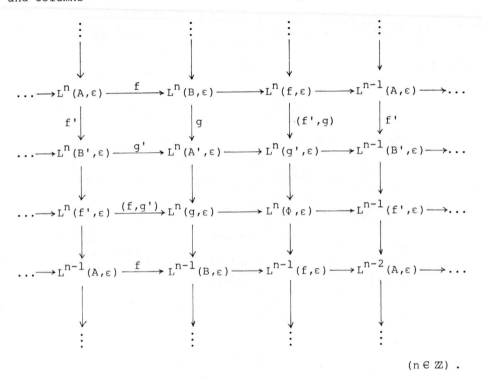

and a Mayer-Vietoris exact sequence of absolute $\begin{cases}\varepsilon\text{-symmetric}\\ \varepsilon\text{-quadratic}\end{cases}$

L-groups

$$\begin{cases} \ldots \longrightarrow L^n(A,\varepsilon) \xrightarrow{\begin{pmatrix} f \\ f' \end{pmatrix}} L^n(B,\varepsilon)\oplus L^n(B',\varepsilon) \xrightarrow{(g \;\; -g')} L^n(A',\varepsilon) \\ \qquad \xrightarrow{\partial} L^{n-1}(A,\varepsilon) \xrightarrow{\begin{pmatrix} f \\ f' \end{pmatrix}} L^{n-1}(B,\varepsilon)\oplus L^{n-1}(B',\varepsilon) \longrightarrow \ldots \\ \ldots \longrightarrow L_n(A,\varepsilon) \xrightarrow{\begin{pmatrix} f \\ f' \end{pmatrix}} L_n(B,\varepsilon)\oplus L_n(B',\varepsilon) \xrightarrow{(g \;\; -g')} L_n(A',\varepsilon) \\ \qquad \xrightarrow{\partial} L_{n-1}(A,\varepsilon) \xrightarrow{\begin{pmatrix} f \\ f' \end{pmatrix}} L_{n-1}(B,\varepsilon)\oplus L_{n-1}(B',\varepsilon) \longrightarrow \ldots \end{cases}$$

$(n \in \mathbb{Z})$

with the connecting maps ∂ given by

$$\begin{cases} \partial : L^n(A',\varepsilon) \longrightarrow L^n(g',\varepsilon) \xrightarrow[\sim]{(f',g)^{-1}} L^n(f,\varepsilon) \longrightarrow L^{n-1}(A,\varepsilon) \\ \partial : L_n(A',\varepsilon) \longrightarrow L_n(g',\varepsilon) \xrightarrow[\sim]{(f',g)^{-1}} L_n(f,\varepsilon) \longrightarrow L_{n-1}(A,\varepsilon) \end{cases} .$$

[]

We shall now interpret the excision condition

$$\begin{cases} L^*(\Phi,\varepsilon) = 0 \\ L_*(\Phi,\varepsilon) = 0 \end{cases}$$

in terms of algebraic Poincaré splittings with respect to Φ of algebraic Poincaré complexes over A'.

A <u>Poincaré splitting (with respect to Φ)</u> (y,y',x) of an n-dimensional $\left\{\begin{array}{l}\varepsilon\text{-symmetric}\\ \varepsilon\text{-quadratic}\end{array}\right.$ Poincaré complex over A' x' $(n \in \mathbb{Z})$ consists of:

i) an $(n-1)$-dimensional $\left\{\begin{array}{l}\varepsilon\text{-symmetric}\\ \varepsilon\text{-quadratic}\end{array}\right.$ Poincaré complex over A x

ii) an n-dimensional $\left\{\begin{array}{l}\varepsilon\text{-symmetric}\\ \varepsilon\text{-quadratic}\end{array}\right.$ Poincaré pair over B $(y, B\otimes_A x)$

iii) an n-dimensional $\left\{\begin{array}{l}\varepsilon\text{-symmetric}\\ \varepsilon\text{-quadratic}\end{array}\right.$ Poincaré pair over B' $(y', B'\otimes_A x)$

iv) a homotopy equivalence

$$A'\otimes_B y \cup_{A'\otimes_A (-x)} A'\otimes_{B'} (-y') \xrightarrow{\sim} x'$$

which will be used as an identification.

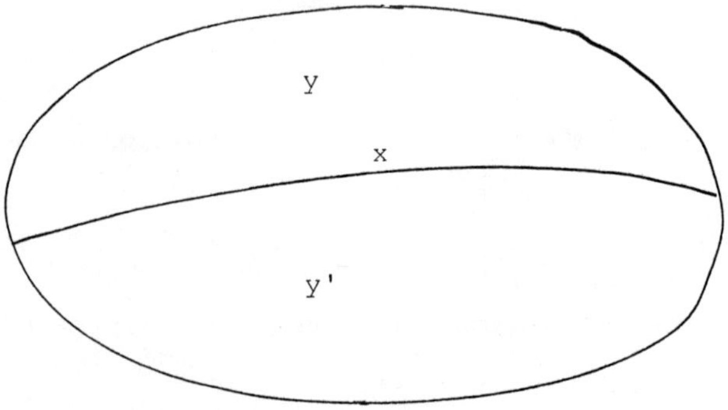

x'

There is also a relative version of Poincaré splitting, as follows.

A <u>Poincaré splitting (with respect to Φ)</u>

$((y,\partial_+y),(y',\partial_+y'),(x,\partial x))$ of an n-dimensional $\left\{\begin{array}{l}\varepsilon\text{-symmetric}\\ \varepsilon\text{-quadratic}\end{array}\right.$ Poincaré pair over A' $(x',\partial x')$ $(n\in\mathbb{Z})$ consists of:

i) an (n-1)-dimensional $\left\{\begin{array}{l}\varepsilon\text{-symmetric}\\ \varepsilon\text{-quadratic}\end{array}\right.$ Poincaré pair over A $(x,\partial x)$

ii) an n-dimensional $\left\{\begin{array}{l}\varepsilon\text{-symmetric}\\ \varepsilon\text{-quadratic}\end{array}\right.$ Poincaré triad over B $(y;\partial_+y,B\boxtimes_A(-x);B\boxtimes_A(-\partial x))$

iii) an n-dimensional $\left\{\begin{array}{l}\varepsilon\text{-symmetric}\\ \varepsilon\text{-quadratic}\end{array}\right.$ Poincaré triad over B' $(y';\partial_+y',B'\boxtimes_A(-x);B'\boxtimes_A(-\partial x))$

iv) a homotopy equivalence of pairs

$$(A'\boxtimes_B y\cup_{A'\boxtimes_A(-x)}A'\boxtimes_{B'}(-y'),A'\boxtimes_B\partial_+y\cup_{A'\boxtimes_A\partial x}A'\boxtimes_{B'}(-\partial_+y'))\xrightarrow{\ \simeq\ }(x',\partial x').$$

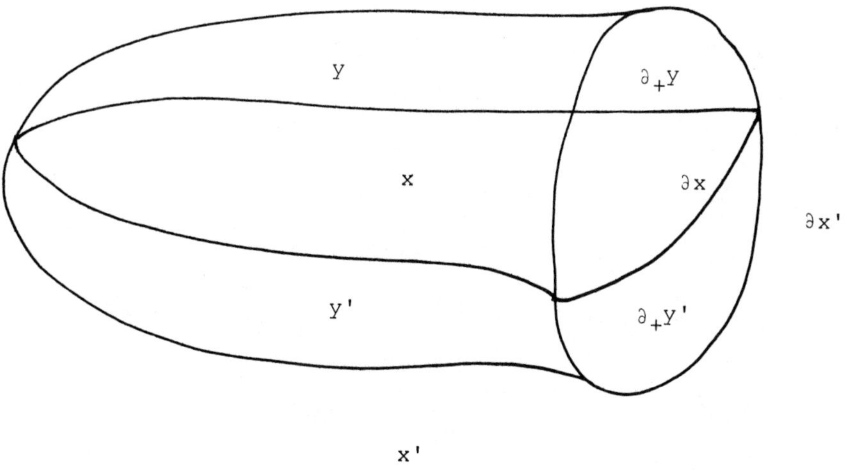

Note that $(\partial_+ y, \partial_+ y', -\partial x)$ is a Poincaré splitting of the boundary (n-1)-dimensional $\begin{cases}\varepsilon\text{-symmetric}\\ \varepsilon\text{-quadratic}\end{cases}$ Poincaré complex over A' $\partial x'$.

An n-dimensional $\begin{cases}\varepsilon\text{-symmetric}\\ \varepsilon\text{-quadratic}\end{cases}$ Poincaré triad over Φ $(x';y,y';x)$ is thus an n-dimensional $\begin{cases}\varepsilon\text{-symmetric}\\ \varepsilon\text{-quadratic}\end{cases}$ Poincaré pair over A' $(x',\partial x')$ together with a Poincaré splitting (y,y',x) of the boundary (n-1)-dimensional $\begin{cases}\varepsilon\text{-symmetric}\\ \varepsilon\text{-quadratic}\end{cases}$ Poincaré complex over A' $\partial x'$, so that

$$\partial x' = A'\otimes_B y \cup_{A'\otimes_A(-x)} A'\otimes_{B'}(-y') \ .$$

A cobordism of such triads $(x_i';y_i,y_i';x_i)$ $(i = 1,2)$ is an (n+1)-dimensional $\begin{cases}\varepsilon\text{-symmetric}\\ \varepsilon\text{-quadratic}\end{cases}$ Poincaré triad over A'

$$(\delta x';\partial_+\delta x', x_1'\oplus -x_2';\partial x_1'\oplus -\partial x_2')$$

such that the n-dimensional $\begin{cases}\varepsilon\text{-symmetric}\\ \varepsilon\text{-quadratic}\end{cases}$ Poincaré pair over A' $(\partial_+\delta x', \partial x_1'\oplus -\partial x_2')$ has a Poincaré splitting

$$((\delta y, y_1\oplus -y_2), (\delta y', y_1'\oplus -y_2'), (\delta x, x_1\oplus -x_2))$$

extending the given Poincaré splitting $(y_1\oplus -y_2, y_1'\oplus -y_2', x_1\oplus -x_2)$ of the boundary (n-1)-dimensional $\begin{cases}\varepsilon\text{-symmetric}\\ \varepsilon\text{-quadratic}\end{cases}$ Poincaré complex over A' $\partial x_1'\oplus -\partial x_2'$.

The n-dimensional $\begin{cases}\varepsilon\text{-symmetric}\\ \varepsilon\text{-quadratic}\end{cases}$ triad L-group $\begin{cases}L^n(\Phi,\varepsilon)\\ L_n(\Phi,\varepsilon)\end{cases}$ $(n\in\mathbb{Z})$

is thus the cobordism group of n-dimensional $\begin{cases}\varepsilon\text{-symmetric}\\ \varepsilon\text{-quadratic}\end{cases}$

Poincaré pairs over A' with a Poincaré split boundary.

Proposition 6.1.2 The $\begin{cases}\varepsilon\text{-symmetric}\\ \varepsilon\text{-quadratic}\end{cases}$ L-theory excision condition

$\begin{cases}L^*(\Phi,\varepsilon)=0\\ L_*(\Phi,\varepsilon)=0\end{cases}$ is satisfied if and only if every $\begin{cases}\varepsilon\text{-symmetric}\\ \varepsilon\text{-quadratic}\end{cases}$

Poincaré pair over A' with a Poincaré split boundary is cobordant to a Poincaré split pair.

In particular, if the excision condition is satisfied

then every $\begin{cases}\varepsilon\text{-symmetric}\\ \varepsilon\text{-quadratic}\end{cases}$ Poincaré complex over A' x' is cobordant

to a Poincaré split complex $A'\otimes_B y\cup_{A'\otimes_A(-x)}A'\otimes_{B'}(-y')$ and the

connecting maps in the $\begin{cases}\varepsilon\text{-symmetric}\\ \varepsilon\text{-quadratic}\end{cases}$ L-theory Mayer-Vietoris

exact sequence of Φ

$$\begin{cases}\partial : L^n(A',\varepsilon)\longrightarrow L^{n-1}(A,\varepsilon)\\ \partial : L_n(A',\varepsilon)\longrightarrow L_{n-1}(A,\varepsilon)\end{cases}\quad (n\in\mathbb{Z})$$

are given by

$$\partial(x') = \partial(A'\otimes_B y\cup_{A'\otimes_A(-x)}A'\otimes_{B'}(-y')) = x\ .$$

[]

While the Poincaré splitting condition for excision of Proposition 6.1.2 has a pleasantly geometric flavour (to which we shall return in §7 below) the following criterion for excision will be found to be of greater use in §6.3 below.

<u>Proposition 6.1.3</u> The induced map of relative ε-symmetric L-groups for some $n \in \mathbb{Z}$

$$(f',g) \; : \; L^n(f,\varepsilon) \longrightarrow L^n(g',\varepsilon)$$

(for a commutative square of rings with involution $\begin{array}{ccc} A & \xrightarrow{f} & B \\ {\scriptstyle f'}\downarrow & \Phi & \downarrow{\scriptstyle g} \\ B' & \xrightarrow{g'} & A' \end{array}$)

is an isomorphism if and only if there exist abelian group morphisms

$$\delta \; : \; L^n(A',\varepsilon) \longrightarrow L^n(f,\varepsilon)$$

$$\hat{\delta} \; : \; L^n(g',\varepsilon) \longrightarrow L^{n-1}(A,\varepsilon)$$

fitting into a commutative diagram

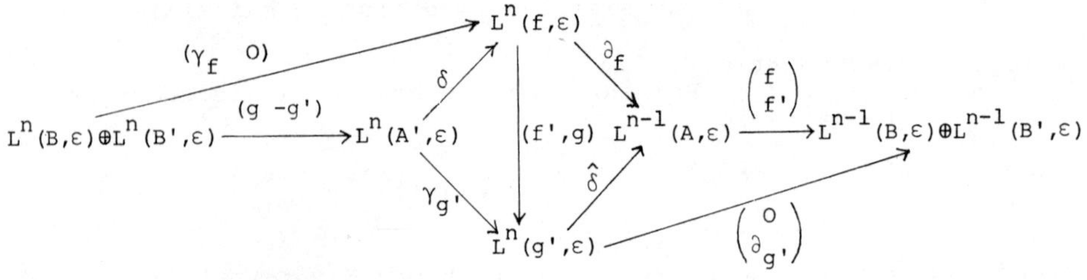

involving the change of rings exact sequences

$$L^n(B,\varepsilon) \xrightarrow{\gamma_f} L^n(f,\varepsilon) \xrightarrow{\partial_f} L^{n-1}(A,\varepsilon) \xrightarrow{f} L^{n-1}(B,\varepsilon)$$

$$L^n(B',\varepsilon) \xrightarrow{g'} L^n(A',\varepsilon) \xrightarrow{\gamma_{g'}} L^n(g',\varepsilon) \xrightarrow{\partial_{g'}} L^{n-1}(B',\varepsilon) \quad .$$

If such morphisms $\delta, \hat{\delta}$ exist there is defined an exact sequence

$$L^n(B,\varepsilon)\oplus L^n(B',\varepsilon) \xrightarrow{(g \;\; -g')} L^n(A',\varepsilon)$$

$$\xrightarrow{\partial} L^{n-1}(A,\varepsilon) \xrightarrow{\begin{pmatrix} f \\ f' \end{pmatrix}} L^{n-1}(B,\varepsilon)\oplus L^{n-1}(B',\varepsilon)$$

with

$$\partial = \gamma_f \delta = \hat{\delta}\gamma_{g'} \; : \; L^n(A',\varepsilon) \longrightarrow L^{n-1}(A,\varepsilon) \quad ,$$

and if $(f',g):L^{n+1}(f,\varepsilon) \longrightarrow L^{n+1}(g',\varepsilon)$ is onto there is an extension of this sequence to the left by an exact sequence

$$L^n(A,\varepsilon) \xrightarrow{\begin{pmatrix} f \\ f' \end{pmatrix}} L^n(B,\varepsilon) \oplus L^n(B',\varepsilon) \xrightarrow{(g \ -g')} L^n(A',\varepsilon) \quad .$$

Similarly for the ε-quadratic L-groups L_*.

<u>Proof</u>: If $(f',g):L^n(f,\varepsilon) \longrightarrow L^n(g',\varepsilon)$ is an isomorphism define

$$\delta = (f',g)^{-1}\gamma_{g'} \ : \ L^n(A',\varepsilon) \longrightarrow L^n(f,\varepsilon)$$

$$\hat{\delta} = \partial_f(f',g)^{-1} \ : \ L^n(g',\varepsilon) \longrightarrow L^{n-1}(A,\varepsilon) \ .$$

Conversely, given $\delta, \hat{\delta}$ we shall verify that (f',g) is an isomorphism by diagram chasing, as follows.

Let $x \in \ker((f',g):L^n(f,\varepsilon) \longrightarrow L^n(g',\varepsilon))$, so that

$$\partial_f(x) = \hat{\delta}(f',g)(x) = 0 \in L^{n-1}(A,\varepsilon)$$

and $x \in \ker(\partial_f:L^n(f,\varepsilon) \longrightarrow L^{n-1}(A,\varepsilon)) = \mathrm{im}(\gamma_f:L^n(B,\varepsilon) \longrightarrow L^n(f,\varepsilon))$. Let $y \in L^n(B,\varepsilon)$ be such that

$$\gamma_f(y) = x \in L^n(f,\varepsilon) \ ,$$

so that

$$\gamma_{g'}g(y) = (f',g)\gamma_f(y) = (f',g)(x) = 0 \in L^n(g',\varepsilon)$$

and $g(y) \in \ker(\gamma_{g'}:L^n(A',\varepsilon) \longrightarrow L^n(g',\varepsilon)) = \mathrm{im}(g':L^n(B',\varepsilon) \longrightarrow L^n(A',\varepsilon))$. Let $z \in L^n(B',\varepsilon)$ be such that

$$g(y) = g'(z) \in L^n(A',\varepsilon) \ ,$$

so that

$$x = \gamma_f(y) = \delta g(y) = \delta g'(z) = 0 \in L^n(f,\varepsilon) \ .$$

Thus $(f',g):L^n(f,\varepsilon) \longrightarrow L^n(g',\varepsilon)$ is one-one.

Given an element $u \in L^n(g',\varepsilon)$ we have

$$f\hat{\delta}(u) = 0 \in L^{n-1}(B,\varepsilon) ,$$

so that

$$\begin{aligned}\hat{\delta}(u) \in \ker(f:L^{n-1}(A,\varepsilon) \longrightarrow L^{n-1}(B,\varepsilon)) \\ = \operatorname{im}(\partial_f:L^n(f,\varepsilon)\longrightarrow L^{n-1}(A,\varepsilon)) .\end{aligned}$$

Let $v \in L^n(f,\varepsilon)$ be such that

$$\hat{\delta}(u) = \partial_f(v) \in L^{n-1}(A,\varepsilon) ,$$

so that

$$\begin{aligned}(u - (f',g)(v)) \in \ker(\hat{\delta}:L^n(g',\varepsilon) \longrightarrow L^{n-1}(A,\varepsilon)) \\ \subseteq \ker(\partial_{g'}:L^n(g',\varepsilon) \longrightarrow L^{n-1}(B',\varepsilon)) \\ = \operatorname{im}(\gamma_{g'}:L^n(A',\varepsilon)\longrightarrow L^n(g',\varepsilon)) .\end{aligned}$$

Let $w \in L^n(A',\varepsilon)$ be such that

$$u - (f',g)(v) = \gamma_{g'}(w) \in L^n(g',\varepsilon) ,$$

so that

$$u = (f',g)(v + \delta(w)) \in \operatorname{im}((f',g):L^n(f,\varepsilon) \longrightarrow L^n(g',\varepsilon)) .$$

Thus $(f',g):L^n(f,\varepsilon)\longrightarrow L^n(g',\varepsilon)$ is onto.

Suppose now that $\delta,\hat{\delta}$ exist.

Given $x \in \ker\left(\begin{pmatrix} f \\ f' \end{pmatrix} :L^{n-1}(A,\varepsilon)\longrightarrow L^{n-1}(B,\varepsilon)\oplus L^{n-1}(B',\varepsilon)\right)$

we have that

$$\begin{aligned}x \in \ker(f:L^{n-1}(A,\varepsilon)\longrightarrow L^{n-1}(B,\varepsilon)) \\ = \operatorname{im}(\partial_f:L^n(f,\varepsilon)\longrightarrow L^{n-1}(A,\varepsilon)) .\end{aligned}$$

Let $y \in L^n(f,\varepsilon)$ be such that

$$x = \partial_f(y) \in L^{n-1}(A,\varepsilon) ,$$

so that

$$\partial_{g'}(f',g)(y) = f'\partial_f(y) = f'(x) = 0 \in L^{n-1}(B',\varepsilon)$$

and there exists $z \in L^n(A',\varepsilon)$ such that

$$(f',g)(y) = \gamma_{g'}(z) \in L^n(g',\varepsilon) .$$

Thus

$$x = \partial_f(y) = \partial_f(f',g)^{-1}\gamma_{g'}(z) = \partial(z) \in L^{n-1}(A,\varepsilon) ,$$

and we have verified the exactness of

$$L^n(A',\varepsilon) \xrightarrow{\partial} L^{n-1}(A,\varepsilon) \xrightarrow{\begin{pmatrix} f \\ f' \end{pmatrix}} L^{n-1}(B,\varepsilon) \oplus L^{n-1}(B',\varepsilon) .$$

Given $x \in \ker(\partial : L^n(A',\varepsilon) \longrightarrow L^{n-1}(A,\varepsilon))$ we have that

$$\gamma_f \delta(x) = 0 \in L^{n-1}(A,\varepsilon) .$$

Let $y \in L^n(B,\varepsilon)$ be such that

$$\delta(x) = \gamma_f(y) \in L^n(f,\varepsilon) ,$$

so that

$$\gamma_{g'}(x - g(y)) = (f',g)(\delta(x) - \gamma_f(y)) = 0 \in L^n(g',\varepsilon) ,$$

and there exists $y' \in L^n(B',\varepsilon)$ such that

$$x = g(y) - g'(y') \in L^n(A',\varepsilon) .$$

This verifies the exactness of

$$L^n(B,\varepsilon) \oplus L^n(B',\varepsilon) \xrightarrow{(g \ -g')} L^n(A',\varepsilon) \xrightarrow{\partial} L^{n-1}(A,\varepsilon).$$

Assume now that $(f',g) : L^{n+1}(f,\varepsilon) \longrightarrow L^{n+1}(g',\varepsilon)$ is onto.

Given $(y,y') \in \ker((g \ -g') : L^n(B,\varepsilon) \oplus L^n(B',\varepsilon) \longrightarrow L^n(A',\varepsilon))$ we have

$$\gamma_f(y) = \delta g(y) = 0 \in L^n(f,\varepsilon) ,$$

so that there exists $z \in L^n(A,\varepsilon)$ such that

$$y = f(z) \in L^n(B,\varepsilon).$$

Now

$$g'(y'-f'(z)) = g(y)-gf(z) = 0 \in L^n(A',\varepsilon) \ ,$$

so that there exists $w \in L^{n+1}(g',\varepsilon)$ such that

$$y'-f'(z) = \partial_{g'}(w) \in L^n(B',\varepsilon) \ .$$

As (f',g) is onto there exists $v \in L^{n+1}(f,\varepsilon)$ such that

$$w = (f',g)(v) \in L^{n+1}(g',\varepsilon) \ .$$

The element

$$x = z+\partial_f(v) \in L^n(A,\varepsilon)$$

is such that

$$(y,y') = (f(x),f'(x)) \in L^n(B,\varepsilon)\oplus L^n(B',\varepsilon) \ ,$$

verifying the exactness of

$$L^n(A,\varepsilon) \xrightarrow{\begin{pmatrix} f \\ f' \end{pmatrix}} L^n(B,\varepsilon)\oplus L^n(B',\varepsilon) \xrightarrow{(g \ \ -g')} L^n(A',\varepsilon) \ .$$

[]

6.2 Change of K-theory

We shall now develop the theory of intermediate triad L-groups, that is the analogues of the triad L-groups of §6.1 for the intermediate L-groups of §1.10. The terminology and and results of §6.1 have obvious intermediate L-theory analogues. In §6.3 the intermediate triad L-groups will be used to obtain excision isomorphisms and Mayer-Vietoris exact sequences in L-theory.

Given a commutative square of rings

$$\begin{array}{ccc} A & \xrightarrow{\ f\ } & B \\ {\scriptstyle f'}\big\downarrow & \Phi & \big\downarrow{\scriptstyle g} \\ B' & \xrightarrow[\ g'\ ]{} & A' \end{array}$$

let

$$\kappa = \left(\begin{array}{ccc} X & \longrightarrow & Y \\ \big\downarrow & & \big\downarrow \\ Y' & \longrightarrow & X' \end{array}\right) \subseteq \left(\begin{array}{ccc} \widetilde{K}_m(A) & \longrightarrow & \widetilde{K}_m(B) \\ \big\downarrow & & \big\downarrow \\ \widetilde{K}_m(B') & \longrightarrow & \widetilde{K}_m(A') \end{array}\right)$$

(m = 0 or 1)

be a commutative square of $*$-invariant subgroups, that is a collection of $*$-invariant subgroups

$$X \subseteq \widetilde{K}_m(A) \quad , \quad Y \subseteq \widetilde{K}_m(B) \quad , \quad Y' \subseteq \widetilde{K}_m(B') \quad , \quad X' \subseteq \widetilde{K}_m(A')$$

such that

$$B \otimes_A X \subseteq Y \quad , \quad B' \otimes_A X \subseteq Y' \quad , \quad A' \otimes_B Y \subseteq X' \quad , \quad A' \otimes_{B'} Y' \subseteq X' \ .$$

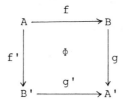

The Tate $\mathbb{Z}_2$-cohomology groups of κ $\hat{H}^*(\mathbb{Z}_2;\kappa)$ are defined by

$$\hat{H}^n(\mathbb{Z}_2;\kappa) = \frac{\ker(d_n : G \longrightarrow G)}{\operatorname{im}(d_{n+1} : G \longrightarrow G)} \quad (n \pmod 2) ,$$

with

$$d_n = \begin{pmatrix} 1-(-)^n T & (-)^n g & -(-)^n g' & 0 \\ 0 & 1+(-)^n T & 0 & (-)^n f \\ 0 & 0 & 1+(-)^n T & (-)^n f' \\ 0 & 0 & 0 & 1-(-)^n T \end{pmatrix}$$

$$: G = X' \oplus Y \oplus Y' \oplus X \longrightarrow G = X' \oplus Y \oplus Y' \oplus X \qquad (d_n d_{n+1} = 0) ,$$

and are such that there is defined a commutative diagram of abelian groups with exact rows and columns

$$\begin{array}{ccccccccc}
 & & \vdots & & \vdots & & \vdots & & \vdots \\
 & & \downarrow & & \downarrow & & \downarrow & & \downarrow \\
\cdots \longrightarrow & & \hat{H}^n(\mathbb{Z}_2;X) & \longrightarrow & \hat{H}^n(\mathbb{Z}_2;Y) & \longrightarrow & \hat{H}^n(\mathbb{Z}_2;X\to Y) & \longrightarrow & \hat{H}^{n-1}(\mathbb{Z}_2;X) \longrightarrow \cdots \\
 & & \downarrow & & \downarrow & & \downarrow & & \downarrow \\
\cdots \longrightarrow & & \hat{H}^n(\mathbb{Z}_2;Y') & \longrightarrow & \hat{H}^n(\mathbb{Z}_2;X') & \longrightarrow & \hat{H}^n(\mathbb{Z}_2;Y'\to X') & \longrightarrow & \hat{H}^{n-1}(\mathbb{Z}_2;Y') \longrightarrow \cdots \\
 & & \downarrow & & \downarrow & & \downarrow & & \downarrow \\
\cdots \longrightarrow & & \hat{H}^n(\mathbb{Z}_2;X\to Y') & \longrightarrow & \hat{H}^n(\mathbb{Z}_2;Y\to X') & \longrightarrow & \hat{H}^n(\mathbb{Z}_2;\kappa) & \longrightarrow & \hat{H}^{n-1}(\mathbb{Z}_2;X\to Y') \longrightarrow \cdots \\
 & & \downarrow & & \downarrow & & \downarrow & & \downarrow \\
\cdots \longrightarrow & & \hat{H}^{n-1}(\mathbb{Z}_2;X) & \longrightarrow & \hat{H}^{n-1}(\mathbb{Z}_2;Y) & \longrightarrow & \hat{H}^{n-1}(\mathbb{Z}_2;X\to Y) & \longrightarrow & \hat{H}^{n-2}(\mathbb{Z}_2;X) \longrightarrow \cdots \\
 & & \downarrow & & \downarrow & & \downarrow & & \downarrow \\
 & & \vdots & & \vdots & & \vdots & & \vdots
\end{array} .$$

Given two squares κ_1,κ_2 such as κ, with $\kappa_1 \subseteq \kappa_2$ (i.e. such that $X_1 \subseteq X_2, Y_1 \subseteq Y_2, Y_1' \subseteq Y_2', X_1' \subseteq X_2'$), there are also defined <u>relative Tate $\mathbb{Z}_2$-cohomology groups</u> $\hat{H}^*(\mathbb{Z}_2;\kappa_2/\kappa_1)$, which fit into an exact sequence

$$\ldots \longrightarrow \hat{H}^n(\mathbb{Z}_2;\kappa_1) \longrightarrow \hat{H}^n(\mathbb{Z}_2;\kappa_2) \longrightarrow \hat{H}^n(\mathbb{Z}_2;\kappa_2/\kappa_1) \longrightarrow \hat{H}^{n-1}(\mathbb{Z}_2;\kappa_1) \longrightarrow \ldots$$
$$(n(\mathrm{mod}\ 2)).$$

Given commutative squares Φ,κ as above define the <u>intermediate</u> $\left\{\begin{array}{l}\underline{\varepsilon\text{-symmetric}}\\ \underline{\varepsilon\text{-quadratic}}\end{array}\right.$ <u>triad L-groups</u> $\left\{\begin{array}{l}L^n_\kappa(\Phi,\varepsilon)\\ L_n^\kappa(\Phi,\varepsilon)\end{array}\right.$ $(n \in \mathbb{Z})$ in exactly the same way as the triad L-groups $\left\{\begin{array}{l}L^*(\Phi,\varepsilon)\\ L_*(\Phi,\varepsilon)\end{array}\right.$ of §6.1

(which are the special case $\kappa = \begin{pmatrix} \tilde{K}_0(A) & \longrightarrow & \tilde{K}_0(B) \\ \downarrow & & \downarrow \\ \tilde{K}_0(B') & \longrightarrow & \tilde{K}_0(A') \end{pmatrix}$),

but using only the $\left\{\begin{array}{l}\varepsilon\text{-symmetric}\\ \varepsilon\text{-quadratic}\end{array}\right.$ Poincaré triads over Φ with K-theory in κ. There are then intermediate versions of Propositions 6.1.1, 6.1.2, 6.1.3.

<u>Proposition 6.2.1</u> Given commutative squares Φ,κ_1,κ_2 such that $\kappa_1 \subseteq \kappa_2$ there is defined an exact sequence of intermediate $\left\{\begin{array}{l}\varepsilon\text{-symmetric}\\ \varepsilon\text{-quadratic}\end{array}\right.$ triad L-groups

$$\left\{\begin{array}{l}\ldots \longrightarrow L^n_{\kappa_1}(\Phi,\varepsilon) \longrightarrow L^n_{\kappa_2}(\Phi,\varepsilon) \longrightarrow \hat{H}^n(\mathbb{Z}_2;\kappa_2/\kappa_1) \longrightarrow L^{n-1}_{\kappa_1}(\Phi,\varepsilon) \longrightarrow \ldots \\ \ldots \longrightarrow L_n^{\kappa_1}(\Phi,\varepsilon) \longrightarrow L_n^{\kappa_2}(\Phi,\varepsilon) \longrightarrow \hat{H}^n(\mathbb{Z}_2;\kappa_2/\kappa_1) \longrightarrow L_{n-1}^{\kappa_1}(\Phi,\varepsilon) \longrightarrow \ldots\ . \end{array}\right. \quad (n \in \mathbb{Z})$$

<u>Proof</u>: By analogy with Proposition 1.10.1.

[]

In §6.3 below we shall prove that for certain Φ,κ $\begin{cases} L^*_\kappa(\Phi,\varepsilon) = 0 \\ L^\kappa_*(\Phi,\varepsilon) = 0 \end{cases}$, thus obtaining a Mayer-Vietoris exact sequence of intermediate $\begin{cases} \varepsilon\text{-symmetric} \\ \varepsilon\text{-quadratic} \end{cases}$ L-groups

$$\begin{cases} \ldots \longrightarrow L^n_X(A,\varepsilon) \longrightarrow L^n_Y(B,\varepsilon)\oplus L^n_{Y'}(B',\varepsilon) \longrightarrow L^n_{X'}(A',\varepsilon) \xrightarrow{\partial} L^{n-1}_X(A,\varepsilon) \longrightarrow \ldots \\ \ldots \longrightarrow L_n^X(A,\varepsilon) \longrightarrow L_n^Y(B,\varepsilon)\oplus L_n^{Y'}(B',\varepsilon) \longrightarrow L_n^{X'}(A',\varepsilon) \xrightarrow{\partial} L_{n-1}^X(A,\varepsilon) \longrightarrow \ldots \end{cases}$$

$$(n \in \mathbb{Z}) .$$

In every such case κ will satisfy the following condition.

The commutative square of $*$-invariant subgroups

$$\kappa = \begin{pmatrix} X & \longrightarrow & Y \\ \downarrow & & \downarrow \\ Y' & \longrightarrow & X' \end{pmatrix} \subseteq \begin{pmatrix} \tilde{K}_m(A) & \longrightarrow & \tilde{K}_m(B) \\ \downarrow & & \downarrow \\ \tilde{K}_m(B') & \longrightarrow & \tilde{K}_m(A') \end{pmatrix} \quad (m = 0 \text{ or } 1)$$

is <u>cartesian</u> if

i) X contains the $*$-invariant subgroup

$$I_m = \ker(\begin{pmatrix} f \\ f' \end{pmatrix} : \tilde{K}_m(A) \longrightarrow \tilde{K}_m(B)\oplus\tilde{K}_m(B')) \subseteq \tilde{K}_m(A)$$

ii) the $\mathbb{Z}[\mathbb{Z}_2]$-module sequence

$$0 \longrightarrow X/I_m \xrightarrow{\begin{pmatrix} f \\ f' \end{pmatrix}} Y\oplus Y' \xrightarrow{(g\ \ -g')} X' \longrightarrow 0$$

is exact.

The Tate $\mathbb{Z}_2$-cohomology groups of such κ are given by

$$\hat{H}^*(\mathbb{Z}_2;\kappa) = \hat{H}^{*-2}(\mathbb{Z}_2;I_m) .$$

In Proposition 6.2.2 below it will be shown that if Φ is such that there is excision in the associated classical algebraic K-groups then the intermediate $\begin{cases} \varepsilon\text{-symmetric} \\ \varepsilon\text{-quadratic} \end{cases}$ triad

L-groups $\begin{cases} L^*_\kappa(\Phi,\varepsilon) \\ L_*^\kappa(\Phi,\varepsilon) \end{cases}$ of cartesian squares κ are in fact independent of κ. In particular, if it can be shown that $\begin{cases} L^*_\kappa(\phi,\varepsilon) = 0 \\ L_*^\kappa(\phi,\varepsilon) = 0 \end{cases}$ for one such κ then this is also the case for all other cartesian κ.

Define $*$-invariant subgroups

$$J_m = \mathrm{im}((g\ g'):\widetilde{K}_m(B)\oplus\widetilde{K}_m(B')\longrightarrow \widetilde{K}_m(A'))\subseteq \widetilde{K}_m(A') \quad (m = 0 \text{ or } 1) .$$

Define abelian group morphisms

$$\Delta : \hat{H}^n(\mathbb{Z}_2;I_0)\longrightarrow \hat{H}^{n+1}(\mathbb{Z}_2;\widetilde{K}_1(A')/J_1) \quad (n(\mathrm{mod}\ 2))$$

as follows. An element $\begin{cases} [P]\in \hat{H}^0(\mathbb{Z}_2;I_0) \\ [P]\in \hat{H}^1(\mathbb{Z}_2;I_0) \end{cases}$ is represented by a f.g. projective A-module P such that for some $q\geqslant 0$ there exist

i) an A-module isomorphism $\begin{cases} h:P\xrightarrow{\sim} P^* \\ h:P\oplus P^*\xrightarrow{\sim} A^{2q} \end{cases}$

ii) a B-module isomorphism $k:B\otimes_A P\xrightarrow{\sim} B^q$

iii) a B'-module isomorphism $k':B'\otimes_A P\xrightarrow{\sim} B'^q$.

Let $\begin{cases} \Delta([P])\in \hat{H}^1(\mathbb{Z}_2;\widetilde{K}_1(A')/J_1) \\ \Delta([P])\in \hat{H}^0(\mathbb{Z}_2;\widetilde{K}_1(A')/J_1) \end{cases}$ be the element represented by

$$\begin{cases} \tau(A'^q \xrightarrow[\sim]{1\otimes_B k^{-1}} A'\otimes_A P \xrightarrow[\sim]{1\otimes_A h} A'\otimes_A P^* \xrightarrow[\sim]{1\otimes_{B'} k'^{*-1}} A'^q)\in \widetilde{K}_1(A') \\ \tau(A'^{2q} = A'^q\oplus A'^q \xrightarrow[\sim]{(1\otimes_B k^{-1})\oplus(1\otimes_{B'} k'^*)} A'\otimes_A(P\oplus P^*) \xrightarrow[\sim]{1\otimes_A h} A'^{2q}) \\ \qquad \in \widetilde{K}_1(A') . \end{cases}$$

Let $\hat{H}^*(\mathbb{Z}_2;\Delta)$ be the relative groups appearing in the exact sequence

$$\ldots\longrightarrow \hat{H}^{n-1}(\mathbb{Z}_2;I_0)\xrightarrow{\Delta}\hat{H}^n(\mathbb{Z}_2;\widetilde{K}_1(A')/J_1)\longrightarrow\hat{H}^n(\mathbb{Z}_2;\Delta)$$
$$\longrightarrow \hat{H}^{n-2}(\mathbb{Z}_2;I_0)\longrightarrow \ldots .$$

The commutative square of rings with involution Φ is $\hat{H}^*$-cartesian if

i) the commutative squares of $*$-invariant subgroups

$$\mathfrak{J}_m = \begin{pmatrix} \tilde{K}_m(A) \longrightarrow \tilde{K}_m(B) \\ \downarrow \qquad\qquad \downarrow \\ \tilde{K}_m(B') \longrightarrow J_m \end{pmatrix} \subseteq \begin{pmatrix} \tilde{K}_m(A) \longrightarrow \tilde{K}_m(B) \\ \downarrow \qquad\qquad \downarrow \\ \tilde{K}_m(B') \longrightarrow \tilde{K}_m(A') \end{pmatrix} \quad (m = 0,1)$$

are cartesian

ii) $\hat{H}^*(\mathbb{Z}_2;\Delta) = 0$.

(See §6.3 below for examples of $\hat{H}^*$-cartesian squares Φ. In particular, if Φ is such that there is defined an algebraic K-theory Mayer-Vietoris exact sequence

$$\tilde{K}_1(A) \longrightarrow \tilde{K}_1(B) \oplus \tilde{K}_1(B') \longrightarrow \tilde{K}_1(A') \xrightarrow{\partial} \tilde{K}_0(A) \longrightarrow \tilde{K}_0(B) \oplus \tilde{K}_0(B') \longrightarrow \tilde{K}_0(A')$$

then Φ is $\hat{H}^*$-cartesian).

Proposition 6.2.2 For an $\hat{H}^*$-cartesian square Φ the intermediate $\begin{cases} \varepsilon\text{-symmetric} \\ \varepsilon\text{-quadratic} \end{cases}$ triad L-groups are such that there are natural identifications

$$\begin{cases} L^*_{\kappa_1}(\Phi,\varepsilon) = L^*_{\kappa_2}(\Phi,\varepsilon) \\ L_*^{\kappa_1}(\Phi,\varepsilon) = L_*^{\kappa_2}(\Phi,\varepsilon) \end{cases}$$

for any two cartesian squares κ_1,κ_2.

Proof: For a fixed Φ and a fixed m (= 0 or 1) the set of cartesian squares $\kappa \subseteq \begin{pmatrix} \tilde{K}_m(A) \longrightarrow \tilde{K}_m(B) \\ \downarrow \qquad\qquad \downarrow \\ \tilde{K}_m(B') \longrightarrow \tilde{K}_m(A') \end{pmatrix}$ is partially ordered by inclusion, with minimal element

$$\mathfrak{T}_m = \begin{pmatrix} I_m & \longrightarrow & 0 \\ \downarrow & & \downarrow \\ 0 & \longrightarrow & 0 \end{pmatrix} \subseteq \begin{pmatrix} \tilde{K}_m(A) & \longrightarrow & \tilde{K}_m(B) \\ \downarrow & & \downarrow \\ \tilde{K}_m(B') & \longrightarrow & \tilde{K}_m(A') \end{pmatrix}$$

and maximal element $\mathfrak{J}_m$. The Tate $\mathbb{Z}_2$-cohomology groups of any cartesian square κ are such that

$$\hat{H}^*(\mathbb{Z}_2;\kappa) = \hat{H}^{*-2}(\mathbb{Z}_2;I_m) = \hat{H}^*(\mathbb{Z}_2;\mathfrak{T}_m) \quad ,$$

so that $\hat{H}^*(\mathbb{Z}_2;\kappa/\mathfrak{T}_m) = 0$ and by the exact sequence of Proposition 6.2.1 there are natural identifications

$$\begin{cases} L^*_\kappa(\Phi,\varepsilon) = L^*_{\mathfrak{T}_m}(\Phi,\varepsilon) \\ L^\kappa_*(\Phi,\varepsilon) = L^{\mathfrak{T}_m}_*(\Phi,\varepsilon) \end{cases} .$$

Thus for fixed Φ,m the intermediate triad L-groups are independent of κ.

In order to identify the intermediate triad L-groups for m = 0 with those for m = 1 consider the commutative diagram of abelian groups with exact rows and columns

$$\begin{array}{ccccccccc}
 & \vdots & & \vdots & & \vdots & & \vdots & \\
 & \downarrow & & \downarrow & & \downarrow & & \downarrow & \\
\dots\to & V^n(A\to B,\varepsilon) & \to & U^n_{I_0}(A\to B,\varepsilon) & \to & \hat{H}^{n-1}(\mathbb{Z}_2;I_0) & \to & V^{n-1}(A\to B,\varepsilon) & \to\dots \\
 & \downarrow & & \downarrow & & \downarrow{\scriptstyle\Delta} & & \downarrow & \\
\dots\to & V^n_{J_1}(B'\to A',\varepsilon) & \to & V^n(B'\to A',\varepsilon) & \to & \hat{H}^n(\mathbb{Z}_2;\tilde{K}_1(A')/J_1) & \to & V^{n-1}_{J_1}(B'\to A',\varepsilon) & \to\dots \\
 & \downarrow & & \downarrow & & \downarrow & & \downarrow & \\
\dots\to & L^n_{\mathfrak{J}_1}(\Phi,\varepsilon) & \to & L^n_{\mathfrak{T}_0}(\Phi,\varepsilon) & \to & \hat{H}^n(\mathbb{Z}_2;\Delta) & \to & L^{n-1}_{\mathfrak{J}_1}(\Phi,\varepsilon) & \to\dots \\
 & \downarrow & & \downarrow & & \downarrow & & \downarrow & \\
\dots\to & V^{n-1}(A\to B,\varepsilon) & \to & U^{n-1}_{I_0}(A\to B,\varepsilon) & \to & \hat{H}^{n-2}(\mathbb{Z}_2;I_0) & \to & V^{n-2}(A\to B,\varepsilon) & \to\dots \\
 & \downarrow & & \downarrow & & \downarrow & & \downarrow & \\
 & \vdots & & \vdots & & \vdots & & \vdots &
\end{array}$$

in which $\hat{H}^*(\mathbb{Z}_2;\Delta) = 0$ (by the hypothesis on Φ). It follows that

$$\begin{cases} L^*_{\mathfrak{J}_1}(\Phi,\varepsilon) = L^*_{\mathfrak{I}_0}(\Phi,\varepsilon) \\ L_*^{\mathfrak{J}_1}(\Phi,\varepsilon) = L_*^{\mathfrak{I}_0}(\Phi,\varepsilon) \end{cases} .$$

[]

The commutative square of rings with involution Φ

is $\begin{cases} L^*\text{-} \\ L_*\text{-} \end{cases}$ <u>cartesian</u> if

i) Φ is $\hat{H}^*$-cartesian

ii) $\begin{cases} L^*_\kappa(\Phi,\varepsilon) = 0 \\ L_*^\kappa(\Phi,\varepsilon) = 0 \end{cases}$ for some cartesian square κ of $*$-invariant

subgroups.

(See §6.3 below for examples of $\begin{cases} L^*\text{-} \\ L_*\text{-} \end{cases}$ cartesian squares Φ).

<u>Proposition 6.2.3</u> Let

$$\begin{array}{ccc} A & \xrightarrow{f} & B \\ {\scriptstyle f'}\big\downarrow & \Phi & \big\downarrow{\scriptstyle g} \\ B' & \xrightarrow{g'} & A' \end{array}$$

be an L^*-cartesian square of rings with involution.

i) For any cartesian square of $*$-invariant subgroups

$$\kappa = \left(\begin{array}{ccc} X & \longrightarrow & Y \\ \big\downarrow & & \big\downarrow \\ Y' & \longrightarrow & X' \end{array}\right) \subseteq \left(\begin{array}{ccc} \tilde{K}_m(A) & \longrightarrow & \tilde{K}_m(B) \\ \big\downarrow & & \big\downarrow \\ \tilde{K}_m(B') & \longrightarrow & \tilde{K}_m(A') \end{array}\right) \quad (m = 0 \text{ or } 1)$$

there are defined excision isomorphisms of relative intermediate ε-symmetric L-groups

$$(f',g)\ :\ L^n_{Y,X}(f:A\longrightarrow B,\varepsilon)\xrightarrow{\sim} L^n_{X',Y'}(g':B'\longrightarrow A',\varepsilon)\qquad (n\in\mathbb{Z})$$

and a Mayer-Vietoris exact sequence of the absolute intermediate ε-symmetric L-groups

$$\ldots\longrightarrow L^n_X(A,\varepsilon)\xrightarrow{\begin{pmatrix} f\\ f'\end{pmatrix}} L^n_Y(B,\varepsilon)\oplus L^n_{Y'}(B',\varepsilon)\xrightarrow{(g\ \ -g')} L^n_{X'}(A',\varepsilon)$$

$$\xrightarrow{\partial} L^{n-1}_X(A,\varepsilon)\xrightarrow{\begin{pmatrix} f\\ f'\end{pmatrix}} L^{n-1}_Y(B,\varepsilon)\oplus L^{n-1}_{Y'}(B',\varepsilon)\longrightarrow\ldots\quad (n\in\mathbb{Z})\ ,$$

in which the connecting maps ∂ are given by the composites

$$\partial\ :\ L^n_{X'}(A',\varepsilon)\longrightarrow L^n_{X',Y'}(g':B'\longrightarrow A',\varepsilon)$$

$$\xrightarrow[\sim]{(f',g)^{-1}} L^n_{Y,X}(f:A\longrightarrow B,\varepsilon)\longrightarrow L^{n-1}_X(A,\varepsilon)\ .$$

ii) For cartesian squares of $*$-invariant subgroups κ_1,κ_2 such that

$$\kappa_1=\begin{pmatrix} X_1 & \longrightarrow & Y_1\\ \downarrow & & \downarrow\\ Y_1' & \longrightarrow & X_1'\end{pmatrix}\subseteq\kappa_2=\begin{pmatrix} X_2 & \longrightarrow & Y_2\\ \downarrow & & \downarrow\\ Y_2' & \longrightarrow & X_2'\end{pmatrix}\subseteq\begin{pmatrix} \tilde{K}_m(A) & \longrightarrow & \tilde{K}_m(B)\\ \downarrow & & \downarrow\\ \tilde{K}_m(B') & \longrightarrow & \tilde{K}_m(A')\end{pmatrix}$$

$(m = 0$ or $1)$

there is defined a commutative diagram of abelian groups with exact rows and columns

$$\begin{array}{ccccccccc}
& & \vdots & & \vdots & & \vdots & & \vdots \\
& & \downarrow & & \downarrow & & \downarrow & & \downarrow \\
\cdots \to & & L^n_{X_1}(A,\varepsilon) & \to & L^n_{Y_1}(B,\varepsilon)\oplus L^n_{Y'_1}(B',\varepsilon) & \to & L^n_{X'_1}(A',\varepsilon) & \xrightarrow{\partial} & L^{n-1}_{X_1}(A,\varepsilon) \to \cdots \\
& & \downarrow & & \downarrow & & \downarrow & & \downarrow \\
\cdots \to & & L^n_{X_2}(A,\varepsilon) & \to & L^n_{Y_2}(B,\varepsilon)\oplus L^n_{Y'_2}(B',\varepsilon) & \to & L^n_{X'_2}(A',\varepsilon) & \xrightarrow{\partial} & L^{n-1}_{X_2}(A,\varepsilon) \to \cdots \\
& & \downarrow & & \downarrow & & \downarrow & & \downarrow \\
\cdots \to & & \hat{H}^n(\mathbb{Z}_2;X_2/X_1) & \to & \hat{H}^n(\mathbb{Z}_2;Y_2/Y_1)\oplus\hat{H}^n(\mathbb{Z}_2;Y'_2/Y'_1) & \to & \hat{H}^n(\mathbb{Z}_2;X'_2/X'_1) & \xrightarrow{\partial} & \hat{H}^{n-1}(\mathbb{Z}_2;X_2/X_1) \to \cdots \\
& & \downarrow & & \downarrow & & \downarrow & & \downarrow \\
\cdots \to & & L^{n-1}_{X_1}(A,\varepsilon) & \to & L^{n-1}_{Y_1}(B,\varepsilon)\oplus L^{n-1}_{Y'_1}(B',\varepsilon) & \to & L^{n-1}_{X'_1}(A',\varepsilon) & \xrightarrow{\partial} & L^{n-2}_{X_1}(A,\varepsilon) \to \cdots \\
& & \downarrow & & \downarrow & & \downarrow & & \downarrow \\
& & \vdots & & \vdots & & \vdots & & \vdots
\end{array}$$

iii) The Mayer-Vietoris exact sequences associated to the cartesian squares of $*$-invariant subgroups $\mathfrak{I}_0$ and $\mathfrak{J}_1$ intertwine in a commutative braid of exact sequences

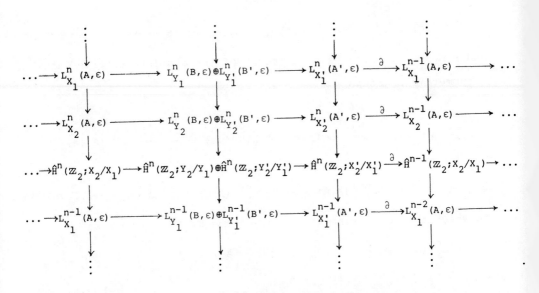

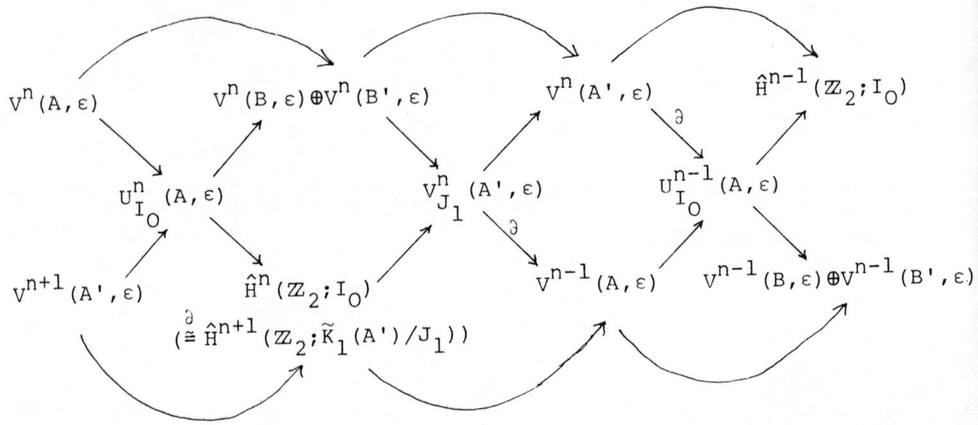

Similarly for L_*-cartesian squares Φ.

Proof: i) It is immediate from Proposition 6.2.1 that

$$L^*_\kappa(\Phi,\varepsilon) = 0 ,$$

which is precisely the condition for there to be excision isomorphisms and a Mayer-Vietoris exact sequence in the intermediate ε-symmetric L-groups associated to κ, by the intermediate version of Proposition 6.1.1 ii).

ii) The only parts of the diagram where commutativity is perhaps not quite obvious are those involving the connecting maps ∂. For those parts consider the more obviously commutative diagram

$$\begin{array}{ccccccc}
\vdots & & \vdots & & \vdots & & \vdots \\
\downarrow & & \downarrow & & \downarrow & & \downarrow \\
L^n_{X'_1}(A',\varepsilon) & \longrightarrow & L^n_{X'_1,Y'_1}(g':B'\to A',\varepsilon) & \xleftarrow[\sim]{(f',g)} & L^n_{Y_1,X_1}(f:A\to B,\varepsilon) & \longrightarrow & L^{n-1}_{X_1}(A,\varepsilon) \\
\downarrow & & \downarrow & & \downarrow & & \downarrow \\
L^n_{X'_2}(A',\varepsilon) & \longrightarrow & L^n_{X'_2,Y'_2}(g':B'\to A',\varepsilon) & \xleftarrow[\sim]{(f',g)} & L^n_{Y_2,X_2}(f:A\to B,\varepsilon) & \longrightarrow & L^{n-1}_{X_2}(A,\varepsilon) \\
\downarrow & & \downarrow & & \downarrow & & \downarrow \\
\hat{H}^n(\mathbb{Z}_2;X'_2/X'_1) & \longrightarrow & \hat{H}^n(\mathbb{Z}_2;Y'_2/Y'_1\to X'_2/X'_1) & \xleftarrow[\sim]{(f',g)} & \hat{H}^n(\mathbb{Z}_2;X_2/X_1\to Y_2/Y_1) & \longrightarrow & \hat{H}^{n-1}(\mathbb{Z}_2;X_2/X_1) \\
\downarrow & & \downarrow & & \downarrow & & \downarrow \\
L^{n-1}_{X'_1}(A',\varepsilon) & \longrightarrow & L^{n-1}_{X'_1,Y'_1}(g':B'\to A',\varepsilon) & \xleftarrow[\sim]{(f',g)} & L^{n-1}_{Y_1,X_1}(f:A\to B,\varepsilon) & \longrightarrow & L^{n-2}_{X_1}(A,\varepsilon) \\
\downarrow & & \downarrow & & \downarrow & & \downarrow \\
\vdots & & \vdots & & \vdots & & \vdots
\end{array}$$

in which the composites of the horizontal maps (inverting the excision isomorphisms (f',g)) are the connecting maps ∂.

iii) By analogy with ii).

[]

Madsen [2,4.11] makes use of a particular case of the naturality property of Proposition 6.2.3 ii), indicating a proof specific to that case.

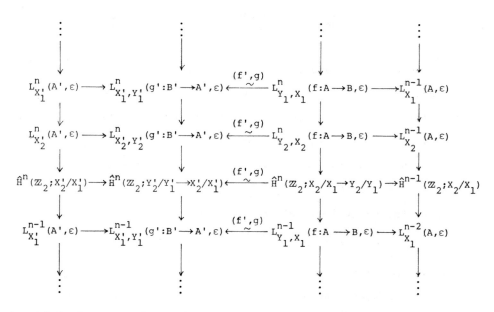

6.3 Cartesian L-theory

A commutative square of rings

$$\begin{array}{ccc} A & \xrightarrow{\ f\ } & B \\ {\scriptstyle f'}\downarrow & \Phi & \downarrow{\scriptstyle g} \\ B' & \xrightarrow{\ g'\ } & A' \end{array}$$

is cartesian if the sequence of abelian groups

$$0 \longrightarrow A \xrightarrow{\begin{pmatrix} f \\ f' \end{pmatrix}} B \oplus B' \xrightarrow{(g\ \ -g')} A' \longrightarrow 0$$

is exact.

The cartesian squares of rings with involution Φ for which we shall obtain excision in L-theory will be such that there is excision in the classical algebraic K-theory of the underlying cartesian square of rings, in the following sense.

A cartesian square of rings Φ is K_*-cartesian if

i) Φ is cartesian

ii) the natural map of relative K-groups

$$(f',g) : K_1(f:A \longrightarrow B) \longrightarrow K_1(g':B' \longrightarrow A')$$

is an isomorphism

iii) the sequences of reduced algebraic K-groups

$$\tilde{K}_m(A) \xrightarrow{\begin{pmatrix} f \\ f' \end{pmatrix}} \tilde{K}_m(B) \oplus \tilde{K}_m(B') \xrightarrow{(g\ \ -g')} \tilde{K}_m(A') \quad (m = 0,1)$$

are exact. (For $m = 0$ this follows from ii)).

In particular, for a K_*-cartesian square Φ there is defined a Mayer-Vietoris exact sequence of reduced algebraic K-groups

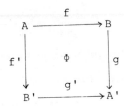

$$\tilde{K}_1(A) \longrightarrow \tilde{K}_1(B) \oplus \tilde{K}_1(B') \longrightarrow \tilde{K}_1(A') \xrightarrow{\partial} \tilde{K}_0(A) \longrightarrow \tilde{K}_0(B) \oplus \tilde{K}_0(B') \longrightarrow \tilde{K}_0(A')$$

with the connecting map ∂ given by

$$\partial : \tilde{K}_1(A') \longrightarrow K_1(g') \xrightarrow[\simeq]{(f',g)^{-1}} K_1(f) \longrightarrow \tilde{K}_0(A) \quad .$$

A cartesian square of rings Φ with $g:B \longrightarrow A'$ (or $g':B' \longrightarrow A'$) onto is K_*-cartesian, by Milnor [4,§4]. The cartesian square

$$\begin{array}{ccc} A & \longrightarrow & B \\ \downarrow & \Phi & \downarrow \\ S^{-1}A & \longrightarrow & T^{-1}B \end{array}$$

associated to a cartesian morphism of rings and multiplicative subsets

$$(A,S) \longrightarrow (B,T)$$

(e.g. $(A,S) \longrightarrow (\hat{A},\hat{S})$ with $\hat{A} = \varprojlim_{s \in S} A/sA$) is K_*-cartesian, by Karoubi [2,App.5] (cf. Proposition 3.1.3 i)).

Let then Φ be a K_*-cartesian square of rings as above. Given

i) a f.g. projective B-module P

ii) a f.g. projective B'-module P'

iii) an isomorphism of the induced f.g. projective A'-modules

$$h : A' \otimes_B P \xrightarrow{\sim} A' \otimes_{B'} P'$$

there are defined

i) the <u>pullback</u> f.g. projective A-module

$(P,h,P') = \{(x,x') \in P \oplus P' \,|\, h(1 \otimes_B x) = 1 \otimes_{B'} x' \in A' \otimes_{B'} P'\}$

$A \times (P,h,P') \longrightarrow (P,h,P')$;

$$(a,(x,x')) \longmapsto (f(a)x, f'(a)x')$$

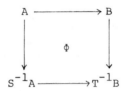

ii) an isomorphism of f.g. projective B-modules

$$i : B\otimes_A(P,h,P') \longrightarrow P \ ; \ b\otimes(x,x') \longmapsto bx$$

iii) an isomorphism of f.g. projective B'-modules

$$i' : B'\otimes_A(P,h,P') \longrightarrow P' \ ; \ b'\otimes(x,x') \longmapsto b'x'$$

such that there is defined a commutative diagram of f.g. projective A'-modules and isomorphisms

$$\begin{array}{ccccc}
 & & A'\otimes_B(B\otimes_A(P,h,P')) & \xrightarrow[\sim]{1\otimes_B i} & A'\otimes_B P \\
 & \overset{1}{\underset{\sim}{\nearrow}} & & & \\
A'\otimes_A(P,h,P') & & & & \wr\big\downarrow h \\
 & \overset{1}{\underset{\sim}{\searrow}} & & & \\
 & & A'\otimes_{B'}(B'\otimes_A(P,h,P')) & \xrightarrow[\sim]{1\otimes_{B'} i'} & A'\otimes_{B'} P'
\end{array} .$$

The isomorphisms i,i' will be used to identify

$$B\otimes_A(P,h,P') = P \quad , \quad B'\otimes_A(P,h,P') = P' \ .$$

The connecting map in the algebraic K-theory Mayer-Vietoris exact sequence of Φ can be expressed in terms of the pullback construction by

$$\partial : \tilde{K}_1(A') \longrightarrow \tilde{K}_0(A) \ ; \ \tau(h:A'^q \xrightarrow{\sim} A'^q) \longmapsto [(B^q,h,B'^q)] \quad (q \geqslant 0) \ .$$

The pullback construction for modules extends to morphisms: if $(P,h,P'),(Q,k,Q')$ are pullback f.g. projective A-modules there is defined a Mayer-Vietoris exact sequence of abelian groups

$$0 \longrightarrow \mathrm{Hom}_A((P,h,P'),(Q,k,Q')) \longrightarrow \mathrm{Hom}_B(P,Q) \oplus \mathrm{Hom}_{B'}(P',Q')$$
$$\longrightarrow \mathrm{Hom}_{A'}(A'\otimes_B P, A'\otimes_{B'} Q') \longrightarrow 0 \ ,$$

so that there is a natural identification

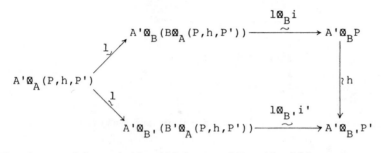

$$\mathrm{Hom}_A((P,h,P'),(Q,k,Q'))$$

$$= \{(e,e') \in \mathrm{Hom}_B(P,Q) \oplus \mathrm{Hom}_{B'}(P',Q')$$

$$| (1\otimes_{B'} e')h = k(1\otimes_B e) \in \mathrm{Hom}_{A'}(A'\otimes_B P, A'\otimes_{B'} Q')\} \ .$$

Let now Φ be a K_*-cartesian square of rings with involution. If (P,h,P') is a pullback f.g. projective A-module there is defined an isomorphism of f.g. projective A-modules

$$(P^*,h^{*-1},P'^*) \xrightarrow{\sim} (P,h,P')^* \ ;$$

$$(e,e') \longmapsto ((x,x') \longmapsto (e(x),e'(x'))$$

$$\in \ker((g\ -g'):B\oplus B' \longrightarrow A') = A)$$

which we shall use to identify

$$(P,h,P')^* = (P^*,h^{*-1},P'^*) \ .$$

It follows that the square

$$\begin{array}{ccc} \tilde{K}_1(A') & \xrightarrow{\partial} & \tilde{K}_0(A) \\ \downarrow * & & \downarrow * \\ \tilde{K}_1(A') & \xrightarrow{\partial} & \tilde{K}_0(A) \end{array}$$

is skew-commutative, that is

$$*\partial = -\partial * \ ,$$

and hence that the abelian group isomorphisms

$$\partial \ : \ \tilde{K}_1(A')/J_1 \xrightarrow{\sim} I_0$$

induces isomorphisms in the Tate $\mathbb{Z}_2$-cohomology groups

$$\partial \ : \ \hat{H}^n(\mathbb{Z}_2;\tilde{K}_1(A')/J_1) \xrightarrow{\sim} \hat{H}^{n-1}(\mathbb{Z}_2;I_0) \qquad (n(\mathrm{mod}\ 2))$$

inverse to the natural maps

$$\Delta \ : \ \hat{H}^{n-1}(\mathbb{Z}_2;I_0) \longrightarrow \hat{H}^n(\mathbb{Z}_2;\tilde{K}_1(A')/J_1) \qquad (n(\mathrm{mod}\ 2))$$

defined in §6.2 above. Thus $\hat{H}^*(\mathbb{Z}_2;\Delta) = 0$, and Φ is $\hat{H}^*$-cartesian in the sense of §6.2.

The first step in proving that a K_*-cartesian square of rings with involution Φ is $\begin{cases} L^*\text{-} \\ L_*\text{-} \end{cases}$ cartesian in the sense of §6.2 would be to extend the pullback construction of f.g. projective A-modules to $\begin{cases} \varepsilon\text{-symmetric} \\ \varepsilon\text{-quadratic} \end{cases}$ forms and formations over A. Unfortunately, such an extension is not always possible. We shall now investigate the extent to which such an extension is in fact possible, in the first instance by considering the behaviour under pullback of the various Q-groups used to define forms in §1.6 above.

Let Φ be a K_*-cartesian square of rings with involution, and let (P,h,P') be a pullback f.g. projective A-module. The split ε-quadratic Q-group

$$\tilde{Q}_\varepsilon((P,h,P')) = \mathrm{Hom}_A((P,h,P'),(P,h,P')^*)$$

fits into an exact sequence

$$0 \longrightarrow \tilde{Q}_\varepsilon((P,h,P')) \longrightarrow \tilde{Q}_\varepsilon(P) \oplus \tilde{Q}_\varepsilon(P') \longrightarrow \tilde{Q}_\varepsilon(A'\otimes_B P) \longrightarrow 0 \ ,$$

which is in fact an exact sequence of $\mathbb{Z}[\mathbb{Z}_2]$-modules with $T \in \mathbb{Z}_2$ acting by the ε-duality involution $T_\varepsilon : \psi \longmapsto \varepsilon\psi^*$ on each $\tilde{Q}$-group. Thus the pullback construction of modules generalizes to split ε-quadratic forms, as detailed further below. The ε-symmetric and ε-quadratic Q-groups

$$Q^\varepsilon(P) = \ker(1-T_\varepsilon : \tilde{Q}_\varepsilon(P) \longrightarrow \tilde{Q}_\varepsilon(P))$$

$$Q_\varepsilon(P) = \mathrm{coker}(1-T_\varepsilon : \tilde{Q}_\varepsilon(P) \longrightarrow \tilde{Q}_\varepsilon(P))$$

are such that there is defined an exact sequence

$$0 \longrightarrow Q^{\varepsilon}((P,h,P')) \longrightarrow Q^{\varepsilon}(P) \oplus Q^{\varepsilon}(P') \longrightarrow Q^{\varepsilon}(A'\otimes_B P)$$
$$\xrightarrow{\partial} Q_{\varepsilon}((P,h,P')) \longrightarrow Q_{\varepsilon}(P) \oplus Q_{\varepsilon}(P') \longrightarrow Q_{\varepsilon}(A'\otimes_B P) \longrightarrow 0$$

with the connecting map ∂ given by

$$\partial : Q^{\varepsilon}(A'\otimes_B P) \longrightarrow Q_{\varepsilon}((P,h,P')) \ ; \ \phi' \longmapsto (\phi_B - \varepsilon\phi_B^*, \phi_{B'} - \varepsilon\phi_{B'}^*) ,$$

expressing ϕ' as

$$\phi' = 1\otimes_B \phi_B - h^*(1\otimes_{B'}\phi_{B'})h \in Q^{\varepsilon}(A'\otimes_B P)$$

for some $\phi_B \in \widetilde{Q}_{\varepsilon}(P)$, $\phi_{B'} \in \widetilde{Q}_{\varepsilon}(P')$. The even ε-symmetric Q-group

$$Q\langle v_0\rangle^{\varepsilon}(P) = \mathrm{im}(1+T_{\varepsilon} : \widetilde{Q}_{\varepsilon}(P) \longrightarrow \widetilde{Q}_{\varepsilon}(P))$$

is such that there is defined an exact sequence

$$0 \to \{\phi \in Q^{\varepsilon}((P,h,P')) \mid \phi(x)(x) \in \mathrm{im}(\hat{\delta} : \hat{H}^1(\mathbb{Z}_2;A',\varepsilon) \longrightarrow \hat{H}^0(\mathbb{Z}_2;A,\varepsilon))$$
$$\text{for all } x \in (P,h,P')\}$$
$$\longrightarrow Q\langle v_0\rangle^{\varepsilon}(P) \oplus Q\langle v_0\rangle^{\varepsilon}(P') \longrightarrow Q\langle v_0\rangle^{\varepsilon}(A'\otimes_B P) \longrightarrow 0 .$$

Thus the pullback construction generalizes to ε-symmetric forms, as detailed further below, but not necessarily to even ε-symmetric and ε-quadratic forms. However, if $\hat{\delta} = 0 : \hat{H}^0(\mathbb{Z}_2;A',\varepsilon) \longrightarrow \hat{H}^1(\mathbb{Z}_2;A,\varepsilon)$ then

$$\partial = 0 : Q^{\varepsilon}(A'\otimes_B P) \longrightarrow Q_{\varepsilon}((P,h,P'))$$

and there are defined short exact sequences

$$0 \longrightarrow Q^{\varepsilon}((P,h,P')) \longrightarrow Q^{\varepsilon}(P) \oplus Q^{\varepsilon}(P') \longrightarrow Q^{\varepsilon}(A'\otimes_B P) \longrightarrow 0$$

$$0 \longrightarrow Q_{\varepsilon}((P,h,P')) \longrightarrow Q_{\varepsilon}(P) \oplus Q_{\varepsilon}(P') \longrightarrow Q_{\varepsilon}(A'\otimes_B P) \longrightarrow 0$$

$$0 \longrightarrow Q\langle v_0\rangle^{-\varepsilon}((P,h,P')) \longrightarrow Q\langle v_0\rangle^{-\varepsilon}(P) \oplus Q\langle v_0\rangle^{-\varepsilon}(P')$$
$$\longrightarrow Q\langle v_0\rangle^{-\varepsilon}(A'\otimes_B P) \longrightarrow 0 ,$$

so that the pullback construction generalizes to ε-quadratic and even $(-\varepsilon)$-symmetric forms. In particular, this is the case for the K_*-cartesian "arithmetic" square associated to a ring with involution A

$$\begin{array}{ccc} A & \longrightarrow & \mathbb{Q}\otimes_{\mathbb{Z}}A \\ \downarrow & & \downarrow \\ \hat{\mathbb{Z}}\otimes_{\mathbb{Z}}A & \longrightarrow & \hat{\mathbb{Q}}\otimes_{\mathbb{Z}}A \end{array}$$

since $\hat{H}^*(\mathbb{Z}_2;\hat{\mathbb{Q}}\otimes_{\mathbb{Z}}A,\varepsilon) = 0$ (on account of $1/2 \in \hat{\mathbb{Q}}$) - the L-theory of such cartesian squares was first studied by Wall [8],[9] as part of his programme for computing the quadratic L-groups $L_*(\mathbb{Z}[\pi])$ of finite groups π. (For torsion-free A, such as $A = \mathbb{Z}[\pi]$ for any group π, this is a localization-completion square as in §3.1, with $S = \mathbb{Z}-\{0\}\subset A$). On the other hand, the K_*-cartesian square of rings with involution

$$\begin{array}{ccc} \mathbb{Z}[\mathbb{Z}_2] & \xrightarrow{e_+} & \mathbb{Z} \\ {\scriptstyle e_-}\downarrow & & \downarrow{\scriptstyle p} \\ \mathbb{Z} & \xrightarrow{p} & \mathbb{Z}_2 \end{array}$$

defined by $\overline{T} = T \in \mathbb{Z}[\mathbb{Z}_2]$ and

$$e_{\pm} : \mathbb{Z}[\mathbb{Z}_2] \longrightarrow \mathbb{Z} \ ; \ a+bT \longmapsto a\pm b$$

$$p = \text{projection} : \mathbb{Z} \longrightarrow \mathbb{Z}_2$$

is such that there is no pullback construction for skew-quadratic forms, since the connecting map

$$\partial : Q^{-1}(\mathbb{Z}_2\otimes_{\mathbb{Z}[\mathbb{Z}_2]}(\mathbb{Z},1,\mathbb{Z})) = \mathbb{Z}_2 \longrightarrow Q_{-1}((\mathbb{Z},1,\mathbb{Z})) = \mathbb{Z}_2$$

$$((\mathbb{Z},1,\mathbb{Z}) = \mathbb{Z}[\mathbb{Z}_2])$$

is non-trivial. (For this example I am indebted to W.Pardon).

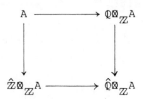

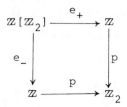

Given

i) $\begin{cases}\text{an } \varepsilon\text{-symmetric} \\ \text{a split } \varepsilon\text{-quadratic}\end{cases}$ form over B $\begin{cases}(M, \phi \in Q^{\varepsilon}(M)) \\ (M, \psi \in \widetilde{Q}_{\varepsilon}(M))\end{cases}$

ii) $\begin{cases}\text{an } \varepsilon\text{-symmetric} \\ \text{a split } \varepsilon\text{-quadratic}\end{cases}$ form over B' $\begin{cases}(M', \phi' \in Q^{\varepsilon}(M')) \\ (M', \psi' \in \widetilde{Q}_{\varepsilon}(M'))\end{cases}$

iii) an isomorphism of the induced $\begin{cases}\varepsilon\text{-symmetric} \\ \text{split } \varepsilon\text{-quadratic}\end{cases}$ forms over A'

$$\begin{cases} h : A'\otimes_B(M,\phi) \xrightarrow{\sim} A'\otimes_{B'}(M',\phi') \\ (h,\chi) : A'\otimes_B(M,\psi) \xrightarrow{\sim} A'\otimes_{B'}(M',\psi') \end{cases}$$

there is defined a <u>pullback</u> $\begin{cases}\varepsilon\text{-symmetric} \\ \text{split } \varepsilon\text{-quadratic}\end{cases}$ form over A

$$\begin{cases} ((M,\phi),h,(M',\phi')) \\ \quad = ((M,h,M'),(\phi,\phi'):(M,h,M') \longrightarrow (M^*,h^{*-1},M'^*) = (M,h,M')^*) \\ ((M,\psi),(h,\chi),(M',\psi')) \\ \quad = ((M,h,M'),(\psi+\chi_B-\varepsilon\chi_B^*,\psi'+\chi_{B'}-\varepsilon\chi_{B'}^*) \\ \qquad\qquad : (M,h,M') \longrightarrow (M^*,h^{*-1},M'^*) = (M,h,M')^*) \end{cases}$$

with $\chi_B \in \mathrm{Hom}_B(M,M^*)$, $\chi_{B'} \in \mathrm{Hom}_{B'}(M',M'^*)$ such that

$$\chi = 1\otimes\chi_B - h^*(1\otimes\chi_{B'})h \in Q_{-\varepsilon}(A'\otimes_B M) .$$

There are natural identifications

$$\begin{cases} B\otimes_A((M,\phi),h,(M',\phi')) = (M,\phi) \\ B\otimes_A((M,\psi),(h,\chi),(M',\psi')) = (M,\psi) \end{cases} ,$$

$$\begin{cases} B'\otimes_A((M,\phi),h,(M',\phi')) = (M',\phi') \\ B'\otimes_A((M,\psi),(h,\chi),(M',\psi')) = (M',\psi') . \end{cases}$$

Given

i) $\begin{cases}\text{an } \varepsilon\text{-symmetric} \\ \text{a split } \varepsilon\text{-quadratic}\end{cases}$ formation over B $\begin{cases}(M,\phi;F,G) \\ (F,(\binom{\gamma}{\mu},\theta)G)\end{cases}$

ii) $\begin{cases}\text{an } \varepsilon\text{-symmetric} \\ \text{a split } \varepsilon\text{-quadratic}\end{cases}$ formation over B' $\begin{cases}(M',\phi';F',G') \\ (F',(\binom{\gamma'}{\mu'}),\theta')G')\end{cases}$

iii) an isomorphism of the induced $\begin{cases}\varepsilon\text{-symmetric} \\ \text{split } \varepsilon\text{-quadratic}\end{cases}$

formations over A'

$$\begin{cases} h : A'\otimes_B(M,\phi;F,G) \xrightarrow{\sim} A'\otimes_{B'}(M',\phi';F',G') \\ (\alpha,\beta,\psi) : A'\otimes_B(F,G) \xrightarrow{\sim} A'\otimes_{B'}(F',G') \end{cases}$$

there is defined a <u>pullback</u> $\begin{cases}\varepsilon\text{-symmetric} \\ \text{split } \varepsilon\text{-quadratic}\end{cases}$ formation over A

$$\begin{cases} ((M,\phi;F,G),h,(M',\phi';F',G')) \\ \qquad = ((M,h,M'),(\phi,\phi');(F,h|,F'),(G,h|,G')) \\ ((F,G),(\alpha,\beta,\psi),(F',G')) \\ \qquad = ((F,\alpha,F'),(\begin{pmatrix}(\gamma+(\psi_B-\varepsilon\psi_B^*)^*\mu,\gamma'+(\psi_{B'}-\varepsilon\psi_{B'}^*)^*\mu') \\ (\mu,\mu')\end{pmatrix}, \\ \qquad\qquad (\theta+\mu^*\psi_B\mu+\chi_B,\theta'+\mu'^*\chi_{B'}\mu'+\chi_{B'})(G,\beta,G')) \end{cases}$$

with $\psi_B \in \mathrm{Hom}_B(F^*,F)$, $\psi_{B'} \in \mathrm{Hom}_{B'}(F'^*,F')$ such that

$$\psi = 1\otimes\psi_B - \alpha^{-1}(1\otimes\psi_{B'})\alpha^{*-1} \in Q_{-\varepsilon}(A'\otimes_B F^*)$$

and $\chi_B \in \mathrm{Hom}_B(G,G^*)$, $\chi_{B'} \in \mathrm{Hom}_{B'}(G',G'^*)$ such that

$$\beta^*(1\otimes_{B'}\theta')\beta - 1\otimes_B\theta - (1\otimes_B\mu^*)\psi(1\otimes_B\mu) \in Q_{-\varepsilon}(A'\otimes_B G) \quad .$$

There are natural identifications

$$\begin{cases} B\otimes_A((M,\phi;F,G),h,(M',\phi';F',G')) = (M,\phi;F,G) \\ B\otimes_A((F,G),(\alpha,\beta,\psi),(F',G')) = (F,G) \end{cases} ,$$

$$\begin{cases} B'\otimes_A((M,\phi;F,G),h,(M',\phi';F',G')) = (M',\phi';F,G') \\ B'\otimes_A((F,G),(\alpha,\beta,\psi),(F',G')) = (F',G') \end{cases} .$$

<u>Proposition 6.3.1</u> Given a cartesian square of rings with involution Φ

$$\begin{array}{ccc} A & \xrightarrow{f} & B \\ {\scriptstyle f'}\downarrow & \Phi & \downarrow{\scriptstyle g} \\ B' & \xrightarrow{g'} & A' \end{array}$$

let κ be a cartesian square of $*$-invariant subgroups

$$\kappa = \begin{pmatrix} X & \longrightarrow & Y \\ \downarrow & & \downarrow \\ Y' & \longrightarrow & X' \end{pmatrix} \subseteq \begin{pmatrix} \tilde{K}_m(A) & \longrightarrow & \tilde{K}_m(B) \\ \downarrow & & \downarrow \\ \tilde{K}_m(B') & \longrightarrow & \tilde{K}_m(A') \end{pmatrix} \quad (m = 0,1) .$$

If either i)$_*$ $g:B \longrightarrow A'$ is onto

or ii)$_*$ Φ is the cartesian square

$$\begin{array}{ccc} A & \longrightarrow & B \\ \downarrow & \Phi & \downarrow \\ S^{-1}A & \longrightarrow & T^{-1}B \end{array}$$

associated to a cartesian morphism of rings with involution and multiplicative subsets

$$(A,S) \longrightarrow (B,T)$$

then Φ is L_*-cartesian

$$L^{\kappa}_*(\Phi,\varepsilon) = 0 ,$$

and there is defined a Mayer-Vietoris exact sequence of intermediate ε-quadratic L-groups

$$\ldots \longrightarrow L_n^X(A,\varepsilon) \xrightarrow{\begin{pmatrix} f \\ f' \end{pmatrix}} L_n^Y(B,\varepsilon) \oplus L_n^{Y'}(B',\varepsilon) \xrightarrow{(g\ \ -g')} L_n^{X'}(A',\varepsilon)$$

$$\xrightarrow{\partial} L_{n-1}^X(A,\varepsilon) \longrightarrow \ldots \quad (n \in \mathbb{Z}) .$$

Define also the conditions

i)* = the maps $g:B \longrightarrow A'$, $g':B' \longrightarrow A'$ and $g:\hat{H}^0(\mathbb{Z}_2;B,\varepsilon) \longrightarrow \hat{H}^0(\mathbb{Z}_2;A',\varepsilon)$, $g':\hat{H}^0(\mathbb{Z}_2;B',\varepsilon) \longrightarrow \hat{H}^0(\mathbb{Z}_2;A',\varepsilon)$ are all onto

ii)* = ii)$_*$ and also

$$\hat{\delta} = 0 : \hat{H}^0(\mathbb{Z}_2;T^{-1}B,\varepsilon) \longrightarrow \hat{H}^1(\mathbb{Z}_2;A,\varepsilon) .$$

Then if $\begin{cases} \text{i)*} \\ \text{ii)*} \end{cases}$ holds Φ is

$$\begin{cases} \text{such that } L_K^n(\Phi,\varepsilon) = 0 \ (n \leqslant 1) \\ L^*\text{-cartesian, with } L_K^*(\Phi,\varepsilon) = 0 \end{cases}$$

and for $\begin{cases} n \leqslant 1 \\ n \in \mathbb{Z} \end{cases}$ there is defined a Mayer-Vietoris exact sequence of intermediate ε-symmetric L-groups

$$\begin{cases} L_Y^1(B,\varepsilon) \oplus L_{Y'}^1(B',\varepsilon) \\ \ldots \end{cases} \longrightarrow \ldots \longrightarrow L_Y^n(B,\varepsilon) \oplus L_{Y'}^n(B',\varepsilon) \xrightarrow{(g\ \ -g')} L_{X'}^n(A',\varepsilon)$$

$$\xrightarrow{\partial} L_X^{n-1}(A,\varepsilon) \xrightarrow{\begin{pmatrix} f \\ f' \end{pmatrix}} L_Y^{n-1}(B,\varepsilon) \oplus L_{Y'}^{n-1}(B',\varepsilon) \longrightarrow \ldots .$$

Proof: i) Consider first the special case

$$\left(\begin{array}{ccc} X & \longrightarrow & Y' \\ \downarrow & \kappa & \downarrow \\ Y' & \longrightarrow & X' \end{array}\right) = \left(\begin{array}{ccc} I_0 & \longrightarrow & 0 \\ \downarrow & \mathcal{I}_0 & \downarrow \\ 0 & \longrightarrow & 0 \end{array}\right) \subseteq \left(\begin{array}{ccc} \tilde{K}_0(A) & \longrightarrow & \tilde{K}_0(B) \\ \downarrow & & \downarrow \\ \tilde{K}_0(B') & \longrightarrow & \tilde{K}_0(A') \end{array}\right) .$$

Assuming condition $\begin{cases} i)^* \\ i)_* \end{cases}$ we shall now define morphisms of $\begin{cases} \varepsilon\text{-symmetric} \\ \varepsilon\text{-quadratic} \end{cases}$ L-groups

$$\begin{cases} \delta : V^n(A',\varepsilon) \longrightarrow U^n_X(f,\varepsilon) \\ \delta : V_n(A',\varepsilon) \longrightarrow U^X_n(f,\varepsilon) \end{cases} ,$$

$$\begin{cases} \hat{\delta} : V^n(g',\varepsilon) \longrightarrow U^{n-1}_X(A,\varepsilon) \\ \hat{\delta} : V_n(g',\varepsilon) \longrightarrow U^X_{n-1}(A,\varepsilon) \end{cases}$$

for $\begin{cases} n \leqslant 1 \\ n \in \mathbb{Z} \end{cases}$ satisfying the hypotheses of the appropriate intermediate version of Proposition 6.1.3, thus obtaining the intermediate $\begin{cases} \varepsilon\text{-symmetric} \\ \varepsilon\text{-quadratic} \end{cases}$ L-theory Mayer-Vietoris exact sequence in the special case $\kappa = \mathcal{I}_0$.

Every element of $\begin{cases} V^0(A',\varepsilon) \\ V_0(A',\varepsilon) \end{cases}$ is represented by a non-singular $\begin{cases} \varepsilon\text{-symmetric} \\ \varepsilon\text{-quadratic} \end{cases}$ form over A' of the type $\begin{cases} (A'^q, \phi' \in Q^{\varepsilon}(A'^q)) \\ (A'^q, \psi' \in Q_{\varepsilon}(A'^q)) \end{cases}$ $(q \geqslant 0)$.

As $g:B \longrightarrow A'$ is onto $\begin{cases} \text{and } g:\hat{H}^0(\mathbb{Z}_2;B,\varepsilon) \longrightarrow \hat{H}^0(\mathbb{Z}_2;A',\varepsilon) \text{ is onto} \\ - \end{cases}$

there exists an $\begin{cases} \varepsilon\text{-symmetric} \\ \varepsilon\text{-quadratic} \end{cases}$ form over B $\begin{cases} (B^q, \phi \in Q^{\varepsilon}(B^q)) \\ (B^q, \psi \in Q_{\varepsilon}(B^q)) \end{cases}$

such that

$$\begin{cases} (A'^q,\phi') = A'\otimes_B(B^q,\phi) \\ (A'^q,\psi') = A'\otimes_B(B^q,\psi) \ . \end{cases}$$

In the ε-quadratic case define

$$\chi' = (\psi'+\varepsilon\psi'^*)^{-1}\psi'(\psi'+\varepsilon\psi'^*)^{-1} \in Q_\varepsilon((A'^q)^*)$$

and let $((B^q)^*,\chi\in Q_\varepsilon((B^q)^*))$ be an ε-quadratic form over B such that

$$((A'^q)^*,\chi') = A'\otimes_B((B^q)^*,\chi) \ .$$

Use the isomorphism of non-singular $\begin{cases}\text{even } (-\varepsilon)\text{-symmetric}\\ \text{split } (-\varepsilon)\text{-quadratic}\end{cases}$ formations over A'

$$\begin{cases} h = \begin{pmatrix} 1 & -\phi'^{-1} \\ 0 & 1 \end{pmatrix} : \\ A'\otimes_B\partial(B^q,\phi) = (A'^q\oplus(A'^q)^*,\begin{pmatrix} 0 & 1 \\ -\varepsilon & 0\end{pmatrix};A'^q, \\ \qquad\qquad \operatorname{im}(\begin{pmatrix} 1 \\ \phi'\end{pmatrix}:A'^q \longrightarrow A'^q\oplus(A'^q)^*)) \\ \xrightarrow{\sim} A'\otimes_{B'}(H^{-\varepsilon}(B'^q);B'^q,(B'^q)^*) \\ \qquad = (A'^q\oplus(A'^q)^*,\begin{pmatrix} 0 & 1 \\ -\varepsilon & 0\end{pmatrix};A'^q,(A'^q)^*) \\ (\alpha,\beta,\sigma) = (1,\psi'+\varepsilon\psi'^*,-\bar{\varepsilon}\chi') : \\ A'\otimes_B\partial(B^q,\psi) = (A'^q,(\begin{pmatrix} 1 \\ \psi'+\varepsilon\psi'^*\end{pmatrix},\psi')A'^q) \\ \qquad \xrightarrow{\sim} A'\otimes_{B'}(B'^q,(B'^q)^*) = (A'^q,(\begin{pmatrix} 0 \\ 1\end{pmatrix},0)(A'^q)^*) \end{cases}$$

to define a pullback non-singular $\begin{cases}\text{even } (-\varepsilon)\text{-symmetric}\\ \text{split } (-\varepsilon)\text{-quadratic}\end{cases}$

formation over A

$$\begin{cases} (N,\nu;F,G) = (\partial(B^q,\phi),h,(H^{-\varepsilon}(B'^q);B'^q,(B'^q)^*)) \\ (F,G) = (\partial(B^q,\psi),(\alpha,\beta,\sigma),(B'^q,(B'^q)^*)) \\ \qquad = ((B^q,1,B'^q),(\begin{pmatrix} (1-(\chi+\varepsilon\chi^*)(\psi+\varepsilon\psi^*),0) \\ (\psi+\varepsilon\psi^*,1) \end{pmatrix}, \\ \qquad\qquad \psi-(\psi+\varepsilon\psi^*)\chi(\psi+\varepsilon\psi^*))(B^q,\psi'+\varepsilon\psi'^*,B'^q)) \\ \qquad ((B^q,1,B'^q) = A^q) \end{cases}$$

with projective class

$$\begin{aligned} [G] - [F^*] &= [(B^q,\phi',B'^q)] - [A^q] \\ &= \partial\tau(\phi':A'^q \longrightarrow (A'^q)^*) \\ &\in X = I_0 = \mathrm{im}(\partial:\tilde{K}_1(A') \longrightarrow \tilde{K}_0(A)) \subseteq \tilde{K}_0(A) \end{aligned}$$

(where $\phi' = \psi' + \varepsilon\psi'^*$ in the ε-quadratic case). The isomorphism of non-singular $\begin{cases}\text{even } (-\varepsilon)\text{-symmetric} \\ \text{split } (-\varepsilon)\text{-quadratic}\end{cases}$ formations over B

$$\begin{cases} 1 : B\otimes_A(N,\nu;F,G) \xrightarrow{\sim} \partial(B^q,\phi) \\ (1,1,\bar{\varepsilon}\chi) : B\otimes_A(F,G) \xrightarrow{\sim} \partial(B^q,\psi) \end{cases}$$

can now be used to define an abelian group morphism

$$\begin{cases} \delta : V^0(A',\varepsilon) \longrightarrow U^0_X(f,\varepsilon) \; ; \; (A'^q,\phi') \longmapsto ((N,\nu;F,G),(B^q,\phi),1) \\ \delta : V_0(A',\varepsilon) \longrightarrow U^X_0(f,\varepsilon) \; ; \; (A'^q,\psi') \longmapsto ((F,G),(B^q,\psi),(1,1,\bar{\varepsilon}\chi)) \end{cases} .$$

The construction of δ in the $(-\varepsilon)$-quadratic case also gives an abelian group morphism

$$\begin{aligned} \delta : V^{-2}(A',\varepsilon) = V\langle v_0\rangle^0(A',-\varepsilon) &\longrightarrow U^{-2}_X(f,\varepsilon) = U\langle v_0\rangle^0_X(f,-\varepsilon) \; ; \\ (A'^q,\psi' - \varepsilon\psi'^* \in Q\langle v_0\rangle^{-\varepsilon}(A'^q)) \\ &\longmapsto ((H_\varepsilon(A^q);A^q,G),(B^q,\psi-\varepsilon\psi^* \in Q\langle v_0\rangle^{-\varepsilon}(B^q)),1) \; , \end{aligned}$$

where $(H_\varepsilon(A^q);A^q,G)$ is the non-singular ε-quadratic formation over A underlying the pullback split ε-quadratic formation $(F,G) = (\partial(B^q,\psi),(\alpha,\beta,\sigma),(B'^q,(B'^q)^*))$.

Every element of $\begin{cases} V^1(A',\varepsilon) \\ V_1(A',\varepsilon) \end{cases}$ is represented by a non-singular

$\begin{cases} \varepsilon\text{-symmetric} \\ \varepsilon\text{-quadratic} \end{cases}$ formation over A' of the type

$\begin{cases} (H^{\varepsilon}((A'^q)^*,\zeta');A'^q,\alpha(A'^q)) \\ (H_{\varepsilon}(A'^q);A'^q,\alpha(A'^q)) \end{cases}$ $(q \geqslant 0)$ for some $\begin{cases} \text{isomorphism} \\ \text{automorphism} \end{cases}$

$$\begin{cases} \alpha \ : \ H^{\varepsilon}((A'^q)^*,\xi') \xrightarrow{\ \sim\ } H^{\varepsilon}((A'^q)^*,\zeta') \\ (\alpha,\chi) \ : \ \widetilde{H}_{\varepsilon}(A'^q) \xrightarrow{\ \sim\ } \widetilde{H}_{\varepsilon}(A'^q) \end{cases}$$

of $\begin{cases} - \\ a \end{cases}$ standard hyperbolic $\begin{cases} \varepsilon\text{-symmetric forms} \\ \text{split } \varepsilon\text{-quadratic form} \end{cases}$ over A,

by Proposition 1.6.2. In the ε-symmetric case we have that the maps $g:B \longrightarrow A'$, $g:\hat{H}^0(\mathbb{Z}_2;B,\varepsilon) \longrightarrow \hat{H}^0(\mathbb{Z}_2;A',\varepsilon)$ are onto, so that there exists an ε-symmetric form over B $((B^q)^*,\xi \in Q^{\varepsilon}((B^q)^*))$ such that

$$((A'^q)^*,\xi') = A'\otimes_B((B^q)^*,\xi) \ ;$$

furthermore, the maps $g':B' \longrightarrow A'$, $g':\hat{H}^0(\mathbb{Z}_2;B',\varepsilon) \longrightarrow \hat{H}^0(\mathbb{Z}_2;A',\varepsilon)$ are onto, so that there also exists an ε-symmetric form over B' $((B'^q)^*,\zeta \in Q^{\varepsilon}((B'^q)^*))$ such that

$$((A'^q)^*,\zeta') = A'\otimes_{B'}((B'^q)^*,\zeta) \ .$$

The pullback non-singular $\begin{cases} \varepsilon\text{-symmetric} \\ \text{split } \varepsilon\text{-quadratic} \end{cases}$ form over A

$$\begin{cases} (M,\phi) = (H^{\varepsilon}((B^q)^*,\xi),\alpha,H^{\varepsilon}((B'^q)^*,\zeta)) \\ (M,\psi) = (\widetilde{H}_{\varepsilon}(B^q),(\alpha,\chi),\widetilde{H}_{\varepsilon}(B'^q)) \end{cases}$$

has projective class

$$[M] = \partial\tau(\alpha:A'^q\oplus(A'^q)^* \longrightarrow A'^q\oplus(A'^q)^*) \in X = I_0 \subseteq \widetilde{K}_0(A) \ .$$

Use the isomorphism of non-singular $\begin{cases}\varepsilon\text{-symmetric}\\ \varepsilon\text{-quadratic}\end{cases}$ forms over B

$$\begin{cases} 1 : B\otimes_A(M,\phi) \xrightarrow{\sim} \partial(H^{\varepsilon}((B^q)^*,\xi);B^q,0) = H^{\varepsilon}((B^q)^*,\xi) \\ 1 : B\otimes_A(M,\psi) \xrightarrow{\sim} \partial(B^q,0) = H_{\varepsilon}(B^q) \end{cases}$$

to define an abelian group morphism

$$\begin{cases} \hat{\delta} : V^1(A',\varepsilon) \longrightarrow U^1_X(f,\varepsilon) \ ; \\ \qquad (H^{\varepsilon}((A'^q)^*,\zeta');A'^q,\alpha(A'^q)) \longmapsto ((M,\phi),(H^{\varepsilon}((B^q)^*,\xi);B^q,0),1) \\ \hat{\delta} : V_1(A',\varepsilon) \longrightarrow U^X_1(f,\varepsilon) \ ; \\ \qquad (H_{\varepsilon}(A'^q);A'^q,\alpha(A'^q)) \longmapsto ((M,\psi),(B^q,0),1) \ . \end{cases}$$

The construction of $\hat{\delta}$ in the $(-\varepsilon)$-quadratic case also gives an abelian group morphism

$$\hat{\delta} : V^{-1}(A',\varepsilon) = V\langle v_0\rangle^1(A',-\varepsilon) \longrightarrow U^{-1}_X(f,\varepsilon) = U\langle v_0\rangle^1_X(f,-\varepsilon) \ ;$$

$$(H^{-\varepsilon}(A'^q);A'^q,\alpha(A'^q)) \longmapsto ((M,\psi-\varepsilon\psi^*),(H^{-\varepsilon}(B^q);B^q,0),1) \ ,$$

with $\alpha:H^{-\varepsilon}(A'^q) \xrightarrow{\sim} H^{-\varepsilon}(A'^q)$ an automorphism of a standard hyperbolic even $(-\varepsilon)$-symmetric form over A' and

$$(M,\psi-\varepsilon\psi^*) = (H^{-\varepsilon}(B^q),\alpha,H^{-\varepsilon}(B'^q))$$

the pullback non-singular even $(-\varepsilon)$-symmetric form over A.

An element of $\begin{cases} V^0(g',\varepsilon) \\ V_0(g',\varepsilon)\end{cases}$ is represented by a non-singular $\begin{cases}\text{even } (-\varepsilon)\text{-symmetric}\\ \text{split } (-\varepsilon)\text{-quadratic}\end{cases}$ formation over B' of the type $\begin{cases}(H^{-\varepsilon}(B'^q);B'^q,G)\\ (B'^q,G)\end{cases}$ (with $G = B'^q$ as a B'-module) together with an $\begin{cases}\varepsilon\text{-symmetric}\\ \varepsilon\text{-quadratic}\end{cases}$ form over A' of the type $\begin{cases}(A'^q,\phi' \in Q^{\varepsilon}(A'^q))\\ (A'^q,\psi' \in Q_{\varepsilon}(A'^q))\end{cases}$

and an isomorphism of $\begin{cases}\text{even } (-\varepsilon)\text{-symmetric} \\ \text{split } (-\varepsilon)\text{-quadratic}\end{cases}$ formations over A'

$$\begin{cases} h : \partial(A'^q,\phi') \xrightarrow{\sim} A'\otimes_{B'}(H^{-\varepsilon}(B'^q);B'^q,G) \\ h : \partial(A'^q,\psi') \xrightarrow{\sim} A'\otimes_{B'}(B'^q,G) \end{cases} .$$

Let $\begin{cases}(B^q,\phi\in Q^{\varepsilon}(B^q)) \\ (B^q,\psi\in Q_{\varepsilon}(B^q))\end{cases}$ be an $\begin{cases}\varepsilon\text{-symmetric} \\ \varepsilon\text{-quadratic}\end{cases}$ form over B such that

$$\begin{cases}(A'^q,\phi') = A'\otimes_B(B^q,\phi) \\ (A'^q,\psi') = A'\otimes_B(B^q,\psi)\end{cases}$$

Use the pullback construction of $\begin{cases}\text{even } (-\varepsilon)\text{-symmetric} \\ \text{split } (-\varepsilon)\text{-quadratic}\end{cases}$ formations over A to define an abelian group morphism

$$\begin{cases} \hat{\delta} : V^0(g',\varepsilon) \longrightarrow U_X^{-1}(A,\varepsilon) = U\langle v_0\rangle_X^1(A,-\varepsilon) \ ; \\ \qquad ((H^{-\varepsilon}(B'^q);B'^q,G),(A'^q,\phi'),h) \longmapsto (\partial(B^q,\phi),h,(H^{-\varepsilon}(B'^q);B'^q,G)) \\ \hat{\delta} : V_0(g',\varepsilon) \longrightarrow U^X_{-1}(A,\varepsilon) = U_1^X(A,-\varepsilon) \ ; \\ \qquad ((B'^q,G),(A'^q,\psi'),h) \longmapsto (\partial(B^q,\psi),h,(B'^q,G)) \ , \end{cases}$$

and define similarly

$$\hat{\delta} : V^{-2}(g',\varepsilon) = V\langle v_0\rangle^0(g',-\varepsilon) \longrightarrow U_X^{-3}(A,\varepsilon) = U_1^X(A,\varepsilon) \ ;$$

$$((H_\varepsilon(B'^q);B'^q,G),(A'^q,\psi'-\varepsilon\psi'^*),$$
$$h:\partial(A'^q,\psi'-\varepsilon\psi'^*) \xrightarrow{\sim} A'\otimes_{B'}(H_\varepsilon(B'^q);B'^q,G))$$
$$\longmapsto (\partial(B^q,\psi-\varepsilon\psi^*),h,(H_\varepsilon(B'^q;B'^q,G))$$

using the pullback construction for ε-quadratic formations over A.

An element of $\begin{cases}V^1(g',\varepsilon) \\ V_1(g',\varepsilon)\end{cases}$ is represented by a non-singular $\begin{cases}\varepsilon\text{-symmetric} \\ \text{split } \varepsilon\text{-quadratic}\end{cases}$ form over B' of the type $\begin{cases}(B'^q,\phi\in Q^{\varepsilon}(B'^q)) \\ (B'^q,\psi\in \tilde{Q}_{\varepsilon}(B'^q))\end{cases}$

together with an isomorphism of $\begin{cases} \varepsilon\text{-symmetric} \\ \text{split } \varepsilon\text{-quadratic} \end{cases}$ forms over A'

$$\begin{cases} h : H^{\varepsilon}((A'^{p})^{*},\xi') \xrightarrow{\sim} A'\otimes_{B'}(B'^{q},\phi) \oplus H^{\varepsilon}((A'^{r})^{*},\zeta') \\ (h,\chi) : \tilde{H}_{\varepsilon}(A'^{p}) \xrightarrow{\sim} A'\otimes_{B'}(B'^{q},\psi) \oplus \tilde{H}_{\varepsilon}(A'^{r}) \ . \end{cases} \quad (2p = q + 2r)$$

In the ε-symmetric case let $((B^{p})^{*},\xi \in Q^{\varepsilon}((B^{p})^{*}))$ be an ε-symmetric form over B such that

$$((A'^{p})^{*},\xi') = A'\otimes_{B}((B^{p})^{*},\xi) \ ,$$

and let $((B'^{r})^{*},\zeta \in Q^{\varepsilon}((B'^{r})^{*}))$ be an ε-symmetric form over B' such that

$$((A'^{r})^{*},\zeta') = A'\otimes_{B'}((B'^{r})^{*},\zeta)$$

Use the pullback construction of $\begin{cases} \varepsilon\text{-symmetric} \\ \text{split } \varepsilon\text{-quadratic} \end{cases}$ forms over A to define an abelian group morphism

$$\begin{cases} \hat{\delta} : V^{1}(g',\varepsilon) \longrightarrow U^{0}_{X}(A,\varepsilon) \ ; \\ \qquad ((B'^{q},\phi),h) \longmapsto (H^{\varepsilon}((B^{p})^{*},\xi),h,(B'^{q},\phi)\oplus H^{\varepsilon}((B'^{r})^{*},\zeta)) \\ \hat{\delta} : V_{1}(g',\varepsilon) \longrightarrow U^{X}_{0}(A,\varepsilon) \ ; \\ \qquad ((B'^{q},\psi),(h,\chi)) \longmapsto (\tilde{H}_{\varepsilon}(B^{p}),(h,\chi),(B'^{q},\psi)\oplus\tilde{H}_{\varepsilon}(B'^{r})) \ , \end{cases}$$

and define similarly

$$\hat{\delta} : V^{-1}(g',\varepsilon) = V\langle v_{0}\rangle^{1}(g',-\varepsilon) \longrightarrow U^{-2}_{X}(A,\varepsilon) = U\langle v_{0}\rangle^{0}_{X}(A,-\varepsilon) \ ;$$

$$((B'^{q},\psi-\varepsilon\psi^{*}),h:H^{-\varepsilon}(A'^{p}) \xrightarrow{\sim} A'\otimes_{B'}(B'^{q},\psi-\varepsilon\psi^{*}) \oplus H^{-\varepsilon}(A'^{r}))$$

$$\longmapsto (H^{-\varepsilon}(B^{p}),h,(B'^{q},\psi-\varepsilon\psi^{*})\oplus H^{-\varepsilon}(B'^{r}))$$

using the pullback construction of even $(-\varepsilon)$-symmetric forms over A.

Having defined maps

$$\begin{cases} \delta : V^n(A',\varepsilon) \longrightarrow U^n_X(f,\varepsilon) \\ \delta : V_n(A',\varepsilon) \longrightarrow U_n^X(f,\varepsilon) \end{cases} , \quad \begin{cases} \hat{\delta} : V^n(g',\varepsilon) \longrightarrow U^{n-1}_X(A,\varepsilon) & (n \leqslant 1) \\ \hat{\delta} : V_n(g',\varepsilon) \longrightarrow U^X_{n-1}(A,\varepsilon) & (n \in \mathbb{Z}) \end{cases}$$

satisfying the hypotheses of the appropriate intermediate version of Proposition 6.1.3 we have from its conclusion that the natural maps of relative intermediate $\begin{cases} \varepsilon\text{-symmetric} \\ \varepsilon\text{-quadratic} \end{cases}$ L-groups

$$\begin{cases} (f',g) : U^n_X(f,\varepsilon) \longrightarrow V^n(g',\varepsilon) & (n \leqslant 1) \\ (f',g) : U^X_n(f,\varepsilon) \longrightarrow V_n(g',\varepsilon) & (n \in \mathbb{Z}) \end{cases}$$

are excision isomorphisms. Thus

$$\begin{cases} L^n_\kappa(\Phi,\varepsilon) = 0 & (n \leqslant 0) \\ L^\kappa_n(\Phi,\varepsilon) = 0 & (n \in \mathbb{Z}) \end{cases}$$

for $\kappa = \mathfrak{X}_0$, and hence by Proposition 6.2.2 also for any other cartesian square of $*$-invariant subgroups κ, giving rise to the Mayer-Vietoris exact sequence of intermediate $\begin{cases} \varepsilon\text{-symmetric} \\ \varepsilon\text{-quadratic} \end{cases}$ L-groups

$$\begin{cases} L^0_Y(B,\varepsilon) \oplus L^0_{Y'}(B',\varepsilon) \longrightarrow L^0_{X'}(A',\varepsilon) \xrightarrow{\partial} L^{-1}_X(A,\varepsilon) \longrightarrow L^{-1}_Y(B,\varepsilon) \oplus L^{-1}_{Y'}(B',\varepsilon) \\ \qquad \longrightarrow \dots \\ \dots \longrightarrow L^X_n(A,\varepsilon) \longrightarrow L^Y_n(B,\varepsilon) \oplus L^{Y'}_n(B',\varepsilon) \longrightarrow L^{X'}_n(A',\varepsilon) \\ \qquad \xrightarrow{\partial} L^X_{n-1}(A,\varepsilon) \longrightarrow \dots \quad (n \in \mathbb{Z}) \end{cases} .$$

In order to extend the ε-symmetric sequence to the left by an exact sequence

$$L^1_Y(B,\varepsilon) \oplus L^1_{Y'}(B',\varepsilon) \longrightarrow L^1_{X'}(A',\varepsilon) \xrightarrow{\partial} L^0_X(A,\varepsilon)$$
$$\longrightarrow L^0_Y(B,\varepsilon) \oplus L^0_{Y'}(B',\varepsilon) \longrightarrow L^0_{X'}(A',\varepsilon)$$

it suffices by Proposition 6.1.3 to prove that the map

$$(f',g) \; : \; L^1_{Y,X}(f,\varepsilon) \longrightarrow L^1_{X',Y'}(g',\varepsilon)$$

is an isomorphism for any cartesian square κ. For $\kappa = \mathcal{I}_0$ this has already been done above, and the construction of the maps $\delta,\hat{\delta}$ used to do this extends to the case $\kappa = \mathcal{I}_1$, so that (f',g) is an isomorphism for $\kappa = \mathcal{I}_1$ also. Any other cartesian square κ is such that $\kappa \supseteq \mathcal{I}_m$ ($m = 0$ or 1), and applying the 5-lemma to the morphism of exact sequences

$$\begin{array}{ccccccccc}
\hat{H}^2(\mathbb{Z}_2;X/I_m \to Y) & \longrightarrow & L^1_{O,I_m}(f,\varepsilon) & \longrightarrow & L^1_{Y,X}(f,\varepsilon) & \longrightarrow & \hat{H}^1(\mathbb{Z}_2;X/I_m \to Y) & \longrightarrow & L^0_{O,I_m}(f,\varepsilon) \\
(f',g)\,\wr\big\downarrow & & (f',g)\,\wr\big\downarrow & & (f',g)\big\downarrow & & (f',g)\,\wr\big\downarrow & & (f',g)\,\wr\big\downarrow \\
\hat{H}^2(\mathbb{Z}_2;Y' \longrightarrow X') & \longrightarrow & L^1_{O,O}(g',\varepsilon) & \longrightarrow & L^1_{X',Y'}(g',\varepsilon) & \longrightarrow & \hat{H}^1(\mathbb{Z}_2:Y' \longrightarrow X') & \longrightarrow & L^0_{O,O}(g',\varepsilon)
\end{array}$$

it is clear that the middle (f',g) is also an isomorphism.

ii) Excision isomorphisms and a Mayer-Vietoris exact sequence for the intermediate $\left\{\begin{array}{l}\varepsilon\text{-symmetric}\\ \varepsilon\text{-quadratic}\end{array}\right.$ L-groups associated to a cartesian square of rings with involution

$$\begin{array}{ccc}
A & \longrightarrow & B \\
\big\downarrow & \Phi & \big\downarrow \\
S^{-1}A & \longrightarrow & T^{-1}B
\end{array}$$

satisfying condition $\left\{\begin{array}{l}\text{ii})^*\\ \text{ii})_*\end{array}\right.$ have already been obtained in Proposition 3.6.3 i), in the special case of the cartesian square of $*$-invariant subgroups

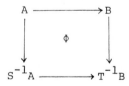

$$\kappa = \left(\begin{array}{ccc} \widetilde{K}_0(A) & \longrightarrow & \widetilde{K}_0(B) \\ \downarrow & & \downarrow \\ \mathrm{im}(\widetilde{K}_0(A)\longrightarrow\widetilde{K}_0(S^{-1}A)) & \longrightarrow & \mathrm{im}(\widetilde{K}_0(B)\longrightarrow\widetilde{K}_0(T^{-1}B)) \end{array}\right)$$

$$\subseteq \left(\begin{array}{ccc} \widetilde{K}_0(A) & \longrightarrow & \widetilde{K}_0(B) \\ \downarrow & & \downarrow \\ \widetilde{K}_0(S^{-1}A) & \longrightarrow & \widetilde{K}_0(T^{-1}B) \end{array}\right) .$$

Thus Φ is $\left\{\begin{array}{l} L^*- \\ L_*- \end{array}\right.$ cartesian in the sense of §6.2, and there are defined excision isomorphisms and a Mayer-Vietoris exact sequence for any cartesian square of $*$-invariant subgroups κ, by Proposition 6.2.3.

[]

It is possible to give an alternative proof of the L-theory Mayer-Vietoris exact sequence of Proposition 6.3.1 ii) (the localization-completion case) which avoids the localization exact sequence of §3, and is closer in spirit to the proof of the Mayer-Vietoris sequences of Proposition 6.3.1 i) (the case of a cartesian square Φ with $g:B\longrightarrow A'$ onto) involving the explicit construction of the maps $\delta,\hat{\delta}$ for $\kappa = \mathfrak{I}_0$. The main idea here is that even though neither $B\longrightarrow T^{-1}B$ nor $S^{-1}A\longrightarrow T^{-1}B$ is onto for every $x\in T^{-1}B$ there exists $t\in T$ such that $tx\in \mathrm{im}(B\longrightarrow T^{-1}B)$, so that every $\left\{\begin{array}{l} \varepsilon\text{-symmetric} \\ \varepsilon\text{-quadratic} \end{array}\right.$ form over $T^{-1}B$ of the type $\left\{\begin{array}{l} (T^{-1}B^q,\phi'\in Q^{\varepsilon}(T^{-1}B^q)) \\ (T^{-1}B^q,\psi'\in Q_{\varepsilon}(T^{-1}B^q)) \end{array}\right.$ is isomorphic to the form induced over $T^{-1}B$ from an $\begin{cases}\varepsilon\text{-symmetric}\\ \varepsilon\text{-quadratic}\end{cases}$ form over B

of the type $\begin{cases}(B^q,\phi\in Q^{\varepsilon}(B^q))\\ (B^q,\psi\in Q_{\varepsilon}(B^q))\end{cases}$ via an isomorphism

$$\begin{cases} t : T^{-1}B\otimes_B(B^q,\phi) = (T^{-1}B^q,\bar{t}\phi' t)\longrightarrow(T^{-1}B^q,\phi') \\ t : T^{-1}B\otimes_B(B^q,\psi) = (T^{-1}B^q,\bar{t}\psi' t)\longrightarrow(T^{-1}B^q,\psi') \end{cases}$$

for some T-isomorphism $t\in \mathrm{Hom}_B(B^q,B^q)$ (e.g. multiplication on the right by an element $t\in T$), and similarly for higher-dimensional $\begin{cases}\varepsilon\text{-symmetric}\\ \varepsilon\text{-quadratic}\end{cases}$ complexes over $T^{-1}B$. Indeed, such was the original approach adopted by Wall [8] in his work on the quadratic L-theory of arithmetic squares.

We refer to Madsen [1,p.249] for an application of the Mayer-Vietoris exact sequence of Proposition 6.3.1 i)$_*$ to a proof of Theorem 13A.4 iii) of Wall [4], that for a finite group π the transfer map $i^!:L_0(\mathbb{Z}[\pi])\longrightarrow L_0(\mathbb{Z})$ induced by the inclusion $i:\mathbb{Z}\longrightarrow\mathbb{Z}[\pi]$ is onto. For another application see Cappell and Shaneson [4].

6.4 Ideal L-theory

Given a ring A and a two-sided ideal $I \triangleleft A$ define the <u>double of A along I</u> $D(A,I)$ to be the ring consisting of ordered pairs (a,b) of elements $a,b \in A$ such that

$$a - b \in I \triangleleft A \ ,$$

with addition and multiplication by

$$(a,b) + (a',b') = (a+a',b+b') \in D(A,I)$$

$$(a,b)(a',b') = (aa',bb') \in D(A,I)$$

$$((a,b),(a',b') \in D(A,I),\ aa' - bb' = (a-b)a' + b(a'-b') \in I \triangleleft A),$$

exactly as in Milnor [4,§4]. There is defined a cartesian square of rings

$$\begin{array}{ccc} D(A,I) & \xrightarrow{\ f\ } & A \\ {\scriptstyle f'}\downarrow & \Phi_I & \downarrow{\scriptstyle g} \\ A & \xrightarrow{\ g'\ } & A/I \end{array}$$

with

$$f : D(A,I) \longrightarrow A \ ;\ (a,b) \longmapsto a$$

$$f' : D(A,I) \longrightarrow A' \ ;\ (a,b) \longmapsto b$$

$$g = g' = \text{projection} : A \longrightarrow A/I \ .$$

The diagonal map

$$\Delta : A \longrightarrow D(A,I) \ ;\ a \longmapsto (a,a)$$

is a ring morphism such that

$$f\Delta = 1 : A \longrightarrow A \ .$$

Thus the <u>relative K-groups of (A,I)</u> $K_m(A,I)$ $(m = 0,1)$ defined by

$$K_m(A,I) = \ker(f:K_m(D(A,I)) \longrightarrow K_m(A))$$

are such that there are natural identifications

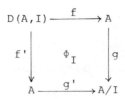

$$K_m(D(A,I)) = K_m(A) \oplus K_m(A,I)$$
$$K_m(A,I) = K_{m+1}(A \longrightarrow A/I) \qquad (m = 0,1)$$

by the excision property of Milnor [4,§4], with an exact sequence

$$K_2(A) \longrightarrow K_2(A/I) \longrightarrow K_1(A,I) \longrightarrow K_1(A) \longrightarrow K_1(A/I)$$
$$\longrightarrow K_0(A,I) \longrightarrow K_0(A) \longrightarrow K_0(A/I) \ .$$

There is also defined a cartesian square of rings

$$\begin{array}{ccc} I^+ & \xrightarrow{F} & \mathbb{Z} \\ {\scriptstyle F'}\downarrow & \Phi_I^+ & \downarrow{\scriptstyle G} \\ A & \xrightarrow{G'} & A/I \end{array}$$

where $I^+ = \mathbb{Z} \oplus I$ is the ring with addition and multiplication by

$$(n,i) + (n',i') = (n+n',i+i') \in I^+$$
$$(n,i)(n',i') = (nn',ni'+n'i+ii') \in I^+$$
$$(n,n' \in \mathbb{Z}\ ,\ i,i' \in I)$$

and

$$F : I^+ \longrightarrow \mathbb{Z}\ ;\ (n,i) \longmapsto n$$
$$F' : I^+ \longrightarrow A\ ;\ (n,i) \longmapsto n1_A + i$$
$$G : \mathbb{Z} \longrightarrow A/I\ ;\ n \longmapsto n1_{A/I}$$
$$G' = \text{projection} : A \longrightarrow A/I \ .$$

The inclusion

$$d : \mathbb{Z} \longrightarrow I^+\ ;\ n \longmapsto (n,0)$$

is such that

$$Fd = 1 : \mathbb{Z} \longrightarrow \mathbb{Z} \ .$$

Thus the <u>algebraic K-groups of I</u> $K_m(I)$ $(m = 0,1)$ defined by

$$K_m(I) = \widetilde{K}_m(I^+) = K_{m+1}(F : I^+ \longrightarrow \mathbb{Z})$$

are such that

$$K_m(I^+) = K_m(I)\oplus K_m(\mathbb{Z}) \quad .$$

The natural map

$$(F',G) : K_1(F) = K_0(I) \longrightarrow K_1(G') = K_0(A,I)$$

is an isomorphism by the excision property of classical algebraic K-theory, cf. Bass [2,IX.1.2]. Swan [1] has constructed examples of pairs (A,I) for which the natural map

$$(F',G) : K_2(F) = K_1(I) \longrightarrow K_2(G') = K_1(A,I)$$

is not an isomorphism, so that excision fails in higher algebraic K-theory.

We shall now investigate analogous results in algebraic L-theory. Roughly speaking, the 4-periodicity of the ε-quadratic L-groups $L_*(A,\varepsilon) = L_{*+4}(A,\varepsilon)$ of a ring with involution A keeps them sufficiently close to being the L-theory analogues of the classical algebraic K-groups $K_0(A), K_1(A)$ for there to be excision with respect to the involution-invariant ideals I of A, as will be shown in Proposition 6.4.1 below. The ε-symmetric L-groups $L^*(A,\varepsilon)$ are closer in spirit to the higher algebraic $K_*(A)$, and in Proposition 6.4.2 we shall give an example of the failure of excision in ε-symmetric L-theory.

Let then A be a ring with involution. A two-sided ideal $I \triangleleft A$ is <u>invariant</u> if

$$\bar{I} = I \triangleleft A \ ,$$

that is $\bar{i} \in I$ for each $i \in I$. The double D(A,I) is then a ring with involution

$$\bar{\ } : D(A,I) \longrightarrow D(A,I) \ ; \ (a,b) \longmapsto (\bar{a},\bar{b}) \ ,$$

and I^+ is a ring with involution

$$\bar{\ } : I^+ \longrightarrow I^+ \ ; \ (n,i) \longmapsto (n,\bar{i}) \ ,$$

so that the cartesian squares Φ_I and Φ_I^+ defined above are in fact cartesian squares of rings with involution. Define the $\left\{\begin{array}{l}\underline{\varepsilon\text{-symmetric}}\\ \underline{\varepsilon\text{-quadratic}}\end{array}\right.$ L-groups of (A,I)

$$\left\{\begin{array}{l} L^n(A,I,\varepsilon) = \ker(f:L^n_{\widetilde{K}_0(A)}(D(A,I),\varepsilon) \longrightarrow L^n(A,\varepsilon)) \\ L_n(A,I,\varepsilon) = \ker(f:L_n^{\widetilde{K}_0(A)}(D(A,I),\varepsilon) \longrightarrow L_n(A,\varepsilon)) \end{array}\right. \quad (n \in \mathbb{Z}) \ .$$

The diagonal map $\Delta : A \longrightarrow D(A,I) ; a \longmapsto (a,a)$ is a morphism of rings with involution such that $f\Delta = 1 : A \longrightarrow A$, so that there are natural identifications

$$\left\{\begin{array}{l} L^n_{\widetilde{K}_0(A)}(D(A,I),\varepsilon) = L^n(A,\varepsilon) \oplus L^n(A,I,\varepsilon) \\ L_n^{\widetilde{K}_0(A)}(D(A,I),\varepsilon) = L_n(A,\varepsilon) \oplus L_n(A,I,\varepsilon) \end{array}\right. \quad (n \in \mathbb{Z}) \ .$$

For $\varepsilon = \pm 1$ define the $\left\{\begin{array}{l}\underline{\varepsilon\text{-symmetric}}\\ \underline{\varepsilon\text{-quadratic}}\end{array}\right.$ L-groups of I

$$\left\{\begin{array}{l} L^n(I,\varepsilon) = \ker(F:L^n(I^+,\varepsilon) \longrightarrow L^n(\mathbb{Z},\varepsilon)) \\ L_n(I,\varepsilon) = \ker(F:L_n(I^+,\varepsilon) \longrightarrow L_n(\mathbb{Z},\varepsilon)) \end{array}\right. \quad (n \in \mathbb{Z}) \ ,$$

so that there are natural identifications

$$\left\{\begin{array}{l} L^n(I^+,\varepsilon) = L^n(I,\varepsilon) \oplus L^n(\mathbb{Z},\varepsilon) \\ L_n(I^+,\varepsilon) = L_n(I,\varepsilon) \oplus L_n(\mathbb{Z},\varepsilon) \end{array}\right. \quad (n \in \mathbb{Z}) \ .$$

Proposition 6.4.1 Given a ring with involution A and an invariant ideal $I \triangleleft A$ let

$$K = \mathrm{im}(\widetilde{K}_0(A) \longrightarrow \widetilde{K}_0(A/I)) \subseteq \widetilde{K}_0(A/I) \ .$$

$\left\{\begin{array}{l}\text{If } \hat{H}^0(\mathbb{Z}_2;A,\varepsilon) \longrightarrow \hat{H}^0(\mathbb{Z}_2;A/I,\varepsilon) \text{ is onto}\\ \text{For all } A,I,\varepsilon\end{array}\right.$ there are natural identifications

$$\begin{cases} L^n(A,I,\varepsilon) = L_K^{n+1}(A \longrightarrow A/I,\varepsilon) & (n \leqslant 0) \\ L_n(A,I,\varepsilon) = L^K_{n+1}(A \longrightarrow A/I,\varepsilon) & (n \in \mathbb{Z}) \end{cases} ,$$

and there is defined an exact sequence of $\begin{cases} \varepsilon\text{-symmetric} \\ \varepsilon\text{-quadratic} \end{cases}$ L-groups

$$\begin{cases} L^1(A,\varepsilon) \longrightarrow L_K^1(A/I,\varepsilon) \longrightarrow L^0(A,I,\varepsilon) \longrightarrow L^0(A,\varepsilon) \longrightarrow L_K^0(A/I,\varepsilon) \longrightarrow \dots \\ \dots \longrightarrow L_n(A,I,\varepsilon) \longrightarrow L_n(A,\varepsilon) \longrightarrow L_n^K(A/I,\varepsilon) \longrightarrow L_{n-1}(A,I,\varepsilon) \longrightarrow \dots \end{cases}$$
$$(n \in \mathbb{Z}) \quad .$$

Furthermore, in the ε-quadratic case for $\varepsilon = \pm 1$ there are also natural identifications

$$L_n(A,I,\varepsilon) = L_n(I,\varepsilon) \quad (n \in \mathbb{Z})$$

and the exact sequence can be written as

$$\dots \longrightarrow L_n(I,\varepsilon) \longrightarrow L_n(A,\varepsilon) \longrightarrow L_n^K(A/I,\varepsilon) \longrightarrow L_{n-1}(I,\varepsilon) \longrightarrow \dots \quad (n \in \mathbb{Z}) \ .$$

<u>Proof</u>: Immediate from Proposition 6.3.1 i) applied to Φ_I and Φ_I^+.

[]

In particular, for the ideal $(2) = 2\mathbb{Z} \triangleleft \mathbb{Z}$ there are defined isomorphisms of rings with involution

$$D(\mathbb{Z},(2)) \longrightarrow \mathbb{Z}[\mathbb{Z}_2] \ ; \ (a,b) \longmapsto \tfrac{1}{2}(a+b) + \tfrac{1}{2}(a-b)T$$

$$(2)^+ \longrightarrow \mathbb{Z}[\mathbb{Z}_2] \ ; \ (n,2i) \longmapsto n + i(1-T)$$

(the involution being the identity in each case), so that the cartesian squares $\Phi_{(2)}$, $\Phi_{(2)}^+$ may be identified with the cartesian square

$$\begin{array}{ccc} \mathbb{Z}[\mathbb{Z}_2] & \xrightarrow{\ e_+\ } & \mathbb{Z} \\ {\scriptstyle e_-}\big\downarrow & \Phi & \big\downarrow{\scriptstyle p} \\ \mathbb{Z} & \xrightarrow{\ p\ } & \mathbb{Z}_2 \end{array}$$

previously defined in §6.3. It follows that for $\varepsilon = \pm 1$ there are natural identifications of $\left\{\begin{matrix}\varepsilon\text{-symmetric}\\ \varepsilon\text{-quadratic}\end{matrix}\right.$ L-groups

$$\left\{\begin{aligned} L^n(\mathbb{Z},(2),\varepsilon) &= L^n((2),\varepsilon) = L^{n+1}(e_+:\mathbb{Z}[\mathbb{Z}_2]\longrightarrow\mathbb{Z},\varepsilon) \\ L_n(\mathbb{Z},(2),\varepsilon) &= L_n((2),\varepsilon) = L_{n+1}(e_+:\mathbb{Z}[\mathbb{Z}_2]\longrightarrow\mathbb{Z},\varepsilon) \end{aligned}\right. \quad (n\in\mathbb{Z}) \quad .$$

Let us write the $\left\{\begin{matrix}\text{skew-symmetric}\\ \text{skew-quadratic}\end{matrix}\right.$ L-groups of a ring with involution A as

$$\left\{\begin{aligned} L^n(A,-1) &= \bar{L}^n(A) \\ L_n(A,-1) &= \bar{L}_n(A) \end{aligned}\right. \quad (n\in\mathbb{Z}) \quad .$$

Proposition 6.4.2 Excision fails for the ε-symmetric L-theory of the cartesian square

$$\begin{array}{ccc} \mathbb{Z}[\mathbb{Z}_2] & \longrightarrow & \mathbb{Z} \\ \downarrow & \Phi & \downarrow \\ \mathbb{Z} & \longrightarrow & \mathbb{Z}_2 \end{array}$$

with $\varepsilon = -1$, since

$$\left\{\begin{aligned} &\bar{L}^0(\mathbb{Z}[\mathbb{Z}_2]\longrightarrow\mathbb{Z}) = 0 \\ &\bar{L}^0(\mathbb{Z}\longrightarrow\mathbb{Z}_2) = \bar{L}^0(\Phi) = \mathbb{Z}_2 \end{aligned}\right. \quad .$$

There is no Mayer-Vietoris exact sequence of ε-symmetric L-groups

$$\bar{L}^0(\mathbb{Z})\oplus\bar{L}^0(\mathbb{Z})\longrightarrow\bar{L}^0(\mathbb{Z}_2)\xrightarrow{\partial}\bar{L}^{-1}(\mathbb{Z}[\mathbb{Z}_2]) \quad ,$$

since

$$\bar{L}^0(\mathbb{Z}) = 0 \ , \ \bar{L}^0(\mathbb{Z}_2) = \mathbb{Z}_2 \ , \ \bar{L}^{-1}(\mathbb{Z}[\mathbb{Z}_2]) = 0 \ .$$

Proof: As $\hat{H}^0(\mathbb{Z}_2;\mathbb{Z}[\mathbb{Z}_2],-1) = 0$ Proposition 1.8.1 identifies

$$\bar{L}^{-1}(\mathbb{Z}[\mathbb{Z}_2]) = L_1(\mathbb{Z}[\mathbb{Z}_2]) \quad .$$

The Mayer-Vietoris exact sequence of quadratic L-groups given by Proposition 6.3.1 i)$_*$

$$L_2(\mathbb{Z})\oplus L_2(\mathbb{Z}) \xrightarrow{(1\ 1)} L_2(\mathbb{Z}[\mathbb{Z}_2]) \xrightarrow{\partial} L_1(\mathbb{Z}[\mathbb{Z}_2]) \longrightarrow L_1(\mathbb{Z})\oplus L_1(\mathbb{Z})$$

shows that $L_1(\mathbb{Z}[\mathbb{Z}_2]) = 0$, since $L_2(\mathbb{Z}) = L_2(\mathbb{Z}_2) = \mathbb{Z}_2$, $L_1(\mathbb{Z}) = 0$ by Proposition 4.3.1.

[]

Anticipating the splitting theorem

$$V^n(A[z,z^{-1}],\varepsilon) = V^n(A,\varepsilon)\oplus U^{n-1}(A,\varepsilon) \quad (n\in\mathbb{Z},\ \bar{z}=z^{-1})$$

conjectured in §I.10 and mentioned in the introduction to §7 below it is possible to extend the failure of excision given by Proposition 6.4.2 to the higher-dimensional ε-symmetric L-groups, as follows. For each $k\geqslant 0$ let Φ^k be the cartesian square of rings with involution

$$\begin{array}{ccc} \mathbb{Z}[\mathbb{Z}_2\times\mathbb{Z}^k] & \xrightarrow{e_+} & \mathbb{Z}[\mathbb{Z}^k] \\ {\scriptstyle e_-}\downarrow & \Phi^k & \downarrow{\scriptstyle p} \\ \mathbb{Z}[\mathbb{Z}^k] & \xrightarrow{p} & \mathbb{Z}_2[\mathbb{Z}^k] \end{array} .$$

Then

$$\bar{L}^0(\Phi^0) = \mathbb{Z}_2 \neq 0$$

$$\bar{L}^k(\Phi^k) = \sum_{j=0}^{k}\binom{k}{j}\bar{L}^j(\Phi^j) \neq 0$$

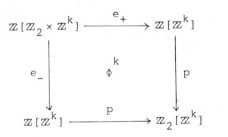

§7. The algebraic theory of codimension q surgery

The Browder-Novikov-Sullivan-Wall surgery theory of topological manifold structures on geometric Poincaré complexes was reformulated in Ranicki [7] in terms of the algebraic Poincaré complex theory of I.,II. and the algebraic theory of surgery classifying spaces. This reformulation is recalled in §7.1, and in §7.2 it is extended to the Browder-Wall surgery theory of topological (manifold, codimension q submanifold) structures on geometric Poincaré (complex, codimension q subcomplex) pairs, with $q \geqslant 1$. In §7.3 the quadratic construction ψ_F on a stable map $F:\Sigma^{\infty}X \longrightarrow \Sigma^{\infty}Y$ of §I.1 is refined to a "spectral quadratic construction" ψ_F on a "semi-stable" map $F:\Sigma^{\infty}X \longrightarrow Y$ (i.e. a map of spectra with domain a suspension spectrum), for use in §7.4 and beyond. In §7.4 we recall and expand the expression due to Quinn of geometric codimension q surgery obstruction theory in terms of geometric Poincaré splittings. The theory is then expressed in terms of algebraic Poincaré splittings in §7.5. The algebraic theory of codimension 1 surgery is developed in §7.6. In §7.7 our methods are extended to surgery with coefficients, such as the Cappell-Shaneson homology surgery obstruction theory. This extension is needed for the algebraic theory of codimension 2 surgery developed in §7.8. Finally, in §7.9 we outline the algebraic theory of knot cobordism (the origin of codimension 2 surgery), giving various algebraic characterizations of the high-dimensional knot cobordism groups C_*.

As noted in the Introduction §7 is only a preliminary account of the algebraic theory of codimension q surgery, just as Ranicki [7] is only a preliminary account of the total surgery obstruction theory, the full account of both to appear as Ranicki [11],[12]. In particular, Ranicki [11] will carry out the programme set out in §7.5 for the algebraic derivation of codimension q splitting theorems for manifolds, such as those of Cappell [i] ($1 \leqslant i \leqslant 9$) for $q = 1$, by proving codimension q splitting theorems for quadratic Poincaré complexes. The algebraic methods should also apply to the symmetric L-groups. For example, the splitting theorem for the quadratic L-groups of the Laurent extension $A[z,z^{-1}]$ ($\bar{z} = z^{-1}$) of Shaneson [1], Novikov [1] and Ranicki [2]

$$V_n(A[z,z^{-1}]) = V_n(A)\oplus U_{n-1}(A) \quad (n \in \mathbb{Z})$$

should be extended to the splitting theorem for the symmetric L-groups conjectured in §I.10

$$V^n(A[z,z^{-1}]) = V^n(A)\oplus U^{n-1}(A) \quad (n \in \mathbb{Z}) \quad .$$

7.1 The total surgery obstruction

We shall now recall the total surgery obstruction theory of Ranicki [7], at the same time extending it to geometric Poincaré complexes which may be disconnected and/or nonorientable. Such complexes arise naturally in codimension q surgery obstruction theory, particularly for $q = 1$. In the first instance, we develop some terminology with which to handle such complexes.

Given a topological space X with a finite number of path-components $X_1, X_2, \ldots, X_m$ define the fundamental groupoid of X $\pi_1(X)$ to be the disjoint union of the fundamental groups of the path-components

$$\pi_1(X) = \pi_1(X_1) \cup \pi_1(X_2) \cup \ldots \cup \pi_1(X_m) .$$

(In dealing with fundamental groups and groupoids we can afford to neglect the effects of the choice of basepoints, since all the algebraic L-functors of groups are such that inner automorphisms induce the identity, at least in the oriented case, cf. Taylor [2] for $L_*(\pi)$). An algebraic Poincaré complex over $\mathbb{Z}[\pi_1(X)]$ x is defined to be a collection $\{x_i \mid 1 \leqslant i \leqslant m\}$ of algebraic Poincaré complexes over $\mathbb{Z}[\pi_1(X_i)]$ x_i.

The $\left\{\begin{matrix}\text{symmetric}\\ \text{quadratic}\end{matrix}\right.$ L-groups of $\mathbb{Z}[\pi_1(X)]$ $\left\{\begin{matrix}L^*(\mathbb{Z}[\pi_1(X)])\\ L_*(\mathbb{Z}[\pi_1(X)])\end{matrix}\right.$ are the cobordism groups of $\left\{\begin{matrix}\text{symmetric}\\ \text{quadratic}\end{matrix}\right.$ Poincaré complexes over $\mathbb{Z}[\pi_1(X)]$, and are such that

$$\begin{cases} L^*(\mathbb{Z}[\pi_1(X)]) = L^*(\mathbb{Z}[\pi_1(X_1)])\oplus L^*(\mathbb{Z}[\pi_1(X_2)])\oplus \\ \qquad\qquad \ldots \oplus L^*(\mathbb{Z}[\pi_1(X_m)]) \\ L_*(\mathbb{Z}[\pi_1(X)]) = L_*(\mathbb{Z}[\pi_1(X_1)])\oplus L_*(\mathbb{Z}[\pi_1(X_2)])\oplus \\ \qquad\qquad \ldots \oplus L_*(\mathbb{Z}[\pi_1(X_m)]) \ . \end{cases}$$

If $\begin{cases} X \\ (f,b):M \longrightarrow X \end{cases}$ is an n-dimensional $\begin{cases} \text{geometric Poincaré complex} \\ \text{normal map} \end{cases}$

the $\begin{cases} \underline{\text{symmetric}} \\ \underline{\text{quadratic}} \end{cases}$ $\underline{\text{signature}}$

$$\begin{cases} \sigma^*(X) \in L^n(\mathbb{Z}[\pi_1(X)]) \\ \sigma_*(f,b) \in L_n(\mathbb{Z}[\pi_1(X)]) \end{cases}$$

is defined exactly as in §1.2, with components

$$\begin{cases} \{\sigma^*(X_i) \in L^n(\mathbb{Z}[\pi_1(X_i)]) \mid 1 \leqslant i \leqslant m\} \\ \{\sigma_*((f_i,b_i) = (f,b)| : M_i = f^{-1}(X_i) \longrightarrow X_i) \in L_n(\mathbb{Z}[\pi_1(X_i)]) \mid 1 \leqslant i \leqslant m\} \ . \end{cases}$$

(The inverse images $M_i = f^{-1}(X_i) \subseteq M$ of the path components X_i of X need not be connected, cf. Ranicki [8]).

In order to deal with nonorientable geometric Poincaré complexes we define generalized homology groups with twisted coefficients in the following manner.

Let (X,w) be a pair consisting of a topological space X and an orientation double covering

$$w : \overline{X} \longrightarrow X \ ,$$

which is classified by a map

$$w : \pi_1(X) \longrightarrow \mathbb{Z}_2 = \{\pm 1\} \ .$$

Let $\underline{M}$ be a spectrum

$$\underline{M} = \{M_k, \Sigma M_k \longrightarrow M_{k+1} \mid k \geqslant 0\}$$

which is equipped with an <u>orientation-reversing</u> involution

$$T : \underline{M} \longrightarrow \underline{M} .$$

The <u>w-twisted M-coefficient</u> $\left\{\begin{array}{l}\underline{\text{cohomology}}\\ \underline{\text{homology}}\end{array}\right.$ <u>groups of X</u> are defined by

$$\left\{\begin{array}{ll} H^n(X,w;\underline{M}) = \varinjlim_k [\Sigma^k\overline{X}_+, M_{n+k}]_{\mathbb{Z}_2} & \\ & (n \in \mathbb{Z}) \\ H_n(X,w;\underline{M}) = \varinjlim_k \pi_{n+k}(\overline{X}_+ \wedge_{\mathbb{Z}_2} M_k) & \end{array}\right. .$$

In particular, for the Eilenberg-MacLane spectrum of $\mathbb{Z}$

$$\underline{K} = \{K(\mathbb{Z},k), \Sigma K(\mathbb{Z},k) \longrightarrow K(\mathbb{Z},k+1) \,|\, k \geqslant 0\}$$

with the orientation-reversing involution $T:\underline{K} \longrightarrow \underline{K}$ induced by

$$T : \mathbb{Z} \longrightarrow \mathbb{Z} \; ; \; z \longmapsto -z$$

the w-twisted $\underline{K}$-coefficient $\left\{\begin{array}{l}\text{cohomology}\\ \text{homology}\end{array}\right.$ groups of X are the

w-twisted integral $\left\{\begin{array}{l}\text{cohomology}\\ \text{homology}\end{array}\right.$ groups of X

$$\left\{\begin{array}{ll} H^n(X,w;\underline{K}) = H^n(X,w) = H_n(\mathrm{Hom}_{\mathbb{Z}[\mathbb{Z}_2]}(C(\overline{X}),\mathbb{Z}^-)) & \\ & (n \in \mathbb{Z}) \\ H_n(X,w;\underline{K}) = H_n(X,w) = H_n(\mathbb{Z}^- \otimes_{\mathbb{Z}[\mathbb{Z}_2]} C(\overline{X})) & \end{array}\right. ,$$

where $\mathbb{Z}^-$ denotes the $\mathbb{Z}[\mathbb{Z}_2]$-module with additive group $\mathbb{Z}$ and $\mathbb{Z}_2$ acting by T.

An n-dimensional geometric Poincaré complex X has a fundamental class $[X] \in H_n(X,w)$ $(\cong \mathbb{Z}^m)$, with orientation map $w = w(X) : \pi_1(X) \longrightarrow \mathbb{Z}_2$. Let $-X$ denote the geometric Poincaré complex with the same underlying CW complex, but with fundamental class

$$[-X] = -[X] \in H_n(X,w) .$$

Homotopy equivalences of geometric Poincaré complexes

$$f : X \longrightarrow X'$$

are required to be orientation-preserving, with

$$f_*([X]) = [X'] \in H_n(X',w') .$$

Every compact n-dimensional topological manifold M is to be equipped with a fundamental class $[M] \in H_n(M,w(M))$, so that it has the structure of a simple n-dimensional geometric Poincaré complex.

Given a pair $(X,w:\overline{X}\longrightarrow X)$ let $\Omega_n^{STOP}(X,w)$ denote the bordism group of maps $f:M\longrightarrow X$ from compact n-dimensional topological manifolds M for which the orientation map factors as

$$w(M) : \pi_1(M) \xrightarrow{f_*} \pi_1(X) \xrightarrow{w} \mathbb{Z}_2 .$$

Then if the spectrum

$$\underline{MSTOP} = \{MSTOP(k), \Sigma MSTOP(k) \longrightarrow MSTOP(k+1) \mid k \geqslant 0\}$$

of the Thom spaces $T(1_k) = MSTOP(k)$ of the universal oriented topological k-disc bundles $1_k:BSTOP(k)\longrightarrow BSTOP(k)$ is given the orientation-reversing involution

$$T : \underline{MSTOP} \longrightarrow \underline{MSTOP}$$

induced by the oppositely oriented bundles $-1_k:BSTOP(k)\longrightarrow BSTOP(k)$ there are natural identifications

$$\Omega_n^{STOP}(X,w) = H_n(X,w;\underline{MSTOP}) \qquad (n \neq 4) .$$

These follow from the identification

$$\overline{X}_+ \wedge_{\mathbb{Z}_2} MSTOP(k) = T(\eta_k)$$

with $T(\eta_k)$ the Thom space of the topological bundle η_k classified by the map η_k appearing in the homotopy-theoretic pullback diagram

$$\begin{array}{ccc} X \times_{K(\mathbb{Z}_2,1)} BTOP(k) & \xrightarrow{\eta_k} & BTOP(k) \\ \downarrow & & \downarrow w_1 \\ X & \xrightarrow{w} & K(\mathbb{Z}_2,1) \end{array}$$

and topological transversality in dimensions $\neq 4$.

The pair $(X, w:\overline{X} \longrightarrow X)$ is untwisted if $\overline{X} = X \sqcup X$ is the trivial double covering of X, that is $w(g) = +1$ for each $g \in \pi_1(X)$, in which case the w-twisted $\underline{M}$-coefficient $\begin{cases}\text{cohomology}\\ \text{homology}\end{cases}$ groups of X are just the usual $\underline{M}$-coefficient $\begin{cases}\text{cohomology}\\ \text{homology}\end{cases}$ groups of X

$$\begin{cases} H^n(X,w;\underline{M}) = \varinjlim_k [\Sigma^k \overline{X}_+, M_{n+k}]_{\mathbb{Z}_2} = \varinjlim_k [\Sigma^k X_+, M_{n+k}] = H^n(X;\underline{M}) \\ H_n(X,w;\underline{M}) = \varinjlim_k \pi_{n+k}(\overline{X}_+ \wedge_{\mathbb{Z}_2} M_k) = \varinjlim_k \pi_{n+k}(X_+ \wedge M_k) = H_n(X;\underline{M}) \end{cases},$$

and the w-twisted integral $\begin{cases}\text{cohomology}\\ \text{homology}\end{cases}$ groups are just the usual integral $\begin{cases}\text{cohomology}\\ \text{homology}\end{cases}$ groups

$$\begin{cases} H^n(X,w) = H_n(\mathrm{Hom}_{\mathbb{Z}[\mathbb{Z}_2]}(C(\overline{X}), \mathbb{Z}^-)) = H_n(\mathrm{Hom}_{\mathbb{Z}}(C(X), \mathbb{Z})) = H^n(X) \\ H_n(X,w) = H_n(\mathbb{Z}^- \otimes_{\mathbb{Z}[\mathbb{Z}_2]} C(\overline{X})) = H_n(C(X)) = H_n(X) \end{cases}.$$

The topological bordism groups of (X,w) are just the usual oriented topological bordism groups of X

$$\Omega_n^{STOP}(X,w) = \Omega_n^{STOP}(X)$$

$$(= H_n(X;\underline{MSTOP}) \text{ for } n \neq 4) .$$

From now on we shall suppress the explicit reference to w in dealing with w-twisted $\left\{\begin{matrix}\text{cohomology}\\ \text{homology}\end{matrix}\right.$ groups, writing $\left\{\begin{matrix}H^*(X,w;\underline{M})\\ H_*(X,w;\underline{M})\end{matrix}\right.$ as $\left\{\begin{matrix}H^*(X;\underline{M})\\ H_*(X;\underline{M})\end{matrix}\right.$, the contributions of the orientation covering $w:\overline{X}\longrightarrow X$ and the orientation-reversing involution $T:\underline{M}\longrightarrow\underline{M}$ being understood.

An <u>s-triangulation</u> of a simple n-dimensional geometric Poincaré complex X is a simple homotopy equivalence

$$f : M \xrightarrow{\sim} X$$

from a compact n-dimensional topological manifold M. A <u>concordance</u> of s-triangulations $f:M\xrightarrow{\sim}X, f':M'\xrightarrow{\sim}X$ is a simple homotopy equivalence of triads

$$(g;f,f') : (N;M,M') \xrightarrow{\sim} (X\times I;X\times 0,X\times 1) \quad (I = [0,1])$$

from a compact (n+1)-dimensional topological manifold triad (N;M,M'). The <u>topological manifold structure set</u> $\mathcal{S}^{TOP}(X)$ of a simple n-dimensional geometric Poincaré complex X is the set (possibly empty) of concordance classes of s-triangulations $f:M\xrightarrow{\sim}X$. For $n \geqslant 5$ the concordance of s-triangulations is also the equivalence relation defined by

$$(f:M\xrightarrow{\sim}X) \sim (f':M'\xrightarrow{\sim}X)$$

if there exist a homeomorphism $h:M\xrightarrow{\sim}M'$ and a homotopy $g : f \simeq f'h : M \longrightarrow X$

by the topological s-cobordism theorem, so that $\mathcal{S}^{TOP}(X)$ is the topological manifold structure set of X in the sense of Sullivan [1] and Wall [4,§10].

An <u>h-triangulation</u> of a finite n-dimensional geometric Poincaré complex X is a homotopy equivalence

$$f : M \xrightarrow{\simeq} X$$

from a compact n-dimensional topological manifold M. <u>Concordance</u> of h-triangulations is defined as for s-triangulations, but using a homotopy equivalence of triads instead of a simple homotopy equivalence.

For the sake of the above application of the s-cobordism theorem we shall be primarily concerned with the s-triangulation theory of simple geometric Poincaré complexes. Accordingly, we shall be dealing with the simple quadratic L-groups of group rings

$$L^s_*(\pi,w) = V_*^{\{\pi\}\subseteq\tilde{K}_1(\mathbb{Z}[\pi])}(\mathbb{Z}[\pi])$$

originally defined by Wall [4] using based $\mathbb{Z}[\pi]$-modules, simple isomorphisms (with $\tau = 0 \in Wh(\pi) = \tilde{K}_1(\mathbb{Z}[\pi])/\{\pi\}$) and the w-twisted involution on $\mathbb{Z}[\pi]$ for some orientation map $w:\pi \longrightarrow \mathbb{Z}_2$. From now on $L_*(\mathbb{Z}[\pi])$ will stand for the simple L-groups $L^s_*(\pi,w)$ rather than the projective L-groups $L^p_*(\pi,w) = U_*(\mathbb{Z}[\pi])$ as heretofore, which will be denoted by $L^p_*(\mathbb{Z}[\pi])$. The free L-groups $L^h_*(\pi,w) = V_*(\mathbb{Z}[\pi])$ are denoted by $L^h_*(\mathbb{Z}[\pi])$.

The s-triangulation theory developed here has of course its counterpart in a parallel theory for the h-triangulation of finite geometric Poincaré complexes, involving $L^h_*(\mathbb{Z}[\pi])$. (Define a <u>p-triangulation</u> of a finitely dominated n-dimensional geometric Poincaré complex X to be an h-triangulation of the finite (n+1)-dimensional geometric Poincaré complex $X \times S^1$.

Following Pedersen and Ranicki [1] there is also a parallel theory for the p-triangulation of finitely dominated geometric Poincaré complexes, involving $L_*^p(\mathbb{Z}[\pi])$).

From now on geometric Poincaré complexes are to be taken as simple (unless specified otherwise), and manifolds are to be taken as compact, topological and triangulable. Similarly for geometric Poincaré pairs and manifolds with boundary.

As in Ranicki [7] let $\begin{cases} \underline{\mathbb{L}}^0 \\ \underline{\mathbb{L}}_0 \\ \hat{\underline{\mathbb{L}}}^0 \end{cases}$ denote the spectrum of

$\begin{cases} \text{symmetric} \\ \text{quadratic} \\ \text{(symmetric,quadratic)} \end{cases}$ Poincaré n-ads over $\mathbb{Z}$ with homotopy groups

$$\begin{cases} \pi_n(\underline{\mathbb{L}}^0) = \begin{cases} L^n(\mathbb{Z}) & \text{if } n \geqslant 0 \\ 0 & \phantom{\text{if }} n \leqslant -1 \end{cases} \\ \pi_n(\underline{\mathbb{L}}_0) = \begin{cases} L_n(\mathbb{Z}) & \text{if } n \geqslant 1 \\ 0 & \phantom{\text{if }} n \leqslant 0 \end{cases} \\ \pi_n(\hat{\underline{\mathbb{L}}}^0) = \begin{cases} \hat{L}^n(\mathbb{Z}) & n \geqslant 1 \\ L^0(\mathbb{Z}) & \text{if } n = 0 \\ 0 & n \leqslant -1 \end{cases} \end{cases},$$

and such that there is defined a fibration sequence

$$\underline{\mathbb{L}}_0 \xrightarrow{1+T} \underline{\mathbb{L}}^0 \xrightarrow{J} \hat{\underline{\mathbb{L}}}^0 \xrightarrow{H} \Sigma\underline{\mathbb{L}}_0 \xrightarrow{1+T} \Sigma\underline{\mathbb{L}}_0 \longrightarrow \dots .$$

(There are defined algebraic $\mathbb{L}$-spectra for any ring with involution A, using algebraic Poincaré n-ads over A, but only the case A = $\mathbb{Z}$ need concern us here. The general theory will be developed in Ranicki [12]).

For any space X equipped with an orientation double covering $w:\overline{X}\longrightarrow X$ there are defined <u>assembly maps</u>

$$\left\{\begin{array}{l} \sigma^* : H_n(X;\underline{\mathbb{L}}^0) \longrightarrow L^n(\mathbb{Z}[\pi_1(X)]) \\ \sigma_* : H_n(X;\underline{\mathbb{L}}_0) \longrightarrow L_n(\mathbb{Z}[\pi_1(X)]) \qquad (n \geqslant 0) \\ \hat{\sigma}^* : H_n(X;\underline{\hat{\mathbb{L}}}^0) \longrightarrow \hat{L}^n(\mathbb{Z}[\pi_1(X)]) \end{array}\right. ,$$

where the homology groups are defined using w-twisted coefficients and the L-groups are defined using the w-twisted involution on the group ring $\mathbb{Z}[\pi_1(X)]$. The assembly maps fit together to define a natural transformation of exact sequences

$$\begin{array}{ccccccccc} \dots\longrightarrow H_n(X;\underline{\mathbb{L}}_0) & \xrightarrow{1+T} & H_n(X;\underline{\mathbb{L}}^0) & \xrightarrow{J} & H_n(X;\underline{\hat{\mathbb{L}}}^0) & \xrightarrow{H} & H_{n-1}(X;\underline{\mathbb{L}}_0) \longrightarrow \dots \\ \downarrow{\scriptstyle \sigma_*} & & \downarrow{\scriptstyle \sigma^*} & & \downarrow{\scriptstyle \hat{\sigma}^*} & & \downarrow{\scriptstyle \sigma_*} \\ \dots\longrightarrow L_n(\mathbb{Z}[\pi]) & \xrightarrow{1+T} & L^n(\mathbb{Z}[\pi]) & \xrightarrow{J} & \hat{L}^n(\mathbb{Z}[\pi]) & \xrightarrow{H} & L_{n-1}(\mathbb{Z}[\pi]) \longrightarrow \dots \end{array}$$

with $\pi = \pi_1(X)$. (The hyperquadratic L-groups $\hat{L}^*(A)$ were defined in §2.

The n-ad version of the Browder-Novikov transversality construction of topological normal maps combined with the n-ad version of the quadratic kernel construction of II. and the computation

$$\pi_n(G/TOP) = L_n(\mathbb{Z}) \quad (n \geqslant 1)$$

give a canonical homotopy equivalence

$$G/TOP \xrightarrow{\sim} \mathbb{L}_0$$

with $\mathbb{L}_0$ the 0th space of the Ω-spectrum $\underline{\mathbb{L}}_0 = \{\mathbb{L}_{-k} = \Omega\mathbb{L}_{-k-1} \mid k \geqslant 0\}$. In fact, the quadratic $\mathbb{L}$-spectrum $\underline{\mathbb{L}}_0$ is homotopy equivalent to the 0-connective cover of the Quinn [1] spectrum of simply-connected surgery problems (= topological normal maps of n-ads $(f,b):M \longrightarrow X$ with $\pi_1(X) = \{1\}$), corresponding to the infinite loop space structure of G/TOP given by the Sullivan characteristic variety addition. The quadratic assembly map $\sigma_*:H_*(X;\underline{\mathbb{L}}_0) \longrightarrow L_*(\mathbb{Z}[\pi_1(X)])$ is the algebraic version of the geometric assembly map of Quinn [2].

Given a (k-1)-spherical fibration $\xi:X \longrightarrow BG(k)$ over a space X let $\overline{X} \longrightarrow X$ be the orientation double covering classified by $w_1(\xi) \in H^1(X;\mathbb{Z}_2)$, so that ξ lifts to a $\mathbb{Z}_2$-equivariant map $\overline{\xi}:\overline{X} \longrightarrow BSG(k)$ classifying an oriented (k-1)-spherical fibration over $\overline{X}$. We shall consider spherical fibrations ξ to be equipped with a choice of lift $\overline{\xi}$, letting $-\xi$ denote the same fibration with the other choice of lift. The base space X will always be taken to be a finitely dominated CW complex.

Let $\underline{R} = \{R_j, \Sigma R_j \longrightarrow R_{j+1} \mid j \geqslant 0\}$ be a ring spectrum, with structure maps

$$\otimes : R_j \wedge R_k \longrightarrow R_{j+k} \quad , \quad 1_j : S^j \longrightarrow R_j \quad (j,k \geqslant 0) \; ,$$

which is equipped with an <u>orientation-reversing</u> involution $T:\underline{R} \longrightarrow \underline{R}$ inducing the additive inverse $T : \pi_*(\underline{R}) \longrightarrow \pi_*(\underline{R})$; $x \longmapsto -x$. An <u>R-orientation</u> of a (k-1)-spherical fibration $\xi:X \longrightarrow BG(k)$ is a $w_1(\xi)$-twisted $\underline{R}$-coefficient Thom class, i.e. an element

$$U_\xi \in \dot{H}^k(T(\xi);\underline{R}) = \varinjlim_j [\Sigma^j T(\overline{\xi}), R_{j+k}]_{\mathbb{Z}_2}$$

such that for each map $i:\{pt.\} \longrightarrow X$ $i^*U_\xi \in \dot{H}^k(T(i^*\xi);\underline{R}) = \pi_0(\underline{R})$ is a unit. $\underline{R}$-orientations are required to be compatible with the choice of lift $\overline{\xi}:\overline{X} \longrightarrow BSG(k)$, so that

$$U_{-\xi} = -U_\xi \in \dot{H}^k(T(\xi);\underline{R}) \; .$$

If $\underline{M}$ is an $\underline{R}$-module spectrum there are defined Thom isomorphisms in $\underline{M}$-coefficient $\begin{cases} \text{cohomology} \\ \text{homology} \end{cases}$

$$\begin{cases} U_\xi \cup - \ : \ H^*(X,w';\underline{M}) \xrightarrow{\ \sim\ } \dot{H}^{*+k}(T(\xi),w'';\underline{M}) \\ U_\xi \cap - \ : \ \dot{H}_*(T(\xi),w'';\underline{M}) \xrightarrow{\ \sim\ } H_{*-k}(X;w';\underline{M}) \end{cases}$$

for any orientation maps $w',w'':\pi_1(X) \longrightarrow \mathbb{Z}_2$ such that $w'w'' = w_1(\xi)$.

The Whitney sum of oriented $\begin{cases}\text{topological bundles}\\ \text{spherical fibrations}\end{cases}$

$$\begin{cases} \oplus \ : \ BSTOP(j) \times BSTOP(k) \longrightarrow BSTOP(j+k) \\ \oplus \ : \ BSG(j) \times BSG(k) \longrightarrow BSG(j+k) \end{cases} \quad (j,k \geqslant 0)$$

induces products in the Thom spaces

$$\begin{cases} \otimes \ : \ MSTOP(j) \wedge MSTOP(k) \longrightarrow MSTOP(j+k) \\ \otimes \ : \ MSG(j) \wedge MSG(k) \longrightarrow MSG(j+k) \end{cases} \quad (j,k \geqslant 0)$$

making $\begin{cases} \underline{MSTOP} = \{MSTOP(j), \Sigma MSTOP(j) \longrightarrow MSTOP(j+1) \mid j \geqslant 0\} \\ \underline{MSG} = \{MSG(j), \Sigma MSG(j) \longrightarrow MSG(j+1) \mid j \geqslant 0\} \end{cases}$ into

a ring spectrum. A $\begin{cases}\text{topological bundle } \xi : X \longrightarrow BTOP(k)\\ \text{spherical fibration } \xi : X \longrightarrow BG(k)\end{cases}$ has

a canonical $\begin{cases}\underline{MSTOP}\text{-}\\ \underline{MSG}\text{-}\end{cases}$ orientation

$$\begin{cases} U_\xi \in \dot{H}^k(T(\xi);\underline{MSTOP}) = \varinjlim_j \, [\Sigma^j T(\overline{\xi}), MSTOP(j+k)]_{\mathbb{Z}_2} \\ U_\xi \in \dot{H}^k(T(\xi);\underline{MSG}) = \varinjlim_j \, [\Sigma^j T(\overline{\xi}), MSG(j+k)]_{\mathbb{Z}_2} \end{cases} ,$$

the element represented by the $\mathbb{Z}_2$-map of Thom spaces

$$\begin{cases} U_\xi \ : \ T(\overline{\xi}) \longrightarrow MSTOP(k) \\ U_\xi \ : \ T(\overline{\xi}) \longrightarrow MSG(k) \end{cases}$$

induced by the classifying map $\begin{cases}\overline{\xi} : \overline{X} \longrightarrow BSTOP(k)\\ \overline{\xi} : \overline{X} \longrightarrow BSG(k)\end{cases}$.

The tensor product of algebraic Poincaré n-ads over $\mathbb{Z}$ gives rise to pairings of spectra

$$\otimes : \underline{\mathbb{L}}^0 \wedge \underline{\mathbb{L}}^0 \longrightarrow \underline{\mathbb{L}}^0$$

$$\otimes : \hat{\underline{\mathbb{L}}}^0 \wedge \hat{\underline{\mathbb{L}}}^0 \longrightarrow \hat{\underline{\mathbb{L}}}^0$$

$$\otimes : \underline{\mathbb{L}}^0 \wedge \underline{\mathbb{L}}_0 \longrightarrow \underline{\mathbb{L}}_0$$

making $\underline{\mathbb{L}}^0$ and $\hat{\underline{\mathbb{L}}}^0$ into ring spectra, and $\underline{\mathbb{L}}_0$ into an $\underline{\mathbb{L}}^0$-module spectrum. The $\begin{cases} \text{symmetric} \\ \text{(symmetric, quadratic)} \end{cases}$ Poincaré n-ads over $\mathbb{Z}$ of oriented $\begin{cases} \text{manifold} \\ \text{normal space} \end{cases}$ n-ads define a map of ring spectra

$$\begin{cases} \sigma^* : \underline{\text{MSTOP}} \longrightarrow \underline{\mathbb{L}}^0 \\ \hat{\sigma}^* : \underline{\text{MSG}} \longrightarrow \hat{\underline{\mathbb{L}}}^0 \end{cases}$$

such that there is defined a commutative square of ring spectra

$$\begin{array}{ccc} \underline{\text{MSTOP}} & \xrightarrow{\sigma^*} & \underline{\mathbb{L}}^0 \\ {\scriptstyle J}\downarrow & & \downarrow{\scriptstyle J} \\ \underline{\text{MSG}} & \xrightarrow{\hat{\sigma}^*} & \hat{\underline{\mathbb{L}}}^0 \end{array} .$$

It follows that a $\begin{cases} \text{topological bundle } \tilde{\xi}:X \longrightarrow \text{BTOP}(k) \\ \text{spherical fibration } \xi:X \longrightarrow \text{BG}(k) \end{cases}$ has a <u>canonical</u> $\begin{cases} \underline{\mathbb{L}}^0\text{-} \\ \hat{\underline{\mathbb{L}}}^0\text{-} \end{cases}$<u>orientation</u> $\begin{cases} U_{\tilde{\xi}} \in \dot{H}^k(T(\xi);\underline{\mathbb{L}}^0) \\ \hat{U}_{\xi} \in \dot{H}^k(T(\xi);\hat{\underline{\mathbb{L}}}^0) \end{cases}$, the image under $\begin{cases} \sigma^*:\dot{H}^k(T(\xi);\underline{\text{MSTOP}}) \longrightarrow \dot{H}^k(T(\xi);\underline{\mathbb{L}}^0) \\ \hat{\sigma}^*:\dot{H}^k(T(\xi);\underline{\text{MSG}}) \longrightarrow \dot{H}^k(T(\xi);\hat{\underline{\mathbb{L}}}^0) \end{cases}$ of the canonical $\begin{cases} \underline{\text{MSTOP}} \\ \underline{\text{MSG}} \end{cases}$-orientation, $\begin{cases} \text{with } \xi = J\tilde{\xi} : X \longrightarrow \text{BG}(k) \\ - \end{cases}$ such that

$$JU_{\tilde{\xi}} = \hat{U}_{\xi} \in \dot{H}^k(T(\xi);\hat{\underline{\mathbb{L}}}^0) .$$

We refer to Rourke and Sanderson [1],[2] for the definition and basic properties of topological block bundles.

A <u>t-triangulation</u> $\tilde{\xi}$ of a spherical fibration $\xi:X \longrightarrow BG(k)$ is a reduction of ξ to a topological block bundle, as defined by a classifying map $\tilde{\xi}:X \longrightarrow B\widetilde{TOP}(k)$ together with a homotopy $b:J\tilde{\xi} \simeq \xi:X \longrightarrow BG(k)$. A <u>concordance</u> of t-triangulations $\tilde{\xi}_0,\tilde{\xi}_1$ is a t-triangulation $\widetilde{(\xi \times I)}:X \times I \longrightarrow B\widetilde{TOP}(k)$ of

$$\xi \times I : X \times I \xrightarrow{\text{projection}} X \xrightarrow{\xi} BG(k)$$

such that

$$\widetilde{(\xi \times I)}|_{X \times \{0\}} = \tilde{\xi}_0,\ \widetilde{(\xi \times I)}|_{X \times \{1\}} = \tilde{\xi}_1 : X \longrightarrow B\widetilde{TOP}(k)$$

(and similarly for the homotopies b_0,b_1), i.e. an isomorphism $\tilde{\xi}_0 \xrightarrow{\sim} \tilde{\xi}_1$ over $1:X \longrightarrow X$ of the $\widetilde{TOP}$ reductions. Define the <u>topological structure set of ξ</u> $\mathcal{T}^{TOP}(\xi)$ to be the set of concordance classes of t-triangulations of ξ. For $k = 1,2$ $B\widetilde{TOP}(k) = BG(k)$ so that $\mathcal{T}^{TOP}(\xi)$ consists of a single element, and for $k \geqslant 3$ $G(k)/\widetilde{TOP}(k) = G/TOP$ so that $\mathcal{T}^{TOP}(\xi)$ is the set of stable equivalence classes of stable reductions of ξ to a genuine topological bundle $\tilde{\xi}:X \longrightarrow BTOP(k')$ (k' large) which is either empty or in unnatural one-one correpondence with $[X,G/TOP]$. In particular, note that a topological block bundle $\tilde{\xi}:X \longrightarrow B\widetilde{TOP}(k)$ has a <u>canonical $\underline{\mathbb{L}}^0$-orientation</u> $U_{\tilde{\xi}} \in \dot{H}^k(T(\xi);\underline{\mathbb{L}}^0)$ such that

$$JU_{\tilde{\xi}} = \hat{U}_{\xi} \in \dot{H}^k(T(\xi);\underline{\hat{\mathbb{L}}}^0)$$

is the canonical $\underline{\hat{\mathbb{L}}}^0$-orientation of the underlying spherical fibration $\xi = J\tilde{\xi}:X \longrightarrow BG(k)$, namely the canonical $\underline{\mathbb{L}}^0$-orientation of any stably equivalent genuine topological bundle.

<u>Proposition 7.1.1</u> Let $\xi : X \longrightarrow BG(k)$ be a $(k-1)$-spherical fibration over a space X.

i) The <u>t-triangulability obstruction of ξ</u>

$$t(\xi) = H(\hat{U}_\xi) \in \dot{H}^{k+1}(T(\xi);\underline{\mathbb{L}}_0)$$

is such that $t(\xi) = 0$ if and only if ξ admits a t-triangulation, i.e. a $\widetilde{TOP}$ reduction $\tilde{\xi} : X \longrightarrow B\widetilde{TOP}(k)$.

ii) Given a subspace $Y \subset X$ and a t-triangulation $\eta : Y \longrightarrow B\widetilde{TOP}(k)$ of the restriction $\xi| : Y \longrightarrow BG(k)$ there is defined a <u>rel∂ t-triangulability obstruction</u>

$$t_\partial(\xi,\eta) \in H^{k+1}(T(\xi),T(\eta);\underline{\mathbb{L}}_0)$$

such that $t_\partial(\xi,\eta) = 0$ if and only if ξ admits a t-triangulation $\tilde{\xi} : X \longrightarrow B\widetilde{TOP}(k)$ such that $\tilde{\xi}| = \eta : Y \longrightarrow B\widetilde{TOP}(k)$. The obstruction has images

$$i^* t_\partial(\xi,\eta) = t(\xi) \in \dot{H}^{k+1}(T(\xi);\underline{\mathbb{L}}_0)$$

$$(1+T)\, t_\partial(\xi,\eta) = \delta U_\eta \in H^{k+1}(T(\xi),T(\eta);\underline{\mathbb{L}}^0)$$

with i = inclusion : $(T(\xi),pt.) \longrightarrow (T(\xi),T(\eta))$, $U_\eta \in \dot{H}^k(T(\eta);\underline{\mathbb{L}}^0)$ the canonical $\underline{\mathbb{L}}^0$-orientation of η and δ the connecting map.

iii) If $k \geqslant 3$ and ξ admits a t-triangulation the topological structure set $\mathcal{J}^{TOP}(\xi)$ carries a natural affine structure with translation group $\dot{H}^k(T(\xi);\underline{\mathbb{L}}_0)$, the <u>difference</u> of two t-triangulations $\tilde{\xi}_0, \tilde{\xi}_1 : X \longrightarrow B\widetilde{TOP}(k)$ being the element

$$t(\tilde{\xi}_0,\tilde{\xi}_1) = t_\partial((\xi \times I, \tilde{\xi}_0 \cup -\tilde{\xi}_1) : (X \times I, X \times \{0,1\}) \longrightarrow (BG(k), B\widetilde{TOP}(k)))$$

$$\in H^{k+1}(T(\xi \times I), T(\xi_0 \cup -\xi_1);\underline{\mathbb{L}}_0) = \dot{H}^{k+1}(\Sigma T(\xi);\underline{\mathbb{L}}_0) = \dot{H}^k(T(\xi);\underline{\mathbb{L}}_0)$$

with image

$$(1+T)\, t(\tilde{\xi}_0,\tilde{\xi}_1) = U_{\tilde{\xi}_0} - U_{\tilde{\xi}_1} \in \dot{H}^k(T(\xi);\underline{\mathbb{L}}^0)\ .$$

A particular choice of t-triangulation $\widetilde{\xi}:X \longrightarrow \widetilde{BTOP}(k)$ determines an isomorphism of abelian groups

$$(U_{\widetilde{\xi}} \cup -)^{-1} : \dot{H}^k(T(\xi);\underline{\mathbb{L}}_0) \xrightarrow{\sim} H^0(X;\underline{\mathbb{L}}_0) = [X,G/TOP]$$

with $U_{\widetilde{\xi}} \in \dot{H}^k(T(\xi);\underline{\mathbb{L}}^0)$ the canonical $\underline{\mathbb{L}}^0$-orientation.

Proof: See Ranicki [7], where it was shown that $\xi:X \longrightarrow BG(k)$ admits a t-triangulation $\widetilde{\xi}:X \longrightarrow \widetilde{BTOP}(k)$ if and only if there exists an $\underline{\mathbb{L}}^0$-orientation $U_{\widetilde{\xi}} \in \dot{H}^k(T(\xi);\underline{\mathbb{L}}^0)$ such that

$$JU_{\widetilde{\xi}} = \hat{U}_{\xi} \in \dot{H}^k(T(\xi);\underline{\hat{\mathbb{L}}}^0)$$

is the canonical $\underline{\hat{\mathbb{L}}}^0$-orientation.

[]

Let X be an n-dimensional geometric Poincaré complex with Spivak normal structure $(\nu_X:X \longrightarrow BG(k),\rho_X:S^{n+k} \longrightarrow T(\nu_X))$ and let $w:\bar{X} \longrightarrow X$ be any double covering of X, so that there is defined an $S\mathbb{Z}_2$-duality map

$$\alpha_X : S^{n+k} \xrightarrow{\rho_X} T(\nu_X) \xrightarrow{\Delta} \bar{X}_+ \wedge_{\mathbb{Z}_2} T(\bar{\nu}_X) \quad .$$

Thus for any coefficient spectrum $\underline{M}$ there are defined $S\mathbb{Z}_2$-duality isomorphisms

$$\begin{aligned}\alpha_X : \dot{H}^r(T(\nu_X);\underline{M}) &= \varinjlim_j [\Sigma^j T(\bar{\nu}_X),M_{j+r}]_{\mathbb{Z}_2} \\ &\xrightarrow{\sim} H_{n+k-r}(X;\underline{M}) = \varinjlim_j \pi_{n+j+k}(\bar{X}_+ \wedge_{\mathbb{Z}_2} M_{j+r}) \ ;\end{aligned}$$

$$(h:\Sigma^j T(\bar{\nu}_X) \longrightarrow M_{j+r})$$

$$\longmapsto (S^{n+j+k} \xrightarrow{\Sigma^j \alpha_X} \bar{X}_+ \wedge_{\mathbb{Z}_2} \Sigma^j T(\bar{\nu}_X) \xrightarrow{1 \wedge h} \bar{X}_+ \wedge_{\mathbb{Z}_2} M_{j+r}) \quad (r \in \mathbb{Z})$$

using w-twisted $\underline{M}$-coefficients and an involution $T:\underline{M} \longrightarrow \underline{M}$.

An $\underline{R}$-orientation of X is an $\underline{R}$-orientation of $\nu_X : X \longrightarrow BG(k)$, or equivalently a w(X)-twisted $\underline{R}$-coefficient fundamental class

$$[X] \in H_n(X;\underline{R})$$

such that $U_{\nu_X} = \alpha_X^{-1}([X]) \in \dot{H}^k(T(\nu_X);\underline{R})$ is an $\underline{R}$-orientation of ν_X. For any $\underline{R}$-module spectrum $\underline{M} = \{M_j, \Sigma M_j \longrightarrow M_{j+1}, R_j \wedge M_k \longrightarrow M_{j+k} \mid j,k \geqslant 0\}$ there are then defined $\underline{R}$-coefficient Poincaré duality isomorphisms

$$[X] \cap - \ : H^*(X,w';\underline{M}) \xrightarrow[\sim]{U_{\nu_X} \cup -} \dot{H}^{*+k}(T(\nu_X),w'';\underline{M})$$

$$\xrightarrow[\sim]{\alpha_X} H_{n-*}(X,w'';\underline{M})$$

for any orientation maps $w',w'' : \pi_1(X) \longrightarrow \mathbb{Z}_2$ such that $w'w'' = w(X)$. In particular, the canonical $\hat{\underline{\mathbb{L}}}^0$-orientation $\hat{U}_{\nu_X} \in \dot{H}^k(T(\nu_X);\hat{\underline{\mathbb{L}}}^0)$ of ν_X determines the canonical $\hat{\underline{\mathbb{L}}}^0$-orientation of X

$$[\hat{X}] = \alpha_X(\hat{U}_{\nu_X}) \in H_n(X;\hat{\underline{\mathbb{L}}}^0) \quad .$$

Proposition 7.1.2 An n-dimensional manifold M has a canonical $\underline{\mathbb{L}}^0$-orientation

$$[M] = \alpha_M(U_{\nu_M}) \in H_n(M;\underline{\mathbb{L}}^0)$$

with $(\nu_M = \nu_{M \subset S^{n+k}} : M \longrightarrow B\widetilde{TOP}(k), \rho_M : S^{n+k} \longrightarrow T(\nu_M))$ the canonical topological normal structure, such that

i) $J([M]) = [\hat{M}] \in H_n(M;\hat{\underline{\mathbb{L}}}^0)$ is the canonical $\hat{\underline{\mathbb{L}}}^0$-orientation

ii) $\sigma^*([M]) = \sigma^*(M) \in L^n(\mathbb{Z}[\pi_1(M)])$, with σ^* the symmetric assembly map and $\sigma^*(M)$ the symmetric signature of M.

Proof: See Ranicki [7].

[]

A <u>t-triangulation</u> of an n-dimensional geometric Poincaré complex X is a topological normal map

$$(f,b) : M \longrightarrow X$$

in the sense of §1.2. A <u>concordance</u> of t-triangulations $(f_0,b_0):M_0 \longrightarrow X$, $(f_1,b_1):M_1 \longrightarrow X$ is a topological normal map of triads

$$((g,c);(f_0,b_0),(f_1,b_1)) : (N;M_0,M_1) \longrightarrow X \times (I;0,1) .$$

The <u>topological normal structure set of X</u> $\mathcal{T}^{TOP}(X)$ is the set of concordance classes of t-triangulations $(f,b):M \longrightarrow X$. (Of course, $\mathcal{T}^{TOP}(X)$ may be empty). In dealing with t-triangulations $(f,b):M \longrightarrow X$ we shall sometimes omit to mention b, writing $f:M \longrightarrow X$ in conformity with the terminology for s-triangulations, even though f does not in general determine b.

<u>Proposition 7.1.3</u> Let X be an n-dimensional geometric Poincaré complex with Spivak normal structure

$$(\nu_X:X \longrightarrow BG(k),\rho_X:S^{n+k} \longrightarrow T(\nu_X)) \quad (k \geqslant 3) .$$

i) The Browder-Novikov transversality construction of topological normal maps defines a natural bijection

$$\alpha_X : \mathcal{T}^{TOP}(\nu_X) \xrightarrow{\simeq} \mathcal{T}^{TOP}(X) ;$$

$$(\tilde{\nu}_X:X \longrightarrow B\widetilde{TOP}(k)) \longmapsto ((f,b):M \longrightarrow X)$$

sending the t-triangulation $\tilde{\nu}_X$ of ν_X to the t-triangulation (f,b) of X obtained by making $\rho_X:S^{n+k} \longrightarrow T(\nu_X)$ topologically transverse at the zero section $X \subset T(\nu_X)$ with respect to $\tilde{\nu}_X$ and setting

$$f = \rho_X| : M = \rho_X^{-1}(X) \longrightarrow X .$$

(For n = 4 M is allowed one singularity, cf. Scharlemann [1]).

ii) The $S\mathbb{Z}_2$-duality isomorphism

$$\alpha_X : \dot{H}^{k+1}(T(\nu_X);\underline{\mathbb{L}}_0) \xrightarrow{\sim} H_{n-1}(X;\underline{\mathbb{L}}_0)$$

sends the t-triangulability obstruction $t(\nu_X)$ of ν_X to the <u>t-triangulability obstruction of X</u>

$$t(X) = \alpha_X(t(\nu_X)) \in H_{n-1}(X;\underline{\mathbb{L}}_0)$$

such that $t(X) = 0$ if and only if X is t-triangulable. The image of $t(X)$ under the quadratic assembly map is

$$\begin{aligned}\sigma_*(t(X)) &= (\Omega C([X]\cap -:C(\tilde{X})^{n-*} \longrightarrow C(\tilde{X})),\psi)\\ &= 0 \in L_{n-1}(\mathbb{Z}[\pi_1(X)])\end{aligned}$$

with $\tilde{X}$ the universal cover of X. (See Proposition 7.4.3 iii) for a generalization of this to normal spaces).

iii) A t-triangulation $(f,b):M \longrightarrow X$ of X determines an $\underline{\mathbb{L}}^0$-orientation of X

$$[X] = f_*([M]) \in H_n(X;\underline{\mathbb{L}}^0)$$

such that $J([X]) = [\hat{X}] \in H_n(X;\hat{\underline{\mathbb{L}}}^0)$ is the canonical $\hat{\underline{\mathbb{L}}}^0$-orientation of X, and such that the surgery obstruction $\sigma_*(f,b) \in L_n(\mathbb{Z}[\pi_1(X)])$ has symmetrization

$$\begin{aligned}(1+T)\sigma_*(f,b) &= \sigma^*(M) - \sigma^*(X)\\ &= \sigma^*([X]) - \sigma^*(X) \in L^n(\mathbb{Z}[\pi_1(X)]) .\end{aligned}$$

iv) If X is t-triangulable the set $\mathcal{J}^{TOP}(X)$ carries a natural affine structure with translation group $H_n(X;\underline{\mathbb{L}}_0)$. the <u>difference</u> of two t-triangulations $(f_0,b_0):M_0 \longrightarrow X$, $(f_1,b_1):M_1 \longrightarrow X$ being the element

$$t(f_0,f_1) = \alpha_X(t((\tilde{\nu}_X)_0,(\tilde{\nu}_X)_1)) \in H_n(X;\underline{\mathbb{L}}_0) ,$$

with $t((\tilde{\nu}_X)_0,(\tilde{\nu}_X)_1) \in \dot{H}^k(T(\nu_X);\underline{\mathbb{L}}_0)$ the difference of the corresponding t-triangulations $(\tilde{\nu}_X)_0,(\tilde{\nu}_X)_1$ of ν_X.

The difference has images

$$(1+T)t(f_0,f_1) = f_{0*}([M_0]) - f_{1*}([M_1]) \in H_n(X;\underline{\mathbb{L}}^0)$$

$$\sigma_*(t(f_0,f_1)) = \sigma_*(f_0,b_0) - \sigma_*(f_1,b_1) \in L_n(\mathbb{Z}[\pi_1(X)]) \quad .$$

Proof: See Ranicki [7], where it was shown that X admits an $\underline{\mathbb{L}}^0$-orientation $[X] \in H_n(X;\underline{\mathbb{L}}^0)$ such that

i) $J([X]) = [\hat{X}] \in H_n(X;\hat{\underline{\mathbb{L}}}^0)$

ii) $\sigma^*([X]) = \sigma^*(X) \in L^n(\mathbb{Z}[\pi_1(X)])$

iii) the relations i) and ii) are compatible on the $\mathbb{L}$-spectrum level

if (and for $n \geqslant 5$ only if) X admits an s-triangulation.

[]

As they stand the conditions i),ii),iii) listed above are just a restatement in algebraic terms of the Browder-Novikov-Sullivan-Wall two-stage obstruction theory for the s-triangulability of an n-dimensional geometric Poincaré complex X, with i) giving a t-triangulation $(f,b):M \longrightarrow X$ and ii) giving a vanishing of the surgery obstruction $\sigma_*(f,b) \in L_n(\mathbb{Z}[\pi_1(X)])$ up to the (8-torsion) difference between the quadratic and symmetric L-groups which is taken care of by iii). However, the three conditions were united and expressed as the vanishing of a single invariant, as follows.

Given a space X with an orientation double covering $w:\overline{X} \longrightarrow X$ define the quadratic $\mathcal{S}$-groups $\mathcal{S}_*(X)$ to be the abelian groups appearing in the exact sequence

$$\dots \longrightarrow H_n(X;\underline{\mathbb{L}}_0) \xrightarrow{\sigma_*} L_n(\mathbb{Z}[\pi_1(X)]) \longrightarrow \mathcal{S}_n(X) \longrightarrow H_{n-1}(X;\underline{\mathbb{L}}_0) \longrightarrow \dots ,$$

in which the homology groups are defined using w-twisted

$\underline{\mathbb{L}}_0$-coefficients and the L-groups are defined using the w-twisted involution on $\mathbb{Z}[\pi_1(X)]$. An orientation-preserving map $f:X \longrightarrow X'$ induces abelian group morphisms

$$f_* : \mathcal{S}_*(X) \longrightarrow \mathcal{S}_*(X')$$

which are isomorphisms if f is a homotopy equivalence.

The <u>total surgery</u> (or <u>s-triangulability</u>) <u>obstruction</u> of an n-dimensional geometric Poincaré complex X is an element

$$s(X) \in \mathcal{S}_n(X)$$

with the following properties.

<u>Proposition 7.1.4</u> i) $s(X) = 0 \in \mathcal{S}_n(X)$ if (and for $n \geqslant 5$ only if) X is s-triangulable, i.e. has the simple homotopy type of a manifold.

ii) The image of $s(X)$ in $H_{n-1}(X;\underline{\mathbb{L}}_0)$ is the t-triangulability obstruction of X

$$[s(X)] = t(X) \in H_{n-1}(X;\underline{\mathbb{L}}_0) \quad .$$

If

$$\begin{aligned} s(X) \in \ker(\mathcal{S}_n(X) &\longrightarrow H_{n-1}(X;\underline{\mathbb{L}}_0)) \\ &= \operatorname{im}(\sigma_*:L_n(\mathbb{Z}[\pi_1(X)]) \longrightarrow \mathcal{S}_n(X)) \subseteq \mathcal{S}_n(X) \end{aligned}$$

(i.e. if X is t-triangulable) the inverse image of $s(X)$ in $L_n(\mathbb{Z}[\pi_1(X)])$ is the coset of the subgroup

$$\begin{aligned} \ker(L_n(\mathbb{Z}[\pi_1(X)]) &\longrightarrow \mathcal{S}_n(X)) \\ &= \operatorname{im}(\sigma_*:H_n(X;\underline{\mathbb{L}}_0) \longrightarrow L_n(\mathbb{Z}[\pi_1(X)])) \subseteq L_n(\mathbb{Z}[\pi_1(X)]) . \end{aligned}$$

consisting of the surgery obstructions $\sigma_*(f,b) \in L_n(\mathbb{Z}[\pi_1(X)])$ of all the t-triangulations $(f,b):M \longrightarrow X$.

iii) If $n \geqslant 5$ and X is s-triangulable the topological manifold structure set $\mathcal{S}^{TOP}(X)$ carries a natural affine structure with translation group $\mathcal{S}_{n+1}(X)$, the <u>difference</u> of two s-triangulations

$f_0:M_0 \xrightarrow{\sim} X$, $f_1:M_1 \xrightarrow{\sim} X$ being an element

$$s(f_0,f_1) \in \mathcal{S}_{n+1}(X)$$

with image

$$[s(f_0,f_1)] = t(f_0,f_1) \in H_n(X;\underline{\mathbb{L}}_0) \quad .$$

(See Proposition 7.1.4 (rel∂) iv) below for the algebraic surgery exact sequence involving $\mathcal{S}^{TOP}(X)$, and for an expression of the difference $s(f_0,f_1)$ as a rel∂ s-triangulability obstruction).

<u>Proof</u>: See Ranicki [7].

[]

(If π is a group equipped with an orientation map $w:\pi \longrightarrow \mathbb{Z}_2$ the Sullivan-Wall homomorphism

$$\theta : \Omega_n^{STOP}(K(\pi,1) \times G/TOP, K(\pi,1) \times *) \longrightarrow L_n(\mathbb{Z}[\pi])$$

(cf. Wall [4,Thm.13.B.3]) factors through the quadratic assembly map as

$$\theta : \Omega_n^{STOP}(K(\pi,1) \times G/TOP, K(\pi,1) \times *) = \dot{H}_n(K(\pi,1);\underline{MSTOP} \wedge G/TOP)$$
$$\longrightarrow\!\!\!\!\!\rightarrow H_n(K(\pi,1);\underline{\mathbb{L}}_0) \xrightarrow{\sigma_*} L_n(\mathbb{Z}[\pi]) ,$$

using the composite map of spectra

$$\underline{MSTOP} \wedge G/TOP = \underline{MSTOP} \wedge \mathbb{L}_0 \xrightarrow{\sigma^* \wedge 1} \underline{\mathbb{L}}^0 \wedge \mathbb{L}_0 \xrightarrow{\otimes} \underline{\mathbb{L}}_0 \quad .$$

Thus if π is finitely presented and $n \geqslant 5$ the subgroup

$$\mathrm{im}(\sigma_*:H_n(K(\pi,1);\underline{\mathbb{L}}_0) \longrightarrow L_n(\mathbb{Z}[\pi])) \subseteq L_n(\mathbb{Z}[\pi])$$

consists of the surgery obstructions $\sigma_*(f,b)$ ot t-triangulations $(f,b):M \longrightarrow X$ of closed n-dimensional manifolds X equipped with a reference map $X \longrightarrow K(\pi,1)$).

There are relative and rel∂ versions of total surgery obstruction theory, which we shall now summarize.

Let (X,Y) be an n-dimensional geometric Poincaré pair $\begin{cases} - \\ \text{such that } Y \text{ is a manifold} \end{cases}$ with Spivak normal structure

$$\begin{cases} ((\nu_X,\nu_Y):(X,Y) \longrightarrow BG(k),\ (\rho_X,\rho_Y):(D^{n+k},S^{n+k-1}) \longrightarrow (T(\nu_X),T(\nu_Y))) \\ ((\nu_X,\nu_Y):(X,Y) \longrightarrow (BG(k),B\widetilde{TOP}(k)), \\ \qquad\qquad (\rho_X,\rho_Y):(D^{n+k},S^{n+k-1}) \longrightarrow (T(\nu_X),T(\nu_Y))). \end{cases}$$

A $\begin{cases} t\text{-} \\ t_\partial\text{-} \end{cases}$ *triangulation* of (X,Y) is a topological normal map of pairs

$$((f,b),(g,c)) \ : \ (M,N) \longrightarrow (X,Y)$$

$\begin{cases} - \\ \text{such that } g:N \longrightarrow Y \text{ is a homeomorphism} \end{cases}$. The $\begin{cases} t\text{-} \\ t_\partial\text{-} \end{cases}$ *triangulation obstruction of* (X,Y)

$$\begin{cases} t(X,Y) = \alpha_X(t(\nu_X)) \in H_{n-1}(X,Y;\underline{\mathbb{L}}_0) \\ t_\partial(X,Y) = \alpha_X(t_\partial(\nu_X,\nu_Y)) \in H_{n-1}(X;\underline{\mathbb{L}}_0) \end{cases}$$

is the image of the $\begin{cases} t\text{-} \\ t_\partial\text{-} \end{cases}$ triangulability obstruction of $\begin{cases} \nu_X \\ (\nu_X,\nu_Y) \end{cases}$ under the $S\mathbb{Z}_2$-duality isomorphism

$$\begin{cases} \alpha_X \ : \ \dot{H}^{k+1}(T(\nu_X);\underline{\mathbb{L}}_0) \xrightarrow{\sim} H_{n-1}(X,Y;\underline{\mathbb{L}}_0) \\ \alpha_X \ : \ \dot{H}^{k+1}(T(\nu_X),T(\nu_Y);\underline{\mathbb{L}}_0) \xrightarrow{\sim} H_{n-1}(X;\underline{\mathbb{L}}_0) \end{cases} .$$

An $\begin{cases} s\text{-} \\ s_\partial\text{-} \end{cases}$ *triangulation* of (X,Y) is a $\begin{cases} t\text{-} \\ t_\partial\text{-} \end{cases}$ triangulation

$$((f,b),(g,c)) \ : \ (M,N) \longrightarrow (X,Y)$$

such that $\begin{cases} (f,g):(M,N) \longrightarrow (X,Y) \\ f:M \longrightarrow X \end{cases}$ is a simple homotopy equivalence.

Let $\begin{cases} \mathcal{S}^{TOP}(X,Y) \\ \mathcal{S}^{TOP}_{\partial}(X,Y) \end{cases}$ be the set of concordance classes of $\begin{cases} s\text{-} \\ s_{\partial}\text{-} \end{cases}$ triangulations of (X,Y).

The <u>rel∂ total surgery</u> (or <u>s_{∂}-triangulability</u>) <u>obstruction</u> of an n-dimensional geometric Poincaré pair (X,Y) with manifold boundary Y is an element

$$s_{\partial}(X,Y) \in \mathcal{S}_n(X)$$

with the following properties.

<u>Proposition 7.1.4</u> (rel∂) i) $s_{\partial}(X,Y) = 0 \in \mathcal{S}_n(X)$ if (and for $n \geq 5$ only if) (X,Y) is s_{∂}-triangulable.

ii) The image of $s_{\partial}(X,Y)$ in $H_{n-1}(X;\underline{\mathbb{L}}_0)$ is the t_{∂}-triangulability obstruction of (X,Y)

$$[s_{\partial}(X,Y)] = t_{\partial}(X,Y) \in H_{n-1}(X;\underline{\mathbb{L}}_0) .$$

iii) If $n \geq 5$ and (X,Y) is s_{∂}-triangulable the structure set $\mathcal{S}^{TOP}_{\partial}(X,Y)$ carries a natural affine structure with translation group $\mathcal{S}_{n+1}(X)$.

iv) For $n \geq 5$ an s-triangulation $f:M \xrightarrow{\sim} X$ of an n-dimensional geometric Poincaré complex X determines an isomorphism between the Sullivan-Wall surgery exact sequence of the manifold M

$$\dots \longrightarrow \mathcal{S}^{TOP}_{\partial}(M \times D^1, M \times S^0) \longrightarrow [M \times D^1, M \times S^0; G/TOP, *]$$

$$\xrightarrow{\theta} L_{n+1}(\mathbb{Z}[\pi_1(M)]) \longrightarrow \mathcal{S}^{TOP}(M) \longrightarrow [M,G/TOP] \xrightarrow{\theta} L_n(\mathbb{Z}[\pi_1(M)])$$

and the exact sequence

$$\dots \longrightarrow \mathcal{S}_{n+2}(X) \longrightarrow H_{n+1}(X;\underline{\mathbb{L}}_0) \xrightarrow{\sigma_*} L_{n+1}(\mathbb{Z}[\pi_1(X)])$$

$$\longrightarrow \mathcal{S}_{n+1}(X) \longrightarrow H_n(X;\underline{\mathbb{L}}_0) \xrightarrow{\sigma_*} L_n(\mathbb{Z}[\pi_1(X)]) .$$

In particular, f determines a bijection

$$f_{\#} : \mathcal{S}^{TOP}(X) \xrightarrow{\sim} \mathcal{S}_{n+1}(X) \; ;$$

$$(f':M' \xrightarrow{\simeq} X) \longmapsto s(f',f) = s_{\partial}(W' \cup_X -W, M' \cup -M)$$

sending f to 0, with W (resp. W') the mapping cylinder of f (resp. f') and using the homotopy invariance of the $\mathcal{S}$-groups to identify $\mathcal{S}_{n+1}(W' \cup_X -W) = \mathcal{S}_{n+1}(X)$. Similarly, f induces bijections

$$f_{\#} : \mathcal{S}_{\partial}^{TOP}(M \times D^k, \partial(M \times D^k)) \xrightarrow{\sim} \mathcal{S}_{n+k+1}(X) \qquad (k \geqslant 0)$$

which are isomorphisms of abelian groups for $k \geqslant 1$.

Proof: See Ranicki [7].

[]

The relative $\mathcal{S}$-groups $\mathcal{S}_*(X,Y)$ of a pair of spaces (X,Y) (equipped with an orientation double covering w) are defined to fit into a commutative diagram of abelian groups with exact rows and columns

$$\begin{array}{ccccccccc}
 & & \vdots & & \vdots & & \vdots & & \vdots \\
 & & \downarrow & & \downarrow & & \downarrow & & \downarrow \\
\cdots \to & & H_n(Y;\underline{\mathbb{L}}_0) & \xrightarrow{\sigma_*} & L_n(\mathbb{Z}[\pi_1(Y)]) & \to & \mathcal{S}_n(Y) & \to & H_{n-1}(Y;\underline{\mathbb{L}}_0) \to \cdots \\
 & & \downarrow & & \downarrow & & \downarrow & & \downarrow \\
\cdots \to & & H_n(X;\underline{\mathbb{L}}_0) & \xrightarrow{\sigma_*} & L_n(\mathbb{Z}[\pi_1(X)]) & \to & \mathcal{S}_n(X) & \to & H_{n-1}(X;\underline{\mathbb{L}}_0) \to \cdots \\
 & & \downarrow & & \downarrow & & \downarrow & & \downarrow \\
\cdots \to & & H_n(X,Y;\underline{\mathbb{L}}_0) & \xrightarrow{\sigma_*} & L_n(\mathbb{Z}[\pi_1(Y)] \to \mathbb{Z}[\pi_1(X)]) & \to & \mathcal{S}_n(X,Y) & \to & H_{n-1}(X,Y;\underline{\mathbb{L}}_0) \to \cdots \\
 & & \downarrow & & \downarrow & & \downarrow & & \downarrow \\
\cdots \to & & H_{n-1}(Y;\underline{\mathbb{L}}_0) & \xrightarrow{\sigma_*} & L_{n-1}(\mathbb{Z}[\pi_1(Y)]) & \to & \mathcal{S}_{n-1}(Y) & \to & H_{n-2}(Y;\underline{\mathbb{L}}_0) \to \cdots \\
 & & \downarrow & & \downarrow & & \downarrow & & \downarrow \\
 & & \vdots & & \vdots & & \vdots & & \vdots
\end{array}$$

The total surgery (or s-triangulability) obstruction of an n-dimensional geometric Poincaré pair (X,Y) is an element

$$s(X,Y) \in \mathcal{S}_n(X,Y)$$

with the following properties.

Proposition 7.1.4 (rel) i) $s(X,Y) = 0 \in \mathcal{S}_n(X,Y)$ if (and for $n \geq 6$ only if) (X,Y) is s-triangulable.

ii) The image of $s(X,Y)$ in $H_{n-1}(X,Y;\underline{\mathbb{L}}_0)$ is the t-triangulability obstruction of (X,Y)

$$[s(X,Y)] = t(X,Y) \in H_{n-1}(X,Y;\underline{\mathbb{L}}_0) .$$

The image of $s(X,Y)$ in $\mathcal{S}_{n-1}(Y)$ is the total surgery obstruction $s(Y)$ of Y. If Y is a manifold $s(X,Y)$ is the image of the rel∂ total surgery obstruction $s_\partial(X,Y) \in \mathcal{S}_n(X)$.

iii) If $n \geq 6$ and (X,Y) is s-triangulable the structure set $\mathcal{S}^{TOP}(X,Y)$ carries a natural affine structure with translation group $\mathcal{S}_{n+1}(X,Y)$. An s-triangulation $f:(M,N) \xrightarrow{\sim} (X,Y)$ determines a bijection

$$f_\# : \mathcal{S}^{TOP}(M,N) \xrightarrow{\sim} \mathcal{S}_{n+1}(X,Y)$$

sending f to 0.

[]

7.2 The geometric theory of codimension q surgery

We shall now extend the total surgery obstruction theory of §7.1 to the problem of simultaneously s-triangulating a geometric Poincaré complex X and a codimension q Poincaré subcomplex $Y \subset X$, that is finding an s-triangulation of X

$$f : M \xrightarrow{\sim} X$$

such that

i) f is topologically transverse at $Y \subset X$ with respect to a t-triangulation $\tilde{\xi} : Y \longrightarrow \widetilde{BTOP}(q)$ of the normal fibration $\xi = \nu_{Y \subset X} : Y \longrightarrow BG(q)$, so that in particular $N = f^{-1}(Y) \subset M$ is a codimension q submanifold

ii) the restriction of f defines an s-triangulation of Y

$$g = f| : N \xrightarrow{\sim} Y$$

iii) the restriction of f to the complements defines a simple homotopy equivalence

$$h = f| : M - N \xrightarrow{\sim} X - Y .$$

(For $q = 2$ there is also a theory for the weaker problem in which f is only required to satisfy i) and ii), so that h need only be a $\mathbb{Z}[\pi_1(X)]$-homology equivalence - see §§7.7,7.8). This problem is closely related to the obstruction theory for deciding whether a particular $\left\{\begin{matrix} s- \\ t- \end{matrix}\right.$ triangulation $f : M \longrightarrow X$ of X is concordant to such a simultaneous s-triangulation of X and Y, i.e. can be "split along $Y \subset X$". Following the solution by Browder [1],[3] of the splitting problem in the simply-connected case Wall [4,§11] developed an obstruction theory for the codimension q splitting problem in the non-simply-connected case.

The obstruction groups $\begin{cases} LS_*(\Phi) \\ LP_*(\Phi) \end{cases}$ were defined geometrically, but shown to depend only on the fundamental group data of the pushout square

$$\begin{array}{ccc} \pi_1(S(\nu_{Y\subset X})) & \longrightarrow & \pi_1(X-Y) \\ \downarrow & \Phi & \downarrow \\ \pi_1(Y) & \longrightarrow & \pi_1(X) \end{array} \quad ,$$

with $\begin{cases} LS_*(\Phi) = L_*(\mathbb{Z}[\pi_1(Y)]) \\ LP_*(\Phi) = L_{*+q}(\mathbb{Z}[\pi_1(X)])\oplus L_*(\mathbb{Z}[\pi_1(Y)]) \end{cases}$ for $q \geqslant 3$. In §7.2 we shall be only concerned with the geometrically defined $\begin{cases} LS- \\ LP- \end{cases}$ groups. In §7.5 we shall give an algebraic definition in the non-trivial cases $q = 1,2$ using quadratic Poincaré complexes. In §7.6 $(q=1)$ and §7.8 $(q=2)$ we shall deal individually with the two cases of the algebraic theory of codimension q surgery

In the first instance, we recall the geometric definition of the $\begin{cases} LS- \\ LP- \end{cases}$ groups, for any $q \geqslant 1$.

A <u>codimension q CW pair</u> (X,Y) is a CW complex X with a decomposition

$$X = E(\xi)\cup_{S(\xi)} Z$$

for some $(q-1)$-spherical fibration $\xi: Y \longrightarrow BG(q)$ over a subcomplex $Y \subset X$, with $Z \subset X$ a disjoint subcomplex and

$$(D^q, S^{q-1}) \longrightarrow (E(\xi), S(\xi)) \longrightarrow Y$$

the associated (D^q, S^{q-1})-fibration. In dealing with geometrically defined L-groups it is assumed that X has a finite 2-skeleton, so that $\pi_1(X)$ is finitely presented - no such restriction is required for the algebraic definitions in §§7.6,7.8 below. Applying the generalized Van Kampen theorem there is obtained an expression for the fundamental groupoid $\pi_1(X)$ as a free product with amalgamation

$$\pi_1(X) = \pi_1(E(\xi)) *_{\pi_1(S(\xi))} \pi_1(Z) \quad ,$$

i.e. there is defined a pushout square in the category of groupoids

$$\begin{array}{ccc} \pi_1(S(\xi)) & \longrightarrow & \pi_1(Z) \\ \big\downarrow & \Phi & \big\downarrow \\ \pi_1(E(\xi)) & \longrightarrow & \pi_1(X) \end{array} \quad .$$

(It is not assumed that the maps in Φ are injective).

Given an orientation map for X

$$w(X) = w : \pi_1(X) \longrightarrow \mathbb{Z}_2$$

define orientation maps for $S(\xi), E(\xi), Z$ using the restrictions

$$w(S(\xi)) = w| : \pi_1(S(\xi)) \longrightarrow \pi_1(X) \xrightarrow{w} \mathbb{Z}_2$$

$$w(E(\xi)) = w| : \pi_1(E(\xi)) \longrightarrow \pi_1(X) \xrightarrow{w} \mathbb{Z}_2$$

$$w(Z) = w| : \pi_1(Z) \longrightarrow \pi_1(X) \xrightarrow{w} \mathbb{Z}_2$$

and give Y the orientation map

$$w(Y) = w(E(\xi))w_1(\xi) : \pi_1(Y) = \pi_1(E(\xi)) \longrightarrow \mathbb{Z}_2$$

with $w_1(\xi): \pi_1(Y) \longrightarrow \mathbb{Z}_2$ the orientation map of $\xi: Y \longrightarrow BG(q)$.

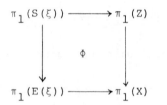

Define the transfer maps in quadratic L-theory induced by (X,Y)

$$p\xi^! : L_n(\mathbb{Z}[\pi_1(Y)]) \longrightarrow L_{n+q}(\mathbb{Z}[\pi_1(Z)] \longrightarrow \mathbb{Z}[\pi_1(X)]) \quad (n \geqslant 0)$$

to be the composites of the transfer maps induced by ξ

$$\xi^! : L_n(\mathbb{Z}[\pi_1(Y)]) \longrightarrow L_{n+q}(\mathbb{Z}[\pi_1(S(\xi))] \longrightarrow \mathbb{Z}[\pi_1(E(\xi))])$$

and the maps naturally induced by Φ

$$p : L_{n+q}(\mathbb{Z}[\pi_1(S(\xi))] \longrightarrow \mathbb{Z}[\pi_1(E(\xi))]) \longrightarrow L_{n+q}(\mathbb{Z}[\pi_1(Z)] \longrightarrow \mathbb{Z}[\pi_1(X)]) \quad ,$$

with $\xi^!$ sending the quadratic signature $\sigma_*(f,b)$ of a normal map of n-dimensional geometric Poincaré complexes

$$(f,b) : M \longrightarrow N$$

equipped with a reference map $g:N \longrightarrow Y$ to the relative quadratic signature $\xi^!\sigma_*(f,b) = \sigma_*((f,b)^!)$ of the induced normal map of (n+q)-dimensional geometric Poincaré pairs

$$(f,b)^! : (E((gf)^*\xi), S((gf)^*\xi)) \longrightarrow (E(g^*\xi), S(g^*\xi))$$

which is equipped with a reference map of pairs $g^!:(E(g^*\xi),S(g^*\xi)) \longrightarrow (E(\xi),S(\xi))$.

Following Wall [4,p.$\begin{Bmatrix}127\\252\end{Bmatrix}$] define the <u>quadratic</u> $\begin{Bmatrix}\text{LS-}\\\text{LP-}\end{Bmatrix}$<u>groups of (X,Y)</u> $\begin{Bmatrix}LS_n(\Phi)\\LP_n(\Phi)\end{Bmatrix}$ $(n \geqslant 0)$ to be the relative groups appearing in the exact sequence

$$\begin{cases} \ldots \longrightarrow L_{n+q+1}(\mathbb{Z}[\pi_1(Z)] \longrightarrow \mathbb{Z}[\pi_1(X)]) \longrightarrow LS_n(\Phi) \longrightarrow L_n(\mathbb{Z}[\pi_1(Y)]) \\ \qquad \xrightarrow{p\xi^!} L_{n+q}(\mathbb{Z}[\pi_1(Z)] \longrightarrow \mathbb{Z}[\pi_1(X)]) \longrightarrow \ldots \\ \ldots \longrightarrow L_{n+q}(\mathbb{Z}[\pi_1(Z)]) \longrightarrow LP_n(\Phi) \longrightarrow L_n(\mathbb{Z}[\pi_1(Y)]) \\ \qquad \xrightarrow{\partial p\xi^!} L_{n+q-1}(\mathbb{Z}[\pi_1(Z)]) \longrightarrow \ldots \end{cases}$$

and satisfying the following properties:

<u>Proposition 7.2.1</u> i) The quadratic $\left\{\begin{matrix}LS\text{-}\\ LP\text{-}\end{matrix}\right.$ groups are 4-periodic

$$\left\{\begin{array}{l} LS_n(\Phi) = LS_{n+4}(\Phi) \\ LP_n(\Phi) = LP_{n+4}(\Phi) \end{array}\right. \quad (n \geqslant 0) \quad .$$

ii) The LS-groups are related to the LP-groups by a commutative braid of exact sequences

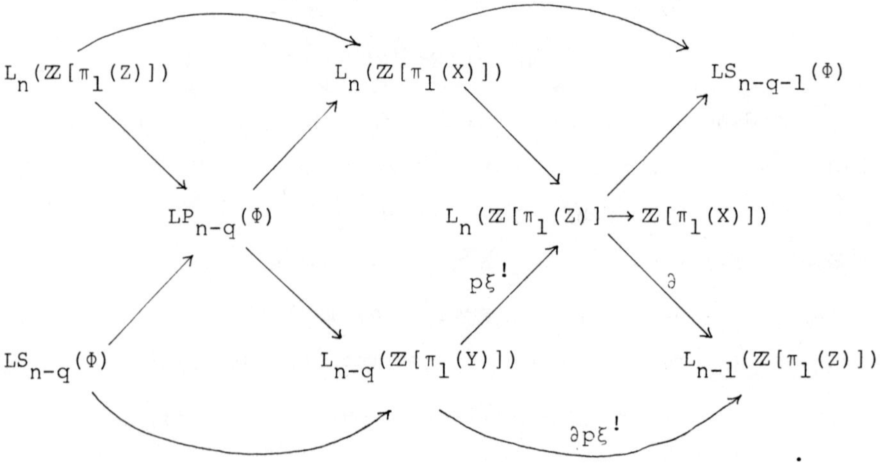

iii) The LS-groups are related to the triad L-groups $L_*(\mathbb{Z}[\Phi])$ by a commutative braid of exact sequences

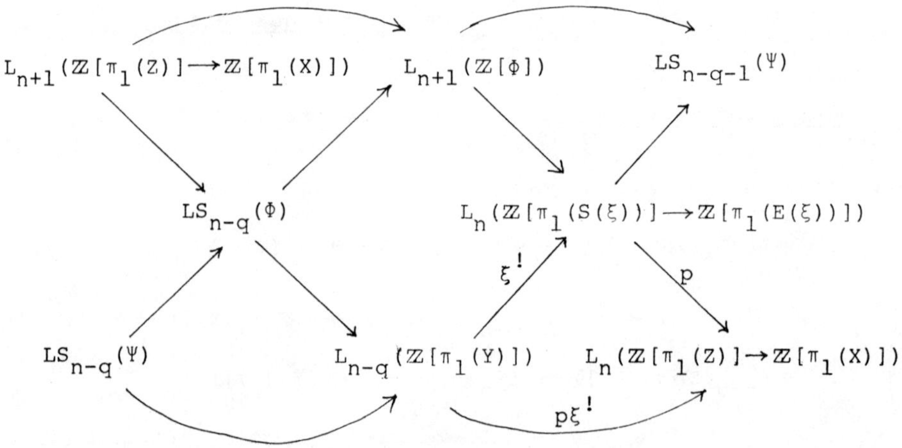

with $LS_*(\Psi)$ the LS-groups of the pushout square of groupoids

$$\begin{array}{ccc} \pi_1(S(\xi)) & \longrightarrow & \pi_1(S(\xi)) \\ \downarrow & \Psi & \downarrow \\ \pi_1(E(\xi)) & \longrightarrow & \pi_1(E(\xi)) \end{array}$$

associated to the codimension q CW pair $(E(\xi),Y)$.

[]

An <u>(n,n-q)-dimensional</u> (or <u>codimension q</u>) <u>geometric Poincaré pair</u> (X,Y) is a codimension q CW pair such that

i) X is an n-dimensional geometric Poincaré complex

ii) Y is an (n-q)-dimensional geometric Poincaré complex

iii) $(Z,S(\xi))$ is an n-dimensional geometric Poincaré pair.

(Actually, iii) implies ii)). Then $Y \subset X$ is a "codimension q Poincaré embedding" with complement Z and normal fibration

$$\nu_{Y \subset X} = \xi : Y \longrightarrow BG(q) .$$

The prescribed Spivak normal structure of X

$$(\nu_X : X \longrightarrow BG(k), \rho_X : S^{n+k} \longrightarrow T(\nu_X))$$

determines a Spivak normal structure of Y

$$(\nu_Y = \xi \oplus \nu_X|_Y : Y \longrightarrow BG(q+k),$$

$$\rho_Y : S^{n+k} \xrightarrow{\rho_X} T(\nu_X) \xrightarrow{\text{collapse}} T(\nu_X)/T(\nu_X|_Z) = T(\nu_Y)) .$$

A <u>normal map</u> of (n,n-q)-dimensional geometric Poincaré pairs

$$((f,b),(g,c)) : (M,N) \longrightarrow (X,Y)$$

is a normal map of n-dimensional geometric Poincaré complexes

$$(f,b) : M \longrightarrow X$$

with a decomposition

$$(f,b) = (g,c)^! \cup (h,d) : M = E(\nu) \cup_{S(\nu)} P \longrightarrow X = E(\xi) \cup_{S(\xi)} Z$$

where $(g,c): N \longrightarrow Y$ is a normal map of $(n-q)$-dimensional geometric Poincaré complexes such that

$$\nu = \nu_{N \subset M} : N \xrightarrow{g} Y \xrightarrow{\xi} BG(k)$$

and $(h,d):(P,S(\nu)) \longrightarrow (Z,S(\xi))$ is a normal map of n-dimensional geometric Poincaré pairs such that

$$(h,d)| = (g,c)^!| : S(\nu) \longrightarrow S(\xi) .$$

<u>Proposition 7.2.2</u> Given a normal map of $(n,n-q)$-dimensional geometric Poincaré pairs

$$((f,b),(g,c)) : (M,N) \longrightarrow (X,Y)$$

$\begin{cases} \text{such that } f:M \longrightarrow X \text{ is a simple homotopy equivalence} \\ - \end{cases}$

there is defined a <u>codimension q quadratic signature</u>

$$\begin{cases} \sigma_*((f,b),(g,c)) \in LS_{n-q}(\Phi) \\ \sigma_*((f,b),(g,c)) \in LP_{n-q}(\Phi) \end{cases}$$

with image $\begin{cases} \sigma_*((f,b),(g,c)) \in LP_{n-q}(\Phi) \\ (\sigma_*(f,b),\sigma_*(g,c)) \in L_n(\mathbb{Z}[\pi_1(X)]) \oplus L_{n-q}(\mathbb{Z}[\pi_1(Y)]) \end{cases}$,

such that $\sigma_*((f,b),(g,c)) = 0$ if $((f,b),(g,c))$ is normal bordant

$\begin{cases} \text{by a geometric Poincaré s-cobordism of } (f,b) \\ - \end{cases}$ to a normal map

of pairs such that the maps $f:M \longrightarrow X$, $g:N \longrightarrow Y$, $h:P \longrightarrow Z$ are all simple homotopy equivalences.

For $q \geqslant 3$ $\pi_1(S(\xi)) = \pi_1(E(\xi))$, $\pi_1(Z) = \pi_1(X)$ and

$$\begin{cases} \sigma_*((f,b),(g,c)) = \sigma_*(g,c) \in LS_{n-q}(\Phi) = L_{n-q}(\mathbb{Z}[\pi_1(Y)]) \\ \sigma_*((f,b),(g,c)) = (\sigma_*(f,b),\sigma_*(g,c)) \\ \qquad \in LP_{n-q}(\Phi) = L_n(\mathbb{Z}[\pi_1(X)]) \oplus L_{n-q}(\mathbb{Z}[\pi_1(Y)]) . \end{cases}$$

<u>Proof</u>: The normal maps of n-dimensional geometric Poincaré pairs $(f,b):(M,\emptyset) \longrightarrow (X,\emptyset)$, $(g,c)^{!}:(E(\nu),S(\nu)) \longrightarrow (E(\xi),S(\xi))$ are normal bordant via the normal map

$$((F,B) = (f,b)\times \mathrm{id}.;(f,b),(g,c)^{!})$$
$$:((M\times I, P\times 1);(M,\emptyset)\times 0,(E(\nu),S(\nu))\times 1)$$
$$\longrightarrow ((X\times I, Z\times 1);(X,\emptyset)\times 0,(E(\xi),S(\xi))\times 1) \quad (I=[0,1]),$$

so that in particular the restriction

$$(F,B)| = (h,d) : (P,S(\nu))\times 1 \longrightarrow (Z,S(\xi))\times 1$$

defines a normal null-bordism of $(g,c)^{!}| : S(\nu) \longrightarrow S(\xi)$.

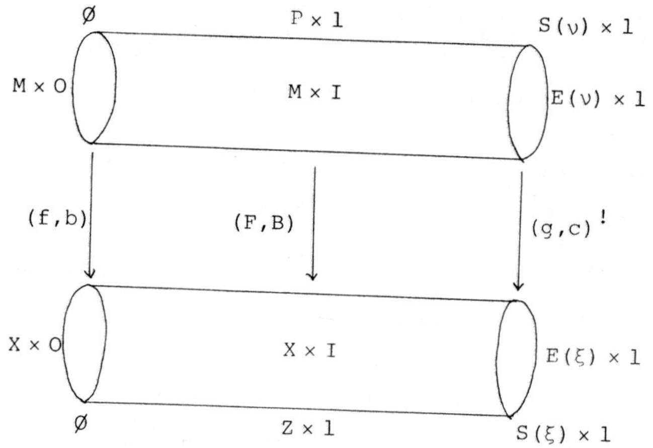

This gives a particular reason for

$$\begin{cases} p\xi^{!}\sigma_*(g,c) = 0 \in L_n(\mathbb{Z}[\pi_1(Z)] \longrightarrow \mathbb{Z}[\pi_1(X)]) \\ \partial p\xi^{!}\sigma_*(g,c) = 0 \in L_{n-1}(\mathbb{Z}[\pi_1(Z)]) \end{cases}$$

and so determines an element $\begin{cases} \sigma_*((f,b),(g,c)) \in LS_{n-q}(\Phi) \\ \sigma_*((f,b),(g,c)) \in LP_{n-q}(\Phi) \end{cases}$

with image $\sigma_*(g,c) \in L_{n-q}(\mathbb{Z}[\pi_1(Y)])$.

[]

An <u>(n,n-q)-dimensional t-normal geometric Poincaré pair</u> $(X,Y,\tilde{\xi})$ is an (n,n-q)-dimensional geometric Poincaré pair (X,Y) together with a choice of t-triangulation $\tilde{\xi}:Y \longrightarrow B\widetilde{TOP}(q)$ of the normal fibration $\xi = \nu_{Y \subset X} : Y \longrightarrow BG(q)$. We shall be primarily concerned with the cases q = 1,2, for which $BG(q) = B\widetilde{TOP}(q)$ so that the t-normal structure $\tilde{\xi}$ is redundant.

An <u>(n,n-q)-dimensional</u> (or <u>codimension q</u>) <u>manifold pair</u> (M,N) is an n-dimensional manifold M together with a locally flat codimension q submanifold $N \subset M$. The normal block bundle

$$\nu = \nu_{N \subset M} : N \longrightarrow B\widetilde{TOP}(q)$$

is such that

$$M = E(\nu) \cup_{S(\nu)} \overline{M \setminus E(\nu)} .$$

In particular, (M,N) has an underlying structure of an (n,n-q)-dimensional t-normal geometric Poincaré pair.

Let $(X,Y,\tilde{\xi})$ be an (n,n-q)-dimensional t-normal geometric Poincaré pair. A <u>topological normal map</u> (or a <u>t-triangulation</u> of $(X,Y,\tilde{\xi})$)

$$((f,b),(g,c)) : (M,N) \longrightarrow (X,Y)$$

is a t-triangulation of X (i.e. a topological normal map)

$$(f,b) : M \longrightarrow X$$

which is topologically transverse at $Y \subset X$ with respect to $\tilde{\xi}$, so that $(M,N = f^{-1}(Y))$ is an (n,n-q)-dimensional manifold pair with normal block bundle

$$\nu : N \xrightarrow{f|} Y \xrightarrow{\tilde{\xi}} B\widetilde{TOP}(q) ,$$

the restriction of (f,b)

$$(f,b)| = (g,c) : N \longrightarrow Y$$

is a t-triangulation of Y, the restriction

$$(f,b) = (h,d)| : (P,S(\nu)) \longrightarrow (Z,S(\xi)) \quad (P = \overline{M \setminus E(\nu)})$$

is a t-triangulation of $(Z,S(\xi))$ such that

$$(h,d)| = (g,c)^!| : S(\nu) \longrightarrow S(\xi) ,$$

and

$$(f,b) = (g,c)^! \cup (h,d) : M = E(\nu) \cup_{S(\nu)} P \longrightarrow X = E(\xi) \cup_{S(\xi)} Z .$$

In particular, $((f,b),(g,c))$ has an underlying structure of a normal map of codimension q geometric Poincaré pairs. Let $\mathcal{T}^{TOP}(X,Y,\tilde{\xi})$ be the set of concordance classes of t-triangulations of $(X,Y,\tilde{\xi})$.

Proposition 7.2.3 The forgetful map

$$\mathcal{T}^{TOP}(X,Y,\tilde{\xi}) \longrightarrow \mathcal{T}^{TOP}(X) \ ; \ ((f,b),(g,c)) \longmapsto (f,b)$$

is a bijection. Thus if X is t-triangulable $\mathcal{T}^{TOP}(X,Y,\tilde{\xi})$ carries a natural affine structure with translation group $H_n(X;\underline{\mathbb{L}}_0)$.

Proof: Topological transversality.

[]

Let $(X,Y,\tilde{\xi})$ be an $(n,n-q)$-dimensional t-normal geometric Poincaré pair. An s-triangulation of $(X,Y,\tilde{\xi})$ is a t-triangulation

$$((f,b),(g,c)) : (M,N) \longrightarrow (X,Y)$$

such that each of the constituent t-triangulations

$$(f,b) : M \longrightarrow X$$

$$(g,c) : N \longrightarrow Y$$

$$(h,d) : (P,S(\nu)) \longrightarrow (Z,S(\xi))$$

is an s-triangulation. Let $\mathcal{S}^{TOP}(X,Y,\tilde{\xi})$ be the set of concordance classes of s-triangulations. The forgetful map

$$\mathcal{S}^{TOP}(X,Y,\tilde{\xi}) \longrightarrow \mathcal{S}^{TOP}(X) \ ; \ ((f,b),(g,c)) \longmapsto (f,b)$$

is in general neither injective nor surjective.

An s-triangulation $f:M \xrightarrow{\sim} X$ is split along $Y \subset X$ if f actually defines an s-triangulation of $(X,Y,\tilde{\xi})$

$$((f,b),(g,c)) : (M,N) \xrightarrow{\sim} (X,Y) .$$

Given an $\left\{\begin{matrix} s- \\ t- \end{matrix}\right.$ triangulation $(f,b):M \longrightarrow X$ make f topologically transverse at $Y \subset X$ with respect to $\tilde{\xi}$, and use the codimension q quadratic signature of the resulting t-triangulation of $(X,Y,\tilde{\xi})$

$$((f,b),(g,c)) : (M,N) \longrightarrow (X,Y)$$

(as given by Proposition 7.2.2) to define the codimension q splitting obstruction of f along $Y \subset X$

$$\left\{\begin{array}{l} s(f,Y) = \sigma_*((f,b),(g,c)) \in LS_{n-q}(\Phi) \\ t(f,Y) = \sigma_*((f,b),(g,c)) \in LP_{n-q}(\Phi) . \end{array}\right.$$

The following is essentially a restatement of the obstruction theory of Wall [4,§11] for the "smoothing of codimension q Poincaré embeddings", by a method of proof going back to Browder [3].

Proposition 7.2.4 The $\left\{\begin{matrix} s- \\ t- \end{matrix}\right.$ triangulation $(f,b):M \longrightarrow X$ is such that $\left\{\begin{array}{l} s(f,Y) = 0 \in LS_{n-q}(\Phi) \\ t(f,Y) = 0 \in LP_{n-q}(\Phi) \end{array}\right.$ if (and for $n-q \geqslant 5$ only if) (f,b) is concordant to an s-triangulation of X which is split along $Y \subset X$.

Proof: The codimension q splitting obstruction $\left\{\begin{array}{l} s(f,Y) \in LS_{n-q}(\Phi) \\ t(f,Y) \in LP_{n-q}(\Phi) \end{array}\right.$ has image

$$\left\{\begin{array}{l} [s(f,Y)] = \sigma_*(g,c) \in L_{n-q}(\mathbb{Z}[\pi_1(Y)]) \\ [t(f,Y)] = \sigma_*(g,c) \in L_{n-q}(\mathbb{Z}[\pi_1(Y)]) , \end{array}\right.$$

the surgery obstruction of the t-triangulation $(g,c):N \longrightarrow Y$.

Now $\sigma_*(g,c) = 0$ if (and for $n-q \geqslant 5$ only if) there exists an $(n-q+1)$-dimensional topological normal map of triads

$$(G,C) : (L;N,N';\emptyset) \longrightarrow Y \times ([1,2];1,2;\emptyset)$$

such that

$$\text{i)}\ (G,C)| = (g,c) : N \longrightarrow Y \times 1$$

$$\text{ii)}\ (G,C)| = (g',c') : N' \longrightarrow Y \times 2 \text{ is an s-triangulation.}$$

Given such an extension (G,C) of (g,c) let

$$(\lambda;\nu,\nu') : (L;N,N') \xrightarrow{G} Y \times ([1,2];1,2) \xrightarrow{\text{projection}} Y \xrightarrow{\tilde{\xi}} \widetilde{BTOP}(q)$$

and define an $(n+1)$-dimensional topological normal map of triads

$$\begin{aligned}
(F',B') &= (F,B) \cup_{(g,c)} !(G,C)^! \\
&: (V;\partial_+V,\partial_-V;\partial\partial_+V) \\
&= (M \times I \cup_{E(\nu) \times 1} E(\lambda); \overline{M \setminus E(\nu)} \times 1 \cup_{S(\nu) \times 1} S(\lambda), M \times 0 \cup E(\nu'); S(\nu')) \\
&\longrightarrow (W;\partial_+W,\partial_-W;\partial\partial_+W) \\
&= (X \times [0,1] \cup_{E(\xi) \times 1} E(\xi) \times [1,2]; Z \times 1 \cup_{S(\xi) \times 1} S(\xi) \times [1,2], \\
&\qquad X \sqcup E(\xi) \times 2; S(\xi) \times 2)
\end{aligned}$$

such that the restriction

$$\begin{cases}
(F',B')|_{\partial_-V} = (f,b) \sqcup (g',c')^! \\
\quad : (\partial_-V,\partial\partial_-V) = (M,\emptyset) \sqcup (E(\nu'),S(\nu')) \\
\qquad\qquad \longrightarrow (\partial_-X,\partial\partial_-X) = (X,\emptyset) \sqcup (E(\xi),S(\xi)) \\
(F',B')|_{\partial\partial_+V} = (g',c')^!| : \partial\partial_+V = \partial\partial_-V = S(\nu') \longrightarrow \partial\partial_+W = S(\xi)
\end{cases}$$

is an s-triangulation, by glueing together topological normal maps of triads as in the picture

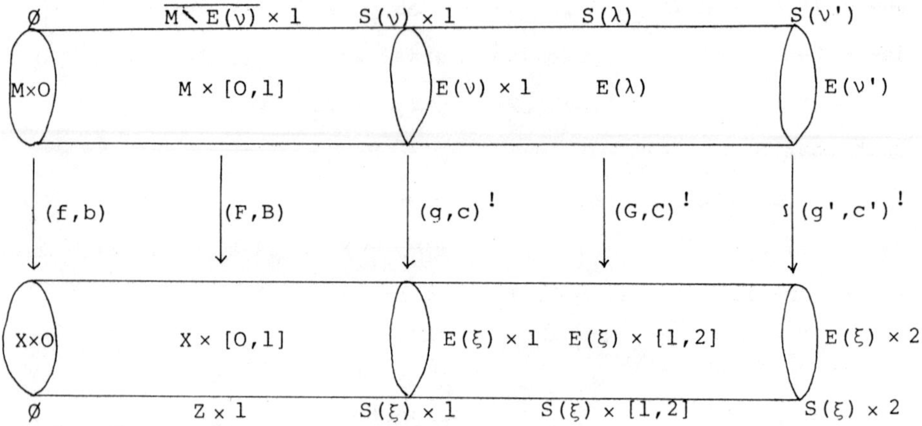

.

The surgery obstruction

$$\begin{cases} \sigma_*(F',B') \in L_{n+1}(\mathbb{Z}[\pi_1(Z)] \longrightarrow \mathbb{Z}[\pi_1(X)]) \\ \sigma_*((F',B')|_{\partial_+ V}) \in L_n(\mathbb{Z}[\pi_1(Z)]) \end{cases}$$

is 0 if (and for $n \geqslant 5$ only if) $\begin{cases} (F',B') \\ (F',B')|_{\partial_+ V} \end{cases}$ is topologically

normal bordant rel $\begin{cases} (F',B')|_{\partial_- V} \\ (F',B')|_{\partial\partial_+ V} \end{cases}$ to an s-triangulation of an

$\begin{cases} (n+1)\text{-} \\ n\text{-} \end{cases}$ dimensional geometric Poincaré $\begin{cases} \text{triad} \\ \text{pair} \end{cases}$

$$\begin{cases} (H,D) : (Q;P,\partial_- V;\partial\partial_+ V) \xrightarrow{\ \simeq\ } (W;\partial_+ W,\partial_- W;\partial\partial_+ W) \\ (\partial_+ H,\partial_+ D) : (P,\partial\partial_+ V) \xrightarrow{\ \simeq\ } (\partial_+ W,\partial\partial_+ W) \\ \text{by an } (n+1)\text{-dimensional topological normal bordism} \\ (H,D) : (Q;\partial_+ V,P;\partial\partial_+ V) \longrightarrow (\partial_+ W \times I;\partial_+ W \times 0,\partial_+ W \times 1;\partial\partial_+ W \times I) \ . \end{cases}$$

Such an (H,D) (if it exists) can be regarded as a concordance

of $\begin{cases} s\text{-} \\ t\text{-} \end{cases}$ triangulations of X

$$(H;f,f') : (Q;M,M') \longrightarrow X \times (I;0,1)$$

with $f' : M' = E(\nu') \cup_{S(\nu')} P \longrightarrow X = E(\xi) \cup_{S(\xi)} Z$ an s-triangulation of X which is split along $Y \subset X$.

Thus if $\sigma_*(g,c) = 0 \in L_{n-q}(\mathbb{Z}[\pi_1(Y)])$ there exists an extension (G,C) of (g,c) satisfying i) and ii), and the corresponding (n+1)-dimensional topological normal map of triads $(F',B'):(V;\partial_+V,\partial_-V;\partial\partial_+V) \longrightarrow (W;\partial_+W,\partial_-W;\partial\partial_+W)$ is such that

$$\begin{cases} s(f,Y) = [\sigma_*(F',B')] \in \ker(LS_{n-q}(\Phi) \longrightarrow L_{n-q}(\mathbb{Z}[\pi_1(Y)])) \\ \qquad = \operatorname{im}(L_{n+1}(\mathbb{Z}[\pi_1(Z)] \longrightarrow \mathbb{Z}[\pi_1(X)]) \longrightarrow LS_{n-q}(\Phi)) \\ t(f,Y) = [\sigma_*((F',B')|_{\partial_+V})] \\ \qquad \in \ker(LP_{n-q}(\Phi) \longrightarrow L_{n-q}(\mathbb{Z}[\pi_1(Y)])) \\ \qquad = \operatorname{im}(L_n(\mathbb{Z}[\pi_1(Z)]) \longrightarrow LP_{n-q}(\Phi)) \end{cases} .$$

If $(f,b):M \longrightarrow X$ is an s-triangulation of X which is split along $Y \subset X$ then $(g,c):N \longrightarrow Y$ is an s-triangulation of Y, so that $\sigma_*(g,c) = 0 \in L_{n-q}(\mathbb{Z}[\pi_1(Y)])$. Taking

$$(G,C) = (g,c) \times \mathrm{id.} : N \times (I;0,1;\emptyset) \longrightarrow Y \times (I;0,1;\emptyset)$$

we have that

$$(F',B') = (f,b) \times \mathrm{id.}_{[0,1]} \cup_{(g,c)^! \times 1} (g,c)^! \times \mathrm{id.}_{[1,2]}$$

$$: (V;\partial_+V,\partial_-V;\partial\partial_+V) \longrightarrow (W;\partial_+W,\partial_-W;\partial\partial_+W)$$

is an s-triangulation of triads, so that $\begin{cases} \sigma_*(F',B') = 0 \\ \sigma_*((F',B')|_{\partial_+V}) = 0 \end{cases}$

and by the above remark $\begin{cases} s(f,Y) = 0 \\ t(f,Y) = 0 \end{cases}$.

Conversely, if $n-q \geqslant 5$ and $\begin{cases} s(f,Y) = 0 \\ t(f,Y) = 0 \end{cases}$ then $\sigma_*(g,c) = 0$ and there exists an extension (G,C) of (g,c) satisfying i) and ii).

Now

$$\begin{cases} \sigma_*(F',B') \in \ker(L_{n+1}(\mathbb{Z}[\pi_1(Z)] \longrightarrow \mathbb{Z}[\pi_1(X)]) \longrightarrow LS_{n-q}(\Phi)) \\ \qquad = \operatorname{im}(p\xi^!: L_{n-q+1}(\mathbb{Z}[\pi_1(Y)]) \longrightarrow L_{n+1}(\mathbb{Z}[\pi_1(Z)] \longrightarrow \mathbb{Z}[\pi_1(X)])) \\ \sigma_*((F',B')|_{\partial_+V}) \in \ker(L_n(\mathbb{Z}[\pi_1(Z)]) \longrightarrow LP_{n-q}(\Phi)) \\ \qquad = \operatorname{im}(\partial p\xi^!: L_{n-q+1}(\mathbb{Z}[\pi_1(Y)]) \longrightarrow L_n(\mathbb{Z}[\pi_1(Z)])) , \end{cases}$$

so that there exists an element $a \in L_{n-q+1}(\mathbb{Z}[\pi_1(Y)])$ such that

$$\begin{cases} \sigma_*(F',B') = p\xi^!(a) \in L_{n+1}(\mathbb{Z}[\pi_1(Z)] \longrightarrow \mathbb{Z}[\pi_1(X)]) \\ \sigma_*((F',B')|_{\partial_+V}) = \partial p\xi^!(a) \in L_n(\mathbb{Z}[\pi_1(Z)]) \quad . \end{cases}$$

By the surgery obstruction realization theorems of Wall [4,§§5,6] there exists an $(n-q+1)$-dimensional topological normal map of triads

$$(G',C') : (L';N',N'';\emptyset) \longrightarrow Y \times (I;0,1;\emptyset)$$

such that

i)' $(G',C')| = (g',c') : N' \longrightarrow Y \times 0$

ii)' $(G',C')| = (g'',c'') : N'' \longrightarrow Y \times 1$ is an s-triangulation

iii)' $\sigma_*(G',C') = -a \in L_{n-q+1}(\mathbb{Z}[\pi_1(Y)])$.

Replacing (G,C) by the extension of (g,c) defined by

$$(G'',C'') = (G,C) \cup_{(g',c')} (G',C')$$
$$: (L \cup_{N'} L';N,N'';\emptyset) \longrightarrow Y \times (I;0,1;\emptyset)$$

we have that (F',B') is replaced by an $(n+1)$-dimensional topological normal map of triads

$$(F'',B'') = (F',B') \cup_{(g',c')^!} (G'',C'')$$
$$: (V';\partial_+V',\partial_-V';\partial\partial_+V') \longrightarrow (W;\partial_+W,\partial_-W;\partial\partial_+W)$$

such that

$$\begin{cases} \sigma_*(F'',B'') = \sigma_*(F',B') + \sigma_*((G',C')^!) \\ \qquad = p\xi^!(a) + p\xi^!(-a) = 0 \in L_{n+1}(\mathbb{Z}[\pi_1(Z)] \longrightarrow \mathbb{Z}[\pi_1(X)]) \end{cases}$$

$$\left\{\begin{aligned}\sigma_*((F'',B'')|_{\partial_+V'}) &= \sigma_*((F',B')|_{\partial_+V}) + \sigma_*((G',C')^!|_{S(\lambda')})\\ &= \partial p\xi^!(a) + \partial p\xi^!(-a) = 0 \in L_n(\mathbb{Z}[\pi_1(Z)])\end{aligned}\right. ,$$

so that the coresponding s-triangulation $f'':M'' \xrightarrow{\sim} X$ is split along $Y \subset X$.

[]

For $q \geqslant 3$ Proposition 7.2.2 gives

$$\left\{\begin{aligned} s(f,Y) &= \sigma_*(g,c) \in LS_{n-q}(\Phi) = L_{n-q}(\mathbb{Z}[\pi_1(Y)])\\ t(f,Y) &= (\sigma_*(f,b),\sigma_*(g,c))\\ &\qquad \in LP_{n-q}(\Phi) = L_n(\mathbb{Z}[\pi_1(X)])\oplus L_{n-q}(\mathbb{Z}[\pi_1(Y)]) .\end{aligned}\right.$$

The following is essentially a restatement of Wall [4,Cor.11.3.1]:

<u>Proposition 7.2.5</u> For $q \geqslant 3$ an $(n,n-q)$-dimensional geometric Poincaré pair (X,Y) is such that the geometric Poincaré complexes X and Y are individually s-triangulable if (and for $n-q \geqslant 5$ only if) (X,Y) admits a t-normal structure $\widetilde{\xi}:Y \longrightarrow B\widetilde{TOP}(q)$ such that $(X,Y,\widetilde{\xi})$ is s-triangulable.

<u>Proof</u>: It is clear that if $(X,Y,\widetilde{\xi})$ is s-triangulable then so are X and Y.

Conversely, suppose that $n-q \geqslant 5$ and that there are given s-triangulations $(f,b):M \xrightarrow{\sim} X$, $(g,c):N \xrightarrow{\sim} Y$. Let

$$(X,\widetilde{\nu}_X:X \longrightarrow B\widetilde{TOP}(k),\rho_X:S^{n+k} \longrightarrow T(\nu_X))$$

$$(Y,\widetilde{\nu}_Y:Y \longrightarrow B\widetilde{TOP}(q+k),\rho_Y:S^{n+k} \longrightarrow T(\nu_Y))$$

be corresponding topological normal structures, with

$$\rho_Y : S^{n+k} \xrightarrow{\rho_X} T(\nu_X) \xrightarrow{\text{projection}} T(\nu_X)/T(\nu_X|_Z) = T(\nu_Y) .$$

The t-triangulations $\widetilde{\nu}_X$ and $\widetilde{\nu}_Y$ of $\nu_X:X \longrightarrow BG(k)$ and $\nu_Y:Y \longrightarrow BG(q+k)$ determine a unique t-triangulation $\widetilde{\xi}$ of

$\xi = \nu_{Y\subset X} : Y \longrightarrow BG(q)$, since

$$\nu_Y = \xi \oplus \nu_X|_Y : Y \longrightarrow BG(q+k) \quad .$$

Making f topologically transverse at $Y \subset X$ with respect to $\tilde{\xi}$ note that the t-triangulation $(f,b) : f^{-1}(Y) \longrightarrow Y$ corresponds to the same topological normal structure $(\tilde{\nu}_Y, \rho_Y)$ as $(g,c) : N \longrightarrow Y$, so that

$$s(f,Y) = \sigma_*(g,c) = 0 \in LS_{n-q}(\Phi) = L_{n-q}(\mathbb{Z}[\pi_1(Y)])$$

and (f,b) is concordant to an s-triangulation of X (also denoted by $(f,b) : M \longrightarrow X)$) which is split along $Y \subset X$, with the restriction $(f,b) : f^{-1}(Y) \longrightarrow Y$ an s-triangulation of Y concordant to $(g,c) : N \longrightarrow Y$. (In fact, the proof of Proposition 7.2.4 gives an embedding $N \subset M$ such that $(f,b)| = (g,c) : f^{-1}(Y) = N \longrightarrow Y)$.

[]

Moreover, Wall [4,Cor.11.3.4] proved that if $(W,\partial W)$ is an n-dimensional manifold with boundary such that W is an h-triangulable (n-q)-dimensional geometric Poincaré complex and $q \geqslant 3$ then every h-triangulation $V \xrightarrow{\simeq} W$ is homotopic to an embedding, the non-simply-connected Browder-Casson-Sullivan theorem.

We shall now extend the total surgery obstruction theory of §7.1 to codimension q t-normal geometric Poincaré pairs $(X,Y,\tilde{\xi})$. (See Levitt and Ranicki [1] for an extension to the s-triangulation theory of "stratified geometric Poincaré complexes" - the pair (X,Y) is the case of one stratum). In the first instance we have to define transfer maps in the $\mathcal{S}$-groups

$$p\tilde{\xi}^! : \mathcal{S}_n(Y) \longrightarrow \mathcal{S}_{n+q}(X,Z) \quad (n \geqslant 0) \quad .$$

A <u>codimension q t-normal CW pair</u> $(X,Y,\widetilde{\xi})$ is a codimension q CW pair (X,Y) together with a t-triangulation $\widetilde{\xi}:Y \longrightarrow \widetilde{BTOP}(q)$ of the normal fibration $\xi:Y \longrightarrow BG(q)$. For example, a codimension q t-normal geometric Poincaré pair is such an object. The composite of the transfer isomorphisms

$$\widetilde{\xi}^! = (U_{\widetilde{\xi}} \cap -)^{-1} : H_*(Y;\underline{\mathbb{L}}_0) \xrightarrow{\sim} H_{*+q}(E(\xi),S(\xi);\underline{\mathbb{L}}_0)$$

(with $U_{\widetilde{\xi}} \in \dot{H}^q(T(\xi);\underline{\mathbb{L}}^0)$ the canonical $\underline{\mathbb{L}}^0$-orientation of $\widetilde{\xi}$) and the excision isomorphisms

$$p : H_{*+q}(E(\xi),S(\xi);\underline{\mathbb{L}}_0) \xrightarrow{\sim} H_{*+q}(X,Z;\underline{\mathbb{L}}_0)$$

define transfer isomorphisms

$$p\widetilde{\xi}^! : H_*(Y;\underline{\mathbb{L}}_0) \xrightarrow{\sim} H_{*+q}(X,Z;\underline{\mathbb{L}}_0) \quad .$$

These are compatible via the assembly maps with the transfer maps in the quadratic L-groups

$$p\xi^! : L_*(\mathbb{Z}[\pi_1(Y)]) \longrightarrow L_{*+q}(\mathbb{Z}[\pi_1(Z)] \longrightarrow \mathbb{Z}[\pi_1(X)]) \quad .$$

Thus there are defined <u>transfer</u> maps in the $\mathcal{S}$-groups

$$p\widetilde{\xi}^! : \mathcal{S}_*(Y) \longrightarrow \mathcal{S}_{*+q}(X,Z) \quad ,$$

which are composites

$$p\widetilde{\xi}^! : \mathcal{S}_*(Y) \xrightarrow{\widetilde{\xi}^!} \mathcal{S}_{*+q}(E(\xi),S(\xi)) \xrightarrow{p} \mathcal{S}_{*+q}(X,Z)$$

and fit into a natural transformation of exact sequences

$$\begin{array}{ccccccccc}
\ldots \to & H_{n-q}(Y;\underline{\mathbb{L}}_0) & \xrightarrow{\sigma_*} & L_{n-q}(\mathbb{Z}[\pi_1(Y)]) & \to & \mathcal{S}_{n-q}(Y) & \to & H_{n-q-1}(Y;\underline{\mathbb{L}}_0) & \to \ldots \\
& \downarrow{\scriptstyle p\widetilde{\xi}^!}\wr & & \downarrow{\scriptstyle p\xi^!} & & \downarrow{\scriptstyle p\widetilde{\xi}^!} & & \downarrow{\scriptstyle p\widetilde{\xi}^!}\wr & \\
\ldots \to & H_n(X,Z;\underline{\mathbb{L}}_0) & \xrightarrow{\sigma_*} & L_n(\mathbb{Z}[\pi_1(Z)] \to \mathbb{Z}[\pi_1(X)]) & \to & \mathcal{S}_n(X,Z) & \to & H_{n-1}(X,Z;\underline{\mathbb{L}}_0) & \to \ldots
\end{array}$$

.

<u>Proposition 7.2.6</u> Let $(X,Y,\tilde{\xi})$ be a codimension q t-normal CW pair with pushout square of fundamental groupoids

$$\begin{array}{ccc} \pi_1(S(\xi)) & \longrightarrow & \pi_1(Z) \\ \big\downarrow & \Phi & \big\downarrow \\ \pi_1(E(\xi)) & \longrightarrow & \pi_1(X) \end{array}$$

i) The LS-groups of Φ are related to the $\mathcal{S}$-groups by a commutative braid of exact sequences

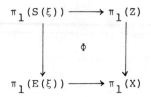

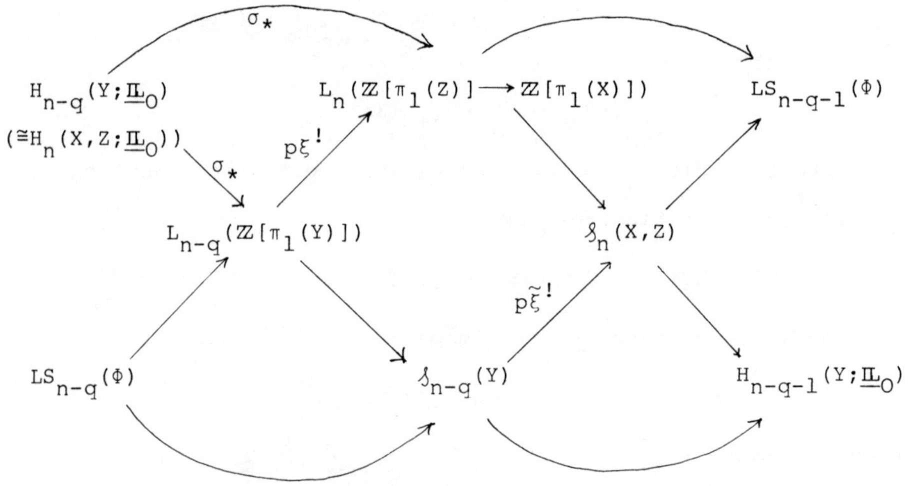

ii) There are defined <u>$\mathcal{S}$-groups</u> $\mathcal{S}_*(X,Y,\tilde{\xi})$ which fit into a commutative braid of exact sequences

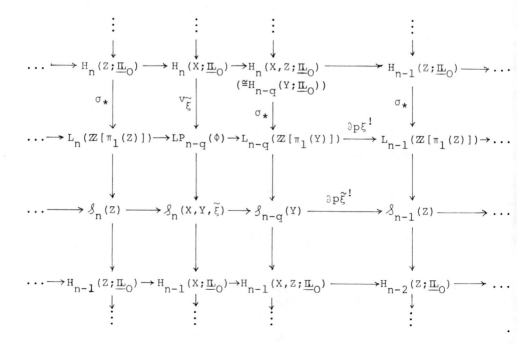

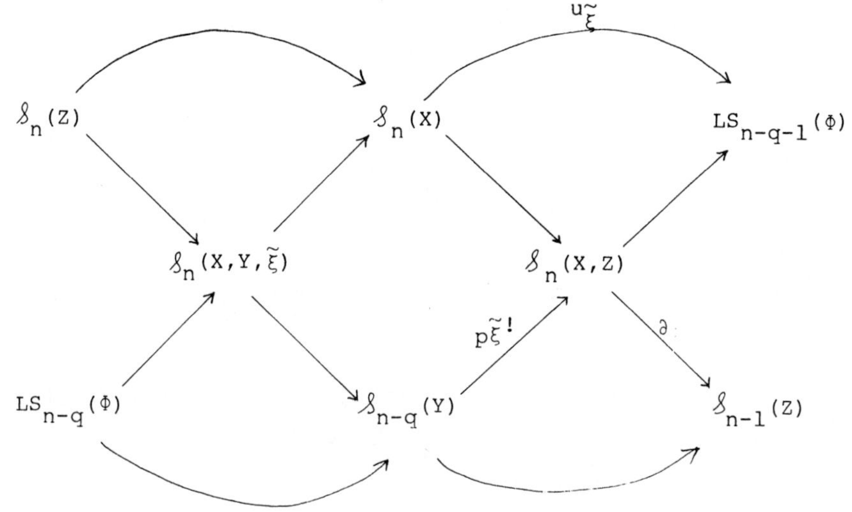

iii) The LP-groups of Φ are related to the $\mathcal{S}$-groups by a commutative diagram with exact rows and columns

$$\begin{array}{ccccccccccc}
 & & \vdots & & \vdots & & \vdots & & \vdots & & \\
 & & \downarrow & & \downarrow & & \downarrow & & \downarrow & & \\
\cdots & \longrightarrow & H_n(Z;\underline{\mathbb{L}}_0) & \longrightarrow & H_n(X;\underline{\mathbb{L}}_0) & \longrightarrow & H_n(X,Z;\underline{\mathbb{L}}_0) & \longrightarrow & H_{n-1}(Z;\underline{\mathbb{L}}_0) & \longrightarrow & \cdots \\
 & & & & & & (\cong H_{n-q}(Y;\underline{\mathbb{L}}_0)) & & & & \\
 & & \downarrow{\scriptstyle\sigma_*} & & \downarrow{\scriptstyle v_{\tilde{\xi}}} & & \downarrow{\scriptstyle\sigma_*} & & \downarrow{\scriptstyle\sigma_*} & & \\
\cdots & \longrightarrow & L_n(\mathbb{Z}[\pi_1(Z)]) & \longrightarrow & LP_{n-q}(\Phi) & \longrightarrow & L_{n-q}(\mathbb{Z}[\pi_1(Y)]) & \xrightarrow{\partial p\xi^!} & L_{n-1}(\mathbb{Z}[\pi_1(Z)]) & \longrightarrow & \cdots \\
 & & \downarrow & & \downarrow & & \downarrow & & \downarrow & & \\
\cdots & \longrightarrow & \mathcal{S}_n(Z) & \longrightarrow & \mathcal{S}_n(X,Y,\tilde{\xi}) & \longrightarrow & \mathcal{S}_{n-q}(Y) & \xrightarrow{\partial p\tilde{\xi}^!} & \mathcal{S}_{n-1}(Z) & \longrightarrow & \cdots \\
 & & \downarrow & & \downarrow & & \downarrow & & \downarrow & & \\
\cdots & \longrightarrow & H_{n-1}(Z;\underline{\mathbb{L}}_0) & \longrightarrow & H_{n-1}(X;\underline{\mathbb{L}}_0) & \longrightarrow & H_{n-1}(X,Z;\underline{\mathbb{L}}_0) & \longrightarrow & H_{n-2}(Z;\underline{\mathbb{L}}_0) & \longrightarrow & \cdots \\
 & & \downarrow & & \downarrow & & \downarrow & & \downarrow & & \\
 & & \vdots & & \vdots & & \vdots & & \vdots & &
\end{array}$$

.

iv) The maps $u_{\tilde{\xi}}, v_{\tilde{\xi}}$ are related to each other by a commutative braid of exact sequences

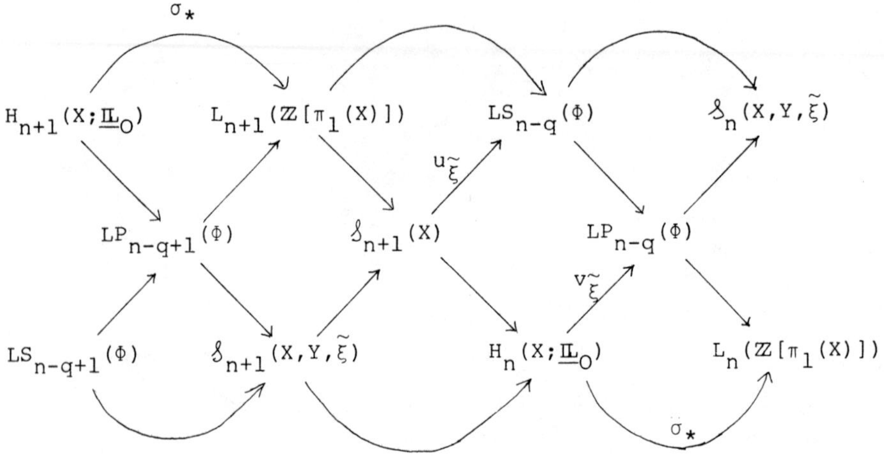

.

[]

The <u>total surgery</u> (or <u>s-triangulability</u>) <u>obstruction</u> of an $(n,n-q)$-dimensional t-normal codimension q geometric Poincaré pair $(X,Y,\tilde{\xi})$ is an element

$$s(X,Y,\tilde{\xi}) \in \mathcal{S}_n(X,Y,\tilde{\xi})$$

with the following properties.

<u>Proposition 7.2.7</u> i) $s(X,Y,\tilde{\xi}) = 0$ if (and for $n-q \geqslant 5$ only if) $(X,Y,\tilde{\xi})$ is s-triangulable.

ii) The obstruction has images

$$[s(X,Y,\tilde{\xi})] = s(X) \in \mathcal{S}_n(X)$$
$$[s(X,Y,\tilde{\xi})] = s(Y) \in \mathcal{S}_{n-q}(Y) \ .$$

iii) If $f_0:M_0 \longrightarrow X$, $f_1:M_1 \longrightarrow X$ are $\begin{cases} s- \\ t- \end{cases}$ triangulations of X with difference $\begin{cases} s(f_0,f_1) \in \mathcal{S}_{n+1}(X) \\ t(f_0,f_1) \in H_n(X;\underline{\mathbb{L}}_0) \end{cases}$ the splitting obstructions along $Y \subset X$ differ by

$$\begin{cases} s(f_0,Y) - s(f_1,Y) = u_{\tilde{\xi}}(s(f_0,f_1)) \in LS_{n-q}(\Phi) \\ t(f_0,Y) - t(f_1,Y) = v_{\tilde{\xi}}(t(f_0,f_1)) \in LP_{n-q}(\Phi) \ . \end{cases}$$

Thus if

$$\begin{cases} s(X,Y,\tilde{\xi}) \in \ker(\mathcal{S}_n(X,Y,\tilde{\xi}) \longrightarrow \mathcal{S}_n(X)) \\ \qquad\qquad = \operatorname{im}(LS_{n-q}(\Phi) \longrightarrow \mathcal{S}_n(X,Y,\tilde{\xi})) \subseteq \mathcal{S}_n(X,Y,\xi) \\ s(X,Y,\tilde{\xi}) \in \ker(\mathcal{S}_n(X,Y,\tilde{\xi}) \longrightarrow H_{n-1}(X;\underline{\mathbb{L}}_0)) \\ \qquad\qquad = \operatorname{im}(LP_{n-q}(\Phi) \longrightarrow \mathcal{S}_n(X,Y,\tilde{\xi})) \subseteq \mathcal{S}_n(X,Y,\tilde{\xi}) \end{cases}$$

(i.e. if X is $\begin{cases} s- \\ t- \end{cases}$ triangulable) the inverse image of $s(X,Y,\tilde{\xi})$ in

$\begin{cases} LS_{n-q}(\Phi) \\ LP_{n-q}(\Phi) \end{cases}$ is the coset of the subgroup

$$\begin{cases} \ker(LS_{n-q}(\Phi) \longrightarrow \mathcal{S}_n(X,Y,\tilde{\xi})) = \operatorname{im}(u_{\tilde{\xi}}:\mathcal{S}_{n+1}(X) \longrightarrow LS_{n-q}(\Phi)) \subseteq LS_{n-q}(\Phi) \\ \ker(LP_{n-q}(\Phi) \longrightarrow \mathcal{S}_n(X,Y,\tilde{\xi})) = \operatorname{im}(v_{\tilde{\xi}}:H_n(X;\underline{\mathbb{L}}_0) \longrightarrow LP_{n-q}(\Phi)) \subseteq LP_{n-q}(\Phi) \end{cases}$$

consisting of the splitting obstructions along $Y \subset X$

$\begin{cases} s(f,Y) \\ t(f,Y) \end{cases}$ of all the $\begin{cases} s- \\ t- \end{cases}$ triangulations $f:M \longrightarrow X$ of X.

iv) If $n-q \geqslant 5$ and (X,Y) is an $(n,n-q)$-dimensional manifold pair there is a natural identification

$$\mathcal{S}^{TOP}(X,Y,\tilde{\xi}) = \mathcal{S}_{n+1}(X,Y,\tilde{\xi})$$

and the commutative exact braid of Proposition 7.2.6 iv) has a natural expression as a braid of surgery exact sequences

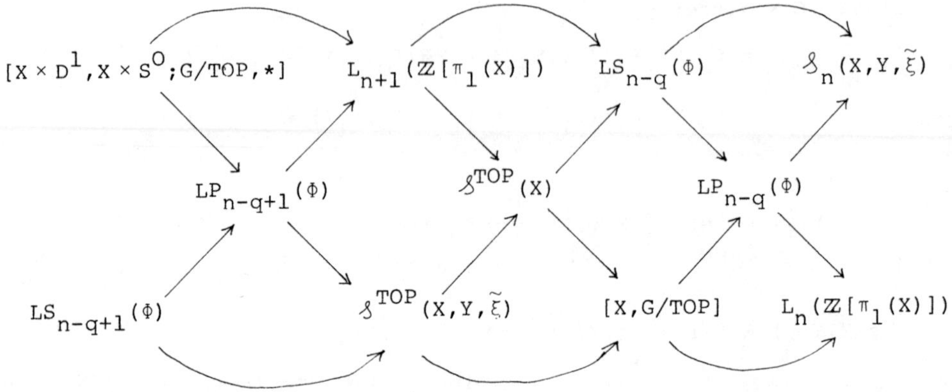

with

$$[X,G/TOP] = \mathcal{S}^{TOP}(X) = \mathcal{S}^{TOP}(X,Y,\tilde{\xi}) = H_n(X;\underline{\mathbb{L}}_0) \quad .$$

[]

(According to Ranicki [7] the topological manifold structure set $\mathcal{S}^{TOP}(X)$ of an n-dimensional geometric Poincaré complex X with $n \geqslant 5$ is in natural one-one correspondence with the set of $\underline{\mathbb{L}}^0$-orientations $[X] \in H_n(X;\underline{\mathbb{L}}^0)$ such that

i) $J([X]) = [\hat{X}] \in H_n(X;\hat{\underline{\mathbb{L}}}^0)$ is the canonical $\hat{\underline{\mathbb{L}}}^0$-orientation

ii) $\sigma^*([X]) = \sigma^*(X) \in L^n(\mathbb{Z}[\pi_1(X)])$

iii) the relations i) and ii) are compatible on the $\mathbb{L}$-spectrum level.

In view of this Proposition 7.2.7 can be interpreted as stating that the structure set $\mathcal{S}^{TOP}(X,Y,\tilde{\xi})$ of an (n,n-q)-dimensional geometric Poincaré pair (X,Y) for $n-q \geqslant 5$ is in natural one-one correspondence with the set of $\underline{\mathbb{L}}^0$-orientations $[X] \in H_n(X;\underline{\mathbb{L}}^0)$ satisfying i),ii),iii) and also

iv) the composite

$$H_n(X;\underline{\mathbb{L}}^0) \longrightarrow H_n(X,Z;\underline{\mathbb{L}}^0) \xrightarrow[\sim]{(p\tilde{\xi}^!)^{-1}} H_{n-q}(Y;\underline{\mathbb{L}}^0)$$

sends [X] to an $\underline{\mathbb{L}}^0$-orientation $[Y] \in H_{n-q}(Y;\underline{\mathbb{L}}^0)$ satisfying analogous conditions i),ii),iii) determining an s-triangulation of the (n-q)-dimensional geometric Poincaré complex Y

v) the composite

$$H_n(X;\underline{\mathbb{L}}^0) \longrightarrow H_n(X,E(\xi);\underline{\mathbb{L}}^0) \xrightarrow[\sim]{(\mathrm{excision})_*^{-1}} H_n(Z,S(\xi);\underline{\mathbb{L}}^0)$$

sends [X] to an $\underline{\mathbb{L}}^0$-orientation $[Z] \in H_n(Z,S(\xi);\underline{\mathbb{L}}^0)$ satisfying analogous conditions i),ii),iii) determining an s-triangulation of the n-dimensional geometric Poincaré pair $(Z,S(\xi))$ which on the boundary is the s-triangulation of $S(\xi)$ induced by $\tilde{\xi}$ from the s-triangulation of Y given by iv)).

In dealing with the geometric theory of codimension q surgery we have only considered the simplest case of geometric Poincaré complexes and closed manifolds. More generally, suppose given

i) an n-dimensional geometric Poincaré pair $(X,\partial X)$

ii) an (n-q)-dimensional geometric Poincaré pair $(Y,\partial Y)$

iii) a geometric Poincaré embedding

$$(Y,\partial Y) \subset (X,\partial X)$$

with normal fibration

$$(\xi,\partial\xi) : (Y,\partial Y) \longrightarrow BG(q) \ ,$$

so that

$$(X,\partial X) = (E(\xi) \cup_{S(\xi)} Z, E(\partial\xi) \cup_{S(\partial\xi)} \partial_+ Z)$$

for some n-dimensional geometric Poincaré triad $(Z;\partial_+ Z,S(\xi);S(\partial\xi))$

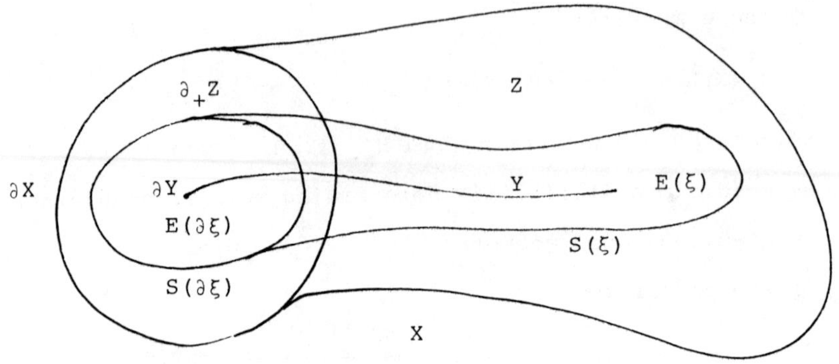

together with a t-triangulation

$$(\tilde{\xi},\widetilde{\partial\xi}) \; : \; (Y,\partial Y) \longrightarrow \widetilde{BTOP}(q) \quad ,$$

iv) an $\begin{cases} s- \\ t- \end{cases}$ triangulation of pairs

$$(f,\partial f) \; : \; (M,\partial M) \longrightarrow (X,\partial X)$$

such that $\partial f:\partial M \longrightarrow \partial X$ is an s-triangulation which is split along $\partial Y \subset \partial X$

there is defined a <u>rel∂ codimension q splitting obstruction along $Y \subset X$</u>

$$\begin{cases} s_\partial(f,Y) \in LS_{n-q}(\Phi) \\ t_\partial(f,Y) \in LP_{n-q}(\Phi) \end{cases}$$

with Φ the pushout square of fundamental groupoids

$$\begin{array}{ccc} \pi_1(S(\xi)) & \longrightarrow & \pi_1(Z) \\ \downarrow & \Phi & \downarrow \\ \pi_1(E(\xi)) & \longrightarrow & \pi_1(X) \end{array} \quad ,$$

such that $\begin{cases} s_\partial(f,Y) = 0 \\ t_\partial(f,Y) = 0 \end{cases}$ if (and for $n-q \geqslant 5$ only if) f is concordant rel ∂f to an s-triangulation of $(X,\partial X)$ which is split along $(Y,\partial Y) \subset (X,\partial X)$. By the realization theorem of Wall [4,§11] every element of $\begin{cases} LP_{n-q}(\Phi) \\ LS_{n-q}(\Phi) \end{cases}$ is a rel∂ codimension q splitting obstruction $\begin{cases} s_\partial(f,Y) \\ t_\partial(f,Y) \end{cases}$. The total surgery obstruction theory for codimension q geometric Poincaré pairs also has a rel∂ version.

7.3 The spectral quadratic construction

The quadratic construction of §II.1 (recalled in §1.2) associates to a stable π-map $F:\Sigma^{\infty}X \longrightarrow \Sigma^{\infty}Y$ of π-spaces X,Y a natural transformation

$$\dot{\psi}_F : \dot{H}_*(X/\pi) \longrightarrow Q_*(\dot{C}(Y))$$

such that

$$(1+T)\dot{\psi}_F = f^{\%}\dot{\phi}_X - \dot{\phi}_Y f_* : \dot{H}_*(X/\pi) \longrightarrow Q^*(\dot{C}(Y)) \quad ,$$

with $f : \dot{C}(X) \longrightarrow \dot{C}(Y)$ a $\mathbb{Z}[\pi]$-module chain map induced by F and $\dot{\phi}_X : \dot{H}_*(X/\pi) \longrightarrow Q^*(\dot{C}(X))$ the symmetric construction on X. The quadratic construction is an equivariant chain level generalization of the functional Steenrod square method used by Browder [6] to define the quadratic function needed to define the Arf invariant of a normal map of even-dimensional geometric Poincaré complexes.

The spectral quadratic construction which we shall now be considering associates to a semi-stable π-map $F:X \longrightarrow \Sigma^{\infty}Y$ of π-spaces X,Y a natural transformation

$$\dot{\psi}_F : \dot{H}_{*+\infty}(X/\pi) \longrightarrow Q_*(C(f))$$

with $f : \Omega^{\infty}\dot{C}(X) \longrightarrow \dot{C}(Y)$ a $\mathbb{Z}[\pi]$-module chain map induced by F. The spectral quadratic construction is an equivariant chain level generalization of the functional Steenrod square method used by Browder [4] to define the quadratic function needed to define the Arf invariant of a Wu-oriented even-dimensional geometric Poincaré complex. The name derives from the use of the spectra of stable homotopy theory, which are only implicit in our terminology.

Let π be a group, and let $w:\pi \longrightarrow \mathbb{Z}_2 = \{\pm 1\}$ be an orientation map.

A <u>semi-stable π-map</u> is a π-map

$$F : X \longrightarrow \Sigma^p Y$$

from a π-space X to the p-fold suspension of a π-space Y, for some $p \geqslant 0$. In the first instance, we shall only be concerned with the case when p is large, which is signified by writing

$$p = \infty \ , \quad F : X \longrightarrow \Sigma^\infty Y \ .$$

The chain level method used to define the quadratic construction in §II.1 (taking into account the correction on p.30) applies equally well to define the <u>spectral quadratic construction</u> on a semi-stable π-map $F:X \longrightarrow \Sigma^\infty Y$ inducing the $\mathbb{Z}[\pi]$-module chain map $f:\Omega^\infty \dot{C}(X) \longrightarrow \dot{C}(Y)$, as abelian group morphisms

$$\psi_F : \dot{H}_{n+\infty}(X/\pi) \longrightarrow Q_n(C(f)) \quad (n \geqslant 0)$$

defined using w-twisted coefficients and the w-twisted involution on $\mathbb{Z}[\pi]$.

<u>Proposition 7.3.1</u> The spectral quadratic construction has the following properties:

i) $(1+T)\psi_F = e^{\%}\dot{\phi}_Y f_* : \dot{H}_{*+\infty}(X/\pi) \longrightarrow Q^*(C(f))$

with $e : \dot{C}(Y) \longrightarrow C(f)$ the inclusion,

ii) $g_{\%}\psi_F = H\dot{\phi}_X : \dot{H}_{*+\infty}(X/\pi) \longrightarrow Q_*(\Omega^{\infty-1}\dot{C}(X))$

with $g : C(f) \longrightarrow \Omega^{\infty-1}\dot{C}(X)$ the projection and

$$H : \hat{Q}^{*+\infty}(\dot{C}(X)) = \hat{Q}^{*+1}(\Omega^{\infty-1}\dot{C}(X)) \longrightarrow Q_*(\Omega^{\infty-1}\dot{C}(X))$$

as in Proposition 1.1.2,

iii) $\psi_{\Sigma F} = \psi_F : \dot{H}_{*+\infty+1}(\Sigma X/\pi) = \dot{H}_{*+\infty}(X/\pi) \longrightarrow Q_*(C(f))$

iv) if $X = \Sigma^\infty X_0$ for some π-space X_0

$$\psi_F : \dot{H}_{*+\infty}(X/\pi) = \dot{H}_*(X_0/\pi) \xrightarrow{\psi_{F_0}} Q_*(\dot{C}(Y)) \xrightarrow{e_\%} Q_*(C(f))$$

with ψ_{F_0} the spectral quadratic construction of §II.1 on the stable π-map

$$F_0 = F : X = \Sigma^\infty X_0 \longrightarrow Y$$

v) if there are given π-spaces X,X',Y,Y' and a commutative diagram of (semi-)stable π-maps

$$\begin{array}{ccc} X & \xrightarrow{F} & \Sigma^\infty Y \\ \downarrow G & & \downarrow H \\ X' & \xrightarrow{F'} & \Sigma^\infty Y' \end{array}$$

inducing the commutative diagram of $\mathbb{Z}[\pi]$-module chain complexes and chain maps

$$\begin{array}{ccccc} \Omega^\infty\dot{C}(X) & \xrightarrow{f} & \dot{C}(Y) & \xrightarrow{e} & C(f) \\ \downarrow g & & \downarrow h & & \downarrow (h,g) \\ \Omega^\infty\dot{C}(X') & \xrightarrow{f'} & \dot{C}(Y') & \xrightarrow{e'} & C(f') \end{array}$$

then

$$\psi_{F'}g_* = (h,g)_\%\psi_F + e'_\%\psi_H f_* : \dot{H}_{*+\infty}(X/\pi) \longrightarrow Q_*(C(f'))$$

with $\psi_F:\dot{H}_{*+\infty}(X/\pi) \longrightarrow Q_*(C(f))$ (resp. $\psi_{F'}:\dot{H}_{*+\infty}(X'/\pi) \longrightarrow Q_*(C(f')))$ the spectral quadratic construction on F (resp. F') and $\psi_H:\dot{H}_*(Y) \longrightarrow Q_*(\dot{C}(Y'))$ the quadratic construction on H.

[]

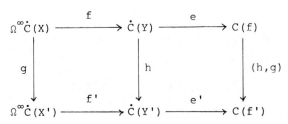

By analogy with the unstable quadratic construction of §II.1 on an unstable π-map $F:\Sigma^p X \longrightarrow \Sigma^p Y$ $(p \geqslant 0)$

$$\psi_F : \dot{H}_*(X/\pi) \longrightarrow Q_*^{[0,p-1]}(\dot{C}(Y))$$

we also have:

Proposition 7.3.2 Given π-spaces X,Y and a semi-stable π-map

$$F : X \longrightarrow \Sigma^p Y$$

for some $p \geqslant 0$ there is defined an unstable spectral quadratic construction

$$\psi_F : \dot{H}_{*+p}(X/\pi) \longrightarrow Q_*^{[0,p-1]}(C(f))$$

with $f:\Omega^p\dot{C}(X) \longrightarrow \dot{C}(Y)$ a $\mathbb{Z}[\pi]$-module chain map induced by F. If $p = 0$ then $\psi_F = 0$.

[]

Given a commutative ring R let the group ring $R[\pi]$ have the w-twisted involution

$$\bar{\ } : R[\pi] \longrightarrow R[\pi] \ ; \ \sum_{g\in\pi} r_g g \longmapsto \sum_{g\in\pi} w(g) r_g g^{-1} \quad (r_g \in R) .$$

Given a π-space X and a ring with involution A equipped with a morphism

$$R[\pi] \longrightarrow A$$

define the A-coefficient chain complex of X to be the A-module chain complex

$$C(X;A) = A\otimes_{R[\pi]} C(X;R)$$

with $C(X;R) = R\otimes_{\mathbb{Z}} C(X)$, and similarly for the reduced complex $\dot{C}(X;A) = A\otimes_{R[\pi]}\dot{C}(X:R)$. Define the A-coefficient symmetric construction on X to be the natural transformation

$$\dot{\phi}_X : \dot{H}_*(X/\pi;R) \longrightarrow Q^*(\dot{C}(X;A))$$

obtained from the $\mathbb{Z}[\pi]$-module chain level symmetric construction

$$\dot{\phi}_X : \dot{C}(X) \longrightarrow \mathrm{Hom}_{\mathbb{Z}[\mathbb{Z}_2]}(W, \dot{C}(X) \otimes_{\mathbb{Z}} \dot{C}(X))$$

(i.e. the underlying diagonal chain approximation) by applying $R\otimes_{\mathbb{Z}[\pi]}-$ and composing with the R-module chain map induced by $R[\pi] \longrightarrow A$

$$\begin{aligned}\dot{C}(X/\pi;R) &= R\otimes_{\mathbb{Z}}\dot{C}(X/\pi) = R\otimes_{\mathbb{Z}[\pi]}\dot{C}(X)\\ &\xrightarrow{1\otimes\dot{\phi}_X} R\otimes_{\mathbb{Z}[\pi]}\mathrm{Hom}_{\mathbb{Z}[\mathbb{Z}_2]}(W,\dot{C}(X)\otimes_{\mathbb{Z}}\dot{C}(X))\\ &= \mathrm{Hom}_{\mathbb{Z}[\mathbb{Z}_2]}(W,\mathrm{Hom}_{R[\pi]}(C(X;R[\pi])^*,C(X;R[\pi])))\\ &\longrightarrow \mathrm{Hom}_{\mathbb{Z}[\mathbb{Z}_2]}(W,\mathrm{Hom}_A(\dot{C}(X;A)^*,\dot{C}(X;A))) \quad .\end{aligned}$$

Define similarly the <u>A-coefficient quadratic construction</u> on a stable π-map $F:\Sigma^\infty X \longrightarrow \Sigma^\infty Y$

$$\dot{\psi}_F : \dot{H}_*(X/\pi;R) \longrightarrow Q_*(\dot{C}(Y;A))$$

and the <u>A-coefficient spectral quadratic construction</u> on a semi-stable π-map $F:X \longrightarrow \Sigma^\infty Y$

$$\psi_F : \dot{H}_{*+\infty}(X/\pi;R) \longrightarrow Q_*(C(f;A))$$

with $C(f;A) = C(f:\Omega^\infty\dot{C}(X;A) \xrightarrow{F} \dot{C}(Y;A))$.

Recall from §II.1 that the $\left\{\begin{matrix}\text{symmetric}\\ \text{quadratic}\end{matrix}\right.$ Wu classes $\left\{\begin{matrix}v_*\\ v^*\end{matrix}\right.$ of the mod 2 (= $\mathbb{Z}_2$-coefficient) $\left\{\begin{matrix}\text{symmetric}\\ \text{quadratic}\end{matrix}\right.$ construction

$$\left\{\begin{matrix}\dot{\phi}_X:\dot{H}_*(X;\mathbb{Z}_2) \longrightarrow Q^*(\dot{C}(X;\mathbb{Z}_2))\\ \dot{\psi}_F:\dot{H}_*(X;\mathbb{Z}_2) \longrightarrow Q_*(\dot{C}(Y;\mathbb{Z}_2))\end{matrix}\right. \text{ on a } \left\{\begin{matrix}\{1\}\text{-space } X\\ \text{stable } \{1\}\text{-map } F:\Sigma^\infty X \longrightarrow \Sigma^\infty Y\end{matrix}\right.$$

have an expression in terms of $\left\{\begin{matrix}-\\ \text{functional}\end{matrix}\right.$ Steenrod squares

$$\left\{\begin{array}{l}
\dot{H}_n(X;\mathbb{Z}_2) \xrightarrow{\dot{\phi}_X} Q^n(\dot{C}(X;\mathbb{Z}_2)) \xrightarrow{v_r} \mathrm{Hom}_{\mathbb{Z}_2}(\dot{H}^{n-r}(X;\mathbb{Z}_2),\mathbb{Z}_2)\ ; \\
\qquad x \longmapsto (y \longmapsto \langle Sq^r(y),x\rangle) \\
\qquad (x \in \dot{H}_n(X;\mathbb{Z}_2),\ y \in \dot{H}^{n-r}(X;\mathbb{Z}_2)) \\
\dot{H}_n(X;\mathbb{Z}_2) \xrightarrow{\dot{\psi}_F} Q_n(\dot{C}(Y;\mathbb{Z}_2)) \xrightarrow{v^r} \mathrm{Hom}_{\mathbb{Z}_2}(\dot{H}^{n-r}(Y;\mathbb{Z}_2),\mathbb{Z}_2)\ ; \\
\qquad x \longmapsto (y \longmapsto \langle Sq_h^{r+1}(\Sigma^\infty \iota),\Sigma^\infty x\rangle) \\
(x \in \dot{H}_n(X;\mathbb{Z}_2),\ y \in \dot{H}^{n-r}(Y;\mathbb{Z}_2) = [Y,K(\mathbb{Z}_2,n-r)], \\
\iota = \text{generator} \in \dot{H}^{n-r}(K(\mathbb{Z}_2,n-r);\mathbb{Z}_2) = \mathbb{Z}_2\ , \\
h = (\Sigma^\infty y)F - \Sigma^\infty(f^*y) \in [\Sigma^\infty X,\Sigma^\infty K(\mathbb{Z}_2,n-r)] \quad (= \{X,K(\mathbb{Z}_2,n-r)\}) \\
f : \dot{C}(X) = \Omega^\infty\dot{C}(\Sigma^\infty X) \xrightarrow{F} \Omega^\infty\dot{C}(\Sigma^\infty Y) = \dot{C}(Y))
\end{array}\right.$$

with $\begin{cases} v_r = 0 \\ v^r = 0 \end{cases}$ for $\begin{cases} n < 2r \\ n > 2r \end{cases}$. The intersection pairing of the complex $(\dot{C}(X;\mathbb{Z}_2),\dot{\phi}_X(x) \in Q^n(\dot{C}(X;\mathbb{Z}_2)))$ is just the evaluation on $x \in \dot{H}_n(X;\mathbb{Z}_2)$ of the cup product

$$\dot{\phi}_X(x)_0 : \dot{H}^r(X;\mathbb{Z}_2) \times \dot{H}^{n-r}(X;\mathbb{Z}_2) \longrightarrow \mathbb{Z}_2\ ;$$
$$(y,z) \longmapsto \langle y \cup z,x\rangle\ ,$$

and it follows from the relation

$$(1+T)\dot{\psi}_F = f^{\%}\dot{\phi}_X - \dot{\phi}_Y f_* : \dot{H}_n(X;\mathbb{Z}_2) \longrightarrow Q^n(\dot{C}(Y;\mathbb{Z}_2))$$

that

$$v^r(\psi_F(x))(y_1+y_2) - v^r(\psi_F(x))(y_1) - v^r(\psi_F(x))(y_2)$$
$$= \begin{cases} \langle f^*y_1 \cup f^*y_2 - f^*(y_1 \cup y_2),x\rangle \\ 0 \end{cases} \text{if } \begin{cases} n = 2r \\ n \neq 2r \end{cases}$$
$$(x \in \dot{H}_n(X;\mathbb{Z}_2),\ y_1,y_2 \in \dot{H}^{n-r}(Y;\mathbb{Z}_2))\quad .$$

<u>Proposition 7.3.3</u> The quadratic Wu classes v^* of the mod 2 spectral quadratic construction $\psi_F : \dot{H}_{*+\infty}(X;\mathbb{Z}_2) \longrightarrow Q_*(C(f;\mathbb{Z}_2))$ on a semi-stable $\{1\}$-map $F : X \longrightarrow \Sigma^\infty Y$ inducing the $\mathbb{Z}$-module chain map $f : \Omega^\infty \dot{C}(X) \longrightarrow \dot{C}(Y)$ are such that

i) the rth quadratic Wu class $v^r(\psi_F)$ has an expression in terms of functional Steenrod squares

$$\dot{H}_{n+\infty}(X;\mathbb{Z}_2) \xrightarrow{\psi_F} Q_n(C(f;\mathbb{Z}_2)) \xrightarrow{v^r} \mathrm{Hom}_{\mathbb{Z}_2}(H^{n-r}(f;\mathbb{Z}_2),\mathbb{Z}_2) ;$$

$$x \longmapsto (y \longmapsto \langle Sq_h^{r+1}(\Sigma^\infty \iota), x\rangle)$$

$$(x \in \dot{H}_n(X;\mathbb{Z}_2),\ y \in H^{n-r}(f;\mathbb{Z}_2),$$

$$h = (\Sigma^\infty(e^*y))F \in [X, \Sigma^\infty K(\mathbb{Z}_2, n-r)],$$

$$e = \text{inclusion} : \dot{C}(Y;\mathbb{Z}_2) \longrightarrow C(f;\mathbb{Z}_2) ,$$

$$e^*y \in \dot{H}^{n-r}(Y;\mathbb{Z}_2) = [Y, K(\mathbb{Z}_2, n-r)]\ )$$

with $v^r(\psi_F) = 0$ if $n > 2r$,

ii) $v^r(\psi_F(x))(y_1+y_2) - v^r(\psi_F(x))(y_1) - v^r(\psi_F(x))(y_2)$

$$= \begin{cases} \langle e^*y_1 \cup e^*y_2, f_*x\rangle & \text{if } n = 2r \\ 0 & \text{if } n \neq 2r \end{cases}$$

$$(x \in \dot{H}_{n+\infty}(X;\mathbb{Z}_2),\ y_1, y_2 \in H^{n-r}(f;\mathbb{Z}_2)) ,$$

iii) $v^r(\psi_F(x))(g^*z) = \langle Sq^{r+1}(z), x\rangle \in \mathbb{Z}_2$

$$(x \in \dot{H}_{n+\infty}(X;\mathbb{Z}_2),\ z \in \dot{H}^{n+\infty-r-1}(X;\mathbb{Z}_2),$$

$$g = \text{projection} : C(f;\mathbb{Z}_2) \longrightarrow \Omega^{\infty-1}\dot{C}(X;\mathbb{Z}_2)) .$$

[]

(The identity of Proposition 7.3.3 $\begin{cases} ii) \\ iii) \end{cases}$ is a direct consequence of the identity $\begin{cases} (1+T)\psi_F = e^{\%}\dot{\phi}_Y f_* \\ g_{\%}\psi_F = H\dot{\phi}_X \end{cases}$ of Proposition 7.3.1 $\begin{cases} i) \\ ii) \end{cases}$).

Recall from §II.9 the <u>hyperquadratic construction</u> θ_X on a π-space X, which is the composite natural transformation

$$\theta_X : \dot{H}^k(X/\pi) \xrightarrow[\sim]{\alpha} \dot{H}_{N-k}(Y/\pi) \xrightarrow{\dot{\phi}_Y} Q^{N-k}(\dot{C}(Y))$$
$$\xrightarrow{\alpha_{\%}} Q^{N-k}(\dot{C}(X)^{N-*}) \xrightarrow{J} \hat{Q}^{N-k}(\dot{C}(X)^{N-*}) = \hat{Q}^{-k}(\dot{C}(X)^{-*}) \quad (k \geqslant 0)$$

defined using any π-space Y $S\pi$-dual to X and any $S\pi$-duality map $\alpha:S^N \longrightarrow X \wedge_\pi Y$ (but which is independent of the choice of Y and α), with J as in Proposition 1.1.2. There is also an <u>A-coefficient hyperquadratic construction</u>

$$\theta_X : \dot{H}^k(X/\pi;R) \longrightarrow \hat{Q}^{-k}(\dot{C}(X;A)^{-*}) \quad (k \geqslant 0) \quad .$$

The hyperquadratic Wu classes $\hat{v}_*$ of the mod 2 hyperquadratic construction $\theta_X:\dot{H}^k(X;\mathbb{Z}_2) \longrightarrow \hat{Q}^{-k}(\dot{C}(X;\mathbb{Z}_2)^{-*})$ on a $\{1\}$-space X have an expression in terms of the dual Steenrod squares

$$\dot{H}^k(X;\mathbb{Z}_2) \xrightarrow{\theta_X} \hat{Q}^{-k}(\dot{C}(X;\mathbb{Z}_2)^{-*}) \xrightarrow{\hat{v}_r} \mathrm{Hom}_{\mathbb{Z}_2}(\dot{H}_{k+r}(X;\mathbb{Z}_2),\mathbb{Z}_2);$$
$$x \longmapsto (y \longmapsto \langle \chi(Sq^r)(x), y\rangle)$$
$$(x \in \dot{H}^k(X;\mathbb{Z}_2), y \in \dot{H}_{k+r}(X;\mathbb{Z}_2)) \quad .$$

Given a spherical fibration $\xi:X \longrightarrow BG(k)$ over a space X we shall say that a covering $\widetilde{X}$ of X is <u>oriented with respect to ξ</u> if the group of covering translations π is equipped with a map $w:\pi \longrightarrow \mathbb{Z}_2$ such that the orientation map of ξ factors as

$$w_1(\xi) : \pi_1(X) \longrightarrow \pi \xrightarrow{w} \mathbb{Z}_2 \quad ,$$

in which case the composite $\tilde{\xi} : \tilde{X} \xrightarrow{\text{projection}} X \xrightarrow{\xi} BG(k)$ is

an oriented spherical fibration over $\tilde{X}$. The <u>formally n-dimensional</u>

<u>hyperquadratic complex of ξ</u> is the pair

$$\hat{\sigma}^*(\xi) = (C(\tilde{X})^{n-*}, \theta_{T\pi(\xi)}(U_\xi) \in \hat{Q}^n(C(\tilde{X})^{n-*}))$$

defined for any $n \in \mathbb{Z}$, with $U_\xi \in \dot{H}^k(T(\xi))$ the w-twisted coefficient

Thom class of ξ and

$$\theta_{T\pi(\xi)} : \dot{H}^k(T(\xi)) \longrightarrow \hat{Q}^{-k}(\dot{C}(T\pi(\xi))^{-*}) = \hat{Q}^n(\dot{C}(T\pi(\xi))^{n+k-*})$$

the hyperquadratic construction on the Thom π-space $T\pi(\xi)$, using

the $\mathbb{Z}[\pi]$-module chain equivalence

$$U_\xi \cap - : \dot{C}(T\pi(\xi)) \xrightarrow{\sim} S^k C(\tilde{X})$$

to identify

$$\hat{Q}^n(\dot{C}(T\pi(\xi))^{n+k-*}) = \hat{Q}^n(C(\tilde{X})^{n-*}) .$$

If A is a ring with involution which is equipped with a morphism

$\mathbb{Z}[\pi] \longrightarrow A$ the <u>A-coefficient Wu classes of ξ</u> $v_*(\xi)$ are defined

to be the hyperquadratic Wu classes of $A \otimes_{\mathbb{Z}[\pi]} \hat{\sigma}^*(\xi)$, the A-module

morphisms

$$v_r(\xi) = \hat{v}_r(\theta_{T\pi(\xi)}(U_\xi)) : H_r(\tilde{X};A) \longrightarrow \hat{H}^r(\mathbb{Z}_2;A) \quad (r \geqslant 0) .$$

The mod 2 Wu classes defined in this way

$$v_r(\xi) : H_r(X;\mathbb{Z}_2) \longrightarrow \hat{H}^r(\mathbb{Z}_2;\mathbb{Z}_2) = \mathbb{Z}_2 \quad (r \geqslant 0)$$

agree with the usual mod 2 Wu classes $v_*(\xi) \in H^*(X;\mathbb{Z}_2)$, which are

characterized by

$$v_r(\xi) \cup U_\xi = \chi(Sq^r)(U_\xi) \in \dot{H}^{k+r}(T(\xi);\mathbb{Z}_2)$$

$$(r \geqslant 0, U_\xi \in \dot{H}^k(T(\xi);\mathbb{Z}_2)) .$$

Let X be an n-dimensional geometric Poincaré complex with Spivak normal structure

$$(\nu_X : X \longrightarrow BG(k), \rho_X : S^{n+k} \longrightarrow T(\nu_X)) \ ,$$

and let $\widetilde{X}$ be an oriented covering of X with group of covering translations π. With A as above there is defined an n-dimensional symmetric Poincaré complex over A

$$\sigma^*(X) = (C(\widetilde{X};A), \phi_{\widetilde{X}}([X]) \in Q^n(C(\widetilde{X};A))) \ ,$$

and as in Proposition II.9.6 it is possible to use the Sπ-duality between $\widetilde{X}_+$ and the Thom π-space $T\pi(\nu_X)$ defined by

$$\alpha_X : S^{n+k} \xrightarrow{\rho_X} T(\nu_X) \xrightarrow{\Delta} \widetilde{X}_+ \wedge_\pi T\pi(\nu_X)$$

and the Poincaré duality A-module chain equivalence

$$\phi_{\widetilde{X}}([X])_0 = [X] \cap - : C(\widetilde{X};A)^{n-*} \xrightarrow{\sim} C(\widetilde{X};A)$$

to identify

$$J\sigma^*(X) = \hat{\sigma}^*(\nu_X) \ .$$

Thus the <u>A-coefficient Wu classes of X</u> defined by

$$v_r(X) : H^{n-r}(\widetilde{X};A) \xrightarrow{v_r(\phi_{\widetilde{X}}([X]))} H^{n-2r}(\mathbb{Z}_2;A,(-)^r) \xrightarrow{J} \hat{H}^r(\mathbb{Z}_2;A)$$

(with J an isomorphism for $n \neq 2r$) can be identified with the A-coefficient Wu classes of ν_X

$$v_r(X) = v_r(\nu_X) : H^{n-r}(\widetilde{X};A) = H_r(\widetilde{X};A) \longrightarrow \hat{H}^r(\mathbb{Z}_2;A) \quad (r \geqslant 0) \ .$$

In particular, for $A = \mathbb{Z}_2$ this recovers the usual identification of the mod 2 Wu classes $v_*(X) \in H^*(X;\mathbb{Z}_2)$ characterized by

$$v_r(X) \cup x = Sq^r(x) \in H^n(X;\mathbb{Z}_2)$$
$$(r \geqslant 0, \ x \in H^{n-r}(X;\mathbb{Z}_2))$$

with the mod 2 Wu classes of $\nu_X : X \longrightarrow BG(k)$

$$v_*(X) = v_*(\nu_X) \in H^*(X;\mathbb{Z}_2) \ .$$

A <u>formally n-dimensional normal space</u> (X,ν_X,ρ_X) (or X for short) consists of

i) a finitely dominated CW complex X

ii) a spherical fibration $\nu_X : X \longrightarrow BG(k)$ over X

iii) a map $\rho_X : S^{n+k} \longrightarrow T(\nu_X)$.

(Normal spaces were introduced by Quinn [3]). The <u>orientation map</u> of X is the orientation map of ν_X

$$w(X) = w_1(\nu_X) : \pi_1(X) \longrightarrow \mathbb{Z}_2 \quad .$$

The <u>fundamental class</u> of X is the w(X)-twisted integral homology class

$$[X] = h(\rho_X) \cap U_{\nu_X} \in H_n(X) \quad ,$$

with $h : \pi_{n+k}(T(\nu_X)) \longrightarrow \dot{H}_{n+k}(T(\nu_X))$ the Hurewicz map and $U_{\nu_X} \in \dot{H}^k(T(\nu_X))$ the w(X)-twisted integral cohomology Thom class of ν_X.

An n-dimensional geometric Poincaré complex X is a formally n-dimensional normal space such that the $\mathbb{Z}[\pi_1(X)]$-module chain level cap product with the fundamental class [X]

$$[X] \cap - : C(\widetilde{X})^{n-*} \xrightarrow{\sim} C(\widetilde{X})$$

is a chain equivalence, with $\widetilde{X}$ the universal cover of X.

A <u>formally n-dimensional degree 1 map</u>

$$f : M \longrightarrow X$$

is a map from an n-dimensional geometric Poincaré complex M to a formally n-dimensional normal space X such that

$$f_*([M]) = [X] \in H_n(X) \quad .$$

A *formally n-dimensional normal map*

$$(f,b) : M \longrightarrow X$$

is a formally n-dimensional degree 1 map $f:M \longrightarrow X$ together with a map of the normal fibrations $b:\nu_M \longrightarrow \nu_X$ covering f such that

$$T(b)_*(\rho_M) = \rho_X \in \pi_{n+k}(T(\nu_X)) \ .$$

Formally n-dimensional normal maps arise in codimension q surgery theory - see §7.5 below, particularly Proposition 7.5.4.

An n-dimensional $\begin{cases} \text{degree 1} \\ \text{normal} \end{cases}$ map $\begin{cases} f:M \longrightarrow X \\ (f,b):M \longrightarrow X \end{cases}$ of geometric Poincaré complexes is a formally n-dimensional $\begin{cases} \text{degree 1} \\ \text{normal} \end{cases}$ map such that X is an n-dimensional geometric Poincaré complex.

We shall now generalize the construction in §1.2 of the $\begin{cases} \text{symmetric} \\ \text{quadratic} \end{cases}$ kernel $\begin{cases} \sigma^*(f) \\ \sigma_*(f,b) \end{cases}$ from an actually to a formally n-dimensional $\begin{cases} \text{degree 1} \\ \text{normal} \end{cases}$ map $\begin{cases} f:M \longrightarrow X \\ (f,b):M \longrightarrow X \end{cases}$.

Let A be a ring with involution.

A *formally n-dimensional* $\begin{cases} \textit{symmetric} \\ \textit{quadratic} \end{cases}$ *complex over A* $\begin{cases} (C,\phi) \\ (C,\psi) \end{cases}$ is a finite chain complex C of f.g. projective A-modules

$$C : \ldots \longrightarrow C_{r+1} \xrightarrow{d} C_r \xrightarrow{d} C_{r-1} \longrightarrow \ldots \quad (r \in \mathbb{Z})$$

together with an element

$$\begin{cases} \phi \in Q^n(C) = H_n(\mathrm{Hom}_{\mathbb{Z}[\mathbb{Z}_2]}(W,\mathrm{Hom}_A(C^*,C))) \\ \psi \in Q_n(C) = H_n(W\otimes_{\mathbb{Z}[\mathbb{Z}_2]}\mathrm{Hom}_A(C^*,C)) \end{cases} .$$

If C is an n-dimensional A-module chain complex (i.e. if $H_r(C) = 0$ for $r < 0$ and $H^r(C) = 0$ for $r > n$) then $\begin{cases}(C,\phi)\\(C,\psi)\end{cases}$ is the same as an n-dimensional $\begin{cases}\text{symmetric}\\\text{quadratic}\end{cases}$ complex over A in the sense of §1.1. The manipulations of finite-dimensional $\begin{cases}\text{symmetric}\\\text{quadratic}\end{cases}$ complexes (such as the algebraic surgery of §1.5) carry over to formally finite-dimensional $\begin{cases}\text{symmetric}\\\text{quadratic}\end{cases}$ complexes.

Given a formally n-dimensional degree 1 map

$$f : M \longrightarrow X$$

and an oriented covering $\widetilde{X}$ of X with group of covering translations π let $\widetilde{M}$ be the induced oriented covering of M, and let $\widetilde{f}:\widetilde{M}\longrightarrow\widetilde{X}$ be a π-equivariant map covering f. The <u>Umkehr chain map of f</u> is the composite $\mathbb{Z}[\pi]$-module chain map

$$f^! : C(\widetilde{X})^{n-*} \xrightarrow{\widetilde{f}^*} C(\widetilde{M})^{n-*} \xrightarrow[\sim]{[M]\cap -} C(\widetilde{M}) .$$

There are natural identifications

$$\begin{cases} H^*(f^!) = H_{n-*+1}(\widetilde{f}) \\ H_*(f^!) = H^{n-*}(\widetilde{f}) \end{cases} ,$$

so that the $\mathbb{Z}[\pi]$-module chain maps

$$e = \text{inclusion} : C(\widetilde{M}) \longrightarrow C(f^!)$$

$$g = \text{projection} : C(f^!) \longrightarrow SC(\widetilde{X})^{n-*}$$

are such that there are defined long exact sequences

$$\begin{cases} \dots \longrightarrow H_{r+1}(\widetilde{M}) \xrightarrow{\widetilde{f}_*} H_{r+1}(\widetilde{X}) \xrightarrow{g^*} H^{n-r}(f^!) \xrightarrow{e^*} H_r(\widetilde{M}) \xrightarrow{\widetilde{f}_*} H_r(\widetilde{X}) \longrightarrow \dots \\ \dots \longrightarrow H^r(\widetilde{X}) \xrightarrow{\widetilde{f}^*} H^r(\widetilde{M}) \xrightarrow{e_*} H_{n-r}(f^!) \xrightarrow{g_*} H^{r+1}(\widetilde{X}) \xrightarrow{\widetilde{f}^*} H^{r+1}(\widetilde{M}) \longrightarrow \dots , \end{cases}$$

identifying $H^*(\widetilde{M}) = H_{n-*}(\widetilde{M})$ by the Poincaré duality of M.

<u>Proposition 7.3.4</u> Given a formally n-dimensional normal space X and an oriented covering $\widetilde{X}$ of X with group of covering translations π the following complexes are defined.

i) The <u>symmetric complex</u> of X is a formally n-dimensional symmetric complex over $\mathbb{Z}[\pi]$

$$\sigma^*(X) = (C(\widetilde{X}), \phi \in Q^n(C(\widetilde{X})))$$

such that

$$\phi_0 = [X] \cap - : C(\widetilde{X})^{n-*} \longrightarrow C(\widetilde{X}) .$$

ii) The <u>quadratic Poincaré complex</u> of X is a formally (n-1)-dimensional quadratic Poincaré complex over $\mathbb{Z}[\pi]$

$$\sigma_*(X) = (\Omega C([X] \cap - : C(\widetilde{X})^{n-*} \longrightarrow C(\widetilde{X})), \psi \in Q_{n-1}(\Omega C([X] \cap -)))$$

such that

$$(1+T)\sigma_*(X) = \partial\sigma^*(X)$$

$$g_\%\sigma_*(X) = H\hat{\sigma}^*(\nu_X) ,$$

where $\partial\sigma^*(X)$ is the boundary of the symmetric complex $\sigma^*(X)$ and

$$g = \text{projection} : \Omega C([X] \cap -) \longrightarrow C(\widetilde{X})^{n-*} .$$

iii) The <u>symmetric kernel</u> of a formally n-dimensional degree 1 map $f: M \longrightarrow X$ is a formally n-dimensional symmetric complex over $\mathbb{Z}[\pi]$

$$\sigma^*(f) = (C(f^!), \phi \in Q^n(C(f^!)))$$

such that there are defined homotopy equivalences

$$h : \partial\sigma^*(f) \xrightarrow{\ \simeq\ } -\partial\sigma^*(X)$$

$$\sigma^*(M) \xrightarrow{\ \simeq\ } \sigma^*(f) \cup_h \sigma^*(X) \ .$$

iv) The <u>quadratic kernel</u> of a formally n-dimensional normal map $(f,b):M \longrightarrow X$ is a formally n-dimensional quadratic complex over $\mathbb{Z}[\pi]$

$$\sigma_*(f,b) = (C(f^!), \psi \in Q_n(C(f^!)))$$

such that

$$(1+T)\sigma_*(f,b) = \sigma^*(f) \ ,$$

and such that there is defined a homotopy equivalence

$$h : \partial\sigma_*(f,b) \xrightarrow{\ \simeq\ } -\sigma_*(X)$$

v) If $\begin{cases} F:M_1 \longrightarrow M_2 \\ (F,B):M_1 \longrightarrow M_2 \end{cases}$ is a $\begin{cases} \text{degree 1} \\ \text{normal} \end{cases}$ map of n-dimensional geometric Poincaré complexes and $\begin{cases} f_i:M_i \longrightarrow X \\ (f_i,b_i):M_i \longrightarrow X \end{cases}$ $(i = 1,2)$ are formally n-dimensional $\begin{cases} \text{degree 1} \\ \text{normal} \end{cases}$ maps such that there is defined a commutative diagram

$$\begin{array}{ccc} M_1 & \xrightarrow{\begin{cases} F \\ (F,B) \end{cases}} & M_2 \\ & \begin{cases} f_1 \\ (f_1,b_1) \end{cases} \searrow \quad \swarrow \begin{cases} f_2 \\ (f_2,b_2) \end{cases} & \\ & X & \end{array}$$

the $\begin{cases} \text{symmetric} \\ \text{quadratic} \end{cases}$ kernel $\begin{cases} \sigma^*(F) \\ \sigma_*(F,B) \end{cases}$ is canonically cobordant to the union formally n-dimensional $\begin{cases} \text{symmetric} \\ \text{quadratic} \end{cases}$ Poincaré complex over $\mathbb{Z}[\pi]$

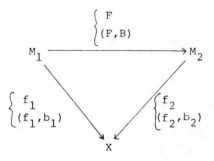

$$\begin{cases} \sigma^*(f_1) \cup_{h_2^{-1}h_1} -\sigma^*(f_2) \\ \sigma_*(f_1,b_1) \cup_{h_2^{-1}h_1} -\sigma_*(f_2,b_2) \end{cases}$$ obtained by glueing along the

composite homotopy equivalence

$$\begin{cases} h_2^{-1}h_1 \ : \ \partial\sigma^*(f_1) \xrightarrow[\simeq]{h_1} -\partial\sigma^*(X) \xrightarrow[\simeq]{h_2^{-1}} \partial\sigma^*(f_2) \\ h_2^{-1}h_1 \ : \ \partial\sigma_*(f_1,b_1) \xrightarrow[\simeq]{h_1} -\sigma_*(X) \xrightarrow[\simeq]{h_2^{-1}} \partial\sigma_*(f_2,b_2) \ . \end{cases}$$

Thus on the L-group level

$$\begin{cases} \sigma^*(F) = \sigma^*(f_1) \cup_{h_2^{-1}h_1} -\sigma^*(f_2) \in L^n(\mathbb{Z}[\pi]) \\ \sigma_*(F,B) = \sigma_*(f_1,b_1) \cup_{h_2^{-1}h_1} -\sigma_*(f_2,b_2) \in L_n(\mathbb{Z}[\pi]) \ . \end{cases}$$

<u>Proof</u>: i) Define $\sigma^*(X) = (C(\widetilde{X}),\phi)$ by

$$\phi = \phi_{\widetilde{X}}([X]) \in Q^n(C(\widetilde{X})) \ .$$

ii) See Proposition 7.4.1 iv) below for the definition of $\sigma_*(X)$.

iii) Define $\sigma^*(f) = (C(f^!),\phi)$ by

$$\phi = e^{\%}\phi_{\widetilde{M}}([M]) \in Q^n(C(f^!))$$

with e = inclusion : $C(\widetilde{M}) \longrightarrow C(f^!)$.

iv) Define $\sigma_*(f,b) = (C(f^!),\psi)$ as follows.

Let $T\pi(\nu_X)^*$ be a π-space $S\pi$-dual to the Thom π-space $T\pi(\nu_X)$.

The $S\pi$-dual of the induced π-map of Thom π-spaces

$$T\pi(b) \ : \ T\pi(\nu_M) \longrightarrow T\pi(\nu_X)$$

is the <u>geometric Umkehr</u> semi-stable π-map

$$T\pi(b)^* \ : \ T\pi(\nu_X)^* \longrightarrow T\pi(\nu_M)^* = \Sigma^\infty \widetilde{M}_+$$

inducing the Umkehr chain map

$$f^! \ : \ C(\widetilde{X})^{n-*} \longrightarrow C(\widetilde{M}) \ .$$

Evaluating the spectral quadratic construction

$$\theta_{T\pi(b)*} : \dot{H}^k(T(\nu_X)) \longrightarrow Q_n(C(f^!))$$

on the Thom class $U_{\nu_X} \in \dot{H}^k(T(\nu_X))$ set

$$\psi = \theta_{T\pi(b)*}(U_{\nu_X}) \in Q_n(C(f^!)) .$$

v) This is a generalization of the sum formula of Proposition $\begin{cases} \text{II.2.5} \\ \text{II.4.3} \end{cases}$ for the composition of $\begin{cases} \text{degree 1} \\ \text{normal} \end{cases}$ maps of geometric Poincaré complexes, and may be proved similarly.

[]

There are evident A-coefficient versions of the constructions of Proposition 7.3.4, for any ring with involution A equipped with a morphism $\mathbb{Z}[\pi] \longrightarrow A$.

A formally n-dimensional topological normal map

$$(f,b) : M \longrightarrow X$$

(or a t-triangulation of X) consists of:

i) an n-dimensional manifold M and an embedding $M \subset S^{n+k}$ with consequent topological normal structure

$$(\nu_M = \nu_{M\subset S^{n+k}} : M \longrightarrow \widetilde{BTOP}(k),$$

$$\rho_M : S^{n+k} \xrightarrow{\text{collapse}} S^{n+k}/\overline{S^{n+k} - E(\nu_M)} = E(\nu_M)/S(\nu_M) = T(\rho_M))$$

ii) a formally n-dimensional normal space X with a topological normal structure

$$(\widetilde{\nu}_X : X \longrightarrow \widetilde{BTOP}(k), \rho_X : S^{n+k} \longrightarrow T(\nu_X))$$

iii) a degree 1 map $f : M \longrightarrow X$

iv) a map of topological block bundles

$$b : \nu_M \longrightarrow \widetilde{\nu}_X$$

covering f, such that

$$T(b)_*(\rho_M) = \rho_X \in \pi_{n+k}(T(\nu_X)) .$$

(The Browder-Novikov transversality construction applies equally well to formally n-dimensional topological normal maps. Thus the set $\mathcal{T}^{TOP}(X)$ of concordance classes of t-triangulations of a formally n-dimensional normal space X is in natural one-one correspondence with the set $\mathcal{T}^{TOP}(\nu_X)$ of concordance classes of t-triangulations $\tilde{\nu}_X : X \longrightarrow \widetilde{BTOP}(k)$ of a normal fibration $\nu_X : X \longrightarrow BG(k)$ $(k \geqslant 3)$, in the non-empty case carrying a natural affine structure with translation group $\dot{H}^k(T(\nu_X);\underline{\mathbb{L}}_0)$, exactly as for the t-triangulations of a geometric Poincaré complex X considered in Proposition 7.1.3). The usual notion of a <u>geometric surgery</u> on an n-dimensional topological normal map

$$((f,b):M \longrightarrow X)$$
$$\longmapsto ((f',b') : M' = \overline{M \backslash S^r \times D^{n-r}} \cup D^{r+1} \times S^{n-r-1} \longrightarrow X)$$

carries over to a formally n-dimensional topological normal map. Indeed, the assertion of Milnor [1,p.46] that every compact, smooth and oriented n-dimensional manifold M is cobordant to one for which the classifying map of the tangent bundle $\tau_M : M \longrightarrow BSO(n)$ induces monomorphisms

$$\tau_{M*} : \pi_r(M) \longrightarrow \pi_r(BSO(n)) \quad (1 \leqslant r \leqslant \tfrac{n}{2} - 1)$$

concerns geometric surgery in the smooth category on the formally n-dimensional topological normal map

$$(\tau_M, b) : M \longrightarrow BSO(n)$$

(replacing BSO(n) by some high-dimensional skeleton).

<u>Proposition 7.3.5</u> A geometric surgery on a formally n-dimensional topological normal map $(f,b):M \longrightarrow X$ determines an algebraic surgery on the quadratic kernel $\sigma_*(f,b)$.

<u>Proof</u>: By analogy with Proposition II.7.3.

[]

If $(f,b):M \longrightarrow X$ is a formally 2i-dimensional normal map then the quadratic kernel over $\mathbb{Z}_2$

$$\sigma_*(f,b) = (C(f^!), \psi \in Q_{2i}(C(f^!)))$$

determines a quadratic self-intersection form over $\mathbb{Z}_2$

$$(H^i(f^!;\mathbb{Z}_2) = H_{i+1}(f;\mathbb{Z}_2),$$
$$\lambda = (1+T)\psi_0 : H_{i+1}(f;\mathbb{Z}_2) \times H_{i+1}(f;\mathbb{Z}_2) \longrightarrow \mathbb{Z}_2,$$
$$\mu = v^i(\psi) : H_{i+1}(f;\mathbb{Z}_2) \longrightarrow \mathbb{Z}_2)$$

such that

i) $\lambda(x,y) = \langle e^*x \cup e^*y, [M]\rangle \in \mathbb{Z}_2$

$(x,y \in H_{i+1}(f;\mathbb{Z}_2),\ e^*x, e^*y \in H_i(M;\mathbb{Z}_2) = H^i(M;\mathbb{Z}_2))$

ii) $\mu(g^*z) = v_{i+1}(\nu_X)(z) \in \mathbb{Z}_2$

$(z \in H_{i+1}(X;\mathbb{Z}_2),\ g^*z \in H_{i+1}(f;\mathbb{Z}_2))$.

This generalizes the functional Steenrod square construction due to Browder [4] of a quadratic self-intersection form over $\mathbb{Z}_2$

$$(\ker(f_*:H_i(M;\mathbb{Z}_2) \longrightarrow H_i(X;\mathbb{Z}_2)), \lambda, \mu)$$

in the case $v_{i+1}(\nu_X) = 0 \in H^{i+1}(X;\mathbb{Z}_2)$. (See Proposition 7.3.7 ii) below for the connection between the two forms). The latter form was given a geometric interpretation by Browder [9] in the case of the formally 2i-dimensional topological normal map $(f,b):M \longrightarrow X$ defined by a framed embedding $f:M^{2i} \subset X^{2i+k}$ $(k \geqslant 0)$ of a 2i-manifold M in a (2i+k)-manifold X (possibly with boundary) such that $v_{i+1}(X) = 0 \in H^{i+1}(X;\mathbb{Z}_2)$.

In general, the above self-intersection forms over $\mathbb{Z}_2$ (K,λ,μ) are singular and $\mu:K \longrightarrow \mathbb{Z}_2$ does not vanish on $\ker(\lambda:K \longrightarrow K^*) \subseteq K$, so that the Arf invariant is not defined for (K,λ,μ). We shall now give an interpretation in terms of our theory of the condition for the Arf invariant to be defined, in the more general context of ε-quadratic forms over any semisimple ring with involution A, extending the results of Browder [8,§2] for $A = \mathbb{Z}_2$.

Let then A be a ring with involution which is semisimple, i.e. 0-dimensional in the sense of §1.2, so that every A-module is projective and every submodule of an A-module is a direct summand.

The radical of an ε-symmetric form over A $(M,\phi \in Q^{\varepsilon}(M))$ is the annihilator of M

$$M^{\perp} = \ker(\phi : M \longrightarrow M^*) \subseteq M .$$

The induced ε-symmetric form on the quotient A-module $M/M^{\perp}$

$$(M/M^{\perp}, \phi/\phi^{\perp} \in Q^{\varepsilon}(M/M^{\perp}))$$

is non-singular and such that

$$(M,\phi) = (M/M^{\perp}, \phi/\phi^{\perp}) \oplus (M^{\perp}, 0)$$

(up to non-canonical isomorphism). The Witt class of (M,ϕ) is defined by

$$\sigma^*(M,\phi) = (M/M^{\perp}, \phi/\phi^{\perp}) \in L^{\varepsilon}(A) .$$

If (M,ϕ) is an even ε-symmetric form then so is $(M/M^{\perp}, \phi/\phi^{\perp})$, allowing the definition

$$\sigma^*(M,\phi) = (M/M^{\perp}, \phi/\phi^{\perp}) \in L\langle v_0 \rangle^{\varepsilon}(A) .$$

An ε-quadratic form over A $(M, \psi \in Q_{\varepsilon}(M))$ is eradicable if

$$\psi|_{M^{\perp}} = 0 \in Q_{\varepsilon}(M^{\perp})$$

with $M^{\perp} = \ker(\psi + \varepsilon\psi^* : M \longrightarrow M^*)$ the radical of the ε-symmetrization $(M, \psi + \varepsilon\psi^* \in Q^{\varepsilon}(M))$, or equivalently if for each $x \in M^{\perp}$

$$\psi(x)(x) = 0 \in Q_{\varepsilon}(A) = A/\{a - \varepsilon\bar{a} \mid a \in A\} .$$

There is induced an ε-quadratic form on the quotient A-module $M/M^{\perp}$

$$(M/M^{\perp},\psi/\psi^{\perp} \in Q_{\varepsilon}(M/M^{\perp}))$$

which is non-singular and such that

$$(M,\psi) = (M/M^{\perp},\psi/\psi^{\perp}) \oplus (M^{\perp},0)$$

(up to non-canonical isomorphism). The <u>Witt class</u> of an eradicable ε-quadratic form over A (M,ψ) is defined by

$$\sigma_*(M,\psi) = (M/M^{\perp},\psi/\psi^{\perp}) \in L_{\varepsilon}(A) \ .$$

We have the following algebraic version of the Novikov additivity property for the signature, involving the glueing of forms defined in §1.7.

<u>Proposition 7.3.6</u> Given $\begin{cases} \text{any } \varepsilon\text{-symmetric} \\ \text{any even } \varepsilon\text{-symmetric} \\ \text{eradicable } \varepsilon\text{-quadratic} \end{cases}$ forms over a semisimple ring with involution A $\begin{cases} (M,\phi) \\ (M,\phi), \\ (M,\psi) \end{cases}$ $\begin{cases} (M',\phi') \\ (M',\phi') \\ (M',\psi') \end{cases}$ and a stable isomorphism of boundary $\begin{cases} \text{even } (-\varepsilon)\text{-symmetric} \\ (-\varepsilon)\text{-quadratic} \\ \text{split } (-\varepsilon)\text{-quadratic} \end{cases}$ formations over A

$$\begin{cases} f : \partial(M,\phi) \xrightarrow{\sim} \partial(M',-\phi') \\ f : \partial(M,\phi) \xrightarrow{\sim} \partial(M',-\phi') \\ f : \partial(M,\psi) \xrightarrow{\sim} \partial(M',-\psi') \end{cases}$$

the Witt class of the union non-singular $\begin{cases} \varepsilon\text{-symmetric} \\ \text{even } \varepsilon\text{-symmetric} \\ \varepsilon\text{-quadratic} \end{cases}$ form over A $\begin{cases} (M,\phi) \cup_f (M',\phi') \\ (M,\phi) \cup_f (M',\phi') \\ (M,\psi) \cup_f (M',\psi') \end{cases}$ is given by

$$\begin{cases} \sigma^*((M,\phi)\cup_f(M',\phi')) = \sigma^*(M,\phi) + \sigma^*(M',\phi') \in L^{\varepsilon}(A) \\ \sigma^*((M,\phi)\cup_f(M',\phi')) = \sigma^*(M,\phi) + \sigma^*(M',\phi') \in L\langle v_0\rangle^{\varepsilon}(A) \\ \sigma_*((M,\psi)\cup_f(M',\psi')) = \sigma_*(M,\psi) + \sigma_*(M',\psi') \in L_{\varepsilon}(A) \ . \end{cases}$$

Proof: As in the proof of Proposition 1.7.1 there is defined

an isomorphism of $\begin{cases} \varepsilon\text{-symmetric} \\ \text{even } \varepsilon\text{-symmetric} \\ \varepsilon\text{-quadratic} \end{cases}$ forms over A

$$\begin{cases} (M,-\phi)\oplus((M,\phi)\cup_f(M',\phi')) \xrightarrow{\ \sim\ } (M',\phi')\oplus(\text{hyperbolic}) \\ (M,-\phi)\oplus((M,\phi)\cup_f(M',\phi')) \xrightarrow{\ \sim\ } (M',\phi')\oplus(\text{hyperbolic}) \\ (M,-\psi)\oplus((M,\psi)\cup_f(M',\psi')) \xrightarrow{\ \sim\ } (M',\psi')\oplus(\text{hyperbolic}) \ . \end{cases}$$

Passing to the quotients by the radicals gives rise to the Witt class sum formula.

[]

It follows from the proof of Proposition 7.3.6 that the eradicability of an ε-quadratic form over A (M,ψ) depends only on the boundary split $(-\varepsilon)$-quadratic formation over A $\partial(M,\psi)$. This dependence has a concise expression in terms of the associated $\pm\varepsilon$-quadratic complexes. For any $i \geqslant 1$ let $(C,\psi \in Q_{2i}(C,(-)^i\varepsilon))$ be the 2i-dimensional $(-)^i\varepsilon$-quadratic complex over A defined by

$$\psi_0 = \psi : C^i = M \longrightarrow C_i = M^* \quad , \quad C_r = 0 \ (r \neq i) \ ,$$

so that the boundary

$$\partial(C,\psi) = (\partial C, \partial\psi \in Q_{2i-1}(C,(-)^i\varepsilon))$$

is the (2i-1)-dimensional $(-)^i\varepsilon$-quadratic Poincaré complex over A defined by

$$d_{\partial C} = \psi + \varepsilon\psi^* : \partial C_i = M \longrightarrow \partial C_{i-1} = M^*$$

$$\partial\psi_0 = \begin{cases} 1 : \partial C^{i-1} = M \longrightarrow \partial C_i = M \\ 0 : \partial C^i = M^* \longrightarrow \partial C_{i-1} = M^* \end{cases}$$

$$\partial\psi_1 = \psi : \partial C^{i-1} = M \longrightarrow \partial C_{i-1} = M^* \quad ,$$

with ith $(-)^i\varepsilon$-quadratic Wu class

$$v^i(\partial\psi) : H^{i-1}(\partial C) = M^\perp \longrightarrow \hat{H}^1(\mathbb{Z}_2;A,\varepsilon) \; ; \; x \longmapsto \psi(x)(x) \; .$$

The ε-quadratic form (M,ψ) is eradicable if and only if $v^i(\partial\psi) = 0$.

Proposition 7.3.7 Let $\begin{cases} f:M \longrightarrow X \\ (f,b):M \longrightarrow X \end{cases}$ be a formally 2i-dimensional $\begin{cases} \text{degree 1} \\ \text{normal} \end{cases}$ map, with $\begin{cases} \text{symmetric} \\ \text{quadratic} \end{cases}$ kernel over A

$$\begin{cases} \sigma^*(f) = (C(f^!;A), \phi \in Q^{2i}(C(f^!;A))) \\ \sigma_*(f,b) = (C(f^!;A), \psi \in Q_{2i}(C(f^!;A))) \end{cases}$$

for some semisimple ring with involution A equipped with a morphism $\mathbb{Z}[\pi_1(X)] \longrightarrow A$.

The $\begin{cases} (-)^i\text{-symmetric intersection} \\ (-)^i\text{-quadratic self-intersection} \end{cases}$ form on $H^i(f^!;A) = H_{i+1}(f;A)$ determined by $\begin{cases} \sigma^*(f) \\ \sigma_*(f,b) \end{cases}$

$$\begin{cases} (H_{i+1}(f;A), \lambda = \phi_0 : H_{i+1}(f;A) \times H_{i+1}(f;A) \longrightarrow A) \\ (H_{i+1}(f;A), \lambda = (1+T)\psi_0 : H_{i+1}(f;A) \times H_{i+1}(f;A) \longrightarrow A, \\ \qquad \mu = v^i(\psi) : H_{i+1}(f;A) \longrightarrow Q_{(-)^i}(A)) \end{cases}$$

has the following properties.

i) The natural A-module morphism

$$e^* = \partial : H_{i+1}(f;A) \longrightarrow H_i(M;A)$$

defines a morphism of $(-)^i$-symmetric forms over A

$$e^* : (H_{i+1}(f;A),\lambda) \longrightarrow (H_i(M;A),\theta)$$

with $(H_i(M;A),\theta = [M]\cap - : H_i(M;A) = H^i(M;A) \xrightarrow{\sim} H_i(M;A) = H^i(M;A)^*)$ the non-singular $(-)^i$-symmetric intersection form over A of M. The radical of $(H_{i+1}(f;A),\lambda)$ is the submodule

$$H_{i+1}(f;A)^{\perp} = \mathrm{im}(H_{i+1}([X]\cap -;A) \longrightarrow H_{i+1}(f;A)) \subseteq H_{i+1}(f;A),$$

and is such that

$$\ker(e^*:H_{i+1}(f;A) \longrightarrow H_i(M;A)) = \mathrm{im}(g^*:H_{i+1}(X;A) \longrightarrow H_{i+1}(f;A))$$

$$\subseteq H_{i+1}(f;A)^{\perp}.$$

The quotient A-module

$$H_{i+1}(f;A)/\ker(e^*:H_{i+1}(f;A) \longrightarrow H_i(M;A))$$

$$= \mathrm{im}(e^*:H_{i+1}(f;A) \longrightarrow H_i(M;A))$$

$$= \ker(f_*:H_i(M;A) \longrightarrow H_i(X;A))$$

supports a $(-)^i$-symmetric form over A induced from $(H_{i+1}(f;A),\lambda)$ which is also a subform of $(H_i(M;A),\theta)$

$$(\ker(f_*:H_i(M;A) \longrightarrow H_i(X;A)),[\lambda] = \theta|)$$

with annihilator

$$\mathrm{im}(f^!:H^i(X;A) \longrightarrow H_i(M;A)) = \ker(e_*:H_i(M;A) \longrightarrow H^{i+1}(f;A))$$

$$\subseteq H_i(M;A)$$

and radical

$$\mathrm{im}(f^!:H^i(X;A) \longrightarrow H_i(M;A)) \cap \ker(f_*:H_i(M;A) \longrightarrow H_i(X;A))$$

$$\subseteq \ker(f_*:H_i(M;A) \longrightarrow H_i(X;A)) .$$

ii) The restriction of $\mu:H_{i+1}(f;A) \longrightarrow Q_{(-)^i}(A)$ to the submodule

$$\mathrm{im}(g^*:H_{i+1}(X;A) \longrightarrow H_{i+1}(f;A)) = \ker(e^*:H_{i+1}(f;A) \longrightarrow H_i(M;A)) \subseteq H_{i+1}(f;A)$$

is given by the (i+1)th A-coefficient Wu class of $\nu_X:X \longrightarrow BG$

$$v_{i+1}(\nu_X) : H_{i+1}(X;A) \xrightarrow{g^*} H_{i+1}(f;A) \longrightarrow Q_{(-)^i}(A) = H_0(\mathbb{Z}_2;A,(-)^i) .$$

Thus there is induced a $(-)^i$-quadratic form over A

$$(\ker(f_*:H_i(M;A) \longrightarrow H_i(X;A)),[\lambda],[\mu])$$

if and only if $v_{i+1}(\nu_X) = 0$. In particular, if $(H_{i+1}(f;A),\lambda,\mu)$ is eradicable then $v_{i+1}(\nu_X) = 0$.

iii) The $(-)^i$-quadratic form over A $(H_{i+1}(f;A),\lambda,\mu)$ is eradicable if and only if the boundary formally (2i-1)-dimensional quadratic Poincaré complex over A

$$\begin{aligned}\partial\sigma_*(f,b) &= -\sigma_*(X)\\ &= (\Omega C([X]\cap -:C(X;A)^{2i-*} \longrightarrow C(X;A)),\xi \in Q_{2i-1}(\Omega C([X]\cap -;A)))\end{aligned}$$

is such that

$$v^i(\xi) = 0 : H_{i+1}([X]\cap -;A) \longrightarrow H_0(\mathbb{Z}_2;A,(-)^i) .$$

In any case, the restriction of $v^i(\xi)$ to the submodule $\mathrm{im}(H_{i+1}(X;A) \longrightarrow H_{i+1}([X]\cap -;A)) \subseteq H_{i+1}([X]\cap -;A)$ is given by

$$v_{i+1}(\nu_X) : H_{i+1}(X;A) \longrightarrow H_{i+1}([X]\cap -;A) \xrightarrow{v^i(\xi)} H_0(\mathbb{Z}_2;A,(-)^i) .$$

<u>Proof</u>: i) Consider the commutative braid of exact sequences of A-modules

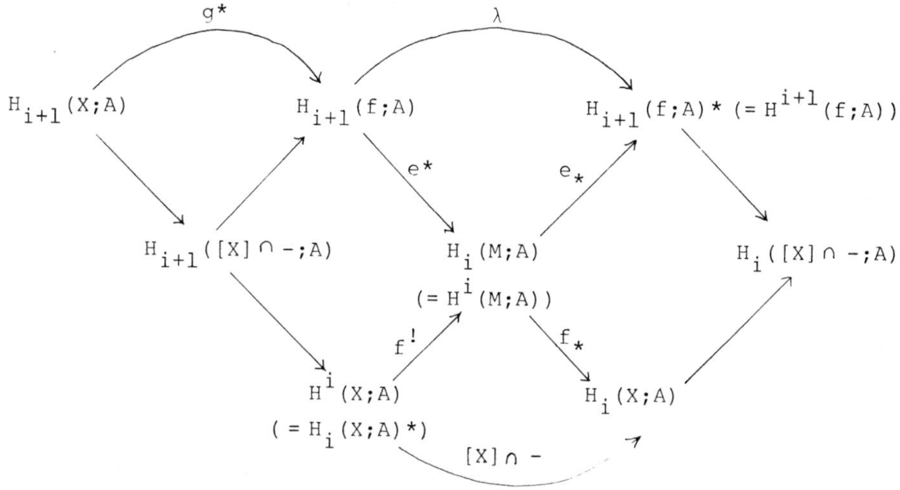

identifying $H^i(X;A) = H_i(X;A)^*$ by the universal coefficient theorem and $H^i(M;A) = H_i(M;A)$ by the Poincaré duality of M. Note that $e_* \in \mathrm{Hom}_A(H_i(M;A), H_{i+1}(f;A)^*)$ has a factorization

$$e_* : H_i(M;A) \longrightarrow \mathrm{im}(e^*: H_{i+1}(f;A) \longrightarrow H_i(M;A))^* \xrightarrow{j^*} H_{i+1}(f;A)^*$$

with j^* the split injection dual to the natural projection

$$j : H_{i+1}(f;A) \longrightarrow \mathrm{im}(e_*: H_{i+1}(f;A) \longrightarrow H_i(M;A)) \ .$$

ii),iii) These follow from Proposition 7.3.4 ii),iv) and the commutative diagram

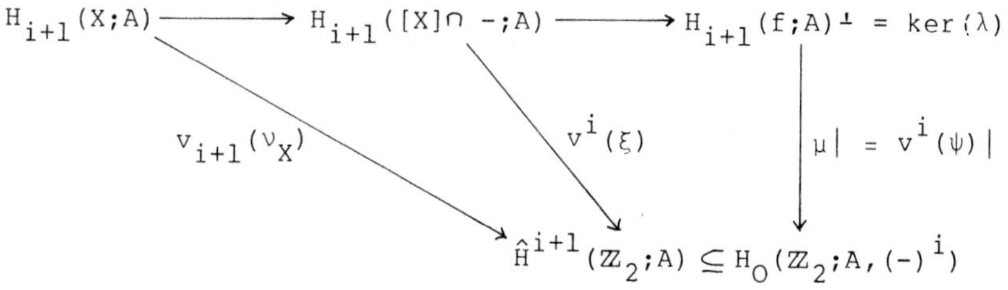

in which the map $H_{i+1}([X] \cap -;A) \longrightarrow H_{i+1}(f;A)^{\perp}$ is onto.

[]

Combining the sum formulae of Propositions 7.3.4 v), 7.3.6 with the eradicability condition of Proposition 7.3.7 there is obtained a sum formula for the quadratic signature over a semisimple ring with involution A of a normal map of 2i-dimensional geometric Poincaré complexes

$$(F,B) : M_1 \longrightarrow M_2$$

which appears in a commutative diagram of formally 2i-dimensional normal maps

$$\begin{array}{ccc} M_1 & \xrightarrow{(F,B)} & M_2 \\ & {\scriptstyle (f_1,b_1)} \searrow \quad \swarrow {\scriptstyle (f_2,b_2)} & \\ & X & \end{array}$$

with the formally 2i-dimensional normal space X such that the quadratic complex over A of X

$$\sigma_*(X) = (\Omega C([X]\cap -:C(X;A)^{2i-*} \longrightarrow C(X;A)), \xi \in Q_{2i-1}(\Omega C([X]\cap -;A)))$$

satisfies

$$v^i(\xi) = 0 : H_{i+1}([X]\cap -;A) \longrightarrow \hat{H}^{i+1}(\mathbb{Z}_2;A) \quad ,$$

namely

$$\begin{aligned} \sigma_*(F,B) &= \sigma_*(f_1,b_1) \cup -\sigma_*(f_2,b_2) \\ &= \sigma_*(f_1,b_1) - \sigma_*(f_2,b_2) \\ &\in L_{2i}(A) = L_{(-)^i}(A) \ . \end{aligned}$$

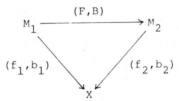

7.4 Geometric Poincaré splitting

Geometric Poincaré surgery is not logically necessary for the development of the algebraic theory of codimension q surgery in §7.5 below. However, it is a convenient halfway point between manifold and algebraic surgery, just as homotopy theory is halfway between geometry and algebra. We refer to Browder [7], Levitt [1], Jones [1], Quinn [3], Lannes, Latour and Morlet [1] and Hodgson [1] for various expositions of geometric Poincaré surgery theory. In particular, Quinn reformulated the codimension q manifold surgery theory in terms of surgery on geometric Poincaré complexes and normal spaces. We shall now recall and extend this reformulation, taking into account the total surgery obstruction theory of Ranicki [7] and replacing geometric Poincaré surgery with algebraic Poincaré surgery as far as possible.

An <u>n-dimensional normal space</u> X is a formally n-dimensional normal space $(X,\nu_X:X\longrightarrow BG(k),\rho_X:S^{n+k}\longrightarrow T(\nu_X))$ in the sense of §7.3 such that X is a finite n-dimensional CW complex. In dealing with normal spaces we shall assume a certain minimal amount of Poincaré duality (which can be achieved by surgery on 0-cells), namely

i) cap product with the fundamental class $[X]\in H_n(X)$ defines a $\mathbb{Z}[\pi_1(X)]$-module epimorphism

$$[X]\cap - : H^n(\widetilde{X}) \longrightarrow\!\!\!\!\to H_0(\widetilde{X})$$

with $\widetilde{X}$ the universal cover of X, so that $\Omega C([X]\cap -:C(\widetilde{X})^{n-*}\longrightarrow C(\widetilde{X}))$ is an (n-1)-dimensional $\mathbb{Z}[\pi_1(X)]$-module chain complex,

ii) slant product with $\alpha_X = \Delta\rho_X \in \pi_{n+k}(\bar{X}_+ \wedge_{\mathbb{Z}_2} T(\bar{\nu}_X))$

defines abelian group isomorphisms

$$\alpha_X : \dot{H}^k(T(\nu_X);\underline{\mathbb{L}}^0) \xrightarrow{\simeq} H_n(X;\underline{\mathbb{L}}^0)$$

$$\alpha_X : \dot{H}^k(T(\nu_X);\hat{\underline{\mathbb{L}}}^0) \xrightarrow{\simeq} H_n(X;\hat{\underline{\mathbb{L}}}^0) \quad ,$$

where the homology and cohomology groups are defined using w(X)-twisted coefficients. It then follows from the commutative diagram of abelian groups with exact rows

$$\begin{array}{ccccc}
\dot{H}^k(T(\nu_X);\underline{\mathbb{L}}^0) & \xrightarrow{J} & \dot{H}^k(T(\nu_X);\hat{\underline{\mathbb{L}}}^0) & \xrightarrow{H} & \dot{H}^{k+1}(T(\nu_X);\underline{\mathbb{L}}_0) \\
\downarrow{\scriptstyle \alpha_X \wr} & & \downarrow{\scriptstyle \alpha_X \wr} & & \downarrow{\scriptstyle \alpha_X} \\
H_n(X;\underline{\mathbb{L}}^0) & \xrightarrow{J} & H_n(X;\hat{\underline{\mathbb{L}}}^0) & \xrightarrow{H} & H_{n-1}(X;\underline{\mathbb{L}}_0)
\end{array}$$

that the restriction

$$\alpha_X| \; : \; \mathrm{im}(H:\dot{H}^k(T(\nu_X);\hat{\underline{\mathbb{L}}}^0) \longrightarrow \dot{H}^{k+1}(T(\nu_X);\underline{\mathbb{L}}_0))$$
$$\longrightarrow \mathrm{im}(H:H_n(X;\hat{\underline{\mathbb{L}}}^0) \longrightarrow H_{n-1}(X;\underline{\mathbb{L}}_0))$$

is an isomorphism. Thus the t-triangulability obstruction of $t(\nu_X) = H(\hat{U}_{\nu_X}) \in \dot{H}^{k+1}(T(\nu_X);\underline{\mathbb{L}}_0)$ is such that $t(\nu_X) = 0$ if and only if $\alpha_X(t(\nu_X)) = 0 \in H_{n-1}(X;\underline{\mathbb{L}}_0)$. The <u>t-triangulability obstruction of X</u> $t(X) = \alpha_X(t(\nu_X)) \in H_{n-1}(X;\underline{\mathbb{L}}_0)$ is therefore such that $t(X) = 0$ if and only if X (i.e. ν_X) is t-triangulable.

An <u>n-dimensional normal pair</u> (X,Y) consists of

i) a finite CW pair (X,Y) with X n-dimensional and Y (n-1)-dimensional

ii) a spherical fibration $\nu_X : X \longrightarrow BG(k)$

iii) a map of pairs

$$(\rho_X,\rho_Y) \; : \; (D^{n+k},S^{n+k-1}) \longrightarrow (T(\nu_X),T(\nu_Y))$$

with $\nu_Y = \nu_X|_Y : Y \longrightarrow BG(k)$, such that (Y,ν_Y,ρ_Y) is an (n-1)-dimensional normal space.

The <u>orientation map</u> of (X,Y) is the orientation map of ν_X

$$w(X) = w_1(\nu_X) : \pi_1(X) \longrightarrow \mathbb{Z}_2$$

and the <u>fundamental class</u> of (X,Y) is the w(X)-twisted integral homology class defined by

$$[X] = h(\rho_X,\rho_Y) \cap U_{\nu_X} \in H_n(X,Y)$$

with $h : \pi_{n+k}(T(\nu_X),T(\nu_Y)) \longrightarrow H_{n+k}(T(\nu_X),T(\nu_Y))$ the Hurewicz map and $U_{\nu_X} \in \dot{H}^k(T(\nu_X))$ the w(X)-twisted integral Thom class of ν_X.

In dealing with normal pairs (X,Y) we shall assume that

i) cap product with $[X] \in H_n(X,Y)$ defines a $\mathbb{Z}[\pi_1(X)]$-module epimorphism

$$[X] \cap - : H^n(\widetilde{X}) \longrightarrow\!\!\!\!\to H_0(\widetilde{X},\widetilde{Y})$$

with $\widetilde{X}$ the universal cover of X and $\widetilde{Y}$ the induced cover of Y, so that $\Omega C([X] \cap - : C(\widetilde{X})^{n-*} \longrightarrow C(\widetilde{X},\widetilde{Y}))$ is an (n-1)-dimensional $\mathbb{Z}[\pi_1(X)]$-module chain complex,

ii) slant product with $\alpha_X = \Delta(\rho_X/\rho_Y) \in \pi_{n+k}(\overline{X}/\overline{Y} \wedge_{\mathbb{Z}_2} T(\overline{\nu}_X))$ defines abelian group isomorphisms

$$\alpha_X : \dot{H}^k(T(\nu_X);\underline{\mathbb{L}}^0) \xrightarrow{\sim} H_n(X,Y;\underline{\mathbb{L}}^0)$$

$$\alpha_X : \dot{H}^k(T(\nu_X);\hat{\underline{\mathbb{L}}}^0) \xrightarrow{\sim} H_n(X,Y;\hat{\underline{\mathbb{L}}}^0)$$

The element $t(X,Y) = \alpha_X(t(\nu_X)) \in H_{n-1}(X,Y;\underline{\mathbb{L}}_0)$ is the <u>t-triangulability obstruction of (X,Y)</u>.

A finite n-dimensional geometric Poincaré pair (X,Y) is an n-dimensional normal pair such that the $\mathbb{Z}[\pi_1(X)]$-module chain map

$$[X] \cap - : C(\widetilde{X})^{n-*} \longrightarrow C(\widetilde{X},\widetilde{Y})$$

is a chain equivalence.

An <u>n-dimensional (normal, geometric Poincaré) pair</u> (X,Y) is an n-dimensional normal pair such that Y is an $(n-1)$-dimensional geometric Poincaré complex.

Given a space K with an orientation double covering $w:\bar{K} \longrightarrow K$ let $\Omega_n^N(K)$ (resp. $\Omega_n^P(K)$, $\Omega_n^{N,P}(K)$) denote the bordism group of n-dimensional normal spaces X (resp. geometric Poincaré complexes X, (normal, geometric Poincaré) pairs (X,Y)) which are equipped with a map $X \longrightarrow K$ such that the orientation map factors as

$$w(X) : \pi_1(X) \longrightarrow \pi_1(K) \xrightarrow{w} \mathbb{Z}_2$$

There is thus defined an exact sequence

$$\ldots \longrightarrow \Omega_n^P(K) \longrightarrow \Omega_n^N(K) \longrightarrow \Omega_n^{N,P}(K) \longrightarrow \Omega_{n-1}^P(K) \longrightarrow \ldots .$$

We shall only be concerned with the case when K is a CW complex with a finite 2-skeleton, for which the Levitt-Jones-Quinn geometric Poincaré surgery theory identifies

$$\Omega_n^{N,P}(K) = L_{n-1}(\mathbb{Z}[\pi_1(K)]) \quad (n \geqslant 5).$$

We shall now use algebraic Poincaré surgery and the spectral quadratic construction to define quadratic signature maps

$$\sigma_* : \Omega_n^{N,P}(K) \longrightarrow L_{n-1}(\mathbb{Z}[\pi_1(K)]) \quad (n \geqslant 1)$$

which the theory implies are isomorphisms for $n \geqslant 5$. (It follows from the surgery obstruction realization theorems of Wall [4] that they are split surjections, at any rate).

<u>Proposition 7.4.1</u> i) Given an n-dimensional (normal, geometric Poincaré) pair (X,Y) and an oriented covering $(\tilde{X},\tilde{Y})$ with group of covering translations π there is defined in a natural way an $(n-1)$-dimensional quadratic Poincaré complex over $\mathbb{Z}[\pi]$, the <u>quadratic Poincaré complex</u> of (X,Y)

$$\sigma_*(X,Y) = (\Omega C([X]\cap - : C(\widetilde{X})^{n-*} \longrightarrow C(\widetilde{X},\widetilde{Y})), \psi \in Q_{n-1}(\Omega C([X]\cap -))) .$$

The <u>quadratic signature</u> of (X,Y) is the cobordism class

$$\sigma_*(X,Y) \in L_{n-1}(\mathbb{Z}[\pi]) .$$

ii) The symmetrization of the quadratic complex $(1+T)\sigma_*(X,Y)$ is canonically cobordant to the symmetric Poincaré complex $\sigma^*(Y)$, so that on the L-group level

$$(1+T)\sigma_*(X,Y) = \sigma^*(Y) \in L^{n-1}(\mathbb{Z}[\pi]) .$$

iii) The $\mathbb{Z}[\pi]$-module chain map

$$g = \text{projection} : \Omega C([X]\cap -) \longrightarrow C(\widetilde{X})^{n-*}$$

is such that

$$g_\%\sigma_*(X,Y) = H\hat{\sigma}^*(\nu_X)$$

where $\hat{\sigma}^*(\nu_X) = (C(\widetilde{X})^{n-*}, \theta_{T\pi(\nu_X)}(U_{\nu_X}) \in \hat{Q}^n(C(\widetilde{X})^{n-*}))$ is the hyperquadratic complex of $\nu_X : X \longrightarrow BG(k)$.

iv) If $Y = \emptyset$ (i.e. given an n-dimensional normal space X) the quadratic Poincaré complex of X is the $(n-1)$-dimensional quadratic Poincaré complex over $\mathbb{Z}[\pi]$

$$\sigma_*(X,\emptyset) = \sigma_*(X) = (\Omega C([X]\cap - : C(\widetilde{X})^{n-*} \longrightarrow C(\widetilde{X})), \psi \in Q_{n-1}(\Omega C([X]\cap -))) .$$

The <u>hyperquadratic signature</u> of X is the element

$$\hat{\sigma}^*(X) \in \hat{L}^n(\mathbb{Z}[\pi])$$

defined by $\sigma_*(X,\emptyset)$ together with the canonical null-cobordism of the symmetrization $(1+T)\sigma_*(X,\emptyset)$. The <u>quadratic signature</u> of X is the quadratic signature of $(X,\emptyset)$, i.e. the element

$$H\hat{\sigma}^*(X) = \sigma_*(X) \in L_{n-1}(\mathbb{Z}[\pi]) .$$

If X is an n-dimensional geometric Poincaré complex then $\sigma_*(X,\emptyset)$ is contractible and

$$\hat{\sigma}^*(X) = J\sigma^*(X) \in \hat{L}^n(\mathbb{Z}[\pi]) , \quad \sigma_*(X) = 0 \in L_{n-1}(\mathbb{Z}[\pi])$$

with $\sigma^*(X) \in L^n(\mathbb{Z}[\pi])$ the symmetric signature.

Proof: Let $(C,\phi \in Q^{n-1}(C))$ be the symmetric Poincaré complex over $\mathbb{Z}[\pi]$ of Y

$$\sigma^*(Y) = (C(\widetilde{Y}),\phi_{\widetilde{Y}}([Y])) = (C,\phi) \ ,$$

and define a $\mathbb{Z}[\pi]$-module chain map

$$f = \text{inclusion} : C = C(\widetilde{Y}) \longrightarrow D = C(\widetilde{X}) \ .$$

The evaluation of the relative symmetric construction of §II.6

$$\phi_{\widetilde{X},\widetilde{Y}} : H_n(X,Y) \longrightarrow Q^n(f)$$

on the fundamental class $[X] \in H_n(X,Y)$ gives a connected n-dimensional symmetric pair over $\mathbb{Z}[\pi]$

$$\begin{aligned}\sigma^*(X,Y) &= (f:C(\widetilde{X}) \longrightarrow C(\widetilde{Y}),\phi_{\widetilde{X},\widetilde{Y}}([X]) \in Q^n(f)) \\ &= (f:C \longrightarrow D,(\delta\phi,\phi) \in Q^n(f)) \ .\end{aligned}$$

Let $(C',\phi' \in Q^{n-1}(C'))$ be the $(n-1)$-dimensional symmetric Poincaré complex over $\mathbb{Z}[\pi]$ obtained from (C,ϕ) by surgery on the pair $(f:C \longrightarrow D,(\delta\phi,\phi))$, so that

$$\begin{aligned}C' &= \Omega C([X] \cap -:C(\widetilde{X})^{n-*} \longrightarrow C(\widetilde{X},\widetilde{Y})) \\ &= \Omega C(g:D^{n-*} \longrightarrow C(f))\end{aligned}$$

with $g:D^{n-*} \longrightarrow C(f)$ the $\mathbb{Z}[\pi]$-module chain map defined by

$$g = \begin{pmatrix} (-)^r\delta\phi_0 \\ (-)^n\phi_0 f^* \end{pmatrix} = [X] \cap - : D^{n-r} \longrightarrow C(f)_r = D_r \oplus C_{r-1} \ .$$

Proposition 1.5.1 ii) (or rather its proof in Proposition I.4.1) gives a canonical symmetric Poincaré cobordism between (C,ϕ) and (C',ϕ'). Let

$$(D',\delta\phi') = (C(f),\delta\phi/\phi \in Q^n(C(f)))$$

be the n-dimensional symmetric complex over $\mathbb{Z}[\pi]$ obtained from the pair $(f:C \longrightarrow D,(\delta\phi,\phi))$ by the algebraic Thom complex construction of §1.4. Equivalently, $(D',\delta\phi')$ is the complex defined by the evaluation on $[X] \in H_n(X,Y) = \dot{H}_n(X/Y)$ of the absolute symmetric construction on $\widetilde{X}/\widetilde{Y}$

$$\dot{\phi}_{\widetilde{X}/\widetilde{Y}} : \dot{H}_n(X/Y) \longrightarrow Q^n(\dot{C}(\widetilde{X}/\widetilde{Y})) \quad ,$$

identifying

$$D' = C(f) = C(\widetilde{X},\widetilde{Y}) = C(\widetilde{X}/\widetilde{Y}) \quad , \quad \delta\phi' = \phi_{\widetilde{X}/\widetilde{Y}}([X]) \quad .$$

The inclusion $C(f) \longrightarrow C(g:D^{n-*} \longrightarrow C(f))$ defines a $\mathbb{Z}[\pi]$-module chain map

$$e : D' = C(f) \longrightarrow C(g) = SC'$$

such that

$$e^{\%}(\delta\phi') = S(\phi') \in Q^n(C') \quad ,$$

where $S : Q^{n-1}(C') \longrightarrow Q^n(SC')$ is the algebraic suspension map of §1.1. Let $T\pi(\nu_X)^*$ be a π-space $S\pi$-dual to the Thom π-space $T\pi(\nu_X)$, so that the $S\pi$-duality theory of §II.3 applied to the composite $\{1\}$-map

$$\alpha_{X/Y} : S^{n+k} = D^{n+k}/S^{n+k-1} \xrightarrow{\rho_X/\rho_Y} T(\nu_X)/T(\nu_Y)$$
$$\xrightarrow{\Delta} T\pi(\nu_X) \wedge_{\pi} \widetilde{X}/\widetilde{Y}$$

gives a semi-stable π-map

$$G : T\pi(\nu_X)^* \longrightarrow \Sigma^{\infty}(\widetilde{X}/\widetilde{Y})$$

inducing the $\mathbb{Z}[\pi]$-module chain map

$$g = [X] \cap - : D^{n-*} = C(\widetilde{X})^{n-*} \longrightarrow C(f) = C(\widetilde{X},\widetilde{Y}) \quad .$$

Evaluating the spectral quadratic construction

$$\psi_G : \dot{H}^k(T(\nu_X)) \longrightarrow Q_n(C(g))$$

on the w-twisted coefficient Thom class $U_{\nu_X} \in \dot{H}^k(T(\nu_X))$ there is obtained an element

$$\psi' = \psi_G(U_{\nu_X}) \in Q_n(C(g))$$

such that

$$(1+T)\psi' = e^{\%}(\delta\phi') \in Q^n(SC') = Q^n(C(g))$$

by Proposition 7.3.1 i). Considering the exact sequence

$$Q_{n-1}(C') \xrightarrow{\begin{pmatrix} 1+T \\ S \end{pmatrix}} Q^{n-1}(C') \oplus Q_n(C') \xrightarrow{(S \quad -(1+T))} Q^n(SC')$$

(or rather the underlying short exact sequence of chain complexes, cf. Proposition I.1.3) there is obtained an element $\psi \in Q_{n-1}(C')$ such that

$$(1+T)\psi = \phi' \in Q^{n-1}(C')$$

$$S\psi = \psi' \in Q_n(SC') ,$$

with $\sigma_*(X,Y) = (C',\psi)$ an $(n-1)$-dimensional quadratic Poincaré complex over $\mathbb{Z}[\pi]$.

[]

If $(f,b):M \longrightarrow X$ is a normal map of n-dimensional geometric Poincaré complexes and W is the mapping cylinder of $f:M \longrightarrow X$ then $(W, M \cup -X)$ is an $(n+1)$-dimensional (normal, geometric Poincaré) pair (cf. Quinn [3]) with quadratic signature

$$\sigma_*(W, M \cup -X) = \sigma_*(f,b) \in L_n(\mathbb{Z}[\pi]) ,$$

that is the quadratic signature of (f,b) in the sense of §1.2.

Proposition 7.4.2 The various signature maps fit together to define a natural transformation of long exact sequences

$$\begin{array}{ccccccccccc}
\cdots \longrightarrow & \Omega^N_{n+1}(K) & \longrightarrow & \Omega^{N,P}_{n+1}(K) & \longrightarrow & \Omega^P_n(K) & \longrightarrow & \Omega^N_n(K) & \longrightarrow \cdots \\
& \downarrow{\scriptstyle \hat{\sigma}^*} & & \downarrow{\scriptstyle \sigma_*} & & \downarrow{\scriptstyle \sigma^*} & & \downarrow{\scriptstyle \hat{\sigma}^*} & \\
\cdots \longrightarrow & \hat{L}^{n+1}(\mathbb{Z}[\pi]) & \xrightarrow{H} & L_n(\mathbb{Z}[\pi]) & \xrightarrow{1+T} & L^n(\mathbb{Z}[\pi]) & \xrightarrow{J} & \hat{L}^n(\mathbb{Z}[\pi]) & \longrightarrow \cdots
\end{array}$$

with $\pi = \pi_1(K)$, $n \geqslant 0$.

[]

As noted above, it follows from the Levitt-Jones-Quinn geometric Poincaré surgery theory that the quadratic signature maps

$$\sigma_* : \Omega^{N,P}_{n+1}(K) \longrightarrow L_n(\mathbb{Z}[\pi_1(K)]) \quad (n \geqslant 4)$$

are isomorphisms. In general, neither the symmetric signature maps

$$\sigma^* : \Omega^P_n(K) \longrightarrow L^n(\mathbb{Z}[\pi_1(K)])$$

nor the hyperquadratic signature maps

$$\hat{\sigma}^* : \Omega^N_n(K) \longrightarrow \hat{L}^n(\mathbb{Z}[\pi_1(K)])$$

are isomorphisms. See Ranicki [7,p.306] for an example in which $\hat{\sigma}^*$ is not onto, and see Proposition 7.6.8 below for an example in which σ^* is not onto.

An <u>(n,n-q)-dimensional</u> (or <u>codimension q</u>) <u>normal pair</u> (X,Y) consists of:

i) an n-dimensional normal space

$$(X, \nu_X : X \longrightarrow BG(k), \rho_X : S^{n+k} \longrightarrow T(\nu_X))$$

ii) an (n-q)-dimensional subcomplex $Y \subset X$

iii) a (q-1)-spherical fibration over Y

$$\nu_{Y\subset X} = \xi : Y \longrightarrow BG(q)$$

and a subcomplex $Z \subset X$ disjoint from Y such that

$$X = E(\xi) \cup_{S(\xi)} Z$$

and such that Y is an (n-q)-dimensional normal space with

$$(\nu_Y = \xi \oplus \nu_X|_Y : Y \longrightarrow BG(q+k),$$

$$\rho_Y : S^{(n-q)+(q+k)} = S^{n+k} \xrightarrow{\rho_X} T(\nu_X) \longrightarrow T(\nu_X)/T(\nu_X|_Z) = T(\nu_Y)) .$$

In particular, (X,Y) is a codimension q CW pair in the sense of §7.2. A codimension q geometric Poincaré pair (X,Y) is a codimension q normal pair with X and Y geometric Poincaré complexes and $(Z,S(\xi))$ a geometric Poincaré pair.

Let (X,Y) be a codimension q CW pair with

$$X = E(\xi) \cup_{S(\xi)} Z \quad , \quad \xi : Y \longrightarrow BG(q) .$$

A map $f:M \longrightarrow X$ from an n-dimensional $\begin{cases} \text{normal space} \\ \text{geometric Poincaré complex} \end{cases}$

M is $\begin{cases} \underline{\text{normal}} \\ \underline{\text{Poincaré}} \end{cases}$ $\underline{\text{transverse at } Y \subset X}$ if $(M, N = f^{-1}(Y) \subset M)$ is an

(n,n-q)-dimensional $\begin{cases} \text{normal} \\ \text{geometric Poincaré} \end{cases}$ pair with

$$\nu_{N \subset M} = \nu : N \xrightarrow{g = f|} Y \xrightarrow{\xi} BG(q) \;,\; M = E(\xi) \cup_{S(\xi)} P \;,\; P = f^{-1}(Z)$$

$$f = g^! \cup h : M = E(\nu) \cup_{S(\nu)} P \longrightarrow X = E(\xi) \cup_{S(\xi)} Z .$$

According to the normal transversality theory of Quinn [3] every map $f:M \longrightarrow X$ from a normal space M is normal transverse at $Y \subset X$, for any codimension q CW pair (X,Y), so that in particular there is an analogue for the normal bordism groups $\Omega_*^N(K)$ of the Pontrjagin-Thom isomorphisms $\Omega_*^{STOP}(K) \cong H_*(K;\underline{MSTOP})$ $(* \neq 4)$ for topological bordism, as follows.

Given an n-dimensional normal space

$$(M,\nu_M : M \longrightarrow BG(k), \rho_M : S^{n+k} \longrightarrow T(\nu_M))$$

use the canonical $\underline{MSG}$-orientation $U_{\nu_M} \in \dot{H}^k(T(\nu_M);\underline{MSG})$ to define the canonical $\underline{MSG}$-fundamental class

$$[M] = \alpha_M(U_{\nu_M}) \in H_n(M;\underline{MSG})$$

using $w(M)$-twisted coefficients. (The cap products

$$[M] \cap - : H^*(M;\underline{MSG}) \longrightarrow H_{n-*}(M;\underline{MSG})$$

are not in general isomorphisms).

The normal space Pontrjagin-Thom isomorphisms are defined by noting that $\overline{K}_+ \wedge_{\mathbb{Z}_2} MSG(k) = T(\eta_k)$ is the Thom space of the $(k-1)$-spherical fibration classified by the map η_k appeaing in the homotopy-theoretic pullback square

$$\begin{array}{ccc} K \times_{K(\mathbb{Z}_2,1)} BG(k) & \xrightarrow{\eta_k} & BG(k) \\ \big\downarrow & & \big\downarrow w_1 \\ K & \xrightarrow{w} & K(\mathbb{Z}_2,1) \end{array}$$

and setting

$$H_n(K;\underline{MSG}) = \varinjlim_k \pi_{n+k}(\overline{K}_+ \wedge_{\mathbb{Z}_2} MSG(k)) \longrightarrow \Omega_n^N(K) \ ;$$

$$(F : S^{n+k} \longrightarrow \overline{K}_+ \wedge_{\mathbb{Z}_2} MSG(k) = T(\eta_k))$$

$$\longmapsto (f : M = F^{-1}(K \times_{K(\mathbb{Z}_2,1)} BG(k)) \xrightarrow{F|} K \times_{K(\mathbb{Z}_2,1)} BG(k) \longrightarrow K) \ ,$$

using normal transversality. The inverse isomorphisms are defined by

$$\Omega_n^N(K) \longrightarrow H_n(K;\underline{MSG}) \ ; \ (f : M \longrightarrow K) \longmapsto f_*([M]) \ ,$$

with $[M] \in H_n(M;\underline{MSG})$ the canonical $\underline{MSG}$-fundamental class.

<u>Proposition 7.4.3</u> An n-dimensional normal space X has a <u>canonical $\hat{\mathbb{L}}^0$-fundamental class</u>

$$[\hat{X}] \in H_n(X;\hat{\underline{\mathbb{L}}}^0)$$

such that

i) the map $H:H_n(X;\hat{\underline{\mathbb{L}}}^0)\longrightarrow H_{n-1}(X;\underline{\mathbb{L}}_0)$ sends $[\hat{X}]$ to the t-triangulability obstruction of X

$$H([\hat{X}]) = t(X) \in H_{n-1}(X;\underline{\mathbb{L}}_0) \quad ,$$

ii) the hyperquadratic assembly map $\hat{\sigma}^*:H_n(X;\hat{\underline{\mathbb{L}}}^0)\longrightarrow \hat{L}^n(\mathbb{Z}[\pi_1(X)])$ sends $[\hat{X}]$ to the hyperquadratic signature of X

$$\hat{\sigma}^*([X]) = \hat{\sigma}^*(X) \in \hat{L}^n(\mathbb{Z}[\pi_1(X)]) \quad ,$$

iii) the quadratic assembly map $\sigma_*:H_{n-1}(X;\underline{\mathbb{L}}_0)\longrightarrow L_{n-1}(\mathbb{Z}[\pi_1(X)])$ sends $H([\hat{X}]) = t(X)$ to the quadratic signature of X

$$\sigma_*(H([\hat{X}])) = \sigma_*(t(X)) = \sigma_*(X) \in L_{n-1}(\mathbb{Z}[\pi_1(X)]) \quad .$$

<u>Proof</u>: Use the canonical $\hat{\underline{\mathbb{L}}}^0$-orientation $\hat{U}_{\nu_X} \in \dot{H}^k(T(\nu_X);\hat{\underline{\mathbb{L}}}^0)$ to define

$$[\hat{X}] = \alpha_X(\hat{U}_{\nu_X}) \in H_n(X;\hat{\underline{\mathbb{L}}}^0)$$

with $(\nu_X:X\longrightarrow BG(k),\rho_X:S^{n+k}\longrightarrow T(\nu_X))$ the normal structure and $\alpha_X = \Delta\rho_X$. Alternatively, regard <u>MSG</u> as the spectrum of oriented normal space n-ads, use the n-ad version of the (symmetric, quadratic) Poincaré complex construction of Proposition 7.4.1 i) to define a morphism of ring spectra

$$\hat{\sigma}^* : \underline{MSG} \longrightarrow \hat{\underline{\mathbb{L}}}^0 \quad ,$$

and use the canonical <u>MSG</u>-fundamental class $[X] \in H_n(X;\underline{MSG})$ to define

$$[\hat{X}] = \hat{\sigma}^*[X] \in H_n(X;\hat{\underline{\mathbb{L}}}^0) \quad .$$

[]

The hyperquadratic signature map on the normal bordism groups factorizes through the hyperquadratic assembly maps

$$\hat{\sigma}^* : \Omega_n^N(K) = H_n(K;\underline{MSG}) \xrightarrow{\hat{\sigma}^*} H_n(K;\underline{\hat{\mathbb{L}}}^0)$$

$$\xrightarrow{\hat{\sigma}^*} \hat{L}^n(\mathbb{Z}[\pi_1(K)]) \quad (n \geqslant 0),$$

by Proposition 7.4.3 i).

A t-triangulable n-dimensional normal space X has quadratic signature

$$\sigma_*(X) = \sigma_*(t(X)) = 0 \in L_{n-1}(\mathbb{Z}[\pi_1(X)]) \quad ,$$

by Proposition 7.4.3 iii). The vanishing of the quadratic signature for t-triangulable normal spaces has a simple geometric interpretation: given a t-triangulation

$$(f,b) : M \longrightarrow X$$

(i.e. a formally n-dimensional topological normal map) note that the mapping cylinder W of $f:M \longrightarrow X$ defines a normal space cobordism $(W;M,X)$ between the manifold M and the normal space X, with a reference map

$$(g;f,1) : (W;M,X) \longrightarrow X,$$

so that

$$\sigma_*(X) = \sigma_*(M) = HJ\sigma^*(M) = 0 \in L_{n-1}(\mathbb{Z}[\pi_1(X)])$$

by Proposition 7.4.2. In fact, the quadratic kernel $\sigma_*(f,b)$ is a connected n-dimensional quadratic complex over $\mathbb{Z}[\pi_1(X)]$ such that the quadratic Poincaré complex $\sigma_*(X)$ used to define the quadratic signature is homotopy equivalent to the boundary $\partial(-\sigma_*(f,b))$, by Proposition 7.3.4 iv).

Given a codimension q CW pair (X,Y) with

$$X = E(\xi)\cup_{S(\xi)} Z \quad , \ \xi : Y \longrightarrow BG(q)$$

there are defined transfer maps in the $\begin{cases}\text{normal space} \\ \text{geometric Poincaré}\end{cases}$ bordism groups

$$\begin{cases} p\xi^! : \Omega_n^N(Y) \xrightarrow{\xi^!} \Omega_{n+q}^N(E(\xi),S(\xi)) \xrightarrow{p} \Omega_{n+q}^N(X,Y) \\ p\xi^! : \Omega_n^P(Y) \xrightarrow{\xi^!} \Omega_{n+q}^P(E(\xi),S(\xi)) \xrightarrow{p} \Omega_{n+q}^P(X,Y) \end{cases} \quad (n \geqslant 0)$$

with

$$\xi^!(f:M \longrightarrow Y) = ((E(f^*\xi),S(f^*\xi)) \longrightarrow (E(\xi),S(\xi)))$$

and p the natural maps induced by the inclusion

$$(E(\xi),S(\xi)) \longrightarrow (X,Z) \ .$$

The normal space bordism transfer maps $p\xi^!$ are isomorphisms, with $\xi^!:\Omega_n^N(Y) \xrightarrow{\sim} \Omega_{n+q}^N(E(\xi),S(\xi))$ the inverses of the MSG-coefficient Thom isomorphisms

$$U_\xi \cap - : \Omega_{n+q}^N(E(\xi),S(\xi)) = H_{n+q}(E(\xi),S(\xi);\underline{MSG}) = \dot{H}_{n+q}(T(\xi);\underline{MSG})$$
$$\xrightarrow{\sim} H_n(X;\underline{MSG}) = \Omega_n^N(X)$$

(with $U_\xi \in \dot{H}^q(T(\xi);\underline{MSG})$ the canonical MSG-orientation of ξ) and

$$p : \Omega_{n+q}^N(E(\xi),S(\xi)) = H_{n+q}(E(\xi),S(\xi);\underline{MSG})$$
$$\xrightarrow{\sim} \Omega_{n+q}^N(X,Z) = H_{n+q}(X,Z;\underline{MSG})$$

the MSG-coefficient homology excision isomorphisms.

Let $\begin{cases}\Omega_n^N(\partial p\xi^!) \\ \Omega_n^P(\partial p\xi^!)\end{cases}$ $(n \geqslant 0)$ be the relative $\begin{cases}\text{normal space} \\ \text{geometric Poincaré}\end{cases}$ bordism groups appearing in the exact sequence

$$\begin{cases} \cdots \longrightarrow \Omega^N_{n-q+1}(Y) \xrightarrow{\partial p\xi^!} \Omega^N_n(Z) \longrightarrow \Omega^N_n(\partial p\xi^!) \longrightarrow \Omega^N_{n-q}(Y) \longrightarrow \cdots \\ \cdots \longrightarrow \Omega^P_{n-q+1}(Y) \xrightarrow{\partial p\xi^!} \Omega^P_n(Z) \longrightarrow \Omega^P_n(\partial p\xi^!) \longrightarrow \Omega^P_{n-q}(Y) \longrightarrow \cdots \end{cases} ,$$

the bordism groups of pairs of maps

$$(g : N \longrightarrow Y \ , \ h : (P,S(f^*\xi)) \longrightarrow (Z,S(\xi)))$$

such that N is an (n-q)-dimensional $\begin{cases}\text{normal space}\\ \text{geometric Poincaré complex}\end{cases}$

and $(P,S(f^*\xi))$ is an n-dimensional $\begin{cases}\text{normal}\\ \text{geometric Poincaré}\end{cases}$ pair,

with

$$h| = g^!| \ : \ S(f^*\xi) \longrightarrow S(\xi) \ .$$

There are defined maps

$$\begin{cases} \Omega^N_n(\partial p\xi^!) \longrightarrow \Omega^N_n(X) \ ; \ (g,h) \longmapsto f \\ \Omega^P_n(\partial p\xi^!) \longrightarrow \Omega^P_n(X) \ ; \ (g,h) \longmapsto f \end{cases} \qquad (n \geqslant 0)$$

with

$$f = g^! \cup h : M = E(f^*\xi) \cup_{S(f^*\xi)} P \longrightarrow X = E(\xi) \cup_{S(\xi)} Z \ .$$

A map $f:M \longrightarrow X$ from an n-dimensional $\begin{cases}\text{normal space}\\ \text{geometric Poincaré complex}\end{cases}$

M is bordant to one which is $\begin{cases}\text{normal}\\ \text{Poincaré}\end{cases}$ transverse at $Y \subset X$ if

and only if

$$\begin{cases} (f:M \longrightarrow X) \in \mathrm{im}(\Omega^N_n(\partial p\xi^!) \longrightarrow \Omega^N_n(X)) \subseteq \Omega^N_n(X) \\ (f:M \longrightarrow X) \in \mathrm{im}(\Omega^P_n(\partial p\xi^!) \longrightarrow \Omega^P_n(X)) \subseteq \Omega^P_n(X) \ . \end{cases}$$

The maps $\Omega^N_*(\partial p\xi^!) \longrightarrow \Omega^N_*(X)$ are isomorphisms, by normal transversality. Recall from §7.2 that the analogously defined relative quadratic L-groups $L_*(\partial p\xi^!)$ appearing in the exact sequence

$$\cdots \longrightarrow L_{n-q+1}(\mathbb{Z}[\pi_1(Y)]) \xrightarrow{\partial p\xi^!} L_n(\mathbb{Z}[\pi_1(Z)])$$

$$\longrightarrow L_n(\partial p\xi^!) \longrightarrow L_{n-q}(\mathbb{Z}[\pi_1(Y)]) \longrightarrow \cdots$$

are the codimension q surgery obstructions $LP_*(\Phi)$ of Wall [4]

$$L_n(\partial p\xi^!) = LP_{n-q}(\Phi) ,$$

and such that the analogously defined maps $LP_{n-q}(\Phi) \longrightarrow L_n(\mathbb{Z}[\pi_1(X)])$ fit into the exact sequence

$$\cdots \longrightarrow LS_{n-q}(\Phi) \longrightarrow LP_{n-q}(\Phi) \longrightarrow L_n(\mathbb{Z}[\pi_1(X)]) \longrightarrow LS_{n-q-1}(\Phi) \longrightarrow \cdots .$$

<u>Proposition 7.4.4</u> Given a codimension q CW pair (X,Y) with the fundamental groupoid pushout square

$$\begin{array}{ccc} \pi_1(E(\xi)) & \longrightarrow & \pi_1(Z) \\ \downarrow & \Phi & \downarrow \\ \pi_1(S(\xi)) & \longrightarrow & \pi_1(X) \end{array}$$

there is defined a commutative braid of exact sequences for n-q ≥ 5

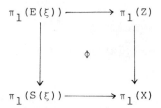

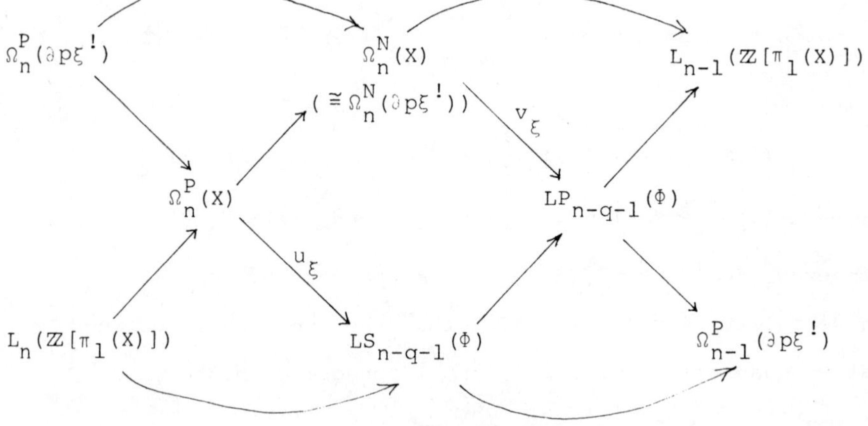

.

Thus $\begin{cases} LS_{n-q-1}(\Phi) \\ LP_{n-q-1}(\Phi) \end{cases}$ $(n-q \geqslant 5)$ is the bordism group of maps

$$(f,\partial f) : (M,\partial M) \longrightarrow X$$

from n-dimensional $\begin{cases} \text{geometric Poincaré} \\ \text{(normal, geometric Poincaré)} \end{cases}$ pairs $(M,\partial M)$ such that $\partial f:\partial M \longrightarrow X$ is Poincaré transverse at $Y \subset X$.

[]

Again, let (X,Y) be a codimension q CW pair.

A map $f:M \longrightarrow X$ from an n-dimensional $\begin{cases} \text{geometric Poincaré complex} \\ \text{normal space} \end{cases}$ M is <u>Poincaré split along $Y \subset X$</u> if f is $\begin{cases} \text{Poincaré} \\ \text{normal} \end{cases}$ bordant to a map $f':M' \longrightarrow X$ from an n-dimensional geometric Poincaré complex M' which is Poincaré transverse at $Y \subset X$. The <u>Poincaré splitting obstruction of f along $Y \subset X$</u> is the element

$$\begin{cases} s^P(f,Y) = u_\xi(f:M \longrightarrow X) \in LS_{n-q-1}(\Phi) \\ t^P(f,Y) = v_\xi(f:M \longrightarrow X) \in LP_{n-q-1}(\Phi) \end{cases}$$

with $\begin{cases} u_\xi \\ v_\xi \end{cases}$ as in Proposition 7.4.4. (For $n-q \leqslant 4$ define $\begin{cases} s^P(f,Y) \\ t^P(f,Y) \end{cases}$ using periodicity, by

$$\begin{cases} s^P(f,Y) = u_\xi(M \times (\mathbb{C}P^2)^k \xrightarrow{\text{projection}} M \xrightarrow{f} X) \\ \qquad \in LS_{n+4k-q-1}(\Phi) = LS_{n-q-1}(\Phi) \\ t^P(f,Y) = v_\xi(M \times (\mathbb{C}P^2)^k \xrightarrow{\text{projection}} M \xrightarrow{f} X) \\ \qquad \in LP_{n+4k-q-1}(\Phi) = LP_{n-q-1}(\Phi) \end{cases}$$

for any $k \geqslant 1$ such that $n+4k-q \geqslant 5$).

The Poincaré splitting obstruction $s^P(f,Y) \in LS_{n-q-1}(\Phi)$ was first obtained by Quinn [3].

For $q \geq 3$ the two Poincaré splitting obstructions along $Y \subset X$ for a map $f:M \longrightarrow X$ from an n-dimensional geometric Poincaré complex M coincide, with

$$s^P(f,Y) \in LS_{n-q-1}(\Phi) = L_{n-q-1}(\mathbb{Z}[\pi_1(Y)])$$

$$t^P(f,Y) = (s^P(f,Y),0) \in LP_{n-q-1}(\Phi) = L_{n-q-1}(\mathbb{Z}[\pi_1(Y)]) \oplus L_{n-1}(\mathbb{Z}[\pi_1(X)]).$$

Proposition 7.4.5 The Poincaré splitting obstruction along $Y \subset X$ of a map $f:M \longrightarrow X$ from an n-dimensional $\left\{\begin{array}{l}\text{geometric Poincaré complex}\\ \text{normal space}\end{array}\right.$ M is such that $\left\{\begin{array}{l} s^P(f,Y) = 0 \in LS_{n-q-1}(\Phi) \\ t^P(f,Y) = 0 \in LP_{n-q-1}(\Phi)\end{array}\right.$ if (and for $n-q \geq 5$ only if) f is Poincaré split along $Y \subset X$.

[]

It is reasonable to expect an expression for the Poincaré splitting obstruction along $Y \subset X$ $\left\{\begin{array}{l} s^P(f,Y) \in LS_{n-q-1}(\Phi) \\ t^P(f,Y) \in LP_{n-q-1}(\Phi)\end{array}\right.$ of a map $f:M \longrightarrow X$ from an n-dimensional $\left\{\begin{array}{l}\text{geometric Poincaré complex}\\ \text{normal space}\end{array}\right.$ M in terms of the t-triangulability obstruction $t(\xi) \in \dot{H}^{q+1}(T(\xi);\underline{\mathbb{L}}_0)$ of $\xi:Y \longrightarrow BG(q)$ and the $\left\{\begin{array}{l} s- \\ t-\end{array}\right.$ triangulability obstruction of M $\left\{\begin{array}{l} s(M) \in \mathcal{S}_n(M) \\ t(M) \in H_{n-1}(M;\underline{\mathbb{L}}_0)\end{array}\right.$, for if ξ is t-triangulable and M is $\left\{\begin{array}{l} s- \\ t-\end{array}\right.$ triangulable then $f:M \longrightarrow X$ is $\left\{\begin{array}{l}\text{Poincaré}\\ \text{normal}\end{array}\right.$ bordant to a map $f':M' \longrightarrow X$ from a manifold M' which is topologically (and a fortiori geometric Poincaré) transverse at $Y \subset X$. We shall obtain such an expression in Proposition 7.4.6 below.

The product of spherical fibrations $\alpha : X \longrightarrow BG(j)$, $\beta : Y \longrightarrow BG(k)$ is the spherical fibration $\alpha\times\beta : X\times Y \longrightarrow BG(j+k)$ defined by

$$(D^j\times D^k, D^j\times S^{k-1}\cup S^{j-1}\times D^k) \longrightarrow (E(\alpha)\times E(\beta), E(\alpha)\times S(\beta)\cup S(\alpha)\times E(\beta))$$
$$(= (D^{j+k}, S^{j+k-1})) \qquad\qquad (= (E(\alpha\times\beta), S(\alpha\times\beta)))$$
$$\longrightarrow X\times Y\ ,$$

with Thom space

$$\begin{aligned} T(\alpha\times\beta) &= E(\alpha\times\beta)/S(\alpha\times\beta) \\ &= (E(\alpha)\times E(\beta))/(E(\alpha)\times S(\beta)\cup S(\alpha)\times S(\beta)) \\ &= (E(\alpha)/S(\alpha))\wedge(E(\beta)/S(\beta)) = T(\alpha)\wedge T(\beta)\ . \end{aligned}$$

The canonical $\hat{\underline{\mathbb{L}}}^0$-orientation of $\alpha\times\beta$ is the product of the canonical $\hat{\underline{\mathbb{L}}}^0$-orientations of α and β

$$\hat{U}_{\alpha\times\beta} = \hat{U}_\alpha \otimes \hat{U}_\beta \in \dot{H}^{j+k}(T(\alpha)\wedge T(\beta);\hat{\underline{\mathbb{L}}}^0)$$

defined using the multiplicative structure of the ring spectrum $\hat{\underline{\mathbb{L}}}^0$. The Whitney sum of spherical fibrations $\alpha : X \longrightarrow BG(j)$, $\beta : X \longrightarrow BG(k)$ over the same base space X is the fibration $\alpha\oplus\beta : X \longrightarrow BG(j+k)$ obtained from the product $\alpha\times\beta$ by pullback along the diagonal map $\Delta : X \longrightarrow X\times X; x \longmapsto (x,x)$

$$\alpha\oplus\beta\ :\ X \longrightarrow X\times X \xrightarrow{\alpha\times\beta} BG(j+k)\ .$$

The canonical $\hat{\underline{\mathbb{L}}}^0$-orientation of $\alpha\oplus\beta$ is the product of the canonical $\hat{\underline{\mathbb{L}}}^0$-orientations of α and β

$$\hat{U}_{\alpha\oplus\beta} = \Delta^*\hat{U}_{\alpha\times\beta} = \hat{U}_\alpha \otimes \hat{U}_\beta \in \dot{H}^{j+k}(T(\alpha\oplus\beta);\hat{\underline{\mathbb{L}}}^0)$$

defined using the multiplicative structure of the ring spectrum $\hat{\underline{\mathbb{L}}}^0$ and the induced map of Thom spaces $\Delta : T(\alpha\oplus\beta) \longrightarrow T(\alpha\times\beta) = T(\alpha)\wedge T(\beta)$. The t-triangulability obstruction of $\alpha\oplus\beta$ is thus given by

$$\begin{aligned} t(\alpha\oplus\beta) &= H(\hat{U}_{\alpha\oplus\beta}) = H(\hat{U}_\alpha \otimes \hat{U}_\beta) \\ &\in \dot{H}^{j+k+1}(T(\alpha\oplus\beta);\underline{\mathbb{L}}_0)\ . \end{aligned}$$

In particular, if $\beta:X \longrightarrow BG(k)$ admits a t-triangulation $\widetilde{\beta}:X \longrightarrow \widetilde{BTOP}(k)$ and $U_{\widetilde{\beta}} \in \dot{H}^k(T(\beta);\underline{\mathbb{L}}^0)$ is the canonical $\underline{\mathbb{L}}^0$-orientation then $\hat{U}_\beta = JU_{\widetilde{\beta}} \in \dot{H}^k(T(\beta);\hat{\underline{\mathbb{L}}}^0)$ and the t-triangulability obstruction of $\alpha\oplus\beta$ is the product

$$t(\alpha\oplus\beta) = t(\alpha) \otimes U_{\widetilde{\beta}} \in \dot{H}^{j+k+1}(T(\alpha\oplus\beta);\underline{\mathbb{L}}_0)$$

defined using the structure of $\underline{\mathbb{L}}_0$ as an $\underline{\mathbb{L}}^0$-module spectrum.

Proposition 7.4.6 Let (X,Y) be a codimension q CW pair, and let $f:M \longrightarrow X$ be a map from an n-dimensional $\begin{cases}\text{geometric Poincaré complex}\\ \text{normal space}\end{cases}$ M which is normal transverse at $Y \subset X$. Let

$$g = f| : N = f^{-1}(M) \longrightarrow Y \ , \ i = \text{inclusion} : N \longrightarrow M$$

and let $\nu_M:M \longrightarrow BG(k)$ be the normal fibration of M.

i) The image of the Poincaré splitting obstruction of f along $Y \subset X$ $\begin{cases} s^P(f,Y) \in LS_{n-q-1}(\Phi) \\ t^P(f,Y) \in LP_{n-q-1}(\Phi)\end{cases}$ in $L_{n-q-1}(\mathbb{Z}[\pi_1(Y)])$ is given by

$$\begin{cases} [s^P(f,Y)] \\ [t^P(f,Y)] \end{cases} = g_*\sigma_*(N) = g_*\sigma_*(t(N)) \in L_{n-q-1}(\mathbb{Z}[\pi_1(Y)])$$

with $\sigma_*(N) \in L_{n-q-1}(\mathbb{Z}[\pi_1(N)])$ the quadratic signature of the $(n-q)$-dimensional normal space N and $t(N) \in H_{n-q-1}(N;\underline{\mathbb{L}}_0)$ the t-triangulability obstruction of N, and hence also of

$$\nu_N = g^*\xi\oplus i^*\nu_M : N \longrightarrow BG(q+k) \ .$$

In particular, for $q \geqslant 3$

$$\begin{cases} s^P(f,Y) = [s^P(f,Y)] \in LS_{n-q-1}(\Phi) = L_{n-q-1}(\mathbb{Z}[\pi_1(Y)]) \\ t^P(f,Y) = ([t^P(f,Y)],0) \\ \qquad \in LP_{n-q-1}(\Phi) = L_{n-q-1}(\mathbb{Z}[\pi_1(Y)])\oplus L_{n-1}(\mathbb{Z}[\pi_1(X)]) \ . \end{cases}$$

ii) If $q = 1$ or 2 and M is $\begin{cases} s- \\ t- \end{cases}$ triangulable then

$$\begin{cases} s^P(f,Y) = 0 \in LS_{n-q-1}(\Phi) \\ t^P(f,Y) = 0 \in LP_{n-q-1}(\Phi) \end{cases}$$

(since $\widetilde{BTOP}(q) = BG(q)$).

If $q \geqslant 3$ and M $\begin{cases} \text{is a manifold with normal bundle} \\ \text{admits a t-triangulation} \end{cases}$ $\tilde{\nu}_M : M \longrightarrow \widetilde{BTOP}(k)$

the Poincaré splitting obstruction of $f:M \longrightarrow X$ along $Y \subset X$ is given by

$$\begin{cases} s^P(f,Y) = g_*\sigma_*(t(N)) = g_*\sigma_*(g^*t(\xi)\otimes i^*U_{\tilde{\nu}_M}) \\ \qquad\qquad \in LS_{n-q-1}(\Phi) = L_{n-q-1}(\mathbb{Z}[\pi_1(Y)]) \\ [t^P(f,Y)] = g_*\sigma_*(t(N)) = g_*\sigma_*(g^*t(\xi)\otimes i^*U_{\tilde{\nu}_M}) \\ \qquad\qquad \in L_{n-q-1}(\mathbb{Z}[\pi_1(Y)]) \end{cases}$$

with $t^P(f,Y) = ([t^P(f,Y)],0)$ as in i) and $U_{\tilde{\nu}_M} \in \dot{H}^k(T(\nu_M);\underline{\mathbb{L}}^0)$ the canonical $\underline{\mathbb{L}}^0$-orientation of $\tilde{\nu}_M$.

iii) If $\xi : Y \longrightarrow BG(q)$ admits a t-triangulation $\tilde{\xi} : Y \longrightarrow \widetilde{BTOP}(q)$ the maps $\begin{cases} u_\xi \\ v_\xi \end{cases}$ appearing in the braid of Proposition 7.4.4 factor as

$$\begin{cases} u_\xi \ : \ \Omega_n^P(X) \longrightarrow \mathcal{S}_n(X) \xrightarrow{u_{\tilde{\xi}}} LS_{n-q-1}(\Phi) \\ v_\xi \ : \ \Omega_n^N(X) \longrightarrow H_{n-1}(X;\underline{\mathbb{L}}_0) \xrightarrow{v_{\tilde{\xi}}} LP_{n-q-1}(\Phi) \end{cases} \qquad (n \geqslant 0)$$

with $\begin{cases} u_{\tilde{\xi}} \\ v_{\tilde{\xi}} \end{cases}$ the maps appearing in the natural transformation of exact sequences given by Proposition 7.2.6 iv)

$$\begin{array}{ccccccccc}
\cdots \longrightarrow & L_n(\mathbb{Z}[\pi_1(X)]) & \longrightarrow & \mathcal{S}_n(X) & \longrightarrow & H_{n-1}(X;\underline{\mathbb{L}}_0) & \xrightarrow{\sigma_*} & L_{n-1}(\mathbb{Z}[\pi_1(X)]) & \longrightarrow \cdots \\
& \| & & \downarrow u_{\tilde{\xi}} & & \downarrow v_{\tilde{\xi}} & & \| & \\
\cdots \longrightarrow & L_n(\mathbb{Z}[\pi_1(X)]) & \longrightarrow & LS_{n-q-1}(\Phi) & \longrightarrow & LP_{n-q-1}(\Phi) & \longrightarrow & L_{n-1}(\mathbb{Z}[\pi_1(X)]) & \longrightarrow \cdots
\end{array}$$

Thus the Poincaré splitting obstruction along $Y \subset X$ of $f:M \longrightarrow X$ is given by

$$\begin{cases} s^P(f,Y) = u_{\tilde{\xi}}(f_* s(M)) \in LS_{n-q-1}(\Phi) \\ t^P(f,Y) = v_{\tilde{\xi}}(f_* t(M)) \in LP_{n-q-1}(\Phi) \end{cases}$$

with $\begin{cases} s(M) \in \mathcal{S}_n(M) \\ t(M) \in H_{n-1}(M;\underline{\mathbb{L}}_0) \end{cases}$ the $\begin{cases} s- \\ t- \end{cases}$ triangulability obstruction of M.

[]

In fact, the t-triangulations $\tilde{\xi}:Y \longrightarrow B\widetilde{TOP}(q)$ of a spherical fibration $\xi:Y \longrightarrow BG(q)$ over a space Y are in a natural one-one correspondence with the geometric Poincaré transversality structures along the zero section $Y \subset T(\xi)$ for maps $f:M \longrightarrow T(\xi)$ from manifolds M, i.e. ways of making them Poincaré transverse at $Y \subset X$ - see Levitt and Morgan [1], Brumfiel and Morgan [1] for the simply-connected case, Levitt and Ranicki [1] for the non-simply-connected case. Dually, the manifold structures on an n-dimensional geometric Poincaré complex M are in a natural one-one correspondence (at least for $n \geqslant 5$) with certain geometric Poincaré transversality structures for maps $f:M \longrightarrow T(\xi)$ to the Thom spaces of topological block bundles $\tilde{\xi}:Y \longrightarrow B\widetilde{TOP}(q)$ - see Levitt and Ranicki [1]. From the point of view of Ranicki [7] such a geometric Poincaré transversality structure on

$$\left\{\begin{array}{l}\text{a } (q-1)\text{-spherical fibration } \xi : Y \longrightarrow BG(q) \\ \text{an } n\text{-dimensional geometric Poincaré complex } M\end{array}\right. \text{ is an}$$

$$\underline{\Omega}^P\text{-orientation} \left\{\begin{array}{l} U_\xi \in \dot{H}^q(T(\xi);\underline{\Omega}^P) \\ [M] \in H_n(M;\underline{\Omega}^P)\end{array}\right. \text{ with image the canonical}$$

$$\underline{\Omega}^N\text{-orientation} \left\{\begin{array}{l} U_\xi \in \dot{H}^q(T(\xi);\underline{\Omega}^N) \\ [M] \in H_n(M;\underline{\Omega}^N)\end{array}\right. , \left\{\begin{array}{l} - \\ \text{with } \sigma^*([M]) = (1:M \longrightarrow M) \in \Omega^P_n(M)\end{array}\right. ,$$

where $\underline{\Omega}^P$ is the spectrum of oriented geometric Poincaré n-ads and $\underline{\Omega}^N = \underline{MSG}$ is the spectrum of oriented normal space n-ads (so that $\pi_*(\underline{\Omega}^P) = \Omega^P_*(pt.)$, $\pi_*(\underline{\Omega}^N) = \pi_*(\underline{MSG}) = \Omega^N_*(pt.)$ and there is defined a cofibration sequence of spectra

$$\underline{\mathbb{L}}_0 \longrightarrow \underline{\Omega}^P \longrightarrow \underline{\Omega}^N \longrightarrow \Sigma^{-1}\underline{\mathbb{L}}_0 \longrightarrow \dots\ ).$$

The geometric Poincaré assembly maps

$$\sigma^* : H_n(X;\underline{\Omega}^P) \longrightarrow \Omega^P_n(X) \qquad (n \geqslant 0)$$

are defined for any space X, and fit into a commutative braid of exact sequences

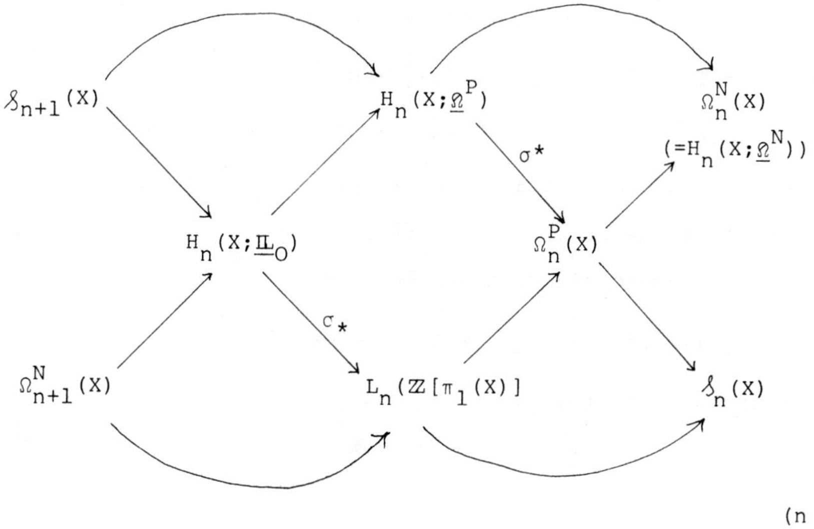

$(n \geqslant 5)$

.

There are evident relative and rel∂ versions of the geometric Poincaré splitting obstruction theory. In particular, given a codimension q CW pair (X,Y) and a map

$$(f,\partial f) : (M,\partial M) \longrightarrow X$$

from an n-dimensional $\begin{cases}\text{geometric Poincaré}\\ \text{(normal, geometric Poincaré)}\end{cases}$ pair $(M,\partial M)$ such that $\partial f:\partial M \longrightarrow X$ is Poincaré transverse at $Y \subset X$ there is defined a <u>rel∂ Poincaré splitting obstruction of f along $Y \subset X$</u>

$$\begin{cases} s_\partial^P(f,Y) \in LS_{n-q-1}(\Phi) \\ t_\partial^P(f,Y) \in LP_{n-q-1}(\Phi) \end{cases}$$

such that the following rel∂ version of Proposition 7.4.5 holds.

<u>Proposition 7.4.7</u> The rel∂ Poincaré splitting obstruction is such that $\begin{cases} s_\partial^P(f,Y) = 0 \in LS_{n-q-1}(\Phi) \\ t_\partial^P(f,Y) = 0 \in LP_{n-q-1}(\Phi)\end{cases}$ if (and for $n-q \geqslant 5$ only if)

there exists a relative $\begin{cases}\text{geometric Poincaré}\\ \text{(normal,geometric Poincaré)}\end{cases}$ bordism

$$(g;f\cup f',\partial_+g;\partial f\cup\partial f') : (W;M\cup -M',\partial_+W;\partial M\cup -\partial M') \longrightarrow X$$

between $(f,\partial f):(M,\partial M) \longrightarrow X$ and a map $(f',\partial f'):(M',\partial M') \longrightarrow X$ from an n-dimensional geometric Poincaré pair $(M',\partial M')$ which is Poincaré transverse at $Y \subset X$, and such that the (n-1)-dimensional geometric Poincaré bordism

$$(\partial_+g;\partial f,\partial f') : (\partial_+W;\partial M,\partial M') \longrightarrow X$$

is Poincaré transverse at $Y \subset X$.

[]

The manifold codimension q splitting obstruction theory described in §7.2 has a natural expression in terms of rel∂ geometric Poincaré splitting obstruction theory, as follows.

Proposition 7.4.8 Let $(X,Y,\widetilde{\xi})$ be an $(n,n-q)$-dimensional t-normal geometric Poincaré pair, and let $f:M\longrightarrow X$ be an $\left\{\begin{matrix} s- \\ t- \end{matrix}\right.$ triangulation of X which is topologically transverse at $Y\subset X$ with respect to $\widetilde{\xi}$. The manifold codimension q splitting obstruction along $Y\subset X$ of f is the rel∂ Poincaré splitting obstruction along $Y\subset X$ of the evident map

$$(g,f\cup 1)\ :\ (W,M\cup -X)\longrightarrow X$$

from the $(n+1)$-dimensional $\left\{\begin{matrix} \text{geometric Poincaré} \\ \text{(normal, geometric Poincaré)} \end{matrix}\right.$ pair $(W,M\cup -X)$ defined by the mapping cylinder W of $f:M\longrightarrow X$

$$\left\{\begin{array}{l} s(f,Y)\ =\ s^P_\partial(g,Y)\ \in LS_{n-q}(\Phi) \\ t(f,Y)\ =\ t^P_\partial(g,Y)\ \in LP_{n-q}(\Phi)\ . \end{array}\right.$$

[]

7.5 Algebraic Poincaré splitting

From now on we shall only be dealing with codimension q surgery theory for $q = 1,2$, since for $q \geqslant 3$ the obstruction groups are just the quadratic L-groups already dealt with in §1.

Let (X,Y) be a codimension q CW pair with

$$X = E(\xi) \cup_{S(\xi)} Z \ , \ \xi : Y \longrightarrow BG(q) = B\widetilde{TOP}(q) \quad (q = 1,2)$$

and let Φ be the corresponding pushout square of fundamental groupoids

$$\begin{array}{ccc} \pi_1(S(\xi)) & \longrightarrow & \pi_1(Z) \\ \downarrow & \Phi & \downarrow \\ \pi_1(E(\xi)) & \longrightarrow & \pi_1(X) \end{array} .$$

We wish to give an algebraic account of the codimension q surgery obstruction groups $\begin{cases} LS_*(\Phi) \\ LP_*(\Phi) \end{cases}$ defined geometrically in §7.2 to fit into the exact sequence

$$\begin{cases} \dots \longrightarrow L_{n-q+1}(\mathbb{Z}[\pi_1(Y)]) \xrightarrow{p\xi^!} L_{n+1}(\mathbb{Z}[\pi_1(Z)] \longrightarrow \mathbb{Z}[\pi_1(X)]) \\ \qquad \longrightarrow LS_{n-q}(\Phi) \longrightarrow L_{n-q}(\mathbb{Z}[\pi_1(Y)]) \longrightarrow \dots \\ \dots \longrightarrow L_{n-q+1}(\mathbb{Z}[\pi_1(Y)]) \xrightarrow{\partial p\xi^!} L_n(\mathbb{Z}[\pi_1(Z)]) \\ \qquad \longrightarrow LP_{n-q}(\Phi) \longrightarrow L_{n-q}(\mathbb{Z}[\pi_1(Y)]) \longrightarrow \dots \ . \end{cases}$$

In §7.6 ($q = 1$) and §7.8 ($q = 2$) we shall construct algebraic transfer functors

$$p\xi^! : (\text{n-dimensional quadratic (Poincaré) complexes over } \mathbb{Z}[\pi_1(Y)])$$
$$\longrightarrow ((n+q)\text{-dimensional quadratic (Poincaré) pairs over } \mathbb{Z}[\pi_1(Z)] \longrightarrow \mathbb{Z}[\pi_1(X)]) \qquad (n \in \mathbb{Z})$$

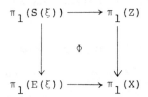

allowing $\left\{\begin{array}{l} LS_*(\Phi) \\ LP_*(\Phi) \end{array}\right.$ to be identified with the relative quadratic L-groups $\left\{\begin{array}{l} L_{*+q+1}(p\xi^!) \\ L_{*+q}(\partial p\xi^!) \end{array}\right.$ defined in the style of §2.3 using appropriately relative quadratic Poincaré cobordism. We shall now assume the existence of such algebraic transfers, leaving the details of the construction to §§7.6,7.8. Instead, we go beyond such a direct algebraic definition of $\left\{\begin{array}{l} LS_*(\Phi) \\ LP_*(\Phi) \end{array}\right.$ to a formulation in terms of algebraic Poincaré splittings with respect to Φ of quadratic Poincaré complexes over $\mathbb{Z}[\pi_1(X)]$, making use of an algebraic analogue of codimension q topological transversality, which we shall need in §§7.6,7.8 to recover the existing algebraic interpretations of the LS-groups in particular cases and also to obtain some new ones. However, the proof of this algebraic formulation will still involve some geometry. A purely algebraic proof will be obtained in Ranicki [11] - this will also apply to rings other than group rings, and also to symmetric L-theory.

In the first instance we extend to pushout squares such as Φ the notion of algebraic Poincaré splitting already developed in §6.1. We continue with the terminology that for $n \in \mathbb{Z}$ an n-dimensional quadratic Poincaré complex x over a ring with involution A is a closed object x of the category $\mathcal{L}_n(A)$ of §1.8. In particular, given an n-dimensional quadratic Poincaré complex y over $\mathbb{Z}[\pi_1(Y)]$ there is defined an (n+q)-dimensional quadratic Poincaré pair $(p\xi^! y, \partial p\xi^! y)$ over $\mathbb{Z}[\pi_1(Z)] \longrightarrow \mathbb{Z}[\pi_1(X)]$.

An <u>n-dimensional quadratic Poincaré splitting over Φ</u> (y,z) consists of

i) an $(n-q)$-dimensional quadratic Poincaré complex y over $\mathbb{Z}[\pi_1(Y)]$

ii) an n-dimensional quadratic Poincaré pair $(z,\partial p\xi^! y)$ over $\mathbb{Z}[\pi_1(Z)]$.

In keeping with the convention that we are considering simple geometric Poincaré complexes (unless specified otherwise) y and $(z,\partial p\xi^! y)$ are to be taken as simple - as usual, there are also free and projective versions of the theory, which are compared to each other in Proposition 7.5.2 below.

It follows from the above that the union

$$p\xi^! y \cup_{\mathbb{Z}[\pi_1(X)]\otimes_{\mathbb{Z}[\pi_1(Z)]}\partial p\xi^! y} \mathbb{Z}[\pi_1(X)]\otimes_{\mathbb{Z}[\pi_1(Z)]} z$$

is a (simple) n-dimensional quadratic Poincaré complex over $\mathbb{Z}[\pi_1(X)]$ which we shall abbreviate to $p\xi^! y \cup z$. The splitting is <u>contractible</u> if the union is contractible.

A <u>Poincaré splitting (with respect to Φ)</u> (y,z) of an n-dimensional quadratic Poincaré complex x over $\mathbb{Z}[\pi_1(X)]$ is an n-dimensional quadratic Poincaré splitting over Φ together with a simple homotopy equivalence

$$p\xi^! y \cup z \xrightarrow{\ \sim\ } x\ .$$

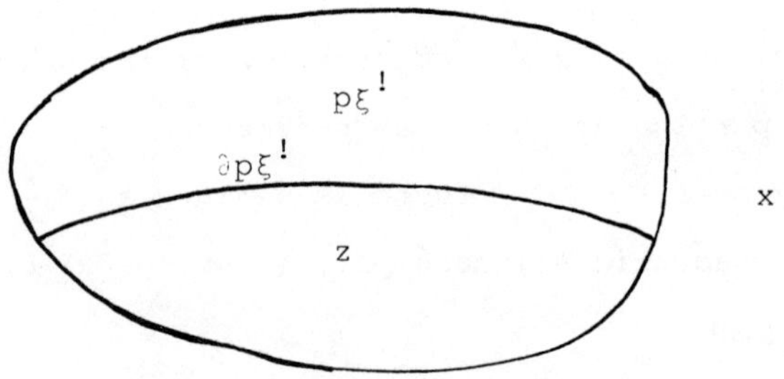

For example, if (X,Y) is an $(n,n-q)$-dimensional geometric Poincaré pair and $(f,b):M \longrightarrow X$ is a normal map from an n-dimensional geometric Poincaré complex M which is Poincaré transverse at $Y \subset X$ so that

$$(f,b) = (g,c)^! \cup (h,d) \; : \; M = E(\nu) \cup_{S(\nu)} P \longrightarrow X = E(\xi) \cup_{S(\xi)} Z$$

then the quadratic kernel $\sigma_*(f,b)$ over $\mathbb{Z}[\pi_1(X)]$ admits a Poincaré splitting, since

$$\sigma_*(f,b) = \sigma_*(g,c)^! \cup \sigma_*(h,d) \; .$$

The splitting is contractible if and only if f is a simple $\mathbb{Z}[\pi_1(X)]$-homology equivalence.

An <u>n-dimensional relative quadratic Poincaré splitting over Φ</u> $((y,\partial y),(z,\partial_+ z))$ consists of

i) an $(n-q)$-dimensional quadratic Poincaré pair $(y,\partial y)$ over $\mathbb{Z}[\pi_1(Y)]$

ii) an n-dimensional quadratic Poincaré triad $(z;\partial_+ z,\partial_+ p\xi^! y;\partial p\xi^! \partial y)$ over $\mathbb{Z}[\pi_1(Z)]$.

It follows that the union

$$(p\xi^! y \cup \mathbb{Z}[\pi_1(X)]\otimes_{\mathbb{Z}[\pi_1(Z)]} \partial_+ p\xi^! y \; \mathbb{Z}[\pi_1(X)]\otimes_{\mathbb{Z}[\pi_1(Z)]} z,$$
$$p\xi^! \partial y \cup \mathbb{Z}[\pi_1(X)]\otimes_{\mathbb{Z}[\pi_1(Z)]} \partial p\xi^! \partial y \; \mathbb{Z}[\pi_1(X)]\otimes_{\mathbb{Z}[\pi_1(Z)]} \partial_+ z)$$

is an n-dimensional quadratic Poincaré pair over $\mathbb{Z}[\pi_1(X)]$ which we shall abbreviate to $(p\xi^! y \cup z, p\xi^! \partial y \cup \partial_+ z)$.

A <u>Poincaré splitting (with respect to Φ)</u> $((y,\partial y),(z,\partial_+ z))$ of an n-dimensional quadratic Poincaré pair $(x,\partial x)$ is an n-dimensional relative quadratic Poincaré splitting over Φ together with a simple homotopy equivalence of pairs

$$(p\xi^! y \cup z, p\xi^! \partial y \cup \partial_+ z) \xrightarrow{\;\simeq\;} (x,\partial x) .$$

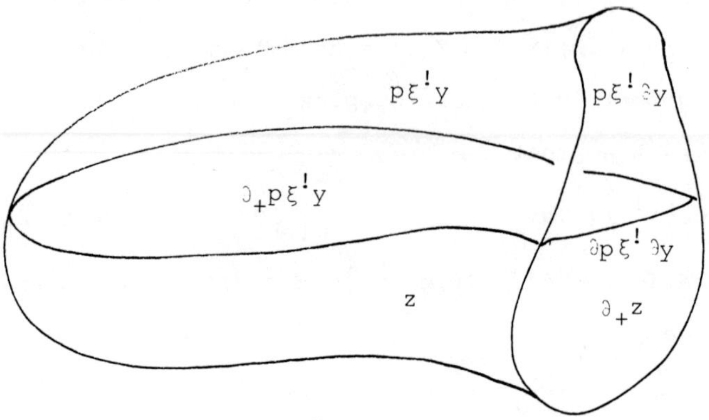

A <u>normal splitting (with respect to Φ)</u> $((y,\partial y),(z,\partial_+z))$ of an n-dimensional quadratic Poincaré complex x over $\mathbb{Z}[\pi_1(X)]$ is a Poincaré splitting of the n-dimensional quadratic Poincaré pair $(x,\partial x)$ over $\mathbb{Z}[\pi_1(X)]$. Note that $(\partial y,\partial_+z)$ is then a contractible Poincaré splitting (of ∂x).

<u>Proposition 7.5.1</u> Let (X,Y) be a codimension q CW pair, and let Φ be the associated pushout square of fundamental groupoids.

i) Every quadratic Poincaré complex x over $\mathbb{Z}[\pi_1(X)]$ cobordant to one which admits a normal splitting $((y,\partial y),(z,\partial_+z))$

ii) The codimension q surgery obstruction group $\begin{cases} LS_{n-q}(\Phi) \\ LP_{n-q}(\Phi) \end{cases}$ $(n \in \mathbb{Z})$

is the cobordism group of $\begin{cases} \text{contractible} \\ - \end{cases}$ n-dimensional

quadratic Poincaré splittings over Φ. The maps appearing in the exact sequence

$$\ldots \longrightarrow LS_{n-q}(\Phi) \longrightarrow LP_{n-q}(\Phi) \longrightarrow L_n(\mathbb{Z}[\pi_1(X)]) \longrightarrow LS_{n-q-1}(\Phi) \longrightarrow \ldots$$

are given by

$$LS_{n-q}(\Phi) \longrightarrow LP_{n-q}(\Phi) \;;\; (y,z) \longmapsto (y,z)$$

$$LP_{n-q}(\Phi) \longrightarrow L_n(\mathbb{Z}[\pi_1(X)]) \;;\; (y,z) \longmapsto p\xi^! y \cup z$$

$$L_n(\mathbb{Z}[\pi_1(X)]) \longrightarrow LS_{n-q-1}(\Phi) \;;\; x \longmapsto (\partial y, \partial_+ z)$$

(if $((y,\partial y),(z,\partial_+ z))$ is a normal splitting of x) .

In particular, the image of an element $x \in L_n(\mathbb{Z}[\pi_1(X)])$ in $LS_{n-q-1}(\Phi)$ is the obstruction to x having a Poincaré splitting.

iii) If $f:M \longrightarrow X$ is map from an n-dimensional $\left\{\begin{array}{l}\text{geometric Poincaré complex}\\ \text{normal space}\end{array}\right.$ M which is normal transverse at $Y \subset X$ with

$$f = g^! \cup h : M = E(\nu) \cup_{S(\nu)} P \longrightarrow X = E(\xi) \cup_{S(\xi)} Z \;,$$

$$\nu : N = f^{-1}(Y) \xrightarrow{g = f|} Y \xrightarrow{\xi} BG(q) \;,$$

$$h = f| : P = f^{-1}(Z) \longrightarrow Z$$

then the codimension q Poincaré splitting obstruction of f along $Y \subset X$ $\left\{\begin{array}{l} s^P(f,Y) \in LS_{n-q-1}(\Phi) \\ t^P(f,Y) \in LP_{n-q-1}(\Phi)\end{array}\right.$ is represented by the $\left\{\begin{array}{l}\text{contractible}\\ -\end{array}\right.$ (n-1)-dimensional quadratic Poincaré splitting over Φ

$$\sigma_*(M) = \sigma_*(N)^! \cup \sigma_*(P,S(\nu))$$

of the (n-1)-dimensional quadratic Poincaré complex over $\mathbb{Z}[\pi_1(X)]$ of M

$$\sigma_*(M) = (\Omega C([M] \cap - : C(\tilde{M})^{n-*} \longrightarrow C(\tilde{M})), \Psi)$$

with

$$\sigma_*(N) = (\Omega C([N] \cap - : C(\tilde{N})^{n-q-*} \longrightarrow C(\tilde{N})), \psi)$$

the (n-q-1)-dimensional quadratic Poincaré complex over $\mathbb{Z}[\pi_1(Y)]$ of the (n-q)-dimensional normal space N and

$$\sigma_*(P,S(\nu)) = (\Omega C([S(\nu)]\cap -:C(\widetilde{S(\nu)})^{n-1-*} \longrightarrow C(\widetilde{S(\nu)}))$$
$$\longrightarrow \Omega C([P]\cap -:C(\widetilde{P},\widetilde{S(\nu)})^{n-*} \longrightarrow C(\widetilde{P})),(\delta\psi,\partial p\xi^!\psi))$$

the (n-1)-dimensional quadratic Poincaré pair over $\mathbb{Z}[\pi_1(Z)]$ of the n-dimensional normal pair $(P,S(\nu))$.

iv) If (X,Y) is an $(n,n-q)$-dimensional geometric Poincaré pair and $(f,b):M \longrightarrow X$ is an $\begin{cases} s- \\ t- \end{cases}$ triangulation of X which is topologically transverse at $Y \subset X$, with

$$(f,b) = (g,c)^! \cup (h,d) : M = E(\nu) \cup_{S(\nu)} P \longrightarrow X = E(\xi) \cup_{S(\xi)} Z$$

$$\nu : N = f^{-1}(Y) \xrightarrow{g = f|} Y \xrightarrow{\xi} BG(q) = \widetilde{BTOP}(q) \quad (q = 1 \text{ or } 2)$$

$$h = f| : P = f^{-1}(Z) \longrightarrow Z ,$$

then the codimension q manifold splitting obstruction of f along $Y \subset X$ $\begin{cases} s(f,Y) \in LS_{n-q}(\Phi) \\ t(f,Y) \in LP_{n-q}(\Phi) \end{cases}$ is represented by the $\begin{cases} \text{contractible} \\ - \end{cases}$ n-dimensional quadratic Poincaré splitting over Φ

$$\sigma_*(f,b) = \sigma_*(g,c)^! \cup \sigma_*(h,d)$$

of the n-dimensional quadratic kernel of (f,b) over $\mathbb{Z}[\pi_1(X)]$, with $\sigma_*(g,c)$ the $(n-q)$-dimensional quadratic kernel of $(g,c):N \longrightarrow Y$ over $\mathbb{Z}[\pi_1(Y)]$ and $\sigma_*(h,d)$ the n-dimensional quadratic kernel of $(h,d):(P,S(\nu)) \longrightarrow (Z,S(\xi))$ over $\mathbb{Z}[\pi_1(Z)]$.

Proof: In the first instance note that ii), iii) and iv) are immediate consequences of i) and its relative version.

To prove i) use the realization theorem of Wall [4] to identify $x \in L_n(\mathbb{Z}[\pi_1(X)])$ with the rel∂ surgery obstruction $\sigma_*(f,b)$ of a t-triangulation of an n-dimensional manifold with boundary $(X_1,\partial X_1)$ $(n \geqslant 5)$

$$((f,b),(\partial f,\partial b)) \ : \ (M,\partial M) \longrightarrow (X_1,\partial X_1)$$

such that $(\partial f,\partial b):\partial M \longrightarrow \partial X_1$ is an s-triangulation of ∂X_1, with respect to a reference map $(r,\partial r):(X_1,\partial X_1) \longrightarrow X$. Make $(r,\partial r)$ topologically transverse at $Y \subset X$, so that

$$(X_1,\partial X_1) = (E(\xi_1) \cup_{S(\xi_1)} Z_1, E(\partial \xi_1) \cup_{S(\partial\xi_1)} \partial_+ Z_1)$$

with

$$(\xi_1,\partial\xi_1) : (Y_1,\partial Y_1) = (r^{-1}(Y),\partial r^{-1}(Y)) \xrightarrow{(r,\partial r)|} Y$$
$$\xrightarrow{\xi} BG(q) = B\widetilde{TOP}(q) \quad (q = 1 \text{ or } 2) \ ,$$
$$(Z_1,\partial_+ Z_1) = (r^{-1}(Z),\partial r^{-1}(Z)) \ .$$

Also, make $(f,\partial f)$ topologically transverse at $(Y_1,\partial Y_1) \subset (X_1,\partial X_1)$, so that

$$((f,b),(\partial f,\partial b)) = ((g,c),(\partial g,\partial c))^! \cup ((h,d),(\partial_+ h,\partial_+ d))$$
$$: (M,\partial M) = (E(\nu) \cup_{S(\nu)} P, E(\partial\nu) \cup_{S(\partial\nu)} \partial_+ P) \longrightarrow (X_1,\partial X_1)$$

with

$$((g,c),(\partial g,\partial c)) = ((f,b),(\partial f,\partial b))|$$
$$: (N,\partial N) = (f^{-1}(Y_1),\partial f^{-1}(\partial Y_1)) \longrightarrow (Y_1,\partial Y_1)$$
$$(\nu,\partial\nu) \ : \ (N,\partial N) \xrightarrow{(g,\partial g)} (Y_1,\partial Y_1) \xrightarrow{(\xi_1,\partial\xi_1)} BG(q) = B\widetilde{TOP}(q)$$
$$((h,d),(\partial_+ h,\partial_+ d)) = ((f,b),(\partial f,\partial b))|$$
$$: (P,\partial_+ P) = (f^{-1}(Z_1),\partial f^{-1}(\partial_+ Z_1)) \longrightarrow (Z,\partial_+ Z) \ .$$

This decomposition of $((f,b),(\partial f,\partial b))$ determines an n-dimensional relative quadratic Poincaré splitting over Φ

$$(\sigma_*((g,c),(\partial g,\partial c)),\sigma_*((h,d),(\partial_+ h,\partial_+ d)))$$

of the n-dimensional quadratic Poincaré pair $(\sigma_*(f,b),\partial\sigma_*(f,b))$ over $\mathbb{Z}[\pi_1(X)]$, i.e. a normal splitting of $x = \sigma_*(f,b)$. Note that in general the s-triangulation $\partial f:\partial M \xrightarrow{\sim} \partial X_1$ is not split along $\partial Y_1 \subset \partial X_1$; in fact, the splitting obstruction

$s(\partial f,\partial Y_1) \in LS_{n-q-1}(\Phi)$ is the image of $x = \sigma_*(f,b) \in L_n(\mathbb{Z}[\pi_1(X)])$ under the canonical map.

Alternatively, it is possible to prove i) using the normal space transversality of Quinn [3]. Consider $x \in L_n(\mathbb{Z}[\pi_1(X)])$ as the quadratic signature $\sigma_*(W,\partial W)$ of an $(n+1)$-dimensional (normal, geometric Poincaré) pair $(W,\partial W)$ equipped with a reference map $(r,\partial r):(W,\partial W) \longrightarrow X$. Making $(r,\partial r)$ normal transverse at $Y \subset X$ note that the constructions of Proposition 7.4.1 translate the consequent normal splitting of $(W,\partial W)$ into a normal splitting of $x = \sigma_*(W,\partial W)$. In general, $\partial r:\partial W \longrightarrow X$ is not Poincaré split along $Y \subset X$; in fact, the splitting obstruction $s^P(\partial r,Y) \in LS_{n-q-1}(\Phi)$ is the image of $x = \sigma_*(W,\partial W) \in L_n(\mathbb{Z}[\pi_1(X)])$ under the canonical map.

(The two methods of proof of i) are related to each other by the mapping cylinder construction, cf. Proposition 7.4.8).

[]

An algebraic proof of Proposition 7.5.1 i) requires an L-theoretic version of the linearization trick of Higman [1] - see the introduction to §7.6 below for a brief survey of the corresponding algebraic K-theory for $q = 1$.

There are evident analogues of Proposition 7.5.1 for the versions of the groups $\begin{cases} LS_*(\Phi) \equiv LS^s_*(\Phi) \\ LP_*(\Phi) \equiv LP^s_*(\Phi) \end{cases}$ appropriate to the free and projective quadratic L-theory, which we denote by $\begin{cases} LS^h_*(\Phi) \\ LP^h_*(\Phi) \end{cases}$ and $\begin{cases} LS^p_*(\Phi) \\ LP^p_*(\Phi) \end{cases}$.

<u>Proposition 7.5.2</u> Let (X,Y) be a codimension q CW pair, and let Φ be the associated pushout square of fundamental groupoids.

The simple groups $\left\{\begin{array}{l} LS_*(\Phi) \\ LP_*(\Phi) \end{array}\right.$ are related to the free groups $\left\{\begin{array}{l} LS^h_*(\Phi) \\ LP^h_*(\Phi) \end{array}\right.$ by a commutative diagram with exact rows and columns

$$\begin{array}{ccccccccccc}
 & \vdots & & \vdots & & \vdots & & \vdots & \\
 & \downarrow & & \downarrow & & \downarrow & & \downarrow & \\
\cdots \longrightarrow & L_{n+1}(\mathbb{Z}[\pi_1(X)]) & \longrightarrow & LP_{n-q}(\Phi) & \longrightarrow & LS_{n-q}(\Phi) & \longrightarrow & L_n(\mathbb{Z}[\pi_1(X)]) & \longrightarrow \cdots \\
 & \downarrow & & \downarrow & & \downarrow & & \downarrow & \\
\cdots \longrightarrow & L^h_{n+1}(\mathbb{Z}[\pi_1(X)]) & \longrightarrow & LP^h_{n-q}(\Phi) & \longrightarrow & LS^h_{n-q}(\Phi) & \longrightarrow & L^h_n(\mathbb{Z}[\pi_1(X)]) & \longrightarrow \cdots \\
 & \downarrow & & \downarrow & & \downarrow & & \downarrow & \\
\cdots \longrightarrow & \hat{H}^{n+1}(\mathbb{Z}_2;Wh(\pi_1(X))) & \longrightarrow & WhP_{n-q}(\Phi) & \longrightarrow & WhS_{n-q}(\Phi) & \longrightarrow & \hat{H}^n(\mathbb{Z}_2;Wh(\pi_1(X))) & \longrightarrow \cdots \\
 & \downarrow & & \downarrow & & \downarrow & & \downarrow & \\
\cdots \longrightarrow & L_n(\mathbb{Z}[\pi_1(X)]) & \longrightarrow & LP_{n-q-1}(\Phi) & \longrightarrow & LS_{n-q-1}(\Phi) & \longrightarrow & L_{n-1}(\mathbb{Z}[\pi_1(X)]) & \longrightarrow \cdots \\
 & \downarrow & & \downarrow & & \downarrow & & \downarrow & \\
 & \vdots & & \vdots & & \vdots & & \vdots &
\end{array}$$

with $\left\{\begin{array}{l} WhS_*(\Phi) \\ WhP_*(\Phi) \end{array}\right.$ the relative groups appearing in the exact sequence

$$\left\{\begin{array}{l}
\cdots \longrightarrow \hat{H}^{n+1}(\mathbb{Z}_2;Wh(\pi_1(Z)) \longrightarrow Wh(\pi_1(X))) \longrightarrow WhS_{n-q}(\Phi) \\
\qquad \longrightarrow \hat{H}^{n-q}(\mathbb{Z}_2;Wh(\pi_1(Y))) \xrightarrow{p\xi^!} \hat{H}^n(\mathbb{Z}_2;Wh(\pi_1(Z)) \longrightarrow Wh(\pi_1(X))) \longrightarrow \cdots \\
\cdots \longrightarrow \hat{H}^n(\mathbb{Z}_2;Wh(\pi_1(Z))) \longrightarrow WhP_{n-q}(\Phi) \longrightarrow \hat{H}^{n-q}(\mathbb{Z}_2;Wh(\pi_1(Y))) \\
\qquad\qquad \xrightarrow{\partial p\xi^!} \hat{H}^{n-1}(\mathbb{Z}_2;Wh(\pi_1(Z))) \longrightarrow \cdots .
\end{array}\right.$$

Similarly for the relation between the free groups $\left\{\begin{array}{l} LS^h_*(\Phi) \\ LP^h_*(\Phi) \end{array}\right.$ and the projective groups $\left\{\begin{array}{l} LS^p_*(\Phi) \\ LP^p_*(\Phi) \end{array}\right.$, with the Whitehead groups

$Wh(\pi)$ replaced by the reduced projective class groups $\tilde{K}_0(\mathbb{Z}[\pi])$.

[]

Furthermore, the splitting theorems for the quadratic L-groups of Shaneson [1], Novikov [1] and Ranicki [2]

$$L_n(\mathbb{Z}[\pi\times\mathbb{Z}]) = L_n(\mathbb{Z}[\pi])\oplus L^h_{n-1}(\mathbb{Z}[\pi]) \qquad (n\in\mathbb{Z})$$
$$L^h_n(\mathbb{Z}[\pi\times\mathbb{Z}]) = L^h_n(\mathbb{Z}[\pi])\oplus L^p_{n-1}(\mathbb{Z}[\pi])$$

extend to the $\left\{\begin{array}{l}LS\text{-}\\ LP\text{-}\end{array}\right.$ groups

$$\left\{\begin{array}{l}LS_n(\Phi\times\mathbb{Z}) = LS_n(\Phi)\oplus LS^h_{n-1}(\Phi)\\ LP_n(\Phi\times\mathbb{Z}) = LP_n(\Phi)\oplus LP^h_{n-1}(\Phi)\end{array}\right. \qquad (n\in\mathbb{Z})$$
$$\left\{\begin{array}{l}LS^h_n(\Phi\times\mathbb{Z}) = LS^h_n(\Phi)\oplus LS^p_{n-1}(\Phi)\\ LP^h_n(\Phi\times\mathbb{Z}) = LP^h_n(\Phi)\oplus LP^p_{n-1}(\Phi)\end{array}\right.$$

with Φ the fundamental groupoid pushout square of a codimension q CW pair (X,Y) and $\Phi\times\mathbb{Z}$ the pushout square of $(X\times S^1, Y\times S^1)$.

The codimension q splitting obstruction theory for t-triangulations (the LP-theory) was developed as a tool for understanding the obstruction theory for s-triangulations (the LS-theory) - from now on we shall be mainly concerned with the latter.

Let (M,N) be an $(n,n-q)$-dimensional manifold pair $(q\geqslant 1)$. Ambient surgery on N inside M is the operation

$$(M,N)\longmapsto(M,N')$$

determined by an embedding

$$e : (D^{r+1},S^r)\times D^{n-q-r}\hookrightarrow (M,N)$$

such that $e^{-1}(N) = S^r\times D^{n-q-r}$, with

$$N' = \overline{N \setminus e(S^r \times D^{n-q-r})} \cup D^{r+1} \times S^{n-q-r-1} \subset M$$

obtained from N by an ordinary surgery. The trace of the surgery on N embeds in $M \times I$ as a codimension q submanifold, defining an ambient cobordism inside $M \times I$ between $N \subset M \times \{0\}$ and $N' \subset M \times \{1\}$

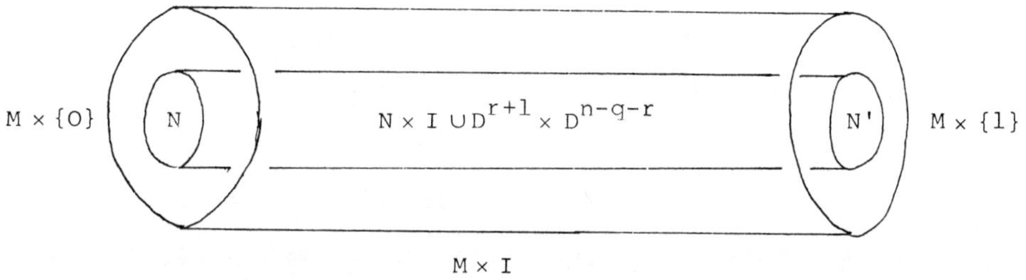

Conversely, every ambient cobordism inside $M \times I$ can be broken up into a finite sequence of ambient surgeries.

A formally (n,n-q)-dimensional normal pair (X,Y) is a codimension q CW pair such that

i) X is a formally n-dimensional normal space (in the sense of §7.3) with normal structure

$$(\nu_X : X \longrightarrow BG(k), \rho_X : S^{n+k} \longrightarrow T(\nu_X))$$

ii) Y is a formally (n-q)-dimensional normal space with normal structure

$$(\nu_Y = \xi \oplus \nu_X|_Y : Y \longrightarrow BG(q+k),$$

$$\rho_Y : S^{n+k} \xrightarrow{\rho_X} T(\nu_X) \longrightarrow T(\nu_X)/T(\nu_X|_Z) = T(\nu_Y)) .$$

Such a pair is t-normal if there is given a t-triangulation $\tilde{\xi} : Y \longrightarrow B\widetilde{TOP}(q)$ of $\xi : Y \longrightarrow BG(q)$. In particular, a formally (n,n-q)-dimensional (t-normal) geometric Poincaré pair (X,Y) is a formally (n,n-q)-dimensional (t-normal) normal pair.

Let (X,Y) be a formally $(n,n-q)$-dimensional t-normal pair. A <u>formally $(n,n-q)$-dimensional topological normal map</u>

$$(f,b) : (M,N) \longrightarrow (X,Y)$$

is a formally n-dimensional topological normal map $(f,b):M \longrightarrow X$ (in the sense of §7.3) which is topologically transverse at $Y \subset X$, so that the restriction

$$(f,b)| = (g,c) : N = f^{-1}(Y) \longrightarrow Y$$

is a formally $(n-q)$-dimensional topological normal map and the restriction

$$(f,b)| = (h,d) : (P,S(\nu)) = f^{-1}(Z,S(\xi)) \longrightarrow (Z,S(\xi))$$

is a formally n-dimensional topological normal map of pairs such that

$$(f,b) = (g,c)^{!} \cup (h,d)$$
$$: M = E(\nu) \cup_{S(\nu)} P \longrightarrow X = E(\xi) \cup_{S(\xi)} Z$$

with

$$\nu = \nu_{N \subset M} : N \xrightarrow{\ g\ } M \xrightarrow{\ \tilde{\xi}\ } \widetilde{BTOP}(q) \ .$$

The notion of ambient surgery on codimension q submanifolds carries over in the obvious way to such topological normal maps. Given a formally $(n,n-q)$-dimensional topological normal map

$$(f,b) : (M,N) \longrightarrow (X,Y)$$

with restriction

$$(f,b)| = (g,c) : N \longrightarrow Y$$

define an <u>ambient surgery on (g,c) inside (f,b)</u> to be a surgery on (g,c) such that the trace normal bordism

$$(G,C) : (N \times I \cup D^{r+1} \times D^{n-q-r};N,N') \longrightarrow Y \times (I;0,1)$$

is a restriction of the normal map of $(n+1)$-dimensional triads

$$(f,b) \times 1 : M \times (I;0,1) \longrightarrow X \times (I;0,1) \ .$$

There is a corresponding algebraic notion of ambient surgery on a pair of the type

(a formally $(n-q)$-dimensional quadratic complex over $\mathbb{Z}[\pi_1(Y)]$ (C,ψ), a formally n-dimensional quadratic pair over $\mathbb{Z}[\pi_1(Z)]$ $(\partial p\xi^! C \longrightarrow D, (\delta\psi, \partial p\xi^!\psi))$ with boundary $\partial p\xi^!(C,\psi)$)

which preserves the (homotopy type of) the union formally n-dimensional quadratic complex over $\mathbb{Z}[\pi_1(X)]$

$$p\xi^!(C,\psi) \cup \mathbb{Z}[\pi_1(X)] \otimes_{\mathbb{Z}[\pi_1(Z)]} (\partial p\xi^! C \longrightarrow D, (\delta\psi, \partial p\xi^!\psi)) \ .$$

Algebraic ambient surgery will be developed further in Ranicki [11]. By analogy with Proposition 7.3.5 (the case $Y = \emptyset$) it is possible to describe the algebraic effect of geometric ambient surgery on the quadratic kernels defined using the spectral quadratic construction:

Proposition 7.5.3 Given a formally $(n,n-q)$-dimensional topological normal map

$$(f,b) : (M,N) \longrightarrow (X,Y)$$

there is defined a quadratic kernel pair

$\sigma_*((g,c),(h,d))$ = (the formally $(n-q)$-dimensional quadratic kernel complex over $\mathbb{Z}[\pi_1(Y)]$ $\sigma_*(g,c) = (C(g^!),\psi)$ of the restriction $(g,c) = (f,b)| : N \longrightarrow Y$,
the formally n-dimensional quadratic kernel pair over $\mathbb{Z}[\pi_1(Z)]$ $\sigma_*(h,d) = (\partial p\xi^! C(g^!) \longrightarrow C(h^!), (\delta\psi, \partial p\xi^!\psi))$ of the restriction $(h,d) = (f,b)| : (P,S(\nu)) \longrightarrow (Z,S(\xi))$ with boundary $\partial p\xi^!(C,\psi)$)

with union

$$p\xi^!\sigma_*(g,c) \cup \sigma_*(h,d) = (C(f^!), p\xi^!\psi \cup \delta\psi) = \sigma_*(f,b)$$

the formally n-dimensional quadratic kernel complex over $\mathbb{Z}[\pi_1(X)]$

of $(f,b):M \longrightarrow X$.

Geometric ambient surgery on (g,c) inside (f,b) has the algebraic effect of ambient surgery on the quadratic kernel $\sigma_*((g,c),(h,d))$, i.e. of algebraic surgery on $\sigma_*(g,c)$ inside $\sigma_*(f,b)$.

[]

In particular, if (X,Y) is an $(n,n-q)$-dimensional t-normal geometric Poincaré pair and $(f,b):(M,N) \longrightarrow (X,Y)$ is an $(n,n-q)$-dimensional normal map such that $f:M \longrightarrow X$ is an s-triangulation of X then the quadratic kernel $\sigma_*((g,c),(h,d))$ is a contractible n-dimensional quadratic Poincaré splitting over the associated pushout square of fundamental groupoids Φ. By Proposition 7.5.1 iv) the splitting obstruction of f along $Y \subset X$ is the cobordism class of this kernel

$$s(f,Y) = \sigma_*((g,c),(h,d)) \in LS_{n-q}(\Phi)$$

(with $s(f,Y) = \sigma_*(g,c) \in LS_{n-q}(\Phi) = L_{n-q}(\mathbb{Z}[\pi_1(Y)])$ for $q \geqslant 3$), which is thus the obstruction to making f concordant to an s-triangulation of (X,Y) by a finite sequence of ambient surgeries on (g,c) inside (f,b).

For $q = 1,2$ ambient surgery on a codimension q submanifold can be related to ambient surgery on a normal map, as follows.

Let $(M,\partial M)$ be an n-dimensional with boundary (which may be empty), and let $N \subset M$ be a codimension q submanifold such that $N \cap \partial M = \emptyset$ with

$$\nu_{N \subset M} = \nu : N \longrightarrow B\widetilde{TOP}(q) = BG(q) \quad , \quad M = E(\nu) \cup_{S(\nu)} P \ ,$$

$$P = \overline{M \setminus E(\nu)} \quad .$$

Let $g:N \longrightarrow M$ be the inclusion, and assume given a factorization of the orientation map of N through $\pi_1(M)$

$$w(N) \ : \ \pi_1(N) \xrightarrow{g_*} \pi_1(M) \xrightarrow{w(N)} \mathbb{Z}_2$$

(which is automatic for $q=1$) so that $\nu:N \longrightarrow BG(q)$ has orientation map

$$w_1(\nu) \ : \ \pi_1(N) \xrightarrow{g_*} \pi_1(M) \xrightarrow{w(M)w(N)} \mathbb{Z}_2 \ .$$

The Poincaré dual of $g_*[N] \in H_{n-q}(M,w(N))$ is an element $\xi \in H^q(M,\partial M,w(M)w(N))$ classifying a $(q-1)$-spherical fibration over M

$$\xi \ : \ M \longrightarrow BG(q)$$

with a section of $\xi|_{\partial M}$ such that

i) $\xi|_N = \nu \ : \ N \longrightarrow BG(q)$

ii) $\xi|_P = \begin{cases} \varepsilon^1 \\ \omega \oplus \varepsilon^1 \end{cases} : P \longrightarrow BG(q) \text{ if } \begin{cases} q=1 \\ q=2 \end{cases}$

with $\omega:P \longrightarrow BG(1)$ the S^0-fibration (= line bundle) over P classified by

$$\omega \ : \ \pi_1(P) \longrightarrow \pi_1(M) \xrightarrow{w(M)w(N)} \mathbb{Z}_2 \ .$$

The inclusion of the zero section $M \subset E(\xi)$ of ξ can be perturbed to define a formally n-dimensional topological normal map of pairs

$$(f,b) \ : \ (M,\partial M) \longrightarrow (E(\xi),S(\xi))$$

such that $f:M \longrightarrow E(\xi)$ is a simple homotopy equivalence, which is topologically transverse at $M \subset E(\xi)$ with $f^{-1}(M) = N \subset M$. The restrictions of (f,b) define a formally $(n-q)$-dimensional topological normal map

$$(f,b)| \ = \ (g,c) \ : \ N \longrightarrow M$$

with $c:\nu_N \longrightarrow \nu_M \oplus \xi$ and a formally n-dimensional normal map of triads

$$(f,b)| = (h,d) : (P;S(\nu),\partial M) \longrightarrow S(\xi) \times (I;0,1) .$$

Proposition 7.5.4 Ambient surgery on N inside M corresponds to ambient surgery on $(g,c):N \longrightarrow M$ inside $(f,b):(M,\partial M) \longrightarrow (E(\xi),S(\xi))$. The algebraic effect is an ambient surgery on the quadratic kernel pair

(the formally (n-q)-dimensional quadratic complex over $\mathbb{Z}[\pi_1(M)]$ $\sigma_*(g,c) = (C(g^!),\psi)$, the formally n-dimensional quadratic triad over $\mathbb{Z}[\pi_1(S(\xi))]$ $\sigma_*(h,d)$ with boundary components $\partial\xi^!\sigma_*(g,c)$ and $\sigma_*(\partial f,\partial b)$)

preserving the union formally n-dimensional quadratic pair over $\mathbb{Z}[\pi_1(M)]$

$$\xi^!\sigma_*(g,c) \cup \mathbb{Z}[\pi_1(M)]\otimes_{\mathbb{Z}[\pi_1(S(\xi))]}\sigma_*(h,d)$$
$$= \sigma_*(f,b) = (0,\mathbb{Z}[\pi_1(M)]\otimes_{\mathbb{Z}[\pi_1(S(\xi))]}\sigma_*(\partial f,\partial b)) .$$

[]

A codimension q spine of an n-dimensional manifold with boundary $(M,\partial M)$ is a codimension q submanifold $N \subset M$ such that the inclusion defines an s-triangulation

$$g : N \xrightarrow{\simeq} M .$$

The problem of finding a codimension q spine is a typical application of ambient surgery obstruction theory. As already noted in the remark following Proposition 7.2.5 for $q \geqslant 3$ an n-dimensional manifold with boundary $(M,\partial M)$ admits a codimension q spine if and only if M is an s-triangulable (n-q)-dimensional geometric Poincaré complex, at least if $n-q \geqslant 5$.

Let $(M,\partial M)$ be an n-dimensional manifold with boundary such that M is an (n-q)-dimensional geometric Poincaré complex ($q = 1$ or 2), and let $N \subset M$ be a codimension q submanifold representing the fundamental class $[M] \in H_{n-q}(M,w)$.

For $\begin{cases} q = 1 \\ q = 2 \end{cases}$ the corresponding n-dimensional topological normal map

$$(f,b) : (M,\partial M) \longrightarrow (E(\xi),S(\xi))$$

is a simple $\begin{cases} \text{homotopy} \\ \mathbb{Z}[\pi_1(M)]\text{-homology} \end{cases}$ equivalence of pairs (assuming $\pi_1(\partial M) = \pi_1(S(\xi)))$ such that the restriction

$$(f,b)| = (g,c) : N \longrightarrow M$$

is an (n-q)-dimensional topological normal map. The quadratic kernel pair $(\sigma_*(g,c),\sigma_*(h,d)/\sigma_*(\partial f,\partial b))$ consists of an (n-q)-dimensional quadratic Poincaré complex over $\mathbb{Z}[\pi_1(M)]$ $\sigma_*(g,c)$ and the n-dimensional quadratic $\begin{cases} \text{Poincaré} \\ \mathbb{Z}[\pi_1(M)]\text{-Poincaré} \end{cases}$ pair over $\mathbb{Z}[\pi_1(S(\xi))]$ $\sigma_*(h,d)/\sigma_*(\partial f,\partial b)$ with boundary $\partial\xi^!\sigma_*(g,c)$ obtained from $\sigma_*(h,d)$ by collapsing $\sigma_*(\partial f,\partial b)$. The union

$$\begin{aligned} &\xi^!\sigma_*(g,c) \cup \mathbb{Z}[\pi_1(M)] \otimes_{\mathbb{Z}[\pi_1(S(\xi))]} (\sigma_*(h,d)/\sigma_*(\partial f,\partial b)) \\ &\qquad = \sigma_*(f,b)/(\mathbb{Z}[\pi_1(M)] \otimes_{\mathbb{Z}[\pi_1(S(\xi))]} \sigma_*(\partial f,\partial b)) \end{aligned}$$

is a contractible n-dimensional quadratic Poincaré complex over $\mathbb{Z}[\pi_1(M)]$. Let Φ be the pushout square of fundamental groups associated to the codimension q CW pair $(E(\xi),M)$

$$\begin{array}{ccc} \pi_1(S(\xi)) & \longrightarrow & \pi_1(S(\xi)) \\ \downarrow & \Phi & \downarrow \\ \pi_1(M) & \longrightarrow & \pi_1(M) \end{array} .$$

For $q = 1$ Proposition 7.5.1 iv) gives the splitting obstruction rel ∂M of f along $M \subset E(\xi)$ to be the element

$$s(f,Y) = (\sigma_*(g,c), \sigma_*(h,d)/\sigma_*(\partial f,\partial b)) \in LS_{n-1}(\Phi) ,$$

so that by Propositions 7.2.4, 7.5.4 $s(f,Y) = 0$ if (and for $n \geqslant 6$ only if) $(M,\partial M)$ admits a codimension 1 spine $N \subset M$.
The map $LS_{n-1}(\Phi) \longrightarrow \mathcal{S}_{n-1}(M)$ appearing in the exact sequence of Proposition 7.2.6 i)

$$\ldots \longrightarrow LS_{n-1}(\Phi) \longrightarrow \mathcal{S}_{n-1}(M) \xrightarrow{\xi^!} \mathcal{S}_n(E(\xi),S(\xi)) \longrightarrow LS_{n-2}(\Phi) \longrightarrow \ldots$$

sends $s(f,Y)$ to $s(M) \in \mathcal{S}_{n-1}(M)$. For $q = 2$ it is necessary to use the algebraic theory of codimension 2 surgery developed in §7.8 below (generalizing the original theory of Cappell and Shaneson [1]) in which only the homology type of the complement of the codimension 2 submanifold is taken into account, not the homotopy type. In terms of that theory Proposition 7.8.6 gives for $q = 2$ the weak splitting obstruction rel ∂M of f along $M \subset E(\xi)$ to be the element

$$ws(f,Y) = (\sigma_*(g,c), \sigma_*(h,d)/\sigma_*(\partial f,\partial b)) \in \Gamma S_{n-2}(\Phi) ,$$

so that by Propositions 7.8.2 i), 7.5.4 $ws(f,Y) = 0$ if (and for $n \geqslant 7$ only if) $(M,\partial M)$ admits a codimension 2 spine $N \subset M$.
The map $\Gamma S_{n-2}(\Phi) \longrightarrow \mathcal{S}_{n-2}(M)$ appearing in the exact sequence of Proposition 7.8.3 i)

$$\ldots \longrightarrow \Gamma S_{n-2}(\Phi) \longrightarrow \mathcal{S}_{n-2}(M) \xrightarrow{\xi^!} \mathcal{S}_n(E(\xi),S(\xi);\mathbb{Z}[\pi_1(M)])$$
$$\longrightarrow \Gamma S_{n-3}(\Phi) \longrightarrow \ldots$$

sends $ws(f,Y)$ to the total surgery obstruction $s(M) \in \mathcal{S}_{n-2}(M)$.

Following Wall [4,p.138] denote the LS-groups of a codimension q CW pair (X,Y) (q = 1 or 2) such that

$$\pi_1(X) = \pi_1(Y) = \pi \ , \quad \pi_1(Z) = \pi_1(S(\xi)) = \pi'$$

$$w(X) = w : \pi_1(X) = \pi \longrightarrow \mathbb{Z}_2$$

by

$$LS_*(\Phi) = LN_*(\pi' \longrightarrow \pi, w) \ .$$

In §7.8 the terminology will be extended to the ΓS-groups, with

$$\Gamma S_*(\Phi) = \Gamma N_*(\pi' \longrightarrow \pi, w) \ .$$

$\left\{\begin{array}{l}\text{Wall [4,§12C]}\\ \text{Matsumoto [1]}\end{array}\right.$ expressed the codimension $\left\{\begin{array}{l}1\\ 2\end{array}\right.$ ambient surgery obstruction groups $\left\{\begin{array}{l}LN_*(\pi' \longrightarrow \pi, w)\\ \Gamma N_*(\pi' \longrightarrow \pi, w)\end{array}\right.$ as the rel∂ obstruction groups for the existence of codimension $\left\{\begin{array}{l}1\\ 2\end{array}\right.$ spines, and obtained an algebraic formulation as a variant of the ordinary surgery obstruction groups $L_*(\mathbb{Z}[\pi])$ by a development of codimension $\left\{\begin{array}{l}1\\ 2\end{array}\right.$ ambient surgery analogous to that of ordinary surgery in §§5,6 of Wall [4]. In $\left\{\begin{array}{l}\text{§7.6}\\ \text{§7.8}\end{array}\right.$ we shall show how the language of algebraic Poincaré splittings can be used to obtain this formulation algebraically, subject only to the (provisional) use of topological transversality in the proof of Proposition 7.5.1 i).

7.6 The algebraic theory of codimension 1 surgery

We start with a brief account of codimension 1 CW surgery and the related algebraic K-theory.

A codimension q CW pair (X,Y) is finite if X is a finite CW complex.

A homotopy equivalence of finite CW complexes $f:M \xrightarrow{\simeq} X$ has a Whitehead torsion $\tau(f) \in Wh(\pi_1(X))$. Two such homotopy equivalences $f:M \xrightarrow{\simeq} X$, $f':M' \xrightarrow{\simeq} X$ are concordant if $f'^{-1}f:M \xrightarrow{\simeq} M'$ is a simple homotopy equivalence, that is if

$$\tau(f) = \tau(f') \in Wh(\pi_1(X)) \ .$$

Let (X,Y) be a finite codimension q CW pair, so that

$$X = E(\xi) \cup_{S(\xi)} Z$$

with $\xi:Y \longrightarrow BG(q)$. A homotopy equivalence $f:M \longrightarrow X$ from a finite CW complex M is split along $Y \subset X$ if f is concordant to a homotopy equivalence (also denoted by f) with a decomposition

$$f = g^! \cup h \ : \ M = E(\nu) \cup_{S(\nu)} P \longrightarrow X = E(\xi) \cup_{S(\xi)} Z$$

such that the restrictions

$$g = f| \ : \ N = f^{-1}(Y) \longrightarrow Y$$
$$h = f| \ : \ P = f^{-1}(Z) \longrightarrow Z$$

are both homotopy equivalences, where

$$\nu \ : \ N \xrightarrow{\ g\ } Y \xrightarrow{\ \xi\ } BG(q) \ ,$$

and such that (M,N) is a finite codimension q CW pair.

A codimension q CW pair (X,Y) is connected if X and Y (but not necessarily Z) are connected CW complexes. We shall be mainly concerned with splitting obstruction theory for connected pairs.

The splitting obstruction theory for finite connected codimension 1 CW pairs (X,Y) divides into three cases:

A) Y is 2-sided in X (i.e. ξ is trivial) and the complement Z is disconnected, with components Z_1, Z_2 say, so that

$$X = Y \times D^1 \cup_{Y \times S^0} (Z_1 \sqcup Z_2)$$

$$(= Z_1 \cup_Y Z_2 \text{ adding collars to } Z_1, Z_2)\ .$$

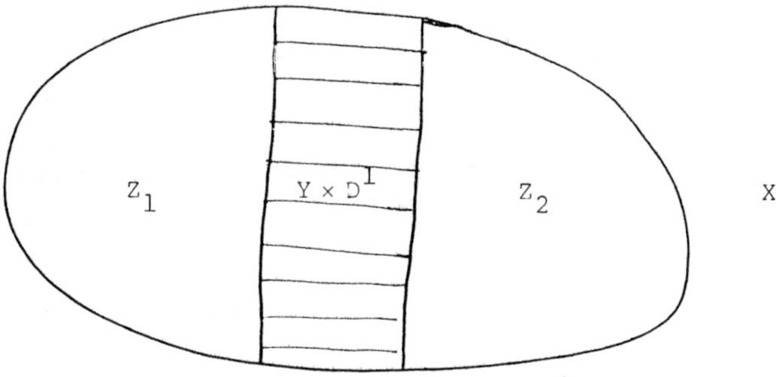

The fundamental group of X is the free product with amalgamation

$$\pi_1(X) = \pi_1(Z_1) *_{\pi_1(Y)} \pi_1(Z_2)$$

determined by the maps $i_1 : \pi_1(Y) \longrightarrow \pi_1(Z_1)$, $i_2 : \pi_1(Y) \longrightarrow \pi_1(Z_2)$ induced by the inclusions $Y \hookrightarrow Z_1$, $Y \hookrightarrow Z_2$.

B) Y is 2-sided in X and the complement Z is connected, so that

$$X = Y \times D^1 \cup_{Y \times S^0} Z\ .$$

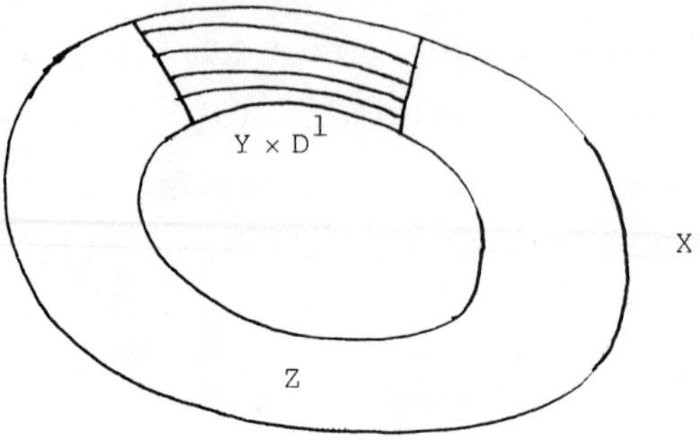

The fundamental group of X is the HNN extension

$$\pi_1(X) = \pi_1(Z) *_{\pi_1(Y)} \{t\}$$

determined by the maps $i_1, i_2 : \pi_1(Y) \longrightarrow \pi_1(Z)$ induced by the inclusions $Y \times \{+1\} \hookrightarrow Z$, $Y \times \{-1\} \hookrightarrow Z$.

C) Y is 1-sided in X (i.e. ξ is non-trivial).

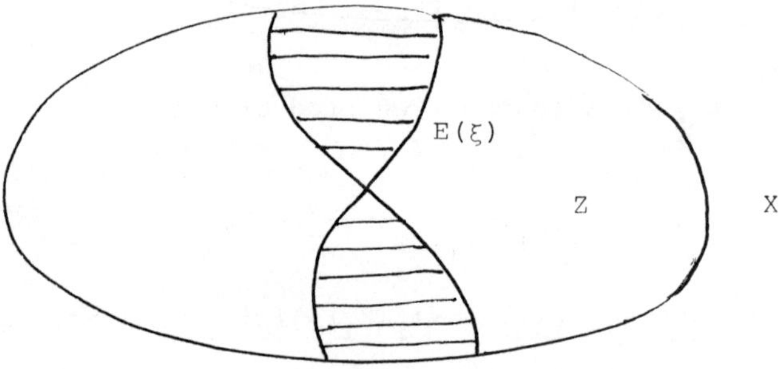

Actually, the codimension 1 CW splitting obstruction theory has only been worked out in the two-sided cases A) and B), under the additional hypothesis that the maps i_1, i_2 are injective. Following the results in special cases of Higman [1], Bass, Heller and Swan [1], Stallings [2], Gersten [1], Farrell and Hsiang [2], Casson [1], Waldhausen [2],[3] obtained a very general splitting theorem in the algebraic K-theory of such cases, as follows. There are defined higher/lower Whitehead groups $Wh_*(X)$ for any space X, to fit into an exact sequence of abelian groups

$$\ldots \longrightarrow H_n(X;\underline{\underline{K}}) \longrightarrow K_n(\mathbb{Z}[\pi_1(X)]) \longrightarrow Wh_n(X) \longrightarrow H_{n-1}(X;\underline{\underline{K}}) \longrightarrow \ldots \quad (n \in \mathbb{Z})$$

with $\underline{\underline{K}}$ the spectrum of the algebraic K-theory of $\mathbb{Z}$, such that $\pi_*(\underline{\underline{K}}) = K_*(\mathbb{Z})$. (Note the analogy with the exact sequence used to define the $\mathcal{S}$-groups $\mathcal{S}_*(X)$ in §7.1). The higher/lower Whitehead groups of a group π are the higher/lower Whitehead groups of the Eilenberg-MacLane space $K(\pi,1)$

$$Wh_*(\pi) = Wh_*(K(\pi,1)) \quad ,$$

with $Wh_0(\pi) = \tilde{K}_0(\mathbb{Z}[\pi])$ the reduced projective class group of the group ring $\mathbb{Z}[\pi]$ and $Wh_1(\pi) = Wh(\pi) = K_1(\mathbb{Z}[\pi])/\{\pm\pi\}$ the usual Whitehead group of π. For a finite connected codimension 1 CW pair (X,Y) of type $\begin{cases} A) \\ B) \end{cases}$ with the maps i_1, i_2 one-one there are defined exotic K-groups $\widetilde{Nil}_*(\Phi)$ of nilpotent objects depending on the pushout square of groupoids

$$\begin{array}{ccc} \pi_1(S(\xi)) = \pi_1(Y) \cup \pi_1(Y) & \xrightarrow{i_1 \cup i_2} & \pi_1(Z) \\ \downarrow & \Phi & \downarrow \\ \pi_1(E(\xi)) = \pi_1(Y) \times \nabla & \longrightarrow & \pi_1(X) \end{array}$$

with ∇ the connected groupoid with two vertices and trivial vertex groups (which is such that $Wh_*(\pi \times \nabla) = Wh_*(\pi)$ for any group π). There are defined split surjections $Wh_*(\pi) \longrightarrow \widetilde{Nil}_*(\Phi)$ which fit into an exact sequence of abelian groups

$$\begin{cases} \dots \longrightarrow Wh_n(\pi_1(Y)) \oplus \widetilde{Nil}_{n+1}(\Phi) \xrightarrow{\begin{pmatrix} i_1 & 0 \\ i_2 & 0 \end{pmatrix}} Wh_n(\pi_1(Z_1)) \oplus Wh(\pi_1(Z_2)) \\ \qquad \longrightarrow Wh_n(\pi_1(X)) \longrightarrow Wh_{n-1}(\pi_1(Y)) \oplus \widetilde{Nil}_n(\Phi) \longrightarrow \dots \\ \dots \longrightarrow Wh_n(\pi_1(Y)) \oplus \widetilde{Nil}_{n+1}(\Phi) \xrightarrow{(i_1 - i_2 \quad 0)} Wh_n(\pi_1(Z)) \\ \qquad \longrightarrow Wh_n(\pi_1(X)) \longrightarrow Wh_{n-1}(\pi_1(Y)) \oplus \widetilde{Nil}_n(\Phi) \longrightarrow \dots . \end{cases}$$

The two main ingredients of the proof of this splitting theorem were:

i) the translation into a generalized Higman linearization trick of the geometric transversality argument in the CW category by which every homotopy equivalence of finite CW complexes $f: M \xrightarrow{\sim} X$ can be made concordant to a map of codimension 1 CW pairs

$$f : (M,N) \longrightarrow (X,Y) ,$$

i.e. such that $(M, N = f^{-1}(Y))$ is a codimension 1 CW pair with

$$f = g^! \cup h : M = E(\nu) \cup_{S(\nu)} P \longrightarrow X = E(\xi) \cup_{S(\xi)} Z ,$$

involving the restrictions

$$g = f| : N \longrightarrow Y$$

$$h = f| : P = f^{-1}(Z) \longrightarrow Z$$

and the pullback

$$\nu : N \xrightarrow{g} Y \xrightarrow{\xi} BG(1) \quad (\xi = \varepsilon)$$

ii) an analysis in terms of nilpotent objects of the obstruction to further deforming the map $f: (M,N) \longrightarrow (X,Y)$ to one for which g and h are homotopy equivalences, i.e. to splitting $f: M \xrightarrow{\sim} X$ along $Y \subset X$, by a finite sequence of "cell exchange" CW surgeries on N inside M.

A homotopy equivalence of finite CW complexes $f: M \xrightarrow{\simeq} X$ (with ξ trivial and i_1, i_2 injective) can be split along $Y \subset X$ if and only if

$$\tau(f) \in \ker(Wh(\pi_1(X)) \longrightarrow \tilde{K}_0(\mathbb{Z}[\pi_1(Y)]) \oplus \widetilde{Nil}_1(\Phi))$$

$$= \operatorname{im}(Wh(\pi_1(Z)) \longrightarrow Wh(\pi_1(X))) \subseteq Wh(\pi_1(X)) \ .$$

If $f = g^! \cup h$ is split along $Y \subset X$ then $\tau(f)$ is the image of $\tau(h) \in Wh(\pi_1(Z))$. In fact, $\widetilde{Nil}_*(\Phi) = 0$ in many cases, and $Wh_*(\pi) = 0$ for any infinite torsion-free group π built up out of the trivial group $\{1\}$ by successive free products with amalgamation and/or HNN extensions (e.g. $\pi = \mathbb{Z}$). In particular, the fundamental groups of irreducible sufficiently large 3-manifolds (the "Haken manifolds") are of this type - it will be recalled from the introduction to Waldhausen [3] that the original motivation for this splitting theorem was the absence of Whitehead torsion in the earlier result of Waldhausen [1] that every homotopy equivalence of such 3-manifolds is homotopic to a homeomorphism.

***

We now turn to the codimension 1 manifold splitting obstruction theory.

Let (X,Y) be a connected $(n,n-1)$-dimensional geometric Poincaré pair. The obstruction theory for splitting s-triangulations $f:M \xrightarrow{\sim} X$ along $Y \subset X$ divides into the same three cases as the codimension 1 CW splitting obstruction theory:

A) Y is 2-sided in X and the complement Z is disconnected, so that

$$Z = Z_1 \sqcup Z_2 \quad , \quad X = Z_1 \cup_Y Z_2 \quad , \quad \pi_1(X) = \pi_1(Z_1) *_{\pi_1(Y)} \pi_1(Z_2) \ .$$

Codimension 1 splitting obstruction theory for A) was first studied by Browder [1] in the simply-connected case

$$\pi_1(X) = \pi_1(Y) = \pi_1(Z_1) = \pi_1(Z_2) = \{1\} \ ,$$

for which every s-triangulation $f:M \xrightarrow{\sim} X$ can be split along $Y \subset X$, at least if $n \geqslant 6$. Lee [1] obtained such a splitting theorem in some further special cases. The expression for the splitting obstruction with arbitrary (X,Y) of type A) as an element

$$s(f,Y) \in LS_{n-1}(\Phi)$$

of a geometrically defined LS-group is due to Wall [4,§11]. Cappell [i] $(1 \leqslant i \leqslant 9)$ has made an extensive study of the obstruction theory for A) in the case when the maps $i_1:\pi_1(Y) \longrightarrow \pi_1(Z_1)$, $i_2:\pi_1(Y) \longrightarrow \pi_1(Z_2)$ are injective, introducing exotic algebraic L-groups $UNil_*(\Phi)$ of nilpotent objects such that

$$LS_{n-1}(\Phi) = \hat{H}^n(\mathbb{Z}_2;I) \oplus UNil_{n+1}(\Phi)$$

with

$$I = \ker\left(\begin{pmatrix} i_1 \\ i_2 \end{pmatrix} : Wh(\pi_1(Y)) \longrightarrow Wh(\pi_1(Z_1)) \oplus Wh(\pi_1(Z_2))\right) \subseteq Wh(\pi_1(Y)) \quad ,$$

and defining split surjections $L_{n+1}(\mathbb{Z}[\pi_1(X)]) \twoheadrightarrow UNil_{n+1}(\Phi)$ (geometrically) to fit into an exact sequence

$$\ldots \longrightarrow L_{n+1}(\mathbb{Z}[\pi_1(X)]) \longrightarrow L_n^I(\mathbb{Z}[\pi_1(Y)]) \oplus UNil_{n+1}(\Phi)$$
$$\xrightarrow{\begin{pmatrix} i_1 & 0 \\ i_2 & 0 \end{pmatrix}} L_n(\mathbb{Z}[\pi_1(Z_1)]) \oplus L_n(\mathbb{Z}[\pi_1(Z_2)]) \longrightarrow L_n(\mathbb{Z}[\pi_1(X)]) \longrightarrow \ldots \ .$$

There is a parallel splitting obstruction theory for h-triangulations, with

$$LS_{n-1}^h(\Phi) = \hat{H}^n(\mathbb{Z}_2; I^h) \oplus UNil_{n+1}^h(\Phi) \quad ,$$

$$I^h = \ker\left(\begin{pmatrix} i_1 \\ i_2 \end{pmatrix} : \tilde{K}_0(\mathbb{Z}[\pi_1(Y)]) \longrightarrow \tilde{K}_0(\mathbb{Z}[\pi_1(Z_1)]) \oplus \tilde{K}_0(\mathbb{Z}[\pi_1(Z_2)])\right)$$
$$\subseteq \tilde{K}_0(\mathbb{Z}[\pi_1(Y)])$$

and the corresponding exact sequence involving L^h-groups. (There is also a parallel splitting obstruction theory for p-triangulations, as usual). In fact, Cappell showed that in many cases $\begin{cases} LS_{*-1}(\Phi) = 0 \\ LS_{*-1}^h(\Phi) = 0 \end{cases}$ by geometrically proving codimension 1 splitting theorems, in which case the above sequences are quadratic L-theory Mayer-Vietoris sequences of the general type considered in §6.2. We shall now use the algebraic characterization of the LS-groups given in §7.5 to provide an algebraic connection between such splitting theorems, Mayer-Vietoris sequences, and the decompositions

$$\begin{cases} LS_{*-1}(\Phi) = \hat{H}^*(\mathbb{Z}_2; I) \oplus UNil_{*+1}(\Phi) \\ LS_{*-1}^h(\Phi) = \hat{H}^*(\mathbb{Z}_2; I^h) \oplus UNil_{*+1}^h(\Phi) \end{cases} .$$

<u>Proposition 7.6.1</u>$_A$ Let (X,Y) be a connected codimension 1 CW pair of type A), with associated pushout square of fundamental groupoids

$$\begin{array}{ccc} \pi_1(Y) \cup \pi_1(Y) & \xrightarrow{i_1 \cup i_2} & \pi_1(Z_1) \cup \pi_1(Z_2) \\ \downarrow & \Phi & \downarrow \\ \pi_1(Y) \times \nabla & \longrightarrow & \pi_1(X) \end{array} ,$$

and let Θ be the pushout square of rings with involution

$$\begin{array}{ccc} \mathbb{Z}[\pi_1(Y)] & \xrightarrow{i_1} & \mathbb{Z}[\pi_1(Z_1)] \\ i_2 \downarrow & \Theta & \downarrow \\ \mathbb{Z}[\pi_1(Z_2)] & \longrightarrow & \mathbb{Z}[\pi_1(X)] \end{array} .$$

i) The LS-groups of Φ are naturally isomorphic to the triad L-groups of $\mathbb{Z}[\Phi]$

$$LS_{n-1}(\Phi) = L_{n+1}(\mathbb{Z}[\Phi]) \quad (n \in \mathbb{Z}) .$$

ii) The LS-groups of Φ are also naturally isomorphic to the triad L-groups of Θ

$$LS_{n-1}(\Phi) = L_{n+1}(\Theta) \quad (n \in \mathbb{Z}) .$$

<u>Proof</u>: i) This identification (which was first observed by Wall [4,Cor.12.4.1]) is immediate from the definition of the LS-groups on noting that the transfer maps are given by

$$\begin{aligned} p\xi^! : L_n(\mathbb{Z}[\pi_1(Y) \times \nabla]) &= L_n(\mathbb{Z}[\pi_1(Y)]) \\ &\longrightarrow L_{n+1}(\mathbb{Z}[\pi_1(Z_1) \cup \pi_1(Z_2)] \longrightarrow \mathbb{Z}[\pi_1(X)]) ; \\ & y \longmapsto (0,(i_1 y, -i_2 y)) \quad (n \in \mathbb{Z}) . \end{aligned}$$

Alternatively, it may be deduced from the braid of Proposition 7.2.1 iii), since $LN_*(\pi \cup \pi \longrightarrow \pi \times \nabla, w) = 0$ $(\pi = \pi_1(Y))$.

ii) This identification (which is also originally due to Wall [4,p.138]) follows from a comparison of the notion of algebraic Poincaré splitting used to define the triad L-groups in §6 with the algebraic Poincaré splitting used to give an algebraic characterization of the LS-groups in §7.5, as follows.

To conform with the terminology of §6 write the square Θ as

$$\begin{array}{ccc} A & \xrightarrow{\quad i_1 \quad} & B_1 \\ {\scriptstyle i_2}\big\downarrow & \Theta & \big\downarrow \\ B_2 & \xrightarrow{\qquad} & A' \end{array} .$$

The triad L-group $L_{n+1}(\Theta)$ was defined in §6.1 to be the cobordism group of (n+1)-dimensional quadratic Poincaré triads $(x;z_1,z_2;y)$ over Θ, consisting of an (n+1)-dimensional quadratic Poincaré pair $(x,\partial x)$ over $A' = \mathbb{Z}[\pi_1(X)]$ such that the boundary ∂x is Poincaré split with respect to Θ

$$x = A'\otimes_{B_1} z_1 \cup_{A'\otimes_A y} A'\otimes_{B_2}(-z_2)$$

for some n-dimensional quadratic Poincaré pairs (z_k,y) $(k = 1,2)$ over $i_k : A = \mathbb{Z}[\pi_1(Y)] \longrightarrow B_k = \mathbb{Z}[\pi_1(Z_k)]$. By the relative version of the algebraic normal transversality of Proposition 7.5.1 i) it can be shown that every such triad is cobordant to one with the pair $(x,\partial x)$ contractible, in which case $(y,z_1 \cup -z_2)$ is a contractible n-dimensional quadratic Poincaré splitting over Φ in the sense of §7.1. Now $LS_{n-1}(\Phi)$ was characterized in Proposition 7.5.1 ii) as the cobordism group of such splittings, so that the natural identification is given by

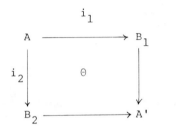

$$LS_{n-1}(\Phi) \longrightarrow L_{n+1}(\Theta)\ ;\ (y,z_1 \cup -z_2) \longmapsto (0,z_1,z_2,y)\ .$$

[]

In particular, Proposition 7.6.1$_A$ ii) identifies the condition $LS_{*-1}(\Phi) = 0$ for there to be type A) codimension 1 splitting with the condition $L_{*+1}(\Theta) = 0$ of Proposition 6.1.1 ii) for there to be a Mayer-Vietoris exact sequence of quadratic L-groups

$$\ldots \longrightarrow L_n(A) \xrightarrow{\begin{pmatrix} i_1 \\ i_2 \end{pmatrix}} L_n(B_1) \oplus L_n(B_2) \longrightarrow L_n(A') \longrightarrow L_{n-1}(A) \longrightarrow \ldots \qquad (n \in \mathbb{Z})\ .$$

Proposition 7.5.1 ii) characterizes the type A) codimension 1 splitting obstruction group for $\begin{cases} s- \\ h- \end{cases}$ triangulations $\begin{cases} LS_{n-1}(\Phi) \\ LS^h_{n-1}(\Phi) \end{cases}$ as the cobordism group of triples

(a $\begin{cases} \text{simple} \\ \text{finite} \end{cases}$ (n-1)-dimensional quadratic Poincaré complex over A

$$(C, \psi \in Q_{n-1}(C))\ ,$$

a $\begin{cases} \text{simple} \\ \text{finite} \end{cases}$ n-dimensional quadratic Poincaré pair over B_1

$$(j_1 : B_1 \otimes_A C \longrightarrow D_1, (\delta\psi_1, 1 \otimes_A \psi) \in Q_n(j_1))\ ,$$

a $\begin{cases} \text{simple} \\ \text{finite} \end{cases}$ n-dimensional quadratic Poincaré pair over B_2

$$(j_2 : B_2 \otimes_A C \longrightarrow D_2, (\delta\psi_2, 1 \otimes_A \psi) \in Q_n(j_2))\ )$$

such that the A'-module chain map

$$\begin{pmatrix} 1 \otimes_{B_1} j_1 \\ 1 \otimes_{B_2} j_2 \end{pmatrix} : A' \otimes_A C \longrightarrow A' \otimes_{B_1} D_1 \oplus A' \otimes_{B_2} D_2$$

is a $\begin{cases} \text{simple} \\ - \end{cases}$ chain equivalence, and similarly for $LS^p_{n-1}(\Phi)$.

Proposition 7.5.1 iii) shows that if M is an $(n+1)$-dimensional geometric Poincaré complex the Poincaré splitting obstruction along $Y \subset X = Z_1 \cup_Y Z_2$ of a map $f:M \longrightarrow X$ normal transverse at $Y \subset X$ is given by

$$s^P(f,Y) = (\sigma_*(f^{-1}(Y)), \sigma_*(f^{-1}(Z_1,Y)), \sigma_*(f^{-1}(Z_2,Y))) \in LS_{n-1}(\Phi) \ .$$

Proposition 7.5.1 iv) shows that if (X,Y) is an $(n,n-1)$-dimensional geometric Poincaré pair (of type A)) the manifold splitting obstruction along $Y \subset X$ of an s-triangulation $f:M \xrightarrow{\simeq} X$ topologically transverse at $Y \subset X$ is given by

$$\begin{aligned} s(f,Y) = (&\sigma_*((f,b)|:f^{-1}(Y) \longrightarrow Y), \sigma_*((f,b)|:f^{-1}(Z_1,Y) \longrightarrow (Z_1,Y)), \\ &\sigma_*((f,b)|:f^{-1}(Z_2,Y) \longrightarrow (Z_2,Y))) \in LS_{n-1}(\Phi) \ . \end{aligned}$$

We shall now interpret in terms of our theory Cappell's decompositions

$$\begin{cases} LS_{n-1}(\Phi) = \hat{H}^n(\mathbb{Z}_2;I) \oplus UNil_{n+1}(\Phi) \\ LS^h_{n-1}(\Phi) = \hat{H}^n(\mathbb{Z}_2;I^h) \oplus UNil^h_{n+1}(\Phi) \end{cases}$$

for a codimension 1 CW pair (X,Y) of type A) with the maps $i_1:\pi_1(Y) \longrightarrow \pi_1(Z_1)$, $i_2:\pi_1(Y) \longrightarrow \pi_1(Z_2)$ injective. As before, let

$$A = \mathbb{Z}[\pi_1(Y)] \ , \quad B_k = \mathbb{Z}[\pi_1(Z_k)] \ (k = 1,2) \ , \quad A' = \mathbb{Z}[\pi_1(X)] \ .$$

The induced morphisms of rings with involution $i_1:A \longrightarrow B_1$, $i_2:A \longrightarrow B_2$ are also injective, and A' is the free product of B_1 and B_2 amalgamated along A

$$A' = B_1 *_A B_2 \ .$$

The (A,A)-bimodules defined by

$$\tilde{B}_k = \mathbb{Z}[\pi_1(Z_k) - i_k\pi_1(Y)] \quad (k = 1,2)$$

are such that

$$B_k = A \oplus \tilde{B}_k \ ,$$

so that A' can be expressed as a direct sum of (A,A)-bimodules

$$A' = A\oplus\widetilde{B}_1\oplus\widetilde{B}_2\oplus(\widetilde{B}_1\otimes_A\widetilde{B}_2)\oplus(\widetilde{B}_2\otimes_A\widetilde{B}_1)\oplus(\widetilde{B}_1\otimes_A\widetilde{B}_2\otimes_A\widetilde{B}_1)\oplus(\widetilde{B}_2\otimes_A\widetilde{B}_1\otimes_A\widetilde{B}_2)\oplus\ldots\ .$$

Before dealing with the elements of $LS_{n-1}(\Phi)$ and $LS^h_{n-1}(\Phi)$ let us consider a triple

$$c = ((C,\psi),(j_1:B_1\otimes_A C \longrightarrow D_1,(\delta\psi_1,1\otimes_A\psi)),(j_2:B_2\otimes_A C \longrightarrow D_2,(\delta\psi_2,1\otimes_A\psi)))$$

consisting of a projective (n-1)-dimensional quadratic Poincaré complex $(C,\psi\in Q_{n-1}(C))$ over A together with projective null-cobordisms $(j_k:B_k\otimes_A C\longrightarrow D_k,(\delta\psi_k,1\otimes_A\ )\in Q_n(j_k))$ over B_k of $B_k\otimes_A(C,\psi)$ $(k=1,2)$ such that the A'-module chain map

$$\begin{pmatrix} 1\otimes_{B_1} j_1 \\ 1\otimes_{B_2} j_2\end{pmatrix} : A'\otimes_A C \longrightarrow A'\otimes_{B_1} D_1 \oplus A'\otimes_{B_2} D_2$$

is a chain equivalence, i.e. a representative of an element $c\in LS^p_{n-1}(\Phi)$ of the projective LS-group. The restriction of the B_k-action to $A\subset B_k$ allows $j_k\in \mathrm{Hom}_{B_k}(B_k\otimes_A C,D_k)$ to be regarded as an A-module morphism

$$i_k^! j_k\ :\ C\oplus(\widetilde{B}_k\otimes_A C) \longrightarrow i_k^! D_k\quad ,$$

and to regard $(j_k:B_k\otimes_A C\longrightarrow(\delta\psi_k,1\otimes_A\psi))$ as an n-dimensional quadratic Poincaré "cobordism" over A from (C,ψ) to $B_k\otimes_A(C,-\psi)$

$$c_k = (i_k^! j_k:C\oplus(\widetilde{B}_k\otimes_A C)\longrightarrow i_k^! D_k,(i_k^!\delta\psi_k,\psi\oplus(1\otimes_A\psi))\in Q_n(i_k^! j_k))\quad (k=1,2)\ .$$

The quotation marks refer to the possibility that $i_k\pi_1(Y)$ may be a subgroup of infinite index in $\pi_1(Z_k)$, in which case $\widetilde{B}_k$ is an infinitely generated free (A,A)-bimodule and the projective A-module chain complexes $\widetilde{B}_k\otimes_A C$, $i_k^! D_k$ are not finite-dimensional in the sense of §1.1 (i.e. not finitely generated). The quadratic Poincaré "cobordism" category defined using possibly infinitely generated projective chain complexes enjoys all the formal properties of the quadratic Poincaré cobordism category of §1 defined using finite-dimensional chain complexes; in particular, the glueing of "cobordisms" may be defined as in §1.7. Use this glueing operation to define two n-dimensional quadratic Poincaré "null-cobordisms" over A of (C,ψ)

$$c_+ = c_2 \cup_{\tilde{B}_2\otimes_A(C,\psi)} \tilde{B}_2\otimes_A c_1 \cup_{\tilde{B}_2\otimes_A\tilde{B}_1\otimes_A(C,\psi)} \tilde{B}_2\otimes_A\tilde{B}_1\otimes_A c_2 \cup \ldots$$

$$= (j_+ : C \longrightarrow D_+, (\delta\psi_+,\psi) \in Q_n(j_+))$$

$$c_- = c_1 \cup_{\tilde{B}_1\otimes_A(C,\psi)} \tilde{B}_1\otimes_A c_2 \cup_{\tilde{B}_1\otimes_A\tilde{B}_2\otimes_A(C,\psi)} \tilde{B}_1\otimes_A\tilde{B}_2\otimes_A c_1 \cup \ldots$$

$$= (j_- : C \longrightarrow D_-, (\delta\psi_-,\psi) \in Q_n(j_-))$$

such that the A-module chain complex

$$C(\begin{pmatrix} j_+ \\ j_- \end{pmatrix} : C \longrightarrow D_+\oplus D_-) = C(\begin{pmatrix} 1\otimes_{B_1} j_1 \\ 1\otimes_{B_2} j_2 \end{pmatrix} : A'\otimes_A C \longrightarrow A'\otimes_{B_1} D_1 \oplus A'\otimes_{B_2} D_2)$$

is chain contractible, restricting the action of A' on the right hand side to $A \subset A'$. It follows that the A-module chain map

$$\begin{pmatrix} j_+ \\ j_- \end{pmatrix} : C \longrightarrow D_+\oplus D$$

is a chain equivalence, and hence that the A-module chain complexes D_+, D_- have the chain homotopy types of n-dimensional A-module chain complexes. Thus up to homotopy equivalence c_+ and c_- are genuine quadratic Poincaré null-cobordisms over A of (C,ψ). The union of c_+ and c_- can be written as

$$c_+ \cup_{(C,\psi)} c_- = A'\otimes_{B_1}(j_1 : C \longrightarrow D_1, (\delta\psi_1, 1\otimes_A\psi))$$

$$\cup_{A'\otimes_A(C,\psi)} A'\otimes_{B_2}(j_2 : C \longrightarrow D_2, (\delta\psi_2, 1\otimes_A\psi)),$$

and so can be regarded as a contractible n-dimensional quadratic Poincaré complex over A'. Also, the union can be expressed as a union of null-cobordisms over B_1 of $B_1 \otimes_A (C,\psi)$

$$c_+ \cup_{(C,\psi)} c_- = c_1 \cup_{B_1 \otimes_A (C,\psi)} c_+ \quad ,$$

so that the B_1-module chain map

$$\begin{pmatrix} j_1 \\ 1\otimes_A j_+ \end{pmatrix} : B_1 \otimes_A C \longrightarrow D_1 \oplus B_1 \otimes_A D_+$$

is a chain equivalence. Similarly, over B_2

$$c_+ \cup_{(C,\psi)} c_- = c_2 \cup_{B_2 \otimes_A (C,\psi)} c_- \quad ,$$

so that the B_2-module chain map

$$\begin{pmatrix} j_2 \\ 1\otimes_A j_- \end{pmatrix} : B_2 \otimes_A C \longrightarrow D_2 \oplus B_2 \otimes_A D_-$$

is also a chain equivalence. Thus in the (reduced) projective class groups

$$\begin{aligned} [C] = [D_+] + [D_-] &= [D_+] + (-)^{n-1}[D_+]^* \\ &= [D_-] + (-)^{n-1}[D_-]^* \in \tilde{K}_0(A) \end{aligned}$$

$$[D_1] = B_1 \otimes_A [D_-] \in \tilde{K}_0(B_1)$$

$$[D_2] = B_2 \otimes_A [D_+] \in \tilde{K}_0(B_2) \quad .$$

Define an (n+1)-quadratic Poincaré relative null-cobordism of the n-dimensional quadratic Poincaré pair over A

$$((1\ 1) : C \oplus C \longrightarrow C, (0, \psi \oplus -\psi) \in Q_n((1\ 1)))$$

(i.e. an (n+1)-dimensional quadratic Poincaré triad over A in the sense of §1.3) by

$$c = \left(\begin{array}{ccc} C\oplus C & \xrightarrow{\ j_+ \oplus j_-\ } & D_+ \oplus D_- \\ {\scriptstyle (1\ 1)}\downarrow & \Gamma_A & \downarrow \\ C & \longrightarrow & 0 \end{array} \quad , \quad (0, 0, \delta\psi_+ \oplus \delta\psi_-, \psi \oplus -\psi) \in Q_{n+1}(\Gamma_A) \right)$$

and define also an $(n+1)$-dimensional quadratic Poincaré null-cobordism of the n-dimensional quadratic Poincaré pair over B_k

$$(j_k : B_k \otimes_A C \longrightarrow D_k, (\delta\psi_k, 1\otimes_A \psi) \in Q_n(j_k)) \quad (k = 1,2)$$

by

$$b_k = \left(j_k \begin{array}{ccc} B_k \otimes_A C & \xrightarrow{1\otimes_A j_{(-)^{k+1}}} & B_k \otimes_A D_{(-)^{k+1}} \\ \downarrow & \Gamma_{B_k} & \downarrow \\ D_k & \longrightarrow & 0 \end{array} , (0, \delta\psi_k, 1\otimes_A \delta\psi_{(-)^{k+1}}, 1\otimes_A \psi) \in Q_{n+1}(\Gamma_{B_k}) \right) .$$

The union of the induced cobordisms over A' is an $(n+1)$-dimensional quadratic Poincaré triad over A' which can be expressed up to homotopy equivalence as

$$A'\otimes_{B_1} b_1 \cup_{A'\otimes_A D_+} A'\otimes_A a \cup_{A'\otimes_A D_-} A'\otimes_{B_2} b_2$$

$$= \left(\begin{array}{ccc} 0 & \longrightarrow & 0 \\ \downarrow & \Gamma_{A'} & \downarrow \\ 0 & \longrightarrow & A'\otimes_A SC \end{array} , (\psi', 0, 0, 0) \in Q_{n+1}(\Gamma_{A'}) \right) ,$$

defining an $(n+1)$-dimensional quadratic Poincaré complex over A'

$$c_L = (A'\otimes_A SC, \psi' \in Q_{n+1}(A'\otimes_A SC)) .$$

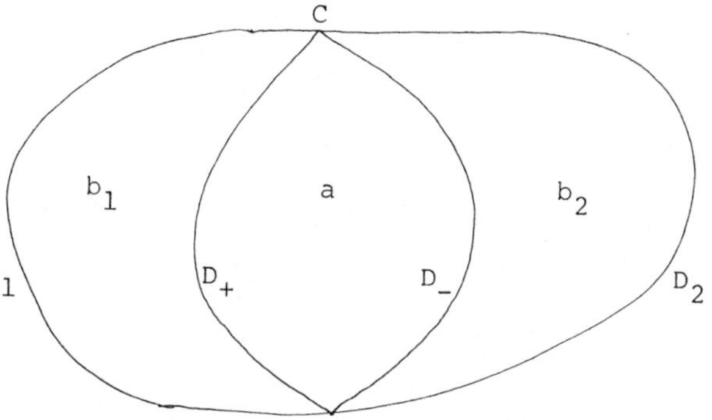

The construction of c_L is the algebraic analogue of the "unitary nilpotent cobordism construction" of Cappell [7,§II.1]. If $[C] = 0 \in \tilde{K}_0(A)$ and $[D_k] = 0 \in \tilde{K}_0(B_k)$ $(k = 1,2)$, so that $c \in LS^h_{n-1}(\Phi)$, then

$$c_L \in L^h_{n+1}(A') \ .$$

Furthermore, the projective class $[D_+] \in \tilde{K}_0(A)$ is such that

$$[D_+] = (-)^n[D_+]^* \in I^h \subseteq \tilde{K}_0(A) \ ,$$

and the element defined by

$$c_K = [D_+] \in \hat{H}^n(\mathbb{Z}_2;I^h)$$

is such that the natural map

$$\hat{H}^{n+1}(\mathbb{Z}_2;Wh(\pi_1(X))) \longrightarrow \hat{H}^n(\mathbb{Z}_2;I^h)\oplus\hat{H}^{n+1}(\mathbb{Z}_2;\widetilde{Nil}_1(\Phi))$$

sends the element

$$\tau(c) = \tau(\begin{pmatrix} 1\otimes_{B_1} j_1 \\ 1\otimes_{B_2} j_2 \end{pmatrix}: A'\otimes_A C \longrightarrow A'\otimes_{B_1} D_1 \oplus A'\otimes_{B_2} D_2) \in Wh(\pi_1(X))$$

(for arbitrary choices of bases for C,D_1,D_2 which may be assumed to be free) to the element $(c_K,\tau(c_L))$.

If C is a based f.g. free A-module chain complex and D_k $(k = 1,2)$ is a based f.g. free B_k-module chain complex, and all the chain equivalences appearing in c are simple so that $c \in LS_{n-1}(\Phi)$, then

$$c_L \in L_{n+1}(A')$$

and there is defined an element

$$c_K = \tau(f_+:C \longrightarrow D_+,(\delta\psi_+,\psi)) \in \hat{H}^n(\mathbb{Z}_2;I) \ .$$

The map

$$LS^q_{n-1}(\Phi) \longrightarrow \hat{H}^n(\mathbb{Z}_2;I^q) \ ; \ c \longmapsto c_K \qquad (q = s,h \quad I^s \equiv I)$$

is a naturally split surjection: the construction of the map $\hat{H}^n(\mathbb{Z}_2;I^q) \longrightarrow L^q_{n-1}(A)$ appearing in the exact sequence of Proposition 1.10.1

$$\ldots \longrightarrow L^{I^q}_n(A) \longrightarrow \hat{H}^n(\mathbb{Z}_2:I^q) \longrightarrow L^q_{n-1}(A) \longrightarrow L^{I^q}_{n-1}(A) \longrightarrow \ldots$$

readily extends to define a natural right inverse

$$\hat{H}^n(\mathbb{Z}_2;I^q) \rightarrowtail LS^q_{n-1}(\Phi) \ .$$

From the present point of view it is convenient to define the UNil-groups of Cappell [4] by

$$UNil^q_{n+1}(\Phi) = \ker(LS^q_{n-1}(\Phi) \twoheadrightarrow \hat{H}^n(\mathbb{Z}_2;I^q)) \quad (q = s,h \ n \in \mathbb{Z}) \ .$$

(In Ranicki [11] the UNil-groups will be expressed as the cobordism groups of quadratic Poincaré complexes with a nilpotent structure, generalizing their original formulation in terms of UNil-forms). The map

$$UNil^q_{n+1}(\Phi) \longrightarrow L^q_{n+1}(A') \ ; \ c \longmapsto c_L$$

is a naturally split surjection: the canonical morphism $L^q_{n+1}(A') \longrightarrow LS^q_{n-1}(\Phi)$ maps onto $UNil^q_{n+1}(\Phi) \subseteq LS^q_{n-1}(\Phi)$, defining a left inverse. Thus every element $c \in LS^q_{n-1}(\Phi)$ can be expressed as

$$c = (c_K, c_L) \in LS^q_{n-1}(\Phi) = \hat{H}^n(\mathbb{Z}_2;I^q) \oplus UNil^q_{n+1}(\Phi)$$
$$(q = s,h) \ .$$

In particular, if $c = s^P(f,Y) \in LS^h_{n-1}(\Phi)$ is the Poincaré splitting obstruction along $Y \subset X$ of a map $f:M \longrightarrow X$ from a finite $(n+1)$-dimensional geometric Poincaré complex M then $c_K \in \hat{H}^n(\mathbb{Z}_2;I^h)$ is the image under the natural map

$$\hat{H}^{n+1}(\mathbb{Z}_2;Wh(\pi_1(X))) \longrightarrow \hat{H}^n(\mathbb{Z}_2;I^h)$$

of the element represented by the Whitehead torsion of M

$$\tau(M) = \tau([M] \cap - : C(\tilde{M})^{n+1-*} \longrightarrow C(\tilde{M})) \in Wh(\pi_1(M))$$

(or rather its image $f_*\tau(M) \in Wh(\pi_1(X))$), with $\tilde{M}$ the universal cover. If (X,Y) is a finite $(n,n-1)$-dimensional geometric Poincaré pair (of type A) with i_1, i_2 injective) and $c = s(f,Y) \in LS^h_{n-1}(\Phi)$ is the splitting obstruction along $Y \subset X$ of an h-triangulation $f: M \xrightarrow{\sim} X$ then $c_K \in \hat{H}^n(\mathbb{Z}_2; I^h)$ is the image under the natural map of the element $\tau(c) = \tau(f) \in \hat{H}^{n+1}(\mathbb{Z}_2; Wh(\pi_1(X)))$ represented by the Whitehead torsion $\tau(f) \in Wh(\pi_1(X))$. Furthermore, in this case $c_K \in \hat{H}^n(\mathbb{Z}_2; I^h)$. is the obstruction to modifying c (resp. f) by a finite sequence of algebraic (resp. geometric) "handle exchanges" to a triple c' (resp. concordant h-triangulation f') which is cobordant to 0 (resp. topologically normal bordant to an h-triangulation of X which is split along $Y \subset X$), and if $c_K = 0$ then $c_L \in UNil^h_{n+1}(\Phi) \subseteq L^h_{n+1}(A')$ is the surgery obstruction of such an algebraic (resp. geometric) cobordism, which in the geometric case is the Cappell unitary nilpotent cobordism. The decomposition $LS^q_{n-1}(\Phi) = \hat{H}^n(\mathbb{Z}_2; I^q) \oplus UNil^q_{n+1}(\Phi)$ $(q = s,h)$ of the LS-groups gives rise to a corresponding decomposition of the L-groups

$$L^q_{n+1}(A') = L^{I^q}_{n+1}(\begin{pmatrix} i_1 \\ i_2 \end{pmatrix} : A \longrightarrow B_1 \oplus B_2) \oplus UNil^q_{n+1}(\Phi) \quad (n \in \mathbb{Z})$$

with $L^{I^q}_*(\begin{pmatrix} i_1 \\ i_2 \end{pmatrix})$ the relative L-groups appearing in the exact sequence

$$\cdots \longrightarrow L^{I^q}_{n+1}(A) \xrightarrow{\begin{pmatrix} i_1 \\ i_2 \end{pmatrix}} L^q_{n+1}(B_1) \oplus L^q_{n+1}(B_2) \longrightarrow L^{I^q}_{n+1}(\begin{pmatrix} i_1 \\ i_2 \end{pmatrix})$$

$$\longrightarrow L^{I^q}_n(A) \longrightarrow \cdots \quad .$$

Proposition 7.6.2$_A$ Let (X,Y) be a finite codimension 1 CW pair of type A) with i_1, i_2 injective. The exact sequence relating the associated free and simple LS-groups

$$\ldots \longrightarrow WhS_n(\Phi) \longrightarrow LS_{n-1}(\Phi) \longrightarrow LS^h_{n-1}(\Phi) \longrightarrow WhS_{n-1}(\Phi) \longrightarrow \ldots \quad (n \in \mathbb{Z})$$

is naturally isomorphic to the direct sum of the exact sequence

$$\ldots \longrightarrow \hat{H}^{n+1}(\mathbb{Z}_2;I^h) \oplus \hat{H}^n(\mathbb{Z}_2;I) \xrightarrow{(0\ 1)} \hat{H}^n(\mathbb{Z}_2;I) \xrightarrow{0} \hat{H}^n(\mathbb{Z}_2;I^h)$$

$$\xrightarrow{\begin{pmatrix}1\\0\end{pmatrix}} \hat{H}^n(\mathbb{Z}_2;I^h) \oplus \hat{H}^{n-1}(\mathbb{Z}_2;I) \longrightarrow \ldots$$

and the exact sequence

$$\ldots \longrightarrow \hat{H}^{n+2}(\mathbb{Z}_2;\widetilde{Nil}_1(\Phi)) \longrightarrow UNil_{n+1}(\Phi) \longrightarrow UNil^h_{n+1}(\Phi)$$

$$\longrightarrow \hat{H}^{n+1}(\mathbb{Z}_2;\widetilde{Nil}_1(\Phi)) \longrightarrow \ldots \ .$$

Similarly for the exact sequence relating the associated projective and free LS-groups, with $\tilde{K}_0$ in place of Wh.

Proof: Immediate from the decompositions of the LS-groups and the comparison exact sequence of Proposition 7.5.2.

[]

B) Y is 2-sided in X and the complement Z is connected, so that

$$X = Y \times D^1 \cup_{Y \times S^0} Z \ , \ \pi_1(X) = \pi_1(Z) *_{\pi_1(Y)} \{t\} \ .$$

Codimension 1 splitting obstruction theory for B) first appeared in the work of Stallings [1], Browder and Levine [1], Farrell [1] and Siebenmann [1] on the characterization of manifolds which fibre over S^1, since a fibre is then a codimension 1 submanifold of type B). A general result was first obtained by Browder [2], who showed that if (X,Y) is an (n,n-1)-dimensional geometric Poincaré pair of type B) with

$$\pi_1(X) = \mathbb{Z} \ , \ \pi_1(Y) = \pi_1(Z) = \{1\}$$

then every s-triangulation $f:M \xrightarrow{\sim} X$ can be split along $Y \subset X$, at least if $n \geqslant 6$. The expression for the splitting obstruction with arbitrary (X,Y) of type B) as an element

$$s(f,Y) \in LS_{n-1}(\Phi)$$

is due to Wall [4,§11], the general obstruction theory being the same for B) as for A).

If the maps $i_1, i_2 : \pi_1(Y) \xrightarrow{\sim} \pi_1(Z)$ are isomorphisms (e.g. if $Y \longrightarrow X \longrightarrow S^1$ is a fibre bundle) the automorphism

$$\alpha = i_2^{-1} i_1 \ : \ \pi_1(Y) \xrightarrow{\sim} \pi_1(Y)$$

is such that the group $\pi_1(X)$ is the α-twisted extension of $\pi_1(Y)$ by $\mathbb{Z}$

$$\pi_1(X) = \pi_1(Y) \times_\alpha \mathbb{Z} \ ,$$

with

$$gt = t\alpha(g) \quad (g \in \pi_1(Y), t = 1 \in \mathbb{Z}) \ .$$

The group ring $\mathbb{Z}[\pi_1(X)]$ is the α-twisted Laurent extension of $\mathbb{Z}[\pi_1(Y)]$

$$\mathbb{Z}[\pi_1(X)] = \mathbb{Z}[\pi_1(Y)]_\alpha[t,t^{-1}] \ ,$$

with

$$at = t\alpha(a) \quad (a \in \mathbb{Z}[\pi_1(Y)])$$

and the $w(X)$-twisted involution

$$\overline{(gt^j)} = w(X)(gt^j)(gt^j)^{-1} = w(Y)(g)t^{-j}g^{-1} \quad (g \in \pi_1(Y), j \in \mathbb{Z}) \ .$$

In this case Wall [4,Thm.12.5] used a generalization of the work of Farrell [1] to identify

$$LS_{n-1}(\Phi) = \hat{H}^n(\mathbb{Z}_2;I)$$

with

$$I = Wh(\pi_1(Y))^\alpha = \ker(1-\alpha : Wh(\pi_1(Y)) \longrightarrow Wh(\pi_1(Y))) \subseteq Wh(\pi_1(Y)) \ .$$

Farrell and Hsiang [1],[3] studied the splitting obstruction theory for h-triangulations in this case, in effect identifying

$$LS^h_{n-1}(\Phi) = \hat{H}^n(\mathbb{Z}_2;I^h)$$

with

$$\begin{aligned} I^h &= \widetilde{K}_0(\mathbb{Z}[\pi_1(Y)])^\alpha \\ &= \ker(1-\alpha:\widetilde{K}_0(\mathbb{Z}[\pi_1(Y)]) \longrightarrow \widetilde{K}_0(\mathbb{Z}[\pi_1(Y)])) \subseteq \widetilde{K}_0(\mathbb{Z}[\pi_1(Y)]) . \end{aligned}$$

In particular, they showed that every s-triangulation $f:M \xrightarrow{\sim} X$ is concordant to one for which $f|:f^{-1}(Y) \longrightarrow Y$ is an h-triangulation, i.e. regarded as an h-triangulation f can be split along $Y \subset X$, at least if $n \geqslant 6$. Shaneson [1] used the Farrell-Hsiang splitting theorem in the case $\alpha = \text{id.} : \pi_1(Y) \longrightarrow \pi_1(Y)$

$$\pi_1(X) = \pi_1(Y) \times \mathbb{Z} \quad , \quad \mathbb{Z}[\pi_1(X)] = \mathbb{Z}[\pi_1(Y)][t,t^{-1}]$$

(e.g. if $X = Y \times S^1$) to give a geometric proof of the splitting theorem for the quadratic L-groups of a Laurent extension

$$L_{n+1}(\mathbb{Z}[\pi \times \mathbb{Z}]) = L_{n+1}(\mathbb{Z}[\pi]) \oplus L^h_n(\mathbb{Z}[\pi]) \quad (\pi = \pi_1(Y)) \ ,$$

which was then proved algebraically by Novikov [1] and Ranicki [2]. For arbitrary α the identification $LS_{n-1}(\Phi) = \hat{H}^n(\mathbb{Z}_2;I)$ is equivalent to the exact sequence

$$\begin{aligned} \dots \longrightarrow L_{n+1}(\mathbb{Z}[\pi_1(X)]) \longrightarrow L^I_n(\mathbb{Z}[\pi_1(Y)]) \xrightarrow{1-\alpha} L_n(\mathbb{Z}[\pi_1(Y)]) \\ \longrightarrow L_n(\mathbb{Z}[\pi_1(X)]) \longrightarrow \dots \quad (n \in \mathbb{Z}) \end{aligned}$$

obtained geometrically by Cappell [1] and algebraically by Ranicki [3].

The codimension 1 splitting obstruction theory of Cappell [i] $(1 \leqslant i \leqslant 9)$ includes the case B) with the maps $i_1, i_2 : \pi_1(Y) \longrightarrow \pi_1(Z)$ injective. As for A) there are defined exotic algebraic L-groups $UNil_*(\Phi)$ of nilpotent objects such that

$$LS_{n-1}(\Phi) = \hat{H}^n(\mathbb{Z}_2;I)\oplus UNil_{n+1}(\Phi)$$

with

$$I = \ker(i_1 - i_2 : Wh(\pi_1(Y)) \longrightarrow Wh(\pi_1(Z))) \subseteq Wh(\pi_1(Y))$$

and there are defined split surjections $L_{n+1}(\mathbb{Z}[\pi_1(X)]) \twoheadrightarrow UNil_{n+1}(\Phi)$ (geometrically) to fit into an exact sequence

$$\ldots \longrightarrow L_{n+1}(\mathbb{Z}[\pi_1(X)]) \longrightarrow L_n^I(\mathbb{Z}[\pi_1(Y)])\oplus UNil_{n+1}(\Phi)$$

$$\xrightarrow{(i_1-i_2\ \ 0)} L_n(\mathbb{Z}[\pi_1(Z)]) \longrightarrow L_n(\mathbb{Z}[\pi_1(X)]) \longrightarrow \ldots .$$

Again, there is a parallel theory for h-triangulations, with

$$LS^h_{n-1}(\Phi) = \hat{H}^n(\mathbb{Z}_2;I^h)\oplus UNil^h_{n+1}(\Phi) ,$$

$$I^h = \ker(i_1-i_2:\tilde{K}_0(\mathbb{Z}[\pi_1(Y)]) \longrightarrow \tilde{K}_0(\mathbb{Z}[\pi_1(Z)]))$$
$$\subseteq \tilde{K}_0(\mathbb{Z}[\pi_1(Y)])$$

and similarly for p-triangulations. If the maps $i_1,i_2:\pi_1(Y) \longrightarrow \pi_1(Z)$ are isomorphisms then $UNil^q_*(\Phi) = 0$ for $q = s,h,p$.

Proposition $7.6.1_B$ Let (X,Y) be a connected codimension 1 CW pair of type B), with associated pushout square of fundamental groupoids

$$\begin{array}{ccc} \pi_1(Y) \cup \pi_1(Y) & \xrightarrow{i_1 \cup i_2} & \pi_1(Z) \\ \downarrow & \Phi & \downarrow \\ \pi_1(Y) \times \nabla & \longrightarrow & \pi_1(X) \end{array} \quad .$$

The LS-groups of Φ are naturally isomorphic to the triad L-groups of $\mathbb{Z}[\Phi]$

$$LS_{n-1}(\Phi) = L_{n+1}(\mathbb{Z}[\Phi]) \quad (n \in \mathbb{Z}) .$$

Proof: By analogy with Proposition $7.6.1_A$ i). The identification (which was also first observed by Wall [4,Cor.12.4.1]) is immediate from the definition of the LS-groups on noting that the transfer maps in this case are given by

$$p\xi^! : L_n(\mathbb{Z}[\pi_1(Y)\times\nabla]) = L_n(\mathbb{Z}[\pi_1(Y)]$$

$$\longrightarrow L_{n+1}(\mathbb{Z}[\pi_1(Z)]\longrightarrow \mathbb{Z}[\pi_1(X)]) ;$$

$$y \longmapsto (0, i_1 y \oplus -i_2 y) \quad (n \in \mathbb{Z}) .$$

Alternatively, it may be deduced from the braid of Proposition 7.2.1 iii), since $LN_*(\pi \cup \pi \longrightarrow \pi \times \nabla, w) = 0$ $(\pi = \pi_1(Y))$.

[]

There is also a case B) version of the identification of Proposition $7.6.1_A$ ii), with an analogous algebraic characterization of the LS-groups, as follows.

Given a codimension 1 CW pair (X,Y) of type B) let

$$A = \mathbb{Z}[\pi_1(Y)] \quad , \quad B = \mathbb{Z}[\pi_1(Z)] \quad , \quad A' = \mathbb{Z}[\pi_1(X)] \quad ,$$

so that $A' = B*_A\{t\}$ is the generalized Laurent extension of B determined by the morphisms of rings with involution $i_1, i_2 : A \longrightarrow B$ induced by the group morphisms $i_1, i_2 : \pi_1(Y) \longrightarrow \pi_1(Z)$, with t an indeterminate over B such that

$$\bar{t} = t^{-1} \quad , \quad i_1(a)t = ti_2(a) \in A' \quad (a \in A) .$$

Let B_k $(k = 1,2)$ be the (B,A)-bimodule with additive group B and

$$B \times B_k \times A \longrightarrow B_k \ ; \ (b,x,a) \longmapsto b.x.i_k(a) .$$

It follows from Proposition $7.6.1_B$ and the algebraic normal transversality of Proposition 7.5.1 ii) that the type B) codimension 1 splitting obstruction group for $\left\{\begin{matrix} s- \\ h- \end{matrix}\right.$ triangulations

$\begin{cases} LS_{n-1}(\Phi) \\ LS^h_{n-1}(\Phi) \end{cases}$ is the cobordism group of pairs

(a $\begin{cases} \text{simple} \\ \text{finite} \end{cases}$ (n-1)-dimensional quadratic Poincaré complex over A,

$(C,\psi \in Q_{n-1}(C))$,

a $\begin{cases} \text{simple} \\ \text{finite} \end{cases}$ n-dimensional quadratic Poincaré pair over B

$((f_1\ f_2) : B_1 \otimes_A C \oplus B_2 \otimes_A C \longrightarrow D, (\delta\psi, 1\otimes_{i_1}\psi \oplus -1\otimes_{i_2}\psi) \in Q_n(f_1\ f_2)))$

such that the A'-module chain map

$$1\otimes f_1 - t\otimes f_2 : A'\otimes_A C \longrightarrow A'\otimes_B D$$

is a $\begin{cases} \text{simple} \\ - \end{cases}$ chain equivalence, regarding A' as an (A',A)-bimodule via the composite $A \xrightarrow{i_1} B \hookrightarrow A'$. Similarly for $LS^p_{n-1}(\Phi)$.

Proposition 7.5.1 iii) shows that if M is an (n+1)-dimensional geometric Poincaré complex the Poincaré splitting obstruction along $Y \subset X = Y \times D^1 \cup_{Y \times S^0} Z$ of a map $f:M \longrightarrow X$ normal transverse at $Y \subset X$ is given by

$$s^P(f,Y) = (\sigma_*(f^{-1}(Y)), \sigma_*(f^{-1}(Z, Y \times S^0))) \in LS_{n-1}(\Phi) \ .$$

Proposition 7.5.1 iv) shows that if (X,Y) is an (n,n-1)-dimensional geometric Poincaré pair (of type B)) the manifold splitting obstruction along $Y \subset X$ of an s-triangulation $f:M \xrightarrow{\sim} X$ topologically transverse at $Y \subset X$ is given by

$$s(f,Y) = (\sigma_*((f,b)| : f^{-1}(Y) \longrightarrow Y), \\ \sigma_*((f,b)| : f^{-1}(Z, Y \times S^0) \longrightarrow (Z, Y \times S^0))) \in LS_{n-1}(\Phi) \ .$$

Note that Proposition $7.6.1_B$ identifies the condition $LS_{*-1}(\Phi) = 0$ for there to be type B) codimension 1 splitting with the condition $L_{*+1}(\mathbb{Z}[\Phi]) = 0$ for there to be a Mayer-Vietoris exact sequence of quadratic L-groups

$$\ldots \longrightarrow L_n(A) \xrightarrow{i_1 - i_2} L_n(B) \longrightarrow L_n(A') \longrightarrow L_{n-1}(A) \longrightarrow \ldots \quad (n \in \mathbb{Z}) .$$

If (X,Y) is a codimension 1 CW pair of type B) such that the maps $i_1, i_2 : \pi_1(Y) \longrightarrow \pi_1(Z)$ are injective there are defined Cappell decompositions

$$\begin{cases} LS_{n-1}(\Phi) = \hat{H}^n(\mathbb{Z}_2; I) \oplus UNil_{n+1}(\Phi) \\ LS^h_{n-1}(\Phi) = \hat{H}^n(\mathbb{Z}_2; I^h) \oplus UNil^h_{n+1}(\Phi) \end{cases} \quad (n \in \mathbb{Z}) ,$$

which may be interpreted in terms of our theory as for the case A) above. In particular, Proposition $7.6.2_A$ carries over word for word to its type B) analogue, Proposition $7.6.2_B$.

A) or B) Y is 2-sided in X (i.e. $\xi = \nu_{Y \subset X}$ is trivial)

Cappell [7] has shown that in many cases $UNil_*(\Phi) = 0$ both for A) and B), by obtaining the equivalent codimension 1 splitting theorems (under the hypothesis $\hat{H}^*(\mathbb{Z}_2; I) = 0$). In effect, Cappell proved that the assembly maps

$$\sigma_* : H_*(K(\pi,1); \underline{\mathbb{L}}_0) \longrightarrow L_*(\mathbb{Z}[\pi])$$

are isomorphisms and hence that $\mathcal{S}_*(K(\pi,1)) = 0$ for any infinite torsion-free group π built up out of the trivial group {1} by successive free products with amalgamation $\pi_1 *_\rho \pi_2$ and/or HNN extensions $\pi_1 *_\rho \{t\}$ along subgroups $\rho \subset \pi_k$ satisfying the "square root closed" condition: if $g \in \pi_k$ is such that $g^2 \in \rho \subset \pi_k$ then $g \in \rho \subset \pi_k$ (e.g. $\pi = \mathbb{Z}$). However, it is not known if $\mathcal{S}_*(K(\pi,1)) = 0$ for the fundamental groups π of irreducible sufficiently large 3-manifolds (except when the square root closed condition is satisfied), although the results of Waldhausen [1],[2],[3] that every homotopy equivalence of such manifolds is homotopic to a homeomorphism and that $Wh_*(\pi) = 0$ do suggest that such ought to be the case.

Let R be a ring such that $\mathbb{Z} \subseteq R \subseteq \mathbb{Q}$. The groups

$$\Gamma_*(\mathbb{Z}[\pi] \longrightarrow R[\pi]) = L_*^S(R[\pi]) \quad (R = S^{-1}\mathbb{Z})$$

are the obstruction groups for surgery on normal maps up to R-homotopy equivalence (see §7.7 below for further details of surgery with coefficients). In particular, for $R = \mathbb{Z}$ this is the ordinary surgery theory up to homotopy equivalence dealt with above. Cappell [4] extended his codimension 1 splitting obstruction theory to surgery with R-coefficients, introducing the appropriate UNil-groups $UNil_*(R[\Phi])$ with all the formal properties of $UNil_*(\Phi) \equiv UNil_*(\mathbb{Z}[\Phi])$. Furthermore, he proved that the groups $UNil_*(R[\Phi])$ are 2-primary for any R, and that $UNil_*(R[\Phi]) = 0$ if $1/2 \in R$. Farrell [2] has shown that the groups $UNil_*(\Phi)$ are in fact of exponent 4 - as pointed out in the introduction to that paper it follows from the localization exact sequence of §3 above that the exponent of $UNil_*(R[\Phi])$ is at most 8 (using the result of Proposition 3.6.4 that the localization maps $L_*(\mathbb{Z}[\pi]) \longrightarrow L_*^S(R[\pi])$ are isomorphisms modulo 8-torsion).

C) Y is 1-sided in X (i.e. $\xi = \nu_{Y \subset X}: Y \longrightarrow BG(1)$ is non-trivial)

Codimension 1 splitting obstruction theory for C) was first studied by Browder and Livesay [1] for the codimension 1 geometric Poincaré pairs $(X,Y) = (\mathbb{R}P^n, \mathbb{R}P^{n-1})$ $(n \geqslant 1)$, in connection with the classification of fixed point free involutions on manifolds which are homotopy spheres. López de Medrano [1],[2] extended the Browder-Livesay theory to fixed point free involutions on arbitrary simply-connected manifolds, thus describing the splitting obstruction theory for codimension 1 geometric Poincaré pairs (X,Y) of type C) with

$$\pi_1(X) = \pi_1(Y) = \mathbb{Z}_2 \ , \quad \pi_1(S(\xi)) = \pi_1(Z) = \{1\} \ .$$

The expression for the splitting obstruction along $Y \subset X$ of an s-triangulation $f: M \xrightarrow{\sim} X$ for an arbitrary $(n,n-1)$-dimensional geometric Poincaré pair (X,Y) of type C) as an element

$$s(f,Y) \in LS_{n-1}(\Phi)$$

is due to Wall [4,§11], the general obstruction theory being the same for C) as for A) and B). Furthermore, in the case $\pi_1(X) = \pi_1(Y)$ Wall [4,§12C] gave an algebraic expression for the obstruction groups $LS_*(\Phi)$, by realizing each element of $LS_{n-1}(\Phi)$ as the rel∂ obstruction to finding a codimension 1 spine $M \subset V$ for an n-dimensional manifold with boundary $(V, \partial V)$ such that V is an (n-1)-dimensional geometric Poincaré complex, and developing a non-simply-connected Browder-Livesay theory. We shall now recover this expression from the general algebraic formulation of the LS-groups in §7.5 above.

We start by extending the formulation of the quadratic L-groups in terms of chain complexes to the quadratic L-groups of rings with antistructure in the sense of Wall [5].

Let A be an associative ring with 1. An <u>antistructure on A</u> (α,ε) consists of a function

$$\alpha : A \longrightarrow A \;;\; a \longmapsto \alpha(a)$$

and a unit $\varepsilon \in A$ such that $\alpha(\varepsilon) = \varepsilon^{-1} \in A$ and also

$$\text{i)}\ \alpha(a+b) = \alpha(a) + \alpha(b)$$

$$\text{ii)}\ \alpha(ab) = \alpha(b)\alpha(a)$$

$$\text{iii)}\ \alpha(1) = 1$$

$$\text{iv)}\ \alpha^2(a) = \varepsilon^{-1}a\varepsilon$$

for all $a,b \in A$. (In particular, if $\varepsilon \in A$ is a central unit then $\alpha : a \longmapsto \bar{a}$ is an involution as in §1.1). Given a f.g. projective A-module M let A act on the dual $M^* = \mathrm{Hom}_A(M,A)$ by

$$A \times M^* \longrightarrow M^* \;;\; (a,f) \longmapsto (x \longmapsto f(x)\alpha(a)) \;,$$

and use the natural A-module isomorphism

$$M \longrightarrow M^{**} \;;\; x \longmapsto (f \longmapsto \alpha(f(x)))$$

to identify $M = M^{**}$. Given also a f.g. projective A-module N define a duality isomorphism

$$\mathrm{Hom}_A(M,N) \longrightarrow \mathrm{Hom}_A(N^*,M^*) \;;\; f \longmapsto (g \longmapsto (x \longmapsto g(f(x)))) \;.$$

Let $T \in \mathbb{Z}_2$ act on $\mathrm{Hom}_A(M,M^*)$ by the <u>(α,ε)-duality involution</u>

$$T_{\alpha,\varepsilon} : \mathrm{Hom}_A(M,M^*) \longrightarrow \mathrm{Hom}_A(M,M^*) \;;$$
$$(\varepsilon\phi^* : x \longmapsto (y \longmapsto \varepsilon\alpha(\phi(y)(x)))) \;.$$

Given a finite-dimensional A-module chain complex C let $T \in \mathbb{Z}_2$ act on $\mathrm{Hom}_A(C^*,C)$ by

$$T_{\alpha,\varepsilon} : \mathrm{Hom}_A(C^p,C_q) \longrightarrow \mathrm{Hom}_A(C^q,C_p) \;;\; \phi \longmapsto (-)^{pq}\varepsilon\phi^* \;,$$

and define the <u>(α,ε)-quadratic Q-groups</u> of C

$$Q_n(C,\varepsilon) = H_n(\mathrm{Hom}_{\mathbb{Z}[\mathbb{Z}_2]}(W,\mathrm{Hom}_A(C^*,C))) \quad (n \in \mathbb{Z}) \;.$$

The <u>(α,ε)-quadratic L-groups of A</u> $L_n(A^{\alpha},\varepsilon)$ $(n \geqslant 0)$ are defined be the cobordism groups of n-dimensional (α,ε)-quadratic Poincaré complexes over A $(C,\psi \in Q_n(C,\varepsilon))$, exactly as in the case of central ε. All the results of §§1-6 in the central case have evident generalizations to rings with antistructure.
(There are also defined (α,ε)-symmetric L-groups $L^*(A^{\alpha},\varepsilon)$ - we shall be mainly concerned with the (α,ε)-quadratic L-groups here). In particular, the (α,ε)-quadratic L-groups are 4-periodic

$$L_n(A^{\alpha},\varepsilon) = L_{n+2}(A^{\alpha},-\varepsilon) = L_{n+4}(A^{\alpha},\varepsilon) \quad (n \geqslant 0)$$

and $L_0(A^{\alpha},\varepsilon)$ is the Witt group of non-singular (α,ε)-quadratic forms over A (M,ψ), as defined by a f.g. projective A-module M together with an element

$$\psi \in Q_{\alpha,\varepsilon}(M) = \operatorname{coker}(1-T_{\alpha,\varepsilon}:\operatorname{Hom}_A(M,M^*) \longrightarrow \operatorname{Hom}_A(M,M^*))$$

such that $\phi = \psi+\varepsilon\psi^* \in \operatorname{Hom}_A(M,M^*)$ is an isomorphism. Note that ϕ can be viewed as an (α,ε)-sesquilinear pairing

$$\phi : M \times M \longrightarrow A \ ; \ (x,y) \longmapsto \phi(x)(y)$$

such that

$$\text{i)} \ \phi(ax,by) = b\phi(x,y)\alpha(a)$$

$$\text{ii)} \ \phi(y,x) = \varepsilon\alpha(\phi(x,y))$$

for all $a,b \in A$, $x,y \in M$.
In keeping with our previous convention we shall now assume that $A = \mathbb{Z}[\pi]$ is a group ring and that the (α,ε)-quadratic L-groups $L_*(A^{\alpha},\varepsilon) \equiv L_*^s(A^{\alpha},\varepsilon)$ are the simple L-groups defined using based A-modules and simple isomorphisms (with $\tau = 0 \in Wh(\pi)$), although there are versions of the theory for arbitrary rings with antistructure and for the free and projective L-groups L_*^h, L_*^p. As we shall be dealing with various antistructures on the same

group ring $\mathbb{Z}[\pi]$ we shall write $M^{*,\alpha}$ for the dual of an A-module M with respect to an antistructure (α,ε), and $Q_*^{\pi,\alpha}(C,\varepsilon)$ for the (α,ε)-quadratic Q-groups of an A-module chain complex C, abbreviating to $Q_*^{\pi,\alpha}(C)$ if $\varepsilon = 1$. If α is the w-twisted involution on $\mathbb{Z}[\pi]$ for some orientation map $w:\pi \longrightarrow \mathbb{Z}_2 = \{\pm 1\}$, that is

$$\alpha(g) = w(g)g^{-1} \in \mathbb{Z}[\pi] \quad (g \in \pi) ,$$

the dual A-module $M^{*,\alpha}$ is denoted by $M^{*,w}$, the Q-groups $Q_*^{\pi,\alpha}(C,\varepsilon)$ are denoted by $Q_*^{\pi,w}(C,\varepsilon)$, and the L-groups $L_*(\mathbb{Z}[\pi]^{\alpha},\varepsilon)$ are denoted by $L_*(\mathbb{Z}[\pi^w],\varepsilon)$, with $Q_*^{\pi,w}(C)$, $L_*(\mathbb{Z}[\pi^w])$ if $\varepsilon = 1$.

Let now (X,Y) be a connected codimension 1 CW pair of type C) with $\pi_1(X) = \pi_1(Y)$. As $\xi:Y \longrightarrow BG(1)$ is non-trivial the double covering $S(\xi)$ of Y is determined by a group extension

$$\{1\} \longrightarrow \pi_1(S(\xi)) \longrightarrow \pi_1(Y) \xrightarrow{w_1(\xi)} \mathbb{Z}_2 \longrightarrow \{1\}$$

which we shall write as

$$\{1\} \longrightarrow \pi' \xrightarrow{p} \pi \xrightarrow{\xi} \mathbb{Z}_2 \longrightarrow \{1\} .$$

Denote the orientation map of X by

$$w(X) = w : \pi_1(X) = \pi \longrightarrow \mathbb{Z}_2 ,$$

so that the other orientation maps are given by

$$w(Y) = w\xi : \pi_1(Y) = \pi \longrightarrow \mathbb{Z}_2 ,$$

$$w(Z) = w(S(\xi)) = w' : \pi_1(Z) = \pi_1(S(\xi)) = \pi' \xrightarrow{p} \pi \xrightarrow{w} \mathbb{Z}_2 .$$

As before, denote the LS-groups of (X,Y)

$$LS_*\begin{pmatrix} \pi_1(S(\xi)) & \xrightarrow{\simeq} & \pi_1(Z) \\ \downarrow & \Phi & \downarrow \\ \pi_1(Y) & \xrightarrow{\simeq} & \pi_1(X) \end{pmatrix}$$

by $LN_*(\pi' \longrightarrow \pi, w)$. By Proposition 7.2.1 ii) the LN-groups fit into the exact sequence

$$\ldots \longrightarrow L_{n+1}(\mathbb{Z}[\pi^w]) \longrightarrow LN_{n-1}(\pi' \longrightarrow \pi, w)$$
$$\longrightarrow L_n(\partial\xi^!:\mathbb{Z}[\pi^{w\xi}] \longrightarrow \mathbb{Z}[\pi'^{w'}]) \longrightarrow L_n(\mathbb{Z}[\pi^w]) \longrightarrow \ldots \ .$$

By Proposition 7.5.1 ii) $LN_{n-1}(\pi' \longrightarrow \pi, w)$ $(n \geqslant 1)$ is the cobordism group of contractible n-dimensional quadratic Poincaré splittings over Φ, i.e. of pairs

$$((C, \psi \in Q_{n-1}^{\pi, w\xi}(C)), (f:\partial\xi^! C \longrightarrow D, (\delta\psi, \partial\xi^!\psi) \in Q_n^{\pi', w'}(f)))$$

consisting of an (n-1)-dimensional quadratic Poincaré complex (C,ψ) over $\mathbb{Z}[\pi^{w\xi}]$ and an n-dimensional quadratic Poincaré pair $(f:\partial\xi^! C \longrightarrow D, (\delta\psi, \partial\xi^!\psi))$ over $\mathbb{Z}[\pi'^{w'}]$ such that the $\mathbb{Z}[\pi]$-module chain map

$$\begin{pmatrix} i \\ 1\otimes f \end{pmatrix} : \mathbb{Z}[\pi]\otimes_{\mathbb{Z}[\pi']}\partial\xi^! C \longrightarrow C \oplus \mathbb{Z}[\pi]\otimes_{\mathbb{Z}[\pi']} D$$

is a (simple) chain equivalence, where i is the $\mathbb{Z}[\pi]$-module chain map appearing in the n-dimensional quadratic Poincaré pair over $\mathbb{Z}[\pi^w]$

$$\xi^!(C,\psi) = (i:\mathbb{Z}[\pi]\otimes_{\mathbb{Z}[\pi']}\partial\xi^! C \longrightarrow C, (\xi^!\psi, 1\otimes\partial\xi^!\psi) \in Q_n^{\pi,w}(i)) \ .$$

We now have to give an algebraic definition of $\xi^!(C,\psi)$.

Choose an element $t \in \pi$ such that $\xi(t) = -1$, so that as a set π is the disjoint union

$$\pi = \pi' \cup t\pi' \ .$$

Given $a = \sum_{g\in\pi} n_g g \in \mathbb{Z}[\pi]$ let

$$[a] = \sum_{g'\in\pi'} n_{g'} g' \in \mathbb{Z}[\pi'] \subset \mathbb{Z}[\pi] \ ,$$

so that

$$a = [a] + [at]t^{-1} \in \mathbb{Z}[\pi]$$

and as an additive group $\mathbb{Z}[\pi]$ has a direct sum decomposition

$$\mathbb{Z}[\pi] = \mathbb{Z}[\pi']\oplus t\mathbb{Z}[\pi'] .$$

Define a ring automorphism

$$\lambda : \mathbb{Z}[\pi] \longrightarrow \mathbb{Z}[\pi] ; a \longmapsto [a] - [at]t^{-1}$$

such that $\lambda^2 = 1$. The induced functor

$$\lambda : \underline{\underline{P}}(\mathbb{Z}[\pi]) = (\text{f.g. projective } \mathbb{Z}[\pi]\text{-modules}) \longrightarrow \underline{\underline{P}}(\mathbb{Z}[\pi]) ;$$
$$M \longmapsto \lambda M$$

sends $M \in |\underline{\underline{P}}(\mathbb{Z}[\pi])|$ to the f.g. projective $\mathbb{Z}[\pi]$-module λM with the same additive group and

$$\mathbb{Z}[\pi] \times \lambda M \longrightarrow \lambda M ; (a,x) \longmapsto \lambda(a)x .$$

The inclusion $p:\pi' \longrightarrow \pi$ induces an inflation functor

$$p_! : \underline{\underline{P}}(\mathbb{Z}[\pi']) \longrightarrow \underline{\underline{P}}(\mathbb{Z}[\pi]) ;$$
$$N \longmapsto p_!N = \mathbb{Z}[\pi]\otimes_{\mathbb{Z}[\pi']}N ,$$

and there is also defined a restriction functor

$$p^! : \underline{\underline{P}}(\mathbb{Z}[\pi]) \longrightarrow \underline{\underline{P}}(\mathbb{Z}[\pi']) ; M \longmapsto p^!M ,$$

sending $M \in |\underline{\underline{P}}(\mathbb{Z}[\pi])|$ to the f.g. projective $\mathbb{Z}[\pi']$-module $p^!M$ with the same additive group and

$$\mathbb{Z}[\pi'] \times p^!M \longrightarrow p^!M ; (a,x) \longmapsto ax .$$

The two functors are related by a short exact sequence

$$0 \longrightarrow \lambda M \xrightarrow{k} p_!p^!M \xrightarrow{i} M \longrightarrow 0$$

which is split (non-canonically), with

$$i : p_!p^!M \longrightarrow M ; a\otimes x \longmapsto ax$$

$$k : \lambda M \longrightarrow p_!p^!M ; x \longmapsto 1\otimes x - t\otimes t^{-1}x .$$

Given $M \in |\underline{\underline{P}}(\mathbb{Z}[\pi])|$ use the $\mathbb{Z}[\pi']$-module isomorphism

$$(p^!M)^{*,w'} \longrightarrow p^!(M^{*,w}) ; f \longmapsto (x \longmapsto f(x) + tf(t^{-1}x))$$

as an identification, and define a $\mathbb{Z}[\mathbb{Z}_2]$-module morphism

$$p^! : \mathrm{Hom}_{\mathbb{Z}[\pi]}(M,M^{*,w}) \longrightarrow \mathrm{Hom}_{\mathbb{Z}[\pi']}(p^!M,(p^!M)^{*,w'}) \ ;$$

$$\phi \longmapsto (p^!\phi : x \longmapsto (y \longmapsto [\phi(x)(y)])) \ ,$$

with $T \in \mathbb{Z}_2$ acting by the duality involution $T:\phi \longmapsto \phi^*$ on both sides. Given $N \in |\underline{\underline{P}}(\mathbb{Z}[\pi'])|$ use the $\mathbb{Z}[\pi]$-module isomorphism

$$p_!(N^{*,w'}) \longrightarrow (p_!N)^{*,w} \ ; \ a\otimes f \longmapsto (b\otimes x \longmapsto bf(x)\bar{a})$$

as an identification, and define a $\mathbb{Z}[\mathbb{Z}_2]$-module morphism

$$p_! : \mathrm{Hom}_{\mathbb{Z}[\pi']}(N,N^{*,w'}) \longrightarrow \mathrm{Hom}_{\mathbb{Z}[\pi]}(p_!N,(p_!N)^{*,w}) \ ;$$

$$\theta \longmapsto (p_!\theta : x \longmapsto (y \longmapsto (1\otimes\theta)(x)(y))) \ .$$

The $\mathbb{Z}[\mathbb{Z}_2]$-module morphisms $p^!, p_!$ can also be defined using the orientation map $w\xi$ instead of w. Note that

$$M^{*,w\xi} = \lambda(M^{*,w}) \ .$$

Given a finite-dimensional $\mathbb{Z}[\pi]$-module chain complex C use the $\mathbb{Z}[\mathbb{Z}_2]$-module chain map

$$p^! : \mathrm{Hom}_{\mathbb{Z}[\pi]}(C^{*,w\xi},C) \longrightarrow \mathrm{Hom}_{\mathbb{Z}[\pi']}((p^!C)^{*,w'},p^!C)$$

to define restriction maps in the Q-groups

$$p^! : Q_*^{\pi,w\xi}(C) \longrightarrow Q_*^{\pi',w'}(p^!C) \ .$$

Given a finite-dimensional $\mathbb{Z}[\pi']$-module chain complex D use the $\mathbb{Z}[\mathbb{Z}_2]$-module chain map

$$p_! : \mathrm{Hom}_{\mathbb{Z}[\pi']}(D^{*,w'},D) \longrightarrow \mathrm{Hom}_{\mathbb{Z}[\pi]}((p_!D)^{*,w},p_!D)$$

to define inflation maps in the Q-groups

$$p_! : Q_*^{\pi',w'}(D) \longrightarrow Q_*^{\pi,w}(p_!D) \ .$$

The short exact sequence of finite-dimensional $\mathbb{Z}[\pi]$-module chain complexes

$$0 \longrightarrow \lambda C \xrightarrow{\ k\ } p_!p^!C \xrightarrow{\ i\ } C \longrightarrow 0$$

is split when regarded as an exact sequence of graded $\mathbb{Z}[\pi]$-modules. Thus applying $\mathrm{Hom}_{\mathbb{Z}[\pi]}(C^{*,w},-)$ there is obtained a short exact sequence of $\mathbb{Z}$-module chain complexes

$$0 \longrightarrow \mathrm{Hom}_{\mathbb{Z}[\pi]}(C^{*,w},\lambda C) \xrightarrow{p^!} \mathrm{Hom}_{\mathbb{Z}[\pi]}(C^{*,w},p_!p^!C) \xrightarrow{i} \mathrm{Hom}_{\mathbb{Z}[\pi]}(C^{*,w},C) \longrightarrow 0 \quad .$$

Using the natural isomorphisms of $\mathbb{Z}$-module chain complexes

$$\mathrm{Hom}_{\mathbb{Z}[\pi]}(C^{*,w\xi},C) \xrightarrow{\simeq} \mathrm{Hom}_{\mathbb{Z}[\pi]}(C^{*,w},\lambda C) \; ; \; \phi \longmapsto (x \longmapsto \phi(x))$$

$$\mathrm{Hom}_{\mathbb{Z}[\pi']}((p^!C)^{*,w'},p^!C) \xrightarrow{\simeq} \mathrm{Hom}_{\mathbb{Z}[\pi]}(C^{*,w},p_!p^!C) \; ;$$
$$\theta \longmapsto (x \longmapsto 1\otimes\theta(x) + t\otimes\theta(t^{-1}x))$$

as identifications there is obtained a sequence

$$0 \longrightarrow \mathrm{Hom}_{\mathbb{Z}[\pi]}(C^{*,w\xi},C) \xrightarrow{p^!} \mathrm{Hom}_{\mathbb{Z}[\pi']}((p^!C)^{*,w'},p^!C) \xrightarrow{i} \mathrm{Hom}_{\mathbb{Z}[\pi]}(C^{*,w},C) \longrightarrow 0$$

which is in fact a short exact sequence of $\mathbb{Z}[\mathbb{Z}_2]$-module chain complexes inducing a long exact sequence of Q-groups

$$\ldots \longrightarrow Q_n^{\pi,w\xi}(C) \xrightarrow{p^!} Q_n^{\pi',w'}(p^!C) \xrightarrow{i} Q_n^{\pi,w}(C) \longrightarrow Q_{n-1}^{\pi,w\xi}(C) \longrightarrow \ldots$$
$$(n \in \mathbb{Z}) \; .$$

Define a $\mathbb{Z}[\mathbb{Z}_2]$-module chain map

$$\xi^! : \mathrm{Hom}_{\mathbb{Z}[\pi]}(C^{*,w\xi},C) \longrightarrow \Omega C(i_{\%}:\mathrm{Hom}_{\mathbb{Z}[\pi]}((p_!p^!C)^{*,w},p_!p^!C) \longrightarrow \mathrm{Hom}_{\mathbb{Z}[\pi]}(C^{*,w},C)) \; ;$$
$$\phi \longmapsto (0,p_!p^!\phi) \; ,$$

so that there is defined a natural transformation of long exact sequences of Q-groups

$$\begin{array}{ccccccccc}
\cdots \longrightarrow & Q_n^{\pi,w\xi}(C) & \xrightarrow{p^!} & Q_n^{\pi',w'}(p^!C) & \xrightarrow{i} & Q_n^{\pi,w}(C) & \longrightarrow & Q_{n-1}^{\pi,w\xi}(C) & \longrightarrow \cdots \\
& \Big\downarrow \xi^! & & \Big\downarrow & & \Big\| & & \Big\downarrow \xi^! & \\
\cdots \longrightarrow & Q_{n+1}^{\pi,w}(i) & \longrightarrow & Q_n^{\pi,w}(p_!p^!C) & \xrightarrow{i_\%} & Q_n^{\pi,w}(C) & \longrightarrow & Q_n^{\pi,w}(i) & \longrightarrow \cdots \ .
\end{array}$$

An element $\psi \in Q_n^{\pi,w\xi}(C)$ is sent by $\xi^!$ to the element

$$\xi^!(\psi) = (\delta\psi, p_!p^!\psi) \in Q_{n+1}^{\pi,w}(i) \ .$$

<u>Proposition 7.6.3</u> The transfer maps in quadratic L-theory associated to a connected codimension 1 CW pair (X,Y) of type C) with $\pi_1(X) = \pi_1(Y) = \pi$ are given algebraically by

$$\xi^! : L_n(\mathbb{Z}[\pi^{w\xi}]) \longrightarrow L_{n+1}(p:\mathbb{Z}[\pi'^{w'}] \longrightarrow \mathbb{Z}[\pi^w]) \ ;$$

$$(C, \psi \in Q_n^{\pi,w\xi}(C)) \longmapsto$$

$$((p^!C, p^!\xi \in Q_n^{\pi',w'}(p^!C)), (i : p_!p^!C \longrightarrow C, \xi^!(\psi) \in Q_{n+1}^{\pi,w}(i)))$$

$$(\pi_1(Z) = \pi_1(S(\xi)) = \pi', \ w(X) = w, \ w' = wp) .$$

[]

(This generalizes the algebraic expression for $\partial\xi^! = p^!$ of Thomas [1]).

Continuing with the previous terminology define

$$\mu : \mathbb{Z}[\pi'] \longrightarrow \mathbb{Z}[\pi'] \ ; \ \sum_{g' \in \pi'} n_{g'} g' \longmapsto \sum_{g' \in \pi'} n_{g'} t^{-1} g' t \ ,$$

an automorphism of the ring $\mathbb{Z}[\pi']$. The induced functor

$$\mu : \underline{\underline{P}}(\mathbb{Z}[\pi']) \longrightarrow \underline{\underline{P}}(\mathbb{Z}[\pi']) \ ; \ N \longmapsto \mu N$$

sends a f.g. projective $\mathbb{Z}[\pi']$-module N to the f.g. projective $\mathbb{Z}[\pi']$-module μN with the same additive group and

$$\mathbb{Z}[\pi'] \times \mu N \longrightarrow \mu N \ ; \ (a,x) \longmapsto \mu(a)x \ .$$

Define an antistructure (β, t^2) on $\mathbb{Z}[\pi']$ by

$$\beta : \mathbb{Z}[\pi'] \longrightarrow \mathbb{Z}[\pi'] \ ;$$

$$\sum_{g' \in \pi'} n_{g'} g' \longmapsto \sum_{g' \in \pi'} w'(g') n_{g'} t^{-1} g'^{-1} t \ .$$

Then for any $N \in |\underline{\underline{P}}(\mathbb{Z}[\pi'])|$ there are defined natural isomorphisms in $\underline{\underline{P}}(\mathbb{Z}[\pi'])$

$$N \oplus \mu N \xrightarrow{\sim} p^! p_! N \ ; \ (x,y) \longmapsto 1 \otimes x + t \otimes y$$

$$N^{*,\beta} \xrightarrow{\sim} \mu(N^{*,w'}) \ ; \ f \longmapsto (x \longmapsto \mu^{-1}(f(x)))$$

which we shall use as identifications. In particular, we have identifications of $\mathbb{Z}$-modules

$$\begin{aligned} \mathrm{Hom}_{\mathbb{Z}[\pi]}(p_!N,(p_!N)^{*,w}) &= \mathrm{Hom}_{\mathbb{Z}[\pi']}(N,p^!p_!(N^{*,w'})) \\ &= \mathrm{Hom}_{\mathbb{Z}[\pi']}(N,N^{*,w'} \oplus \mu(N^{*,w'})) \\ &= \mathrm{Hom}_{\mathbb{Z}[\pi']}(N,N^{*,w'}) \oplus \mathrm{Hom}_{\mathbb{Z}[\pi']}(N,N^{*,\beta}) \ , \end{aligned}$$

and hence an identification of $\mathbb{Z}[\mathbb{Z}_2]$-modules

$$\mathrm{Hom}_{\mathbb{Z}[\pi]}(p_!N,(p_!N)^{*,w}) = \mathrm{Hom}_{\mathbb{Z}[\pi']}(N,N^{*,w'}) \oplus \mathrm{Hom}_{\mathbb{Z}[\pi']}(N,N^{*,\beta})$$

with $T \in \mathbb{Z}_2$ acting by the duality involution $T:\phi \longmapsto \phi^*$ on the left and by $T \oplus T_{w(t)t^2}$ on the right.

Given a finite-dimensional $\mathbb{Z}[\pi']$-module chain complex D we thus have an identification of $\mathbb{Z}[\mathbb{Z}_2]$-module chain complexes

$$\mathrm{Hom}_{\mathbb{Z}[\pi]}((p_!D)^{*,w},p_!D) = \mathrm{Hom}_{\mathbb{Z}[\pi']}(D^{*,w'},D) \oplus \mathrm{Hom}_{\mathbb{Z}[\pi']}(D^{*,\beta},D) \ ,$$

so that

$$Q_*^{\pi,w}(p_!D) = Q_*^{\pi',w'}(D) \oplus Q_*^{\pi',\beta}(D,w(t)t^2)$$

Replacing w by $w\xi$ we also have

$$Q_*^{\pi,w\xi}(p_!D) = Q_*^{\pi',w'}(D) \oplus Q_*^{\pi',\beta}(D,-w(t)t^2)$$

since $\xi(t) = -1$.

Assume now that the underlying codimension 1 CW pair (X,Y) is a formally $(n,n-1)$-dimensional normal pair (in the sense of §7.5) and that there is given a formally $(n,n-1)$-dimensional topological normal map

$$(f,b) : (M,N) \longrightarrow (X,Y) \quad ,$$

denoting the restriction normal maps by

$$(f,b)| = (g,c) : N = f^{-1}(Y) \longrightarrow Y$$

$$(f,b)| = (h,d) : (P,S(\nu)) = f^{-1}(Z,S(\xi)) \longrightarrow (Z,S(\xi)) \ .$$

According to Proposition 7.5.4 ambient surgery on (g,c) inside (f,b) has the algebraic effect of surgery on the quadratic kernel pair $(\sigma_*(g,c),\sigma_*(h,d))$ preserving the union $\xi^!\sigma_*(g,c) \cup p\sigma_*(h,d) = \sigma_*(f,b)$. We shall now associate to the pair $(\sigma_*(g,c),\sigma_*(h,d))$ a formally $(n-1)$-dimensional $(\beta,-w(t)t^2)$-quadratic complex over $\mathbb{Z}[\pi']$ $\sigma_*^!(f,b)$ such that surgery on the pair determines surgery on the complex, and such that if $f:M \longrightarrow X$ is an s-triangulation algebraic surgery determines geometric surgery, generalizing the treatment of type C) codimension 1 surgery due to Wall [4,§12C].

$$\text{The quadratic kernel of } \begin{cases} (f,b):M \longrightarrow X \\ (g,c):N \longrightarrow Y \\ (h,d):(P,S(\nu)) \longrightarrow (Z,S(\xi)) \end{cases}$$

$$\text{is the formally } \begin{cases} n- \\ (n-1)- \\ n- \end{cases} \text{dimensional quadratic } \begin{cases} \text{complex} \\ \text{complex} \\ \text{pair} \end{cases}$$

$$\text{over } \begin{cases} \mathbb{Z}[\pi^w] \\ \mathbb{Z}[\pi^{w\xi}] \\ \mathbb{Z}[\pi'^{w'}] \end{cases}$$

$$\begin{cases} \sigma_*(f,b) = (C(f^!),\Psi \in Q_n^{\pi,w}(C(f^!))) \\ \sigma_*(g,c) = (C(g^!),\psi \in Q_{n-1}^{\pi,w\xi}(C(g^!))) = (C,\psi) \\ \sigma_*(h,d) = (e:p^!C(g^!) \longrightarrow C(h^!),(\delta\psi,p^!\psi) \in Q_n^{\pi',w'}(e)) \\ \qquad = (e:p^!C \longrightarrow D,(\delta\psi,p^!\psi)) \end{cases}$$

with $\begin{cases} C(f^!) \\ C = C(g^!) \\ D = C(h^!) \end{cases}$ the algebraic mapping cone of the $\begin{cases} \mathbb{Z}[\pi]- \\ \mathbb{Z}[\pi]\text{-module} \\ \mathbb{Z}[\pi']- \end{cases}$

Umkehr chain map

$$\begin{cases} f^! : C(\tilde{X})^{n-*,w} \xrightarrow{\tilde{f}^*} C(\tilde{M})^{n-*,w} \xrightarrow[\sim]{[M]\cap -} C(\tilde{M}) \\ g^! : C(\tilde{Y})^{n-1-*,w\xi} \xrightarrow{\tilde{g}^*} C(\tilde{N})^{n-1-*,w\xi} \xrightarrow[\sim]{[N]\cap -} C(\tilde{N}) \\ h^! : C(\tilde{Z},\widetilde{S(\xi)})^{n-*,w'} \xrightarrow{\tilde{h}^*} C(\tilde{P},\widetilde{S(\nu)})^{n-*,w'} \xrightarrow[\sim]{[P]\cap -} C(\tilde{P}) \end{cases} ,$$

where $\begin{cases} \tilde{X} \\ \tilde{Y} \\ (\tilde{Z},\widetilde{S(\xi)}) \end{cases}$ is the universal cover of $\begin{cases} X \\ Y \\ (Z,S(\xi)) \end{cases}$ and

$\begin{cases} \tilde{M} \\ \tilde{N} \\ (\tilde{P},\widetilde{S(\nu)}) \end{cases}$ is the induced cover of $\begin{cases} M \\ N \\ (P,S(\nu)) \end{cases}$. It follows from

the geometric decomposition of the normal map

$$(f,b) = (g,c)^! \cup (h,d) : M = E(\nu) \cup_{S(\nu)} P \longrightarrow X = E(\xi) \cup_{S(\xi)} Z$$

that there is an algebraic decomposition of the quadratic kernel

$$\sigma_*(f,b) = \xi^!\sigma_*(g,c) \cup p\sigma_*(h,d) .$$

Define the <u>antiquadratic kernel</u> of (f,b) to be the formally $(n-1)$-dimensional $(\beta,-w(t)t^2)$-quadratic complex over $\mathbb{Z}[\pi']$

$$\sigma_*^!(f,b) = (D,e_\% p^! t(\psi) \in Q_{n-1}^{\pi',\beta}(D,-w(t)t^2)) .$$

The function $(\sigma_*(g,c),\sigma_*(h,d)) \longmapsto \sigma_*^!(f,b)$ will now be used to recover the expression of the type C) codimension 1 LN-groups $LN_*(\pi' \longrightarrow \pi, w)$ as the L-groups of a ring with antistructure due to Wall [4,Thm.12.9]. The identification of an exact sequence characterizing the LN-groups with a relative L-theory exact sequence was first obtained by Wall [4,Cor.12.9.2] in the special case $\pi = \pi' \times \mathbb{Z}_2^-$ and by Cappell and Shaneson [3] (implicitly) and Hambleton [1] in general.

<u>Proposition 7.6.4</u> Given a group extension

$$\{1\} \longrightarrow \pi' \xrightarrow{\;p\;} \pi \xrightarrow{\;\xi\;} \mathbb{Z}_2 \longrightarrow \{1\}$$

and an orientation map $w:\pi \longrightarrow \mathbb{Z}_2$ there is defined a natural isomorphism of exact sequences of abelian groups

$$\begin{array}{ccccc}
\cdots \longrightarrow & L_{n+1}(\mathbb{Z}[\pi^w]) & \longrightarrow & LN_{n-1}(\pi' \longrightarrow \pi, w) & \\
& \Big\downarrow t \,\wr & & \Big\downarrow \wr & \\
\cdots \longrightarrow & L_{n+1}(\mathbb{Z}[\pi]^{\alpha}, w(t)t^2) & \xrightarrow{\;p^!\;} & L_{n+1}(\mathbb{Z}[\pi']^{\beta}, w(t)t^2) &
\end{array}$$

$$\begin{array}{ccccc}
\longrightarrow & L_n(p^!:\mathbb{Z}[\pi^{w\xi}] \longrightarrow \mathbb{Z}[\pi'^{w'}]) & \longrightarrow & L_n(\mathbb{Z}[\pi^w]) & \longrightarrow \cdots \\
& \Big\downarrow \wr & & \Big\downarrow t\,\wr & \\
\longrightarrow & L_{n+1}(p^!:\mathbb{Z}[\pi]^{\alpha} \longrightarrow \mathbb{Z}[\pi']^{\beta}, w(t)t^2) & \longrightarrow & L_n(\mathbb{Z}[\pi]^{\alpha}, w(t)t^2) & \longrightarrow \cdots ,
\end{array}$$

with $t \in \pi$ an element such that $\xi(t) = -1$, (α, t^2) the antistructure on $\mathbb{Z}[\pi]$ defined by

$$\alpha : \mathbb{Z}[\pi] \longrightarrow \mathbb{Z}[\pi] \; ; \; \sum_{g \in \pi} n_g g \longmapsto \sum_{g \in \pi} w(g) n_g t^{-1} g^{-1} t$$

and (β, t^2) the antistructure on $\mathbb{Z}[\pi']$ defined by

$$\beta = \alpha| \; : \; \mathbb{Z}[\pi'] \longrightarrow \mathbb{Z}[\pi'] \; .$$

<u>Proof</u>: Given a f.g. projective $\mathbb{Z}[\pi]$-module M use the scaling isomorphism of the dual $\mathbb{Z}[\pi]$-modules

$$t : M^{*,w} \xrightarrow{\sim} M^{*,\alpha} \; ; \; f \longmapsto (x \longmapsto f(x)t)$$

to define a scaling isomorphism of $\mathbb{Z}[\mathbb{Z}_2]$-modules

$$t : (\mathrm{Hom}_{\mathbb{Z}[\pi]}(M,M^{*,w}),T_{w,1}) \xrightarrow{\sim} (\mathrm{Hom}_{\mathbb{Z}[\pi]}(M,M^{*,\alpha}),T_{\alpha,w(t)t^2}) \; .$$

Given a finite-dimensional $\mathbb{Z}[\pi]$-module chain complex C there is thus defined a scaling isomorphism of $\mathbb{Z}[\mathbb{Z}_2]$-module chain complexes

$$t : (\mathrm{Hom}_{\mathbb{Z}[\pi]}(C^{*,w},C),T_{w,1}) \xrightarrow{\sim} (\mathrm{Hom}_{\mathbb{Z}[\pi]}(C^{*,\alpha},C),T_{\alpha,w(t)t^2})$$

inducing scaling isomorphisms of Q-groups

$$t : Q_n^{\pi,w}(C) \xrightarrow{\sim} Q_n^{\pi,\alpha}(C,w(t)t^2)$$

and hence also of L-groups

$$t : L_n(\mathbb{Z}[\pi^w]) \xrightarrow{\sim} L_n(\mathbb{Z}[\pi]^\alpha,w(t)t^2) \; ; \; (C,\psi) \longmapsto (C,t\psi) \quad (n \geqslant 0) \; .$$

Note that the morphisms induced in the Q-groups by the $\mathbb{Z}[\pi]$-module chain map

$$\lambda k : C \longrightarrow p_!p^!C \; ; \; x \longmapsto 1\otimes x + t\otimes t^{-1}x$$

are given by

$$\begin{pmatrix} p^! \\ p^!t \end{pmatrix} : Q_n^{\pi,w}(C)$$

$$\longrightarrow Q_n^{\pi,w}(p_!p^!C) = Q_n^{\pi',w'}(p^!C) \oplus Q_n^{\pi',\beta}(p^!C,w(t)t^2) \; ,$$

sending the element

$$\psi = \{\psi_s : C^{r,w} \times C^{n-r-s,w} \longrightarrow \mathbb{Z}[\pi] \,|\, r \in \mathbb{Z}, s \geqslant 0\} \in Q_n^{\pi,w}(C)$$

to the element

$$(p^! \psi, p^! t\psi) = (\{p^! \psi_s : C^{r,w} \times C^{n-r-s,w} \xrightarrow{\psi_s} \mathbb{Z}[\pi] \xrightarrow{p^!} \mathbb{Z}[\pi']\},$$

$$\{p^! t\psi_s : C^{r,w} \times C^{n-r-s,w} \xrightarrow{\psi_s} \mathbb{Z}[\pi] \xrightarrow{p^! t} \mathbb{Z}[\pi']\})$$

$$\in Q_n^{\pi',w'}(p^!C) \oplus Q_n^{\pi',\beta}(p^!C, w(t)t^2) \qquad ,$$

defined using the abelian group morphisms

$$p^! : \mathbb{Z}[\pi] \longrightarrow \mathbb{Z}[\pi'] \ ; \ a \longmapsto [a]$$

$$p^! t : \mathbb{Z}[\pi] \longrightarrow \mathbb{Z}[\pi'] \ ; \ a \longmapsto [at]$$

and identifying

$$\mathrm{Hom}_{\mathbb{Z}[\pi]}(C_r, \mathbb{Z}[\pi]) = C^{r,w} = (p^!C)^{r,w'} = (p^!C)^{r,\alpha} = \mathrm{Hom}_{\mathbb{Z}[\pi']}(p^!C_r, \mathbb{Z}[\pi'])$$

as abelian groups.

Next, we shall define the natural isomorphisms

$$LN_{n-1}(\pi' \longrightarrow \pi, w) \xrightarrow{\sim} L_{n-1}(\mathbb{Z}[\pi']^{\beta}, -w(t)t^2) \ ; \ s \longmapsto s' \qquad (n \geqslant 1)$$

(identifying $L_{n-1}(\mathbb{Z}[\pi']^{\beta}, -w(t)t^2) = L_{n+1}(\mathbb{Z}[\pi']^{\beta}, w(t)t^2)$ by the skew-suspension isomorphisms).

Given an element

$$s = ((C, \psi \in Q_{n-1}^{\pi,w}(C)), (e : p^!C \longrightarrow D, (\delta\psi, p^!\psi) \in Q_n^{\pi',w'}(e)))$$

$$\in LN_{n-1}(\pi' \longrightarrow \pi, w)$$

observe that the composite $\mathbb{Z}[\pi]$-module chain map

$$j : C \xrightarrow{\lambda k} p_! p^! C \xrightarrow{\lambda p_! e} p_! D$$

is a (simple) chain equivalence, since it is given that the $\mathbb{Z}[\pi]$-module chain map

$$\begin{pmatrix} i \\ p^! e \end{pmatrix} : p_! p^! C \longrightarrow C \oplus p_! D$$

is a (simple) chain equivalence, $p_! p^! C$ fits into the short exact sequence of $\mathbb{Z}[\pi]$-module chain complexes

$$0 \longrightarrow C \xrightarrow{\ k\ } p_! p^! C \xrightarrow{\ \lambda i\ } \lambda C \longrightarrow 0$$

and the ring morphisms $p:\mathbb{Z}[\pi'] \longrightarrow \mathbb{Z}[\pi]$, $\lambda:\mathbb{Z}[\pi] \longrightarrow \mathbb{Z}[\pi]$ are such that $\lambda p = p$. The induced isomorphism of Q-groups

$$j_{\%} = \begin{pmatrix} e_{\%}p^! \\ e_{\%}p^! t \end{pmatrix} :$$

$$Q_{n-1}^{\pi,w\xi}(C) \xrightarrow{\ \sim\ } Q_{n-1}^{\pi,w\xi}(p_!D) = Q_{n-1}^{\pi',w'}(D) \oplus Q_{n-1}^{\pi',\beta}(D,-w(t)t^2)$$

sends $\psi \in Q_{n-1}^{\pi,w\xi}(C)$ to the element

$$j_{\%}(\psi) = (0, e_{\%}p^! t(\psi)) .$$

Now $(D, e_{\%}p^! t(\psi) \in Q_{n-1}^{\pi',\beta}(D,-w(t)t^2))$ is an $(n-1)$-dimensional $(\beta,-w(t)t^2)$-quadratic Poincaré complex over $\mathbb{Z}[\pi']$, and the element corresponding to s is

$$s' = (D, e_{\%}p^! t(\psi)) \in L_{n-1}(\mathbb{Z}[\pi']^{\beta}, -w(t)t^2) .$$

Conversely, suppose given an element

$$s' = (D, \psi') \in L_{n-1}(\mathbb{Z}[\pi']^{\beta}, -w(t)t^2) .$$

Define an $(n-1)$-dimensional quadratic Poincaré complex (C,ψ) over $\mathbb{Z}[\pi^{w\xi}]$ by

$$C = p_!D \ , \ \psi = (0,\psi') \in Q_{n-1}^{\pi,w\xi}(C) = Q_{n-1}^{\pi',w'}(D) \oplus Q_{n-1}^{\pi',\beta}(D,-w(t)t^2) .$$

Define a $\mathbb{Z}[\pi']$-module chain map

$$e = (1\ 0) : p^! C = D \oplus \mu D \longrightarrow D$$

and let

$$(\delta\psi, p^!\psi) = (0, (1+T_{\beta,-w(t)t^2})\psi_0')$$

$$\in Q_n^{\pi',w'}(e) = Q_{n-1}^{\pi',w'}(D) \oplus H_{n-1}(\mathrm{Hom}_{\mathbb{Z}[\pi']}(D^{*,\beta}, D)) .$$

(The expression for $Q_n^{\pi',w'}(e)$ follows from the identification of the exact sequence

$$\ldots \longrightarrow Q_n^{\pi,w}(C) \longrightarrow Q_{n-1}^{\pi,w}(C) \xrightarrow{p^!} Q_{n-1}^{\pi',w'}(p^!C) \xrightarrow{i} Q_{n-1}^{\pi,w}(C) \longrightarrow \ldots$$

with direct sum of the exact sequence

$$\ldots \longrightarrow Q_n^{\pi',w'}(D) \xrightarrow{0} Q_{n-1}^{\pi',w'}(D) \xrightarrow{\begin{pmatrix}1\\0\end{pmatrix}} Q_{n-1}^{\pi',w'}(D) \oplus Q_{n-1}^{\pi',w'}(\mu D)$$

$$\xrightarrow{(0\ \ \iota)} Q_{n-1}^{\pi',w'}(D) \longrightarrow \ldots \qquad (*)$$

and the exact sequence

$$\ldots \longrightarrow Q_n^{\pi',\beta}(D,w(t)t^2) \xrightarrow{S} Q_{n-1}^{\pi',\beta}(D,-w(t)t^2)$$

$$\longrightarrow H_{n-1}(\mathrm{Hom}_{\mathbb{Z}[\pi']}(D^{*,\beta},D)) \longrightarrow Q_{n-1}^{\pi',\beta}(D,w(t)t^2) \longrightarrow \ldots \qquad (**)\ .$$

The isomorphisms $\iota : Q_*^{\pi',w'}(\mu D) \xrightarrow{\sim} Q_*^{\pi',w'}(D)$ appearing in (*) are those induced by the isomorphism of $\mathbb{Z}[\mathbb{Z}_2]$-module chain complexes

$$\iota : \mathrm{Hom}_{\mathbb{Z}[\pi']}((\mu D)^{*,w'},\mu D) \xrightarrow{\sim} \mathrm{Hom}_{\mathbb{Z}[\pi']}(D^{*,w'},D)\ ;\ h \longmapsto \mu^{-1}h\ ,$$

using the automorphism $\mu : \mathbb{Z}[\pi'^{w'}] \longrightarrow \mathbb{Z}[\pi'^{w'}]$ of the ring $\mathbb{Z}[\pi']$ with the w'-twisted involution to identify $(\mu D)^{*,w'} = \mu(D^{*,w'})$. The sequence (**) is a special case of the sequence of Proposition 1.1.3 ($p = 1$), using the skew-suspension isomorphisms

$$\overline{S} : Q_{n-2}^{\pi',\beta}(\Omega D,-w(t)t^2) \xrightarrow{\sim} Q_n^{\pi',\beta}(D,w(t)t^2)$$

as identifications). The element corresponding to s' is defined by

$$s = ((C,\psi),(e:p^!C \longrightarrow D,(\delta\psi,p^!\psi))) \in LN_{n-1}(\pi' \longrightarrow \pi,w)\ .$$

In order to verify that the diagram

$$\begin{array}{ccc}
L_{n+1}(\mathbb{Z}[\pi^w]) & \longrightarrow & LN_{n-1}(\pi' \longrightarrow \pi, w) \\
\Big\downarrow{\scriptstyle t\,\wr} & & \Big\downarrow{\scriptstyle \wr} \\
L_{n+1}(\mathbb{Z}[\pi]^w, w(t)t^2) & \xrightarrow{\;p^!\;} & L_{n+1}(\mathbb{Z}[\pi']^\beta, w(t)t^2)
\end{array}$$

commutes consider an $(n+1)$-dimensional quadratic Poincaré complex over $\mathbb{Z}[\pi^w]$ $(C, \psi \in Q_{n+1}^{\pi,w}(C))$. The composite

$$L_{n+1}(\mathbb{Z}[\pi^w]) \xrightarrow[\sim]{t} L_{n+1}(\mathbb{Z}[\pi]^\alpha, w(t)t^2) \xrightarrow{p^!} L_{n+1}(\mathbb{Z}[\pi']^\beta, w(t)t^2)$$

sends (C,ψ) to $(p^!C, p^!t\psi \in Q_{n+1}^{\pi',\beta}(p^!C, w(t)t^2)))$. Let

$$((g:\partial D \longrightarrow D, (\theta,\partial\theta) \in Q_n^{\pi,w\xi}(g)),$$

$$\left(\begin{array}{ccc}
p^!\partial D & \xrightarrow{\;p^!g\;} & p^!D \\
\Big\downarrow{\scriptstyle \partial_+ f} & T & \Big\downarrow{\scriptstyle f} \\
\partial_+ E & \xrightarrow{\;h\;} & E
\end{array}, (\upsilon, \partial_+\upsilon, p^!\theta, p^!\partial\theta) \in Q_{n+1}^{\pi',w'}(T)\right))$$

be a normal splitting of (C,ψ), as given by Proposition 7.5.1 i), so that up to (simple) homotopy equivalence

$$(C,\psi) = (D \cup_{p_!p^!D} p_!E, \xi^!\theta \cup p_!\upsilon) \quad .$$

The composite

$$L_{n+1}(\mathbb{Z}[\pi^w]) \longrightarrow LN_{n-1}(\pi' \longrightarrow \pi, w) \xrightarrow{\;\sim\;} L_{n+1}(\mathbb{Z}[\pi']^\beta, w(t)t^2)$$

sends (C,ψ) to $\bar{S}(\partial_+E, \partial_+f_\% p^!t(\partial\theta) \in Q_{n-1}^{\pi',\beta}(\partial_+E, -w(t)t^2))$. Define a $\mathbb{Z}[\pi]$-module chain map $j: p^!C \longrightarrow E$ by

$$j : p^!C_r = p^!D_r \oplus p^!p_!p^!D_{r-1} \oplus p^!p_!E_r \longrightarrow E_r \ ;$$

$$(x,y,a\otimes z) \longmapsto az \qquad (a \in \mathbb{Z}[\pi],\ z \in E_r) \ .$$

Surgery on $(p^!C, p^!t\psi)$ by the connected $(n+2)$-dimensional $(\beta, w(t)t^2)$-quadratic pair over $\mathbb{Z}[\pi']$

$$(j:p^!C \longrightarrow E,(0,p^!t\psi) \in Q^{\pi',\beta}_{n+2}(j,w(t)t^2))$$

results in the $(n+1)$-dimensional $(\beta,w(t)t^2)$-quadratic Poincaré complex over $\mathbb{Z}[\pi']$ $\overline{S}(\partial_+E,\partial_+f_\circ p^!t(\partial\theta))$, so that the above diagram does indeed commute. Moreover, if (C,ψ) is such that it admits a Poincaré splitting it is possible to set $\partial D = 0$, $\partial_+E = 0$, and the above procedure defines abelian group morphisms

$$L_{n+1}(p^!:\mathbb{Z}[\pi^{w\xi}] \longrightarrow \mathbb{Z}[\pi'^{w'}]) \longrightarrow L_{n+2}(p^!:\mathbb{Z}[\pi]^\alpha \longrightarrow \mathbb{Z}[\pi']^\beta,w(t)t^2) ;$$

$$((D,\theta),(f:p^!D \longrightarrow E,(\upsilon,p^!\theta)))$$

$$\longmapsto ((C,t\psi),(j:p^!C \longrightarrow E,(0,p^!t\psi)))$$

such that there are defined commutative diagrams

$$\begin{array}{ccc}
L_{n+1}(p^!:\mathbb{Z}[\pi^{w\xi}] \longrightarrow \mathbb{Z}[\pi'^{w'}]) & \longrightarrow & L_{n+1}(\mathbb{Z}[\pi^w]) \\
\downarrow & & \downarrow \wr\, t \\
L_{n+2}(p^!:\mathbb{Z}[\pi]^\alpha \longrightarrow \mathbb{Z}[\pi']^\beta,w(t)t^2) & \longrightarrow & L_{n+1}(\mathbb{Z}[\pi]^\alpha,w(t)t^2)
\end{array}$$

$$\begin{array}{ccc}
LN_n(\pi' \longrightarrow \pi,w) & \longrightarrow & L_{n+1}(p^!:\mathbb{Z}[\pi^{w\xi}] \longrightarrow \mathbb{Z}[\pi'^{w'}]) \\
\downarrow \wr & & \downarrow \\
L_{n+2}(\mathbb{Z}[\pi']^\beta,w(t)t^2) & \longrightarrow & L_{n+2}(p^!:\mathbb{Z}[\pi]^\alpha \longrightarrow \mathbb{Z}[\pi']^\beta,w(t)t^2) .
\end{array}$$

It now follows from a 5-lemma argument that these morphisms are also isomorphisms.

[]

(The natural isomorphism of exact sequences of Proposition 7.6.4 extends to a natural isomorphism of commutative braids of exact sequences, from

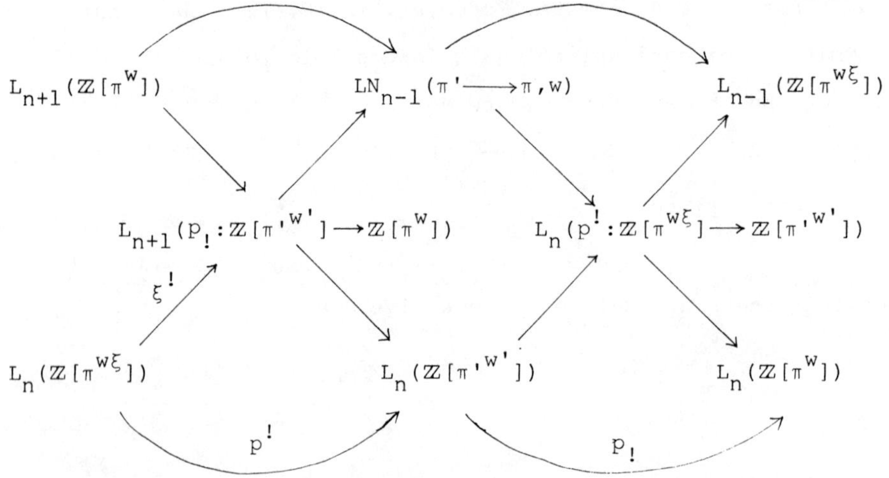

to

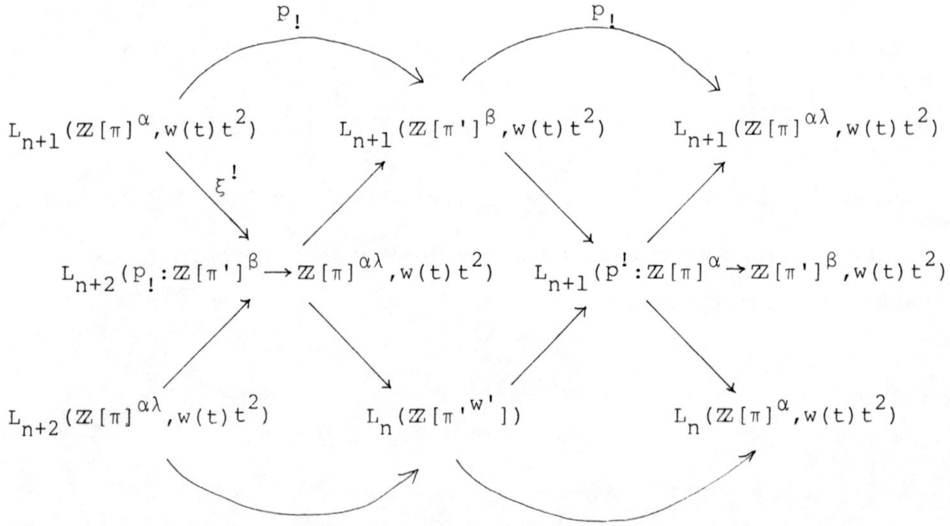

The isomorphism of exact sequences obtained by Hambleton [1] is the one involving the sequences 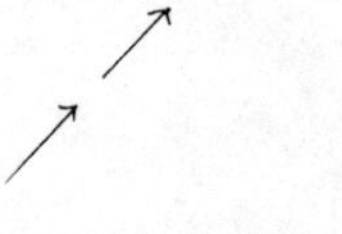).

The LN-groups have been used by Cappell and Shaneson [3],[4] and Hambleton [1] to detect which elements of the quadratic L-groups $L_*(\mathbb{Z}[\pi])$ of finite groups π are the surgery obstructions of topological normal maps of closed manifolds (i.e. belong to $\mathrm{im}(\sigma_*:H_*(K(\pi,1);\underline{\mathbb{L}}_0)\longrightarrow L_*(\mathbb{Z}[\pi]))\subseteq L_*(\mathbb{Z}[\pi]))$. In effect, they were making use of a special case of the natural transformation of exact sequences given by Proposition 7.4.6 iii)

$$\begin{array}{ccccc}
\ldots\longrightarrow \mathcal{S}_{n+1}(K(\pi,1)) & \longrightarrow & H_n(K(\pi,1),w;\underline{\mathbb{L}}_0) \\
\Big\downarrow u_\xi & & \Big\downarrow v_\xi \\
\ldots\longrightarrow LN_{n-1}(\pi'\longrightarrow\pi,w) & \longrightarrow & L_n(i^!:\mathbb{Z}[\pi^{w\xi}]\longrightarrow\mathbb{Z}[\pi'^{w'}]) \\
(=LS_{n-1}(\Phi)) & & (=LP_{n-1}(\Phi)) \\
\xrightarrow{\sigma_*} L_n(\mathbb{Z}[\pi^w]) & \longrightarrow & \mathcal{S}_n(K(\pi,1)) \longrightarrow \ldots \\
\Big\| & & \Big\downarrow u_\xi \\
\longrightarrow L_n(\mathbb{Z}[\pi^w]) & \longrightarrow & LN_{n-2}(\pi'\longrightarrow\pi,w)\longrightarrow\ldots
\end{array},$$

which is defined for any group extension

$$\{1\}\longrightarrow\pi'\xrightarrow{p}\pi\xrightarrow{\xi}\mathbb{Z}_2\longrightarrow\{1\}$$

(at least if the groups are finitely presented) and any orientation map $w:\pi\longrightarrow\mathbb{Z}_2$, with $w' = wp : \pi'\longrightarrow\mathbb{Z}_2$ as before. In connection with their work on pseudo-free group actions Cappell and Shaneson [3] associated to a non-trivial line bundle $\xi:M\longrightarrow BG(1)$ over an n-dimensional manifold M a surgery exact sequence

$$\ldots \longrightarrow [M \times D^1, M \times S^0; G/TOP, *] \oplus LN_{n+1}(\pi' \longrightarrow \pi, w)$$
$$\longrightarrow L_{n+1}(\mathbb{Z}[\pi^{w\xi}]) \longrightarrow \mathcal{S}^{TOP}(E(\xi), S(\xi))$$
$$\longrightarrow [M, G/TOP] \oplus LN_n(\pi' \longrightarrow \pi, w) \longrightarrow L_n(\mathbb{Z}[\pi^{w\xi}])$$

with

$$(\pi_1(M), w(M)) = (\pi, w\xi) \quad , \quad \pi_1(S(\xi)) = \pi' \ .$$

From the point of view of §7.2 this is just a part of the Mayer-Vietoris exact sequence

$$\ldots \longrightarrow H_{n+1}(M, w\xi; \underline{\mathbb{L}}_0) \oplus LN_{n+1}(\pi' \longrightarrow \pi, w)$$
$$\longrightarrow L_{n+1}(\mathbb{Z}[\pi^{w\xi}]) \longrightarrow \mathcal{S}_{n+2}(E(\xi), S(\xi))$$
$$\longrightarrow H_n(M, w\xi; \underline{\mathbb{L}}_0) \oplus LN_n(\pi' \longrightarrow \pi, w) \longrightarrow L_n(\mathbb{Z}[\pi^{w\xi}]) \longrightarrow \ldots$$

associated to the commutative braid of exact sequences of abelian groups

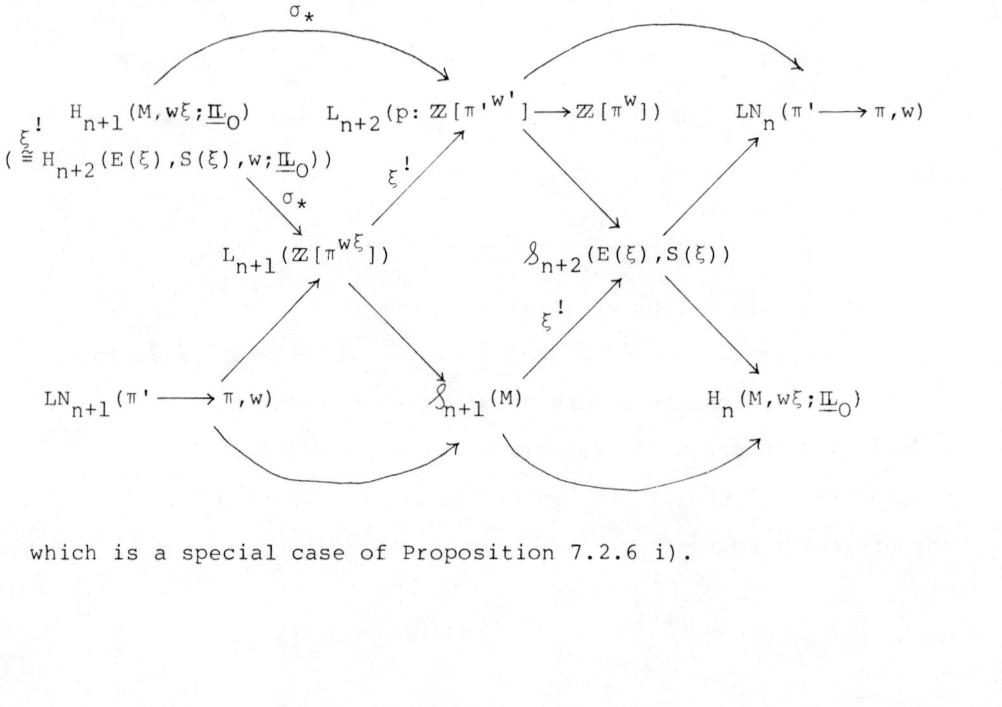

which is a special case of Proposition 7.2.6 i).

The proof of Proposition 7.6.4 can be used to relate the type C) codimension 1 splitting obstruction theory of Browder and Livesay [1] and López de Medrano [1] in the special case $\pi_1(X) = \pi_1(Y) = \mathbb{Z}_2$ to the theory of Wall [4,§12C] in the more general case $\pi_1(X) = \pi_1(Y)$, by expressing both in terms of our algebraic theory, as follows.

In the first instance, let us recall from II. the connection between the quadratic construction and the self-intersections of immersed manifolds. Let $f:M^m \looparrowright N^n$ $(m \leqslant n)$ be an oriented immersion in general position of an m-manifold M in an n-manifold N. Let $\tilde{N}$ be an oriented cover of N with group of covering translations π and orientation map $w:\pi \longrightarrow \mathbb{Z}_2$, such that the pullback of $\tilde{N}$ along f is the trivial cover $\tilde{M} = \pi \times M$ of M, and let $\tilde{f}:\tilde{M} \looparrowright \tilde{N}$ be a π-equivariant lift of f. The double point set of f

$$\begin{aligned} S_2(f) &= \{(x,y) \in M \times M \mid f(x) = f(y),\ x \neq y\}/\Sigma_2 \\ &= \{(\tilde{x},\tilde{y}) \in \tilde{M} \times_\pi \tilde{M} \mid \tilde{f}(\tilde{x}) = \tilde{f}(\tilde{y}),\ \tilde{x} \neq \tilde{y}\}/\Sigma_2 \end{aligned}$$

is then a (2m-n)-dimensional manifold (which may be empty, e.g. if $2m < n$), with $T \in \Sigma_2$ acting by $T:(\tilde{x},\tilde{y}) \longmapsto (\tilde{y},\tilde{x})$. Let $S_2(f)'$ be the evident double cover of $S_2(f)$, and let $c':S_2(f)' \longrightarrow E\Sigma_2$ be a Σ_2-equivariant map lifting a classifying map $c:S_2(f) \longrightarrow B\Sigma_2$, so that $S_2(f)$ is equipped with a map

$$\gamma : S_2(f) \longrightarrow E\Sigma_2 \times_{\Sigma_2} (\tilde{M} \times_\pi \tilde{M}) \ ;$$
$$[\tilde{x},\tilde{y}] \longmapsto [c'(\tilde{x},\tilde{y}),\tilde{x},\tilde{y}] \ .$$

Let $\nu_f:M \longrightarrow B\widetilde{STOP}(n-m)$ be the normal block bundle of the immersion f, so that applying the Pontrjagin-Thom construction to an embedding $f_\infty:M \hookrightarrow N \times \mathbb{R}^\infty$ approximating f there is defined a stable π-map

$$F : \Sigma^{\infty}\widetilde{N}_{+} = \widetilde{N} \times \mathbb{R}^{\infty}/\widetilde{N} \times S^{\infty-1} \xrightarrow{\text{collapse}} \widetilde{N} \times \mathbb{R}^{\infty}/\overline{\widetilde{N} \times \mathbb{R}^{\infty} - \text{nbhd. of } \widetilde{f}_{\infty}(\widetilde{M})}$$
$$= \Sigma^{\infty}T\pi(\nu_f)$$

inducing the Umkehr $\mathbb{Z}[\pi]$-module chain map

$$f^{!} : C(\widetilde{N}) \simeq C(\widetilde{N})^{n-*,w} \xrightarrow{\widetilde{f}^{*}} C(\widetilde{M})^{n-*,w} \simeq S^{n-m}C(\widetilde{M}) \simeq \dot{C}(T\pi(\nu_f)) \ .$$

The quadratic construction on F

$$\psi_F : H_n(N,w) \longrightarrow Q_n^{\pi,w}(\dot{C}(T\pi(\nu_f))) = Q_n^{\pi,w}(S^{n-m}C(\widetilde{M}))$$

sends the fundamental class $[N] \in H_n(N,w)$ to the image under γ of the fundamental class $[S_2(f)] \in H_{2m-n}(S_2(f))$ defined using the appropriately twisted coefficients

$$\psi_F([N]) = \gamma_*([S_2(f)]) \in Q_n^{\pi,w}(S^{n-m}C(\widetilde{M})) = H_{2m-n}(E\Sigma_2 \times_{\Sigma_2} (\widetilde{M} \times_{\pi} \widetilde{M}),w)$$

by the argument outlined on pp.279 - 282 of II. (The reference there to the work of Koschorke and Sanderson should be augmented by a reference to the earlier work of Vogel [1] on the interpretation of the approximation theorem $\Omega^{\infty}\Sigma^{\infty}X = (\bigsqcup_{k \geqslant 1} E\Sigma_k \times_{\Sigma_k} (\prod_k X))/\sim$ for connected spaces X in terms of immersion theory).

In particular, if $n = 2m$, $M^m = S^m$ then

$$\psi_F([N]) = \mu(f) \in H_0(\mathbb{Z}_2;\mathbb{Z}[\pi^w],(-)^m)$$

is the self-intersection number defined geometrically by Wall [4,§5] for an immersion $f:S^m \looparrowright N^{2m}$, with $\mu(f) = 0$ if (and for $m \geqslant 3, \pi = \pi_1(N)$ only if) f is regularly homotopic to an embedding $f:S^m \hookrightarrow N^{2m}$.

In the subsequent applications we shall be concerned with a double covering $p = \text{projection} : N' = \widetilde{N}/\pi' \longrightarrow N$ defined by a subgroup $\pi' \subset \pi$ of index 2, so that as before there is defined a group extension

$$\{1\} \longrightarrow \pi' \xrightarrow{\ p\ } \pi \xrightarrow{\ \xi\ } \mathbb{Z}_2 \longrightarrow \{1\}$$

and there is an element $t \in \pi$ such that $\xi(t) = -1$. Furthermore, the immersion $f: M \looparrowright N$ will be assumed to lift to an embedding $f': M \hookrightarrow N'$, in which case the double point set of f can be expressed as

$$S_2(f) = \{(x,y) \in M \times M \mid f'(x) = Tf'(y)\}/\Sigma_2$$

with $T: N' \longrightarrow N'$ the covering translation. The commutative diagram

$$\begin{array}{ccc} & & N' \\ & \overset{f'}{\nearrow} & \\ M & & \Big\downarrow p \\ & \underset{f}{\searrow} & \\ & & N \end{array}$$

gives rise to a commutative diagram of stable π-maps

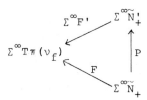

$$\begin{array}{ccc} & & \Sigma^\infty \widetilde{N}'_+ \\ & \overset{\Sigma^\infty F'}{\swarrow} & \\ \Sigma^\infty T\pi(\nu_f) & & \Big\uparrow P \\ & \underset{F}{\nwarrow} & \\ & & \Sigma^\infty \widetilde{N}_+ \end{array}$$

inducing the Umkehr $\mathbb{Z}[\pi]$-module chain maps

$$\begin{array}{ccc} & & C(\widetilde{N}') \\ & \overset{f'^!}{\swarrow} & \\ S^{n-m}C(\widetilde{M}) & & \Big\uparrow p^! \\ & \underset{f^!}{\nwarrow} & \\ & & C(\widetilde{N}) \end{array}$$

with $\widetilde{N}'$ the pullback of $\widetilde{N}$ along $p: N' \longrightarrow N$ and

$$p^! : C(\widetilde{N}) \longrightarrow C(\widetilde{N}') = p_! p^! C(\widetilde{N}) \ ; \ x \longmapsto 1 \otimes x + t \otimes t^{-1}x \ .$$

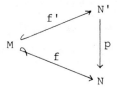

The stable π-map $P:\Sigma^\infty\widetilde{N}_+ \longrightarrow \Sigma^\infty\widetilde{N}'_+$ can be defined in one of two (equivalent) ways, either in the same way as F using a framed embedding $N' \hookrightarrow N\times\mathbb{R}^\infty$ such that

$$p : N' \hookrightarrow N\times\mathbb{R}^\infty \xrightarrow{\text{projection}} N\ ,$$

lifting to the covers and applying the Pontrjagin-Thom construction

$$\begin{aligned} P : \Sigma^\infty\widetilde{N}_+ = \widetilde{N}\times\mathbb{R}^\infty/\widetilde{N}\times S^{\infty-1} &\xrightarrow{\text{collapse}} \widetilde{N}\times\mathbb{R}^\infty/\overline{\widetilde{N}\times\mathbb{R}^\infty - \text{nbhd. of } \widetilde{N}'} \\ &= \widetilde{N}'\times\mathbb{R}^\infty/\widetilde{N}'\times S^{\infty-1} = \Sigma^\infty\widetilde{N}'_+ \end{aligned}$$

(i.e. using the fact that $p:N'\longrightarrow N$ is an immersion with normal bundle $\nu_p = 0 : N' \longrightarrow BG(O)$, although the construction of P is valid for any double covering $p:N'\longrightarrow N$), or using the Dyer-Lashof map $\mathcal{D}$ (cf. Brumfiel and Milgram [1]) with the adjoint of P given by

$$\mathrm{adj}(P) : \widetilde{N}_+ \longrightarrow (E\Sigma_2\times_{\Sigma_2}(\widetilde{N}'\times\widetilde{N}'))_+ \xrightarrow{\ \mathcal{D}\ } \Omega^\infty\Sigma^\infty\widetilde{N}'_+$$

with

$$\widetilde{N} \longrightarrow E\Sigma_2\times_{\Sigma_2}(\widetilde{N}'\times\widetilde{N}')\ ;\ [\widetilde{x}'] \longmapsto [p'(x'),\widetilde{x}',T\widetilde{x}']$$

defined using a Σ_2-equivariant lift $p':N'\longrightarrow E\Sigma_2$ of a classifying map $p:N\longrightarrow B\Sigma_2$. The quadratic construction on P

$$\psi_P:H_*(N,w)\longrightarrow Q_*^{\pi,w}(C(\widetilde{N}')) = Q_*^{\pi',w'}(p^!C(\widetilde{N}))\oplus Q_*^{\pi',\beta}(p^!C(\widetilde{N}),w(t)t^2)$$

has symmetrization

$$\begin{aligned}(1+T)\psi_P &= \phi_{\widetilde{N}'}p^! - p^{!\%}\phi_{\widetilde{N}} \\ &: H_*(N,w) \longrightarrow Q^*_{\pi,w}(C(\widetilde{N}')) = Q^*_{\pi',w'}(p^!C(\widetilde{N}))\oplus Q^*_{\pi',\beta}(p^!C(\widetilde{N}),w(t)t^2)\ .\end{aligned}$$

Let $\phi^!_{\widetilde{N}} : H_*(N',w')\longrightarrow Q^*_{\pi',w'}(p^!C(\widetilde{N}))$ be the symmetric construction on $\widetilde{N}$ with respect to the restriction of the π-action on $\widetilde{N}$ to $\pi'\subset\pi$ (noting that $\widetilde{N}/\pi' = N'$), so that the symmetric construction on $\widetilde{N}'$ with respect to the π-action is given by

$$\phi_{\tilde{N}'} = \begin{pmatrix} \phi^!_{\tilde{N}} \\ 0 \end{pmatrix}$$

$$: H_*(N',w') \longrightarrow Q^*_{\pi,w}(C(\tilde{N}')) = Q^*_{\pi',w'}(p^!C(\tilde{N})) \oplus Q^*_{\pi',\beta}(p^!C(\tilde{N}),w(t)t^2)$$

since $\tilde{p}:\tilde{N}' \longrightarrow \tilde{N}$ is a trivial double cover. It now follows from the chain level commutativity of the diagram

$$\begin{array}{ccc} H_*(N,w) & \xrightarrow{\phi_{\tilde{N}}} & Q^*_{\pi,w}(C(\tilde{N})) \\ \downarrow p^! & & \downarrow p^! \\ H_*(N',w') & \xrightarrow{\phi_{\tilde{N}'}} & Q^*_{\pi',w'}(p^!C(\tilde{N})) \end{array}$$

that the first component of ψ_P is 0

$$\psi_P = \begin{pmatrix} 0 \\ \psi'_P \end{pmatrix}$$

$$: H_*(N,w) \longrightarrow Q^{\pi,w}_*(C(\tilde{N}')) = Q^{\pi',w'}_*(p^!C(\tilde{N})) \oplus Q^{\pi',\beta}_*(p^!C(\tilde{N}),w(t)t^2) \ .$$

The second component defines a natural transformation

$$\psi'_P : H_*(N,w) \longrightarrow Q^{\pi',\beta}_*(p^!C(\tilde{N}),w(t)t^2) \quad ,$$

the <u>antiquadratic construction</u> associated to the double covering $p:N' \longrightarrow N$, such that

$$-(1+T_{w(t)t^2})\psi'_P = p^!t\phi_{\tilde{N}}$$

$$: H_*(N,w) \xrightarrow{\phi_{\tilde{N}}} Q^*_{\pi,w}(C(\tilde{N})) \xrightarrow[\sim]{t} Q^*_{\pi,\alpha}(C(\tilde{N}),w(t)t^2)$$

$$\xrightarrow{p^!} Q^*_{\pi',\beta}(p^!C(\tilde{N}),w(t)t^2) \ .$$

The quadratic construction on F in this case is given by

$$\psi_F = f^!_{\%}\psi_P = \begin{pmatrix} 0 \\ f^{!\,!}_{\%}\psi'_P \end{pmatrix} : H_n(N,w) \longrightarrow Q^{\pi,w}_n(S^{n-m}C(\pi \times M))$$

$$= Q^{\pi',w'}_n(S^{n-m}C(\pi' \times M)) \oplus Q^{\pi',\beta}_n(S^{n-m}C(\pi' \times M),w(t)t^2) \ ,$$

and

$$\gamma_*([S_2(f)]) = \psi_F([N]) = (0,f^{!\,!}_{\%}\psi'_P([N])) \ .$$

In particular, if $n = 2m$ the non-trivial component is the <u>equivariant self-intersection</u> of $(f:M^m \looparrowright N^{2m}, f':M^m \hookrightarrow N'^{2m})$

$$\mu_0(f) = f'_{\%}\psi'_P([N]) \in Q_{2m}^{\pi',\beta}(S^m C(\pi' \times M), w(t)t^2)$$
$$= H_0(\mathbb{Z}_2; \mathbb{Z}[\pi']^{\beta}, (-)^m w(t)t^2)$$

such that $\mu(f) = \mu_0(f)t^{-1}$, measuring the number of pairs of points in the intersection $f'(M) \cap Tf'(M) \subset N'$, which is a 0-dimensional manifold with a free Σ_2-action. (If (α,ε) is an antistructure on a ring A then $H_0(\mathbb{Z}_2; A, \varepsilon) = A/\{a - \varepsilon\alpha(a) \mid a \in A\}$, by definition). Wall [4,§12C] defined $\mu_0(f)$ geometrically. In the original work of Browder and Livesay [1] $\mu_0(f)$ was expressed in the case $\pi' = \{1\}$ in terms of a mod2 cohomology operation, which was expressed as a functional Steenrod square in Ranicki [8] and which has been extensively studied by Conner and Miller [1].

Let (M,N) be an $(n,n-1)$-dimensional manifold pair of type C), i.e. such that the normal bundle $\nu = \nu_{N \subset M} : N \longrightarrow BG(1)$ is non-trivial. Let $p_N : N' = S(\nu) \longrightarrow N$ be the associated double cover of N, and let $P = \overline{M \setminus E(\nu)}$ so that

$$M = E(\nu) \cup_{S(\nu)} P \quad .$$

The double covering of M

$$p_M : M' = P_+ \cup_{N'} P_- \longrightarrow M$$

defined using two copies P_+, P_- interchanged by the covering translation $T:M' \longrightarrow M'$ is such that

$$p_M| = p_N : p_M^{-1}(N) = N' \longrightarrow N \quad .$$

A <u>one-sided handle exchange</u> on N inside M is the ambient surgery

$$(M,N) \longmapsto (M,\overline{N \setminus S^r \times D^{n-r-1}} \cup D^{r+1} \times S^{n-r-2})$$

determined by an embedding $(D^{r+1},S^r) \times D^{n-r-1} \subset (M,N)$ which lifts to an embedding $(D^{r+1},S^r) \times D^{n-r-1} \subset (P_+,N')$ such that

$$(S^r \times D^{n-r-1}) \cap T(S^r \times D^{n-r-1}) = \emptyset \subset N' .$$

This operation is equivalent to an <u>equivariant handle exchange</u> on N' inside M'

$$(M',N') \longmapsto (M',\overline{N' \setminus (S^r \times D^{n-r-1} \sqcup T(S^r \times D^{n-r-1})}$$

$$\cup (D^{r+1} \times S^{n-r-2} \sqcup T(D^{r+1} \times S^{n-r-2}))$$

(cf. López de Medrano [1,§I.1.2]).

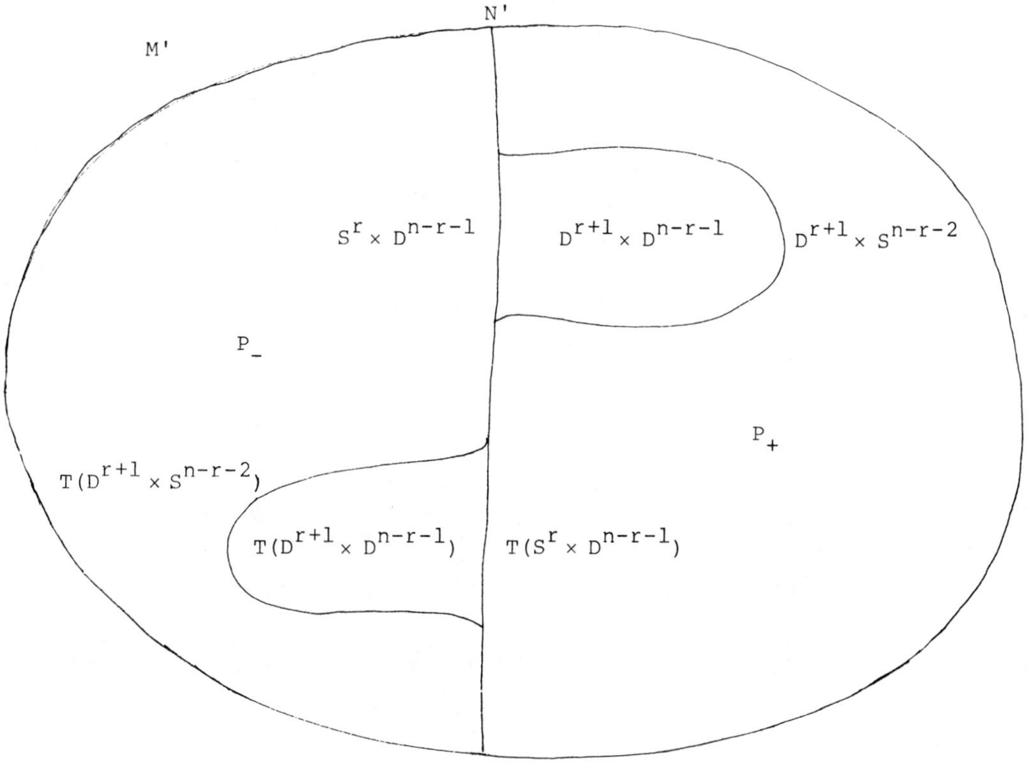

Proposition 7.6.5 Let (X,Y) be a formally $(n,n-1)$-dimensional normal pair of type C) with

$$\pi_1(X) = \pi_1(Y) = \pi \ , \ \pi_1(Z) = \pi_1(S(\xi)) = \pi' \ , \ w(X) = w : \pi \longrightarrow \mathbb{Z}_2 \ .$$

i) If $(f,b):(M,N) \longrightarrow (X,Y)$ is a formally $(n,n-1)$-dimensional topological normal map a one-sided handle exchange on $(g,c) = (f,b)|:N \longrightarrow Y$ inside $(f,b):M \longrightarrow X$ has the algebraic effect of surgery on the antiquadratic kernel $\sigma^!_*(f,b)$.

ii) If (X,Y) is an $(n,n-1)$-dimensional geometric Poincaré pair and $(f,b):(M,N) \longrightarrow (X,Y)$ is an $(n,n-1)$-dimensional topological normal map such that $f:M \longrightarrow X$ is an s-triangulation of X the antiquadratic kernel $\sigma^!_*(f,b)$ is an $(n-1)$-dimensional $(-w(t)t^2)$-quadratic Poincaré complex over $\mathbb{Z}[\pi']^\beta$. The splitting obstruction of f along $Y \subset X$ is given by

$$s(f,Y) = \sigma^!_*(f,b) \in LN_{n-1}(\pi' \longrightarrow \pi,w) = L_{n-1}(\mathbb{Z}[\pi']^\beta,-w(t)t^2) \ .$$

Proof: i) The antiquadratic kernel $\sigma^!_*(f,b)$ was defined using the quadratic kernel $\sigma_*(g,c)$. We shall now obtain it using the antiquadratic construction associated to the double covering of N classified by $\nu = \nu_{N \subset M} : N \longrightarrow BG(1)$, thus relating $\sigma^!_*(f,b)$ to the equivariant self-intersections which are the obstructions to individual one-sided handle exchanges.

Let

$$p_X : X' = Z_+ \cup_{Y'} Z_- \longrightarrow X$$

be the double covering of X defined using two copies Z_+,Z_- of Z interchanged by the covering translation $T:X' \longrightarrow X'$, with

$$p_X| = p_Y : p_X^{-1}(Y) = Y' = S(\xi) \longrightarrow Y$$

the double covering of Y associated to $\xi:Y \longrightarrow BG(1)$.

The formally n-dimensional topological normal map $(f,b):M \longrightarrow X$ has a decomposition

$$(f,b) = (g,c)^! \cup (h,d) \; : \; M = E(\nu) \cup_{S(\nu)} P \longrightarrow X = E(\xi) \cup_{S(\xi)} Z$$

with

$$(g,c) = (f,b)| \; : \; N = f^{-1}(Y) \longrightarrow Y$$

$$(h,d) = (f,b)| \; : \; (P,S(\nu)) = f^{-1}(Z,S(\xi)) \longrightarrow (Z,S(\xi))$$

$$\nu = \nu_{N\subset M} \; : \; N \xrightarrow{\;g\;} Y \xrightarrow{\;\xi\;} BG(1) \; .$$

Let $\widetilde{Y}$ be the universal cover of Y, and let $\widetilde{g}:\widetilde{N} \longrightarrow \widetilde{Y}$ be a π-equivariant lift of $g:N \longrightarrow Y$, so that the Umkehr $\mathbb{Z}[\pi]$-module chain map of g is defined by

$$g^! \; : \; C(\widetilde{Y})^{n-1-*,w\xi} \xrightarrow{\;\widetilde{g}^*\;} C(\widetilde{N})^{n-1-*,w\xi} \xrightarrow[\sim]{\;[N]\cap -\;} C(\widetilde{N}) \; .$$

The quadratic kernel of the formally $(n-1)$-dimensional topological normal map $(g,c):N \longrightarrow Y$ is the formally $(n-1)$-dimensional quadratic complex over $\mathbb{Z}[\pi^{w\xi}]$

$$\sigma_*(g,c) = (C(g^!), \psi_G([Y]) \in Q_{n-1}^{\pi,w\xi}(C(g^!))) = (C,\psi)$$

with $\psi_G:H_{n-1}(Y,w\xi) \longrightarrow Q_{n-1}^{\pi,w\xi}(C(g^!))$ the spectral quadratic construction on the geometric Umkehr semi-stable π-map $G:T\pi(\nu_Y)^* \longrightarrow \Sigma^\infty \widetilde{N}_+$ obtained by equivariant S-duality as in §7.3. The quadratic kernel of the formally $(n-1)$-dimensional topological normal induced from $(g,c):N \longrightarrow Y$ by $p_Y:Y' \longrightarrow Y$

$$(g',c') \; : \; N' \longrightarrow Y'$$

is the restriction of the quadratic kernel of (g,c)

$$\sigma_*(g',c') = p^!\sigma_*(g,c)$$

with $p : \pi' = \pi_1(Y') \longrightarrow \pi = \pi_1(Y)$ the inclusion.

The quadratic kernel of the formally n-dimensional topological normal map of pairs

$$(h,d) : (P,N') \longrightarrow (Z,Y')$$

is a formally n-dimensional quadratic pair over $\mathbb{Z}[\pi'^{w'}]$

$$\sigma_*(h,d) = (e:p^!C \longrightarrow D,(\delta\psi,p^!\psi) \in Q_n^{\pi',w'}(e))$$

defined using the relative spectral quadratic construction, with $D = C(h^!)$ the algebraic mapping cone of the Umkehr $\mathbb{Z}[\pi']$-module chain map

$$h^! : C(\tilde{Z},\tilde{Y})^{n-*,w'} \xrightarrow{\tilde{h}^*} C(\tilde{P},\tilde{N})^{n-*,w'} \xrightarrow{[P]\cap -} C(\tilde{P})$$

with $\tilde{Z}$ the universal cover of Z and $\tilde{h}:(\tilde{P},\tilde{N}) \longrightarrow (\tilde{Z},\tilde{Y})$ a π'-equivariant lift of $h:(P,N) \longrightarrow (Z,Y)$. Let $\tilde{p}_Y:\tilde{Y}' \longrightarrow \tilde{Y}$ be the π-equivariant (trivial) double cover of $\tilde{Y}$ obtained from $p_Y:Y' \longrightarrow Y$ by pullback along the covering projection $\tilde{Y} \longrightarrow Y$, inducing the $\mathbb{Z}[\pi]$-module chain map

$$\tilde{p}_Y : C(\tilde{Y}') = p_!p^!C(\tilde{Y}) \longrightarrow C(\tilde{Y}) \ ; \ a\otimes x \longmapsto ax$$
$$(a \in \mathbb{Z}[\pi],\ x \in C(\tilde{Y}))\ ,$$

and similarly for N. The $\mathbb{Z}[\pi]$-module chain map

$$i : p_!p^!C(g^!) = p_!p^!C \longrightarrow C \ ; \ a\otimes x \longmapsto ax$$

fits into a commutative diagram

$$\begin{array}{ccccc}
C(\tilde{Y}')^{n-1-*,w\xi} & \xrightarrow{g'^!} & C(\tilde{N}') & \longrightarrow & p_!p^!C \\
\downarrow{\scriptstyle \tilde{p}_Y^{!*}} & & \downarrow{\scriptstyle \tilde{p}_N} & & \downarrow{\scriptstyle i} \\
C(\tilde{Y})^{n-1-*,w\xi} & \xrightarrow{g^!} & C(\tilde{N}) & \longrightarrow & C
\end{array}\quad .$$

The $\mathbb{Z}[\pi]$-module chain map

$$p_!e : p_!p^!C \longrightarrow p_!D$$

fits into a commutative diagram

$$\begin{array}{ccccc}
C(\widetilde{Y}')^{n-1-*,w\xi} & \xrightarrow{g'^!} & C(\widetilde{N}') & \longrightarrow & p_!p^!C \\
\downarrow & & \downarrow \widetilde{q}_N & & \downarrow p_!e \\
C(\widetilde{Z}_+ \sqcup \widetilde{Z}_-, \widetilde{Y}')^{n-*,w\xi} & \xrightarrow{p_!h^!} & C(\widetilde{P}_+ \sqcup \widetilde{P}_-) & \longrightarrow & p_!D \\
(= p_!(C(\widetilde{Z},\widetilde{Y})^{n-*,w'})) & & (= p_!C(\widetilde{P})) & &
\end{array}$$

with $q_N : N \longrightarrow P$ the inclusion. It is thus possible to identify

$$C\left(\begin{pmatrix} i \\ p_!e \end{pmatrix} : p_!p^!C \longrightarrow C\oplus p_!D\right) = \xi^!C(g^!) \cup p_!C(h^!)$$

$$= C(f^! : C(\widetilde{X})^{n-*,w} \longrightarrow C(\widetilde{M})) \ .$$

The quadratic kernel of the formally n-dimensional topological normal map $(f,b):M \longrightarrow X$ is the formally n-dimensional quadratic complex over $\mathbb{Z}[\pi^w]$

$$\sigma_*(f,b) = (C(f^!), \psi_F([X]) \in Q_n^{\pi,w}(C(f^!)))$$

$$= \xi^!\sigma_*(g,c) \cup p_!\sigma_*(h,d)$$

with $\psi_F : H_n(X,w) \longrightarrow Q_n^{\pi,w}(C(f^!))$ the spectral quadratic construction on a geometric Umkehr semi-stable π-map $F:T\pi(\nu_X)^* \longrightarrow \Sigma^\infty \widetilde{M}_+$ inducing the Umkehr $\mathbb{Z}[\pi]$-module chain map

$$f^! : C(\widetilde{X})^{n-*,w} \xrightarrow{\widetilde{f}^*} C(\widetilde{M})^{n-*,w} \xrightarrow[\sim]{[M]\cap -} C(\widetilde{M})$$

with $\widetilde{X}$ the universal cover of X and $\widetilde{f}:\widetilde{M} \longrightarrow \widetilde{X}$ a π-equivariant lift of $f:M \longrightarrow X$. Define a $\mathbb{Z}[\pi]$-module chain map

$$j : C \xrightarrow{\lambda k} p_!p^!C \xrightarrow{p_!e} p_!D \ ,$$

with

$$\lambda k : C \longrightarrow p_!p^!C \ ; \ x \longmapsto 1\otimes x + t\otimes t^{-1}x$$

the $\mathbb{Z}[\pi]$-module chain map appearing in the short exact sequence

$$0 \longrightarrow C \xrightarrow{\lambda k} p_! p^! C \xrightarrow{\lambda i} \lambda C \longrightarrow 0 \quad .$$

(Recall that k was defined by

$$k : \lambda C \longrightarrow p_! p^! C \ ; \ x \longmapsto 1 \otimes x - t \otimes t^{-1} x \quad) \ .$$

It follows that the $\mathbb{Z}[\pi]$-module chain map

$$(\begin{pmatrix} 0 \\ 1 \end{pmatrix}, \lambda k) \ : \ C(j) \longrightarrow C(\begin{pmatrix} \lambda i \\ p_! e \end{pmatrix}) \ = \ \lambda C(f^!)$$

appearing in the commutative diagram

$$\begin{array}{ccccc} C & \xrightarrow{\quad j \quad} & p_! D & \longrightarrow & C(j) \\ \Big\downarrow \lambda k & & \Big\downarrow \begin{pmatrix} 0 \\ 1 \end{pmatrix} & & \Big\downarrow (\begin{pmatrix} 0 \\ 1 \end{pmatrix}, \lambda k) \\ p_! p^! C & \xrightarrow{\begin{pmatrix} \lambda i \\ p_! e \end{pmatrix}} & \lambda C \oplus p_! D & \longrightarrow & C(\begin{pmatrix} \lambda i \\ p_! e \end{pmatrix}) \end{array}$$

is a simple chain equivalence. The $\mathbb{Z}[\pi]$-module chain map $\lambda k : C \longrightarrow p_! p^! C$ fits into a commutative diagram

$$\begin{array}{ccccc} C(\widetilde{Y})^{n-1-*, w\xi} & \xrightarrow{\quad g^! \quad} & C(\widetilde{N}) & \longrightarrow & C \\ \Big\downarrow & & \Big\downarrow p_N^! & & \Big\downarrow \lambda k \\ C(\widetilde{Y}')^{n-1-*, w\xi} & \xrightarrow{\quad g'^! \quad} & C(\widetilde{N}') & \longrightarrow & p_! p^! C \end{array} \quad .$$

The double covering of $(f,b) : M \longrightarrow X$ induced from the double covering $p_X : X' \longrightarrow X$ is a formally n-dimensional topological normal map

$$\begin{aligned} (f',b') &= (h_+, d_+) \cup_{(g',c')} (h_-, d_-) \\ &: \ M' = P_+ \cup_{N'} P_- \longrightarrow X' = Z_+ \cup_{Y'} Z_- \end{aligned}$$

for two copies $(h_+,d_+),(h_-,d_-)$ of (h,d). Regarded as a $\mathbb{Z}[\pi']$-module chain map j can be written as

$$p^! j = \begin{pmatrix} j_+ \\ j_- \end{pmatrix} : p^! C = C(p^! g^! : p^!(C(\widetilde{Y})^{n-1-*,w\xi}) \longrightarrow p^! C(\widetilde{N}))$$

$$\longrightarrow p^! p_! D = C(h_+^!) \oplus C(h_-^!)$$

with $j_+ : p^! C \longrightarrow D$ a copy of the inclusion $p^! C(g^!) \longrightarrow C(h^!)$ and $j_- = \mu j_+$.

The antiquadratic kernel of $(f,b):(M,N) \longrightarrow (X,Y)$ is the formally (n-1)-dimensional $(\beta,-w(t)t^2)$-quadratic complex over $\mathbb{Z}[\pi']$

$$\sigma^!_*(f,b) = (D, e_\% p^! t(\psi) \in Q^{\pi',\beta}_{n-1}(D)) \quad .$$

In order to relate this to the antiquadratic construction $\psi^!_{P_N} : H_*(N,w\xi) \longrightarrow Q^{\pi',\beta}_*(p^! C(\widetilde{N}),-w(t)t^2)$ on the geometric Umkehr π-map $P_N : \Sigma^\infty N_+ \longrightarrow \Sigma^\infty N'_+$ of the double covering $p_N : N' \longrightarrow N$ consider the commutative diagram of normal maps

$$\begin{array}{ccc} N' & \xrightarrow{(g',c')} & Y' \\ \big\downarrow{\scriptstyle (p_N,b_N)} & & \big\downarrow{\scriptstyle (p_Y,b_Y)} \\ N & \xrightarrow{(g,c)} & Y \end{array} \quad .$$

The equivariant S-dual of the induced diagram of maps of Thom π-spaces is a commutative diagram of (semi-)stable π-maps

$$\begin{array}{ccc} \Sigma^\infty \widetilde{N}'_+ & \xleftarrow{G'} & T\pi(\nu_{Y'})^* \\ \big\uparrow{\scriptstyle P_N} & & \big\uparrow{\scriptstyle T\pi(b_Y)^*} \\ \Sigma^\infty \widetilde{N}_+ & \xleftarrow{G} & T\pi(\nu_Y)^* \end{array} \quad .$$

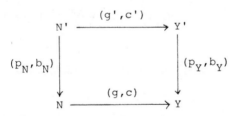

By the sum formula for the spectral quadratic construction of Proposition 7.3.1 v)

$$\psi_{G'}p^!_Y = (\lambda k)_\%\psi_G + (p_!p^!a)_\%\psi_{P_N}g^!$$

$$: H_{n-1}(Y,w\xi) \longrightarrow Q^{\pi,w\xi}_{n-1}(C(g'^!))$$

where a = inclusion : $C(\tilde{N}) \longrightarrow C(g^!) = C$ and

$$Q^{\pi,w\xi}_{n-1}(C(g'^!)) = Q^{\pi,w\xi}_{n-1}(p_!p^!C) = Q^{\pi',w'}_{n-1}(p^!C)\oplus Q^{\pi',\beta}_{n-1}(p^!C,-w(t)t^2) \ .$$

The antiquadratic construction ψ'_{P_N} is such that $\psi_{P_N} = \begin{pmatrix} 0 \\ \psi'_{P_N}\end{pmatrix}$

and $\psi_{G'}p^!_Y = \begin{pmatrix} p^!\psi_G \\ 0 \end{pmatrix}$ (since the double covering $\tilde{p}_Y:\tilde{Y}'\longrightarrow\tilde{Y}$ is trivial), so that

$$p^!t(\psi) = p^!t\psi_G([Y]) = -(p^!a)_\%\psi'_{P_N}([N]) \in Q^{\pi',\beta}_{n-1}(p^!C,-w(t)t^2)$$

and

$$\begin{aligned}\sigma^!_*(f,b) &= (D,e_\% p^!t(\psi)) \\ &= (D,-j_{+\%}(p^!a)_\%\psi'_{P_N}([N]) \in Q^{\pi',\beta}_{n-1}(D,-w(t)t^2)) \ .\end{aligned}$$

The verification that a one-sided handle exchange on $(g,c):N\longrightarrow Y$ inside $(f,b):M\longrightarrow X$ determines an algebraic surgery on $\sigma^!_*(f,b)$ now proceeds as for ordinary surgery in Proposition II.7.3, with the equivariant self-intersection μ_O playing the role of μ.

ii) The natural isomorphism of Proposition 7.6.4

$$LN_{n-1}(\pi'\longrightarrow\pi,w) \xrightarrow{\sim} L_{n-1}(\mathbb{Z}[\pi']^\beta,-w(t)t^2)$$

sends the splitting obstruction $s(f,Y) = (\sigma_*(g,c),\sigma_*(h,d))$ to the cobordism class of the antiquadratic kernel $\sigma^!_*(f,b)$.

[]

In the original example of Browder and Livesay [1] $(X,Y) = (\mathbb{R}P^n, \mathbb{R}P^{n-1})$ and the splitting obstruction along $Y \subset X$ of an s-triangulation $f: M \xrightarrow{\sim} X$ (for $n \geqslant 5$) is an element

$$s(f,Y) \in LN_{n-1}(\{1\} \longrightarrow \mathbb{Z}_2, (-)^{n-1}) = L_{n-1}(\mathbb{Z}, (-)^n) .$$

Thus the obstruction is 0 if $n \equiv 0 \pmod 2$, and is

$$\begin{cases} \text{the Arf invariant} \\ \frac{1}{8}(\text{the signature}) \end{cases} \text{of a non-singular} \begin{cases} \text{skew-quadratic} \\ \text{quadratic} \end{cases} \text{form}$$

over $\mathbb{Z}$ if $\begin{cases} n \equiv 1 \pmod 4 \\ n \equiv 3 \pmod 4 \end{cases}$. López de Medrano [1] studied the

splitting obstruction theory for arbitrary type C) $(n,n-1)$-dimensional geometric Poincaré pairs (X,Y) with $\pi_1(X) = \pi_1(Y) = \mathbb{Z}_2$, for which the splitting obstruction along $Y \subset X$ of an s-triangulation $f: M \xrightarrow{\sim} X$ is an element

$$s(f,Y) \in LN_{n-1}(\{1\} \longrightarrow \mathbb{Z}_2, w(X)(t)) = L_{n+1}(\mathbb{Z}, \varepsilon)$$

with $\varepsilon = w(X)(t) = +1$ if $T: X' \longrightarrow X'$ is orientation-preserving (i.e. if X is orientable) and $\varepsilon = -1$ otherwise.

If $(M, \partial M)$ is an n-dimensional manifold with boundary such that M is an $(n-1)$-dimensional geometric Poincaré complex the Poincaré-Lefschetz dual of the mod2 fundamental class $[M] \in H_{n-1}(M; \mathbb{Z}_2)$ is an element $\xi \in H^1(M, \partial M; \mathbb{Z}_2)$ classifying an S^0-fibration $\xi: M \longrightarrow BG(1)$ such that

    i) $\xi|_N = \nu = \nu_{N \subset M}: N \longrightarrow BG(1)$ is the non-trivial normal bundle of a codimension 1 submanifold $N \subset M$ such that $N \cap \partial M = \emptyset$ and $g_*[N] = [M] \in H_{n-1}(M, w\xi)$ $(w = w(M, \partial M))$, with $g: N \longrightarrow M$ the inclusion

    ii) $\xi|_P = \varepsilon: P \longrightarrow BG(1)$, with $P = \overline{M \setminus E(\nu)}$.

Let $(f,b):(M,\partial M) \longrightarrow (E(\xi),S(\xi))$ be an s-triangulation of the n-dimensional geometric Poincaré pair $(E(\xi),S(\xi))$ topologically transverse at the zero section $M \subset E(\xi)$ with $f^{-1}(M) = N \subset M$, as in the discussion at the end of §7.5. The rel ∂M splitting obstruction of f along $M \subset E(\xi)$

$$s_\partial(f,M) = \sigma_*^!(f,b) = (D,\psi' \in Q_{n-1}^{\pi',\beta}(D,-w(t)t^2))$$
$$\in LN_{n-1}(\pi' \longrightarrow \pi,w) = L_{n-1}(\mathbb{Z}[\pi']^\beta,-w(t)t^2)$$

is the obstruction to the existence of a codimension 1 spine $N \subset M$ obtained by Wall [4,§12C]. In particular, if $n-1 = 2m$ and $(f,b)| = (g,c) : N \longrightarrow M$ is $(m-1)$-connected (as can be achieved by preliminary one-sided handle exchanges below the middle dimension) it is possible to represent every element

$$x \in H_m(D) = H_m(h_+^!) = \ker(H_m(\tilde{N}) \longrightarrow H_m(\tilde{M})) \cap \ker(H_m(\tilde{N}) \longrightarrow H_m(\tilde{P}_+))$$

by a framed immersion $x_1 : S^m \looparrowright N^{2m}$ with a lift to an embedding $x_1' = x_2 : S^m \hookrightarrow N'^{2m}$ in the double cover $N' = S(\nu)$ which extends to a framed embedding $(x_3,x_2):(D^{m+1},S^m) \hookrightarrow (P_+,N')$, together with a null-homotopy $(x_4,gx_1):(D^{m+1},S^m) \longrightarrow M$ of gx_1. The antiquadratic kernel $\sigma_*^!(f,b)$ is given in this case by the non-singular $(\beta,(-)^{m+1}w(t)t^2)$-quadratic form over $\mathbb{Z}[\pi']$

$$(H_m(h_+^!),\lambda_0 = (1+T_{(-)^{m+1}w(t)t^2})\psi_0' : H_m(h_+^!) \times H_m(h_+^!) \longrightarrow \mathbb{Z}[\pi'],$$
$$\mu_0 = v^m(\psi') : H_m(h_+^!) \longrightarrow H_0(\mathbb{Z}_2;\mathbb{Z}[\pi']^\beta,(-)^{m+1}w(t)t^2))$$

defined geometrically by Wall [4,§12C], with $\mu_0(x)$ the equivariant self-intersection of (x_1,x_1'). An element $x \in H_m(h_+^!)$ is such that $\mu_0(x) = 0$ if (and for $m \geqslant 3$ only if) it can be killed by a one-sided handle exchange on $(g,c):N \longrightarrow M$ inside $(f,b):(M,\partial M) \longrightarrow (E(\xi),S(\xi))$.

If (X,Y) is a codimension 1 CW pair of type C) with

$$\pi_1(X) = \pi_1(Y) = \pi \ , \ \pi_1(Z) = \pi_1(S(\xi)) = \pi' \ , \ w(X) = w : \pi \longrightarrow \mathbb{Z}_2$$

and $f:M \longrightarrow X$ is a map from an n-dimensional geometric Poincaré complex M the Poincaré splitting obstruction of f along $Y \subset X$ is given by Propositions 7.5.1 iii), 7.6.4 to be an element

$$s^P(f,Y) \in LN_{n-2}(\pi' \longrightarrow \pi, w) = L_n(\mathbb{Z}[\pi']^\beta, w(t)t^2) \ .$$

As it stands the construction of this invariant requires f to be normal transverse at $Y \subset X$. However, in the case $\pi = \mathbb{Z}_2$, $\pi' = \{1\}$, $n = 2m$ Hambleton and Milgram [1] identified this Poincaré splitting obstruction with the Arf invariant of the non-singular quadratic form over $\mathbb{Z}_2$

$$(H^m(M';\mathbb{Z}_2), \lambda_0 : H^m(M';\mathbb{Z}_2) \times H^m(M';\mathbb{Z}_2) \longrightarrow \mathbb{Z}_2; (x,y) \longmapsto \langle x \cup Ty, [M'] \rangle,$$

$$\mu_0 = v^m(\psi'_{p_M}([M])) : H^m(M';\mathbb{Z}_2) \longrightarrow \mathbb{Z}_2) \quad ,$$

with $p_M : M' \longrightarrow M$ the double cover of M induced along f from the double cover $p_X : X' = Z_+ \cup_{S(\xi)} Z_- \longrightarrow X$ of X, which is defined without normal transversality. We shall now use the antiquadratic construction to express the non-simply-connected Poincaré splitting obstruction in terms of a generalization of this form, which is also defined intrinsically (i.e. without appealing to normal transversality).

Let M be an n-dimensional normal space, and let $\widetilde{M}$ be an oriented covering of M with group of covering translations π and orientation map $w:\pi \longrightarrow \mathbb{Z}_2$ such that π is equipped with a subgroup $\pi' \subset \pi$ of index 2, so that

$$p = \text{projection} : \widetilde{M}/\pi' = M' \longrightarrow M$$

is a non-trivial double covering of M with a geometric Umkehr stable π-map

$$P : \Sigma^{\infty}\widetilde{M}_+ \longrightarrow \Sigma^{\infty}\widetilde{M}'_+ \quad .$$

As before, write the group extension as

$$\{1\} \longrightarrow \pi' \xrightarrow{\ p\ } \pi \xrightarrow{\ \xi\ } \mathbb{Z}_2 \longrightarrow \{1\} \ ,$$

choose an element $t \in \pi - \pi'$ and define an antistructure (β, t^2) on $\mathbb{Z}[\pi']$ by

$$\beta : \mathbb{Z}[\pi'] \longrightarrow \mathbb{Z}[\pi'] \ ; \ \sum_{g' \in \pi'} n_{g'} g' \longmapsto \sum_{g' \in \pi'} w'(g') n_{g'} t^{-1} g'^{-1} t$$

$$(w' = wp : \pi' \longrightarrow \mathbb{Z}_2) \quad .$$

Use the antiquadratic construction on P

$$\psi'_P : H_n(M,w) \longrightarrow Q_n^{\pi',\beta}(p^! C(\widetilde{M}), w(t)t^2)$$

to define the <u>antiquadratic complex</u> of (M,p), the n-dimensional $(\beta, w(t)t^2)$-quadratic complex over $\mathbb{Z}[\pi']$

$$\sigma_*(M,p) = (p^! C(\widetilde{M}), \psi'_P([M]))$$

with antisymmetrization

$$\begin{aligned}(1+T_{w(t)t^2})\sigma_*(M,p) \\ &= (p^! C(\widetilde{M}), p^! t\phi_{\widetilde{M}}([M]) \in Q^n_{\pi',\beta}(p^! C(\widetilde{M}), w(t)t^2)) \\ &= p^! t\sigma^*(M) \ .\end{aligned}$$

Proposition 7.6.6 i) The antiquadratic complex $\sigma_*(M,p)$ is such that

$$(1+T_{w(t)t^2})\sigma_*(M,p) = p^! t\sigma^*(M)$$

$$g_\%\sigma_*(M,p) = Sp^! t\sigma_*(M) \quad ,$$

where $\sigma^*(M) = (C(\widetilde{M}),\phi_{\widetilde{M}}([M]) \in Q^n_{\pi,w}(C(\widetilde{M})))$ is the symmetric complex of M, $\sigma_*(M) = (\Omega C([M]\cap -),\psi \in Q^{\pi,w}_{n-1}(\Omega C([M]\cap -)))$ is the quadratic Poincaré complex of M and

$$g = \text{inclusion} : p^! C(\widetilde{M}) \longrightarrow p^! C([M]\cap -:C(\widetilde{M})^{n-*,w} \longrightarrow C(\widetilde{M}))\ .$$

ii) If M is an n-dimensional geometric Poincaré complex then $\sigma_*(M,p)$ is an n-dimensional $(\beta,w(t)t^2)$-quadratic Poincaré complex over $\mathbb{Z}[\pi']$. The antiquadratic signature of (M,p) is the cobordism class

$$\sigma_*(M,p) \in L_n(\mathbb{Z}[\pi']^\beta,w(t)t^2)\ .$$

iii) The antiquadratic signature vanishes if (M,p) is the boundary of an (n+1)-dimensional geometric Poincaré pair $(\delta M,M)$ equipped with a double cover $(\delta p,p):(\delta M',M') \longrightarrow (\delta M,M)$ such that $\widetilde{M}$ extends to a cover $(\widetilde{\delta M},\widetilde{M})$ of $(\delta M,M)$, in which case the antiquadratic complex $\sigma_*(M,p)$ is the boundary of the (n+1)-dimensional (β,t^2)-quadratic Poincaré pair over $\mathbb{Z}[\pi']$

$$\sigma_*(\delta M,\delta p) = (f:p^! C(\widetilde{M}) \longrightarrow p^! C(\widetilde{\delta M}),\psi^!_{\delta p,p}([\delta M]) \in Q^{\pi',\beta}_{n+1}(f))\ ,$$

so that

$$\sigma_*(M,p) = 0 \in L_n(\mathbb{Z}[\pi']^\beta,w(t)t^2)\ .$$

iv) The antiquadratic signature vanishes if (M,p) admits a characteristic geometric Poincaré subcomplex, that is if the classifying map $p:M \longrightarrow BG(1) = \mathbb{RP}^\infty$ is Poincaré transverse at $\mathbb{RP}^{\infty-1} \subset \mathbb{RP}^\infty$, so that $(M,N = p^{-1}(\mathbb{RP}^{\infty-1}))$ is an (n,n-1)-dimensional geometric Poincaré pair of type C) with $\widetilde{M} = \widetilde{P}_+ \cup_{\widetilde{N}} \widetilde{P}_-$.

More precisely, the antiquadratic signature vanishes

$$\sigma_*(M,p) = 0 \in L_n(\mathbb{Z}[\pi']^{\beta}, w(t)t^2) \quad ,$$

since the antiquadratic complex $\sigma_*(M,p)$ is the boundary of the $(n+1)$-dimensional $(\beta, w(t)t^2)$-quadratic Poincaré pair over $\mathbb{Z}[\pi']$

$$\sigma_*(M,N) = (j_+ : p^! C(\widetilde{M}) \longrightarrow C(\widetilde{M}, \widetilde{P}_+), (0, \psi_P'([M])) \in Q_{n+1}^{\pi',\beta}(j_+)) \ .$$

<u>Proof</u>: i) Let $(\nu_M : M \longrightarrow BG(k), \rho_M : S^{n+k} \longrightarrow T(\nu_M))$ be the normal structure of M. As in the proof of Proposition 7.4.1 i) let $F : T\pi(\nu_M)^* \longrightarrow \Sigma^{\infty}\widetilde{M}_+$ be a semi-stable π-map inducing the $\mathbb{Z}[\pi]$-module chain map

$$[M] \cap - \ : \ C(\widetilde{M})^{n-*,w} \longrightarrow C(\widetilde{M}) \ .$$

The projection $p : M' \longrightarrow M$ is covered by a map of $(k-1)$-spherical fibrations $b : p^*\nu_M \longrightarrow \nu_M$, with a commutative diagram

$$\begin{array}{ccc} S(p^*\nu_M) & \xrightarrow{S(b)} & S(\nu_M) \\ \downarrow & & \downarrow \\ M' & \xrightarrow{p} & M \end{array}$$

in which both p and S(b) are double coverings. The geometric Umkehr {1}-maps P and S(B) induced by p and S(b) fit into the commutative diagram

$$\begin{array}{ccc} \Sigma^{\infty}S(p^*\nu_M)_+ & \xleftarrow{S(B)} & \Sigma^{\infty}S(\nu_M)_+ \\ \downarrow & & \downarrow \\ \Sigma^{\infty}M'_+ & \xleftarrow{P} & \Sigma^{\infty}M_+ \end{array}$$

so that there is induced a stable {1}-map of Thom spaces

$$T(B) = P/S(B)$$

$$: \Sigma^\infty T(\nu_M) = \Sigma^\infty M_+/\Sigma^\infty S(\nu_M)_+ \longrightarrow \Sigma^\infty T(p^*\nu_M) = \Sigma^\infty M'_+/\Sigma^\infty S(p^*\nu_M)_+ \ .$$

Define a normal structure for the double covering M' of M by

$$(\nu_{M'} = p^*\nu_M \oplus \varepsilon^\infty : M' \longrightarrow BG(k+\infty),$$

$$\rho_{M'} : S^{n+k+\infty} \xrightarrow{\Sigma^\infty \rho_M} \Sigma^\infty T(\nu_M) \xrightarrow{T(B)} \Sigma^\infty T(p^*\nu_M) = T(\nu_{M'})) \ ,$$

so that the corresponding semi-stable π-map $F' : T\pi(\nu_{M'})^* \longrightarrow \Sigma^\infty \widetilde{M}'_+$ fits into the commutative diagram of (semi-)stable π-maps

$$\begin{array}{ccc} T\pi(\nu_M)^* & \xrightarrow{F} & \Sigma^\infty \widetilde{M}_+ \\ {\scriptstyle T\pi(b)^*}\downarrow & & \downarrow{\scriptstyle P} \\ T\pi(\nu_{M'})^* & \xrightarrow{F'} & \Sigma^\infty \widetilde{M}'_+ \end{array}$$

inducing the commutative diagram of $\mathbb{Z}[\pi]$-module chain maps

$$\begin{array}{ccc} C(\widetilde{M})^{n-*,w} & \xrightarrow{[M]\cap -} & C(\widetilde{M}) \\ {\scriptstyle p^*}\downarrow & & \downarrow{\scriptstyle p^!} \\ C(\widetilde{M}')^{n-*,w} & \xrightarrow{[M']\cap -} & C(\widetilde{M}') \end{array} \ .$$

By the sum formula for the spectral quadratic construction of Proposition 7.3.1 v)

$$\psi_{F'} T(b)^* = (p^!, p^*)_\% \psi_F + (p_! p^! g)_\% \psi_P([M]\cap -)$$

$$: \dot{H}^k(T(\nu_M), w) = H^0(M) \longrightarrow Q_n^{\pi,w}(C([M']\cap -))$$

$$= Q_n^{\pi',w'}(p^! C([M]\cap -)) \oplus Q_n^{\pi',\beta}(p^! C([M]\cap -), w(t)t^2)$$

with

$$g = \text{inclusion} : C(\widetilde{M}) \longrightarrow C([M]\cap -)$$

$$p_! p^! g = \text{inclusion} : p_! p^! C(\widetilde{M}) = C(\widetilde{M}')$$
$$\longrightarrow p_! p^! C([M]\cap -) = C([M']\cap -)$$

$$(p^!, p^*) = \lambda k : C([M]\cap -)$$
$$\longrightarrow C([M']\cap -) = p_! p^! C([M]\cap -) \quad .$$

As $\widetilde{p}:\widetilde{M}' \longrightarrow \widetilde{M}$ is a trivial double covering $p^! t\psi_{F'} = 0$, and $p^!\psi_P = 0$, so that

$$\psi_{F'}T(b)^* = \begin{pmatrix} p^!\psi_{F'}T(b)^* \\ 0 \end{pmatrix} = \begin{pmatrix} \psi_{p^!F} \\ p^! t\psi_F + g_\%\psi'_P([M]\cap -) \end{pmatrix}$$

$$: \dot{H}^k(T(\nu_M),w) \longrightarrow Q_n^{\pi,w}(C([M']\cap -))$$
$$= Q_n^{\pi',w'}(p^! C([M]\cap -)) \oplus Q_n^{\pi',\beta}(p^! C([M]\cap -), w(t)t^2) .$$

Evaluating the second component on the Thom class $U_{\nu_M} \in \dot{H}^k(T(\nu_M),w)$ and rearranging we thus have

$$-g_\%\psi'_P([M]) = p^! t\psi_F(U_{\nu_M}) \in Q_n^{\pi',\beta}(p^! C([M]\cap -), w(t)t^2)$$

and

$$g_\%\sigma_*(M,p) = p^! tS\sigma_*(M)$$

(with $S\sigma_*(M)$ the suspension of the (n-1)-dimensional quadratic Poincaré complex over $\mathbb{Z}[\pi^w]$

$$\sigma_*(M) = (\Omega C([M]\cap -), \psi \in Q_{n-1}^{\pi,w}(\Omega C([M]\cap -)))$$

defined in Proposition 7.4.1 i)).

ii),iii),iv) are direct consequences of the definitions.

[]

(The definition in Proposition 7.6.6 ii) of the antiquadratic signature $\sigma_*(M,p) \in L_n(\mathbb{Z}[\pi']^\beta, w(t)t^2)$ of an n-dimensional geometric Poincaré complex M with a non-trivial double covering $p:M' \longrightarrow M$ corrects the definition of $\sigma_*(M,p)$ in Ranicki [8,p.566]).

In particular, it follows from Proposition 7.6.5 iii) that for any space X equipped with an orientation map $w : \pi_1(X) = \pi \longrightarrow \mathbb{Z}_2$ and a non-trivial map $\xi : \pi \longrightarrow \mathbb{Z}_2$ the antiquadratic signature defines abelian group morphisms

$$\sigma_*^{\xi} : \Omega_n^P(X,w) \longrightarrow L_n(\mathbb{Z}[\pi']^{\beta}, w(t)t^2) \; ;$$

$$(f:M \longrightarrow X) \longmapsto \sigma_*(M,p_M) \qquad (n \geqslant 0)$$

with $\pi' = \ker(\xi:\pi \longrightarrow \mathbb{Z}_2)$, $p_M = f^*\xi : M \longrightarrow BG(1)$.
If $p_M:M' \longrightarrow M$ is a trivial double covering then $\sigma_*(M,p_M) = 0$, by a special case of Proposition 7.6.6 iv) (with $N = \emptyset$).
The antiquadratic signature maps σ_*^{ξ} are related to the symmetric signature maps σ^* by a commutative diagram

$$\begin{array}{ccc} \Omega_n^P(X,w) & \xrightarrow{\sigma_*^{\xi}} & L_n(\mathbb{Z}[\pi']^{\beta}, w(t)t^2) \\ \downarrow{\scriptstyle \sigma^*} & & \downarrow{\scriptstyle 1+T_{w(t)t^2}} \\ L^n(\mathbb{Z}[\pi^w]) & \xrightarrow{p^!t} & L^n(\mathbb{Z}[\pi']^{\beta}, w(t)t^2) \end{array} \; .$$

<u>Proposition 7.6.7</u> Let (X,Y) be a codimension 1 CW pair of type C) with

$$\pi_1(X) = \pi_1(Y) = \pi \;,\; \pi_1(Z) = \pi_1(S(\xi)) = \pi' \;,\; w(X) = w \; .$$

i) The Poincaré splitting obstruction function $u_{\xi}:f \longmapsto s^P(f,Y)$ on $\Omega_*^P(X,w)$ coincides with the antiquadratic signature map

$$u_{\xi} = \sigma_*^{\xi} : \Omega_n^P(X,w) \longrightarrow LN_{n-2}(\pi' \longrightarrow \pi, w) = L_n(\mathbb{Z}[\pi']^{\beta}, w(t)t^2) \; ;$$

$$(f:M \longrightarrow X) \longmapsto s^P(f,Y) = \sigma_*(M,p_M) \; .$$

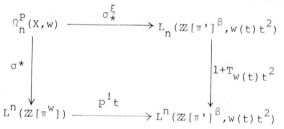

ii) The hyperquadratic signature map $\hat{\sigma}^*$ on $\Omega_*^N(X,w)$ is such that

$$p^!t\hat{\sigma}^* = 0 : \Omega_n^N(X,w) \xrightarrow{\hat{\sigma}^*} \hat{L}^n(\mathbb{Z}[\pi^w]) \xrightarrow{p^!t} \hat{L}^n(\mathbb{Z}[\pi']^\beta, w(t)t^2) .$$

iii) If $(f,\partial f):(M,\partial M) \longrightarrow X$ is a map from an n-dimensional geometric Poincaré pair $(M,\partial M)$ which is (normal,Poincaré) transverse at $Y \subset X$ with $(f,\partial f)^{-1}(Y) = (N,\partial N) \subset (M,\partial M)$ then the rel∂ Poincaré splitting obstruction of f along $Y \subset X$ is given by

$$s_\partial^P(f,Y) = \sigma_*(M,p_M) \cup_{\sigma_*(\partial M,p_{\partial M})} -\sigma_*(\partial M,\partial N)$$

$$\in LN_{n-2}(\pi' \longrightarrow \pi,w) = L_n(\mathbb{Z}[\pi']^\beta, w(t)t^2) .$$

Proof: i) In the first instance we shall combine Propositions 7.5.1 iii), 7.6.4 to give an explicit description of the Poincaré splitting obstruction of f along $Y \subset X$

$$s^P(f,Y) \in LN_{n-2}(\pi' \longrightarrow \pi,w) = L_n(\mathbb{Z}[\pi']^\beta, w(t)t^2) ,$$

assuming that $f:M \longrightarrow X$ is normal transverse at $Y \subset X$. Let

$$f = g^! \cup h : M = E(\nu) \cup_{S(\nu)} P \longrightarrow X = E(\xi) \cup_{S(\xi)} Z ,$$

with

$$g = f| : N = f^{-1}(Y) \longrightarrow Y$$

$$h = f| : P = f^{-1}(Z) \longrightarrow Z$$

$$\nu : N \xrightarrow{g} Y \xrightarrow{\xi} BG(1) .$$

Let $\widetilde{M}$ be the covering of M obtained from the universal covering $\widetilde{X}$ of X by pullback along f, so that $\widetilde{N} \subset \widetilde{M}$ is the covering of N obtained from the universal covering $\widetilde{Y}$ of Y by pullback along g, and

$$\widetilde{M} = \widetilde{P}_+ \cup_{\widetilde{N}} \widetilde{P}_-$$

for two copies $\widetilde{P}_+, \widetilde{P}_-$ of the covering $\widetilde{P}$ of P obtained from the universal covering $\widetilde{Z}$ of Z by pullback along h. The construction of Proposition 7.4.1 i) associates to the (n-1)-dimensional normal space N an (n-2)-dimensional quadratic Poincaré complex over $\mathbb{Z}[\pi^{w\xi}]$

$$\begin{aligned}\sigma_*(N) &= (C,\psi)\\ &= (\Omega C([N]\cap -:C(\widetilde{N})^{n-1-*,w\xi} \longrightarrow C(\widetilde{N})),\psi \in Q_{n-2}^{\pi,w\xi}(C))\ .\end{aligned}$$

Denote the double covering $\widetilde{N}/\pi' = S(\nu)$ of N by N', so that (P,N') is an n-dimensional normal pair. The relative version of the construction of Proposition 7.4.1 i) associates to (P,N') an (n-1)-dimensional quadratic Poincaré pair over $\mathbb{Z}[\pi'^{w'}]$

$$\begin{aligned}\sigma_*(P,N') &= (e:p^!C \longrightarrow D,(\delta\psi,p^!\psi))\\ &= (e : \Omega C([N]\cap - : p^!C(\widetilde{N})^{n-1-*,w'} \longrightarrow p^!C(\widetilde{N}))\\ &\qquad \longrightarrow C([P]\cap - : C(\widetilde{P},\widetilde{N})^{n-*,w'} \longrightarrow C(\widetilde{P})),\\ &\qquad (\delta\psi,p^!\psi) \in Q_{n-1}^{\pi',w'}(e))\ .\end{aligned}$$

Define a $\mathbb{Z}[\pi]$-module chain map

$$i : p_!p^!C \longrightarrow C\ ;\ g\otimes x \longmapsto gx\quad (g\in\pi)\ ,$$

so that

$$C(\begin{pmatrix} i \\ p_!e\end{pmatrix}: p_!p^!C \longrightarrow C\oplus p_!D) = C([M]\cap - : C(\widetilde{M})^{n-*,w} \xrightarrow{\sim} C(\widetilde{M}))$$

is a simple chain contractible (based) $\mathbb{Z}[\pi]$-module chain complex and $\begin{pmatrix} i \\ p_!e\end{pmatrix}$ is a simple chain equivalence. The expression for the Poincaré splitting obstruction given by Proposition 7.5.1 iii) is

$$\begin{aligned} s^P(f,Y) &= (\sigma_*(N),\sigma_*(P,N')) \\ &= ((C,\psi),(e:p^!C \longrightarrow D,(\delta\psi,p^!\psi)) \\ &\in LS_{n-2}(\Phi) = LN_{n-2}(\pi' \longrightarrow \pi,w) \ . \end{aligned}$$

As in the proof of Proposition 7.6.4 we have that the composite $\mathbb{Z}[\pi]$-module chain map

$$j : C \xrightarrow{\lambda k} p_!p^!C \xrightarrow{p_!e} p_!D$$

is a simple chain equivalence, with

$$\lambda k : C \longrightarrow p_!p^!C \ ; \ x \longmapsto 1\otimes x + t\otimes t^{-1}x \ .$$

The restriction of the $\mathbb{Z}[\pi]$-action to $\mathbb{Z}[\pi'] \subset \mathbb{Z}[\pi]$ defines a simple $\mathbb{Z}[\pi']$-module chain equivalence

$$p^!j = \begin{pmatrix} j_+ \\ j_- \end{pmatrix} : p^!C \xrightarrow{\sim} p^!p_!D = D_+\oplus D_-$$

with D_+ a copy of D and $D_- = \mu D_+$. It now follows from Proposition 7.6.4 that the Poincaré splitting obstruction of f along $Y \subset X$ is given by

$$\begin{aligned} s^P(f,Y) &= (D_+,\psi') \\ &\in LN_{n-2}(\pi' \longrightarrow \pi,w) = L_{n-2}(\mathbb{Z}[\pi']^\beta,-w(t)t^2) \end{aligned}$$

with ψ' defined by

$$\psi' = j_{+\%}p^!t(\psi) \in Q_{n-2}^{\pi',\beta}(D_+,-w(t)t^2) \ .$$

As it stands $\psi \in Q_{n-2}^{\pi,w\xi}(C)$ is defined using the spectral quadratic construction ψ_F on a semi-stable π-map $F:T\pi(\nu_N)^* \longrightarrow \Sigma^\infty \tilde{N}_+$ inducing the $\mathbb{Z}[\pi]$-module chain map

$$[N]\cap - \ : \ C(\tilde{N})^{n-1-*,w\xi} \longrightarrow C(\tilde{N}) \ .$$

Working as in the proof of Proposition 7.6.6 i) it is possible to express $p^!t(\psi) \in Q_{n-2}^{\pi',\beta}(p^!C,-w(t)t^2)$ and hence also ψ' in terms of the antiquadratic construction ψ'_{P_N}. It follows from this expression that surgery on the connected (n+1)-dimensional $(\beta,w(t)t^2)$-quadratic pair over $\mathbb{Z}[\pi']$

$$(g_+ = \text{projection} : p^!C(\widetilde{M}) \longrightarrow C(\widetilde{M},\widetilde{P}_+),$$
$$(0,\psi'_{P_M}([M])) \in Q^{\pi',\beta}_{n+1}(g_+,w(t)t^2))$$

results in the skew-suspension $\overline{S}(D_+,\psi')$ of (D_+,ψ'). Thus the skew-suspension isomorphism

$$\overline{S} : L_{n-2}(\mathbb{Z}[\pi']^\beta,-w(t)t^2) \xrightarrow{\sim} L_n(\mathbb{Z}[\pi']^\beta,w(t)t^2)$$

sends the Poincaré splitting obstruction $s^P(f,Y) = (D_+,\psi')$ to the antiquadratic signature $\overline{S}(D_+,\psi') = \sigma_*(M,p_M) = (p^!C(\widetilde{M}),\psi'_{P_M}([M]))$.

ii) Working as in the proof of Proposition 7.6.4 it may be verified that the composite

$$\hat{L}^n(p^!:\mathbb{Z}[\pi^{w\xi}] \longrightarrow \mathbb{Z}[\pi'^{w'}]) \longrightarrow \hat{L}^n(\mathbb{Z}[\pi^w]) \xrightarrow{p^!t} \hat{L}^n(\mathbb{Z}[\pi']^\beta,w(t)t^2)$$

is 0. (This does not require any algebraic transversality). If $f:M \longrightarrow X$ is a map from an n-dimensional normal space M which is normal transverse at $Y \subset X$, with $N = f^{-1}(Y)$, $P = f^{-1}(Z)$, $N' = f^{-1}(Y')$ as in i), then the hyperquadratic signature $\hat{\sigma}^*(M) \in \hat{L}^n(\mathbb{Z}[\pi^w])$ is the image of $(\hat{\sigma}^*(N),\hat{\sigma}^*(P,N')) \in \hat{L}^n(i^!)$. It follows that $p^!t\hat{\sigma}^*(M) = 0$.

iii) This is a direct generalization of i), and may be proved similarly.

[]

The expression for the rel∂ Poincaré splitting obstruction as a union given by Proposition 7.6.7 iii)

$$s^P_\partial(f,Y) = \sigma_*(M,p_M) \cup_{\sigma_*(\partial M,p_{\partial M})} -\sigma_*(\partial M,\partial N)$$

can be combined with the sum formula of Proposition 7.3.6 to recover the result of Mann and Miller [1] that for $n \equiv 0 \pmod 2$ the Arf invariant of $\mathbb{Z}_2 \otimes \sigma_*(M, p_M)$ is defined (i.e. the middle-dimensional self-intersection form is eradicable) if and only if the Arf invariant of $\mathbb{Z}_2 \otimes \sigma_*(\partial M, \partial N)$ is defined, and that if such is the case the difference of Arf invariants is the mod2 rel∂ Poincaré splitting obstruction $\mathbb{Z}_2 \otimes s_\partial^P(f,Y)$.

Proposition 7.6.8 Let $k \geqslant 0$, and let

$$x_k = \bar{S}^k(\mathbb{Z}[\mathbb{Z}_2^{(-)^k}], t) = (C, \phi \in Q^{2k}_{\mathbb{Z}_2,(-)^k}(C))$$

be the 2k-dimensional symmetric Poincaré complex over $\mathbb{Z}[\mathbb{Z}_2^{(-)^k}]$ (= the group ring $\mathbb{Z}[\mathbb{Z}_2]$ with the involution $\bar{t} = (-)^k t$) defined by the k-fold skew-suspension of the non-singular $(-)^k$-symmetric form $(\mathbb{Z}[\mathbb{Z}_2], t)$ over $\mathbb{Z}[\mathbb{Z}_2^{(-)^k}]$. Then

i) $x_k \notin \operatorname{im}(\sigma^* : \Omega^P_{2k}(K(\mathbb{Z}_2,1),(-)^k) \longrightarrow L^{2k}(\mathbb{Z}[\mathbb{Z}_2^{(-)^k}]))$

ii) $Jx_k \notin \operatorname{im}(\hat{\sigma}^* : \Omega^N_{2k}(K(\mathbb{Z}_2,1),(-)^k) \longrightarrow \hat{L}^{2k}(\mathbb{Z}[\mathbb{Z}_2^{(-)^k}]))$.

Proof: i) Let $\xi : K(\mathbb{Z}_2,1) \longrightarrow BG(1)$ ($= K(\mathbb{Z}_2,1)$) be the universal line bundle, and consider the commutative diagram

$$\begin{array}{ccc}
\Omega^P_{2k}(K(\mathbb{Z}_2,1),(-)^k) & \xrightarrow{\ \sigma^\xi_*\ } & L_{2k}(\mathbb{Z},(-)^k) = L_0(\mathbb{Z}) = \mathbb{Z} \\
\downarrow{\scriptstyle \sigma^*} & & \downarrow{\scriptstyle 1+T\ =\ 8} \\
L^{2k}(\mathbb{Z}[\mathbb{Z}_2^{(-)^k}]) & \xrightarrow{\ p^!t\ } & L^{2k}(\mathbb{Z},(-)^k) = L^0(\mathbb{Z}) = \mathbb{Z}
\end{array} .$$

The 2k-dimensional $(-)^k$-symmetric Poincaré complex over $\mathbb{Z}$

$$p^!t(x_k) = \bar{S}^k(\mathbb{Z} \oplus \mathbb{Z}, \begin{pmatrix} 1 & 0 \\ 0 & 1 \end{pmatrix})$$

is the k-fold skew-suspension of the non-singular symmetric

form over $\mathbb{Z}$ $p^!t(\mathbb{Z}[\mathbb{Z}_2^{(-)^k}],t) = (\mathbb{Z}\oplus\mathbb{Z}, \begin{pmatrix} 1 & 0 \\ 0 & 1 \end{pmatrix})$ of signature 2. As $2 \not\equiv 0 \pmod 8$ it follows that $x_k \notin \mathrm{im}(\sigma^*)$.

ii) By i)

$$p^!tJx_k = Jp^!tx_k = 2 \neq 0 \in \hat{L}^{2k}(\mathbb{Z},(-)^k) = \hat{L}^0(\mathbb{Z}) = \mathbb{Z}_8 ,$$

so that $Jx_k \notin \mathrm{im}(\hat{\sigma}^*)$ by Proposition 7.6.6 ii).

[]

In conclusion, we shall use the LN-groups to give a geometric interpretation of the exact sequence of Proposition 5.2.2 for the simple ε-quadratic L-groups $L_*(A_\alpha[x,x^{-1}],\varepsilon)$ $(\varepsilon = \pm 1)$ of the α-twisted Laurent polynomial extension $A_\alpha[x,x^{-1}]$ $(ax = x\alpha(a))$ of a group ring $A = \mathbb{Z}[\pi]$ with the involution extended by $\bar{x} = x$

$$\dots \longrightarrow L_n(A,\varepsilon)\oplus L_n(A^\alpha,\varepsilon) \longrightarrow L_n(A_\alpha[x,x^{-1}],\varepsilon)$$
$$\longrightarrow \widetilde{LNil}_n(A,\alpha,\varepsilon)\oplus\hat{H}^n(\mathbb{Z}_2;Wh(\pi)^\alpha)\oplus\widetilde{LNil}_n(A,\alpha,\varepsilon)$$
$$\longrightarrow L_{n-1}(A,\varepsilon)\oplus L_{n-1}(A^\alpha,\varepsilon) \longrightarrow \dots$$

with $\alpha: A \longrightarrow A$ the ring automorphism induced by a group automorphism $\alpha: \pi \longrightarrow \pi$ such that $\alpha^2 = \mathrm{id.}$, $Y = \{\pi\} \subseteq \tilde{K}_1(A)$, $Z_\pm = \{0\} \subseteq \widetilde{Nil}_1(A,\alpha^{\pm 1})$. The key idea here is due to Tom Farrell and Sylvain Cappell (independently) - I am particularly indebted to the former for a helpful letter. The idea is to express the infinite dihedral group

$$D_\infty = \{x,y \mid (xy)^2 = y^2 = 1\}$$

in two different ways:

i) as the free product of two copies of $\mathbb{Z}_2$

$$D_\infty = \mathbb{Z}_2 * \mathbb{Z}_2 = \{t_1,t_2 \mid t_1^2 = t_2^2 = 1\}$$

with generators $t_1 = xy$, $t_2 = y$

ii) as an extension of $\mathbb{Z}$ by $\mathbb{Z}_2$

$$\{1\} \longrightarrow \mathbb{Z} \xrightarrow{\;p\;} D_\infty \xrightarrow{\;\xi\;} \mathbb{Z}_2 \longrightarrow \{1\}$$

with $p(1) = x \in D_\infty$, $\xi(x) = 1$, $\xi(y) = -1 \in \mathbb{Z}_2 = \{\pm 1\}$,

and to compare the codimension 1 splitting obstruction theory of type A) associated to i) with the codimension 1 splitting obstruction theory of type C) associated to ii). This can be done using either the manifold splitting theory of §7.2, or the geometric Poincaré splitting theory of §7.4, or the algebraic Poincaré splitting theory of §7.5 - we shall stick to manifolds.

Let then π be a finitely presented group which is equipped with an orientation map $w:\pi \longrightarrow \mathbb{Z}_2$ and an automorphism $\alpha:\pi \longrightarrow \pi$ such that

$$\alpha^2 = \mathrm{id.} : \pi \longrightarrow \pi \ , \ w\alpha = w : \pi \longrightarrow \mathbb{Z}_2 \ .$$

Give the group ring $A = \mathbb{Z}[\pi]$ the w-twisted involution

$$\bar{\ } : A \longrightarrow A \ ; \ \sum_{g\in\pi} n_g g \longmapsto \sum_{g\in\pi} w(g) n_g g^{-1} \quad (n_g \in \mathbb{Z})$$

and note that the automorphism

$$\alpha : A \longrightarrow A \ ; \ \sum_{g\in\pi} n_g g \longmapsto \sum_{g\in\pi} n_g \alpha(g)$$

is such that

$$\overline{\alpha(a)} = \alpha(\bar{a}) = \alpha^{-1}(\bar{a}) \in A \quad (a \in A) \ ,$$

so that A,α satisfy the hypotheses of §5.1 and the α-twisted polynomial extensions $A_\alpha[x]$, $A_\alpha[x,x^{-1}]$ of A are defined as rings with involution ($ax = x\alpha(a)$, $\bar{x} = x$). For each element $s \in L_n(A_\alpha[x,x^{-1}],\varepsilon)$ ($\varepsilon = \pm 1$) write the image of s as

$$\begin{aligned}[s] &= ([s]_1,[s]_2,[s]_3) \\ &\in \widetilde{LNil}_n(A,\alpha,\varepsilon) \oplus \hat{H}^n(\mathbb{Z}_2;Wh(\pi)^\alpha) \oplus \widetilde{LNil}_n(A,\alpha,\varepsilon) \ ,\end{aligned}$$

and if $[s] = 0$ denote an inverse image of s by

$$[[s]] = ([[s]]_1, [[s]]_2) \in L_n(A,\varepsilon) \oplus L_n(A^\alpha,\varepsilon) .$$

We seek a geometric interpretation of these decompositions.

Let $\pi \times_\alpha D_\infty$ be the extension of π by D_∞ defined by

$$gt_1 = t_1 g \quad , \quad gt_2 = t_2\alpha(g) \quad (g \in \pi) .$$

As for D_∞ above (the special case $\pi = \{1\}$) there are two different ways of expressing $\pi \times_\alpha D_\infty$:

i) as the free product with amalgamation

$$D_\infty = (\pi \times \mathbb{Z}_2) *_\pi (\pi \times_\alpha \mathbb{Z}_2)$$

with $\pi \times_\alpha \mathbb{Z}_2$ the extension of π by $\mathbb{Z}_2$ defined by $gt_2 = t_2\alpha(g)$

ii) as an extension of $\pi \times_\alpha \mathbb{Z}$ by $\mathbb{Z}_2$

$$\{1\} \longrightarrow \pi \times_\alpha \mathbb{Z} \xrightarrow{\ p\ } \pi \times_\alpha D_\infty \xrightarrow{\ \xi\ } \mathbb{Z}_2 \longrightarrow \{1\}$$

with $i(g) = g$, $p(1) = x$, $\xi(g) = 1$, $\xi(x) = 1$, $\xi(y) = -1$.

Note that $x = t_1 t_2$ (by definition) so that

$$gx = gt_1t_2 = t_1gt_2 = t_1t_2\alpha(g) = x\alpha(g) \in \pi \times_\alpha D_\infty$$

Fix numbers $k \geqslant 3$, $n \geqslant k+7$ and let $(M^{n-k-2}, \partial M^{n-k-3})$ be an $(n-k-2)$-dimensional manifold with boundary such that

$$\pi_1(M) = \pi_1(\partial M) = \pi,\ w(M) = w : \pi \longrightarrow \mathbb{Z}_2 .$$

Let $P_1^k \# P_2^k$ be the connected sum of two copies P_1^k, P_2^k of the real projective k-space $\mathbb{R}P^k$, and let

$$(D^1, S^0) \longrightarrow (N^{k+1}, S^{k-1} \times S^1) \xrightarrow{\ p\ } P_1^k \# P_2^k$$

be the (D^1, S^0)-bundle over $P_1^k \# P_2^k$ classified by

$$\xi : \pi_1(P_1^k \# P_2^k) = \mathbb{Z}_2 * \mathbb{Z}_2 = D_\infty \longrightarrow \mathbb{Z}_2 .$$

Let $(Q_i^{n-2}; \partial_+ Q_i^{n-3}, \partial_- Q_i^{n-3})$ $(i = 1,2)$ be $(n-2)$-dimensional manifold triads such that there are defined fibre bundles

$$(M,\partial M) \longrightarrow (Q_i,\partial_+Q_i) \xrightarrow{q_i} \overline{P_i^k - D^k}$$

with

$$\partial_-Q_i = q_i^{-1}(S^{k-1}) = M \times S^{k-1}$$

$$\pi_1(Q_1) = \pi_1(\partial_+Q_1) = \pi \times \mathbb{Z}_2$$

$$\pi_1(Q_2) = \pi_1(\partial_+Q_2) = \pi \times_\alpha \mathbb{Z}_2$$

(e.g. $(Q_1;\partial_+Q_1,\partial_-Q_1) = (M \times N;\partial M \times N, M \times S^{k-1})$). The $(n-2)$-dimensional manifold with boundary defined by

$$(Q^{n-2},\partial Q^{n-3}) = (Q_1 \cup_{M \times S^{k-1}} Q_2, \partial_+Q_1 \cup_{M \times S^{k-1}} \partial_+Q_2)$$

is then such that

$$\pi_1(Q) = \pi_1(\partial Q) = \pi_1(Q_1) *_{\pi_1(M)} \pi_1(Q_2) = \pi \times_\alpha D_\infty ,$$

and there is defined a fibre bundle $q = q_1 \cup q_2$ over $P_1^k \# P_2^k = \overline{P_1^k - D^k} \cup_{S^{k-1}} \overline{P_2^k - D^k}$

$$(M,\partial M) \longrightarrow (Q,\partial Q) \xrightarrow{(q,\partial q)} P_1^k \# P_2^k \quad .$$

Define also an $(n-1)$-dimensional manifold triad $(X^{n-1};\partial_+X^{n-2},\partial_-X^{n-2})$ by

$$X = \{(u,v) \in N \times Q \mid p(u) = q(v) \in P_1^k \# P_2^k\}$$

$$\partial_+X = \{(u,v) \in X \mid v \in \partial Q \subset Q\}$$

$$\partial_-X = \{(u,v) \in X \mid u \in S^{k-1} \times S^1 = \partial N \subset N\} \quad ,$$

so that (X,∂_-X) is the total space of a (D^1,S^0) bundle ξ over Q

$$(D^1,S^0) \longrightarrow (X,\partial_-X) \xrightarrow{(\xi,\partial_-\xi)} Q$$

(namely the pullback of p along q, with classifying map the group morphism $\xi : \pi_1(Q) = \pi \times_\alpha D_\infty \longrightarrow \mathbb{Z}_2$ defined above), and $(\partial_+X,\partial\partial_+X)$ is the total space of the restriction of ξ to a (D^1,S^0)-bundle $\partial_+\xi$ over ∂Q

$$(D^1,S^0) \longrightarrow (\partial_+X,\partial\partial_+X) \xrightarrow{\partial_+\xi = \xi|} \partial Q$$

(namely the pullback of p along ∂q). Define also the manifold triads and manifolds with boundary

$$(X_i^{n-1};\partial_+X_i^{n-2},\partial_-X_i^{n-2})$$
$$= (\xi^{-1}(Q_i);\xi^{-1}(\partial_-Q_i)\cup\partial_+\xi^{-1}(\partial_+Q_i),\partial_-\xi^{-1}(Q_i)) \quad (i=1,2)$$

$$(Y^{n-2},\partial Y^{n-3}) = \text{the zero section of } (\xi,\partial_+\xi) = (Q,\partial Q)\times 0 \subset (X,\partial_+X)$$

$$(Y_i^{n-2};\partial_+Y_i^{n-3},\partial_-Y_i^{n-3})$$
$$= (X_i\cap Y;\partial_+X_i\cap Y,\partial_+X_1\cap\partial_+X_2\cap Y) = (Q_i;\partial_+Q_i,\partial_-Q_i)\times 0 \quad (i=1,2)$$

$$(Z^{n-2};\partial_+Z^{n-3},\partial_-Z^{n-3})$$
$$= (\xi^{-1}(M\times S^{k-1});\partial_+\xi^{-1}(\partial M\times S^{k-1}),\partial_-\xi^{-1}(M\times S^{k-1}))$$
$$= (X_1\cap X_2;\partial_+X\cap\partial_+X_1\cap\partial_+X_2,\partial_-X_1\cap\partial_-X_2)$$
$$= (M\times S^{k-1}\times D^1;\partial M\times S^{k-1}\times D^1,M\times S^{k-1}\times S^0)$$

$$(W^{n-3},\partial W^{n-4}) = (Y\cap Z,\partial Y\cap Z) = (M\times S^{k-1},\partial M\times S^{k-1})\times 0$$

such that

$$(X;\partial_+X,\partial_-X) = (X_1\cup_Z X_2;\overline{\partial_+X_1\setminus Z}\cup_{\partial_+Z}\overline{\partial_+X_2\setminus Z},\partial_-X_1\cup_{\partial_-Z}\partial_-X_2)$$

$$(Y,\partial Y) = (Y_1\cup_W Y_2,\partial_+Y_1\cup_{\partial W}\partial_+Y_2) \ .$$

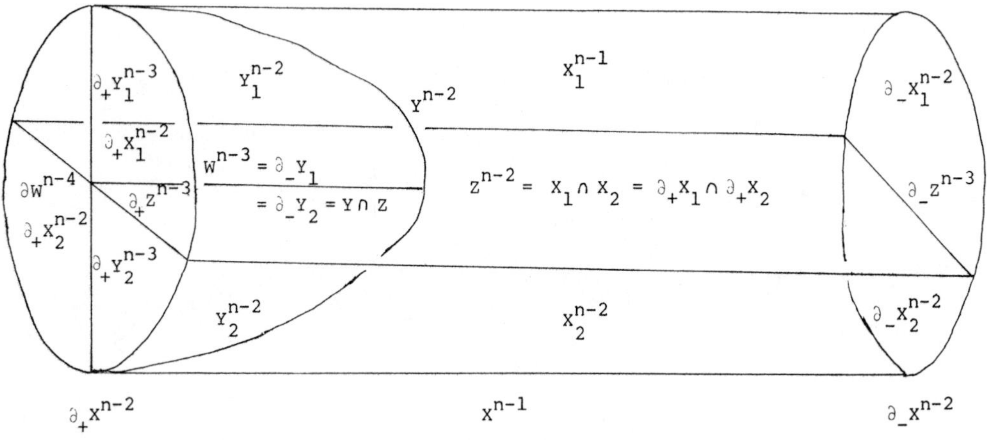

For $\varepsilon = (-)^{k+1}$ let $\mathbb{Z}_2^{\varepsilon}$ (resp. $D_{\infty}^{\varepsilon} = \mathbb{Z}_2^{\varepsilon} * \mathbb{Z}_2^{\varepsilon}$) denote the group $\mathbb{Z}_2$ (resp. D_{∞}) with the orientation map $w_{\varepsilon}(t) = \varepsilon$ (resp. $w_{\varepsilon}(t_i) = \varepsilon$ $(i = 1,2)$), and write $LN_*(\pi \longrightarrow \pi\times_{\alpha}\mathbb{Z}_2, w\times w_{\varepsilon})$ (resp. $LN_*(\pi\times_{\alpha}\mathbb{Z} \longrightarrow \pi\times_{\alpha}D_{\infty}, w\times w_{\varepsilon})$) as $LN_*(\pi \longrightarrow \pi\times_{\alpha}\mathbb{Z}_2^{\varepsilon})$ (resp. $LN_*(\pi\times_{\alpha}\mathbb{Z} \longrightarrow \pi\times_{\alpha}D_{\infty}^{\varepsilon})$).

By Wall [4,Thms.11.7,12.9] every element

$$s \in LN_{n-2}(\pi \times_{\alpha}\mathbb{Z} \longrightarrow \pi \times_{\alpha}D_{\infty}^{\varepsilon})$$

is the rel∂ splitting obstruction along $Y \subset X$ of an s-triangulation of the (n-1)-dimensional manifold triad $(X;\partial_+X,\partial_-X)$

$$f : (V^{n-1};\partial_+V^{n-2},\partial_-V^{n-2}) \xrightarrow{\ \sim\ } (X;\partial_+X,\partial_-X)$$

such that $\partial_+f = f| : \partial_+V \xrightarrow{\ \sim\ } \partial_+X$ is split along $\partial Y \subset \partial_+X$, and the LN-group may be expressed as an L-group

$$LN_{n-2}(\pi \times_{\alpha}\mathbb{Z} \longrightarrow \pi \times_{\alpha}D_{\infty}^{\varepsilon}) = L_n(A_{\alpha}[x,x^{-1}],\varepsilon)$$

with $t = t_2 \in \pi\times_{\alpha}D_{\infty}$ here (cf. Proposition 7.6.4). The image of $s = s_{\partial}(f,Y) \in L_n(A_{\alpha}[x,x^{-1}],\varepsilon)$ in $L_{n-2}(\mathbb{Z}[\pi\times_{\alpha}D^{-\varepsilon}])$ is the rel surgery obstruction $\sigma_*(g,c)$ of the (n-2)-dimensional normal map of pairs

$$(g,c) = (f,b) \ : f^{-1}(Y,\partial Y) \longrightarrow (Y,\partial Y)$$

which restricts to an s-triangulation of ∂Y. The image of $\sigma_*(g,c) \in L_{n-2}(\mathbb{Z}[\pi\times_{\alpha}D^{-\varepsilon}])$ in the group

$$LS_{n-4}\begin{pmatrix} \pi & \longrightarrow & \pi\times_{\alpha}\mathbb{Z}_2^{-\varepsilon} \\ \downarrow & \Phi^{-\varepsilon} & \downarrow \\ \pi\times\mathbb{Z}_2^{-\varepsilon} & \longrightarrow & \pi\times_{\alpha}D_{\infty}^{-\varepsilon} \end{pmatrix} = UNil_{n-2}(\Phi^{-\varepsilon})$$

is the splitting obstruction $s(\partial g,\partial W)$ along $\partial W^{n-4} \subset \partial Y^{n-3}$ of the s-triangulation

$$\partial g = g| : f^{-1}(\partial Y) \xrightarrow{\ \sim\ } \partial Y \ .$$

It is thus possible to identify

$$[s]_1 = s(\partial g, \partial W) \in \widetilde{LNil}_n(A,\alpha,\varepsilon) = UNil_{n-2}(\Phi^{-\varepsilon}) \ .$$

By the unitary nilpotent cobordism construction of Cappell [7] it is possible to replace f by a normal bordant s-triangulation of $(X;\partial_+X,\partial_-X)$ with ∂g split along $\partial W \subset \partial Y$. By the Browder-Wall π-π theorem it is possible to extend this splitting of ∂g to a splitting of the s-triangulation $\partial_+f : \partial_+V \xrightarrow{\sim} \partial_+X$ along $\partial_+Z \subset \partial_+X$. The obstruction to extending the splitting of $\partial\partial_+f = \partial\partial_-f : \partial\partial_-V \xrightarrow{\sim} \partial\partial_-X$ along $\partial\partial_+Z = \partial\partial_-Z \subset \partial\partial_-X$ to a splitting of the s-triangulation $\partial_-f : \partial_-V \xrightarrow{\sim} \partial_-X$ along $\partial_-Z \subset \partial_-X$ is the element

$$s_\partial(\partial_-f,\partial_-Z) = [s]_2 \in LS_{n-3}\left(\begin{array}{ccc} \pi\cup\pi\cup\pi\cup\pi & \longrightarrow & \pi\cup\pi \\ \downarrow & & \downarrow \\ (\pi\cup\pi)\times\nabla & \longrightarrow & \pi\times_\alpha\mathbb{Z} \end{array}\right) = \hat{H}^n(\mathbb{Z}_2;Wh(\pi)^\alpha) \ .$$

Applying the unitary nilpotent cobordism construction again, it is possible to replace f by a normal bordant s-triangulation of $(X;\partial_+X,\partial_-X)$ with ∂_-f split along $\partial_-Z \subset \partial_-X$. The obstruction to extending the splitting of the s-triangulation

$$\partial f = \partial_+f \cup \partial_-f : \partial V = \partial_+V \cup \partial_-V \xrightarrow{\sim} \partial X = \partial_+X \cup \partial_-X$$

along $\partial Z = \partial_+Z \cup \partial_-Z \subset \partial X$ to a splitting of f along $Z \subset X$ is the element

$$[s]_3 = s_\partial(f,Z) \in \widetilde{LNil}_n(A,\alpha,\varepsilon) = LS_{n-2}(\Phi^\varepsilon) = UNil_n(\Phi^\varepsilon) \ .$$

Applying the unitary nilpotent cobordism construction once more, it is possible to replace f by a normal bordant s-triangulation of $(X;\partial_+X,\partial_-X)$ which is split along $(Z;\partial_+Z,\partial_-Z) \subset (X;\partial_+X,\partial_-X)$, so that f restricts to s-triangulations

$$f_i = f| : (V_i^{n-1}; \partial_+ V_i^{n-2}, \partial_- V_i^{n-2}) = f^{-1}(X_i; \partial_+ X_i, \partial_- X_i) \longrightarrow (X_i; \partial_+ X_i, \partial_- X_i) \quad (i = 1,2)$$

such that $\partial_+ f_i = f_i| : \partial_+ V_i \xrightarrow{\sim} \partial_+ X_i$ is split along $\partial Y_i = \partial_+ Y_i \cup_{\partial W} W \subset \partial_+ X_i$. Thus if

$$[s] = ([s]_1, [s]_2, [s]_3) = 0 \in \widetilde{LNil}_n(A,\alpha,\varepsilon) \oplus \hat{H}^n(\mathbb{Z}_2; Wh(\pi)^\alpha) \oplus \widetilde{LNil}_n(A,\alpha,\varepsilon)$$

the original s-triangulation f of $(X;\partial_+ X, \partial_- X)$ is concordant to one which is split along $(Z;\partial_+ Z, \partial_- Z) \subset (X;\partial_+ X, \partial_- X)$, in which case a choice of concordance (which is unique up to $\hat{H}^{n+1}(\mathbb{Z}_2; Wh(\pi)^\alpha)$) determines an inverse image of $s \in L_n(A_\alpha[x,x^{-1}],\varepsilon)$

$$[[s]] = ([[s]]_1, [[s]]_2) \in L_n(A,\varepsilon) \oplus L_n(A^\alpha,\varepsilon) \quad .$$

The obstruction to extending the splitting of $\partial_+ f_1$ along $\partial Y_1 \subset \partial_+ X_1$ to a splitting of f_1 along $Y_1 \subset X_1$ is

$$s_\partial(f_1, Y_1) = [[s]]_1 \in LN_{n-2}(\pi \longrightarrow \pi \times \mathbb{Z}_2^\varepsilon) = L_n(A,\varepsilon) \quad ,$$

and the obstruction to extending the splitting of $\partial_+ f_2$ along $\partial Y_2 \subset \partial_+ X_2$ to a splitting of f_2 along $Y_2 \subset X_2$ is

$$s_\partial(f_2, Y_2) = [[s]]_2 \in LN_{n-2}(\pi \longrightarrow \pi \times {}_\alpha\mathbb{Z}_2^\varepsilon) = L_n(A^\alpha,\varepsilon) \quad .$$

From the point of view of the algebraic theory of codimension 1 surgery the above decompositions of the elements $s \in L_n(A_\alpha[x,x^{-1}],\varepsilon)$ may be deduced from the following commutative diagram of exact sequences of abelian groups, in which the horizontal sequences are of types A) and B), the vertical sequences are of type C), and $L_*(\pi) \equiv L_*(\mathbb{Z}[\pi])$ (as usual).

$$
\begin{array}{ccccccccc}
 & & \vdots & & \vdots & & \vdots & & \vdots \\
 & & \downarrow & & \downarrow & & \downarrow & & \downarrow \\
\cdots \to & & \begin{matrix} L_n(\pi \longrightarrow \pi\times\mathbb{Z}_2^{\varepsilon}) \\ \oplus \\ L_n(\pi \longrightarrow \pi\times_{\alpha}\mathbb{Z}_2^{\varepsilon}) \end{matrix} & \longrightarrow & \begin{matrix} LN_{n-2}(\pi \longrightarrow \pi\times\mathbb{Z}_2^{\varepsilon}) \\ \oplus \\ LN_{n-2}(\pi \longrightarrow \pi\times_{\alpha}\mathbb{Z}_2^{\varepsilon}) \\ (= L_n(A,\varepsilon)\oplus L_n(A^{\alpha},\varepsilon)) \end{matrix} & \longrightarrow & \begin{matrix} L_{n-2}(\pi\times\mathbb{Z}_2^{-\varepsilon}) \\ \oplus \\ L_{n-2}(\pi\times_{\alpha}\mathbb{Z}_2^{-\varepsilon}) \end{matrix} & \longrightarrow & \begin{matrix} L_{n-1}(\pi \longrightarrow \pi\times\mathbb{Z}_2^{\varepsilon}) \\ \oplus \\ L_{n-1}(\pi \longrightarrow \pi\times_{\alpha}\mathbb{Z}_2^{\varepsilon}) \end{matrix} \longrightarrow \cdots \\
 & & \downarrow & & \downarrow & & \downarrow & & \downarrow \\
\cdots \to & & L_n(\pi\times_{\alpha}\mathbb{Z} \longrightarrow \pi\times_{\alpha}D_{\infty}^{\varepsilon}) & \longrightarrow & \begin{matrix} LN_{n-2}(\pi\times_{\alpha}\mathbb{Z} \longrightarrow \pi\times_{\alpha}D_{\infty}^{\varepsilon}) \\ (= L_n(A_{\alpha}[x,x^{-1}],\varepsilon)) \end{matrix} & \longrightarrow & L_{n-2}(\pi\times_{\alpha}D_{\infty}^{-\varepsilon}) & \longrightarrow & L_{n-1}(\pi\times_{\alpha}\mathbb{Z} \to \pi\times_{\alpha}D_{\infty}^{\varepsilon}) \longrightarrow \cdots \\
 & & \downarrow & & \downarrow & & \downarrow & & \downarrow \\
\cdots \to & & \begin{matrix} UNil_n(\Phi^{\varepsilon}) \\ \oplus \\ L_{n-2}^{Wh(\pi)^{\alpha}}(\pi) \end{matrix} & \longrightarrow & \begin{matrix} UNil_n(\Phi^{\varepsilon})\oplus UNil_{n-2}(\Phi^{-\varepsilon}) \\ \oplus \\ \hat{H}^{n-2}(\mathbb{Z}_2;Wh(\pi)^{\alpha}) \end{matrix} & \longrightarrow & \begin{matrix} UNil_{n-2}(\Phi^{-\varepsilon}) \\ \oplus \\ L_{n-3}(\pi) \end{matrix} & \longrightarrow & \begin{matrix} UNil_{n-1}(\Phi^{\varepsilon}) \\ \oplus \\ L_{n-3}^{Wh(\pi)^{\alpha}}(\pi) \end{matrix} \longrightarrow \cdots \\
 & & \multicolumn{5}{c}{(= \widetilde{LNil}_n(A,\alpha,\varepsilon)\oplus\hat{H}^{n}(\mathbb{Z}_2;Wh(\pi)^{\alpha})\oplus\widetilde{LNil}_n(A,\alpha,\varepsilon))} & & \\
 & & \downarrow & & \downarrow & & \downarrow & & \downarrow \\
\cdots \to & & \begin{matrix} L_{n-1}(\pi \longrightarrow \pi\times\mathbb{Z}_2^{\varepsilon}) \\ \oplus \\ L_{n-1}(\pi \longrightarrow \pi\times_{\alpha}\mathbb{Z}_2^{\varepsilon}) \end{matrix} & \longrightarrow & \begin{matrix} LN_{n-3}(\pi \longrightarrow \pi\times\mathbb{Z}_2^{\varepsilon}) \\ \oplus \\ LN_{n-3}(\pi \longrightarrow \pi\times_{\alpha}\mathbb{Z}_2^{\varepsilon}) \\ (= L_{n-1}(A,\varepsilon)\oplus L_{n-1}(A^{\alpha},\varepsilon)) \end{matrix} & \longrightarrow & \begin{matrix} L_{n-3}(\pi\times\mathbb{Z}_2^{-\varepsilon}) \\ \oplus \\ L_{n-3}(\pi\times_{\alpha}\mathbb{Z}_2^{-\varepsilon}) \end{matrix} & \longrightarrow & \begin{matrix} L_{n-2}(\pi \longrightarrow \pi\times\mathbb{Z}_2^{\varepsilon}) \\ \oplus \\ L_{n-2}(\pi \longrightarrow \pi\times_{\alpha}\mathbb{Z}_2^{\varepsilon}) \end{matrix} \longrightarrow \cdots \\
 & & \downarrow & & \downarrow & & \downarrow & & \downarrow \\
 & & \vdots & & \vdots & & \vdots & & \vdots
\end{array}
$$

.

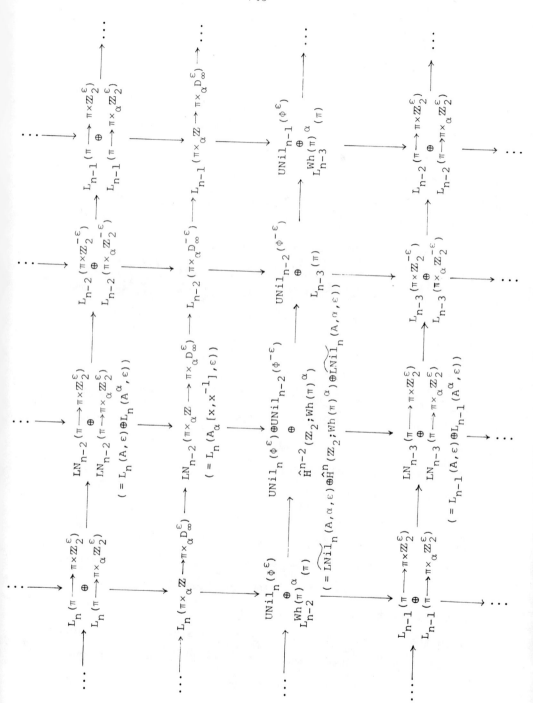

7.7 Surgery with coefficients

In the original theory of Wall [4] quadratic L-groups $L_*(A)$ were defined for all rings with involution A, but only the quadratic L-groups $L_*(\mathbb{Z}[\pi]) = L_*(\pi)$ of integral group rings $\mathbb{Z}[\pi]$ were given a geometric interpretation as surgery obstruction groups. Since then many authors have developed analogues of the theory for surgery with various types of coefficients, giving geometric interpretations of the quadratic L-groups $L_*(S^{-1}\mathbb{Z}[\pi])$ of the localizations away from appropriate multiplicative subsets $S \subset \mathbb{Z}[\pi]$. We shall now list these analogues, after which we shall develop the algebraic theory of the Cappell-Shaneson homology surgery which is needed for the algebraic theory of codimension 2 surgery of §7.8.

I) $\mathbb{Q}$-coefficients

Even prior to the theory of Wall [4] it was clear from the work of Kervaire and Milnor [1] and Wall [2] that quadratic linking forms over $(\mathbb{Z}[\pi],\mathbb{Z}-\{0\})$ play an important role in surgery obstruction theory, in the first instance as a computational tool for finite groups π. Later, Passman and Petrie [1] and Connolly [1] obtained special cases of the localization exact sequence

$$\ldots \longrightarrow L_n(\mathbb{Z}[\pi]) \longrightarrow L_n^S(\mathbb{Q}[\pi]) \longrightarrow L_n(\mathbb{Z}[\pi],S) \longrightarrow L_{n-1}(\mathbb{Z}[\pi]) \longrightarrow \ldots$$
$$(S = \mathbb{Z}-\{0\} \subset \mathbb{Z}[\pi],\ n \in \mathbb{Z})$$

using a mixture of geometry and algebra. Pardon [1],[2],[3] obtained the sequence in general, purely algebraically (at least for finite π), and interpreted $L_n(\mathbb{Z}[\pi],S)$ as the obstruction group for the problem of making an (n-1)-dimensional topological normal map $(f,b):M \longrightarrow X$ which is a rational homotopy equivalence $(\pi_*(f)\otimes\mathbb{Q} = 0)$ normal bordant to a homotopy equivalence $(f',b'):M' \longrightarrow X$ $(\pi_*(f') = 0)$ by a normal bordism $(g,c):(N;M,M') \longrightarrow X \times (I;0,1)$ which is also a rational homotopy equivalence.

II) $\underline{\mathbb{Z}_P\text{-coefficients}}$

Let $P \subseteq \{\text{all primes in } \mathbb{Z}\}$ be a subset (possibly empty), so that there is defined a multiplicative subset

$$S_P = \{q_1^{i_1} q_2^{i_2} \ldots q_k^{i_k} \mid q_1, q_2, \ldots, q_k \in \{\text{all primes in } \mathbb{Z}\} - P\,,\ i_1, i_2, \ldots, i_k \geqslant 0\} \subset \mathbb{Z}$$

and the localization of $\mathbb{Z}$ at P

$$\mathbb{Z}_P = S_P^{-1}\mathbb{Z} \subseteq \mathbb{Q}$$

is defined as usual. Every ring R such that $\mathbb{Z} \subseteq R \subseteq \mathbb{Q}$ is of the type $R = \mathbb{Z}_P$ for some P. A map of finite CW complexes $f:M \longrightarrow X$ such that $f_*:\pi_1(M) \xrightarrow{\sim} \pi_1(X)$ is a P-local homotopy equivalence $(\pi_*(f)\otimes\mathbb{Z}_P = 0)$ if and only if it is a $\mathbb{Z}_P[\pi_1(X)]$-coefficient homology equivalence $(H_*(f;\mathbb{Z}_P[\pi_1(X)]) = 0)$, by the P-local Whitehead theorem. The theory of Wall [4] is the case $P = \{\text{all primes in } \mathbb{Z}\}$, when $\mathbb{Z}_P = \mathbb{Z}$; the theory of I) is the case $P = \emptyset$, when $\mathbb{Z}_P = \mathbb{Q}$.

Surgery on topological normal maps up to $\mathbb{Z}_P[\pi]$-coefficient homology equivalence was first studied by Jones [1], in connection with his work on the fixed point sets of semi-free actions of cyclic groups on manifolds (Smith theory). In particular, the theory of Wall [4] was extended there to surgery on formally n-dimensional topological normal maps $(f,b):M \longrightarrow X$ to n-dimensional $\mathbb{Z}_P[\pi_1(X)]$-coefficient geometric

Poincaré complexes X. It was shown that such a map is normal bordant to a P-local homotopy equivalence if (and for $n \geq 5$ only if) $\sigma_*(f,b) = 0 \in L_n(\mathbb{Z}_P[\pi_1(X)])$. Quinn [4] extended this theory to surgery on $\mathbb{Z}_P$-homology manifolds.

The original application of surgery to the classification of smooth manifolds which are homotopy spheres due to Kervaire and Milnor [1] was generalized by Barge, Lannes, Latour and Vogel [1] to the classification of manifolds which are $\mathbb{Z}_P$-homology spheres.

G.A.Anderson [1] developed an analogue of the Browder-Novikov-Sullivan-Wall theory (the special case P = {all primes in $\mathbb{Z}$}) for the classification of spaces with the P-local homotopy types of manifolds. The theory was reformulated by Taylor and Williams [2], and applied there to the classification of embeddings of manifolds in P-local homotopy spheres, the P-local version of some of the results of Browder [3]. This theory deals with P-local Spivak normal structure; we shall only be concerned with normal spaces $(X, \nu_X : X \longrightarrow BG(k), \rho_X : S^{n+k} \longrightarrow T(\nu_X))$ with P-local Poincaré duality, i.e. such that the $\mathbb{Z}_P[\pi_1(X)]$-module chain map

$$[X] \cap - \ : \ C(\tilde{X};\mathbb{Z}_P)^{n-*} \longrightarrow C(\tilde{X};\mathbb{Z}_P)$$

is a chain equivalence with $\tilde{X}$ the universal cover of X and

$$C(\tilde{X};\mathbb{Z}_P) = \mathbb{Z}_P \otimes_{\mathbb{Z}} C(\tilde{X}) = \mathbb{Z}_P[\pi_1(X)] \otimes_{\mathbb{Z}[\pi_1(X)]} C(\tilde{X}) \ .$$

Pardon [4] used local surgery theory to extend the work of Madsen, Thomas and Wall on the classification of free actions of finite groups on spheres ("the topological spherical space form problem") to the classification of free actions of finite groups on manifolds which are $\mathbb{Z}_P$-homology spheres.

III) Λ-coefficients

Cappell and Shaneson [1] developed an obstruction theory for the problem of making a topological normal map $(f,b):M \longrightarrow X$ to an n-dimensional Λ-coefficient geometric Poincaré complex X normal bordant to a Λ-coefficient homology equivalence, for any locally epic morphism of rings with involution $\mathbb{Z}[\pi_1(X)] \longrightarrow \Lambda$, in connection with their work on codimension 2 surgery. In particular, the theory introduced the Γ-groups Γ_* and the Λ-coefficient homology surgery obstruction was expressed as an element $\sigma_*(f,b) \in \Gamma_n(\mathbb{Z}[\pi_1(X)] \longrightarrow \Lambda)$. The homology surgery theory of I) (resp. II) is essentially the special case $\Lambda = \mathbb{Q}[\pi_1(X)]$ (resp. $\Lambda = \mathbb{Z}_p[\pi_1(X)]$), with $\Gamma_*(\mathbb{Z}[\pi_1(X)] \longrightarrow \Lambda) = L_*(\Lambda)$.

As already noted in §3.2 above Smith [1] expressed the Γ-groups $\Gamma_*(\mathbb{Z}[\pi] \longrightarrow \mathbb{Z}[\rho])$ for certain surjective group morphisms $\pi \longrightarrow \rho$ as the L-groups $L_*(S^{-1}\mathbb{Z}[\pi])$ of the localization of $\mathbb{Z}[\pi]$ away from the multiplicative subset

$$S = \{1+i \mid i \in \ker(\mathbb{Z}[\pi] \longrightarrow \mathbb{Z}[\rho])\} \subset \mathbb{Z}[\pi] ,$$

and that more generally Vogel [1] has expressed the Γ-groups $\Gamma_*(\mathbb{Z}[\pi] \longrightarrow \Lambda)$ of any locally epic morphism $\mathbb{Z}[\pi] \longrightarrow \Lambda$ as the L-groups $L_*(A)$ of an appropriate ring with involution A. Furthermore, Smith [2] developed an obstruction theory for the problem of making a topological normal map $(f,b):M \longrightarrow X$ which is a Λ-coefficient homology equivalence normal bordant to a homotopy equivalence by a normal bordism which is also a Λ-coefficient homology equivalence, expressing the obstruction as an element

$$\sigma_*(f,b) \in \Gamma_{n+1}\begin{pmatrix} \mathbb{Z}[\pi_1(X)] & \longrightarrow & \mathbb{Z}[\pi_1(X)] \\ \downarrow & & \downarrow \\ \mathbb{Z}[\pi_1(X)] & \longrightarrow & \Lambda \end{pmatrix} \quad (n = \dim X).$$

(This theory will be described and generalized further below).

IV) $\mathbb{Z}_m$-coefficients

As already noted in §2.3 above there is an obstruction theory for surgery on $\mathbb{Z}_m$-manifolds (= manifolds with $\mathbb{Z}_m$-type singularities), going back to Sullivan [2], for which the obstruction groups are the relative L-groups $L_*(\mathbb{Z}[\pi];\mathbb{Z}_m)$ appearing in the exact sequence

$$\ldots \longrightarrow L_n(\mathbb{Z}[\pi]) \xrightarrow{m} L_n(\mathbb{Z}[\pi]) \longrightarrow L_n(\mathbb{Z}[\pi];\mathbb{Z}_m) \longrightarrow L_{n-1}(\mathbb{Z}[\pi]) \longrightarrow \ldots .$$

Again, we refer to Morgan and Sullivan [1], Wall [13], Jones [2], Taylor and Williams [1] for applications of surgery on $\mathbb{Z}_m$-manifolds to ordinary surgery.

***

An <u>n-dimensional geometric Λ-Poincaré complex</u> X is an n-dimensional normal space such that the $\mathbb{Z}[\pi_1(X)]$-module chain level cap product

$$[X] \cap - : C(\widetilde{X})^{n-*} \longrightarrow C(\widetilde{X})$$

is a Λ-equivalence for some locally epic morphism of rings with involution $\mathbb{Z}[\pi_1(X)] \longrightarrow \Lambda$, with $\widetilde{X}$ the universal cover. (For id. : $\mathbb{Z}[\pi_1(X)] \longrightarrow \Lambda = \mathbb{Z}[\pi_1(X)]$ this is just the usual notion of a geometric Poincaré complex). If X is a finite n-dimensional CW complex with a fundamental class $[X] \in H_n(X)$ such that $[X] \cap - : H^{n-*}(X;\Lambda) \longrightarrow H_*(X;\Lambda)$ is an isomorphism for some locally epic morphism $\mathbb{Z}[\pi_1(X)] \longrightarrow \Lambda$ and there exists a ring morphism $\Lambda \longrightarrow \mathbb{Z}$ then X is an n-dimensional geometric Λ-Poincaré complex. In keeping with our previous conventions we shall assume that the geometric Λ-Poincaré complexes X we are dealing with are finite and such that $[X] \cap - : C(X;\Lambda)^{n-*} \longrightarrow C(X;\Lambda) = \Lambda \otimes_{\mathbb{Z}[\pi_1(X)]} C(\widetilde{X})$ is a simple

Λ-module chain equivalence. (As usual, there are also versions for finite and finitely-dominated complexes).

Recall from Proposition 2.4.6 that the relative quadratic Γ-group $\Gamma_n\begin{pmatrix} A \longrightarrow A \\ \downarrow \quad \downarrow \\ A \longrightarrow B \end{pmatrix}$ is the cobordism group of (n-1)-dimensional B-acyclic quadratic Poincaré complexes over A, for any locally epic morphism $A \longrightarrow B$. In particular, if X is an n-dimensional geometric Λ-Poincaré complex the construction of Proposition 7.4.1 i) associates to X an (n-1)-dimensional (simple) Λ-acyclic quadratic Poincaré complex over $\mathbb{Z}[\pi_1(X)]$

$$\sigma_*^{\Lambda}(X) = H\hat{\sigma}^*(X) = (\Omega C([X] \cap -:C(\tilde{X})^{n-*} \longrightarrow C(\tilde{X})), \psi \in Q_{n-1}(\Omega C([X] \cap -)))$$

such that $(1+T)\sigma_*^{\Lambda}(X) = \partial\sigma^*(X)$, representing the <u>quadratic signature</u>

$$\sigma_*^{\Lambda}(X) \in \Gamma_n\begin{pmatrix} \mathbb{Z}[\pi_1(X)] \longrightarrow \mathbb{Z}[\pi_1(X)] \\ \Big\downarrow \qquad\qquad \Big\downarrow \\ \mathbb{Z}[\pi_1(X)] \longrightarrow \Lambda \end{pmatrix}.$$

An <u>(n+1)-dimensional (normal,Λ-Poincaré) pair</u> (X,Y) is an (n+1)-dimensional normal pair such that the boundary Y is an n-dimensional geometric Λ-Poincaré complex, with respect to a locally epic morphism $\mathbb{Z}[\pi_1(X)] \longrightarrow \Lambda$. Proposition 7.4.1 i) associates to such a pair (X,Y) an n-dimensional quadratic Λ-Poincaré complex over $\mathbb{Z}[\pi_1(X)]$

$$\sigma_*^{\Lambda}(X,Y) = (\Omega C([X] \cap -:C(\tilde{X})^{n+1-*} \longrightarrow C(\tilde{X},\tilde{Y}), \psi \in Q_n(\Omega C([X] \cap -)))$$

such that $(1+T)\sigma_*^{\Lambda}(X,Y)$ is symmetric Λ-Poincaré cobordant to

$$\sigma_{\Lambda}^*(Y) = (C(\tilde{Y}), \phi_{\tilde{Y}}([Y]) \in Q^n(C(\tilde{Y}))) \quad ,$$

representing the <u>quadratic signature</u> of (X,Y)

$$\sigma_*^{\Lambda}(X,Y) \in \Gamma_n(\mathbb{Z}[\pi_1(X)] \longrightarrow \Lambda) \quad ,$$

with $\tilde{X}$ the universal cover of X and $\tilde{Y}$ the induced cover of Y.

Given a space K and a locally epic morphism of rings with involution $\mathbb{Z}[\pi_1(K)] \longrightarrow \Lambda$ let $\Omega_n^{\Lambda P}(K)$ denote the bordism group of maps $f:X \longrightarrow K$ from n-dimensional geometric Λ-Poincaré complexes X, and let $\Omega_n^{N,\Lambda P}(K)$ denote the bordism group of maps $f:(X,Y) \longrightarrow K$ from n-dimensional (normal,Λ-Poincaré) pairs (X,Y), so that there is defined a long exact sequence

$$\ldots \longrightarrow \Omega_n^{\Lambda P}(K) \longrightarrow \Omega_n^N(K) \longrightarrow \Omega_n^{N,\Lambda P}(K) \longrightarrow \Omega_{n-1}^{\Lambda P}(K) \longrightarrow \ldots .$$

It is tacitly assumed that K is equipped with an orientation map $w : \pi_1(K) \longrightarrow \mathbb{Z}_2$, so that $\mathbb{Z}[\pi_1(K)]$ is given the w-twisted involution, and $w(X) = wf_*$. Also, it is assumed that K is a CW complex with a finite 2-skeleton, so that in particular $\pi_1(K)$ is finitely presented. In the special case $\mathrm{id.} : \mathbb{Z}[\pi_1(K)] \longrightarrow \Lambda = \mathbb{Z}[\pi_1(K)]$

$$\Omega_*^{\Lambda P}(K) = \Omega_*^P(K) \quad , \quad \Omega_*^{N,\Lambda P}(K) = \Omega_*^{N,P}(K) \quad ,$$

using the geometric Poincaré surgery of Browder [7] to ensure that $f_*:\pi_1(X) \longrightarrow \pi_1(K)$ is an isomorphism.

Proposition 7.7.1 The various quadratic signature maps fit together to define a natural transformation of long exact sequences

$$\begin{array}{ccccccc}
\ldots \longrightarrow \Omega_{n+1}^N(K) & \longrightarrow & \Omega_{n+1}^{N,\Lambda P}(K) & \longrightarrow & \Omega_n^{\Lambda P}(K) & \longrightarrow & \Omega_n^N(K) \longrightarrow \ldots \\
\big\downarrow \sigma_* = H\hat{\sigma}^* & & \big\downarrow \sigma_*^{\Lambda} & & \big\downarrow \sigma_*^{\Lambda} & & \big\downarrow \sigma_* \\
\ldots \longrightarrow L_n(\mathbb{Z}[\pi]) & \longrightarrow & \Gamma_n(\mathbb{Z}[\pi] \longrightarrow \Lambda) & \longrightarrow & \Gamma_n\left(\begin{array}{ccc} \mathbb{Z}[\pi] & \longrightarrow & \mathbb{Z}[\pi] \\ \downarrow & & \downarrow \\ \mathbb{Z}[\pi] & \longrightarrow & \Lambda \end{array}\right) & \longrightarrow & L_{n-1}(\mathbb{Z}[\pi]) \longrightarrow \ldots
\end{array}$$

where $\pi = \pi_1(K)$.

[]

In dealing with geometric Λ-Poincaré complexes we shall assume that the Λ-coefficient analogue of the Levitt-Jones-Quinn geometric Poincaré surgery sequence

$$\ldots \longrightarrow \Omega^N_{n+1}(K) \longrightarrow \Gamma_n(\mathbb{Z}[\pi_1(K)] \longrightarrow \Lambda) \longrightarrow \Omega^{\Lambda P}_n(K) \longrightarrow \Omega^N_n(K) \longrightarrow \ldots$$

is exact (at least for $n \geqslant 5$, which we shall also assume). It follows that the quadratic signature maps

$$\sigma^\Lambda_* : \Omega^{N,\Lambda P}_{n+1}(K) \longrightarrow \Gamma_n(\mathbb{Z}[\pi_1(K)] \longrightarrow \Lambda)$$

are isomorphisms. Note that if $1/2 \in \Lambda$ the composites

$$1+T : \Gamma_n(\mathbb{Z}[\pi_1(K)] \longrightarrow \Lambda) \longrightarrow \Omega^{\Lambda P}_n(K) \xrightarrow{\sigma^*_\Lambda} \Gamma^n(\mathbb{Z}[\pi_1(K)] \longrightarrow \Lambda)$$

are isomorphisms, so that there are defined natural direct sum decompositions

$$\Omega^{\Lambda P}_n(K) = \Gamma_n(\mathbb{Z}[\pi_1(K)] \longrightarrow \Lambda) \oplus \Omega^N_n(K) \quad .$$

An <u>n-dimensional geometric (Λ,Λ')-Poincaré pair</u> (X,Y) is an n-dimensional normal pair such that the $\mathbb{Z}[\pi_1(X)]$-module chain level cap product

$$[X] \cap - : C(\widetilde{X})^{n-*} \longrightarrow C(\widetilde{X},\widetilde{Y})$$

is a Λ-equivalence, and such that Y is an $(n-1)$-dimensional geometric Λ'-Poincaré complex, for some locally epic morphisms $\mathbb{Z}[\pi_1(X)] \longrightarrow \Lambda'$, $\Lambda' \longrightarrow \Lambda$. Given a space K and locally epic morphisms $\mathbb{Z}[\pi_1(K)] \longrightarrow \Lambda'$, $\Lambda' \longrightarrow \Lambda$ let $\Omega^{\Lambda P,\Lambda' P}_n(K)$ denote the bordism group of maps $f:(X,Y) \longrightarrow K$ from n-dimensional geometric (Λ,Λ')-Poincaré pairs, so that there is defined an exact sequence

$$\ldots \longrightarrow \Omega^{\Lambda' P}_n(K) \longrightarrow \Omega^{\Lambda P}_n(K) \longrightarrow \Omega^{\Lambda P,\Lambda' P}_n(K) \longrightarrow \Omega^{\Lambda' P}_{n-1}(K) \longrightarrow \ldots \; .$$

Recall from Proposition 2.4.6 that the relative quadratic Γ-group $\Gamma_n\begin{pmatrix} A & \longrightarrow & A \\ \downarrow & & \downarrow \\ B' & \longrightarrow & B \end{pmatrix}$ is the cobordism group of (n-1)-dimensional B-acyclic quadratic B'-Poincaré complexes over A, for any locally epic morphisms $A \longrightarrow B'$, $B' \longrightarrow B$. In particular, if (X,Y) is an n-dimensional geometric (Λ,Λ')-Poincaré pair the construction of Proposition 7.4.1 i) gives an (n-1)-dimensional Λ-acyclic quadratic Λ'-Poincaré complex over $\mathbb{Z}[\pi_1(X)]$

$$\sigma_*^{\Lambda,\Lambda'}(X,Y) = (\Omega C([X]\cap -:C(\tilde{X})^{n-*} \longrightarrow C(\tilde{X},\tilde{Y})),\psi \in Q_{n-1}(\Omega C([X]\cap -)))$$

representing the <u>quadratic signature</u>

$$\sigma_*^{\Lambda,\Lambda'}(X,Y) \in \Gamma_n\begin{pmatrix} \mathbb{Z}[\pi_1(X)] & \longrightarrow & \mathbb{Z}[\pi_1(X)] \\ \downarrow & & \downarrow \\ \Lambda' & \longrightarrow & \Lambda \end{pmatrix}.$$

<u>Proposition 7.7.2</u> i) The quadratic signature maps

$$\Omega_n^{\Lambda P,\Lambda' P}(K) \longrightarrow \Gamma_n\begin{pmatrix} \mathbb{Z}[\pi] & \longrightarrow & \mathbb{Z}[\pi] \\ \downarrow & \Phi & \downarrow \\ \Lambda' & \longrightarrow & \Lambda \end{pmatrix} ; \ (X,Y) \longmapsto \sigma_*^{\Lambda,\Lambda'}(X,Y)$$

are isomorphisms, where $\pi = \pi_1(K)$.

ii) There is defined a natural transformation of exact sequences

$$\begin{array}{ccccccccc} \dots \longrightarrow & \Omega_n^{\Lambda' P}(K) & \longrightarrow & \Omega_n^{\Lambda P}(K) & \longrightarrow & \Gamma_n(\Phi) & \longrightarrow & \Omega_{n-1}^{\Lambda' P}(K) & \longrightarrow \dots \\ & \downarrow{\scriptstyle \sigma^*_{\Lambda'}} & & \downarrow{\scriptstyle \sigma^*_{\Lambda}} & & \downarrow{\scriptstyle 1+T} & & \downarrow{\scriptstyle \sigma^*_{\Lambda'}} & \\ \dots \longrightarrow & \Gamma^n(\mathbb{Z}[\pi]\longrightarrow\Lambda') & \longrightarrow & \Gamma^n(\mathbb{Z}[\pi]\longrightarrow\Lambda) & \longrightarrow & \Gamma^n(\Phi) & \longrightarrow & \Gamma^{n-1}(\mathbb{Z}[\pi]\longrightarrow\Lambda') & \longrightarrow \dots \end{array}$$

.

iii) There is defined a commutative braid of exact sequences

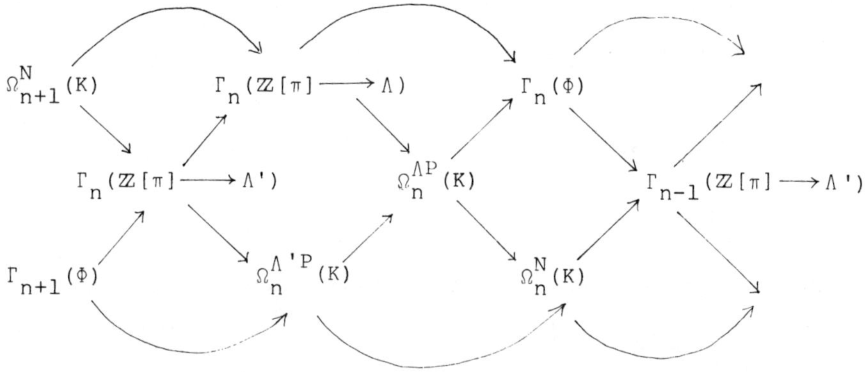

.

[]

Given a formally n-dimensional normal map

$$(f,b) : M \longrightarrow X$$

from an n-dimensional geometric Poincaré complex M to an n-dimensional geometric Λ-Poincaré complex X there is defined a quadratic kernel as in §7.3

$$\sigma^\Lambda_*(f,b) = (C(f^!), \psi_F([X]) \in Q_n(C(f^!)))$$

using the spectral quadratic construction $\psi_F : H_n(X) \longrightarrow Q_n(C(f^!))$ on the geometric Umkehr semi-stable π-map

$$F = T\pi(b)^* : T\pi(\nu_X)^* \longrightarrow T\pi(\nu_M)^* = \Sigma^\infty \widetilde{M}_+ \quad (\pi = \pi_1(X))$$

inducing the Umkehr $\mathbb{Z}[\pi]$-module chain map

$$f^! : C(\widetilde{X})^{n-*} \xrightarrow{\widetilde{f}^*} C(\widetilde{M})^{n-*} \xrightarrow[\sim]{[M] \cap -} C(\widetilde{M}) \quad .$$

The quadratic kernel is an n-dimensional quadratic Λ-Poincaré complex over $\mathbb{Z}[\pi]$ representing the <u>quadratic signature</u>

$$\sigma^\Lambda_*(f,b) \in \Gamma_n(\mathbb{Z}[\pi_1(X)] \longrightarrow \Lambda)$$

such that

$$(1+T)\sigma^\Lambda_*(f,b) = \sigma^*(M) - \sigma^*_\Lambda(X) \in \Gamma^n(\mathbb{Z}[\pi_1(X)] \longrightarrow \Lambda) \quad .$$

The quadratic signature of (f,b) can be interpreted as the quadratic signature $\sigma_*^{\Lambda}(W, M \cup -X)$ of the (n+1)-dimensional (normal,Λ-Poincaré) pair $(W, M \cup -X)$ defined by the mapping cylinder W of $f: M \longrightarrow X$.

The quadratic signature of a formally n-dimensional topological normal map

$$(f,b) : M \longrightarrow X$$

from a manifold M to a geometric Λ-Poincaré complex X is the <u>Λ-homology surgery obstruction</u>

$$\sigma_*^{\Lambda}(f,b) \in \Gamma_n(\mathbb{Z}[\pi_1(X)] \longrightarrow \Lambda)$$

as originally defined by Cappell and Shaneson [1] by a direct generalization of the method of Wall [4], which is the special case $\Lambda = \mathbb{Z}[\pi_1(X)]$. The obstruction vanishes $\sigma_*^{\Lambda}(f,b) = 0$ if (and for $n \geqslant 5$ only if) (f,b) is bordant to a simple Λ-homology equivalence.

More generally, given a formally n-dimensional topological normal map of pairs

$$(f,b) : (M, \partial M) \longrightarrow (X, \partial X)$$

such that $\partial f = f| : \partial M \longrightarrow \partial X$ is a simple Λ-homology equivalence there is defined a <u>rel∂ Λ-homology surgery obstruction</u>

$$\sigma_*^{\Lambda}(f,b) \in \Gamma_n(\mathbb{Z}[\pi_1(X)] \longrightarrow \Lambda)$$

such that $\sigma_*^{\Lambda}(f,b) = 0$ if (and for $n \geqslant 5$ only if) (f,b) is bordant rel ∂f to a simple Λ-homology equivalence of pairs. In this case the quadratic kernel is an n-dimensional quadratic Λ-Poincaré pair over $\mathbb{Z}[\pi_1(X)]$

$$\sigma_*^{\Lambda}(f,\partial f;b,\partial b) = (i: C(\partial f^!) \longrightarrow C(f^!), (\psi_F([X]), \psi_{\partial F}([\partial X])) \in Q_n(i))$$

$$(i = \text{inclusion})$$

such that the boundary (n-1)-dimensional quadratic Λ-Poincaré complex

$$\sigma_*^{\Lambda}(\partial f,\partial b) = (C(\partial f^!),\psi_{\partial F}([\partial X]) \in Q_{n-1}(C(\partial f^!)))$$

is simple Λ-acyclic. The rel∂ Λ-homology surgery obstruction is the cobordism class of the n-dimensional quadratic Λ-Poincaré complex over $\mathbb{Z}[\pi_1(X)]$

$$\sigma_*^{\Lambda}(f,b) = (C(i),\psi_F([X])/\psi_{\partial F}([\partial X]) \in Q_n(C(i)))$$

obtained from $\sigma_*^{\Lambda}(f,\partial f;b,\partial b)$ by applying the algebraic Thom complex construction of §1.2 to collapse the boundary $\sigma_*^{\Lambda}(\partial f,\partial b)$.

Given a commutative square of locally epic morphisms of rings with involution

$$\begin{array}{ccc} \mathbb{Z}[\pi'] & \longrightarrow & \mathbb{Z}[\pi] \\ \downarrow & \Phi & \downarrow \\ \Lambda' & \longrightarrow & \Lambda \end{array}$$

there are defined relative quadratic Γ-groups $\Gamma_*(\Phi)$ to fit into the exact sequence

$$\ldots \longrightarrow \Gamma_n(\mathbb{Z}[\pi'] \longrightarrow \Lambda') \longrightarrow \Gamma_n(\mathbb{Z}[\pi] \longrightarrow \Lambda)$$

$$\longrightarrow \Gamma_n(\Phi) \longrightarrow \Gamma_{n-1}(\mathbb{Z}[\pi'] \longrightarrow \Lambda') \longrightarrow \ldots$$

either geometrically as in Cappell and Shaneson [1] or algebraically as in §2.4 above. Given an n-dimensional geometric (Λ,Λ')-Poincaré triad $(X;X',Y;Y')$ with $\pi_1(X) = \pi$, $\pi_1(X') = \pi'$ and a topological normal map of triads

$$(f,b) : (M;M',N;N') \longrightarrow (X;X',Y;Y')$$

(with $\partial M' = \partial N = N'$, $\partial M = M' \cup_{N'} -N$) such that the restriction

$$(g,c) = (f,b)| : (N,N') \longrightarrow (Y,Y')$$

is a simple (Λ,Λ')-homology equivalence of pairs there is defined a <u>relative (Λ,Λ')-homology surgery obstruction</u>

$$\sigma_*^{\Lambda,\Lambda'}(f,b) \in \Gamma_n(\Phi)$$

such that $\sigma_*^{\Lambda,\Lambda'}(f,b) = 0$ if (and for $n \geqslant 6$ only if) (f,b) is normal bordant rel (g,c) to a simple (Λ,Λ')-homology equivalence of triads. The image of $\sigma_*^{\Lambda,\Lambda'}(f,b) \in \Gamma_n(\Phi)$ in $\Gamma_{n-1}(\mathbb{Z}[\pi'] \longrightarrow \Lambda')$ is the rel∂ Λ'-homology surgery obstruction of the restriction $(f,b)| : (M',N') \longrightarrow (X',Y')$.

In the applications of homology surgery theory to codimension 2 surgery due to Cappell and Shaneson [1] the obstruction groups arising are actually the relative Γ-groups $\Gamma_*(\Phi_0)$ of commutative squares of the type

$$\begin{array}{ccc} \mathbb{Z}[\pi] & \longrightarrow & \mathbb{Z}[\pi] \\ \downarrow & \Phi_0 & \downarrow \\ \mathbb{Z}[\pi] & \longrightarrow & \Lambda \end{array}$$

(cf. the algebraic theory of codimension 2 surgery of §7.8). Pardon [1] (for $\Lambda = \mathbb{Q}[\pi]$) and Smith [2] (for arbitrary Λ) have interpreted the groups $\Gamma_*(\Phi_0)$ as the obstruction groups for making a topological normal map $(f,b):M \longrightarrow X$ $(\pi_1(X) = \pi)$ which is a simple Λ-homology equivalence normal bordant to an s-triangulation of X by a normal bordism which is also a simple Λ-homology equivalence - we shall generalize this interpretation in Proposition 7.7.3 below. In particular, for $\Lambda = \mathbb{Q}[\pi]$ the Γ-group exact sequence

$$\ldots \longrightarrow \Gamma_n(1:\mathbb{Z}[\pi] \longrightarrow \mathbb{Z}[\pi]) \longrightarrow \Gamma_n(\mathbb{Z}[\pi] \longrightarrow \mathbb{Q}[\pi])$$
$$\longrightarrow \Gamma_n(\Phi_0) \longrightarrow \Gamma_{n-1}(1:\mathbb{Z}[\pi] \longrightarrow \mathbb{Z}[\pi]) \longrightarrow \ldots$$

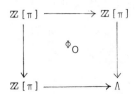

coincides with the quadratic L-theory localization exact sequence for the multiplicative subset $S = \mathbb{Z}-\{0\} \subset \mathbb{Z}[\pi]$

$$\ldots \longrightarrow L_n(\mathbb{Z}[\pi]) \longrightarrow L_n^S(\mathbb{Q}[\pi]) \longrightarrow L_n(\mathbb{Z}[\pi],S) \longrightarrow L_{n-1}(\mathbb{Z}[\pi]) \longrightarrow \ldots$$

obtained by Pardon [1],[2],[3] and in §3 above.

Let

$$\begin{array}{ccc} \mathbb{Z}[\pi] & \longrightarrow & \mathbb{Z}[\pi] \\ \downarrow & \Phi_0 & \downarrow \\ \Lambda' & \longrightarrow & \Lambda \end{array}$$

be a commutative square of locally epic morphisms of rings with involution. Given an n-dimensional geometric (Λ,Λ)-Poincaré pair $(X,\partial X)$ with $\pi_1(X) = \pi$ and a topological normal map of pairs

$$(f,b) : (M,M') \longrightarrow (X,\partial X)$$

such that $f:M \longrightarrow X$ is a simple Λ-homology equivalence and $\partial f = f| : M' \longrightarrow \partial X$ is a simple Λ'-homology equivalence, so that the quadratic kernel $\sigma_*(f,\partial f;b,\partial b) = (i:C(\partial f^!) \longrightarrow C(f^!),(\psi,\partial\psi))$ is an n-dimensional Λ-acyclic quadratic Λ'-Poincaré pair over $\mathbb{Z}[\pi]$ with a Λ'-acyclic boundary $\sigma_*(\partial f,\partial b) = (C(\partial f^!),\partial\psi)$. Collapsing the boundary by the algebraic Thom complex construction (as before) there is obtained an n-dimensional Λ-acyclic quadratic Λ'-Poincaré complex over $\mathbb{Z}[\pi]$ $\sigma_*^{\Lambda,\Lambda'}(f,b) = (C(i),\psi/\partial\psi)$ representing the <u>quadratic signature</u>

$$\sigma_*^{\Lambda,\Lambda'}(f,b) \in \Gamma_{n+1}(\Phi_0) .$$

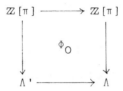

Proposition 7.7.3 i) The quadratic signature is such that $\sigma_*^{\Lambda,\Lambda'}(f,b) = 0 \in \Gamma_{n+1}(\Phi_0)$ if (and for $n \geq 5$ only if) (f,b) extends to a topological normal map of $(n+1)$-dimensional triads

$$(F,B) : (N;N',M;M') \longrightarrow (X \times I; X \times 0, X \times 1; \partial X \times I) \quad (I = [0,1])$$

such that $F:N \longrightarrow X \times I$ is a simple Λ-homology equivalence and $F|:N' \longrightarrow X \times 0$ is a simple Λ'-homology equivalence.

ii) Let $(F,B):(N;N',M;M') \longrightarrow (X \times I; X \times 0, X \times 1; \partial X \times I)$ be a topological normal map of $(n+1)$-dimensional triads such that the restriction $(f,b) = (F,B)| : (M,M') \longrightarrow (X \times 1, \partial X \times I)$ $(\simeq (X,\partial X))$ is a simple (Λ,Λ')-homology equivalence of pairs. Then the relative (Λ,Λ')-homology surgery obstruction of (F,B) is the quadratic signature of (f,b)

$$\sigma_*^{\Lambda,\Lambda'}(F,B) = \sigma_*^{\Lambda,\Lambda'}(f,b) \in \Gamma_{n+1}(\Phi_0) .$$

Proof: i) The image of $\sigma_*^{\Lambda,\Lambda'}(f,b) \in \Gamma_{n+1}(\Phi_0)$ in $\Gamma_n(\mathbb{Z}[\pi] \longrightarrow \Lambda')$ is the obstruction to extending (f,b) to a topological normal map of $(n+1)$-dimensional triads

$$(F,B) : (N;N',M;M') \longrightarrow (X \times I; X \times 0, X \times 1; \partial X \times I)$$

such that $F|:N' \longrightarrow X \times 0$ is a simple Λ'-homology equivalence. Such an extension (F,B) determines a quadratic Λ'-Poincaré null-cobordism of the quadratic kernel $\sigma_*(f,b)$, and conversely every such null-cobordism determines such an extension (F,B). The null-cobordism is Λ-acyclic precisely when $F:N \longrightarrow X \times I$ is a (simple) Λ-homology equivalence.

ii) There is a canonical topological normal bordism rel $f:M \longrightarrow X$ of topological normal maps of $(n+1)$-dimensional triads, from $(F,B):(N;N';M;M') \longrightarrow (X \times I; X \times 0, X \times 1; \partial X \times I)$ to

$$(f,b) \times 1 : (M \times I; M \times 0, M \times 1; M' \times I) \longrightarrow (X \times I; X \times 0, X \times 1; \partial X \times I)$$

(essentially given by $N \times I$, involving a copy of N from N' to $M \times 0$) so that

$$\sigma_*^{\Lambda,\Lambda'}(F,B) = \sigma_*^{\Lambda,\Lambda'}((f,b) \times 1) \in \Gamma_{n+1}(\Phi_0)$$

The relative (Λ,Λ')-homology surgery obstruction $\sigma_*^{\Lambda,\Lambda'}((f,b) \times 1) \in \Gamma_{n+1}^{\Lambda,\Lambda'}(\Phi_0)$ is represented by the pair

(the n-dimensional quadratic Λ'-Poincaré complex over $\mathbb{Z}[\pi]$ $\sigma_*(f,b)$, the quadratic Λ-Poincaré null-cobordism of $\sigma_*(f,b)$ determined by its Λ-acyclicity) ,

which is just the Λ-acyclic quadratic Λ'-Poincaré cobordism class $\sigma_*^{\Lambda,\Lambda'}(f,b) \in \Gamma_{n+1}(\Phi_0)$ appearing in i).

[]

The total surgery obstruction theory of §7.1 extends to homology surgery as follows.

Given a topological space X (equipped with an orientation map $w:\pi_1(X) \longrightarrow \mathbb{Z}_2$) and a locally epic morphism of rings with involution $\mathbb{Z}[\pi_1(X)] \longrightarrow \Lambda$ define the <u>Λ-coefficient $\mathcal{S}$-groups</u> $\mathcal{S}_*(X;\Lambda)$ to fit into the exact sequence

$$\ldots \longrightarrow H_n(X;\underline{\mathbb{L}}_0) \xrightarrow{\sigma_*^\Lambda} \Gamma_n(\mathbb{Z}[\pi_1(X)] \longrightarrow \Lambda) \longrightarrow \mathcal{S}_n(X;\Lambda) \longrightarrow H_{n-1}(X;\underline{\mathbb{L}}_0) \longrightarrow \ldots ,$$

where σ_*^Λ is the composite

$$\sigma_*^\Lambda : H_n(X;\underline{\mathbb{L}}_0) \xrightarrow{\sigma_*} L_n(\mathbb{Z}[\pi_1(X)]) \longrightarrow \Gamma_n(\mathbb{Z}[\pi_1(X)] \longrightarrow \Lambda) .$$

For id. : $\mathbb{Z}[\pi_1(X)] \longrightarrow \Lambda = \mathbb{Z}[\pi_1(X)]$ these are just the $\mathcal{S}$-groups of §7.1

$$\mathcal{S}_*(X;\mathbb{Z}[\pi_1(X)]) = \mathcal{S}_*(X) .$$

<u>Proposition 7.7.4</u> Given a space X and a commutative square of locally epic morphisms

$$\begin{array}{ccc} \mathbb{Z}[\pi] & \longrightarrow & \mathbb{Z}[\pi] \\ \downarrow & \Phi_0 & \downarrow \\ \Lambda' & \longrightarrow & \Lambda \end{array} \qquad (\pi = \pi_1(X))$$

there is defined a commutative braid of exact sequences

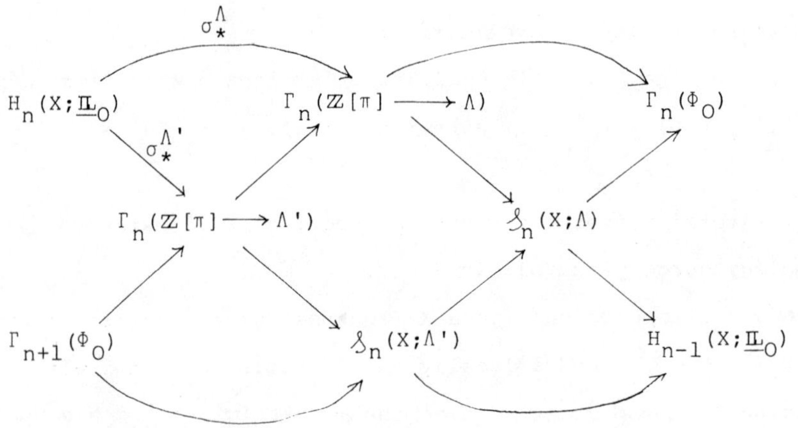

.

[]

An <u>s^{Λ}-triangulation</u> of an n-dimensional geometric Λ-Poincaré complex X is a formally n-dimensional topological normal map

$$(f,b) : M \longrightarrow X$$

such that $f:M \longrightarrow X$ is a simple Λ-homology equivalence. (If there exists a ring morphism $\Lambda \longrightarrow \mathbb{Z}$ every Λ-homology equivalence $f:M \xrightarrow{\simeq} X$ from a manifold M can be given the structure of an s^{Λ}-triangulation $(f,b):M \longrightarrow X$, since integral homology equivalences induce isomorphisms in topological K-theory). Let $\mathcal{S}^{TOP}(X;\Lambda)$ denote the set of concordance classes of s^Λ-triangulations of X.

The <u>total Λ-homology surgery</u> (or <u>s^Λ-triangulability</u>) <u>obstruction</u> of an n-dimensional geometric Λ-Poincaré complex X is an element

$$s^\Lambda(X) \in \mathcal{S}_n(X;\Lambda)$$

with the following properties.

<u>Proposition 7.7.5</u> i) $s^\Lambda(X) = 0 \in \mathcal{S}_n(X;\Lambda)$ if (and for $n \geqslant 5$ only if) X is s^Λ-triangulable.

ii) The image of $s^\Lambda(X)$ in $H_{n-1}(X;\underline{\mathbb{L}}_0) \oplus \Gamma_n\begin{pmatrix}\mathbb{Z}[\pi] \longrightarrow \mathbb{Z}[\pi] \\ \downarrow \qquad\quad \downarrow \\ \mathbb{Z}[\pi] \longrightarrow \Lambda\end{pmatrix}$ $\quad(\pi = \pi_1(X))$

is given by

$$[s^\Lambda(X)] = (\text{the t-triangulability obstruction } t(X), \text{ the quadratic signature } \sigma_*^\Lambda(X)) \ .$$

If

$$s^\Lambda(X) \in \ker(\mathcal{S}_n(X;\Lambda) \longrightarrow H_{n-1}(X;\underline{\mathbb{L}}_0)) = \mathrm{im}(\Gamma_n(\mathbb{Z}[\pi_1(X)] \longrightarrow \Lambda) \longrightarrow \mathcal{S}_n(X;\Lambda))$$

(i.e. if X is t-triangulable) the inverse image of $s^\Lambda(X)$ in $\Gamma_n(\mathbb{Z}[\pi_1(X)] \longrightarrow \Lambda)$ is the coset of the subgroup

$$\ker(\Gamma_n(\mathbb{Z}[\pi_1(X)] \longrightarrow \Lambda) \longrightarrow \mathcal{S}_n(X;\Lambda)) = \mathrm{im}(\sigma_*^\Lambda : H_n(X;\underline{\mathbb{L}}_0) \longrightarrow \Gamma_n(\mathbb{Z}[\pi_1(X)] \longrightarrow \Lambda))$$

consisting of the Λ-homology surgery obstructions $\sigma_*^\Lambda(f,b) \in \Gamma_n(\mathbb{Z}[\pi_1(X)] \longrightarrow \Lambda)$ of all the topological normal maps $(f,b) : M \longrightarrow X$.

iii) If $n \geqslant 5$ and X is s^Λ-triangulable the structure set $\mathcal{S}^{TOP}(X;\Lambda)$ carries a natural affine structure with translation group $\mathcal{S}_{n+1}(X;\Lambda)$. If X is an n-dimensional manifold there is a natural identification

$$\mathcal{S}^{TOP}(X;\Lambda) = \mathcal{S}_{n+1}(X;\Lambda)$$

and the Λ-homology surgery exact sequence

$$\ldots \longrightarrow \Gamma_{n+1}(\mathbb{Z}[\pi_1(X)] \longrightarrow \Lambda) \longrightarrow \mathcal{S}^{TOP}(X;\Lambda)$$
$$\longrightarrow [X,G/TOP] \longrightarrow \Gamma_n(\mathbb{Z}[\pi_1(X)] \longrightarrow \Lambda)$$

can be identified with the exact sequence of abelian groups

$$\ldots \longrightarrow \Gamma_{n+1}(\mathbb{Z}[\pi_1(X)] \longrightarrow \Lambda) \longrightarrow \mathcal{S}_{n+1}(X;\Lambda)$$
$$\longrightarrow H_n(X;\underline{\mathbb{L}}_0) \xrightarrow{\sigma_*^\Lambda} \Gamma_n(\mathbb{Z}[\pi_1(X)] \longrightarrow \Lambda) \quad .$$

Furthermore, if $\mathbb{Z}[\pi_1(X)] \longrightarrow \Lambda$ factors through a locally epic morphism $\mathbb{Z}[\pi_1(X)] \longrightarrow \Lambda'$ the canonical map

$$\mathcal{S}^{TOP}(X;\Lambda) = \mathcal{S}_{n+1}(X;\Lambda) \longrightarrow \Gamma_{n+1}\begin{pmatrix} \mathbb{Z}[\pi_1(X)] & \longrightarrow & \mathbb{Z}[\pi_1(X)] \\ \downarrow & \Phi_0 & \downarrow \\ \Lambda' & \longrightarrow & \Lambda \end{pmatrix}$$

sends an s^Λ-triangulation $(f,b):M \xrightarrow{\sim} X$ to the obstruction $\sigma_*^{\Lambda,\Lambda'}(f,b) \in \Gamma_{n+1}(\Phi_0)$ to making (f,b) concordant (= topologically normal bordant by an s^Λ-triangulation of triads $((F,B);(f,b),(f',b')) : (N;M,M') \xrightarrow{\sim} X \times (I;0,1))$ to an $s^{\Lambda'}$-triangulation $(f',b'):M' \longrightarrow X$.

[]

The total Λ-homology surgery obstruction defines abelian group morphisms

$$s^\Lambda : \Omega_n^{\Lambda P}(K) \longrightarrow \mathcal{S}_n(K;\Lambda) \ ; \ (f:X \longrightarrow K) \longmapsto f_* s^\Lambda(X)$$

for any space K equipped with a locally epic morphism $\mathbb{Z}[\pi_1(K)] \longrightarrow \Lambda$, which fit into a commutative braid of exact sequences

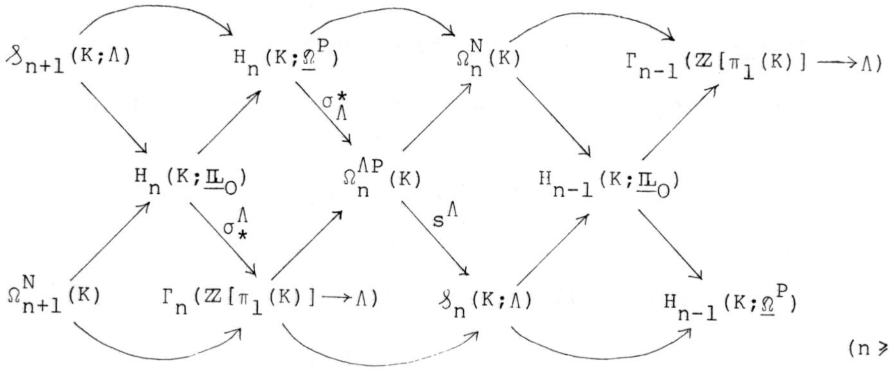

$(n \geqslant 5)$

involving the geometric Λ-Poincaré assembly maps

$$\sigma_\Lambda^* \;:\; H_n(K;\underline{\Omega}^P) \xrightarrow{\sigma^*} \Omega_n^P(K) \longrightarrow \Omega_n^{\Lambda P}(K) \quad .$$

with $\underline{\Omega}^P$ the geometric Poincaré bordism spectrum $(\pi_*(\underline{\Omega}^P) = \Omega_*^P(\mathrm{pt.}))$. Thus an n-dimensional geometric Λ-Poincaré complex X admits an $\underline{\Omega}^P$-orientation $[X] \in H_n(X;\underline{\Omega}^P)$ such that

$$\sigma_\Lambda^*([X]) = (1:X \longrightarrow X) \in \Omega_n^{\Lambda P}(X)$$

if (and for $n \geqslant 5$ only if) X is s^Λ-triangulable.

If $1/2 \in \Lambda$ then $\Gamma_*(\mathbb{Z}[\pi_1(X)] \longrightarrow \Lambda) = \Gamma^*(\mathbb{Z}[\pi_1(X)] \longrightarrow \Lambda)$ and

$$\begin{aligned}(1:X \longrightarrow X) &= ((1:X \longrightarrow X), \sigma_\Lambda^*(X)) \\ &\in \Omega_n^{\Lambda P}(X) = \Omega_n^N(X) \oplus \Gamma^n(\mathbb{Z}[\pi_1(X)] \longrightarrow \Lambda) \quad .\end{aligned}$$

Also, in this case the Λ-coefficient $\mathcal{S}$-groups $\mathcal{S}_*(X;\Lambda)$ fit into the exact sequence

$$\ldots \longrightarrow H_n(X;\underline{\mathbb{L}}^0) \xrightarrow{\begin{pmatrix} J \\ \sigma_\Lambda^* \end{pmatrix}} H_n(X;\hat{\underline{\mathbb{L}}}^0) \oplus \Gamma^n(\mathbb{Z}[\pi_1(X)] \longrightarrow \Lambda) \longrightarrow \mathcal{S}_n(X;\Lambda)$$
$$\longrightarrow H_{n-1}(X;\underline{\mathbb{L}}^0) \longrightarrow \ldots$$

and $s^\Lambda(X) \in \mathcal{S}_n(X;\Lambda)$ is the image of $([\hat{X}], \sigma_\Lambda^*(X))$, so that

X admits an $\underline{\mathbb{L}}^{O}$-orientation $[X] \in H_n(X;\underline{\mathbb{L}}^{O})$ such that

i) $J([X]) = [\hat{X}] \in H_n(X;\hat{\underline{\mathbb{L}}}^{O})$ is the canonical $\hat{\underline{\mathbb{L}}}^{O}$-orientation of X

ii) $\sigma^*_\Lambda([X]) = \sigma^*_\Lambda(X) \in \Gamma^n(\mathbb{Z}[\pi_1(X)] \longrightarrow \Lambda)$ is the symmetric signature of X

if (and for $n \geqslant 5$ only if) X is s^Λ-triangulable.

Thus if $1/2 \in \Lambda$ it is not necessary to consider the delicate $\underline{\mathbb{L}}$-spectrum level compatibility condition needed for $\Lambda = \mathbb{Z}[\pi_1(X)]$.

7.8 The algebraic theory of codimension 2 surgery

Codimension 2 surgery goes back to knot theory, which is the classification theory of embeddings $k:S^1 \subset S^3$. The application of surgery methods to the classification of high-dimensional knots $k:S^n \subset S^{n+2}$ $(n \geq 1)$ was initiated by Kervaire [1] and Levine [2],[4]. (We shall discuss high-dimensional knot theory in §7.9 below). Non-simply-connected codimension 2 surgery first occurred in the work of Browder [5] on free $\mathbb{Z}_p$-actions on homotopy spheres. The general codimension q surgery obstruction theory of Wall [4,§11] applies equally well for q = 2, provided it is generalized to take into account "the general philosophy for dealing with surgery problems in codimension 2: do not insist on obtaining homotopy equivalences when you are doing surgery on the complement of a submanifold, be happy if you can obatin the correct homology conditions" suggested by López de Medrano [2], and the homology surgery theory developed for that purpose by Cappell and Shaneson [1]. Codimension 2 surgery has also been studied by Matsumoto [1] and Freedman [1], by considering ambient surgery on the submanifold instead of homology surgery on the complement. We shall now specialize the algebraic theory of codimension q surgery of §7.5 to the case q = 2, making use of the algebraic homology surgery theory of §7.7. In particular, this permits the comparison to each other of the previous approaches.

To start with we shall modify the geometric theory of codimension q surgery of §7.2 for q = 2 so as to only take the homology type of the complement into account.

A <u>weak (n,n-2)-dimensional</u> (or <u>codimension 2</u>) <u>geometric Poincaré pair</u> (X,Y) is a codimension 2 CW pair (X,Y) such that

i) X is an n-dimensional geometric Poincaré complex

ii) Y is an $(n-2)$-dimensional geometric Poincaré complex.

Then $\pi_1(Z) \longrightarrow \pi_1(X)$ is onto, and $(Z,S(\xi))$ is an n-dimensional geometric $\mathbb{Z}[\pi_1(X)]$-Poincaré pair, with $X = E(\xi) \cup_{S(\xi)} Z$, $\xi : Y \longrightarrow BG(2) = B\widetilde{TOP}(2)$.

A <u>t-triangulation</u> of a weak codimension 2 geometric Poincaré pair (X,Y)

$$((f,b),(g,c)) : (M,N) \longrightarrow (X,Y)$$

is a t-triangulation $(f,b):M \longrightarrow X$ of X (i.e. a topological normal map) which is topologically transverse at $Y \subset X$, so that $(M,N = f^{-1}(Y))$ is a codimension 2 manifold pair with normal block bundle

$$\nu : N \xrightarrow{\ f| \ } Y \xrightarrow{\ \xi\ } B\widetilde{TOP}(2) \quad .$$

The restrictions of (f,b) define topological normal maps

$$(f,b)| = (g,c) : N \longrightarrow Y$$

$$(f,b)| = (h,d) : (P,S(\nu)) \longrightarrow (Z,S(\xi)) \qquad (P = \overline{M \setminus E(\nu)}) \ ,$$

so that

$$(f,b) = (g,c)^{!} \cup (h,d) : M = E(\nu) \cup_{S(\nu)} P \longrightarrow X = E(\xi) \cup_{S(\xi)} Z \ .$$

As for a strong codimension 2 geometric Poincaré pair (X,Y) (Proposition 7.2.3) the set of concordance classes $\mathcal{T}^{TOP}(X,Y,\xi)$ of t-triangulations of (X,Y) is naturally identified with $\mathcal{T}^{TOP}(X)$, and hence also with $\mathcal{T}^{TOP}(\nu_X)$, by topological transversality.

A <u>weak s-triangulation</u> of a weak codimension 2 geometric Poincaré pair (X,Y) is a t-triangulation

$$((f,b),(g,c)) : (M,N) \longrightarrow (X,Y)$$

such that $(f,b):M \longrightarrow X$ is an s-triangulation of X and $(g,c):N \longrightarrow Y$ is an s-triangulation of Y, in which case

$$(h,d) = (f,b)| : (P,S(\nu)) \longrightarrow (Z,S(\xi))$$

is an s^{Λ}-triangulation of $(Z,S(\xi))$ with $\Lambda = \mathbb{Z}[\pi_1(X)]$. Let $\mathcal{WS}^{TOP}(X,Y,\xi)$ denote the set of concordance classes of weak s-triangulations of (X,Y). An s-triangulation $(f,b):M \xrightarrow{\sim} X$ of X is <u>weakly split along $Y \subset X$</u> if f actually defines a weak s-triangulation of (X,Y).

The weak splitting obstruction theory for $\begin{Bmatrix} \text{s-} \\ \text{t-} \end{Bmatrix}$ triangulations involves the following analogues of the $\begin{Bmatrix} \text{LS-} \\ \text{LP-} \end{Bmatrix}$ groups appearing in the strong splitting obstruction theory of §7.2.

Given a (connected) codimension 2 CW pair (X,Y) let Φ denote the associated pushout square of fundamental groups

$$\begin{array}{ccc} \pi_1(S(\xi)) & \longrightarrow & \pi_1(Z) \\ \downarrow & \Phi & \downarrow \\ \pi_1(E(\xi)) & \longrightarrow & \pi_1(X) \end{array} ,$$

and let Φ_Z, Φ_X denote the commutative squares of group rings

$$\begin{array}{ccc} \mathbb{Z}[\pi_1(Z)] & \longrightarrow & \mathbb{Z}[\pi_1(Z)] \\ \downarrow & \Phi_Z & \downarrow \\ \mathbb{Z}[\pi_1(Z)] & \longrightarrow & \mathbb{Z}[\pi_1(X)] \end{array} , \quad \begin{array}{ccc} \mathbb{Z}[\pi_1(Z)] & \longrightarrow & \mathbb{Z}[\pi_1(X)] \\ \downarrow & \Phi_X & \downarrow \\ \mathbb{Z}[\pi_1(X)] & \longrightarrow & \mathbb{Z}[\pi_1(X)]) \end{array} .$$

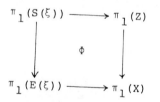

Define the <u>transfer</u> maps in quadratic Γ-theory induced by (X,Y) to be the composites

$$p\xi^! : L_n(\mathbb{Z}[\pi_1(Z)]) \xrightarrow{p\xi^!} L_{n+2}(\mathbb{Z}[\pi_1(Z)] \to \mathbb{Z}[\pi_1(X)]) \longrightarrow \Gamma_{n+2}(\Phi_X) \ .$$

Define the <u>quadratic</u> $\left\{\begin{matrix}\Gamma S\text{-}\\ \Gamma P\text{-}\end{matrix}\right.$ <u>groups of (X,Y)</u> $\left\{\begin{matrix}\Gamma S_n(\Phi)\\ \Gamma P_n(\Phi)\end{matrix}\right.$ $(n \geqslant 0)$ to be the relative groups appearing in the exact sequence

$$\left\{\begin{array}{l} \dots \longrightarrow \Gamma_{n+3}(\Phi_X) \longrightarrow \Gamma S_n(\Phi) \longrightarrow L_n(\mathbb{Z}[\pi_1(Y)]) \xrightarrow{p\xi^!} \Gamma_{n+2}(\Phi_X) \longrightarrow \dots \\ \dots \longrightarrow \Gamma_{n+2}(\mathbb{Z}[\pi_1(Z)] \to \mathbb{Z}[\pi_1(X)]) \longrightarrow \Gamma P_n(\Phi) \\ \qquad \longrightarrow L_n(\mathbb{Z}[\pi_1(Y)]) \xrightarrow{\partial p\xi^!} \Gamma_{n+1}(\mathbb{Z}[\pi_1(Z)] \to \mathbb{Z}[\pi_1(X)]) \longrightarrow \dots \ . \end{array}\right.$$

<u>Proposition 7.8.1</u> i) The quadratic $\left\{\begin{matrix}\Gamma S\text{-}\\ \Gamma P\text{-}\end{matrix}\right.$ groups are 4-periodic

$$\left\{\begin{array}{l} \Gamma S_n(\Phi) = \Gamma S_{n+4}(\Phi) \\ \Gamma P_n(\Phi) = \Gamma P_{n+4}(\Phi) \end{array}\right. \quad (n \geqslant 0) \quad .$$

ii) The ΓS-groups are related to the LS-groups by the commutative braid of exact sequences

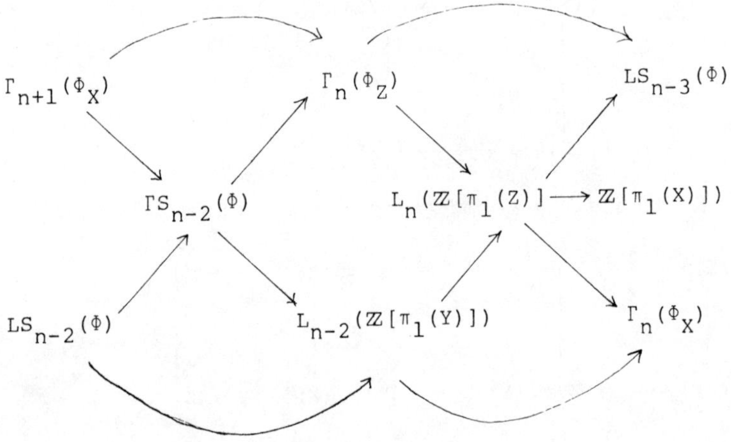

iii) The ΓP-groups are related to the LP-groups by the commutative braid of exact sequences

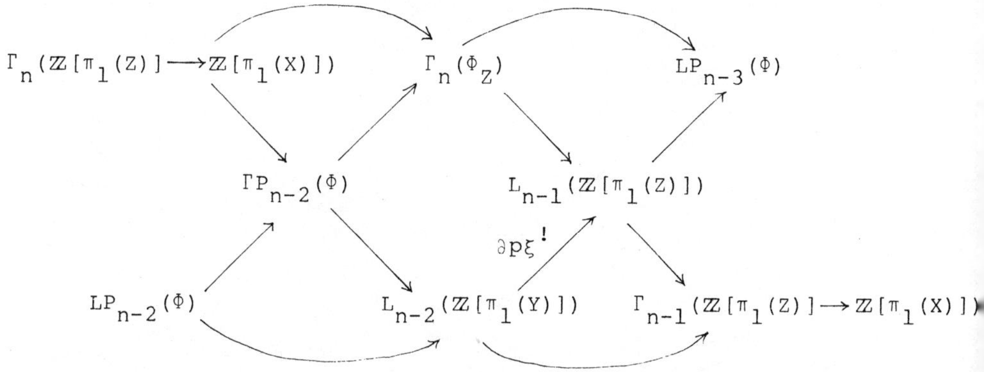

.

iv) The ΓP-groups are related to the ΓS-groups by the commutative braid of exact sequences

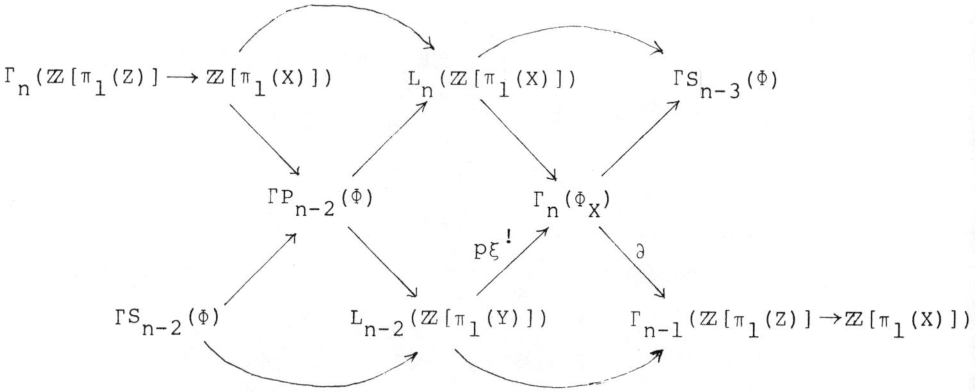

.

v) The LS-,ΓS-,LP-,ΓP-groups are related to each other by the commutative braid of eaxct sequences

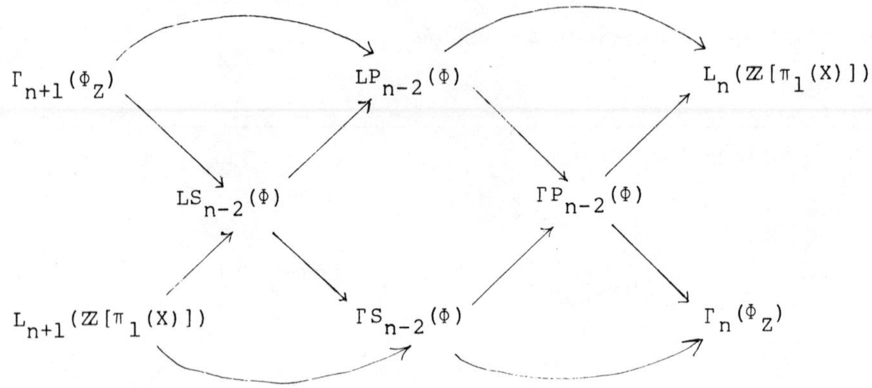

.

[]

Given a weak (n,n-2)-dimensional geometric Poincaré pair (X,Y) and an $\begin{cases} s- \\ t- \end{cases}$ triangulation $(f,b):M \longrightarrow X$ topologically transverse at $Y \subset X$ there is defined a <u>weak codimension 2 splitting obstruction along $Y \subset X$</u>

$$\begin{cases} ws(f,Y) \in \Gamma S_{n-2}(\Phi) \\ wt(f,Y) \in \Gamma P_{n-2}(\Phi) \end{cases}$$

with image the surgery obstruction $\sigma_*(g,c) \in L_{n-2}(\mathbb{Z}[\pi_1(Y)])$ of the (n-2)-dimensional topological normal map

$$(g,c) = (f,b)| : N = f^{-1}(Y) \longrightarrow Y ,$$

by analogy with the strong splitting obstruction of §7.2. The canonical map $\Gamma S_{n-2}(\Phi) \longrightarrow \Gamma P_{n-2}(\Phi)$ sends ws(f,Y) to wt(f,Y).

<u>Proposition 7.8.2</u> i) The $\begin{cases} s- \\ t- \end{cases}$ triangulation $(f,b):M \longrightarrow X$ is such that $\begin{cases} ws(f,Y) = 0 \in \Gamma S_{n-2}(\Phi) \\ wt(f,Y) = 0 \in \Gamma P_{n-2}(\Phi) \end{cases}$ if (and for $n \geqslant 7$ only if) (f,b) is concordant to an s-triangulation of X which is weakly split along $Y \subset X$.

ii) The canonical map $\Gamma P_{n-2}(\Phi) \longrightarrow \Gamma_n(\Phi_Z)$ sends $wt(f,Y)$ to the cobordism class of the $(n-1)$-dimensional $\mathbb{Z}[\pi_1(X)]$-acyclic quadratic Poincaré complex over $\mathbb{Z}[\pi_1(Z)]$ obtained from the quadratic kernel $\sigma_*((f,b)| = \partial(g,c)^! : S(\nu) \longrightarrow S(\xi))$ by surgery on the n-dimensional quadratic $\mathbb{Z}[\pi_1(X)]$-Poincaré pair over $\mathbb{Z}[\pi_1(Z)]$ $\sigma_*((f,b)| = (h,d) : (P,S(\nu)) \longrightarrow (Z,S(\xi)))$, with

$$\nu = \nu_{N \subset M} : N \xrightarrow{g} Y \xrightarrow{\xi} \widetilde{BTOP}(2) \ , \ P = f^{-1}(Z) = \overline{M \setminus E(\nu)} \ .$$

[]

The weak codimension 2 splitting obstruction of Cappell and Shaneson [1,§8] for an s-triangulation $(f,b):M \xrightarrow{\sim} X$ such that $\sigma_*(g,c) = 0 \in L_{n-2}(\mathbb{Z}[\pi_1(Y)])$ (for some weak $(n,n-2)$-dimensional geometric Poincaré pair (X,Y), with $(g,c) = (f,b)| : N = f^{-1}(Y) \longrightarrow Y$) is the preimage of

$$\begin{aligned} ws(f,Y) \in \ & \ker(\Gamma S_{n-2}(\Phi) \longrightarrow L_{n-2}(\mathbb{Z}[\pi_1(Y)])) \\ & = \operatorname{im}(\Gamma_{n+1}(\Phi_X) \longrightarrow \Gamma S_{n-2}(\Phi)) \end{aligned}$$

in the group

$$\begin{aligned} & \Gamma_{n+1}(\Phi_X)/\ker(\Gamma_{n+1}(\Phi_X) \longrightarrow \Gamma S_{n-2}(\Phi)) \\ & \qquad = \operatorname{coker}(p\xi^! : L_{n-1}(\mathbb{Z}[\pi_1(Y)]) \longrightarrow \Gamma_{n+1}(\Phi_X)) , \end{aligned}$$

as is clear from the exact sequence

$$\begin{aligned} \ldots \longrightarrow L_{n-1}(\mathbb{Z}[\pi_1(Y)]) \xrightarrow{p\xi^!} \Gamma_{n+1}(\Phi_X) \longrightarrow \Gamma S_{n-2}(\Phi) \\ \longrightarrow L_{n-2}(\mathbb{Z}[\pi_1(Y)]) \longrightarrow \ldots \ . \end{aligned}$$

The total codimension q surgery obstruction theory of §7.2 extends to weak codimension 2 geometric Poincaré pairs as follows.

Given a codimension 2 CW pair (X,Y) define the <u>$\mathcal{WS}$-groups</u> $\mathcal{WS}_*(X,Y,\xi)$ by analogy with the groups $\mathcal{S}_*(X,Y,\xi)$ of §7.2, to fit into the commutative braid of exact sequences

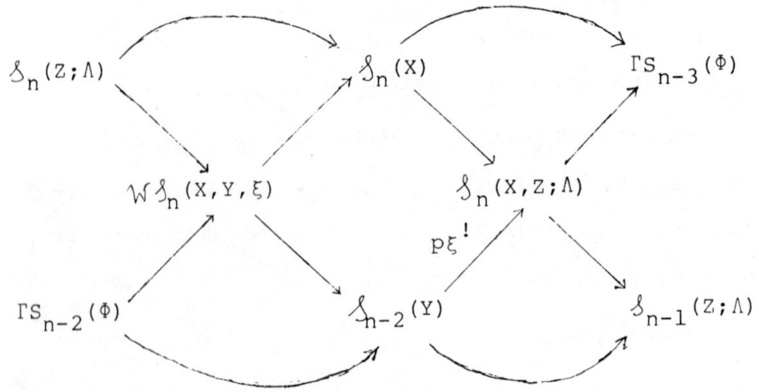

with $\Lambda = \mathbb{Z}[\pi_1(X)]$, $p\xi^!$ the composite

$$p\xi^! : \mathcal{S}_{n-2}(Y) \xrightarrow{p\xi^!} \mathcal{S}_n(X,Z) \longrightarrow \mathcal{S}_n(X,Z;\Lambda) ,$$

$\mathcal{S}_*(X,Z;\Lambda)$ the relative Λ-coefficient $\mathcal{S}$-groups appearing in the exact sequence

$$\ldots \longrightarrow \mathcal{S}_n(Z;\Lambda) \longrightarrow \mathcal{S}_n(X) \longrightarrow \mathcal{S}_n(X,Z;\Lambda) \longrightarrow \mathcal{S}_{n-1}(Z;\Lambda) \longrightarrow \ldots$$

and $\mathcal{S}_*(Z;\Lambda)$ the Λ-coefficient $\mathcal{S}$-groups appearing in the exact sequence

$$\ldots \longrightarrow H_n(Z;\underline{\mathbb{L}}_0) \xrightarrow{\sigma_*^\Lambda} \Gamma_n(\mathbb{Z}[\pi_1(Z)] \longrightarrow \Lambda)$$

$$\longrightarrow \mathcal{S}_n(Z;\Lambda) \longrightarrow H_{n-1}(Z;\underline{\mathbb{L}}_0) \longrightarrow \ldots .$$

<u>Proposition 7.8.3</u> i) The ΓS-groups are related to the $\mathcal{S}$-groups by the commutative braid of exact sequences

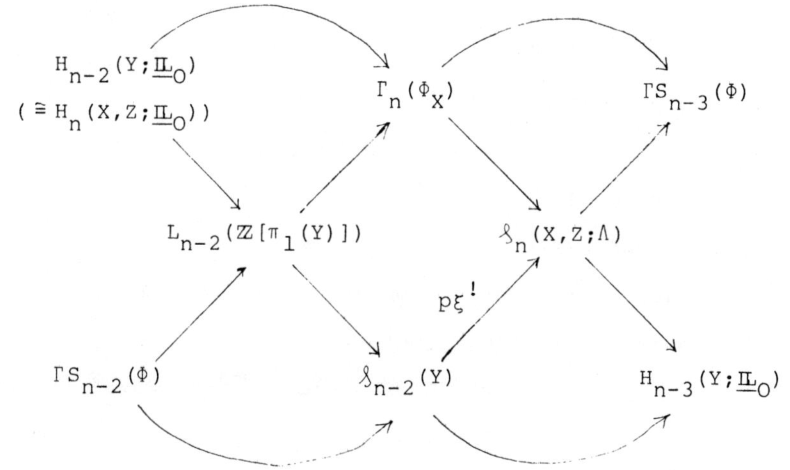

ii) The $\mathcal{WS}$-groups of (X,Y,ξ) are related to the $\mathcal{S}$-groups by the commutative braid of exact sequences

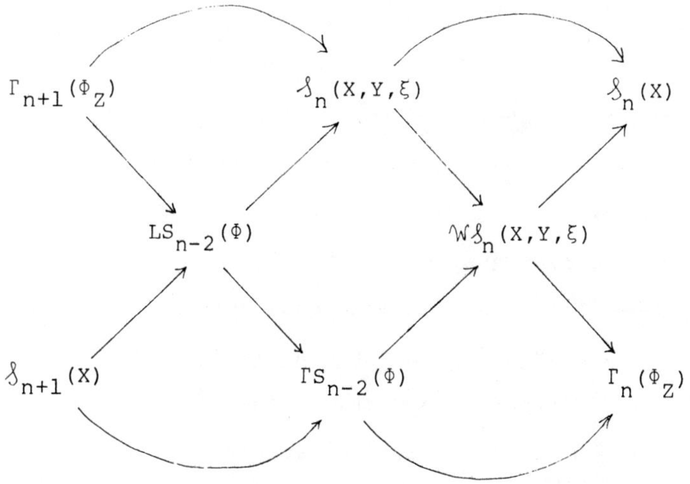

[]

The <u>total weak surgery obstruction</u> of a weak (n,n-2)-dimensional geometric Poincaré pair (X,Y) is an element

$$ws(X,Y,\xi) \in \mathcal{WS}_n(X,Y,\xi)$$

with the following properties.

<u>Proposition 7.8.4</u> i) $ws(X,Y,\xi) = 0$ if (and for $n \geqslant 7$ only if) (X,Y) is weakly s-triangulable.

ii) The obstruction has image

$$\begin{aligned}[ws(X,Y,\xi)] = (&\text{the total surgery obstructions } (s(X),s(Y)),\\ &\text{the quadratic signature}\\ &\sigma_*^{\Lambda}(Z,S(\xi)) = (\Omega C([Z]\cap - : C(\widetilde{Z})^{n-*} \longrightarrow C(\widetilde{Z},\widetilde{S(\xi)})),\psi))\\ \in\ & \mathcal{S}_n(X) \oplus \mathcal{S}_{n-2}(Y) \oplus \Gamma_n(\Phi_Z) \qquad (\Lambda = \mathbb{Z}[\pi_1(X)])\ .\end{aligned}$$

iii) If $n \geqslant 7$ and (X,Y) is an (n,n-2)-dimensional manifold pair there is a natural identification

$$\mathcal{WS}^{TOP}(X,Y,\xi) = \mathcal{WS}_{n+1}(X,Y,\xi)\ .$$

[]

The codimension q geometric Poincaré splitting theory of §7.4 extends to weak codimension 2 geometric Poincaré splitting, as follows.

Let (X,Y) be a codimension 2 CW pair.

A map $f:M \longrightarrow X$ from an n-dimensional geometric Poincaré complex M is <u>weakly Poincaré transverse at $Y \subset X$</u> if $(M,N = f^{-1}(Y))$ is a weak (n,n-2)-dimensional geometric Poincaré pair.

Let $\Omega_n^{\Lambda P,P}(\partial p\xi^!)$ $(n \geqslant 0)$ be the relative geometric Poincaré bordism groups appearing in the exact sequence

$$\dots \longrightarrow \Omega_{n-1}^P(Y) \xrightarrow{\partial p\xi^!} \Omega_n^{\Lambda P}(Z) \longrightarrow \Omega_n^{\Lambda P,P}(\partial p\xi^!) \longrightarrow \Omega_{n-2}^P(Y) \longrightarrow \dots$$
$$(\Lambda = \mathbb{Z}[\pi_1(X)])\ .$$

There are defined maps

$$\Omega_n^{\Lambda P,P}(\partial p\xi^!) \longrightarrow \Omega_n^P(X) \ ;$$

$$((h,g):(P,N) \longrightarrow (Z,Y))$$

$$\longmapsto (f = g^! \cup h : M = E(g^*\xi) \cup_{S(g^*\xi)} P \longrightarrow X = E(\xi) \cup_{S(\xi)} Z) \ .$$

A map $f:M \longrightarrow X$ from an n-dimensional geometric Poincaré complex M is bordant to one which is weakly Poincaré transverse at $Y \subset X$ if and only if

$$(f:M \longrightarrow X) \in \mathrm{im}(\Omega_n^{\Lambda P,P}(\partial p\xi^!) \longrightarrow \Omega_n^P(X)) \subseteq \Omega_n^P(X) \ .$$

Proposition 7.8.5 Given a codimension 2 CW pair (X,Y) there is defined a commutative braid of exact sequences

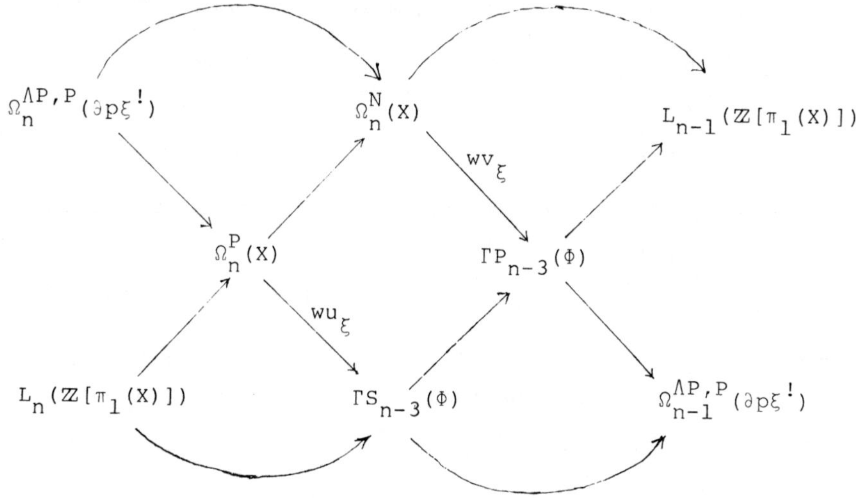

$(n \geqslant 7)$.

Thus $\begin{cases} \Gamma S_{n-3}(\Phi) \\ \Gamma P_{n-3}(\Phi) \end{cases}$ is the bordism group of maps $(f,\partial f):(M,\partial M) \longrightarrow X$ from n-dimensional $\begin{cases} \text{geometric Poincaré} \\ \text{(normal, geometric Poincaré)} \end{cases}$ pairs such that $\partial f:\partial M \longrightarrow \partial X$ is weakly Poincaré transverse at $Y \subset X$.

[]

If $f:M \longrightarrow X$ is a map from an n-dimensional $\begin{cases}\text{geometric Poincaré complex} \\ \text{normal space}\end{cases}$ M the <u>weak Poincaré splitting obstruction of along $Y \subset X$</u>

$$\begin{cases} ws^P(f,Y) = wu_\xi(f) \in \Gamma S_{n-3}(\Phi) \\ wt^P(f,Y) = wv_\xi(f) \in \Gamma P_{n-3}(\Phi) \end{cases}$$

is therefore such that $\begin{cases} ws^P(f,Y) = 0 \\ wt^P(f,Y) = 0 \end{cases}$ if (and for $n \geqslant 7$ only if)

f is $\begin{cases}\text{Poincaré} \\ \text{normal}\end{cases}$ bordant to a map from an n-dimensional geometric Poincaré complex which is weakly Poincaré transverse at $Y \subset X$.

The codimension q algebraic Poincaré splitting theory of §7.5 also extends to weak codimension 2 algebraic Poincaré splitting, as follows.

A <u>weak n-dimensional quadratic Poincaré splitting over Φ</u> (y,z) consists of

i) an (n-2)-dimensional quadratic Poincaré complex y over $\mathbb{Z}[\pi_1(Y)]$

ii) an n-dimensional quadratic $\mathbb{Z}[\pi_1(X)]$-Poincaré pair $(z,\partial p\xi^! y)$ over $\mathbb{Z}[\pi_1(Z)]$.

The union

$$p\xi^! y \cup_{\mathbb{Z}[\pi_1(X)]\otimes_{\mathbb{Z}[\pi_1(Z)]}\partial p\xi^! y} \mathbb{Z}[\pi_1(X)]\otimes_{\mathbb{Z}[\pi_1(Z)]} z$$

is an n-dimensional quadratic Poincaré complex over $\mathbb{Z}[\pi_1(X)]$, which we shall abbreviate to $p\xi^! y \cup z$. The splitting is <u>contractible</u> if the union is contractible.

The normal splitting over Φ of quadratic Poincaré complexes over $\mathbb{Z}[\pi_1(X)]$ given by Proposition 7.5.1 i) provides the following extension to $\begin{cases}\Gamma S_*(\Phi)\\ \Gamma P_*(\Phi)\end{cases}$ of the expression for $\begin{cases}LS_*(\Phi)\\ LP_*(\Phi)\end{cases}$ in terms of $\begin{cases}\text{contractible}\\ -\end{cases}$ quadratic Poincaré splittings over Φ given by Proposition 7.5.1 ii).

Proposition 7.8.6 The weak codimension 2 surgery obstruction group $\begin{cases}\Gamma S_{n-2}(\Phi)\\ \Gamma P_{n-2}(\Phi)\end{cases}$ $(n \in \mathbb{Z})$ is the cobordism group of $\begin{cases}\text{contractible}\\ -\end{cases}$ weak n-dimensional quadratic Poincaré splittings over Φ. The maps appearing in the exact sequence

$$\ldots \longrightarrow \Gamma S_{n-2}(\Phi) \longrightarrow \Gamma P_{n-2}(\Phi) \longrightarrow L_n(\mathbb{Z}[\pi_1(X)]) \longrightarrow \Gamma S_{n-3}(\Phi) \longrightarrow \ldots$$

are given by

$$\Gamma S_{n-2}(\Phi) \longrightarrow \Gamma P_{n-2}(\Phi)\ ;\ (y,z) \longmapsto (y,z)$$

$$\Gamma P_{n-2}(\Phi) \longrightarrow L_n(\mathbb{Z}[\pi_1(X)])\ ;\ (y,z) \longmapsto p\xi^! y \cup z$$

$$L_n(\mathbb{Z}[\pi_1(X)]) \longrightarrow \Gamma S_{n-3}(\Phi)\ ;\ x \longmapsto (\partial y, \partial_+ z)$$

(if $((y,\partial y),(z,\partial_+ z))$ is a normal splitting of x).

In particular, the image of an element $x \in L_n(\mathbb{Z}[\pi_1(X)])$ in $\Gamma S_{n-3}(\Phi)$ is the obstruction to x having a weak Poincaré splitting.

[]

Moreover, the weak codimension 2 geometric Poincaré splitting obstruction $\begin{cases}ws^P(f,Y) \in \Gamma S_{n-3}(\Phi)\\ wt^P(f,Y) \in \Gamma P_{n-3}(\Phi)\end{cases}$ of a map $f:M \longrightarrow X$ from an n-dimensional $\begin{cases}\text{geometric Poincaré complex}\\ \text{normal space}\end{cases}$ M

(resp. the weak codimension 2 manifold splitting obstruction $\begin{cases} ws(f,Y) \in \Gamma S_{n-2}(\Phi) \\ wt(f,Y) \in \Gamma P_{n-2}(\Phi) \end{cases}$ of an $\begin{cases} s- \\ t- \end{cases}$ triangulation $(f,b):M \longrightarrow X$ if (X,Y) is a weak $(n,n-2)$-dimensional geometric Poincaré pair) can be expressed in terms of a $\begin{cases} \text{contractible} \\ - \end{cases}$ weak quadratic Poincaré splitting over Φ using normal (resp. topological) transversality, exactly as was done for the strong case in Proposition 7.5.1 iii) (resp. iv)).

For a codimension 2 CW pair (X,Y) with

$$\pi_1(X) = \pi_1(Y) = \pi \ , \ \pi_1(Z) = \pi_1(S(\xi)) = \pi' \ , \ w(X) = w$$

the LS-groups are written

$$LS_*(\Phi) = LN_*(\pi' \longrightarrow \pi, w)$$

as before, and the ΓS-groups are written

$$\Gamma S_*(\Phi) = \Gamma N_*(\pi' \longrightarrow \pi, w) \ .$$

We shall now investigate the algebraic properties of the $\begin{cases} - \\ \text{weak} \end{cases}$ codimension 2 surgery obstruction groups $\begin{cases} LN_*(\pi' \longrightarrow \pi, w) \\ \Gamma N_*(\pi' \longrightarrow \pi, w) \end{cases}$, by analogy with the algebraic investigation in §7.6 of the type C) codimension 1 surgery obstruction groups $LN_*(\pi' \longrightarrow \pi, w)$. In the first instance we have to give an algebraic description of the transfer maps induced in quadratic L-theory

$$\xi^! \ : \ L_n(\mathbb{Z}[\pi]) \longrightarrow L_{n+2}(\mathbb{Z}[\pi'] \longrightarrow \mathbb{Z}[\pi]) \quad (n \geqslant 0)$$

by the S^1-fibration $\xi : Y \longrightarrow BG(2)$ $(= BO(2))$.

Let (Π,ω) be a pair consisting of

i) an exact sequence of groups

$$\Pi : \mathbb{Z} \xrightarrow{\ i\ } \pi' \xrightarrow{\ p\ } \pi \longrightarrow \{1\}$$

ii) a group morphism

$$\omega : \pi \longrightarrow \mathbb{Z}_2 = \{\pm 1\}$$

such that

$$g'tg'^{-1} = t^{\omega'(g')} \quad (g' \in \pi')$$

where $t = i(1) \in \pi'$, and $\omega' = \omega p : \pi' \longrightarrow \mathbb{Z}_2$.

Wall [4,Prop.11.4] has shown that associated to (Π,ω) there is an S^1-fibration

$$S^1 \longrightarrow X' \xrightarrow{\ p_X\ } X$$

with fundamental group exact sequence

$$\Pi_X : \pi_1(S^1) \longrightarrow \pi_1(X') \xrightarrow{\ p_X\ } \pi_1(X) \longrightarrow \{1\}$$

isomorphic to Π, $X' = K(\pi',1)$ and with orientation map

$$w_1(p_X) = \omega : \pi_1(X) = \pi \longrightarrow \mathbb{Z}_2$$

which has the following universal property: given an S^1-fibration

$$S^1 \longrightarrow Y' \xrightarrow{\ p_Y\ } Y$$

and a map of exact sequences of groups

$$\begin{array}{ccccccccc}
\Pi_Y & : & \pi_1(S^1) & \longrightarrow & \pi_1(Y') & \xrightarrow{\ p_Y\ } & \pi_1(Y) & \longrightarrow & \{1\} \\
\downarrow f & & \| & & \downarrow f' & & \downarrow f & & \\
\Pi & : & \mathbb{Z} & \longrightarrow & \pi' & \xrightarrow{\ p\ } & \pi & \longrightarrow & \{1\}
\end{array}$$

such that

$$w_1(p_Y) : \pi_1(Y) \xrightarrow{\ f\ } \pi \xrightarrow{\ \omega\ } \mathbb{Z}_2$$

there exists a map of S^1-fibrations

$$\begin{array}{ccccc} S^1 & \longrightarrow & Y' & \xrightarrow{p_Y} & Y \\ \Vert & & \downarrow{\scriptstyle f'} & & \downarrow{\scriptstyle f} \\ S^1 & \longrightarrow & X' & \xrightarrow{p_X} & X \end{array}$$

inducing the morphism

$$f : (\Pi_Y, w_1(p_Y)) \longrightarrow (\Pi_X, w_1(p_X)) = (\Pi, \omega) .$$

An S^1-fibration $S^1 \longrightarrow Y' \xrightarrow{p_Y} Y$ has <u>data</u> (Π,ω) if it is equipped with such a morphism $f : (\Pi_Y, w_1(p_Y)) \longrightarrow (\Pi,\omega)$. Note that p_Y is orientable if (and for $f:\pi_1(Y) \xrightarrow{\sim} \pi$ only if) $t \in \pi'$ is a central element. We shall be only concerned with S^1-fibrations p_Y over finite CW complexes Y, although the theory may easily be extended to S^1-fibrations over finitely dominated CW complexes.

Given an S^1-fibration $S^1 \longrightarrow Y' \xrightarrow{p} Y$ with data $(\Pi: \mathbb{Z} \longrightarrow \pi' \longrightarrow \pi \longrightarrow \{1\}, \omega: \pi \longrightarrow \mathbb{Z}_2)$ let $S^1 \longrightarrow X' \longrightarrow X$ be the universal S^1-fibration with data (Π,ω). We shall now introduce the category of "pseudo chain complexes over (Π,ω)" to help explain the relationship between the $\mathbb{Z}[\pi]$-module chain complex $C(\tilde{Y})$ of the cover $\tilde{Y}$ of Y classified by the composite $Y \longrightarrow X \longrightarrow K(\pi,1)$ and the $\mathbb{Z}[\pi']$-module chain complex $C(\tilde{Y}')$ of the cover $\tilde{Y}'$ of Y' classified by $Y' \longrightarrow X' = K(\pi',1)$. (In the special case of the universal data $(\Pi_Y: \mathbb{Z} \longrightarrow \pi_1(Y') \xrightarrow{p} \pi_1(Y) \longrightarrow \{1\}, w_1(p): \pi_1(Y) \longrightarrow \mathbb{Z}_2)$ $\tilde{Y}$ is the universal cover of Y and $\tilde{Y}'$ is the universal cover of Y'). This relationship is considerably more complicated than that between the universal covers of the total and base space of an S^0-fibration (= double cover), for in that case the universal covers coincide as spaces.

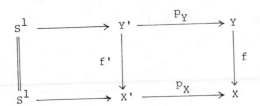

Given data (Π,ω) define ring automorphisms

$$\alpha : \mathbb{Z}[\pi] \longrightarrow \mathbb{Z}[\pi] \; ; \; \sum_{g\in\pi} n_g g \longmapsto \sum_{g\in\pi} n_g \omega(g) g$$

$$\alpha' : \mathbb{Z}[\pi'] \longrightarrow \mathbb{Z}[\pi'] \; ;$$

$$\sum_{g'\in\pi'} n_{g'} g' \longmapsto \sum_{\substack{g'\in\pi' \\ \omega'(g') = +1}} n_{g'} g' - \sum_{\substack{g'\in\pi' \\ \omega'(g') = -1}} n_{g'} g' t$$

$$(n_g, n_{g'} \in \mathbb{Z}) \; .$$

Given a $\mathbb{Z}[\pi]$-module M let αM denote the $\mathbb{Z}[\pi]$-module with the same additive group and $\mathbb{Z}[\pi]$ acting by

$$\mathbb{Z}[\pi] \times \alpha M \longrightarrow \alpha M \; ; \; (a,x) \longmapsto \alpha(a)x \; .$$

Similarly, given a $\mathbb{Z}[\pi']$-module M' let $\alpha' M'$ denote the $\mathbb{Z}[\pi']$-module with the same additive group and $\mathbb{Z}[\pi']$ acting by

$$\mathbb{Z}[\pi'] \times \alpha' M' \longrightarrow \alpha' M' \; ; \; (a,x) \longmapsto \alpha'(a)x \; .$$

For any such M' there is defined a $\mathbb{Z}[\pi']$-module morphism

$$1-t : \alpha' M' \longrightarrow M' \; ; \; x \longmapsto x - tx \; ,$$

which is natural in the sense that for any $\mathbb{Z}[\pi']$-module morphism $f' \in \mathrm{Hom}_{\mathbb{Z}[\pi']}(M',N')$ there is defined a commutative diagram

$$\begin{array}{ccc} \alpha' M' & \xrightarrow{\;1-t\;} & M' \\ {\scriptstyle \alpha' f'} \big\downarrow & & \big\downarrow {\scriptstyle f'} \\ \alpha' N' & \xrightarrow{\;1-t\;} & N' \end{array}$$

with $\alpha' f' \in \mathrm{Hom}_{\mathbb{Z}[\pi']}(\alpha' M', \alpha' N')$ defined by

$$\alpha' f' : \alpha' M' \longrightarrow \alpha' N' \; ; \; x \longmapsto f'(x) \; .$$

Also, for any such M' there is defined a natural $\mathbb{Z}[\pi']$-module isomorphism

$$t : \alpha'^2 M' \longrightarrow M' \; ; \; x \longmapsto tx \; .$$

A pseudo chain complex over $\mathbb{Z}[\Pi]$ $\mathcal{C} = (C',d',e')$ is a collection of $\mathbb{Z}[\pi']$-modules

$$C' = \{C'_r \mid r \in \mathbb{Z}\}$$

and $\mathbb{Z}[\pi']$-module morphisms

$$d' = \{d' \in \mathrm{Hom}_{\mathbb{Z}[\pi']}(C'_r, C'_{r-1}) \mid r \in \mathbb{Z}\}$$

$$e' = \{e' \in \mathrm{Hom}_{\mathbb{Z}[\pi']}(C'_r, \alpha' C'_{r-2}) \mid r \in \mathbb{Z}\}$$

such that

$$\left.\begin{array}{ll} \text{i)} & d'^2 = (1-t)e' : C'_r \longrightarrow C'_{r-2} \\ \text{ii)} & (\alpha' d')e' = e'd' : C'_r \longrightarrow \alpha' C'_{r-3} \end{array}\right. \qquad (r \in \mathbb{Z}) .$$

The projection of $\mathcal{C}$ is the (genuine) $\mathbb{Z}[\pi]$-module chain complex C defined by

$$d_C = 1 \otimes d' : C_r = \mathbb{Z}[\pi] \otimes_{\mathbb{Z}[\pi']} C'_r \longrightarrow C_{r-1} = \mathbb{Z}[\pi] \otimes_{\mathbb{Z}[\pi']} C'_{r-1} .$$

The induction of $\mathcal{C}$ is the $\mathbb{Z}[\pi']$-module chain complex $C^!$ defined by

$$d_{C^!} = \begin{pmatrix} d' & (-)^r(1-t) \\ (-)^r e' & \alpha' d' \end{pmatrix}$$

$$: C^!_r = C'_r \oplus \alpha' C'_{r-1} \longrightarrow C^!_{r-1} = C'_{r-1} \oplus \alpha' C'_{r-2} .$$

The projection and induction are related by the identity

$$\mathbb{Z}[\pi] \otimes_{\mathbb{Z}[\pi']} C^! = C(e : \Omega C \longrightarrow \alpha SC)$$

with e the $\mathbb{Z}[\pi]$-module chain map

$$e = 1 \otimes e' : \Omega C_r = \mathbb{Z}[\pi] \otimes_{\mathbb{Z}[\pi']} C'_{r+1} \longrightarrow \alpha SC_r = \mathbb{Z}[\pi] \otimes_{\mathbb{Z}[\pi']} \alpha' C'_{r-1} .$$

The pseudo chain complex $\mathcal{C}$ is n-dimensional if $C'_r = 0$ for $r < 0$ and $r > n$, and each C'_r $(0 \leqslant r \leqslant n)$ is a f.g. free $\mathbb{Z}[\pi']$-module, in which case C is an n-dimensional $\mathbb{Z}[\pi]$-module chain complex and $C^!$ is an $(n+1)$-dimensional $\mathbb{Z}[\pi']$-module chain complex. The pseudo chain complex $\mathcal{C}$ is contractible if the projection C is chain contractible.

A *pseudo chain map* of pseudo chain complexes over $\mathbb{Z}[\Pi]$

$$\mathcal{F} = (f',g') : \mathcal{C} = (C',d'_C,e'_C) \longrightarrow \mathcal{D} = (D',d'_D,e'_D)$$

consists of collections of $\mathbb{Z}[\pi']$-module morphisms

$$f' = \{f' \in \mathrm{Hom}_{\mathbb{Z}[\pi']}(C'_r,D'_r) \mid r \in \mathbb{Z}\}$$

$$g' = \{g' \in \mathrm{Hom}_{\mathbb{Z}[\pi']}(C'_r,\alpha' D'_{r-1}) \mid r \in \mathbb{Z}\}$$

such that

i) $d'_D f' - f' d'_C = (1-t)g' : C'_r \longrightarrow D'_{r-1}$

ii) $e'_D f' - (\alpha' f')e'_C = (\alpha' d'_D)g' + g' d'_C : C'_r \longrightarrow \alpha' D'_{r-2}$

$(r \in \mathbb{Z})$.

The *projection* of $\mathcal{F}$ is the $\mathbb{Z}[\pi]$-module chain map

$$f = 1 \otimes f' : C = \mathbb{Z}[\pi]\otimes_{\mathbb{Z}[\pi']} C' \longrightarrow D = \mathbb{Z}[\pi]\otimes_{\mathbb{Z}[\pi']} D' .$$

The *induction* of $\mathcal{F}$ is the $\mathbb{Z}[\pi']$-module chain map

$$f^! : C^! \longrightarrow D^!$$

defined by

$$f^! = \begin{pmatrix} f' & 0 \\ (-)^{r-1}g' & \alpha' f' \end{pmatrix} : C^!_r = C'_r \oplus \alpha' C'_{r-1} \longrightarrow D^!_r = D'_r \oplus \alpha' D'_{r-1} .$$

The pseudo chain map $\mathcal{F} : \mathcal{C} \longrightarrow \mathcal{D}$ is an *equivalence* if the projection $f : C \longrightarrow D$ is a chain equivalence.

A pseudo chain complex $\mathcal{C} = (C',d',e')$ is *untwisted* if $e' = 0$, so that C' is a $\mathbb{Z}[\pi']$-module chain complex and

$$C^! = C(1-t : \alpha' C' \longrightarrow C') .$$

Similarly, a pseudo chain map $\mathcal{F} = (f',g') : \mathcal{C} \longrightarrow \mathcal{D}$ is *untwisted* if $\mathcal{C},\mathcal{D}$ are untwisted and $g' = 0$, so that $f' : C' \longrightarrow D'$ is a $\mathbb{Z}[\pi']$-module chain map.

Proposition 7.8.7 i) The induction $C^!$ of a finite-dimensional pseudo chain complex $\mathcal{C}$ is chain contractible if (and for untwisted $\mathcal{C}$ only if) $\mathcal{C}$ is contractible (= C is chain contractible).
ii) The induction $f^!:C^! \longrightarrow D^!$ of a pseudo chain map $\mathcal{F}:\mathcal{C} \longrightarrow \mathcal{D}$ of finite-dimensional pseudo chain complexes is a chain equivalence if (and for untwisted $\mathcal{F}$ only if) $\mathcal{F}$ is an equivalence (= $f:C \longrightarrow D$ is a chain equivalence).
iii) A finite-dimensional $\mathbb{Z}[\pi']$-module chain complex C' is $\mathbb{Z}[\pi]$-acyclic if and only if $1-t:\alpha'C' \longrightarrow C'$ is a chain equivalence.

Proof: i) If the projection C of $\mathcal{C}$ is chain contractible let

$$\Delta = \{\Delta \in \mathrm{Hom}_{\mathbb{Z}[\pi]}(C_r, C_{r+1}) \,|\, r \in \mathbb{Z}\}$$

be a chain contraction, so that

$$d\Delta + \Delta d = 1 : C_r \longrightarrow C_r \quad .$$

As $p:\mathbb{Z}[\pi'] \longrightarrow \mathbb{Z}[\pi]$ is onto with $\ker(p) = \mathrm{im}(1-t:\mathbb{Z}[\pi'] \longrightarrow \mathbb{Z}[\pi'])$ there exist $\Delta' \in \mathrm{Hom}_{\mathbb{Z}[\pi']}(C'_r, C'_{r+1})$, $E' \in \mathrm{Hom}_{\mathbb{Z}[\pi']}(C'_r, \alpha'C'_r)$ such that $1\otimes\Delta' = \Delta$ and

$$d'\Delta' + \Delta'd' - 1 = (1-t)E' : C'_r \longrightarrow C'_r \quad (r \in \mathbb{Z}) .$$

The $\mathbb{Z}[\pi']$-module morphisms

$$\Delta^! = \begin{pmatrix} \Delta' & 0 \\ (-)^r E' & \alpha'\Delta' \end{pmatrix} : C^!_r = C'_r \oplus \alpha' C'_{r-1} \longrightarrow C^!_{r+1} = C'_{r+1} \oplus \alpha' C'_r$$

are such that for each $r \in \mathbb{Z}$

$$d^!\Delta^! + \Delta^! d^! = \begin{pmatrix} 1 & 0 \\ (-)^r((\alpha'd')E' - E'd' + (\alpha'\Delta')e' - e'\Delta') & 1 \end{pmatrix}$$

$$: C^!_r = C'_r \oplus \alpha' C'_{r-1} \longrightarrow C^!_r = C'_r \oplus \alpha' C'_{r-1}$$

is a (simple) $\mathbb{Z}[\pi']$-module automorphism. Thus the automorphism of the induction $C^!$ of $\mathcal{C}$ defined by

$$d^! \Delta^! + \Delta^! d^! : C^! \longrightarrow C^!$$

is null chain homotopic, and $C^!$ is chain contractible.

ii) Define the <u>algebraic mapping cone</u> of the pseudo chain map $\mathfrak{F} = (f',g') : \mathcal{C} \longrightarrow \mathcal{D}$ to be the pseudo chain complex

$$C(\mathfrak{F}) = (D'_r \oplus C'_{r-1}, \begin{pmatrix} d'_D & (-)^{r-1}f' \\ 0 & d'_C \end{pmatrix}, \begin{pmatrix} e'_D & (-)^{r-1}g' \\ 0 & e'_C \end{pmatrix})$$

with projection $C(f:C \longrightarrow D)$ and induction $C(f)^! = C(f^!:C^! \longrightarrow D^!)$. Now apply i) to $C(\mathfrak{F})$.

iii) Apply i) to the untwisted pseudo chain complex $(C',d',0)$.

[]

Pseudo chain complexes arise from S^1-fibrations as follows.

<u>Proposition 7.8.8</u> Given an S^1-fibration

$$S^1 \longrightarrow Y' \xrightarrow{\ p\ } Y$$

with data $(\Pi:\mathbb{Z} \longrightarrow \pi' \longrightarrow \pi \longrightarrow \{1\}, \omega:\pi \longrightarrow \mathbb{Z}_2)$ over a finite n-dimensional CW complex Y there is defined in a natural way an n-dimensional pseudo chain complex over (Π,ω)

$$\mathcal{C}(\tilde{Y},p) = (C',d',e')$$

such that

i) the projection $C = C(\tilde{Y})$ is the n-dimensional $\mathbb{Z}[\pi]$-module chain complex of the cover $\tilde{Y}$ of Y classified by $Y \longrightarrow K(\pi,1)$

ii) the induction $C^! = C(\tilde{Y}')$ is the (n+1)-dimensional $\mathbb{Z}[\pi']$-module chain complex of the cover $\tilde{Y}'$ of Y' classified by $Y' \longrightarrow K(\pi',1)$

iii) the $\mathbb{Z}[\pi]$-module chain map

$$e : \Omega C = \Omega C(\tilde{Y}) \longrightarrow \alpha SC = \alpha SC(\tilde{Y})$$

is the chain level cap product $e = e(p) \cap -$ with the ω-twisted

Euler class $e(p) \in H^2(Y,\omega)$ (which together with the orientation map $w_1(p) : \pi_1(Y) \longrightarrow \pi \xrightarrow{\omega} \mathbb{Z}_2$ classifies $p:Y \longrightarrow BG(2)$)

iv) if $p = \omega \oplus \varepsilon : Y \longrightarrow BG(2)$ (i.e. if $e(p) = 0 \in H^2(Y,\omega)$)

then $\mathcal{C}(\widetilde{Y},p) = (C',d',0)$ is the untwisted pseudo chain complex defined by the n-dimensional $\mathbb{Z}[\pi']$-module chain complex $(C',d') = C(\overline{Y})$ of the covering $\overline{Y}$ of Y classified by the composite $Y \xrightarrow{s} Y' \longrightarrow K(\pi',1)$ with s the section of p given by ε.

Furthermore, a map of S^1-fibrations with data (Π,ω)

$$\begin{array}{ccc} S^1 & = & S^1 \\ \downarrow & & \downarrow \\ Y_1' & \xrightarrow{F'} & Y_2' \\ p_1 \downarrow & & \downarrow p_2 \\ Y_1 & \xrightarrow{F} & Y_2 \end{array}$$

determines a pseudo chain map

$$\mathcal{F} = (f',g') \;:\; \mathcal{C}(\widetilde{Y}_1,p_1) \longrightarrow \mathcal{C}(\widetilde{Y}_2,p_2)$$

with projection

$$f = \widetilde{F} : C_1 = C(\widetilde{Y}_1) \longrightarrow C_2 = C(\widetilde{Y}_2)$$

and induction

$$f^! = \widetilde{F}' : C_1^! = C(\widetilde{Y}_1') \longrightarrow C_2^! = C(\widetilde{Y}_2') \quad .$$

<u>Proof</u>: The construction of $\mathcal{C}(\widetilde{Y},p)$ is by induction on the number of cells in Y, starting with $Y = \emptyset$ for which $\mathcal{C}(\widetilde{Y},p) = 0$. Assume inductively that $\mathcal{C}(\widetilde{Y},p)$ has already been defined for $p:Y \longrightarrow BG(2)$ and let

$$p_1 = p \cup q : Y_1 = Y \cup_f D^r \longrightarrow BG(2)$$

be an extension of p to the complex Y_1 obtained from Y by attaching an r-cell. The attaching map $f : S^{r-1} \longrightarrow Y$ defines a map of S^1-fibrations

$$\begin{array}{ccc}
S^1 & = & S^1 \\
\downarrow & & \downarrow \\
S^{r-1} \times S^1 & \xrightarrow{f'} & Y' \\
{\scriptstyle \varepsilon}\downarrow & & \downarrow{\scriptstyle p} \\
S^{r-1} & \xrightarrow{f} & Y
\end{array} \quad ,$$

using the trivialization of $f^*p : S^{r-1} \xrightarrow{f} Y \xrightarrow{p} BG(2)$ given by $q : D^r \longrightarrow BG(2)$ to identify it with the trivial S^1-fibration ε = projection : $S^{r-1} \times S^1 \longrightarrow S^{r-1}$ over S^{r-1}. The covering $\bar{S}^{r-1}$ of S^{r-1} classified by the composite

$$S^{r-1} \hookrightarrow S^{r-1} \times S^1 \xrightarrow{f'} Y' \longrightarrow K(\pi',1)$$

is trivial, so that $C(\bar{S}^{r-1}) = \mathbb{Z}[\pi'] \otimes_{\mathbb{Z}} C(S^{r-1})$. The $\mathbb{Z}[\pi']$-module chain map

$$g' = \begin{pmatrix} 1 \\ 0 \end{pmatrix} : S^{r-1}\mathbb{Z}[\pi'] \longrightarrow C(\bar{S}^{r-1}) = S^{r-1}\mathbb{Z}[\pi'] \oplus \mathbb{Z}[\pi']$$

determines an (untwisted) pseudo chain map

$$\mathcal{G} = (g',0) : (S^{r-1}\mathbb{Z}[\pi'],0,0) \longrightarrow \mathcal{C}(\tilde{S}^{r-1},\varepsilon) = (C(\bar{S}^{r-1}),0,0) \ .$$

Define the pseudo chain complex $\mathcal{C}(\tilde{Y}_1,p_1)$ to be the algebraic mapping cone of the composite pseudo chain map

$$\mathcal{F}\mathcal{G} : (S^{r-1}\mathbb{Z}[\pi'],0,0) \xrightarrow{\mathcal{G}} \mathcal{C}(\tilde{S}^{r-1},\varepsilon) \xrightarrow{\mathcal{F}} \mathcal{C}(\tilde{Y},p) \ ,$$

that is

$$\mathcal{C}(\tilde{Y}_1,p_1) = C(\mathcal{F}\mathcal{G}) \ .$$

[]

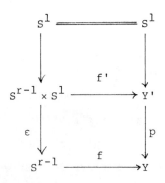

Given an S^1-fibration $S^1 \longrightarrow Y' \xrightarrow{p} Y$ with data $(\Pi:\mathbb{Z} \longrightarrow \pi' \longrightarrow \pi \longrightarrow \{1\}, \omega:\pi \longrightarrow \mathbb{Z}_2)$ there is defined a transfer map in the Whitehead groups

$$p^! : Wh(\pi) \longrightarrow Wh(\pi') ;$$
$$\tau(f:M \xrightarrow{\sim} X) \longmapsto \tau(f':M' \xrightarrow{\sim} X')$$

sending the Whitehead torsion $\tau(f)$ of a homotopy equivalence of finite CW complexes $f:M \xrightarrow{\sim} X$ equipped with a reference map $X \longrightarrow Y$ to the Whitehead torsion $\tau(f')$ of the homotopy equivalence $f':M' \xrightarrow{\sim} X'$ of the total spaces of the S^1-fibrations induced from p. This transfer was first defined geometrically by D.R.Anderson [1], and an algebraic description was first obtained by Munkholm and Pedersen [1]. From the point of view of Proposition 7.8.8 the transfer map sends the Whitehead torsion $\tau(C) \in Wh(\pi)$ of a based acyclic f.g. free $\mathbb{Z}[\pi]$-module chain complex C which is the projection of a based contractible pseudo chain complex $\mathscr{C} = (C',d',e')$ over (Π,ω) to the Whitehead torsion $p^!\tau(C) = \tau(C^!) \in Wh(\pi')$ of the induction $C^!$. In particular, if C is 1-dimensional ($C_r = 0$ for $r \neq 0,1$) then any lift of the $\mathbb{Z}[\pi]$-module isomorphism $d:C_1 \xrightarrow{\sim} C_0$ to a $\mathbb{Z}[\pi']$-module morphism $d':C'_1 \longrightarrow C'_0$ defines a based pseudo chain complex $\mathscr{C} = (C',d',0)$ with projection C, and in this case the induction

$$C^! : \ldots \longrightarrow 0 \longrightarrow \alpha'C'_1 \xrightarrow{\begin{pmatrix} 1-t \\ \alpha'd' \end{pmatrix}} C'_1 \oplus \alpha'C'_0 \xrightarrow{(d' \ -(1-t))} C'_0 \longrightarrow 0 \longrightarrow \ldots$$

is the algebraic description of $p^!\tau(C) = \tau(C^!)$ due to Munkholm and Pedersen [1].

Next, we shall construct the Q-groups of pseudo chain complexes, allowing the definition of "algebraic Poincaré pseudo complexes".

Given data $\xi = (\Pi:\mathbb{Z} \longrightarrow \pi' \xrightarrow{p} \pi \longrightarrow \{1\}, \omega:\pi \longrightarrow \mathbb{Z}_2)$ and an orientation map $w:\pi \longrightarrow \mathbb{Z}_2$ define orientation maps by

$$w^{\xi} = w\omega : \pi \longrightarrow \mathbb{Z}_2$$

$$w' = wp : \pi' \xrightarrow{p} \pi \xrightarrow{w} \mathbb{Z}_2$$

$$w'^{\xi} = w^{\xi}p : \pi' \xrightarrow{p} \pi \xrightarrow{w^{\xi}} \mathbb{Z}_2 \quad .$$

Define the <u>w-twisted dual</u> $\mathscr{C}^{n-*,w}$ of an n-dimensional pseudo chain complex $\mathscr{C} = (C',d',e')$ to be the n-dimensional pseudo chain complex $\mathscr{C}^{n-*,w} = (D',d'_D,e'_D)$ with

$$d'_D = (-)^r\alpha'd'^* : D'_r = \alpha'C'^{n-r,w'^{\xi}} \longrightarrow D'_{r-1} = \alpha'C'^{n-r+1,w'^{\xi}}$$

$$e'_D = t^{-1}e'^* : D'_r = \alpha'C'^{n-r,w'^{\xi}} \xrightarrow{e'^*} C'^{n-r+1,w'^{\xi}} \underset{\sim}{\xrightarrow{t^{-1}}} \alpha'^2C'^{n-r+1,w'^{\xi}} = \alpha'D'_{r-1} \ .$$

The projection D of $\mathscr{C}^{n-*,w}$ is the w-twisted dual of the projection C of $\mathscr{C}$

$$D = C^{n-*,w} \quad .$$

The induction $D^!$ of $\mathscr{C}^{n-*,w}$ may be identified with the w'^{ξ}-twisted dual of the induction $C^!$ of $\mathscr{C}$

$$D^! = (C^!)^{n+1-*,w'^{\xi}} \quad ,$$

since the $\mathbb{Z}[\pi']$-module isomorphisms

$$\begin{pmatrix} 0 & t \\ 1 & 0 \end{pmatrix} : D^!_r = \alpha'C'^{n-r,w'^{\xi}} \oplus \alpha'^2C'^{n-r+1,w'^{\xi}} \xrightarrow{\sim} ((C^!)^{n+1-*,w'^{\xi}})_r = C'^{n-r+1,w'^{\xi}} \oplus \alpha'C'^{n-r,w'^{\xi}}$$

define a canonical isomorphism of $\mathbb{Z}[\pi']$-module chain complexes

$$D^! \xrightarrow{\sim} (C^!)^{n+1-*,w'^{\xi}} .$$

Given a finite-dimensional pseudo chain complex $\mathcal{C} = (C',d',e')$ define a $\mathbb{Z}[\mathbb{Z}_2]$-module chain complex $\mathrm{Hom}_{\Pi}(\mathcal{C}^{*,w},\mathcal{C})$ by

$$\begin{aligned} d : \mathrm{Hom}_{\Pi}(\mathcal{C}^{*,w},\mathcal{C})_r &= \sum_{p+q=r} \mathrm{Hom}_{\mathbb{Z}[\pi']}(\alpha' C'^{p,w'^{\xi}}, C'_q) \oplus \mathrm{Hom}_{\mathbb{Z}[\pi']}(\alpha' C'^{p,w'^{\xi}}, \alpha' C'_{q-1}) \\ &\longrightarrow \mathrm{Hom}_{\Pi}(\mathcal{C}^{*,w},\mathcal{C})_{r-1} ; \end{aligned}$$

$$\begin{aligned} (\phi',\theta') \longmapsto (&d'\phi' + (-)^q \phi'(\alpha' d')^* + (-)^r (1-t)\theta', \\ &(\alpha' d')\theta' + (-)^{q-1}\theta'(\alpha' d')^* + (-)^r (e'\phi' - (\alpha'\phi')t^{-1}e'^*)) \end{aligned}$$

$$T : \mathrm{Hom}_{\Pi}(\mathcal{C}^{*,w},\mathcal{C})_r \longrightarrow \mathrm{Hom}_{\Pi}(\mathcal{C}^{*,w},\mathcal{C})_r ;$$

$$(\phi',\theta') \longmapsto (T_t\phi', T\theta') = ((-)^{pq} t(\alpha'\phi')^*, (-)^{p(q-1)}\theta'^*) .$$

Define the $\begin{cases} \underline{\text{pseudosymmetric}} \\ \underline{\text{pseudoquadratic}} \end{cases}$ $\underline{\text{w-twisted Q-groups of } \mathcal{C}}$ $\begin{cases} Q^*_{\Pi,w}(\mathcal{C}) \\ Q^{\Pi,w}_*(\mathcal{C}) \end{cases}$ by

$$\begin{cases} Q^n_{\Pi,w}(\mathcal{C}) = H_n(\mathrm{Hom}_{\mathbb{Z}[\mathbb{Z}_2]}(W, \mathrm{Hom}_{\Pi}(\mathcal{C}^{*,w},\mathcal{C}))) \\ Q^{\Pi,w}_n(\mathcal{C}) = H_n(W \otimes_{\mathbb{Z}[\mathbb{Z}_2]} \mathrm{Hom}_{\Pi}(\mathcal{C}^{*,w},\mathcal{C})) \end{cases} \quad (n \in \mathbb{Z}) .$$

An element $\begin{cases} (\phi',\theta') \in Q^n_{\Pi,w}(\mathcal{C}) \\ (\psi',\chi') \in Q^{\Pi,w}_n(\mathcal{C}) \end{cases}$ is represented by a collection of chains

$$\begin{cases} \{(\phi'_s,\theta'_s) \in \mathrm{Hom}_{\mathbb{Z}[\pi']}(\alpha' C'^{n-r+s,w'^{\xi}}, C'_r) \oplus \mathrm{Hom}_{\mathbb{Z}[\pi']}(\alpha' C'^{n-r+s,w'^{\xi}}, \alpha' C'_{r-1}) \\ \qquad\qquad |\, r \in \mathbb{Z}, s \geqslant 0\} \\ \{(\psi'_s,\chi'_s) \in \mathrm{Hom}_{\mathbb{Z}[\pi']}(\alpha' C'^{n-r-s,w'^{\xi}}, C'_r) \oplus \mathrm{Hom}_{\mathbb{Z}[\pi']}(\alpha' C'^{n-r-s,w'^{\xi}}, \alpha' C'_{r-1}) \\ \qquad\qquad |\, r \in \mathbb{Z}, s \geqslant 0\} \end{cases}$$

$$\left\{\begin{array}{l}\left\{\begin{array}{l}d'\phi'_s + (-)^r\phi'_s(\alpha'd')^* + (-)^{n+s-1}(\phi'_{s-1} + (-)^s T_t\phi'_{s-1}) + (-)^{n+s}(1-t)\theta'_s \\ \qquad = 0 : \alpha'C'^{n-r+s-1,w'^\xi} \longrightarrow C'_r \\ (\alpha'd')\theta'_s + (-)^r\theta'_s(\alpha'd')^* + (-)^{n+s}(\theta'_{s-1} + (-)^s T\theta'_{s-1}) \\ \qquad + (-)^{n+s-1}(e'\phi'_s - (\alpha'\phi'_s)t^{-1}e'^*) \end{array}\right. \\ \qquad = 0 : \alpha'C'^{n-r+s-2,w'^\xi} \longrightarrow \alpha'C'_r \qquad (s \geqslant 0,\ \phi'_{-1} = 0,\ \theta'_{-1} = 0) \\ \left\{\begin{array}{l}d'\psi'_s + (-)^r\psi'_s(\alpha'd')^* + (-)^{n-s-1}(\psi'_{s+1} + (-)^{s+1}T_t\psi'_{s+1}) + (-)^{n-s}(1-t)\chi'_s \\ \qquad = 0 : \alpha'C'^{n-r-s-1,w'^\xi} \longrightarrow C'_r \\ (\alpha'd')\chi'_s + (-)^r\chi'_s(\alpha'd')^* + (-)^{n-s}(\chi'_{s+1} + (-)^{s+1}T\chi'_{s+1}) \\ \qquad + (-)^{n-s-1}(e'\psi'_s - (\alpha'\psi'_s)t^{-1}e'^*) \\ \qquad = 0 : \alpha'C'^{n-r-s-2,w'^\xi} \longrightarrow \alpha'C'_r \qquad (s \geqslant 0)\ . \end{array}\right.\end{array}\right.$$

The $\mathbb{Z}[\mathbb{Z}_2]$-module chain map

$$p_! : \mathrm{Hom}_\Pi(\mathcal{C}^{*,w},\mathcal{C}) \longrightarrow \mathrm{Hom}_{\mathbb{Z}[\pi]}(C^{*,w},C)\ ;$$
$$(\phi',\theta') \longmapsto \phi = 1\otimes\phi'$$

defines the <u>projection</u> maps in the Q-groups

$$\left\{\begin{array}{l} p_! : Q^n_{\Pi,w}(\mathcal{C}) \longrightarrow Q^n_{\pi,w}(C)\ ;\ (\phi',\theta') \longmapsto \phi \\ p_! : Q_n^{\Pi,w}(\mathcal{C}) \longrightarrow Q_n^{\pi,w}(C)\ ;\ (\psi',\chi') \longmapsto \psi \end{array}\right.$$

(If $t \in \pi'$ is of infinite order and $e' = 0$ these maps are isomorphisms). The $\mathbb{Z}[\mathbb{Z}_2]$-module chain map

$$p^! : \mathrm{Hom}_\Pi(\mathcal{C}^{*,w},\mathcal{C}) \longrightarrow \Omega\mathrm{Hom}_{\mathbb{Z}[\pi']}((C^!)^{*,w'^\xi},C^!)\ ;$$
$$(\phi',\theta') \longmapsto \phi^!$$

with

$$\phi^! = \begin{pmatrix} 0 & \phi' \\ (\alpha'\phi')t^{-1} & (-)^p\theta' \end{pmatrix}$$

$$: C'^{p,w'^{\xi}} \oplus \alpha' C'^{p-1,w'^{\xi}} \longrightarrow C'_q \oplus \alpha' C'_{q-1}$$

defines the <u>induction</u> maps in the Q-groups

$$\begin{cases} p^! : Q^n_{\Pi,w}(\mathscr{C}) \longrightarrow Q^{n+1}_{\pi',w'^{\xi}}(C^!) \ ; \ (\phi',\theta') \longmapsto \phi^! \\ p^! : Q_n^{\Pi,w}(\mathscr{C}) \longrightarrow Q_{n+1}^{\pi',w'^{\xi}}(C^!) \ ; \ (\psi',\chi') \longmapsto \psi^! \end{cases} .$$

Replacing w by w^{ξ} there are also defined w^{ξ}-twisted Q-groups $\begin{cases} Q^*_{\Pi,w^{\xi}}(\mathscr{C}) \\ Q_*^{\Pi,w^{\xi}}(\mathscr{C}) \end{cases}$ with projection maps

$$\begin{cases} p_! : Q^*_{\Pi,w^{\xi}}(\mathscr{C}) \longrightarrow Q^*_{\pi,w^{\xi}}(C) \\ p_! : Q_*^{\Pi,w^{\xi}}(\mathscr{C}) \longrightarrow Q_*^{\pi,w^{\xi}}(C) \end{cases}$$

and induction maps

$$\begin{cases} p^! : Q^*_{\Pi,w^{\xi}}(\mathscr{C}) \longrightarrow Q^{*+1}_{\pi',w'}(C^!) \\ p^! : Q_*^{\Pi,w^{\xi}}(\mathscr{C}) \longrightarrow Q_{*+1}^{\pi',w'}(C^!) \end{cases} .$$

(Both the w-twisted and the w^{ξ}-twisted Q-groups arise in the applications).

A pseudo chain map of finite dimensional pseudo chain complexes over $\mathbb{Z}[\Pi]$

$$\mathscr{F} = (f',g') : \mathscr{C} = (C',d'_C,e'_C) \longrightarrow \mathscr{D} = (D',d'_D,e'_D)$$

induces the $\mathbb{Z}[\mathbb{Z}_2]$-module chain map

$$\mathrm{Hom}_{\Pi}(\mathscr{F}^{*,w},\mathscr{F}) : \mathrm{Hom}_{\Pi}(\mathscr{C}^{*,w},\mathscr{C}) \longrightarrow \mathrm{Hom}_{\Pi}(\mathscr{D}^{*,w},\mathscr{D})$$

defined by

$$\mathrm{Hom}_{\Pi}(\mathcal{F}^{*,w},\mathcal{F}) \; : \; \mathrm{Hom}_{\Pi}(\mathcal{C}^{*,w},\mathcal{C})_r$$

$$= \sum_{p+q=r} \mathrm{Hom}_{\mathbb{Z}[\pi']}(\alpha' C'^{p,w'^{\xi}}, C'_q) \oplus \mathrm{Hom}_{\mathbb{Z}[\pi']}(\alpha' C'^{p,w'^{\xi}}, \alpha' C'_{q-1})$$

$$\longrightarrow \mathrm{Hom}_{\Pi}(\mathcal{D}^{*,w},\mathcal{D})_r \; ;$$

$$(\phi',\theta') \longmapsto$$

$$(f'\phi'(\alpha'f')^*, (\alpha'f')^*\theta'(\alpha'f') + (-)^{q-1}(g'\phi'(\alpha'f')^* - \alpha'(f'\phi'g'^*)t^{-1})) \; .$$

Thus there are induced morphisms in the Q-groups

$$\begin{cases} \mathcal{F}^{\%} \; : \; Q^n_{\Pi,w}(\mathcal{C}) \longrightarrow Q^n_{\Pi,w}(\mathcal{D}) \\ \mathcal{F}_{\%} \; : \; Q^{\Pi,w}_n(\mathcal{C}) \longrightarrow Q^{\Pi,w}_n(\mathcal{D}) \end{cases} \qquad (n \in \mathbb{Z})$$

(which are isomorphisms if $\mathcal{F}: \mathcal{C} \longrightarrow \mathcal{D}$ is an equivalence) which are compatible with the projection and induction maps in the Q-groups.

An <u>n-dimensional</u> $\begin{cases} \text{pseudosymmetric} \\ \text{pseudoquadratic} \end{cases}$ <u>complex over</u> $\underline{\mathbb{Z}[\Pi^w]}$

$\begin{cases} (\mathcal{C},(\phi',\theta')) \\ (\mathcal{C},(\psi',\chi')) \end{cases}$ is an n-dimensional pseudo chain complex over $\mathbb{Z}[\Pi]$

$\mathcal{C} = (C',d',e')$ together with an element $\begin{cases} (\phi',\theta') \in Q^n_{\Pi,w}(\mathcal{C}) \\ (\psi',\chi') \in Q^{\Pi,w}_n(\mathcal{C}) \end{cases}$.

Such a complex is <u>Poincaré</u> if the projection $\begin{cases} (C,\phi \in Q^n_{\pi,w}(C)) \\ (C,\psi \in Q^{\pi,w}_n(C)) \end{cases}$

is an n-dimensional $\begin{cases} \text{symmetric} \\ \text{quadratic} \end{cases}$ Poincaré complex over $\mathbb{Z}[\pi^w]$,

that is if the pseudo chain map

$$\begin{cases} (\phi'_0,\theta'_0) \; : \mathcal{C}^{n-*,w} \longrightarrow \mathcal{C} \\ (1+T)(\psi'_0,\chi'_0) \; : \mathcal{C}^{n-*,w} \longrightarrow \mathcal{C} \end{cases}$$

is an equivalence, since it has projection

$$\begin{cases} \phi_0 : C^{n-*,w} \longrightarrow C \\ (1+T)\psi_0 : C^{n-*,w} \longrightarrow C \end{cases} ,$$

in which case the induction $\begin{cases} (C^!, \phi^! \in Q^{n+1}_{\pi',w',\xi}(C^!)) \\ (C^!, \psi^! \in Q_{n+1}^{\pi',w',\xi}(C^!)) \end{cases}$ is an

(n+1)-dimensional $\begin{cases} \text{symmetric} \\ \text{quadratic} \end{cases}$ Poincaré complex over $\mathbb{Z}[\pi'^{w',\xi}]$,

by Proposition 7.8.7 ii).

Define the $\begin{cases} \underline{\text{pseudosymmetric}} \\ \underline{\text{pseudoquadratic}} \end{cases}$ $\underline{\text{L-groups of } \mathbb{Z}[\Pi^w]}$ $\begin{cases} L^n(\mathbb{Z}[\Pi^w]) \\ L_n(\mathbb{Z}[\Pi^w]) \end{cases}$ $(n \geqslant 0)$

to be the cobordism groups of n-dimensional $\begin{cases} \text{pseudosymmetric} \\ \text{pseudquadratic} \end{cases}$

Poincaré complexes over $\mathbb{Z}[\Pi^w]$. Pseudoquadratic surgery below the middle dimension gives the periodicity

$$L_n(\mathbb{Z}[\Pi^w]) = L_{n+4}(\mathbb{Z}[\Pi^w]) \quad (n \geqslant 0) ,$$

and identifies $L_{2i}(\mathbb{Z}[\Pi^w])$ (resp. $L_{2i+1}(\mathbb{Z}[\Pi^w])$) with the Witt group of non-singular $(-)^i$-pseudoquadratic forms (resp. formations) over $\mathbb{Z}[\Pi^w]$, by analogy with the usual quadratic L-groups $L_*(\mathbb{Z}[\pi^w])$. Every $(-)^i$-quadratic form (resp. formation) over $\mathbb{Z}[\pi^w]$ lifts to a $(-)^i$-pseudoquadratic form (resp. formation) over $\mathbb{Z}[\Pi^w]$, and the projection maps in the quadratic L-groups

$$p_! : L_n(\mathbb{Z}[\Pi^w]) \longrightarrow L_n(\mathbb{Z}[\pi^w]) \ ; \ (\mathscr{C}, (\psi', \chi')) \longmapsto (C, \psi) \quad (n \geqslant 0)$$

are isomorphisms, which we shall use as identifications.

(It is not clear if the projection maps in the symmetric L-groups

$$p_! : L^n(\mathbb{Z}[\Pi^w]) \longrightarrow L^n(\mathbb{Z}[\pi^w]) \ ; \ (\mathscr{C}, (\phi', \theta')) \longmapsto (C, \phi) \quad (n \geqslant 0)$$

are also isomorphisms, except in special cases, e.g. if $t \in \pi'$ is of infinite order so that $1-t : \mathbb{Z}[\pi'] \longrightarrow \mathbb{Z}[\pi']$ is injective, or if there exists a group morphism $s : \pi \longrightarrow \pi'$ such that $ps = 1 : \pi \longrightarrow \pi$).

There are defined induction maps in the $\begin{cases}\text{(pseudo) symmetric}\\ \text{(pseudo) quadratic}\end{cases}$ L-groups

$$\begin{cases} p^! : L^n(\mathbb{Z}[\Pi^w]) \longrightarrow L^{n+1}(\mathbb{Z}[\pi'^{w'^{\xi}}]) \ ; \ (\mathcal{C},(\phi',\theta')) \longmapsto (C^!,\phi^!) \\ p^! : L_n(\mathbb{Z}[\Pi^w]) \longrightarrow L_{n+1}(\mathbb{Z}[\pi'^{w'^{\xi}}]) \ ; \ (\mathcal{C},(\psi',\chi')) \longmapsto (C^!,\psi^!) \\ \quad (= L_n(\mathbb{Z}[\pi^w])) \end{cases} \quad (n \geqslant 0)$$

.

In terms of forms and formations the induction (= transfer) map in quadratic L-theory $\begin{cases} p^! : L_{2i}(\mathbb{Z}[\pi^w]) \longrightarrow L_{2i+1}(\mathbb{Z}[\pi'^{w'^{\xi}}]) \\ p^! : L_{2i+1}(\mathbb{Z}[\pi^w]) \longrightarrow L_{2i+2}(\mathbb{Z}[\pi'^{w'^{\xi}}]) \end{cases}$ (i(mod 2))

sends a non-singular $(-)^i$-quadratic $\begin{cases}\text{form}\\ \text{formation}\end{cases}$ over $\mathbb{Z}[\pi^w]$

$\begin{cases} (M,\psi) \\ (H_{(-)^i}(F);F,\mathrm{im}(\begin{pmatrix}\gamma\\ \mu\end{pmatrix}:G \longrightarrow F\oplus F^{*,w})) \end{cases}$ to the non-singular

$\begin{cases} (-)^i\text{-} \\ (-)^{i+1}\text{-} \end{cases}$ quadratic $\begin{cases}\text{formation}\\ \text{form}\end{cases}$ over $\mathbb{Z}[\pi'^{w'^{\xi}}]$

$$\begin{cases} (H_{(-)^i}(M');M',\mathrm{im}(\begin{pmatrix} 1-t \\ \alpha'\psi'+(-)^i t(\alpha'\psi')^* \end{pmatrix} : \alpha'M' \longrightarrow M'\oplus M'^{*,w'^{\xi}})) \\ (\mathrm{coker}(\begin{pmatrix} 1-t \\ \alpha'\gamma' \\ \alpha'\mu' \end{pmatrix} : \alpha'G' \longrightarrow G'\oplus\alpha'F'\oplus F'^{*,w'^{\xi}}), \begin{pmatrix} \chi' & 0 & 0 \\ t^{-1}\mu' & 0 & 0 \\ \gamma' & 1-t & 0 \end{pmatrix}) \end{cases}$$

with $\begin{cases} \psi' \in \mathrm{Hom}_{\mathbb{Z}[\pi']}(M',M'^{*,w'}) \\ \begin{pmatrix}\gamma'\\ \mu'\end{pmatrix} \in \mathrm{Hom}_{\mathbb{Z}[\pi']}(G',F'\oplus F'^{*,w'}),\ \chi' \in \mathrm{Hom}_{\mathbb{Z}[\pi']}(G',G'^{*,w'^{\xi}}) \end{cases}$

such that

$$\begin{cases} 1\otimes\psi' = \psi \in \mathrm{Hom}_{\mathbb{Z}[\pi]}(M,M^{*,w}) \\ \begin{cases} 1\otimes\begin{pmatrix}\gamma'\\ \mu'\end{pmatrix} = \begin{pmatrix}\gamma\\ \mu\end{pmatrix} \in \mathrm{Hom}_{\mathbb{Z}[\pi]}(G,F\oplus F^{*,w}) \ . \\ \gamma'^*(\alpha'\mu') = \theta' + (-)^i t(\alpha'\theta')^* + (-)^{i+1}\chi'(1-t) \in \mathrm{Hom}_{\mathbb{Z}[\pi']}(\alpha'G',G'^{*,w'^{\xi}}) \\ \text{for some } \theta' \in \mathrm{Hom}_{\mathbb{Z}[\pi']}(\alpha'G',G'^{*,w'^{\xi}}) \ . \end{cases} \end{cases}$$

(Munkholm and Pedersen have also obtained the algebraic description of the transfer map $p^!:L_*(\mathbb{Z}[\pi^w]) \longrightarrow L_{*+1}(\mathbb{Z}[\pi'^{w',\xi}])$ in terms of forms and formations, extending their algebraic description of $p^!:Wh(\pi) \longrightarrow Wh(\pi'))$. For $\pi' = \pi\times\mathbb{Z}$ $p^!:L_*(\mathbb{Z}[\pi]) \longrightarrow L_{*+1}(\mathbb{Z}[\pi\times\mathbb{Z}])$ is just the splitting map $\overline{B} = \sigma^*(S^1)\otimes -$ appearing in the splitting theorem of Novikov [1] and Ranicki [2]. We shall now relate the algebraic L-theory induction maps $p^!$ to geometric transfer maps.

Let

$$S^1 \longrightarrow X' \xrightarrow{\quad p \quad} X$$

be an S^1-fibration over a finite CW complex X with data $(\Pi:\mathbb{Z} \xrightarrow{i} \pi' \xrightarrow{p} \pi \longrightarrow \{1\}, \omega:\pi \longrightarrow \mathbb{Z}_2)$. Let $\mathscr{C}(\widetilde{X},p) = (C',d',e')$ be the finite-dimensional pseudo chain complex over $\mathbb{Z}[\Pi]$ associated to $p:X \longrightarrow BG(2)$ by Proposition 7.8.8, with projection $C = C(\widetilde{X})$ the chain complex of the cover $\widetilde{X}$ of X classified by $X \longrightarrow K(\pi,1)$ and induction $C^! = C(\widetilde{X}')$ the chain complex of the cover $\widetilde{X}'$ of X' classified by $X' \longrightarrow K(\pi',1)$. Let $\widetilde{X}'/t$ be the quotient of $\widetilde{X}'$ by the action of $i(\mathbb{Z})\subset\pi'$, which is the cover of X' classified by $X' \longrightarrow K(\pi',1) \xrightarrow{\quad p \quad} K(\pi,1)$.
Define the <u>Umkehr</u> $\mathbb{Z}[\pi]$-module chain map

$$p^! : \alpha SC(\widetilde{X}) \longrightarrow C(\widetilde{X}'/t)$$

by

$$p^! = \begin{pmatrix} 0 \\ 1 \end{pmatrix} : \alpha SC(\widetilde{X})_r = C_{r-1} \longrightarrow C(\widetilde{X}'/t)_r = (\mathbb{Z}[\pi]\otimes_{\mathbb{Z}[\pi']}C^!)_r = C_r \oplus \alpha C_{r-1}.$$

(There is also defined a geometric Umkehr stable π-map $P:\Sigma^\infty T\pi(\omega) \longrightarrow \Sigma^\infty(\widetilde{X}'/t)_+$ inducing $p^!$ on the chain level).
The definition of $\mathscr{C}(\widetilde{X},p)$ by the algebraic glueing of the untwisted pseudo chain complexes of the restrictions of $p:X \longrightarrow BG(2)$ to the cells of X extends to the symmetric construction:

<u>Proposition 7.8.9</u> There is defined a natural transformation of abelian groups, the <u>pseudosymmetric construction on (X,p)</u>

$$\phi_{\widetilde{X},p} : H_n(X,w) \longrightarrow Q^n_{\Pi,w}(\mathcal{C}(\widetilde{X},p)) \qquad (n \geqslant 0)$$

for any orientation map $w:\pi \longrightarrow \mathbb{Z}_2$, which is related to the symmetric constructions on $\widetilde{X}$ and $\widetilde{X}'$ by a commutative diagram

$$\begin{array}{ccc}
H_{n+1}(X',w'^{\xi}) & \xrightarrow{\phi_{\widetilde{X}'}} & Q^{n+1}_{\pi',w'^{\xi}}(C(\widetilde{X}')) \\
\uparrow{\scriptstyle p^!} & & \uparrow{\scriptstyle p^!} \\
H_n(X,w) & \xrightarrow{\phi_{\widetilde{X},p}} & Q^n_{\Pi,w}(\mathcal{C}(\widetilde{X},p)) \\
& \searrow{\scriptstyle \phi_{\widetilde{X}}} & \downarrow{\scriptstyle p_!} \\
& & Q^n_{\pi,w}(C(\widetilde{X}))
\end{array}$$

[]

Given an n-dimensional geometric Poincaré complex M and a map $f:M \longrightarrow X$ such that $w(M) = f^*w$ the pullback S^1-fibration $f^*p:M \xrightarrow{f} X \xrightarrow{p} BG(2)$ over M

$$S^1 \longrightarrow M' \xrightarrow{f^*p} M$$

has total space M' an (n+1)-dimensional geometric Poincaré complex with orientation map $w(M') = f'^*w'^{\xi}$ and fundamental class $[M'] = p^![M] \in H_{n+1}(M',w'^{\xi})$ $([M] \in H_n(M,w))$. The n-dimensional pseudosymmetric Poincaré complex over $\mathbb{Z}[\pi^w]$

$$\sigma^*(M,p) = (\mathcal{C}(\widetilde{M},f^*p), \phi_{M,f^*p}([M]) \in Q^n_{\Pi,w}(\mathcal{C}(\widetilde{M},f^*p)))$$

has projection the n-dimensional symmetric Poincaré complex over $\mathbb{Z}[\pi^w]$

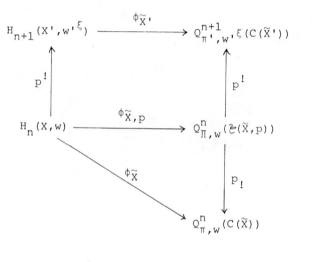

$$p_!\sigma^*(M,p) = \sigma^*(M) = (C(\widetilde{M}),\phi_{\widetilde{M}}([M]) \in Q^n_{\pi,w}(C(\widetilde{M})))$$

and induction the (n+1)-dimensional symmetric Poincaré complex over $\mathbb{Z}[\pi'^{w'^{\xi}}]$

$$p^!\sigma^*(M,p) = \sigma^*(M') = (C(\widetilde{M}'),\phi_{\widetilde{M}'}([M']) \in Q^{n+1}_{\pi',w'^{\xi}}(C(\widetilde{M}'))) .$$

The <u>pseudosymmetric signature</u> map

$$\sigma^*_p : \Omega^P_n(X,w) \longrightarrow L^n(\mathbb{Z}[\Pi^W]) \ ; \ (f:M \longrightarrow X) \longmapsto \sigma^*(M,p) \qquad (n \geqslant 0)$$

fits into a commutative diagram

$$\begin{array}{ccc}
\Omega^P_{n+1}(X',w'^{\xi}) & \xrightarrow{\ \sigma^*\ } & L^{n+1}(\mathbb{Z}[\pi'^{w'^{\xi}}]) \\
\uparrow p^! & & \uparrow p^! \\
\Omega^P_n(X,w) & \xrightarrow{\ \sigma^*_p\ } & L^n(\mathbb{Z}[\Pi^W]) \\
& \searrow^{\sigma^*} & \downarrow p_! \\
& & L^n(\mathbb{Z}[\pi^W])
\end{array}$$

with

$$p^! : \Omega^P_n(X,w) \longrightarrow \Omega^P_{n+1}(X',w'^{\xi}) \ ; \ (f:M \longrightarrow X) \longmapsto (f':M' \longrightarrow X') .$$

The pullback of a formally n-dimensional normal map $(f,b):M \longrightarrow X$ along an S^1-fibration $p:X \longrightarrow BG(2)$ is a formally (n+1)-dimensional normal map $(f',b'):M' \longrightarrow X'$. The Umkehr $\mathbb{Z}[\pi]$-module chain map of f

$$f^! : C(\widetilde{X})^{n-*,w} \xrightarrow{\ \widetilde{f}^*\ } C(\widetilde{M})^{n-*,w} \xrightarrow[\ \simeq\ ]{[M]\cap -} C(\widetilde{M})$$

is the projection of the Umkehr pseudo chain map

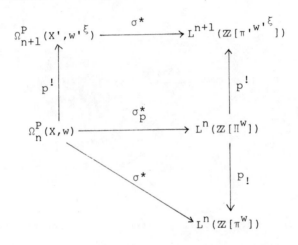

$$\mathcal{F}^! : \mathcal{C}(\tilde{X},p)^{n-*,w} \xrightarrow{\mathcal{F}^*} \mathcal{C}(\tilde{M},f^*p)^{n-*,w} \xrightarrow[\simeq]{\phi_{\tilde{M},p}([M])_0} \mathcal{C}(\tilde{M},f^*p) ,$$

and the Umkehr $\mathbb{Z}[\pi']$-module chain map of f'

$$f'^! : C(\tilde{X}')^{n+1-*,w'} \xrightarrow{\tilde{f}'^*} C(\tilde{M}')^{n+1-*,w} \xrightarrow[\simeq]{[M'] \cap -} C(\tilde{M}')$$

is the induction of $\mathcal{F}^!$. The quadratic kernel of (f,b) is the formally n-dimensional quadratic complex over $\mathbb{Z}[\pi^w]$

$$\sigma_*(f,b) = (C(f^!),\psi_F([X])) \in Q_n^{\pi,w}(C(f^!)))$$

with $\psi_F : H_n(X,w) \longrightarrow Q_n^{\pi,w}(C(f^!))$ the spectral quadratic construction on a geometric Umkehr semi-stable π-map $F: T\pi(\nu_X)^* \longrightarrow \Sigma^\infty \tilde{M}_+$ inducing $f^!$. The quadratic kernel of (f',b') is the formally $(n+1)$-dimensional quadratic complex over $\mathbb{Z}[\pi'^{w'^\xi}]$

$$\sigma_*(f',b') = (C(f'^!),\psi_{F'}([X'])) \in Q_{n+1}^{\pi',w'^\xi}(C(f'^!)))$$

with $\psi_{F'} : H_{n+1}(X',w'^\xi) \longrightarrow Q_{n+1}^{\pi',w'^\xi}(C(f'^!))$ the spectral quadratic construction on a geometric Umkehr semi-stable π'-map $F': T\pi'(\nu_{X'})^* \longrightarrow \Sigma^\infty \tilde{M}'_+$ inducing $f'^!$, with $\nu_{X'} = p^*(\nu_X \oplus \omega)$ (involving the orientation line bundle $\omega = w_1(p) : X \longrightarrow BG(1)$ of $p: X \longrightarrow BG(2)$). The definition of $\mathcal{C}(\tilde{X},p)$ also extends to the quadratic construction:

<u>Proposition 7.8.10</u> There is defined a natural transformation of abelian groups, the <u>spectral pseudoquadratic construction</u>

$$\psi_{F,p} : H_n(X,w) \longrightarrow Q_n^{\Pi,w}(C(\mathcal{F}^!))$$

with symmetrization

$$(1+T)\psi_{F,p} = \mathcal{E}^\% \phi_{\tilde{M}',p} f^! : H_n(X,w) \longrightarrow Q^n_{\Pi,w}(C(\mathcal{F}^!))$$

$$(\mathcal{E} = \text{projection} : \mathcal{C}(\tilde{M},f^*p) \longrightarrow C(\mathcal{F}^!)) ,$$

and such that there is defined a commutative diagram

$$
\begin{array}{ccc}
H_{n+1}(X',w'^{\xi}) & \xrightarrow{\ \psi_{F'}\ } & Q^{\pi',w'^{\xi}}_{n+1}(C(f'^{!})) \\
\uparrow{\scriptstyle p^{!}} & & \uparrow{\scriptstyle p^{!}} \\
H_n(X,w) & \xrightarrow{\ \psi_{F,p}\ } & Q^{\Pi,w}_{n}(C(\mathcal{F}^{!})) \\
& \searrow{\scriptstyle \psi_F} & \downarrow{\scriptstyle p_{!}} \\
& & Q^{\pi,w}_{n}(C(f^{!}))
\end{array}
\quad .
$$

[]

The <u>pseudoquadratic kernel of $((f,b),p)$</u> is the formally n-dimensional pseudoquadratic complex over $\mathbb{Z}[\Pi^w]$

$$\sigma^p_*(f,b) = (C(\mathcal{F}^!),\psi_{F,p}([X]) \in Q^{\Pi,w}_n(C(\mathcal{F}^!)))$$

with projection the quadratic kernel of (f,b)

$$p_!\sigma^p_*(f,b) = \sigma_*(f,b)$$

and induction the quadratic kernel of (f',b')

$$p^!\sigma^p_*(f,b) = \sigma_*(f',b') \ .$$

If $(f,b):M \longrightarrow X$ is a genuine normal map, that is if X is an n-dimensional geometric Poincaré complex, then $\sigma^p_*(f,b)$ is an n-dimensional pseudoquadratic Poincaré complex over $\mathbb{Z}[\Pi^w]$ with cobordism class

$$\sigma^p_*(f,b) = \sigma_*(f,b) \in L_n(\mathbb{Z}[\Pi^w]) = L_n(\mathbb{Z}[\pi^w]) \ .$$

(Moreover, in this case the spectral pseudoquadratic construction is a composite

$$\psi_{F,p} : H_n(X,w) \xrightarrow{\ \psi_{F,p}\ } Q^{\Pi,w}_n(\mathcal{C}(\tilde{M},f^*p)) \xrightarrow{\ \mathcal{E}_{\%}\ } Q^{\Pi,w}_n(C(\mathcal{F}^!)) \quad).$$

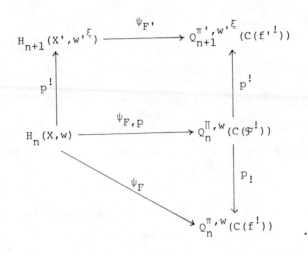

Let (X,Y) be a connected codimension 2 CW pair with

$$\pi_1(X) = \pi_1(Y) \ , \ \pi_1(Z) = \pi_1(S(\xi))$$

and let Φ be the pushout square of fundamental groups

$$\begin{array}{ccc} \pi_1(S(\xi)) & \xrightarrow{\sim} & \pi_1(Z) \\ \downarrow & \Phi & \downarrow \\ \pi_1(E(\xi)) & \xrightarrow{\sim} & \pi_1(X) \end{array} .$$

Denote the universal data of $\xi : Y \longrightarrow BG(2)$

$$(\pi_1(S^1) \longrightarrow \pi_1(S(\xi)) \longrightarrow \pi_1(Y) \longrightarrow \{1\} \ , \ w_1(\xi) : \pi_1(Y) \longrightarrow \mathbb{Z}_2)$$

by

$$(\Pi : \mathbb{Z} \longrightarrow \pi' \xrightarrow{p} \pi \longrightarrow \{1\} \ , \ \omega : \pi \longrightarrow \mathbb{Z}_2)$$

and write the orientation map of X as

$$w(X) = w : \pi_1(X) = \pi \longrightarrow \mathbb{Z}_2 \ ,$$

so that the other orientation maps are given by

$$w(Y) = w^{\xi} : \pi_1(Y) = \pi \longrightarrow \mathbb{Z}_2 \ ,$$

$$w(Z) = w(S(\xi)) = w' : \pi_1(Z) = \pi_1(S(\xi)) = \pi' \longrightarrow \mathbb{Z}_2 \ .$$

By Proposition $\left\{\begin{array}{l} 7.2.1 \text{ ii)} \\ 7.8.1 \text{ iv)} \end{array}\right.$ the $\left\{\begin{array}{l} LS\text{-} \\ \Gamma S\text{-} \end{array}\right.$ groups of (X,Y)

$\left\{\begin{array}{l} LS_*(\Phi) = LN_*(\pi' \longrightarrow \pi, w) \\ \Gamma S_*(\Phi) = \Gamma N_*(\pi' \longrightarrow \pi, w) \end{array}\right.$ fit into the exact sequence

$$\left\{\begin{array}{l} \dots \longrightarrow L_{n+1}(\mathbb{Z}[\pi^w]) \longrightarrow LN_{n-2}(\pi' \longrightarrow \pi, w) \\ \qquad \longrightarrow L_n(\partial\xi^! : \mathbb{Z}[\pi^{w^{\xi}}] \longrightarrow \mathbb{Z}[\pi'^{w'}]) \longrightarrow L_n(\mathbb{Z}[\pi^w]) \longrightarrow \dots \end{array}\right.$$

$$\begin{cases} \ldots \longrightarrow L_{n+1}(\mathbb{Z}[\pi^w]) \longrightarrow \Gamma N_{n-2}(\pi' \longrightarrow \pi, w) \\ \longrightarrow \Gamma_n(\partial\xi^!: \begin{pmatrix} \mathbb{Z}[\pi^{w^\xi}] \\ \downarrow \xi \\ \mathbb{Z}[\pi^w] \end{pmatrix} \longrightarrow \begin{pmatrix} \mathbb{Z}[\pi'^{w'}] \\ \downarrow \\ \mathbb{Z}[\pi^w] \end{pmatrix}) \longrightarrow L_n(\mathbb{Z}[\pi^w]) \longrightarrow \ldots \end{cases}$$

By Proposition $\begin{cases} 7.5.1 \text{ ii)} \\ 7.8.6 \end{cases}$ the codimension 2 $\begin{cases} - \\ \text{weak} \end{cases}$ splitting obstruction group $\begin{cases} LN_{n-2}(\pi' \longrightarrow \pi, w) \\ \Gamma N_{n-2}(\pi' \longrightarrow \pi, w) \end{cases}$ $(n \geqslant 2)$ is the cobordism group of contractible $\begin{cases} - \\ \text{weak} \end{cases}$ n-dimensional quadratic Poincaré splittings over Φ, i.e. of pairs

$$((C, \psi \in Q_{n-2}^{\pi, w^\xi}(C)), (f: \partial\xi^! C \longrightarrow D, (\delta\psi, \partial\xi^!\psi) \in Q_n^{\pi', w'}(f)))$$

consisting of an (n-2)-dimensional quadratic Poincaré complex (C, ψ) over $\mathbb{Z}[\pi^{w^\xi}]$ and a $\begin{cases} - \\ \text{weak} \end{cases}$ n-dimensional quadratic Poincaré pair $(f: \partial\xi^! C \longrightarrow D, (\delta\psi, \partial\xi^!\psi))$ over $\mathbb{Z}[\pi'^{w'}]$ such that the $\mathbb{Z}[\pi]$-module chain map

$$\begin{pmatrix} i \\ p_! f \end{pmatrix} : \mathbb{Z}[\pi] \otimes_{\mathbb{Z}[\pi']} \partial\xi^! C \longrightarrow C \oplus \mathbb{Z}[\pi] \otimes_{\mathbb{Z}[\pi']} D$$

is a (simple) chain equivalence, where i is the $\mathbb{Z}[\pi]$-module chain map appearing in the n-dimensional quadratic Poincaré pair over $\mathbb{Z}[\pi^w]$

$$\xi^!(C, \psi) = (i: \mathbb{Z}[\pi] \otimes_{\mathbb{Z}[\pi']} \partial\xi^! C \longrightarrow C, (\xi^!\psi, 1 \otimes \partial\xi^!\psi) \in Q_n^{\pi, w}(i)) .$$

We now have to give an algebraic definition of $\xi^!(C, \psi)$.

Proposition 7.8.11 The transfer maps in quadratic L-theory associated to a codimension 2 CW pair (X,Y) with $\pi_1(X) = \pi_1(Y) = \pi$, $\pi_1(Z) = \pi_1(S(\xi)) = \pi'$, $w(X) = w : \pi \longrightarrow \mathbb{Z}_2$ are given algebraically by

$$\xi^! : L_n(\mathbb{Z}[\pi^{w^\xi}]) \longrightarrow L_{n+2}(p:\mathbb{Z}[\pi'^{w'}] \longrightarrow \mathbb{Z}[\pi^w]) \ ;$$

$$(C,\psi) \longmapsto ((C^!,\psi^!),(i:\mathbb{Z}[\pi]\otimes_{\mathbb{Z}[\pi']}C^! \longrightarrow C,(0,1\otimes\psi^!))) \quad (n \geqslant 0)$$

with $(C,\psi \in Q_n^{\pi,w^\xi}(C))$ the projection of an n-dimensional pseudoquadratic Poincaré complex $(\mathcal{C},(\psi',\chi') \in Q_n^{\Pi,w^\xi}(\mathcal{C}))$ over $\mathbb{Z}[\Pi^{w^\xi}]$ and $(i:\mathbb{Z}[\pi]\otimes_{\mathbb{Z}[\pi']}C^! \longrightarrow C,(0,1\otimes\psi^!) \in Q_{n+2}^{\pi,w}(i))$ the $(n+2)$-dimensional quadratic Poincaré pair over $\mathbb{Z}[\pi^w]$ with

$$i = (1\ 0) : (\mathbb{Z}[\pi]\otimes_{\mathbb{Z}[\pi']}C^!)_r = C_r \oplus \alpha C_{r-1} \longrightarrow C_r \ .$$

Proof: Immediate from the pseudoquadratic kernel construction.

[]

Continuing with the previous terminology define an antistructure (β,t) on $\mathbb{Z}[\pi']$ by

$$\beta : \mathbb{Z}[\pi'] \longrightarrow \mathbb{Z}[\pi'] \ ;$$

$$\sum_{g'\in\pi'} n_{g'}g' \longmapsto \sum_{\substack{g'\in\pi' \\ \omega'(g')=+1}} n_{g'}w'(g')g'^{-1} - \sum_{\substack{g'\in\pi' \\ \omega'(g')=-1}} n_{g'}w'(g')g'^{-1}t \ ,$$

so that there is defined a morphism of rings with antistructure

$$p : (\mathbb{Z}[\pi'],\beta,t) \longrightarrow (\mathbb{Z}[\pi],w^\xi,1) \ .$$

In the oriented case $w_1(\xi) = \omega = +1$ the unit $t \in \mathbb{Z}[\pi']$ is central and $\beta:\mathbb{Z}[\pi'] \longrightarrow \mathbb{Z}[\pi']$ is the w'-twisted involution.

Assume now that the underlying codimension 2 CW pair (X,Y) is a formally $(n,n-2)$-dimensional normal pair (in the sense of §7.5) and that there is given a formally $(n,n-2)$-dimensional topological normal map

$$(f,b) : (M,N) \longrightarrow (X,Y) ,$$

denoting the restriction normal maps by

$$(f,b)| = (g,c) : N = f^{-1}(Y) \longrightarrow Y$$

$$(f,b)| = (h,d) : (P,S(\nu)) = f^{-1}(Z,S(\xi)) \longrightarrow (Z,S(\xi)) ,$$

with

$$\nu : N \xrightarrow{\ g\ } Y \xrightarrow{\ \xi\ } BG(2) .$$

According to Proposition 7.5.4 ambient surgery on (g,c) inside (f,b) has the algebraic effect of surgery on the (pseudo)quadratic kernel pair $(\sigma_*^p(g,c),\sigma_*(h,d))$ preserving the union $\xi^!\sigma_*^p(g,c) \cup p\sigma_*(h,d) = \sigma_*(f,b)$. We shall now associate to the pair $(\sigma_*^p(g,c),\sigma_*(h,d))$ a formally $(n-2)$-dimensional (β,t)-quadratic complex over $\mathbb{Z}[\pi']$ $\sigma_*^!(f,b)$ such that surgery on the pair determines surgery on the complex, and such that if $f:M \longrightarrow X$ is an s-triangulation algebraic surgery determines geometric surgery, generalizing the treatment of codimension 2 surgery due to Matsumoto [1] and Freedman [1].

The pair $(\sigma_*^p(g,c),\sigma_*(h,d))$ consists of a formally $(n-2)$-dimensional pseudoquadratic complex over $\mathbb{Z}[\Pi^{w^\xi}]$ which we shall write as

$$\sigma_*^p(g,c) = (C(\mathcal{G}^!),\psi_{G,p}([Y])) = (\mathcal{C}= (C',d',e'),(\psi',\chi') \in Q_{n-2}^{\Pi,w^\xi}(\mathcal{C}))$$

and a formally n-dimensional quadratic pair over $\mathbb{Z}[\pi'^{w'}]$ which we shall write as

$$\begin{aligned}\sigma_*(h,d) &= (f_0 : C(g'^{!}) \longrightarrow C(h^{!}), \psi_H([P])) \\ &= (f_0 : C^{!} \longrightarrow \alpha' SD', (\delta\psi', \psi^{!}) \in Q_n^{\pi',w'}(f_0))\end{aligned}$$

with

$$f_0 = (\alpha' k \ \ \alpha' j) : (C^{!})_r = C'_r \oplus \alpha' C'_{r-1} \longrightarrow \alpha' D'_{r-1} \ .$$

Define an untwisted pseudo chain complex over $\mathbb{Z}[\Pi]$

$$\mathcal{D} = (D', 0) \ ,$$

and note that j,k define a pseudo chain map

$$\mathcal{J} = (j,k) : \mathcal{C} \longrightarrow \mathcal{D} \ .$$

As $\mathcal{D}$ is untwisted the $\mathbb{Z}[\mathbb{Z}_2]$-module chain complex $(\mathrm{Hom}_{\Pi}(\mathcal{D}^{*,w^{\xi}}, \mathcal{D}), T)$ is the algebraic mapping cone of the $\mathbb{Z}[\mathbb{Z}_2]$-module chain map

$$1-t : (\mathrm{Hom}_{\mathbb{Z}[\pi']}(\alpha' D'^{*,w'}, \alpha' D'), T_{-1}) \longrightarrow (\mathrm{Hom}_{\mathbb{Z}[\pi']}(D'^{*,\beta}, D'), T_t)$$

and $\mathcal{J}$ induces a natural transformation of exact sequences of abelian groups

$$\begin{array}{ccccccc} \dots \longrightarrow & Q_n^{\pi',w'}(\alpha' SD') & \longrightarrow & Q_n(f_0 p^{!}) & \longrightarrow & & \\ & \bar{S}^{-1} \downarrow \wr & & \mathcal{J}'_{\%} \downarrow & & & \\ \dots \longrightarrow & Q_{n-2}^{\pi',w'}(\alpha' D', -1) & \xrightarrow{1-t} & Q_{n-2}^{\pi',\beta}(D', t) & \longrightarrow & & \\ & & & & & & \\ & Q_{n-2}^{\Pi,w^{\xi}}(\mathcal{C}) & \xrightarrow{f_{0\%} p^{!}} & Q_{n-1}^{\pi',w'}(\alpha' SD') & \longrightarrow \dots & & \\ & \mathcal{J}_{\%} \downarrow & & \bar{S}^{-1} \downarrow \wr & & & \\ & Q_{n-2}^{\Pi,w^{\xi}}(\mathcal{D}) & \longrightarrow & Q_{n-3}^{\pi',w'}(\alpha' D', -1) & \longrightarrow \dots & & . \end{array}$$

Define the <u>antiquadratic kernel</u> of $(f,b) : (M,N) \longrightarrow (X,Y)$ to be the formally $(n-2)$-dimensional (β,t)-quadratic complex over $\mathbb{Z}[\pi']$

$$\sigma'_*(f,b) = (D', \mathcal{J}'_{\%}(\delta\psi', (\psi', \chi'))) \in Q_{n-2}^{\pi',\beta}(D', t))$$

$$(D' = \alpha' \Omega C(h^{!})) \quad .$$

Proposition 7.8.12 Given data

$$(\Pi:\mathbb{Z} \longrightarrow \pi' \xrightarrow{\ p\ } \pi \longrightarrow \{1\}, \omega:\pi \longrightarrow \mathbb{Z}_2)$$

and an orientation map $w:\pi \longrightarrow \mathbb{Z}_2$ there are defined natural isomorphisms of abelian groups

$$LN_{n-2}(\pi' \longrightarrow \pi, w) \xrightarrow{\ \sim\ } L_{n-2}(\mathbb{Z}[\pi']^{\beta}, t) \ ;$$
$$(\sigma_*^p(g,c), \sigma_*(h,d)) \longmapsto \sigma_*^!(f,b)$$

$$\Gamma N_{n-2}(\pi' \longrightarrow \pi, w) \xrightarrow{\ \sim\ } \Gamma_{n-2}(p:\mathbb{Z}[\pi']^{\beta} \longrightarrow \mathbb{Z}[\pi^{w^{\xi}}], t)$$
$$(\sigma_*^p(g,c), \sigma_*(h,d)) \longmapsto \sigma_*^!(f,b)$$

$$\Gamma_n\left(\begin{array}{ccc} \mathbb{Z}[\pi'^{w'}] & \longrightarrow & \mathbb{Z}[\pi'^{w'}] \\ \downarrow & \Theta^{w'} & \downarrow \\ \mathbb{Z}[\pi'^{w'}] & \longrightarrow & \mathbb{Z}[\pi^{w}] \end{array}\right) \xrightarrow{\ \sim\ } \Gamma_{n-2}\left(\begin{array}{ccc} \mathbb{Z}[\pi']^{\beta} & \longrightarrow & \mathbb{Z}[\pi']^{\beta} \\ \downarrow & \Theta^{\beta} & \downarrow \\ \mathbb{Z}[\pi']^{\beta} & \longrightarrow & \mathbb{Z}[\pi^{w^{\xi}}] \end{array}, t\right) ;$$

$$(C', \psi' \in Q_{n-1}^{\pi',w'}(C')) \longmapsto (\alpha'\Omega C', \overline{S}^{-1}(1-t)\psi' \in Q_{n-3}^{\pi',\beta}(\alpha'\Omega C', t))$$
$$(H_*(\mathbb{Z}[\pi] \otimes_{\mathbb{Z}[\pi']} C') = 0)$$

which fit together to define a natural isomorphism of exact sequences

$$\begin{array}{ccccc} \dots \longrightarrow & \Gamma_{n+1}(\Theta^{w'}) & \longrightarrow & LN_{n-2}(\pi' \longrightarrow \pi, w) & \longrightarrow \\ & \downarrow \wr & & \downarrow \wr & \\ \dots \longrightarrow & \Gamma_{n-1}(\Theta^{\beta}, t) & \longrightarrow & L_{n-2}(\mathbb{Z}[\pi']^{\beta}, t) & \longrightarrow \end{array}$$

$$\begin{array}{ccccc} \Gamma N_{n-2}(\pi' \longrightarrow \pi, w) & \longrightarrow & \Gamma_n(\Theta^{w'}) & \longrightarrow & \dots \\ \downarrow \wr & & \downarrow \wr & & \\ \Gamma_{n-2}(p:\mathbb{Z}[\pi']^{\beta} \longrightarrow \mathbb{Z}[\pi^{w^{\xi}}], t) & \longrightarrow & \Gamma_{n-2}(\Theta^{\beta}, t) & \longrightarrow & \dots \ . \end{array}$$

Proof: Given an (n-2)-dimensional (β,t)-quadratic $\begin{cases}\text{Poincaré}\\ \mathbb{Z}[\pi]\text{-Poincaré}\end{cases}$ complex over $\mathbb{Z}[\pi']$ $(D',\psi'\in Q^{\pi',\beta}_{n-2}(D',t))$ define an (n-2)-dimensional pseudoquadratic complex over $\mathbb{Z}[\Pi^{w^\xi}]$ $(\mathcal{D}=(D',0),(\psi',0)\in Q^{\Pi,w^\xi}_{n-2}(\mathcal{D}))$ with projection $\mathbb{Z}[\pi]\otimes_{\mathbb{Z}[\pi']}(D',\psi')=(D,\psi)$, and define also an n-dimensional quadratic $\begin{cases}\text{Poincaré}\\ \mathbb{Z}[\pi]\text{-Poincaré}\end{cases}$ pair over $\mathbb{Z}[\pi'^{w'}]$ $(f_0:D^!\longrightarrow\alpha'SD',(0,\psi^!)\in Q^{\pi',w'}_n(f_0))$ with

$$f_0=(0\ 1)\ :\ (D^!)_r=D'_r\oplus\alpha'D'_{r-1}\longrightarrow\alpha'D'_{r-1}\ ,$$

such that $\begin{pmatrix} i\\ p_!f_0\end{pmatrix}:\mathbb{Z}[\pi]\otimes_{\mathbb{Z}[\pi']}D^!\longrightarrow D\oplus\mathbb{Z}[\pi]\otimes_{\mathbb{Z}[\pi']}\alpha'SD'$ is a simple $\mathbb{Z}[\pi]$-module chain equivalence (the identity in fact). The corresponding abelian group morphisms

$$\begin{cases} L_{n-2}(\mathbb{Z}[\pi']^{\beta},t)\longrightarrow LN_{n-2}(\pi'\longrightarrow\pi,w)\ ;\\ \qquad (D',\psi')\longmapsto((\mathcal{D},(\psi',0)),(f_0:D^!\longrightarrow\alpha'SD',(0,\psi^!)))\\ \Gamma_{n-2}(p:\mathbb{Z}[\pi']^{\beta}\longrightarrow\mathbb{Z}[\pi^{w^\xi}],t)\longrightarrow\Gamma N_{n-2}(\pi'\longrightarrow\pi,w)\ ;\\ \qquad (D',\psi')\longmapsto((\mathcal{D},(\psi',0)),(f_0:D^!\longrightarrow\alpha'SD',(0,\psi^!)))\end{cases}$$

are the isomorphisms inverse to the morphisms defined above.

Given a $\mathbb{Z}[\pi]$-acyclic (n-2)-dimensional (β,t)-quadratic Poincaré complex over $\mathbb{Z}[\pi']$ $(D',\psi'\in Q^{\pi',\beta}_{n-2}(D',t))$ we have that the $\mathbb{Z}[\pi']$-module chain map $1-t:\alpha'D'\longrightarrow D'$ is a chain equivalence, by Proposition 7.8.7 iii), so that there is induced an isomorphism

$$1-t\ :\ Q^{\pi',w'}_{n-2}(\alpha'D',-1)\xrightarrow{\ \sim\ }Q^{\pi',\beta}_{n-2}(D',t)$$

and $(\alpha'SD',\bar{S}(1-t)^{-1}\psi'\in Q^{\pi',w'}_n(\alpha'SD'))$ is a $\mathbb{Z}[\pi]$-acyclic n-dimensional quadratic Poincaré complex over $\mathbb{Z}[\pi'^{w'}]$.

The corresponding abelian group morphisms

$$\Gamma_{n-1}(\Theta^{\beta},t) \longrightarrow \Gamma_{n+1}(\Theta^{w'}) \; ; \; (D',\psi') \longmapsto (\alpha' SD', \bar{S}(1-t)^{-1}\psi')$$

are the isomorphisms inverse to the morphisms defined above.

[]

If (X,Y) is a $\begin{cases} - \\ \text{weak} \end{cases}$ $(n,n-2)$-dimensional geometric Poincaré pair (such that $\pi_1(X) = \pi_1(Y) = \pi$, $\pi_1(S(\xi)) = \pi_1(Z) = \pi'$, $w(X) = w$) and $(f,b):(M,N) \longrightarrow (X,Y)$ is an $(n,n-2)$-dimensional topological normal map such that $(f,b):M \longrightarrow X$ is an s-triangulation of X the antiquadratic kernel $\sigma_*^!(f,b)$ is an $(n-2)$-dimensional (β^{ξ},t)-quadratic $\begin{cases} \text{Poincaré} \\ \mathbb{Z}[\pi]\text{-Poincaré} \end{cases}$ complex over $\mathbb{Z}[\pi']$.

<u>Proposition 7.8.13</u> The $\begin{cases} - \\ \text{weak} \end{cases}$ splitting obstruction of f along $Y \subset X$ is given by

$$\begin{cases} s(f,Y) = \sigma_*^!(f,b) \in LN_{n-2}(\pi' \longrightarrow \pi, w) = L_{n-2}(\mathbb{Z}[\pi']^{\beta},t) \\ ws(f,Y) = \sigma_*^!(f,b) \in \Gamma N_{n-2}(\pi' \longrightarrow \pi, w) \\ \qquad = \Gamma_{n-2}(p:\mathbb{Z}[\pi']^{\beta} \longrightarrow \mathbb{Z}[\pi^{w^{\xi}}],t) \ . \end{cases}$$

<u>Proof</u>: Immediate from Proposition 7.8.12.

[]

Matsumoto [1] and Freedman [1] (independently) analyzed ambient surgery on codimension 2 submanifolds in terms of a geometrically defined t-quadratic form analogous to the self-intersection form of Wall [4,§5] and the equivariant self-intersection form needed for codimension 1 surgery (cf. §7.6). The antiquadratic kernel $\sigma_*^!(f,b)$ is evidently a homological version of this t-quadratic form.

If $(M,\partial M)$ is an n-dimensional manifold with boundary and $U \subset \partial M$ is a codimension 0 submanifold (which may be empty) of the boundary such that (M,U) is an (n-2)-dimensional geometric Poincaré pair, and such that there is given a codimension 2 spine $K \subset U$, then the obstruction $\sigma_*(M,K) \in P_{n-2}(\Pi,w)$ obtained by Matsumoto [1] (in the oriented case $\omega = w = +1$) for the existence of a codimension 2 spine $(N,K) \subset (M,U)$ is the rel∂ weak splitting obstruction along the zero section $M \subset E(\xi)$ of an $s^{\mathbb{Z}[\pi]}$-triangulation (defined as in Proposition 7.5.4)

$$(f,b) : (M;U,X;\partial U) \longrightarrow (E(\xi);E(\xi|_U),S(\xi);S(\xi|_U)) \quad (X = \overline{\partial M \setminus U})$$

topologically transverse at the zero section $(M,U) \subset (E(\xi),E(\xi|_U))$ with $f^{-1}(U) = K \subset U$

$$\sigma_*(M,K) = ws_\partial(f,M) = \sigma'_*(f,b)$$
$$\in P_{n-2}(\Pi,w) = \Gamma N_{n-2}(\pi' \longrightarrow \pi,w) = \Gamma_{n-2}(p:\mathbb{Z}[\pi']^\beta \longrightarrow \mathbb{Z}[\pi^{w^\xi}],t)$$

with $\xi : M \longrightarrow BG(2)$ an S^1-fibration over M extending $\nu_{K \subset U} : V \longrightarrow BG(2)$, $\pi = \pi_1(M)$, $\pi' = \pi_1(S(\xi))$, $w = w(M,\partial M)$. The obstruction to the existence of a codimension 2 spine obtained by Cappell and Shaneson [2] is the relative $(\mathbb{Z}[\pi'],\mathbb{Z}[\pi])$-homology surgery obstruction of the n-dimensional topological normal map of triads

$$(h,d) = (f,b)| : (P;S(\nu),X;\partial U)$$
$$\longrightarrow (S(\xi) \times I;S(\xi) \times 0,S(\xi) \times 1;S(\xi|_U) \times I)$$

with $\nu = \nu_{N \subset M} : N = f^{-1}(M) \xrightarrow{f|} M \longrightarrow BG(2)$, $P = \overline{M \setminus E(\nu)}$, which is the image $\sigma_*(h,d) \in \Gamma_n(\Theta^{w'})$ under the canonical map of $ws_\partial(f,M) \in \Gamma N_{n-2}(\pi' \longrightarrow \pi,w)$. By Proposition 7.7.3 ii) this obstruction is the rel∂ quadratic signature of the $s^{\mathbb{Z}[\pi]}$-triangulation

$$(\partial_+ f, \partial_+ b) = (f,b)| : (X, \partial U) \longrightarrow (S(\xi), S(\xi|_U)) \quad ,$$

that is

$$\sigma_*(h,d) = \sigma_*(\partial_+ f, \partial_+ b) \in \Gamma_n(\Theta^{w'}) \ .$$

We shall now consider the codimension 2 surgery obstruction theory in the case when the normal S^1-fibration admits a section, e.g. if it is trivial (the situation arising in knot theory). We shall develop non-simply-connected analogues of various knot invariants, which will be related to their origins in knot theory in §7.9 below. For example, the above expression for $\sigma_*(h,d)$ is a generalization of the expression of the knot cobordism class of a (high-dimensional) knot in terms of the Blanchfield pairing in the homology of the infinite cyclic cover of the knot complement.

Let then (X,Y) be a codimension 2 CW pair such that

$$\xi = \omega \oplus \varepsilon : Y \longrightarrow BG(2)$$

for some line bundle $\omega : Y \longrightarrow BG(1)$, and such that

$$\pi_1(X) = \pi_1(Y) = \pi \ , \ \pi_1(Z) = \pi_1(S(\xi)) = \pi' \ ,$$

in which case $\pi' = \pi \rtimes_\omega \mathbb{Z}$ is the semidirect product of π and $\mathbb{Z}$ determined by the orientation map $\omega : \pi \longrightarrow \mathbb{Z}_2 = \mathrm{Aut}(\mathbb{Z})$

$$\{1\} \longrightarrow \mathbb{Z} \longrightarrow \pi' \xrightarrow{\ p\ } \pi \longrightarrow \{1\}$$

with $\pi' = \{gt^j | g \in \pi, j \in \mathbb{Z}\}$ as a set and

$$(gt^j)(ht^k) = (gh)t^{\omega(h)j+k} \in \pi' \quad (g,h \in \pi,\ j,k \in \mathbb{Z})$$

$$p(gt^j) = g \in \pi \quad .$$

(If ω is trivial $\pi' = \pi \times \mathbb{Z}$). Denote the orientation map of X by

$$w(X) = w : \pi \longrightarrow \mathbb{Z}_2 \ ,$$

so that

$$w(Y) = w^{\xi} : \pi \longrightarrow \mathbb{Z}_2 \ ; \ g \longmapsto w(g)\omega(g)$$

$$w(Z) = w' : \pi' \longrightarrow \mathbb{Z}_2 \ ; \ gt^j \longmapsto w(g) \quad .$$

A mild generalization of the splitting theorem of Shaneson [1] (the trivial case $\omega = +1$) identifies

$$L_n(\mathbb{Z}[\pi'^{w'}]) = L_n(\mathbb{Z}[\pi^w]) \oplus L^h_{n-1}(\mathbb{Z}[\pi^{w^\xi}]) \ ,$$

so that the transfer map in quadratic L-theory associated to (X,Y)

$$\xi^! : L_n(\mathbb{Z}[\pi^{w^\xi}]) \longrightarrow L_{n+2}(p:\mathbb{Z}[\pi'^{w'}] \longrightarrow \mathbb{Z}[\pi^w]) = L^h_n(\mathbb{Z}[\pi^{w^\xi}])$$

is just the forgetful map appearing in the Rothenberg exact sequence. The resulting identification

$$LN_n(\pi' \longrightarrow \pi, w) = \hat{H}^{n+1}(\mathbb{Z}_2; Wh(\pi)^{w^\xi})$$

was first obtained by Wall [4, Prop.13A.10] (for $\omega = +1$). It now follows from the exact sequence given by Proposition 7.8.1 i)

$$\ldots \longrightarrow \hat{H}^{n+1}(\mathbb{Z}_2; Wh(\pi)^{w^\xi}) \longrightarrow \Gamma N_n(\pi' \longrightarrow \pi, w)$$

$$\longrightarrow \Gamma_{n+2}\begin{pmatrix} \mathbb{Z}[\pi'^{w'}] & \longrightarrow & \mathbb{Z}[\pi'^{w'}] \\ \big\downarrow & \Theta^{w'} & \big\downarrow \\ \mathbb{Z}[\pi'^{w'}] & \longrightarrow & \mathbb{Z}[\pi^w] \end{pmatrix} \longrightarrow \hat{H}^n(\mathbb{Z}_2; Wh(\pi)^{w^\xi}) \longrightarrow \ldots$$

that $\Gamma N_n(\pi' \longrightarrow \pi, w)$ can be identified with the cobordism group of $\mathbb{Z}[\pi]$-acyclic (n+1)-dimensional quadratic Poincaré complexes over $\mathbb{Z}[\pi'^{w'}]$ (C,ψ) such that C is based, $\tau(\mathbb{Z}[\pi]\otimes_{\mathbb{Z}[\pi']}C) = 0 \in Wh(\pi)$, $\tau((1+T)\psi_0 : C^{n+1-*,w'} \longrightarrow C) = 0 \in Wh(\pi')$, and such that an invariant in the second Whitehead group $Wh_2(\pi)$ is 0. We shall use this expression for ΓN_* in §7.9 below in the special case $\pi = \{1\}$ (when all the Whitehead groups are 0) to describe the high-dimensional knot cobordism groups C_* as the cobordism groups of $\mathbb{Z}$-acyclic algebraic Poincaré complexes over $\mathbb{Z}[\mathbb{Z}]$, generalizing Blanchfield duality. In §7.9 there will also be given a description of C_* as the cobordism groups of "ultraquadratic" Poincaré complexes over $\mathbb{Z}$, generalizing the Seifert form, which motivates the following expression for $\Gamma N_*(\pi\times\mathbb{Z}\longrightarrow\pi,w)$.

Let A be a ring with involution, and let C be a finite-dimensional A-module chain complex. The case p = 1 of Proposition 1.1.3 gives the exact sequence

$$\ldots\longrightarrow Q^{n+2}(SC,\varepsilon)\longrightarrow H_n(\mathrm{Hom}_A(C^*,C))\xrightarrow{1+T_\varepsilon}Q^n(C,\varepsilon)\xrightarrow{S}Q^{n+1}(SC,\varepsilon)\longrightarrow\ldots,$$

with $\varepsilon\in A$ a central unit such that $\bar{\varepsilon}=\varepsilon^{-1}\in A$ and

$$1+T_\varepsilon\ :\ H_n(\mathrm{Hom}_A(C^*,C))\longrightarrow Q^n(C,\varepsilon)\ ;$$

$$\hat{\psi}\longmapsto(1+T_\varepsilon)\hat{\psi}=\{(1+T_\varepsilon)\hat{\psi}_s=\begin{cases}(1+T_\varepsilon)\hat{\psi} & s=0\\ 0 & s\geqslant 1\end{cases}\}\ .$$

The $\mathbb{Z}[\mathbb{Z}_2]$-module defined by the abelian group $H_n(\mathrm{Hom}_A(C^*,C))$ with $T\in\mathbb{Z}_2$ acting by the ε-duality involution $T_\varepsilon:\hat{\psi}\longmapsto(-)^{pq}\varepsilon\hat{\psi}^*$ $(\hat{\psi}\in\mathrm{Hom}_A(C^p,C_q))$ is denoted by $\hat{Q}_n(C,\varepsilon)$. An element

$$\hat{\psi}\in\hat{Q}_n(C,\varepsilon)=H_n(\mathrm{Hom}_A(C^*,C))$$

is a chain homotopy class of A-module chain maps

$$\hat{\psi}\ :\ C^{n-*}\longrightarrow C\ .$$

An <u>n-dimensional ε-ultraquadratic complex over A</u> $(C,\hat{\psi})$ is an n-dimensional A-module chain complex C together with an element $\hat{\psi}\in\hat{Q}_n(C,\varepsilon)$. Such a complex is <u>Poincaré</u> if

$$(1+T_\varepsilon)\hat{\psi}\ :\ C^{n-*}\longrightarrow C$$

is a chain homotopy class of chain equivalences. Similarly for pairs. Define the <u>n-dimensional ε-ultraquadratic L-group of A</u> $\hat{L}_n(A,\varepsilon)$ $(n\geqslant 0)$ to be the cobordism group of n-dimensional

ε-ultraquadratic Poincaré complexes over A. The ε-ultraquadratic version of the algebraic surgery of §1.5 shows that the skew-suspension maps

$$\bar{S} : \hat{L}_n(A,\varepsilon) \longrightarrow \hat{L}_{n+2}(A,-\varepsilon) \ ; \ (C,\hat{\psi}) \longmapsto (SC,S\hat{\psi}) \quad (n \geqslant 0)$$

are isomorphisms, just as for the ε-quadratic L-groups $L_*(A,\varepsilon)$. There are defined forgetful maps

$$\hat{L}_n(A,\varepsilon) \longrightarrow L_n(A,\varepsilon) \ ;$$
$$(C,\hat{\psi} \in \hat{Q}_n(C,\varepsilon)) \longmapsto (C,\psi \in Q_n(C,\varepsilon)) \quad (n \geqslant 0)$$

with $\psi_0 = \hat{\psi}$, $\psi_s = 0$ $(s \geqslant 1)$.

An <u>ε-ultraquadratic form over A</u> $(M,\hat{\psi})$ is a f.g. projective A-module M together with an element $\hat{\psi} \in \mathrm{Hom}_A(M,M^*)$. Such a form is <u>non-singular</u> if the A-module morphism

$$\hat{\psi}+\varepsilon\hat{\psi}^* : M \longrightarrow M^*$$

is an isomorphism. A <u>morphism</u> (resp. <u>isomorphism</u>) of such forms

$$f : (M,\hat{\psi}) \longrightarrow (M',\hat{\psi}')$$

is an A-module morphism (resp. isomorphism) $f \in \mathrm{Hom}_A(M,M')$ such that

$$f^*\hat{\psi}'f = \hat{\psi} \in \mathrm{Hom}_A(M,M^*) \ .$$

A <u>sublagrangian</u> of an ε-ultraquadratic form $(M,\hat{\psi})$ is a direct summand L of M such that the inclusion $j \in \mathrm{Hom}_A(L,M)$ defines a morphism

$$j : (L,0) \longrightarrow (M,\hat{\psi})$$

and such that $j^*(\hat{\psi}+\varepsilon\hat{\psi}^*) \in \mathrm{Hom}_A(M,L^*)$ is onto. A <u>lagrangian</u> is a sublagrangian for which

$$L = \ker(j^*(\hat{\psi}+\varepsilon\hat{\psi}^*) : M \longrightarrow L^*) \ .$$

An <u>ε-ultraquadratic formation over A</u> $(M,\hat{\psi};F,G)$ is a non-singular ε-ultraquadratic form over A over A $(M,\hat{\psi})$ together with a lagrangian F and a sublagrangian G. Such a formation is non-singular if G is a lagrangian. The ε-ultraquadratic version of the theory of §1.6 identifies the homotopy equivalence classes of $\begin{cases}\text{0-dimensional}\\ \text{connected 1-dimensional}\end{cases}$ ε-ultraquadratic (Poincaré) complexes over A with the $\begin{cases}\text{isomorphism}\\ \text{stable isomorphism}\end{cases}$ classes of (non-singular) ε-ultraquadratic $\begin{cases}\text{forms}\\ \text{formations}\end{cases}$ over A, and also identifies

$$\begin{cases}\hat{L}_0(A,\varepsilon)\\ \hat{L}_1(A,\varepsilon)\end{cases} = \text{the Witt group of non-singular } \varepsilon\text{-ultraquadratic} \begin{cases}\text{forms}\\ \text{formations}\end{cases} \text{over } A\ .$$

The forgetful map $\begin{cases}\hat{L}_0(A,\varepsilon) \longrightarrow L_0(A,\varepsilon)\\ \hat{L}_1(A,\varepsilon) \longrightarrow L_1(A,\varepsilon)\end{cases}$ is $\begin{cases}\text{onto}\\ \text{one-one}\end{cases}$.

By analogy with the the intermediate ε-quadratic L-groups $L_*^X(A,\varepsilon)$ $(X \subseteq \tilde{K}_m(A),\ m = 0,1)$ of §1.10 there are defined intermediate ε-ultraquadratic L-groups $\hat{L}_*^X(A,\varepsilon)$, with an exact sequence

$$\ldots \longrightarrow \hat{H}^{n+1}(\mathbb{Z}_2;Y/X) \longrightarrow \hat{L}_n^X(A,\varepsilon) \longrightarrow \hat{L}_n^Y(A,\varepsilon) \longrightarrow \hat{H}^n(\mathbb{Z}_2;Y/X) \longrightarrow \hat{L}_{n-1}^X(A,\varepsilon) \longrightarrow \ldots$$

for $X \subseteq Y \subseteq \tilde{K}_m(A)$.

For $\varepsilon = 1 \in A$ the terminology is contracted in the usual fashion

$$\hat{Q}_n(C,1) = \hat{Q}_n(C)\ ,\ \hat{L}_n(A,1) = \hat{L}_n(A)$$

$$\text{1-ultraquadratic} = \text{ultraquadratic}\ .$$

In dealing with the ultraquadratic L-groups in the topological context we shall be working with $\hat{L}_*^{\{\pi\}\subseteq \tilde{K}_1(\mathbb{Z}[\pi])}(\mathbb{Z}[\pi])$, and from now on $\hat{L}_*(\mathbb{Z}[\pi])$ will denote these $\hat{L}$-groups.

Ultraquadratic complexes arise in topology by applying the unstable spectral quadratic construction of Proposition 7.3.2 to a π-map $F:X \longrightarrow \Sigma Y$, to obtain a natural transformation

$$\hat{\psi}_F : \dot{H}_{n+1}(X/\pi) \longrightarrow \hat{Q}_n^{[0,0]}(C(f)) = \hat{Q}_n(C(f)) \quad (n \geqslant 0)$$

with $C(f)$ the algebraic mapping cone of the induced $\mathbb{Z}[\pi]$-module chain map $f : \Omega\dot{C}(X) \longrightarrow \dot{C}(Y)$. We shall call $\hat{\psi}_F$ the <u>ultraquadratic construction</u> on F. If $X = \Sigma X_0$ is the suspension of a π-space X_0 then $\hat{\psi}_F$ is the composite

$$\hat{\psi}_F : \dot{H}_{n+1}(X/\pi) = \dot{H}_n(X_0/\pi) \xrightarrow{\psi_F} \hat{Q}_n(\dot{C}(Y)) \xrightarrow{\hat{e}_\%} \hat{Q}_n(C(f))$$

with ψ_F the unstable quadratic construction on $F:\Sigma X_0 \longrightarrow \Sigma Y$ in the sense of §II.1 and $e:\dot{C}(Y) \longrightarrow C(f)$ the inclusion. For connected Y it is possible to construct $\hat{\psi}_F$ by means of the adjoint π-map $\mathrm{adj}(F) : X_0 \longrightarrow \Omega\Sigma Y$ and the approximation theorem

$$\Omega\Sigma Y = \bigcup_{k\geqslant 1} (\prod_k Y)/\sim$$

due to James [1], with

$$\hat{\psi}_F : \dot{H}_n(X_0/\pi) \xrightarrow{\mathrm{adj}(F)_*} \dot{H}_n(\Omega\Sigma Y/\pi) = \bigoplus_{k=1}^{\infty} \dot{H}_n(\bigwedge_k Y/\pi)$$
$$\xrightarrow{\text{projection}} \dot{H}_n(Y \wedge_\pi Y) = \hat{Q}_n(\dot{C}(Y)) \ .$$

Similarly for disconnected Y of the type $(Y')_+$ for some space with π-action Y', using $\Omega\Sigma Y = \Omega B(\coprod_{k\geqslant 1} (\prod_k Y'))$ and $\dot{H}_n(\Omega\Sigma Y/\pi) = \mathbb{Z}[\mathbb{Z}]\otimes_{\mathbb{Z}[\mathbb{N}]}\bigoplus_{k=1}^{\infty} H_n((\prod_k Y')/\pi)$, i.e. the group completion version of the James construction.

Let X be an n-dimensional geometric Poincaré complex with $(\pi_1(X),w(X)) = (\pi,w)$, and let $\mathbb{Z}[\pi]$ have the w-twisted involution. Given an $s^{\mathbb{Z}[\pi]}$-triangulation

$$(g,c) : W \longrightarrow X \times S^1$$

topologically transverse at $X \times pt. \subset X \times S^1$ there is defined an n-dimensional topological normal map

$$(f,b) = (g,c)| : M = g^{-1}(X \times pt.) \longrightarrow X .$$

We shall call normal maps arising in this way ultranormal. The ultraquadratic kernel of (f,b) is the n-dimensional ultraquadratic Poincaré complex over $\mathbb{Z}[\pi]$

$$\hat{\sigma}_*(f,b) = (C(f^!),\hat{\psi} \in \hat{Q}_n(C(f^!)))$$

refining the quadratic kernel $\sigma_*(f,b) = (C(f^!),\psi \in Q_n(C(f^!)))$, which is defined as follows. Let $\tilde{X}$ be the universal cover of X, let $\tilde{W}$ be the pullback of $\tilde{X} \times S^1$ along g, and let $\tilde{g}:\tilde{W} \longrightarrow \tilde{X} \times S^1$ be a π-equivariant homology equivalence covering g. The embedding $M \times D^1 = g^{-1}(X \times D^1) \subset W$ lifts to a π-equivariant embedding $M \times D^1 \subset W$, where $X \times D^1 \subset X \times S^1$ is a normal D^1-bundle of $X \times pt. \subset X \times S^1$. Applying the Pontrjagin-Thom construction there is obtained a π-map

$$H : \tilde{W}_+ \xrightarrow{\text{collapse}} \tilde{W}/\overline{\tilde{W} - \tilde{M} \times D^1} = \tilde{M} \times D^1/\tilde{M} \times S^0 = \Sigma\tilde{M}_+$$

inducing the $\mathbb{Z}[\pi]$-module chain map

$$h = (0\ f^!) : \Omega C(\tilde{W}) = \Omega C(\tilde{X} \times S^1) = \Omega C(\tilde{X}) \oplus C(\tilde{X}) \longrightarrow C(\tilde{M}) .$$

Define a $\mathbb{Z}[\pi]$-module chain map

$$e = (0\ 1) : C(h) = C(\tilde{X}) \oplus C(f^!) \longrightarrow C(f^!) .$$

The composite

$$H_{n+1}(W) \xrightarrow{\hat{\psi}_H} \hat{Q}_n(C(h)) \xrightarrow{\hat{e}_\%} \hat{Q}_n(C(f^!))$$

sends the fundamental class $[W] \in H_{n+1}(W)$ to the element

$$\hat{\psi} = e_{\%}\hat{\psi}_H([W]) \in \hat{Q}_n(C(f^!))$$

appearing in $\hat{\sigma}_*(f,b)$. The cobordism class

$$\hat{\sigma}_*(f,b) \in \hat{L}_n(\mathbb{Z}[\pi])$$

is the <u>ultraquadratic signature</u> of (f,b).

<u>Proposition 7.8.14</u> Given a (finitely presented) group π there are defined natural isomorphisms of abelian groups

$$\hat{L}_n(\mathbb{Z}[\pi]) \xrightarrow{\sim} \Gamma N_n(\pi\times\mathbb{Z} \longrightarrow \pi) = \Gamma_n(\mathbb{Z}[\pi\times\mathbb{Z}] \longrightarrow \mathbb{Z}[\pi], z) ;$$

$$(C,\hat{\psi}) \longmapsto (C[z,z^{-1}],\psi) \quad (n \geqslant 0)$$

with $z = (1,1) \in \pi\times\mathbb{Z}$, $C[z,z^{-1}] = \mathbb{Z}[\pi\times\mathbb{Z}]\otimes_{\mathbb{Z}[\pi]}C$ and $\psi_0 = \hat{\psi}$, $\psi_s = 0$ $(s \geqslant 1)$.

<u>Proof</u>: By the theory of Matsumoto [1] every element of $\Gamma N_n(\pi\times\mathbb{Z} \longrightarrow \pi)$ (at least for $n \geqslant 5$) is the obstruction $\sigma_*(M,K)$ to extending a codimension 2 spine $K \subset U$ to a codimension 2 spine $(N,K) \subset (M,U)$, for some $(n+2)$-dimensional manifold with boundary $(M,\partial M)$ and codimension 0 submanifold $U \subset \partial M$, such that (M,U) is an n-dimensional geometric Poincaré pair with $\pi_1(M) = \pi$ and $H_n(M,U) \longrightarrow H_n(M,\partial M) = H^2(M)$; $[M] \longmapsto 0$. The associated $s^{\mathbb{Z}[\pi]}$-triangulation of triads

$$(g,c) : (M;U,X;\partial U) \longrightarrow (M\times D^2;U\times D^2,M\times S^1;U\times S^1) \quad (X = \overline{\partial M\setminus U})$$

restricts to an n-dimensional ultranormal map of pairs

$$(f,b) = (g,c)| : (N,K) = g^{-1}((M,U)\times pt.) \longrightarrow (M,U)$$

such that $(\partial f,\partial b) : K \longrightarrow U$ is an s-triangulation of U. The ultraquadratic signature defines the inverse isomorphisms

$$\Gamma N_n(\pi\times\mathbb{Z} \longrightarrow \pi) \longrightarrow \hat{L}_n(\mathbb{Z}[\pi]) ;$$

$$\sigma_*(M,K) = ws_\partial(g,M) \longmapsto \hat{\sigma}_*(f,b) .$$

[]

Let $A[z,z^{-1}]$ be the ring of finite Laurent polynomials $\sum_{j=-\infty}^{\infty} a_j z^j$ $(a_j \in A)$ in a central invertible indetermineate z over a ring with involution A, extending the involution by $\bar{z} = z^{-1}$. The projection

$$A[z,z^{-1}] \longrightarrow A \; ; \; \sum_{j=-\infty}^{\infty} a_j z^j \longmapsto \sum_{j=-\infty}^{\infty} a_j$$

is a morphism of rings with involution. Define the <u>covering</u> of an n-dimensional ε-ultraquadratic Poincaré complex over A $(C,\hat{\psi} \in \hat{Q}_n(C,\varepsilon))$ to be the A-acyclic (n+1)-dimensional ε-quadratic Poincaré complex over $A[z,z^{-1}]$

$$\beta(C,\hat{\psi}) = (D,\theta \in Q_{n+1}(D,\varepsilon))$$

given by

$$d_D = \begin{pmatrix} d_C & (-)^{r-1}(1+T_{\varepsilon z})\hat{\psi} \\ 0 & (-)^{r-1}d_C^* \end{pmatrix}$$

$$: D_r = C_r[z,z^{-1}] \oplus C^{n-r+1}[z,z^{-1}] \longrightarrow D_{r-1} = C_{r-1}[z,z^{-1}] \oplus C^{n-r+2}[z,z^{-1}]$$

$$\theta_0 = \begin{pmatrix} 0 & (-)^{r(n-r)}z \\ (-)^{n-r} & 0 \end{pmatrix}$$

$$: D^{n-r+1} = C^{n-r+1}[z,z^{-1}] \oplus C_r[z,z^{-1}] \longrightarrow D_r = C_r[z,z^{-1}] \oplus C^{n-r+1}[z,z^{-1}]$$

$$\theta_s = 0 \quad (s \geqslant 1) \; ,$$

where $C[z,z^{-1}] = A[z,z^{-1}] \otimes_A C$. If $(C,\hat{\psi})$ is projective (resp. free) then $\beta(C,\hat{\psi})$ is free (resp. simple). If $A = \mathbb{Z}[\pi]$ then $A[z,z^{-1}] = \mathbb{Z}[\pi \times \mathbb{Z}]$.

<u>Proposition 7.8.15</u> Given a (finitely presented) group π there are defined natural isomorphisms of abelian groups

$$\beta : \hat{L}^h_n(\mathbb{Z}[\pi]) \xrightarrow{\sim} \Gamma_{n+2}\left(\begin{array}{ccc} \mathbb{Z}[\pi\times\mathbb{Z}] & \longrightarrow & \mathbb{Z}[\pi\times\mathbb{Z}] \\ \big\downarrow & \Theta & \big\downarrow \\ \mathbb{Z}[\pi\times\mathbb{Z}] & \longrightarrow & \mathbb{Z}[\pi] \end{array}\right) ;$$

$$(C,\hat{\psi}) \longmapsto \beta(C,\hat{\psi}) \qquad (n \geqslant 0) ,$$

where $\hat{L}^h_n(\mathbb{Z}[\pi])$ is the cobordism group of free n-dimensional ultraquadratic Poincaré complexes over $\mathbb{Z}[\pi]$.

<u>Proof</u>: This follows from Proposition 7.8.14 and a 5-lemma argument applied to the natural transformation of exact sequences

$$\begin{array}{ccccc} \ldots \longrightarrow & \hat{H}^{n+1}(\mathbb{Z}_2;Wh(\pi)) & \longrightarrow & \hat{L}_n(\mathbb{Z}[\pi]) & \longrightarrow \\ & \big\downarrow \wr & & \big\downarrow \wr & \\ \ldots \longrightarrow & LN_n(\pi\times\mathbb{Z} \longrightarrow \pi) & \longrightarrow & \Gamma N_n(\pi\times\mathbb{Z} \longrightarrow \pi) & \longrightarrow \end{array}$$

$$\begin{array}{ccccc} \hat{L}^h_n(\mathbb{Z}[\pi]) & \longrightarrow & \hat{H}^n(\mathbb{Z}_2;Wh(\pi)) & \longrightarrow & \ldots \\ \big\downarrow \beta & & \big\downarrow \wr & & \\ \Gamma_{n+2}(\Theta) & \longrightarrow & LN_{n-1}(\pi\times\mathbb{Z} \longrightarrow \pi) & \longrightarrow & \ldots \; . \end{array}$$

[]

In the full account (Ranicki [11]) we shall be obliged to obtain the identifications $\Gamma N_*(\pi\times\mathbb{Z} \longrightarrow \pi) = \hat{L}_*(\mathbb{Z}[\pi])$, $\Gamma_{*+2}(\Theta) = \hat{L}^h_*(\mathbb{Z}[\pi])$ of Propositions 7.8.14,7.8.15 algebraically, using an appropriate Higman linearization trick to replace the codimension 1 transversality.

7.9 The algebraic theory of knot cobordism

We shall now illustrate the various approaches of the algebraic theory of surgery to codimension 2 embeddings by giving various L-theoretic interpretations of the high-dimensional knot cobordism groups C_*, as well as defining some isotopy invariants of knots.

We refer to Kervaire and Weber [1] for a survey of high-dimensional knot theory.

Given a (locally flat) topological knot $k:S^n \subset S^{n+2}$ $(n \geqslant 1)$ let $U = S^n \times D^2 \subset S^{n+2}$ be a closed regular neighbourhood of $k(S^n) = S^n \times 0 \subset S^{n+2}$. The knot complement is the $(n+2)$-dimensional manifold with boundary

$$(X, \partial X) = (\overline{S^{n+2} - U}, S^n \times S^1) \quad ,$$

with X a deformation retract of the actual complement $S^{n+2} - k(S^n)$. The generator $1 \in \pi^1(S^{n+2} - k(S^n)) = \mathbb{Z}$ is represented by an $(n+2)$-dimensional topological normal map of pairs

$$(g,c) \; : \; (X, \partial X) \longrightarrow (D^{n+1} \times S^1, S^n \times S^1)$$

which is a $\mathbb{Z}$-homology equivalence with $g| = \text{id.} : \partial X \longrightarrow S^n \times S^1$, i.e. an $s^{\mathbb{Z}}_{\partial}$-triangulation of $(D^{n+1} \times S^1, S^n \times S^1)$.

The Blanchfield complex of a knot $k:S^n \subset S^{n+2}$ is the $\mathbb{Z}$-acyclic $(n+2)$-dimensional quadratic Poincaré complex over $\mathbb{Z}[z,z^{-1}]$

$$\sigma_*(k) = \sigma_*(g,c)$$

defined by the quadratic kernel of (g,c). The chain complex involved in $\sigma_*(k) = (C(g^!), \psi \in Q_{n+2}(C(g^!)))$ is the algebraic mapping cone $C(g^!)$ of the $\mathbb{Z}[z,z^{-1}]$-module Umkehr chain map

$$g^! \; : \; C(D^{n+1} \times \mathbb{R}) \xrightarrow[\simeq]{([D^{n+1} \times S^1] \cap -)^{-1}} C(D^{n+1} \times \mathbb{R}, S^n \times \mathbb{R})^{n+1-*}$$

$$\xrightarrow{\tilde{g}^*} C(\tilde{X}, \widetilde{\partial X})^{n+2-*} \xrightarrow[\simeq]{([X] \cap -)} C(\tilde{X})$$

with $\widetilde{X}$ the infinite cyclic covering of X. The non-trivial homology kernel $\mathbb{Z}[z,z^{-1}]$-modules of g are the <u>knot modules</u> of k

$$H_*(C(g^!)) = H_*(\widetilde{X}) \quad (* \neq 0) \ .$$

<u>Proposition 7.9.1</u> The homotopy equivalence class of the Blanchfield complex $\sigma_*(k)$ is an isotopy invariant of the knot $k:S^n \subset S^{n+2}$ $(n \geqslant 1)$.

[]

Define a multiplicative subset

$$P = \{\sum_{j=-\infty}^{\infty} p_j z^j \in \mathbb{Z}[z,z^{-1}] \mid \sum_{j=-\infty}^{\infty} p_j = 1 \in \mathbb{Z}\} \subset \mathbb{Z}[z,z^{-1}] \ .$$

<u>Proposition 7.9.2</u> i) The following conditions on a finite-dimensional $\mathbb{Z}[z,z^{-1}]$-module chain complex C are equivalent:

a) C is $\mathbb{Z}$-acyclic, i.e. $H_*(\mathbb{Z}\otimes_{\mathbb{Z}[z,z^{-1}]}C) = 0$,

b) C is P-acyclic, i.e. $pH_*(C) = 0$ for some $p \in P$,

c) $1-z : C \longrightarrow C$ is a $\mathbb{Z}[z,z^{-1}]$-module chain equivalence,

i.e. $1-z : H_*(C) \longrightarrow H_*(C)$ is a an automorphism.

ii) If C is a finite-dimensional $\mathbb{Z}$-acyclic $\mathbb{Z}[z,z^{-1}]$-module chain complex the ε-symmetrization maps in the Q-groups

$$1+T_\varepsilon \ : \ Q_*(C,\varepsilon) \longrightarrow Q^*(C,\varepsilon)$$

are isomorphisms, for any unit $\varepsilon \in \mathbb{Z}[z,z^{-1}]$.

<u>Proof</u>: i) a) $\Leftrightarrow$ c) is immediate from the short exact sequence of $\mathbb{Z}$-module chain complexes

$$0 \longrightarrow C \xrightarrow{1-z} C \longrightarrow \mathbb{Z}\otimes_{\mathbb{Z}[z,z^{-1}]}C \longrightarrow 0 \quad .$$

b) $\Rightarrow$ a) by the factorization of the projection

$$\mathbb{Z}[z,z^{-1}] \longrightarrow P^{-1}\mathbb{Z}[z,z^{-1}] \longrightarrow \mathbb{Z} \ .$$

c) $\Rightarrow$ b) The homology $H_*(C)$ is a f.g. $\mathbb{Z}[z,z^{-1}]$-module such that $1-z:H_*(C) \longrightarrow H_*(C)$ is an automorphism. We now use the argument of Levine [5,Cor.1.3]. Let $x_1,x_2,\ldots,x_m \in H_*(C)$ be a finite set of $\mathbb{Z}[z,z^{-1}]$-module generators, so that $x_i = (1-z)y_i$ for some $y_i = \sum_{j=1}^{m} a_{ij}x_j \in H_*(C)$ $(a_{ij} \in \mathbb{Z}[z,z^{-1}],\ 1 \leqslant i \leqslant m)$. Define an $m \times m$ matrix over $\mathbb{Z}[z,z^{-1}]$ $B = (b_{ij})_{1 \leqslant i,j \leqslant m}$ by

$$b_{ij} = \begin{cases} (z-1)a_{ij} & \text{if } i \neq j \\ 1 + (z-1)a_{ii} & \text{if } i = j \ , \end{cases}$$

so that $\sum_{j=1}^{m} b_{ij}x_j = 0 \in H_*(C)$ $(1 \leqslant i \leqslant m)$. Now $p = \det(B) \in P$ and there exists an $m \times m$ matrix over $\mathbb{Z}[z,z^{-1}]$ $B' = (b'_{ij})_{1 \leqslant i,j \leqslant m}$ such that $B'B = pI$ (with $B^{-1} = p^{-1}B'$ over $P^{-1}\mathbb{Z}[z,z^{-1}]$), so that

$$px_i = \sum_{j=1}^{m} b'_{ij}\left(\sum_{k=1}^{m} b_{jk}x_k\right) = 0 \in H_*(C) \quad (1 \leqslant i \leqslant m)$$

and $pH_*(C) = 0$.

(This is the special case $\pi' = \mathbb{Z}$, $\pi = \{1\}$ of the result of Smith [1,Prop.2.3] that if $\pi' \longrightarrow \pi$ is a surjection of groups such that

$$P = \{p \in \mathbb{Z}[\pi'] \mid p - 1 \in \ker(\mathbb{Z}[\pi'] \longrightarrow \mathbb{Z}[\pi])\} \subset \mathbb{Z}[\pi']$$

is a multiplicative subset then a finite-dimensional $\mathbb{Z}[\pi']$-module chain complex C is $\mathbb{Z}[\pi]$-acyclic if and only if it is P-acyclic. If the surjection is part of data

$$(\Pi : \mathbb{Z} \xrightarrow{\ i\ } \pi' \longrightarrow \pi \longrightarrow \{1\}, \omega : \pi \longrightarrow \mathbb{Z}_2)$$

with $\omega = +1$ and $i(1) = z \in \pi'$ then Proposition 7.8.7 iii) shows that a finite-dimensional $\mathbb{Z}[\pi']$-module chain complex C is $\mathbb{Z}[\pi]$-acyclic if and only if $1-z : H_*(C) \longrightarrow H_*(C)$ is a $\mathbb{Z}[\pi']$-module isomorphism. Note however that the result of Smith [1,Cor.3] is false: if $\ker(\pi' \longrightarrow \pi) = \mathbb{Z}$ and

$$P_0 = \{\sum_{j=-\infty}^{\infty} p_j z^j \mid \sum_{j=-\infty}^{\infty} p_j = 1,\ p_j \in \mathbb{Z}\} \subsetneq P \subset \mathbb{Z}[\pi']$$

it is not in general true that a finite-dimensional $\mathbb{Z}[\pi]$-acyclic $\mathbb{Z}[\pi']$-module chain complex C is P_0-acyclic. The error in the proof arises in assuming that if M is a f.g. $\mathbb{Z}[\pi']$-module such that $1-z:M \longrightarrow M$ is an automorphism and M_0 is the f.g. $\mathbb{Z}[z,z^{-1}]$-submodule of M generated by a finite set of $\mathbb{Z}[\pi']$-module generators then the restriction $1-z:M_0 \longrightarrow M_0$ is also an automorphism).

ii) By Proposition 3.2.1 i) an (n+1)-dimensional P-acyclic $\mathbb{Z}[z,z^{-1}]$-module chain complex C is the resolution of an n-dimensional $(\mathbb{Z}[z,z^{-1}],P)$-module chain complex D, with $H_*(C) = H_*(D)$ and $1-z:D \longrightarrow D$ an automorphism by i). By the exact sequence of Vogel [2,2.4] (cf. the discussion in §3.1)

$$\ldots \longrightarrow \bar{Q}_n^P(D,\varepsilon) \longrightarrow Q_n^P(D,\varepsilon) \longrightarrow Q_{n+1}(C,-\varepsilon) \longrightarrow \bar{Q}_{n-1}^P(D,\varepsilon) \longrightarrow \ldots ,$$

its ε-symmetric analogue

$$\ldots \longrightarrow \bar{Q}_P^n(D,\varepsilon) \longrightarrow Q_P^n(D,\varepsilon) \longrightarrow Q^{n+1}(C,-\varepsilon) \longrightarrow \bar{Q}_P^{n-1}(D,\varepsilon) \longrightarrow \ldots$$

and a 5-lemma argument it suffices to show that the ε-symmetrization maps in the Q-groups of D

$$\begin{cases} 1+T_\varepsilon \ : \ \bar{Q}_*^P(D,\varepsilon) \longrightarrow \bar{Q}_P^*(D,\varepsilon) \\ 1+T_\varepsilon \ : \ Q_*^P(D,\varepsilon) \longrightarrow Q_P^*(D,\varepsilon) \end{cases}$$

are isomorphisms. The automorphism

$$u = (1-z)^{-1} : D \longrightarrow D$$

is such that $\begin{cases} \phi = u\phi + \phi u^\wedge \\ \phi = (1\otimes u)\phi + (u\otimes 1)\phi \end{cases}$ for any $\begin{cases} \phi \in \mathrm{Hom}_{\mathbb{Z}[z,z^{-1}]}(D^\wedge,D) \\ \phi \in D\otimes_{\mathbb{Z}[z,z^{-1}]}D \end{cases}$,

since $U = (1-z)^{-1} \in \mathbb{Z}[z,z^{-1},(1-z)^{-1}]$ is such that $U + \bar{U} = 1$.

Use $u:D \longrightarrow D$ to define isomorphisms inverse to the ε-symmetrization maps

$$\begin{cases} \bar{Q}^*_P(D,\varepsilon) = H_*(\mathrm{Hom}_{\mathbb{Z}[\mathbb{Z}_2]}(W,\mathrm{Hom}_{\mathbb{Z}[z,z^{-1}]}(D^\wedge,D))) \\ \qquad \longrightarrow \bar{Q}^P_*(D,\varepsilon) = H_*(W\otimes_{\mathbb{Z}[\mathbb{Z}_2]}\mathrm{Hom}_{\mathbb{Z}[z,z^{-1}]}(D^\wedge,D)) \ ; \\ \qquad \phi = \{\phi_s \mid s \geqslant 0\} \longmapsto \psi = \{\psi_s = \begin{cases} u\phi_0 & \text{if } s=0 \\ 0 & \text{otherwise} \end{cases}\} \\ Q^*_P(D,\varepsilon) = H_*(\mathrm{Hom}_{\mathbb{Z}[\mathbb{Z}_2]}(W,D\otimes_{\mathbb{Z}[z,z^{-1}]}D)) \\ \qquad \longrightarrow Q^P_*(D,\varepsilon) = H_*(W\otimes_{\mathbb{Z}[\mathbb{Z}_2]}(D\otimes_{\mathbb{Z}[z,z^{-1}]}D)) \ ; \\ \qquad \phi = \{\phi_s \mid s \geqslant 0\} \longmapsto \psi = \{\psi_s = \begin{cases} (1\otimes u)\phi_0 & \text{if } s=0 \\ 0 & \text{otherwise} \end{cases}\} \ . \end{cases}$$

(The isomorphism $1+T_\varepsilon : Q_*(C,\varepsilon) \xrightarrow{\ \sim\ } Q^*(C,\varepsilon)$ is a generalization of the result of Levine [5,Prop.12.3] that the symmetric Blanchfield pairing on a knot module admits a quadratic refinement. The use of the automorphism $u=(1-z)^{-1}$ was suggested by Neal Stoltzfus).

[]

The <u>Alexander polynomial</u> $p \in P$ of a finite-dimensional $\mathbb{Z}$-acyclic $\mathbb{Z}[z,z^{-1}]$-module chain complex C is the generator (unique up to unit) of the maximal principal ideal contained in the order ideal $\{s \in \mathbb{Z}[z,z^{-1}] \mid sH_*(C) = 0\} \triangleleft \mathbb{Z}[z,z^{-1}]$. Thus if $M(\lambda)$ denotes the cyclic $\mathbb{Q}[z,z^{-1}]$-module of order $\lambda \in \mathbb{Z}[z,z^{-1}]$ and $\mathbb{Q}\otimes_{\mathbb{Z}}H_*(C) = \bigoplus_\lambda M(\lambda)$ is the decomposition of the induced $\mathbb{Q}[z,z^{-1}]$-module $\mathbb{Q}\otimes_{\mathbb{Z}}H_*(C)$ as a direct sum of irreducible cyclic modules then $p \in P$ is the lowest common multiple (l.c.m.) of the polynomials $\lambda \in \mathbb{Z}[z,z^{-1}]$.

The Alexander polynomial $p \in P$ of a knot $k:S^n \subset S^{n+2}$ $(n \geqslant 1)$ is the Alexander polynomial of the $(n+2)$-dimensional P-acyclic $\mathbb{Z}[z,z^{-1}]$-module chain complex $C(g^!)$. For $n = 1$ this is just the polynomial originally defined by Alexander [1]. For $n \geqslant 1$ it is the l.c.m. of the knot polynomials defined by Levine [1]. The linking pairing of the Blanchfield complex $\sigma_*(k) = (C(g^!),\psi)$

$$(1+T)\psi_0^P : H^r(\tilde{X}) \times H^{n-r+3}(\tilde{X}) \longrightarrow P^{-1}\mathbb{Z}[z,z^{-1}]/\mathbb{Z}[z,z^{-1}] \ ;$$

$$(x,y) \longmapsto \frac{1}{p}(1+T)\psi_0(x)(w)$$

$$(r \neq 0, n+3 \ \ x \in C(\tilde{X})^r, \ y \in C(\tilde{X})^{n-r+3}, \ w \in C(\tilde{X})^{n-r+2}, \ d^*w = py)$$

agrees via the Poincaré duality $H^{n+2-*}(\tilde{X}) = H_*(\tilde{X})$ $(* \neq n,n+2)$ with the pairing originally defined by Blanchfield [1]

$$H_{n+2-r}(\tilde{X}) \times H_{r-1}(\tilde{X}) \longrightarrow P^{-1}\mathbb{Z}[z,z^{-1}]/\mathbb{Z}[z,z^{-1}]$$

using geometric linking numbers of homology classes. The knot module parings have been studied more recently by Levine [5].

A Seifert surface for a knot $k:S^n \subset S^{n+2}$ $(n \geqslant 1)$ is a codimension 1 framed submanifold $M^{n+1} \subset S^{n+2}$ with boundary $\partial M = k(S^n)$. Given a knot $k:S^n \subset S^{n+2}$ make the $s_\partial^{\mathbb{Z}}$-triangulation $(g,c):(X,\partial X) \longrightarrow (D^{n+1} \times S^1, S^n \times S^1)$ topologically transverse at $(D^{n+1},S^n) \times pt. \subset (D^{n+1},S^n) \times S^1$, thus obtaining a Seifert surface $M = g^{-1}(D^{n+1} \times pt.) \subset X \subset S^{n+2}$ (with a collar removed) together with an $(n+1)$-dimensional ultranormal map

$$(f,b) = (g,c)| \ : \ (M,\partial M) = g^{-1}((D^{n+1},S^n) \times pt.) \longrightarrow (D^{n+1},S^n)$$

such that $f| = id. : \partial M = k(S^n) \longrightarrow S^n$. Conversely, every Seifert surface M determines an ultranormal map $(f,b):(M,\partial M) \longrightarrow (D^{n+1},S^n)$, by the method recalled in the proof of Proposition 7.9.3 below. In the original work of

Seifert [1] M was obtained for $k:S^1 \subset S^3$ using the knot projection.

The <u>Seifert complex</u> $\hat{\sigma}_*(k,M)$ of a pair (k,M) consisting of a knot $k:S^n \subset S^{n+2}$ $(n \geqslant 1)$ and a Seifert surface $M^{n+1} \subset S^{n+2}$ for k is the (n+1)-dimensional ultraquadratic Poincaré complex over $\mathbb{Z}$

$$\hat{\sigma}_*(k,M) = \hat{\sigma}_*(f,b)$$

defined by the ultraquadratic kernel of the associated ultranormal map $(f,b):(M,\partial M) \longrightarrow (D^{n+1},S^n)$. The chain complex appearing in $\hat{\sigma}_*(f,b) = (C(f^!),\hat{\psi} \in \hat{Q}_{n+1}(C(f^!)))$ is the algebraic mapping cone $C(f^!)$ of the $\mathbb{Z}$-module Umkehr chain map

$$f^! : C(D^{n+1}) \xrightarrow[\simeq]{([D^{n+1}] \cap -)^{-1}} C(D^{n+1},S^n)^{n+1-*}$$

$$\xrightarrow{f^*} C(M,\partial M)^{n+1-*} \xrightarrow[\simeq]{([M] \cap -)} C(M) ,$$

so that there is an identification

$$C(f^!) = \dot{C}(M) .$$

Identifying $\dot{C}(M) = \dot{C}(M)^{n+1-*}$ by the Poincaré-Lefschetz duality of the (n+2)-dimensional manifold triad $(M^{n+1};k(D^n_+),k(D^n_-);k(S^{n-1}))$, and $\dot{C}(S^{n+2}-M) = \dot{C}(M)$ by Alexander duality, note that the ultraquadratic structure $\hat{\psi} \in \hat{Q}_{n+1}(C(f^!)) = H_{n+1}(\mathrm{Hom}_{\mathbb{Z}}(C(f^!)^*,C(f^!)))$ can be identified with the chain homotopy class of the $\mathbb{Z}$-module chain map

$$\hat{\psi} : C(f^!)^{n+1-*} = \dot{C}(M)^{n+1-*} = \dot{C}(M) \xrightarrow{\nu_+} \dot{C}(S^{n+2}-M) = \dot{C}(M)$$

induced by the map $\nu_+:M \longrightarrow S^{n+2}-M \,;\, (x,0) \longmapsto (x,1)$ pushing $M = M \times 0 \subset S^{n+2}$ off itself along the positive normal direction determined by the framing of the normal bundle $\nu_{M \subset S^{n+2}}$, with $E(\nu_{M \subset S^{n+1}}) = M \times [-1,1] \subset S^{n+2}$. In particular, for $n = 2i-1$ the pairing

$$\hat{\psi} : H_i(M)/\text{torsion} \times H_i(M)/\text{torsion} \longrightarrow \mathbb{Z}$$

is the usual Seifert form of $(k:S^{2i-1} \subset S^{2i+1}, M^{2i})$.

The Blanchfield and Seifert complexes of a knot are related by the covering operation of §7.8

$$\beta : \{(n+1)\text{-dimensional ultraquadratic Poincaré complexes over } \mathbb{Z}\}$$
$$\longrightarrow \{\mathbb{Z}\text{-acyclic } (n+2)\text{-dimensional quadratic Poincaré complexes over } \mathbb{Z}[z,z^{-1}]\} \quad .$$

Define $(n+1)$-dimensional ultraquadratic Poincaré complexes over $\mathbb{Z}$ $(C,\hat{\psi})$, $(C',\hat{\psi}')$ to be <u>S-equivalent</u> if their coverings $\beta(C,\hat{\psi})$, $\beta(C',\hat{\psi}')$ are homotopy equivalent. S-equivalence is an equivalence relation such that

$$\text{homotopy equivalence} \Longrightarrow \text{S-equivalence} \Longrightarrow \text{cobordism} \ .$$

We shall relate this notion of S-equivalence with the usual S-equivalence of Seifert matrices further below.

<u>Proposition 7.9.3</u> Let $k:S^n \subset S^{n+2}$ $(n \geqslant 1)$ be a knot, and let $M^{n+1} \subset S^{n+2}$ be a Seifert surface for k. Then

i) $\beta\hat{\sigma}_*(k,M) = \sigma_*(k)$, up to homotopy equivalence

ii) the S-equivalence class of the Seifert complex $\hat{\sigma}_*(k,M)$ is an isotopy invariant of k, namely the homotopy equivalence class of $\sigma_*(k)$.

<u>Proof</u>: There is a standard way of constructing the infinite cyclic covering $\tilde{X}$ of the knot complement X from a Seifert surface M : cut X along M to obtain an $(n+2)$-dimensional manifold triad $(N^{n+2};M,zM;k(S^n))$ involving a copy zM of M, and set

$$\tilde{X} = \sum_{j=-\infty}^{\infty} z^j N \quad .$$

Accordingly, the $(n+2)$-dimensional topological normal map

$(g,c):(X,\partial X) \longrightarrow (D^{n+1} \times S^1, S^n \times S^1)$ used to define $\sigma_*(k) = \sigma_*(g,c)$ may be constructed from the $(n+1)$-dimensional ultranormal map $(f,b):(M,\partial M) \longrightarrow (D^{n+1}, S^n)$ used to define $\hat{\sigma}_*(k,M) = \hat{\sigma}_*(f,b)$: glue together $\mathbb{Z}$ copies of an $(n+2)$-dimensional topological normal map of triads

$$(G,C) : (N;M,zM;k(S^n)) \longrightarrow (D^{n+1} \times I; D^{n+1} \times 0, D^{n+1} \times 1; S^n \times I)$$

and quotient out the free $\mathbb{Z}$-action to obtain $(G,C)/\mathbb{Z} = (g,c)$. Passing to algebra it follows that $\sigma_*(k)$ may be constructed from $\hat{\sigma}_*(k,M)$ in the same way, using the algebraic glueing operation of §1.7, which in this case gives the covering operation $\beta:\hat{\sigma}_*(k,M) \longmapsto \beta\hat{\sigma}_*(k,M) = \sigma_*(k)$.

[]

An $\left\{\begin{array}{l}\text{odd-}\\ \text{even-}\end{array}\right.$ dimensional knot $\left\{\begin{array}{l}k:S^{2i-1} \subset S^{2i+1}\\ k:S^{2i} \subset S^{2i+2}\end{array}\right.$ $(i \geqslant 1)$ is <u>simple</u> if $\pi_r(X) = \pi_r(S^1)$ for $r < i$, that is if the $s^{\mathbb{Z}}_{\partial}$-triangulation $(g,c):(X,\partial X) \longrightarrow (D^{n+1} \times S^1, S^n \times S^1)$ is $(i-1)$-connected. The Blanchfield complex of a simple knot k is the i-fold skew-suspension

$$\sigma_*(k) = \bar{S}^i \sigma_i(k)$$

of a P-acyclic $\left\{\begin{array}{l}1-\\ 2-\end{array}\right.$ dimensional $(-)^i$-quadratic Poincaré complex over $\mathbb{Z}[z,z^{-1}]$ $\sigma_i(k)$.

The <u>Blanchfield linking</u> $\left\{\begin{array}{l}\underline{\text{form}}\\ \underline{\text{formation}}\end{array}\right.$ of a simple knot $\left\{\begin{array}{l}k:S^{2i-1} \subset S^{2i+1}\\ k:S^{2i} \subset S^{2i+2}\end{array}\right.$ is the non-singular $(-)^{i+1}$-symmetric linking

$\begin{cases} \text{form} \\ \text{formation} \end{cases}$ over $(\mathbb{Z}[z,z^{-1}],P)$ associated to $(1+T_{(-)}i)\sigma_i(k)$ by Proposition $\begin{cases} 3.4.1 \\ 3.5.2 \end{cases}$. There is no loss of structure in passing from $\sigma_i(k)$ to $(1+T_{(-)}i)\sigma_i(k)$, since it follows from Proposition 7.9.2 ii) that there are natural identifications of categories

$$(\varepsilon\text{-quadratic linking} \begin{cases} \text{forms} \\ \text{formations} \end{cases} \text{over } (\mathbb{Z}[z,z^{-1}],P))$$

$$= (\varepsilon\text{-symmetric linking} \begin{cases} \text{forms} \\ \text{formations} \end{cases} \text{over } (\mathbb{Z}[z,z^{-1}],P))$$

$$(\varepsilon = \pm 1) \ .$$

As a special case of Proposition 7.9.1 we have that the $\begin{cases} \text{isomorphism} \\ \text{stable equivalence} \end{cases}$ class of the Blanchfield linking $\begin{cases} \text{form} \\ \text{formation} \end{cases}$ is an isotopy invariant of a simple knot. (Indeed, the linking formation is only defined up to stable equivalence).

A Seifert surface $\begin{cases} M^{2i} \subset S^{2i+1} \\ M^{2i+1} \subset S^{2i+2} \end{cases}$ of an $\begin{cases} \text{odd-} \\ \text{even-} \end{cases}$ dimensional knot $\begin{cases} k:S^{2i-1} \subset S^{2i+1} \\ k:S^{2i} \subset S^{2i+2} \end{cases}$ $(i \geqslant 1)$ is <u>simple</u> if M is (i-1)-connected, that is $\pi_r(M) = 0$ for $r < i$. A knot is simple if and only if it admits a simple Seifert surface. If M is a simple Seifert surface the Seifert complex is the i-fold skew-suspension

$$\hat{\sigma}_*(k,M) = \bar{S}^i\hat{\sigma}_i(k,M)$$

of a $\begin{cases} 0\text{-} \\ 1\text{-} \end{cases}$ dimensional $(-)^i$-ultraquadratic Poincaré complex over $\mathbb{Z}$ $\hat{\sigma}_i(k,M)$.

The <u>Seifert</u> $\left\{\begin{matrix}\underline{\text{form}}\\ \underline{\text{formation}}\end{matrix}\right.$ of a simple (knot, Seifert surface) pair $\left\{\begin{matrix}(k:S^{2i-1}\subset S^{2i+1},M^{2i})\\ (k:S^{2i}\subset S^{2i+2},M^{2i+1})\end{matrix}\right.$ $(i\geqslant 1)$ is the non-singular $(-)^i$-ultraquadratic $\left\{\begin{matrix}\text{form}\\ \text{formation}\end{matrix}\right.$ over $\mathbb{Z}$ $\left\{\begin{matrix}(Q,\hat{\psi}\in \mathrm{Hom}_{\mathbb{Z}}(Q,Q^*))\\ (Q,\hat{\psi};F,G)\end{matrix}\right.$ associated to $\hat{\sigma}_i(k,M)$, with $\left\{\begin{matrix}Q = H_i(M)\\ G = C_{i+1}(M)\xrightarrow{d} Q/F = C_i(M)\end{matrix}\right.$.

This is the Seifert $\left\{\begin{matrix}\text{form}\\ \text{formation}\end{matrix}\right.$ used by $\left\{\begin{matrix}\text{Levine [4]}\\ \text{Kearton [3]}\end{matrix}\right.$ in the isotopy classification of simple $\left\{\begin{matrix}\text{odd-}\\ \text{even-}\end{matrix}\right.$ dimensional knots $(i\geqslant 3)$.

The Seifert form was originally defined by Seifert [2] for classical knots $k:S^1\subset S^3$. An "ε-form" $(Q,\hat{\psi};F,G;\phi)$ in the sense of Kearton [3] is a non-singular ε-ultraquadratic formation over $\mathbb{Z}$ $(Q,\hat{\psi};F,G)$ together with an exact sequence of abelian groups

$$0\longrightarrow (Q/(F+G))\otimes\mathbb{Z}_2\xrightarrow{\ g\ }\pi\xrightarrow{\ h\ } F\cap G\longrightarrow 0$$

and a bilinear pairing $\phi:\pi\times\pi\longrightarrow\mathbb{Z}_2$ such that

$$\phi(a,gb) = \hat{\psi}(ha,b)\ ,\ \phi(gb,a) = \hat{\psi}(b,ha)\in\mathbb{Z}\quad (a,b\in Q)\ .$$

The $(-)^i$-form associated to a simple pair $(k:S^{2i}\subset S^{2i+2},M^{2i+1})$ consists of the Seifert formation $(Q,\hat{\psi};F,G)$ and the homotopy theoretic analogue on $\pi = \pi_{i+1}(M)$ of the Seifert pairing

$$\begin{aligned}\phi:\pi_{i+1}(M)\times\pi_{i+1}(M)&\xrightarrow{1\times\nu_+}\pi_{i+1}(M)\times\pi_{i+1}(S^{2i+2}-M) = \pi_{i+1}(M)\times\pi^i(M)\\ &\xrightarrow{\text{composition}}\pi_{i+1}(S^i) = \mathbb{Z}_2\quad (i\geqslant 3)\end{aligned}$$

with $h:\pi=\pi_{i+1}(M)\longrightarrow F\cap G = H_{i+1}(M)$ the Hurewicz map.

For a simple odd-dimensional knot $k:S^{2i-1}\subset S^{2i+1}$ $(i\geq 1)$ Proposition 7.9.3 i) reiterates the well-known relationship between the Blanchfield linking form

$$(1+T_{(-)^i})\sigma_i(k)$$
$$= (H_i(\widetilde{X}),\theta:H_i(\widetilde{X})\times H_i(\widetilde{X})\longrightarrow P^{-1}\mathbb{Z}[z,z^{-1}]/\mathbb{Z}[z,z^{-1}])$$

and the Seifert form of a simple Seifert surface $M^{2i}\subset S^{2i+1}$ for k

$$\hat{\sigma}_i(k,M) = (H_i(M),\hat{\psi}:H_i(M)\times H_i(M)\longrightarrow \mathbb{Z})$$

with

$$H_i(\widetilde{X}) = \mathrm{coker}(\hat{\psi}+(-)^i z\hat{\psi}^*:H_i(M)[z,z^{-1}]\longrightarrow H_i(M)^*[z,z^{-1}])$$

$$\theta : H_i(\widetilde{X})\times H_i(\widetilde{X}) \longrightarrow P^{-1}\mathbb{Z}[z,z^{-1}]/\mathbb{Z}[z,z^{-1}] \ ;$$
$$(x,y)\longmapsto \frac{(1-z)}{p}x(w)$$

$$(x,y\in H_i(M)^*[z,z^{-1}],\ w\in H_i(M)[z,z^{-1}],\ p\in P$$
$$(\hat{\psi}+(-)^i z\hat{\psi}^*)(w) = py\in H_i(M)^*[z,z^{-1}]) \ .$$

A <u>Seifert matrix of type ε</u> $(\varepsilon=\pm 1)$ is a square matrix V with entries in $\mathbb{Z}$ such that $V+\varepsilon V'$ is invertible, where V' is the transpose of V $(v'_{ij} = v_{ji})$. There is an evident one-one correspondence between such matrices and non-singular ε-ultraquadratic forms over $\mathbb{Z}$ $(Q,\hat{\psi})$ with a choice of base for the f.g. free $\mathbb{Z}$-module Q. Trotter [1] and Murasugi [1] introduced the S-equivalence relation on Seifert matrices of type $\varepsilon = -1$, using congruences and elementary enlargements, corresponding to elementary ambient surgeries on a Seifert surface $M^2\subset S^3$ of a knot $k:S^1\subset S^3$. Levine [4] extended this to $\varepsilon = +1$, and used the results of Kervaire [1] on the classification of high-dimensional knots to identify (isotopy classes of simple odd-dimensional knots $k:S^{2i-1} \subset S^{2i+1}$)

= (S-equivalence classes of Seifert matrices of type $\varepsilon = (-)^i$) $(i \geqslant 3)$.

Trotter [2],[3] (algebraically) and Kearton [1] (geometrically) then used the Blanchfield linking form to identify

(S-equivalence classes of Seifert matrices of type $\varepsilon = (-)^i$)

= (isomorphism classes of non-singular $(-)^{i+1}$-symmetric linking forms over $(\mathbb{Z}[z,z^{-1}],P)$) .

Thus our notion of S-equivalence for 0-dimensional ε-ultraquadratic Poincaré complexes over $\mathbb{Z}$ is the same as the S-equivalence of Seifert matrices of type ε. Kearton [3] used elementary operations to define a T-equivalence relation on ε-forms (= Seifert formations with a homotopy pairing) and used the results of Kervaire [2] and Levine [4] to identify

(isotopy classes of simple even-dimensional knots $k:S^{2i} \subset S^{2i+2}$ such that $\pi_i(X)$ has no 2-torsion)

= (T-equivalence classes of $(-)^i$-forms $(Q,\hat{\psi};F,G;\phi)$ such that $Q/(F+G)$ has no 2-torsion) $(i \geqslant 3)$.

(See Kearton [4] and Richter [1] for some preliminary results expressing this set in terms of Blanchfield linking formations with a homotopy pairing). In the full account of codimension 1 splitting theorems in Ranicki [11] there will also be included an ultraquadratic version, in particular expressing the S-equivalence relation on Seifert complexes in terms of elementary operations, and using the covering operation β to identify

(S-equivalence classes of n-dimensional ε-ultraquadratic Poincaré complexes over $\mathbb{Z}$)

= (homotopy equivalence classes of $\mathbb{Z}$-acyclic (n+1)-dimensional ε-quadratic Poincaré complexes over $\mathbb{Z}[z,z^{-1}]$) $(n \geqslant 0, \varepsilon = \pm 1)$.

For $n = 1$ this will identify the part of the T-equivalence relation concerning the Seifert formation $(Q,\hat{\psi};F,G)$ with the S-equivalence relation defined above. Farber [1],[2] has extended the classification of high-dimensional simple knots in terms of stable homotopy theory to the metastable range, identifying the isotopy classes of knots $k:S^n \subset S^{n+2}$ such that $\pi_r(X) = \pi_r(S^1)$ $(r \leqslant \frac{1}{3}(n+1), n \geqslant 5)$ with "R-equivalence" classes of homotopy Seifert pairings. As for T-equivalence, the chain level part of R-equivalence is the S-equivalence of ultraquadratic Poincaré complexes over $\mathbb{Z}$. In particular, Farber [2] completed the classification due to Kearton [3] of simple even-dimensional knots $k:S^{2i} \subset S^{2i+2}$ $(i \geqslant 3)$ in terms of stable algebra, including the case when $\pi_i(X)$ has 2-torsion.

A knot $k:S^n \subset S^{n+2}$ $(n \geqslant 1)$ is <u>fibred</u> if the canonical map $g:S^{n+2}-k(S^n) \longrightarrow S^1$ is a fibre bundle, in which case the closure of the fibre is a Seifert surface $M^{n+1} \subset S^{n+2}$ for k. The corresponding Seifert surface $\hat{\sigma}_*(k,M) = (\dot{C}(M),\hat{\psi} \in \hat{Q}_{n+1}(\dot{C}(M)))$ is such that

$$\hat{\psi} : \dot{C}(M)^{n+1-*} \longrightarrow \dot{C}(M)$$

is a chain homotopy class of $\mathbb{Z}$-module chain equivalences (the <u>monodromy</u> of k). Simple fibred knots are of interest in the study of algebraic singularities, cf. Milnor [2]. Odd-dimensional simple fibred knots have been classified by Durfee [1] in terms of non-singular Seifert matrices. Kojima [1] has obtained a partial classification of even-dimensional simple fibred knots in terms of Seifert formations with a homotopy pairing (the same as the one of Kearton [3]).

Following the work of Fox and Milnor [1] on C_1 Kervaire [1] defined the cobordism groups C_n of knots $k:S^n \subset S^{n+2}$ $(n \geqslant 1)$.

Proposition 7.9.4 The high-dimensional knot cobordism groups C_n $(n \geqslant 4)$ have natural identifications

$$\text{i)}\quad C_n = \hat{L}_{n+1}(\mathbb{Z})$$

$$\text{ii)}\quad C_n = \Gamma_{n+3}\begin{pmatrix} \mathbb{Z}[z,z^{-1}] & \longrightarrow & \mathbb{Z}[z,z^{-1}] \\ \big\downarrow & \Theta & \big\downarrow \\ \mathbb{Z}[z,z^{-1}] & \longrightarrow & \mathbb{Z} \end{pmatrix}$$

$$\text{iii)}\quad C_n = \Gamma_{n+1}(\mathbb{Z}[z,z^{-1}] \longrightarrow \mathbb{Z}, z)$$

$$\text{iv)}\quad C_n = \Gamma_{n+1}(\Theta, z)$$

$$\text{v)}\quad C_n = \Gamma N_{n+1}(\mathbb{Z} \longrightarrow \{1\})$$

$$\text{vi)}\quad C_n = L_{n+3}(\mathbb{Z}[z,z^{-1}], P)$$

$$\text{vii)}\quad C_n = \mathcal{L}_{n+3}(S^1;\mathbb{Z}) \quad .$$

Proof: i) The ultranormal maps $(f,b):(M,\partial M) \longrightarrow (D^{n+1},S^n)$ associated to the various Seifert surfaces $M^{n+1} \subset S^{n+2}$ of a knot $k:S^n \subset S^{n+2}$ are ultranormal bordant. More generally, the ultranormal maps associated to Seifert surfaces of cobordant knots are ultranormal bordant. Thus the Seifert complexes of cobordant knots are cobordant, and the ultraquadratic signature defines abelian group morphisms

$$\hat{\sigma}_* : C_n \longrightarrow \hat{L}_{n+1}(\mathbb{Z}) \ ; \ (k:S^n \subset S^{n+2}) \longmapsto \hat{\sigma}_*(k,M) \quad (n \geqslant 1) \ .$$

Kervaire [1] showed that $C_{2i} = 0$ $(i \geqslant 2)$. It follows from Proposition 1.6.5 iii) that the forgetful maps

$$\hat{L}_{2i+1}(A,\varepsilon) \longrightarrow L_{2i+1}(A,\varepsilon)$$

are one-one, for any ring with involution A. In particular, $\hat{L}_{2i+1}(\mathbb{Z}) \subseteq L_{2i+1}(\mathbb{Z}) = 0$, so that $\hat{L}_{2i+1}(\mathbb{Z}) = 0 = C_{2i}$ $(i \geqslant 2)$.

The odd-dimensional knot cobordism groups C_{2i-1} ($i \geqslant 3$) were identified by Levine [2] with the Witt groups of non-singular $(-)^i$-ultraquadratic forms over $\mathbb{Z}$ (i.e. Seifert forms), so that $C_{2i-1} = \hat{L}_0(\mathbb{Z},(-)^i) = \hat{L}_{2i}(\mathbb{Z})$ and $\hat{\sigma}_*$ is an isomorphism in this case also.

ii) The identification $C_n = \Gamma_{n+3}(\Theta)$ was first obtained geometrically by Cappell and Shaneson [1], as a special case of their theory of "semi-local knots".

iii) The identification $C_n = \Gamma_{n+1}(\mathbb{Z}[z,z^{-1}] \longrightarrow \mathbb{Z},z)$ was obtained by Matsumoto [1],[2],[3] both geometrically and algebraically. The cobordism class of a knot $k:S^n \subset S^{n+2}$ is identified with the obstruction to extending the inclusion $k(S^n) \subset U$ of $k(S^n)$ in a closed regular neighbourhood $U = S^n \times D^2 \subset S^{n+2}$ to a codimension 2 spine $(M,k(S^n)) \subset (D^{n+3},U)$.

iv) Immediate from iii) and the exact sequence

$$\ldots \longrightarrow L_n(\mathbb{Z}[z,z^{-1}],z) \longrightarrow \Gamma_n(\mathbb{Z}[z,z^{-1}] \longrightarrow \mathbb{Z},z)$$
$$\longrightarrow \Gamma_n(\Theta,z) \longrightarrow L_{n-1}(\mathbb{Z}[z,z^{-1}],z) \longrightarrow \ldots ,$$

since $L_*(\mathbb{Z}[z,z^{-1}],z) = 0$.

v) Immediate from iii) and Proposition 7.8.12.

The cobordism class of a knot $k:S^n \subset S^{n+2}$ is interpreted as the rel∂ obstruction $ws_\partial(h,Y) \in \Gamma N_{n+1}(\mathbb{Z} \longrightarrow \{1\})$ to a weak codimension 2 splitting of an $s^{\mathbb{Z}}$-triangulation of the $(n+3)$-dimensional geometric Poincaré triad $(D^{n+3},U) \times (D^2,S^1)$

$$h : (D^{n+3};U,X;\partial X) \longrightarrow (D^{n+3} \times D^2;U \times D^2,D^{n+3} \times S^1;U \times S^1)$$

along $(Y,\partial Y) = (D^{n+3},U) \times \mathrm{pt.} \subset (D^{n+3} \times D^2,U \times D^2)$.

vi) The quadratic signature of the Blanchfield complex defines abelian group morphisms

$$\sigma_* : C_n \longrightarrow L_{n+3}(\mathbb{Z}[z,z^{-1}],P) ;$$
$$(k:S^n \subset S^{n+2}) \longmapsto \sigma_*(k) \quad (n \geqslant 1) .$$

The expression of the odd-dimensional knot-cobordism groups C_{2i-1} $(i \geqslant 3)$ as the Witt groups of non-singular $(-)^{i+1}$-quadratic linking forms over $(\mathbb{Z}[z,z^{-1}],P)$ (i.e. Blanchfield forms) is due to Kearton [2]. The actual identification $C_n = L_{n+3}(\mathbb{Z}[z,z^{-1}],P)$ is due to Pardon [1] and Smith [2].

vii) According to the theory of §7.7 for $n \geqslant 3$

$$\mathcal{S}_{n+3}(S^1;\mathbb{Z}) = \mathcal{S}^{TOP}_\partial(D^{n+1} \times S^1, S^n \times S^1;\mathbb{Z})$$

is the set of concordance classes of $s^{\mathbb{Z}}_\partial$-triangulations of $(D^{n+1} \times S^1, S^n \times S^1)$, which (by definition) are $(n+2)$-dimensional topological normal maps

$$(g,c) : (X,\partial X) \longrightarrow (D^{n+1} \times S^1, S^n \times S^1)$$

such that $g:X \longrightarrow D^{n+1} \times S^1$ is a $\mathbb{Z}$-homology equivalence and $g| : \partial X \longrightarrow S^n \times S^1$ is a homeomorphism. In particular, the knot complement $(X,\partial X)$ of a knot $k:S^n \subset S^{n+2}$ determines such an $s^{\mathbb{Z}}_\partial$-triangulation, so that there are defined abelian group morphisms

$$C_n \longrightarrow \mathcal{S}_{n+3}(S^1;\mathbb{Z}) ; k \longmapsto (g,c) .$$

The inverse isomorphisms are defined by associating to an $s^{\mathbb{Z}}_\partial$-triangulation (g,c) with $\pi_1(X) = \pi_1(S^1)$ the cobordism class of the knot

$$k : S^n \times 0 \subset S^n \times D^2 \cup_{S^n \times S^1} X = S^{n+2} .$$

[]

For any ring with involution A and any multiplicative subset $S \subset A$ define <u>∂-equivalence</u> to be the equivalence relation on $S^{-1}A$-non-singular $\begin{cases} \varepsilon\text{-symmetric} \\ \varepsilon\text{-quadratic} \end{cases}$ forms over A given by

$$X \sim X' \text{ if there exists an isomorphism } X \oplus Y \xrightarrow{\sim} X' \oplus Y'$$
$$\text{for some non-singular forms } Y, Y' .$$

Combining Proposition 1.7.1 with the results on linking forms of §3.4 we have that the boundary operation defines a natural one-one correspondence

$$\partial : \{\partial\text{-equivalence classes of } S^{-1}A\text{-non-singular } \begin{cases} \varepsilon\text{-symmetric} \\ \varepsilon\text{-quadratic} \end{cases} \text{forms over } A\}$$

$$\xrightarrow{\sim} \{\text{isomorphism classes of non-singular } \begin{cases} \text{even } \varepsilon\text{-symmetric} \\ \text{split } \varepsilon\text{-quadratic} \end{cases}$$

linking forms over (A,S) which are null-cobordant regarded as non-singular $\begin{cases} \text{even } (-\varepsilon)\text{-symmetric} \\ (-\varepsilon)\text{-quadratic} \end{cases}$ formations over A} .

For linking forms over $(\mathbb{Z}[z,z^{-1}],P)$

ε-symmetric = even ε-symmetric = ε-quadratic = split ε-quadratic ,

by Proposition 7.9.2 ii). A non-singular $(-)^{i+1}$-symmetric linking form over $(\mathbb{Z}[z,z^{-1}],P)$ (M,λ) is null-cobordant regarded as a non-singular $\begin{cases} \text{even } (-)^{i+1}\text{-symmetric} \\ (-)^{i+1}\text{-quadratic} \end{cases}$ formation over $\mathbb{Z}[z,z^{-1}]$ if and only if $\begin{cases} \sigma^*(M,\lambda) = 0 \\ \sigma_*(M,\lambda) = 0 \end{cases}$, where $\begin{cases} \sigma^*(M,\lambda) \\ \sigma_*(M,\lambda) \end{cases}$ denotes the image of the linking form cobordism class

$$\begin{cases} (M,\lambda) \in L^0(\mathbb{Z}[z,z^{-1}],P,(-)^{i+1}) = C_{2i-1} \\ (M,\lambda) \in L_0(\mathbb{Z}[z,z^{-1}],P,(-)^{i+1}) = C_{2i-1} \end{cases} \text{ in}$$

$$\begin{cases} L^{-1}(\mathbb{Z}[z,z^{-1}],(-)^{i+1}) = L\langle v_0\rangle^0(\mathbb{Z},(-)^i) \\ L_{-1}(\mathbb{Z}[z,z^{-1}],(-)^{i+1}) = L_0(\mathbb{Z},(-)^i) \end{cases} \text{ under the map}$$

appearing in the localization exact sequence

$$\begin{cases} L^0(P^{-1}\mathbb{Z}[z,z^{-1}],(-)^{i+1}) \longrightarrow L^0(\mathbb{Z}[z,z^{-1}],P,(-)^{i+1}) \\ (=\Gamma^0(\mathbb{Z}[z,z^{-1}]\longrightarrow \mathbb{Z},(-)^{i+1})) \\ \qquad\qquad \longrightarrow L^{-1}(\mathbb{Z}[z,z^{-1}],(-)^{i+1}) \\ L_0(P^{-1}\mathbb{Z}[z,z^{-1}],(-)^{i+1}) \longrightarrow L_0(\mathbb{Z}[z,z^{-1}],P,(-)^{i+1}) \\ (=\Gamma_0(\mathbb{Z}[z,z^{-1}]\longrightarrow \mathbb{Z},(-)^{i+1})) \\ \qquad\qquad \longrightarrow L_{-1}(\mathbb{Z}[z,z^{-1}],(-)^{i+1}) . \end{cases}$$

If $(N,\hat{\psi})$ is a Seifert form for the Blanchfield linking form (M,λ), that is a non-singular $(-)^i$-ultraquadratic form over $\mathbb{Z}$ such that $\beta(N,\hat{\psi}) = (M,\lambda)$, then $\begin{cases} \sigma^*(M,\lambda) \in L\langle v_0\rangle^0(\mathbb{Z},(-)^i) \\ \sigma_*(M,\lambda) \in L_0(\mathbb{Z},(-)^i) \end{cases}$ is $\frac{1}{8}$(the signature) of $(N,\hat{\psi}+\hat{\psi}^*)$ if $i \equiv 0 \pmod 2$ and is $\begin{cases} 0 \\ \text{the Arf invariant of } (N,\hat{\psi}) \end{cases}$ if $i \equiv 1 \pmod 2$. Applying the above special case of Proposition 1.7.1 we have:

<u>Proposition 7.9.5</u> The boundary operation defines a natural one-one correspondence

$$\partial : \{\partial\text{-equivalence classes of } \mathbb{Z}\text{-non-singular } \begin{cases} (-)^{i+1}\text{-symmetric} \\ (-)^{i+1}\text{-quadratic} \end{cases} \text{ forms over } \mathbb{Z}[z,z^{-1}]\}$$

$$\xrightarrow{\sim} \{\text{isomorphism classes of non-singular } (-)^{i+1}\text{-symmetric linking forms over } (\mathbb{Z}[z,z^{-1}],P)\ (M,\lambda) \text{ with } \begin{cases} \sigma^*(M,\lambda) = 0 \\ \sigma_*(M,\lambda) = 0 \end{cases}\} .$$

[]

The result that $\mathbb{Z}$-non-singular ε-quadratic forms over $\mathbb{Z}[z,z^{-1}]$ have isomorphic boundary ε-symmetric linking forms over $(\mathbb{Z}[z,z^{-1}],P)$ if and only if they are ∂-equivalent was first obtained by Stoltzfus [2,Prop.5.5], by a generalization of the method of Wall [10].

The computation $C_{2i} = 0$ $(i \geqslant 2)$ can be used to express the stable equivalence classes of non-singular $(-)^{i+1}$-symmetric $(= (-)^{i+1}$-quadratic) linking formations over $(\mathbb{Z}[z,z^{-1}],P)$ in terms of $(-)^{i}$-symmetric linking forms over $(\mathbb{Z}[z,z^{-1}],P)$, relating non-singular Blanchfield linking formations to singular Blanchfield linking forms as follows.

For any ring with involution A and multiplicative subset $S \subset A$ define <u>∂-equivalence</u> to be the equivalence relation on ε-symmetric linking forms over (A,S) (M,λ) generated by the elementary operations:

i) $(M,\lambda) \longmapsto (M',\lambda')$ is (M',λ') is isomorphic to (M,λ)

ii) $(M,\lambda) \longmapsto (L^{\perp}/L,\lambda^{\perp}/\lambda)$ if L is a sublagrangian of (M,λ)

iii) $(M,\lambda) \longmapsto (M,\lambda)\oplus(M',\lambda')$ if (M',λ') is non-singular.

A special case of the S-acyclic analogue of Proposition 1.8.3 shows that the boundary operation defines a natural one-one correspondence

$$\begin{aligned}\partial : \{&\partial\text{-equivalence classes of } \varepsilon\text{-symmetric linking forms over } (A,S)\}\\ \xrightarrow{\sim} \{&\text{stable equivalence classes of null-cobordant non-singular even } (-\varepsilon)\text{-symmetric linking formations over } (A,S)\}\end{aligned}$$

(cf. Proposition 3.5.4). Now $C_{2i} = L_1(\mathbb{Z}[z,z^{-1}],P,(-)^{i+1}) = 0$, so that every non-singular $(-)^{i+1}$-symmetric linking formation over $(\mathbb{Z}[z,z^{-1}],P)$ is null-cobordant, and consequently:

Proposition 7.9.6 The boundary operation defines a natural one-one correspondence

$$\partial : \{\partial\text{-equivalence classes of } (-)^i\text{-symmetric linking forms over } (\mathbb{Z}[z,z^{-1}],P)\}$$
$$\xrightarrow{\sim} \{\text{stable equivalence classes of non-singular } (-)^{i+1}\text{-symmetric linking formations over } (\mathbb{Z}[z,z^{-1}],P)\}\ .$$

[]

In conclusion, it should perhaps be pointed out that the various characterizations of the odd-dimensional knot cobordism groups C_{2i+1} $(i \geqslant 2)$ given by Proposition 7.9.4 have little computational significance. The actual computations use the "isometric structures" of Milnor [3] - see Levine [3], Kervaire [2] and Stoltzfus [1].

References

J.W.Alexander [1] Topological invariants of knots and links
Trans. A.M.S. 30, 275 - 306 (1928)

D.R.Anderson [1] The Whitehead torsion of a fiber-homotopy equivalence Mich. Math. J. 21, 171 - 180 (1974)

G.A.Anderson [1] Surgery with coefficients
Springer Lecture Notes 591 (1977)

Č.Arf [1] Untersuchungen über quadratische Formen in Körpern der Charakteristik 2 I.
J. reine angew. Math. 183, 148 - 167 (1941)

A.Bak [1] K-theory of forms
Ann. of Math. Study 98 (1981)

[2] Surgery and K-theory groups of quadratic forms over finite groups and orders preprint

and M.Kolster

[1] The computation of odd dimensional projective surgery groups of all finite groups
to appear in Topology

and W.Scharlau

[1] Grothendieck and Witt groups of orders and finite groups Invent. Math. 23, 207 - 240 (1974)

J.Barge, J.Lannes, F.Latour and P.Vogel

[1] Λ-sphères
Ann. Sc. Éc. Norm. Sup. (4) 4, 463 - 506 (1974)

H.Bass [1] K-theory and stable algebra
Publ. Math. IHES 22, 5 - 60 (1964)

[2] Algebraic K-theory Benjamin (1968)

H.Bass (contd.) [3] Unitary algebraic K-theory
Proc. 1972 Battelle Seattle Conference on Algebraic K-theory, Vol. III., Springer Lecture Notes 343, 57 - 265 (1973)

A.Heller and R.G.Swan
[1] The Whitehead group of a polynomial extension
Publ. Math. IHES 22, 61 - 80 (1964)

R.C.Blanchfield [1] Intersection theory of manifolds with operators with applications to knot theory
Ann. of Math. 65, 340 - 356 (1957)

W.Browder [1] Embedding 1-connected manifolds
Bull. A.M.S. 72, 225 - 231 (1966)

[2] Manifolds with $\pi_1 = \mathbb{Z}$
ibid., 238 - 244 (1966)

[3] Embedding smooth manifolds
Proc. 1966 ICM, Moscow, 712 - 719 (1968)

[4] The Kervaire invariant of framed manifolds and its generalization
Ann. of Math. 90, 157 - 186 (1969)

[5] Free $\mathbb{Z}_p$-actions on homotopy spheres
Proc. 1969 Georgia Conference on the Topology of Manifolds, Markham, 217 - 226 (1970)

[6] Surgery on simply-connected manifolds
Springer (1972)

[7] Poincaré spaces, their normal fibrations and surgery Invent. Math. 17, 191 - 202 (1972)

W.Browder (contd.)

[8] Cobordism invariants, the Kervaire invariant and fixed point free involutions
Trans. A.M.S. 178, 193 - 225 (1973)

[9] Complete intersections and the Kervaire invariant Proc. 1978 Århus Conference on Algebraic Topology, Springer Lecture Notes 763, 88 - 108 (1979)

and J.Levine

[1] Fibering manifolds over a circle
Comm. Math. Helv. 40, 153 - 160 (1966)

and G.R.Livesay

[1] Fixed point free involutions on homotopy spheres Tôhoku J. Math. 25, 69 - 88 (1973)

G.Brumfiel and J.Milgram

[1] Normal maps, covering spaces and quadratic functions
Duke Math. J. 44, 663 - 694 (1977)

G.Brumfiel and J.Morgan

[1] Homotopy-theoretic consequences of N.Levitt's obstruction theory to transversality for spherical fibrations
Pac. J. Math. 67, 1 - 100 (1976)

S.Cappell [1] A splitting theorem for manifolds and surgery groups Bull. A.M.S. 77, 281 - 286 (1971)

[2] Mayer-Vietoris sequences in hermitian K-theory
Proc. 1972 Battelle Seattle Conference on Algebraic K-theory, Vol. III., Springer Lecture Notes 343, 478 - 512 (1973)

S.Cappell (contd.)

[3] Splitting obstructions for hermitian forms and manifolds with $\mathbb{Z}_2 \subset \pi_1$
Bull. A.M.S. 79, 909 - 913 (1973)

[4] Unitary nilpotent groups and hermitian K-theory I.
ibid. 80, 1117 - 1122 (1974)

[5] Manifolds with fundamental group a generalized free product ibid., 1193 - 1198 (1974)

[6] On connected sums of manifolds
Topology 13, 395 - 400 (1974)

[7] A splitting theorem for manifolds
Invent. Math. 33, 69 - 170 (1976)

[8] On homotopy invariance of higher signatures
ibid., 171 - 179 (1976)

[9] Submanifolds of small codimension
Proc. 1978 ICM, Helsinki, 455 - 462 (1980)

and J.Shaneson

[1] The codimension two placement problem and homology equivalent manifolds
Ann. of Math. 99, 277 - 348 (1974)

[2] Totally spineless manifolds
Ill. J. Math. 21, 231 - 239 (1977)

[3] Pseudo-free group actions I.
Proc. 1978 Århus Conference on Algebraic Topology, Springer Lecture Notes 763, 395 - 447 (1979)

[4] A counterexample on the oozing problem for closed manifolds ibid., 627 - 634 (1979)

G.Carlsson [1] Desuspension in the symmetric L-groups
ibid., 175 - 197 (1979)

[2] Wu invariants of hermitian forms
J. of Algebra 65, 188 - 205 (1980)

and J.Milgram

[1] Torsion Witt rings for orders and finite groups
Proc. 1977 Evanston Conference on the Geometric Applications of Homotopy Theory, Vol.I., Springer Lecture Notes 657, 85 - 105 (1978)

[2] The structure of odd L-groups
Proc. 1978 Waterloo Conference on Algebraic Topology, Springer Lecture Notes 741, 1 - 72 (1979)

[3] Some exact sequences in the theory of hermitian forms J. Pure and App. Alg. 18, 233 - 252 (1980)

D.Carter [1] Localization in lower algebraic K-theory
Comm. in Alg. 8, 603 - 622 (1980)

J.W.S.Cassels [1] Rational quadratic forms Academic Press (1978)

A.J.Casson [1] Whitehead groups of free products with amalgamation Proc. 1972 Battelle Seattle Conference on Algebraic K-theory, Vol. II., Springer Lecture Notes 342, 144 - 154 (1973)

P.E.Conner and E.Y.Miller

[1] Equivariant self-intersection notes

F.X.Connolly [1] Linking forms and surgery
Topology 12, 389 - 409 (1973)

G.deRham [1] Sur l'analysis situs des varietes à n dimensions
J. de mat. pures X, 115 - 200 (1931)

A.Durfee [1] Fibered knots and algebraic singularities
Topology 13, 47 - 59 (1974)

[2] Bilinear and quadratic forms on torsion modules
Adv. in Math. 25, 133 - 164 (1977)

M.S.Farber [1] Isotopy types of knots of codimension two
Trans. A.M.S. 261, 185 - 209 (1980)

[2] A classification of some knots of codimension two Uspekhi Math. 213, 105 - 111 (1980)

F.T.Farrell [1] The obstruction to fibering a manifold over a circle Indiana Univ. J. 21, 315 - 346 (1971)

[2] The exponent of UNil Topology 18, 305 - 312 (1979)

and W.C.Hsiang

[1] A geometric interpretation of the Künneth formula for algebraic K-theory
Bull. A.M.S. 74, 548 - 553 (1968)

[2] A formula for $K_1R_\alpha[T]$
Proc. Symp. Pure Math. A.M.S. 17, 192 - 218 (1970)

[3] Manifolds with $\pi_1 = G \times_\alpha T$
Amer. J. Math. 95, 813 - 845 (1973)

R.Fox and J.Milnor

[1] Singularities of 2-spheres in 4-space and cobordism of knots
Osaka J. Math. 3, 257 - 267 (1966)

M.Freedman [1] Surgery on codimension two submanifolds
A.M.S. Memoir 191 (1977)

A.Fröhlich [1] On the K-theory of unimodular forms over rings of algebraic integers
Quart. J. Math. Oxf. (2) 22, 401 - 423 (1971)

S.M.Gersten [1] On class groups of free products
Ann. of Math. 87, 392 - 398 (1968)

[2] On the spectrum of algebraic K-theory
Bull. A.M.S. 78, 216 - 219 (1972)

D.Grayson [1] Higher algebraic K-theory II. (after D.Quillen)
Proc. 1976 Evanston Conference on Algebraic K-theory, Springer Lecture Notes 551, 217 - 240 (1976)

[2] K-theory and localization of noncommutative rings J. Pure and App. Alg. 18, 125 - 127 (1980)

I.Hambleton [1] Projective surgery obstructions for closed manifolds preprint

and J.Milgram

[1] Poincaré transversality for double covers
Can. J. Math. XXX, 1319 - 1330 (1978)

[2] The surgery obstruction groups for finite 2-groups Invent. Math. 61, 33 - 52 (1980)

G.Higman [1] Units in group rings
Proc. Lond. Math. Soc. (2) 46, 231 - 248 (1940)

J.P.E.Hodgson [1] Surgery on Poincaré complexes
preprint

I.M.James [1] Reduced product spaces
Ann. of Math. 62, 170 - 197 (1955)

L.Jones [1] The converse to a fixed point theorem of P.A.Smith II.
Indiana Univ. Math. J. 22, 309 - 325 (1972)

[2] The nonsimply connected characteristic variety theorem Proc. Symp. Pure Math. A.M.S. 32, 131 - 140 (1978)

M.Karoubi [1] Periodicité de la K-theorié hermitienne
Proc. 1972 Battelle Seattle Conference on Algebraic K-theory, Vol. III., Springer Lecture Notes 343, 301 - 411 (1973)

[2] Localisation de formes quadratiques I.
Ann. Sci. Éc. Norm. Sup. (4) 7, 359 - 404 (1974)

[3] Localisation de formes quadratiques II.
ibid. 8, 99 - 155 (1975)

C.Kearton [1] Blanchfield duality and simple knots
Trans. A.M.S. 202, 141 - 160 (1975)

[2] Cobordism of knots and Blanchfield duality
J. Lond. Math. Soc. (2) 10, 406 - 408 (1975)

[3] An algebraic classification of some even-dimensional knots
Topology 15, 363 - 373 (1976)

[4] An algebraic classification of certain simple even-dimensional knots preprint

M.Kervaire [1] Les noeuds de dimensions supérieures
Bull. Soc. Math. France 93, 225 - 271 (1965)

[2] Knot cobordism in codimension two
Proc. 1970 Amsterdam Conference on Manifolds, Springer Lecture Notes 197, 83 - 105 (1970)

and J.Milnor

[1] Groups of homotopy spheres I.
Ann. of Math. 77, 504 - 537 (1963)

and C.Weber

[1] A survey of multidimensional knots
Proc. 1977 Plans Conference on Knots, Springer Lecture Notes 685, 61 - 134 (1977)

M.Knebusch and W.Scharlau

[1] Quadratische Formen und quadratische Reziprozitätsgesetze über algebraischen Zahlkörpern Math. Z. 121, 346 - 368 (1971)

M.Kneser and D.Puppe

[1] Quadratische Formen und Verschlingungsinvarianten von Knoten Math. Z. 58, 376 - 384 (1953)

S.Kojima [1] A classification of some even-dimensional fibered knots
J. Fac. Sci. Tokyo (IA) 24, 671 - 683 (1977)

M.Kolster [1] Computations of Witt groups of finite groups
Math. Ann. 241, 129 - 158 (1979)

[2] Even dimensional projective surgery groups of finite groups preprint

J.Lannes [1] Formes quadratiques d'enlacement sur l'anneau des entiers d'un corps de nombres
Ann. Sci. Éc. Norm. Sup. (4) 8, 535 - 579 (1975)

, F.Latour and C.Morlet

[1] Geométrie des complexes de Poincaré et chirurgie IHES notes

R.Lee [1] Splitting a manifold into two parts IAS notes (196

J.Levine [1] Polynomial invariants of knots of codimension two Ann. of Math. 84, 537 - 554 (1966)

[2] Knot cobordism in codimension two
Comm. Math. Helv. 44, 229 - 244 (1969)

[3] Invariants of knot cobordism
Invent. Math. 8, 98 - 110 (1969)

J.Levine (contd.)

[4] An algebraic classification of some knots of codimension two
Comm. Math. Helv. 45, 185 - 198 (1970)

[5] Knot modules I.
Trans. A.M.S. 229, 1 - 50 (1977)

N.Levitt [1] Poincaré duality cobordism
Ann. of Math. 96, 211 - 244 (1972)

and J.Morgan

[1] Transversality structures and PL structures on spherical fibrations
Bull. A.M.S. 78, 1064 - 1068 (1978)

and A.A.Ranicki

[1] Intrinsic transversality structures
preprint

S.López de Medrano

[1] Involutions on manifolds Springer (1971)

[2] Invariant knots and surgery in codimension 2
Proc. 1970 ICM, Nice, Vol. 2, 99 - 112 (1971)

I.Madsen [1] Smooth spherical space forms
Proc. 1977 Evanston Conference on the Geometric Applications of Homotopy theory, I., Springer Lecture Notes 657, 301 - 352 (1978)

[2] Spherical space forms in the period dimension I.
preprint

B.M.Mann and E.Y.Miller

[1] A formula for the Browder-Livesay invariant of an involution J. Lond. Math. Soc.

Y.Matsumoto [1] Knot cobordism groups and surgery in codimension two
J. Fac. Sci. Tokyo (IA) 20, 253 - 317 (1973)

[2] Some relative notions in the theory of hermitian forms
Proc. Jap. Acad. 49, 583 - 587 (1973)

[3] On the equivalence of algebraic formulations of knot cobordism
Jap. J. Math. 3, 81 - 104 (1977)

S.Maumary [1] Proper surgery groups and Wall-Novikov groups
Proc. 1972 Battelle Seattle Conference on Algebraic K-theory, Vol. III, Springer Lecture Notes 343, 526 - 539 (1973)

J.Milnor [1] A procedure for killing the homotopy groups of differentiable manifolds
Proc. A.M.S. Symp. Pure Math. III, 39 - 55 (1961)

[2] Singular points of complex hypersurfaces
Ann. of Math. Study 61 (1968)

[3] Isometries of inner product spaces
Invent. Math. 8, 83 - 97 (1969)

[4] Introduction to algebraic K-theory
Ann. of Math. Study 72 (1971)

and D.Husemoller

[1] Symmetric bilinear forms Springer (1972)

A.S.Mishchenko [1] Homotopy invariants of non-simply-connected manifolds III. Higher signatures Izv. Akad. Nauk SSSR, ser. mat. 35, 1316 - 1355 (1971)

J.Morgan and D.Sullivan

[1] The transversality characteristic class and linking cycles in surgery theory Ann. of Math. 99, 463 - 544 (1974)

H.J.Munkholm and E.K.Pedersen

[1] Whitehead transfers for S^1-bundles: an algebraic description preprint

K.Murasugi [1] On a certain numerical invariant of link types Trans. A.M.S. 114, 377 - 383 (1965)

O.T.O'Meara [1] Introduction to quadratic forms Springer (1962)

S.P.Novikov [1] The algebraic construction and properties of hermitian analogues of K-theory for rings with involution, from the point of view of the hamiltonian formalism. Some applications to differential topology and the theory of characteristic classes Izv. Akad. Nauk SSSR, ser. mat. 34, 253 - 288, 478 - 500 (1970)

W.Pardon [1] Local surgery and the theory of quadratic forms Bull. A.M.S. 82, 131 - 133 (1976)

[2] The exact sequence of a localization for Witt groups Proc. 1976 Evanston Conference on Algebraic K-theory, Springer Lecture Notes 551, 336 - 379 (1976)

W.Pardon (contd.)

[3] Local surgery and the exact sequence of a localization for Witt groups A.M.S. Memoir 196 (19

[4] Mod2 semi-characteristics and the converse to a theorem of Milnor
Math. Z. 171, 247 - 268 (1980)

[5] The exact sequence of a localization for Witt groups II. Numerical invariants of odd-dimensional surgery obstructions
to appear in Pac. J. Math.

[6] A "Gersten Conjecture" for Witt groups to appear i
Proc. 1980 Oberwolfach Algebraic K-theory Conferen

[7] The map of the Witt group of a regular local ring to the Witt group of its field of fractions prepri

D.S.Passman and T.Petrie

[1] Surgery with coefficients in a field
Ann. of Math. 95, 385 - 405 (1972)

E.K.Pedersen and A.A.Ranicki

[1] Projective surgery theory
Topology 19, 239 - 254 (1980)

D.G.Quillen

[1] Higher K-theory for categories with exact sequences Proc. 1972 Oxford Conference "New Developments in Topology"
LMS Lecture Notes 11, 95 - 104 (1974)

[2] Higher algebraic K-theory I.
Proc. 1972 Battelle Seattle Conference on Algebraic K-theory, Vol. I., Springer Lecture Notes 341, 85 - 147 (1973)

F.Quinn

[1] A geometric formulation of surgery
Proc. 1969 Georgia Conference on the Topology of Manifolds, Markham, 500 - 512 (1970)

[2] $\underline{B}_{(\underline{TOP})_n^{\sim}}$ and the surgery obstruction I.
Bull. A.M.S. 77, 596 - 600 (1971)

[3] Surgery on Poincaré and normal spaces
ibid. 78, 262 - 267 (1972)

[4] Semifree group actions and surgery on PL homology manifolds
Proc. 1974 Conference on Geometric Topology, Park City, Utah, Springer Lecture Notes 438, 395 - 414 (1975)

A.A.Ranicki

[1] Algebraic L-theory I. Foundations
Proc. Lond. Math. Soc. (3) 27, 101 - 125 (1973)

[2] II. Laurent extensions
ibid., 126 - 158 (1973)

[3] III. Twisted Laurent extensions
Proc. 1972 Battelle Seattle Conference on Algebraic K-theory, Vol. III., Springer Lecture Notes 343, 412 - 463 (1973)

[4] IV. Polynomial extension rings
Comm. Math. Helv. 49, 137 - 167 (1974)

[5] An algebraic formulation of surgery
Trinity College fellowship dissertation (1972)

[6] Localization in quadratic L-theory
Proc. 1978 Waterloo Conference on Algebraic Topology, Springer Lecture Notes 741, 102 - 157 (1979)

A.A.Ranicki (contd.)

[7] The total surgery obstruction
Proc. 1978 Århus Conference on Algebraic Topology, Springer Lecture Notes 763, 275 - 316 (1979)

[8] The surgery obstruction of a disjoint union
J. Lond. Math. Soc. 20, 559 - 566 (1979)

[9] The algebraic theory of surgery I. Foundations
Proc. Lond. Math. Soc. (3) 40, 87 - 192 (1980)

[10] II. Applications to topology
ibid., 193 - 283 (1980)

[11] Splitting theorems in the algebraic theory of surgery in preparation

[12] Classifying spaces in the algebraic theory of surgery in preparation

W.Richter [1] Simple even-dimensional knots and Blanchfield duality preprint

C.P.Rourke and B.J.Sanderson

[1] Block bundles I.
Ann. of Math. 87, 1 - 28 (1968)

[2] On topological neighbourhoods
Comp. Math. 22, 387 - 424 (1970)

M.G.Scharlemann [1] Transversality theories at dimension 4
Invent. Math. 33, 1 - 14 (1976)

H.Seifert [1] Verschlingungsinvarianten
Sit. Preuß. Akad. Wiss. 26, 811 - 823 (1933)

[2] Über das Geschlecht von Knoten
Math. Ann. 110, 571 - 592 (1934)

J.Shaneson [1] Wall's surgery obstruction groups for $G \times \mathbb{Z}$ Ann. of Math. 90, 296 - 334 (1969)

R.W.Sharpe [1] On the structure of the unitary Steinberg group ibid. 96, 444 - 479 (1972)

[2] Surgery on compact manifolds: the bounded even-dimensional case ibid. 98, 187 - 209 (1973)

L.Siebenmann [1] A total Whitehead torsion obstruction to fibering over the circle Comm. Math. Helv. 45, 1 - 48 (1970)

J.Smith [1] Acyclic localizations J. Pure and App. Alg. 12, 117 - 127 (1978)

[2] Complements of codimension two submanifolds III. Cobordism theory to appear in Pac. J. Math.

J.Stallings [1] On fibering certain 3-manifolds Proc. 1961 Georgia Conference on the Topology of 3-manifolds, Prentice-Hall, 95 - 100 (1962)

[2] Whitehead torsion of free products Ann. of Math. 82, 354 - 363 (1965)

B.Stenström [1] Rings of quotients Springer (1975)

N.Stoltzfus [1] Unraveling the integral knot concordance group A.M.S. Memoir 192 (1977)

[2] Equivariant concordance of invariant knots Trans. A.M.S. 254, 1 - 45 (1979)

D.Sullivan [1] Triangulating homotopy equivalences Princeton Ph.D. thesis (1965)

D.Sullivan (contd.)

[2] Geometric periodicity and the invariants of manifolds
Proc. 1970 Amsterdam Conference on Manifolds, Springer Lecture Notes 197, 44 - 75 (1970)

R.G.Swan [1] Excision in algebraic K-theory
J. Pure and App. Alg. 1, 221 - 252 (1971)

L.R.Taylor [1] Surgery on paracompact manifolds
Berkeley Ph.D. thesis (1972)

[2] Surgery groups and inner automorphisms
Proc. 1972 Battelle Seattle Conference on Algebraic K-theory, Vol. III., Springer Lecture Notes 343, 471 - 477 (1973)

and B.Williams

[1] Surgery spaces: formulae and structure
Proc. 1978 Waterloo Conference on Algebraic Topology, Springer Lecture Notes 741, 170 - 195 (1979)

[2] Local surgery: foundations and applications
Proc. 1978 Århus Conference on Algebraic Topology, Springer Lecture Notes 763, 673 - 695 (1979)

[3] Surgery on closed manifolds preprint

C.B.Thomas [1] Frobenius reciprocity for hermitian forms
J. Alg. 18, 237 - 244 (1971)

H.F.Trotter [1] Homology of group systems with applications to knot theory
Ann. of Math. 76, 464 - 498 (1962)

[2] On S-equivalence of Seifert matrices
Invent. Math. 20, 173 - 207 (1973)

[3] Knot modules and Seifert matrices
Proc. 1977 Plans Conference on Knots, Springer Lecture Notes 685, 291 - 299 (1978)

P.Vogel [1] Cobordisme d'immersions
Ann. Sci. Éc. Norm. Sup. (4) 7, 317 - 358 (1974)

[2] Localization in algebraic L-theory
Proc. 1979 Siegen Topology Symposium, Springer Lecture Notes 788, 482 - 495 (1980)

[3] On the obstruction group in homology surgery
preprint

F.Waldhausen [1] On irreducible 3-manifolds which are sufficiently large Ann. of Math. 87, 56 - 88 (1968)

[2] Whitehead groups of generalized free products
notes (1969)

[3] Algebraic K-theory of generalized free products
Ann. of Math. 108, 135 - 256 (1978)

C.T.C.Wall [1] Quadratic forms on finite groups and related topics Topology 2, 281 - 298 (1963)

[2] Surgery on non-simply-connected manifolds
Ann. of Math. 84, 217 - 276 (1966)

[3] Poincaré complexes I. ibid. 86, 213 - 245 (1970)

[4] Surgery on compact manifolds
Academic Press (1970)

C.T.C.Wall (contd.)

[5] On the axiomatic foundations of the theory of Hermitian forms
Proc. Camb. Phil. Soc. 67, 243 - 250 (1970)

[6] On the classification of hermitian forms I. Rings of algebraic integers
Comp. Math. 22, 425 - 451 (1970)

[7] III. Complete semilocal rings
Invent. Math. 19, 59 - 71 (1973)

[8] V. Global rings ibid. 23, 261 - 288 (1974)

[9] VI. Group rings
Ann. of Math. 103, 1 - 80 (1976)

[10] Quadratic forms on finite groups II.
Bull. Lond. Math. Soc. 4, 156 - 160 (1972)

[11] Foundations of algebraic L-theory
Proc. 1972 Battelle Seattle Conference on Algebraic K-theory, Vol.III., Springer Lecture Notes 343, 266 - 300 (1973)

[12] Periodicity in algebraic L-theory
Proc. 1973 Tokyo Conference on Manifolds, 57 - 68 (1974)

[13] Formulae for surgery obstructions
Topology 15, 189 - 210 (1976)

H.Weyl

[1] The algebraic theory of numbers
Ann. of Math. Study 1 (1940)

O.Zariski and P.Samuel

[1] Commutative algebra
Van Nostrand (1958), Springer (1975)

INDEX OF PROPOSITIONS